JEEP

CJ/SCRAMBLER/WRANGLER 1971-90 REPAIR MANUAL

CHILTON'S

Sr. Vice President	Ronald A. Hoxter
Publisher and Editor-In-Chief	Kerry A. Freeman, S.A.E.
Managing Editors	Peter M. Conti, Jr. □ W. Calvin Settle, Jr., S.A.E.
Assistant Managing Editor	Nick D'Andrea
Senior Editors	Richard J. Rivele, S.A.E. □ Ron Webb
Director of Manufacturing	Mike D'Imperio
Manager of Manufacturing	John F. Butler

CHILTON BOOK COMPANY

ONE OF THE DIVERSIFIED PUBLISHING COMPANIES,
A PART OF CAPITAL CITIES/ABC, INC.

Manufactured in USA
© 1990 Chilton Book Company
Chilton Way Radnor, Pa. 19089
ISBN 0–8019–8034–8
Library of Congress Catalog Card No. 90–055441

3 4 5 6 7 8 9 0   1 0 9 8 7 6 5 4 3 2

Contents

Contents

SAFETY NOTICE

Proper service and repair procedures are vital to the safe, reliable operation of all motor vehicles, as well as the personal safety of those performing repairs. This manual outlines procedures for servicing and repairing vehicles using safe, effective methods. The procedures contain many NOTES, CAUTIONS and WARNINGS which should be followed along with standard safety procedures to eliminate the possibility of personal injury or improper service which could damage the vehicle or compromise its safety.

It is important to note that the repair procedures and techniques, tools and parts for servicing motor vehicles, as well as the skill and experience of the individual performing the work vary widely. It is not possible to anticipate all of the conceivable ways or conditions under which vehicles may be serviced, or to provide cautions as to all of the possible hazards that may result. Standard and accepted safety precautions and equipment should be used when handling toxic or flammable fluids, and safety goggles or other protection should be used during cutting, grinding, chiseling, prying, or any other process that can cause material removal or projectiles.

Some procedures require the use of tools specially designed for a specific purpose. Before substituting another tool or procedure, you must be completely satisfied that neither your personal safety, nor the performance of the vehicle will be endangered

Although information in this manual is based on industry sources and is complete as possible at the time of publication, the possibility exists that some car manufacturers made later changes which could not be included here. While striving for total accuracy, Chilton Book Company cannot assume responsibility for any errors, changes or omissions that may occur in the compilation of this data.

PART NUMBERS

Part numbers listed in this reference are not recommendations by Chilton for any product by brand name. They are references that can be used with interchange manuals and aftermarket supplier catalogs to locate each brand supplier's discrete part number.

SPECIAL TOOLS

Special tools are recommended by the vehicle manufacturer to perform their specific job. Use has been kept to a minimum, but where absolutely necessary, they are referred to in the text by the part number of the tool manufacturer. These tools can be purchased under the appropriate part number, from your Jeep dealer or regional distributor or an equivalent tool can be purchased locally from a tool supplier or parts outlet. Before substituting any tool for the recommended one, read the SAFETY NOTICE at the top of this page.

ACKNOWLEDGMENTS

The Chilton Book Company expresses its appreciation to the Jeep Corporation, A Division of Chrysler Corporation, Detroit, Michigan for their generous assistance in the preparation of this book.

General Information *and* Maintenance

1

QUICK REFERENCE INDEX

GENERAL INDEX

HOW TO USE THIS BOOK

This book covers all 1971-90 CJ-5, CJ-6, CJ-7, Scrambler and Wrangler models.

The first two sections will be the most used, since they contain maintenance and tune-up information and procedures. Studies have shown that a properly tuned and maintained Jeep can get at least 10% better gas mileage (which translates into lower operating costs) and periodic maintenance will catch minor problems before they turn into major repair bills. The other sections deal with the more complex systems of your Jeep. Operating systems from engine through brakes are covered to the extent that the average do-it-yourselfer becomes mechanically involved. This book will not explain such things as rebuilding the differential for the simple reason that the expertise required and the investment in special tools make this task impractical and uneconomical. It will give you the detailed instructions to help you change your own brake pads and shoes, tune-up the engine, replace spark plugs and filters, and do many more jobs that will save you money, give you personal satisfaction and help you avoid expensive problems.

A secondary purpose of this book is a reference guide for owners who want to understand their Jeep and/or their mechanics better. In this case, no tools at all are required. Knowing just what a particular repair job requires in parts and labor time will allow you to evaluate whether or not you're getting a fair price quote and help decipher itemized bills from a repair shop.

Before attempting any repairs or service on your Jeep, read through the entire procedure outlined in the appropriate section. This will give you the overall view of what tools and supplies will be required. There is nothing more frustrating than having to walk to the bus stop on Monday morning because you were short one gasket on Sunday afternoon. So read ahead and plan ahead. Each operation should be approached logically and all procedures thoroughly understood before attempting any work. Some special tools that may be required can often be rented from local automotive jobbers or places specializing in renting tools and equipment. Check the yellow pages of your phone book.

All sections contain adjustments, maintenance, removal and installation procedures, and overhaul procedures. When overhaul is not considered practical, we tell you how to remove the failed part and then how to install the new or rebuilt replacement. In this way, you at least save the labor costs. Backyard overhaul of some components (such as the alternator or water pump) is just not practical, but the removal and installation procedure is often simple and well within the capabilities of the average Jeep owner.

Two basic mechanic's rules should be mentioned here. First, whenever the LEFT side of the Jeep or engine is referred to, it is meant to specify the DRIVER'S side of the Jeep. Conversely, the RIGHT side of the Jeep means the PASSENGER'S side. Second, all screws and bolts are removed by turning counterclockwise, and tightened by turning clockwise.

Safety is always the most important rule. Constantly be aware of the dangers involved in working on or around an automobile and take proper precautions to avoid the risk of personal injury or damage to the vehicle. See the part in this section, Servicing Your Vehicle Safely, and the SAFETY NOTICE on the acknowledgment page before attempting any service procedures and pay attention to the instructions provided. There are 3 common mistakes in mechanical work:

1. Incorrect order of assembly, disassembly or adjustment. When taking something apart or putting it together, doing things in the wrong order usually just costs you extra time; however it CAN break something. Read the entire procedure before beginning disassembly. Do everything in the order in which the instructions say you should do it, even if you can't immediately see a reason for it. When you're taking apart something that is very intricate (for example a carburetor), you might want to draw a picture of how it looks when assembled at one point in order to make sure you get everything back in its proper position. We will supply exploded views whenever possible, but sometimes the job requires more attention to detail than an illustration provides. When making adjustments (especially tune-up adjustments), do them in order. One adjustment often affects another and you cannot expect satisfactory results unless each adjustment is made only when it cannot be changed by any other.

2. Overtorquing (or undertorquing) nuts and bolts. While it is more common for overtorquing to cause damage, undertorquing can cause a fastener to vibrate loose and cause serious damage, especially when dealing with aluminum parts. Pay attention to torque specifications and utilize a torque wrench in assembly. If a torque figure is not available remember that, if you are using the right tool to do the job, you will probably not have to strain yourself to get a fastener tight enough. The pitch of most threads is so slight that the tension you put on the wrench will be multiplied many times in actual force on what you are tightening. A good example of how critical torque is can be seen in the case of spark plug installation, especially where you are putting the plug into an aluminum cylinder head. Too little torque can fail to crush the gasket, causing leakage of combustion gases and consequent overheating of the plug and engine parts. Too much torque can damage the threads or distort the plug, which changes the spark gap at the electrode. Since more and more manufacturers are using aluminum in their engine and chassis parts to save weight, a torque wrench should be in any serious do-it-yourselfer's tool box.

here are many commercial chemical products available for ensuring that fasteners won't come loose, even if they are not torqued just right (a very common brand is Loctite®). If you're worried about getting something together tight enough to hold, but loose enough to avoid mechanical damage during assembly, one of these products might offer substantial insurance. Read the label on the package and make sure the product is compatible with the materials, fluids, etc. involved before choosing one.

3. Crossthreading. This occurs when a part such as a bolt is screwed into a nut or casting at the wrong angle and forced, causing the threads to become damaged. Crossthreading is more likely to occur if access is difficult. It helps to clean and lubricate fasteners, and to start threading with the part to be installed going straight in, using your fingers. If you encounter resistance, unscrew the part and start over again at a different angle until it can be inserted and turned several times without much effort. Keep in mind that many parts, especially spark plugs, use tapered threads so that gentle turning will automatically bring the part you're threading to the proper angle if you don't force it or resist a change in angle. Don't put a wrench on the part until it's been turned in a couple of times by hand. If you suddenly encounter resistance and the part has not seated fully, don't force it. Pull it back out and make sure it's clean and threading properly.

Always take your time and be patient; once you have some experience, working on your Jeep will become an enjoyable hobby.

TOOLS AND EQUIPMENT

Without the proper tools and equipment it is impossible to properly service your vehicle. It would be impossible to catalog each tool that you would need to perform each or every operation in this book. It would also be unwise for the amateur to rush out and buy an expensive set of tools an the theory that he may need one or more of them at sometime.

The best approach is to proceed slowly, gathering together a good quality set of those tools that are used most frequently. Don't be misled by the low cost of bargain tools. It is far better to spend a little more for better quality. Forged wrenches, 6- or 12-point sockets and fine tooth ratchets are by far preferable to their less expensive counterparts. As any good mechanic can tell you, there are few worse experiences than trying to work on a Jeep with bad tools. Your monetary savings will be far outweighed by frustration and mangled knuckles.

Certain tools, plus a basic ability to handle tools, are required to get started. A basic mechanics tool set, a torque wrench, and a Torx® bits set. Torx® bits are hexlobular drivers which fit both inside and outside on special Torx® head fasteners used in various places on Jeep vehicles.

A special wheel bearing nut socket would be helpful when removing the front wheel bearings.

Begin accumulating those tools that are used most frequently; those associated with routine maintenance and tune-up.

In addition to the normal assortment of screwdrivers and pliers you should have the following tools for routine maintenance jobs (your Jeep, depending on the model year, uses both SAE and metric fasteners):

1. SAE/Metric wrenches, sockets and combination open end/box end wrenches in sizes from 1/8 in. (3mm) to 3/4 in. (19mm), and a spark plug socket (13/16 in. or 5/8 in.). If possible, buy various length socket drive extensions. One break in this department is that the metric sockets available in the U.S. will all fit the ratchet handles and extensions you may already have (1/4, 3/8, and 1/2 in. drive).
2. Jackstands for support.
3. Oil filter wrench.
4. Oil filter spout for pouring oil.
5. Grease gun for chassis lubrication.
6. Hydrometer for checking the battery.
7. A container for draining oil.
8. Many rags for wiping up the inevitable mess.

In addition to the above items there are several others that are not absolutely necessary, but handy to have around. These include oil-dry, a transmission funnel and the usual supply of lubricants, antifreeze and fluids, although these can be purchased as needed. This is a basic list for routine maintenance, but only your personal needs and desires can accurately determine your list of necessary tools.

The second list of tools is for tune-ups. While the tools involved here are slightly more sophisticated, they need not be outrageously expensive. There are several inexpensive tach/dwell meters on the market that are every bit as good for the average mechanic as a $100.00 professional model. Just be sure that it goes to at least 1,200-1,500 rpm on the tach scale and that it works on 4, 6 and 8 cylinder engines. A basic list of tune-up equipment could include:

1. Tach-dwell meter.
2. Spark plug wrench.
3. Timing light (a DC light that works from the Jeep's battery is best, although an AC light that plugs into 110V house current will suffice at some sacrifice in brightness).
4. Wire spark plug gauge/adjusting tools.
5. Set of feeler blades.

Here again, be guided by your own needs.

In addition to these basic tools, there are several other tools and gauges you may find useful. These include:

1. A compression gauge. The screw-in type is slower to use, but eliminates the possibility of a faulty reading due to escaping pressure.
2. A manifold vacuum gauge.
3. A test light.
4. An induction meter. This is used for determining whether or not there is current in a wire. These are handy for use if a wire is broken somewhere in a wiring harness.

As a final note, you will probably find a torque wrench necessary for all but the most basic work. The beam type models are perfectly adequate, although the newer click (breakaway) type are more precise, and you don't have to crane your neck to see a torque reading in awkward situations. The breakaway torque wrenches are more expensive and should be recalibrated periodically.

Torque specification for each fastener will be given in the procedure in any case that a specific torque value is required. If no torque specifications are given, use the following values as a guide, based upon fastener size:

Bolts marked 6T
6mm bolt/nut - 5-7 ft. lbs.
8mm bolt/nut - 12-17 ft. lbs.
10mm bolt/nut - 23-34 ft. lbs.
12mm bolt/nut - 41-59 ft. lbs.
14mm bolt/nut - 56-76 ft. lbs.

Bolts marked 8T
6mm bolt/nut - 6-9 ft. lbs.
8mm bolt/nut - 13-20 ft. lbs.
10mm bolt/nut - 27-40 ft. lbs.
12mm bolt/nut - 46-69 ft. lbs.
14mm bolt/nut - 75-101 ft. lbs.

Special Tools

Normally, the use of special factory tools is avoided for repair procedures, since these are not readily available for the do-it-yourself mechanic. When it is possible to perform the job with more commonly available tools, it will be pointed out, but occa-

TWO-WIRE CONDUCTOR THIRD WIRE GROUNDING THE CASE

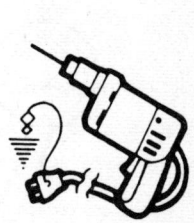
THREE-WIRE CONDUCTOR GROUNDING THRU A CIRCUIT

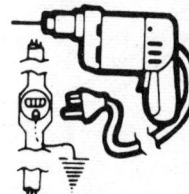
THREE-WIRE CONDUCTOR ONE WIRE TO A GROUND

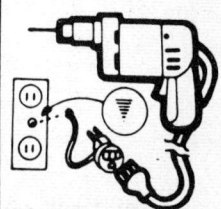
THREE-WIRE CONDUCTOR GROUNDING THRU AN ADAPTER PLUG

When using electric tools make sure they are properly grounded

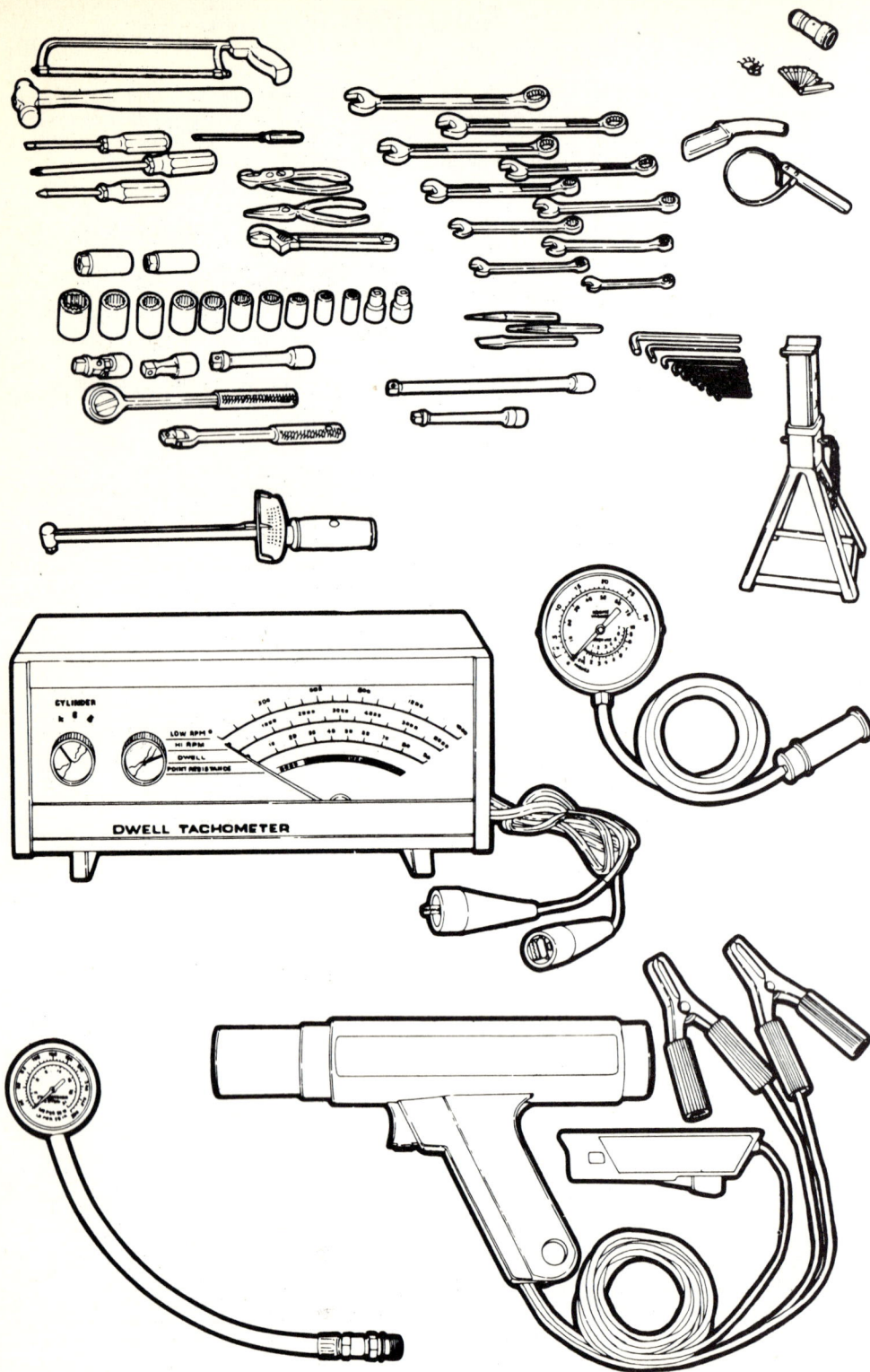

You need only a basic assortment of hand tools for most maintenance and repair jobs

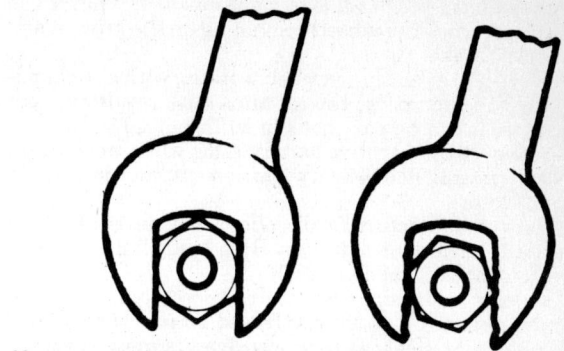

Keep screwdriver tips in good shape. They should fit the slot as shown in "A". If they look like those in "B", they need grinding or replacing

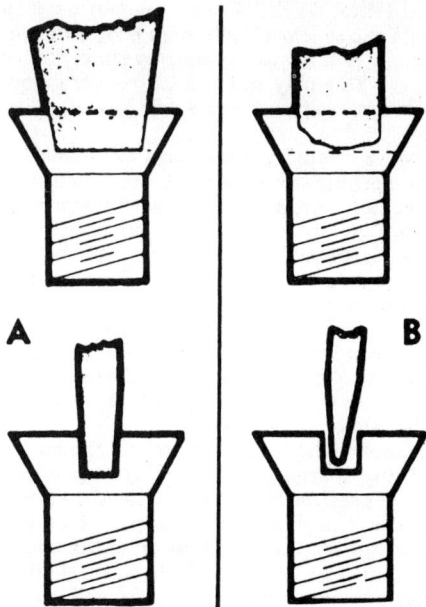

If you're using an open end wrench, use the correct size, and position it properly on the nut or bolt

sionally, a special tool was designed to perform a specific function and should be used. Before substituting another tool, you should be convinced that neither your safety nor the performance of the vehicle will be compromised.

Some special tools are available through your Jeep dealer or major tool manufacturers, such as:

Service Tool Division
Kent-Moore
29784 Little Mack
Roseville, MI 48066-2298

Miller Special Tools
Utica Tool Co.
32615 Park La.
Garden City, MI 48135

Owatonna Tool Co
Owatonna, MN 55060

Robert Bosch Corp.
2800 S.25th St.
Broadview, IL 60153

Utica Tool Co.
32615 Park La.
Garden City, MI 48135

Equivalent tools may be purchased at most independent tool dealers or auto parts stores.

SERVICING YOUR JEEP SAFELY

It is virtually impossible to anticipate all of the hazards involved with automotive maintenance and service, but care and common sense will prevent most accidents.

The rules of safety for mechanics range from "don't smoke around gasoline," to "use the proper tool for the job." The trick to avoiding injuries is to develop safe work habits and take every possible precaution.

Do's

• Do keep a fire extinguisher and first aid kit within easy reach.

• Do wear safety glasses or goggles when cutting, drilling or prying, even if you have 20-20 vision. If you wear glasses for the sake of vision, they should be made of hardened glass that can also serve as safety glasses, or wear safety goggles over your regular glasses.

• Do shield your eyes whenever you work around the battery. Batteries contain sulphuric acid. In case of contact with the eyes or skin, flush the area with water or a mixture of water and baking soda and get medical attention immediately.

• Do use safety stands for any under-Jeep service. Jacks are for raising vehicles; safety stands are for making sure the vehicle stays raised until you want it to come down. Whenever the vehicle is raised, block the wheels remaining on the ground and set the parking brake.

- Do use adequate ventilation when working with any chemicals. Like carbon monoxide, the asbestos dust resulting from brake lining wear can be poisonous in sufficient quantities.
- Do disconnect the negative battery cable when working on the electrical system. The primary ignition system can contain up to 40,000 volts.
- Do follow manufacturer's directions whenever working with potentially hazardous materials. Both brake fluid and antifreeze are poisonous if taken internally.
- Do properly maintain your tools. Loose hammerheads, mushroomed punches and chisels, frayed or poorly grounded electrical cords, excessively worn screwdrivers, spread wrenches (open end), cracked sockets, slipping ratchets, or faulty droplight sockets can cause accidents.
- Do use the proper size and type of tool for the job being done.
- Do when possible, pull on a wrench handle rather than push on it, and adjust your stance to prevent a fall.
- Do be sure that adjustable wrenches are tightly adjusted on the nut or bolt and pulled so that the face is on the side of the fixed jaw.
- Do select a wrench or socket that fits the nut or bolt. The wrench or socket should sit straight, not cocked.
- Do strike squarely with a hammer-avoid glancing blows.
- Do set the parking brake and block the drive wheels if the work requires that the engine be running.

Don'ts

- Don't run an engine in a garage or anywhere else without proper ventilation-EVER! Carbon monoxide is poisonous. It takes a long time to leave the human body and you can build up a deadly supply of it in your system by simply breathing in a little every day. You may not realize you are slowly poisoning yourself. Always use power vents, windows, fans or open the garage doors.
- Don't work around moving parts while wearing a necktie or other loose clothing. Short sleeves are much safer than long, loose sleeves and hard-toed shoes with neoprene soles protect your toes and give a better grip on slippery surfaces. Jewelry such as watches, fancy belt buckles, beads or body adornment of any kind is not safe working around a Jeep. Long hair should be hidden under a hat or cap.
- Don't use pockets for toolboxes. A fall or bump can drive a screwdriver deep into you body. Even a wiping cloth hanging from the back pocket can wrap around a spinning shaft or fan.
- Don't smoke when working around gasoline, cleaning solvent or other flammable material.
- Don't smoke when working around the battery. When the battery is being charged, it gives off explosive hydrogen gas.
- Don't use gasoline to wash your hands! There are excellent soaps available. Gasoline may contain lead, and lead can enter the body through a cut, accumulating in the body until you are very ill. Gasoline also removes all the natural oils from the skin so that bone dry hands will suck up oil and grease.
- Don't service the air conditioning system unless you are equipped with the necessary tools and training. The refrigerant, R-12, is extremely cold and when exposed to the air, will instantly freeze any surface it comes in contact with, including your eyes. Although the refrigerant is normally non-toxic, R-12 becomes a deadly poisonous gas in the presence of an open flame. One good whiff of the vapors from burning refrigerant can be fatal.

HISTORY AND MODEL IDENTIFICATION

American Motors purchased the Jeep division of Willys Motors in 1971. The CJ line, at that time, was made up of the CJ-5 and CJ-6. The engine line-up included the F4-134 Willys engine and V6-225 Buick engine.

In 1972, both of these engines were dropped and replaced by AMC 6-cylinder engines of 232 and 258 cid. The AMC 304 cid V8 was also offered.

In 1972 the wheelbase of the CJ-5 was lengthened from 81 in. (2,057mm) to 84 in. (2,134mm) to accommodate the larger American Motors engines. The CJ-6 also had its wheelbase elongated to 104 in. (2,642mm) to accommodate the larger American Motors engines.

For the 1976 model year the CJ-6 was discontinued in the U.S. and Canada, although still exported. A new model, the CJ-7, featuring an optional one piece removable plastic hardtop, automatic transmission, steel side doors with roll up windows and the full time 4WD system, Quadra-Trac® was introduced. The CJ-7 has a wheelbase of 93.5 in. (2,375mm).

In mid-year 1981 Jeep introduced its newest model, the Scrambler. Designed for rugged dependability and good fuel economy, the Scrambler is both a work and recreational vehicle. The standard engine is a GM built 151 cid 4-cylinder with an American Motors 6-258 an an option. A manual 4-speed transmission is standard with automatic as an option. The standard transfer case is the 2-speed Dana 300.

For 1984 the CJ-5 was discontinued. The 4-151 was replaced by an AMC 4-150.

In 1987, the CJ series was discontinued entirely, and Chrysler Corp., Jeep's new owners, introduced the Wrangler series. The Wrangler utilizes Dana axles, an AISIN or Peugeot 5-speed as standard equipment, the Chrysler 999 automatic and the 4-150 as the standard engine, with the 6-258 as the option.

SERIAL NUMBER IDENTIFICATION

Vehicle

When American Motors Corporation took over the Jeep Corporation, the numbering system was changed to the American Motors 13 digit (through 1980) or 17 digit (1981 and later) alpha-numerical Vehicle Identification Number (VIN).

This number is stamped on a metal plate on the left side of the firewall.

Engine

4-134

The engine serial number for the Kaiser/Willys built F-Head 4-cylinder engine is located on the water pump boss at the front of the engine. It consists of a 5 or 6 digit number. The engine code prefix for the F-Head is 4J.

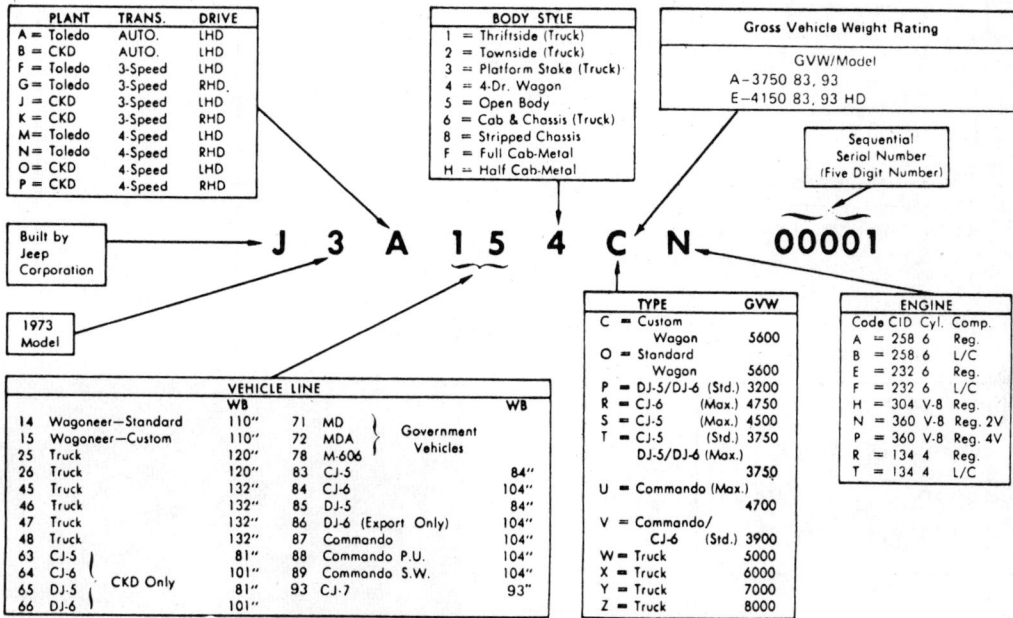

PLANT	TRANS.	DRIVE
A = Toledo	AUTO.	LHD
B = CKD	AUTO.	LHD
F = Toledo	3-Speed	LHD
G = Toledo	3-Speed	RHD
J = CKD	3-Speed	LHD
K = CKD	3-Speed	RHD
M = Toledo	4-Speed	LHD
N = Toledo	4-Speed	RHD
O = CKD	4-Speed	LHD
P = CKD	4-Speed	RHD

BODY STYLE
1 = Thriftside (Truck)
2 = Townside (Truck)
3 = Platform Stake (Truck)
4 = 4-Dr. Wagon
5 = Open Body
6 = Cab & Chassis (Truck)
8 = Stripped Chassis
F = Full Cab-Metal
H = Half Cab-Metal

Gross Vehicle Weight Rating

	GVW/Model
A	3750 83, 93
E	4150 83, 93 HD

Sequential Serial Number (Five Digit Number)

Built by Jeep Corporation

1973 Model

J 3 A 1 5 4 C N 00001

	TYPE	GVW
C =	Custom Wagon	5600
O =	Standard Wagon	5600
P =	DJ-5/DJ-6 (Std.)	3200
R =	CJ-6 (Max.)	4750
S =	CJ-5 (Max.)	4500
T =	CJ-5 (Std.) 3750 DJ-5/DJ-6 (Max.)	3750
U =	Commando (Max.)	4700
V =	Commando/ CJ-6 (Std.)	3900
W =	Truck	5000
X =	Truck	6000
Y =	Truck	7000
Z =	Truck	8000

ENGINE

Code	CID	Cyl.	Comp.
A =	258	6	Reg.
B =	258	6	L/C
E =	232	6	Reg.
F =	232	6	L/C
H =	304	V-8	Reg.
N =	360	V-8	Reg. 2V
P =	360	V-8	Reg. 4V
R =	134	4	Reg.
T =	134	4	L/C

VEHICLE LINE

		WB					WB
14	Wagoneer—Standard	110"	71	MD			
15	Wagoneer—Custom	110"	72	MDA	Government Vehicles		
25	Truck	120"	78	M-606			
26	Truck	120"	83	CJ-5			84"
45	Truck	132"	84	CJ-6			104"
46	Truck	132"	85	DJ-5			84"
47	Truck	132"	86	DJ-6 (Export Only)			104"
48	Truck	132"	87	Commando			104"
63	CJ-5	81"	88	Commando P.U.			104"
64	CJ-6	101"	89	Commando S.W.			104"
65	DJ-5	81"	93	CJ-7			93"
66	DJ-6	101"					

(64 CKD Only, 65 CKD Only)

Serial number system for 1971–74

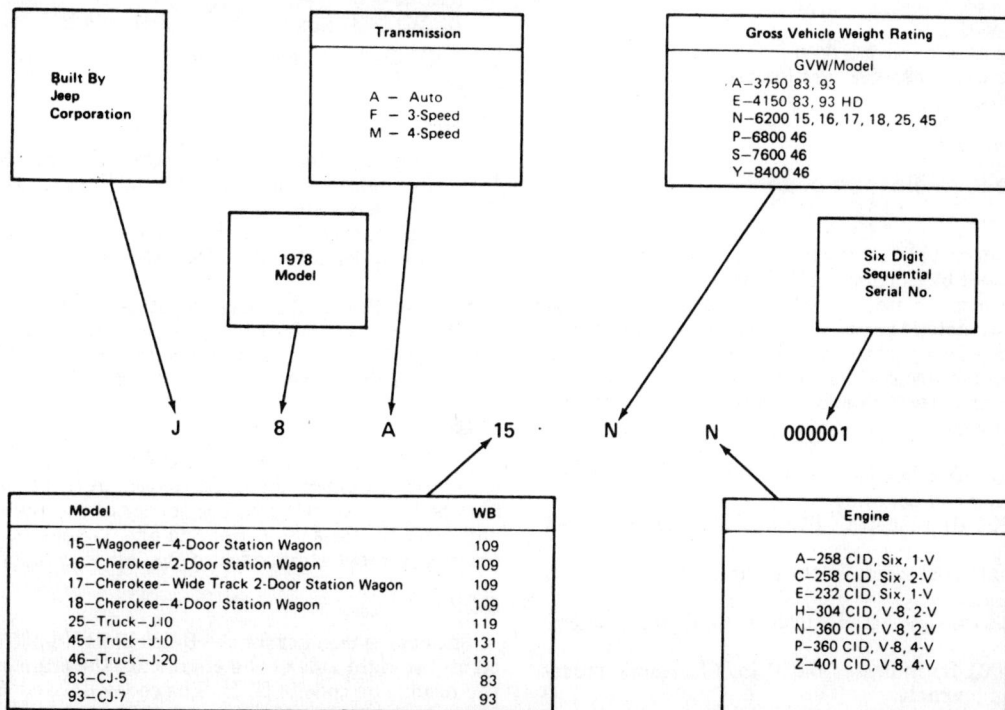

Built By Jeep Corporation

Transmission
A — Auto
F — 3-Speed
M — 4-Speed

Gross Vehicle Weight Rating

	GVW/Model
A	3750 83, 93
E	4150 83, 93 HD
N	6200 15, 16, 17, 18, 25, 45
P	6800 46
S	7600 46
Y	8400 46

1978 Model

Six Digit Sequential Serial No.

J 8 A 15 N N 000001

Model

	Model	WB
15	Wagoneer—4-Door Station Wagon	109
16	Cherokee—2-Door Station Wagon	109
17	Cherokee—Wide Track 2-Door Station Wagon	109
18	Cherokee—4-Door Station Wagon	109
25	Truck—J-10	119
45	Truck—J-10	131
46	Truck—J-20	131
83	CJ-5	83
93	CJ-7	93

Engine

	Engine
A	258 CID, Six, 1-V
C	258 CID, Six, 2-V
E	232 CID, Six, 1-V
H	304 CID, V-8, 2-V
N	360 CID, V-8, 2-V
P	360 CID, V-8, 4-V
Z	401 CID, V-8, 4-V

Serial number code system for 1975–80

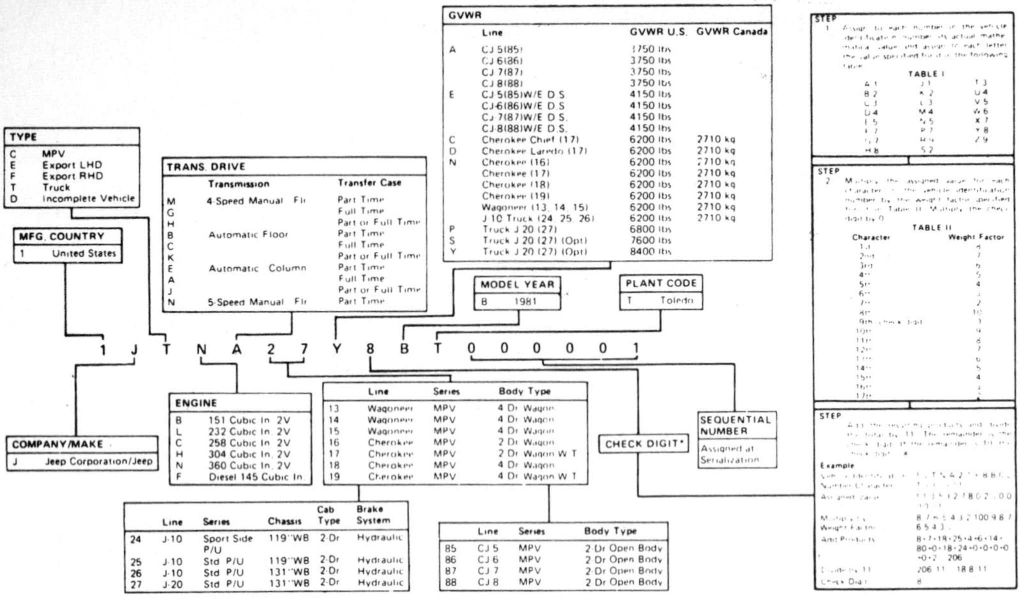

1981–86 VIN decoding chart

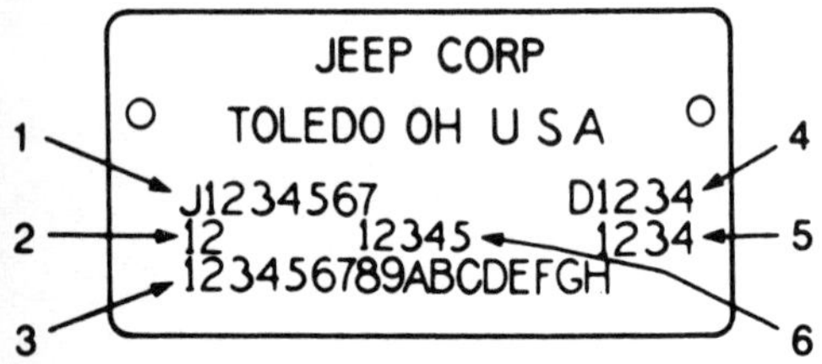

A metal identification plate is riveted to the driver side of the dash panel in the engine compartment.

1. Order number
2. Paint gun number
3. Vehicle identification number (VIN)
4. Vehicle deviation or special sales request and order (SSR & O)
5. Trim option number
6. Paint option number

1981–86 vehicle identification plate

It is sometimes necessary to machine oversize or undersize clearances for cylinder blocks and crankshafts. If your engine is equipped with oversized or undersized parts, it is necessary to order parts that will match the old parts. To find out if your engine is one with odd-sized parts, check the engine code letter or the engine code number itself-which in some cases is followed by a letter or a series of letters. The following chart explains just what the letters indicate:

● Letter A (10001-A) indicates 0.010 in. (0.25mm) undersized main and connecting rod bearings.
● Letter B (10001-B) indicates 0.010 in. (0.25mm) oversized cylinder bore.
● Letter AB (10001-AB) indicates the combination of A and B above.
● Letter C (10001-C) indicates 0.002 in. (0.051mm) undersized piston pin.
● Letter D (10001-D) indicates 0.010 in. (0.25mm) undersized main bearing journals.
● Letter E (10001-E) indicates 0.010 in. (0.25mm) undersized connecting rod bearing journals.

CJ-5 CJ-6, CJ-7 and Scrambler serial number plate location

4-150

The engine serial number for the American Motors built 4-150 is located on a machined pad at the rear right side of the block, just below the head.

Also on the block, just above the oil filter, is the oversized/undersized component code. The codes are explained as follows:

● Letter B: cylinder bores 0.010 in. (0.25mm) over
● Letter C: camshaft bearing bores 0.010 in. (0.25mm) over
● Letter M: main bearing journals 0.010 in. (0.25mm) under
● Letter P: connecting rod journals 0.010 in. (0.25mm) under

4-151

In 1980, a General Motors 151 cid, 4-cylinder engine became standard equipment on all CJ models. A three character code is stamped into the left rear top corner of the block. Additionally, engines built for sale in Georgia and Tennessee have a non-repeating number stamped into the left rear block flange.

6-225

The engine number for the Buick built V6-225 engine is located on the right side of the engine, on the crankcase, just below the head. The code is KLH. The codes RU and RV, included in the engine number of 1965 and 1966 engines, indicate manual or automatic transmission, respectively.

TRANSMISSION - TRANSFER CASE
E—3 SPEED AUTO. — PART TIME
L—5 SPEED MANUAL — PART TIME

TRIM TYPE AND GVWR
1—BASE 4001-5000 LBS.
    (1815-2268 kg.)
3—SPORT 4001-5000 LBS.
    (1815-2268 kg.)
4—LAREDO 4001-5000 LBS.
    (1815-2268 kg.)
J—BASE 3001-4000 LBS.
    (1361-1814 kg.)
K—SPORT 3001-4000 LBS.
    (1361-1814 kg.)
L—LAREDO 3001-4000 LBS.
    (1361-1814 kg.)

C-MPV

MANUFACTURING PLANT
B—BRAMPTON

MANUFACTURING COUNTRY
2—CANADA

2   B   C   C   E   8 1   K   X   H   B   500   001

CHECK DIGIT

MANUFACTURER
B—JEEP CORP.
CANADA

MODEL YEAR
H—1987

ENGINE TYPE
C—258 CUBIC IN.
    (4.2L), 2V, I-6
    AMC - GASOLINE

H—150 CUBIC IN.
    (2.46L), FUEL
    INJECTION I-4
    AMC - GASOLINE

SERIES
81—2 DR. MPV
    (WRANGLER/YJ-S)
82—2 DR. MPV
    (WRANGLER/YJ-L)

SERIAL NUMBER
SIX DIGIT SEQUENTIAL
SERIAL NUMBER

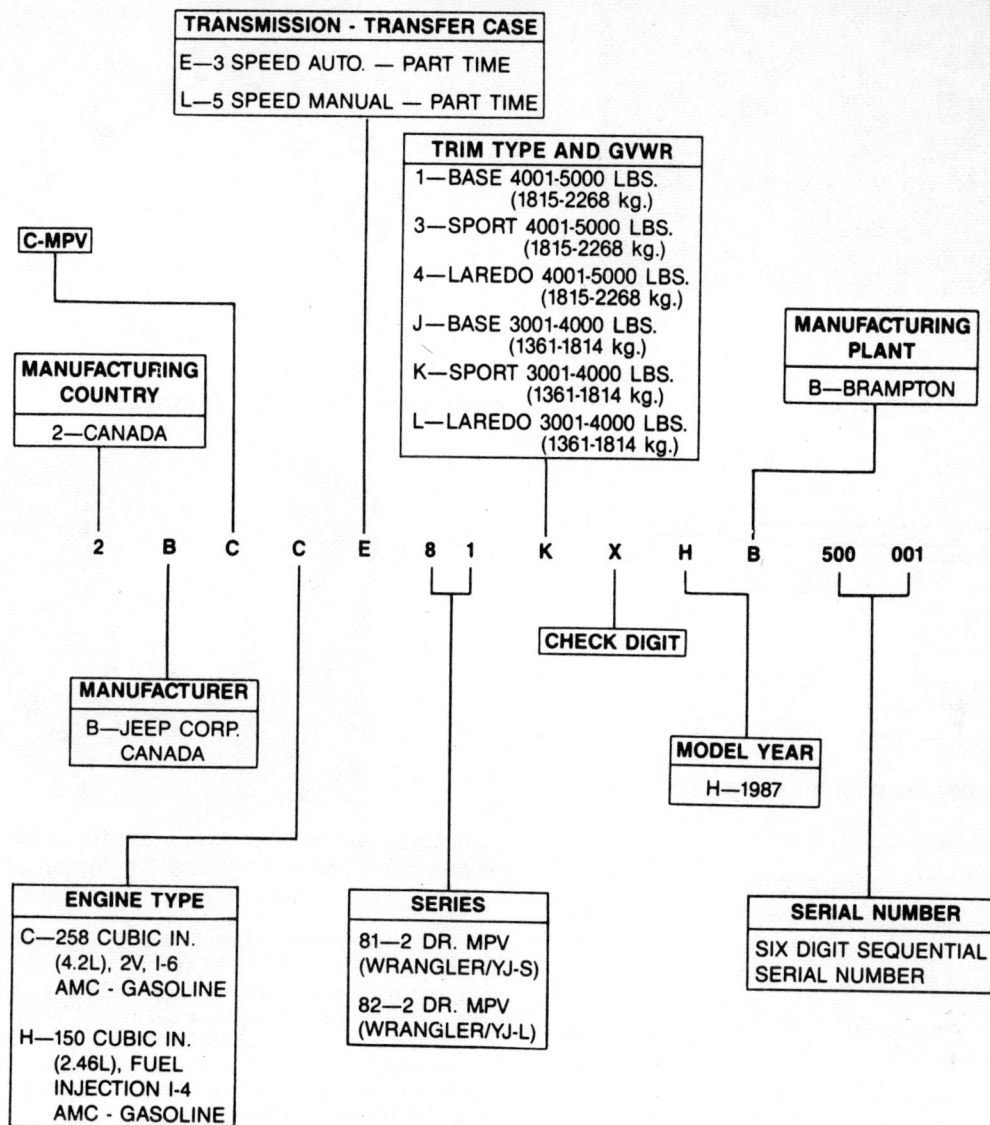

1987–90 VIN decoding chart

ENGINE APPLICATION

Engine	cu. In.	Actual Displacement cc	Liters	Type	Manufacturer	Years	Models
4-134	134.21	2,1990.0	2.2	F-Head	Kaiser	1971	CJ-5, CJ-6
4-150	150.45	2,465.4	2.5	OHV	Jeep	1984–90	CJ-5, CJ-7, Scrambler, Wrangler
4-151	150.78	2,470.8	2.5	OHV	Chevrolet	1980–83	CJ-5, CJ-7 Scrambler
6-225	225.29	3,691.8	3.7	OHV	Buick	1971	CJ-5, CJ-6
6-232	231.91	3,800.3	3.8	OHV	Jeep	1972–78	CJ-5, CJ-6
6-258	258.08	4,229.2	4.2	OHV	Jeep	1972–90	CJ-5, CJ-6, CJ-7, Scrambler, Wrangler
8-304	303.92	4,980.3	5.0	OHV	Jeep	1976–81	CJ-5, CJ-6

4–134 engine serial number

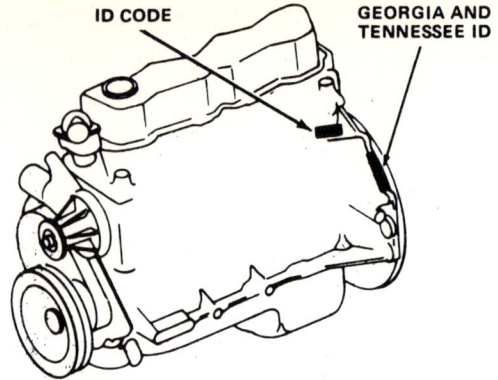

1980–83 4–151 engine ID number

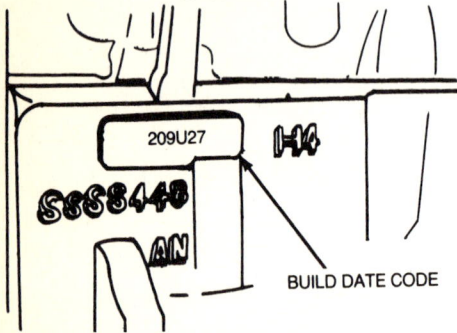

KENOSHA-BUILT
"E-1197277" OR
"W-1207177"

209U27

BUILD DATE CODE

Engine serial number location for the 4–150

The parts size letter code is on the boss directly above the oil filter on inline 6-cylinder engines

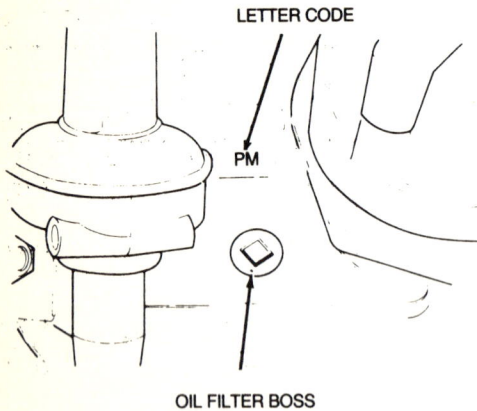

LETTER CODE

PM

OIL FILTER BOSS

4–150 oversized/undersized component code location

It is sometimes necessary to machine oversize or undersize clearances for cylinder blocks and crankshafts. If your engine is equipped with oversized or undersized parts, it is necessary to order parts that will match the old parts. To find out if your engine is one with odd-sized parts, check the engine code letter or the engine code number itself-which in some cases is followed by a letter or a series of letters. The following chart explains just what the letters indicate:

- Letter A (10001-A) indicates 0.010 in. (0.25mm) undersized main and connecting rod bearings.
- Letter B (10001-B) indicates 0.010 in. (0.25mm) oversized cylinder bore.
- Letter AB (l0001-AB) indicates the combination of A and B above.
- Letter C (10001-C) indicates 0.002 in. (0.051mm) undersized piston pin.
- Letter D (10001-D) indicates 0.010 in. (0.25mm) undersized main bearing journals.
- Letter E (10001-E) indicates 0.010 in. (0.25mm) undersized connecting rod bearing journals.

6-232 and 6-258

The engine code is found in the identification plate on the firewall. The second location is on a machined surface of the block between number 2 and 3 spark plugs. For further identification, the displacement is cast into the side of the block. The letter in the code identifies the engine by displacement (cu. in.), carburetor type and compression ratio.

All of the engines have the same undersize/oversize letter codes, located on the boss directly above the oil filter. The parts size code is as follows:

6–225 engine serial number location

6–232, 6–258 engine serial number location

- Letter B indicates 0.010 in. (0.25mm) oversized cylinder bore.
- Letter M indicates 0.010 in. (0.25mm) undersized main bearings.
- Letter P indicates 0.010 in. (0.25mm) undersized connecting rod bearings.
- Letter C indicates 0.010 in. (0.25mm) oversized camshaft block bores.

8-304

On the American Motors built 8-304 engines, the number is located on a tag attached to the right valve cover. For further identification, the displacement is cast into the side of the block. The letter in the code identifies the engine by displacement (cu. in.), carburetor type and compression ratio.

All of the engines made after 1971 have the same undersize/oversize letter codes, located on a tag next to the engine number. The parts size code is as follows:

- Letter B indicates 0.010 in. (0.25mm) oversized cylinder bore.
- Letter M indicates 0.010 in. (0.25mm) undersized main bearings.

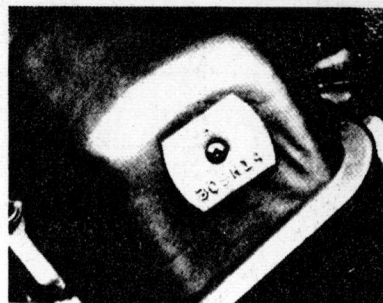

8–304 engine serial number location

- Letter P indicates 0.010 in. (0.25mm) undersized connecting rod bearings.
- Letters PM indicate a combination of the above specifications for P and M.
- Letter C indicates 0.010 in. (0.25mm) oversized camshaft block bores.

Transmission/Transfer Case/Axle Identification

There is a tag attached to the housing that identifies the manufacturer and model. It is necessary to have the information on this tag before ordering parts. When reassembling the unit, be sure that this tag is replaced on the case so identification can be made in the future.

In some cases, the transmission identification number may be embossed on the transmission housing.

MANUAL TRANSMISSION APPLICATION

Transmission Types	Years	Models
AISIN AX5 5-speed ①	1987–90	Standard on some models
Peugeot BA 10/5 5-speed ①	1987–90	Standard on some models
Tremec T-150 3-speed	1976–79	Standard on all models
Tremec T-176 4-speed	1980	All CJ-5 and CJ-7 with 8-304
	1981	Some CJ-5, CJ-7 with 6-258
		All models with 8-304
	1982–83	All models with 6-258
	1984–86	Standard on some models with 6-258
Warner SR-4 4-speed	1980	CJ-7 with 6-258
	1981	All CJ-5, CJ-7 with 4-151
		Some CJ-5 and CJ-7 with 6-258
Warner T-4 4-speed	1982–83	Standard on all models with 4-151
	1984–86	Standard on all models with 4-150
		Standard on some models with 6-258
Warner T-5 5-speed	1982–83	Optional on CJ-5 with 4-151
		Optional on CJ-7 and Scrambler with 6-258
	1984–86	Optional on all models
Warner T-14A 3-speed	1972–75	Standard on all models
Warner T-18 4-speed	1971–75	Optional on all models
Warner T-18A 4-speed	1976–79	Optional on all models
Warner T-86AA 3-speed	1971	Standard on all V6 models

MANUAL TRANSMISSION APPLICATION (Continued)

Transmission Types	Years	Models
Warner T-90C 3-speed	1971	Standard on all models
Warner T-98A 4-speed	1971	Optional on all models

①Which transmission is in your vehicle is determined by availability at the time of production.

TRANSFER CASE APPLICATION

Transfer Case Type	Years	Models
Dana 300	1980–86	All models
New Process NP-207	1987	All models
New Process NP-231	1988–90	All models
Spicer 18	1971	All models
Spicer 20	1972–79	All models
Warner Quadra-Trac®	1976–79	CJ-7 w/Auto. Trans.

AUTOMATIC TRANSMISSION APPLICATION

Transmission Type	Years	Models
Turbo Hydra-Matic 400	1976–79	CJ-7
Chrysler 904	1981	CJ-7 with 4-151
Chrysler 999	1980–81	CJ-7 with 6-258 and 8-304
	1982–86	CJ-7 and Scrambler with 6-258
	1987–90	Wranger with 6-258

FRONT DRIVE AXLE APPLICATION

Axle Type	Years	Models
Dana 27	1971	CJ-5, CJ-6
Dana 27A	1971	CJ-5, CJ-6
Dana 30	1971–90	All models

REAL AXLE APPLICATION

Axle Type	Years	Models
AMC 7⁹/₁₆ in. ring gear	1984–86	All models
AMC 8⁷/₈ in. ring gear	1973–83	All models
Dana 35C	1987–90	All models
Dana/Spicer 44	1971–72	CJ-5, CJ-6

ROUTINE MAINTENANCE

See the Maintenance Intervals Chart in this section for the recommended maintenance intervals for the components covered here.

Air Cleaner

SERVICE

Oil Bath Type

4-134

To service the oil bath type air cleaner on the 4-134, first unscrew the oil cup clamp and remove the oil cup from the cleaner body. Remove the oil from the cup and scrape out all the dirt inside, on the bottom. Wash the cup with a safe solvent. Refill the oil cup and replace it on the air cleaner body. Use the same viscosity of oil as you use in the engine crankcase.

To service the air cleaner body (less the oil cup), loosen the hose clamp and remove the hose form the cleaner. Detach the breather hose from the fitting on the cleaner. Remove the two wing nuts and lift the cleaner from the vehicle. Agitate the cleaner body thoroughly in a cleaning solution to clean the filtering element and then dry the element with compressed air. Reinstall the air cleaner body and replace the oil cup. The air cleaner should be serviced every 2,000 miles.

6-225

To service the oil bath type air cleaner on V6 engines, first remove the air cleaner from the engine by unscrewing the wing nut on top of the air cleaner. Remove the oil cup from the body of the air cleaner and remove all of the oil from the oil cup. Remove all of the dirt from the inside of the coil cup with a safe solvent. Wash the filter element in solvent, air dry it, and then fill the oil cup to the indicated level with clean oil. Assemble the air cleaner element to the oil cup, making sure that the gasket is in place between the two pieces. Mount the air cleaner assembly in the carburetor, making sure that the gasket between the air cleaner and the carburetor is in place and making a good seal. Secure the air cleaner to the carburetor with the wing nut.

Paper Element Type

Remove the wing nut or hex nuts on top of the cover. On 4-

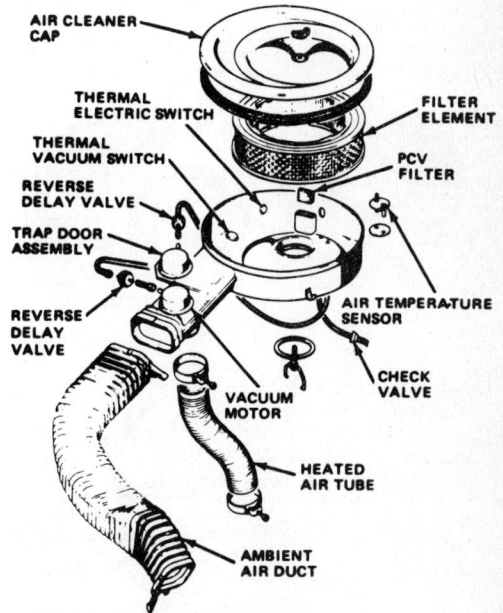

6–232, 6–258 air cleaner assemly

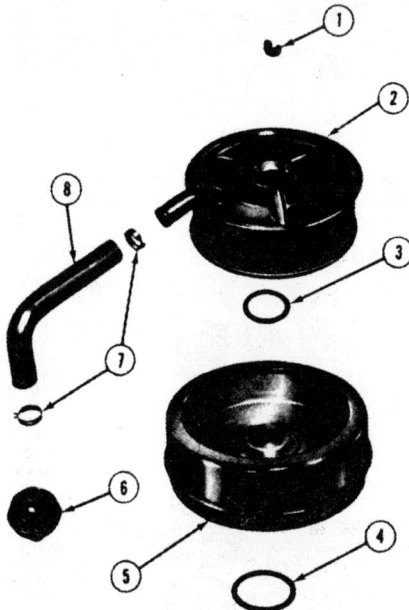

1. Wing nut
2. Cover
3. Rubber gasket
4. Cork gasket
5. Oil cup
6. Breather
7. Clamps
8. Vent tube

Exploded view of an oil bath air cleaner for the 6–225

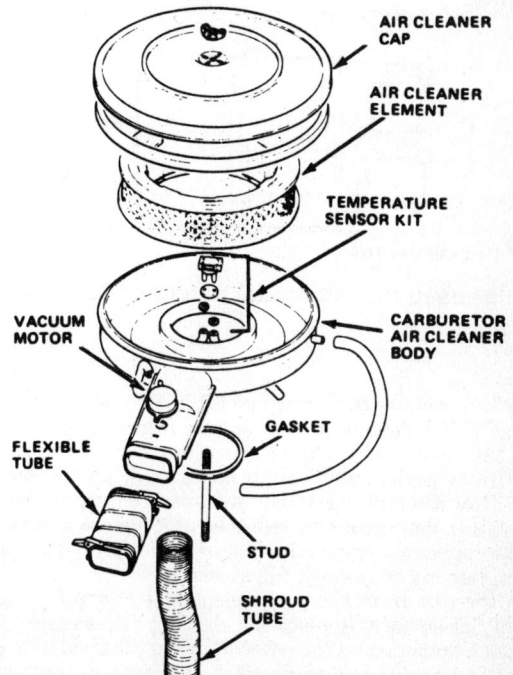

8-cylinder air cleaner assemly

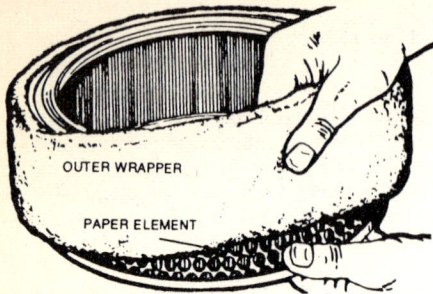

Polyurethane and paper element air cleaner

- OUTER WRAPPER
- PAPER ELEMENT

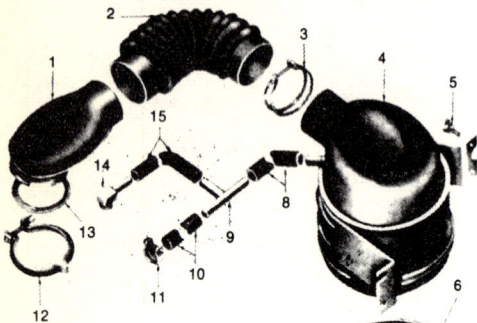

1. Horn
2. Flexible connector
3. Hose clamp
4. Body
5. Wing nut

11. Hose clamp
12. Clamp
13. Gasket
14. Elbow
15. Hose

Exploded view of an oil bath air cleaner for the 4–134

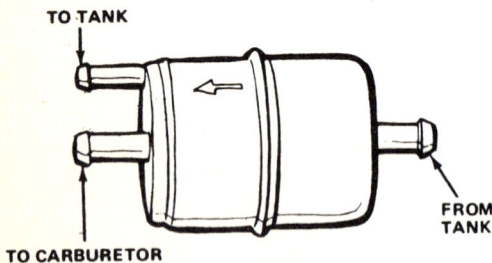

- TO TANK
- TO CARBURETOR
- FROM TANK

Fuel filter used on late model American Motors built engines

151 and V8s, remove the cover and lift out the element. On the 6-232 and 6-258, detach the rubber hose from the engine rocker arm (valve) cover and set the cover aside, being careful not to damage the large diameter hose or hoses to the air cleaner inlet.

If the filter element has a foam wrapper, remove the wrapper and wash it in detergent or a safe solvent. Squeeze and blot dry. Wet the wrapper in engine oil and squeeze it tightly in an absorbent towel or rag to remove the excess.

Clean the dirt from the paper element by rapping it gently against a flat surface. Replace the element as necessary.

Clean the housing and the cover. Replace the oiled wrapper, if any, on the element and reinstall the element in the housing, placing it 180° from its original position.

NOTE: The oiled foam wrapper element is a factory option for some years. It should be available through Jeep parts. There are also aftermarket variations on this, both dry and oiled.

Fuel Filter

REPLACEMENT

4-134
1984-86 4-150
6-225
6-232
6-258
8-304

All these engines have a throwaway cartridge filter in the line between the fuel pump and the carburetor. To replace it:
1. Remove the air cleaner as necessary.
2. Put an absorbent rag under the filter to catch spillage.
3. Remove the hose clamps.
4. Remove the filter and short attaching hoses.
5. Remove the hoses if they are to be reused.
6. Assemble the new filter and hoses.

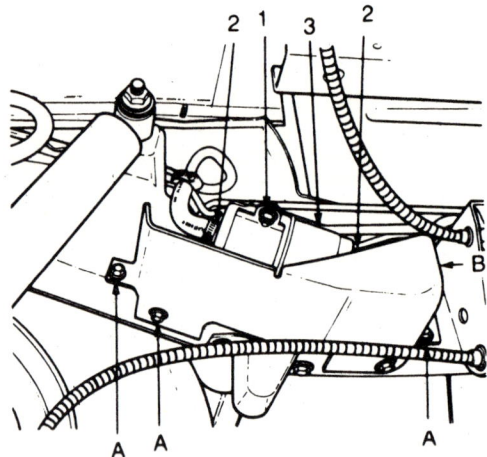

1. Filter retaining bolt
2. Hose clamps
3. Filter

a. Filter shield bolts
b. Filter shield

1987–90 4–150 fuel filter

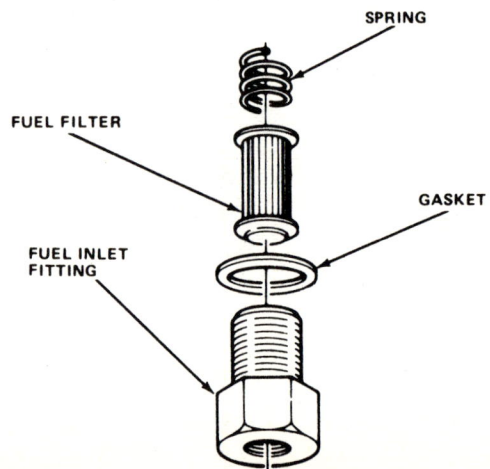

- SPRING
- FUEL FILTER
- GASKET
- FUEL INLET FITTING

4–151 fuel filter

NOTE: **The original equipment wire hose clamps should be replaced with screw type band clamps for the best results.**

7. 1975 V8 and all 1976 and later filters have two outlets. The extra one is to return fuel vapors and bubbles to the tank so as to prevent vapor lock. The tank line outlet must be up.

8. Install the filter, tighten the clamps, start the engine, and check for leaks. Discard the rag and old filter safely.

1987-90 4-150

The filter is located behind a protective shield on the left frame rail, just in front of the shock absorber.

--- CAUTION ---
Wear protective goggles to prevent fuel from spraying into your eyes. Have the new filter handy to install immediately.

1. Raise and support the rear end on jackstands.
2. Remove the protective shield bolts (A) and remove the shield.
3. Remove the filter retaining strap bolt, (1).
4. Clamp shut the hose on the inlet side of the filter to prevent fuel from draining once the filter is removed.
5. Remove the hose clamps, (2).
6. Remove the filter.
7. Installation is the reverse of removal.

4-151

The filter is located behind the large inlet nut in the carburetor. It is a small, paper, throwaway type.

--- CAUTION ---
DO NOT perform a filter change on a hot engine!

1. Place an absorbent rag under the carburetor inlet nut.
2. Hold the inlet nut with one wrench, while loosening the fuel line fitting with another. Remove the fuel line.
3. Unscrew the inlet nut and remove the nut, washer, spring and filter.
4. Installation is the reverse of removal. It's always best to use a new washer. Coat the threads of the inlet nut with a non-hardening, fuel-proof gasket cement. DO NOT OVERTIGHTEN THE INLET NUT! The threads are easily stripped in the carburetor. Hold the inlet nut with a wrench while tightening the fuel line fitting.

PCV Valve

The PCV valve, which is the heart of the positive crankcase ventilation system, should be free of dirt and residue and in working order. As long as the valve is kept clean and is not showing signs of becoming damaged or gummed up, it should work properly. When the valve cannot be cleaned sufficiently or becomes sticky and will not operate freely, it should be replaced.

The PCV valve is used to control the rate at which crankcase vapors are returned to the intake manifold. The action of the valve plunger is controlled by intake manifold vacuum and the spring. During deceleration and idle, when manifold vacuum is high, it overcomes the tension of the valve spring and the plunger bottoms in the manifold end of the valve housing. Because of the valve construction, it reduces, but dies not stop, the passage of vapors to the intake manifold. When the engine is lightly accelerated or operated at constant speed, spring tension matches intake manifold vacuum pull and the plunger takes a mid-position in the valve body, allowing more vapors to flow into the manifold.

SERVICE

An inoperative PCV system will cause rough idling, sludge and

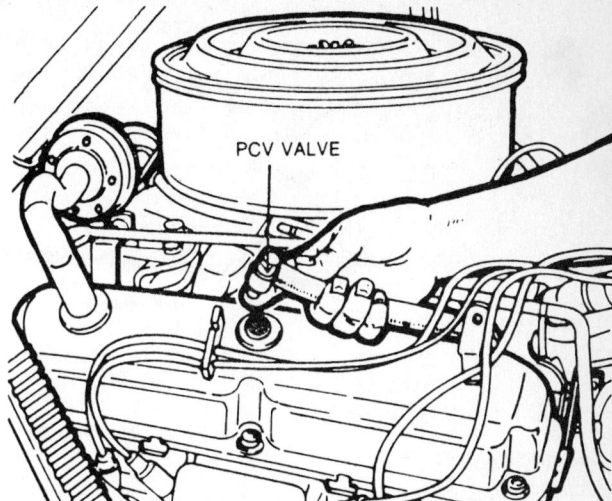

Removing the PCV valve on 6–225 engines

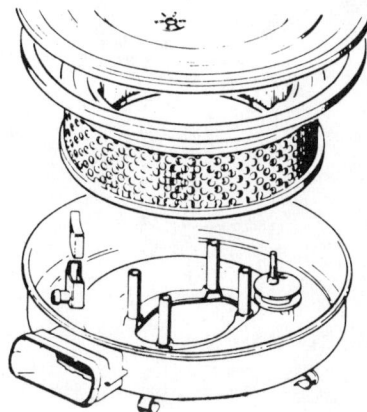

6–232, 6–258 PCV air filter

oil dilution. In the event erratic idle, never attempt to compensate by disconnecting the PCV system. Disconnecting the PCV system will adversely affect engine ventilation. It could also shorten engine life through the buildup of sludge.

On 4-151 and V8 engines, the air being drawn into the PCV system passes through a polyurethane foam filter located in the oil filler cap. The filler cap is vented only by a hose connected to the air cleaner. The foam filter in the oil filler cap should be cleaned with safe solvent.

The PCV valve is in the right rocker arm (valve) cover on the V6, in the intake manifold on the 4-cylinder, in the intake manifold behind the carburetor on the V8, and in the rocker arm cover on the 6-232 and 6-258.

To inspect the PCV valve, proceed as follows:

1. With the engine idling, remove the PCV valve from the rocker cover. If the valve is not plugged, a hissing sound will be heard. A strong vacuum should be felt when you place your finger over the valve.
2. Reinstall the PCV valve and allow about a minute for pressure to drop.
3. Remove the crankcase intake air cleaner. Cover the opening in the rocker cover with a piece of stiff paper. The paper should be sucked against the opening with noticeable force.
4. With the engine stopped, remove the PCV valve and shake it. A rattle or clicking should be heard to indicate that the valve is free.

PCV valve location on the 6-232, 6-258

Typical V8 PCV system

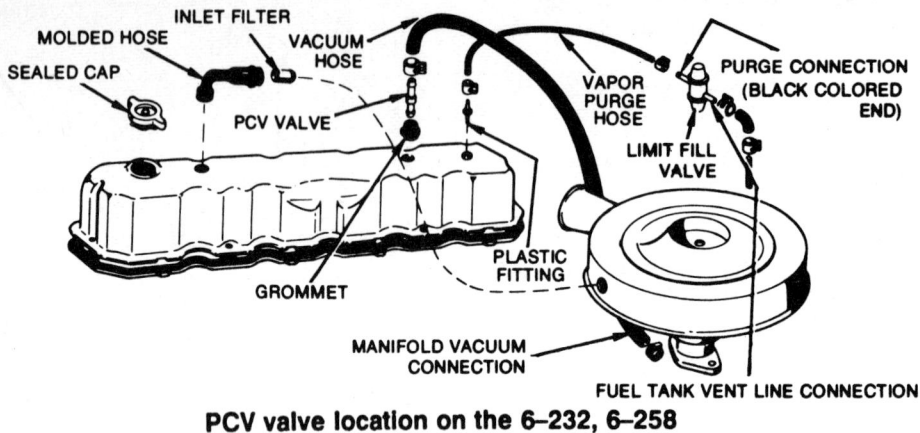

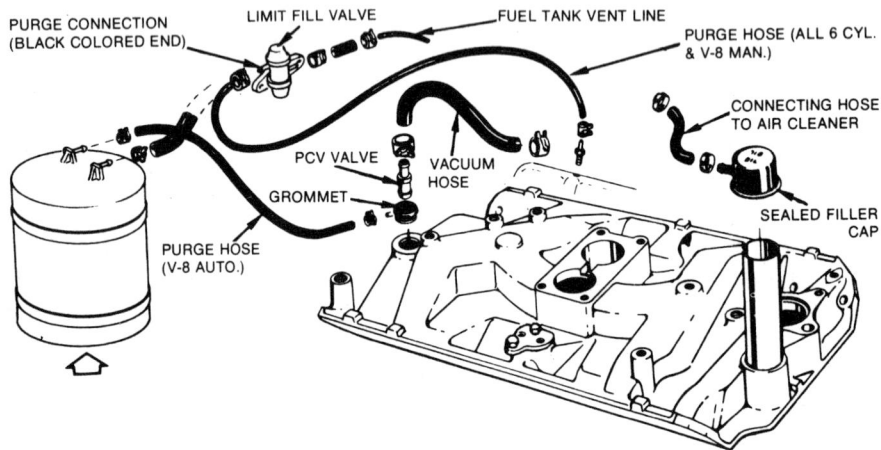

5. If the system meets the tests in Steps 1, 2, 3, and 4 (above), no further service is required, unless replacement is specified in the Maintenance Intervals Chart. If the system does not meet the tests, the valve should be replaced with a new one.

NOTE: Do not attempt to clean a PCV valve.

6. With a new PCV valve installed, if the paper is not sucked against the crankcase air intake opening (see Step 2), it will be necessary to clean the PCV valve hose and the passage in the lower part of the carburetor.

7. Clean the line with Combustion Chamber Conditioner or similar solvent. Do not leave the hoses in solvent for more than ½ hour. Allow the line to air dry.

8. Remove the carburetor or throttle body and HAND turn a ¼″ drill through the passages to dislodge solid particles and blow clean.

NOTE: It is not necessary to disassembly the carburetor for this operation. If necessary, use a smaller drill, so that no metal is removed.

9. After checking and/or servicing the Crankcase Ventilation System, any components that do not allow passage or air to the intake manifold should be replaced.

Heat Riser

The heat riser is a thermostatically operated valve in the exhaust manifold. It closes when the engine is cold, to direct hot exhaust gases to the intake manifold, in order to preheat the incoming fuel/air mixture. If it sticks closed, the result will be a rough idle after the engine warms up. If it sticks open, there will be frequent stalling during warmup, especially in cold and damp weather.

On the V6 and V8, the valve is between the exhaust manifold and the exhaust pipe. On the 6-232 and 6-258, it is an integral part of the exhaust manifold. The heat riser counterweight should move freely. If it sticks, apply Jeep Heat Valve Lubricant or something similar (engine cool) to the ends of the shaft. Sometimes rapping the end of the shaft sharply with a hammer (engine hot) will break it loose. If this fails, parts must be removed for repair or replacement.

Evaporative Canister

All models equipped with V8s, and 1973 and later 4- and 6-cylinder engines, have fuel evaporative emission control systems which include an evaporative storage canister. The purpose of this charcoal canister is to store gasoline vapors until they can be drawn into the engine and burned along with the air/fuel mixture. The air filter in the bottom of the canister should be replaced every 15,000 miles.

Battery

MAINTENANCE

Loose, dirty, or corroded battery terminals are a major cause

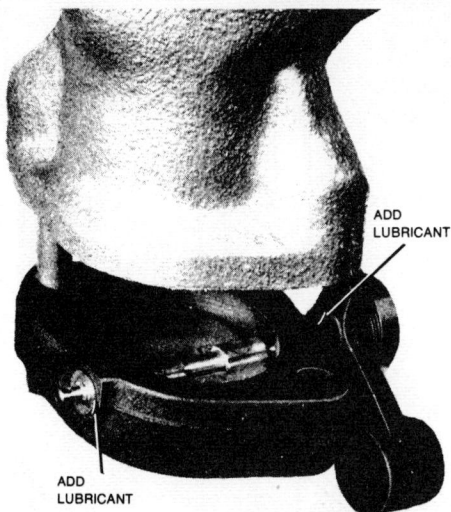

ADD LUBRICANT

ADD LUBRICANT

ADD LUBRICANT

ADD LUBRICANT

ADD LUBRICANT

V-8 ENGINE

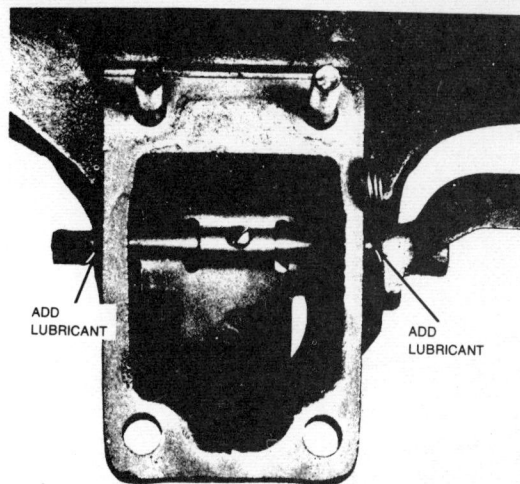

SIX-CYLINDER ENGINE

Heat riser lubrication points (shown detached)

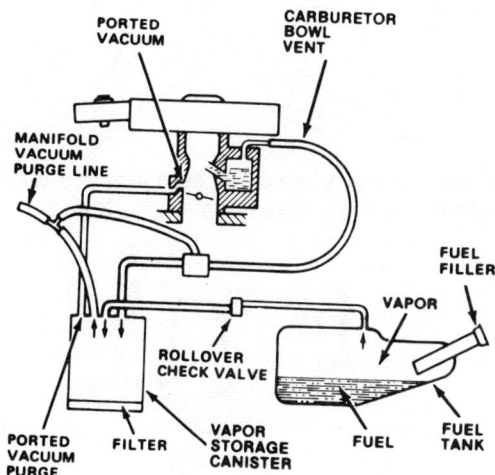

PORTED VACUUM

CARBURETOR BOWL VENT

MANIFOLD VACUUM PURGE LINE

FUEL FILLER

VAPOR

ROLLOVER CHECK VALVE

PORTED VACUUM PURGE

FILTER

VAPOR STORAGE CANISTER

FUEL

FUEL TANK

Typial fuel vapor control system

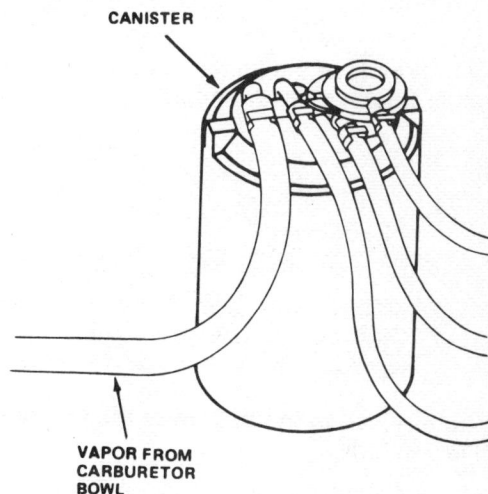

CANISTER

VAPOR FROM CARBURETOR BOWL

Fuel vapor storage canister and hoses

of "no-start." Every 3 months or so, remove the battery terminals and clean them, giving them a light coating of petroleum jelly when you are finished. This will help to retard corrosion.

Check the battery cables for signs of wear or chafing and replace any cable or terminal that looks marginal. Battery terminals can be easily cleaned and inexpensive terminal cleaning tools are an excellent investment that will pay for themselves many times over. They can usually be purchased from any well-equipped auto store or parts department. Side terminal batteries require a different tool to clean the threads in the battery case. The accumulated white powder and corrosion can be cleaned from the top of the battery with an old toothbrush and a solution of baking soda and water.

Unless you have a maintenance-free battery, check the electrolyte level (see Battery under Fluid Level Checks in this section) and check the specific gravity of each cell. Be sure that the vent holes in each cell cap are not blocked by grease or dirt. The vent holes allow hydrogen gas, formed by the chemical reaction in the battery, to escape safely.

REPLACEMENT BATTERIES

The cold power rating of a battery measures battery starting

performance and provides an approximate relationship between battery size and engine size. The cold power rating of a replacement battery should match or exceed your engine size in cubic inches.

ELECTROLYTE LEVEL

The correct level should be at the bottom of the well inside each cell opening. The surface of the electrolyte should appear distorted, not flat. Only colorless, odorless, preferably distilled, water should be added. It is a good idea to add the water with a squeeze bulb to avoid splashing and spills. If water is frequently needed, the most likely cause is overcharging, caused by voltage regulator problems. If any acid should escape, it can be neutralized with a baking soda and water solution.

Avoid sparks and smoking around the battery! It gives off explosive hydrogen gas. If you get acid on your skin or eyes, rinse it off immediately with lots of water. See a doctor immediately if you get acid in your eyes! In winter, add water only before driving to prevent the battery from freezing and cracking.

NOTE: Original equipment batteries with the ganged caps are often chronically wet on top, causing a lot of

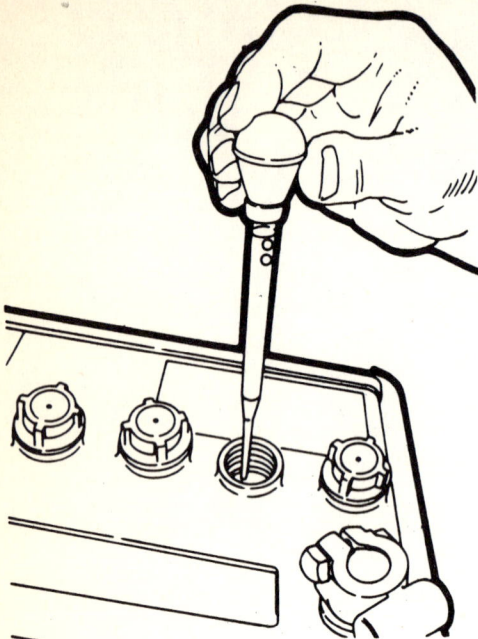

The specific gravity of the battery can be checked with a simple float-type hydrometer

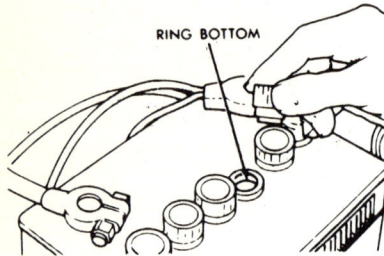

RING BOTTOM

Fill each battery cell to the bottom of the splitring with distilled water

Battery State of Charge at Room Temperature

Specific Gravity Reading	Charged Condition
1.260–1.280	Fully Charged
1.230–1.250	¾ Charged
1.200–1.220	½ Charged
1.170–1.190	¼ Charged
1.140–1.160	Almost no Charge
1.110–1.130	No Charge

Special tools are available for cleaning the terminals and cable clamps on side terminal batteries

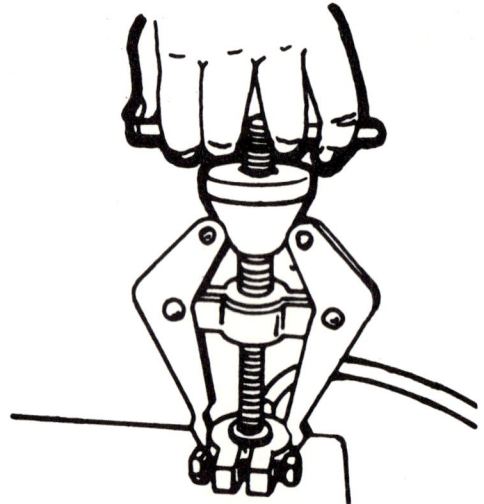

Use a small puller to remove the battery cables

The terminals can be cleaned with a stiff wire brush or with a terminal cleaner made for the purpose. These are inexpensive and can be purchased in most any decently equipped parts store.

corrosion in the battery tray. The problem is insufficient venting. Solve it by removing the caps and drilling a tiny vent hole for each cell through the top of the cap.

If water is added during freezing weather, the truck should be driven several miles to allow the water to mix with the electrolyte. Otherwise the battery could freeze.

At least once a year check the specific gravity of the battery. It should be between 1.20-1.26 in. at room temperature. Clean and tighten the terminal clamps and apply a thin coating of petroleum jelly to the terminals. This will help to retard corrosion.

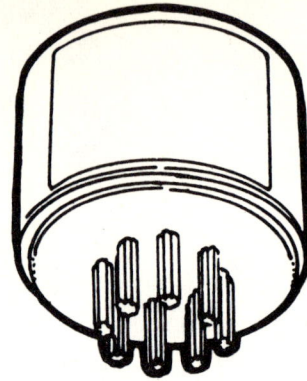

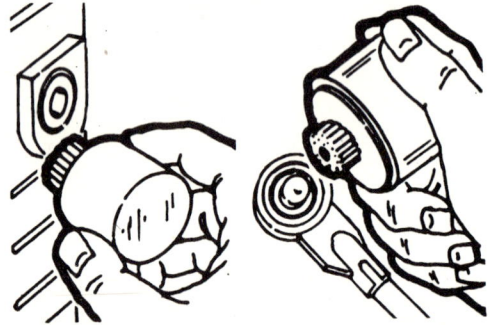

SPECIFIC GRAVITY (EXCEPT MAINTENANCE FREE BATTERIES)

At least once a year, check the specific gravity of the battery. It should be between 1.20 in.Hg and 1.26 in.Hg at room temperature.

The specific gravity can be check with the use of an hydrome-

CABLES AND CLAMPS

Once a year, the battery terminals and the cable clamps should be cleaned. Loosen the clamps and remove the cables, negative cable first. On batteries with posts on top, the use of a puller specially made for the purpose is recommended. These are inexpensive, and available in auto parts stores. Side terminal battery cables are secured with a bolt.

Clean the cable lamps and the battery terminal with a wire brush, until all corrosion, grease, etc., is removed and the metal is shiny. It is especially important to clean the inside of the clamp thoroughly, since a small deposit of foreign material or oxidation there will prevent a sound electrical connection and inhibit either starting or charging. Special tools are available for cleaning these parts, one type for conventional batteries and another type for side terminal batteries.

Before installing the cables, loosen the battery holddown clamp or strap, remove the battery and check the battery tray. Clear it of any debris, and check it for soundness. Rust should be wire brushed away, and the metal given a coat of anti-rust paint. Replace the battery and tighten the holddown clamp or strap securely, but be careful not to overtighten, which will crack the battery case.

After the clamps and terminals are clean, reinstall the cables, negative cable last; do not hammer on the clamps to install. Tighten the clamps securely, but do not distort them. Give the clamps and terminals a thin external coat of grease after installation, to retard corrosion.

Check the cables at the same time that the terminals are cleaned. If the cable insulation is cracked or broken, or if the ends are frayed, the cable should be replaced with a new cable of the same length and gauge.

CAUTION

Keep flame or sparks away from the battery; it gives off explosive hydrogen gas. Battery electrolyte contains sulphuric acid. If you should splash any on your skin or in your eyes, flush the affected area with plenty of clear water. If it lands in your eyes, get medical help immediately.

CHARGING

Battery chargers are relatively inexpensive, easy to use and available at any retail outlet that sell auto parts. For the average vehicle owner, a 6 amp charger is more than sufficient. Always follow the manufacturer's instructions when using the charger.

CAUTION

Keep flame or sparks away from the battery! It gives off explosive hydrogen gas, while it is being charged.

Belts

Once a year or at 12,000 mile intervals, the tension (and condition) of the alternator, power steering (if so equipped), air conditioning (if so equipped), and Thermactor air pump drive belts should be checked, and, if necessary, adjusted. Loose accessory drive belts can lead to poor engine cooling and diminish alternator, power steering pump, air conditioning compressor or air pump output. A belt that is too tight places a severe strain on the water pump, alternator, power steering pump, compressor or air pump bearings.

Replace any belt that is so glazed, worn or stretched that it cannot be tightened sufficiently.

NOTE: The material used in late model drive belts is such that the belts do not show wear. Replace belts at least every three years.

On vehicles with matched belts, replace both belts. New ½", ⅜" and ¹⁵⁄₃₂" wide belts are to be adjusted to a tension of 140 lbs.; ¼" wide belts are adjusted to 80 lbs., measured on a belt tension

Cleaning the inside of the cable end

Cleaning the battery terminal

ter, an inexpensive instrument available from many sources, including auto parts stores. The hydrometer has a squeeze bulb at one end and a nozzle at the other. Battery electrolyte is sucked into the hydrometer until the float is lifted from its seat. The specific gravity is then read by noting the position of the float. Generally, if after charging, the specific gravity between any two cells varies more than 50 points (0.50), the battery is bad and should be replaced.

It is not possible to check the specific gravity in this manner on sealed (maintenance free) batteries. Instead, the indicator built into the top of the case must be relied on to display any signs of battery deterioration. If the indicator is dark, the battery can be assumed to be OK. If the indicator is light, the specific gravity is low, and the battery should be charged or replaced.

JUMP STARTING A DEAD BATTERY

The chemical reaction in a battery produces explosive hydrogen gas. This is the safe way to jump start a dead battery, reducing the chances of an accidental spark that could cause an explosion.

Jump Starting Precautions

1. Be sure both batteries are of the same voltage.
2. Be sure both batteries are of the same polarity (have the same grounded terminal).
3. Be sure the vehicles are not touching.
4. Be sure the vent cap holes are not obstructed.
5. Do not smoke or allow sparks around the battery.
6. In cold weather, check for frozen electrolyte in the battery. Do not jump start a frozen battery.
7. Do not allow electrolyte on your skin or clothing.
8. Be sure the electrolyte is not frozen.

CAUTION: Make certin that the ignition key, in the vehicle with the dead battery, is in the OFF position. Connecting cables to vehicles with on-board computers will result in computer destruction if the key is not in the OFF position.

Jump Starting Procedure

1. Determine voltages of the two batteries; they must be the same.
2. Bring the starting vehicle close (they must not touch) so that the batteries can be reached easily.
3. Turn off all accessories and both engines. Put both vehicles in Neutral or Park and set the handbrake.
4. Cover the cell caps with a rag—do not cover terminals.
5. If the terminals on the run-down battery are heavily corroded, clean them.
6. Identify the positive and negative posts on both batteries and connect the cables in the order shown.
7. Start the engine of the starting vehicle and run it at fast idle. Try to start the car with the dead battery. Crank it for no more than 10 seconds at a time and let it cool for 20 seconds in between tries.
8. If it doesn't start in 3 tries, there is something else wrong.
9. Disconnect the cables in the reverse order.
10. Replace the cell covers and dispose of the rags.

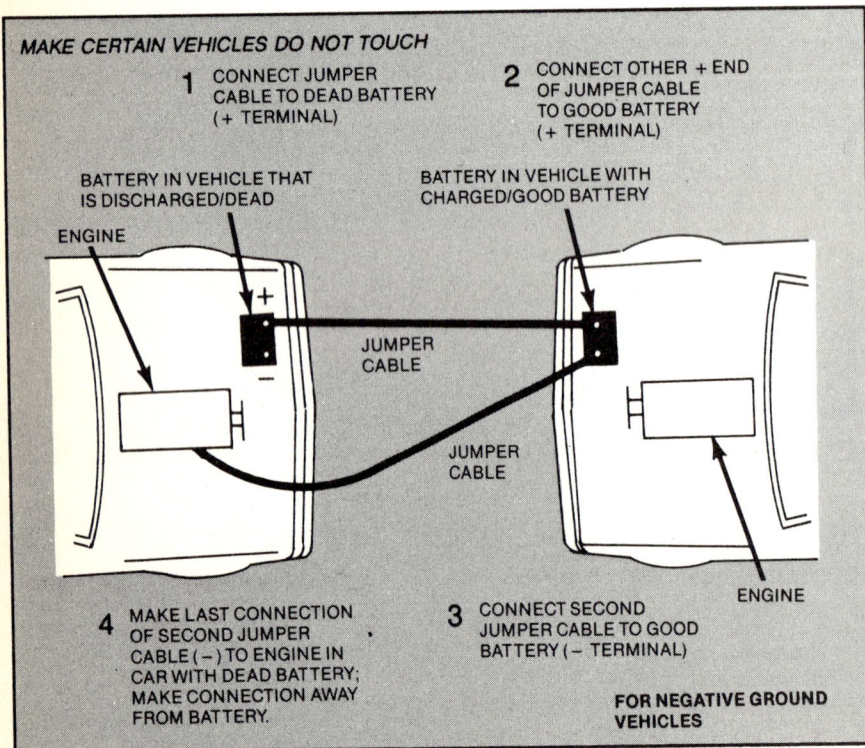

MAKE CERTAIN VEHICLES DO NOT TOUCH

1 CONNECT JUMPER CABLE TO DEAD BATTERY (+ TERMINAL)

2 CONNECT OTHER + END OF JUMPER CABLE TO GOOD BATTERY (+ TERMINAL)

BATTERY IN VEHICLE THAT IS DISCHARGED/DEAD

ENGINE

BATTERY IN VEHICLE WITH CHARGED/GOOD BATTERY

JUMPER CABLE

JUMPER CABLE

ENGINE

4 MAKE LAST CONNECTION OF SECOND JUMPER CABLE (−) TO ENGINE IN CAR WITH DEAD BATTERY; MAKE CONNECTION AWAY FROM BATTERY.

3 CONNECT SECOND JUMPER CABLE TO GOOD BATTERY (− TERMINAL)

FOR NEGATIVE GROUND VEHICLES

Side terminal batteries occasionally pose a problem when connecting jumper cables. There frequently isn't enough room to clamp the cables without touching sheet metal. Side terminal adaptors are available to alleviate this problem and should be removed after use

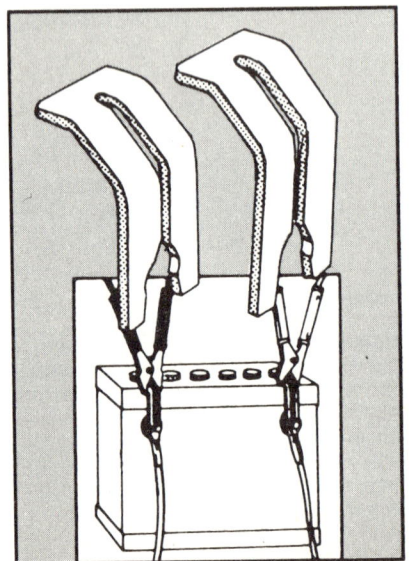

HOW TO SPOT WORN V-BELTS

V–Belts are vital to efficient engine operation—they drive the fan, water pump and other accessories. They require little maintenance (occasional tightening) but they will not last forever. Slipping or failure of the V–belt will lead to overheating. If your V–belt looks like any of these, it should be replaced.

Cracking or Weathering

This belt has deep cracks, which cause it to flex. Too much flexing leads to heat build–up and premature failure. These cracks can be caused by using the belt on a pulley that is too small. Notched belts are available for small diameter pulleys.

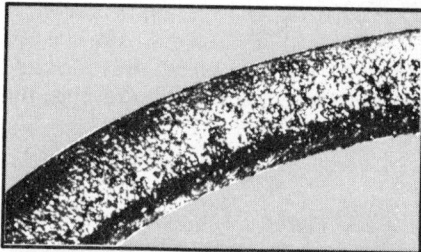

Softening (Grease and Oil)

Oil and grease on a belt can cause the belt's rubber compounds to soften and separate from the reinforcing cords that hold the belt together. The belt will first slip, then finally fail altogether.

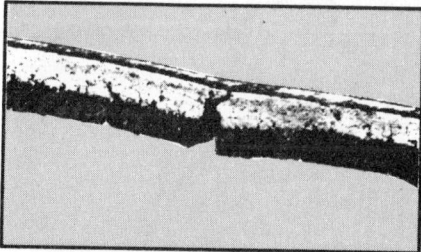

Glazing

Glazing is caused by a belt that is slipping. A slipping belt can cause a run-down battery, erratic power steering, overheating or poor accessory performance. The more the belt slips, the more glazing will be built up on the surface of the belt. The more the belt is glazed, the more it will slip. If the glazing is light, tighten the belt.

Worn Cover

The cover of this belt is worn off and is peeling away. The reinforcing cords will begin to wear and the belt will shortly break. When the belt cover wears in spots or has a rough jagged appearance, check the pulley grooves for roughness.

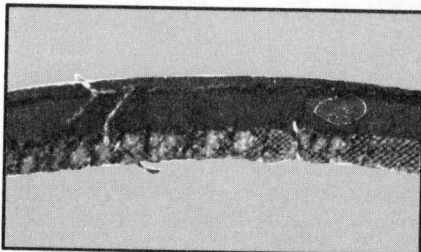

Separation

This belt is on the verge of breaking and leaving you stranded. The layers of the belt are separating and the reinforcing cords are exposed. It's just a matter of time before it breaks completely.

gauge. Any belt that has been operating for a minimum of 10 minutes is considered a used belt. In the first 10 minutes, the belt should stretch to its maximum extent. After 10 minutes, stop the engine and recheck the belt tension. Belt tension for a used belt should be maintained at 110 lbs. (all except ¼" wide belts) or 60 lbs. (¼" wide belts). If a belt tension gauge is not available, the following procedures may be used.

ADJUSTMENTS FOR ALL V-BELTS

— CAUTION —

On models equipped with an electric cooling fan, disconnect the negative battery cable or fan motor wiring harness connector before replacing or adjusting drive belts. The fan may come on, under certain circumstances, even though the ignition is off.

Alternator (Fan Drive) Belt

1. Position the ruler perpendicular to the drive belt at its longest straight run. Test the tightness of the belt by pressing it firmly with your thumb. The deflection should not exceed ¼".
2. If the deflection exceeds ¼", loosen the alternator mounting and adjusting arm bolts.
3. Place a 1" open-end or adjustable wrench on the adjusting ridge cast on the body, and pull on the wrench until the proper tension is achieved.
4. Holding the alternator in place to maintain tension, tighten the adjusting arm bolt. Recheck the belt tension. When the belt is properly tensioned, tighten the alternator mounting bolt.

Power Steering Drive Belt

1. Hold a ruler perpendicularly to the drive belt at its longest run, test the tightness of the belt by pressing it firmly with your thumb. The deflection should not exceed ¼".
2. To adjust the belt tension, loosen the adjusting and mounting bolts on the front face of the steering pump cover plate (hub side).
3. Using a pry bar or broom handle on the pump hub, move the power steering pump toward or away from the engine until the proper tension is reached. Do not pry against the reservoir as it is relatively soft and easily deformed.
4. Holding the pump in place, tighten the adjusting arm bolt and then recheck the belt tension. When the belt is properly tensioned tighten the mounting bolts.

Air Conditioning Compressor Drive Belt

1. Position a ruler perpendicular to the drive belt at its longest run. Test the tightness of the belt by pressing it firmly with your thumb. The deflection should not exceed ¼".
2. If the engine is equipped with an idler pulley, loosen the idler pulley adjusting bolt, insert a pry bar between the pulley and the engine (or in the idler pulley adjusting slot), and adjust the tension accordingly. If the engine is not equipped with an idler pulley, the alternator must be moved to accomplish this adjustment, as outlined under Alternator (Fan Drive) Belt.
3. When the proper tension is reached, tighten the idler pulley adjusting bolt (if so equipped) or the alternator adjusting and mounting bolts.

Air Pump Drive Belt

1. Position a ruler perpendicular to the drive belt at its longest run. Test the tightness of the belt by pressing it firmly with your thumb. The deflection should be about ¼".
2. To adjust the belt tension, loosen the adjusting arm bolt slightly. If necessary, also loosen the mounting belt slightly.
3. Using a pry bar or broom handle, pry against the pump rear cover to move the pump toward or away from the engine as necessary.

— CAUTION —

Do not pry against the pump housing itself, as damage to the housing may result.

4. Holding the pump in place, tighten the adjusting arm bolt and recheck the tension. When the belt is properly tensioned, tighten the mounting bolt.

SERPENTINE (SINGLE) DRIVE BELT MODELS

Some models feature a single, wide, ribbed V-belt that drives

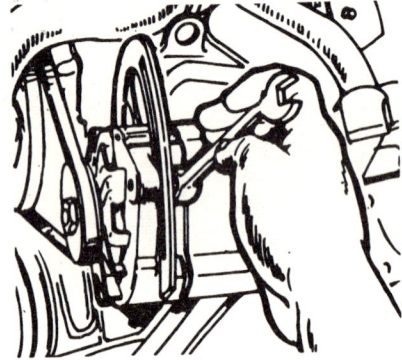

To adjust belt tension or to replace belts, first loosen the component's mounting and adjusing bolts slightly

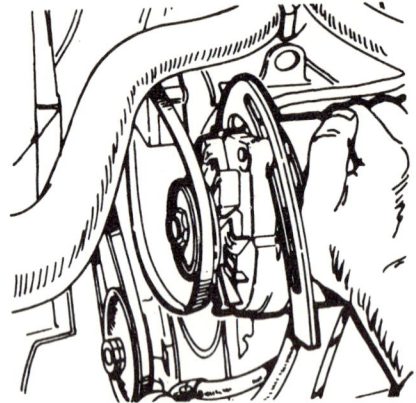

Push the component toward the engine and slip off the belt

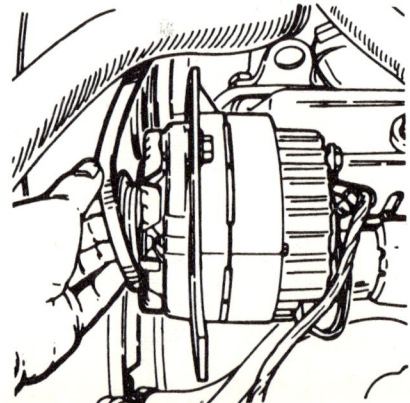

Slip the new belt over the pulley

the water pump, alternator, and (on some models) the air conditioner compressor. To install a new belt, loosen the bracket lock bolt, retract the belt tensioner with a pry bar and slide the old belt off of the pulleys. Slip on a new belt and release the tensioner and tighten the lock bolt. The spring powered tensioner eliminates the need for periodic adjustments.

WARNING: Check to make sure that the V-ribbed belt is located properly in all drive pulleys before applying tensioner pressure.

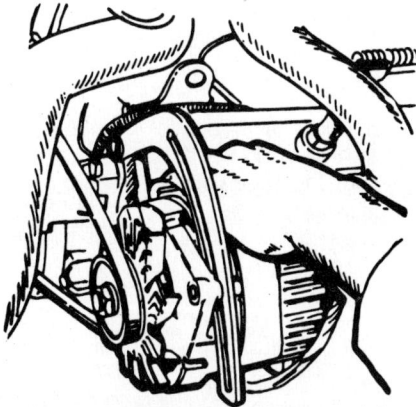

Pull outward on the component and tighten the mounting bolts

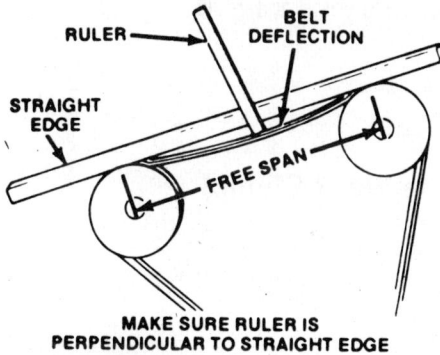

Measuring belt deflection

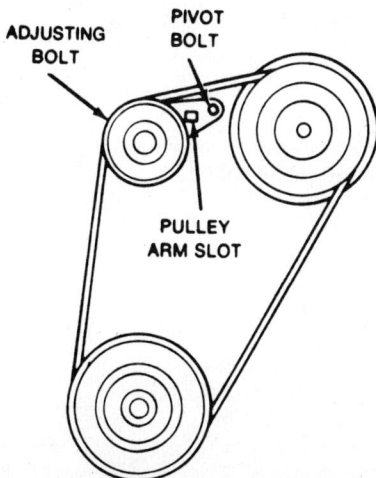

Some pulleys have a rectangular slot to aid in moving the accessories to be tightened

Hoses

REMOVAL AND INSTALLATION

Radiator hoses are generally of two constructions, the preformed (molded) type, which is custom made for a particular application, and the spring-loaded type, which is made to fit several different applications. Heater hoses are all of the same general construction.

REPLACEMENT

Inspect the condition of the radiator and heater hoses periodically. Early spring and at the beginning of the fall or winter, when you are performing other maintenance, are good times. Make sure the engine and cooling system are cold. Visually inspect for cracking, rotting or collapsed hoses, replace as necessary. Run your hand along the length of the hose. If a weak or swollen spot is noted when squeezing the hose wall, replace the hose.

1. Drain the cooling system into a suitable container (if the coolant is to be reused).

――――――――― **CAUTION** ―――――――――

When draining the coolant, keep in mind that cats and dogs are attracted by the ethylene glycol antifreeze, and are quite likely to drink any that is left in an uncovered container or in puddles on the ground. This will prove fatal in sufficient quantity. Always drain the coolant into a sealable container. Coolant should be reused unless it is contaminated or several years old.

――――――――――――――――――――――――――――

2. Loosen the hose clamps at each end of the hose that requires replacement.
3. Twist, pull and slide the hose off the radiator, water pump, thermostat or heater connection.
4. Clean the hose mounting connections. Position the hose clamps on the new hose.
5. Coat the connection surfaces with a water resistant sealer and slide the hose into position. Make sure the hose clamps are located beyond the raised bead of the connector (if equipped) and centered in the clamping area of the connection.
6. Tighten the clamps to 20-30 in. lbs. Do not overtighten.
7. Fill the cooling system.
8. Start the engine and allow it to reach normal operating temperature. Check for leaks.

Windshield Wipers

For maximum effectiveness and longest element life, the windshield and wiper blades should be kept clean. Dirt, tree sap, road tar and so on will cause streaking, smearing and blade deterioration if left on the windshield. It is advisable to wash the windshield carefully with a commercial glass cleaner at least once a month. Wipe off the rubber blades with a wet rag afterwards. Do not attempt to move the wipers back and forth by hand! Damage to the motor and drive mechanism will result.

If the blades are found to be cracked, broken or torn they should be replaced immediately. Replacement intervals will vary with usage, although ozone deterioration usually limits blade lift to about one year. If the wiper pattern is smeared or streaked, or if the blade chatters across the glass, the blades should be replaced. It is easiest and most sensible to replace them in pairs.

There are basically three different types of wiper blade refills, which differ in their method of replacement. One type has two release buttons, approximately ⅓ of the way up from the ends of the blade frame. Pushing the buttons down releases a lock and

HOW TO SPOT BAD HOSES

Both the upper and lower radiator hoses are called upon to perform difficult jobs in an inhospitable environment. They are subject to nearly 18 psi at under hood temperatures often over 280°F, and must circulate nearly 7500 gallons of coolant an hour—3 good reasons to have good hoses.

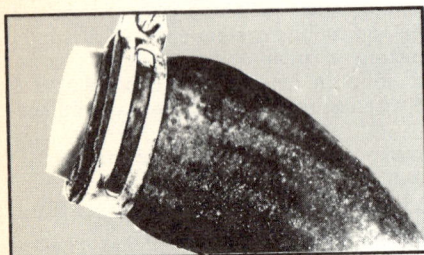

Swollen Hose

A good test for any hose is to feel it for soft or spongy spots. Frequently these will appear as swollen areas of the hose. The most likely cause is oil soaking. This hose could burst at any time, when hot or under pressure.

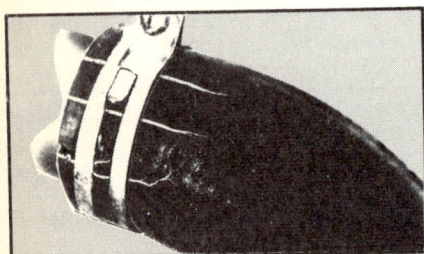

Cracked Hose

Cracked hoses can usually be seen but feel the hoses to be sure they have not hardened; a prime cause of cracking. This hose has cracked down to the reinforcing cords and could split at any of the cracks.

Frayed Hose End (Due to Weak Clamp)

Weakened clamps frequently are the cause of hose and cooling system failure. The connection between the pipe and hose has deteriorated enough to allow coolant to escape when the engine is hot.

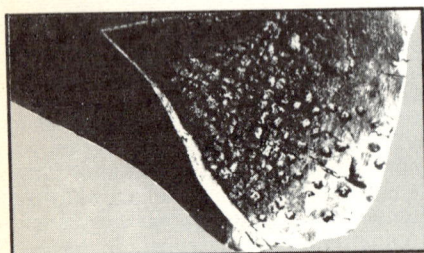

Debris In Cooling System

Debris, rust and scale in the cooling system can cause the inside of a hose to weaken. This can usually be felt on the outside of the hose as soft or thinner areas.

TRICO

BLADE FRAME LEVER

RUBBER BLADE ELEMENT ASSY.

SQUEEZE SIDES OF RETAINER

LEVER JAWS

LATCH LOCK RELEASE

METAL BACKING IS WIDER

RETAINING TABS

HOLD FRAME FROM TWISTING

METAL BACKING STRIP

METAL BACKING STRIP

FRAME

INSERT SCREWDRIVER BEHIND TAB AND PUSH HANDLE DOWN.

ANCO

LATCH-PIN

YOKE JAWS

RUBBER BLADE ELEMENT ASSY.

YOKE JAWS

POLYCARBONATE

UNLOCKED

LOCKED

TRIDON

PLASTIC BACKING STRIP

NOTCH

FRAME

PULL UP & TWIST

PRESSURE DOWN

RUBBER BLADE

RETAINING TABS

FIRM SURFACE

16

16.5

THE LENGTH OF THE 16" AND 16.5" TRIDON BLADES ARE MOLDED IN EACH END. REPLACE ONLY WITH IDENTICAL BLADES OR REFILLS.

FRAME

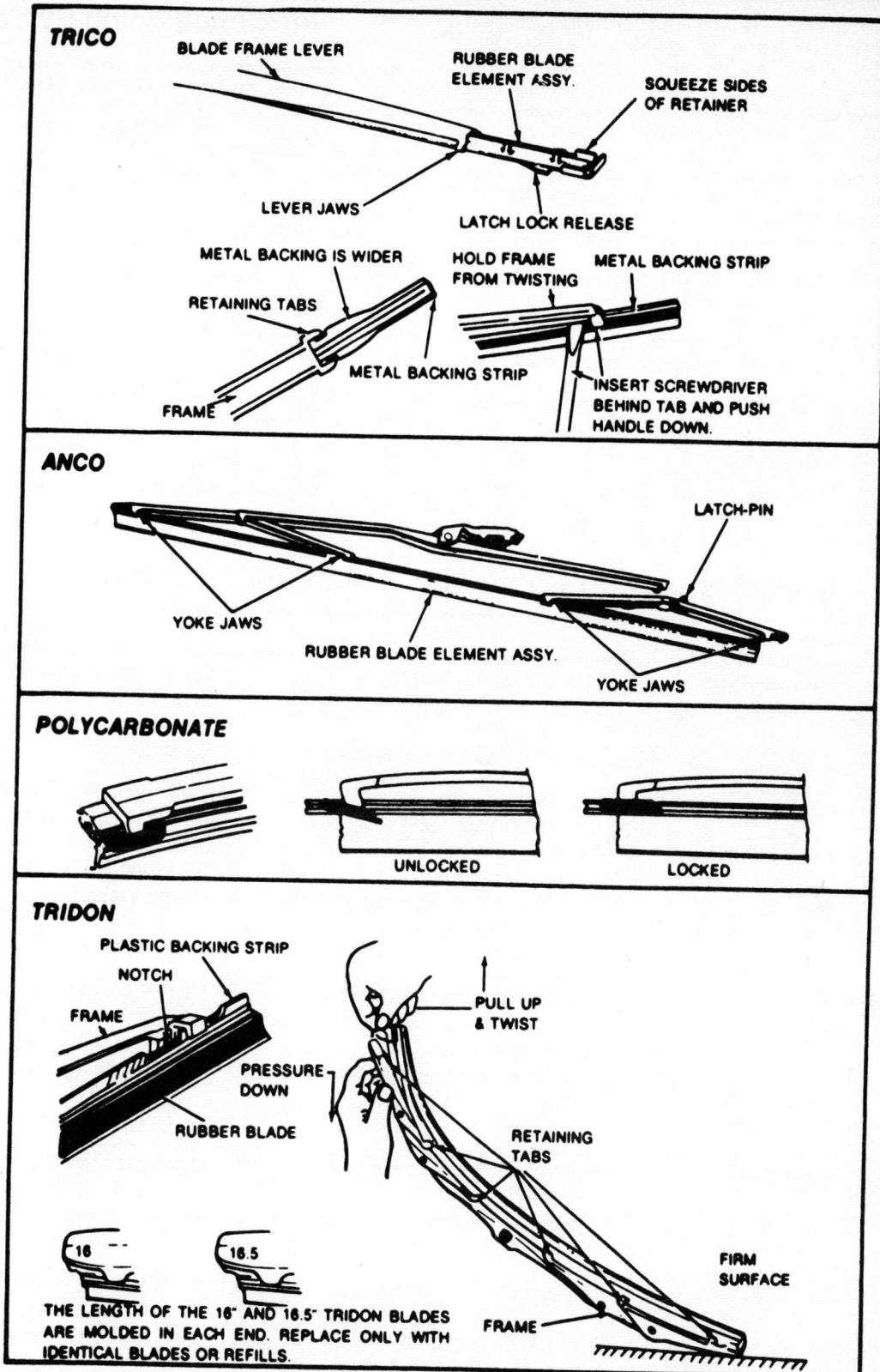

Popular styles of wiper refills

allows the rubber blade to be removed from the frame. The new blade slides back into the frame and locks in place.

The second type of refill has two metal tabs which are unlocked by squeezing them together. The rubber blade can then be withdrawn from the frame jaws. A new one is installed by inserting it into the front frame jaws and sliding it rearward to engage the remaining frame jaws. There are usually four jaws. Be certain when installing that the refill is engaged in all of them. At the end of its travel, the tabs will lock into place on the front jaws of the wiper blade frame.

The third type is a refill made from polycarbonate. The refill has a simple locking device at one end which flexes downward out of the groove into which the jaws of the holder fit, allowing easy release. By sliding the new refill through all the jaws and pushing through the slight resistance when it reaches the end of its travel, the refill will lock into position.

Regardless of the type of refill used, make sure that all of the frame jaws are engaged as the refill is pushed into place and locked. The metal blade holder and frame will scratch the glass if allowed to touch it.

Air Conditioning System

Air conditioning was first offered on CJ models in 1977. This system remained unchanged through 1980. The compressor used was the vertical, 2-cylinder Tecumseh model.

For the 1981 model year, all Jeep vehicles built for sale in California, and equipped with the 6-258 engine, utilized a Japanese made Sankyo, 5-cylinder axial compressor. All non-California 6-258 engines and all 8-304 engines utilized the Tecumseh compressor.

In 1982, the Sankyo became the only compressor used on CJ models.

All systems utilize a sight glass for system inspection.

NOTE: This book contains simple testing and charging procedures for your Jeep's air conditioning system. More comprehensive testing, diagnosis and service procedures may be found in CHILTON'S GUIDE TO AIR CONDITIONING SERVICE AND REPAIR, book part number 7580, available at your local retailer.

Troubleshooting Basic Air Conditioning Problems

Problem	Cause	Solution
There's little or no air coming from the vents (and you're sure it's on)	• The A/C fuse is blown • Broken or loose wires or connections • The on/off switch is defective	• Check and/or replace fuse • Check and/or repair connections • Replace switch
The air coming from the vents is not cool enough	• Windows and air vent wings open • The compressor belt is slipping • Heater is on • Condenser is clogged with debris • Refrigerant has escaped through a leak in the system • Receiver/drier is plugged	• Close windows and vent wings • Tighten or replace compressor belt • Shut heater off • Clean the condenser • Check system • Service system
The air has an odor	• Vacuum system is disrupted • Odor producing substances on the evaporator case • Condensation has collected in the bottom of the evaporator housing	• Have the system checked/repaired • Clean the evaporator case • Clean the evaporator housing drains
System is noisy or vibrating	• Compressor belt or mountings loose • Air in the system	• Tighten or replace belt; tighten mounting bolts • Have the system serviced

Sight glass condition
  Constant bubbles, foam or oil streaks
  Clear sight glass, but no cold air
  Clear sight glass, but air is cold
  Clouded with milky fluid
	• Undercharged system
	
	• No refrigerant at all
	• System is OK
	• Receiver drier is leaking dessicant
		• Charge the system
		
		• Check and charge the system
		
		• Have system checked

Large difference in temperature of lines	• System undercharged	• Charge and leak test the system

Troubleshooting Basic Air Conditioning Problems (cont.)

Problem	Cause	Solution
Compressor noise	• Broken valves • Overcharged • Incorrect oil level • Piston slap • Broken rings • Drive belt pulley bolts are loose	• Replace the valve plate • Discharge, evacuate and install the correct charge • Isolate the compressor and check the oil level. Correct as necessary. • Replace the compressor • Replace the compressor • Tighten with the correct torque specification
Excessive vibration	• Incorrect belt tension • Clutch loose • Overcharged • Pulley is misaligned	• Adjust the belt tension • Tighten the clutch • Discharge, evacuate and install the correct charge • Align the pulley
Condensation dripping in the passenger compartment	• Drain hose plugged or improperly positioned • Insulation removed or improperly installed	• Clean the drain hose and check for proper installation • Replace the insulation on the expansion valve and hoses
Frozen evaporator coil	• Faulty thermostat • Thermostat capillary tube improperly installed • Thermostat not adjusted properly	• Replace the thermostat • Install the capillary tube correctly • Adjust the thermostat
Low side low—high side low	• System refrigerant is low • Expansion valve is restricted	• Evacuate, leak test and charge the system • Replace the expansion valve
Low side high—high side low	• Internal leak in the compressor—worn	• Remove the compressor cylinder head and inspect the compressor. Replace the valve plate assembly if necessary. If the compressor pistons, rings or
Low side high—high side low (cont.)	 • Cylinder head gasket is leaking • Expansion valve is defective • Drive belt slipping	cylinders are excessively worn or scored replace the compressor • Install a replacement cylinder head gasket • Replace the expansion valve • Adjust the belt tension
Low side high—high side high	• Condenser fins obstructed • Air in the system • Expansion valve is defective • Loose or worn fan belts	• Clean the condenser fins • Evacuate, leak test and charge the system • Replace the expansion valve • Adjust or replace the belts as necessary
Low side low—high side high	• Expansion valve is defective • Restriction in the refrigerant hose	• Replace the expansion valve • Check the hose for kinks—replace if necessary

Troubleshooting Basic Air Conditioning Problems (cont.)

Problem	Cause	Solution
Low side low—high side high	• Restriction in the receiver/drier • Restriction in the condenser	• Replace the receiver/drier • Replace the condenser
Low side and high normal (inadequate cooling)	• Air in the system • Moisture in the system	• Evacuate, leak test and charge the system • Evacuate, leak test and charge the system

GENERAL SERVICING PROCEDURES

The most important aspect of air conditioning service is the maintenance of pure and adequate charge of refrigerant in the system. A refrigeration system cannot function properly if a significant percentage of the charge is lost. Leaks are common because the severe vibration encountered in an automobile can easily cause a sufficient cracking or loosening of the air conditioning fittings. As a result, the extreme operating pressures of the system force refrigerant out.

The problem can be understood by considering what happens to the system as it is operated with a continuous leak. Because the expansion valve regulates the flow of refrigerant to the evaporator, the level of refrigerant there is fairly constant. The receiver-drier stores any excess of refrigerant, and so a loss will first appear there as a reduction in the level of liquid. As this level nears the bottom of the vessel, some refrigerant vapor bubbles will begin to appear in the stream of liquid supplied to the expansion valve. This vapor decreases the capacity of the expansion valve very little as the valve opens to compensate for its presence. As the quantity of liquid in the condenser decreases, the operating pressure will drop there and throughout the high side of the system. As the R-12 continues to be expelled, the pressure available to force the liquid through the expansion valve will continue to decrease, and, eventually, the valve's orifice will prove to be too much of a restriction for adequate flow even with the needle fully withdrawn.

At this point, low side pressure will start to drop, and severe reduction in cooling capacity, marked by freeze-up of the evaporator coil, will result. Eventually, the operating pressure of the evaporator will be lower than the pressure of the atmosphere surrounding it, and air will be drawn into the system wherever there are leaks in the low side.

Because all atmospheric air contains at least some moisture, water will enter the system and mix with the R-12 and the oil. Trace amounts of moisture will cause sludging of the oil, and corrosion of the system. Saturation and clogging of the filter-drier, and freezing of the expansion valve orifice will eventually result. As air fills the system to a greater and greater extend, it will interfere more and more with the normal flows of refrigerant and heat.

A list of general precautions that should be observed while doing this follows:

1. Keep all tools as clean and dry as possible.
2. Thoroughly purge the service gauges and hoses of air and moisture before connecting them to the system. Keep them capped when not in use.
3. Thoroughly clean any refrigerant fitting before disconnecting it, in order to minimize the entrance of dirt into the system.
4. Plan any operation that requires opening the system beforehand in order to minimize the length of time it will be exposed to open air. Cap or seal the open ends to minimize the entrance of foreign material.
5. When adding oil, pour it through an extremely clean and dry tube or funnel. Keep the oil capped whenever possible. Do not use oil that has not been kept tightly sealed.
6. Use only refrigerant 12. Purchase refrigerant intended for use in only automotive air conditioning system. Avoid the use of refrigerant 12 that may be packaged for another use, such as cleaning, or powering a horn, as it is impure.
7. Completely evacuate any system that has been opened to replace a component, other than when isolating the compressor, or that has leaked sufficiently to draw in moisture and air. This requires evacuating air and moisture with a good vacuum pump for at least one hour.

NOTE: If a system has been open for a considerable length of time it may be advisable to evacuate the system for up to 12 hours (overnight).

8. Use a wrench on both halves of a fitting that is to be disconnected, so as to avoid placing torque on any of the refrigerant lines.

Antifreeze

In order to prevent heater core freeze-up during A/C operation, it is necessary to maintain permanent type antifreeze protection of $+15°F$ ($-9°C$) or lower. A reading of $-15°F$ ($-26°C$) is ideal since this protection also supplies sufficient corrosion inhibitors for the protection of the engine cooling system.

NOTE: The same antifreeze should not be used longer than the manufacturer specified.

Radiator Cap

For efficient operation of an air conditioned car's cooling system, the radiator cap should have a holding pressure which meets manufacturer's specifications. A cap which fails to hold these pressure should be replaced.

Condenser

Any obstruction of or damage to the condenser configuration will restrict the air flow which is essential to its efficient operation. It is therefore, a good rule to keep this unit clean and in proper physical shape.

NOTE: Bug screens are regarded as obstructions.

Condensation Drain Tube

This single molded drain tube expels the condensation, which accumulates on the bottom of the evaporator housing, into the engine compartment.

If this tube is obstructed, the air conditioning performance can be restricted and condensation buildup can spill over onto the vehicle's floor.

SAFETY PRECAUTIONS

Because of the importance of the necessary safety precautions

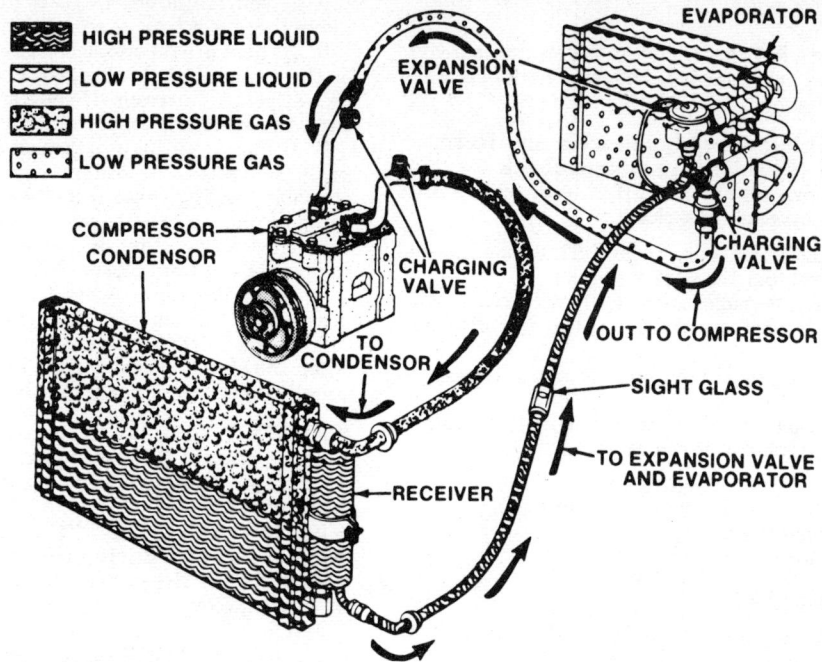

HIGH PRESSURE LIQUID
LOW PRESSURE LIQUID
HIGH PRESSURE GAS
LOW PRESSURE GAS

EVAPORATOR
EXPANSION VALVE
COMPRESSOR CONDENSOR
CHARGING VALVE
CHARGING VALVE
TO CONDENSOR
OUT TO COMPRESSOR
SIGHT GLASS
TO EXPANSION VALVE AND EVAPORATOR
RECEIVER

Basic air conditioning system components and flow diagram, showing the Tecumseh 2-cylinder compressor

that must be exercised when working with air conditioning systems and R-12 refrigerant, a recap of the safety precautions are outlined.

1. Avoid contact with a charged refrigeration system, even when working on another part of the air conditioning system or vehicle. If a heavy tool comes into contact with a section of copper tubing or a heat exchanger, it can easily cause the relatively soft material to rupture.

2. When it is necessary to apply force to a fitting which contains refrigerant, as when checking that all system couplings are securely tightened, use a wrench on both parts of the fitting involved, if possible. This will avoid putting torque on refrigerant tubing. (It is advisable, when possible, to use tube or line wrenches when tightening these flare nut fittings.).

3. Do not attempt to discharge the system by merely loosening a fitting, or removing the service valve caps and cracking these valves. Precise control is possibly only when using the service gauges. Place a rag under the open end of the center charging hose while discharging the system to catch any drops of liquid that might escape. Wear protective gloves when connecting or disconnecting service gauge hoses.

4. Discharge the system only in a well ventilated area, as high concentrations of the gas can exclude oxygen and act as an anesthesia. When leak testing or soldering, this is particularly important, as toxic gas is formed when R-12 contacts any flame.

5. Never start a system without first verifying that both service valves are backseated, if equipped, and that all fittings are throughout the system are snugly connected.

6. Avoid applying heat to any refrigerant line or storage vessel. Charging may be aided by using water heated to less than 125°F (52°C) to warm the refrigerant container. Never allow a refrigerant storage container to sit out in the sun, or near any other source of heat, such as a radiator.

7. Always wear goggles when working on a system to protect the eyes. If refrigerant contacts the eye, it is advisable in all cases to see a physician as soon as possible.

8. Frostbite from liquid refrigerant should be treated by first gradually warming the area with cool water, and then gently applying petroleum jelly. A physician should be consulted.

9. Always keep refrigerant can fittings capped when not in use. Avoid sudden shock to the can which might occur from dropping it, or from banging a heavy tool against it. Never carry a can in the passenger compartment of a car.

10. Always completely discharge the system before painting the vehicle (if the paint is to be baked on), or before welding anywhere near the refrigerant lines.

TEST GAUGES

Most of the service work performed in air conditioning requires the use of a set of two gauges, one for the high (head) pressure side of the system, the other for the low (suction) side.

The low side gauge records both pressure and vacuum. Vacuum readings are calibrated from 0 to 30 inches and the pressure graduations read from 0 to no less than 60 psi.

The high side gauge measures pressure from 0 to at last 600 psi.

Both gauges are threaded into a manifold that contains two hand shut-off valves. Proper manipulation of these valves and the use of the attached test hoses allow the user to perform the following services:

1. Test high and low side pressures.
2. Remove air, moisture, and contaminated refrigerant.
3. Purge the system (of refrigerant).
4. Charge the system (with refrigerant).

The manifold valves are designed so that they have no direct effect on gauge readings, but serve only to provide for, or cut off, flow of refrigerant through the manifold. During all testing and hook-up operations, the valves are kept in a close position to avoid disturbing the refrigeration system. The valves are opened only to purge the system or refrigerant or to charge it.

INSPECTION

The compressed refrigerant used in the air conditioning system expands into the atmosphere at a temperature of −21.7°F (−29.8°C) or lower. This will freeze any surface, including your eyes, that it contacts. In addition, the refrigerant decomposes into a poisonous gas in the presence of a flame. Do not open or disconnect any part of the air conditioning system.

Sight Glass Check

You can safely make a few simple checks to determine if your air conditioning system needs service. The tests work best if the temperature is warm (about 70°F [21.1°C]).

NOTE: If your vehicle is equipped with an aftermarket air conditioner, the following system check may not apply. You should contact the manufacturer of the unit for instructions on systems checks.

1. Place the automatic transmission in Park or the manual transmission in Neutral. Set the parking brake.
2. Run the engine at a fast idle (about 1,500 rpm) either with the help of a friend or by temporarily readjusting the idle speed screw.
3. Set the controls for maximum cold with the blower on High.
4. Locate the sight glass in one of the system lines. Usually it is on the left alongside the top of the radiator.

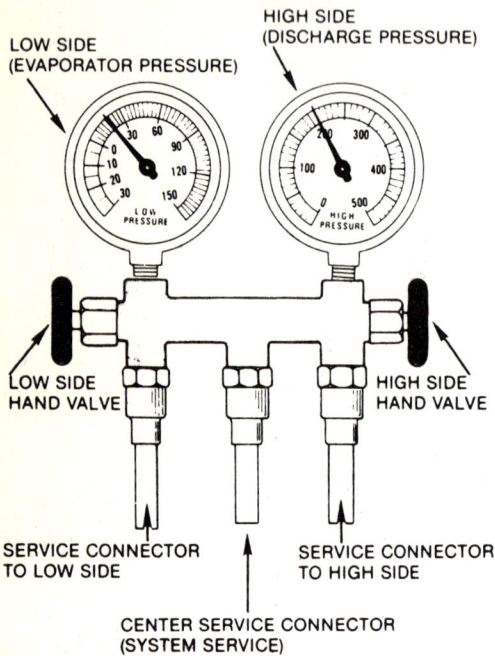

Typical manifold gauge set

5. If you see bubbles, the system must be recharged. Very likely there is a leak at some point.
6. If there are no bubbles, there is either no refrigerant at all or the system is fully charged. Feel the two hoses going to the belt-driven compressor. If they are both at the same temperature, the system is empty and must be recharged.
7. If one hose (high-pressure) is warm and the other (low-pressure) is cold, the system may be all right. However, you are probably making these tests because you think there is something wrong, so proceed to the next step.
8. Have an assistant in the car turn the fan control on and off to operate the compressor clutch. Watch the sight glass.
9. If bubbles appear when the clutch is disengaged and disappear when it is engaged, the system is properly charged.
10. If the refrigerant takes more than 45 seconds to bubble when the clutch is disengaged, the system is overcharged. This usually causes poor cooling at low speeds.

WARNING: If it is determined that the system has a leak, it should be corrected as soon as possible. Leaks may allow moisture to enter and cause a very expensive rust problem.

Exercise the air conditioner for a few minutes, every two weeks or so, during the cold months. This avoids the possibility of the compressor seals drying out from lack of lubrication.

TESTING THE SYSTEM

1. Connect a gauge set.
2. Close (clockwise) both gauge set valves.
3. Mid-position both service valves.
4. Park the Jeep in the shade. Start the engine, set the parking brake, place the transmission in NEUTRAL and establish an idle of 1,500 rpm.
5. Run the air conditioning system for full cooling, but NOT in the MAX or COLD mode.
6. Insert a thermometer into the center air outlet.
7. Use the accompanying performance chart for a specifications reference. If pressures are abnormal, refer to the accompanying Pressure Diagnosis Chart.

ISOLATING THE COMPRESSOR

It is not necessary to discharge the system for compressor removal. The compressor can be isolated from the rest of the system, eliminating the need for recharging.

1. Connect a manifold gauge set.

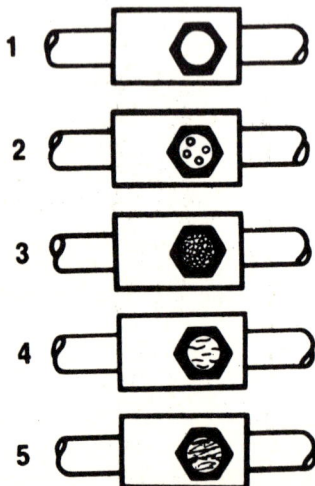

1 Clear sight glass — system correctly charged or overcharged

2 Occasional bubbles — refrigerant charge slightly low

3 Oil streaks on sight glass — total lack of refrigerant

4 Heavy stream of bubbles — serious shortage of refrigerant

5 Dark or clouded sight glass — contaminent present

Sight glass inspection

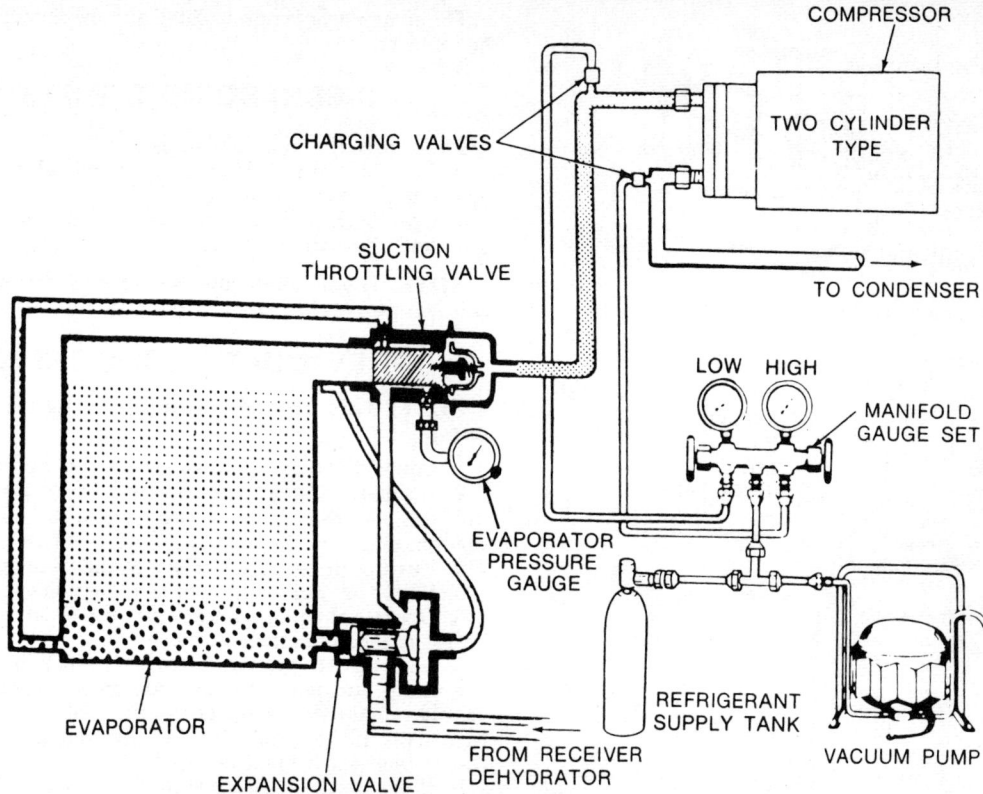

Gauge connections on the Tecumseh compressor

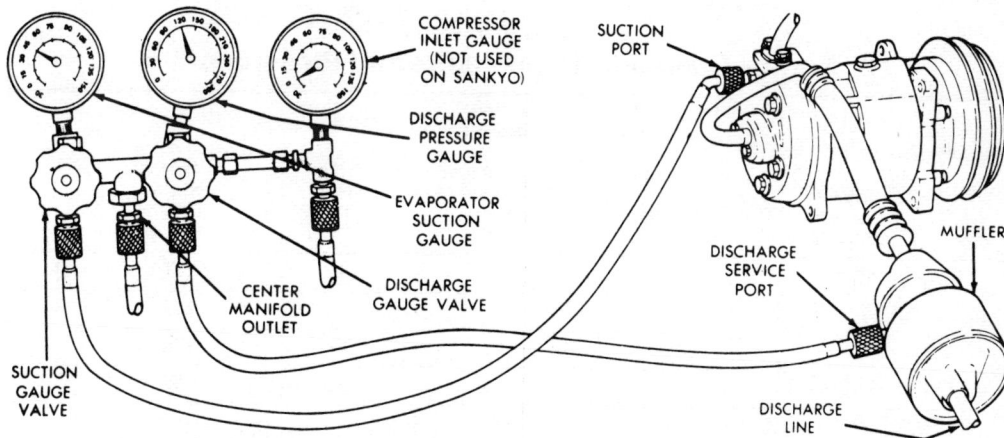

Gauge connections on the Sankyo compressor

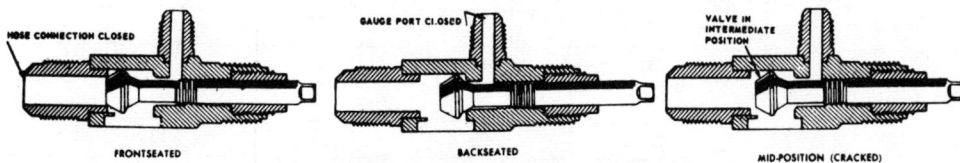

Manual service valve positions

2. Close both gauge hand valves and mid-position (crack) both compressor service valves.

3. Start the engine and turn on the air conditioning.

4. Turn the compressor suction valve slowly clockwise towards the front-seated position. When the suction pressure drops to zero, stop the engine and turn off the air conditioning. Quickly front-seat the valve completely.

5. Front-seat the discharge service valve.

6. Loosen the oil level check plug to remove any internal pressure.

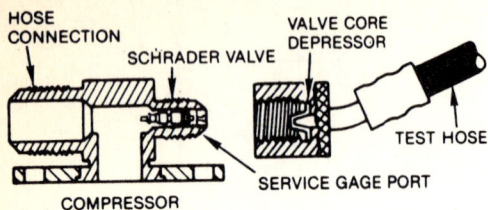

Schrader valve

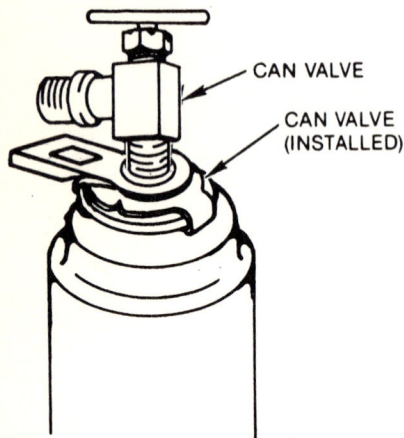

One pound R-12 can with opener valve connected

The compressor is now isolated and the service valves can now be removed.

DISCHARGING THE SYSTEM

1. Connect the manifold gauge set.
2. Turn both manifold gauge set hand valves to the full open (counterclockwise) position.
3. Open both service valve slightly, from the backseated position, and allow the refrigerant to discharge slowly.

NOTE: If you allow the refrigerant to rush out, it will take some refrigerant oil with it!

EVACUATING THE SYSTEM

NOTE: This procedure requires the use of a vacuum pump.

1. Connect the manifold gauge set.
2. Discharge the system.
3. Connect the center service hose to the inlet fitting of the vacuum pump.
4. Turn both gauge set valves to the wide open position.
5. Start the pump and note the low side gauge reading.
6. Operate the pump for a minimum of 30 minutes after the lowest observed gauge reading.
7. Leak test the system. Close both gauge set valves. Turn off the pump and note the low side gauge reading. The needle should remain stationary at the point at which the pump was turned off. If the needle drops to zero rapidly, there is a leak in the system which must be repaired.
8. If the needle remains stationary for 3 to 5 minutes, open the gauge set valves and run the pump for at least 30 minutes more.

Normal Operating Temperature and Pressures*

Relative Humidity (percent)	Surrounding Air Temperature (°F)	Maximum Desirable Center Register Discharge Air Temp. (°F)	Suction Pressure PSI (REF)	Head Pressure PSI (+25 PSI)
20	70	40	11	177
	80	41	15	208
	90	42	20	226
	100	43	23	255
30	70	40	12	181
	80	41	16	214
	90	42	22	234
	100	44	26	267
40	70	40	13	185
	80	42	18	220
	90	43	23	243
	100	44	26	278
50	70	40	14	189
	80	42	19	226
	90	44	25	251
	100	46	27	289

Normal Operating Temperature and Pressures*

Relative Humidity (percent)	Surrounding Air Temperature (°F)	Maximum Desirable Center Register Discharge Air Temp. (°F)	Suction Pressure PSI (REF)	Head Pressure PSI (+25 PSI)
60	70	41	15	193
	80	43	21	233
	90	45	25	259
	100	46	28	300
70	70	41	16	198
	80	43	22	238
	90	45	26	267
	100	46	29	312
80	70	42	18	202
	80	44	23	244
	90	47	27	277
	100	—	—	—
90	70	42	19	206
	80	47	24	250
	90	48	28	284
	100	—	—	—

*Operate engine with transmission in neutral. Keep vehicle out of direct sunlight.

Pressure Diagnosis

Condition	Possible Cause	Correction
Low side low—High side low	System refrigerant low	Evacuate, leak test, and charge system
Low side high—High side low	Internal leak in compressor—worn	Remove compressor cylinder head and inspect compressor. Replace valve plate assembly if necessary. If compressor pistons, rings, or cylinders are excessively worn or scored, replace compressor.
	Head gasket leaking	Install new cylinder head gasket
	Expansion valve	Replace expansion valve
	Drive belt slipping	Set belt tension
Low side high—High side high	Clogged condenser fins	Clean out condenser fins
	Air in system	Evacuate, leak test, and charge system
	Expansion valve	Replace expansion valve
	Loose or worn fan belts	Adjust or replace belts as necessary
Low side low—High side high	Expansion valve	Replace expansion valve
	Restriction in liquid line	Check line for kinks—replace if necessary
	Restriction in receiver	Replace receiver
	Restriction in condenser	Replace condenser
Low side and high side normal (inadequate cooling)	Air in system	Evacuate, leak test, and charge system
	Moisture in system	Evacuate, leak test, and charge system.

9. Close both gauge set valves, stop the pump and disconnect the gauge set. The system is now ready for charging.

LEAK TESTING

Some leak tests can be performed with a soapy water solution. There must be at least a ½ lb. charge in the system for a leak to be detected. The most extensive leak tests are performed with either a Halide flame type leak tester or the more preferable electronic leak tester.

In either case, the equipment is expensive, and, the use of a Halide detector can be extremely hazardous!

CHARGING THE SYSTEM

---- CAUTION ----

NEVER OPEN THE HIGH PRESSURE SIDE WITH A CAN OF RE-FRIGERANT CONNECTED TO THE SYSTEM! OPENING THE HIGH PRESSURE SIDE WILL OVERPRESSURIZE THE CAN, CAUSING IT TO EXPLODE!

Systems With Sight Glass

In this procedure the refrigerant enters the suction side of the system as a vapor while the compressor is running. Before proceeding, the system should be in a partial vacuum after adequate evacuation. Both hand valves on the gauge manifold should be closed.

1. Attach both test hoses to their respective service valve ports. Mid-position manually operated service valves, if present.
2. Install the dispensing valve (closed position) on the refrigerant container. (Single and multiple refrigerant manifolds are available to accommodate one to four 15 oz. cans.)
3. Attach the center charging hose to the refrigerant container valve.
4. Open dispensing valve on the refrigerant valve.
5. Loosen the center charging hose coupler where it connect to the gauge manifold to allow the escaping refrigerant to purge the hose of contaminants.
6. Tighten the center charging hose connector.
7. Purge the low pressure test hose at the gauge manifold.
8. Start the truck engine, roll down the truck windows and adjust the air conditioner to maximum cooling. The truck engine should be at normal operating temperature before proceeding. The heated environment helps the liquid vaporize more efficiently.
9. Crack open the low side hand valve on the manifold. Manipulate the valve so that the refrigerant that enters the system does not cause the low side pressure to exceed 40 psi. Too sudden a surge may permit the entrance of unwanted liquid to the compressor. Since liquids cannot be compressed, the compressor will suffer damage if compelled to attempt it. If the suction side of the system remains in a vacuum the system is blocked. Locate and correct the condition before proceeding any further.

NOTE: Placing the refrigerant can in a container of warm water (no hotter than +125°F [+51.6°C]) will speed the charging process. Slight agitation of the can is helpful too, but be careful not to turn the can upside down.

Systems Without a Sight Glass

1. Connect the gauge set.
2. Close (clockwise) both gauge set valves.
3. Connect the center hose to the refrigerant can opener valve.
4. Make sure the can opener valve is closed, that is, the needle is raised, and connect the valve to the can. Open the valve, puncturing the can with the needle.
5. Loosen the center hose fitting at the pressure gauge, allowing refrigerant to purge the hose of air. When the air is bled, tighten the fitting.

---- CAUTION ----

IF THE LOW PRESSURE GAUGE SET HOSE IS NOT CONNECTED TO THE ACCUMULATOR/DRIER, KEEP THE CAN IN AN UPRIGHT POSITION!

6. Disconnect the wire harness snap-lock connector from the clutch cycling pressure switch and install a jumper wire across the two terminals of the connector.
7. Open the low side gauge set valve and the can valve.
8. Allow refrigerant to be drawn into the system.
9. When no more refrigerant is drawn into the system, start the engine and run it at about 1,500 rpm. Turn on the system and operate it at the full high position. The compressor will operate and pull refrigerant gas into the system.

NOTE: To help speed the process, the can may be placed, upright, in a pan of warm water, not exceeding 125°F (52°C).

10. If more than one can of refrigerant is needed, close the can valve and gauge set low side valve when the can is empty and connect a new can to the opener. Repeat the charging process until no more refrigerant is drawn into the system. The frost line on the outside of the can will indicate what portion of the can has been used.

---- CAUTION ----

NEVER ALLOW THE HIGH PRESSURE SIDE READING TO EXCEED 240 psi.

11. When the charging process has been completed, close the gauge set valve and can valve. Remove the jumper wire and reconnect the cycling clutch wire. Run the system for at least five minutes to allow it to normalize. Low pressure side reading should be 4-25 psi; high pressure reading should be 120-210 psi at an ambient temperature of 70-90°F (21-32°C).
12. Loosen both service hoses at the gauges to allow any refrigerant to escape. Remove the gauge set and install the dust caps on the service valves.

NOTE: Multi-can dispensers are available which allow a simultaneous hook-up of up to four 1 lb. cans of R-12.

Capacities

Never exceed the recommended maximum charge for the system! The maximum charge for systems using the 2-cylinder Tecumseh compressor is 2½ lb.; 2 lb. for those systems using the 5-cylinder Sankyo compressor.

Front Hub and Wheel Bearings

ADJUSTMENT

NOTE: Sodium-based grease is not compatible with lithium-based grease. Read the package labels and be careful not to mix the two types. If there is any doubt as to the type of grease used, completely clean the old grease from the bearing and hub before replacing.

Before handling the bearings, there are a few things that you should remember to do and not to do.

Remember to DO the following:
• Remove all outside dirt from the housing before exposing the bearing.
• Treat a used bearing as gently as you would a new one.
• Work with clean tools in clean surroundings.
• Use clean, dry canvas gloves, or at least clean, dry hands.
• Clean solvents and flushing fluids are a must.
• Use clean paper when laying out the bearings to dry.

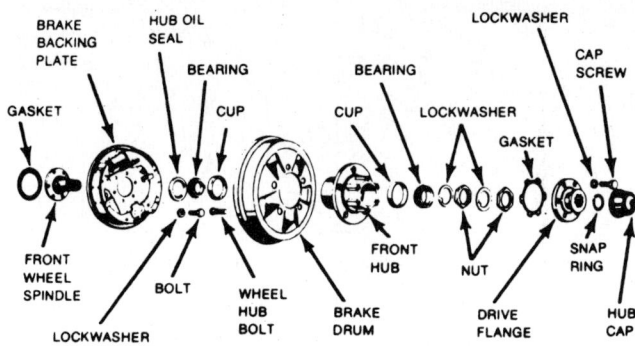

Typical front wheel hub and bearings, with drum brakes

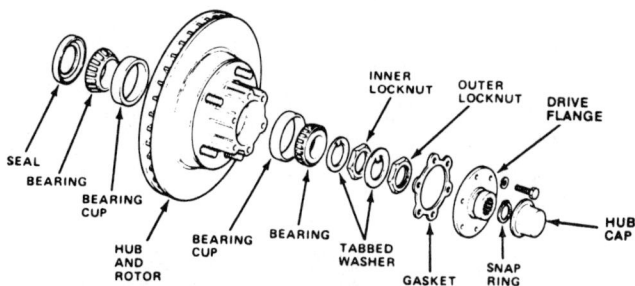

Hub and wheel bearings on models with disc brakes through 1979

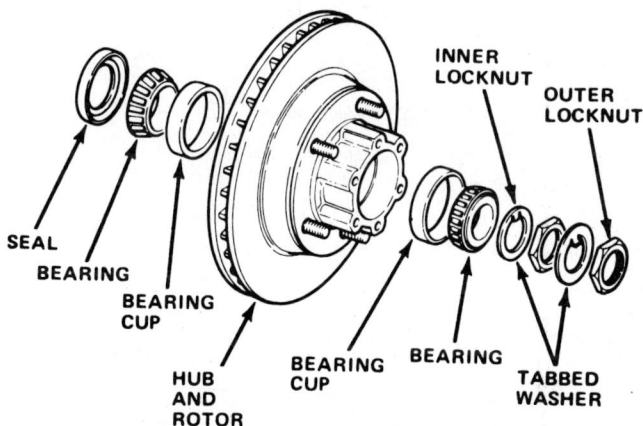

Hub and wheel bearings on 1980–86 models

- Protect disassembled bearings from rust and dirt. Cover them up.
- Use clean rags to wipe bearings.
- Keep the bearings in oil-proof paper when they are to be stored or are not in use.
- Clean the inside of the housing before replacing the bearing.

Do NOT do the following:

- Don't work in dirty surroundings.
- Don't use dirty, chipped or damaged tools.
- Try not to work on wooden work benches or use wooden mallets.
- Don't handle bearings with dirty or moist hands.
- Do not use gasoline for cleaning. Use a safe solvent.
- Do not spin-dry bearings with compressed air. They will be damaged.

- Do not spin dirty bearings.
- Avoid using cotton waste or dirty cloths to wipe bearings.
- Try not to scratch or nick bearing surfaces.
- Do not allow the bearing to come in contact with dirt or rust at any time.

1971-86

1. Raise the front of the vehicle and place jackstands under the axle.
2. Remove the wheel.
3. Remove the front hub grease cap and driving hub snapring. On models equipped with locking hubs, remove the retainer knob hub ring, agitator knob, snapring, outer clutch retaining ring and actuating cam body.
4. Remove the splined driving hub and the pressure spring. This may require slight prying with a screwdriver.
5. Remove the external snapring from the spindle shaft and remove the hub shaft drive gear.
6. Remove the wheel bearing locknut, lockring, adjusting nut and inner lockring.
7. On vehicles with drum brakes, remove the hub and drum assembly. This may require that the brake adjusting wheel be backed off a few turns. The outer wheel bearing and spring retainer will come off with the hub.
8. On vehicles with disc brakes, remove the caliper and suspend it out of the way by hanging it from a suspension or frame member with a length of wire. Do not disconnect the brake hose, and be careful to avoid stretching the hose. Remove the rotor and hub assembly. The outer wheel bearing and, on vehicles with locking hubs, the spring collar, will come off with the hub.
9. Carefully drive out the inner bearing and seal from the hub, using a wood block.
10. Inspect the bearing races for excessive wear, pitting or grooves. If they are cracked or grooved, or if pitting and excess wear is present, drive them out with a drift or punch.
11. Check the bearing for excess wear, pitting or cracks, or excess looseness.

NOTE: If it is necessary to replace either the bearing or the race, replace both. Never replace just a bearing or a race. These parts wear in a mating pattern. If just one is replaced, premature failure of the new part will result.

12. If the old parts are retained, thoroughly clean them in a safe solvent and allow them to dry on a clean towel. Never spin dry them with compressed air.
13. On vehicles with drum brakes, cover the spindle with a cloth and thoroughly brush all dirt from the brakes. Never blow the dirt off the brakes, due to the presence of asbestos in the dirt, which is harmful to your health when inhaled.
14. Remove the cloth and thoroughly clean the spindle.
15. Thoroughly clean the inside of the hub.
16. Pack the inside of the hub with EP wheel bearing grease. Add grease to the hub until it is flush with the inside diameter of the bearing cup.
17. Pack the bearing with the same grease. A needle-shaped wheel bearing packer is best for this operation. If one is not available, place a large amount of grease in the palm of your hand and slide the edge of the bearing cage through the grease to pick up as much as possible, then work the grease in as best you can with your fingers.
18. If a new race is being installed, very carefully drive it into position until it bottoms all around, using a brass drift. Be careful to avoid scratching the surface.
19. Place the inner bearing in the race and install a new grease seal.
20. Place the hub assembly onto the spindle and install the inner lockring and outer bearing. Install the wheel bearing nut and torque it to 50 ft. lbs. while turning the wheel back and

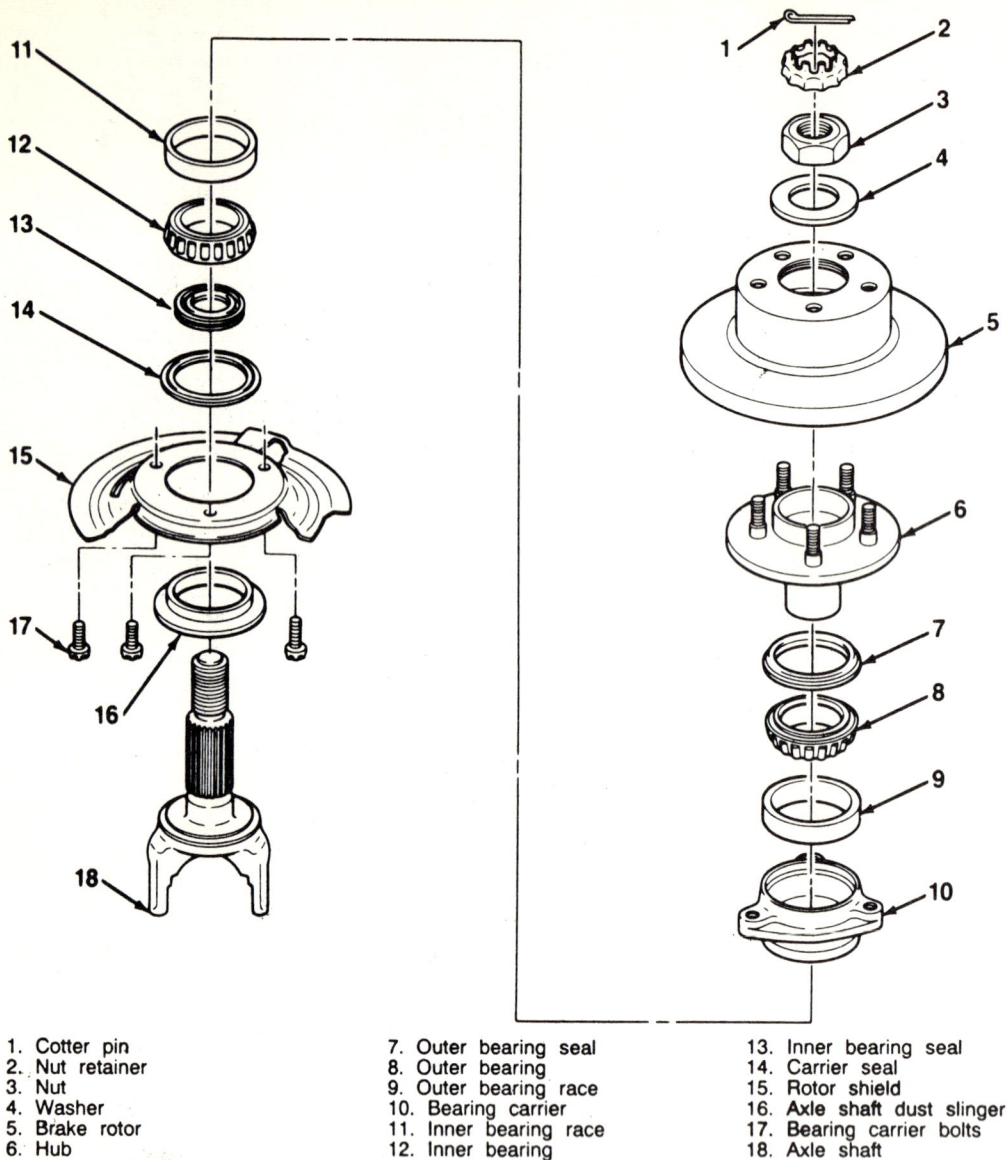

1. Cotter pin
2. Nut retainer
3. Nut
4. Washer
5. Brake rotor
6. Hub
7. Outer bearing seal
8. Outer bearing
9. Outer bearing race
10. Bearing carrier
11. Inner bearing race
12. Inner bearing
13. Inner bearing seal
14. Carrier seal
15. Rotor shield
16. Axle shaft dust slinger
17. Bearing carrier bolts
18. Axle shaft

1987-90 front hub and bearings

forth to seat the bearings. Back off the nut about ¼ turn (90°) maximum.

21. Install the lockwasher with the tab aligned with the keyway in the spindle and turn the inner wheel bearing adjusting nut until the peg on the nut engages the nearest hole in the lockwasher.

22. Install the outer locknut and torque it to 50 ft. lbs.

23. Install the spring collar, drive flange, snapring, pressure spring, and hub cap.

24. Install the caliper over the rotor.

1987-90

1. Raise and support the front end on jackstands.
2. Remove the wheel.
3. Dismount the caliper and suspend it out of the way.
4. Remove the rotor.
5. Remove the hub nut pin, cap and nut.
6. Remove the hub.
7. The hub and bearings are usually replaced as a unit. The hub and bearing carrier may, however, be disassembled and the

bearings replaced as a set. Once the hub and bearing carrier have been separated, the bearings should not be reused.

8. Pack the hub cavity and bearings with wheel bearing grease and install the hub on the axle shaft. If the carrier was separated from the hub, make sure you install a new carrier seal and inner bearing seal.

9. Install the hub washer and nut. Torque the nut to 175 ft. lbs. and instal the cap and new cotter pin.

10. Install the rotor, caliper and wheel.

Locking Hub Service

Jeep vehicles through 1979 were not factory equipped with locking hubs. Locking hubs were a dealer installed option or installed by the owner after purchase. Beginning in 1980, factory installed hubs were offered.

1971-79

The following is a general service procedure that should apply to all types. Locking hubs should be lubricated at least once a year and as soon as possible if running for extended periods sub-

RETAINING
RING

WEAR
WASHER

HUB SHAFT

BEARING
HUB

RETAINING
RING

COMPRESSOR
SPRING

RING
CLUTCH

RETAINING
RING

DIAL
SCREW

NUT
CLUTCH

O-RING

CLUTCH
CUP

COMPRESSOR
SPRING

HUB

SEAL

DIAL
DETENT

CONTROL
DIAL

SCREW

LABEL

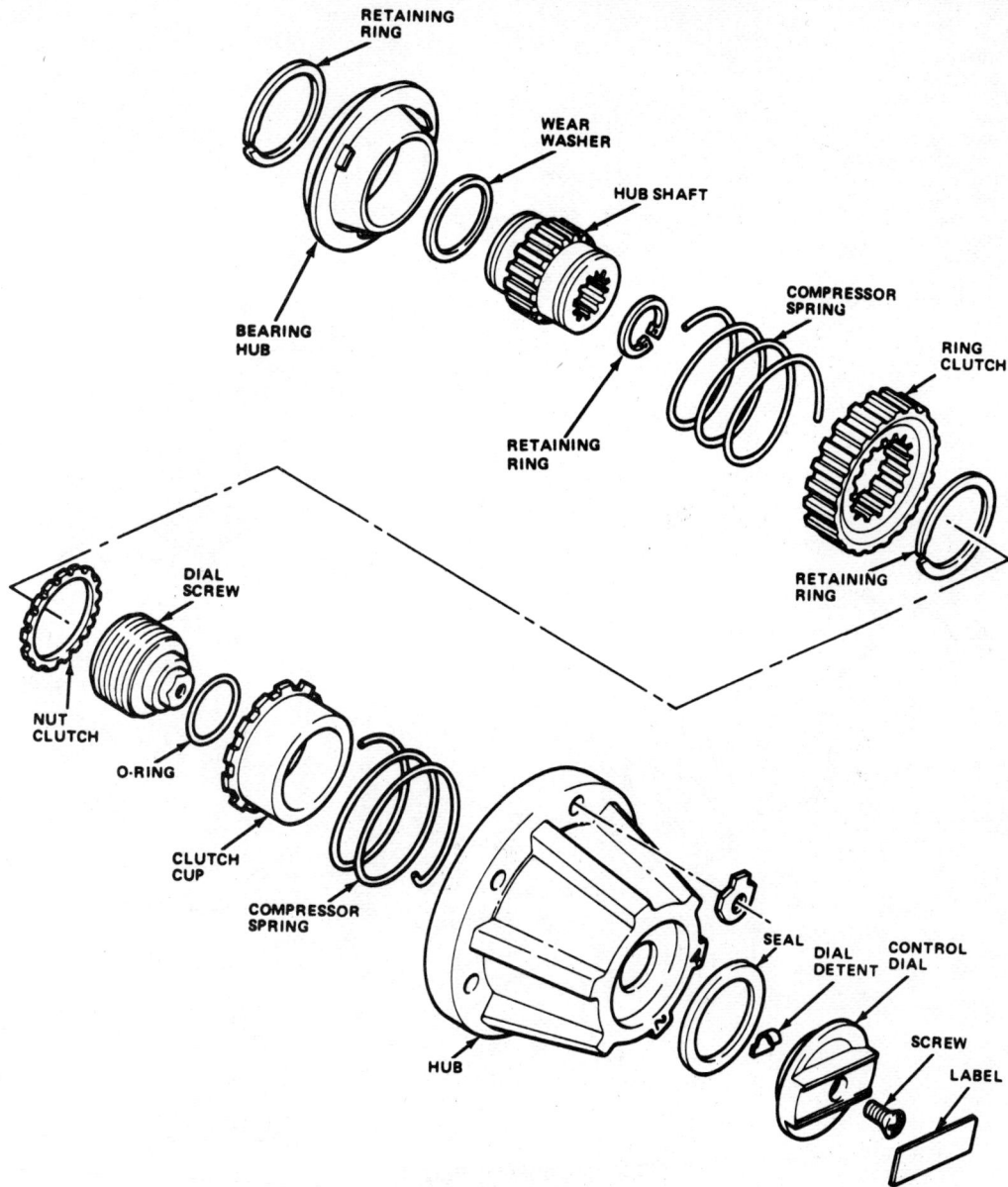

1980–81 locking hub

merged in water. The same type of grease should be used in the locking hubs as is used on the wheel bearings. EP lithium based chassis lube is preferred.

1. Remove the lockout screws and washers.
2. Remove the hub ring and knob.
3. Remove the internal snapring from the groove in the hub.
4. Remove the cam body ring and clutch retainer from the hub and disassemble the parts.
5. Remove the axle shaft snapring. It may be necessary to push in on the gear and pull out on the axle with a bolt to make the snapring removal easier.
6. Remove the drive gear and clutch gear. A slight rocking of the hub may make them slide out easier.
7. Remove the coil spring and spring retainer.
8. Clean all the components in a safe solvent. Wipe out the hub with a clean cloth.
9. Grease the inside of the hub liberally.
10. Install the spring retainer ring with the undercut area facing inwards. Be sure it seats against the bearing.

11. Install the coil spring with the large end going in first.
12. Install the axle shaft sleeve and ring and the inner clutch ring with the teeth of both components meshed together in a locked position. It may be necessary to rock the hub to mesh the splines of the axle with those of the axle shaft sleeve and ring. Keep the two gears locked in position.
13. Install the axle shaft snapring. Push in on the gear and pull out on the axle with a bolt to allow the snapring to go into the groove.
14. Install the actuating cam body ring into the outer clutch retaining ring and install them in the hub.
15. Install the internal snapring.
16. Apply a small amount of Lubriplate® grease to the ears of the cam.
17. Assemble the knob in the hub ring and assemble them to the axle with the knob in the locked position. Tighten the screws and washers evenly and alternately, making sure the retainer ring is not cocked in the hub.
18. Torque the screws to 40 in. lbs.

1. Retaining ring
2. Bearing hub
3. Wear washer
4. Hub shaft
5. Retaining ring
6. Compressor spring
7. Ring clutch
8. Retaining ring
9. Nut clutch
10. Dial screw
11. O-Ring
12. Clutch cup
13. Compressor spring
14. Hub
15. Control dial
16. Screw

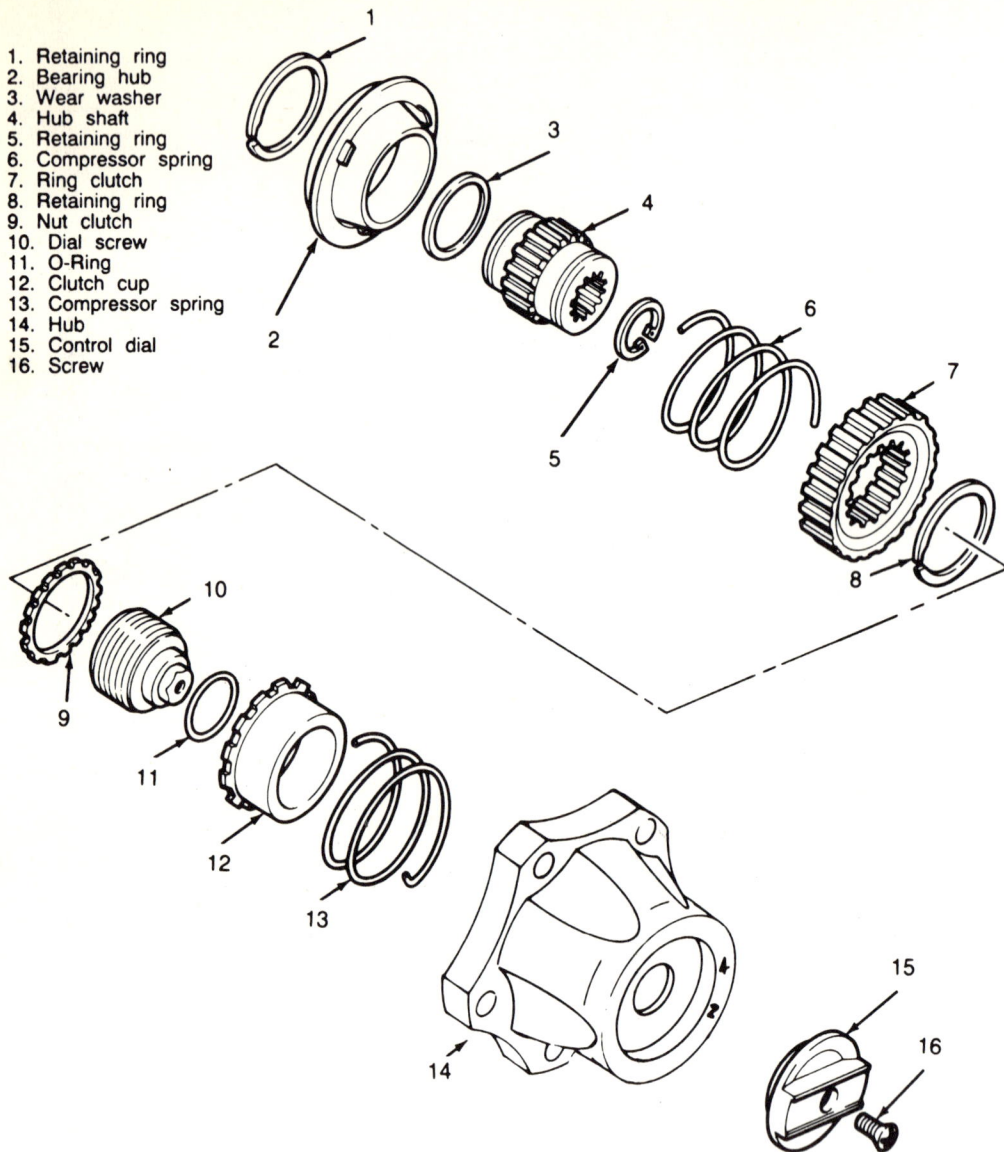

1982–86 locking hub

1980-86

1. Remove the bolts and lockwashers attaching the hub body to the axle.

2. Remove the hub body and discard the gasket.

WARNING: Do not turn the hub control dial after removed!

3. Remove the retaining ring from the axle shaft.

4. Remove the hub clutch and bearing assembly.

5. Clean and inspect all parts.

6. Lubricate all parts with chassis lubricant. DO NOT PACK THE HUB WITH GREASE!

7. Install in reverse of removal. Torque the retaining nuts to 30 ft. lbs.

Tires and Wheels

Inspect the tire treads for cuts, bruises and other damage. Check the air valves to be sure that they are tight. Replace any missing valve caps.

The tires should be checked frequently for proper air pressure. A chart in the glove compartment or on the driver's door pillar gives the recommended inflation pressure. Pressures can increase as much as 6 psi due to heat buildup. It is a good idea to have your own accurate gauge, and to check pressures weekly. Not all gauges on service station air pumps can be trusted.

Inspect tires for uneven wear that might indicate the need for front end alignment or tire rotation. Tires should be replaced when a tread wear indicator appears as a solid band across the tread.

When you buy new tires, give some thought to these points, especially if you are switching to larger tires or to another profile series (50, 60, 70, 78):

1. All four tires should be the same. Four wheel drive requires that all tires be the same size, type, and tread pattern to provide even traction on loose surfaces, to prevent driveline bind when conventional part time four wheel drive is used, and to prevent excessive wear on the center differential with full time four wheel drive.

2. The wheels must be the correct width for the tire. Tire dealers have charts of tire and rim compatibility. A mismatch

Troubleshooting Basic Wheel Problems

Problem	Cause	Solution
The car's front end vibrates at high speed	• The wheels are out of balance • Wheels are out of alignment	• Have wheels balanced • Have wheel alignment checked/adjusted
Car pulls to either side	• Wheels are out of alignment • Unequal tire pressure • Different size tires or wheels	• Have wheel alignment checked/adjusted • Check/adjust tire pressure • Change tires or wheels to same size
The car's wheel(s) wobbles	• Loose wheel lug nuts • Wheels out of balance • Damaged wheel • Wheels are out of alignment • Worn or damaged ball joint • Excessive play in the steering linkage (usually due to worn parts) • Defective shock absorber	• Tighten wheel lug nuts • Have tires balanced • Raise car and spin the wheel. If the wheel is bent, it should be replaced • Have wheel alignment checked/adjusted • Check ball joints • Check steering linkage • Check shock absorbers
Tires wear unevenly or prematurely	• Incorrect wheel size • Wheels are out of balance • Wheels are out of alignment	• Check if wheel and tire size are compatible • Have wheels balanced • Have wheel alignment checked/adjusted

Troubleshooting Basic Tire Problems

Problem	Cause	Solution
The car's front end vibrates at high speeds and the steering wheel shakes	• Wheels out of balance • Front end needs aligning	• Have wheels balanced • Have front end alignment checked
The car pulls to one side while cruising	• Unequal tire pressure (car will usually pull to the low side) • Mismatched tires • Front end needs aligning	• Check/adjust tire pressure • Be sure tires are of the same type and size • Have front end alignment checked
Abnormal, excessive or uneven tire wear See "How to Read Tire Wear"	• Infrequent tire rotation • Improper tire pressure • Sudden stops/starts or high speed on curves	• Rotate tires more frequently to equalize wear • Check/adjust pressure • Correct driving habits
Tire squeals	• Improper tire pressure • Front end needs aligning	• Check/adjust tire pressure • Have front end alignment checked

Tire Size Comparison Chart

| "Letter" sizes | | | Inch Sizes | Metric-inch Sizes | | |
"60 Series"	"70 Series"	"78 Series"	1965–77	"60 Series"	"70 Series"	"80 Series"
		Y78-12	5.50-12, 5.60-12 6.00-12	165/60-12	165/70-12	155-12
		W78-13	5.20-13	165/60-13	145/70-13	135-13
		Y78-13	5.60-13	175/60-13	155/70-13	145-13
			6.15-13	185/60-13	165/70-13	155-13, P155/80-13
A60-13	A70-13	A78-13	6.40-13	195/60-13	175/70-13	165-13
B60-13	B70-13	B78-13	6.70-13	205/60-13	185/70-13	175-13
			6.90-13			
C60-13	C70-13	C78-13	7.00-13	215/60-13	195/70-13	185-13
D60-13	D70-13	D78-13	7.25-13			
E60-13	E70-13	E78-13	7.75-13			195-13
			5.20-14	165/60-14	145/70-14	135-14
			5.60-14	175/60-14	155/70-14	145-14
			5.90-14			
A60-14	A70-14	A78-14	6.15-14	185/60-14	165/70-14	155-14
	B70-14	B78-14	6.45-14	195/60-14	175/70-14	165-14
	C70-14	C78-14	6.95-14	205/60-14	185/70-14	175-14
D60-14	D70-14	D78-14				
E60-14	E70-14	E78-14	7.35-14	215/60-14	195/70-14	185-14
F60-14	F70-14	F78-14, F83-14	7.75-14	225/60-14	200/70-14	195-14
G60-14	G70-14	G77-14, G78-14	8.25-14	235/60-14	205/70-14	205-14
H60-14	H70-14	H78-14	8.55-14	245/60-14	215/70-14	215-14
J60-14	J70-14	J78-14	8.85-14	255/60-14	225/70-14	225-14
L60-14	L70-14		9.15-14	265/60-14	235/70-14	
	A70-15	A78-15	5.60-15	185/60-15	165/70-15	155-15
B60-15	B70-15	B78-15	6.35-15	195/60-15	175/70-15	165-15
C60-15	C70-15	C78-15	6.85-15	205/60-15	185/70-15	175-15
	D70-15	D78-15				
E60-15	E70-15	E78-15	7.35-15	215/60-15	195/70-15	185-15
F60-15	F70-15	F78-15	7.75-15	225/60-15	205/70-15	195-15
G60-15	G70-15	G78-15	8.15-15/8.25-15	235/60-15	215/70-15	205-15
H60-15	H70-15	H78-15	8.45-15/8.55-15	245/60-15	225/70-15	215-15
J60-15	J70-15	J78-15	8.85-15/8.90-15	255/60-15	235/70-15	225-15
	K70-15		9.00-15	265/60-15	245/70-15	230-15
L60-15	L70-15	L78-15, L84-15	9.15-15			235-15
	M70-15	M78-15				255-15
		N78-15				

NOTE: Every size tire is not listed and many size comaprisons are approximate, based on load ratings. Wider tires than those supplied new with the vehicle should always be checked for clearance

can cause sloppy handling and rapid tread wear. The old rule of thumb is that the tread width should match the rim width (inside bead to inside bead) within an inch. For radial tires, the rim width should be 80% or less of the tire (not tread) width.

3. The height (mounted diameter) of the new tires can greatly change speedometer accuracy, engine speed at a given road speed, fuel mileage, acceleration, and ground clearance. Tire makers furnish full measurement specifications. Speedometer drive gears are available from Jeep parts for correction.

NOTE: Dimensions of tires marked the same size may vary significantly, even among tires from the same maker.

4. The spare tire should be usable, at least for low speed operation, with the new tires. You will probably have to remove the side mounted spare for clearance. This is especially true on 1972 and later models, since they have a wider tread and minimal tire-to-spare clearance.

5. There shouldn't be any body interference when loaded, on bumps, or in turning.

The only sure way to avoid problems with these points is to stick to tire and wheel sizes available as factory options.

HOW TO READ TIRE WEAR

The way your tires wear is a good indicator of other parts of your car. Abnormal wear patterns are often caused by the need for simple tire maintenance, or for front end alignment.

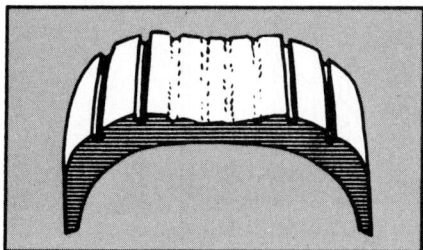

Over-Inflation

Excessive wear at the center of the tread indicates that the air pressure in the tire is consistently too high. The tire is riding on the center of the tread and wearing it prematurely. Occasionally, this wear pattern can result from outrageously wide tires on narrow rims. The cure for this is to replace either the tires or the wheels.

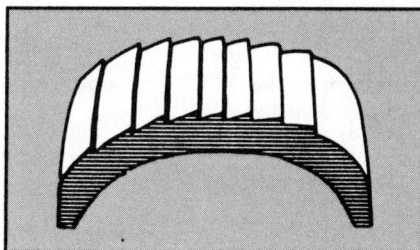

Feathering

Feathering is a condition when the edge of each tread rib develops a slightly rounded edge on one side and a sharp edge on the other. By running your hand over the tire, you can usually feel the sharper edges before you'll be able to see them. The most common causes of feathering are incorrect toe–in setting or deteriorated bushings in the front suspension.

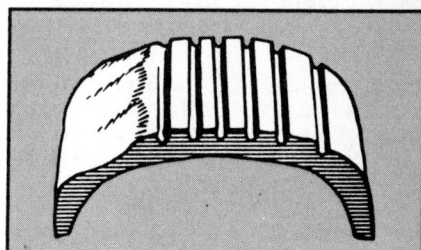

Cupping

Cups or scalloped dips appearing around the edge of the tread almost always indicate worn (sometimes bent) suspension parts. Adjustment of wheel alignment alone will seldom cure the problem. Any worn component that connects the wheel to the vehicle can cause this type of wear. Occasionally, wheels that are out of balance will wear like this, but wheel imbalance usually shows up as bald spots between the outside edges and center of the tread.

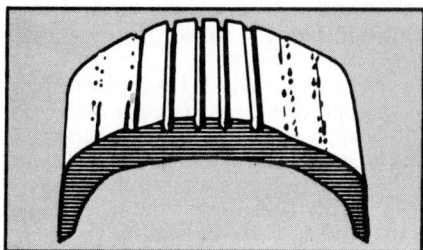

Under-Inflation

This type of wear usually results from consistent under–inflation. When a tire is under inflated, there is too much contact with the road by the outer threads, which wear prematurely. When this type of wear occurs, and the tire pressure is known to be consistently correct, a bent or worn steering component or the need for wheel alignment could be indicated.

One Side Wear

When an inner or outer rib wears faster than the rest of the tire, the need for wheel alignment is indicated. There is excessive camber in the front suspension, causing the wheel to lean too much, putting excessive load on one side of the tire. Misalignment could also be due to sagging springs, worn ball joints, or worn control arm bushings. Be sure the vehicle is loaded the way it's normally driven when you have the wheels aligned.

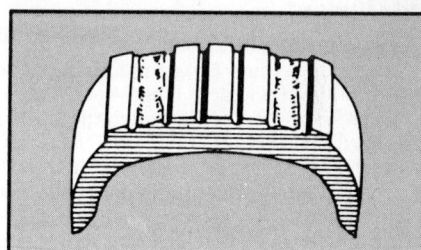

Second-Rib Wear

Second-rib wear is normally found only in radial tires, and appears where the steel belts end in relation to the tread. Normally, it can be kept to a minimum by paying careful attention to tire pressure and frequently rotation the tires. This is often considered normal wear but excessive amounts indicate that the tires are too wide for the wheels.

TIRE ROTATION

Tire rotation is recommended to obtain maximum tread wear. The pattern you use depends on personal preference, and whether or not you have a usable spare. Radial tires should not be cross-switched. They last longer if their direction of rotation is not changed. Truck type tires sometimes have directional tread indicated by arrows on the sidewalls. The arrow shows the direction of rotation. They will wear very rapidly if reversed. Studded snow tires will lose their studs if their rotation direction is reversed.

NOTE: Mark the wheel position or direction of rotation on radial, or studded snow tires before removing them.

Avoid overtightening the lug nuts to prevent damage to the brake disc or drum. Alloy wheels can also be cracked by overtightening. Use of a torque wrench is highly recommended. Tighten the lug nuts in a criss-cross sequence shown to 85 ft. lbs.

TIRE USAGE

The tires on your truck were selected to provide the best all around performance for normal operation when inflated as specified. Oversize tires will not increase the maximum carrying capacity of the vehicle, although they will provide an extra margin of tread life. Be sure to check overall height before using larger size tires which may cause interference with suspension components or wheel wells. When replacing conventional tire sizes with other tire size designations, be sure to check the manufacturer's recommendations. Interchangeability is not always possible because of differences in load ratings, tire dimensions, wheel well clearances, and rim size. Also due to differences in handling characteristics, 70 Series and 60 Series tires should be used only in pairs on the same axle; radial tires should be used only in sets of four.

NOTE: Many states have vehicle height restrictions; some states prohibit the lifting of vehicles beyond their design limits.

The wheels must be the correct width for the tire. Tire dealers have charts of tire and rim compatibility. A mismatch can cause sloppy handling and rapid tread wear. The old rule of thumb is that the tread width should match the rim width (inside bead to inside bead) within an inch. For radial tires, the rim width should be 80% or less of the tire (not tread) width.

The height (mounted diameter) of the new tires can greatly change speedometer accuracy, engine speed at a given road speed, fuel mileage, acceleration, and ground clearance. Tire manufacturers furnish full measurement specifications. Speedometer drive gears are available for correction.

NOTE: Dimensions of tires marked the same size may vary significantly, even among tires from the same manufacturer.

The spare tire should be of the same size, construction and design as the tires on the vehicle. It's not a good idea to carry a spare of a different contstruction.

TIRE DESIGN

For maximum satisfaction, tires should be used in sets of five.

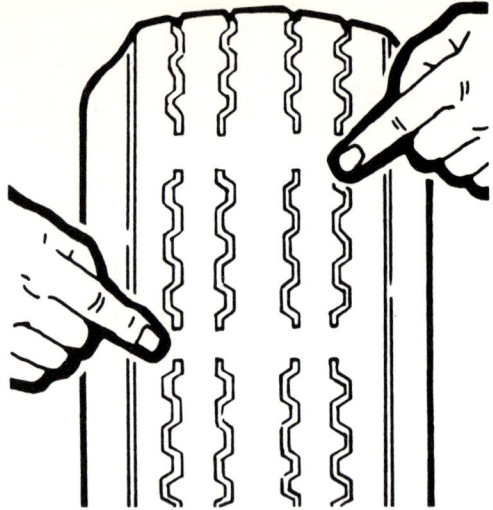

Tread wear indicators are built into all new tires. When they appear, it's time to replace the tires

Tread depth can be checked with a penny; when the top of Lincoln's head is visible, it's time for new tires

Mixing or different types (radial, bias-belted, fiberglass belted) should be avoided. Conventional bias tires are constructed so that the cords run bead-to-bead at an angle. Alternate plies run at an opposite angle. This type of construction gives rigidity to both tread and sidewall. Bias-belted tires are similar in construction to conventional bias ply tires. Belts run at an angle and also at a 90° angle to the bead, as in the radial tire. Tread life is improved considerably over the conventional bias tire. The radial tire differs in construction, but instead of the carcass plies running at an angle of 90° to each other, they run at an angle of 90° to the bead. This gives the tread a great deal of rigidity and the sidewall a great deal of flexibility and accounts for the characteristic bulge associated with radial tires.

When radial tires are used, tire sizes and wheel diameters should be selected to maintain ground clearance and tire load capacity equivalent to the minimum specified tire. Radial tires should always be used in sets of five, but in an emergency, radial tires can be used with caution on the rear axle only. If this is done, both tires on the rear should be of radial design.

WARNING: Radial tires should never be used on only the front axle!

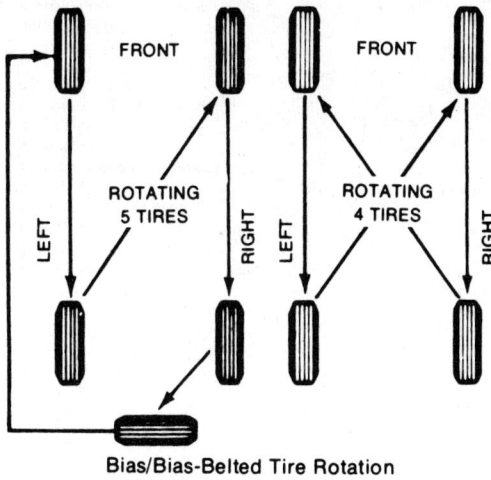

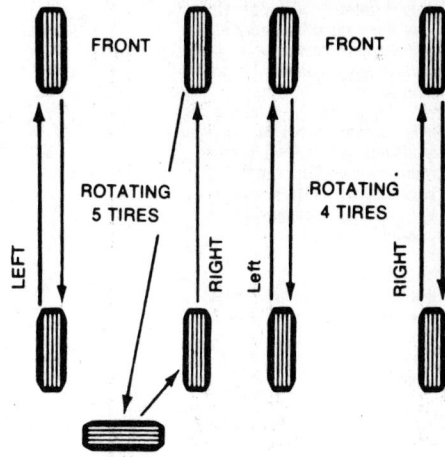

Tire rotation

Tread depth can also be checked with an inexpensive gauge made for the purpose

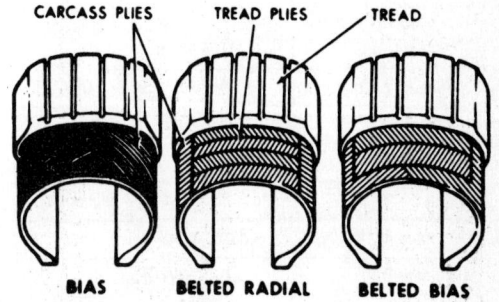

Types of tire construction

FLUIDS AND LUBRICANTS

Fuel and Oil Recommendations

FUEL

All models through 1974 are designed to use a regular grade of gasoline. All 1975 and later models must use unleaded gasoline.

ENGINE OIL

Many factors help to determine the proper oil for your Jeep. The big question is what viscosity to use and when. The whole question of viscosity revolves around the lowest anticipated ambient temperature to be encountered before your next oil change. The recommended viscosity ratings for temperatures ranging from below 0°F (−18°C) to above +32°F (0°C) are listed in the accompanying chart. They are broken down into multiviscosities and single viscosities. Multi-viscosity oils are recommended because of their wider range of acceptable temperatures and driving conditions.

The SAE grade number indicates the viscosity of the engine oil, or its ability to lubricate under a given temperature. The lower the SAE grade number, the lighter the oil. The lower the viscosity, the easier it is to crank the engine in cold weather.

This is the oil's SAE viscosity grade. The numbers followed by a 'W' indicate an oil with low temperature performance characteristics and the 'non-W' numbers describe an oil with high temperature characteristics. If there is one number, it is a single grade. Two or more numbers indicate a 'multi-viscosity' oil which has both low and high temperature characteristics.

This is the manufacturer's brand name.

These letters generally mean that the oil meets or exceeds established standards for use in gasoline (indicated by 'S' and a following letter) and diesel and commercial engines (indicated by 'C' and a following letter). These designations replace the older classifications which may be called for in some owners' manuals. The SF rating is the highest standard for gasoline automobiles.

This means that the oil will protect expensive engine components. Even if your car is no longer under warranty, it indicates that the oil is of good quality.

ALL-CLIMATE HEAVY DUTY
SAE
10W-20W-40
PART NO. 141
ENGINE CAR MANUFACTURERS
WARRANTY REQUIREMENTS
API SERVICES
SC, SD, SE
CA, CB, CC

The top of the oil can will tell you all you need to know about the oil

Lowest Air Temperature Anticipated	Multiviscosity Engine Oil
Above 40°F	SAE 10W-30, 40, 50 or 20W-40, 50
Above 32°F	SAE 10W-30 or 10W-40
Above 0°F	SAE 10W-30 or 10W-40
Below 0°F	SAE 5W-20 or 5W-30
	Single-Viscosity Engine Oil
Above 40°F	SAE 30 or 40
Above 32°F	SAE 20W-20
Above 0°F	SAE 20
Below 0°F	SAE 10W

Engine oil viscosity selection chart

The API (American Petroleum Institute) designation indicates the classification of engine oil for use under given operating conditions. For gasoline engines, only oils designated for Service SF/SG, or just SF, should be used. You can find the SF or SG marking either on the top or on the side of the container. The viscosity rating should be in the same place. Select the viscosity rating to be used by your type of driving and the temperature range anticipated before the next oil change.

The multi-viscosity oils offer the advantage of being adaptable to temperature extremes. They allow easy starts at low temperatures, yet still give good protection at high speeds and warm temperatures.

Engine

OIL LEVEL CHECK

Make sure that your vehicle is on a level surface to ensure an accurate reading. Then, raise the hood, position the holdup rod, and measure the oil with the dipstick which is on the left side of 4-cylinder and V8 engines and on the right of 6-cylinder engines. Add oil through the filler pipe on the right side of 4-134 engines. Add oil through the valve cover filler hole on 4-150, 4-151, 6-225, 6-232 and 6-258, and through the filler pipe at the front of the engine on V8s.

If the oil is below the ADD mark, add a quart of oil, then recheck the level. If the level is still not reading full, add only a half of a quart at a time, until the dipstick reads FULL. Do not overfill the engine. When you check the oil, make sure that you allow

sufficient time for all of the oil to drain back into the crankcase after stopping the engine. A minute or so should be enough time.

OIL AND FILTER CHANGE

CAUTION

The EPA warns that prolonged contact with used engine oil may cause a number of skin disorders, including cancer! You should make every effort to minimize your exposure to used engine oil. Protective gloves should be worn when changing the oil. Wash your hands and any other exposed skin areas as soon as possible after exposure to used engine oil. Soap and water, or waterless hand cleaner should be used.

The engine oil is to be changed every 4,000 miles. The oil should be changed more frequently, however, under conditions such as:
- Driving in dusty conditions
- Continuous trailer pulling or RV use
- Extensive or prolonged idling
- Extensive short trip operation in freezing temperatures (when the engine is not thoroughly warmed up)
- Frequent long runs at high speeds and high ambient temperatures
- Stop-and-go service, such as delivery trucks,

the oil change interval and filter replacement interval should be cut in half. Operation of the engine in severe conditions, such as a dust storm, volcanic ash or deep water, may require an immediate oil and filter change.

Before draining the oil, make sure that the engine is at operating temperature. Hot oil will hold more impurities in suspension and will flow better, removing more oil and dirt.

NOTE: Drain the oil into a suitable receptacle. Waste oil may be disposed of at a garage or service facility which accepts waste oil for recycling.

After the drain plug is loosened, unscrew the plug with your fingers, using a rag to shield your fingers from the heat. Push in on the plug as you unscrew it so you can feel when all of the screw threads are out of the hole. You can then remove the plug quickly with the minimum amount of oil running down your arm. You will also have the plug in your hand and not in the bottom of a pan of hot oil.

Change the oil filter every time you change the oil. The engine should be at operating temperature.

The oil filter is the spin-on type.

To replace the filter, you will need an oil filter wrench. Loosen the filter with the filter wrench. With a rag wrapped around the filter, unscrew the filter from the oil pump housing. Be careful

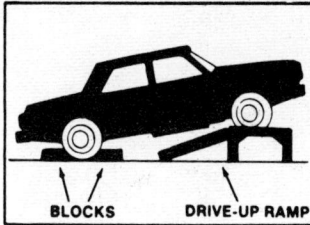

1. Warm the car up before changing your oil. Raise the front end of the car and support it on drive-on ramps or jackstands.

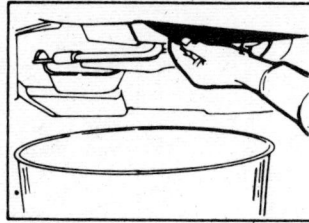

BLOCKS DRIVE-UP RAMP

2. Locate the drain plug on the bottom of the oil pan and slide a low flat pan of sufficient capacity under the engine to catch the oil. Loosen the plug with a wrench and turn it out the last few turns by hand. Keep a steady inward pressure on the plug to avoid hot oil from running down your arm.

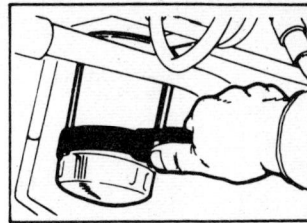

3. Remove the oil filter with a filter wrench. The filter can hold more than a quart of oil, which will be hot. Be sure the gasket comes off with the filter and clean the mounting base on the engine.

4. Lubricate the gasket on the new filter with clean engine oil. A dry gasket may not make a good seal and will allow the filter to leak.

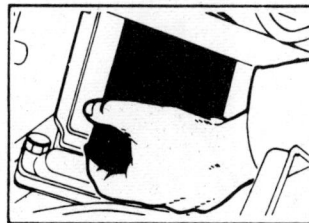

5. Position a new filter on the mounting base and spin it on by hand. Do not use a wrench. When the gasket contacts the engine, tighten it another ½–1 turn by hand.

6. Using a rag, clean the drain plug and the area around the drain hole in the oil pan.

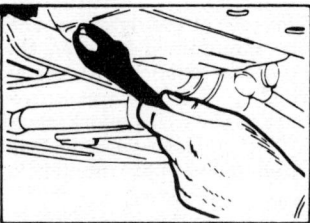

7. Install the drain plug and tighten it finger-tight. If you feel resistance, stop and be sure you are not cross-threading the plug. Finally, tighten the plug with a wrench.

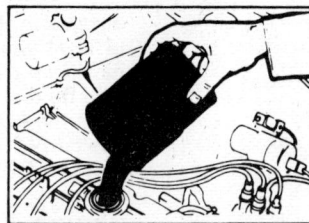

8. Locate the oil cap on the valve cover. An oil spout is the easiest way to add oil, but a funnel will do just as well.

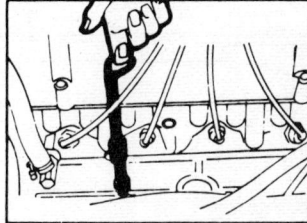

9. Start the engine and check for leaks. The oil pressure warning light will remain on for a few seconds; when it goes out, stop the engine and check the level on the dipstick.

Follow these 9 easy and safe steps to change your engine oil

of hot oil that might run down the side of the filter, especially on the sixes and V8s.

On the F4-134 engines, the filter is mounted with the open side facing downward so you won't have to worry about oil running down on your hand.

Make sure that you have a pan under the filter before you start to remove it from the engine so you won't make a mess and, if some of the hot oil does happen to get on you, you will have a place to dump the filter in a hurry. Wipe the base of the mounting plate with a clean, dry cloth. When you install the new filter, smear a small amount of oil on the gasket with your finger, just enough to coat the entire surface where it comes in contact with the mounting plate. When you tighten the filter, turn it only a quarter of a turn after it comes in contact with the mounting plate.

Manual Transmission

FLUID LEVEL CHECK

The level of lubricant in the transmission should be main-

tained at the filler hole on all manual transmissions. This hole is on the right side. When you check the level in the transmission, make sure that the vehicle is level so that you get a true reading. When you remove the filler plug, lubricant should run out of the hole. Replace the plug quickly for a minimum loss of lubricant. If lubricant does not run out of the hole when the plug is removed, lubricant should be added until it does. Replace the plug as soon as the lubricant reaches the level of the hole.

FLUID CHANGE

Remove the drain plug which is at the bottom of the transmission or else on the side near the bottom. Allow all the lubricant to run out before replacing the plug. Replace the case with the correct viscosity oil. All manual transmissions, except the T4 & T5, use SAE 80W-90 gear oil. The T4 & T5 use Dexron®II automatic transmission fluid.

Automatic Transmission

FLUID LEVEL CHECK

The fluid level in automatic transmissions is checked with a

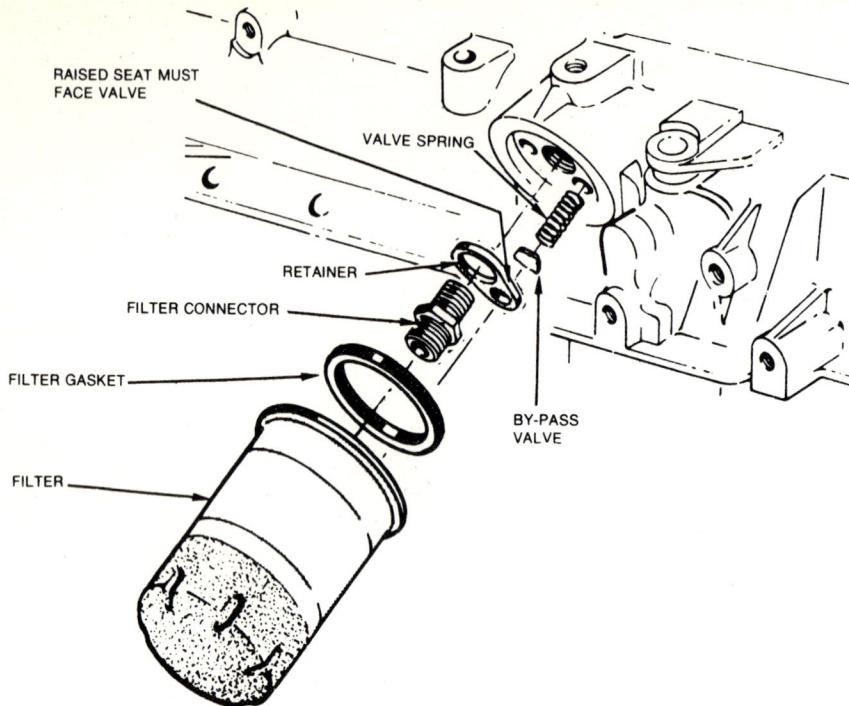

Spin-on oil filter on inline 6-cylinder and V8

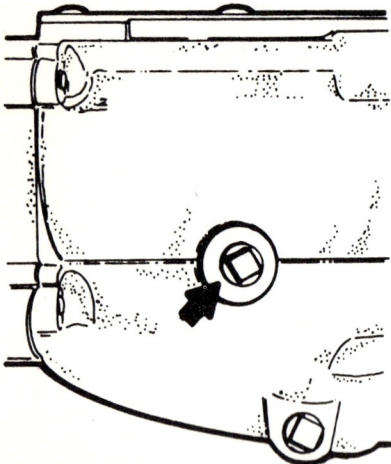

Manual transmission fill and drain plugs with the drain plug at the bottom center

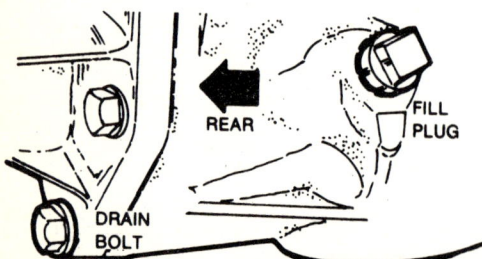

Manual transmission fill and drain plugs, using a tailshaft bolt as the drain plug

dipstick in the filler pipe at the right rear of the engine. The fluid level should be maintained between the ADD and FULL marks on the end of the dipstick with the automatic transmission fluid at normal operating temperatures. To raise the level from the ADD mark to the FULL mark, requires the addition of one pint of fluid. The fluid level with the fluid at room temperature (75°F [24°C]) should be approximately 1/4 in. below the ADD mark.

NOTE: In checking the automatic transmission fluid, insert the dipstick in the filler tube with the markings toward the center of the vehicle. Also, remember that the FULL mark on the dipstick is calibrated for normal operating temperature. This temperature is obtained only after at least 15 miles of expressway driving or the equivalent of city driving.

1. With the transmission in Park, the engine running at idle speed, the foot brake applied and the vehicle resting on level ground, move the transmission gear selector through each of the gear positions, including Reverse, allowing time for the transmission to engage. Return the shift selector to the Park position and apply the parking brake. Do not turn the engine off, but leave it running at idle speed.

2. Clean all dirt from around the transmission dipstick cap and the end of the filler tube.

3. Pull the dipstick out of the tube, wipe it off with a clean cloth, and push it back into the tube all the way, making sure that it seats completely.

4. Pull the dipstick out of the tube again and read the level of the fluid on the stick. The level should be between the ADD and FULL marks. If fluid must be added, add enough fluid through the tube to raise the level to between the ADD and FULL marks. Do not overfill the transmission because this will cause foaming and loss of fluid through the vent.

NOTE: Use only Dexron®II transmission fluid.

DRAIN, FILTER SERVICE AND REFILL

WARNING: If, when the transmission fluid level is

checked, the fluid is noticed to be discolored from a clear red to brown, has a burned smell, or contains water, it should be changed immediately!

1. Drive the vehicle for at least 20 minutes at expressway speeds or the equivalent to raise the temperature of the fluid to its normal operating range.

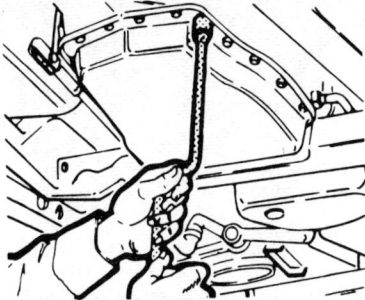

Many late model vehicles have no drain plug. Loosen the pan bolts and allow one corner of the pan to hang, so that the fluid will drain out

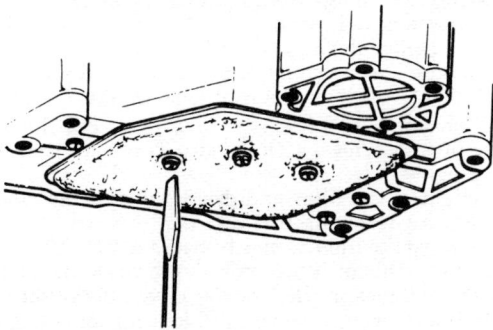

Removing automatic transmission filter

Clean the pan thoroughly with a safe solvent and allow it to air dry

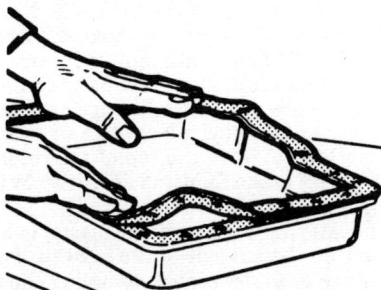

Install a new pan gasket

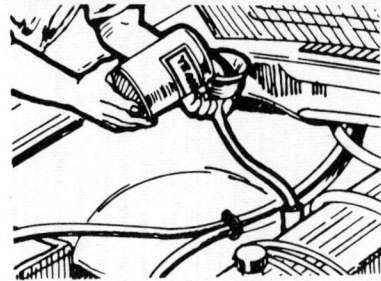

Fill the transmission with the required amount of fluid. Do not overfill. Start the engine and run the selector through all the shift points. Check the fluid and add as necessary

2. Drain the automatic transmission fluid into an appropriate container before it has cooled. The fluid is drained by loosening the transmission pan and allowing the fluid to run out around the edges. It is best to loosen only one corner of the pan and allow most of the fluid to drain out.

3. Remove the remaining pan screws, and remove the pan and pan gasket.

4. Remove the strainer and discard it.

5. Remove the O-ring seal from the pickup pipe and discard it.

6. Install a new O-ring seal on the pickup pipe and install the new strainer and pipe assembly.

7. Thoroughly clean the bottom pan and position a new gasket on the pan mating surface.

8. Install the pan and tighten the attaching screws to 10-13 ft. lbs.

9. Pour about 5 qts. of Dexron®II automatic transmission fluid down the dipstick tube. Make sure that the funnel, container, hose or any other item used to assist in filling the transmission is clean.

10. Start the engine with the transmission in Park. Do NOT race it. Allow the engine to idle for a few minutes.

11. After the engine has been running for a few minutes, move the selector lever through all of the gears.

12. With the selector lever in Park, check the transmission fluid level and adjust as necessary. Remember the transmission fluid must be warm when at the Full mark.

NOTE: On some 1977-79 models, fluid may overflow through the filler pipe, or vent tube. If this condition occurs:

1. Insert a length of stiff wire into the vent tube. If the tube is restricted, clean of replace it.

2. If the tube is not restricted, make sure that the fluid level is correct. If the fluid level is correct perform a road test. If the fluid overflows during the road test.

3. Raise and support the vehicle on jackstands.

4. Loosen the vacuum modulator adapter retaining bolt. Pull the modulator outward about ½-1 in. (13-25mm). Drain off about 1 pint of fluid. Seat the modulator and tighten the attaching bolt. Lower the vehicle and road test. Check the fluid level and file a new mark on the dipstick for the new fill level.

Part Time Transfer Case

FLUID LEVEL CHECK

The transfer case should be checked in the same manner as the manual transmission. The level should be up to the filler hole. Use the same viscosity oil as in the transmission. The filler hole is on the right side. Check the oil level at the top hole. The bottom one is for draining.

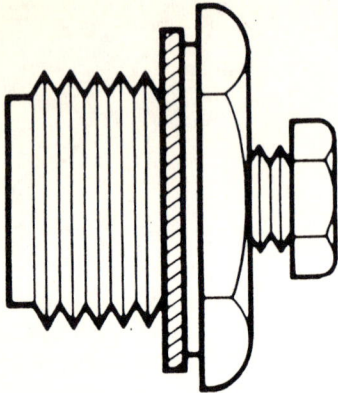

2-piece NP-207 drain plug

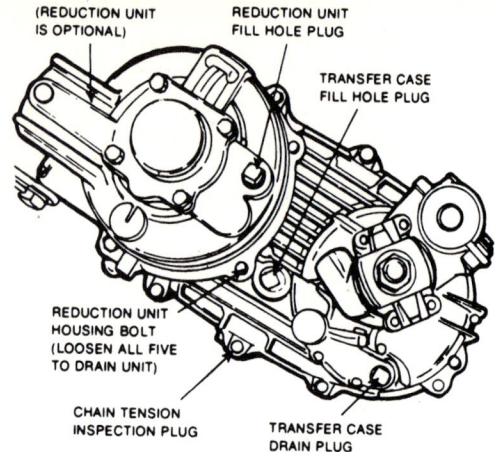

Typical Quadra-Trac® external components

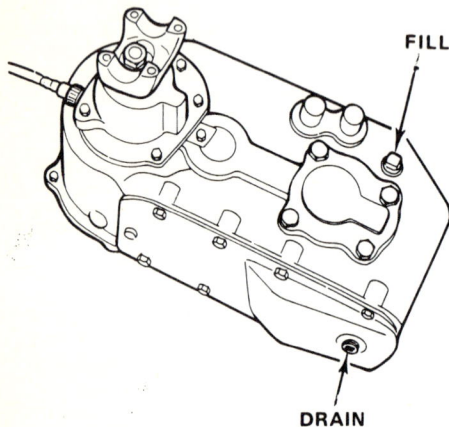

Dana 300 transfer case drain and fill plugs

NOTE: Some NP-207 models have a two piece fill plug, i.e., a small threaded plug inside a larger one. If your transfer case is so equipped, the fluid level is checked at the bottom of the smaller plug. To remove the smaller plug, hold the larger one firmly with a wrench while removing the smaller plug.

DRAIN AND REFILL

All manual transfer cases are to be serviced at the same time and in the same manner as the manual transmissions. The transfer case has its own drain plug which should be opened. Do not rely on the transmission drain plug to completely drain the transfer case, even if they are interconnected. Once the transfer case has been drained, replace the drain plug, remove the fill plug and fill the transfer case. The Dana/Spicer 18 and 20 used SAE 80W-90 gear oil. The NP-207 and 231 use Dexron®II. The Dana 300 uses SAE 85W-90 gear oil. Replace the fill plug.

Quadra-Trac® Full Time Transfer Case

FLUID LEVEL CHECK

Fluid levels in the Quadra-Trac® transfer case and low range reduction unit, if so equipped, should be checked at the same time. The lubricant levels are checked at the filler plug holes. The filler plug holes are located on the rear side of the transfer case assembly, just below center in the middle of the case housing and to the right of center of the reduction unit housing. The lubricant should be level with each filler plug hole. If not, replenish with Quadra-Trac® lubricant.

DRAIN AND REFILL

Remove the filler plugs from both the transfer case and the optional low range reduction unit. Remove the transfer case drain plug and allow the unit to drain. Replace the plug. Loosen the five bolts on the reduction unit (it has no drain plug) and pull it out slightly. Add one pint of Quadra-Trac® lubricant to the reduction unit and install the filler plug. Fill the transfer case to the filler hole level with Quadra-Trac® lubricant and replace the plug.

Don't overtighten the filler and drain plugs in the aluminum case. The correct torque is 10-25 ft. lbs.

After changing the fluid, it may be necessary to drive the vehicle in figure 8s for about fifteen minutes to work the fresh lubricant into the clutches in the transfer case differential.

Don't hold the steering wheel on full lock for more than about five seconds at a time during these maneuvers. The power steering fluid can overheat and cause pump and gear damage.

NOTE: The Quadra-Trac® in some Jeep vehicles may develop a low frequency, pulsating noise or grating or rasping which sometimes occurs in low speed cornering or parking. This is caused by the brake cones releasing suddenly after sticking. The condition is known as stick-slip. As a remedy, a lubricant, part number Jeep 8130444, which must be used in a drain and refill procedure. However, this new lubricant is especially prone to water contamination, so a new vent kit must be installed. The kit is part number 81030445. On vehicles that do not exhibit this problem, the original lubricant, part number 5358652 may be used. Directions for installing the vent kit are supplied with the kit.

Brake and Clutch Master Cylinders

On models through 1971, the master cylinder is located under the floor. To check the level of the brake fluid remove the floor plate. Clean the area of all dirt so that, when you remove the cover, no dirt will fall in and contaminate the brake fluid. Dirt in the hydraulic system could score the inside of the master cylinder or wheel cylinders and cause leakage or brake failure. Unscrew the lid of the master cylinder with a wrench. The fluid level should be within ½" from the top of the reservoir chamber. Use only heavy duty brake fluid and keep it away from any other fluids or vapors that could contaminate it.

On 1972 and later Jeep vehicles, the master cylinder is located under the hood, on the left side of the firewall. To check the fluid, use a screwdriver to pry off the retaining clip from the lid of the reservoir. The fluid level should be within ¼ in. (6mm) of the top of the reservoir.

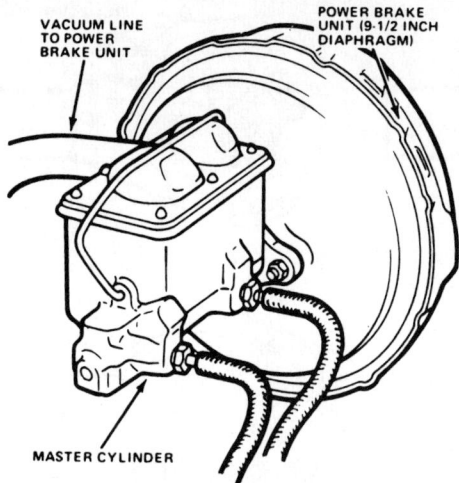

Typical 1972 and later master cylinder mounting

If the master cylinder is less than half full, there is probably a leak somewhere in the hydraulic system. Investigate the problem before driving the vehicle.

Coolant

COOLANT LEVEL CHECK

The coolant level should be maintained 1½-2 in. (38-52mm) below the bottom of the filler cap when the engine is cold. Since operating temperatures reach as high as +205°F (+96°C) for the 4-cylinder and 6-232/258, and +190°F (+88°C) for the V6 and V8s, coolant could be forced out of the radiator if it is filled too high. The radiator coolant level should be checked regularly, such as every time you fill the vehicle with gas. Never open the radiator cap of an engine that hasn't had sufficient time to cool or the pressure can blow off the cap and send out a spray of scalding water.

On systems with a coolant recovery tank, maintain the coolant level at the level marks on the recovery bottle.

For best protection against freezing and overheating, maintain an approximate 50% water and 50% ethylene glycol antifreeze mixture in the cooling system. Do not mix different brands of antifreeze to avoid possible chemical damage to the cooling system.

Avoid using water that is known to have a high alkaline content or is very hard, except in emergency situations. Drain and flush the cooling system as soon as possible after using such water.

Cover the radiator cap with a thick cloth before removing it from a radiator in a vehicle that is hot. Turn the cap counterclockwise slowly until pressure can be heard escaping. Allow all pressure to escape from the radiator before completely removing the radiator cap. It is best to allow the engine to cool if possible, before removing the radiator cap.

NOTE: Never add cold water to an overheated engine while the engine is not running.

After filling the radiator, run the engine until it reaches normal operating temperature, to make sure that the thermostat has opened and all the air is bled from the system.

DRAINING, FLUSHING AND REFILLING

—————————— CAUTION ——————————

When draining the coolant, keep in mind that cats and dogs are attracted by the ethylene glycol antifreeze, and are quite likely to drink any that is

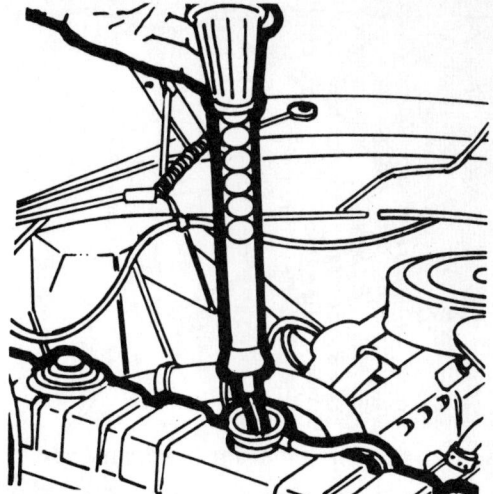

Coolant protection can be checked with a simple float-type tester

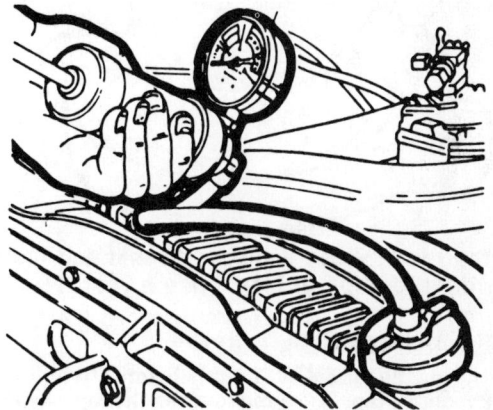

The system should be pressure tested once a year

left in an uncovered container or in puddles on the ground. This will prove fatal in sufficient quantity. Always drain the coolant into a sealable container. Coolant should be reused unless it is contaminated or several years old.

To drain the cooling system, allow the engine to cool down BEFORE ATTEMPTING TO REMOVE THE RADIATOR CAP. Then turn the cap until it hisses. Wait until all pressure is off the cap before removing it completely.

To avoid burns and scalding, always handle a warm radiator cap with a heavy rag.

1. At the dash, set the heater TEMP control lever to the fully HOT position.
2. With the radiator cap removed, drain the radiator by loosening the petcock at the bottom of the radiator. Locate any drain plugs in the block and remove them. Flush the radiator with water until the fluid runs clear.
3. Close the petcock and replace the plug(s), then refill the system with a 50/50 mix of ethylene glycol antifreeze. Fill the system to ¾-1¼ in. (19-32mm) from the bottom of the filler neck. Reinstall the radiator cap.

NOTE: If equipped with a fluid reservoir tank, fill it up to the MAX level.

4. Operate the engine at 2,000 rpm for a few minutes and check the system for signs of leaks.

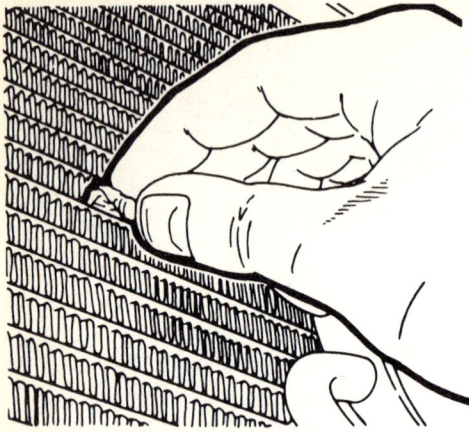

Keep the radiator fins clear for maximum cooling

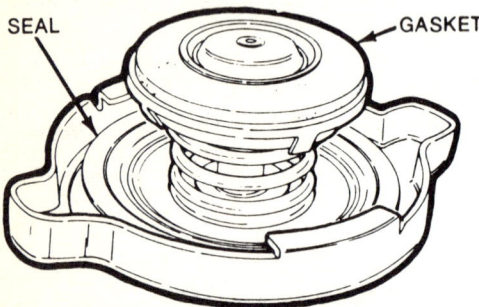

SEAL   GASKET

Check the radiator cap's rubber gasket and metal seal for deterioration at least once a year

Radiator Cap Inspection

Allow the engine to cool sufficiently before attempting to remove the radiator cap. Use a rag to cover the cap, then remove by pressing down and turning counterclockwise to the first stop. If any hissing is noted (indicating the release of pressure), wait until the hissing stops completely, then press down again and turn counterclockwise until the cap can be removed.

― **CAUTION** ―

DO NOT attempt to remove the radiator cap while the engine is hot. Severe personal injury from steam burns can result.

Check the condition of the radiator cap gasket and seal inside of the cap. The radiator cap is designed to seal the cooling system under normal operating conditions which allows the build up of a certain amount of pressure (this pressure rating is stamped or printed on the cap). The pressure in the system raises the boiling point of the coolant to help prevent overheating. If the radiator cap does not seal, the boiling point of the coolant is lowered and overheating will occur. If the cap must be replaced, purchase the new cap according to the pressure rating which is specified for your vehicle.

Prior to installing the radiator cap, inspect and clean the filler neck. If you are reusing the old cap, clean it thoroughly with clear water. After turning the cap on, make sure the arrows align with the overflow hose.

Axles

FLUID LEVEL CHECK

The standard front and rear axle differentials use SAE 80W/90 gear oil. Either is acceptable for use in the differential hous-

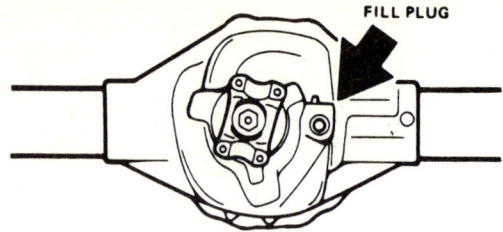

FILL PLUG

AMC axle fill plug location

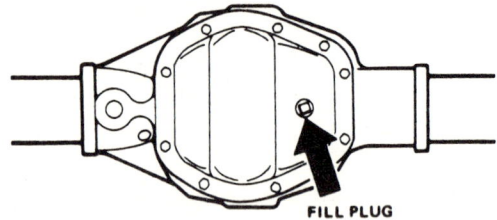

FILL PLUG

Dana/Spicer axle fill plug location

ing. Powr-Lok® differentials use only Jeep Powr-Lok® Lubricant or its equivalent. In Trac-Lok® axles, use any limited slip gear oil meeting SAE 75W/90, 80W/90 or 85W/90 specifications. Check the level of the oil in the differential housing every 5,000 miles under normal driving conditions and every 3,000 miles if the vehicle is used in severe driving conditions. The level should be up to the filler hole. When you remove the filler plug, the oil should start to run out. If it does not, replenish the supply until it does.

The lubricant should be changed every 30,000 miles. If running in deep water, change the lubricant daily.

DRAIN AND REFILL

1. Remove the axle differential housing cover and allow the lubricant to drain out into a proper container.
2. Install the differential housing cover and a new gasket.
3. Tighten the cover attaching bolts to 15-25 ft. lbs.
4. Remove the fill plug and add new lubricant to the fill hole level.
5. Replace the fill plug.

NOTE: Trac-Lok® (limited-slip) differentials may be cleaned only by disassembling the unit and wiping with clean, lint-free rags.

Manual Steering Gear

FLUID LEVEL CHECK

1971

There is a fill plug on top of the steering gear box. The level should be maintained at the bottom of the fill plug hole. The correct lubricant is SAE 80W/90 gear oil.

1972 and Later

These models use a steering box which is packed with grease. There is normally no need to add lubricant. However, the cover bolt opposite the adjuster may be removed for filling.

Power Steering Reservoir

The level of the fluid should be at the correct point on the dipstick attached to the inside of the lid of the power steering pump. Replenish the supply with DEXRON®II automatic transmission fluid.

Steering Knuckle

The axle shaft universal joints on 1971 models are located in the steering knuckle and are bathed in oil as they turn. To check the fluid level in the steering knuckle, remove the filler plug from the inside of the knuckle. The fluid should be at the level of the hole. If it is not, replenish the supply. Examine the knuckle for leaks if the level is abnormally low. A leak should be readily visible.

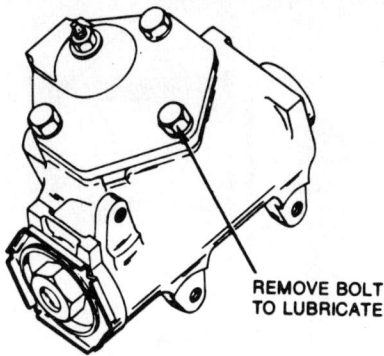

Manual steering gear fill hole

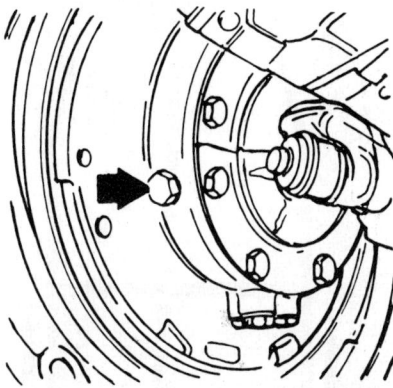

Front axle steering knuckle fill plug location

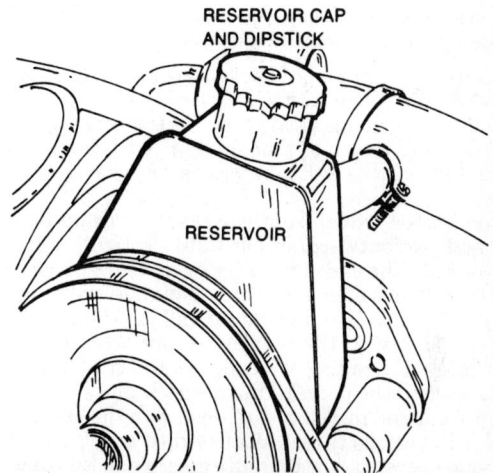

Power steering pump dipstick location

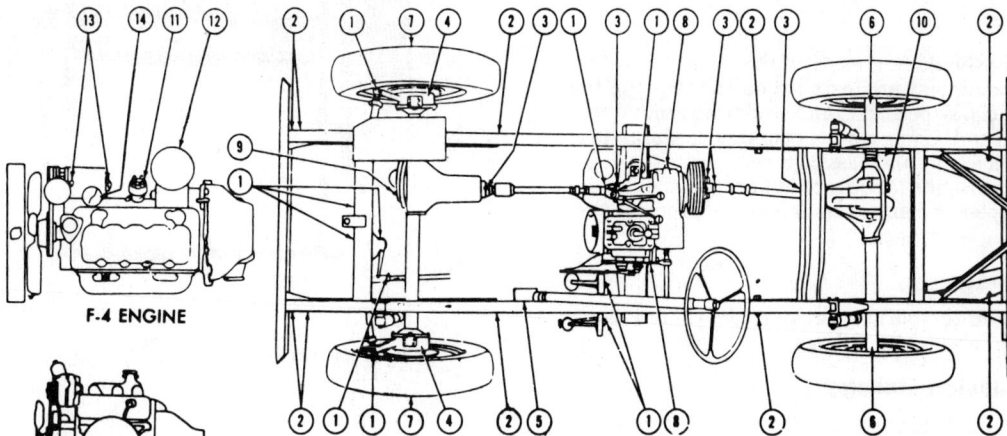

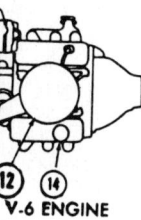

1. Chasis bearings
2. Spring shackle bushings
   Spring pivot bolt bushings
   Universal joints
3. Driveshafts
4. Front axle shaft
5. Steering gear housing
6. Rear wheel bearings
7. Front wheel bearings
8. 3 Speed transmission and transfer case
   4 Speed transmission and transfer case[1]

9. Differentials Front, Rear
10. Speedometer cable
11. Distributor
    Oiler
    Wick
    Picot
    Cam
12. Air cleaner
13. Generator
14. Engine

1971 lubrication chart

Chassis Greasing

The lubrication chart indicates where the grease fittings are located. The vehicle should be greased according to the intervals in the Preventive Maintenance Schedule at the end of this section.

Water resistant EP chassis lubricant (grease) should be used for all chassis grease points.

Every year or 7,500 miles the front suspension ball points, both upper and lower on each side of the truck, must be greased. Most trucks covered in this guide should be equipped with grease nipples on the ball joints, although some may have plugs which must be removed and nipples fitted.

WARNING: Do not pump so much grease into the ball joint that excess grease squeezes out of the rubber boot. This destroys the watertight seal.

Jack up the front end of the truck and safely support it with jackstands. Block the rear wheels and firmly apply the parking brake. If the truck has been parked in temperatures below 20°F (−6°C) for any length of time, park it in a heated garage for an hour or so until the ball joints loosen up enough to accept the grease.

Depending on which front wheel you work on first, turn the wheel and tire outward, either full-lock right or full-lock left. You now have the ends of the upper and lower suspension control arms in front of you; the grease nipples are visible pointing up (top ball joint) and down (lower ball joint) through the end of each control arm. If the nipples are not accessible enough, remove the wheel and tire. Wipe all dirt and crud from the nipples or from around the plugs (if installed). If plugs are on the truck, remove them and install grease nipples in the holes (nipples are available in various thread sizes at most auto parts stores). Using a hand operated, low pressure grease gun loaded with a quality chassis grease, grease the ball joint only until the rubber joint boot begins to swell out.

Steering Linkage

The steering linkage should be greased at the same interval as the ball joints. Grease nipples are installed on the steering tie rod ends on most models. Wipe all dirt and crud from around the nipples at each tie rod end. Using a hand operated, low pressure grease gun loaded with a suitable chassis grease, grease the linkage until the old grease begins to squeeze out around the tie rod ends. Wipe off the nipples and any excess grease. Also grease the nipples on the steering idler arms.

Parking Brake Linkage

Use chassis grease on the parking brake cable where it contacts the cable guides, levers and linkage.

Automatic Transmission Linkage

Apply a small amount of clean engine oil to the kickdown and shift linkage points at 7,500 mile intervals.

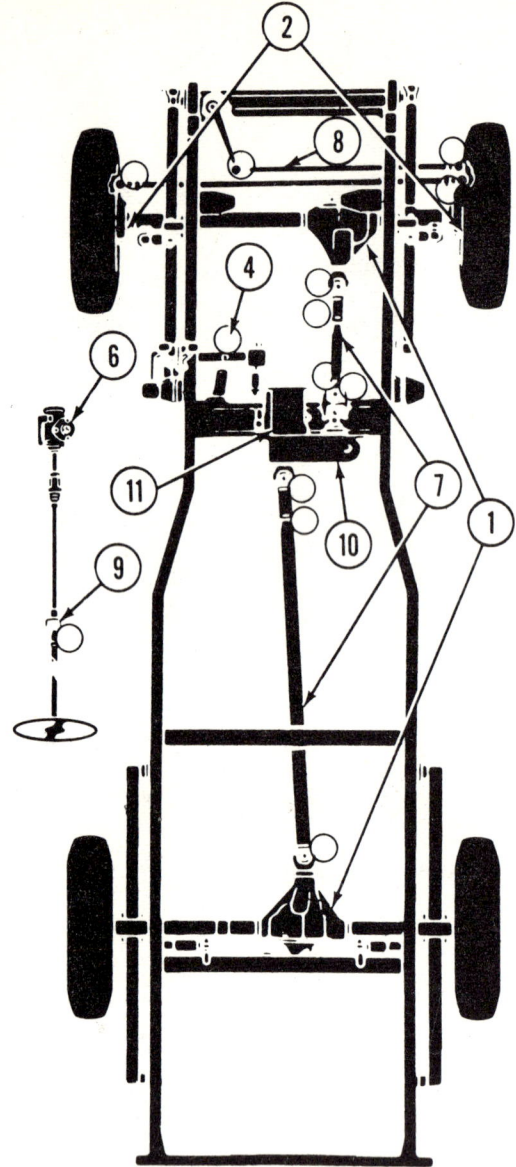

⊙ LUBRICATION POINTS

1. Differentials
2. Front wheel bearings
3. Not used
4. Clutch lever and linkage
5. Not used
6. Manual steering gear
7. Driveshafts
8. Steering linkage
9. Steering shaft U-joint
10. Transfer case
11. Transmission

1972 and later lubrication chart

OUTSIDE VEHICLE MAINTENANCE

Lock Cylinders

Apply graphite lubricant sparingly through the key slot. Insert the key and operate the lock several times to be sure that the lubricant is worked into the lock cylinder.

Door Hinges and Hinge Checks

Spray a silicone lubricant on the hinge pivot points to eliminate any binding conditions. Open and close the door several times to be sure that the lubricant is evenly and thoroughly distributed.

Tailgate or Liftgate

Spray a silicone lubricant on all of the pivot and friction surfaces to eliminate any squeaks or binds. Work the tailgate to distribute the lubricant

Body Drain Holes

Be sure that the drain holes in the doors and rocker panels are cleared of obstruction. A small screwdriver can be used to clear them of any debris.

PUSHING AND TOWING

To push start your vehicle, (manual transmission only) follow the procedures below. Check to make sure that the bumpers of both vehicles are aligned so neither will be damaged. Be sure that all electrical system components are turned off (headlights, heater blower, etc.). Turn on the ignition switch. Place the shift lever in second or third and push in the clutch pedal. At about 15 mph (24 kmh), signal the driver of the pushing vehicle to fall back, depress the accelerator pedal, and release the clutch pedal slowly. The engine should start.

When you are doing the pushing or pulling, make sure that the two bumpers match so you won't damage the vehicle you are to push. Another good idea is to put an old tire in between the two vehicles. If the bumpers don't match, perhaps you should tow the other vehicle. Try to keep your Jeep right up against the other vehicle while you are pushing. If the two vehicles do separate, stop and start over again instead of trying to catch up and ramming the other vehicle. Also try, as much as possible, to avoid riding or slipping the clutch.

If you have to tow the other vehicle, make sure that the tow chain or rope is sufficiently long and strong, and that it is attached securely to both vehicles at a strong place. Attach the chain at a point on the frame or as close to it as possible. Once again, go slowly and tell the other driver to do the same. Warn the other driver not to allow too much slack in the line when he gains traction and can move under his own power. Otherwise he may run over the tow line and damage both vehicles.

If your Jeep must be towed, follow these guidelines:

1. A Jeep with a manual transmission can be towed with either all four wheels or either axle on the ground for any distance at a safe speed with both the transmission and transfer case in Neutral.

2. To tow a Jeep with an automatic transmission and Quadra-Trac®, the driveshaft to the axle(s) remaining on the ground must be disconnected. Be sure to index mark the driveshafts and yoke flanges for alignment upon assembly. Also, the driveshafts must be tied securely up out of the way or removed completely while the vehicle is being towed.

3. A Jeep equipped with an automatic transmission and Quadra-Trac® with the optional low range reduction unit can be towed with all four wheels on the ground without disconnecting the driveshafts. Place the transmission shift lever in Park, the low range reduction unit shift lever in Neutral, and the emergency drive control knob in the Normal position. If the emergency drive system was engaged when the engine was shut down, it will have to be restarted and the emergency drive control knob turned to the Normal position to disengage the system since the control mechanism is vacuum operated.

WARNING: Never tow the Jeep with the emergency drive system engaged or the reduction unit in low range!

In all cases, unnecessary wear and tear can be avoided by disconnecting the driveshafts at the differentials and either tying them up out of the way or removing them altogether. Be sure to index mark the driveshafts and yoke flanges for proper alignment during assembly. If the Jeep is equipped with free running front hubs (manual transmission only), there is no need to remove the front driveshaft, simply disengage the hubs.

JACKING

Scissors jacks or hydraulic jacks are recommended for all Jeep vehicles. To change a tire, place the jack beneath the spring plate, below the axle, near the wheel to be changed.

Make sure that you are on level ground, that the transmission is in Reverse or with automatic transmissions, Park; the park-ing brake is set, and the tire diagonally opposite to the one to be changed is blocked so that it will not roll. Loosen the lug nuts before you jack the wheel to be changed completely free of the ground.

TRAILER TOWING

Jeep vehicles have long been popular as trailer towing vehicles. Their strong construction, 4-wheel drive and wide range of engine/transmission combinations make them ideal for towing campers, boat trailers and utility trailers.

Factory trailer towing packages are available on most Jeep vehicles. However, if you are installing a trailer hitch and wiring on your Jeep, there are a few thing that you ought to know.

Trailer Weight

Trailer weight is the first, and most important, factor in determining whether or not your vehicle is suitable for towing the trailer you have in mind. The horsepower-to-weight ratio should be calculated. The basic standard is a ratio of 35:1. That is, 35 pounds of GVW for every horsepower.

To calculate this ratio, multiply you engine's rated horsepower by 35, then subtract the weight of the vehicle, including passengers and luggage. The resulting figure is the ideal maximum trailer weight that you can tow. One point to consider: a numerically higher axle ratio can offset what appears to be a low trailer weight. If the weight of the trailer that you have in mind is somewhat higher than the weight you just calculated, you might consider changing your rear axle ratio to compensate.

Hitch Weight

There are three kinds of hitches: bumper mounted, frame mounted, and load equalizing.

Bumper mounted hitches are those which attach solely to the vehicle's bumper. Many states prohibit towing with this type of hitch, when it attaches to the vehicle's stock bumper, since it subjects the bumper to stresses for which it was not designed. Aftermarket rear step bumpers, designed for trailer towing, are acceptable for use with bumper mounted hitches.

Frame mounted hitches can be of the type which bolts to two or more points on the frame, plus the bumper, or just to several points on the frame. Frame mounted hitches can also be of the tongue type, for Class I towing, or, of the receiver type, for classes II and III.

Load equalizing hitches are usually used for large trailers. Most equalizing hitches are welded in place and use equalizing bars and chains to level the vehicle after the trailer is hooked up.

The bolt-on hitches are the most common, since they are relatively easy to install.

Check the gross weight rating of your trailer. Tongue weight is usually figured as 10% of gross trailer weight. Therefore, a trailer with a maximum gross weight of 2,000 lb. will have a maximum tongue weight of 200 lb. Class I trailers fall into this category. Class II trailers are those with a gross weight rating of 2,000-3,500 lb., while Class III trailers fall into the 3,500-6,000 lb. category. Class IV trailers are those over 6,000 lb. and are for use with fifth wheel trucks, only.

When you've determined the hitch that you'll need, follow the manufacturer's installation instructions, exactly, especially when it comes to fastener torques. The hitch will subjected to a lot of stress and good hitches come with hardened bolts. Never substitute an inferior bolt for a hardened bolt.

Wiring

Wiring the car for towing is fairly easy. There are a number of good wiring kits available and these should be used, rather than trying to design your own. All trailers will need brake lights and turn signals as well as tail lights and side marker lights. Most states require extra marker lights for overwide trailers. Also, most states have recently required back-up lights for trailers, and most trailer manufacturers have been building trailers with back-up lights for several years.

Additionally, some Class I, most Class II and just about all Class III trailers will have electric brakes.

Add to this number an accessories wire, to operate trailer internal equipment or to charge the trailer's battery, and you can have as many as seven wires in the harness.

Determine the equipment on your trailer and buy the wiring kit necessary. The kit will contain all the wires needed, plus a plug adapter set which included the female plug, mounted on the bumper or hitch, and the male plug, wired into, or plugged into the trailer harness.

When installing the kit, follow the manufacturer's instructions. The color coding of the wires is standard throughout the industry.

One point to note, some domestic vehicles, and most imported vehicles, have separate turn signals. On most domestic vehicles, the brake lights and rear turn signals operate with the same bulb. For those vehicles with separate turn signals, you can pur-

Recommended Equipment Checklist

Equipment	Class I Trailers Under 2,000 pounds	Class II Trailers 2,000-3,500 pounds	Class III Trailers 3,500-6,000 pounds	Class IV Trailers 6,000 pounds and up
Hitch	Frame or Equalizing	Equalizing	Equalizing	Fifth wheel Pick-up truck only
Tongue Load Limit**	Up to 200 pounds	200-350 pounds	350-600 pounds	600 pounds and up
Trailer Brakes	Not Required	Required	Required	Required
Safety Chain	3/16″ diameter links	1/4″ diameter links	5/16″ diameter links	—
Fender Mounted Mirrors	Useful, but not necessary	Recommended	Recommended	Recommended
Turn Signal Flasher	Standard	Constant Rate or heavy duty	Constant Rate or heavy duty	Constant Rate or heavy duty
Coolant Recovery System	Recommended	Required	Required	Required
Transmission Oil Cooler	Recommended	Recommended	Recommended	Recommended
Engine Oil Cooler	Recommended	Recommended	Recommended	Recommended
Air Adjustable Shock Absorbers	Recommended	Recommended	Recommended	Recommended
Flex or Clutch Fan	Recommended	Recommended	Recommended	Recommended
Tires	***	***	***	***

NOTE: The information in this chart is a guide. Check the manufacturer's recommendations for your car if in doubt.

*Local laws may require specific equipment such as trailer brakes or fender mounted mirrors. Check your local laws. Hitch weight is usually 10-15% of trailer gross weight and should be measured with trailer loaded.

**Most manufacturer's do not recommend towing trailers of over 1,000 pounds with compacts. Some intermediates cannot tow Class III trailers

***Check manufacturer's recommendations for your specific car : trailer combination.

—Does not apply

chase an isolation unit so that the brake lights won't blink whenever the turn signals are operated, or, you can go to your local electronics supply house and buy four diodes to wire in series with the brake and turn signal bulbs. Diodes will isolate the brake and turn signals. The choice is yours. The isolation units are simple and quick to install, but far more expensive than the diodes. The diodes, however, require more work to install properly, since they require the cutting of each bulb's wire and soldering in place of the diode.

One, final point, the best kits are those with a spring loaded cover on the vehicle mounted socket. This cover prevent dirt and moisture from corroding the terminals. Never let the vehicle socket hang loosely. Always mount it securely to the bumper or hitch.

Cooling

ENGINE

One of the most common, if not THE most common, problems associated with trailer towing is engine overheating.

With factory installed trailer towing packages, a heavy duty cooling system is usually included. Heavy duty cooling systems are available as optional equipment on most Jeep vehicles, with or without a trailer package. If you have one of these extra-capacity systems, you shouldn't have any overheating problems.

If you have a standard cooling system, without an expansion tank, you'll definitely need to get an aftermarket expansion tank kit, preferably one with at least a 2 quart capacity. These kits are easily installed on the radiator's overflow hose, and come with a pressure cap designed for expansion tanks.

Another helpful accessory is a Flex Fan. These fan are large diameter units are designed to provide more airflow at low speeds, with blades that have deeply cupped surfaces. The blades then flex, or flatten out, at high speed, when less cooling air is needed. These fans are far lighter in weight than stock fans, requiring less horsepower to drive them. Also, they are far quieter than stock fans.

If you do decide to replace your stock fan with a flex fan, note that if your Jeep has a fan clutch, a spacer between the flex fan and water pump hub will be needed.

Aftermarket engine oil coolers are helpful for prolonging engine oil life and reducing overall engine temperatures. Both of these factors increase engine life.

While not absolutely necessary in towing Class I and some Class II trailers, they are recommended for heavier Class II and all Class III towing.

Engine oil cooler systems consist of an adapter, screwed on in place of the oil filter, a remote filter mounting and a multi-tube, finned heat exchanger, which is mounted in front of the radiator or air conditioning condenser.

TRANSMISSION

An automatic transmission is usually recommended for trailer towing. Modern automatics have proven reliable and, of course, easy to operate, in trailer towing.

The increased load of a trailer, however, causes an increase in the temperature of the automatic transmission fluid. Heat is the worst enemy of an automatic transmission. As the temperature of the fluid increases, the life of the fluid decreases.

It is essential, therefore, that you install an automatic transmission cooler.

The cooler, which consists of a multi-tube, finned heat exchanger, is usually installed in front of the radiator or air conditioning compressor, and hooked inline with the transmission cooler tank inlet line. Follow the cooler manufacturer's installation instructions.

Select a cooler of at least adequate capacity, based upon the combined gross weights of the Jeep and trailer.

Cooler manufacturers recommend that you use an aftermarket cooler in addition to, and not instead of, the present cooling tank in your Jeep radiator. If you do want to use it in place of the

radiator cooling tank, get a cooler at least two sizes larger than normally necessary.

NOTE: A transmission cooler can, sometimes, cause slow or harsh shifting in the transmission during cold weather, until the fluid has a chance to come up to normal operating temperature. Some coolers can be purchased with or retrofitted with a temperature bypass valve which will allow fluid flow through the cooler only when the fluid has reached operating temperature, or above.

CAPACITIES

Model	Engine	Crankcase Incl. Filter (qt.)	Transmission (pt.) ▲				Transfer Case (pt.)	Drive Axle (pt.)		Fuel Tank (gal.)	Cooling System (qt.)	
			3-sp	4-sp	5-sp	Auto		Front	Rear		w/AC	wo/AC
CJ-5	4-134	5.0	3.0	6.75	—	—	3.50	2.5	2.50	10.5	—	12.0
	4-151	3.0	—	3.00	—	—	4.00	2.5	4.8	14.8	7.8	7.8
	6-225	4.0	3.0	6.75	—	—	3.50	2.5	2.5	10.5	10.0	—
	6-232	5.0	2.5①	6.5	—	—	3.25	2.5	②	16.0	10.5	—
	6-258	5.0	2.5①	③	4.5	—	④	2.5	②	⑤	10.5	12.0
	8-304	5.0	2.5①	③	4.5	—	④	2.5	②	⑤	14.0	15.5
CJ-6	4-134	5.0	3.0	6.75	—	—	3.50	2.5	2.5	10.5	12.0	—
	6-225	4.0	3.0	6.75	—	—	3.50	2.5	2.5	10.5	12.0	—
	6-232	5.0	2.5①	6.5	—	—	3.25	2.5	②	16.0	10.5	—
	6-258	5.0	2.5①	③	4.5	—	④	2.5	②	⑤	10.5	12.0
	8-304	5.0	2.75	③	—	—	3.25	2.5	2.5	16.0	10.5	12.0
CJ-7	4-150	4.0	—	3.0	4.5	—	4.00	2.5	4.8	20.0	7.8	7.8
	4-151	3.0	—	3.0	4.5	17.0	4.00	2.5	4.8	14.8	7.8	7.8
	6-232	5.0	2.8	6.5	—	—	3.25	2.5	4.8	16.0	10.5	—
	6-258	5.0	2.8	③	4.5	⑥	⑦	2.5	4.8	⑤	10.5	12.0
	8-304	5.0	2.8	③	—	17.0	⑦	2.5	4.8	⑤	13.0	14.5
Scrambler	4-150	4.0	—	3.9	4.5	17.0	4.00	2.5	4.8	20.0	7.8	7.8
	4-151	3.0	—	③	4.5	17.0	4.00	2.5	4.8	16.0	7.8	7.8
	6-258	5.0	—	③	4.5	17.0	4.00	2.5	4.8	⑧	10.5	12.0
Wrangler	4-150	4.0	—	—	⑨	—	⑭	2.5	2.5	⑩	9.0	9.0
	6-258	6.0	—	—	⑨	17.0	⑭	2.5	2.5	⑩	10.5	10.5

▲ The automatic transmission figure is for total capacity. For drain and refill only, use about one half this amount, then run the engine until the fluid is warm and add as much as necessary to bring the level to the full mark on the dipstick.

① 1976 and later: 3.0
② Dana rear: 3.0
   AMC 7⁹/₁₆ in. rear: 4.0
   AMC 8⁷/₈ in. rear: 4.8
③ T-18: 6.5
   SR-4: 3.0
   T-176: 3.5
   T-4: 3.9
④ 1972–79: 3.25
   1980–83: 4.0
⑤ 1972–79: 16.0
   1980 and later: 14.8
⑥ 1976–79: 22.0
   1980 and later: 17.0

⑦ Except Quadra-Trac®: 3.25
   Quadra-Trac® without reduction unit: 3.5
   Quadra-Trac® with reduction unit: 4.5
⑧ 1981–83: 16.0
   1984–86: 20.0
⑨ AISIN AX5: 7.0
   Peugeot BA10/5: 4.5
⑩ Standard: 15.0
   Optional: 20.0
⑪ Transmission and transfer case are filled together, sharing a common sump.
   Total capacity for the 3-speed with transfer case is 6.5 pts.
⑫ Transfer case capacity on vehicles with the optional 4-speed transmission is 3.5 pts.
⑬ Dana 27: 2.5
   Dana 44: 3.0
⑭ NP-207: 4.5
   NP-231: 3.3

PREVENTIVE MAINTENANCE
1971–73

Interval	Item	Service
Every 6,000 miles	Engine oil and filter	Change
	Steering gear	Check level
	Differentials	Check level
	Manual transmission	Check level
	Transfer case	Check level
	Drive belts	Check
	Air cleaner	Change filter
Every 12,000 miles	All chassis lube fittings	EP chassis lube
	Front and rear wheel bearings	Clean and repack
	U-joints	EP chassis lube
	Fuel filter	Replace
	PCV valve	Replace
	Oil filler cap	Clean
	Timing and dwell	Check
	Heat riser	Lubricate
	Point, condenser, rotor	Replace
	Spark plugs	Replace
Every 30,000 miles	Differentials	Change fluid
	Manual transmission	Change fluid
	Spark plugs wires	Change
	Transfer case	Change fluid

PREVENTIVE MAINTENANCE
1974–76

Interval	Item	Service
Every 5,000 miles	Engine oil and filter	Change
	Steering gear	Check level
	Power steering reservoir	Check level
	Heat riser	Lubricate
	Differentials	Check level
	Manual transmission	Check level
	Transfer case	Check level
	Automatic transmission	Check level
	All chassis lube fittings	EP chassis lube
	Drive belts	Check
	Air cleaner	Change filter
Every 10,000 miles	Driveshaft splines	EP chassis lube
Every 15,000 miles	Front and rear wheel bearings	Clean and repack
	U-joints	EP chassis lube
	Fuel filter	Replace
	PCV valve	Replace
	Oil filler cap	Clean
	Timing and dwell	Check
	Point, condenser, rotor	Replace

PREVENTIVE MAINTENANCE
1974–76

Interval	Item	Service
Every 15,000 miles	Spark plugs	Replace
	EGR valve port	Clean
Every 25,000 miles	Automatic transmission	Change fluid and filter
Every 30,000 miles	Differentials	Change fluid
	Manual transmission	Change fluid
	Spark plug wires	Change
	Model 20 transfer case	Change fluid

PREVENTIVE MAINTENANCE
1977–79

Interval	Item	Service
Every 5,000 miles	Engine oil and filter	Change
	Steering gear	Check level
	Power steering reservoir	Check level
	Heat riser	Lubricate
	Differentials	Check level
	Manual transmission	Check level
	Transfer case	Check level
	Automatic transmission	Check level
	Drive belts	Check
	Air cleaner	Change filter
Every 10,000 miles	Driveshaft splines	EP chassis lube
Every 15,000 miles	All chassis lube fittings	EP chassis lube
	U-joints	EP chassis lube
	Fuel filter	Replace
	PCV valve	Replace
	Oil filler cap	Clean
	Timing and dwell	Check
	Point, condenser, rotor	Replace
	Spark plugs	Replace
	EGR valve port	Clean
Every 25,000 miles	Automatic transmission	Change fluid and filter
Every 30,000 miles	Differentials	Change fluid
	Manual transmission	Change fluid
	Model 20 transfer case	Change fluid
	Spark plug wires	Change
	Front wheel bearings	Clean and repack

PREVENTIVE MAINTENANCE
1980–86

Interval	Item	Service
Every 5,000 miles	Engine oil and filter	Change
	Steering gear	Check level
	Power steering reservoir	Check level

PREVENTIVE MAINTENANCE
1980–86

Interval	Item	Service
Every 5,000 miles	Heat riser	Lubricate
	Differentials	Check level
	Manual transmission	Check level
	Transfer case	Check level
	Automatic transmission	Check level
	Air cleaner	Change filter
	Drive belts	Check
Every 10,000 miles	Driveshaft splines	EP chassis lube
Every 15,000 miles	All chassis lube fittings	EP chassis lube
	U-joints	EP chassis lube
	Fuel filter	Replace
	PCV valve	Replace
	Oil filler cap	Clean
	Spark plugs	Replace
	EGR valve port	Clean
Every 25,000 miles	Automatic transmission	Change fluid and filter
Every 30,000 miles	Differentials	Change fluid
	Manual transmission	Change fluid
	Spark plug wires	Change
	Front wheel bearings	Clean and repack
	Transfer case	Change fluid

PREVENTIVE MAINTENANCE
1987–90

Interval	Item	Service
Every 5,000 miles	Engine oil and filter	Change
	Steering gear	Check level
	Power steering reservoir	Check level
	Differentials	Check level
	Manual transmission	Check level
	Transfer case	Check level
	Automatic transmission	Check level
	Air cleaner	Change filter
	Drive belts	Check
Every 15,000 miles	All chassis lube fittings	EP chassis lube
	U-joints	EP chassis lube
	Fuel filter	Replace
	PCV valve	Replace
	Oil filler cap	Clean
	Spark plugs	Replace
Every 30,000 miles	Spark plug wires	Change
	Front wheel bearings	Clean and repack

PREVENTIVE MAINTENANCE
1987–90

Interval	Item	Service
Every 48,000 miles	Manual transmission	Change fluid
	Differentials	Change fluid
	Automatic transmission	Change fluid and filter
	Transfer case	Change fluid

Engine Performance and Tune-Up

2

QUICK REFERENCE INDEX

GENERAL INDEX

TUNE-UP PROCEDURES

TUNE-UP SPECIFICATIONS

Engine	Years	Spark Plugs Type	Gap (in.)	Distributor Point Gap (in.)	Dwell (deg.)	Ignition Timing (deg.) Man. Trans.	Auto. Trans.	Valve Clearance ▲ In.	Exh.	Idle Speed Man. Trans.	Auto. Trans.
4-134	1971	J-8	0.030	①	②	5B	—	0.018	0.016	600	—
4-150	1984–86	RFN-14LY	0.035	—Electronic—		③	—	Hyd.	Hyd.	750	—
	1987–90	RC-12LYC	0.035	—Electronic—		④	—	Hyd.	Hyd.	⑤	—
4-151	1980	R44TSX	0.06	—Electronic—		10B	—	Hyd.	Hyd.	900	—
	1981	R44TSX	0.060	—Electronic—		10B	12B	Hyd.	Hyd.	900	700
	1982–83	R44TSX	0.060	—Electronic—		12B	—	Hyd.	Hyd.	900	—
6-225	1971	44S	0.035	0.016	30	5B	—	Hyd.	Hyd.	550	—
6-232	1972	N-12Y	0.035	0.016	32	5B	—	Hyd.	Hyd.	675	—
	1973	N-12Y	0.035	0.016	32	5B	—	Hyd.	Hyd.	700	—
	1974	N-12Y	0.035	0.016	32	5B	—	Hyd.	Hyd.	600	—
	1975	N-12Y	0.035	—Electronic—		5B	—	Hyd.	Hyd.	700	—
	1976	N-12Y	0.035	—Electronic—		8B	—	Hyd.	Hyd.	600	—
	1977	N-12Y	0.035	—Electronic—		⑥	—	Hyd.	Hyd.	⑦	—
	1978	N-13L	0.035	—Electronic—		⑥	—	Hyd.	Hyd.	⑦	—
6-258	1972–73	N-12Y	0.035	0.016	32	3B	—	Hyd.	Hyd.	700	—
	1974	N-12Y	0.035	0.016	32	3B	—	Hyd.	Hyd.	600	—
	1975	N-12Y	0.035	—Electronic—		3B	—	Hyd.	Hyd.	600	—
	1976	N-12Y	0.035	—Electronic—		6B	8B	Hyd.	Hyd.	600	⑧
	1977	N-12Y	0.035	—Electronic—		⑨	⑩	Hyd.	Hyd.	⑪	⑧
	1978	N-13L	0.035	—Electronic—		⑫	⑩	Hyd.	Hyd.	⑪	550
	1979	N-13L	0.035	—Electronic—		6B	4B	Hyd.	Hyd.	700	600
	1980	N-14LY	0.035	—Electronic—		⑬	⑭	Hyd.	Hyd.	700	600
	1981	RFN-14LY	0.035	—Electronic—		⑮	⑬	Hyd.	Hyd.	650	550
	1982–83	RFN-14LY	0.035	—Electronic—		⑯	⑯	Hyd.	Hyd.	⑰	⑱
	1984–90	RFN-14LY	0.035	—Electronic—		⑲	⑲	Hyd.	Hyd.	⑳	㉑
8-304	1972–74	N-12Y	0.035	0.016	32	5B	—	Hyd.	Hyd.	750	—
	1975	N-12Y	0.035	—Electronic—		5B	—	Hyd.	Hyd.	750	—
	1976	N-12Y	0.035	—Electronic—		5B	10B	Hyd.	Hyd.	750	700
	1977–78	N-12Y	0.035	—Electronic—		5B	㉒	Hyd.	Hyd.	750	700
	1979	N-12Y	0.035	—Electronic—		5B	8B	Hyd.	Hyd.	㉓	600
	1980	N-12Y	0.035	—Electronic—		㉔	㉕	Hyd.	Hyd.	700	600
	1981	N-12Y	0.035	—Electronic—		㉖	10B	Hyd.	Hyd.	㉗	600

NOTE: The specifications on the underhood sticker often reflect changes made during production. If the specifications on your vehicle's sticker disagree with the specifications in this chart, use the sticker specifications.

▲ Valve clearance is set on a cold engine.
① Autolite distributor: 0.020
   Delco distributor: 0.022
② Autolite distributor: 42
   Delco distributor: 25–34
③ All except high altitude: 12B
   High altitude: 19B
④ Set by computer; not adjustable
⑤ Not adjustable
⑥ Except high altitude: 5B
   High altitude: 10B

⑦ Except high altitude: 850
   High altitude: 600
⑧ Except Calif.: 550
   Calif.: 700
⑨ Except Calif. and high altitude: 3B
   Calif.: 6B
   High altitude: 10B
⑩ Except high altitude: 8B
   High altitude: 10B
⑪ Except high altitude: 850
   High altitude: 600

⑫ Except Calif and high altitude: 3B
   Calif.: 8B
   High altitude: 10B
⑬ Except Calif.: 8B
   Calif.: 6B
⑭ Except Calif.: 10B
   Calif.: 8B
⑮ Except Calif.: 8B
   Calif.: 4B
⑯ Except Calif.: 6B
   Calif.: 13B

TUNE-UP SPECIFICATIONS
(Footnotes, continued)

⑰ Except Calif.: 600
Calif.: 650
⑱ Except Calif.: 500
Calif.: 550
⑲ Except high altitude: 9B
High altitude: 16B
⑳ Except high altitude: 680
High altitude: 700 ± 70

㉑ Except high altitude: 600
High altitude: 650 ± 70
㉒ Except Calif.: 10B
Calif.: 5B
㉓ Except Calif.: 700
Calif.: 750
㉔ Except Calif.: 8B
Calif.: 5B

㉕ Except Calif.: 12B
Calif.: 10B
㉖ Except high altitude: 8B
High altitude: 12B
㉗ Except high altitude: 600
High altitude: 700

In order to extract the full measure of performance and economy from your engine it is essential that it be properly tuned at regular intervals. A regular tune-up will keep your vehicle's engine running smoothly and will prevent the annoying minor breakdowns and poor performance associated with an untuned engine.

A complete tune-up should be performed every 12,000 miles or twelve months, whichever comes first. This interval should be halved if the vehicle is operated under severe conditions, such as trailer towing, prolonged idling, continual stop and start driving, or if starting or running problems are noticed. It is assumed that the routine maintenance described in Section 1 has been kept up, as this will have a decided effect on the results of a tune-up. All of the applicable steps of a tune-up should be followed in order, as the result is a cumulative one.

If the specifications on the tune-up sticker in the engine compartment disagree with the Tune-Up Specifications chart in this section, the figures on the sticker must be used. The sticker often reflects changes made during the production run.

Spark Plugs

Spark plugs ignite the air and fuel mixture in the cylinder as the piston reaches the top of the compression stroke. The controlled explosion that results forces the piston down, turning the crankshaft and the rest of the drive train.

The average life of a spark plug is dependent on a number of factors; the mechanical condition of the engine; the type of fuel; driving conditions; and the driver.

When you remove the spark plugs, check their condition. They are a good indicator of the condition of the engine.

A small deposit of light tan or gray material on a spark plug that has been used for any period of time is to be considered normal. Additives in unleaded fuels may give a number of unusual color indications; for instance, MMT (a manganese anti-knock compound) will cause rust red deposits.

The gap between the center electrode and the side or ground electrode can be expected to increase not more than 0.001 in. (0.025mm) every 1,000 miles under normal conditions.

When a spark plug is functioning normally or, more accurately, when the plug is installed in an engine that is functioning properly, the plugs can be taken out, cleaned, regapped, and reinstalled in the engine without doing the engine any harm.

When, and if, a plug fouls and begins to misfire, you will have to investigate, correct the cause of the fouling, and either clean or replace the plug.

There are several reasons why a spark plug will foul and you can learn which reason by just looking at the plug. The two most common problems are oil fouling and pre-ignition/detonation.

Oil fouling is easily noticed as dark, wet oily deposits on the plug's electrodes. Oil fouling is caused by internal engine problems, the most common of which are worn valve seals or guides and worn or damaged piston rings. These problems can be corrected only by engine repairs.

Pre-ignition or detonation problems are characterized by extensive burning and/or damage to the plug's electrodes. The problem is caused by incorrect ignition timing or faulty spark control. Check the timing and/or diagnose the spark control system.

SPARK PLUG HEAT RANGE

Spark plug heat range is the ability of the plug to dissipate heat. The longer the insulator (or the farther it extends into the engine), the hotter the plug will operate; the shorter the insulator the cooler it will operate. A plug that absorbs little heat and remains too cool will quickly accumulate deposits of oil and carbon since it is not hot enough to burn them off. This leads to plug fouling and consequently to misfiring. A plug that absorbs too much heat will have no deposits, but, due to the excessive heat, the electrodes will burn away quickly and in some instances, preignition may result. Preignition takes place when plug tips get so hot that they glow sufficiently to ignite the fuel/air mixture before the actual spark occurs. This early ignition will usually cause a pinging during low speeds and heavy loads.

The general rule of thumb for choosing the correct heat range when picking a spark plug is: if most of your driving is long distance, high speed travel, use a colder plug; if most of your driving is stop and to, use a hotter plug. Original equipment plugs are compromise plugs, but most people never have occasion to change their plugs from the factory-recommended heat range.

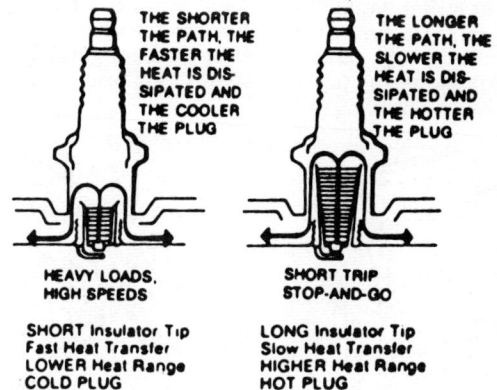

THE SHORTER THE PATH, THE FASTER THE HEAT IS DISSIPATED AND THE COOLER THE PLUG

THE LONGER THE PATH, THE SLOWER THE HEAT IS DISSIPATED AND THE HOTTER THE PLUG

HEAVY LOADS, HIGH SPEEDS

SHORT TRIP STOP-AND-GO

SHORT Insulator Tip
Fast Heat Transfer
LOWER Heat Range
COLD PLUG

LONG Insulator Tip
Slow Heat Transfer
HIGHER Heat Range
HOT PLUG

Spark plug heat range

REMOVAL

1. Remove the wires one at a time and number them so you won't cross them when you replace them.
2. Remove the wire from the end of the spark plug by grasp-

ing the wire by the rubber boot. If the boot sticks to the plug, remove it by twisting and pulling at the same time. Do not pull the wire itself or you will most certainly damage the core, or tear the connector.

3. Use a spark plug socket to loosen all of the plugs about two turns.

4. If compressed air is available, blow off the area around the spark plug holes. Otherwise, use a rag or a brush to clean the area. Be careful not to allow any foreign material to drop into the spark plug holes.

5. Remove the plugs by unscrewing them the rest of the way from the engine.

INSPECTION

Check the plugs for deposits and wear. If they are not going to be replaced, clean the plugs thoroughly. Remember that any kind of deposit will decrease the efficiency of the plug. Plugs can be cleaned on a spark plug cleaning machine, which can sometimes be found in service stations, or you can do an acceptable job of cleaning with a stiff brush.

Check spark plug gap before installation. The ground electrode must be aligned with the center electrode and the specified size wire gauge should pass through the gap with a slight drag. If the electrodes are worn, it is possible to file them level.

INSTALLATION

1. Insert the plugs in the spark plug hole and tighten them hand tight. Take care not to crossthread them.

2. Tighten the plugs to 11 ft. lbs. on the 4-151; 25-30 ft. lbs. on all other engines.

3. Install the spark plug wires on their plugs. Make sure that each wire is firmly connected to each plug.

CHECKING AND REPLACING SPARK PLUG CABLES

Visually inspect the spark plug cables for burns, cuts, or breaks in the insulation. Check the spark plug boots and the nipples on the distributor cap and coil. Replace any damages wiring. If no physical damage is obvious, the wires can be checked with an ohmmeter for excessive resistance.

When installing a new set of spark plug cables, replace the cables one at a time so there will be no mixup. Start by replacing the longest cable first. Install the boot firmly over the spark plug. Route the wire exactly the same as the original. Insert the nipple firmly into the tower on the distributor cap. Repeat the process for each cable.

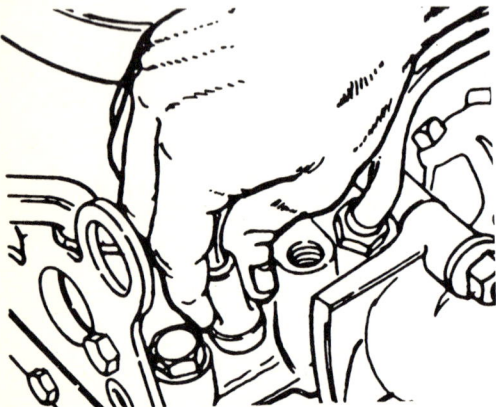

Plugs in good condition can be filled and reused

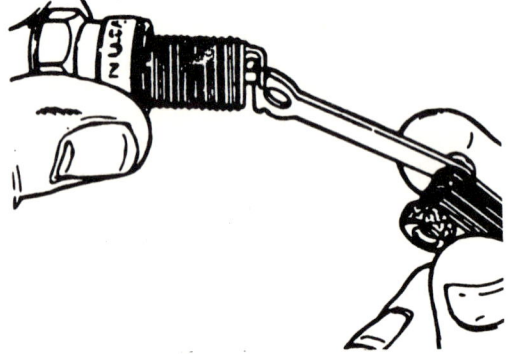

Adjust electrode gap by bending the side electrode

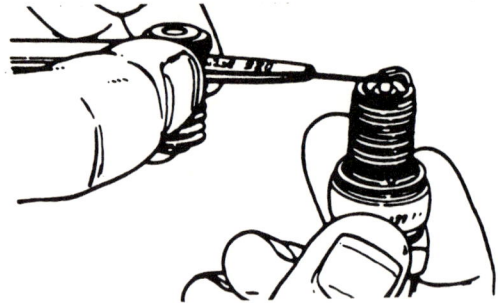

Always use a wire gauge to check the electrode gap

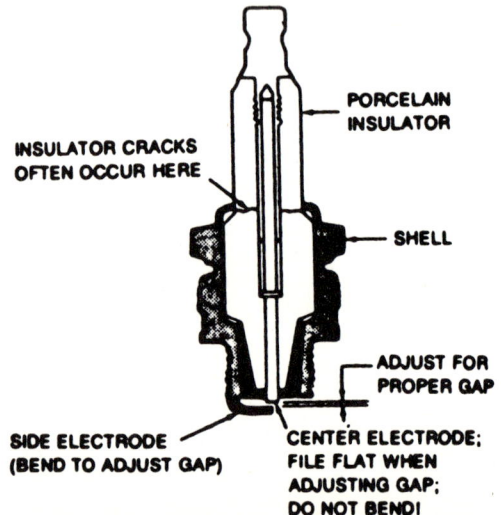

Twist and pull on the rubber boot to remove the spark plug wires; never pull on the wire itself

PORCELAIN INSULATOR

INSULATOR CRACKS OFTEN OCCUR HERE

SHELL

ADJUST FOR PROPER GAP

SIDE ELECTRODE (BEND TO ADJUST GAP)

CENTER ELECTRODE; FILE FLAT WHEN ADJUSTING GAP; DO NOT BEND!

Cross section of a spark plug

HEI PLUG WIRE RESISTANCE

Wire Length (inches)	Minimum Ohms	Maximum Ohms
Up to 15	3,000	10,000
15–25	4,000	15,000
25-35	6,000	20,000
Over 35	8,000	25,000

FIRING ORDERS

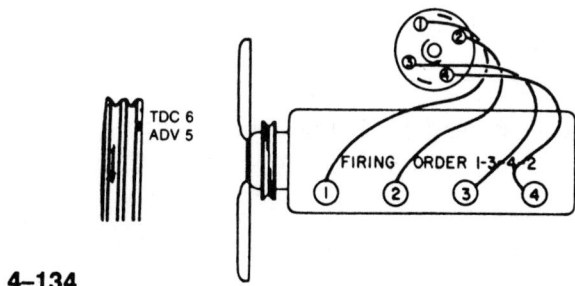

4–134

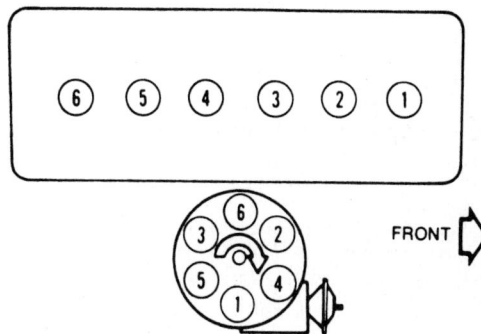

CLOCKWISE ROTATION
1-5-3-6-2-4

SIX-CYLINDER ENGINES

Inline 6-cylinder engines starting 1975 (electronic ignition)

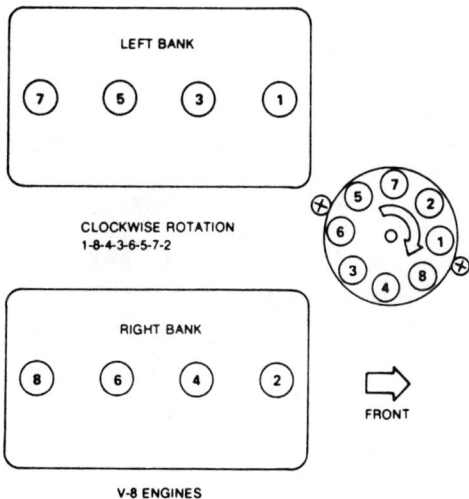

V8 engines through 1974 (point-type ignition)

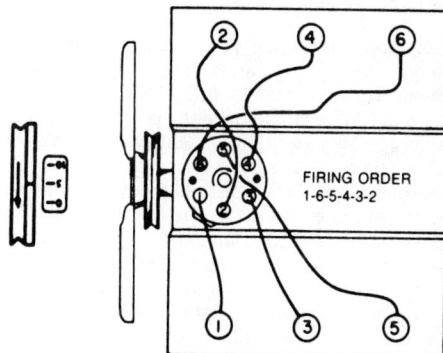

V6

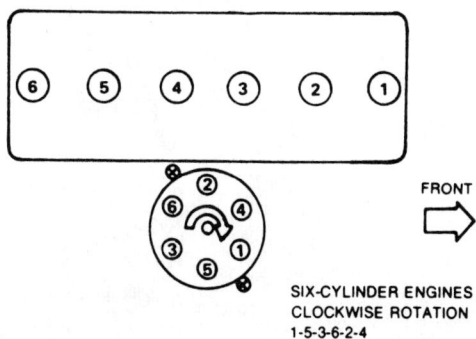

Inline 6-cylinder engines through 1974 (point-type ignition)

FIRING ORDERS

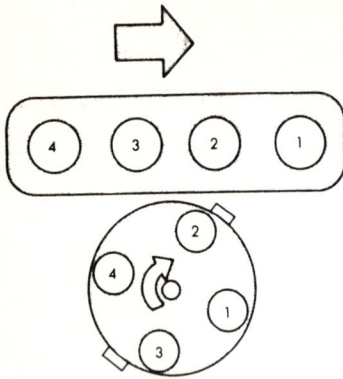

Distributor cap and firing order for the 4–150

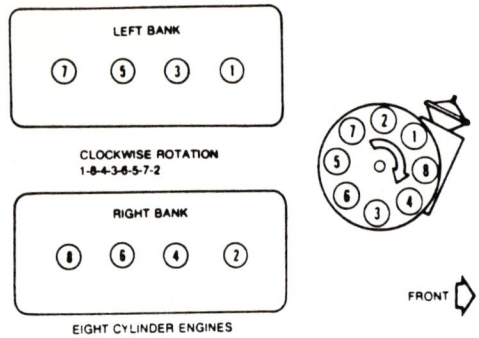

V8 engines starting 1975 (electronic ignition)

POINT-TYPE IGNITION

Breaker Points and Condenser

NOTE: When you replace a set of points, always replace the condenser at the same time.

When you change the point gap or the dwell, you will also have changed the ignition timing. So, if the point gap or dwell is changed, the ignition timing must be adjusted.

There are two ways to check the breaker point gap; it can be done with a feeler gauge or a dwell meter. Either way you set the amount of time that the points remain closed or open. The time is measured in degrees of gap between the breaker points with a feeler gauge, you are setting the maximum amount the points will open when the rubbing block on the points is on a high point of the distributor cam. When you adjust the points with a dwell meter, you are adjusting the number of degrees that the points will remain closed before they start to open as a high point of the distributor cam approaches the rubbing block.

INSPECTION OF THE POINTS

1. Disconnect the high tension wire from the top of the distributor and the coil, and unsnap the distributor retaining caps.

2. Remove the distributor cap by prying off the spring clips on the 4-134, or depressing and turning the holddown screws on the side of the cap on all other engines.

3. Remove the rotor from the distributor shaft by pulling it straight up. On the 304 cu in. V8 and 225 cu. in. V6, the rotor is attached to the distributor shaft by screws. Remove the screws to remove the rotor. Examine the condition of the rotor. If it is cracked or the metal tip is excessively worn or burned, it should be replaced.

4. Pry open the contact points with a screwdriver and check the condition of the contacts. If they are excessively worn, burned, or pitted, they should be replaced.

5. If the points are in good condition, adjust them, and replace the rotor and the distributor cap. If the points need to be replaced, follow the replacement procedure below.

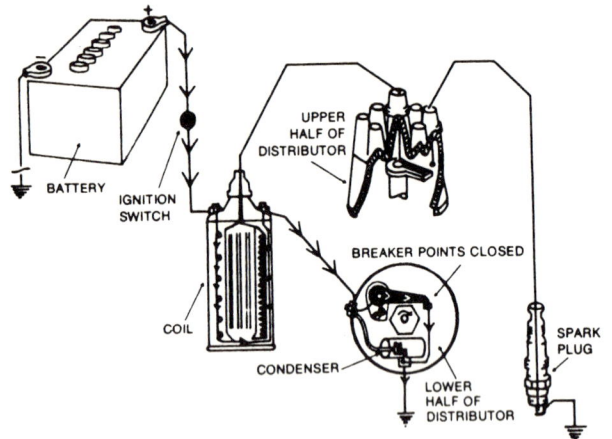

Diagram of a point type primary ignition circuit

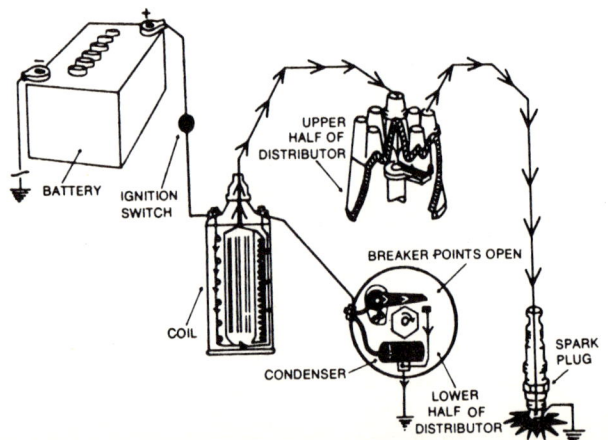

Diagram of a point type secondary ignition circuit

TROUBLESHOOTING BASIC POINT-TYPE IGNITION SYSTEM PROBLEMS

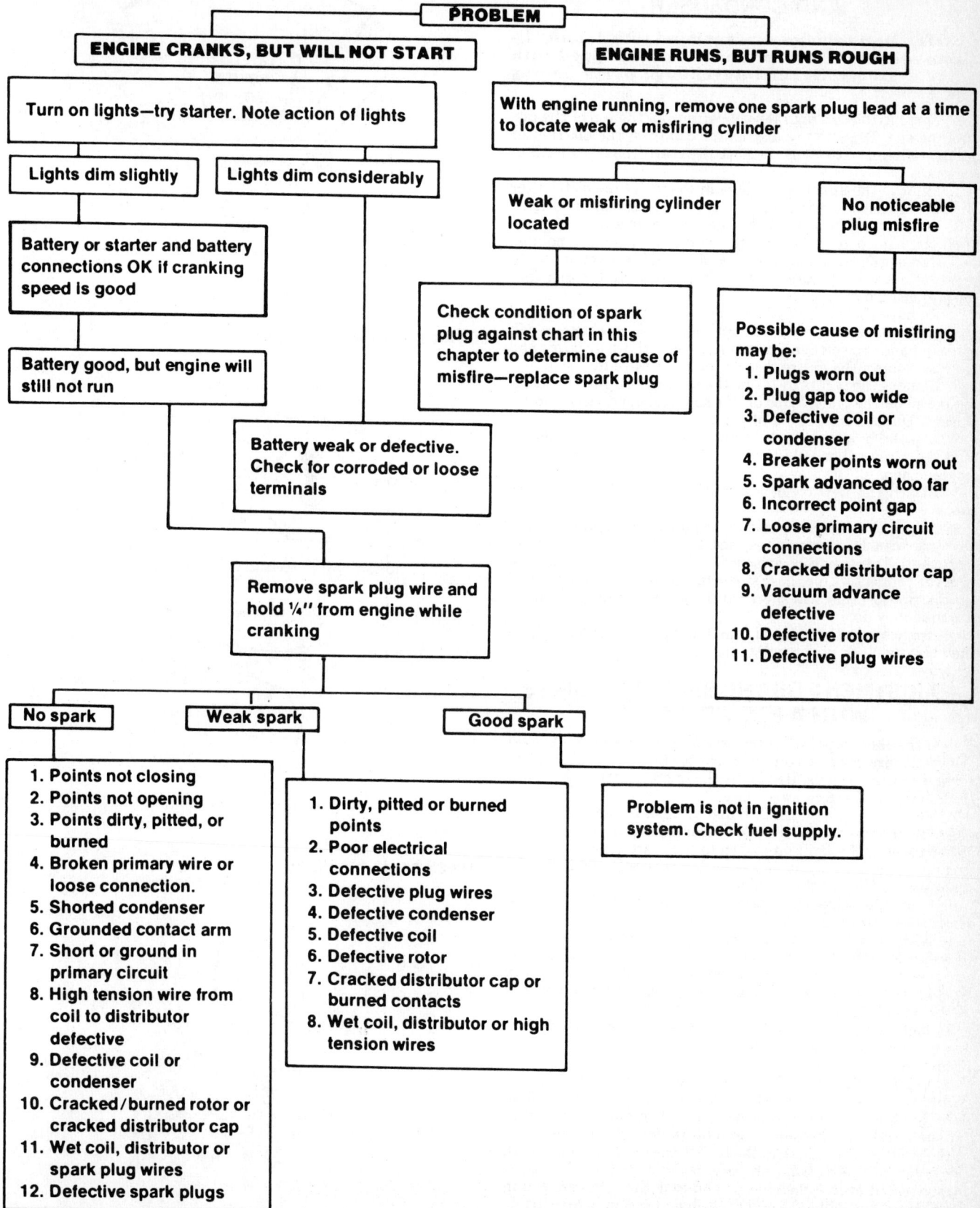

PROBLEM

ENGINE CRANKS, BUT WILL NOT START

ENGINE RUNS, BUT RUNS ROUGH

Turn on lights—try starter. Note action of lights

With engine running, remove one spark plug lead at a time to locate weak or misfiring cylinder

Lights dim slightly

Lights dim considerably

Weak or misfiring cylinder located

No noticeable plug misfire

Battery or starter and battery connections OK if cranking speed is good

Check condition of spark plug against chart in this chapter to determine cause of misfire—replace spark plug

Possible cause of misfiring may be:
1. Plugs worn out
2. Plug gap too wide
3. Defective coil or condenser
4. Breaker points worn out
5. Spark advanced too far
6. Incorrect point gap
7. Loose primary circuit connections
8. Cracked distributor cap
9. Vacuum advance defective
10. Defective rotor
11. Defective plug wires

Battery good, but engine will still not run

Battery weak or defective. Check for corroded or loose terminals

Remove spark plug wire and hold ¼" from engine while cranking

No spark

Weak spark

Good spark

1. Points not closing
2. Points not opening
3. Points dirty, pitted, or burned
4. Broken primary wire or loose connection.
5. Shorted condenser
6. Grounded contact arm
7. Short or ground in primary circuit
8. High tension wire from coil to distributor defective
9. Defective coil or condenser
10. Cracked/burned rotor or cracked distributor cap
11. Wet coil, distributor or spark plug wires
12. Defective spark plugs

1. Dirty, pitted or burned points
2. Poor electrical connections
3. Defective plug wires
4. Defective condenser
5. Defective coil
6. Defective rotor
7. Cracked distributor cap or burned contacts
8. Wet coil, distributor or high tension wires

Problem is not in ignition system. Check fuel supply.

REPLACEMENT OF THE BREAKER POINTS AND CONDENSER

NOTE: Most vehicles were equipped with Autolite ignition systems. However, some were equipped with Delco systems. Never interchange parts during replacement.

1. Remove the coil high tension wire from the top of the distributor cap. Remove the distributor cap from the distributor and place it out of the way. Remove the rotor from the distributor shaft.

2. Remove the dust cover that is in the top of the distributor on some models, covering the points. It is pressed in handtight.

3. Loosen the screw that holds the condenser lead to the body of the breaker points. Remove the condenser from the points.

4. Remove the screw that holds and grounds the condenser to the distributor body. Remove the condenser from the distributor and discard it.

5. Remove the points assembly attaching screws and adjustment lockscrews. A screwdriver with a holding mechanism will come in handy so you don't drop a screw into the distributor and have to remove the entire distributor to retrieve it.

6. Remove the points by lifting them straight up off the locating dowel on the plate. Wipe off the cam and apply new cam lubricant. Discard the old set of points.

7. Slip the new set of points onto the locating dowel and install the screws that hold the assembly onto the plate. Do not tighten them all the way.

8. Attach the new condenser to the plate with the ground screw.

9. Attach the condenser lead to the points at the proper place. On American Motors engines, and the V6, the primary wire from the coil must now be attached to the points also. Make sure that the connectors for these two wires do not touch the body of the distributor; they will short out the primary circuit of the ignition if they do.

10. Apply a small amount of cam lubricant to the shaft where the rubbing block of the points touches.

ADJUSTMENT OF THE BREAKER POINTS WITH A FEELER GAUGE

1. If the contact points of the assembly are not parallel, bend the stationary contact so they make contact across the entire surface of the contacts. Bend only the bracket part of the point assembly, not the contact surface.

2. Turn the engine until the rubbing block of the points is on one of the high points of the distributor cam. You can do this by either turning the ignition switch to the start position and releasing it quickly or by using a wrench on the bolt that holds the crankshaft pulley to the crankshaft.

3. Place the correct size feeler gauge between the contacts. Make sure it is parallel with the contact surfaces.

4. With your free hand, insert a screwdriver into the notch provided for adjustment or into the eccentric adjusting screw, and then twist the screwdriver to either increase or decrease the gap to the proper setting. V6 and V8 engines have to be adjusted at the adjusting screw with an allen wrench.

5. Tighten the adjustment lockscrew and recheck the contact gap to make sure that it didn't change when the lockscrew was tightened.

6. Replace the rotor, distributor cap, and the high tension wire that connects the top of the distributor and the coil. Make sure that the rotor if firmly seated all the way onto the distributor shaft and that the tab of the rotor is aligned with the notch in the shaft. Align the tab in the base of the distributor cap with the notch in the distributor body. Make sure that the cap is firmly seated on the distributor and that the retainers are in place. Make sure that the end of the high tension wire is firmly placed in the top of the distributor and the coil.

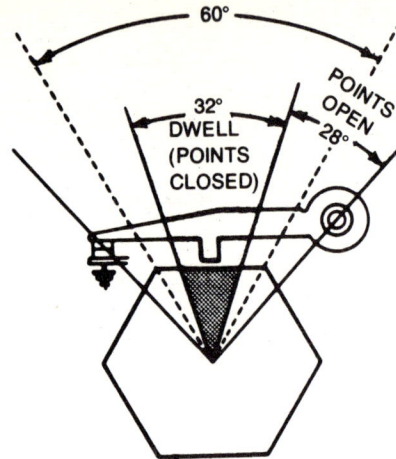

Typical breaker point dwell

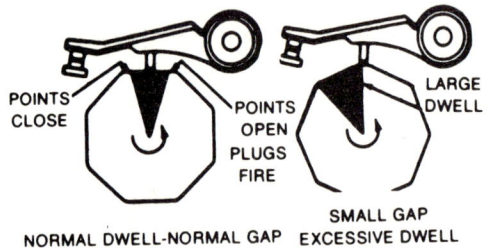

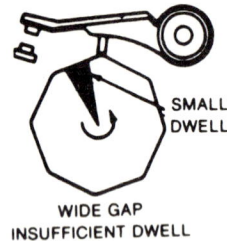

Dwell angle functions

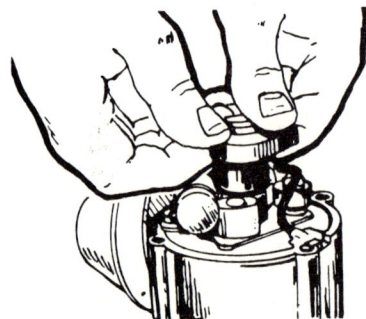

Removing the rotor, except Delco V6 and V8

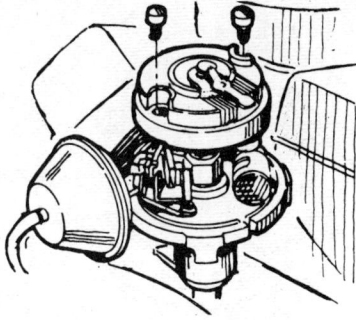

Removing the rotor from Delco V6 or V8

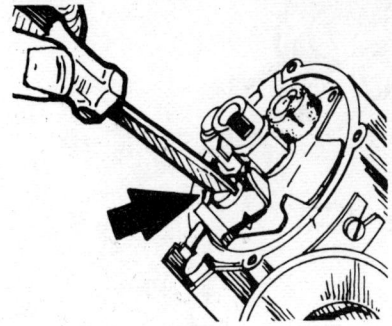

Removing the point set holddown screws from all but Delco V6 or V8

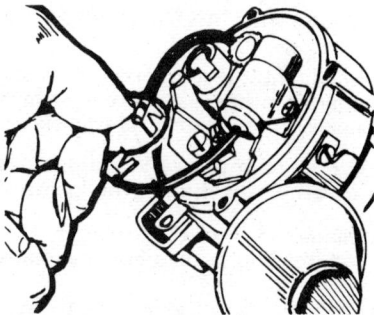

Removing the wires from the points on all but Delco V6 or V8

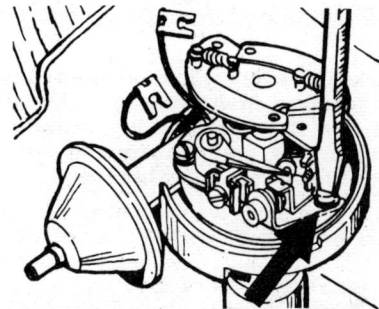

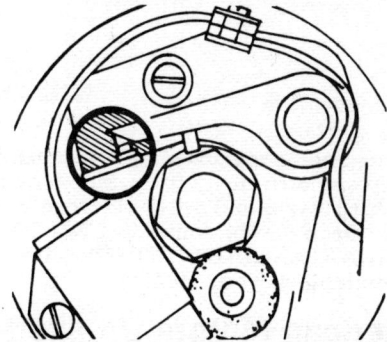

On the Delco V6 and V8 distributors, loosen, but don't remove the point set holddown screws. Slide the point set out from under the screws

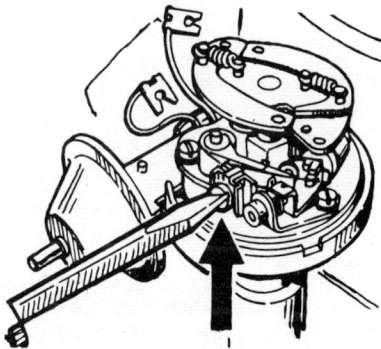

Removing the wires from the points on Delco V6 or V8

Once the points are installed, make certain that the contact surfaces are properly aligned. If there is misalignment, correct it by bending the STATIONARY arm, NOT THE MOVING ARM! Use a pair of needle-nosed pliers to bend the arm

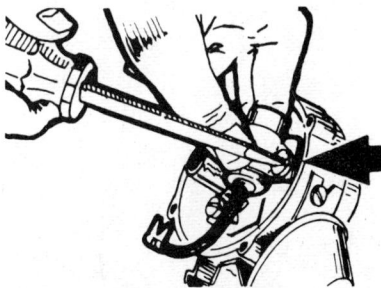

ADJUSTMENT OF THE BREAKER POINTS WITH A DWELL METER

1. Adjust the points with a feeler gauge as described above.
2. Connect the dwell meter to the ignition circuit as according to the manufacturer's instructions. One lead of the meter is to be connected to a ground and the other lead is to be connected to the distributor post on the coil. An adapter is usually provided for this purpose.
3. If the dwell meter has a set line on it, adjust the meter to zero the indicator.
4. Start the engine.

Removing the condenser from all but the Delco V6 or V8

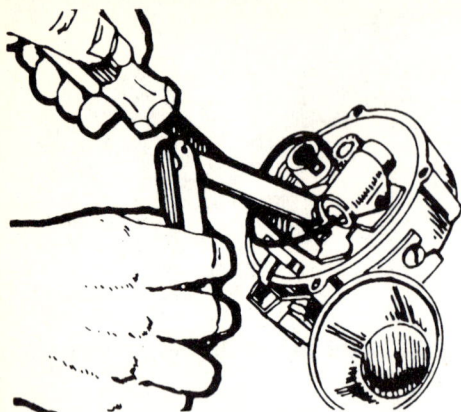

Adjusting the points on all but Delco V6 and V8

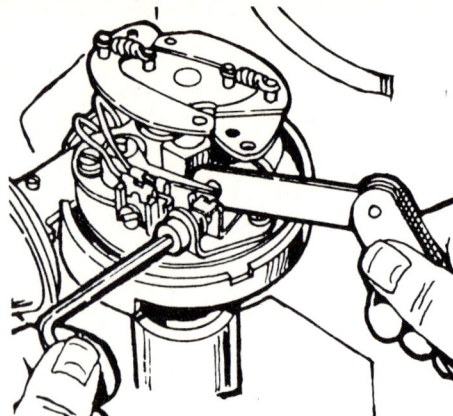

Adjusting the points on the Delco V6 and V8

NOTE: Be careful when working on any vehicle while the engine is running. Make sure that the transmission is in neutral and that the parking brake is on. Keep hands, clothing, tools, and the wires of the test instruments clear of the rotating fan blades.

5. Observe the reading on the dwell meter. If the meter does not have a scale for 4-cylinder engines, multiply the 8-cylinder reading by two. If the reading is within the specified range, turn off the engine and remove the dwell meter.

6. If the reading is above the specified range, the breaker point gap is too small. If the reading is below the specified range, the gap is too large. In either case, the engine must be stopped and the gap adjusted in the manner previously covered.

NOTE: On the V6 engine and V8 engines, it is possible to adjust the dwell while the engine is running.

7. Start the engine and check the reading on the dwell meter. When the correct reading is obtained, disconnect the dwell meter.

8. Check the adjustment of the ignition timing.

ELECTRONIC IGNITION

NOTE: This book contains simple testing procedures for your Jeep's electronic ignition. More comprehensive testing on this system and other electronic control systems on your Jeep can be found in CHILTON'S GUIDE TO ELECTRONIC ENGINE CONTROLS, book part number 7535, available at your local retailer.

American Motors Breakerless Inductive Discharge Ignition System

During the years 1975 through 1977, all American Motors built engines were equipped with the Breakerless Inductive Discharge (BID) ignition system. The system consists of an electronic ignition control unit, a standard type ignition coil, a distributor that contains an electronic sensor and trigger wheel instead of a cam, breaker points and condenser, and the usual high tension wires and spark plugs. There are no contacting (and thus wearing) surfaces between the trigger wheel and the sensor. The dwell angle remains the same and never requires adjustment. The dwell angle is determined by the control unit and the angle between the trigger wheel spokes.

COMPONENTS

The AMC breakerless inductive discharge (BID) ignition system consists of five components:
- Control unit
- Coil
- Breakerless distributor
- Ignition cables
- Spark plugs

The control unit is a solid state, epoxy sealed module with waterproof connectors. The control unit has a built-in current regulator, so no separate ballast resistor or resistance wire is needed in the primary circuit. Battery voltage is supplied to the ignition coil positive (+) terminal when the ignition key is turned to the **ON** or **START** position; low voltage coil primary current is also supplied by the control unit.

In place of the points, cam, and condenser, the distributor has a sensor and trigger wheel. The sensor is a small coil which generates an electromagnetic field when excited by the oscillator in the control unit. This system was last used in 1977.

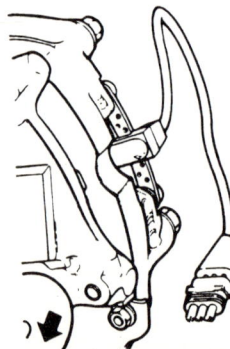

BID distributor sensor

OPERATION

When the ignition switch is turned on, the control unit is activated. The control unit then sends an oscillating signal to the sensor, which cause the sensor to generate a magnetic field. When one of the trigger wheel teeth enters this field, the strength of the oscillation in the sensor is reduced. One the strength drops to a predetermined level, a demodulator circuit operates the control unit's switching transistor. The switching transistor is wired in series with the coil primary circuit; it switches the circuit off, inducing high voltage in the coil secondary winding when it gets the demodulator signal. From this point on, the BID ignition system works in the same manner as a conventional system.

SYSTEM TEST

1. Check all the BID ignition system electrical connections.
2. Disconnect the coil-to-distributor high tension lead from the distributor cap.
3. Using insulated pliers and a heavy glove, hold the end of the lead ½ in. (13mm) away from a ground. Crank the engine. If there is a spark, the trouble is not in the ignition system. Check the distributor cap, rotor, and wires.
4. Replace the spark plug lead. Turn the ignition switch off and disconnect the coil high tension cable from the center tower on the distributor cap. Place a paper clip around the cable ½-¾ in. (13-19mm) from the metal end. Ground the paper clip to the engine. Crank the engine. If there is spark, the distributor cap or rotor may be at fault.
5. Turn the ignition switch off and replace the coil wire. Make the spark test of Step 3 again. If there is no spark, check the coil high tension wire with an ohmmeter. It should show 5,000-10,000Ω resistance. If not, replace it and repeat the spark test.
6. Detach the distributor sensor lead wire plug. Check the wire connector by trying a no. 16 (0.177 in.) drill bit for a snug fit in the female terminals. Apply a light coat of Silicone Dielectric Compound or its equivalent to the male terminals. Fill the female cavities ¼ full. Reconnect the plug.
7. Repeat the test of Step 4.
8. If there was a spark in Step 7, detach the sensor lead plug and try a replacement sensor. Try the test again. If there is a spark, the sensor was defective.
9. Connect a multitester with a volt scale, between the coil positive terminal and an engine ground. With the ignition switch on, the volt scale should read battery voltage. If it is lower, there is a high resistance between the battery (through the ignition switch) and the coil.
10. Connect the multitester between the coil negative terminal and an engine ground. With the ignition switch on, the voltage should be 5-8. If not, replace the coil. If you get a battery voltage reading, crank the engine slightly to move the trigger wheel tooth away from the sensor; voltage should drop to 5-8.
11. Check the sensor resistance by connecting an ohmmeter to its leads. Resistance should be 1.6-2.4Ω.

TESTING

Test the coil with a conventional coil checker or an ohmmeter. Primary resistance should be 1.25-1.40Ω and secondary resistance should be 9-12KΩ. The open output circuit should be more than 20 kilovolts. Replace the coil if it doesn't meet specifications.

DISTRIBUTOR OVERHAUL

NOTE: If you must remove the sensor from the distributor for any reason, it will be necessary to have the special sensor positioning gauge in order to align it properly during installation.

Disassembly

1. Scribe matchmarks on the distributor housing, rotor, and engine block. Disconnect the leads and vacuum lines from the distributor. Remove the distributor. Unless the cap is to be replaced, leave it connected to the spark plug cables and position it out of the way.
2. Remove the rotor and dust cap.
3. Place a small gear puller over the trigger wheel, so that its jaws grip the inner shoulders of the wheel and not its arms. Place a thick washer between the gear puller and the distributor shaft to act as a spacer; do not press against the smaller inner shaft.
4. Loosen the sensor holddown screw with a small pair of needlenosed pliers; it has a tamper proof head. Pull the sensor lead grommet out of the distributor body and pull out the leads from around the spring pivot pin.
5. Release the sensor securing spring by lifting it. Make sure that it clears the leads. Slide the sensor off the bracket. Remember, a special gauge is required for sensor installation.
6. Remove the vacuum advance unit securing screw. Slide the vacuum unit out of the distributor. Remove it only if it is to be replaced.
7. Clean the vacuum unit and sensor brackets. Lubrication of these parts is not necessary.

Assembly

1. Install the vacuum unit, if it was removed.
2. Assemble the sensor, sensor guide, flat washer, and retaining screw. Tighten the screw only far enough to keep the assembly together; don't allow the screw to project below the bottom of the sensor.

NOTE: Replacement sensors come with a slotted head screw to aid in assembly. If the original sensor is being used, replace the tamperproof screw with a conventional one. Use the original washer.

3. Secure the sensor on the vacuum advance unit bracket, making sure that the tip of the sensor is placed in the notch on the summing bar.
4. Position the spring on the sensor and route the leads around the spring pivot pin. Fit the sensor lead grommet into the slot on the distributor body. Be sure that the lead can't get caught in the trigger wheel.
5. Place the special sensor positioning gauge over the distributor shaft, so that the flat on the shaft is against the large notch on the gauge. Move the sensor until the sensor core fits into the small notch on the gauge. Tighten the sensor securing screw with the gauge in place (through the round hole in the gauge).
6. It should be possible to remove and install the gauge without any side movement of the sensor. Check this and remove the gauge.
7. Position the trigger wheel on the shaft. Check to see that the sensor core is centered between the trigger wheel legs and that the legs don't touch the core.
8. Bend a piece of 0.050 in. (1.3mm) gauge wire, so that it has a 90° angle and one leg ½ in. (13mm) long. Use the gauge to measure the clearance between the trigger wheel legs and the sensor boss. Press the trigger wheel on the shaft until it just touches the gauge. Support the shaft during this operation.
9. Place 3 to 5 drops of SAE 20 oil on the felt lubricator wick.
10. Install the dust shield and rotor on the shaft.
11. Install the distributor on the engine using the matchmarks made during removal and adjust the timing. Use a new distributor mounting gasket.

American Motors Solid State Ignition (SSI) System

AMC introduced Solid State Ignition (SSI) as a running change on some 1977 Canadian models. It is standard equipment on all 1978 and later American Motors built engines, except the 1987-90 4-150.

The system consists of a sensor and toothed trigger wheel inside the distributor, and a permanently sealed electronic control unit which determines dwell, in addition to the coil, ignition wires, and spark plugs.

The trigger wheel rotates on the distributor shaft. As one of its teeth nears the sensor magnet, the magnetic field shifts toward the tooth. When the tooth and sensor are aligned, the magnetic field is shifted to its maximum, signaling the electronic control unit to switch off the coil primary current. This starts an electronic timer inside the control unit, which allows the primary current to remain off only long enough for the spark plug to fire. The timer adjusts the amount of time primary current is off according to conditions, thus automatically adjusting dwell. There is also a special circuit within the control unit to detect and ignore spurious signals. Spark timing is adjusted by both mechanical (centrifugal) and vacuum advance.

A wire of 1.35Ω resistance is spliced into the ignition feed to reduce voltage to the coil during running conditions. The resistance wire is bypassed when the engine is being started so that full battery voltage may be supplied to the coil. Bypass is accomplished by the I-terminal on the solenoid.

SECONDARY CIRCUIT TEST

1. Disconnect the coil wire from the center of the distributor cap.

NOTE: Twist the rubber boot slightly in either direction, then grasp the boot and pull straight up. Do not pull on the wire, and do not use pliers.

2. Hold the wire ½ in. (13mm) from a ground with a pair of insulated pliers and a heavy glove. As the engine is cranked, watch for a spark.
3. If a spark appears, reconnect the coil wire. Remove the wire from one spark plug, and test for a spark as above.

WARNING: Do not remove the spark plug wires from cylinder 3 on the 4-150, or cylinder 3 or 5 on a 1977-79 6-258 or 1 or 5 on a 1980 and later 6-258, or cylinders 3 or 4 of an 8-304, when performing this test, as sensor damage could occur!

3. If a spark occurs, the problem is in the fuel system or ignition timing. If no spark occurs, check for a defective rotor, cap, or spark plug wires.
4. If no spark occurs from the coil wire in Step 2, test the coil wire resistance with an ohmmeter. It should be 7,700-9,300Ω at +75°F (24°C) or 12,000Ω maximum at +93°F (34°C).

COIL PRIMARY CIRCUIT TEST

1. Turn the ignition On. Connect a multitester to the coil positive (+) terminal and a ground. If the voltage is 5.5-6.5 volts, go to Step 2. If above 7 volts, go to Step 4. If below 5.5 volts, disconnect the condenser lead and measure. If the voltage is now 5.5-6.5 volts, replace the condenser. If not, go to Step 6.
2. With the multitester connected as in Step 1, read the voltage with the engine cranking. If battery voltage is indicated, the circuit is okay. If not, go to Step 3.
3. Check for a short or open in the starter solenoid I-terminal wire. Check the solenoid for proper operation.
4. Disconnect the wire from the starter solenoid I-terminal, with the ignition On and the multitester connected as in Step 1.

If the voltage drops to 5.5-6.5 volts, replace the solenoid. If not, connect a jumper between the coil negative (−) terminal and a ground. If the voltage drops to 5.5-6.5 volts, go to Step 5. If not, repair the resistance wire.

5. Check for continuity between the coil negative (−) terminal and D4, and D1 to ground. If the continuity is okay, replace the control unit. If not, check for an open wire and go back to Step 2.
6. Turn ignition Off. Connect an ohmmeter between the + coil terminal and dash connector AV. If above 1.40Ω, repair the resistance wire.
7. With the ignition Off, connect the ohmmeter between connector AV and ignition switch terminal 11. If less than 0.1Ω, replace the ignition switch or repair the wire, whichever is the cause. If above 0.1Ω, check connections, and check for defective wiring.

COIL TEST

1. Check the coil for cracks, carbon tracks, etc., and replace as necessary.
2. Connect an ohmmeter across the coil + and − terminals, with the coil connector removed. If 1.13-1.23Ω @ 75°F (24°C), the coil is okay. If not, replace it.

CONTROL UNIT AND SENSOR TEST

1. With the ignition On, remove the coil high tension wire from the distributor cap and hold ½ in. (13mm) from ground with insulated pliers. Disconnect the 4-wire connector at the control unit. If a spark occurs (normal), go to Step 2. If not, go to Step 5.
2. Connect an ohmmeter to D2 and D3. If the resistance is 400-800Ω (normal), go to Step 6. If not, go to Step 3.
3. Disconnect and reconnect the 3-wire connector at distributor. If the reading is now 400-800Ω, go to Step 6. If not, disconnect the 3-wire connector and go to Step 4.
4. Connect the ohmmeter across B2 and B3. If 300-800Ω, repair the harness between the 3-wire and 4-wire connectors. If not, replace the sensor.
5. Connect the ohmmeter between D1 and the battery negative terminal. If the reading is 0 (0.002 or less), go to Step 2. If above 0.002Ω, there is a bad ground in the cable or at the distributor. Repair the ground and retest.
6. Connect a multitester across D2 and D3. Crank the engine. If the needle fluctuates, the system is okay. If not, either the trigger wheel is defective, or the distributor is not turning. Repair or replace as required.

IGNITION FEED TO CONTROL UNIT TEST

NOTE: Do not perform this test without first performing the Coil Primary Circuit Test.

1. With the ignition On, unplug the 2-wire connector at the module. Connect a multitester between F2 and ground. If the reading is battery voltage, replace the control unit and go to Step 3. If not, go to Step 2.
2. Repair the cause of the voltage reduction: either the ignition switch or a corroded dash connector. Check for a spark at the coil wire. If okay, stop. If not, replace the control unit and check for proper operation.
3. Reconnect the 2-wire connector at the control unit, and unplug the 4-wire connector at the control unit. Connect an ammeter between C1 and ground. If it reads 0.9-1.1 amps, the system is okay. If not, replace the module.

American Motors Solid State (Renix) Ignition System for 1987-90 4-150 Engines

These engines are equipped with electronically controlled fuel injection. Therefore, the electronic ignition system is different from that used on carbureted engines.

The system consists of:
- a solid state ignition control module (ICM)
- an electronic control module (ECU)
- a forty tooth rotor in the distributor
- TDC sensor mounted at the rear of the engine on the flywheel housing

The control module consists of a solid state ignition circuit and an integrated ignition coil each of which can be removed and serviced separately. Spark timing control is determined by the ignition control module. Signals from the ECU relay information about engine load and other driving conditions to both the ICM and fuel injection system electronic control components.

Electrical feed to the ICM is through terminal **A** of connector 1 (see illustration). Electrical feed occurs only when the ignition switch is in the **START** and **RUN** positions. Terminal **B** of connector 1 is grounded at the engine oil dipstick bracket, along with the ECU ground wire and the O₂ sensor ground.

DIAGNOSIS

Primary System

Primary system diagnosis is made through the diagnostic connector, using the appropriate diagnostic computer. Primary circuit tests are made at **(D1-2) B+** after ignition; tachometer voltage is at **D1-1**; vehicle ground is at **D1-3**.

Secondary System

1. Remove the center wire from the distributor cap.
2. Using insulated pliers, hold the terminal end about ½ in. (13mm) from the engine head and crank the engine.
3. If a spark jumps from the wire to the head, reconnect the

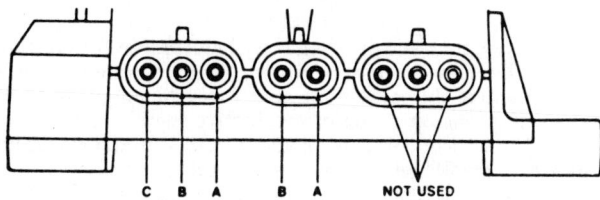

CONNECTOR 1:
A - Ignition (+)
B - Ground (−)
C - Tach Signal Diagnostic Connector
D1 - Pin 1

CONNECTOR 2:
A - Not Used
B - ECU Square Wave Output
Ignition Coil Interface

4–150 ignition control module

wire and remove a wire from one of the spark plugs.
4. Make a metal extension to insert in the spark plug wire boot, and, holding the wire and extension about ½ in. (13mm) from the head, crank the engine.
5. If a spark occurs, check ECU sensors using tester MS 1700, or equivalent. If the sensors check out okay, the problem is probably in the fuel system.
6. If no spark occurs, The rotor, distributor cap or spark plug wires are defective.

Delco High Energy Ignition (HEI) System 4-151

The General Motors HEI system is a pulse triggered, transistor controlled, inductive discharge ignition system. The entire HEI system is contained within the distributor cap.

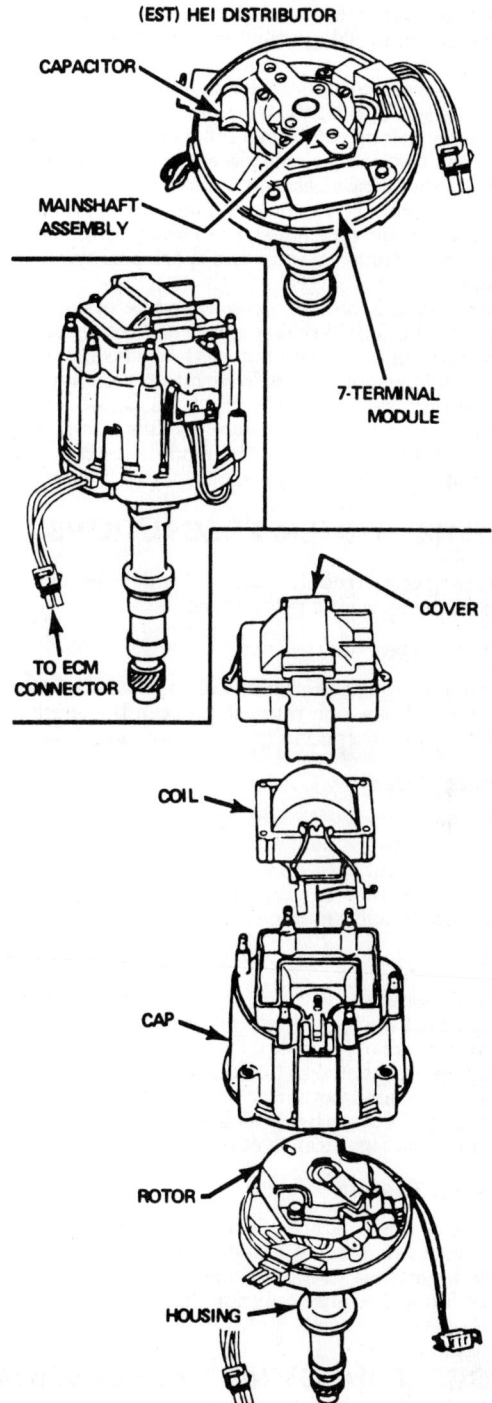

(EST) HEI DISTRIBUTOR

CAPACITOR
MAINSHAFT ASSEMBLY
7-TERMINAL MODULE
TO ECM CONNECTOR
COVER
COIL
CAP
ROTOR
HOUSING

H.E.I. (EST) distributor

The distributor, in addition to housing the mechanical and vacuum advance mechanisms, contains the ignition coil, the electronic control module, and the magnetic triggering device. The magnetic pick-up assembly contains a permanent magnet, a pole piece with internal teeth, and a pick-up coil (not to be confused with the ignition coil).

In the HEI system, as in other electronic ignition systems, the breaker points have been replaced with an electronic switch-a transistor, which is located within the control module. This switching transistor performs the same function the points did in a conventional ignition system; it simply turns coil primary current on and off at the correct time. Essentially then, electronic and conventional ignition systems operate on the same principle.

The module which houses the switching transistor is controlled (turned on and off) by a magnetically generated impulse induced in the pick-up coil. When the teeth of the rotating timer align with the teeth of the pole piece, the induced voltage in the pick-up coil signals the electronic module to open the coil primary circuit. The primary current then decreases, and a high voltage is induced in the ignition coil secondary windings, which is then directed through the rotor and spark plug wires to fire the spark plugs.

In essence, then, the pick-up coil module system simply replaces the conventional breaker points and condenser. The condenser found within the distributor is for radio suppression purposes only and has nothing to do with the ignition process. The module automatically controls the dwell period, increasing it with increasing engine speed. Since dwell is automatically controlled, it cannot be adjusted. The module itself is non-adjustable and non-repairable and must bereplaced if found defective.

HEI SYSTEM PRECAUTIONS

Before going on to troubleshooting, it might be a good idea to take note of the following precautions.

Timing Light Use

Inductive pick-up timing lights are the best kind to use with HEI. Timing lights which connect between the spark plug and the spark plug wire occasionally (not always) give false readings.

Spark Plug Wires

The plug wires used with HEI systems are of a different construction than conventional wires. When replacing them, make sure you get the correct wires, since conventional wires won't carry the voltage. Also handle them carefully to avoid cracking or splitting them and never pierce them.

Tachometer Use

Not all tachometers will operate or indicate correctly when used on an HEI system. While some tachometers may give a reading, this does not necessarily mean the reading is correct. In addition, some tachometers hook up differently from others. If you can't figure out whether or not your tachometer will work on your truck, check with the tachometer manufacturer. Dwell readings have no significance at all.

HEI System Testers

Instruments designed specifically for testing HEI systems are available from several tool manufacturers. Some of these will even test the module itself. However, the test given in the following section will require only a multitester with volt and ohm scales.

TROUBLESHOOTING THE HEI SYSTEM

The symptoms of a defective component within the HEI system are exactly the same as those you would encounter in a con-ventional system. Some of these symptoms are:

- Hard or no starting
- Rough idle
- Poor fuel economy

Engine Misses Under Load or While Accelerating

If you suspect a problem in the ignition system, there are certain preliminary checks which you should carry out before you begin to check the electronic portions of the system.

First, it is extremely important to make sure that the vehicle's battery is in good condition. A defective or poorly charged battery will cause the various components of the ignition system to read incorrectly when tested.

Second, make sure all of the wiring connections are clean and tight, not only at the battery, but also at the distributor cap, coil and module.

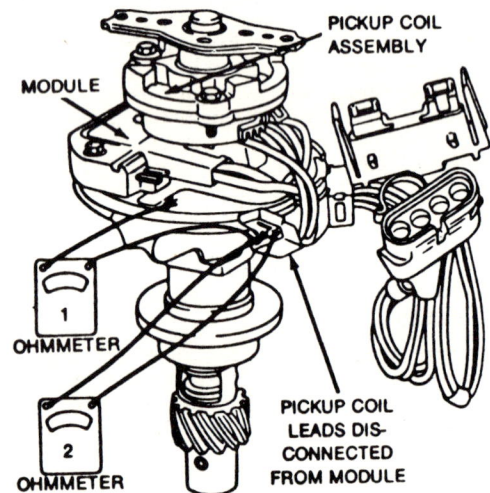

Ohmmeter 1 shows the connections for testing the pick-up coil. Ohmmeter 2 shows the connections for testing the pick-up continuity

Since the major difference between electronic and point type ignition systems is in the distributor area, it is imperative to check the secondary ignition wires first. If the secondary system checks out okay, then the problem is probably not in the ignition system. To check the secondary system, perform a simple spark test. Remove on of the spark plug wires from the plug and insert a makeshift extension made of conductive metal, in the wire boot. Hold the wire and extension about ¼ in. (6mm) away from the block and crank the engine. If a normal spark occurs, then the problem is most likely not in the ignition system. Check for fuel system problems, or fouled spark plugs.

If, however, there is no spark or a weak spark, then further ignition system testing will have to be done. Troubleshooting techniques fall into two categories, depending on the nature of the problem. The categories are (1) Engine cranks, but won't start, and (2) Engine runs, but runs rough or cuts out.

Engine Fails to Start

If the engine won't start, perform a spark test as described earlier. If no spark occurs, check for the presence of normal battery voltage at the battery (BAT) terminal in the distributor cap. The ignition switch must be in the on position for this test. Either a multitester or a test light may be used for this test. Connect the test light wire to ground and the probe end to the BAT terminal at the distributor. If the light comes on, you have voltage to the distributor. If the light fails to come on, this indicates an open circuit in the ignition primary wiring leading to the distributor. In this case, you will have to check wiring continuity back to the ignition switch using test light. If there is battery voltage at the BAT terminal, but no spark at the plugs, then the problem lies within the distributor assembly. Go on to the distributor components test section.

Engine Runs, but Runs Roughly or Cuts Out

1. Make sure the plug wires are in good shape first. There should be no obvious cracks or breaks. You can check the plug wires with an ohmmeter, but do not pierce the wires with a probe. Check the chart for the correct plug wire resistance.

2. If the plug wires are okay, remove the cap assembly, and check for moisture, cracks, chips, or carbon tracks, or any other high voltage leaks or failures. Replace the cap if you find any defects. Make sure the timer wheel rotates when the engine is cranked. If everything is all right so far, go on to the distributor components test section.

Distributor Components Testing

If the trouble has been narrowed down to the units within the distributor, the following tests can help pinpoint the defective component. An ohmmeter with both high and low ranges should be used. These tests are made with the cap assembly removed and the battery wire disconnected.

1. Connect an ohmmeter between the TACH and BAT terminals in the distributor cap. The primary coil resistance should be less than one ohm (zero or nearly zero).

2. To check the coil secondary resistance, connect an ohmmeter between the rotor button and the BAT terminal. Then connect the ohmmeter between the ground terminal and the rotor button. The resistance in both cases should be between 6,000 and 30,000Ω.

3. Replace the coil only if the readings in steps 1 and 2 are infinite.

NOTE: These resistance checks will not disclose shorted coil windings. This condition can be detected only with scope analysis or a suitably designed coil tester. If these instruments are unavailable, replace the coil with a known good coil as a final coil test.

4. To test the pick-up coil, first disconnect the white and green module leads. Set the ohmmeter on the high scale and connect it between a ground and either the white or green lead. Any resistance measurement less than infinity requires replacement of the pick-up coil.

5. Pick-up coil continuity is tested by connecting the ohmmeter (on low range) between the white and green leads. Normal resistance is between 500 and 1500Ω. Move the vacuum advance arm while performing this test. This will detect any break in coil continuity. Such a condition can cause intermittent misfiring. Replace the pick-up coil if the reading is outside the specific limits.

6. If no defects have been found at this time, and you still have a problem, then the module will have to be checked. If you do not have access to a module tester, the only possible alternative is a substitution test. If the module fails the substitution test, replace it.

COMPONENT REPLACEMENT

Integral Ignition Coil

1. Disconnect the feed and module wire terminal connectors from the distributor cap.

2. Remove the ignition set retainer.

3. Remove the 4 coil cover-to-distributor cap screws and coil cover.

4. Remove the 4 coil-to-distributor cap screws.

5. Using a blunt drift, press the coil wire spade terminals up out of distributor cap.

6. Lift the coil up out of the distributor cap.

7. Remove and clean the coil spring, rubber seal washer and coil cavity of the distributor cap.

8. Coat the rubber seal with a dielectric lubricant furnished in the replacement ignition coil package.

9. Reverse the above procedures to install.

Distributor Cap

1. Remove the feed and module wire terminal connectors from the distributor cap.

2. Remove the retainer and spark plug wires from the cap.

3. Depress and release the 4 distributor cap-to-housing retainers and lift off the cap assembly.

4. Remove the 4 coil cover screws and cover.

5. Using a finger or a blunt drift, push the spade terminals up out of the distributor cap.

6. Remove all 4 coil screws and lift the coil, coil spring, and rubber seal washer out of the cap coil cavity.

7. Using a new distributor cap, reverse the above procedures to assembly, being sure to clean and lubricate the rubber seal washer with dielectric lubricant.

Rotor

1. Disconnect the feed and module wire connectors from the distributor.

2. Depress and release the 4 distributor cap to housing retainers and lift off the cap assembly.

3. Remove the two rotor attaching screws and rotor.

4. Reverse the above procedure to install.

Vacuum Advance

1. Remove the distributor cap and rotor as previously described.

2. Disconnect the vacuum hose from the vacuum advance unit.

3. Remove the two vacuum advance retaining screws, pull the advance unit outward, rotate, and disengage the operating rod from its tang.

4. Reverse the above procedure to install.

Module

1. Remove the distributor cap and rotor as previously described.

2. Disconnect the harness connector and pick-up coil spade connectors from the module. Be careful not to damage the wires when removing the connector.

3. Remove the two screws and module from the distributor housing.

4. Coat the bottom of the new module with dielectric lubricant supplied with the new module. Reverse the above procedure to install.

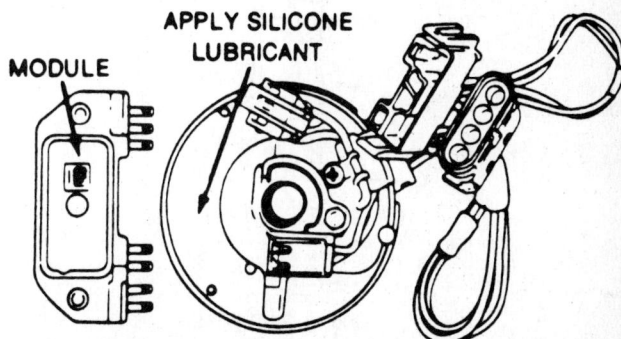

Module replacement; be sure to coat the mating surfaces with silicone lubricant

IGNITION TIMING

Ignition timing is the measurement, in degrees of crankshaft rotation, of the point at which the spark plugs fire in each of the cylinders. It is measured in degrees before or after Top Dead Center (TDC) of the compression stroke. Ignition timing is controlled by turning the distributor in the engine.

Ideally, the air/fuel mixture in the cylinder will be ignited by the spark plug just as the piston passes TDC of the compression stroke. If this happens, this piston will be beginning the power stroke just as the compressed and ignited air/fuel mixture starts to expand. The expansion of the air/fuel mixture then forces the piston down on the power stroke and turns the crankshaft.

Because it takes a fraction of a second for the spark plug to ignite the gases in the cylinder, the spark plug must fire a little before the piston reaches TDC. Otherwise, the mixture will not be completely ignited as the piston TDC and the full benefit of the explosion will not be used by the engine. The timing measurement is given in degrees of crankshaft rotation before the piston reaches TDC (BTDC). If the setting for the ignition timing is 5 degrees BTDC, the spark plug must fire 5 degrees before that piston reaches TDC. This only holds true, however, when the engine is at idle speed.

As the engine speed increases, the pistons go faster. The spark plugs have to ignite the fuel even sooner if it is to be completely ignited when the piston reaches TDC. To do this, the distributor has a means to advance the timing of the spark as the engine speed increases. In some Jeep vehicles that were made before 1972, the advancing of the spark in the distributor was accomplished by weights alone. Others have a vacuum diaphragm to assist the weights. It is necessary to disconnect the vacuum line to the distributor when the engine is being timed.

If the ignition is set too far advanced (BTDC), the ignition and expansion of the fuel in the cylinder will occur too soon and tend to force the piston down while it is still traveling up. This causes engine ping. If the engine is too far retarded after TDC (ATDC), the piston will have already passed TDC and started on its way down when the fuel is ignited. This will cause the piston to be forced down for only a portion of its travel. This will result in poor engine performance and lack of power.

The timing is best checked with a timing light. This device is connected in series with the no. 1 spark plug. The current that fires the spark plug also causes the light to flash.

There is a notch on the front of the crankshaft pulley on the 4-134 engine. There are also marks to indicate TDC and 5° BTDC on the timing gear cover that will assist you in setting ignition timing.

On the 6-232 and 6-258, there is a mark on the crankshaft pulley and a scale divided into degrees. The 8-304 has the same mark and scale arrangement.

The 6-225 has the scale on the crankshaft pulley and the pointer mark on the engine.

When the engine is running, the timing light is aimed at the marks on the engine and crankshaft pulley.

There are three basic types of timing lights available. The first is a simple neon bulb with two wire connections. One wire connects to the spark plug terminal and the other plugs into the end of the spark plug wire for the No. 1 cylinder, thus connecting the light in series with the spark plug. This type of light is pretty dim and must be held very close to the timing marks to be seen. Sometimes a dark corner has to be sought out to see the flash at all. This type of light is very inexpensive. The second type operates from the car battery-two alligator clips connect to the battery terminals, while an adapter enables a third clip to be connected to the No. 1 spark plug and wire. This type is a bit more expensive, but it provides a nice bright flash that you can see even in bright sunlight. It is the type most often seen in professional shops. The third type replaces the battery power source with 110 volt current.

NOTE: Connect a tachometer to the BID or SSI ignition system in the conventional way; to the negative (distributor) side of the coil and to a ground. HEI distributor caps have a Tach terminal. Some tachometers may not work with a BID, SSI, or HEI ignition system and there is a possibility that some could be damaged. Check with the manufacturer of the tachometer to make sure it can be used.

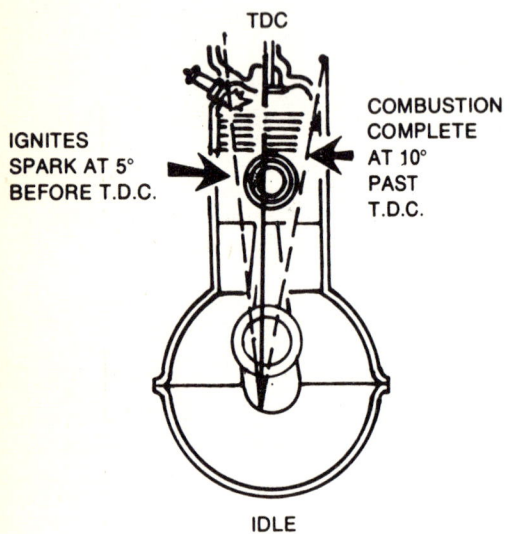

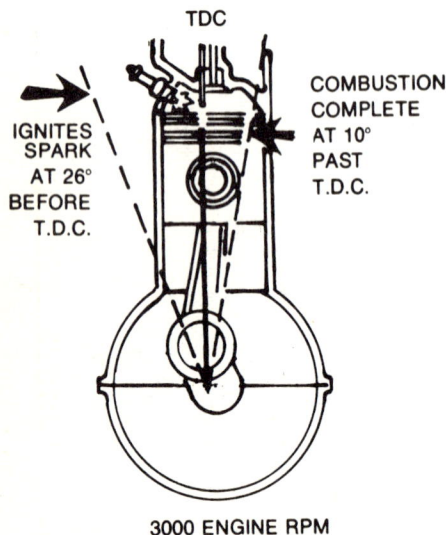

TDC

IGNITES SPARK AT 5° BEFORE T.D.C.

COMBUSTION COMPLETE AT 10° PAST T.D.C.

IDLE

TDC

IGNITES SPARK AT 26° BEFORE T.D.C.

COMBUSTION COMPLETE AT 10° PAST T.D.C.

3000 ENGINE RPM

Ignition timing at idle and at 3,000 rpm

Timing should be checked at each tune-up and any time the points are adjusted or replaced. The timing marks consist of a notch on the rim of the crankshaft pulley and a graduated scale attached to the engine front (timing) cover. A stroboscopic flash (dynamic) timing light must be used, as a static light is too inaccurate for emission controlled engines.

IGNITION TIMING ADJUSTMENT

Point Type Ignition

1. Locate the timing marks on the pulley and on the front of the engine.
2. Clean off the timing marks so you can see them.
3. Mark the timing marks with a piece of chalk or white paint. Mark the one on the engine that will indicate correct timing when it is aligned with the mark on the pulley or flywheel.
4. Attach a tachometer to the engine.
5. Attach a timing light according to the manufacturer's instructions. If the timing light has three wires, one is attached to the no. 1 spark plug lead with an adapter. The other two are connected to the battery. The red one goes to the positive side of the battery and the black one to the negative terminal.
6. Disconnect the vacuum line to the distributor at the distributor. Plug the end of the hose.
7. Check to make sure that all of the wires clear the fan and then start the engine.
8. If there is an idle speed solenoid, disconnect it.
9. Aim the timing light at the timing marks. If the marks that you put on the pulley and the engine are aligned, the timing is correct. Turn off the engine and remove the tachometer and the timing light. If the marks are not in alignment, proceed to the following steps.
10. Turn off the engine.
11. Loosen the distributor lockbolt just enough so that the distributor can be turned with a little effort.
12. Start the engine. Keep the cords of the timing light clear of the fan.
13. With the timing light aimed at the pulley and the marks on the engine, turn the distributor in the direction of rotor rotation to retard the spark, and in the opposite direction of rotor rotation to advance the spark. Line up the marks on the pulley and the engine.
14. When the marks are aligned, tighten the distributor lockbolt and recheck the timing with the timing light to make sure that the distributor did not move when you tightened the distributor lockbolt.
15. Turn off the engine and remove the timing light.

Electronic Ignition

1. Warm up the engine to normal operating temperature. Stop the engine and connect the timing light to the No. 1 (left front on V8, front on 4- or 6-cylinder) spark plug wire. Clean off the timing marks and mark the pulley notch and timing scale with white chalk.
2. Disconnect and plug the vacuum line at the distributor. This is done to prevent any distributor vacuum advance.
3. Start the engine and adjust the idle to 500 rpm with the carburetor idle speed screw on 1975-77 Jeep vehicles. On 1978 and later models, set the idle speed to the figure shown on the underhood sticker. This is done to prevent any distributor centrifugal advance. If there is a throttle stop solenoid, disconnect it electrically.
4. Aim the timing light at the pointer marks. Be careful not to touch the fan, because it may appear to be standing still. If the pulley notch isn't aligned with the proper timing mark (refer to the Tune-Up Specifications chart), the timing will have to be adjusted.

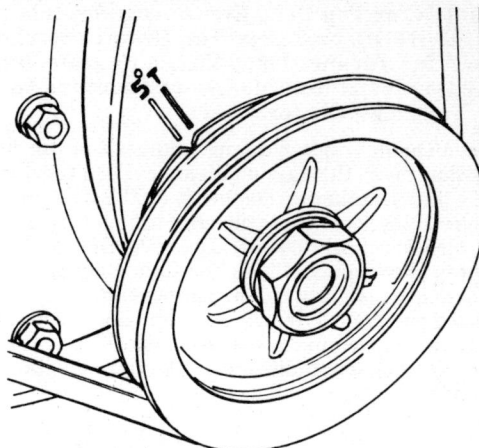

4–134 timing marks

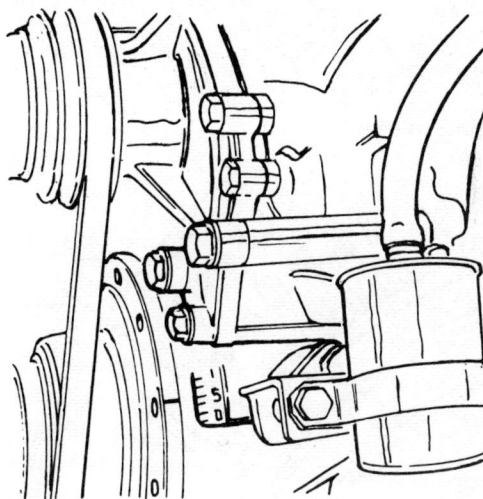

V6–225 timing marks

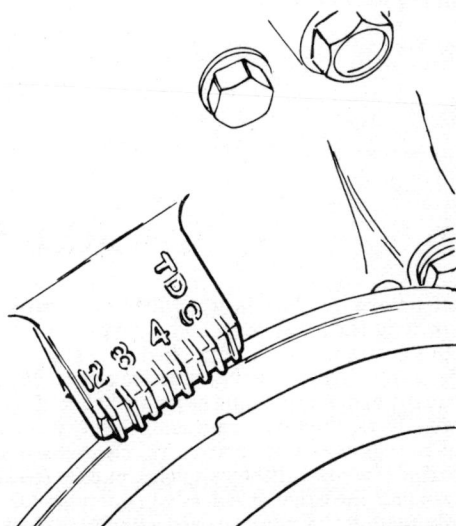

6–232, 6–258 timing marks with point type ignition

NOTE: TDC or Top Dead Center corresponds to 0 degrees. B, or BTDC, or Before Top Dead Center, may be shown as A for Advanced on a V8 timing scale. R on a V8 timing scale means Retarded, corresponding to ATDC, or After Top Dead Center.

5. Loosen the distributor clamp locknut. You can buy trick wrenches that make this task a lot easier. Turn the distributor slowly to adjust the timing, holding it by the base and not the cap. Turn counterclockwise to advance timing (toward BTDC), and clockwise to retard (toward TDC or ATDC).

6. Tighten the locknut. Check the timing again, in case the distributor moved slightly as you tightened it.

7. Replace the distributor vacuum line and correct the idle speed to that specified in the Tune-Up Specifications chart.

8. Stop the engine and disconnect the timing light.

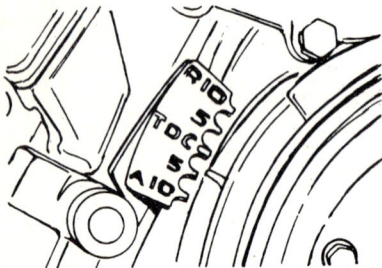

4–151 timing marks

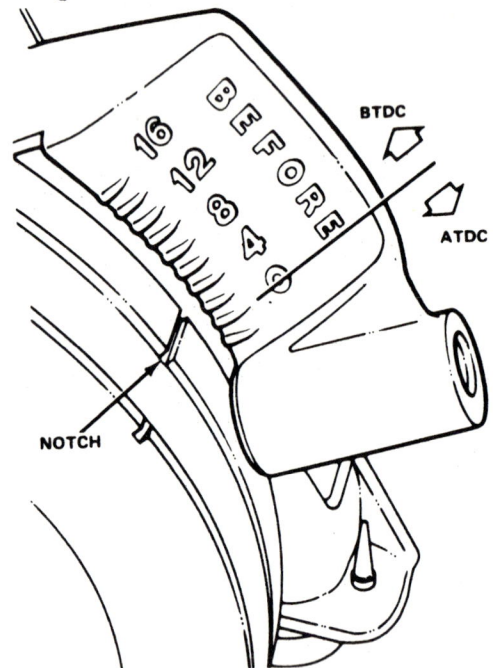

8–304 timing marks

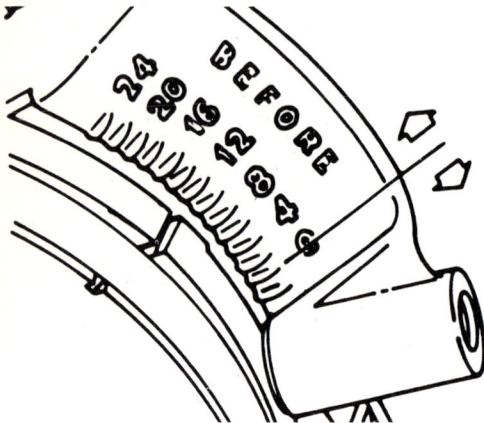

4–150 timing marks

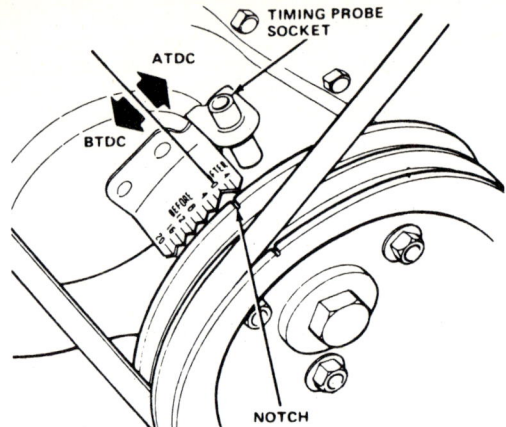

6–258 timing marks with electronic ignition

VALVE LASH ADJUSTMENT

Valve lash determines how far the valves enter into the cylinder and how long they stay open and closed.

If the valve clearance is too large, part of the lift of the camshaft will be used in removing the excessive clearance. The valve will, consequently, not be opening as far as it should. This condition has two effects; the valve train components will emit a tapping sound as they take up the excessive clearance and the engine will perform poorly. If the valve clearance is too small, the intake valves and the exhaust valves will open too far and they will not fully seat on the cylinder head when they close. When a valve seats itself on the cylinder head, it does two things; it seals the combustion chamber so that none of the gases in the cylinder escape and it cools itself by transferring some of the heat it absorbs from the combustion in the cylinder to the cylinder head and to the engine's cooling system. If the valve clearance is too small, the engine will run poorly because of the gases escaping from the combustion chamber. The valves will also become overheated and will warp, since they cannot transfer heat unless they are touching the valve seat in the cylinder head.

NOTE: While all valve adjustments must be made as accurately as possible, it is better to have the valve adjustment slightly loose than slightly tight, as burned valves may result from overly tight adjustments.

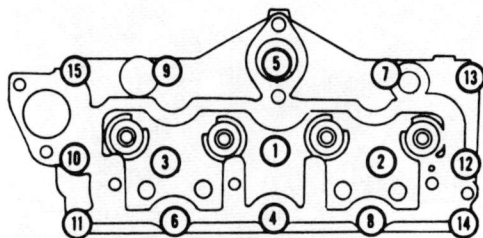

4–134 head bolt torque sequence

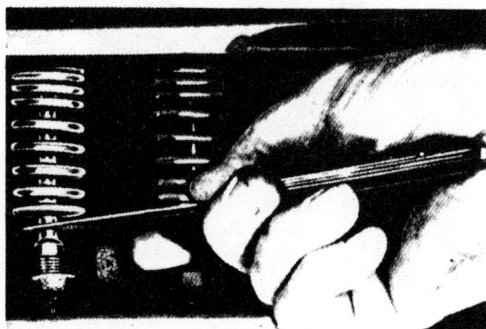

4–134 exhaust valve adjustment

The 4-134 has adjustable valves. All other engines have hydraulic valve lifters which maintain a zero clearance.

4-134 Engine

NOTE: The engine must be cold when the valves are adjusted.

1. Remove the valve cover. Check all the cylinder head bolts to make sure they are tightened to the correct torque specifications.
2. Remove the valve side cover.
3. Turn the engine until the lifter for the front intake valve is down as far as it will go. The lifter should be resting on the center of the heel (back) of the cam lobe for that valve. You can observe the position of the lifter by looking through the side valve spring cover opening. Put the correct size feeler gauge between the rocker arm and the valve stem. There should be a very slight drag on the feeler gauge when it is pulled through the gap. If there is a slight drag, the valve is at the correct setting. If the feeler gauge cannot pass between the rocker arm and the valve

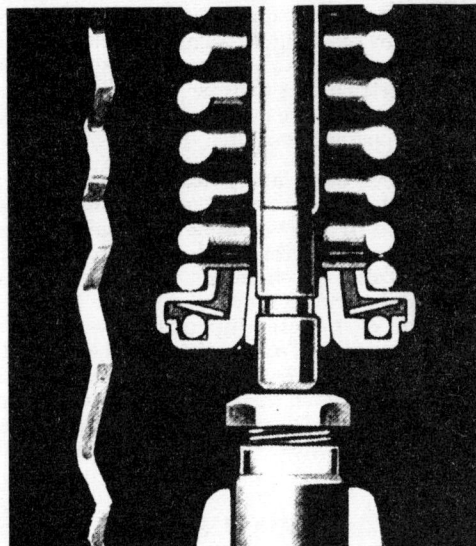

4–134 exhaust valve adjustment screw

stem, the gap between them is too small and must be increased. If the gauge can be passed through the gap without any drag, the gap is too large and must be decreased. Loosen the locknut on the top of the rocker arm (pushrod side) by turning it counterclockwise. Turn the adjusting screw clockwise to lessen the gap and counterclockwise to increase the gap. When the gap is correct, turn the locknut clockwise to lock the adjusting screw. Follow this procedure for all of the intake valves, making sure that the lifter is all the way down for each adjustment.

4. Turn the engine so that the first exhaust valve is completely closed and the lifter that operates that particular valve is all of the way down and on the heel of the cam lobe that operates it.
5. Insert the correct size feeler gauge between the valve stem of the exhaust valve and the adjusting screw. This is done through the side of the engine in the space that is exposed when the side valve spring cover is removed. If there is a slight drag on the feeler gauge, you can assume that the gap is correct. If there is too much drag or not enough, turn the adjusting screw clockwise to increase the gap and counterclockwise to decrease the gap.
6. When all of the valves have been adjusted to the proper clearance, replace the covers with new gaskets.

FUEL SYSTEM

This section contains only tune-up adjustment procedures for fuel systems. Descriptions, adjustments, and overhaul procedures for fuel system components can be found in Section 5.

IDLE SPEED ADJUSTMENT 1971-74

1. Start the engine and run it until it reaches operating temperature.
2. If it hasn't already been done, check and adjust the ignition timing. After you have set the timing, turn off the engine.
3. Attach a tachometer to the engine.
4. Leave the air cleaner on. Turn on the headlights to high beam.

5. Start the engine and, with the transmission in Neutral or Park, check the idle speed on the tachometer. If the reading on the tachometer is correct, turn off the engine and remove the tachometer. If it is not correct, proceed to the following steps.
6. Turn the idle adjusting screw at the bottom of the carburetor with a screwdriver-clockwise to increase idle speed and counterclockwise to decrease it.

MIXTURE ADJUSTMENT 1971-74

The idle mixture screw is located at the very bottom of the carburetor.

1. Turn the screw until it is all the way in. Do not force the

screw in any further because it is very easy to damage the needle valve and its seat by screwing the adjusting screw in too tightly.

2. Turn the screw out ¾ to 1¾ turns. This should be the normal adjustment setting. For a richer mixture, turn the screw out. The ideal setting for the mixture adjustment screw results in the maximum engine rpm.

NOTE: Limiter caps are installed on all engines. These caps limit the amount of adjustment that can be made and should not be removed, if possible. If a satisfactory idle cannot be obtained, however, they can be removed by installing a sheet metal screw in the center of the screw and turning clockwise. After removing the caps, adjust the carburetor in the same manner as was used without the caps. There are special service limiter caps available to replace the ones removed. Install the service limiter caps with the ears positioned against the full rich stops. Be careful not to disturb the idle setting while installing the caps. Press the caps squarely and firmly into place.

IDLE SPEED AND MIXTURE ADJUSTMENTS

1975-78

WARNING: On vehicles equipped with a catalytic converter, do not idle the engine over three minutes at a time! If the adjustments are not completed within three minutes, run the engine at 2000 rpm for one minute.

1. Turn the idle screw(s) to the full rich position. Note the position of the screw head slot inside the limiter cap slots.

2. Remove the limiter cap(s) carefully with a pair of needlenosed pliers. Reset the idle speed screws to the approximate position before cap removal.

3. Connect an accurate tachometer to the engine according to the manufacturer's instructions.

4. Run the engine to operating temperature.

5. Adjust the idle to 30 rpm above the recommended idle speed.

NOTE: On V8 engines with automatic transmissions, the throttle stop solenoid is used to adjust the idle speed. Use the following procedure for these vehicles:

a. With the solenoid wire connected, loosen the locknut and turn the solenoid in or out to obtain the specified rpm.

b. Tighten the solenoid bracket.

c. Disconnect the solenoid wire and adjust the idle speed screw to 500 rpm. Connect the wire.

6. Starting from the full rich stop position (established before the limiters were removed), turn the mixture screws clockwise (leaner) until a slight rpm drop is indicated.

7. Turn the mixture screws counterclockwise until the highest rpm reading is obtained at the best lean idle setting. On carburetors with two screws, turn them evenly in alternating equal increments.

NOTE: If the idle speed changed more than 30 rpm during the adjustment, reset it to 30 rpm above the specified rpm and repeat the adjustment.

1979-80

The procedure for adjusting the idle speed and mixture is called the lean drop procedure and is made with the engine operating at normal operating temperature and the air cleaner in place as follows:

1. Turn the mixture screws to the full rich position with the tabs on the limiters against the stops. Note the position of the screw head slot inside the limiter cap slots.

2. Remove the idle limiter caps by threading a sheet metal screw in center of the cap and turning clockwise. Discard the limiter caps.

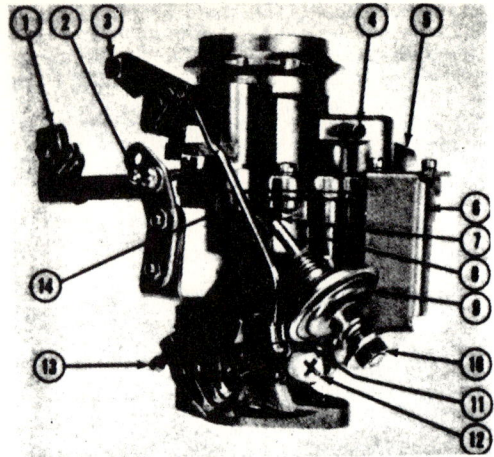

1. Choke cable bracket
2,7. Throttle lever
3. Choke shaft
4. Bowl vent
5. Fuel inlet
6. Dashpot bracket
8,9. Dashpot plunger
10. Locknut
11. Stop pin
12. Idle mixture limiter
13. Idle speed screw
14. Fast idle rod

Late Model Carter YF used on the 4–134 engine

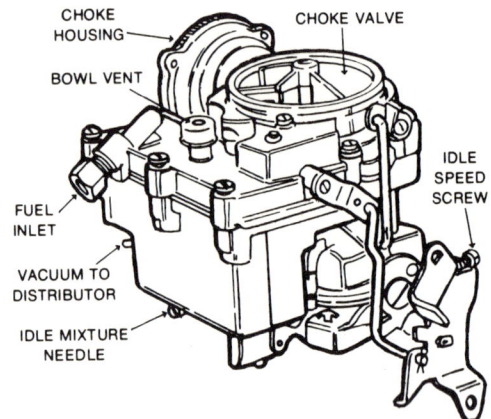

Rochester 2GC carburetor adjustment points

3. Reset the adjustment screws to the same position noted before the limiter caps were removed.

4. Start the engine and allow it to reach normal operating temperature.

5. Adjust the idle speed to 30 rpm above the specified rpm. See the Tune-Up Specifications chart. On 6-cylinder engines with a throttle stop solenoid, turn the solenoid in or out to obtain the specified rpm. On V8 engines with a throttle stop solenoid, turn the hex screw on the throttle stop solenoid carriage to obtain the specified rpm. This is done with the solenoid wire connected. Tighten the solenoid locknut, if so equipped. Disconnect the solenoid wire and adjust the curb idle speed screw to obtain an idle speed of 500 rpm. Reconnect the solenoid wire.

6. Starting from the full rich stop position, as was determined before the limiter caps were removed, turn the mixture adjusting screws clockwise (leaner) until a loss of engine speed is noticed.

7. Turn the screws counterclockwise (richer) until the highest rpm reading is obtained at the best lean idle setting. The best lean idle setting is on the lean side of the highest rpm setting without changing rpm.

8. If the idle speed changed more than 30 rpm during the

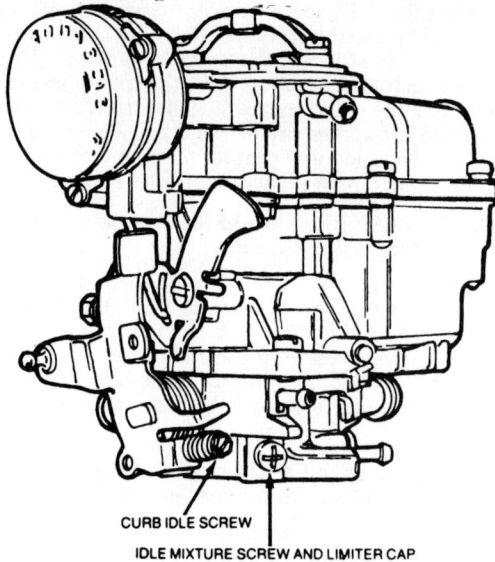

Carter YF used on the 6–232 and 6–258

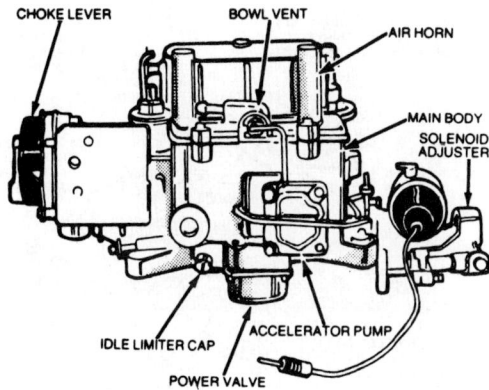

CURB IDLE SCREW
IDLE MIXTURE SCREW AND LIMITER CAP

CHOKE LEVER
BOWL VENT
AIR HORN
MAIN BODY SOLENOID ADJUSTER
IDLE LIMITER CAP
ACCELERATOR PUMP
POWER VALVE

Autolite/Motorcraft 2100 right side

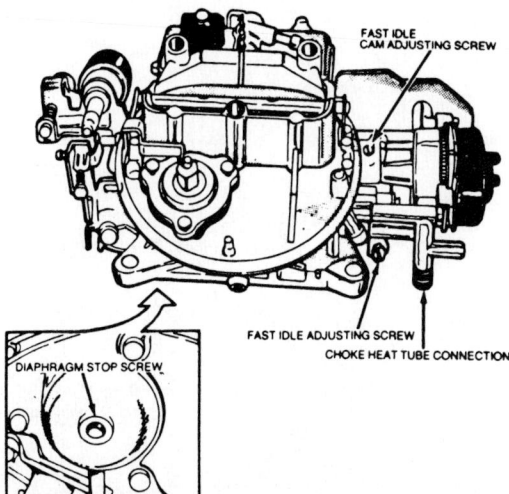

FAST IDLE CAM ADJUSTING SCREW
DIAPHRAGM STOP SCREW
FAST IDLE ADJUSTING SCREW
CHOKE HEAT TUBE CONNECTION

Autolite/Motorcarft 2100 left side

mixture adjustment procedure, reset the idle speed to 30 rpm above the specified rpm with the idle speed adjusting screw or the throttle stop solenoid and repeat the mixture adjustment.

9. Install new limiter caps over the mixture adjusting screws with the tabs positioned against the full rich stops. Be careful not to disturb the idle mixture setting while installing the caps.

1981-82

Idle mixture screws on these carburetors are sealed with plugs or dowel pins. A mixture adjustment must be undertaken ONLY when the carburetor is overhauled, the throttle body replaced, or the engine does not meet required emission standards. Since expensive testing equipment is needed to properly set the mixture, only the idle speed adjusting procedure is given below.

NOTE: The adjustment is made with the manual transmission in neutral and the automatic in drive. Therefore, make certain that the vehicle's parking brake is set firmly, and that the wheels are blocked. It may be a good idea to have someone in the vehicle with their foot on the brake.

1. Connect a tachometer, start engine and warm it to normal operating temperature. The choke and intake manifold heater (6-cylinder engine only) must be off.

2. If the engine speed is not within the OK range, turn the curb idle adjustment screw to obtain the specified curb idle rpm.

3. For the 6-cylinder engine (BBD carburetor): Disconnect the vacuum hose from the vacuum actuator and holding solenoid wire connector. Adjust the curb (slow) idle speed adjustment screw to obtain the specified curb (slow) idle rpm, if it is not within the OK range. Refer to the Emission Control Information label, and the Tune-Up Specifications Chart. Apply a direct source of vacuum to the vacuum actuator. Turn the vacuum actuator adjustment screw on the throttle lever until the specified rpm is obtained (900 rpm for manual transmissions, and 800 rpm for automatic transmissions). Disconnect the manifold vacuum source from the vacuum actuator. With the jumper wire, apply battery voltage (12v) to energize the holding solenoid. Turn the A/C on, if equipped.

NOTE: The throttle must be opened manually to allow the Sol-Vac throttle positioner to be extended.

With the Sol-Vac throttle positioner extended, the idle speed should be 650 rpm for automatic transmission equipped vehicles and 750 rpm for manual transmission equipped vehicles. If the idle speed is not within tolerance, adjust the Sol-Vac (hexhead adjustment screw) to obtain the specified rpm. Remove the jumper wire from the Sol-Vac holding solenoid wire connector. Connect the Sol-Vac holding solenoid wire connector. Connect the original hose to the vacuum actuator.

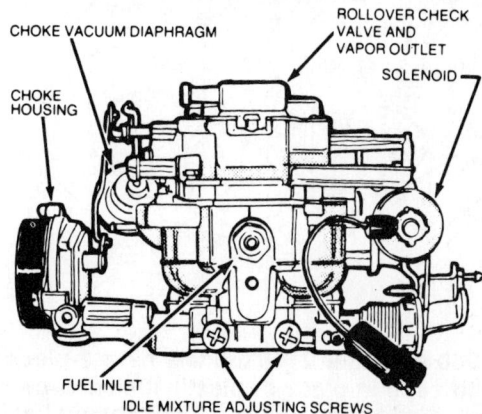

CHOKE VACUUM DIAPHRAGM
ROLLOVER CHECK VALVE AND VAPOR OUTLET
SOLENOID
CHOKE HOUSING
FUEL INLET
IDLE MIXTURE ADJUSTING SCREWS

BBD 2-bbl adjustments

4. For four and eight cylinder engines (2SE, E2SE or 2150 carburetor), turn the nut on the solenoid plunger or the hex screw on the solenoid carriage to obtain the specified idle rpm. Tighten the locknut, if equipped. Disconnect the solenoid wire connector and adjust the curb idle screw to obtain a 500 rpm idle speed. Connect the solenoid wire connector. If the model 2150 carburetor (8-cylinder engine), is equipped with a dashpot, fully depress the dashpot stem with the throttle at the curb idle position, and measure the clearance between the stem and throttle lever. The clearance should be 0.032 in. (0.8mm). Adjust it by loosening the locknut and turning the dashpot.

1983-84 4-150

1. Fully warm up the engine.
2. Check the choke fast idle adjustment: Disconnect and plug the EGR valve vacuum hose. Position the fast idle adjustment screw on the second step of the fast idle cam with the transmission in neutral. Adjust the fast idle speed to 2,000 rpm for manual transmission and 2,300 rpm for automatic transmission. Allow the throttle to return to normal curb idle and reconnect the EGR vacuum hose.

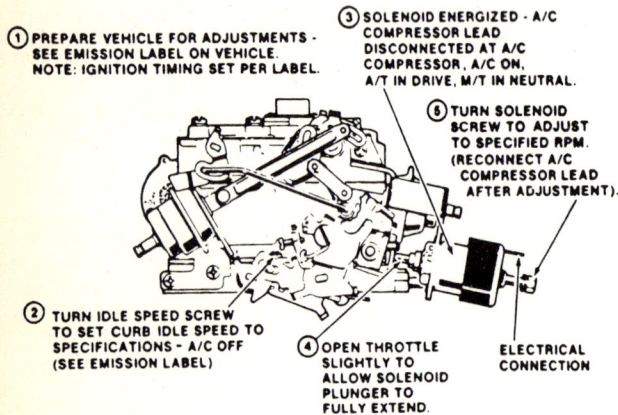

Idle speed adjustment, without air conditioning for the E2SE

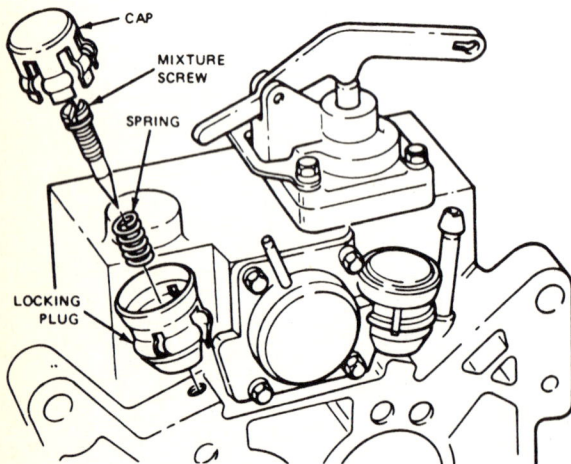

Some 1980 and later 2150 models have 2-piece metal plugs and caps in place of plastic limiter caps on the idle mixture adjusting screws. They should be carefully removed before attempting any adjustments

3. To adjust the Sol-Vac Vacuum Actuator: Remove the vacuum hose from the vacuum actuator and plug the hose. Connect an external vacuum source to the actuator and apply 10-15 inches Hg of vacuum to the actuator. Shift the transmission to Neutral. Adjust the idle speed to the following rpm using the vacuum actuator adjustment screw on the throttle lever: 850 rpm for automatic transmission 950 rpm for manual transmission. The adjustment is made with all accessories turned off.

NOTE: The curb idle should always be adjusted after vacuum actuator adjustment.

4. To adjust the curb idle: Remove the vacuum hose from the Sol-Vac vacuum actuator and plug the hose. Shift the transmission into Neutral. Adjust the curb idle using the ¼ in. (6mm) hex-head adjustment screw on the end of the Sol-Vac unit. Set the speed to 750 rpm for manual transmission, 700 rpm for automatic transmission. Reconnect the vacuum hose to the vacuum actuator.

NOTE: Engine speed will vary 10-30 rpm during this mode due to the closed loop fuel control.

5. To adjust the TRC (Anti-Diesel): The TRC screw is preset at the factory and should not require adjustment. However, to check adjustment, the screw should be ¾ turn from closed throttle position.

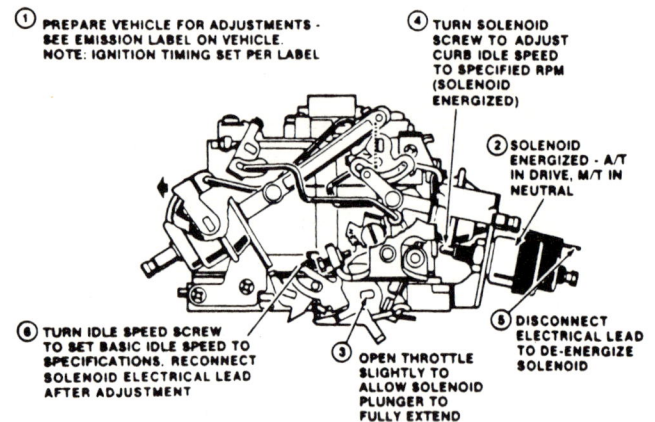

Idle speed adjustment, with air conditioning for the E2SE

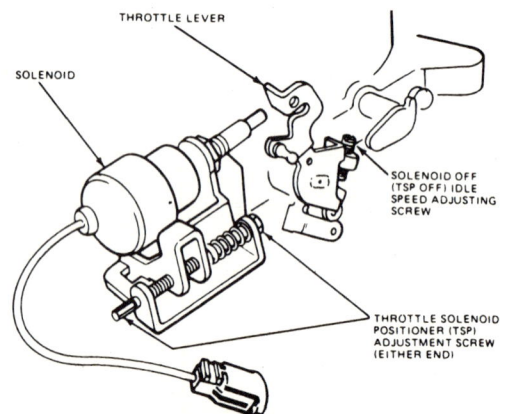

Location of the idle speed adjustment on the Motorcraft/Autolite 2100, 2150

1983-84 6-258

SOL-VAC VACUUM ACTUATOR ADJUSTMENT

1. Disconnect and plug the vacuum hose to the Sol-Vac vacuum actuator.
2. Disconnect the Sol-Vac electrical connector. Connect an external vacuum source to the vacuum actuator and apply 10-15 inches Hg of vacuum.
3. Open throttle for at least 3.0 seconds (1200 rpm); then close throttle.
4. Set the speed using the vacuum actuator adjustment screw on the throttle lever to obtain specified rpm.
5. Disconnect the external vacuum source. Reconnect the Sol-Vac vacuum hose and electrical connector.

SOL-VAC HOLDING SOLENOID ADJUSTMENT

NOTE: The Sol-Vac vacuum actuator adjustment should always precede the Sol-Vac solenoid adjustment.

1. Disconnect and plug the vacuum hose at the Sol-Vac vacuum actuator.
2. Disconnect the Sol-Vac electrical connector.
3. Energize the Sol-Vac holding solenoid with either of the two following methods:
   a. Apply battery voltage (12v) to the solenoid, or,
   b. Reconnect the Sol-Vac electrical connector and turn on the rear window defogger or turn on the air conditioner with the compressor disconnected.
4. Open throttle for at least 3.0 seconds (1,200 rpm) to allow the Sol-Vac holding solenoid to fully extend.
5. Set the speed using the ¼ in. (6mm) hex-head adjustment screw on the end of the Sol-Vac unit to obtain the specified rpm.
6. Reopen the throttle above 1,200 rpm to insure the correct holding position and reset the speed if necessary. Reconnect the vacuum hose to the Sol-Vac actuator. Reconnect the Sol-Vac electrical connector if disconnected.

1985-86

4-150 with YFA Carburetor

1. The TRC (anti-Diesel) adjustment screw is statically set at ¾ of turn from the throttle valve closed position during factory assembly and does not normally require readjustment. Should this adjustment be required, turn the adjustment screw counterclockwise to the throttle plate closed position and then turn the screw clockwise ¾ turn.
2. Connect a tachometer to the ignition coil TACH wire connector.
3. Place the transmission in NEUTRAL and lock the parking brake.
4. Start the engine and allow it to reach normal operating temperature.
5. Connect an external vacuum source to the Sol-Vac vacuum actuator and apply 10-15 in.Hg of vacuum. Plug the engine vacuum hose.
6. Adjust the vacuum actuator until an engine speed of approximately 1,000 rpm is achieved.

NOTE: Refer to the Vehicle Emission Control Information Label for the latest specifications for the particular engine being adjusted.

7. Remove the vacuum source from the vacuum actuator and retain the plug in the vacuum hose from the engine.
8. Turn the hex-head curb idle speed adjustment screw until the speed of 500 rpm is obtained.

NOTE: Refer to the Vehicle Emission Control Label for the latest specifications for the particular engine being adjusted.

9. Stop the engine and connect the engine vacuum hose to the vacuum actuator.
10. Remove the tachometer from the engine.

4-150 with Throttle Body Fuel Injection

Adjustments are not possible on this unit, as all functions are computer controlled.

1985-90 6-258

NOTE: The carburetor choke and intake manifold heater must be off. This occurs when the engine coolant heats to approximately +160°F (71°C).

1. Have the engine at normal operating temperature. Connect a tachometer to the ignition coil negative (TACH) terminal.
2. Remove the vacuum hose from the Sol-Vac vacuum actuator unit. Plug the vacuum hose. Disconnect the holding solenoid wire connector.
3. Adjust the curb (slow) idle speed screw to obtain the correct curb idle speed. Refer to the specifications under Idle Speed or refer to the Emission Information label, under the hood, for the correct curb idle engine rpm.
4. Apply a direct source of vacuum to the vacuum actuator, using a hand vacuum pump or its equivalent. When the Sol-Vac throttle positioner is fully extended, turn the vacuum actuator adjustment screw on the throttler lever until the specified engine rpm is obtained. Disconnect the vacuum source from the vacuum actuator.
5. With a jumper wire, apply battery voltage (12v) to energize the holding solenoid.

NOTE: The holding wire connector can be installed and either the rear window defroster or the air conditioner (with the compressor clutch wire disconnected) can be turned on to energize the holding solenoid.

6. Hold the throttle open manually to allow the throttle positioner to fully extend.

NOTE: Without the vacuum actuator, the throttle must be opened manually to allow the Sol-Vac throttle positioner to fully extend.

7. If the holding solenoid idle speed is not within specifications, adjust the idle using the ¼ in. (6mm) hex-headed adjustment screw on the end of the Sol-Vac unit. Adjust to specifications.
8. Disconnect the jumper wire from the Sol-Vac holding solenoid wire connector, if used. Connect the wire connector to the Sol-Vac unit, if not connected. Install the original vacuum hose to the vacuum actuator.
9. Remove the tachometer and if disconnected, connect the compressor clutch wire. Install any other component that was previously removed.

Troubleshooting Engine Performance

Problem	Cause	Solution
Hard starting (engine cranks normally)	• Binding linkage, choke valve or choke piston	• Repair as necessary
	• Restricted choke vacuum diaphragm	• Clean passages
	• Improper fuel level	• Adjust float level
	• Dirty, worn or faulty needle valve and seat	• Repair as necessary
	• Float sticking	• Repair as necessary
	• Faulty fuel pump	• Replace fuel pump
	• Incorrect choke cover adjustment	• Adjust choke cover
	• Inadequate choke unloader adjustment	• Adjust choke unloader
	• Faulty ignition coil	• Test and replace as necessary
	• Improper spark plug gap	• Adjust gap
	• Incorrect ignition timing	• Adjust timing
	• Incorrect valve timing	• Check valve timing; repair as necessary
Rough idle or stalling	• Incorrect curb or fast idle speed	• Adjust curb or fast idle speed
	• Incorrect ignition timing	• Adjust timing to specification
	• Improper feedback system operation	• Refer to Chapter 4
	• Improper fast idle cam adjustment	• Adjust fast idle cam
	• Faulty EGR valve operation	• Test EGR system and replace as necessary
	• Faulty PCV valve air flow	• Test PCV valve and replace as necessary
	• Choke binding	• Locate and eliminate binding condition
	• Faulty TAC vacuum motor or valve	• Repair as necessary
	• Air leak into manifold vacuum	• Inspect manifold vacuum connections and repair as necessary
	• Improper fuel level	• Adjust fuel level
	• Faulty distributor rotor or cap	• Replace rotor or cap
	• Improperly seated valves	• Test cylinder compression, repair as necessary
	• Incorrect ignition wiring	• Inspect wiring and correct as necessary
	• Faulty ignition coil	• Test coil and replace as necessary
	• Restricted air vent or idle passages	• Clean passages
	• Restricted air cleaner	• Clean or replace air cleaner filler element
	• Faulty choke vacuum diaphragm	• Repair as necessary
Faulty low-speed operation	• Restricted idle transfer slots	• Clean transfer slots
	• Restricted idle air vents and passages	• Clean air vents and passages
	• Restricted air cleaner	• Clean or replace air cleaner filter element
	• Improper fuel level	• Adjust fuel level
	• Faulty spark plugs	• Clean or replace spark plugs

Troubleshooting Engine Performance

Problem	Cause	Solution
Faulty low-speed operation (cont.)	• Dirty, corroded, or loose ignition secondary circuit wire connections	• Clean or tighten secondary circuit wire connections
	• Improper feedback system operation	• Refer to Chapter 4
	• Faulty ignition coil high voltage wire	• Replace ignition coil high voltage wire
	• Faulty distributor cap	• Replace cap
Faulty acceleration	• Improper accelerator pump stroke	• Adjust accelerator pump stroke
	• Incorrect ignition timing	• Adjust timing
	• Inoperative pump discharge check ball or needle	• Clean or replace as necessary
	• Worn or damaged pump diaphragm or piston	• Replace diaphragm or piston
	• Leaking carburetor main body cover gasket	• Replace gasket
	• Engine cold and choke set too lean	• Adjust choke cover
	• Improper metering rod adjustment (BBD Model carburetor)	• Adjust metering rod
	• Faulty spark plug(s)	• Clean or replace spark plug(s)
	• Improperly seated valves	• Test cylinder compression, repair as necessary
	• Faulty ignition coil	• Test coil and replace as necessary
	• Improper feedback system operation	• Refer to Chapter 4
Faulty high speed operation	• Incorrect ignition timing	• Adjust timing
	• Faulty distributor centrifugal advance mechanism	• Check centrifugal advance mechanism and repair as necessary
	• Faulty distributor vacuum advance mechanism	• Check vacuum advance mechanism and repair as necessary
	• Low fuel pump volume	• Replace fuel pump
	• Wrong spark plug air gap or wrong plug	• Adjust air gap or install correct plug
	• Faulty choke operation	• Adjust choke cover
	• Partially restricted exhaust manifold, exhaust pipe, catalytic converter, muffler, or tailpipe	• Eliminate restriction
	• Restricted vacuum passages	• Clean passages
	• Improper size or restricted main jet	• Clean or replace as necessary
	• Restricted air cleaner	• Clean or replace filter element as necessary
	• Faulty distributor rotor or cap	• Replace rotor or cap
	• Faulty ignition coil	• Test coil and replace as necessary
	• Improperly seated valve(s)	• Test cylinder compression, repair as necessary
	• Faulty valve spring(s)	• Inspect and test valve spring tension, replace as necessary
	• Incorrect valve timing	• Check valve timing and repair as necessary

Troubleshooting Engine Performance (cont.)

Problem	Cause	Solution
Faulty high speed operation (cont.)	• Intake manifold restricted	• Remove restriction or replace manifold
	• Worn distributor shaft	• Replace shaft
	• Improper feedback system operation	• Refer to Chapter 4
Misfire at all speeds	• Faulty spark plug(s)	• Clean or replace spark plug(s)
	• Faulty spark plug wire(s)	• Replace as necessary
	• Faulty distributor cap or rotor	• Replace cap or rotor
	• Faulty ignition coil	• Test coil and replace as necessary
	• Primary ignition circuit shorted or open intermittently	• Troubleshoot primary circuit and repair as necessary
	• Improperly seated valve(s)	• Test cylinder compression, repair as necessary
	• Faulty hydraulic tappet(s)	• Clean or replace tappet(s)
	• Improper feedback system operation	• Refer to Chapter 4
	• Faulty valve spring(s)	• Inspect and test valve spring tension, repair as necessary
	• Worn camshaft lobes	• Replace camshaft
	• Air leak into manifold	• Check manifold vacuum and repair as necessary
	• Improper carburetor adjustment	• Adjust carburetor
	• Fuel pump volume or pressure low	• Replace fuel pump
	• Blown cylinder head gasket	• Replace gasket
	• Intake or exhaust manifold passage(s) restricted	• Pass chain through passage(s) and repair as necessary
	• Incorrect trigger wheel installed in distributor	• Install correct trigger wheel
Power not up to normal	• Incorrect ignition timing	• Adjust timing
	• Faulty distributor rotor	• Replace rotor
	• Trigger wheel loose on shaft	• Reposition or replace trigger wheel
	• Incorrect spark plug gap	• Adjust gap
	• Faulty fuel pump	• Replace fuel pump
	• Incorrect valve timing	• Check valve timing and repair as necessary
	• Faulty ignition coil	• Test coil and replace as necessary
	• Faulty ignition wires	• Test wires and replace as necessary
	• Improperly seated valves	• Test cylinder compression and repair as necessary
	• Blown cylinder head gasket	• Replace gasket
	• Leaking piston rings	• Test compression and repair as necessary
	• Worn distributor shaft	• Replace shaft
	• Improper feedback system operation	• Refer to Chapter 4
Intake backfire	• Improper ignition timing	• Adjust timing
	• Faulty accelerator pump discharge	• Repair as necessary
	• Defective EGR CTO valve	• Replace EGR CTO valve
	• Defective TAC vacuum motor or valve	• Repair as necessary

Troubleshooting Engine Performance (cont.)

Problem	Cause	Solution
Intake backfire (cont.)	• Lean air/fuel mixture	• Check float level or manifold vacuum for air leak. Remove sediment from bowl
Exhaust backfire	• Air leak into manifold vacuum	• Check manifold vacuum and repair as necessary
	• Faulty air injection diverter valve	• Test diverter valve and replace as necessary
	• Exhaust leak	• Locate and eliminate leak
Ping or spark knock	• Incorrect ignition timing	• Adjust timing
	• Distributor centrifugal or vacuum advance malfunction	• Inspect advance mechanism and repair as necessary
	• Excessive combustion chamber deposits	• Remove with combustion chamber cleaner
	• Air leak into manifold vacuum	• Check manifold vacuum and repair as necessary
	• Excessively high compression	• Test compression and repair as necessary
	• Fuel octane rating excessively low	• Try alternate fuel source
	• Sharp edges in combustion chamber	• Grind smooth
	• EGR valve not functioning properly	• Test EGR system and replace as necessary
Surging (at cruising to top speeds)	• Low carburetor fuel level	• Adjust fuel level
	• Low fuel pump pressure or volume	• Replace fuel pump
	• Metering rod(s) not adjusted properly (BBD Model Carburetor)	• Adjust metering rod
	• Improper PCV valve air flow	• Test PCV valve and replace as necessary
	• Air leak into manifold vacuum	• Check manifold vacuum and repair as necessary
	• Incorrect spark advance	• Test and replace as necessary
	• Restricted main jet(s)	• Clean main jet(s)
	• Undersize main jet(s)	• Replace main jet(s)
	• Restricted air vents	• Clean air vents
	• Restricted fuel filter	• Replace fuel filter
	• Restricted air cleaner	• Clean or replace air cleaner filter element
	• EGR valve not functioning properly	• Test EGR system and replace as necessary
	• Improper feedback system operation	• Refer to Chapter 4

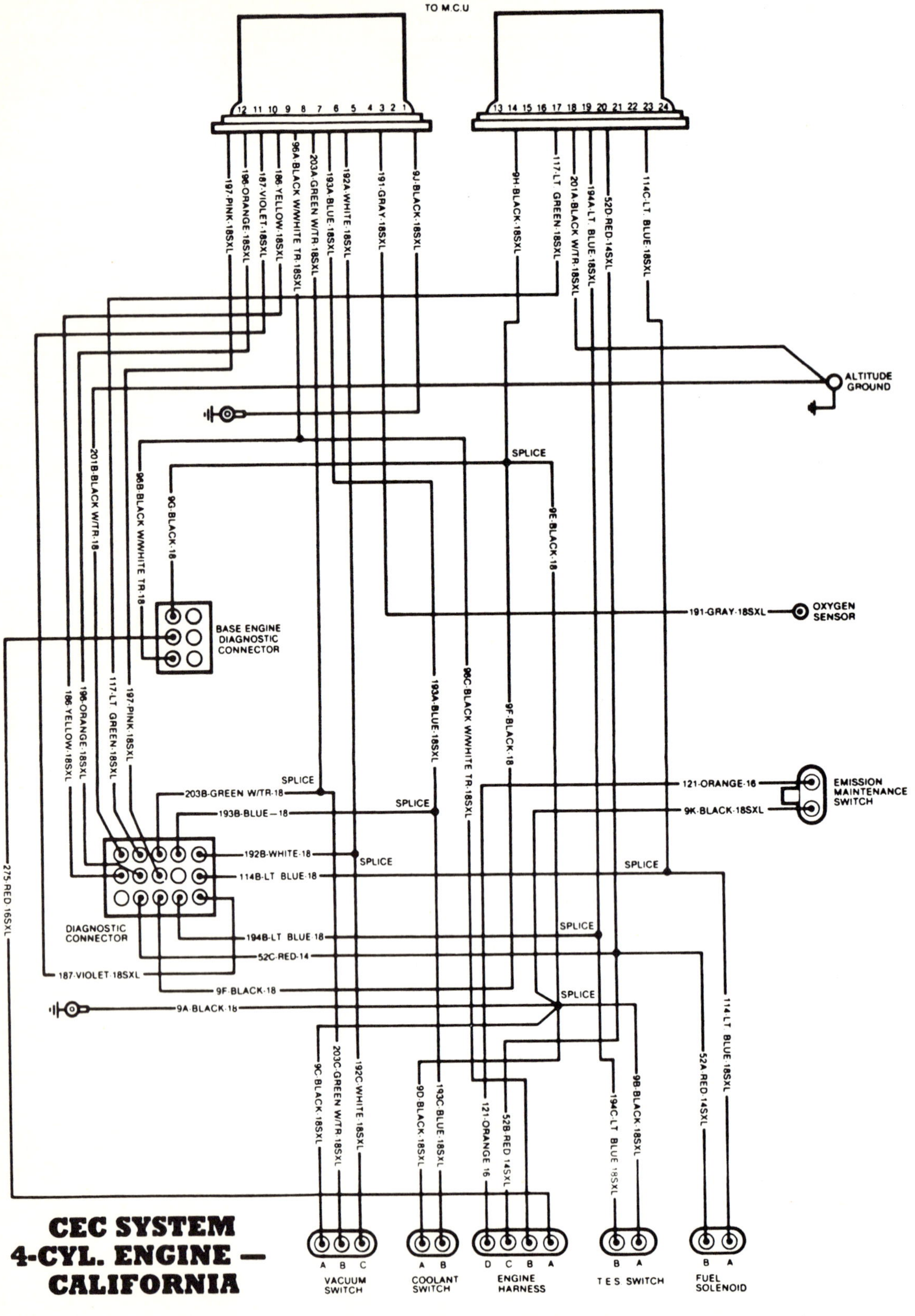

TO M.C.U

12 11 10 9 8 7 6 5 4 3 2 1

13 14 15 16 17 18 19 20 21 22 23 24

197-PINK-18SXL
196-ORANGE-18SXL
187-VIOLET-18SXL
186-YELLOW-18SXL
96A-BLACK W/WHITE TR-18SXL
203A-GREEN W/TR-18SXL
192A-WHITE-18SXL
193A-BLUE-18SXL
191-GRAY-18SXL
9J-BLACK-18SXL

9H-BLACK-18SXL
117-LT GREEN-18SXL
201A-BLACK W/TR-18SXL
194A-LT BLUE-18SXL
52D-RED-14SXL
114C-LT BLUE-18SXL

ALTITUDE GROUND

SPLICE

201B-BLACK W/TR-18

96B-BLACK W/WHITE TR-18
9G-BLACK-18

BASE ENGINE DIAGNOSTIC CONNECTOR

9E-BLACK-18

191-GRAY-18SXL ⊙ OXYGEN SENSOR

186-YELLOW-18SXL
196-ORANGE-18SXL
117-LT GREEN-18SXL
197-PINK-18SXL

193A-BLUE-18SXL
96C-BLACK W/WHITE TR-18SXL
9F-BLACK-18

121-ORANGE-16   EMISSION
9K-BLACK-18SXL   MAINTENANCE SWITCH

203B-GREEN W/TR-18   SPLICE
193B-BLUE—18   SPLICE
192B-WHITE-18   SPLICE
114B-LT BLUE-18   SPLICE

DIAGNOSTIC CONNECTOR

194B-LT BLUE-18   SPLICE
52C-RED-14

187-VIOLET-18SXL
9F-BLACK-18
9A-BLACK-18   SPLICE

275-RED-16SXL

9C-BLACK-18SXL
203C-GREEN W/TR-18SXL
192C-WHITE-18SXL
90-BLACK-18SXL
193C-BLUE-18SXL
121-ORANGE-16
52B-RED-14SXL
194C-LT BLUE-18SXL
9B-BLACK-18SXL
52A-RED-14SXL
114-LT BLUE-18SXL

**CEC SYSTEM
4-CYL. ENGINE —
CALIFORNIA**

A B C
VACUUM SWITCH

A B
COOLANT SWITCH

D C B A
ENGINE HARNESS

B A
T E S SWITCH

B A
FUEL SOLENOID

Engine and Engine Overhaul

3

QUICK REFERENCE INDEX

GENERAL INDEX

ENGINE ELECTRICAL

ALTERNATOR AND REGULATOR SPECIFICATIONS

Engine	Year	Manufacturer	Alternator Field Current @ 12v (amps)	Output (amps)	Regulator Manufacturer	Volts @ 75°F
4-134	1971	Motorola	1.2–1.7	35	Motorola	14.2–14.6
4-150	1984–90	Delco-Remy	4.0–5.0	56 ①	Delco-Remy	13.9–14.9
4-151	1980–83	Delco-Remy	4.0-5.0	42 ②	Delco-Remy	12.0–15.5
6-225	1971	Motorola	1.2–1.7	35	Delco-Remy	14.2–14.6
6-232, 258	1972–74	Motorola	1.8–2.5	37	Motorola	13.7–14.2
6-232, 258	1975	Delco-Remy	1.8–2.5	37 ③	Delco-Remy	13.7–14.2
	1976–79	Delco-Remy	4.0–5.0	37 ②	Delco-Remy	12.0–15.5
6-258	1980–90	Delco-Remy	4.0–5.0	42 ②	Delco-Remy	13.9–14.9
8-304	1972–74	Motorola	1.8–2.5	37 ④	Motorola	13.7–14.2
	1975	Motorola	1.8–2.5	37 ④	Motorola	12.7–15.3
	1976–77	Motorcraft	2.5–3.0	40 ⑤	Motorcraft	13.1–14.8
	1978–79	Delco-Remy	4.0–5.0	37 ②	Delco-Remy	12.0–15.5
	1980–81	Delco-Remy	4.0–5.0	42 ②	Delco-Remy	12.0–15.5

① Optional 68 and 78
② Optional 56, 63, 78 and 85
③ Optional 55 and 63
④ Optional 51 and 62
⑤ Optional 60
⑥ Optional 40 amp

STARTER SPECIFICATIONS

Engine	Year	Manufacturer	Lock Test Amps	Volts	Torque (ft. lbs.)	No-Load Test Amps	Volts	rpm	Brush Spring Tension (oz.)
4-134	1971	Autolite	①	4.0	②	50	10.0	4,400	31–47
		Delco	435	5.8	1.5	75	10.3	6,900	24 min.
		Prestolite	405	N.A.	9.0	50	10.0	5,300	32–40
4-150	1984	Delco	—Not Recommended—			67	12.0	8,500	30–40
	1985–86	Motorcraft	—Not Recommended—			67	12.0	8,368	N.A.
	1987–90	Bosch	120	9.6	N.A.	75	12.5	2,900	N.A.
6-225	1971	Delco	—Not Recommended—			75	10.6	6,200	32–40
6-232, 6-258	1972–78	Autolite	600	3.4	13.0	65	12.0	9,250	35–40
6-258	1979–81	Motorcraft	—Not Recommended—			77	12.0	9,250	35–40
	1982–90	Motorcraft	Not Recommended			67	12.0	7,868	35–40
8-304	1972–81	Autolite	600	3.4	13.0	65	12.0	9,250	35–40

N.A.: Information Not Available
min.: minimum
① Starter No. MDU7004: 280
   All others: 170
② Starter No. MDU7004: 6.2
   All others: 1.5

GENERAL ENGINE SPECIFICATIONS

Engine	Years	Fuel System Type	SAE net Horsepower @ rpm	SAE net Torque ft. lbs. @ rpm	Bore × Stroke	Comp. Ratio	Oil Press. (psi.) @ 2000 rpm
4-134	1971	1-bbl	75 @ 4,000	114 @ 2,000	3.125 × 4.375	6.7:1	35
4-150	1984–86	1-bbl	83 @ 4,200	116 @ 2,600	3.876 × 3.188	9.2:1	40
	1987–90	TBI	117 @ 5,000	135 @ 3,000	3.876 × 3.188	9.2:1	40
4-151	1980–83	2-bbl	87 @ 4,400	128 @ 2,400	4.000 × 3.000	8.3:1	38
6-225	1971	2-bbl	160 @ 4,200	235 @ 3,500	3.750 × 3.400	9.0:1	33
6-232	1972–78	1-bbl	100 @ 3,600	185 @ 1,800	3.750 × 3.500	8.0:1	50
6-258	1972–76	1-bbl	110 @ 3,500	195 @ 2,000	3.750 × 3.895	8.0:1	50
	1977–86	2-bbl	114 @ 3,600	196 @ 2,000	3.750 × 3.895	8.0:1	50
	1987–90	2-bbl	112 @ 3,000	210 @ 2,000	3.750 × 3.895	8.6:1	50
8-304	1972–81	2-bbl	150 @ 4,200	245 @ 2,500	3.750 × 3.753	8.4:1	50

TBI: Throttle Body Injection
1-bbl: one barrel carburetor
2-bbl: two barrel carburetor

VALVE SPECIFICATIONS

Engine	Years	Seat Angle (deg)	Face Angle (deg)	Spring Test Pressure (lbs. @ in.)	Spring Installed Height (in.)	Stem to Guide Clearance (in.) Intake	Stem to Guide Clearance (in.) Exhaust	Stem Diameter (in.) Intake	Stem Diameter (in.) Exhaust
4-134	1971	45	45	①	1.660	0.0007–0.0022	0.0025–0.0045	0.3733–0.3738	0.3710–0.3720
4-150	1984–86	44.5	44	212 @ 1.203	1.625	0.0010–0.0030	0.0010–0.0030	0.3110–0.3120	0.3110–0.3120
	1987–90	44.5	45	200 @ 1.216	1.640	0.0010–0.0030	0.0010–0.0030	0.3110–0.3120	0.3110–0.3120
4-151	1980–83	46	45	176 @ 1.250	1.660	0.0010–0.0027	0.0010–0.0027	0.3422	0.3422
6-225	1971	45	45	168 @ 1.260	1.640	0.0012–0.0032	0.0015–0.0035	0.3415–0.3427	0.3402–0.3412
6-232	1972–78	②	③	④	⑤	0.0010–0.0030	0.0010–0.0030	0.3715–0.3725	0.3715–0.3725
6-258	1972–90	②	③	⑥	⑦	0.0010–0.0030	0.0010–0.0030	0.3715–0.3725	0.3715–0.3725
8-304	1976–81	②	③	218 @ 1.359	1.812	0.0010–0.0030	0.0010–0.0030	0.3715–0.3725	0.3715–0.3725

① Intake: 153 @ 1.400
Exhaust: 120 @ 1.750
② Intake: 30
Exhaust: 44.5
③ Intake: 29
Exhaust: 44
④ With rotators: 218 @ 1.875
Without rotators: 195 @ 1.437
⑤ Intake: 1.786
Exhaust: 2.110

⑥ 1972–76: with rotators: 218 @ 1.875
without rotators: 195 @ 1.437
1977–78: 20B @ 1.386
1979–87: Intake, 195 @ 1.411
Exhaust, 220 @ 1.188
1988–90: All, 195 @ 1.411
⑦ 1972–78: 1.786
1979–84 Intake, 1.786
Exhaust, 1.625
1985–90: 1.786

CAMSHAFT SPECIFICATIONS

(All specifications in inches)

Engine	Journal Diameter 1	2	3	4	5	Bearing Clearance	Lobe Lift Intake	Exhaust	End Play
4-134	2.1860–2.1855	2.1225–2.1215	2.0600–2.0590	1.6230–1.6225	—	0.0010–0.0025	0.2600	0.3510	0.004–0.007
4-150	2.0300–2.0290	2.0200–2.0190	2.0100–2.0009	2.0000–1.9990	—	0.0010–0.0030	0.2650	0.2650	0
4-151	1.8690	1.8690	1.8690	—	—	0.0007–0.0027	0.3980	0.3980	0.0015–0.0050
6-225	1.7560–1.7550	1.7260–1.7250	1.6960–1.6950	1.6660–1.6650	—	0.0015–0.0040	N.A.	N.A.	N.A.
6-232	2.0300–2.0290	2.0200–2.0190	2.0100–2.0090	2.0000–1.9990	—	0.0010–0.0030	0.2540	0.2540	0
6-258	2.0300–2.0290	2.0200–2.0190	2.0100–2.0090	2.0000–1.9990	—	0.0010–0.0030	0.2540 ①	0.2540 ①	0
8-304	2.1205–2.1195	2.0905–2.0895	2.0605–2.0595	2.0305–2.0295	2.0005–1.9995	0.0010–0.0030	0.2660	0.2660	0

N.A.: Information not available
①1985–90: 0.2531

CRANKSHAFT AND CONNECTING ROD SPECIFICATIONS

(All specifications in inches)

Engine	Years	Crankshaft Main Bearing Journal Diameter	Main Bearing Oil Clearance	Shaft End Play	Thrust on No.	Connecting Rod Journal Diameter	Oil Clearance	Side Clearance
4-134	1971	2.331–2.3341	0.0003–0.0029	0.0040–0.0060	1	1.9375–1.9383	0.0001–0.0019	0.004–0.010
4-150	1984–90	2.4996–2.5001	0.0010–0.0025	0.0015–0.0065	2	2.0934–2.0955	0.0010–0.0030	0.010–0.019
4-151	1980–83	2.2988	0.0005–0.0022	0.0035–0.0085	5	1.8690	0.0007–0.0027	0.006–0.022
6-225	1971	2.4993–2.4997	0.0005–0.0021	0.0040–0.0080	2	1.9998–2.002	0.0020–0.0023	0.006–0.017
6-232	1972–78	2.4986–2.5001	0.0010–0.0020	0.0015–0.0065	3	2.0934–2.0955	0.0010–0.0020	0.005–0.014
6-258	1972–90	⑥	①	0.0015–0.0065	3	2.0934–2.0955	②	③
8-304	1976–81	④	⑤	0.0030–0.0080	3	2.0934–2.0955	0.0010–0.0020	0.006–0.018

①1972–73: 0.0010–0.0020
  1974–80: 0.0010–0.0030
  1981: No. 1 0.0005–0.0026
      No. 2, 3, 4, 5, 6 0.0005–0.0030
      No. 7 0.0011–0.0035
  1982–90: 0.0010–0.0025
②1972–73: 0.0010–0.0020
  1974–76: 0.0010–0.0030
  1977–81: 0.0010–0.0025
  1982–90: 0.0010–0.0030

③1972–80: 0.005–0.014
  1981–90: 0.010–0.019
④No. 1, 2, 3, 4: 2.7474–2.7489
  No. 5: 2.7464–2.7479
⑤No. 1, 2, 3, 4 0.0010–0.0020
  No. 5 0.0020–0.0030
⑥1972–86: 2.4986–2.5001
  1987–90: 2.4996–2.5001

PISTON AND RING SPECIFICATIONS
(All specifications in inches)

| Engine | Years | Ring Gap | | | Ring Side Clearance | | | Piston▲ Clearance |
		#1 Compr.	#2 Compr.	Oil Control	#1 Compr.	#2 Compr.	Oil Control	
4-134	1971	0.0070–0.0170	0.0070–0.0170	0.0070–0.0170	0.0020–0.0040	0.0015–0.0035	0.0010–0.0025	0.0025–0.0045
4-150	1984–87	0.0100–0.0200	0.0100–0.0200	0.0100–0.0250	0.0017–0.0032	0.0017–0.0032	0.0010–0.0080	0.0009–0.0017
	1988–90	0.0100–0.0200	0.0100–0.0200	0.0150–0.0550	0.0010–0.0032	0.0010–0.0032	0.0010–0.0085	0.0013–0.0021
4-151	1980–83	0.0027–0.0033	0.0090–0.0190	0.0150–0.0550	0.0025–0.0033	0.0025–0.0033	0.0025–0.0033	0.0025–0.0033
6-225	1971	0.0100–0.0200	0.0100–0.0200	0.0150–0.0350	0.0020–0.0035	0.0030–0.0050	0.0015–0.0085	0.0005–0.0011
6-232	1972–78	0.0100–0.0200	0.0100–0.0200	0.0150–0.0550	0.0015–0.0030	0.0015–0.0030	0.0010–0.0080	0.0009–0.0017
6-258	1972–90	0.0100–0.0200	0.0100–0.0200	①	②	②	0.0010–0.0080	0.0009–0.0017
8-304	1976–81	0.0100–0.0200	0.0100–0.0200	0.0100–0.0250	0.0015–0.0030	0.0015–0.0030	0.0011–0.0080	0.0010–0.0018

▲ Measured at the skirt
① 1972–73: 0.0150–0.0550
  1974–90: 0.0100–0.0250
② 1972–80: 0.0015–0.0030
  1981–90: 0.0017–0.0032

TORQUE SPECIFICATIONS
(All specifications in ft. lbs.)

| Engine | Years | Cyl. Head | Conn. Rod | Main Bearing | Crankshaft Damper | Flywheel | Manifold | |
							Intake	Exhaust
4-134	1971	60–70	35–45	65–75	65–75	35–41	29–35	29–35
4-150	1984–87	80–90	30–35	75–85	75–85	50 ①	20–25	20–25
	1988–90	④	30–35	75–85	75–85	50 ①	20–25	20–25
4-151	1980–83	93–97	28–32	63–67	155–165	53–57	Bolt: 40 Nut: 30	Bolt: 40 Nut: 30
6-225	1971	65–85	30–40	85–95	140–150	50–65	45–55	14–20
6-232	1972–78	100–110	25–30	75–85	50–60	100–110	40–45	23–28
6-258	1972–90	②	30–35	75–85	75–85	100–110	20–25	③
8-304	1976	100–110	25–30	95–105	53–58	100–110	40–45	23–27

① Plus a 60° turn
② 1972–80: 105
  1981–90: 85
③ 1972–79: 23
  1980 and later (See illustration in text):
    Bolts No. 1 thru 11: 23
    Bolts No. 12 & 13: 50
④ See the illustration in the text
    Nos. 1, 2, 3, 4, 5, 7, 9, 10: 110
    No. 8: 100

Understanding the Engine Electrical System

The engine electrical system can be broken down into three separate and distinct systems:

1. The starting system.
2. The charging system.
3. The ignition system.

BATTERY AND STARTING SYSTEM

Basic Operating Principles

The battery is the first link in the chain of mechanisms which work together to provide cranking of the automobile engine. In most modern cars, the battery is a lead/acid electrochemical device consisting of six 2v subsections connected in series so the unit is capable of producing approximately 12v of electrical pressure. Each subsection, or cell, consists of a series of positive and negative plates held a short distance apart in a solution of sulfuric acid and water. The two types of plates are of dissimilar metals. This causes a chemical reaction to be set up, and it is this reaction which produces current flow from the battery when its positive and negative terminals are connected to an electrical appliance such as a lamp or motor. The continued transfer of electrons would eventually convert the sulfuric acid in the electrolyte to water, and make the two plates identical in chemical composition. As electrical energy is removed from the battery, its voltage output tends to drop. Thus, measuring battery voltage and battery electrolyte composition are two ways of checking the ability of the unit to supply power. During the starting of the engine, electrical energy is removed from the battery. However, if the charging circuit is in good condition and the operating conditions are normal, the power removed from the battery will be replaced by the generator (or alternator) which will force electrons back through the battery, reversing the normal flow, and restoring the battery to its original chemical state.

The battery and starting motor are linked by very heavy electrical cables designed to minimize resistance to the flow of current. Generally, the major power supply cable that leaves the battery goes directly to the starter, while other electrical system needs are supplied by a smaller cable. During starter operation, power flows from the battery to the starter and is grounded through the car's frame and the battery's negative ground strap.

The starting motor is a specially designed, direct current electric motor capable of producing a very great amount of power for its size. One thing that allows the motor to produce a great deal of power is its tremendous rotating speed. It drives the engine through a tiny pinion gear (attached to the starter's armature), which drives the very large flywheel ring gear at a greatly reduced speed. Another factor allowing it to produce so much power is that only intermittent operation is required of it. This, little allowance for air circulation is required, and the windings can be built into a very small space.

The starter solenoid is a magnetic device which employs the small current supplied by the starting switch circuit of the ignition switch. This magnetic action moves a plunger which mechanically engages the starter and electrically closes the heavy switch which connects it to the battery. The starting switch circuit consists of the starting switch contained within the ignition switch, a transmission neutral safety switch or clutch pedal switch, and the wiring necessary to connect these in series with the starter solenoid or relay.

A pinion, which is a small gear, is mounted to a one-way drive clutch. This clutch is splined to the starter armature shaft. When the ignition switch is moved to the **start** position, the solenoid plunger slides the pinion toward the flywheel ring gear via a collar and spring. If the teeth on the pinion and flywheel match properly, the pinion will engage the flywheel immediate-

ly. If the gear teeth butt one another, the spring will be compressed and will force the gears to mesh as soon as the starter turns far enough to allow them to do so. As the solenoid plunger reaches the end of its travel, it closes the contacts that connect the battery and starter and then the engine is cranked.

As soon as the engine starts, the flywheel ring gear begins turning fast enough to drive the pinion at an extremely high rate of speed. At this point, the one-way clutch begins allowing the pinion to spin faster than the starter shaft so that the starter will not operate at excessive speed. When the ignition switch is released from the starter position, the solenoid is de-energized, and a spring contained within the solenoid assembly pulls the gear out of mesh and interrupts the current flow to the starter.

Some starter employ a separate relay, mounted away from the starter, to switch the motor and solenoid current on and off. The relay thus replaces the solenoid electrical switch, buy does not eliminate the need for a solenoid mounted on the starter used to mechanically engage the starter drive gears. The relay is used to reduce the amount of current the starting switch must carry.

THE CHARGING SYSTEM

Basic Operating Principles

The automobile charging system provides electrical power for operation of the vehicle's ignition and starting systems and all the electrical accessories. The battery services as an electrical surge or storage tank, storing (in chemical form) the energy originally produced by the engine driven generator. The system also provides a means of regulating generator output to protect the battery from being overcharged and to avoid excessive voltage to the accessories.

The storage battery is a chemical device incorporating parallel lead plates in a tank containing a sulfuric acid/water solution. Adjacent plates are slightly dissimilar, and the chemical reaction of the two dissimilar plates produces electrical energy when the battery is connected to a load such as the starter motor. The chemical reaction is reversible, so that when the generator is producing a voltage (electrical pressure) greater than that produced by the battery, electricity is forced into the battery, and the battery is returned to its fully charged state.

The vehicle's generator is driven mechanically, through V-belts, by the engine crankshaft. It consists of two coils of fine wire, one stationary (the stator), and one movable (the rotor). The rotor may also be known as the armature, and consists of fine wire wrapped around an iron core which is mounted on a shaft. The electricity which flows through the two coils of wire (provided initially by the battery in some cases) creates an intense magnetic field around both rotor and stator, and the interaction between the two fields creates voltage, allowing the generator to power the accessories and charge the battery.

There are two types of generators: the earlier is the direct current (DC) type. The current produced by the DC generator is generated in the armature and carried off the spinning armature by stationary brushes contacting the commutator. The commutator is a series of smooth metal contact plates on the end of the armature. The commutator is a series of smooth metal contact plates on the end of the armature. The commutator plates, which are separated from one another by a very short gap, are connected to the armature circuits so that current will flow in one directions only in the wires carrying the generator output. The generator stator consists of two stationary coils of wire which draw some of the output current of the generator to form a powerful magnetic field and create the interaction of fields which generates the voltage. The generator field is wired in series with the regulator.

Newer automobiles use alternating current generators or alternators, because they are more efficient, can be rotated at higher speeds, and have fewer brush problems. In an alternator, the field rotates while all the current produced passes only through the stator winding. The brushes bear against continuous slip rings rather than a commutator. This causes the current produced to periodically reverse the direction of its flow. Diodes (electrical one-way switches) block the flow of current from traveling in the wrong direction. A series of diodes is wired together to permit the alternating flow of the stator to be converted to a pulsating, but unidirectional flow at the alternator output. The alternator's field is wired in series with the voltage regulator.

The regulator consists of several circuits. Each circuit has a core, or magnetic coil of wire, which operates a switch. Each switch is connected to ground through one or more resistors. The coil of wire responds directly to system voltage. When the voltage reaches the required level, the magnetic field created by the winding of wire closes the switch and inserts a resistance into the generator field circuit, thus reducing the output. The contacts of the switch cycle open and close many times each second to precisely control voltage.

While alternators are self-limiting as far as maximum current is concerned, DC generators employ a current regulating circuit which responds directly to the total amount of current flowing through the generator circuit rather than to the output voltage. The current regulator is similar to the voltage regulator except that all system current must flow through the energizing coil on its way to the various electrical components.

Ignition Coil

REMOVAL AND INSTALLATION

All Except the 4–151 and 1987-90 4–150

1. Disconnect the battery ground.
2. Disconnect the two small and one large wire from the coil.
3. Disconnect the condenser connector from the coil, if equipped.
4. Unbolt and remove the coil.
5. Installation is the reverse of removal.

4–151 1980–81

1. Remove the distributor cap.
2. Remove the three coil cover attaching screws and lift off the cover.
3. Remove the four coil attaching screws and lift off the coil.
4. Installation is the reverse of removal.

4-151 1982–83

1. Disconnect the harness at the coil.
2. Pulling on the boot, only, pull the coil-to-distributor cap wire from the coil.
3. Remove the three coil mounting screws and lift off the coil.
4. Installation is the reverse of removal.

4-150 1987-90

The coil is an integral part of the Ignition Control Module (ICM), mounted to the left of the battery on the firewall. The coil can, however, be removed for separate replacement.

Ignition Module

REMOVAL AND INSTALLATION

The ignition module is mounted next to the battery on all models. It is a sealed, weatherproof unit on all models, except the 1987-90 4–150. The 1987-90 4–150 incorporates the coil in the control module. The coil, on these models, can be removed separately.

Removing the module, on all models, is a matter of simply removing the fasteners that attach it to the fender or firewall and

pulling apart the connectors. When unplugging the connectors, pull them apart with a firm, straight pull. NEVER PRY THEM APART! To pry them will cause damage. When reconnecting them, coat the mating ends with silicone dielectric grease to wa-

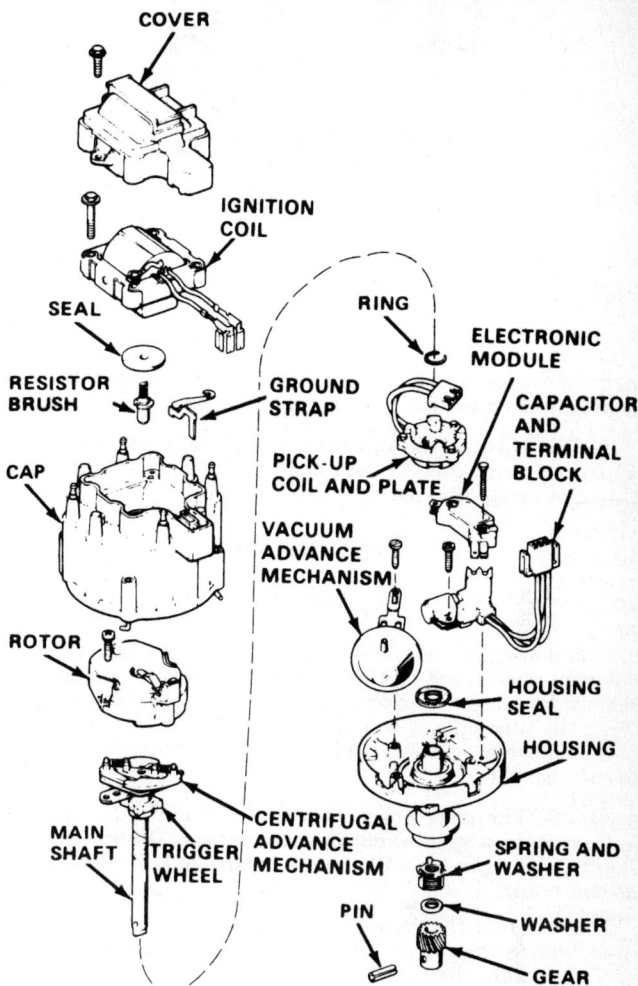

On the 1980–81 4–151, the coil is in the distributor cap

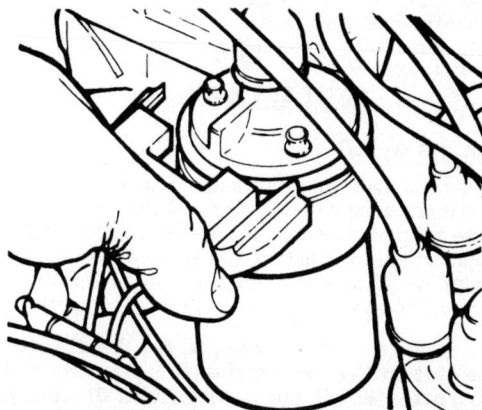

Coil connections used on all late model engines except the 4–151 and 1987-90 4–150 engines. Earlier engines had separate wires rather than the slip-on connector

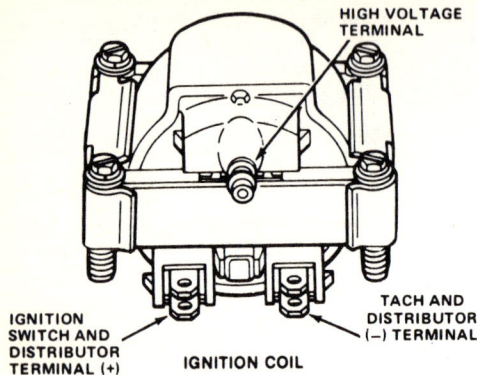

HIGH VOLTAGE
TERMINAL

IGNITION
SWITCH AND
DISTRIBUTOR
TERMINAL (+)

IGNITION COIL

TACH AND
DISTRIBUTOR
(−) TERMINAL

1982–83 4–151 coil

terproof the connection. Press the connectors together firmly to overcome any vacuum lock caused by the grease.

NOTE: If the locking tabs weaken or break, don't replace the unit. Just secure the connection with electrical tape or tie straps.

Distributor

REMOVAL

1. Remove the high-tension wires from the distributor cap terminal towers, noting their positions to assure correct reassembly. For diagrams of firing orders and distributor wiring, refer to the tune-up and troubleshooting section.
2. Remove the primary lead from the terminal post at the side of the distributor.

NOTE: The wire connector on 1978 and later models will contain a special conductive grease. Do not remove it. The same grease will also be found on the metal parts of the rotor.

3. Disconnect the vacuum line if there is one.
4. Remove the two distributor cap retaining hooks or screws and remove the distributor cap.
5. Note the position of the rotor in relation to the base. Scribe a mark on the base of the distributor and on the engine block to facilitate reinstallation. Align the marks with the direction the metal tip of the rotor is pointing.
6. Remove the bolt that holds the distributor to the engine.
7. Lift the distributor assembly from the engine.

INSTALLATION

All Except the 4–150

1. Insert the distributor shaft and assembly into the engine. Line up the mark on the distributor and the one on the engine with the metal tip of the rotor. Make sure that the vacuum advance diaphragm is pointed in the same direction as it pointed originally. This will be done automatically if the marks on the engine and the distributor are line up with the rotor.

NOTE: On the 6–225 and F4–134, the distributor shaft fits into a slot in the end of the oil pump shaft. Therefore, the rotor won't turn when the distributor is pressed into place.

2. Install the distributor holddown bolt and clamp. Leave the screw loose enough so that you can move the distributor with heavy hand pressure.
3. Connect the primary wire to the distributor side of the coil.

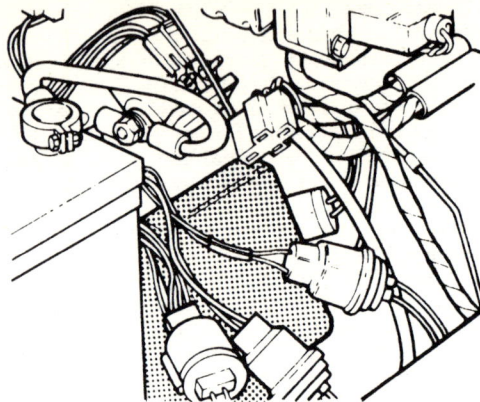

Control module on all except the 4–151 and 1987–90 4–150 engines

Install the distributor cap on the distributor housing. Secure the distributor cap with the spring clips or the screw type retainers, whichever is used.

4. Install the spark plug wires. Make sure that the wires are pressed all of the way into the top of the distributor cap and firmly onto the spark plugs.
5. Adjust the point cam dwell and set the ignition timing. Refer to the tune-up section.

If the engine was turned while the distributor was removed, or if the marks were not drawn, it will be necessary to initially time the engine. Follow the procedure below.

NOTE: Design of the V6 engine requires a special form of distributor cam. The distributor may be serviced in the regular way and should cause no more problems than any other distributor, if the firing plan is thoroughly understood. The distributor cam is not ground to standard 6-cylinder indexing intervals. This particular form requires that the original pattern of spark plug wiring be used. The engine will not run in balance if number one spark plug wire is inserted into number 6 distributor cap tower, even though each wire in the firing sequence is advanced to the next distributor tower. There is a difference between the firing intervals of each succeeding cylinder through the 720° engine cycle.

INSTALLATION, ENGINE ROTATED

All Except the 4–150

1. If the engine has been rotated while the distributor was out, you'll have to first put the engine on No. 1 cylinder at Top Dead Center firing position. You can either remove the valve cover or No. 1 spark plug to determine engine position. Rotate the engine with a socket wrench on the nut at the center of the front pulley in the normal direction of rotation. Either feel for air being expelled forcefully through the spark plug hole or watch for the engine to rotate up to the Top Center mark without the valves moving (both valves will be closed). Stop turning F4–134 engines when either the 5° mark on the flywheel is in the middle of the flywheel inspection opening, or the marks on the crankshaft pulley and the timing gear cover are in alignment. If the valves are moving as you approach TDC or there is no air being expelled through the plug hole, turn the engine another full turn until you get the appropriate indication as the engine approaches TDC position.
2. Start the distributor into the engine with the matchmarks between the distributor body and the engine lined up. Turn the rotor slightly until the matchmarks on the bottom of the distrib-

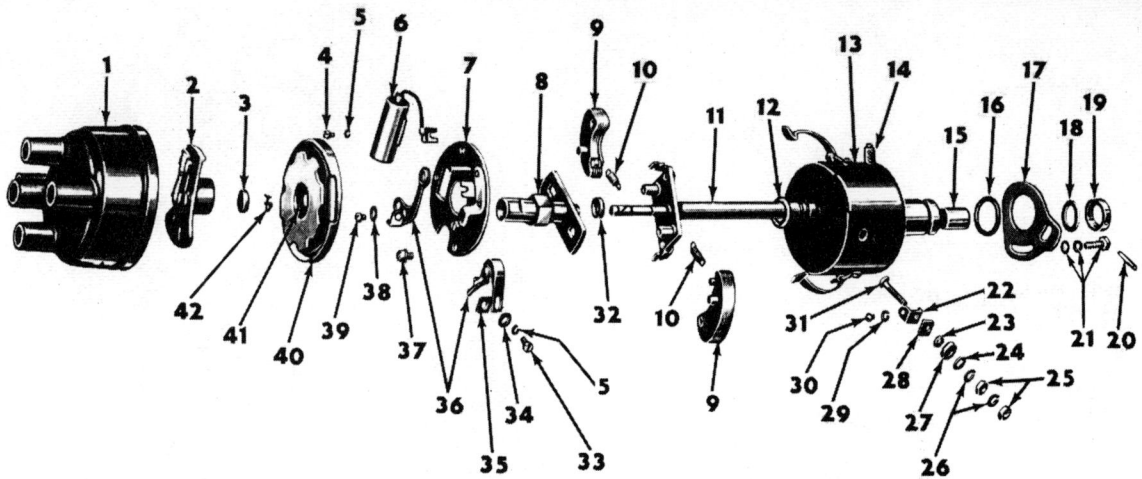

4–134 distributor

1. Cap
2. Rotor
3. Cam oiling wick
4. Condenser mounting screw
5. Lockwasher
6. Condenser
7. Breaker plate
8. Cam and stop plate
9. Governor weight
10. Governor spring
11. Driveshaft
12. Thrust washer
13. Base
14. Oiler

15. Bearing
16. Rubber O-ring
17. Advance arm
18. Lower thrust washer
19. Driveshaft collar
20. Collar rivet
21. Screw and washers
22. Connector
23. Bushing
24. Terminal washer
25. Terminal nut
26. Terminal lockwasher
27. Insulating washer
28. Terminal insulation

29. Connector lockwasher
30. Connector screw
31. Terminal post
32. Cam spacer
33. Breaker arm spring clip screw
34. Spring clip screw washer
35. Spring clip
36. Distributor points
37. Breaker plate screw
38. Washer
39. Locking screw
40. Plate seal
41. Felt washer
42. Snapring

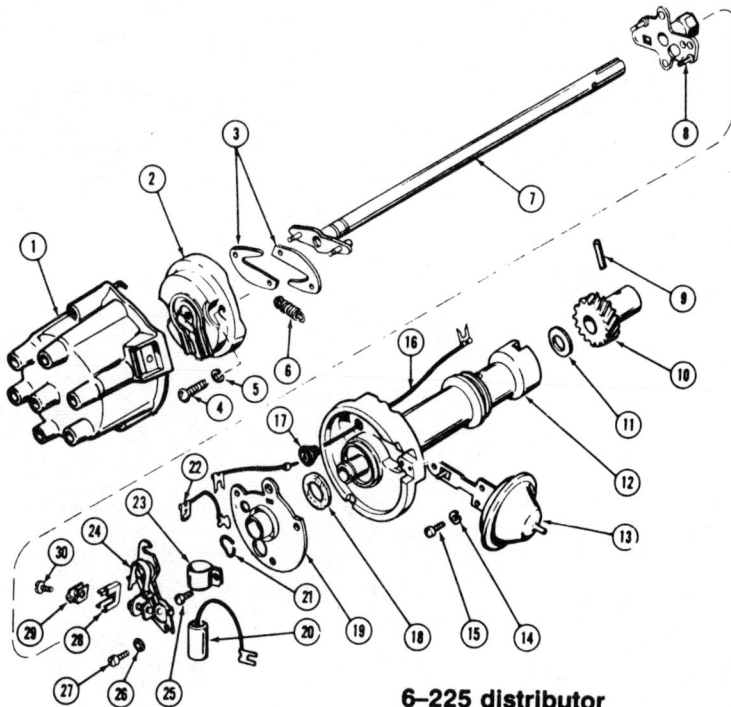

6–225 distributor

1. Cap assembly
2. Rotor
3. Governor weight
4. Rotor mounting screw
5. Lockwasher
6. Weight spring (governor)
7. Shaft
8. Cam assembly
9. Gear pin
10. Drive gear
11. Spacer washer
12. Housing
13. Vacuum control
14. Lockwasher
15. Control mounting screw
16. Primary lead
17. Lead grommet
18. Washer
19. Breaker plate
20. Condenser
21. Retaining spring
22. Ground lead
23. Condenser clamp
24. Contact set
25. Clamp screw
26. Lockwasher
27. Contact screw
28. Insulator
29. Spring clip
30. Screw

utor body and the bottom of the distributor shaft near the gear are aligned.

NOTE: On the 4–134 and 6–225, the distributor shaft indexes with the oil pump driveshaft.

Then, insert the distributor all the way into the engine. If you have trouble getting the distributor and camshaft gears to mesh, turn the rotor back and forth very slightly until the distributor can be inserted easily. If the rotor is not now lined up with the position of No. 1 plug terminal, you'll have to pull the distributor back out slightly, shift the position of the rotor appropriately, and then reinstall it.

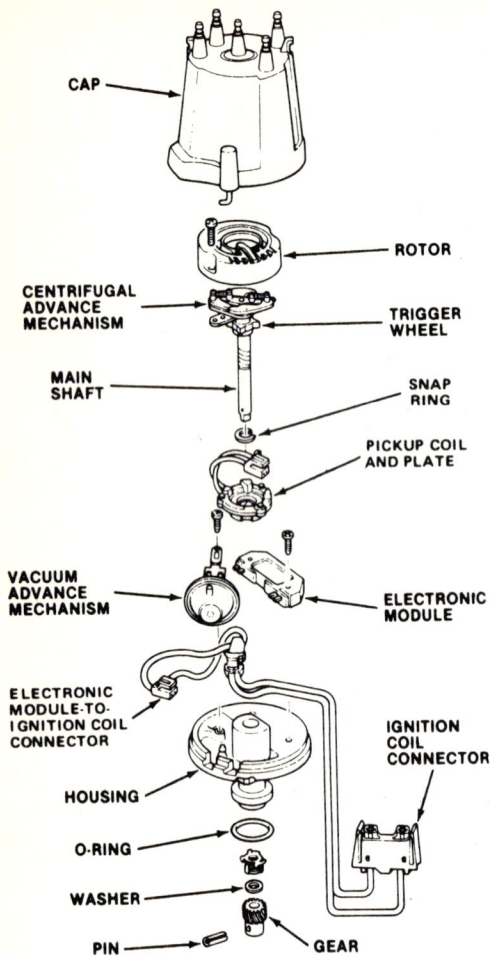

CAP

ROTOR

CENTRIFUGAL
ADVANCE
MECHANISM

TRIGGER
WHEEL

MAIN
SHAFT

SNAP
RING

PICKUP COIL
AND PLATE

VACUUM
ADVANCE
MECHANISM

ELECTRONIC
MODULE

ELECTRONIC
MODULE-TO-
IGNITION COIL
CONNECTOR

IGNITION
COIL
CONNECTOR

HOUSING

O-RING

WASHER

PIN

GEAR

1982–83 4–151 distributor

3. Align the matchmarks between the distributor and engine. Install the distributor mounting bolt and tighten it finger tight. Reconnect the vacuum advance line and distributor wiring connector, and reinstall the gasket and cap. Reconnect the negative battery cable. Adjust the ignition timing as described in Section 2. Then, tighten the distributor mounting bolt securely.

INSTALLATION

4–150

1. Rotate the engine until the No.1 piston is at TDC compression.
2. Using a flat bladed screwdriver, in the distributor hole, rotate the oil pump gear so that the slot in the oil pump shaft is slightly past the 3:00 o'clock position, relative to the length of the engine block.
3. With the distributor cap removed, install the distributor with the rotor at the 5:00 o'clock position, relative to the oil pump gear shaft slot. When the distributor is completely in place, the rotor should be at the 6:00 o'clock position. If not, remove the distributor and perform the entire procedure again.
4. Tighten the lockbolt.

Alternator

ALTERNATOR PRECAUTIONS

To prevent damage to the alternator and regulator, the fol-

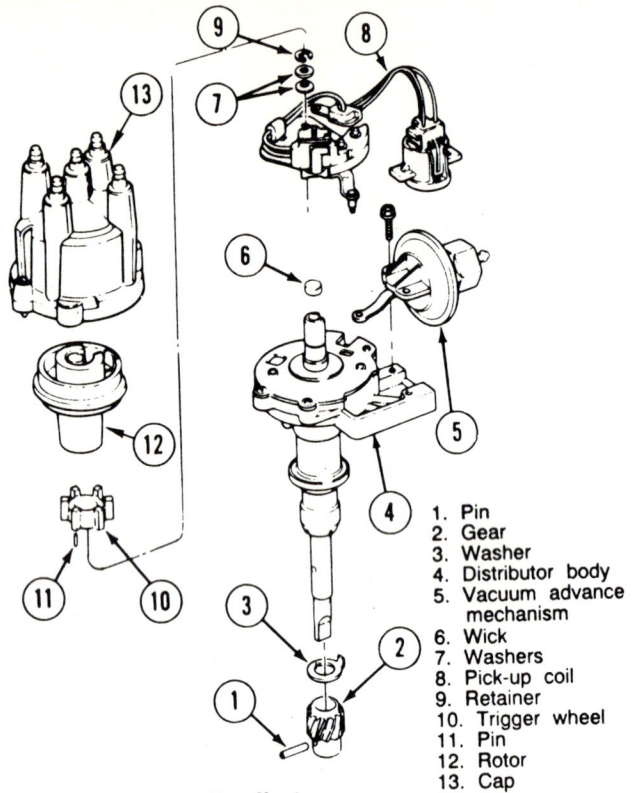

1. Pin
2. Gear
3. Washer
4. Distributor body
5. Vacuum advance mechanism
6. Wick
7. Washers
8. Pick-up coil
9. Retainer
10. Trigger wheel
11. Pin
12. Rotor
13. Cap

1984–86 4–150 distributor

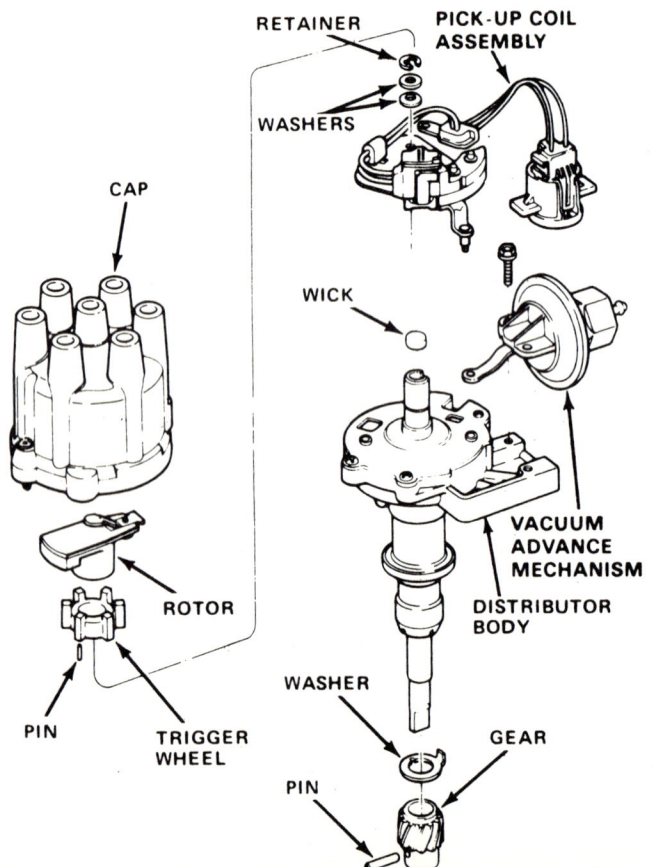

RETAINER

PICK-UP COIL
ASSEMBLY

WASHERS

CAP

WICK

ROTOR

VACUUM
ADVANCE
MECHANISM

DISTRIBUTOR
BODY

PIN

TRIGGER
WHEEL

WASHER

GEAR

PIN

SSI distributor, 6-cylinder is shown; the V8 is similar

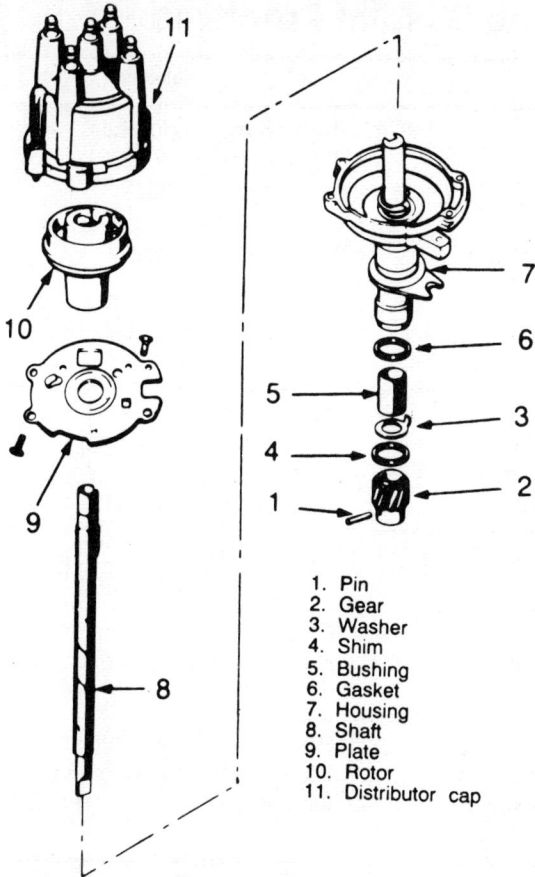

1. Pin
2. Gear
3. Washer
4. Shim
5. Bushing
6. Gasket
7. Housing
8. Shaft
9. Plate
10. Rotor
11. Distributor cap

1987–90 4–150 engine distributor

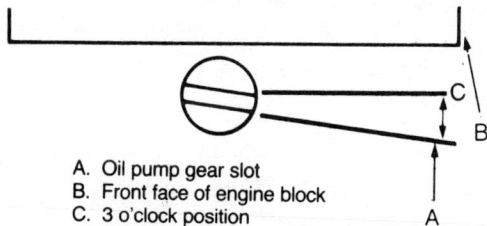

1. Front face of engine block
2. Rotor pre-positioned
3. 5 o'clock position (approx.)

Positioning the distributor rotor and shaft for installation, on the 4–150

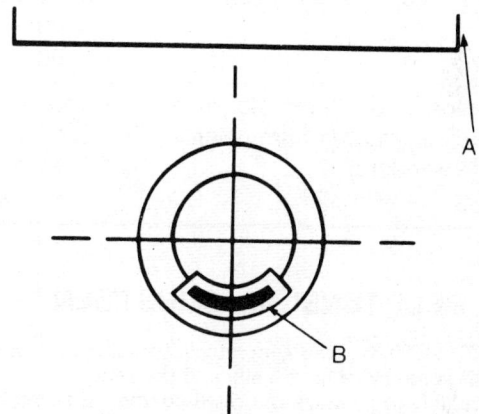

A. Front face of engine block
B. Rotor position when properly installed

Rotor position with the distributor properly installed on the 4–150

A. Oil pump gear slot
B. Front face of engine block
C. 3 o'clock position

Positioning the oil pump shaft for distributor installation on the 4–150

lowing precautionary measures must be taken when working with the electrical system.

1. Never reverse battery connections. Always check the battery polarity visually. This is to be done before any connections are made to be sure that all of the connections correspond to the battery ground polarity of the Jeep.

2. Booster batteries for starting must be connected properly. Make sure that the positive cable of the booster battery is connected to the positive terminal of the battery that is getting the boost. This applies to both negative and ground cables.

3. Disconnect the battery cables before using a fast charger. The charger has a tendency to force current through the diodes in the opposite direction for which they were designed. This burns out the diodes.

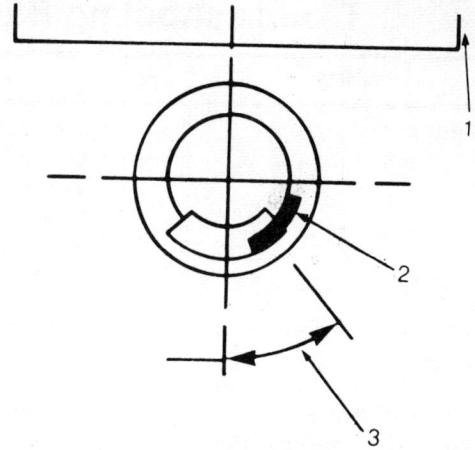

4. Never use a fast charger as a booster for starting the vehicle.

5. Never disconnect the voltage regulator while the engine is running.

6. Do not ground the alternator output terminal.

7. Do not operate the alternator on an open circuit with the field energized.

8. Do not attempt to polarize an alternator.

REMOVAL AND INSTALLATION

1. Remove all of the electrical connections from the alternator or generator. Label all of the wires so that you can install them correctly.

2. Remove all of the attaching nuts, bolts and washers noting different sized threads or nuts and bolts that go in certain holes.

3. Remove the alternator carefully.

4. To install, reverse the above procedure and adjust the belt as described below. Torque the mounting bolts to 25–30 ft. lbs.; the sliding adjuster bolt to 20 ft. lbs.

Troubleshooting Basic Charging System Problems

Problem	Cause	Solution
Noisy alternator	• Loose mountings • Loose drive pulley • Worn bearings • Brush noise • Internal circuits shorted (High pitched whine)	• Tighten mounting bolts • Tighten pulley • Replace alternator • Replace alternator • Replace alternator
Squeal when starting engine or accelerating	• Glazed or loose belt	• Replace or adjust belt
Indicator light remains on or ammeter indicates discharge (engine running)	• Broken fan belt • Broken or disconnected wires • Internal alternator problems • Defective voltage regulator	• Install belt • Repair or connect wiring • Replace alternator • Replace voltage regulator
Car light bulbs continually burn out—battery needs water continually	• Alternator/regulator overcharging	• Replace voltage regulator/alternator
Car lights flare on acceleration	• Battery low • Internal alternator/regulator problems	• Charge or replace battery • Replace alternator/regulator
Low voltage output (alternator light flickers continually or ammeter needle wanders)	• Loose or worn belt • Dirty or corroded connections • Internal alternator/regulator problems	• Replace or adjust belt • Clean or replace connections • Replace alternator or regulator

BELT TENSION ADJUSTMENT

The fan belt drives the generator/alternator and the water pump. If it is too loose, it will slip and the generator/alternator will not be able to produce the rated current. if the belt is too loose, the water pump would not be driven and the engine could overheat. Check the tension of the fan belt by pushing your thumb down on the longest span of belt midway between the pulleys. If the belt flexes more than ½ in. (13mm), it should be tightened. Loosen the bolt on the adjusting bracket and pivot bolt and move the alternator or generator away from the engine to tighten the belt. Do not apply pressure to the rear of the case aluminum housing of an alternator; it might break. Tighten the adjusting bolts when the proper tension is reached.

Regulator

The voltage regulators that are used with alternators are transistorized and cannot be serviced. If the voltage regulator is not operating properly, it must be replaced.

Starter

REMOVAL AND INSTALLATION

The starter on the 4–134 engine can be removed from the top of the engine. The starter motor on all other engines must be removed from beneath the vehicle.
1. Disconnect the battery ground.
2. Raise and support the vehicle on jackstands.
3. Remove all wires from the starter and tag them for installation.

4. Remove all but one upper attaching bolt, support the starter (it's heavier than it looks) and remove the last bolt.
5. Pull the starter from the engine.
6. Installation is the reverse of removal. Torque the mounting bolts to:
● 4–134, 6–225, 6–232, 6–258, 8–304: 25 ft. lbs.
● 4–150: 17 ft. lbs.

1. Auxiliary terminal
2. Output terminal
3. Auxiliary terminal
4. Field terminal
5. Ground terminal
6. Ground terminal

1971 Motorola alternator

BRUSH TERMINAL

VOLTAGE REGULATOR GROUND TERMINAL

REGULATOR TERMINAL

NEGATIVE DIODES

OUTPUT TERMINAL

POSITIVE DIODES

7 VOLT AC TERMINAL

.5 MFD CAPACITOR

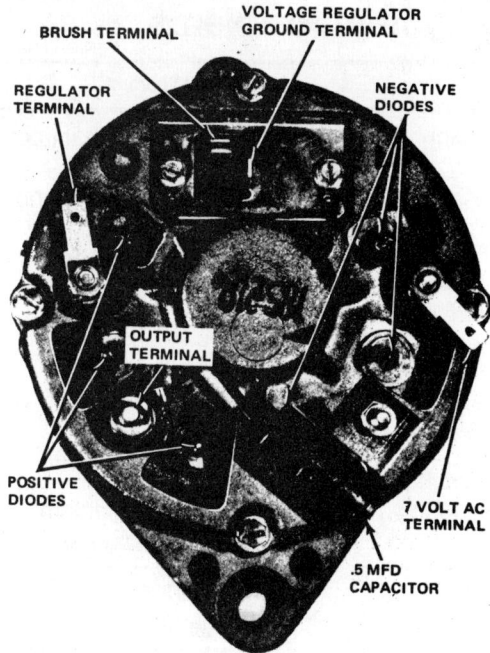

1972–75 Prestolite alternator

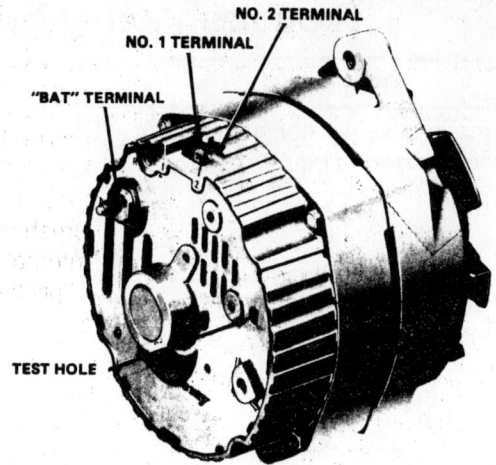

NO. 2 TERMINAL

NO. 1 TERMINAL

"BAT" TERMINAL

TEST HOLE

Delcotron alternator used on 1976 and later Jeep

Troubleshooting Basic Starting System Problems

Problem	Cause	Solution
Starter motor rotates engine slowly	• Battery charge low or battery defective	• Charge or replace battery
	• Defective circuit between battery and starter motor	• Clean and tighten, or replace cables
	• Low load current	• Bench-test starter motor. Inspect for worn brushes and weak brush springs.
	• High load current	• Bench-test starter motor. Check engine for friction, drag or coolant in cylinders. Check ring gear-to-pinion gear clearance.
Starter motor will not rotate engine	• Battery charge low or battery defective	• Charge or replace battery
	• Faulty solenoid	• Check solenoid ground. Repair or replace as necessary.
	• Damage drive pinion gear or ring gear	• Replace damaged gear(s)
	• Starter motor engagement weak	• Bench-test starter motor
	• Starter motor rotates slowly with high load current	• Inspect drive yoke pull-down and point gap, check for worn end bushings, check ring gear clearance
	• Engine seized	• Repair engine

Troubleshooting Basic Starting System Problems

Problem	Cause	Solution
Starter motor drive will not engage (solenoid known to be good)	• Defective contact point assembly	• Repair or replace contact point assembly
	• Inadequate contact point assembly ground	• Repair connection at ground screw
	• Defective hold-in coil	• Replace field winding assembly
Starter motor drive will not disengage	• Starter motor loose on flywheel housing	• Tighten mounting bolts
	• Worn drive end busing	• Replace bushing
	• Damaged ring gear teeth	• Replace ring gear or driveplate
	• Drive yoke return spring broken or missing	• Replace spring
Starter motor drive disengages prematurely	• Weak drive assembly thrust spring	• Replace drive mechanism
	• Hold-in coil defective	• Replace field winding assembly
Low load current	• Worn brushes	• Replace brushes
	• Weak brush springs	• Replace springs

STARTER DRIVE REPLACEMENT

Autolite

1. Remove the cover of the starter drive's actuating lever arm. Remove the through bolts, starter drive gear housing, and the return spring of the driver gear's actuating lever.
2. Remove the pivot pin which retains the starter gear actuating lever and remove the lever and armature.
3. Remove the stopring retainer. Remove and discard the stopring which holds the drive gear to the armature shaft and then remove the drive gear assembly.

To install the unit:

1. Lightly Lubriplate® the armature shaft splines and install the starter drive gear assembly on the shaft. Install a new stopring and stopring retainer.
2. Position the starter drive gear actuating lever to the frame and starter drive assembly. Install the pivot pin.
3. Fill the starter drive gear housing one quarter full of grease.
4. Position the drive actuating lever return spring and the drive gear housing to the frame, then install and tighten the through bolts. Be sure that the stopring retainer is properly seated in the drive housing.

Delco-Remy

1. Remove the through bolts.
2. Remove the starter drive housing.
3. Slide the two piece thrust collar off the end of the armature shaft.
4. Slide a standard ½ in. pipe coupling, or other spacer, onto the shaft so the end of the coupling butts against the edge of the retainer.
5. Tap the end of the coupling with a hammer, driving the retainer toward the armature end of the snapring.
6. Remove the snapring from its groove in the shaft with pliers. Slide the retainer and the starter drive from the armature.

To install the unit:

1. Lubricate the drive end of the shaft with silicone lubricant.
2. Slide the drive gear assembly onto the shaft, with the gear facing outward.
3. Slide the retainer onto the shaft with the cupped surface facing away from the gear.

4. Stand the whole starter assembly on a block of wood with the snapring positioned on the upper end of the shaft. Drive the snapring down with a small block of wood and a hammer. Slide the snapring into its groove.
5. Install the thrust collar onto the shaft with the shoulder next to the snapring.
6. With the retainer on one side of the snapring and the thrust collar on the other side, squeeze them together with a pair of pliers until the ring seats in the retainer. On models without a thrust collar, use a washer. Remember to remove the washer before installing the starter in the engine.

Prestolite

1. Slide the thrust collar off the armature shaft.
2. Using a standard ½ in. pipe connector, drive the snapring retainer off the shaft.
3. Remove the snapring from the groove, and then remove the drive assembly.

To install the unit:

1. Lubricate the drive end and splines with Lubriplate®.
2. Install the clutch assembly onto the shaft.
3. Install the snapring retainer with the cupped surface facing toward the end of the shaft.
4. Install the snapring into the groove. Use a new snapring if necessary.
5. Install the thrust collar onto the shaft with the shoulder against the snapring.
6. Force the retainer over the snapring in the same manner as was used for the Delco-Remy starters.

SOLENOID OR RELAY REPLACEMENT

Autolite

1. Disconnect the battery ground.
2. Remove all of the leads to the solenoid.
3. Remove the connecting lever.
4. Remove the attaching bolts that hold the solenoid assembly to the starter housing.
5. Remove the solenoid assembly from the starter housing.
6. To install the solenoid assembly, reverse the above procedure.

Delco-Remy

Remove the leads from the solenoid. Remove the drive housing of the starter motor. Remove the shift lever pin and bolt from the shift lever. Remove the attaching bolts that hold the solenoid assembly to the housing of the starter motor. Remove the starter solenoid from the starter housing. To install the solenoid, reverse the above procedure.

Prestolite

1. Remove the leads from the solenoid assembly.
2. Remove the attaching bolts that hold the solenoid to the starter housing.
3. Remove the bolt form the shift lever.
4. Remove the solenoid assembly from the starter housing.
5. Reverse the procedure for installation.

STARTER OVERHAUL

Autolite/Motorcraft
DISASSEMBLY

1. Remove the cover screw, the cover through-bolts, the starter drive end housing and the starter drive plunger lever return spring.
2. Remove the starter gear plunger lever pivot pin, the lever and the armature. Remove the stop ring retainer and the stop ring from the armature shaft (discard the ring), then the starter drive gear assembly.

3. Remove the brush end plate, the insulator assembly and the brushes from the plastic holder, then lift out the brush holder. For reassembly, note the position of the brush holder with respect to the end terminal.
4. Remove the two ground brush-to-frame screws.
5. Bend up the sleeve's edges which are inserted in the frame's rectangular hole, then remove the sleeve and the retainer. Detach the field coil ground wire from the copper tab.
6. Remove the three coil retaining screws. Cut the field coil connection at the switch post lead, then remove the pole shoes and the coils from the frame.
7. Cut the positive brush leads from the field coils (as close to the field connection point as possible).
8. Check the armature and the armature windings for broken or burned insulation, open circuits or grounds.
9. Check the commutator for runout. If it is rough, has flat spots or is more than 0.005 in. (0.13mm) out of round, reface the commutator face.
10. Inspect the armature shaft and the two bearings for scoring and excessive wear, then replace (if necessary).
11. Inspect the starter drive. If the gear teeth are pitted, broken or excessively worn, replace the starter drive.

NOTE: The factory brush length is ½ in. (13mm); the wear limit is ¼ in. (6mm).

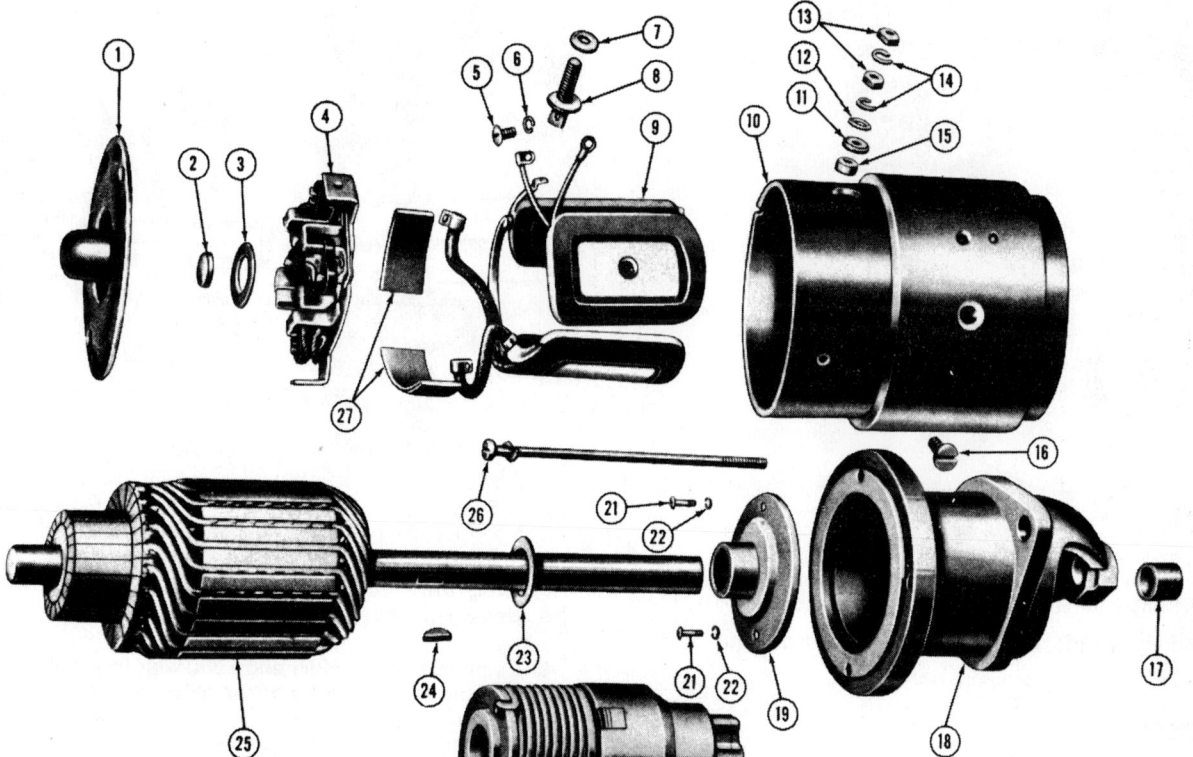

1. End plate
2. Plug
3. Thrust washer
4. Brush plate assembly
5. Screw
6. Lockwasher
7. Insulating washer
8. Terminal
9. Field coil and pole shoe set
10. Frame
11. Insulating washer
12. Washer
13. Nut
14. Lockwasher
15. Insulating bushing
16. Pole shoe screw
17. Sleeve bearing
18. Drive end frame
19. Intermediate bearing
20. Bendix drive
21. Screw
22. Lockwasher
23. Thrust washer
24. Key
25. Armature
26. Thru-bolt
27. Insulator

4–134 starter motor

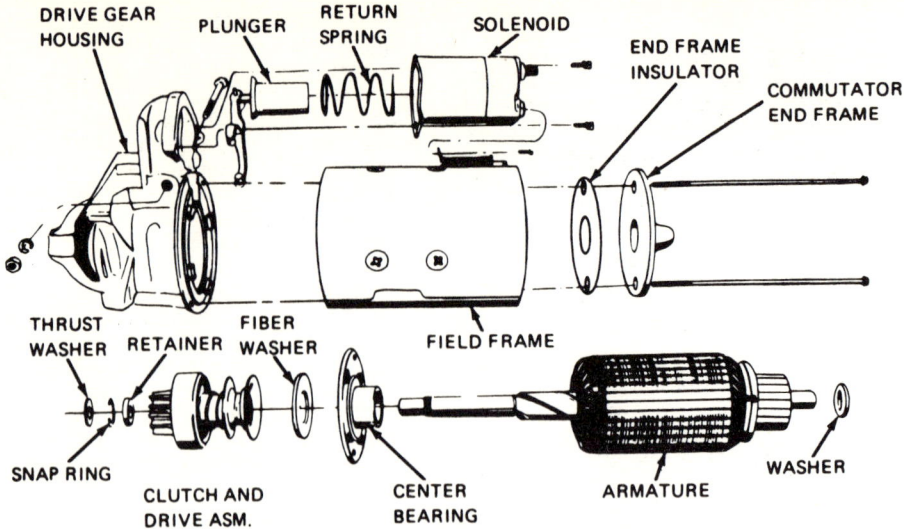

Delco starter used on the 6–225

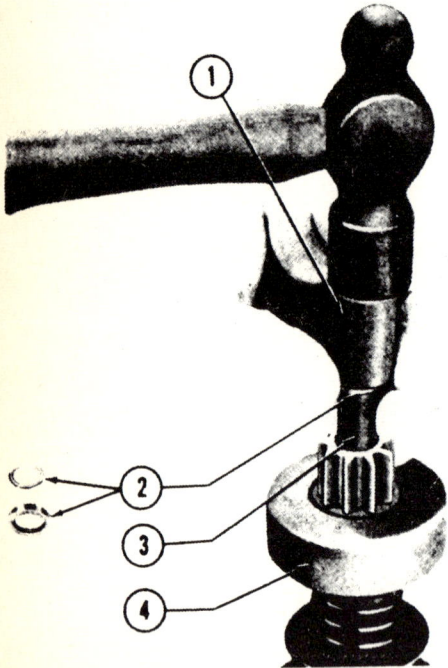

1. ½ in. pipe coupling
2. Snap ring and retainer
3. Armature shaft
4. Drive assembly

Removing the starter drive assembly from the armature shaft

1. Retainer
2. Snap ring
3. Thrust collar
4. Drive assembly
5. Retainer
6. Groove in the armature shaft
7. Snap ring

Installing the pinion stop retainer and thrust collar on the armature shaft

ASSEMBLY

1. Install the starter terminal, the insulator, the washers and the nut in the frame.

NOTE: Be sure to position the screw slot perpendicular to the frame end surface.

2. Position the coils and the pole pieces, with the coil leads in the terminal screw slot, then install the screws. When tightening the pole screws, strike the frame with several sharp hammer blows to align the pole shoes, then stake the screws.

3. Install the solenoid coil and the retainer, then bend the tabs to hold the coils to the frame.

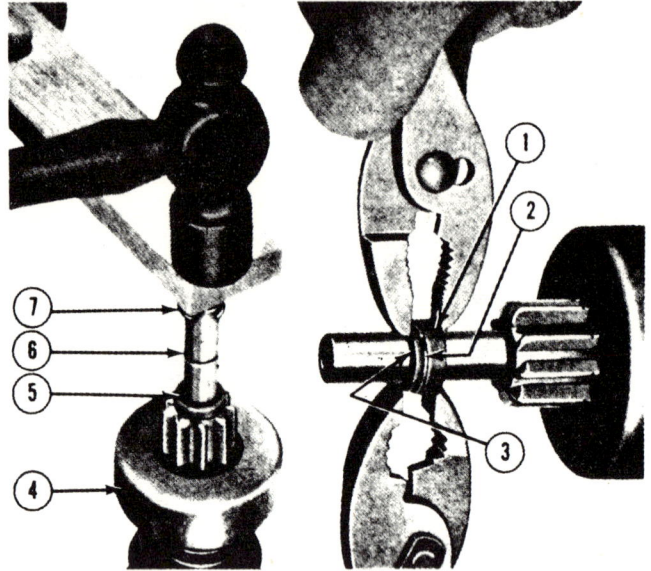

4. Using resin-core solder and a 300 watt iron, solder the field coils and the solenoid wire to the starter terminal. Check for continuity and ground connections of the assembled coils.

5. Position the solenoid coil ground terminal over the nearest ground screw hole and the ground brushes-to-starter frame, then install the screws.

6. Apply a thin coating of Lubriplate® on the armature shaft splines. Install the starter motor drive gear assembly-to-armature shaft, followed by a new stop ring and retainer. Install the armature in the starter frame.

7. Position the starter drive gear plunger lever to the frame and the starter drive assembly, then install the pivot pin. Place some grease into the end housing bore. Fill it about ¼ full, then position the drive end housing to the frame.

8. Install the brush holder and the brush springs. The posi-

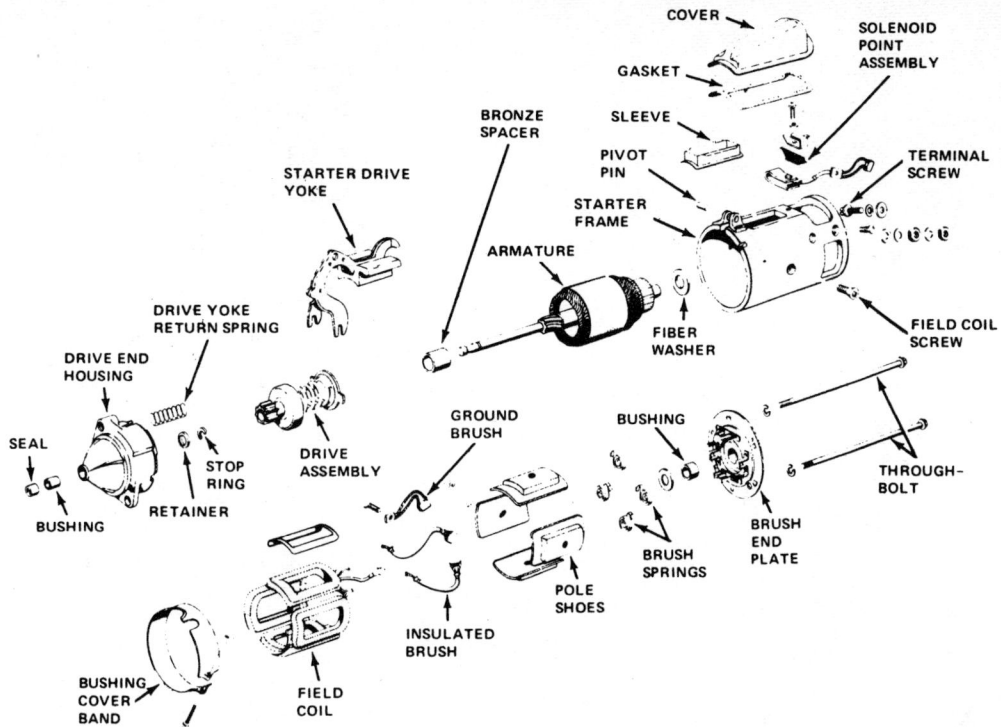

1972–77 starter motor

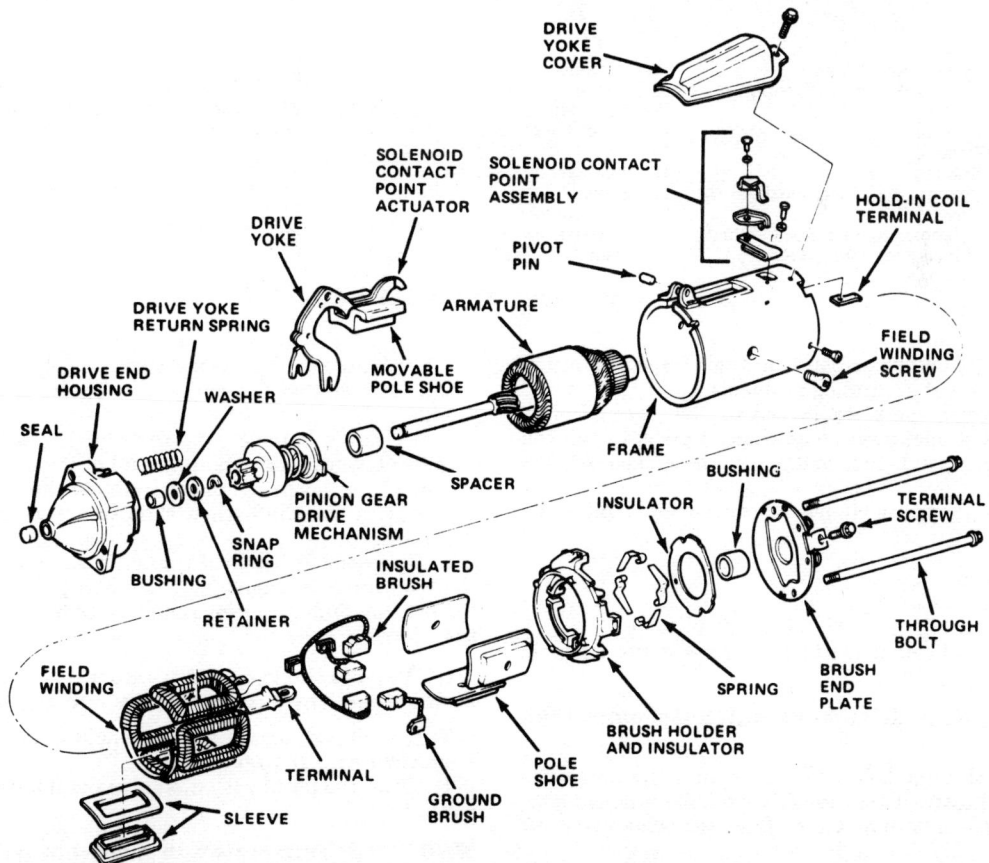

Starter motor used on all 1978–86 engines, except the 4–150 and 4–151

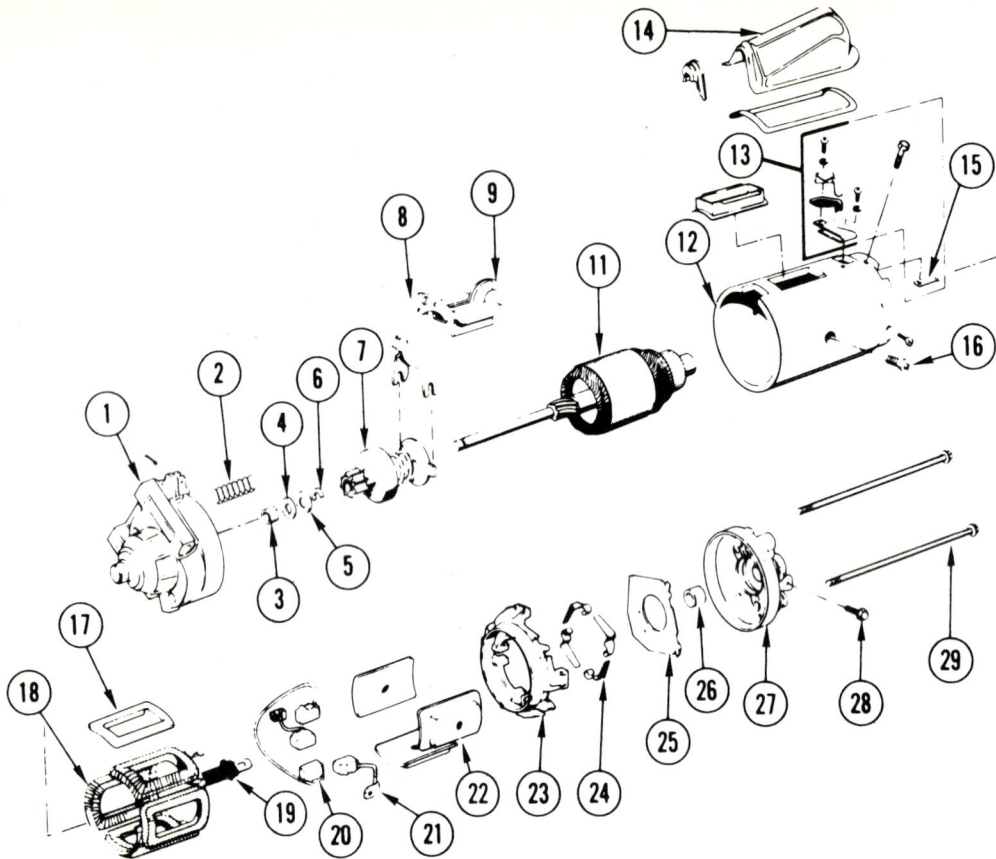

1. Drive end housing
2. Drive yoke return spring
3. Bushing
4. Washer
5. Retainer
6. Snapring
7. Pinion gear drive mechanism
8. Drive yoke
9. Solenoid contact point actuator
10. Moveable pole shoe
11. Armature
12. Frame
13. Solenoid contact point assembly
14. Drive yoke cover
15. Hold-in coil terminal
16. Field winding screw
17. Sleeve
18. Field winding
19. Terminal
20. Insulated brush
21. Ground brush
22. Pole shoe
23. Brush holder and insulator
24. Springs
25. Insulator
26. Bushing
27. Brush end plate
28. Terminal screw
29. Through bolt

Starter motor used on the 1984–86 4–150

tive brush leads should be positioned in their respective brush holder slots, to prevent grounding problems.

9. Install the brush end plate. Be certain that the end plate insulator is in the proper position on the end plate. Install the two starter frame through-bolts and torque them to 55–75 inch lbs.

10. Install the starter drive plunger lever cover and tighten the retaining screw.

Delco-Remy

DISASSEMBLY

1. Detach the field coil connectors from the motor solenoid terminal.

NOTE: If equipped, remove solenoid mounting screws.

2. Remove the through-bolts, the commutator end frame, the field frame and the armature assembly from drive housing.

3. Remove the overrunning clutch from the armature shaft as follows:

    a. Slide the two piece thrust collar off the end of the armature shaft.

    b. Slide a standard ½ in. pipe coupling or other spacer onto the shaft, so that the coupling end butts against the retainer edge.

    c. Using a hammer, tap the coupling end, driving the retainer towards the armature end of the snapring.

    d. Using snapring pliers, remove the snapring from its groove in the shaft, then slide the retainer and the clutch from the shaft.

4. Disassemble the field frame brush assembly by releasing the V-spring and removing the support pin. The brush holders, the brushes and the springs can now be pulled out as a unit and the leads disconnected.

NOTE: On the integral frame units, remove the brush holder from the brush support and the brush screw.

5. If equipped, separate the solenoid from the lever housing.
Cleaning And Inspection

1. Clean the parts with a rag. Do not immerse the parts in a solvent.

WARNING: Immersion in a solvent will dissolve the grease that is packed in the clutch mechanism. It will damage the armature and the field coil insulation!

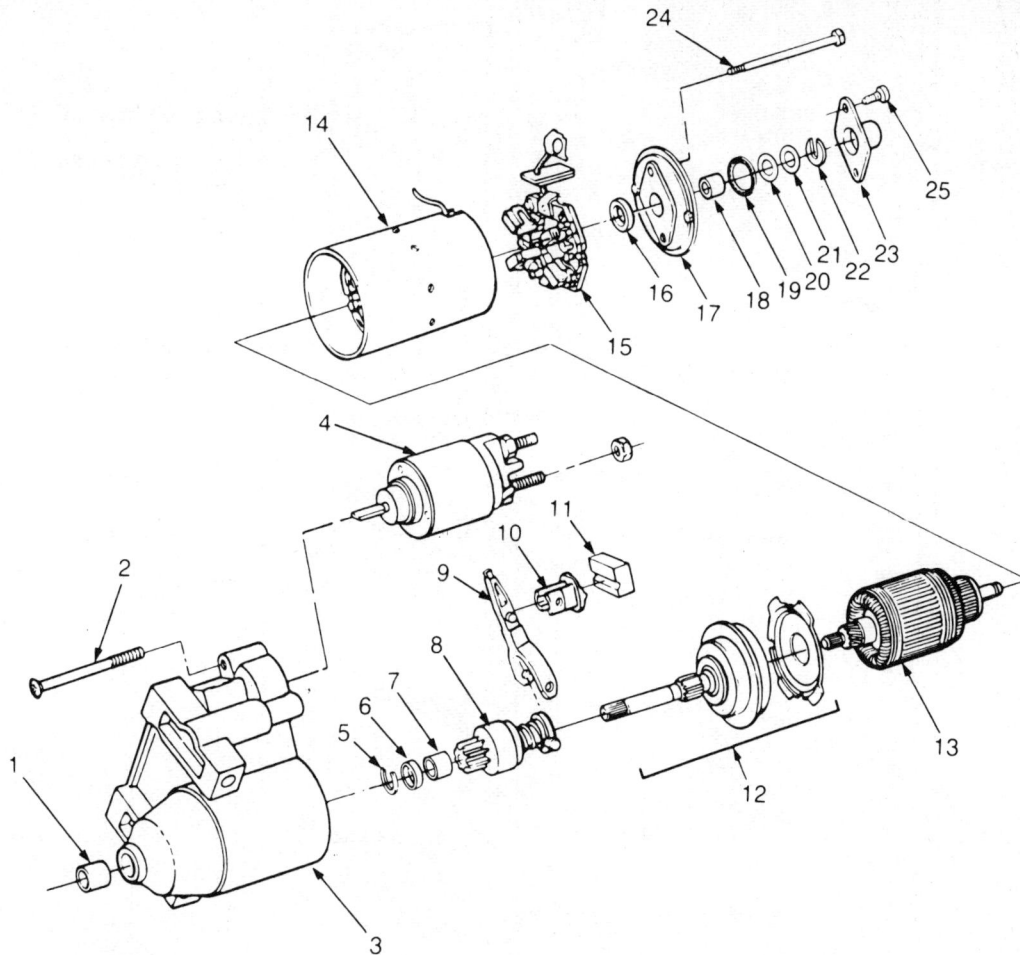

1. Bushing
2. Screw
3. Shield
4. Solenoid switch
5. Retainer
6. Stop ring
7. Bushing
8. Overrunning clutch drive
9. Fork
10. Bearing pedestal
11. Sealing rubber
12. Planetary gear system
13. Armature
14. Stator frame
15. Brush holder
16. Gasket
17. Commutator end shield
18. Bushing
19. Seal ring
20. Shim
21. Shim
22. Retaining washer
23. Closure cap
24. Hexagon screw
25. Screw

Starter used on the 1987–90 4–150 engine

2. Test the overrunning clutch action. The pinion should turn freely in the overrunning direction but must not slip in the cranking direction. Check that the pinion teeth have not been chipped, cracked or excessively worn. Replace the unit (if necessary).

3. Inspect the armature commutator. If the commutator is rough or out of round, it should be machined and undercut.

NOTE: Undercut the insulation between the commutator bars by $1/32$ in. (0.8mm). The undercut must be the full width of the insulation and flat at the bottom. A triangular groove will not be satisfactory. Most late model starter motor use a molded armature commutator design. No attempt to undercut the insulation should be made or serious damage may result to the commutator.

ASSEMBLY

1. Install the brushes into the holders, then install solenoid (if equipped).

2. Assemble the insulated and the grounded holder together. Using the V-spring, position and assemble the unit on the support pin. Push the holders and the spring to bottom of the support, then rotate the spring to engage the slot in the support. Attach the ground wire to the grounded brush and the field lead wire to the insulated brush, then repeat this procedure for other brush sets.

3. Assemble the overrunning clutch to the armature shaft as follows:

   a. Lubricate the drive end of the shaft with silicone lubricant.

   b. Slide the clutch assembly onto the shaft with the pinion outward.

   c. Slide the retainer onto the shaft with the cupped surface facing away from the pinion.

   d. Stand the armature up on a wood surface with the commutator downward. Position the snapring on the upper end of the shaft and drive it onto the shaft with a small block of wood and a hammer, then slide the snapring into groove.

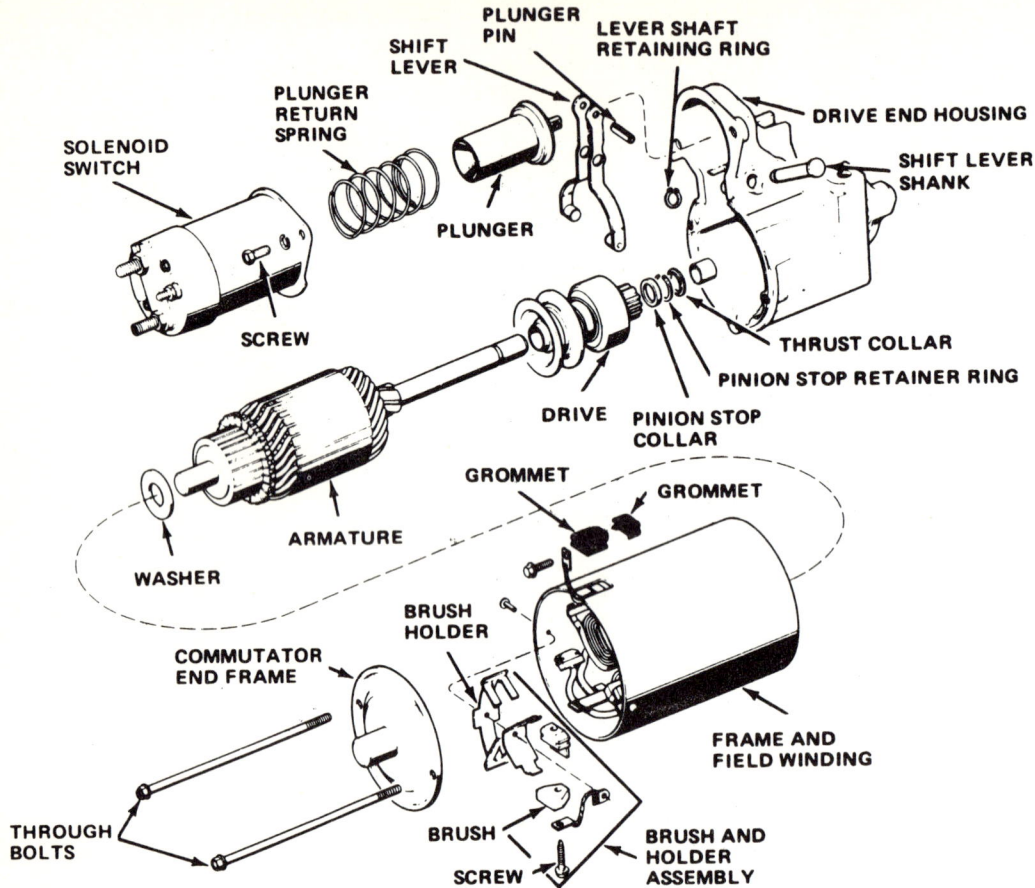

Starter motor used on the 4–151

e. Install the thrust collar onto the shaft with the shoulder next to snapring.

f. With the retainer on one side of the snapring and the thrust collar on the other side, squeeze two sets together (with pliers) until the ring seats in the retainer. On models without a thrust collar use a washer. Remember to remove the washer before continuing.

4. Lubricate the drive end bushing with silicone lubricant, then slide the armature and the clutch assembly into place, while engaging the shift lever with the clutch.

NOTE: On the non-integral starters, the shift lever may be installed in the drive gear housing first.

5. Position the field frame over the armature and apply sealer (silicone) between the frame and the solenoid case. Position the frame against the drive housing, making sure the brushes are not damaged in the process.

6. Lubricate the commutator end bushing with silicone lubricant, place a washer on the armature shaft and slide the commutator end frame onto the shaft. Install the through-bolts and tighten.

7. Reconnect the field coil connections to the solenoid motor terminal. Install the solenoid mounting screws (if equipped).

8. Check the pinion clearance. It should be 0.010–0.140 in. (0.25–3.56mm) with the pinion in the cranking position, on all models.

Prestolite

DISASSEMBLY AND ASSEMBLY

1. To remove the solenoid, remove the screw from the field coil connector and solenoid mounting screws. Rotate the solenoid 90° and remove it along with the plunger return spring.

2. For further service, remove the two through-bolts, then remove the commutator end frame and washer.

3. To replace the clutch and drive assembly proceed as follows:

a. Remove the thrust washer or the collar from the armature shaft.

b. Slide a 5⁄8 in. deep socket or a piece of pipe of suitable size over the shaft and against the retainer as a driving tool. Tap the tool to remove the retainer off the snapring.

c. Remove the snapring from the groove in the shaft. Check and make sure the snapring isn't distorted. If it is, it will be necessary to replace it with a new one upon reassembly.

d. Remove the retainer and clutch assembly from the armature shaft.

4. The shift lever may be disconnected from the plunger at this time by removing the roll pin.

5. On models with the standard starter, the brushes may be removed by removing the brush holder pivot pin which positions one insulated and one grounded brush. Remove the brush and spring and replace the brushes as necessary.

6. On models with the smaller 5MT starter, remove the brush and holder from the brush support, then remove the screw from the brush holder and separate the brush and holder. Replace the brushes as necessary.

7. Installation is the reverse of removal. Assemble the armature and clutch and drive assembly as follows:

a. Lubricate the drive end of the armature shaft and slide the clutch assembly onto the armature shaft with the pinion away from the armature.

b. Slide the retainer onto the shaft with the cupped side facing the end of the shaft.

c. Install the snapring into the groove on the armature shaft.

d. Install the thrust washer on the shaft.

e. Position the retainer and thrust washer with the snapring in between. Using two pliers, grip the retainer and thrust washer or collar and squeeze until the snapring is forced into the retainer and is held securely in the groove in the armature shaft.

f. Lubricate the drive gear housing bushing.

g. Engage the shift lever yoke with the clutch and slide the complete assembly into the drive gear housing.

NOTE: When the starter motor has been disassembled or the solenoid has been replaced, it is necessary to check the pinion clearance. Pinion clearance must be correct to prevent the buttons on the shift lever yoke from rubbing on the clutch collar during cranking.

CHECKING PINION CLEARANCE

1. Disconnect the motor field coil connector from the solenoid motor terminal and insulate it carefully.

2. Connect one 12 volt battery lead to the solenoid switch terminal and the other to the starter frame.

3. Flash a jumper lead momentarily from the solenoid motor terminal to the starter frame. This will shift the pinion into cranking position and it will remain there until the battery is disconnected.

4. Push the pinion back as far as possible to take up any movement, and check the clearance with a feeler gauge. The clearance should be 0.010–0.140 in. (0.25–3.56mm).

5. There is no means for adjusting pinion clearance on the starter motor. If clearance does not fall within the limits, check for improper installation and replace all worn parts.

Bosch

DISASSEMBLY

1. Disconnect the field coil wire from the solenoid terminal.

2. Remove the solenoid and work the plunger off the shift fork.

3. Remove the two end shield bearing cap screws, the cap and the washers.

4. Remove the two commutator end frame cover through-bolts, the cover, the two brushes and the brush plate.

5. Slide the field frame off over the armature. Remove the shift lever pivot bolt, the rubber gasket and the metal plate.

6. Remove the armature assembly and the shift lever from the drive end housing. Press the stop collar off the snapring, then remove the snapring, the clutch assembly, the clutch assembly and the drive end housing from the armature.

INSPECTION AND SERVICE

1. The brushes that are worn more than ½ the length of new brushes or are oil-soaked, should be replaced. The new brushes are $^{11}/_{16}$ in. (17.5mm) long.

2. Do not immerse the starter clutch unit in cleaning solvent. Solvent will wash the lubricant from the clutch.

3. Place the drive unit on the armature shaft, then, while holding the armature, rotate the pinion.

NOTE: The drive pinion should rotate smoothly in one direction only. The pinion may not rotate easily but as long as it rotates smoothly it is in good condition. If the clutch unit does not function properly or if the pinion is worn, chipped or burred, replace the unit.

ASSEMBLY

1. Lubricate the armature shaft and the splines with SAE 10W or 30W oil.

2. Fit the drive end housing onto the armature, then install the clutch, the stop collar and the snapring onto the armature.

3. Install the shift fork pivot bolt, the rubber gasket and the metal plate. Slide the field frame into position and install the brush holder and the brushes.

4. Position the commutator end frame cover and the through-bolts.

5. Install the shim and the armature shaft lock. Check the endplay, which should be 0.002–0.012 in. (0.05–0.30mm), then install the bearing cover.

6. Assemble the plunger to the shift fork, then install the solenoid with its mounting bolts. Connect the field wire to the solenoid.

Battery

REMOVAL AND INSTALLATION

1. Remove the holddown screws from the battery box. Loosen the nuts that secure the cable ends to the battery terminals. Lift the battery cables from the terminals with a twisting motion.

2. If there is a battery cable puller available, make use of it. Lift the battery from the vehicle.

3. Before installing the battery in the vehicle, make sure that the battery terminals are clean and free from corrosion. Use a battery terminal cleaner on the terminals and on the inside of the battery cable ends. If a cleaner is not available, use a heavy sandpaper to remove the corrosion. A mixture of baking soda and water will neutralize any acid. Place the battery in the vehicle. Install the cables on the terminals. Tighten the nuts on the cable ends. Smear a light coating of grease on the cable ends and the tops of the terminals. This will prevent buildup of oxidized acid on the terminals and the cable ends. Install and tighten the nuts of the battery box.

Troubleshooting Engine Mechanical Problems

Problem	Cause	Solution
External oil leaks	• Fuel pump gasket broken or improperly seated	• Replace gasket
	• Cylinder head cover RTV sealant broken or improperly seated	• Replace sealant; inspect cylinder head cover sealant flange and cylinder head sealant surface for distortion and cracks
	• Oil filler cap leaking or missing	• Replace cap

Troubleshooting Engine Mechanical Problems (cont.)

Problem	Cause	Solution
External oil leaks	• Oil filter gasket broken or improperly seated	• Replace oil filter
	• Oil pan side gasket broken, improperly seated or opening in RTV sealant	• Replace gasket or repair opening in sealant; inspect oil pan gasket flange for distortion
	• Oil pan front oil seal broken or improperly seated	• Replace seal; inspect timing case cover and oil pan seal flange for distortion
	• Oil pan rear oil seal broken or improperly seated	• Replace seal; inspect oil pan rear oil seal flange; inspect rear main bearing cap for cracks, plugged oil return channels, or distortion in seal groove
	• Timing case cover oil seal broken or improperly seated	• Replace seal
	• Excess oil pressure because of restricted PCV valve	• Replace PCV valve
	• Oil pan drain plug loose or has stripped threads	• Repair as necessary and tighten
	• Rear oil gallery plug loose	• Use appropriate sealant on gallery plug and tighten
	• Rear camshaft plug loose or improperly seated	• Seat camshaft plug or replace and seal, as necessary
	• Distributor base gasket damaged	• Replace gasket
Excessive oil consumption	• Oil level too high	• Drain oil to specified level
	• Oil with wrong viscosity being used	• Replace with specified oil
	• PCV valve stuck closed	• Replace PCV valve
	• Valve stem oil deflectors (or seals) are damaged, missing, or incorrect type	• Replace valve stem oil deflectors
	• Valve stems or valve guides worn	• Measure stem-to-guide clearance and repair as necessary
	• Poorly fitted or missing valve cover baffles	• Replace valve cover
	• Piston rings broken or missing	• Replace broken or missing rings
	• Scuffed piston	• Replace piston
	• Incorrect piston ring gap	• Measure ring gap, repair as necessary
	• Piston rings sticking or excessively loose in grooves	• Measure ring side clearance, repair as necessary
	• Compression rings installed upside down	• Repair as necessary
	• Cylinder walls worn, scored, or glazed	• Repair as necessary
	• Piston ring gaps not properly staggered	• Repair as necessary
	• Excessive main or connecting rod bearing clearance	• Measure bearing clearance, repair as necessary

Troubleshooting Engine Mechanical Problems (cont.)

Problem	Cause	Solution
No oil pressure	• Low oil level	• Add oil to correct level
	• Oil pressure gauge, warning lamp or sending unit inaccurate	• Replace oil pressure gauge or warning lamp
	• Oil pump malfunction	• Replace oil pump
	• Oil pressure relief valve sticking	• Remove and inspect oil pressure relief valve assembly
	• Oil passages on pressure side of pump obstructed	• Inspect oil passages for obstruction
	• Oil pickup screen or tube obstructed	• Inspect oil pickup for obstruction
	• Loose oil inlet tube	• Tighten or seal inlet tube
Low oil pressure	• Low oil level	• Add oil to correct level
	• Inaccurate gauge, warning lamp or sending unit	• Replace oil pressure gauge or warning lamp
	• Oil excessively thin because of dilution, poor quality, or improper grade	• Drain and refill crankcase with recommended oil
	• Excessive oil temperature	• Correct cause of overheating engine
	• Oil pressure relief spring weak or sticking	• Remove and inspect oil pressure relief valve assembly
	• Oil inlet tube and screen assembly has restriction or air leak	• Remove and inspect oil inlet tube and screen assembly. (Fill inlet tube with lacquer thinner to locate leaks.)
	• Excessive oil pump clearance	• Measure clearances
	• Excessive main, rod, or camshaft bearing clearance	• Measure bearing clearances, repair as necessary
High oil pressure	• Improper oil viscosity	• Drain and refill crankcase with correct viscosity oil
	• Oil pressure gauge or sending unit inaccurate	• Replace oil pressure gauge
	• Oil pressure relief valve sticking closed	• Remove and inspect oil pressure relief valve assembly
Main bearing noise	• Insufficient oil supply	• Inspect for low oil level and low oil pressure
	• Main bearing clearance excessive	• Measure main bearing clearance, repair as necessary
	• Bearing insert missing	• Replace missing insert
	• Crankshaft end play excessive	• Measure end play, repair as necessary
	• Improperly tightened main bearing cap bolts	• Tighten bolts with specified torque
	• Loose flywheel or drive plate	• Tighten flywheel or drive plate attaching bolts
	• Loose or damaged vibration damper	• Repair as necessary

Troubleshooting Engine Mechanical Problems (cont.)

Problem	Cause	Solution
Connecting rod bearing noise	• Insufficient oil supply	• Inspect for low oil level and low oil pressure
	• Carbon build-up on piston	• Remove carbon from piston crown
	• Bearing clearance excessive or bearing missing	• Measure clearance, repair as necessary
	• Crankshaft connecting rod journal out-of-round	• Measure journal dimensions, repair or replace as necessary
	• Misaligned connecting rod or cap	• Repair as necessary
	• Connecting rod bolts tightened improperly	• Tighten bolts with specified torque
Piston noise	• Piston-to-cylinder wall clearance excessive (scuffed piston)	• Measure clearance and examine piston
	• Cylinder walls excessively tapered or out-of-round	• Measure cylinder wall dimensions, rebore cylinder
	• Piston ring broken	• Replace all rings on piston
	• Loose or seized piston pin	• Measure piston-to-pin clearance, repair as necessary
	• Connecting rods misaligned	• Measure rod alignment, straighten or replace
	• Piston ring side clearance excessively loose or tight	• Measure ring side clearance, repair as necessary
	• Carbon build-up on piston is excessive	• Remove carbon from piston
Valve actuating component noise	• Insufficient oil supply	• Check for: (a) Low oil level (b) Low oil pressure (c) Plugged push rods (d) Wrong hydraulic tappets (e) Restricted oil gallery (f) Excessive tappet to bore clearance
	• Push rods worn or bent	• Replace worn or bent push rods
	• Rocker arms or pivots worn	• Replace worn rocker arms or pivots
	• Foreign objects or chips in hydraulic tappets	• Clean tappets
	• Excessive tappet leak-down	• Replace valve tappet
	• Tappet face worn	• Replace tappet; inspect corresponding cam lobe for wear
	• Broken or cocked valve springs	• Properly seat cocked springs; replace broken springs
	• Stem-to-guide clearance excessive	• Measure stem-to-guide clearance, repair as required
	• Valve bent	• Replace valve
	• Loose rocker arms	• Tighten bolts with specified torque
	• Valve seat runout excessive	• Regrind valve seat/valves
	• Missing valve lock	• Install valve lock
	• Push rod rubbing or contacting cylinder head	• Remove cylinder head and remove obstruction in head
	• Excessive engine oil (four-cylinder engine)	• Correct oil level

Troubleshooting the Cooling System

Problem	Cause	Solution
High temperature gauge indication—overheating	• Coolant level low • Fan belt loose • Radiator hose(s) collapsed • Radiator airflow blocked • Faulty radiator cap • Ignition timing incorrect • Idle speed low • Air trapped in cooling system • Heavy traffic driving • Incorrect cooling system component(s) installed • Faulty thermostat • Water pump shaft broken or impeller loose • Radiator tubes clogged • Cooling system clogged • Casting flash in cooling passages • Brakes dragging • Excessive engine friction • Antifreeze concentration over 68% • Missing air seals • Faulty gauge or sending unit • Loss of coolant flow caused by leakage or foaming • Viscous fan drive failed	• Replenish coolant • Adjust fan belt tension • Replace hose(s) • Remove restriction (bug screen, fog lamps, etc.) • Replace radiator cap • Adjust ignition timing • Adjust idle speed • Purge air • Operate at fast idle in neutral intermittently to cool engine • Install proper component(s) • Replace thermostat • Replace water pump • Flush radiator • Flush system • Repair or replace as necessary. Flash may be visible by removing cooling system components or removing core plugs. • Repair brakes • Repair engine • Lower antifreeze concentration percentage • Replace air seals • Repair or replace faulty component • Repair or replace leaking component, replace coolant • Replace unit
Low temperature indication—undercooling	• Thermostat stuck open • Faulty gauge or sending unit	• Replace thermostat • Repair or replace faulty component
Coolant loss—boilover	• Overfilled cooling system • Quick shutdown after hard (hot) run • Air in system resulting in occasional "burping" of coolant • Insufficient antifreeze allowing coolant boiling point to be too low • Antifreeze deteriorated because of age or contamination • Leaks due to loose hose clamps, loose nuts, bolts, drain plugs, faulty hoses, or defective radiator	• Reduce coolant level to proper specification • Allow engine to run at fast idle prior to shutdown • Purge system • Add antifreeze to raise boiling point • Replace coolant • Pressure test system to locate source of leak(s) then repair as necessary

Troubleshooting the Cooling System (cont.)

Problem	Cause	Solution
Coolant loss—boilover	• Faulty head gasket • Cracked head, manifold, or block • Faulty radiator cap	• Replace head gasket • Replace as necessary • Replace cap
Coolant entry into crankcase or cylinder(s)	• Faulty head gasket • Crack in head, manifold or block	• Replace head gasket • Replace as necessary
Coolant recovery system inoperative	• Coolant level low • Leak in system • Pressure cap not tight or seal missing, or leaking • Pressure cap defective • Overflow tube clogged or leaking • Recovery bottle vent restricted	• Replenish coolant to FULL mark • Pressure test to isolate leak and repair as necessary • Repair as necessary • Replace cap • Repair as necessary • Remove restriction
Noise	• Fan contacting shroud • Loose water pump impeller • Glazed fan belt • Loose fan belt • Rough surface on drive pulley • Water pump bearing worn • Belt alignment	• Reposition shroud and inspect engine mounts • Replace pump • Apply silicone or replace belt • Adjust fan belt tension • Replace pulley • Remove belt to isolate. Replace pump. • Check pulley alignment. Repair as necessary.
No coolant flow through heater core	• Restricted return inlet in water pump • Heater hose collapsed or restricted • Restricted heater core • Restricted outlet in thermostat housing • Intake manifold bypass hole in cylinder head restricted • Faulty heater control valve • Intake manifold coolant passage restricted	• Remove restriction • Remove restriction or replace hose • Remove restriction or replace core • Remove flash or restriction • Remove restriction • Replace valve • Remove restriction or replace intake manifold

NOTE: *Immediately after shutdown, the engine enters a condition known as heat soak. This is caused by the cooling system being inoperative while engine temperature is still high. If coolant temperature rises above boiling point, expansion and pressure may push some coolant out of the radiator overflow tube. If this does not occur frequently it is considered normal.*

Troubleshooting the Serpentine Drive Belt

Problem	Cause	Solution
Tension sheeting fabric failure (woven fabric on outside circumference of belt has cracked or separated from body of belt)	• Grooved or backside idler pulley diameters are less than minimum recommended • Tension sheeting contacting (rubbing) stationary object • Excessive heat causing woven fabric to age • Tension sheeting splice has fractured	• Replace pulley(s) not conforming to specification • Correct rubbing condition • Replace belt • Replace belt
Noise (objectional squeal, squeak, or rumble is heard or felt while drive belt is in operation)	• Belt slippage • Bearing noise • Belt misalignment • Belt-to-pulley mismatch • Driven component inducing vibration • System resonant frequency inducing vibration	• Adjust belt • Locate and repair • Align belt/pulley(s) • Install correct belt • Locate defective driven component and repair • Vary belt tension within specifications. Replace belt.
Rib chunking (one or more ribs has separated from belt body)	• Foreign objects imbedded in pulley grooves • Installation damage • Drive loads in excess of design specifications • Insufficient internal belt adhesion	• Remove foreign objects from pulley grooves • Replace belt • Adjust belt tension • Replace belt
Rib or belt wear (belt ribs contact bottom of pulley grooves)	• Pulley(s) misaligned • Mismatch of belt and pulley groove widths • Abrasive environment • Rusted pulley(s) • Sharp or jagged pulley groove tips • Rubber deteriorated	• Align pulley(s) • Replace belt • Replace belt • Clean rust from pulley(s) • Replace pulley • Replace belt
Longitudinal belt cracking (cracks between two ribs)	• Belt has mistracked from pulley groove • Pulley groove tip has worn away rubber-to-tensile member	• Replace belt • Replace belt
Belt slips	• Belt slipping because of insufficient tension • Belt or pulley subjected to substance (belt dressing, oil, ethylene glycol) that has reduced friction • Driven component bearing failure • Belt glazed and hardened from heat and excessive slippage	• Adjust tension • Replace belt and clean pulleys • Replace faulty component bearing • Replace belt
"Groove jumping" (belt does not maintain correct position on pulley, or turns over and/or runs off pulleys)	• Insufficient belt tension • Pulley(s) not within design tolerance • Foreign object(s) in grooves	• Adjust belt tension • Replace pulley(s) • Remove foreign objects from grooves

Troubleshooting the Serpentine Drive Belt (cont.)

Problem	Cause	Solution
"Groove jumping" (belt does not maintain correct position on pulley, or turns over and/or runs off pulleys)	• Excessive belt speed • Pulley misalignment • Belt-to-pulley profile mismatched • Belt cordline is distorted	• Avoid excessive engine acceleration • Align pulley(s) • Install correct belt • Replace belt
Belt broken (Note: identify and correct problem before replacement belt is installed)	• Excessive tension • Tensile members damaged during belt installation • Belt turnover • Severe pulley misalignment • Bracket, pulley, or bearing failure	• Replace belt and adjust tension to specification • Replace belt • Replace belt • Align pulley(s) • Replace defective component and belt
Cord edge failure (tensile member exposed at edges of belt or separated from belt body)	• Excessive tension • Drive pulley misalignment • Belt contacting stationary object • Pulley irregularities • Improper pulley construction • Insufficient adhesion between tensile member and rubber matrix	• Adjust belt tension • Align pulley • Correct as necessary • Replace pulley • Replace pulley • Replace belt and adjust tension to specifications
Sporadic rib cracking (multiple cracks in belt ribs at random intervals)	• Ribbed pulley(s) diameter less than minimum specification • Backside bend flat pulley(s) diameter less than minimum • Excessive heat condition causing rubber to harden • Excessive belt thickness • Belt overcured • Excessive tension	• Replace pulley(s) • Replace pulley(s) • Correct heat condition as necessary • Replace belt • Replace belt • Adjust belt tension

ENGINE MECHANICAL

Design

F-Head 4-Cylinder

The F4–134, 4-cylinder engine is of a combination valve-in-head and valve-in-block construction. The intake valves are mounted in the head and are operated by pushrods through rocker arms. The intake manifold is cast as an integral part of the cylinder head and is completely water jacketed. This type of construction transfers heat from the cooling system to the intake passages and assists in vaporizing the fuel when the engine is cold. Therefore, there is no heat control valve (heat riser) needed in the exhaust manifold.

The exhaust valves are mounted in the block with thorough water jacketing to provide effective cooling of the valves.

The engine is pressure lubricated. An oil pump which is driven by the camshaft forces the lubricant through oil channels and drilled passages in the crankshaft to efficiently lubricate the main and connecting rod bearings. Lubricant is also force fed to the camshaft bearings, rocker arms, and timing gears. Cylinder walls and piston pins are lubricated from spurt holes in the "follow" side of the connecting rods.

The engine is equipped with a fully counterbalanced crankshaft that is supported by three main bearings. The counterweights of the crankshaft are independently forged and are permanently attached to the crankshaft with dowels and cap screws that are tack welded. Crankshaft endplay is adjusted by placing shims between the crankshaft thrust washer and the shoulder on the crankshaft.

The pistons have an extra groove directly above the top ring which acts as a heat dam or insulator.

The engine was available in compression ratios ranging from 6.3:1 to 7.8:1. This permits the use of regular octane gas. The displacement of the F4–134 engine is 134.2 cu. in.

4–150

The 4–150 engine used in 1984 and later models is a new design, developed from existing technology at work in the venerable 6–258. It is a four cylinder, overhead valve configuration with cast iron head and block. The crankshaft rides in 5 main bearings. Both manifolds are on the left side of the engine. The engine is thoroughly conventional in all respects. 1987-90 models utilize a throttle body fuel injection system, replacing the 2-bbl carburetor.

4–151

The 151 cid General Motors built, overhead valve, four cylinder engine has a crossflow cylinder head, five main crankshaft bearings, hydraulic lifters, conventional ball socket rocker arms, exceptionally long pushrods, a gear driven camshaft and a coolant heated aluminum intake manifold.

V6–225

The V6 engine has a displacement of 225 cu. in. and a compression ratio of 9.0:1 which permits the use of regular octane gas.

It has two banks of three cylinders each of which are opposed to one another at a 90° angle. The left bank of cylinders, as viewed from the driver's seat, is set forward of the right bank so that the connecting rods of opposite pairs of pistons and rods can be attached to the same crankpin.

6–232, 258

The American Motors 6-cylinder engines are inline sixes with overhead intake and exhaust valves. None of the rocker arms are adjustable. The 232 was last used in the 1978 model year.

8–304

The 304 V8 has two banks of four cylinders each which are opposed to each other at a 90° angle. The camshaft is located above the crankshaft, between the two banks. It operates the valves through the use of hydraulic lifters, pushrods, and rocker arms mounted in pairs on bridged pivots. A two barrel carburetor is used.

Engine Overhaul Tips

Most engine overhaul procedures are fairly standard. In addition to specific parts replacement procedures and complete specifications for your individual engine, this section also is a guide to accept rebuilding procedures. Examples of standard rebuilding practice are shown and should be used along with specific details concerning your particular engine.

Competent and accurate machine shop services will ensure maximum performance, reliability and engine life.

In most instances it is more profitable for the do-it-yourself mechanic to remove, clean and inspect the component, buy the necessary parts and deliver these to a shop for actual machine work.

On the other hand, much of the rebuilding work (crankshaft, block, bearings, piston rods, and other components) is well within the scope of the do-it-yourself mechanic.

TOOLS

The tools required for an engine overhaul or parts replacement will depend on the depth of your involvement. With a few exceptions, they will be the tools found in a mechanic's tool kit (see Section 1). More in-depth work will require any or all of the following:
- a dial indicator (reading in thousandths) mounted on a universal base
- micrometers and telescope gauges
- jaw and screw-type pullers
- scraper
- valve spring compressor
- ring groove cleaner
- piston ring expander and compressor
- ridge reamer
- cylinder hone or glaze breaker
- Plastigage®
- engine stand

The use of most of these tools is illustrated in this section. Many can be rented for a one-time use from a local parts jobber or tool supply house specializing in automotive work.

Occasionally, the use of special tools is called for. See the information on Special Tools and Safety Notice in the front of this book before substituting another tool.

INSPECTION TECHNIQUES

Procedures and specifications are given in this section for inspecting, cleaning and assessing the wear limits of most major components. Other procedures such as Magnaflux® and Zyglo® can be used to locate material flaws and stress cracks. Magnaflux® is a magnetic process applicable only to ferrous materials. The Zyglo® process coats the material with a fluorescent dye penetrant and can be used on any material Check for suspected surface cracks can be more readily made using spot check dye. The dye is sprayed onto the suspected area, wiped off and the area sprayed with a developer. Cracks will show up brightly.

OVERHAUL TIPS

Aluminum has become extremely popular for use in engines, due to its low weight. Observe the following precautions when handling aluminum parts:

● Never hot tank aluminum parts (the caustic hot tank solution will eat the aluminum.

● Remove all aluminum parts (identification tag, etc.) from engine parts prior to the tanking.

● Always coat threads lightly with engine oil or antiseize compounds before installation, to prevent seizure.

● Never overtorque bolts or spark plugs especially in aluminum threads.

Stripped threads in any component can be repaired using any of several commercial repair kits (Heli-Coil®, Microdot®, Keenserts®, etc.).

When assembling the engine, any parts that will be frictional contact must be prelubed to provide lubrication at initial start-up. Any product specifically formulated for this purpose can be used, but engine oil is not recommended as a prelube.

When semi-permanent (locked, but removable) installation of bolts or nuts is desired, threads should be cleaned and coated with Loctite® or other similar, commercial non-hardening sealant.

REPAIRING DAMAGED THREADS

Several methods of repairing damaged threads are available. Heli-Coil® (shown here), Keenserts® and Microdot® are among the most widely used. All involve basically the same principle—drilling out stripped threads, tapping the hole and installing a prewound insert—making welding, plugging and oversize fasteners unnecessary.

Two types of thread repair inserts are usually supplied: a standard type for most Inch Coarse, Inch Fine, Metric Course and Metric Fine thread sizes and a spark lug type to fit most spark plug port sizes. Consult the individual manufacturer's catalog to determine exact applications. Typical thread repair kits will contain a selection of prewound threaded inserts, a tap (corresponding to the outside diameter threads of the insert) and an installation tool. Spark plug inserts usually differ because they require a tap equipped with pilot threads and a combined reamer/tap section. Most manufacturers also supply blister-packed thread repair inserts separately in addition to a master kit containing a variety of taps and inserts plus installation tools.

Before effecting a repair to a threaded hole, remove any snapped, broken or damaged bolts or studs. Penetrating oil can be used to free frozen threads. The offending item can be removed with locking pliers or with a screw or stud extractor. After the hole is clear, the thread can be repaired, as follows:

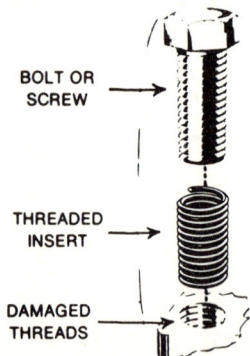

Damaged bolt holes can be repaired with thread repair inserts

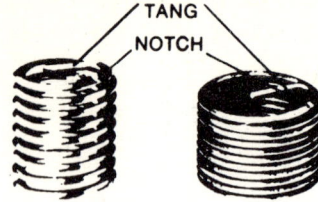

Standard thread repair insert (left) and spark plug thread insert (right)

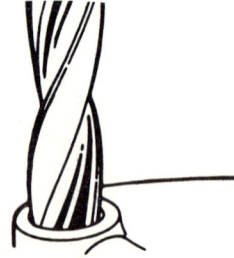

Drill out the damaged threads with specified drill. Drill completely through the hole or to the bottom of the blind hole

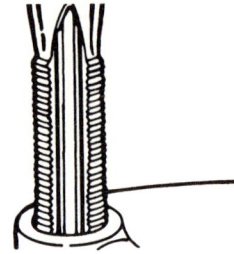

With the tap supplied, tap the hole to receive the thread insert. Keep the tap well oiled and back it out frequently to avoid clogging the threads

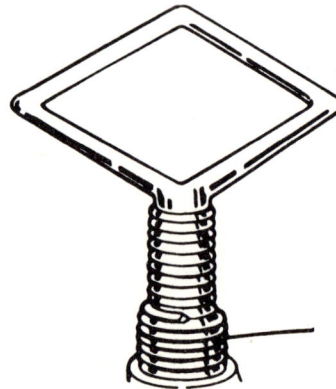

Screw the threaded insert onto the installation tool until the tang engages the slot. Screw the insert into the tapped hole until it is ¼-½ turn below the top surface. After installation break off the tang with a hammer and punch

Standard Torque Specifications and Fastener Markings

In the absence of specific torques, the following chart can be used as a guide to the maximum safe torque of a particular size/grade of fastener.
- There is no torque difference for fine or coarse threads.
- Torque values are based on clean, dry threads. Reduce the value by 10% if threads are oiled prior to assembly.
- The torque required for aluminum components or fasteners is considerably less.

U.S. Bolts

SAE Grade Number	1 or 2			5			6 or 7		
Number of lines always 2 less than the grade number.									
Bolt Size (Inches)—(Thread)	Maximum Torque			Maximum Torque			Maximum Torque		
	Ft./Lbs.	Kgm	Nm	Ft./Lbs.	Kgm	Nm	Ft./Lbs.	Kgm	Nm
¼ — 20	5	0.7	6.8	8	1.1	10.8	10	1.4	13.5
— 28	6	0.8	8.1	10	1.4	13.6			
5/16 — 18	11	1.5	14.9	17	2.3	23.0	19	2.6	25.8
— 24	13	1.8	17.6	19	2.6	25.7			
⅜ — 16	18	2.5	24.4	31	4.3	42.0	34	4.7	46.0
— 24	20	2.75	27.1	35	4.8	47.5			
7/16 — 14	28	3.8	37.0	49	6.8	66.4	55	7.6	74.5
— 20	30	4.2	40.7	55	7.6	74.5			
½ — 13	39	5.4	52.8	75	10.4	101.7	85	11.75	115.2
— 20	41	5.7	55.6	85	11.7	115.2			
9/16 — 12	51	7.0	69.2	110	15.2	149.1	120	16.6	162.7
— 18	55	7.6	74.5	120	16.6	162.7			
⅝ — 11	83	11.5	112.5	150	20.7	203.3	167	23.0	226.5
— 18	95	13.1	128.8	170	23.5	230.5			
¾ — 10	105	14.5	142.3	270	37.3	366.0	280	38.7	379.6
— 16	115	15.9	155.9	295	40.8	400.0			
⅞ — 9	160	22.1	216.9	395	54.6	535.5	440	60.9	596.5
— 14	175	24.2	237.2	435	60.1	589.7			
1 — 8	236	32.5	318.6	590	81.6	799.9	660	91.3	894.8
— 14	250	34.6	338.9	660	91.3	849.8			

Metric Bolts

Relative Strength Marking	4.6, 4.8			8.8		
Bolt Markings						
Bolt Size Thread Size x Pitch (mm)	Maximum Torque			Maximum Torque		
	Ft./Lbs.	Kgm	Nm	Ft./Lbs.	Kgm	Nm
6 x 1.0	2–3	.2–.4	3–4	3–6	.4–.8	5–8
8 x 1.25	6–8	.8–1	8–12	9–14	1.2–1.9	13–19
10 x 1.25	12–17	1.5–2.3	16–23	20–29	2.7–4.0	27–39
12 x 1.25	21–32	2.9–4.4	29–43	35–53	4.8–7.3	47–72
14 x 1.5	35–52	4.8–7.1	48–70	57–85	7.8–11.7	77–110
16 x 1.5	51–77	7.0–10.6	67–100	90–120	12.4–16.5	130–160
18 x 1.5	74–110	10.2–15.1	100–150	130–170	17.9–23.4	180–230
20 x 1.5	110–140	15.1–19.3	150–190	190–240	26.2–46.9	160–320
22 x 1.5	150–190	22.0–26.2	200–260	250–320	34.5–44.1	340–430
24 x 1.5	190–240	26.2–46.9	260–320	310–410	42.7–56.5	420–550

Standard Torque Specifications and Fastener Markings

Checking Engine Compression

A noticeable lack of engine power, excessive oil consumption and/or poor fuel mileage measured over an extended period are all indicators of internal engine war. Worn piston rings, scored or worn cylinder bores, blown head gaskets, sticking or burnt valves and worn valve seats are all possible culprits here. A check of each cylinder's compression will help you locate the problems.

As mentioned in the Tools and Equipment section of Section 1, a screw-in type compression gauge is more accurate that the type you simply hold against the spark plug hole, although it takes slightly longer to use. It's worth it to obtain a more accurate reading. Follow the procedures below.

1. Warm up the engine to normal operating temperature.
2. Remove all spark plugs.
3. Disconnect the high tension lead from the ignition coil.
4. On fully open the throttle either by operating the carburetor throttle linkage by hand or by having an assistant floor the accelerator pedal.
5. Screw the compression gauge into the no.1 spark plug hole until the fitting is snug.

NOTE: Be careful not to crossthread the plug hole. On aluminum cylinder heads use extra care, as the threads in these heads are easily ruined.

6. Ask an assistant to depress the accelerator pedal fully on both carbureted and fuel injected Jeep vehicles. Then, while you read the compression gauge, ask the assistant to crank the engine two or three times in short bursts using the ignition switch.
7. Read the compression gauge at the end of each series of cranks, and record the highest of these readings. Repeat this procedure for each of the engine's cylinders. Compare the highest reading of each cylinder to the compression pressure specification in the Tune-Up Specifications chart in Section 2. The specs in this chart are maximum values.

NOTE: A cylinder's compression pressure is usually acceptable if it is not less than 80% of maximum. The difference between any two cylinders should be no more than 12–14 pounds.

8. If a cylinder is unusually low, pour a tablespoon of clean engine oil into the cylinder through the spark plug hole and repeat the compression test. If the compression comes up after adding the oil, it appears that the cylinder's piston rings or bore are damaged or worn. If the pressure remains low, the valves may not be seating properly (a valve job is needed), or the head gasket may be blown near that cylinder. If compression in any two adjacent cylinders is low, and if the addition of oil doesn't help the compression, there is leakage past the head gasket. Oil and coolant water in the combustion chamber can result from this problem. There may be evidence of water droplets on the engine dipstick when a head gasket has blown.

Engine

REMOVAL AND INSTALLATION

F4–134

1. Drain the cooling system by opening the draincocks at the bottom of the radiator and the lower right side of the cylinder block.

——————— CAUTION ———————

When draining the coolant, keep in mind that cats and dogs are attracted by the ethylene glycol antifreeze, and are quite likely to drink any that is left in an uncovered container or in puddles on the ground. This will prove fatal in sufficient quantity. Always drain the coolant into a sealable container. Coolant should be reused unless it is contaminated or several years old.

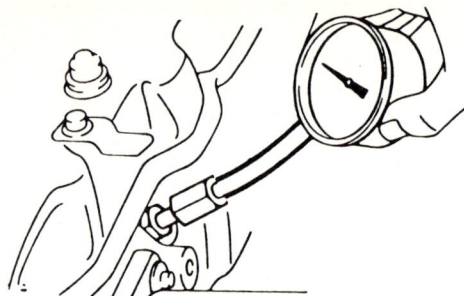

The screw-in type compression gauge is more accurate

2. Disconnect the battery at the positive terminal to avoid the possibility of a short circuit.
3. Remove the air cleaner horn from the carburetor and disconnect the breather hose at the oil filler pipe.
4. Disconnect the carburetor choke and throttle controls by loosening the clamp bolts and setscrews.
5. Disconnect the fuel tank-to-fuel pump line at the fuel pump by unscrewing the connecting nut.
6. Plug the fuel line to prevent leakage. Disconnect the windshield wiper vacuum hose at the fuel pump.
7. Remove the upper and lower radiator hoses. Remove the heater hoses, if so equipped, from the water pump and the rear of the cylinder head.
8. Remove the fan hub and fan blades.
9. Remove the four radiator attaching screws and remove the radiator and shroud as one unit.
10. Remove the starter motor cables and remove the starter motor.
11. Disconnect the wires from the alternator or the generator. Disconnect the ignition primary wire at the ignition coil.
12. Disconnect the oil pressure and temperature sending unit wires at the units.
13. Disconnect the exhaust pipe at the exhaust manifold by removing the stud nuts.
14. Remove the spark plug wires from the cable bracket that is mounted to the rocker arm cover. Remove the cable bracket by removing the stud nuts.
15. Remove the rocker arm cover by removing the attaching stud nuts.
16. Attach a lifting bracket to the engine using the head bolts. Be sure that the bolts selected will hold the engine with the weight balanced. Attach the lifting bracket to a boom hoist, or other lifting device, and take up all of the slack.
17. Remove the two nuts and bolts from each front engine support. Disconnect the engine ground strap.
18. Remove the engine supports. Lower the engine slightly to permit access to the two top bolts on the flywheel housing.
19. Remove the bolts that attach the flywheel housing to the engine.
20. Pull the engine forward, or roll the vehicle backward, until the clutch clears the flywheel housing. Lift the engine from the vehicle.

To install:

21. Lower the engine into the vehicle. Push the engine backwards until the clutch enters the bell housing.
22. Bolt the engine to the bell housing.
23. Let the engine down, onto the two front supports.
24. Bolt the engine to the supports.
25. Install the rocker arm cover.
26. Install the spark plug wires on the cable bracket that is mounted to the rocker arm cover. Install the cable bracket by installing the stud nuts.
27. Connect the exhaust pipe at the exhaust manifold by installing the stud nuts.

28. Connect the oil pressure and temperature sending unit wires at the units.
29. Connect the wires from the alternator or the generator. Connect the ignition primary wire at the ignition coil.
30. Install the starter motor and install the starter motor cables.
31. Install the radiator and shroud as one unit.
32. Install the fan hub and fan blades.
33. Install the upper and lower radiator hoses. Install the heater hoses, if so equipped, on the water pump and the rear of the cylinder head.
34. Connect the windshield wiper vacuum hose at the fuel pump.
35. Connect the fuel tank-to-fuel pump line at the fuel pump.
36. Connect the carburetor choke and throttle controls.
37. Install the air cleaner horn on the carburetor and connect the breather hose at the oil filler pipe.
38. Connect the battery at the positive terminal.
39. Fill the cooling system.

4-150

1. Disconnect the battery ground cable.
2. Remove the air cleaner.
3. Remove the hood.
4. Drain the coolant.

CAUTION

When draining the coolant, keep in mind that cats and dogs are attracted by the ethylene glycol antifreeze, and are quite likely to drink any that is left in an uncovered container or in puddles on the ground. This will prove fatal in sufficient quantity. Always drain the coolant into a sealable container. Coolant should be reused unless it is contaminated or several years old.

5. Remove the lower radiator hose.
6. Remove the upper radiator hose.
7. Disconnect the coolant recovery hose.
8. Remove the fan shroud.
9. Disconnect the automatic transmission coolant lines.
10. Discharge the refrigerant system. See Section 1.

CAUTION

Do this CAREFULLY, or let someone with experience do it for you. GREAT PERSONAL INJURY CAN OCCUR WHEN MISHANDLING REFRIGERANT GAS!

11. Disconnect and remove the condenser. Cap all openings at once!
12. Remove the radiator.
13. Remove the fan and install a $\frac{5}{16}$ in. x $\frac{1}{2}$ in. capscrew through the pulley and into the water pump flange to maintain the pulley-to-pump alignment.

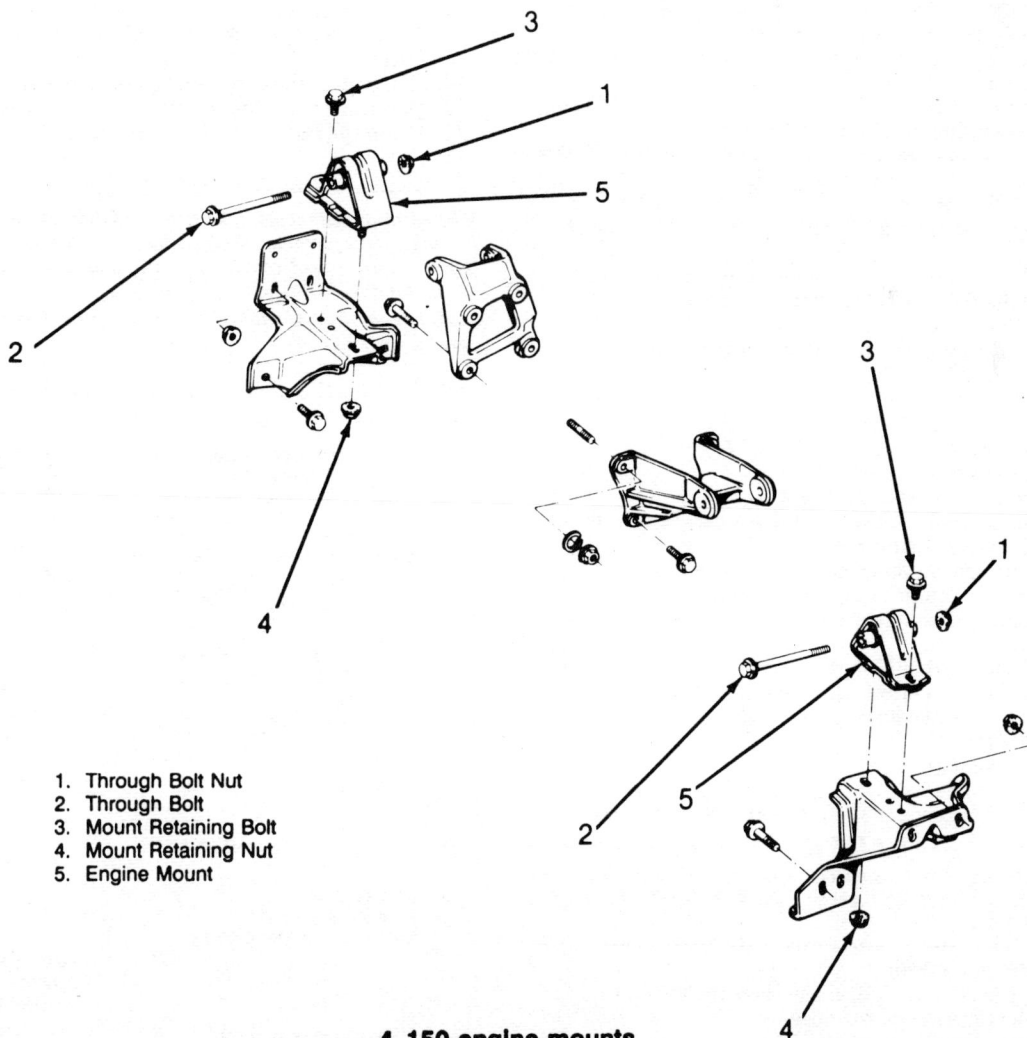

1. Through Bolt Nut
2. Through Bolt
3. Mount Retaining Bolt
4. Mount Retaining Nut
5. Engine Mount

4-150 engine mounts

14. Disconnect the heater hoses.
15. Disconnect the throttle linkage.
16. Disconnect the cruise control linkage.
17. Disconnect the oil pressure sending unit wire.
18. Disconnect the temperature sending unit wire.
19. Disconnect and tag all vacuum hoses connected to the engine.
20. Remove the air conditioning compressor.
21. Remove the power steering hoses at the gear.
22. Drain the power steering reservoir.
23. Remove the power brake vacuum check valve from the booster.
24. Raise and support the front end on jackstands.
25. Disconnect and tag the starter wires.
26. Remove the starter.
27. Disconnect the exhaust pipe at the manifold.
28. Remove the flywheel housing access cover.
29. On vehicles equipped with automatic transmission, matchmark the converter and flywheel and remove the attaching bolts.
30. Remove the upper flywheel housing-to-engine bolts and loosen the lower ones.
31. Remove the engine mount cushion-to-engine compartment bolts.
32. Attach a shop crane to the lifting eyes on the engine.
33. Raise the engine off the front supports.
34. Place a floor jack under the flywheel housing.
35. Remove the remaining flywheel housing bolts.
36. Lift the engine out of the vehicle.
37. Mount the engine on a work stand or cradle. Never let it rest on the oil pan.

To install the engine:

38. Lower the engine into place in the vehicle.
39. Lubricate the manual transmission input shaft with chassis lube before insertion into clutch splines.
40. Install the flywheel housing bolts. Torque the top flywheel housing-to-engine bolts to 27 ft. lbs. and the bottom ones to 43 ft. lbs.
41. Install the engine mount cushion-to-engine compartment bolts. Torque the front bracket support bolts to 33 ft. lbs.
42. Remove the shop crane.
43. On vehicles equipped with automatic transmission, install the converter attaching bolts.
44. Install the flywheel housing access cover.
45. Connect the exhaust pipe at the manifold.
46. Install the starter.
47. Connect the wires.
48. Raise and support the front end on jackstands.
49. Install the power brake vacuum check valve on the booster.
50. Fill the power steering reservoir.
51. Connect the power steering hoses at the gear.
52. Install the air conditioning compressor.
53. Connect all vacuum hoses.
54. Connect all wires.
55. Connect the throttle linkage.
56. Connect the cruise control linkage.
57. Connect the heater hoses.
58. Install the fan.
59. Install the radiator.
60. Install the condenser.
61. Evacuate and charge the refrigerant system. See Section 1.

──────── **CAUTION** ────────

Do this CAREFULLY, or let someone with experience do it for you. GREAT PERSONAL INJURY CAN OCCUR WHEN MISHANDLING REFRIGERANT GAS!

62. Install the fan shroud.
63. Connect the automatic transmission coolant lines.
64. Install the upper radiator hose.
65. Connect the coolant recovery hose.
66. Install the lower radiator hose.
67. Fill the cooling system.
68. Install the hood.
69. Install the air cleaner.
70. Connect the battery ground cable.

4-151

1. Disconnect the battery.
2. Remove the air cleaner.
3. Jack up the vehicle and support it on jackstands.
4. Disconnect the exhaust pipe from the manifold.
5. Disconnect the oxygen sensor.
6. Disconnect the wires from the starter.
7. Unbolt the starter and remove it from the vehicle.
8. Disconnect the wires from the distributor and oil pressure sending unit.
9. Remove the engine mount nuts.
10. On vehicles with manual transmission, remove the clutch slave cylinder and flywheel inspection plate.
11. Remove the clutch or converter housing-to-engine bolts.
12. On vehicle with automatic transmission, disconnect the converter from the drive plate.
13. Lower the vehicle.
14. Support the transmission with a jack.
15. Tag all hoses at the carburetor and remove them.
16. Disconnect the mixture control solenoid wire from the carburetor, (not all vehicles have these).
17. Disconnect the wires from the alternator.
18. Disconnect the throttle cable from the bracket and the carburetor.
19. Disconnect the choke and solenoid wires at the carburetor.
20. Disconnect the temperature sender wire.
21. Drain the radiator at the drain cock, then remove the lower hose.

──────── **CAUTION** ────────

When draining the coolant, keep in mind that cats and dogs are attracted by the ethylene glycol antifreeze, and are quite likely to drink any that is left in an uncovered container or in puddles on the ground. This will prove fatal in sufficient quantity. Always drain the coolant into a sealable container. Coolant should be reused unless it is contaminated or several years old.

22. Remove the upper radiator hose and the heater hoses.
23. Remove the fan shroud, and radiator.

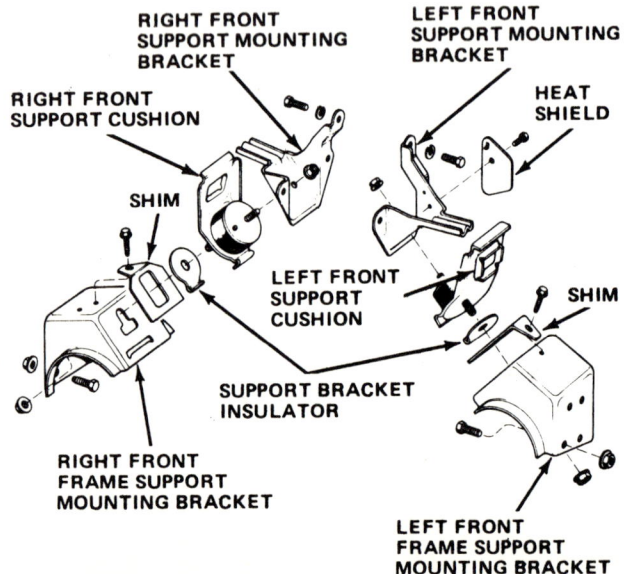

4-151 engine mounts

24. Remove the power steering hoses at the pump.
25. Attach a shop crane to the engine and lift it out of the vehicle.

NOTE: The manual transmission may have to be raised slightly to allow a smooth separation.

To install:
26. Lower the engine into the truck.
27. On vehicle with automatic transmission, Connect the converter to the drive plate.
28. Install the clutch or converter housing-to-engine bolts. Torque the bolts to 35 ft. lbs.
29. On vehicles with manual transmission, Install the clutch slave cylinder and flywheel inspection plate. Torque the slave cylinder bolts to 18 ft. lbs.
30. Install the engine mount nuts. Torque them to 34 ft. lbs.
31. Connect the wires to the distributor and oil pressure sending unit.
32. Install the starter. Torque the mounting bolts to 27 ft. lbs.; the bracket nut to 40 inch lbs.
33. Connect the wires to the starter.
34. Connect the oxygen sensor.
35. Connect the exhaust pipe at the manifold. Torque the nuts to 35 ft. lbs.
36. Install the power steering hoses at the pump.
37. Install the fan shroud, and radiator.
38. Install the upper radiator hose and the heater hoses.
39. Connect the temperature sender wire.
40. Connect the choke and solenoid wires at the carburetor.
41. Connect the throttle cable at the bracket and the carburetor.
42. Connect the wires to the alternator.
43. Connect the mixture control solenoid wire at the carburetor, (not all vehicles have these).
44. Install all hoses at the carburetor.
45. Lower the vehicle.
46. Install the air cleaner.
47. Connect the battery.
48. Fill the cooling system.

6-225

1. Remove the hood.
2. Disconnect the battery ground cable from the engine and the battery.
3. Remove the air cleaner.
4. Drain the coolant from the radiator and engine.

━━━━━━━━━━ **CAUTION** ━━━━━━━━━━
When draining the coolant, keep in mind that cats and dogs are attracted by the ethylene glycol antifreeze, and are quite likely to drink any that is left in an uncovered container or in puddles on the ground. This will prove fatal in sufficient quantity. Always drain the coolant into a sealable container. Coolant should be reused unless it is contaminated or several years old.

5. Disconnect the alternator wiring harness from the connector at the regulator.
6. Disconnect the upper and lower radiator hoses from the engine.
7. Remove the right and left radiator support bars.
8. Remove the radiator from the vehicle.
9. Disconnect the engine wiring harnesses from the connectors which are located on the firewall.
10. Disconnect the battery cable and wiring from the engine starter assembly.
11. Remove the starter assembly from the engine.
12. Disconnect the engine fuel hoses from the fuel lines at the right fame rails.
13. Plug the fuel lines.
14. Disconnect the throttle linkage and the choke cable from

the carburetor and remove the cable support bracket that is mounted on the engine.
15. Disconnect the exhaust pipes from the right and left sides of the engine.
16. Place a jack under the transmission and support the weight of the transmission.
17. Remove the bolts that secure the engine to the front motor mounts.
18. Attach a suitable sling to the engine lifting eyes and, using a hoist, lift the engine just enough to support its weight.
19. Remove the bolts that secure the engine to the flywheel housing.
20. Raise the engine slightly and slide the engine forward to remove the transmission main shaft form the clutch plate splines.

NOTE: The engine and the transmission must be raised slightly to release the spline from the clutch plate while sliding the engine forward.

21. When the engine is free of the transmission shaft, raise the engine and remove it from the vehicle.
To install:
22. Lower the engine into the vehicle.
23. Slide the engine rearward to install the transmission main shaft in the clutch plate splines.
24. Install the bolts that secure the engine to the flywheel housing. Torque the bolts to 30–40 ft. lbs.
25. Install the bolts that secure the engine to the front motor mounts. Torque the bolts to 75 ft. lbs.
26. Connect the exhaust pipes to the right and left sides of the engine.
27. Connect the throttle linkage and the choke cable to the carburetor and Install the cable support bracket that is mounted on the engine.
28. Connect the engine fuel hoses to the fuel lines at the right fame rails.
29. Install the starter assembly on the engine. Torque the starter-to-block bolts to 30–40 ft. lbs.; the bracket bolts to 10–12 ft. lbs.
30. Connect the battery cable and wiring to the engine starter assembly.
31. Connect the engine wiring harnesses to the connectors which are located on the firewall.
32. Install the radiator.
33. Install the right and left radiator support bars.
34. Connect the upper and lower radiator hoses to the engine.
35. Connect the alternator wiring harness to the connector at the regulator.
36. Fill the cooling system.
37. Install the air cleaner.
38. Connect the battery ground cable to the engine and the battery.
39. Install the hood.

6-232, 6-258

1. Remove the air cleaner.
2. Drain the cooling system.

━━━━━━━━━━ **CAUTION** ━━━━━━━━━━
When draining the coolant, keep in mind that cats and dogs are attracted by the ethylene glycol antifreeze, and are quite likely to drink any that is left in an uncovered container or in puddles on the ground. This will prove fatal in sufficient quantity. Always drain the coolant into a sealable container. Coolant should be reused unless it is contaminated or several years old.

3. Disconnect the upper and lower radiator hoses.
4. If equipped with an automatic transmission, disconnect the cooler lines from the radiator.
5. Remove the radiator and the fan. If the Jeep has air condi-

tioning, evacuate the system, remove the condenser and cap all openings immediately. See Section 1.

6. If so equipped, remove the power steering pump and the drive belt, and place the unit aside. Do not remove the power steering hoses.

7. If the vehicle has air conditioning, remove the compressor.

8. Disconnect all wires, lines, linkage, and hoses that are connected to the engine. Remove the oil filter.

Engine mounts for the 6–258 through 1986

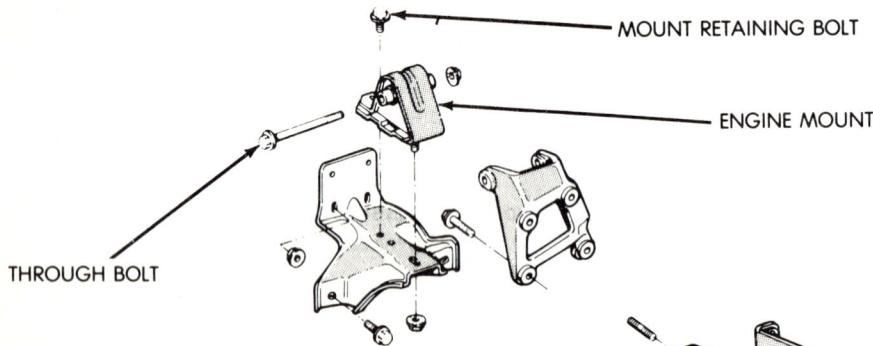

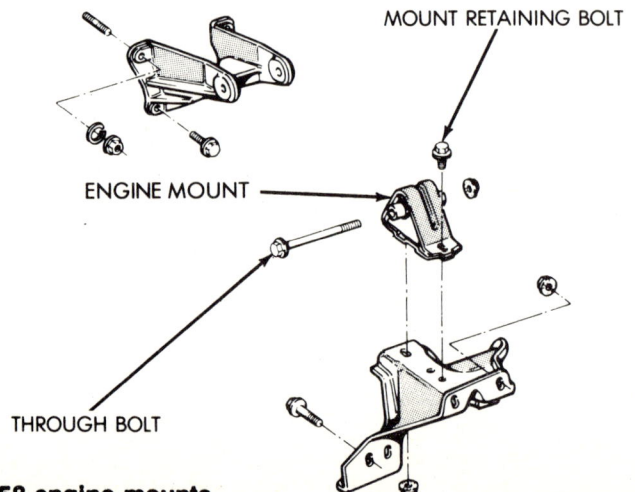

9. Remove both of the engine front support cushion-to-frame retaining nuts.

10. Disconnect the exhaust pipe at the support bracket and exhaust manifold.

11. Support the weight of the engine with a lifting device.

12. Remove the front support cushion and bracket assemblies from the engine.

13. Remove the transfer case shift lever boot and the transmission access cover.

14. If equipped with an automatic transmission, remove the upper bolts securing the transmission bellhousing to the engine. If equipped with a manual transmission, remove the upper bolts that secure the clutch housing to the engine.

15. Remove the starter motor.

16. If the vehicle is equipped with an automatic transmission:

   a. Remove the engine to transmission adapter plate inspection covers.

   b. Mark the assembled position of the converter and flex plate and remove the converter-to-flex plate retaining screws.

   c. Remove the remaining bolts securing the transmission bellhousing to the engine.

17. If equipped with a manual transmission, remove the lower cover of the clutch housing and the remaining bolts that secure the clutch housing to the engine.

18. Support the transmission with a floor jack.

19. Attach a suitable sling to the engine and using a hoist, lift the engine upward and forward at the same time, removing it from the vehicle.

To install:

20. Lower the engine into the vehicle and slide it rearward to engage the transmission.

21. If equipped with an automatic transmission, install the upper bolts securing the transmission bellhousing to the engine. Torque them to 27 ft. lbs. If equipped with a manual transmission, install the upper bolts that secure the clutch housing to the engine. Torque them to 27 ft. lbs.

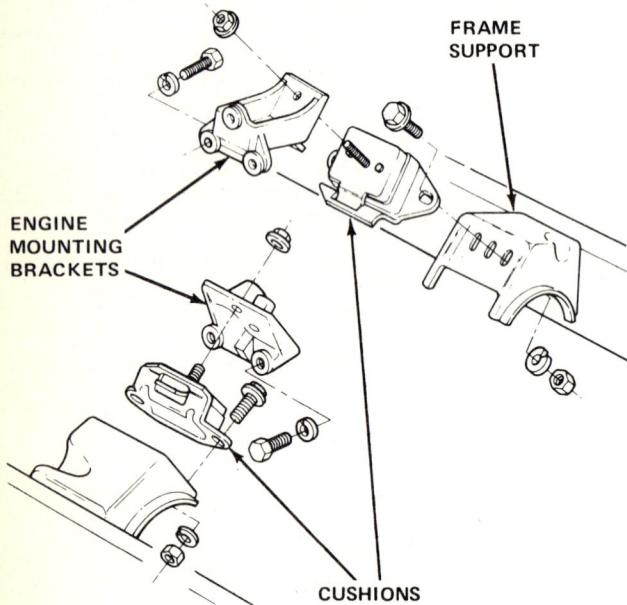

1987–90 6–258 engine mounts

22. If equipped with a manual transmission, install the lower cover of the clutch housing and the remaining bolts that secure the clutch housing to the engine. Torque the clutch housing spacer-to-bolts to 12–15 ft. lbs. Torque the clutch housing lower bolts to 43 ft. lbs.

23. If the vehicle is equipped with an automatic transmission:

    a. Install the remaining bolts securing the transmission bellhousing to the engine. Torque the bellhousing lower bolts to 43 ft. lbs.

    b. Mark the assembled position of the converter and flex plate and install the converter-to-flex plate retaining screws. Torque the bolts to 20–25 ft. lbs.

    c. Install the engine to transmission adapter plate inspection covers.

24. Install the starter motor. Torque the mounting bolts to 18 ft. lbs.

25. Install the transfer case shift lever boot and the transmission access cover.

26. Install the front support cushion and bracket assemblies from the engine.

27. Remove the shop crane.

28. Connect the exhaust pipe at the support bracket and exhaust manifold. Torque the nuts to 20 ft. lbs.

29. Install both of the engine front support cushion-to-frame retaining nuts. Torque them to 35 ft. lbs.

30. Connect all wires, lines, linkage, and hoses that are connected to the engine.

31. Install the oil filter.

32. If so equipped, install the power steering pump and the drive belt.

NOTE: If the vehicle has air conditioning, mount and connect the air conditioning compressor. See Section 1.

34. Install the radiator and the fan, and the condenser.

35. Evacuate, charge and leak test the refrigerant system.

36. If equipped with an automatic transmission, connect the cooler lines to the radiator.

37. Connect the upper and lower radiator hoses.

38. Fill the cooling system.

39. Install the air cleaner.

8–304

1. Remove the air cleaner.
2. Drain the cooling system.

────────────── **CAUTION** ──────────────

When draining the coolant, keep in mind that cats and dogs are attracted by the ethylene glycol antifreeze, and are quite likely to drink any that is left in an uncovered container or in puddles on the ground. This will prove fatal in sufficient quantity. Always drain the coolant into a sealable container. Coolant should be reused unless it is contaminated or several years old.

3. Disconnect the upper and lower radiator hoses.

4. If equipped with an automatic transmission, disconnect the cooler lines from the radiator.

5. Remove the radiator and the fan. If the Jeep has air conditioning, evacuate the system, remove the condenser and cap all openings immediately. See Section 1.

6. If so equipped, remove the power steering pump and the drive belt, and place the unit aside. Do not remove the power steering hoses.

7. If the vehicle has air conditioning, remove the compressor.

8. Disconnect all wires, lines, linkage, and hoses that are connected to the engine.

9. Remove both of the engine front support cushion-to-frame retaining nuts.

10. Disconnect the exhaust pipes at the support bracket and exhaust manifold.

11. Support the weight of the engine with a lifting device.

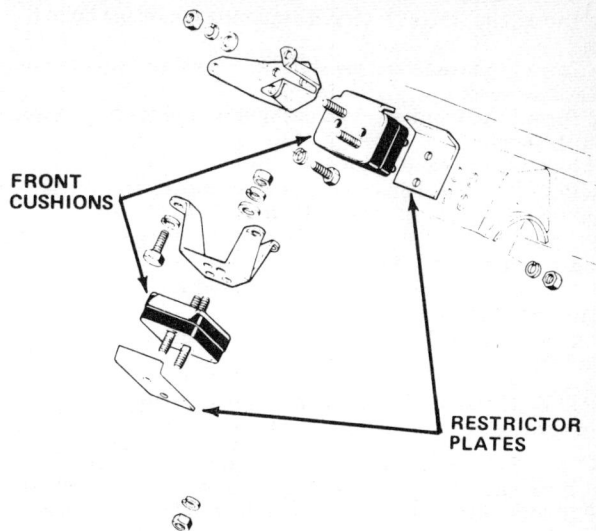

FRONT CUSHIONS

RESTRICTOR PLATES

8–304 engine mounts

12. Remove the front support cushion and bracket assemblies from the engine.

13. Remove the transfer case shift lever boot and the transmission access cover.

14. If equipped with an automatic transmission, remove the upper bolts securing the transmission bellhousing to the engine. If equipped with a manual transmission, remove the upper bolts that secure the clutch housing to the engine.

15. Remove the starter motor.

16. If the vehicle is equipped with an automatic transmission:

    a. Remove the engine to transmission adapter plate inspection covers.

    b. Mark the assembled position of the converter and flex plate and remove the converter-to-flex plate retaining screws.

    c. Remove the remaining bolts securing the transmission bellhousing to the engine.

17. If equipped with a manual transmission, remove the lower cover of the clutch housing and the remaining bolts that secure the clutch housing to the engine.

18. Support the transmission with a floor jack.

19. Attach a suitable sling to the engine and using a hoist, lift the engine upward and forward at the same time, removing it from the vehicle.

To install:

20. Lower the engine into the vehicle and slide it rearward to engage the transmission.

21. If equipped with an automatic transmission, install the upper bolts securing the transmission bellhousing to the engine. Torque them to 27 ft. lbs. If equipped with a manual transmission, install the upper bolts that secure the clutch housing to the engine. Torque them to 27 ft. lbs.

22. If equipped with a manual transmission, install the lower cover of the clutch housing and the remaining bolts that secure the clutch housing to the engine. Torque the clutch housing spacer-to-bolts to 12–15 ft. lbs. Torque the clutch housing lower bolts to 43 ft. lbs.

23. If the vehicle is equipped with an automatic transmission:

    a. Install the remaining bolts securing the transmission bellhousing to the engine. Torque the bellhousing lower bolts to 43 ft. lbs.

    b. Mark the assembled position of the converter and flex plate and install the converter-to-flex plate retaining screws. Torque the bolts to 20–25 ft. lbs.

    c. Install the engine to transmission adapter plate inspection covers.

24. Install the starter motor. Torque the mounting bolts to 18 ft. lbs.

25. Install the transfer case shift lever boot and the transmission access cover.

26. Install the front support cushion and bracket assemblies from the engine.

27. Remove the shop crane.

28. Connect the exhaust pipes at the support bracket and exhaust manifold. Torque the nuts to 20 ft. lbs.

29. Install both of the engine front support cushion-to-frame retaining nuts. Torque them to 35 ft. lbs.

30. Connect all wires, lines, linkage, and hoses that are connected to the engine.

31. If so equipped, install the power steering pump and the drive belt.

NOTE: If the vehicle has air conditioning, mount and connect the air conditioning compressor. See Section 1.

32. Install the radiator and the fan, and the condenser.

33. Evacuate, charge and leak test the refrigerant system.

34. If equipped with an automatic transmission, connect the cooler lines to the radiator.

35. Connect the upper and lower radiator hoses.

36. Fill the cooling system.

37. Install the air cleaner.

Rocker Shafts and Rocker Studs

REMOVAL AND INSTALLATION

F4-134

1. Remove the rocker arm cover attaching bolts and remove the rocker arm cover.

2. Remove the nuts from the rocker arm shaft support studs.

3. Remove the intake valve pushrods from the engine.

4. Install in the reverse order. Tighten the rocker arm retaining bolts to 30–33 ft. lbs.

4-150

1. Remove the rocker arm cover. The cover seal is RTV sealer. Break the seal with a clean putty knife or razor blade. Don't at-

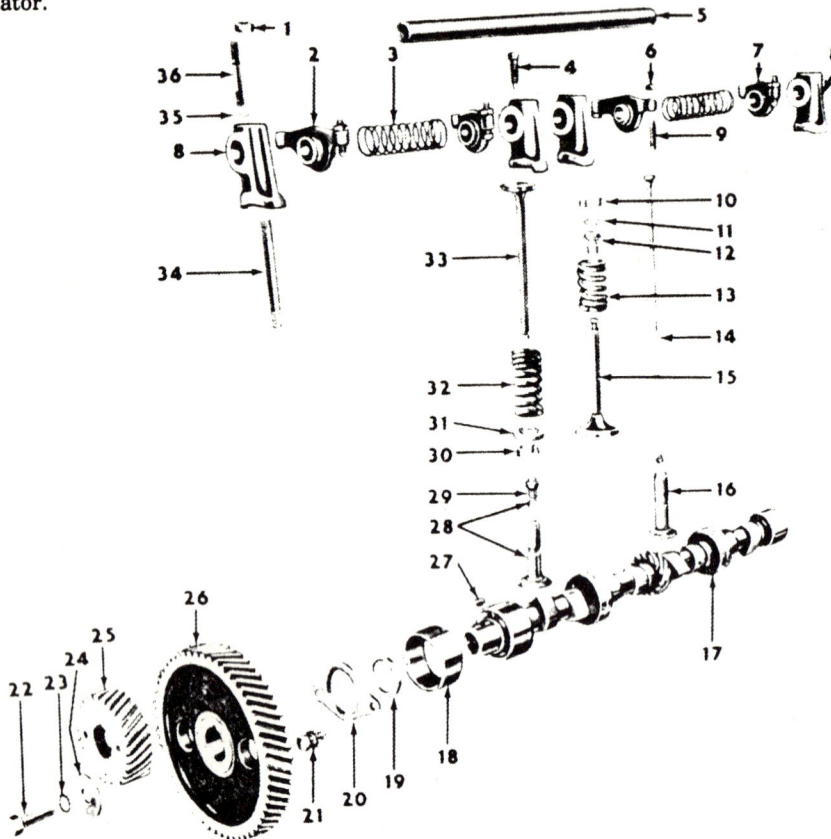

1. Nut
2. Left rocker arm
3. Rocker arm shaft spring
4. Rocker shaft lock screw
5. Rocker shaft
6. Nut
7. Right rocker arm
8. Rocker arm shaft bracket
9. Intake valve tappet adjusting screw
10. Intake valve upper retainer lock
11. Oil seal
12. Intake valve spring upper retainer
13. Intake valve spring
14. Intake valve push rod
15. Intake valve
16. Intake valve tappet
17. Camshaft
18. Camshaft front bearing
19. Camshaft thrust plate spacer
20. Camshaft thrust plate
21. Bolt and lock washer
22. Bolt
23. Lockwasher
24. Camshaft gear washer
25. Crankshaft gear
26. Camshaft gear
27. Woodruff key No. 9
28. Exhaust valve tappet
29. Tappet adjusting screw
30. Spring retainer lock
31. Roto cap assembly
32. Exhaust valve spring
33. Exhaust valve
34. Rocker shaft support stud
35. Washer
36. Rocker arm cover stud

F4-134 valve train

tempt to remove the cover until the seal is broken. To remove the cover, pry where indicated at the bolt holes.

2. Remove the two capscrews at each bridge and pivot assembly. It's best to remove the capscrews alternately, a little at a time each to avoid damage to the bridge.

3. Remove the bridges, pivots and rocker arms. Keep them in order.

4. Install the rocker arms and bridge and pivot assemblies. Tighten the capscrews to 19 ft. lbs.

5. Clean the mating surfaces of the cover and head.

6. Run a ⅛ in. (3mm) bead of RTV sealer around the mating surface of the head. Install the cover within 10 minutes! Don't allow any of the RTV material to drop into the engine! In the engine it will form and set and possibly block and oil passage. Torque the cover bolts to 36–60 inch lbs.

4–151

1. Remove the rocker arm cover.
2. Remove the rocker arm capscrew and ball.
3. Remove the rocker arm.
4. Install rocker arms. Tighten the capscrews to 20 ft. lbs. DO NOT OVERTORQUE!
5. Clean the mating surfaces of the cover and head.
6. Install the cover. Tighten the bolts to 36–60 inch lbs.

6–225

1. Remove the crankcase ventilator valve from the right side valve cover.
2. Remove the four attaching bolts from the right and left side valve covers and remove both of the valve covers.
3. Unscrew, but do not remove, the bolts that attach the rocker arm assemblies to the cylinder heads.
4. Remove the rocker arm assemblies, with the bolts in place, from the cylinder heads.
5. Mark each of the pushrods so that they can be installed in their original positions.
6. Remove the pushrods.
To install:
7. Install the pushrods.

8. Install the rocker arm assemblies. Tighten the bolts to 30 ft. lbs., a little at a time.
9. Clean the mating surfaces of the covers and heads.
10. Install the valve covers using new gaskets. Torque the bolts to 36–60 in. lbs.
11. Install the crankcase ventilator valve from the right side valve cover.
12. Install the crankcase ventilator valve from the right side valve cover.

1972 and 1974 6–232, 6–258

1. Remove the valve cover by removing the 6 valve cover attaching screws.
2. Loosen, but do not remove, the 6 bolts that attach the rocker arm assembly to the cylinder head.
3. Lift the whole rocker arm assembly off the head with the bolts in place.
4. Identify each of the pushrods so that they can be replaced in their original positions.
5. Remove the pushrods.
6. Install the rocker arm assembly. Tighten bolts, working evenly, from the center outward, to 22 ft. lbs.
7. Clean the mating surfaces of the valve cover and head.
8. Install the valve cover, using a new gasket. Torque the bolts to 36–60 ft. lbs.

1973, 1975–78 6–232
1975–87 6–258
1973–81 8–304

On these engines the rocker arms pivot on a bridged pivot that is secured with two capscrews. The bridged pivots maintain proper rocker arm-to-valve tip alignment.
1. Remove the rocker cover and gasket.
2. Remove the two capscrews at each bridged pivot, backing

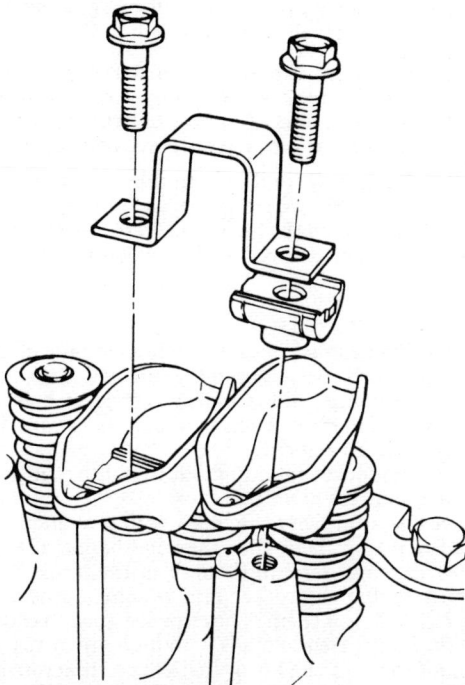

4–150 rocker arm assembly

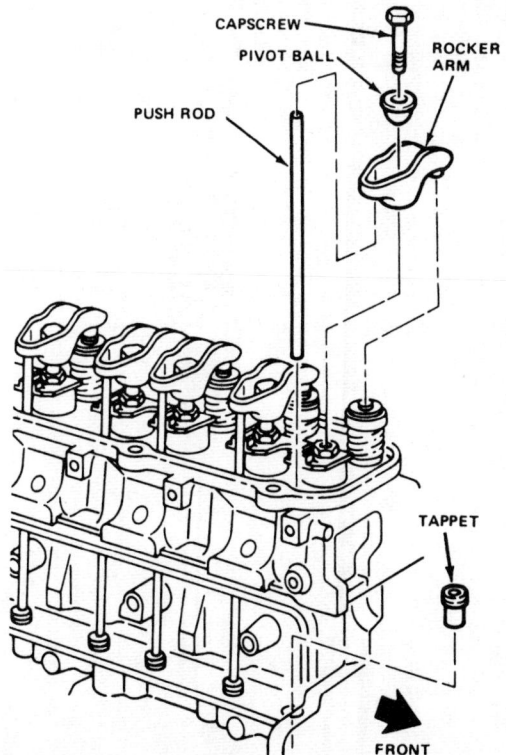

4–151 rocker arm and pushrod removal and installation

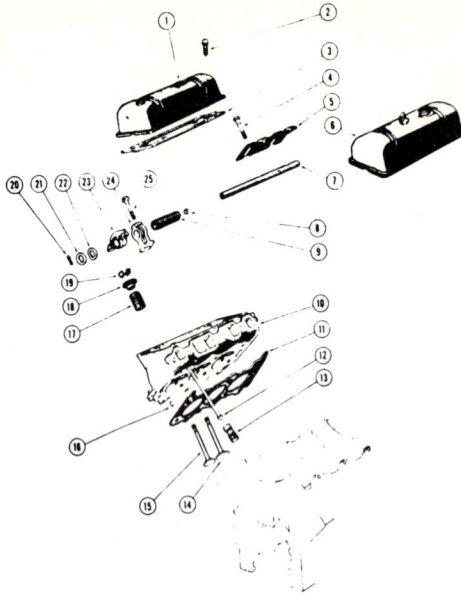

1. Right rocker arm cover
2. Rocker arm cover bolt
3. Gasket
4. Bolt
5. Baffle
6. Left rocker arm cover
7. Rocker arm shaft
8. Plug
9. Rocker arm spring
10. Cylinder head
11. Head gasket
12. Pushrod
13. Valve lifter
14. Intake valve
15. Exhaust valve
16. Dowel pin
17. Valve spring
18. Valve spring cap
19. Valve spring cap key
20. Cotter pin
21. Rocker arm shaft end washer
22. Rocker arm shaft spring
23. Rocker arm
24. Rocker arm shaft bracket
25. Bolt

6–225 valve train

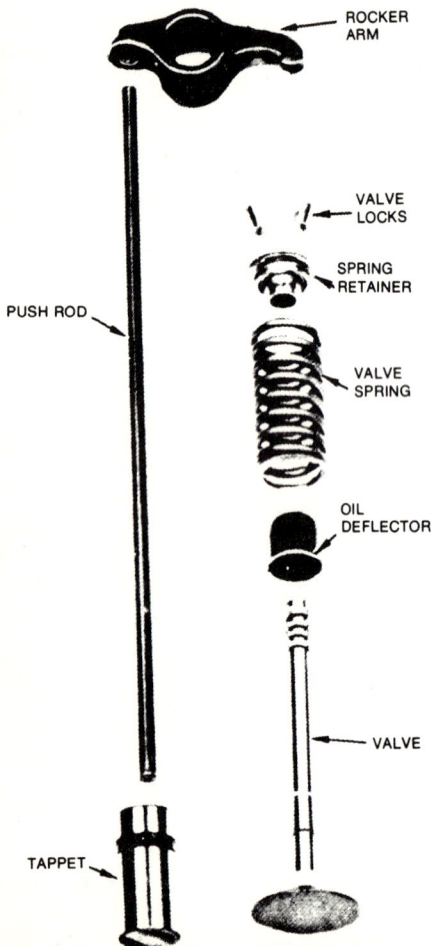

1972 and 1974 6-cylinder engine valve train

off each capscrew one turn at a time to avoid breaking the bridge.

3. Remove each bridged pivot and corresponding pair of rocker arms and place them on a clean surface in the same order as they are removed.

NOTE: Bridged pivots, capscrews, rockers, and pushrods must all be reinstalled in their original positions.

4. Clean all the parts in a suitable solvent and use compressed air to blow out the oil passages in the pushrods and the rocker arms. Replace any excessively worn parts.

5. Install rocker arms, pushrods and bridged pivots in the same positions from which they were removed.

NOTE: Be sure that the bottom end of each pushrod is centered in the plunger which they were removed. Be sure that the bottom end of each pushrod is centered in the plunger cap of each hydraulic valve tappet.

6. Install the capscrews and tighten them one turn at a time, alternating between the two screws on each bridge. Tighten the capscrews to 21 ft. lbs. on the Sixes and 10 ft. lbs. on the 304 V8.

7. Install the rocker cover(s) with new gasket(s).

1972 8–304

The 1972 8–304 has each rocker arm individually mounted on a separate stud. Each rocker assembly consists of the following: a rocker arm retaining stud, a rocker arm pivot ball, a rocker arm, and a retaining stud. Each assembly is removed and installed separately. To remove:

1. Unscrew the rocker retaining nut from the stud and lift off the rocker arm and its pivot ball.

2. Remove the stud from the block with a wrench.

3. Label the pushrods so that they can be installed in their original positions and remove them from the block.

4. When installing the rocker arm retaining studs, use caution not to cross thread them. They are designed to cause an interference fit. Lubricate the studs with high pressure grease before installing them in the head. Install the rocker arm assemblies in the reverse order of removal. Tighten the rocker arm retaining nuts to 23 ft. lbs.

Thermostat

REMOVAL AND INSTALLATION

When draining the coolant, keep in mind that cats and dogs are attracted by the ethylene glycol antifreeze, and are quite likely to drink any that is left in an uncovered container or in

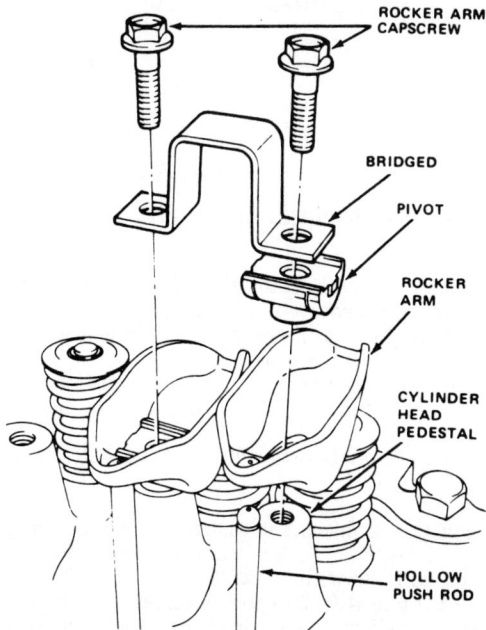

Rocker arm assembly used on the 1873, 1975–78 6–232; 1975–90 6–258; 1973–81 8–304

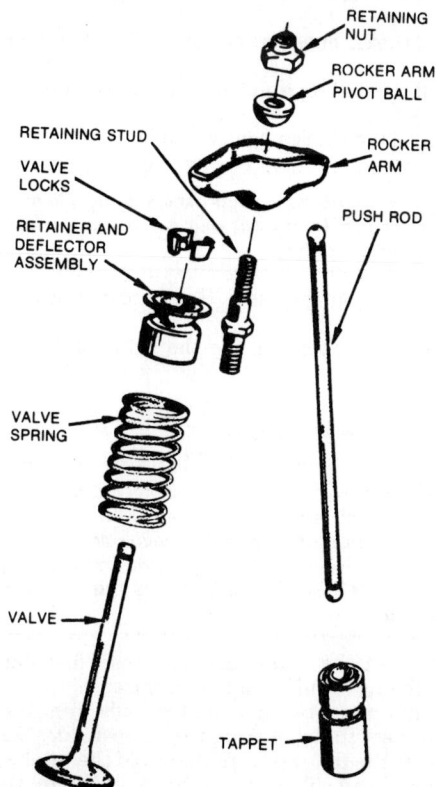

1972 V8 valve train

puddles on the ground. This will prove fatal in sufficient quantity. Always drain the coolant into a sealable container. Coolant should be reused unless it is contaminated or several years old.

The thermostat is located in the water outlet housing at the front or on top of the engine. On the V6 and the 304 V8 the water outlet housing is located in the front of the intake manifold.

To remove the thermostats from all of these engines, first drain the cooling system. It is not necessary to disconnect or remove any of the hoses. Remove the two attaching screws and lift the housing from the engine. Remove the thermostat and the gasket. To install, place the thermostat in the housing with the spring inside the engine. Install a new gasket with a small amount of sealing compound applied to both sides. Install the water outlet and tighten the attaching bolts to 30 ft. lbs. Refill the cooling system.

Intake Manifold

REMOVAL AND INSTALLATION

F4–134

On the F4–134 engine the intake manifold is cast as an integral part of the head.

4–150

NOTE: It may be necessary to remove the carburetor or the throttle body from the intake manifold before the manifold is removed.

1. Disconnect the negative battery cable. Drain the radiator.

——————— **CAUTION** ———————

When draining the coolant, keep in mind that cats and dogs are attracted by the ethylene glycol antifreeze, and are quite likely to drink any that is left in an uncovered container or in puddles on the ground. This will prove fatal in sufficient quantity. Always drain the coolant into a sealable container. Coolant should be reused unless it is contaminated or several years old.

2. Remove the air cleaner. Disconnect the fuel pipe. Remove the carburetor or the throttle body, as required.
3. Disconnect the coolant hoses from the intake manifold.
4. Disconnect the throttle cable from the bellcrank.
5. Disconnect the PCV valve vacuum hose from the intake manifold.
6. If equipped, remove the vacuum advance CTO valve vacuum hoses.
7. Disconnect the system coolant temperature sender wire connector (located on the intake manifold). Disconnect the air temperature sensor wire, if equipped.
8. Disconnect the vacuum hose from the EGR valve.
9. On vehicles equipped with power steering remove the power steering pump and its mounting bracket. Do not detach the power steering pump hoses.
10. Disconnect the intake manifold electric heater wire connector, as required.
11. Disconnect the throttle valve linkage, if equipped with automatic transmission.
12. Disconnect the EGR valve tube from the intake manifold.
13. Remove the intake manifold attaching screws, nuts and clamps. Remove the intake manifold. Discard the gasket.
14. Clean the mating surfaces of the manifold and cylinder head.

To install:

NOTE: If the manifold is being replaced, ensure all fittings, etc., are transferred to the replacement manifold.

15. Install the intake manifold. Install the intake manifold attaching screws, nuts and clamps. Torque manifold fasteners to 23 ft. lbs.

3 ENGINE AND ENGINE OVERHAUL

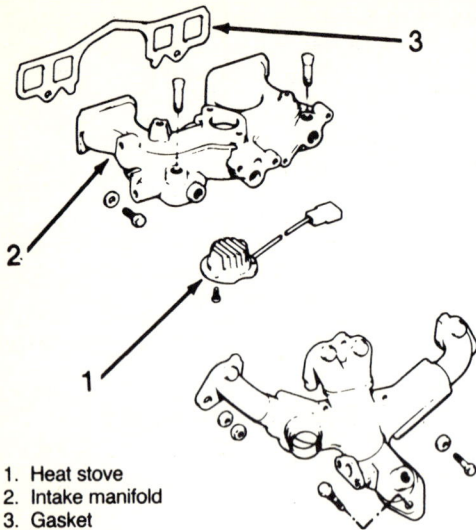

1. Heat stove
2. Intake manifold
3. Gasket

4–150 manifolds

16. Connect the EGR valve tube to the intake manifold.
17. Connect the throttle valve linkage, if equipped with automatic transmission.
18. Connect the intake manifold electric heater wire connector, as required.
19. On vehicles equipped with power steering install the power steering pump and its mounting bracket.
20. Connect the vacuum to from the EGR valve.
21. Connect the system coolant temperature sender wire connector (located on the intake manifold). Connect the air temperature sensor wire, if equipped.
22. If equipped, install the vacuum advance CTO valve vacuum hoses.
23. Install the carburetor or the throttle body. Torque the carburetor or throttle body mounting bolts to 14 ft. lbs.
24. Connect the PCV valve vacuum hose to the intake manifold.
25. Connect the throttle cable to the bellcrank.
26. Connect the fuel pipe.
27. Connect the coolant hoses to the intake manifold.
28. Install the air cleaner.
29. Connect the negative battery cable.
30. Fill the cooling system.

4–151

1. Remove the negative cable.
2. Remove the air cleaner and the PCV valve hose.

— CAUTION —
DO NOT remove the block drain plugs or loosen the radiator draincock the with system hot and under pressure because serious burns from the coolant can occur.

3. Drain the cooling system.

— CAUTION —
When draining the coolant, keep in mind that cats and dogs are attracted by the ethylene glycol antifreeze, and are quite likely to drink any that is left in an uncovered container or in puddles on the ground. This will prove fatal in sufficient quantity. Always drain the coolant into a sealable container. Coolant should be reused unless it is contaminated or several years old.

4. Tag and remove the vacuum hoses (ensure the distributor vacuum advance hose is removed).
5. Disconnect the fuel pipe and electrical wire connections from the carburetor.

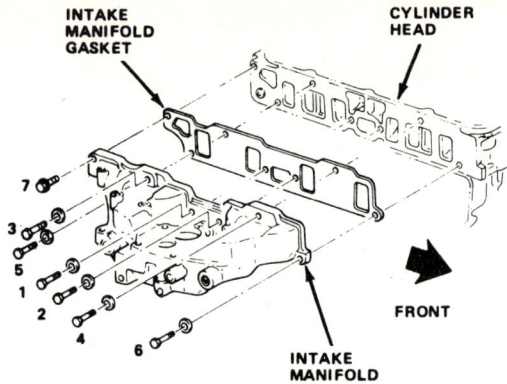

4–151 Intake manifold bolt tightening sequence

6. Disconnect the carburetor throttle linkage. Remove the carburetor and the carburetor spacer.
7. Remove the bellcrank and throttle linkage brackets and move them to one side for clearance.
8. Remove the heater hose at the intake manifold.
9. Remove the alternator. Note the position of spacers for installation.
10. Remove the manifold-to-cylinder head bolts and remove the manifold.
To install:
11. Position the replacement gasket and install the replacement manifold on the cylinder head. Start all bolts.
12. Tighten all bolts with 37 ft. lbs.
13. Connect the heater hose to the intake manifold.
14. Install the bellcrank and throttle linkage brackets.
15. Connect the carburetor throttle linkage to the brackets and bellcrank.
16. Install the carburetor spacer and tighten the bolts with 15 ft. lbs.
17. Install the carburetor and gasket. Tighten the nuts with 15 ft. lbs.
18. Install the fuel pipe and electrical wire connections. Install the vacuum hoses.
19. Install the battery negative cable.

— CAUTION —
Use extreme caution when the engine is operating! Do not stand in a direct line with the fan! Do not put your hands near pulleys, belts or fan! Do not wear loose clothing!

20. Refill the cooling system. Start the engine and inspect for leaks.
21. Install the air cleaner and the PCV valve hose.

6–225

1. Drain the cooling system.

— CAUTION —
When draining the coolant, keep in mind that cats and dogs are attracted by the ethylene glycol antifreeze, and are quite likely to drink any that is left in an uncovered container or in puddles on the ground. This will prove fatal in sufficient quantity. Always drain the coolant into a sealable container. Coolant should be reused unless it is contaminated or several years old.

2. Disconnect the crankcase vent hose, distributor vacuum hose, and the fuel line from the carburetor.
3. Disconnect the two distributor leads from the coil.
4. Disconnect the wire from the temperature sending unit.
5. Remove the ten cap bolts that hold the intake manifold to the cylinder head. They must be replaced in their original location.

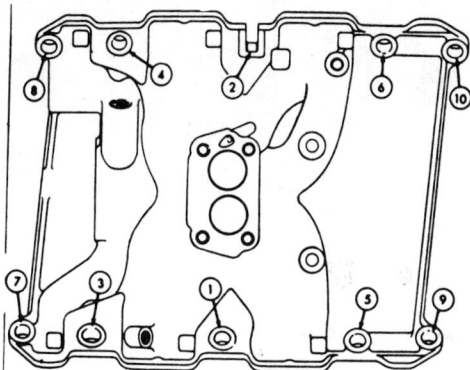

6–225 Intake manifold tightening sequence

6. Remove the intake manifold assembly and gasket from the engine.

7. Reverse the above procedure for installation. Tighten the bolts to the correct torque, and in the proper sequence.

6–232, 6–258

The intake manifold and exhaust manifold are mounted externally on the left side of the engine and are attached to the cylinder head. The intake and exhaust manifolds are removed as a unit. On some engines, an exhaust gas recirculation valve is mounted on the side of the intake manifold.

1. Remove the air cleaner and carburetor.
2. Disconnect the accelerator cable from the accelerator bellcrank.
3. Disconnect the PCS vacuum hose from the intake manifold.
4. Disconnect the distributor vacuum hose and electrical wires at the TCS solenoid vacuum valve.
5. Remove the TCS solenoid vacuum valve and bracket from the intake manifold. In some cases it might not be necessary to remove the TCS unit.
6. If so equipped, disconnect the EGR valve vacuum hoses.
7. Remove the power steering mounting bracket and pump and set it aside without disconnecting the hoses.
8. Remove the EGR valve, if so equipped.
9. Disconnect the exhaust pipe from the manifold flange. Disconnect the spark CTO hoses and remove the oxygen sensor.
10. Remove the manifold attaching bolts, nuts and clamps.
11. Separate the intake manifold and exhaust manifold from the engine as an assembly. Discard the gasket.
12. If either manifold is to be replaced, they should be separated at the heat riser area.

To install:

13. Clean the mating surfaces of the manifolds and the cylinder head before replacing the manifolds. Replace them in reverse order of the above procedure with a new gasket. Tighten the bolts and nuts to the specified torque in the proper sequence.
14. Connect the exhaust pipe to the manifold flange. Torque the nuts to 20 ft. lbs. Connect the spark CTO hoses and install the oxygen sensor.
15. Install the EGR valve, if so equipped.
16. Install the power steering mounting bracket and pump.
17. If so equipped, connect the EGR valve vacuum hoses.
18. Install the TCS solenoid vacuum valve and bracket to the intake manifold.
19. Connect the distributor vacuum hose and electrical wires at the TCS solenoid vacuum valve.
20. Connect the PCS vacuum hose to the intake manifold.
21. Connect the accelerator cable to the accelerator bellcrank.
22. Install the air cleaner and carburetor.

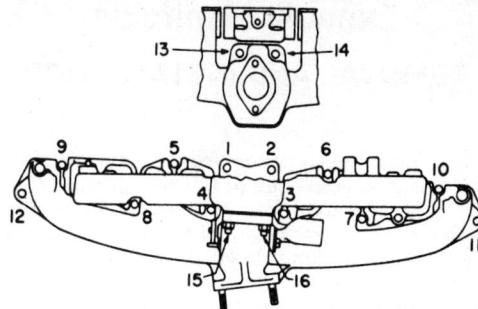

6–232, 6–258 Intake manifold tightening sequence

8–304

1. Drain the coolant from the radiator.

CAUTION

When draining the coolant, keep in mind that cats and dogs are attracted by the ethylene glycol antifreeze, and are quite likely to drink any that is left in an uncovered container or in puddles on the ground. This will prove fatal in sufficient quantity. Always drain the coolant into a sealable container. Coolant should be reused unless it is contaminated or several years old.

2. Remove the air cleaner assembly.
3. Disconnect the spark plug wires. Remove the spark plug wire brackets from the valve covers, and the bypass valve bracket.
4. Disconnect the upper radiator hose and the by-pass hose from the intake manifold. Disconnect the heater hose from the rear of the manifold.
5. Disconnect the ignition coil bracket and lay the coil aside.
6. Disconnect the TCS solenoid vacuum valve from the right side valve cover.
7. Disconnect all lines, hoses, linkages and wires from the carburetor and intake manifold and TCS components as required.
8. Disconnect the air delivery hoses at the air distribution manifolds.
9. Disconnect the air pump diverter valve and lay the valve and the bracket assembly, including the hoses, forward of the engine.
10. Remove the intake manifold after removing the cap bolts that hold it in place. Remove and discard the side gaskets and the end seals.

To install:

11. Clean the mating surfaces of the intake manifold and the cylinder head before replacing the intake manifold. Use new gaskets and tighten the cap bolts to the correct torque. Install in reverse order of the above procedure.

NOTE: There is no specified tighten sequence for this intake manifold. Start at the center bolts and work outward.

12. Connect the air pump diverter valve.
13. Connect the air delivery hoses at the air distribution manifolds.
14. Connect all lines, hoses, linkages and wires to the carburetor and intake manifold and TCS components as required.
15. Install the TCS solenoid vacuum valve to the right side valve cover.
16. Install the ignition coil bracket.
17. Connect the upper radiator hose and the by-pass hose to the intake manifold. Connect the heater hose to the rear of the manifold.
18. Connect the spark plug wires. Install the spark plug wire brackets on the valve covers, and the bypass valve bracket.
19. Install the air cleaner assembly.
20. Fill the cooling system.

Exhaust Manifolds

REMOVAL AND INSTALLATION

F4–134

1. Remove the air delivery hose from the air injection tube assembly if the engine is so equipped. If not, proceed to step two.
2. Remove the five nuts from the manifold studs.
3. Pull the manifold from the mounting studs. Be careful not to damage the air injection tubes if the engine is equipped with an air pump.
4. Remove the gaskets from the cylinder block.
5. If the exhaust manifold is to be replaced it will be necessary to remove the air injection tubes from the exhaust manifold. The application of heat may be necessary to aid removal.
6. Use new gaskets when replacing the exhaust manifold. Make sure that the cylinder head are clean. Tighten the attaching nuts to the correct torque specification.
7. If the exhaust manifold was be replaced, install the air injection tubes. The application of heat may be necessary to aid installation.
8. Install the air delivery hose on the air injection tube assembly if the engine is so equipped.

4–150

1. Remove the intake manifold.
2. Disconnect the EGR tube.
3. Disconnect the exhaust pipe at the manifold.
4. Disconnect the oxygen sensor wire.
5. Support the manifold and remove the nuts from the studs.
6. If a new manifold is being installed, transfer the oxygen sensor. Torque the sensor to 35 ft. lbs.
7. Thoroughly clean the gasket mating surfaces of the manifold and head.
8. Install the manifold, using a new gasket. Torque the nuts to 23 ft. lbs.
9. Connect the oxygen sensor wire.
10. Connect the exhaust pipe at the manifold.
11. Connect the EGR tube.
12. Install the intake manifold.

NOTE: On some 1984–85 4–150 engines, the manifold end studs can be bent or broken if the manifold is misaligned. To remedy this:

    a. Remove the manifold.
    b. Replace any bent or broken studs.
    c. Using a straightedge, check the flatness of the manifold mating surface. If a 0.015 in. (0.38mm) flat feeler gauge can be inserted between the straightedge and the manifold, at any point, replace the manifold.
    d. Modify the original or replacement manifold by grinding the mounting flanges as shown in the accompanying illustration.
    e. Install the manifold

4–151

1. Remove the air cleaner and heated air tube.
2. Remove the engine oil dipstick tube attaching bolt.
3. Remove the oxygen sensor, if equipped.
4. Raise the vehicle and disconnect the exhaust pipe from the manifold. Lower the vehicle.
5. Remove the exhaust manifold bolts and remove the manifold and gasket.
6. Install the replacement gasket and the exhaust manifold on the cylinder head. Tighten all bolts to 39 ft. lbs.
7. Install the dipstick tube attaching bolts.

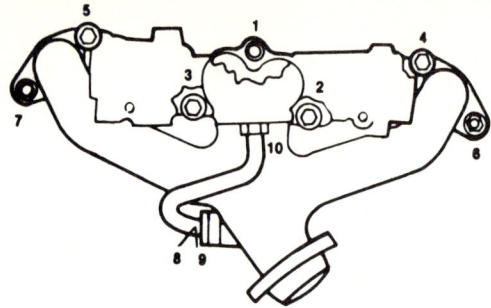

4–150 exhaust manifold torque sequence

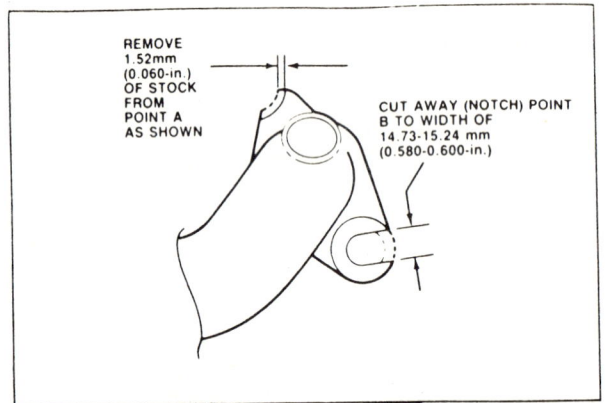

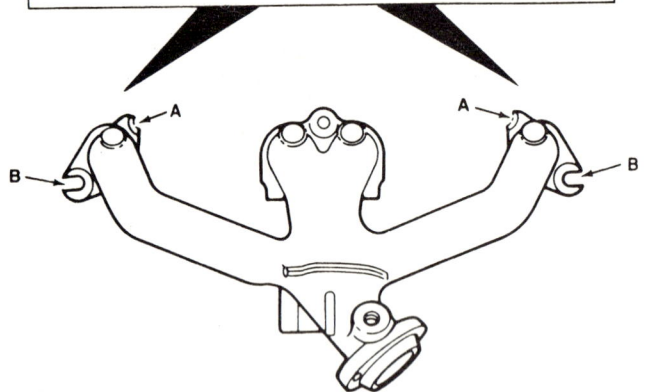

REMOVE 1.52mm (0.060-in.) OF STOCK FROM POINT A AS SHOWN

CUT AWAY (NOTCH) POINT B TO WIDTH OF 14.73-15.24 mm (0.580-0.600-in.)

Exhaust manifold modification on the 4–150

6–225

1. Remove the five attaching screws, one nut, and exhaust manifold from the side of the cylinder head.
2. Use a new gasket when replacing the exhaust manifolds. Make sure that the mating surfaces of the manifold and the cylinder head are clean. Tighten the manifold nuts and bolts to the correct torque.

6–232, 6–258

The intake and exhaust manifolds of the 232 and 258 cu. in. Sixes must be removed together. See the procedure for removing and installing the intake manifold.

8–304

1. Disconnect the spark plug wires.
2. Disconnect the air delivery hose at the distribution manifold.
3. Remove the air distribution manifold and the injection tubes.

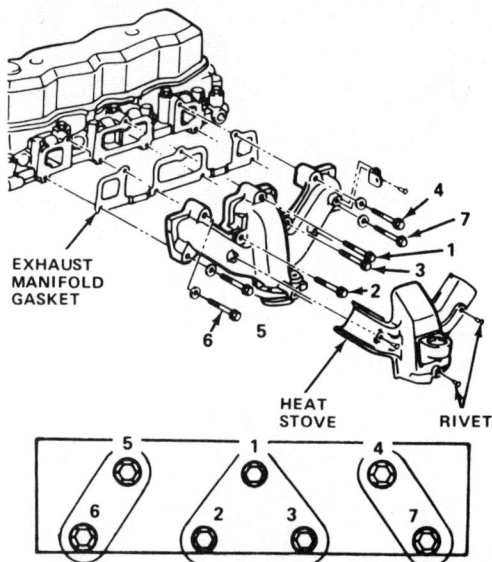

4-151 exhaust manifold bolt tightening sequence

6-225 exhaust manifold bolts

4. Disconnect the exhaust pipe at the manifold.

5. Remove the exhaust manifold attaching bolts and washers along with the spark plug shields.

6. Separate the exhaust manifold from the cylinder head.

7. Install in reverse order of the above procedure. Clean the mating surfaces and tighten the attaching bolts to the correct torque.

Air Conditioning Compressor

REMOVAL AND INSTALLATION

1987-90 4-150

1. Isolate the compressor. See Section 1.
2. Disconnect the battery ground.
3. Remove the discharge and suction hoses from the compressor and cap all openings immediately.
4. Loosen the alternator and remove the drive belt.
5. Remove the alternator from its brackets and set it out of the way.
6. Unbolt and remove the compressor.
7. For installation, install the compressor and alternator, connect the hoses and install and tension the drive belts. Connect the compressor clutch wire. Open the valves.

6-258

1977-80 ALL
1981, EXCEPT CALIF.

1. Isolate the compressor. See Section 1.

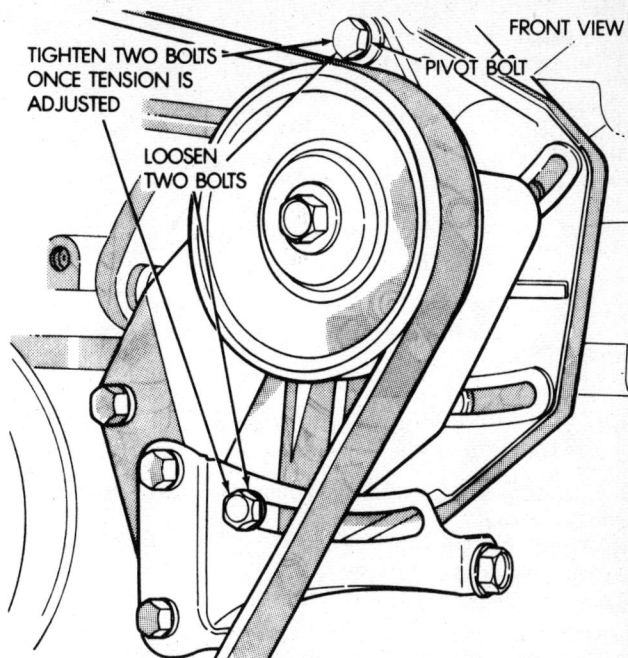

4-150 compressor mounting, front view

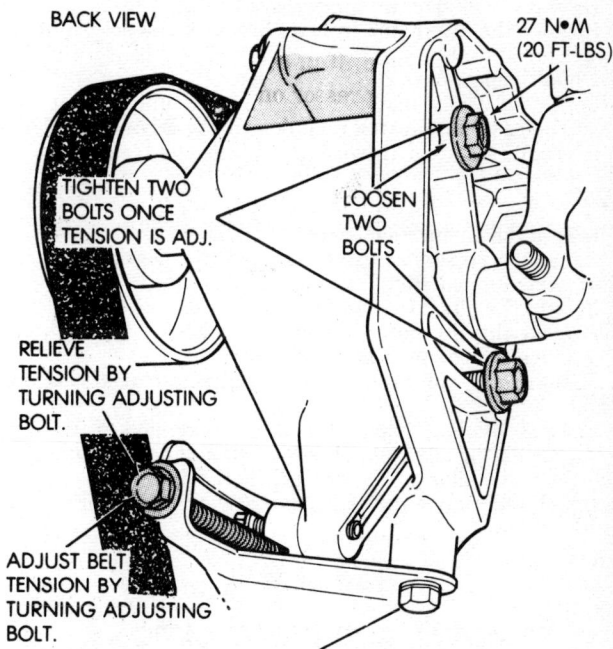

4-150 compressor mounting, rear view

2. Remove both service valves and cap the valves and compressor ports immediately.
3. Remove the alternator belt and adjusting bolt.
4. Remove the upper alternator mounting bolt and loosen the lower mounting bolt.
5. Remove the idler pulley.
6. Disconnect the compressor clutch wire. Remove the compressor mounting nuts and lift out the compressor. BE CAREFUL; IT'S HEAVY!
7. For installation, install the compressor, alternator and belt, idler pulley, and adjust the drive belts.

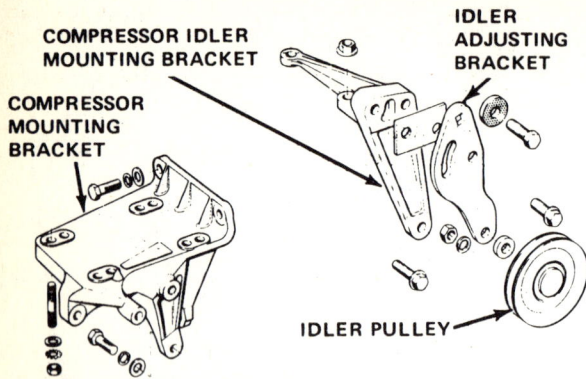

Compressor mounting on the 6–258 for 1977–80 all, and 1981, except Calif

8. Install the service valves and purge the compressor of air. Open the valves.
9. Connect the clutch wire.

1981 CALIFORNIA MODELS
1982–90 ALL MODELS

1. Isolate the compressor. See Section 1.
2. Disconnect the battery ground.
3. Remove the discharge and suction hoses from the compressor and cap all openings immediately.
4. Loosen the alternator and remove the drive belts.
5. Remove the alternator from its brackets and set it out of the way.
6. Unbolt and remove the compressor.
7. For installation, install the compressor and alternator, connect the hoses and install and tension the drive belts. Connect the compressor clutch wire. Open the valves.

8–304

1. Isolate the compressor. See Section 1.
2. Remove both service valves and cap the valves and compressor ports immediately.
3. Loosen the alternator and remove the drive belts.
4. Remove the alternator mounting bracket.
5. Remove the compressor clutch wire.
6. Remove the compressor and mounting bracket as an assembly.
7. For installation, install the compressor and mounting bracket, install the alternator bracket, connect the clutch wire, install the belts, idler pulley, and adjust the drive belts.
8. Install the service valves and purge the compressor of air. Open the valves.

Radiator

REMOVAL AND INSTALLATION

1. Drain the radiator by opening the drain cock and removing the radiator pressure cap.

――――――――― **CAUTION** ―――――――――
When draining the coolant, keep in mind that cats and dogs are attracted by the ethylene glycol antifreeze, and are quite likely to drink any that is left in an uncovered container or in puddles on the ground. This will prove fatal in sufficient quantity. Always drain the coolant into a sealable container. Coolant should be reused unless it is contaminated or several years old.
―――――――――――――――――――――――――――――――

2. Remove the upper and lower hose clamps and hoses at the radiator.
3. Disconnect the automatic transmission oil cooler lines at

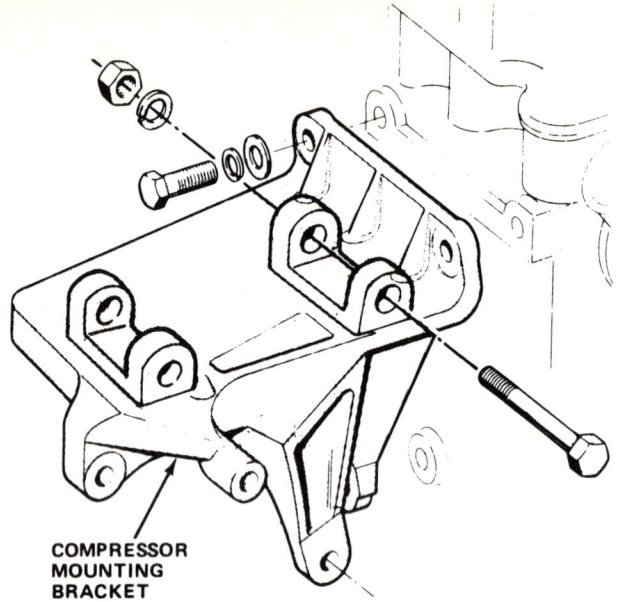

Compressor mounting used on the 1981 California 6–258 and all 1982–90 models

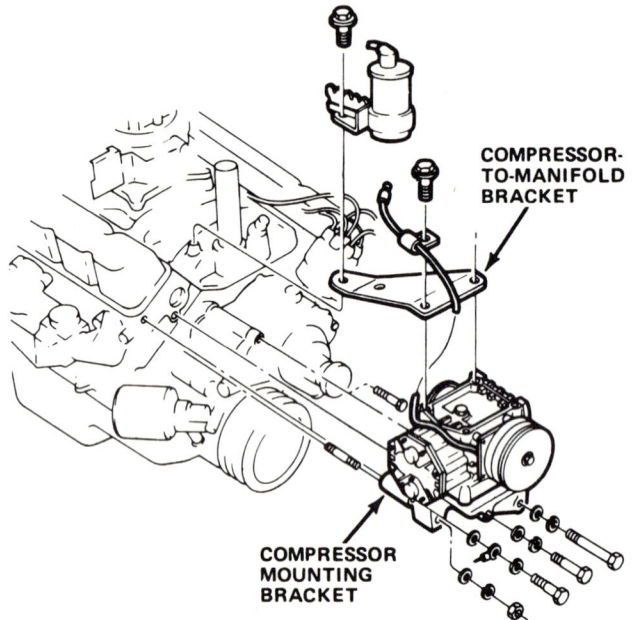

8–304 compressor mounting

the radiator, if so equipped. Remove the radiator shroud from the radiator, if so equipped.
4. Remove all attaching screws that secure the radiator to the radiator body support.
5. Remove the radiator.
6. Replace in reverse order of the above procedure.

Water Pump

REMOVAL AND INSTALLATION

4–134

1. Drain the cooling system.

prove fatal in sufficient quantity. Always drain the coolant into a sealable container. Coolant should be reused unless it is contaminated or several years old.

2. Disconnect the hoses at the pump.
3. Remove the drive belts.
4. Remove the power steering pump bracket.
5. Remove the fan and shroud.
6. Unbolt and remove the pump.

To install:

7. Clean the mating surfaces thoroughly.
8. Using a new gasket, install the pump and torque the bolts to 13 ft. lbs.
9. Install the fan and shroud.
10. Install the power steering pump bracket.
11. Install and adjust the drive belts.
12. Connect the hoses at the pump.
13. Fill the cooling system.

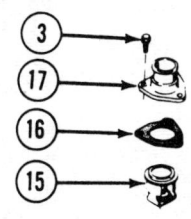

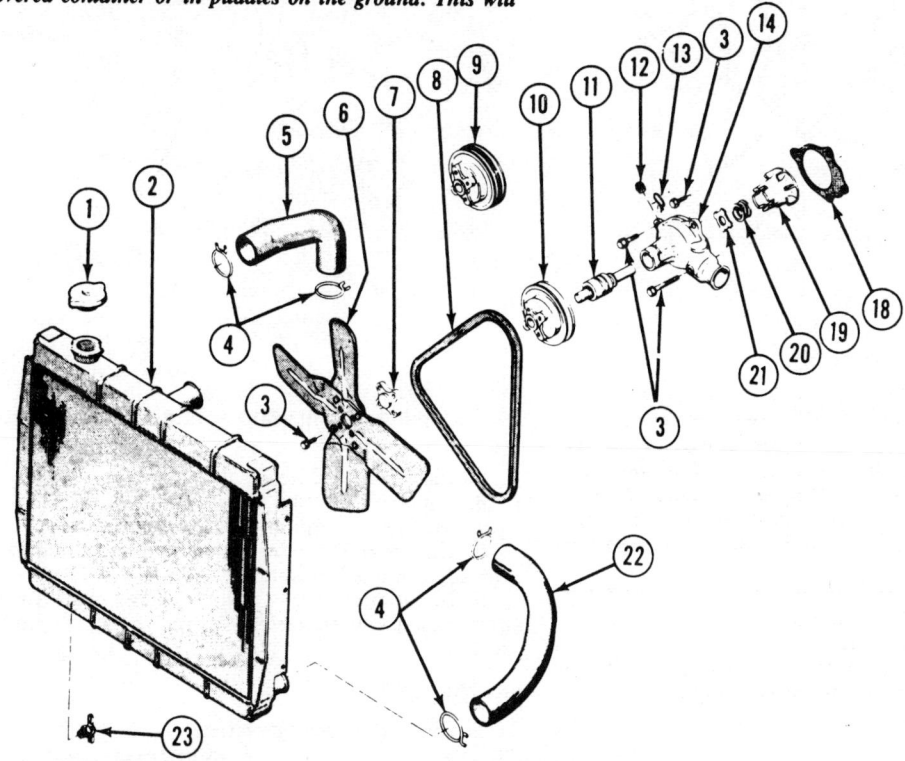

CAUTION

When draining the coolant, keep in mind that cats and dogs are attracted by the ethylene glycol antifreeze, and are quite likely to drink any that is left in an uncovered container or in puddles on the ground. This will prove fatal in sufficient quantity. Always drain the coolant into a sealable container. Coolant should be reused unless it is contaminated or several years old.

2. Disconnect the hoses at the pump.
3. Remove the fan belt.
4. Unbolt the fan and hub assembly.
5. Unbolt and remove the pump.
6. Installation is the reverse of removal. Torque the pump bolts to 17 ft. lbs. Always use a new gasket coated with sealer.

4–150

NOTE: **Some 4–150 engines with air conditioning are equipped with a serpentine drive belt and have a reverse rotating water pump coupled with a viscous fan drive assembly. The components are identified by the words REVERSE stamped on the cover of the viscous drive and on the inner side of the fan. The word REV is also cast into the body of the water pump.**

1. Drain the cooling system.

CAUTION

When draining the coolant, keep in mind that cats and dogs are attracted by the ethylene glycol antifreeze, and are quite likely to drink any that is left in an uncovered container or in puddles on the ground. This will

1. Radiator pressure cap	9. Pulley (double groove)	17. Water outlet fitting
2. Radiator	10. Pulley (single groove)	18. Gasket
3. Bolt	11. Bearing and shaft	19. Impeller
4. Hose clamp	12. Pipe plug	20. Pump seal
5. Upper hose	13. Bearing retainer spring	21. Seal washer
6. Fan	14. Pump body	22. Lower hose
7. Fan spacer	15. Thermostat	23. Drain cock
8. Fan and alternator belt	16. Gasket	

4–134 cooling system components

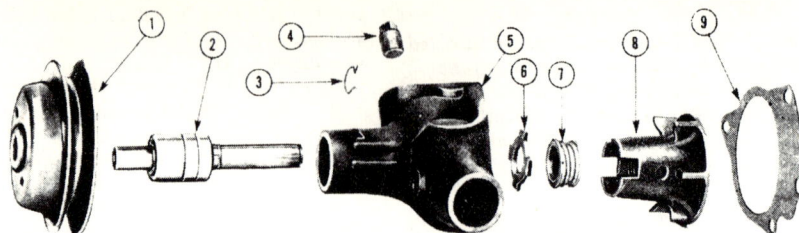

1. Fan and pump pulley
2. Bearing and shaft
3. Bearing retainer spring
4. Pipe plug
5. Pump body
6. Seal washer
7. Pump seal
8. Impeller
9. Gasket

4–134 water pump

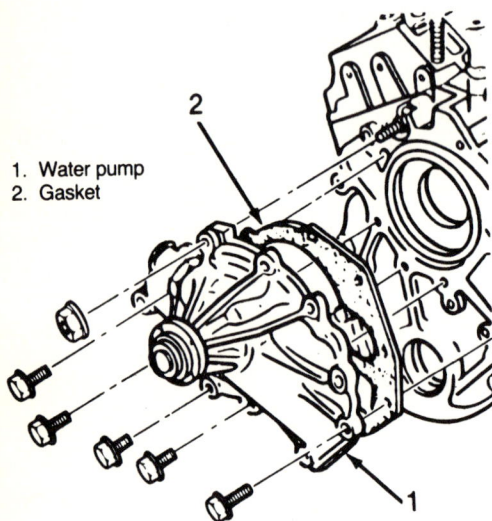

1. Water pump
2. Gasket

4–150 water pump

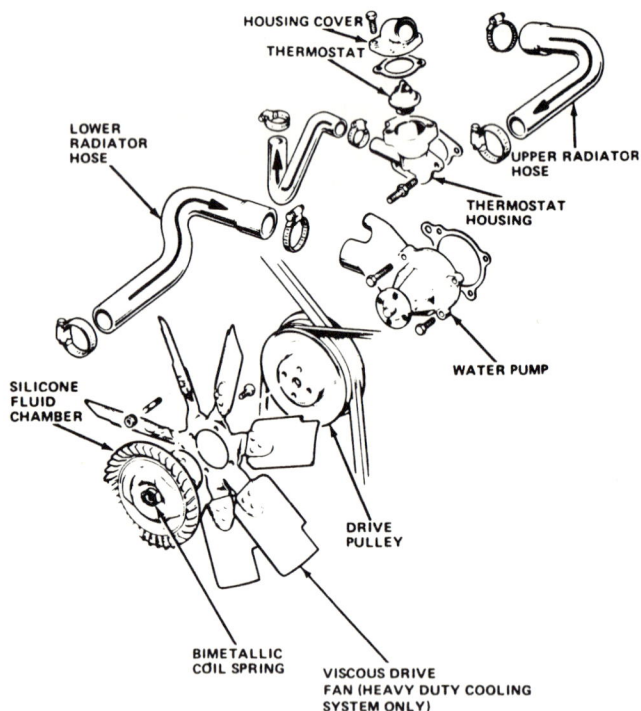

4–151 water pump and related components

4–151

1. Remove the fan belt.
2. Remove the fan and hub assembly.
3. Drain the cooling system.

─────── CAUTION ───────

When draining the coolant, keep in mind that cats and dogs are attracted by the ethylene glycol antifreeze, and are quite likely to drink any that is left in an uncovered container or in puddles on the ground. This will prove fatal in sufficient quantity. Always drain the coolant into a sealable container. Coolant should be reused unless it is contaminated or several years old.

4. Disconnect the hoses at the pump.
5. Unbolt and remove the pump.
6. Installation is the reverse of removal. Always use a new gasket coated with sealer. Torque the water pump bolts to 25 ft. lbs. Tighten the fan and hub bolts to 18 ft. lbs.

6–225

1. Drain the cooling system.

─────── CAUTION ───────

When draining the coolant, keep in mind that cats and dogs are attracted by the ethylene glycol antifreeze, and are quite likely to drink any that is left in an uncovered container or in puddles on the ground. This will prove fatal in sufficient quantity. Always drain the coolant into a sealable container. Coolant should be reused unless it is contaminated or several years old.

2. Disconnect all hoses at the pump.
3. Remove the drive belts.
4. Remove the fan and hub assembly.
5. Unbolt and remove the water pump along with the alternator adjustment bracket.
6. Installation is the reverse of removal. Always use a new gasket coated with sealer. Torque the water pump bolts to 6–8 ft. lbs.

6–232, 258

1. Drain the cooling system.

─────── CAUTION ───────

When draining the coolant, keep in mind that cats and dogs are attracted by the ethylene glycol antifreeze, and are quite likely to drink any that is left in an uncovered container or in puddles on the ground. This will prove fatal in sufficient quantity. Always drain the coolant into a sealable container. Coolant should be reused unless it is contaminated or several years old.

2. Disconnect all hoses at the pump.
3. Remove the drive belts.

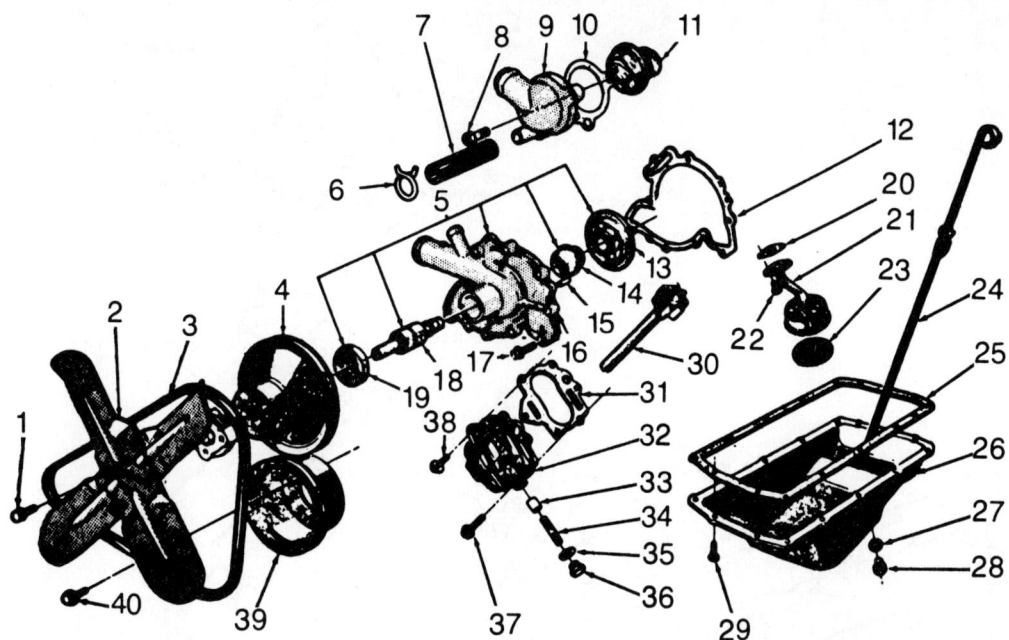

1. Bolt and lock washer
2. Fan assembly
3. Fan and alternator belt
4. Fan driven pulley
5. Water pump assembly
6. Hose clamp
7. Thermostat by-pass hose
8. Hex head bolt
9. Water outlet elbow
10. Water outlet elbow gasket
11. Thermostat
12. Water pump gasket
13. Impeller and insert, water pump
14. Water pump seal

15. Dowel pin
16. Water pump cover
17. Bolt
18. Water pump shaft and bearing
19. Fan hub
20. Oil suction pipe gasket
21. Oil suction housing, pipe and flange
22. Bolt
23. Oil pump screen
24. Oil dipstick
25. Oil pan gasket
26. Oil pan assembly
27. Drain plug gasket
28. Drain plug

29. Screw and lockwasher
30. Oil pump shaft and gear
31. Oil pump cover gasket
32. Valve by-pass and cover assembly
33. Oil pressure valve
34. Valve by-pass spring
35. Oil pressure valve cap gasket
36. Oil pressure valve cap
37. Screw
38. Screw
39. Fan driving pulley
40. Hex head bolt

6–225 cooling system components

4. Remove the fan shroud attaching screws.
5. Unbolt the fan and fan drive assembly and remove along with the shroud. On some models it may be easier to turn the shroud ½ turn.
6. Unbolt and remove the pump.

NOTE: Engines built for sale in California having a single, serpentine drive belt and viscous fan drive, have a reverse rotating pump and drive. These components are identified by the word REVERSE stamped on the drive cover and inner side of the fan, and REV cast into the water pump body. Never interchange standard rotating parts with these.

7. Installation is the reverse of removal. Always use a new gasket coated with sealer. Torque the water pump bolts to 13 ft. lbs.; the fan bolts to 18 ft. lbs.

8–304

1. Drain the cooling system.

─────────────── **CAUTION** ───────────────

When draining the coolant, keep in mind that cats and dogs are attracted by the ethylene glycol antifreeze, and are quite likely to drink any that is left in an uncovered container or in puddles on the ground. This will prove fatal in sufficient quantity. Always drain the coolant into a sealable container. Coolant should be reused unless it is contaminated or several years old.

2. Disconnect all hoses at the pump.
3. Loosen all drive belts.
4. Remove the shroud, but reinsert one bolt to hold the radiator.
5. Remove the fan and hub.
6. If the vehicle is equipped with A/C install a double nut on the compressor bracket-to-water pump stud and remove the stud.
7. Remove, but do not disconnect the alternator and bracket.
8. If so equipped, remove the nuts that attach the power steering pump to the rear half of the pump bracket.
9. Remove the two bolts that attach the front half to the rear half of the bracket.
10. Remove the remaining upper screw from the inner air pump support bracket, loosen the lower bolt and drop the bracket away from the power steering front bracket.
11. Remove the front half of the power steering bracket from the water pump mounting stud.
12. Unbolt and remove the water pump.

To install:
13. Install the pump. Always use a new gasket coated with sealer. Torque the pump-to-timing case bolts to 48 inch lbs. and the pump-to-block bolts to 25 ft. lbs.
14. Install the front half of the power steering bracket to the water pump mounting stud. Torque the power steering pulley nut to 60 ft. lbs.

UPPER RADIATOR HOSE

THERMOSTAT

HOSE (FROM HEATER)

THERMOSTAT HOUSING

DRIVE PULLEY

WATER PUMP

LOWER RADIATOR HOSE

RADIATOR

SILICONE FLUID CHAMPER

BIMETALLIC COIL SPRINGS

VISCOUS DRIVE FAN (HEAVY DUTY COOLING SYSTEM ONLY)

SHROUD

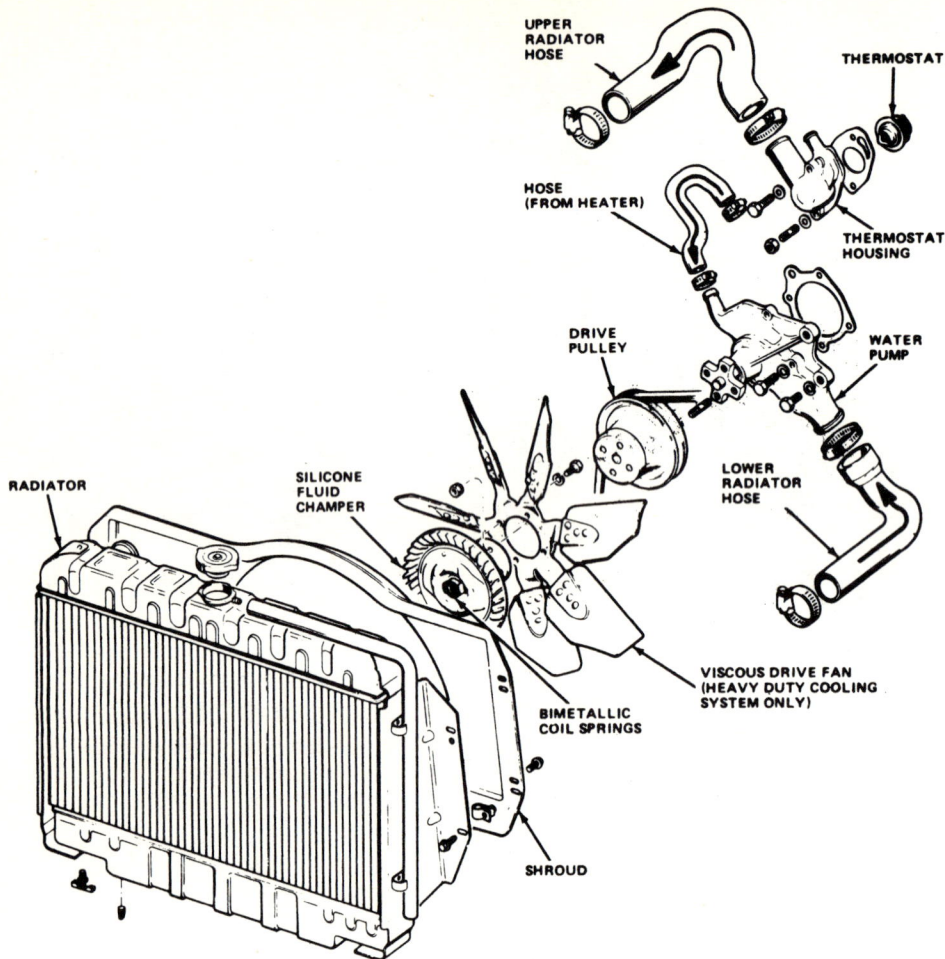

6–232, 6–258 cooling system components

15. Install the upper screw from the inner air pump support bracket, and tighten the lower bolt.

16. If so equipped, install the nuts that attach the power steering pump to the rear half of the pump bracket.

17. Install the two bolts that attach the front half to the rear half of the bracket.

18. Install the alternator and bracket.

19. If the vehicle is equipped with A/C install a double nut on the compressor bracket-to-water pump stud and install the stud.

20. Install the fan and hub.

21. Install the shroud.

22. Adjust all drive belts.

23. Connect all hoses at the pump.

24. Fill the cooling system.

Cylinder Head

REMOVAL AND INSTALLATION

NOTE: It is important to note that each engine has its own head bolt tightening sequence and torque. Incorrect tightening procedure may cause head warpage and compression loss. Correct sequence and torque for each engine model is shown in this section.

4–134

1. Drain the coolant.

——— CAUTION ———

When draining the coolant, keep in mind that cats and dogs are attracted by the ethylene glycol antifreeze, and are quite likely to drink any that is left in an uncovered container or in puddles on the ground. This will prove fatal in sufficient quantity. Always drain the coolant into a sealable container. Coolant should be reused unless it is contaminated or several years old.

2. Remove the upper radiator hose.

3. Remove the carburetor.

4. Remove the rocker arm cover.

5. Remove the rocker arm attaching stud nuts and rocker arm shaft assembly.

6. Remove the cylinder head bolts. One of the bolts is located below the carburetor mounting, inside the intake manifold.

7. Lift off the cylinder head.

8. Thoroughly clean the gasket mating surfaces. Remove all traces of old gasket material. Remove all carbon deposits from the combustion chambers. Lay a straightedge across the head and check for flatness. Total deviation should not exceed 0.001 in. (0.025mm).

To install:

9. Position the head on the block.

10. Tighten the head bolts first to 40 ft. lbs. then to the specified torque in the correct sequence.

11. Install the rocker arm shaft assembly.

12. Install the rocker arm cover.

13. Install the carburetor.

1. Radiator upper hose
2. Thermostat housing cover
3. Gasket
4. Thermostat
5. Gasket
6. Drive pulley
7. Viscous drive fan
8. Coolant recovery bottle
9. Radiator
10. Shroud
11. Bimetallic coil spring
12. Silicone fluid chamber
13. Stud
14. Hose (from heater)
15. Water pump
16. Bypass hose

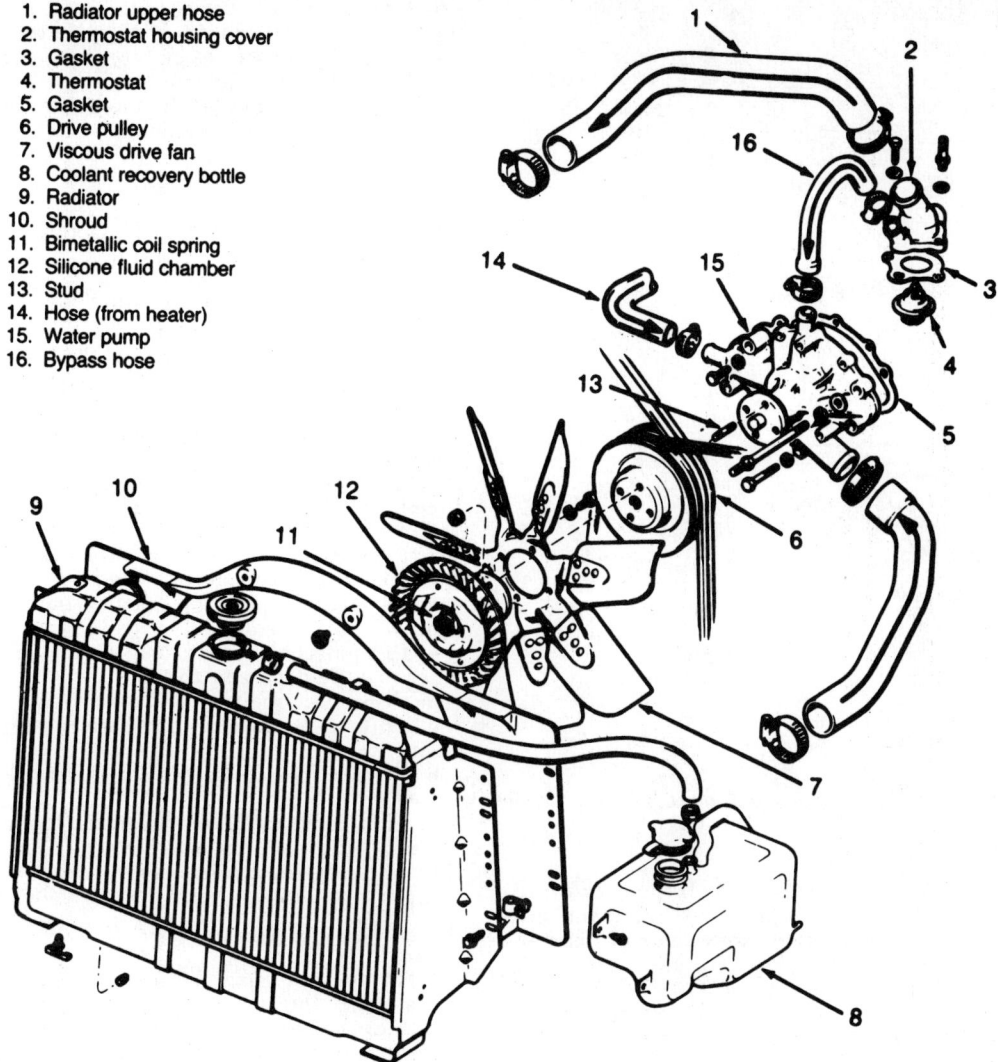

8–304 cooling system components

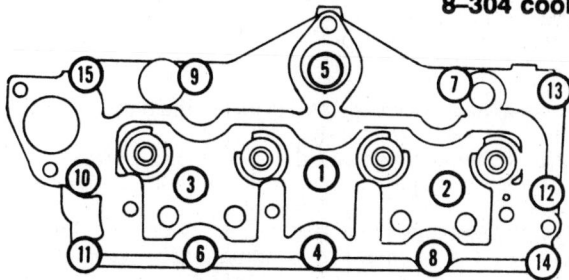

F4–134 cylinder head torque sequence

14. Install the upper radiator hose.
15. Fill the coolaning system.

4–150

1. Disconnect the battery ground.
2. Drain the cooling system.

────────── **CAUTION** ──────────

When draining the coolant, keep in mind that cats and dogs are attracted by the ethylene glycol antifreeze, and are quite likely to drink any that is left in an uncovered container or in puddles on the ground. This will

prove fatal in sufficient quantity. Always drain the coolant into a sealable container. Coolant should be reused unless it is contaminated or several years old.

3. Disconnect the hoses at the thermostat housing.
4. Remove the air cleaner.
5. Remove the rocker arm cover. The cover seal is RTV sealer. Break the seal with a clean putty knife or razor blade. Don't attempt to remove the cover until the seal is broken. To remove the cover, pry where indicated at the bolt holes.
6. Remove the rocker arms. Keep them in order!
7. Remove the pushrods. Keep them in order!
8. Remove the power steering pump bracket.
9. Suspend the pump out of the way.
10. Remove the intake and exhaust manifolds.
11. Remove the air conditioning compressor drive belt.
12. Loosen the alternator drive belt.
13. Remove the compressor/alternator bracket mounting bolt.
14. Unbolt the compressor and suspend it out of the way. DO NOT DISCONNECT THE REFRIGERANT LINES!
15. Remove the spark plugs.
16. Disconnect the temperature sending unit wire.
17. Remove the head bolts.

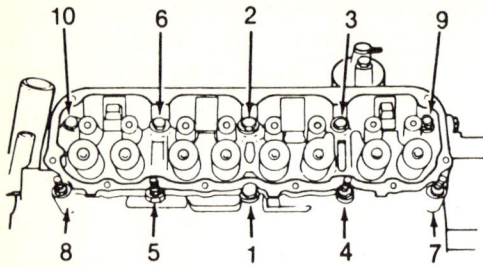

4–150 head bolt tightening sequence

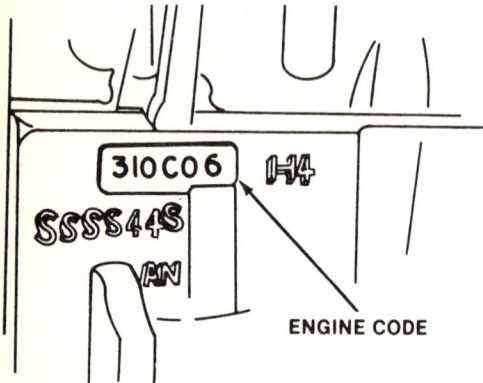

Engine code location

18. Lift the head off the engine and place it on a clean workbench.

19. Remove the head gasket.

CHILTON TIP: Some head bolts used on the spark plug side of the 1984 4–150 were improperly hardened and may break under the head during service or at head installation while torquing the bolts. Engines with the defective bolts are serial numbers 310U06 through 310U14. Whenever a broken bolts is found, replace all bolts on the spark plug side of the head with bolt #400 6593.

To install:
20. Thoroughly clean the gasket mating surfaces. Remove all traces of old gasket material. Remove all carbon deposits from the combustion chambers. Lay a straightedge across the head and check for flatness. Total deviation should not exceed 0.001 in. (0.025mm).

21. Do not apply sealant to the head or block. Coat both sides of the gasket with sealer. The gasket should be stamped **TOP** for installation. Place the gasket on the block.

22. Install the head on the block.

23. Coat the bolt labeled 8 in the torque sequence illustration, with Permatex #2, or equivalent. Install all the bolts. Tighten the bolts in three equal steps, in the sequence shown, to 85 ft. lbs. Torque #8 to 75 ft. lbs.

24. Connect the temperature sending unit wire.
25. Install the spark plugs.
26. Install the compressor.
27. Install the compressor/alternator bracket mounting bolt.
28. Adjust the alternator drive belt.
29. Install the air conditioning compressor drive belt.
30. Install the intake and exhaust manifolds.
31. Install the power steering pump bracket.
32. Install the pushrods. Keep them in order!
33. Install the rocker arms. Keep them in order!
34. Thoroughly clean the mating surfaces of the head and rocker cover. Run a ⅛ in. (3mm) bead of RTV sealer along the

length of the sealing surface of the head. Position the cover on the head within 10 minutes of applying the sealer. Torque the cover bolts, in a crisscross pattern, to 55 inch lbs.

35. Install the air cleaner.
36. Connect the hoses at the thermostat housing.
37. Connect the battery ground.
38. Fill the cooling system.

4–151

NOTE: The 4–151 rocker cover is sealed with RTV silicone gasket material. Do not use a conventional gasket.

1. Drain the cooling system and disconnect the hoses at the thermostat housing.

— CAUTION —
When draining the coolant, keep in mind that cats and dogs are attracted by the ethylene glycol antifreeze, and are quite likely to drink any that is left in an uncovered container or in puddles on the ground. This will prove fatal in sufficient quantity. Always drain the coolant into a sealable container. Coolant should be reused unless it is contaminated or several years old.

2. Remove the cylinder head cover (valve cover), the gasket, the rocker arm assembly, and the pushrods.
3. Remove the intake and exhaust manifold from the cylinder head.
4. Disconnect the spark plug wires and remove the spark plugs to avoid damaging them.
5. Disconnect the temperature sending unit wire, ignition coil and bracket assembly and battery ground cable from the engine.
6. Remove the cylinder head bolts, the cylinder head and gasket from the block.
7. Thoroughly clean the gasket mating surfaces. Remove all traces of old gasket material. Remove all carbon deposits from the combustion chambers. Lay a straightedge across the head and check for flatness. Total deviation should not exceed 0.001 in. (0.025mm).

To install:
8. Thoroughly clean the mating surfaces of the head and rocker cover. Run a ⅛ in. (3mm) bead of RTV sealer along the length of the sealing surface of the head, inboard of the bolt holes. Position the cover on the head within 10 minutes of applying the sealer. Torque the cover bolts, in a crisscross pattern, to 55 inch lbs.

9. Do not apply sealant to the head or block. Coat both sides of the gasket with sealer. The gasket should be stamped TOP for installation. Place the gasket on the block.

10. Install the head on the block. Insert the bolts and tighten them, in sequence, to the proper torque.
11. Install the pushrods and the rocker arm assembly.
12. Install the intake and exhaust manifold on the cylinder head.
13. Install the spark plugs and connect the spark plug wires.
14. Connect the temperature sending unit wire, ignition coil and bracket assembly and battery ground cable at the engine.
15. Connect the hoses at the thermostat housing.
16. Fill the cooling system.

6–225

1. Drain the cooling system.

— CAUTION —
When draining the coolant, keep in mind that cats and dogs are attracted by the ethylene glycol antifreeze, and are quite likely to drink any that is left in an uncovered container or in puddles on the ground. This will prove fatal in sufficient quantity. Always drain the coolant into a sealable container. Coolant should be reused unless it is contaminated or several years old.

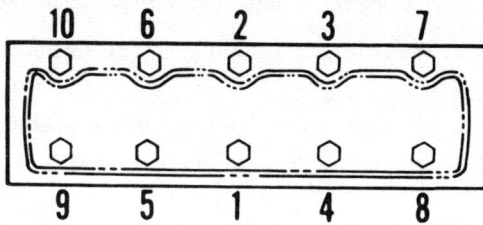

4–151 cylinder head torque sequence

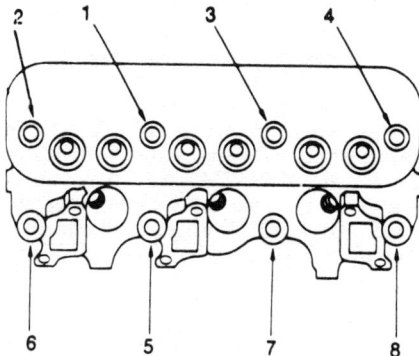

V6–225 cylinder head torque sequence

2. Remove the intake manifold.
3. Remove the rocker cover.
4. Remove the exhaust pipes at the flanges.
5. Remove the alternator in order to remove the right head.
6. Remove the dipstick and power steering pump, if so equipped, in order to remove the left head.
7. Remove the valve cover and the rocker assemblies. Mark these parts so that they can be reinstalled in exactly the same positions.
8. Unbolt the head bolts and lift off the cylinder head(s). It is very important that the inside of the engine be protected from dirt. The hydraulic lifters are particularly susceptible to being damaged by dirt.
9. Thoroughly clean the gasket mating surfaces. Remove all traces of old gasket material. Remove all carbon deposits from the combustion chambers. Lay a straightedge across the head and check for flatness. Total deviation should not exceed 0.001 in. (0.025mm).
To install:
10. Do not apply sealant to the head or block. Coat both sides of the gasket with sealer. The gasket should be stamped TOP for installation. Place the gasket on the block.
11. Install the head on the block. Insert the bolts and tighten them, in sequence, to the 65–85 ft. lbs., in three progressive passes.
12. Install the rocker assemblies and the valve covers. Torque the rocker arm assemblies to 25–35 ft. lbs.; the valve covers to 36–60 in. lbs.
13. Install the dipstick and power steering pump, if so equipped.
14. Install the alternator. Torque the bracket-to-head bolts to 30–40 ft. lbs.; the bracket to water pump bolts to 18–25 ft. lbs.
15. Install the exhaust pipes at the flanges. Torque the nuts to 20 ft. lbs.
16. Install the intake manifold. Torque the bolts to 45–55 ft. lbs.
17. Fill the cooling system.

6–232, 6–258

1. Drain the cooling system and disconnect the hoses at the thermostat housing.

2. Remove the cylinder head cover (valve cover), the gasket, the rocker arm assembly, and the pushrods.

NOTE: The pushrods must be replaced in their original positions.

3. Remove the intake and exhaust manifold from the cylinder head.
4. Disconnect the spark plug wires and the spark plugs to avoid damaging them.
5. Disconnect the temperature sending unit wire, ignition coil and bracket assembly from the engine.
6. Remove the cylinder head bolts, the cylinder head and gasket from the block.
7. Thoroughly clean the gasket mating surfaces. Remove all traces of old gasket material. Remove all carbon deposits from the combustion chambers. Lay a straightedge across the head and check for flatness. Total deviation should not exceed 0.001 in. (0.025mm).
To install:
8. Do not apply sealant to the head or block. Coat both sides of the gasket with sealer. The gasket should be stamped TOP for installation. Place the gasket on the block.
9. Install the head on the block. Insert the bolts and tighten them, in sequence, to the proper torque, in three progressive passes. Torque all bolts to 85 ft. lbs., except #11 in the torque sequence, which should be coated with sealer and torqued to 75 ft. lbs.

CHILTON TIP: Some head bolts used on the spark plug side of the 1984 6–258 were improperly hardened and may break under the head during service or at head installation while torquing the bolts. Engines with the defective bolts are serial numbers 310C06 through 310C14. Whenever a broken bolts is found, replace all seven bolts on the spark plug side of the head with bolt #400 6593.

10. Connect the temperature sending unit wire, ignition coil and bracket assembly at the engine.
11. Install the spark plugs and connect the spark plug wires.
12. Install the intake and exhaust manifold on the cylinder head.
13. Install the the rocker arm assembly, and the pushrods.

NOTE: The pushrods must be replaced in their original positions.

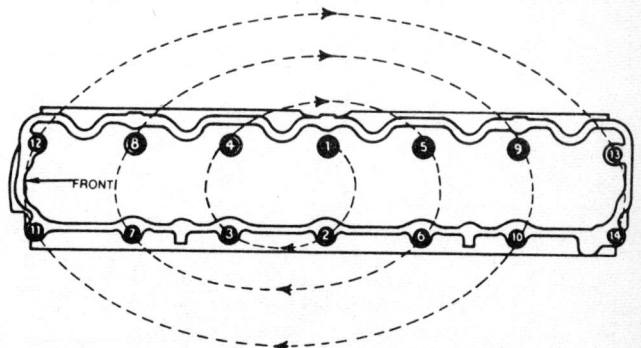

6–232, 6–258 head torque sequence

14. Install the cylinder head cover (valve cover) and the gasket.
15. Connect the hoses at the thermostat housing.
16. Fill the cooling system

8–304

1. Drain the cooling system and cylinder block.

──────────────── CAUTION ────────────────

When draining the coolant, keep in mind that cats and dogs are attracted by the ethylene glycol antifreeze, and are quite likely to drink any that is left in an uncovered container or in puddles on the ground. This will prove fatal in sufficient quantity. Always drain the coolant into a sealable container. Coolant should be reused unless it is contaminated or several years old.

──

2. When removing the right cylinder head, it may be necessary to remove the heater core housing from the firewall.
3. Remove the valve cover(s) and gasket(s).
4. Remove the rocker arm assemblies and the pushrods.

NOTE: The valve train components must be replaced in their original positions.

5. Remove the spark plugs to avoid damaging them.
6. Remove the intake manifold with the carburetor still attached.
7. Remove the exhaust pipes at the flange of the exhaust manifold. When replacing the exhaust pipes it is advisable to install new gaskets at the flange.
8. Loosen all of the drive belts.
9. Disconnect the battery ground cable and alternator bracket from the right cylinder head.
10. Disconnect the air pump and power steering pump brackets from the left cylinder head.
11. Remove the cylinder head bolts and lift the head(s) from the cylinder block.
12. Remove the cylinder head gasket from the head or the block.
13. Thoroughly clean the gasket mating surfaces. Remove all traces of old gasket material. Remove all carbon deposits from the combustion chambers. Lay a straightedge across the head and check for flatness. Total deviation should not exceed 0.001 in. (0.025mm).
To install:
14. Do not apply sealant to the head or block. Coat both sides of the gasket with sealer. The gasket should be stamped TOP for installation. Place the gasket on the block.
15. Install the head on the block. Insert the bolts and tighten them, in sequence, to the proper torque, in three progressive passes. Torque all bolts to 80 ft. lbs.
16. Connect the air pump and power steering pump brackets to the left cylinder head.
17. Connect the battery ground cable and alternator bracket to the right cylinder head.
18. Adjust all of the drive belts.
19. Install the exhaust pipes at the flange of the exhaust mani-

fold. When replacing the exhaust pipes it is advisable to install new gaskets at the flange.
20. Install the intake manifold with the carburetor still attached.
21. Install the spark plugs.
22. Install the rocker arm assemblies and the pushrods.

NOTE: The valve train components must be replaced in their original positions.

23. Install the valve cover(s) and gasket(s).
24. Install the heater core housing on the firewall.
25. Fill the cooling system and cylinder block.

Valves and Springs

NOTE: Fabricate a valve arrangement board to use when you remove the valves, which will indicate the port in which each valve was originally installed (and which cylinder head on V6 and V8 models). Also note that the valve keys, rotators, caps, etc. should be arranged in a manner which will allow you to install them on the valve on which they were originally used.

REMOVAL

All Except the F4–134 Exhaust Valves

1. The head must be removed from the engine. In all but the 4–134, all the valves are in the head. In the 4–134, just the intakes are in the head. The exhaust valves are in the block.
2. Remove the rocker arm assemblies.
3. Using a spring compressor, compress the valve springs and remove the keepers (locks). Relax the compressor and remove the washers or rotators, the springs, and the lower washers (on some engines). Keep all parts in order.

F4–134 Exhaust Valves

1. Remove the attaching bolts from the side valve spring cover. Remove the side valve spring cover and gasket.
2. Use rags to block off the three holes in the exhaust chamber to prevent the valve retaining locks from falling into the crankcase, should they be accidentally dropped.
3. Using a valve spring compressor, compress the valve springs only on those valves which are in the closed position (valve seated against the head). Remove the valve spring retainer locks, the retainer, and the exhaust valve spring. Close the other valves by rotating the camshaft and repeat the above operation for the remaining valves.
4. Lift all of the valves from the cylinder block. If the valve cannot be removed from the block, pull the valve upward as far as possible and remove the spring. Lower the valve and remove any carbon deposits from the valve stem. This will permit removal of the valve.

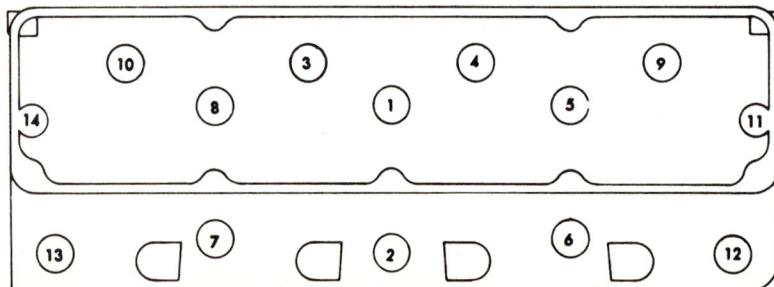

V8 cylinder head torque sequence

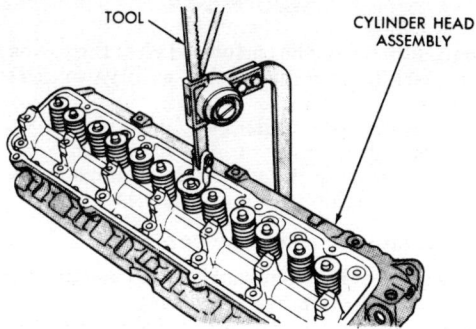

Using a valve spring compressor

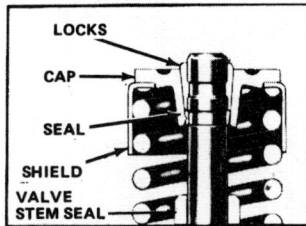

1 & 3 CHECK DIAGONALLY
2 CHECK ACROSS CENTER

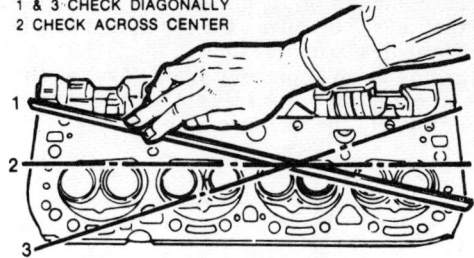

Check the cylinder head for warpage

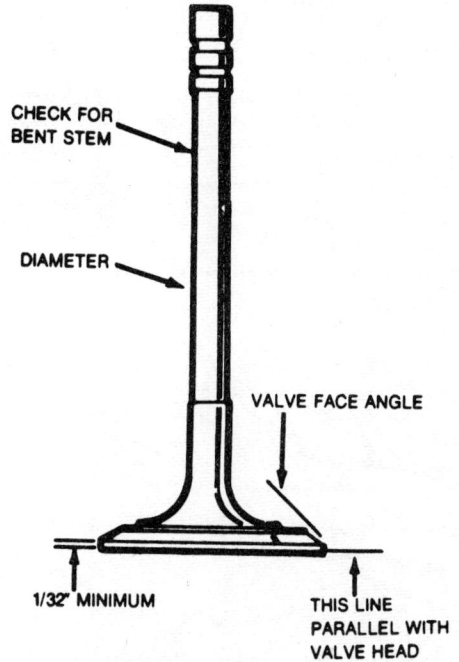

Typical upper valve train components

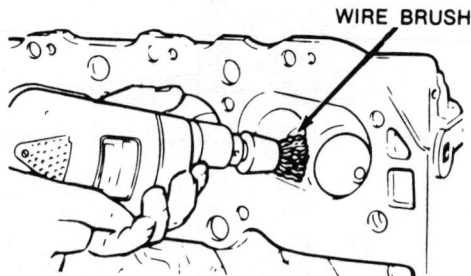

CHECK FOR BENT STEM

DIAMETER

VALVE FACE ANGLE

1/32" MINIMUM

THIS LINE PARALLEL WITH VALVE HEAD

Critical valve dimensions

Remove the carbon from the cylinder head with a wire brush and electric drill

INSPECTION AND REFACING

1. Clean the valves with a wire wheel.
2. Inspect the valves for warping, cracks or wear.
3. The valves may be refaced if not worn or pitted excessively.
4. Using a valve guide cleaner chucked into a drill, clean all of the valve guides. Check the valve stem diameter and the guide diameter with micrometers. Valve guides on the 4–134 are re-placeable. On the other engines, the guide must be reamed and an insert pressed in, or they may be knurled to bring up interior metal, restoring their diameter. Oversized valve stems are available to compensate for wear.
5. Install each valve into its respective port (guide) of the cylinder head.
6. Mount a dial indicator so that the stem is at 90° to the valve stem, as close to the valve guide as possible.
7. Move the valve off its seat, and measure the valve guide-to-stem clearance by rocking the stem back and forth to actuate the dial indicator.
8. In short, the refacing of valves and other such head work is most easily done at a machine shop. The quality and time saved easily justifies the cost.
9. Inspect the springs for obvious signs of wear. Check their installed height and tension using the values in the Valve Specifications Chart in this section.

REFACING

Using a valve grinder, resurface the valves according to specifications in this section.

NOTE: All machine work should be performed by a competent, professional machine shop.
Valve face angle is not always identical to valve seat angle.

A minimum margin of $\frac{1}{32}$ in. (0.8mm) should remain after grinding the valve. The valve stem top should also be squared and resurfaced, by placing the stem in the V-block of the grinder, and turning it while pressing lightly against the grinding wheel. Be sure to chamfer the edge of the tip so that the squared edges don't dig into the rocker arm or cam.

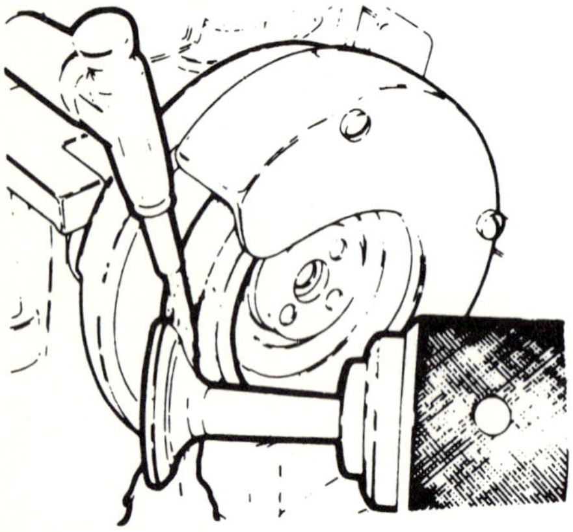

Refacing a valve. Any well equipped machine shop can do this job

Lapping the valves by hand

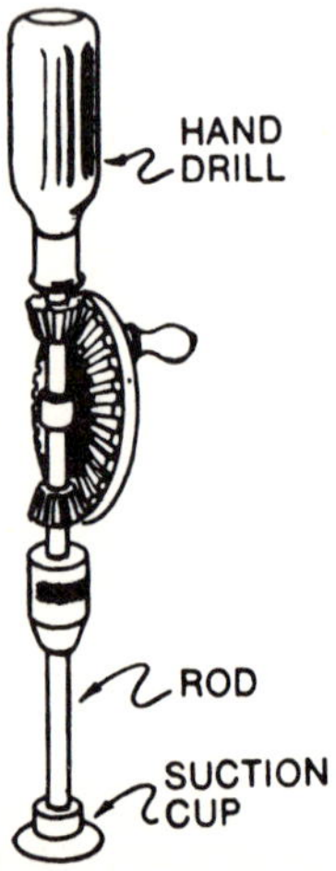

Home-made valve lapping tool

LAPPING

This procedure should be performed after the valves and seats have been machined, to insure that each valve mates to each seat precisely.

1. Invert the cylinder head, lightly lubricate the valve stems, and install the valves in the head as numbered.
2. Coat valve seats with fine grinding compound, and attach the lapping tool suction cup to a valve head.

NOTE: Moisten the suction cup.

3. Rotate the tool between your palms, changing position and lifting the tool often to prevent grooving.
4. Lap the valve until a smooth, polished seat is evident.

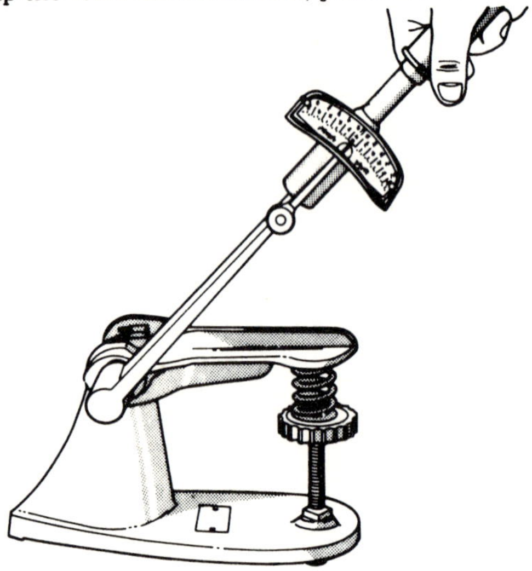

Testing a valve spring

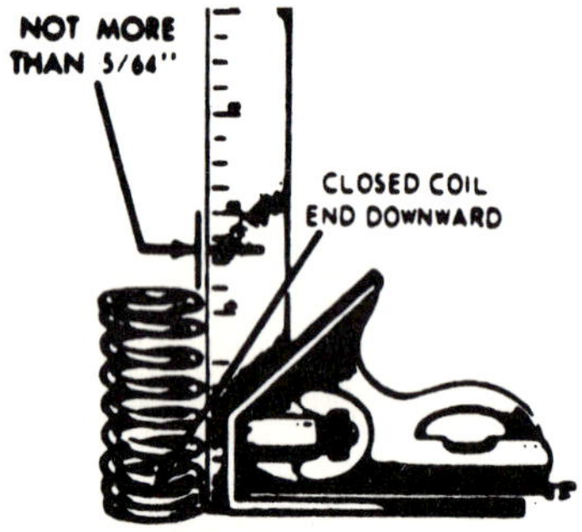

Check the valve spring free length and squareness

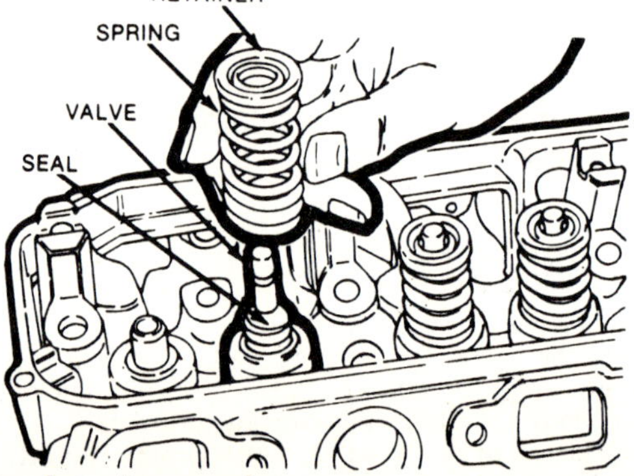

Install valve stem seals

Check valve spring installed height:

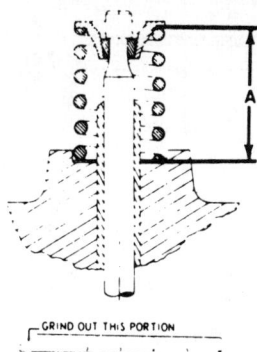

Valve spring installed height (A)

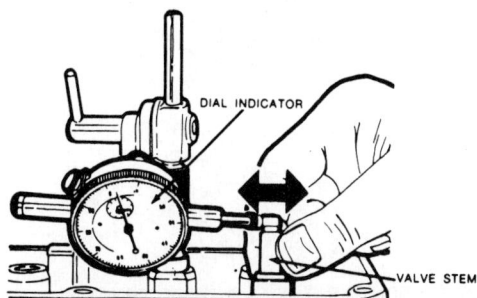

Check the valve stem-to-guide clearance

5. Remove the valve and tool, and rinse away all traces of grinding compound.

VALVE SPRING TESTING

Place the spring on a flat surface next to a square. Measure the height of the spring, and rotate it against the edge of the square to measure distortion. If spring height varies (by comparison) by more than $\frac{1}{16}$ in. (1.6mm) or if distortion exceeds $\frac{1}{16}$ in. (1.6mm), replace the spring.

In addition to evaluating the spring as above, test the spring pressure at the installed and compressed (installed height minus valve lift) height using a valve spring tester. Spring pressure should be ± 1 lb. of all other springs in either position.

INSTALLATION

1. Coat all parts with clean engine oil. Install all parts in their respective locations. The spring is installed with the closely wound coils toward the valve head. Always use new valve seals.
2. Use a spring compressor to install the keepers and slowly release the compressor after the keepers are in place.
3. Release the spring compressor. Tap the end of the stem with a wood mallet to insure that the keepers are securely in place.
4. Install all other parts in reverse order of removal.

Valve Guides

REMOVAL AND INSTALLATION

4–134

NOTE: A press is used for removal and installation of guides.

1. Place the head in the press and press the old guide out

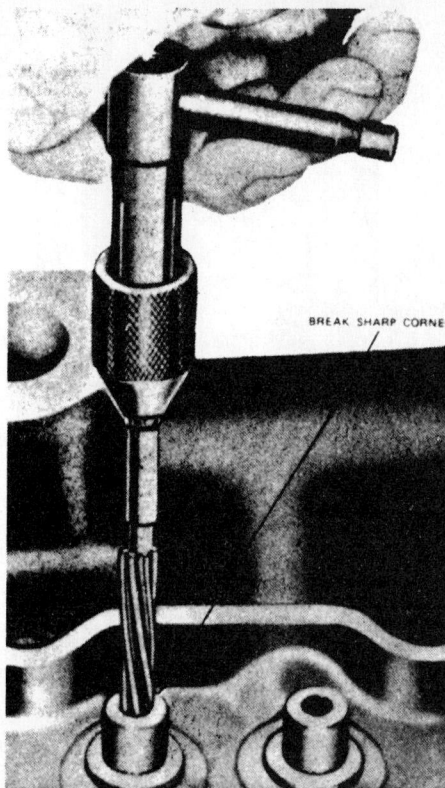

Close-up of a hand reamer

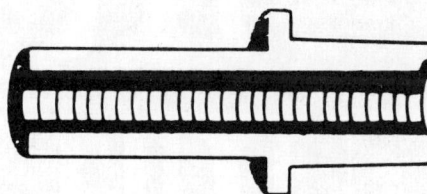

Cross section of a knurled valve guide

through the bottom and the new one in through the top. The only guides in the head are the intakes.
2. Once the new guide is in place, check its protrusion. The guide should be flush with the head surface.
3. The exhaust valve guides are also replaceable. These guides are located in the block. A press is not necessary for this procedure. The engine should be removed from the vehicle. The head, oil pan and crankshaft should be removed. The guides are driven out of their bores from above, with a driver made for the purpose. The new guides are then driven in with the same tool, until their tops are 1 inch below the block surface.
4. After the guides are in place they must be reamed:
● Intake, 0.3740–0.3760 in. (9.4996–9.5504mm)
● Exhaust, 0.3735–0.3765 in. (9.4869–9.5631mm)

Valve Seats

INSPECTION AND REFACING

The exhaust valve seats on the 4–134 and 6–226 are replaceable. All others have integral seats. Check the condition of the seats for excessive wear, pitting or cracks. Remove all traces of deposits from the seats. The seats may be refaced with a special

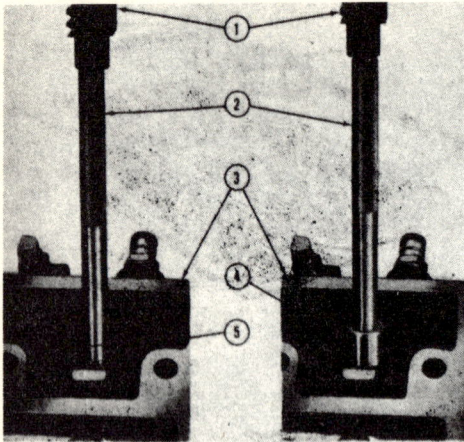

1. Arbor press   4. Stop
2. Driver   5. Valve guide
3. Cylinder head

Removing and installing a valve guide

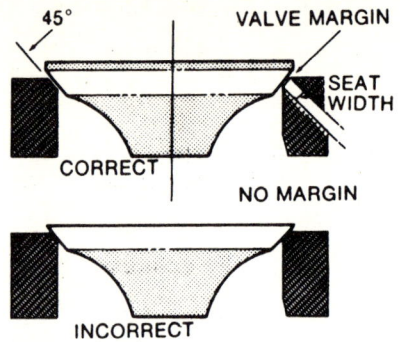

Valve seat width and centering

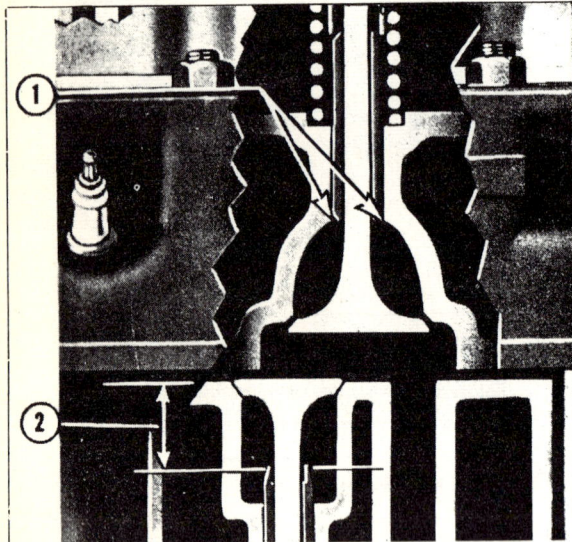

1. Flush at this point   2. One inch [25.4 mm.]

4-134 valves

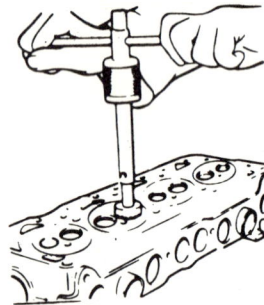

Reaming a valve seat with a head reamer

Checking valve seat concentricity with a dial gauge

grinding tool, to the dimensions shown in the Valve Specifications Chart.

You can replace the exhaust valve seats by driving them out from the bottom with a driver made for the purpose. This requires removal of the engine from the truck. Removal of the head, crankshaft, camshaft and pistons and rods. The new seats should be chilled with dry ice and installed immediately. They are driven into place with a seat installation tool. After installation, they should be ground to a 45° angle.

Oil Pan

REMOVAL AND INSTALLATION

4-134
6-225

To remove the oil pan on these engines, remove the oil pan at-

taching bolts and remove the oil pan. Clean all of the attaching surfaces and install new gaskets.

4-150

1. Disconnect the battery ground.
2. Raise and support the truck on jackstands.
3. Drain the oil.

— CAUTION —
The EPA warns that prolonged contact with used engine oil may cause a number of skin disorders, including cancer! You should make every effort to minimize your exposure to used engine oil. Protective gloves should be worn when changing the oil. Wash your hands and any other exposed skin areas as soon as possible after exposure to used engine oil. Soap and water, or waterless hand cleaner should be used.

4. Disconnect the exhaust pipe at the manifold.

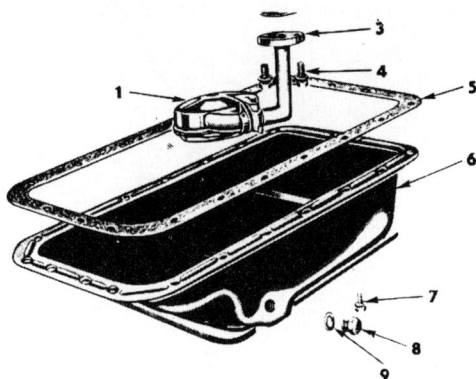

1. Floating oil intake
2. Gasket
3. Oil float support
4. Screw and lockwasher
5. Oil pan gasket
6. Oil pan
7. Bolt and lockwasher
8. Drain plug
9. Drain plug gasket

4–134 oil pan

5. Remove the starter.
6. Remove the bellhousing access plate.
7. Unbolt and remove the oil pan.
To install:
8. Clean the gasket surfaces thoroughly.
9. Install a replacement seal at the bottom of the timing case cover and at the rear bearing cap.
10. Using new gaskets coated with sealer, install the pan and torque the bolts to 10 ft. lbs.
11. Install all other parts in reverse order of removal.

4–151

1. Disconnect the battery ground.
2. Raise the vehicle and support it on jackstands.
3. Drain the oil.

— **CAUTION** —

The EPA warns that prolonged contact with used engine oil may cause a number of skin disorders, including cancer! You should make every effort to minimize your exposure to used engine oil. Protective gloves should be worn when changing the oil. Wash your hands and any other exposed skin areas as soon as possible after exposure to used engine oil. Soap and water, or waterless hand cleaner should be used.

4. Remove the starter.
5. Unbolt and remove the oil pan.
To install:
6. Clean all gasket surfaces, and remove all sludge and deposits from the pan.
7. Install the rear pan gasket in the main bearing cap and apply a small amount of RTV sealant in the depressions where the pan gasket contacts the block.
8. Position the gasket on the pan. Apply a ⅛ x ¼ in. (3mm x 6mm) bead of RTV sealant at the split lines of the front and side gaskets.
9. Position the pan on the block carefully to avoid gasket misalignment. Install the bolts and tighten them to 45 inch lbs.
10. Install the starter. Tighten the bolts to 17 ft. lbs.; the nut to 40 inch lbs.
11. Connect the starter cables, lower the vehicle, fill the crankcase and run the engine to operating temperature, checking for leaks.

6–232, 6–258, 8–304

1. Raise the vehicle and drain the engine oil.

— **CAUTION** —

The EPA warns that prolonged contact with used engine oil may cause a number of skin disorders, including cancer! You should make every effort to minimize your exposure to used engine oil. Protective gloves should be worn when changing the oil. Wash your hands and any other exposed skin areas as soon as possible after exposure to used engine oil. Soap and water, or waterless hand cleaner should be used.

2. Remove the starter motor.
3. Place a jack under the transmission bell housing. Disconnect the engine right support cushion bracket from the block and raise the engine to allow sufficient clearance for oil pan removal.
4. Remove the oil pan attaching bolts and remove the oil pan.
To install:
5. Remove the oil pan front and rear neoprene oil seals and the side gaskets. Thoroughly clean the gasket surfaces of the oil pan and the engine block. Remove all of the sludge and dirt from the oil pan sump.
6. Apply a generous amount of RTV silicone to the end tabs of a new oil pan front seal and install the seal to the timing case cover.
7. Cement new oil pan side gaskets into position on the engine block and apply a generous amount of RTV silicone to the side gasket contacting surface of the seal end tabs.
8. Install the seal in the recess of the rear main bearing cap, making sure that it is fully seated.
9. Coat the oil pan contacting surface of the front and rear oil pan seals with engine oil.
10. Install the oil pan and assemble the engine mount in the reverse order of removal.

Oil Pump

REMOVAL AND INSTALLATION

4-134

1. Set number one piston at TDC in order to reinstall the oil pump without disturbing the ignition timing.
2. Remove the distributor cover and note the position of the rotor. Keep the rotor in that position when the oil pump is installed.
3. Remove the cap screws and lockwashers that attach the oil pump to the cylinder block. Carefully slide the oil pump and its driveshaft out of the cylinder block.
The oil pump is driven by the camshaft by means of a spiral gear. The distributor in turn is driven by the oil pump by means of a tongue on the end of the distributor shaft which engages a slot in the end of the oil pump shaft. Because the tongue and the slot are both machined off center, the two shafts can be meshed in only one position. Since the position of the distributor shaft determines the timing of the engine, and is controlled by the oil pump shaft, the position of the oil pump shaft with respect to the camshaft is important. If only the oil pump has been removed, install it so that the slot in the end of the shaft lines up with the tip of the distributor shaft and allows that shaft to slip into it without disturbing the the original position of the distributor. If the engine has been disturbed or both the distributor and the oil pump have been removed, follow the procedure given below.
1. Turn the crankshaft to align the timing marks on the crankshaft and camshaft timing gears.
2. Install the oil pump gasket on the pump.
3. With the wider side of the slot on top, start the oil pump driveshaft into the opening in the cylinder keeping the mounting holes in the body of the pump in alignment with the holes in the cylinder block.
4. Insert a long blade screwdriver into the distributor shaft opening in the side of the cylinder block and engage the slot in

1. Oil dipstick
2. Oil pan baffle
3. Oil pan gasket
4. Oil pan
5. Drain plug gasket
6. Drain plug
7. Oil pump screen
8. Oil suction housing, pipe and flange
9. Oil suction pipe gasket
10. Oil pump idler gear
11. Valve by-pass and cover assembly
12. Oil pressure valve
13. Spring
14. Gasket
15. Oil pressure valve cap
16. Oil filter
17. Oil pump cover gasket
18. Oil pump shaft and gear

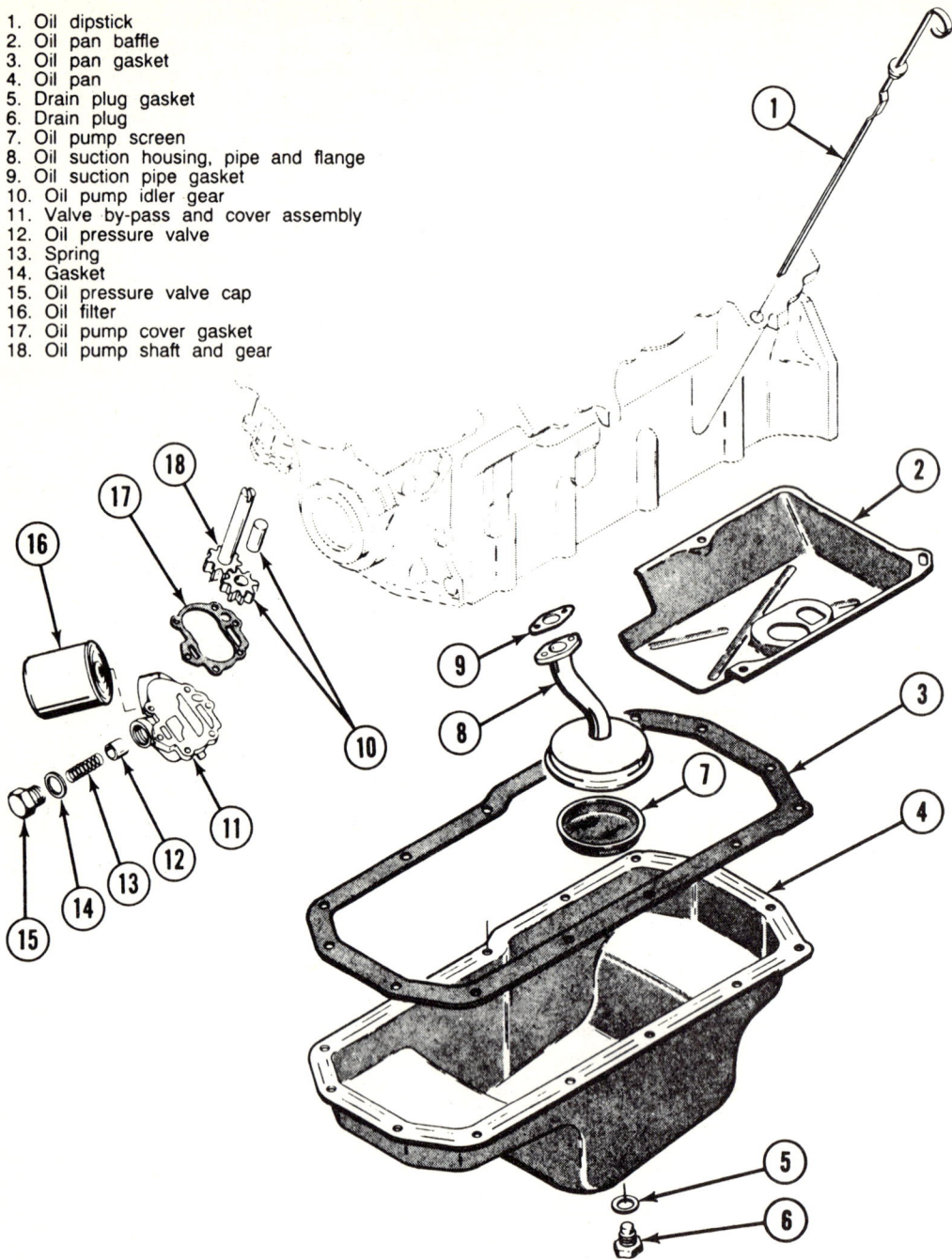

6–225 oil pan and pump

the oil pump shaft. Turn the shaft so that the slot is positioned at what would be roughly the nine thirty position on a clock face.

5. Remove the screwdriver and observe the position of the slot in the end of the oil pump shaft to make certain it is properly positioned.

6. Replace the screwdriver and, while turning the screwdriver clockwise to guide the oil pump driveshaft gear into engagement with the camshaft gear, press against the oil pump to force it into position.

7. Remove the screwdriver and again observe the position of the slot. If installation was properly made, the slot will be in a position roughly equivalent to the 11 o'clock position on the face of a clock, with the wider side of the slot still on the top. If the slot is improperly positioned, remove the oil pump and repeat the operation.

8. Coat the threads of the capscrews with gasket cement and secure the oil pump in place.

4–150

1. Disconnect the battery ground.
2. Raise and support the Jeep on jackstands.
3. Drain the oil.

1. Front seal
2. Rear seal
3. Gaskets

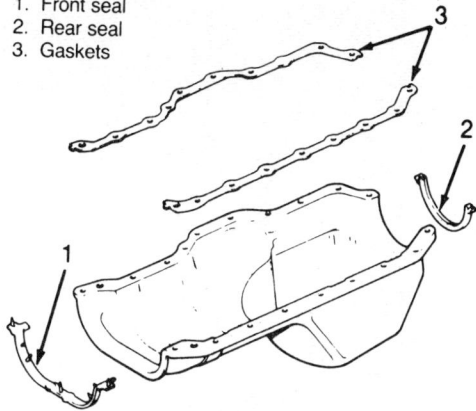

4–150 oil pan gasket and seal positioning

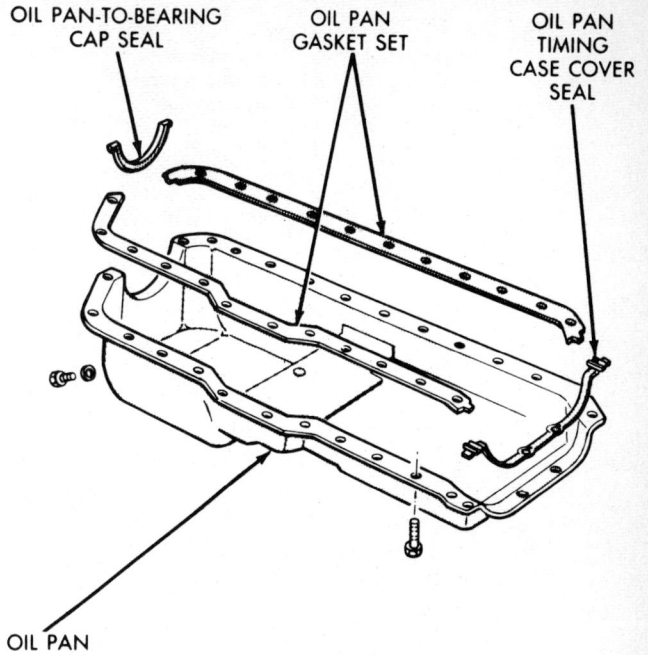

OIL PUMP DRIVESHAFT

OIL PUMP PICKUP TUBE AND STRAINER

OIL PUMP BODY

OIL PUMP GEARS

PRESSURE RELIEF AND SPRING

OIL PUMP COVER

OIL PAN GASKET

OIL PAN

OIL PAN BOLTS

OIL PAN DRAIN

OIL PAN REINFORCEMENTS

WIRE HARNESS PROTECTOR

4–151 oil pan and oil pump

OIL PAN-TO-BEARING CAP SEAL

OIL PAN GASKET SET

OIL PAN TIMING CASE COVER SEAL

OIL PAN

6–232, 6–258 oil pan and gaskets

11. Install all other parts in reverse order of removal.

4–151

1. Drain the oil and remove the oil pan.

CAUTION

The EPA warns that prolonged contact with used engine oil may cause a number of skin disorders, including cancer! You should make every effort to minimize your exposure to used engine oil. Protective gloves should be worn when changing the oil. Wash your hands and any other exposed skin areas as soon as possible after exposure to used engine oil. Soap and water, or waterless hand cleaner should be used.

2. Remove the oil pump retaining screws and separate the oil pump and gasket from the engine block.
3. Install in reverse order of the above procedure.

6–225

1. Remove the oil filter.
2. Disconnect the wire from the oil pressure indicator switch in the filter by-pass pump cover assembly to the timing chain cover.
3. Remove the screws that attach the oil pump cover assembly to the timing chain cover.
4. Remove the cover assembly and slide out the oil pump.
5. Install in reverse order of the above procedure.

6–232, 6–258

1. Drain the oil and remove the oil pan.

CAUTION

The EPA warns that prolonged contact with used engine oil may cause a number of skin disorders, including cancer! You should make every effort to minimize your exposure to used engine oil. Protective gloves should be worn when changing the oil. Wash your hands and any other exposed skin areas as soon as possible after exposure to used engine oil. Soap and water, or waterless hand cleaner should be used.

2. Remove the oil pump retaining screws and separate the oil pump and gasket from the engine block.

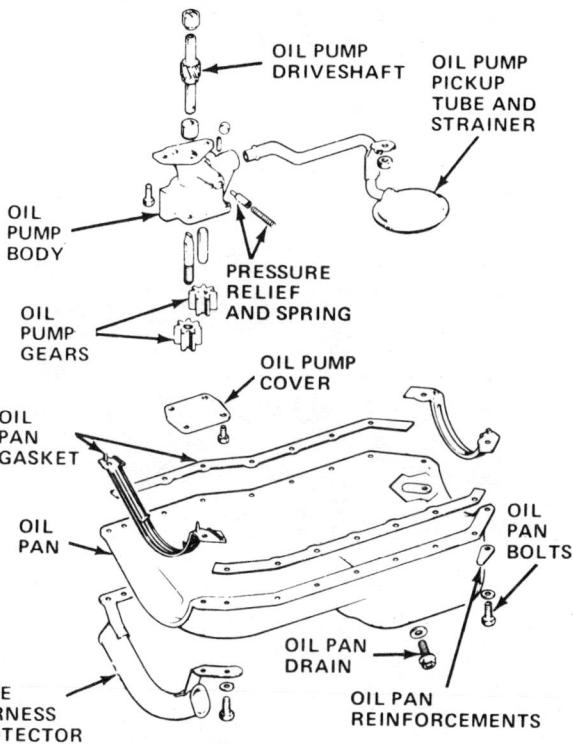

CAUTION

The EPA warns that prolonged contact with used engine oil may cause a number of skin disorders, including cancer! You should make every effort to minimize your exposure to used engine oil. Protective gloves should be worn when changing the oil. Wash your hands and any other exposed skin areas as soon as possible after exposure to used engine oil. Soap and water, or waterless hand cleaner should be used.

4. Disconnect the exhaust pipe at the manifold.
5. Remove the starter.
6. Remove the bellhousing access plate.
7. Unbolt and remove the oil pan.
To install:
8. Clean the gasket surfaces thoroughly.
9. Install a replacement seal at the bottom of the timing case cover and at the rear bearing cap.
10. Using new gaskets coated with sealer, install the pan and torque the bolts to 10 ft. lbs.

1. Cover screw
2. Cover
3. Cover gasket
4. Shaft and rotors
5. Body assembly
6. Driven gear
7. Pump gasket
8. Gear retaining pin
9. Relief valve retainer
10. Relief valve retainer gasket
11. Relief valve spring
12. Relief valve plunger

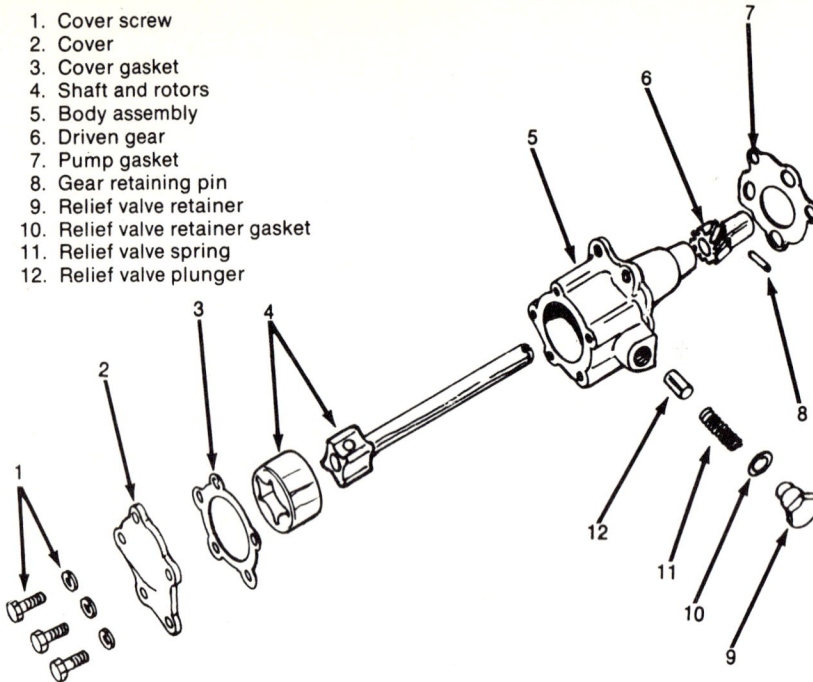

4–134 oil pump

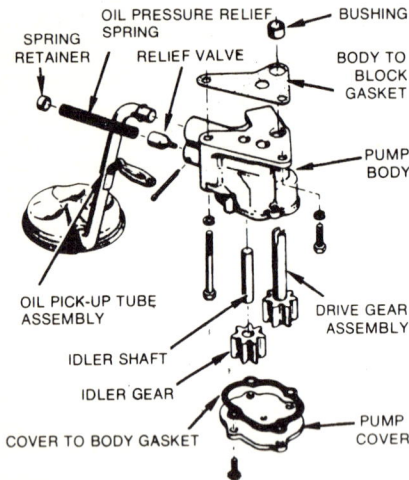

4–150 and 6–232, 6–258 oil pump

4–151 oil pump

NOTE: Do not disturb the position of the oil pick-up tube and screen assembly in the pump body. If the tube is moved within the pump body, a new assembly must be installed to assure an airtight seal.

3. Installation is the reverse of removal. Torque the short bolts to 10 ft. lbs.; the long bolts to 17 ft. lbs.

8–304

Remove the retaining screws and separate the oil pump cover, gasket and oil filter as an assembly from the pump body (timing chain cover). Install in reverse order with a new filter and gasket.

Crankshaft Pulley (Vibration Damper)

REMOVAL AND INSTALLATION

1. Remove the fan shroud, as required. If necessary, drain the cooling system and remove the radiator. Remove drive belts from pulley.

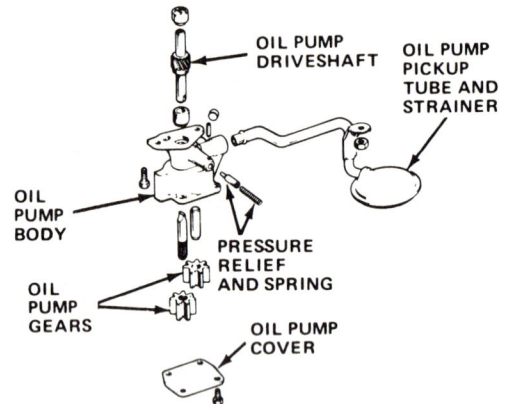

— CAUTION —

When draining the coolant, keep in mind that cats and dogs are attracted by the ethylene glycol antifreeze, and are quite likely to drink any that is left in an uncovered container or in puddles on the ground. This will prove fatal in sufficient quantity. Always drain the coolant into a sealable container. Coolant should be reused unless it is contaminated or several years old.

2. On those engines with a separate pulley, remove the retaining bolts and separate the pulley from the vibration damper.

3. Remove the vibration damper/pulley retaining bolt from the crankshaft end.

4. Using a puller, remove the damper/pulley from the crankshaft.

5. Upon installation, align the key slot of the pulley hub to the crankshaft key. Complete the assembly in the reverse order of removal. Torque the retaining bolts to specifications.

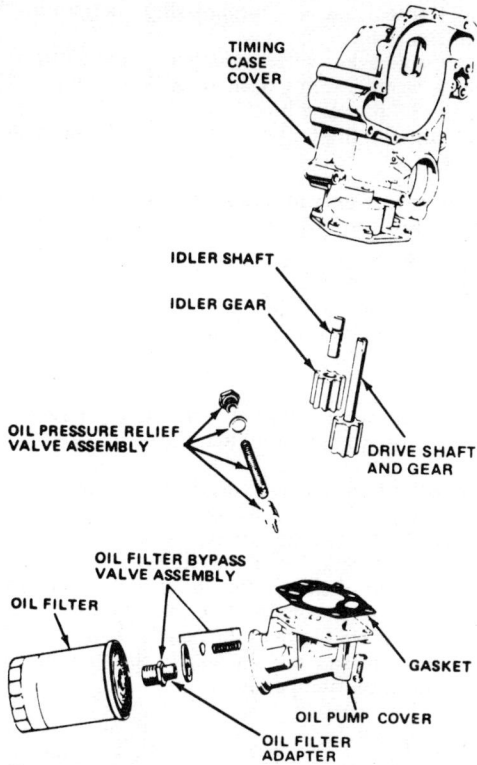

TIMING CASE COVER

IDLER SHAFT

IDLER GEAR

OIL PRESSURE RELIEF VALVE ASSEMBLY

DRIVE SHAFT AND GEAR

OIL FILTER BYPASS VALVE ASSEMBLY

OIL FILTER

GASKET

OIL PUMP COVER

OIL FILTER ADAPTER

8–304 oil pump

J-21791

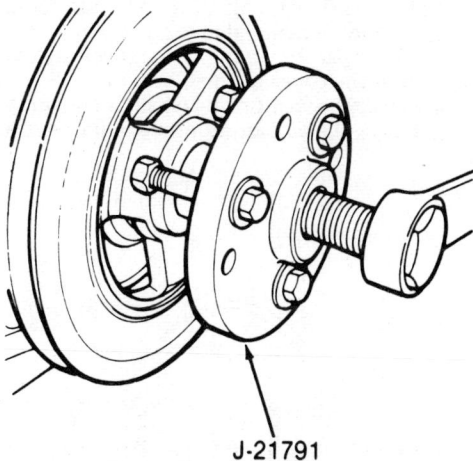

Using a puller to remove the crankshaft damper

Timing Gear Cover

TIMING GEAR COVER AND OIL SEAL REPLACEMENT

4–134

1. Remove the drive belts and crankshaft pulley.
2. Remove the attaching bolts, nuts and lock washers that hold the timing gear cover to the engine.
3. Remove the timing gear cover.
4. Remove the timing pointer.
5. Remove the timing gear cover gasket.
6. Remove and discard the crankshaft oil seal from the timing gear cover.

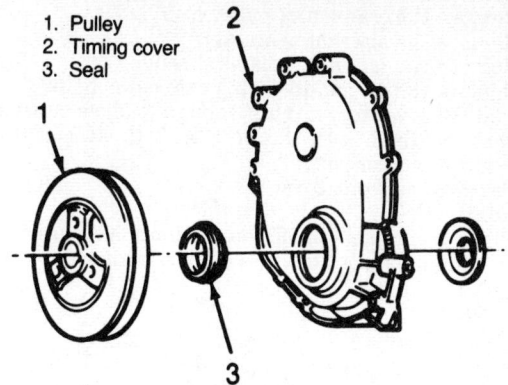

1. Pulley
2. Timing cover
3. Seal

4–150 timing cover assembly

7. Replace in reverse order of the above procedure. Replace the crankshaft oil seal. Use a new timing gear cover gasket.

4–150

NOTE: **Special tools are needed for this job.**

1. Remove the drive belts and fan shroud.
2. Unscrew the vibration damper bolts and washer.
3. Using a puller, remove the vibration damper.
4. Remove the fan assembly. If the fan is equipped with a fan clutch DO NOT LAY IT DOWN! If you lay it down, the fluid will leak out of the clutch and irreversibly damage the fan.
5. Disconnect the battery ground.
6. Remove the air conditioning compressor/alternator bracket assembly and lay it out of the way. DO NOT DISCONNECT THE REFRIGERANT LINES!
7. Unbolt the cover from the block and oil pan. Remove the cover and front seal.
8. Cut off the oil pan side gasket end tabs and oil pan front seal tabs.
9. Clean all gasket mating surfaces thoroughly.
10. Remove the seal from the cover.

To install:

11. Apply sealer to both sides of the new case cover gasket and position it on the block.
12. Cut the end tabs off the new oil pan side gaskets corresponding to those cut off the original gasket and attach the tabs to the oil pan with gasket cement.
13. Coat the front cover seal end tab recesses generously with RTV sealant and position the side seal in the cover.
14. Apply engine oil to the seal-to-pan contact surface.
15. Position the cover on the block.
16. Insert alignment tool J-22248 into the crankshaft opening in the cover.
17. Install the cover bolts. Tighten the cover-to-block bolts to 5 ft. lbs.; the cover-to-pan bolts to 11 ft. lbs.
18. Remove the alignment tool and position the new front seal on the tool with the seal lip facing outward. Apply a light film of sealer to the outside diameter of the seal. Lightly coat the crankshaft with clean engine oil.
19. Position the tool and seal over the end of the crankshaft and insert the Draw Screw J-9163-2 into the installation tool.
20. Tighten the nut until the tool just contacts the cover.
21. Remove the tools and apply a light film of engine oil on the vibration damper hub contact surface of the seal.
22. With the key inserted in the keyway in the crankshaft, install the vibration damper, washer and bolt. Lubricate the bolt and tighten it to 108 ft. lbs.
23. Install all other parts in reverse order of removal.

4–151

1. Disconnect the battery ground.

2. Remove the crankshaft pulley hub.
3. Remove the alternator bracket.
4. Remove the fan and radiator shroud.
5. Remove the oil pan-to-timing case cover bolts.
6. Pull the cover forward just enough to allow cutting the oil pan front seal flush with the block on both sides of the cover. Use a sharp knife or razor.
7. Remove the front cover.

To install:
8. Clean the gasket surface on the block and cover.
9. Cut the tabs from the new oil pan front seal.

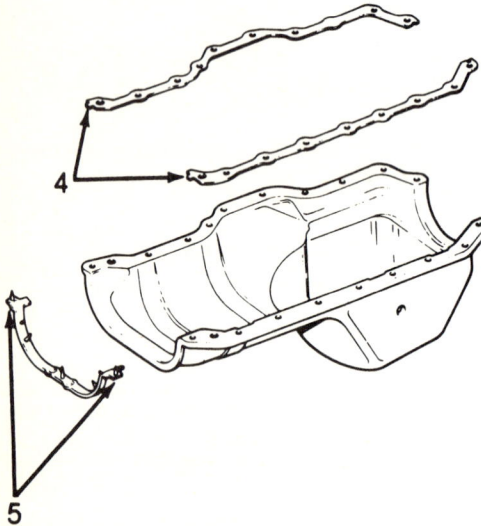

4–150 oil pan and gaskets

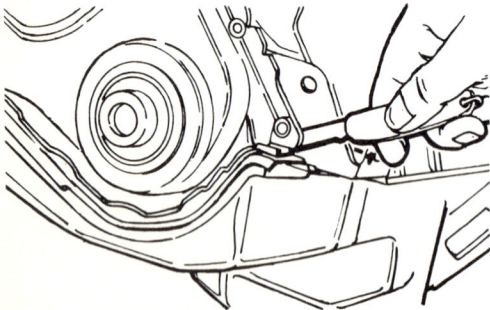

Cutting the pan gasket

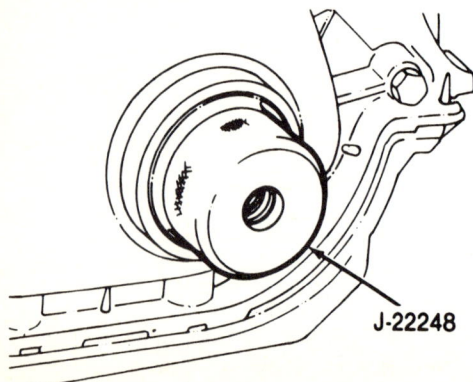

Timing cover centering tool installed on a 4–150

10. Install the seal on the cover, pressing the tips into the holes provided in the cover.
11. Coat a new gasket with sealer and place on the cover.
12. Apply a ⅛ in. (3mm) bead of RTV sealant to the joint formed at the oil pan and block.
13. Install an aligning tool such as tool J-23042 in the timing case cover seal.

NOTE: It is important that an aligning tool is used to avoid seal damage and to ensure a tight, even seal fit.

14. Position the cover on the block and partially retighten the two oil pan-to-cover bolts.
16. Install all other parts in reverse order of removal. Torque the fan assembly bolts to 18 ft. lbs.

6–225

1. Remove the water pump and crankshaft pulley.
2. Remove the two bolts that attach the oil pan to the timing chain cover.
3. Remove the five bolts that attach the timing chain cover to the engine block.
4. Remove the cover and gasket.
5. Remove the crankshaft front oil seal.
6. From the rear of the timing chain cover, coil new packing around the crankshaft hole in the cover so that the ends of the packing are at the top. Drive in the new packing with a punch. It will be necessary to ream out the hole to obtain clearance for the crankshaft vibration damper hub.

6–232, 6–258

COVER REMOVED

1. Remove the drive belts, engine fan and hub assembly, the accessory pulley, and vibration damper.
2. Remove the oil pan to timing chain cover screws and the screws that attach the cover to the block.
3. Raise the timing chain cover just high enough to detach the retaining nibs of the oil pan neoprene seal from the bottom

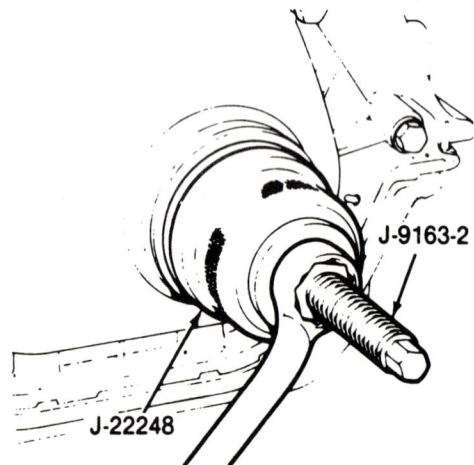

Oil seal installation tool

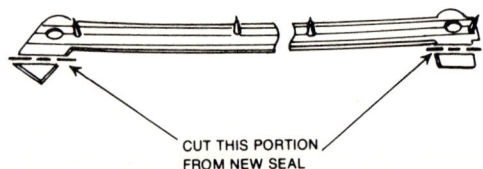

CUT THIS PORTION
FROM NEW SEAL

4–151 oil pan seal modification

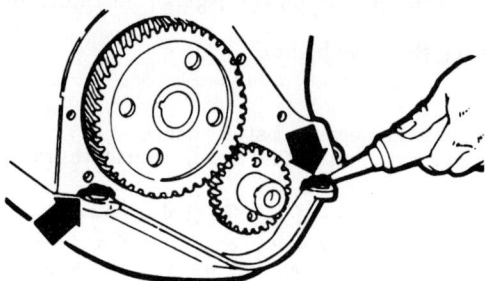
Applying RTV sealant on 4–151 engines

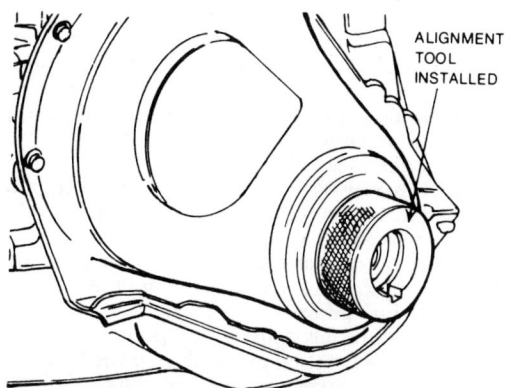
Timing case cover alignment tool installed on 4–151 engines

Water pump and timing chain cover bolts location 6–225

Trim the timing gear cover gasket as indicated before installation on 6–232, 6–258

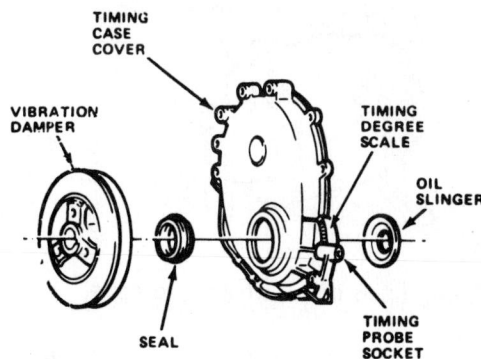
6–232, 6–258 timing cover and seal

side of the cover. This must be done to prevent pulling the seal end tabs away from the tongues of the oil pan gaskets which would cause a leak.

4. Remove the timing chain cover and gasket from the engine.

To install:

5. Use a razor blade to cut off the oil pan seal end tabs flush with the front face of the cylinder block and remove the seal. Clean the timing chain cover, oil pan, and cylinder block surfaces.

6. Apply seal compound (Perfect Seal, or equivalent) to both sides of the replacement timing case cover gasket and position the gasket on cylinder block.

7. Cut the end tabs off the replacement oil pan gasket corresponding to the pieces cut off the original gasket. Cement these pieces on the oil pan.

8. Coat the oil pan seal end tabs generously with Permatex® No. 2 or equivalent, and position the seal on the timing case cover.

9. Position the timing case cover on the engine. Place Timing Case Cover Alignment Tool and Seal Installer J-22248 in the crankshaft opening of the cover.

10. Install the cover-to-block screws and oil pan-to-cover screws. Tighten the cover-to-block screws to 60 inch lbs. and the oil pan-to-cover screws to 11 ft. lbs.

11. Remove the cover aligning tool and position the replacement oil seal aligning tool in the case. Position the replacement oil seal on the tool with the seal lip facing outward. Apply a light film of Perfect Seal, or equivalent, on the outside diameter of the seal.

12. Insert the draw screw from Tool J-9163 into the seal installing tool. Tighten the nut against the tool until the tool contacts the cover.

13. Remove the tools and apply a light film of engine oil to the seal lip.

14. Install the vibration damper and tighten the retaining screw to 80 ft. lbs.

15. Install the damper pulley. Tighten the capscrews to 20 ft. lbs.

16. Install the engine fan and hub assembly.

17. Install the drive belt(s).

COVER INSTALLED

1. Remove the drive belts.
2. Remove the vibration damper pulley.
3. Remove the vibration damper.
4. Remove the oil seal with Tool J-9256.
5. Position the replacement oil seal on the Timing Case Cover Alignment Tool and Seal Installer J-22248 with the seal lip fac-

ing outward. Apply a light film of Perfect Seal, or equivalent, to the outside diameter of the seal.

6. Insert the draw screw from Tool J-9163 into the seal installing tool. Tighten the nut against the tool until the tool contacts the cover.

7. Remove the tools. Apply a light film of engine oil to the seal lip.

8. Install the vibration damper and tighten the retaining bolt to 80 ft. lbs.

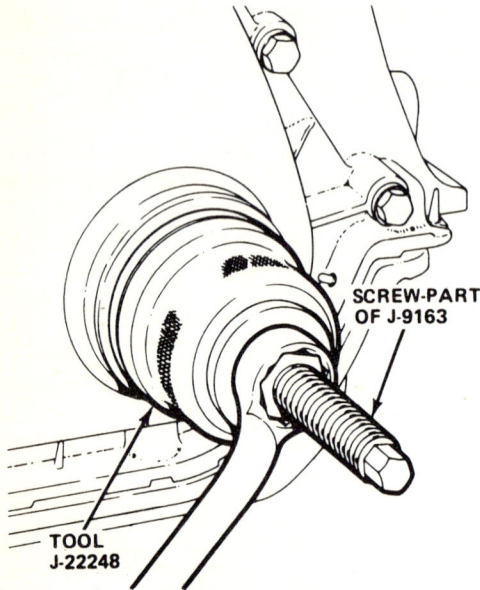

Timing case oil seal installation, 6–258 engines

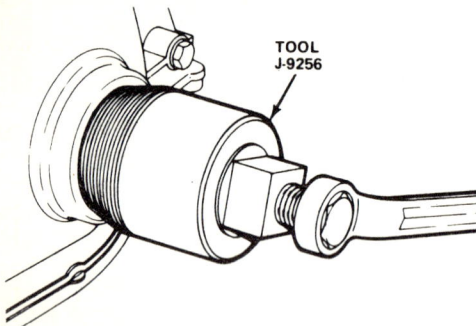

6–258 timing case cover oil seal removal

9. Install the damper pulley. Tighten the capscrews to 20 ft. lbs.

10. Install the drive belt(s).

8–304

1. Remove the negative battery cable.
2. Drain the cooling system and disconnect the radiator hoses and by-pass hose.

--- **CAUTION** ---

When draining the coolant, keep in mind that cats and dogs are attracted by the ethylene glycol antifreeze, and are quite likely to drink any that is left in an uncovered container or in puddles on the ground. This will prove fatal in sufficient quantity. Always drain the coolant into a sealable container. Coolant should be reused unless it is contaminated or several years old.

---

3. Remove all of the drive belts and the fan and spacer assembly.

4. Remove the alternator and the front portion of the alternator bracket as an assembly.

5. Disconnect the heater hose.

6. Remove the power steering pump and/or the air pump, and the mounting bracket as an assembly. Do not disconnect the power steering hoses.

7. Remove the distributor cap and note the position of the rotor. Remove the distributor. See the Engine Electrical Section.

8. Remove the fuel pump.

9. Remove the vibration damper and pulley.

10. Remove the two front oil pan bolts and the bolts which secure the timing chain cover to the engine block.

NOTE: The timing gear cover retaining bolts vary in length and must be installed in the same locations from which they were removed.

11. Remove the cover by pulling forward until it is free of the locating dowel pins.

12. Clean the gasket surface of the cover and the engine block.

13. Pry out the original seal from inside the timing chain cover and clean the seal bore.

To install:

14. Drive the new seal into place from the inside with a block of wood until it contacts the outer flange of the cover.

15. Apply a light film of motor oil to the lips of the new seal.

16. Before reinstalling the timing gear cover, remove the lower locating dowel pin from the engine block. The pin is required for correct alignment of the cover and must either be reused or a replacement dowel pin installed after the cover is in position.

17. Cut both sides of the oil pan gasket flush with the engine block with a razor blade.

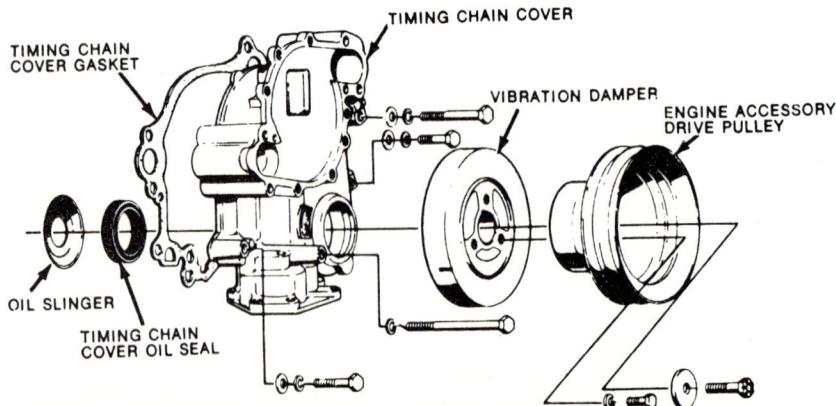

8–304 timing case, cover and seal. The unit for engines through 1976 is shown. 1977 and later engines are identical, except for the location of the seal

18. Trim a new gasket to correspond to the amount cut off at the oil pan.

19. Apply sealer to both sides of the new gasket and install the gasket on the timing case cover.

20. Install the new front oil pan seal.

21. Align the tongues of the new oil pan gasket pieces with the oil pan seal and cement them into place on the cover.

22. Apply a bead of sealer to the cutoff edges of the original oil pan gaskets.

23. Place the timing case cover into position and install the front oil pan bolts. Tighten the bolts slowly and evenly until the cover aligns with the upper locating dowel.

24. Install the lower dowel through the cover and drive it into the corresponding hole in the engine block.

25. Install the cover retaining bolts in the same locations from which they were removed. Tighten to 25 ft. lbs.

26. Assemble the remaining components in the reverse order of removal.

1. Puller   2. Camshaft gear

Pulling the 4–134 valve timing gears for engines after serial No. 3175402

Timing Chain or Gears and Tensioner

REMOVAL AND INSTALLATION

4–134

1. Remove the timing gear cover.

2. Use a puller to remove both the crankshaft and the camshaft gear from the engine after removing all attaching nuts and bolts.

3. Remove the Woodruff keys.

Installation is as follows:

1. Install the Woodruff key in the longer of the two keyways on the front end of the crankshaft.

2. Install the crankshaft timing gear on the front end of the crankshaft with the timing mark facing away from the cylinder block.

3. Align the keyway in the gear with the Woodruff key and then drive or press the gear onto the crankshaft firmly against the thrust washer.

4. Turn the camshaft or the crankshaft as necessary so that the timing marks on the two gears will be together after the camshaft gear is installed.

5. Install the Woodruff key in the keyway on the front of the camshaft.

6. Start the large timing gear on the camshaft with the timing mark facing out.

NOTE: Do not drive the gear onto the camshaft as the camshaft may drive the plug out of the rear of the engine and cause an oil leak.

7. Install the camshaft retaining screw and torque it to 30–40 ft. lbs. This will draw the gear onto the camshaft as the screw is tightened. Standard running tolerance between the timing gears is 0-0.002 in. (0.05mm).

8. Install the timing gears with the marks aligned as shown.

9. Set the intake valve clearance to 0.020 in. (0.5mm) on the #1 cylinder.

10. Rotate the crankshaft until the #1 cylinder intake valve is ready to open as indicated by the IO mark on the flywheel. The mark should be centered in the hole.

NOTE: Some models do not have an IO mark. TC and 5° are the only marks. On these engines, the intake valve opens at 9° BTC. To estimate valve opening, measure the distance between TC and 5° and measure about that distance further on.

4–134 timing gear alignment for engines after serial No. 3175402

4–150

1. Remove the timing case cover.

2. Rotate the crankshaft so that the timing marks on the cam and crank sprockets align next to each other, as illustrated.

3. Remove the oil slinger from the crankshaft.

4. Remove the cam sprocket retaining bolt and remove the sprocket and chain. The crank sprocket may also be removed at this time. If the tensioner is to be removed, the oil pan must be removed. first.

To install:

5. Prior to installation, turn the tensioner lever to the unlock (down) position.

6. Pull the tensioner block toward the tensioner to compress the spring. Hold the block and turn the tensioner lever to the lock (up) position. The camshaft sprocket bolt should be torqued to 50 ft. lbs. on 1984–86 models; 80 ft. lbs. on 1987-90 models.

7. Install the sprockets and chain together, as a unit. Make sure the timing marks are aligned.

To verify that the timing chain is correctly installed:

a. Turn the crankshaft to place the camshaft timing mark at approximately the one o'clock position.

b. At this point, there should be a tooth on the crankshaft sprocket meshed with the chain at the three o'clock position.

c. Count the number of timing chain pins between the two timing marks on the right side (your right, facing the engine). There should be 20 pins.

8. Install the oil pan, slinger and timing cover.

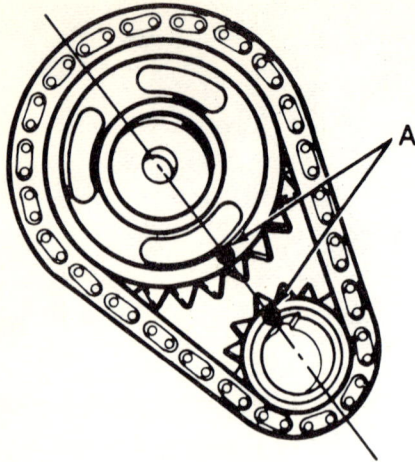

4–150 valve timing mark alignment

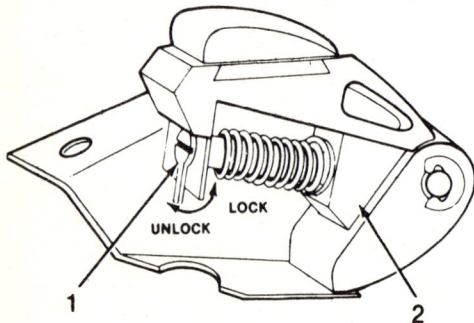

4–150 timing chain tensioner. 1 is the tensioner lever, 2 is the block

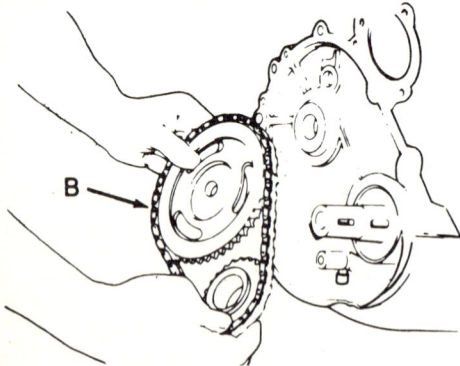

Installing the timing chain and sprockets on the 4–150

4–151

NOTE: Removal of the camshaft gear requires a special adapter #J-971 and the use of a press. Camshaft removal is necessary.

1. Place the adapter on the press and place the camshaft through the opening.
2. Press the shaft out of the gear using a socket or other suitable tool.

WARNING: The thrust plate must be in position so that the woodruff key does not damage the gear when the shaft is pressed out.

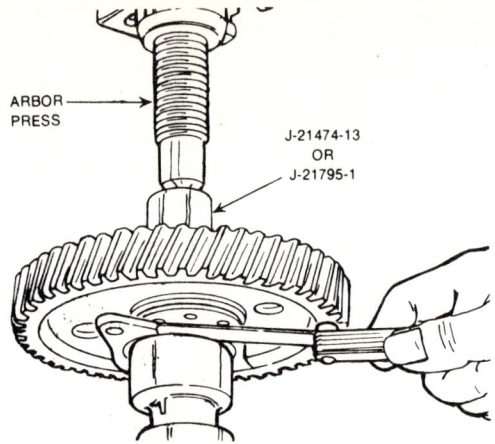

Installing 4–151 camshaft timing gear and measuring thrust plate end clearance

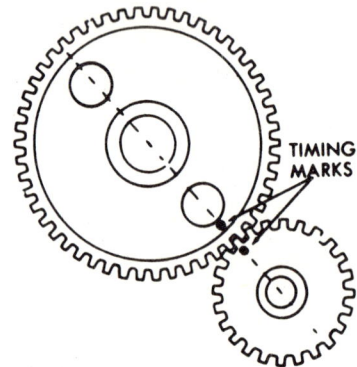

4–151 timing mark alignment

3. To install the gear firmly support the shaft at the back of the front journal in an arbor press using pressplate adapters J-21474-13 or J-21795-1.
4. Place the gear spacer ring and thrust plate over the end of the shaft, and install the woodruff key in the shaft keyway.
5. Install the camshaft gear and press it onto the shaft until it bottoms against the gear spacer ring. The end clearance of the thrust plate should be 0.0015–0.0050 in. (0.038–0.13mm). If less than 0.0015 in. (0.038mm), the spacer ring should be replaced.

6–225

1. Remove the timing chain cover.
2. Make sure that the timing marks on the crankshaft and the camshaft sprockets are aligned. This will make installing the parts easier.

NOTE: It is not necessary to remove the timing chain dampers (tensioners) unless they are worn or damaged and require replacement.

3. Remove the front crankshaft oil slinger.
4. Remove the bolt and the special washer that hold the camshaft distributor drive gear and fuel pump eccentric at the forward end of the camshaft. Remove the eccentric and the gear from the camshaft.
5. Alternately pry forward the camshaft sprocket and then the crankshaft sprocket until the camshaft sprocket is pried from the camshaft.
6. Remove the camshaft sprocket, sprocket key, and timing chain from the engine.

V6 valve timing sprocket alignment

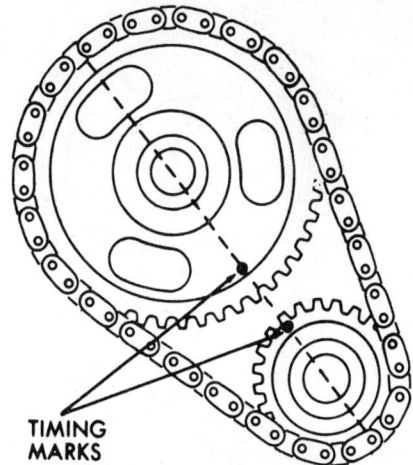

6–232, 6–258 timing mark alignment

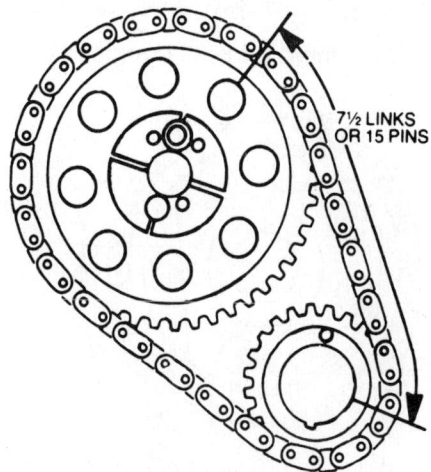

6–232, 6–258 alignment verification

7. Pry the crankshaft sprocket from the crankshaft.
Install as follows:
1. If the engine has not been disturbed proceed to step Number 4 for installation procedures.
2. If the engine has been disturbed turn the crankshaft so that number one piston is at top dead center.
3. Temporarily install the sprocket key and the camshaft sprocket on the camshaft. Turn the camshaft so that the index mark of the sprocket is downward. Remove the key and sprocket from the camshaft.
4. Assemble the timing chain and sprockets. Install the keys, sprockets, and chain assembly on the camshaft and crankshaft so that the index marks of both the sprockets are aligned.

NOTE: It will be necessary to hold the spring loaded timing chain damper out of the way while installing the timing chain and sprocket assembly.

5. Install the front oil slinger on the crankshaft with the inside diameter against the sprocket (concave side toward the front of the engine).
6. Install the fuel pump eccentric on the camshaft and the key, with the oil groove of the eccentric forward.
7. Install the distributor drive gear on the camshaft. Secure the gear and eccentric to the camshaft with the retaining washer and bolt.
8. Torque the bolt to 40–55 ft. lbs.

6–232, 6–258

1. Remove the drive belts, engine fan and hub assembly, accessory pulley, vibration damper and timing chain cover.
2. Remove the oil seal from the timing chain cover.
3. Remove the camshaft sprocket retaining bolt and washer.
4. Rotate the crankshaft until the timing mark on the crankshaft sprocket is closest to and in a center line with the timing pointer of the camshaft sprocket.
5. Remove the crankshaft sprocket, camshaft sprocket and timing chain as an assembly. Disassemble the chain and sprockets.
Installation is as follows:
1. Assemble the timing chain, crankshaft sprocket and camshaft sprocket with the timing marks aligned.
2. Install the assembly to the crankshaft and the camshaft.
3. Install the camshaft sprocket retaining bolt and washer and tighten to 45–55 ft. lbs.
4. Install the timing chain cover and a new oil seal.
5. Install the vibration damper, accessory pulley, engine fan and hub assembly and drive belts. Tighten the belts to the proper tension.

8–304

1. Remove the timing chain cover and gasket.
2. Remove the crankshaft oil slinger.
3. Remove the camshaft sprocket retaining bolt. and washer, distributor drive gear and fuel pump eccentric.
4. Rotate the crankshaft until the timing mark on the crankshaft sprocket is adjacent to, and on a center line with, the timing mark on the camshaft sprocket.
5. Remove the crankshaft sprocket, camshaft sprocket and timing chain as an assembly.
6. Clean all of the gasket surfaces.
Installation is as follows:
1. Assemble the timing chain, crankshaft sprocket and camshaft sprocket with the timing marks on both sprockets aligned.
2. Install the assembly to the crankshaft and the camshaft.
3. Install the fuel pump eccentric, distributor drive gear, washer and retaining bolt. Tighten the bolt to 25–35 ft. lbs.
4. Install the crankshaft oil slinger.
5. Install the timing chain cover using a new gasket and oil seal.

NOTE: In mid-year 1979, a new timing chain, camshaft sprocket and crankshaft sprocket were phased into production on all 8–304 engines. These are offered as replacement parts for older engines. When installing any one of these parts on an older engine, all three parts

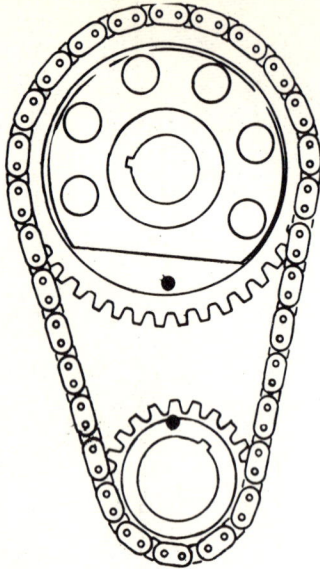

Timing gear alignment on 8–304 engines

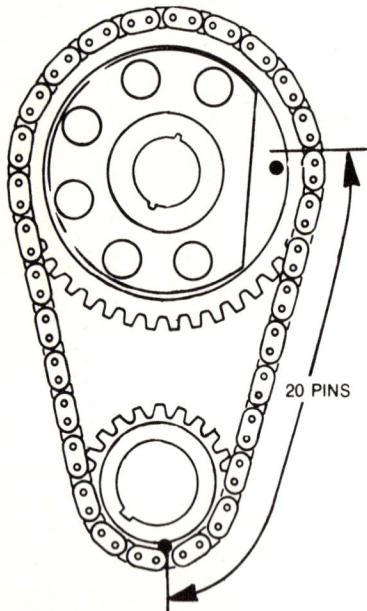

Correct timing chain installation verification, 8–304

must be installed. None of the new parts is usable in conjunction with the older parts. They must be installed as a set. To determine the necessity for a replacement of an older chain, perform the following deflection test:

1. Remove the timing case cover.
2. Rotate the sprockets until all slack is removed from the right side of the chain.
3. Locate the dowel on the lower left side of the engine and measure up ¾ in. (19mm). Make a mark.
4. Measure across the chain with a straightedge from the mark to a point at the bottom of the camshaft sprocket.
5. Grab the chain at the point where the straightedge crosses it. Push the chain left (inward) as far as it will go. Make a mark on the block at this point. Push the chain to the right as far as it will go. Make another mark. Measure between the two marks. Total deflection should not exceed ⅞ in. (22mm).

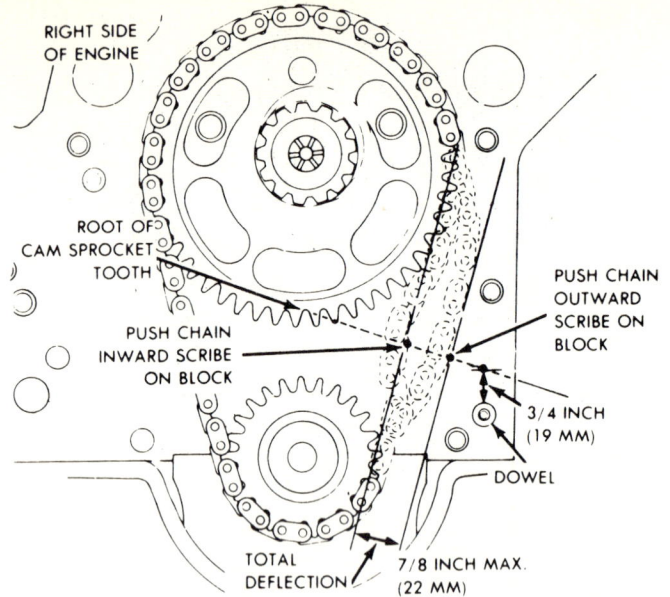

Measuring timing chain deflection on 8–304 engines

6. Replace the chain and sprockets if deflection is not within specifications.
7. Replace the timing case cover.

Valve Timing

4–134

1. Install the timing gears with the marks aligned as shown.
2. Set the intake valve clearance to 0.020 in. (0.5mm) on the #1 cylinder.
3. Rotate the crankshaft until the #1 cylinder intake valve is ready to open as indicated by the IO mark on the flywheel. The mark should be centered in the hole.

NOTE: Some models do not have an IO mark. TC and 5° are the only marks. On these engines, the intake valve opens at 9° BTC. To estimate valve opening, measure the distance between TC and 5° and measure about that distance further on.

Camshaft

REMOVAL AND INSTALLATION

NOTE: Caution must be taken when performing this procedure. Camshaft bearings are coated with babbit material, which can be damaged by scraping the cam lobes across the bearing.

4–134

1. Remove the engine.
2. Remove the exhaust manifold.
3. Remove the oil pump and the distributor.
4. Remove the crankshaft pulley.
5. Remove the cylinder head.
6. Remove the exhaust valves.
7. Remove the timing gear cover and the crankshaft and camshaft timing gears.
8. Remove the front end plate.
9. Push the lifters into the cylinder block as far as possible so that the ends of the lifters are not in contact with the camshaft.

1. Nut
2. Left Rocker Arm
3. Rocker Arm Shaft Spring
4. Rocker Shaft Lock Screw
5. Rocker Shaft
6. Nut
7. Right Rocker Arm
8. Rocker Arm Shaft Bracket
9. Intake Valve Tappet Adjusting Screw
10. Intake Valve Upper Retainer Lock
11. Oil Seal
12. Intake Valve Spring Upper Retainer
13. Intake Valve Spring
14. Intake Valve Push Rod
15. Intake Valve
16. Intake Valve Tappet
17. Camshaft
18. Camshaft Front Bearing
19. Camshaft Thrust Plate Spacer
20. Camshaft Thrust Plate
21. Bolt and Lockwasher
22. Bolt
23. Lockwasher
24. Camshaft Gear Washer
25. Crankshaft Gear
26. Camshaft Gear
27. Woodruff Key No. 9
28. Exhaust Valve Tappet
29. Tappet Adjusting Screw
30. Spring Retainer Lock
31. Roto Cap Assembly
32. Exhaust Valve Spring
33. Exhaust Valve
34. Rocker Shaft Support Stud
35. Washer
36. Rocker Arm Cover Stud

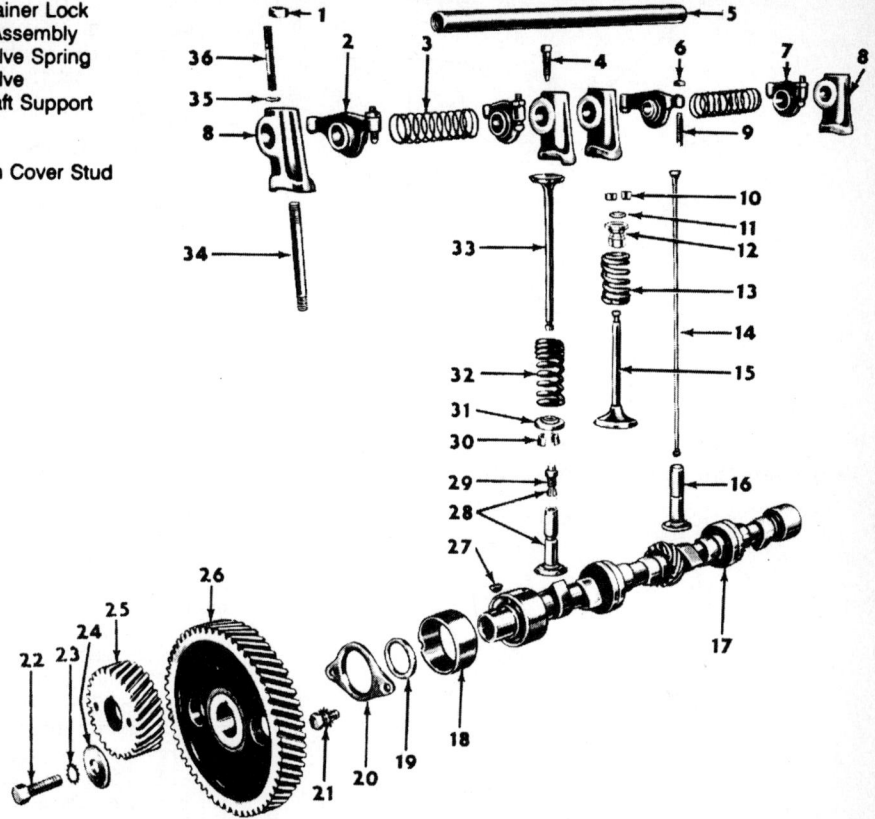

4–134 camshaft and related parts

10. Secure each tappet in the raised position by installing a clip type clothes pin on the shank of each tappet or tie them up in the plate and spacer.

11. Remove the crankshaft thrust plate attaching screws. Remove the camshaft thrust plate and spacer.

12. Pull the camshaft forward out of the cylinder block being careful to prevent damage to the camshaft bearing surfaces.

To install:

13. Slide the camshaft into the cylinder block being careful to prevent damage to the camshaft bearing surfaces.

14. Install the crankshaft thrust plate attaching screws. Install the camshaft thrust plate and spacer.

15. Remove the clothes pins.

16. Install the front end plate.

17. Install the timing gear cover and the crankshaft and camshaft timing gears.

18. Install the exhaust valves.

19. Install the cylinder head.

20. Install the crankshaft pulley.

21. Install the oil pump and the distributor.

22. Install the exhaust manifold.

23. Install the engine.

4–150

———————— CAUTION ————————

To remove perform this procedure the air conditioning system must be discharged. Mishandling of refrigerant gas can cause severe personal injury! If you are not completely familiar with the handling of refrigerant systems, have the system discharged by someone who is.

1. Disconnect the battery ground.
2. Drain the cooling system.

———————— CAUTION ————————

When draining the coolant, keep in mind that cats and dogs are attracted by the ethylene glycol antifreeze, and are quite likely to drink any that is left in an uncovered container or in puddles on the ground. This will prove fatal in sufficient quantity. Always drain the coolant into a sealable container. Coolant should be reused unless it is contaminated or several years old.

3. Remove the radiator, discharge the refrigerant system (See Section 1) and remove the condenser.

4. Remove the fuel pump.

5. Matchmark the distributor and engine for installation. Note the rotor position by marking it on the distributor body. Unbolt and remove the distributor and wires.

6. Remove the rocker arm cover.

7. Remove the rocker arm assemblies.

8. Remove the pushrods.

NOTE: Keep everything in order for installation.

9. Using a tool J-21884, or equivalent, remove the hydraulic lifters.

10. Remove the pulley, vibration damper and timing case cover. Remove the crankshaft oil slinger.

NOTE: If the camshaft sprocket appears to have been rubbing against the cover, check the oil pressure relief holes in the rear cam journal for debris.

11. Remove the timing chain and sprockets.

12. Slide the camshaft from the engine.

13. Inspect all parts for wear and damage. Lubricate all moving parts with engine oil supplement.

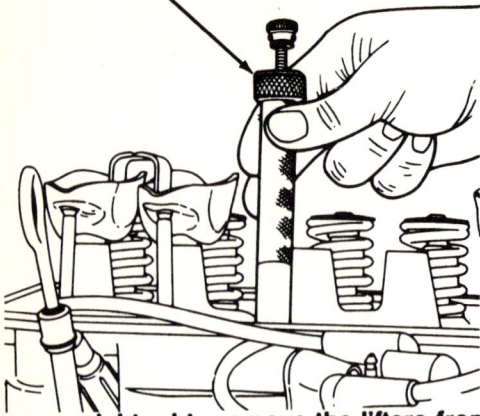

Using a special tool to remove the lifters from a 4-150

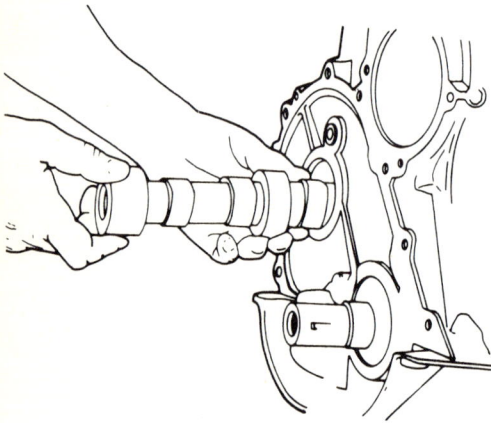

Removing the 4-150 camshaft

To install:

14. Slide the camshaft into the engine, carefully, to avoid damage to the bearing surfaces.

15. Install the timing chain and sprockets. Make sure that all camshaft timing marks align. Torque the camshaft sprocket bolt to 50 ft. lbs. on 1984–86 models; 80 ft. lbs. on 1987–90 models.

16. Install the timing case cover.

17. Using a tool J-21884, or equivalent, install the hydraulic lifters.

18. Install the pushrods.

19. Install the rocker arm assemblies.

20. Install the rocker arm cover.

21. Install the distributor and wires. When installing the distributor, make sure that all matchmarks align. It may be necessary to rotate the oil pump drive tang with a long-bladed screwdriver to facilitate installation of the distributor.

For 1987-90 engines: Position the engine at the number 1 cylinder TDC location. Rotate the gear slot on the oil pump shaft to a point slightly past the 3 o'clock position. Install the distributor with the rotor at the 5 o'clock position. When fully engaged, the rotor should be at the 6 o'clock position.

NOTE: If the distributor is not installed correctly, or removed later, the complete installation procedure must be done again.

22. Install the fuel pump.

23. Install the radiator.

24. Install the condenser and evacuate and charge the refrigerant system. (See Section 1).

25. Fill the cooling system.

26. Connect the battery ground.

4-151

1. Remove the air cleaner.

2. Drain the cooling system.

————————— CAUTION —————————

When draining the coolant, keep in mind that cats and dogs are attracted by the ethylene glycol antifreeze, and are quite likely to drink any that is left in an uncovered container or in puddles on the ground. This will prove fatal in sufficient quantity. Always drain the coolant into a sealable container. Coolant should be reused unless it is contaminated or several years old.

3. Remove the timing gear cover.

4. Disconnect the radiator hoses at the radiator. Remove the radiator.

5. Remove the two camshaft thrust plate screws through the holes in the camshaft gear.

6. Remove the tappets.

7. Remove the distributor, oil pump drive and fuel pump.

8. Remove the camshaft and gear assembly by pulling out through front of the block. Support the shaft carefully when removing it to prevent damaging the camshaft bearings.

To install:

9. Thoroughly coat the camshaft journals with a high quality engine oil supplement such as STP or its equivalent.

10. Install the camshaft assembly in engine block. Use care to prevent damaging the bearings or the camshaft.

11. Turn the crankshaft and camshaft so that the valve timing marks on the gear teeth are aligned. The engine is now in number four cylinder firing position. Install the camshaft thrust plate-to-block screws and tighten to 75 inch lbs.

12. Install the timing gear cover and gasket.

13. Line up the keyway in the hub with the key on crankshaft and slide the hub onto the shaft. Install the center bolt and tighten to 160 ft. lbs.

14. Install the valve tappets, pushrods, pushrod cover, oil pump shaft and gear assembly and the fuel pump. Install the distributor according to the following procedure:

    a. Turn the crankshaft 360° to the firing position of the number one cylinder (number one exhaust and intake valve tappets both on base circle [heel] of the camshaft and the timing notch on the vibration damper is indexed with the top dead center mark TDC on the timing degree scale).

    b. Install the distributor and align the shaft so that the rotor arm points toward the number one cylinder spark plug contact.

15. Install the rocker arms and pivot balls over the pushrods. With the tappets on the base circle (heel) of camshaft, tighten the rocker arm capscrews to 20 ft. lbs. Do not overtighten.

16. Install the cylinder head cover.

17. Install the intake manifold.

18. Install the radiator and lower radiator hose.

19. Install the belt, fan and shroud. Tighten the fan bolts to 18 ft. lbs.

20. Install the upper radiator hose.

21. Tighten the belts.

6-225

1. Remove the engine.

————————— CAUTION —————————

When draining the coolant, keep in mind that cats and dogs are attracted by the ethylene glycol antifreeze, and are quite likely to drink any that is left in an uncovered container or in puddles on the ground. This will prove fatal in sufficient quantity. Always drain the coolant into a sealable container. Coolant should be reused unless it is contaminated or several years old.

Removing 4–151 camshaft thrust plate screws

2. Remove the intake manifold and carburetor assembly.
3. Remove the distributor.
4. Remove the fuel pump.
5. Remove the alternator, drive belts, cooling fan, fan pulley and water pump.
6. Remove the crankshaft pulley and the vibration damper.
7. Remove the oil pump.
8. Remove the timing chain cover.
9. Remove the timing chain and the camshaft sprocket, along with the distributor drive gear and the fuel pump eccentric.
10. Remove the rocker arm assemblies.

NOTE: The pushrods need not be removed. But if they are, be sure that they are replaced in their original positions.

11. Lift the tappets up so that they are not in contact with the camshaft. Use wire clips or clip type pins to hold the tappets up.
12. Carefully guide the camshaft forward out of the engine. Avoid marring the bearing surfaces.
To install:
13. Coat all moving parts with an engine oil supplement, such as STP or its equivalent.
14. Slide the camshaft, carefully, into position. Avoid marring the bearing surfaces.
15. Drop the tappets back onto the camshaft.
16. Install the rocker arm assemblies.
17. Install the timing chain and the camshaft sprocket, along with the distributor drive gear and the fuel pump eccentric.
18. Install the timing chain cover.
19. Install the oil pump.
20. Install the crankshaft pulley and the vibration damper.
21. Install the alternator, drive belts, cooling fan, fan pulley and water pump.
22. Install the fuel pump.
23. Install the distributor.
24. Install the intake manifold and carburetor assembly.
25. Install the engine.

6–232, 6–258

1. Drain the cooling system and remove the radiator. With air conditioning, remove the condenser and receiver assembly as a unit, without disconnecting any lines or discharging the system.

─────────── **CAUTION** ───────────
When draining the coolant, keep in mind that cats and dogs are attracted by the ethylene glycol antifreeze, and are quite likely to drink any that is left in an uncovered container or in puddles on the ground. This will prove fatal in sufficient quantity. Always drain the coolant into a sealable container. Coolant should be reused unless it is contaminated or several years old.

2. Remove the valve cover and gasket, the rocker assemblies, pushrods, cylinder head and gasket and the lifters.

NOTE: The valve train components must be replaced in their original locations.

3. Remove the drive belts, cooling fan, fan hub assembly, vibration damper and the timing chain cover.
4. Remove the fuel pump and distributor assembly, including the spark plug wires.
5. Rotate the crankshaft until the timing mark of the crankshaft sprocket is adjacent to, and on a center line with, the timing mark of the camshaft sprocket.
6. Remove the crankshaft sprocket, camshaft sprocket, and the timing chain as an assembly.
7. Remove the front bumper or grille as required and carefully slide out the camshaft.
To install:
8. Coat all parts with engine oil supplement.
9. Slide the camshaft into place.
10. Install the front bumper and/or grille.
11. Install the crankshaft sprocket, camshaft sprocket, and the timing chain as an assembly.
12. Rotate the crankshaft until the timing mark of the crankshaft sprocket is adjacent to, and on a center line with, the timing mark of the camshaft sprocket.
13. Install the fuel pump and distributor assembly, including the spark plug wires.
14. Install the drive belts.
15. Install the cooling fan and fan hub assembly.
16. Install the vibration damper.
17. Install the cylinder head and gasket and the lifters.
18. Install the rocker assemblies and pushrods.
19. Install the valve cover and gasket.
20. Install the the timing chain cover.
21. With air conditioning, install the condenser and receiver.

NOTE: The valve train components must be replaced in their original locations.

22. Fill the cooling system and install the radiator.

8–304

1. Disconnect the battery cables.
2. Drain the radiator and both banks of the block. Remove the lower hose at the radiator, the by-pass hose at the pump, the thermostat housing and the radiator. With air conditioning, remove the condenser and receiver assembly as a unit, without disconnecting any lines or discharging the system.

─────────── **CAUTION** ───────────
When draining the coolant, keep in mind that cats and dogs are attracted by the ethylene glycol antifreeze, and are quite likely to drink any that is left in an uncovered container or in puddles on the ground. This will prove fatal in sufficient quantity. Always drain the coolant into a sealable container. Coolant should be reused unless it is contaminated or several years old.

3. Remove the distributor, all wires, and the coil from the manifold.
4. Remove the intake manifold as an assembly.
5. Remove the valve covers, rocker arms and pushrods.
6. Remove the lifters.

NOTE: The valve train components must be replaced in their original locations.

7. Remove the cooling fan and hub assembly, fuel pump, and heater hose at the water pump.
8. Remove the alternator and bracket as an assembly. Just move it aside, do not disconnect the wiring.
9. Remove the crankshaft pulley and the damper. Remove the lower radiator hose at the water pump.

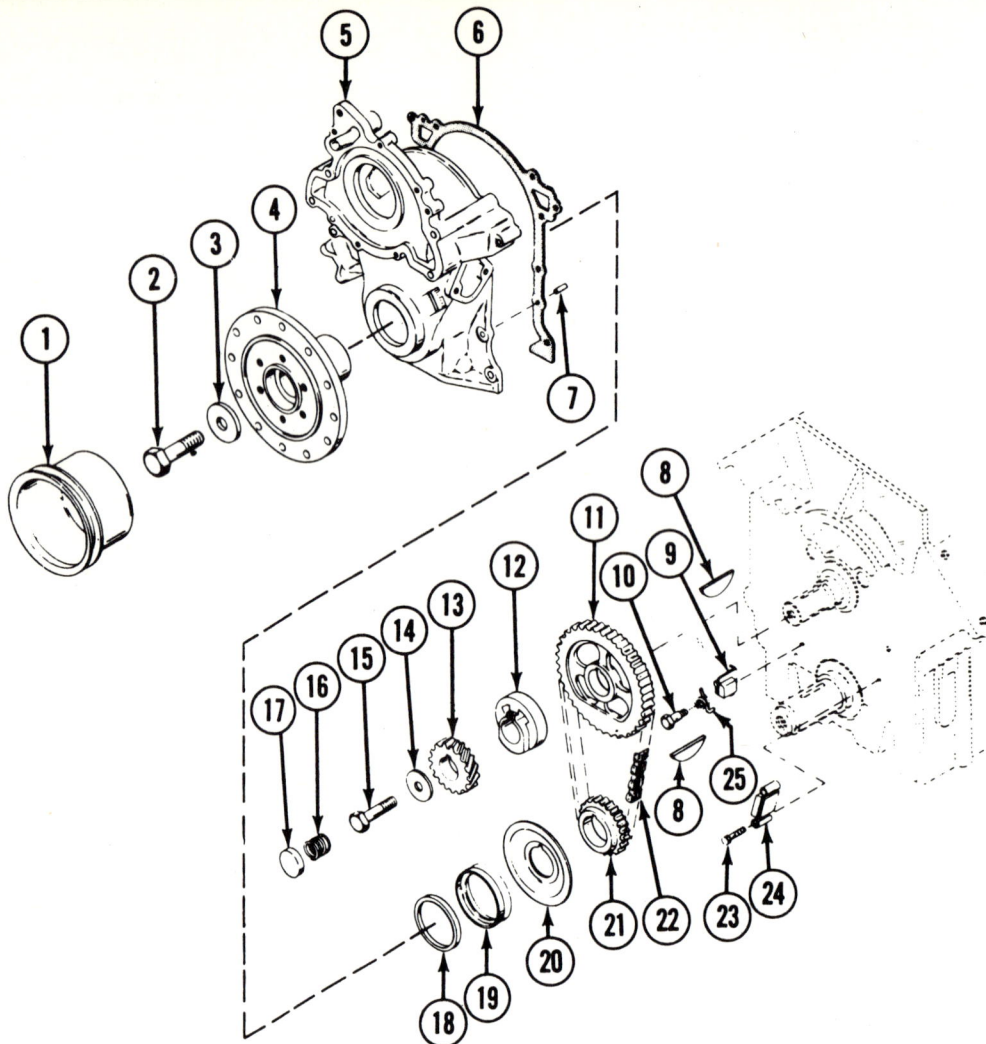

1. Crankshaft pulley
2. Crankshaft pulley bolt
3. Washer
4. Vibration damper
5. Timing gear cover
6. Gasket
7. Dowel pin
8. Woodruff key
9. Timing chain damper (right)
10. Damper bolt
11. Camshaft sprocket
12. Fuel pump eccentric
13. Distributor camshaft gear
14. Washer
15. Special bolt
16. Thrust spring
17. Thrust button
18. Oil shedder (crankshaft)
19. Crankshaft packing (front)
20. Crankshaft slinger
21. Crankshaft sprocket
22. Timing chain
23. Damper bolt
24. Timing chain damper (left)
25. Spring

6–225 timing chain, sprockets and related parts

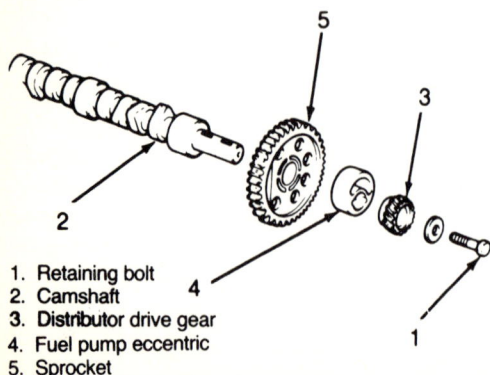

1. Retaining bolt
2. Camshaft
3. Distributor drive gear
4. Fuel pump eccentric
5. Sprocket

8–304 camshaft and related parts

10. Remove the timing chain cover.
11. Remove the distributor/oil pump drive gear, fuel pump eccentric, sprockets and the timing chain.
12. Remove the grille.
13. Remove the camshaft carefully by sliding it forward out of the engine.

To install:
14. Coat all parts with engine oil supplement.
15. Slide the camshaft, carefully, into the engine.
16. Install the grille.
17. Install the distributor/oil pump drive gear, fuel pump eccentric, sprockets and the timing chain.
18. Install the timing chain cover.
19. Install the crankshaft pulley and the damper.
20. Install the lower radiator hose at the water pump.
21. Install the alternator and bracket as an assembly.

22. Install the cooling fan and hub assembly.
23. Install the fuel pump.
24. Install the heater hose at the water pump.
25. Install the lifters.

NOTE: The valve train components must be replaced in their original locations.

26. Install the valve covers, rocker arms and pushrods.
27. Install the intake manifold as an assembly.
28. Install the distributor, all wires, and the coil on the manifold.
29. Install the radiator.
30. With air conditioning, install the condenser and receiver assembly.
31. Install the thermostat and housing.
32. Install the by-pass hose at the pump.
33. Install the lower hose at the radiator.
34. Fill the cooling system.
35. Connect the battery cables.

CHECKING CAMSHAFT

Camshaft Lobe Lift

Check the lift of each lobe in consecutive order and make a note of the reading.
1. Remove the fresh air inlet tube and the air cleaner. Remove the heater hose and crankcase ventilation hoses. Remove valve rocker arm cover(s).
2. Remove the rocker arm stud nut or fulcrum bolts, fulcrum seat and rocker arm.
3. Make sure the pushrod is in the valve tappet socket. Install a dial indicator so that the actuating point of the indicator is in the pushrod socket (or the indicator ball socket adaptor is on the end of the pushrod) and in the same plane as the push rod movement.
4. Install an auxiliary starter switch Crank the engine with the ignition switch off. Turn the crankshaft over until the tappet is on the base circle of the camshaft lobe. At this position, the pushrod will be in its lowest position.
5. Zero the dial indicator. Continue to rotate the crankshaft slowly until the pushrod is in the fully raised position.
6. Compare the total lift recorded on the dial indicator with the specification shown on the Camshaft Specification chart.
To check the accuracy of the original indicator reading, continue to rotate the crankshaft until the indicator reads zero. If the left on any lobe is below specified wear limits listed, the camshaft and the valve tappet operating on the worn lobe(s) must be replaced.
7. Install the dial indicator and auxiliary starter switch.
8. Install the rocker arm, fulcrum seat and stud nut or fulcrum bolts. Check the valve clearance. Adjust if required (refer to procedure in this section).
9. Install the valve rocker arm cover(s) and the air cleaner.

Camshaft End Play

NOTE: On engines with an aluminum or nylon camshaft sprocket, prying against the sprocket, with the valve train load on the camshaft, can break or damage the sprocket. Therefore, the rocker arm adjusting nuts must be backed off, or the rocker arm and shaft assembly must be loosened sufficiently to free the camshaft. After checking the camshaft end play, check the valve clearance. Adjust if required (refer to procedure in this section).

1. Push the camshaft toward the rear of the engine. Install a dial indicator so that the indicator point is on the camshaft sprocket attaching screw.
2. Zero the dial indicator. Position a prybar between the cam-

Check the camshaft for straightness

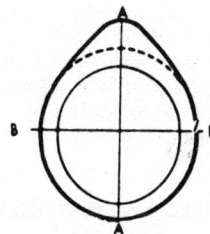

Camshaft lobe measurement

shaft gear and the block. Pull the camshaft forward and release it. Compare the dial indicator reading with the specifications.
3. If the end play is excessive, check the spacer for correct installation before it is removed. If the spacer is correctly installed, replace the thrust plate.
4. Remove the dial indicator.

CAMSHAFT BEARING REPLACEMENT

1. Remove the engine following the procedures in this chapter and install it on a work stand.
2. Remove the camshaft, flywheel and crankshaft, following the appropriate procedures. Push the pistons to the top of the cylinder.
3. Remove the camshaft rear bearing bore plug. Remove the camshaft bearings with a bearing removal tool.
4. Select the proper size expanding collet and back-up nut and assemble on the mandrel. With the expanding collet collapsed, install the collet assembly in the camshaft bearing and tighten the back-up nut on the expanding mandrel until the collet fits the camshaft bearing.
5. Assemble the puller screw and extension (if necessary) and install on the expanding mandrel. Wrap a cloth around the threads of the puller screw to protect the front bearing or journal. Tighten the pulling nut against the thrust bearing and pulling plate to remove the camshaft bearing. Be sure to hold a wrench on the end of the puller screw to prevent it from turning.
6. To remove the front bearing, install the puller from the rear of the cylinder block.

7. Position the new bearings at the bearing bores, and press them in place. Be sure to center the pulling plate and puller screw to avoid damage to the bearing. Failure to use the correct expanding collet can cause severe bearing damage. Align the oil holes in the bearings with the oil holes in the cylinder block before pressing bearings into place.

8. Install the camshaft rear bearing bore plug.

9. Install the camshaft, crankshaft, flywheel and related parts, following the appropriate procedures.

10. Install the engine in the truck, following procedures described earlier in this section.

Pistons and Connecting Rods

REMOVAL

NOTE: In most cases, this procedure is easier with the engine out of the vehicle.

1. Remove the head(s).
2. Remove the oil pan.
3. Rotate the engine to bring each piston, in turn, to the bottom of its stroke. With the piston bottomed, use a ridge reamer to remove the ridge at the top of the cylinder. DO NOT CUT TOO DEEPLY!
4. Matchmark the rods and caps. If the pistons are to be removed from the connecting rod, mark the cylinder number on the piston with a silver pencil or quick drying paint for proper cylinder identification and cap-to-rod location. Remove the connecting rod capnuts and lift off the rod caps, keeping them in order. Install a guide hose over the threads of the rod bolts. This is to prevent damage to the bearing journal and rod bolt threads.
5. Using a hammer handle, push the piston and rod assemblies up out of the block.

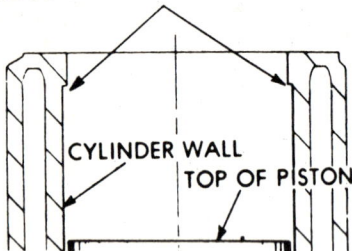
Cylinder bore ridge

PISTON PIN REMOVAL AND INSTALLATION

Use care at all times when handling and servicing connecting rods and pistons. To prevent possible damage to these units, do not clamp the rod or piston in a vise since they may become distorted. Do not allow the pistons to strike against one another, against hard objects or bench surfaces, since distortion of the piston contour or nicks in the soft aluminum material may result.

1. Remove the piston rings using a suitable piston ring remover.
2. Remove the piston pin lockring, if used. Install the guide bushing of the piston pin removing and installing tool.
3. Install the piston and connecting rod assembly on a support, and place the assembly in an arbor press. Press the pin out of the connecting rod, using the appropriate piston pin tool.
4. Assembly is the reverse of disassembly. Use new lockrings where needed.

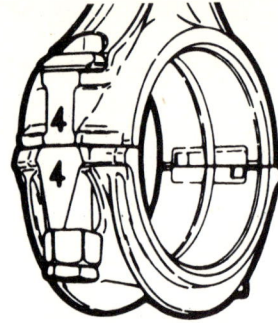
Number each rod and cap accordingly

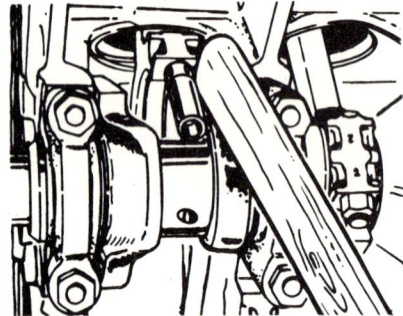
Push the piston out with a hammer handle

Remove the piston rings

Install the piston pin lock-rings (if used)

INSPECTION

Cylinder Block

Check the cylinder walls for evidence of rust, which would indicate a cracked block. Check the block face for distortion with a straightedge. Maximum distortion variance is 0.005 in. (0.13mm). The block cannot be planed, so it will have to be replaced if too distorted. Using a micrometer, check the cylinders for out-of-roundness.

Connecting Rods and Bearings

Wash connecting rods in cleaning solvent and dry with com-

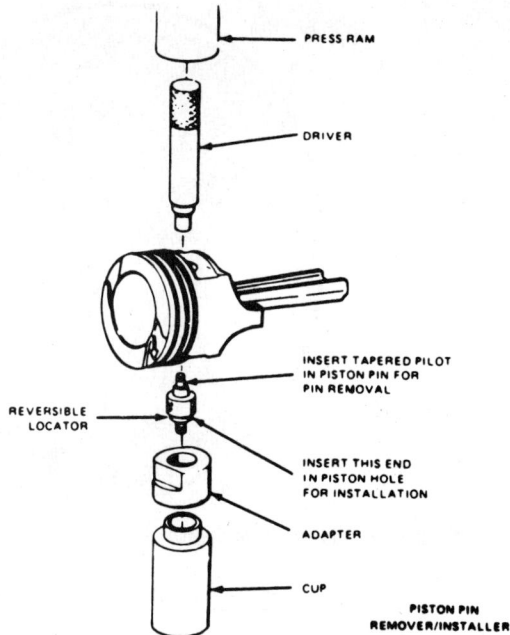

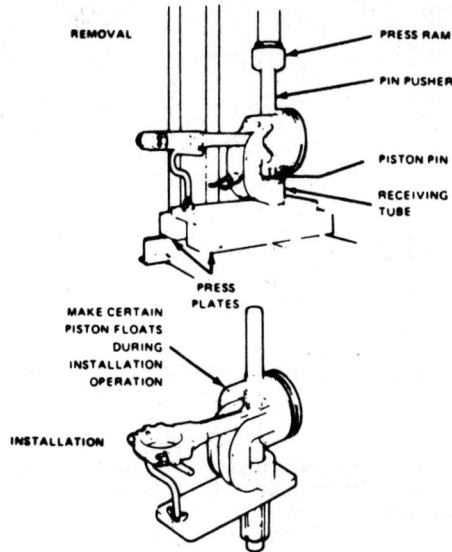

Piston pins must be pressed in with an arbor press

pressed air. Check for twisted or bent rods and inspect for nicks or cracks. Replace connecting rods that are damaged.

Inspect journals for roughness and wear. Slight roughness may be removed with a fine grit polishing cloth saturated with engine oil. Burrs may be removed with a fine oil stone by moving the stone on the journal circumference. Do not move the stone back and forth across the journal. If the journals are scored or ridged, the crankshaft must be replaced.

The connecting rod journals should be checked for out-of-round and correct size with a micrometer.

NOTE: Crankshaft rod journals will normally be standard size. If any undersized bearings are used, the size will be stamped on a counterweight.

If plastic gauging material is to be used:

1. Clean oil from the journal bearing cap, connecting rod and outer and inner surfaces of the bearing inserts. Position the insert so that the tang is properly aligned with the notch in the rod and cap.

2. Place a piece of plastic gauging material in the center of lower bearing shell.

3. Remove the bearing cap and determine the bearing clearances by comparing the width of the flattened plastic gauging material at its widest point with the graduation on the container. The number within the graduation on the envelope indicates the clearance in thousandths of an inch or millimeters. If this clearance is excessive, replace the bearing and recheck the clearance with the plastic gauging material. Lubricate the bearing with engine oil before installation. Repeat the procedure on the remaining connecting rod bearings. All rods must be connected to their journals when rotating the crankshaft, to prevent engine damage.

Pistons

Clean varnish from piston skirts and pins with a cleaning solvent. DO NOT WIRE BRUSH ANY PART OF THE PISTON. Clean the ring grooves with a groove cleaner and make sure oil ring holes and slots are clean.

Inspect the piston for cracked ring lands, skirts or pin bosses, wavy or worn ring lands, scuffed or damaged skirts, eroded areas at the top of the piston. Replace pistons that are damaged

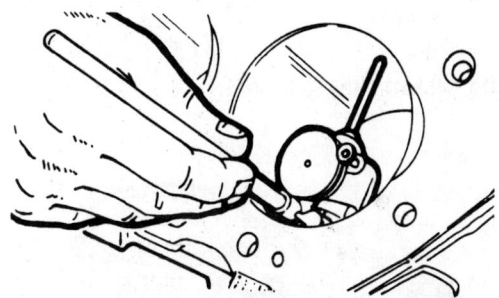

Measure the cylinder bore with a dial gauge

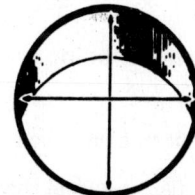

A—AT RIGHT ANGLE TO CENTERLINE OF ENGINE
B—PARALLEL TO CENTERLINE OF ENGINE

Cylinder bore measuring points. Take the top measurement 1/2 inch below the top, the bottom measurement 1/2 inch above the top of the piston at BDC

or show signs of excessive wear. Inspect the grooves for nicks or burrs that might cause the rings to hang up.

Measure piston skirt (across center line of piston pin) and check piston clearance.

MEASURING THE OLD PISTONS

Check used piston-to-cylinder bore clearance as follows:
1. Measure the cylinder bore diameter with a telescope gauge.
2. Measure the piston diameter. When measuring the pistons

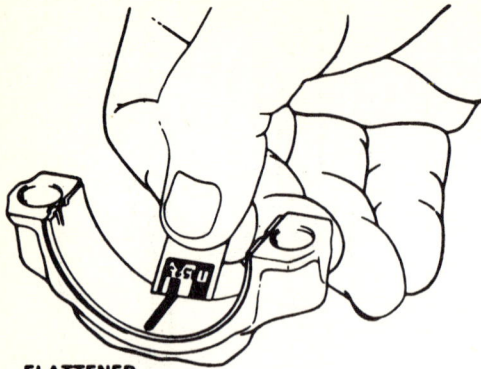

FLATTENED GAGING PLASTIC

Checking the connecting rod bearing clearance with Plastigage®

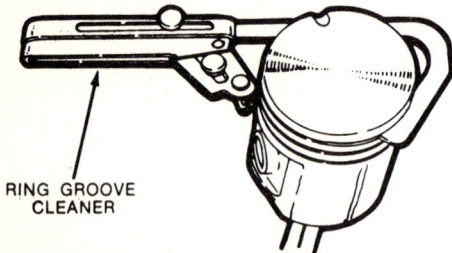

RING GROOVE CLEANER

Clean the piston ring grooves

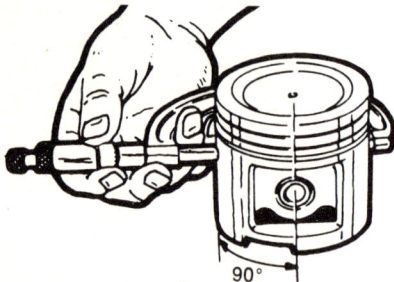

90°

Measure the piston prior to fitting

for size or taper, measurements must be made with the piston pin removed.

3. Subtract the piston diameter from the cylinder bore diameter to determine piston-to-bore clearance.

4. Compare the piston-to-bore clearances obtained with those clearances recommended. Determine if the piston-to-bore clearance is in the acceptable range.

5. When measuring taper, the largest reading must be at the bottom of the skirt.

SELECTING NEW PISTONS

1. If the used piston is not acceptable, check the service piston size and determine if a new piston can be selected. (Service pistons are available in standard, high limit and standard oversize.

2. If the cylinder bore must be reconditioned, measure the new piston diameter, then hone the cylinder bore to obtain the preferred clearance.

3. Select a new piston and mark the piston to identify the cylinder for which it was fitted. On some vehicles, oversize pistons may be found. These pistons will be 0.010 in. oversize.

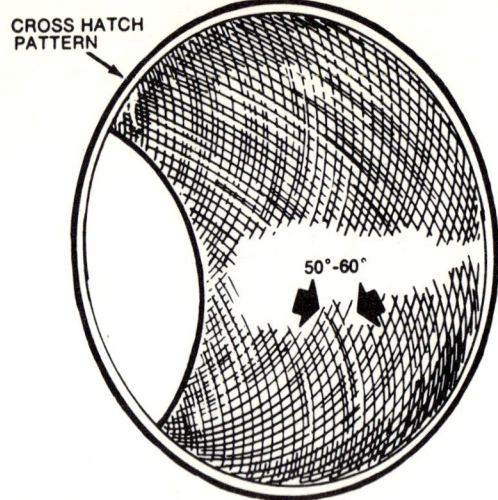

CROSS HATCH PATTERN

50°–60°

Cylinder bore after honing

CYLINDER HONING

1. When cylinders are being honed, follow the manufacturer's recommendations for the use of the hone.

2. Occasionally, during the honing operation, the cylinder bore should be thoroughly cleaned and the selected piston checked for correct fit.

3. When finish-honing a cylinder bore, the hone should be moved up and down at a sufficient speed to obtain a very fine uniform surface finish in a cross-hatch pattern of approximately 45–65° included angle. The finish marks should be clean but not sharp, free from imbedded particles and torn or folded metal.

4. Permanently mark the piston for the cylinder to which it has been fitted and proceed to hone the remaining cylinders.

NOTE: Handle the pistons with care. Do not attempt to force the pistons through the cylinders until the cylinders have been honed to the correct size. Pistons can be distorted through careless handling.

5. Thoroughly clean the bores with hot water and detergent. Scrub well with a stiff bristle brush and rinse thoroughly with hot water. It is extremely essential that a good cleaning operation be performed. If any of the abrasive material is allowed to remain in the cylinder bores, it will rapidly wear the new rings and cylinder bores. The bores should be swabbed several times with light engine oil and a clean cloth and then wiped with a clean dry cloth. CYLINDERS SHOULD NOT BE CLEANED WITH KEROSENE OR GASOLINE. Clean the remainder of the cylinder block to remove the excess material spread during the honing operation.

CHECKING CYLINDER BORE

Cylinder bore size can be measured with inside micrometers or a cylinder gauge. The most wear will occur at the top of the ring travel.

Reconditioned cylinder bores should be held to not more than 0.001 in. taper.

If the cylinder bores are smooth, the cylinder walls should not be deglazed. If the cylinder walls are scored, the walls may have to be honed before installing new rings. It is important that reconditioned cylinder bores be thoroughly washed with a soap and water solution to remove all traces of abrasive material to eliminate premature wear.

RING TOLERANCES

When installing new rings, ring gap and side clearance should be checked as follows:

Piston Ring and Rail Gap

Each ring and rail gap must be measured with the ring or rail positioned squarely and at the bottom of the ring travel area of the bore.

Side Clearance

Each ring must be checked for side clearance in its respective piston groove by inserting a feeler gauge between the ring and its upper land. The piston grooves must be cleaned before checking the ring for side clearance specifications. To check oil ring side clearance, the oil rings must be installed on the piston.

RING INSTALLATION

For service ring specifications and detailed installation productions, refer to the instructions furnished with the parts package.

PISTON ASSEMBLY AND INSTALLATION

1. Using a ring expander, install new rings in the grooves, with their gaps staggered to be 270° apart.
2. Using a straightedge, check the rods for straightness. Check, also, for cracks. Before assembling the block, it's a good idea to have the block checked for cracks with Magnaflux® or its equivalent.
3. Install the pins and retainers.
4. Coat the pistons with clean engine oil and apply a ring compressor. Position the assembly over the cylinder bore and slide the piston into the cylinder slowly, taking care to avoid nicking the walls. The pistons will have a mark on the crown, such as a

groove or notch or stamped symbol. This mark indicates the side of the piston which should face front. Lower the piston slowly, until it bottoms on the crankshaft. A good idea is to cover the rod studs with length of rubber hose to avoid nicking the crank journals. Assemble the rod caps at this time. Check the rod bear-

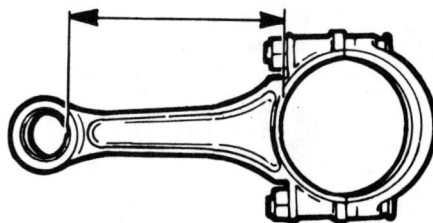

Check the connecting rod length (arrow)

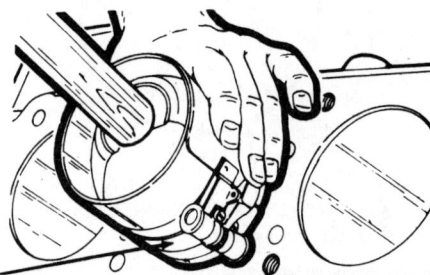

RING COMPRESSOR

Install the piston using a ring compressor

USE A SHORT PIECE OF 3/8" HOSE AS A GUIDE

Use lengths of vacuum hose or rubber tubing to protect the crankshaft journals and cylinder walls during piston installation

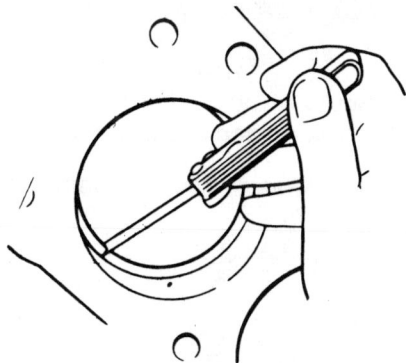

Check the piston ring end gap

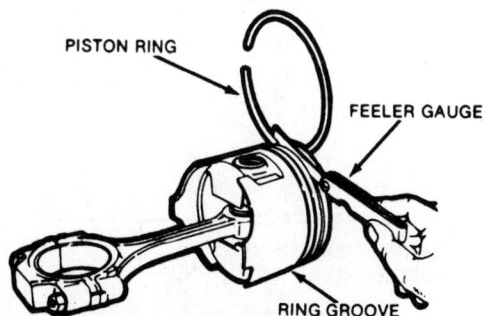

PISTON RING

FEELER GAUGE

RING GROOVE

Check the piston ring side clearance

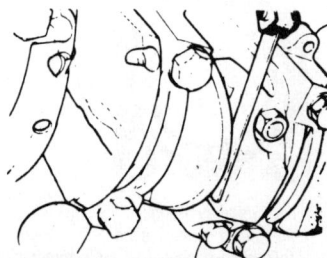

Check the connecting rod side clearance with a feeler gauge

ing clearances using Plastigage®, going by the instructions on the package.

5. Install the bearing caps with the stamped numbers matched. Torque the caps to the figure shown in the Torque Specifications Chart. See the accompanying illustrations for proper piston and rod installation.

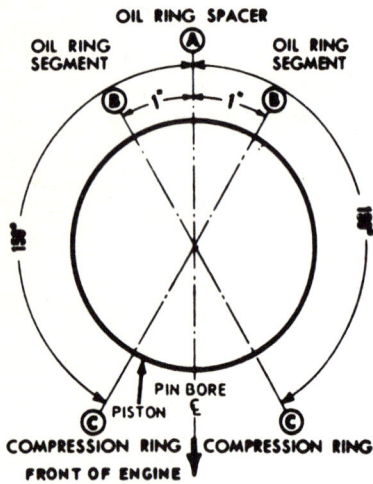

Proper ring gap spacing

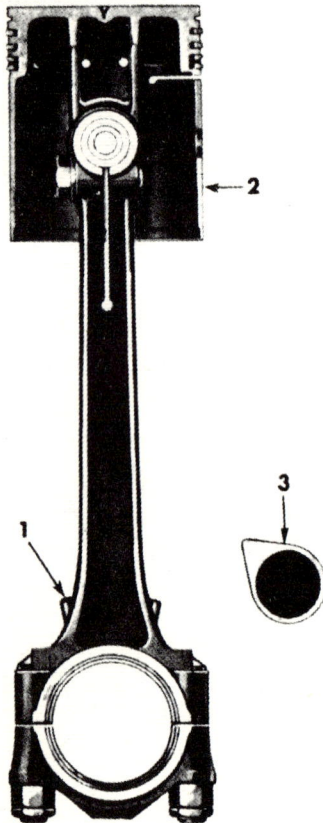

1. Oil spray hole
2. Piston skirt T-slot
3. Relative position of camshaft

4–134 piston and rod assembly

Rear Main Oil Seal

REPLACEMENT

4–134

1. Raise and support the vehicle on jackstands.
2. Remove the oil pan.
3. Remove the rear main bearing cap.
4. Using a center punch, drive the upper seal out of its groove just far enough to grasp it with a pliers, and pull it the rest of the way.

To install:

5. Apply a light film of chassis grease to the lower seal and install it in the cap.
6. Install the rubber packings in the upper crankcase half. The packings are of a predetermined length to allow about ¼ in. (6mm) protrusion. This protrusion is necessary for a positive seal. DO NOT TRIM THESE SEALS!

V6–225 left bank piston and rod assembly

V6–225 right bank piston and rod assembly

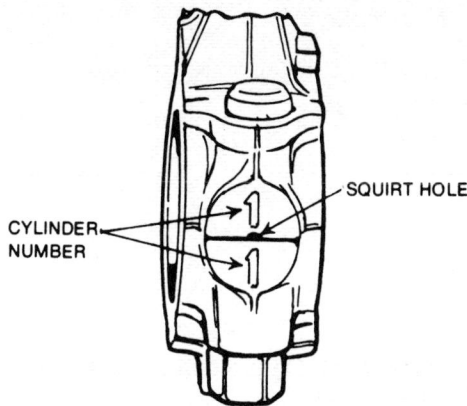

6–232, 6–258 connecting rod numbering

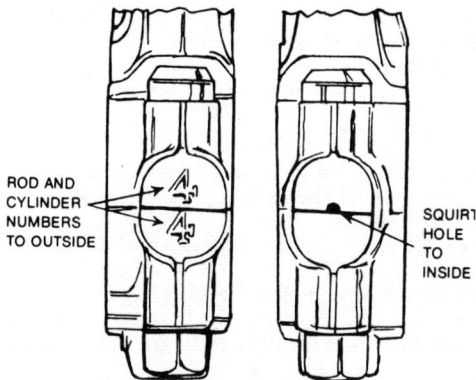

4–151 rod number and squirt hole

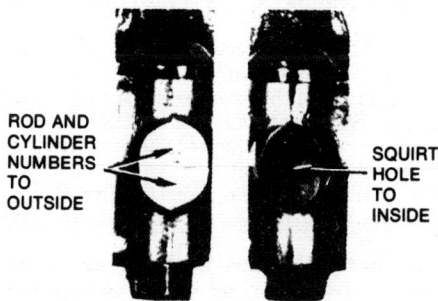

8–304 connecting rod and cap mating

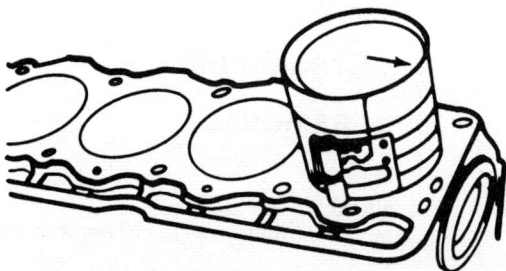

On the 4-150, the arrow on the piston crown faces front

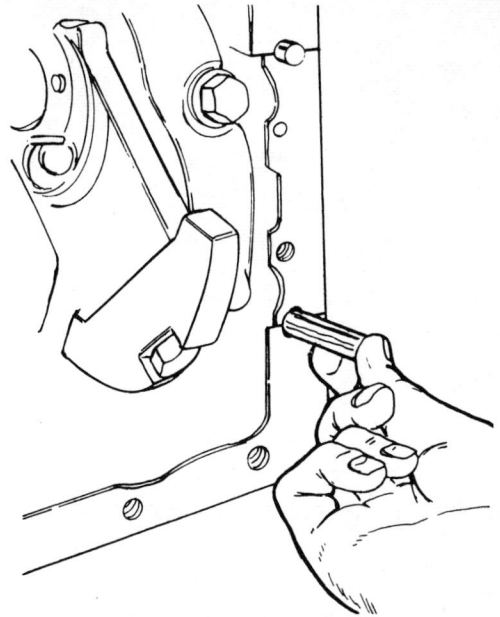

4–134 rear main bearing cap packing

7. Apply a small amount of RTV sealant to both sides and face of the bearing cap and install it.

8. Torque the cap to the figure shown in the torque specifications chart.

9. Install the oil pan.

6–225

NOTE: For removal of both upper and lower seals, the crankshaft must be removed from the engine. Crankshaft removal is easiest with the engine out of the vehicle.

1. Remove engine.

2. Remove the timing case and gears, and any spacers and shims from the front of the crankshaft.

3. Remove the oil pan and oil pickup.

4. Note the mating of crankshaft and main caps. Match mark them with an indelible inker.

5. Unbolt the #1 main bearing cap. Using a pry bar, carefully lift the cap from the dowels. Take great care to avoid damaging the dowels. Any bent dowels must be replaced.

6. In the same manner, remove the next two caps.

7. To remove the rear cap, rear main bearing bolt remover, special tool W-323 or its equivalent must be used.

8. Match mark the connecting rod caps and remove them.

9. Lift out the crankshaft.

10. Remove the braided seal from the inner groove of the cap and the neoprene seals from the outer grooves.

11. Using a center punch, drive the block seal out just far enough to grasp with a pliers and pull out.

To install:

12. Dip a new block seal in engine oil and force it into place in the block.

13. Insert the new braided cap seal into the groove in the cap and coat it with engine oil.

14. Cut the ends of the braided seals flush with the cap and block surfaces.

15. The new neoprene seals are installed after the cap is torqued in place. These seals are supposed to project about $\frac{1}{16}$ in. (1.6mm) above the cap. DO NOT CUT THESE SEALS FLUSH WITH THE CAP! Before installation, dip the neoprene seals in kerosene for about 1–2 minutes. After installation squirt some more kerosene on the protruding ends of the seals.

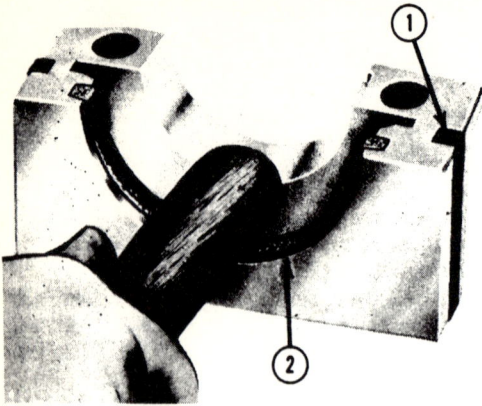

1. Neoprene seal
2. Fabric seal

Installing the V6-225 crankshaft rear oil seal

Then, peen the ends of the seals with a hammer to make sure of a tight seal at the upper parting line between the cap and block.

16. Installation of the crankshaft and remaining parts is the reverse of removal. Torque the main cap bolts evenly, one side then the other, a little at a time until the torque value is reached.

NOTE: Whenever the second cap is removed, the thrust surfaces must be aligned. To do this, pry the crankshaft back and forth several times throughout its end travel with the cap bolts of the second main cap only finger tight.

4-150

1. Remove the transmission.
2. Remove the flywheel.
3. Pry out the seal from around the crankshaft flange.
To install:
4. Coat the inner lip of the new seal with clean engine oil.
5. Gently tap the new seal into place, flush with the block, using a rubber or plastic mallet.
6. Install the flywheel.
7. Install the transmission.

4-151

NOTE: The seal is a one piece unit that can be removed and installed without removing the oil pan or crankshaft.

1. Raise and support the vehicle on jackstands.
2. Remove the transmission and transfer case as an assembly.
3. Disconnect and remove the starter.
4. On manual transmission vehicles, remove the flywheel inspection plate, and clutch slave cylinder.
5. Remove the flywheel or drive plate housing.
6. On manual transmission, remove the clutch assembly by backing out the bolts evenly around the pressure plate.
7. Remove the flywheel or drive plate. It is a good idea to match mark the flywheel location for assembly.
8. Using a small bladed screwdriver, pry the rear seal out from around the crankshaft hub. Be careful to avoid damaging the sealing groove or hub.
To install:
9. With a light hammer, tap the seal into position with the lip facing the front of the engine.
10. Install the flywheel or drive plate.

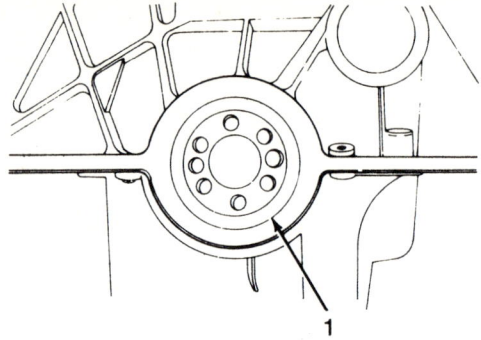

4-150 rear main seal. (1) indicates the actual seal around the crankshaft

11. On manual transmission, install the clutch assembly by tightening the bolts evenly around the pressure plate.
12. Install the flywheel or drive plate housing.
13. On manual transmission vehicles, install the flywheel inspection plate, and clutch slave cylinder.
14. Install the starter.
15. Install the transmission and transfer case as an assembly.

6-232, 6-258, 8-304

This seal is a two piece neoprene type with a single lip.
1. Raise and support the vehicle on jackstands.
2. Remove the oil pan.
3. Remove the rear main bearing cap and discard the lower seal.
4. Loosen all remaining main bearing caps.
5. Using a center punch, carefully drive the upper half of the seal out of the block just far enough to graph with a pliers and pull out.
6. Remove the oil pan front and rear seals and the side gaskets.
To install:
7. Clean all gasket surfaces.
8. Wipe clean the sealing surface of the crankshaft and coat it lightly with engine oil.
9. Coat the lip of the upper seal with engine oil and install it in the block. The lip faces forward.
10. Coat both end tabs of the lower seal with RTV silicone sealer. Do not get any RTV sealer on the seal lip.
11. Coat the outer curved surface of the seal with liquid soap. Coat the seal lip with engine oil.
12. Install the seal into the cap, pressing firmly.
13. Coat both chamferred edges of the cap with RTV sealer.

WARNING: Do not allow any RTV sealer to get on the mating surfaces of the cap or block as this will affect bearing clearance!

14. Install the rear main cap.
15. Tighten all main bearing cap bolts gradually to 80 ft. lbs.
16. Replace the oil pan.

Crankshaft

REMOVAL

4-134

1. Remove the engine from the vehicle.
2. Remove the fan, hub, timing cover, pulleys and timing gears from the engine.
3. Slide the thrust washers and adjusting shims from the front end of the crankshaft. Be careful to avoid losing or mismatching the washers and shims!

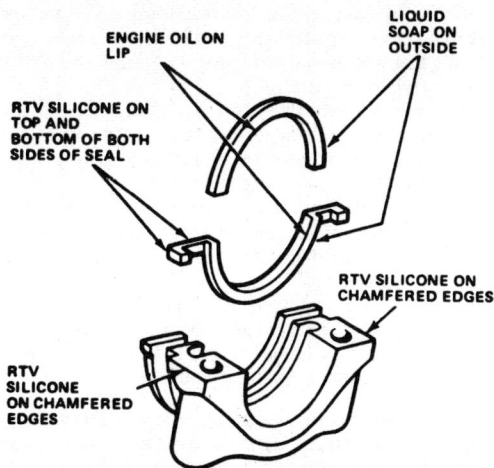

Rear main seal installation for the 6-232, 6-258 and 8-304

Rear main seal installation for the 6-232, 6-258 and 8-304

4. Remove the oil pan and pickup.
5. Pull the two pieces of the rear main cap packing out from the sides of the cap.
6. Note the match marks on the bearing caps and remove the nuts and washers from the cap dowels.
7. Using a pry bar under the ends of each main cap, carefully lift the caps off the dowels. Take great care to avoid damage to the caps or dowels, as bent dowels must be replaced.
8. Note the match marks on the connecting rod caps and remove the bearing caps from the rods.
9. Lift the crankshaft from the block.

4-151

1. Remove the engine and mount it on a work stand.
2. Remove the spark plugs.
3. Remove the fan and pulley.
4. Remove the vibration damper and hub.
5. Remove the oil pan and oil pump.
6. Remove the timing case cover.
7. Remove the crankshaft timing gear.
8. Remove the connecting rod bearing caps. Mark each for reassembly.
9. Remove the main bearing caps, marking each for reassembly.
10. Remove the crankshaft.

6-225

1. Remove the engine and mount it on a work stand.
2. Remove the flywheel.
3. Remove the fan and hub.
4. Remove the crankshaft pulley and vibration damper.
5. Remove the timing chain and sprocket.
6. Remove the oil pan, and pickup.
7. Remove the connecting rod caps, marking them for reassembly.
8. Mark the main bearing caps.
9. Remove the two bolts from the cap and carefully lift it off with the aid of a pry bar. Similarly remove the next two caps.
10. To remove the last cap, a special tool, main bearing bolt remover W-323 is necessary.
11. Lift the crankshaft from the block.

4-150, 6-232, 6-258 and 8-304

1. Remove the engine from the vehicle and mount it on a work stand.
2. Drain the oil.

3. Remove the flywheel or torque converter, match marking the pieces for installation.
4. Remove all drive belts.
5. Remove the fan and hub assembly.
6. Remove the crankshaft pulley and vibration damper.
7. Remove the timing case cover.
8. Remove the oil pan.
9. Remove the oil pump and pickup.
10. Remove the rod bearing caps, marking them for installation.
11. Remove the main bearing caps, marking them for installation.
12. Lift out the crankshaft.

NOTE: A replacement oil pickup tube must be used. Do not attempt to install the original. Make sure the plastic button is inserted in the bottom of the pickup screen. Always use a new rear main seal. If new bearings are installed, check clearances with Plastigage®.

INSPECTION

1. Check the crankshaft for wear or damage to the bearing surfaces of the journals. Crankshafts that are damaged can be reconditioned by a professional machine shop.
2. Using a dial indicator, check the crankshaft journal runout. Measure the crankshaft journals with a micrometer to determine the correct size rod and main bearings to be used. Whenever a new or reconditioned crankshaft is installed, new connecting rod bearings and main bearings should be installed.
3. Clean all oil passages in the block (and crankshaft if it is being reused).

NOTE: A new rear main seal should be installed any time the crankshaft is removed or replaced.

4. Wipe the oil from the crankshaft journal and the outer and inner surfaces of the bearing shell.
5. Place a piece of plastic gauging material in the center of the bearing.
6. Use a floor jack or other means to hold the crankshaft against the upper bearing shell. This is necessary to obtain accurate clearance readings when using plastic gauging material.
7. Install the bearing cap and bearing. Place engine oil on the cap bolts and install. Torque the bolts to specification.
8. Remove the bearing cap and determine the bearing clearance by comparing the width of the flattened plastic gauging material at its widest point with the graduations on the gauging material container. The number within the graduation on the envelope indicates the clearance in millimeters or thousandths of an inch. If the clearance is greater than allowed, REPLACE BOTH BEARING SHELLS AS A SET. Recheck the clearance after replacing the shells.

INSTALLATION

NOTE: Main bearing clearances must be corrected by the use of selective upper and lower shells. UNDER NO CIRCUMSTANCES should the use of shims behind the shells to compensate for wear be attempted!

1. Install new bearing upper halves in the block. If the crankshaft has been turned to resurface the journals, undersized

bearing must be used to compensate. Lay the crankshaft in the block.

2. Install the lower bearing halves in the caps.
3. Use Plastigage® to check bearing fit.
4. When the bearings are properly fitted, install and torque the bearing caps.
5. Check crankshaft endplay to determine the need for thrust washers.
6. While you're at it, it's a good idea to replace the rear main seal at this time.
7. Install sufficient oil pan bolts in the block to align with the connecting rod bolts. Use rubber bands between the bolts to position the connecting rods as required. Connecting rod position

can be adjusted by increasing the tension on the rubber bands with additional turns around the pan bolts or thread protectors.

8. Position the upper half of main bearings in the block and lubricate them with engine oil.
9. Position crankshaft keyway in the same position as removed and lower it into block. The connecting rods will follow the crank pins into the correct position as the crankshaft is lowered.
10. Lubricate the thrust flanges with clean engine oil or engine rebuilding oil. Install caps with the lower half of the bearings lubricated with engine oil. Lubricate the cap bolts with engine oil and install, but do not tighten.
11. With a block of wood, bump the shaft in each direction to align the thrust flanges of the main bearing. After bumping the shaft in each direction, wedge the shaft to the front and hold it while torquing the thrust bearing cap bolts.

NOTE: In order to prevent the possibility of cylinder block and/or main bearing cap damage, the main bearing caps are to be tapped into their cylinder block cavity using a wood or rubber mallet before the bolts are in-

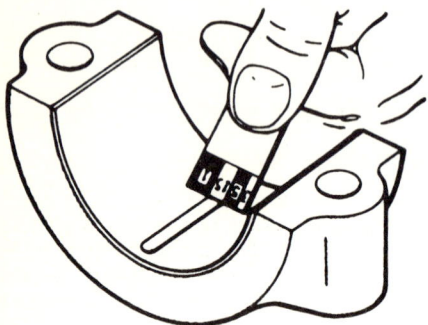
Checking main bearing oil clearance with Plastigage®

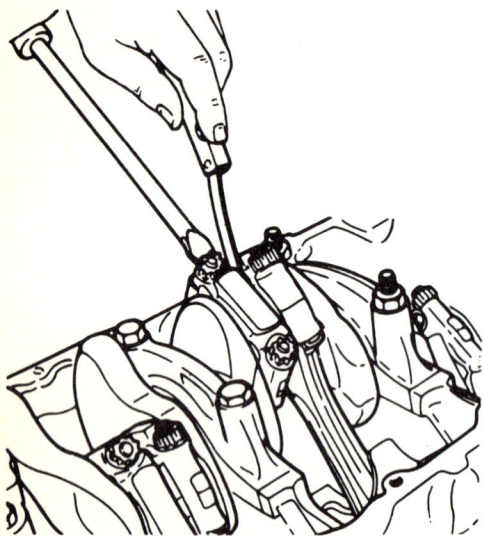
Checking rod side clearance with a flat feeler gauge. Use a small prybar to spread the rods

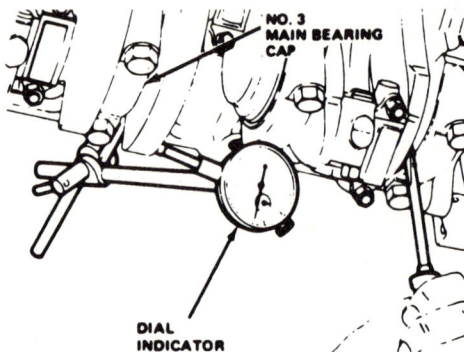
Checking crankshaft endplay with a dial indicator

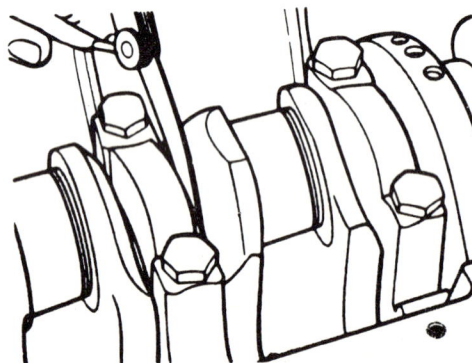
Checking crankshaft endplay with a feeler gauge

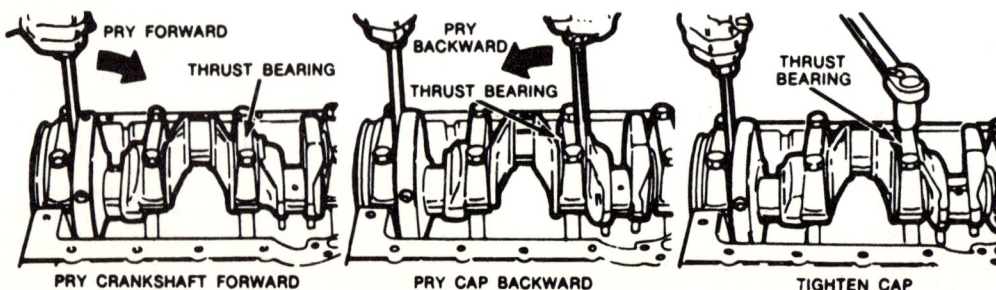
Crankshaft thrust bearing alignment

stalled. Do not use attaching bolts to pull the main bearing caps into their seats. Failure to observe this information may damage the cylinder block or a bearing cap.

12. Torque all main bearing caps to specification. Check crankshaft endplay, using a flat feeler gauge.
13. Remove the connecting rod bolt thread protectors and lubricate the connecting rod bearings with engine oil.
14. Install the connecting rod bearing caps in their original position. Torque the nuts to specification.
15. Install all parts in reverse order of removal. See related procedures in this section for component installation.

Flywheel/Flex Plate and Ring Gear

NOTE: Flex plate is the term for a flywheel mated with an automatic transmission.

REMOVAL AND INSTALLATION

All Engines

NOTE: The ring gear is replaceable only on engines mated with a manual transmission. Engines with automatic transmissions have ring gears which are welded to the flex plate.

1. Remove the transmission and transfer case.
2. Remove the clutch, if equipped, or torque converter from the flywheel. The flywheel bolts should be loosened a little at a time in a cross pattern to avoid warping the flywheel. On Jeep vehicles with manual transmission, replace the pilot bearing in the end of the crankshaft if removing the flywheel.
3. The flywheel should be checked for cracks and glazing. It can be resurfaced by a machine shop.
4. If the ring gear is to be replaced, drill a hole in the gear between two teeth, being careful not to contact the flywheel surface.
5. Using a cold chisel at this point, crack the ring gear and remove it.
6. Polish the inner surface of the new ring gear and heat it in an oven to about 600°F (316°C). Quickly place the ring gear on the flywheel and tap it into place, making sure that it is fully seated.

NOTE: Never heat the ring gear past 800°F (426°C), or the tempering will be destroyed.

7. Position the flywheel on the end of the crankshaft. Torque the bolts a little at a time, in a cross pattern, to the torque figure shown in the Torque Specifications Chart.
8. Install the clutch or torque converter.
9. Install the transmission and transfer case.

EXHAUST SYSTEM

— CAUTION —
When working on exhaust systems, ALWAYS wear protective goggles! Avoid working on a hot exhaust system!

Muffler

REMOVAL AND INSTALLATION

NOTE: The following applies to exhaust systems using clamped joints. Most later model, original equipment systems use welded joints at the muffler. These joints will, of course, have to be cut.

1. Raise and support the rear end on jackstands, placed under the frame, so that the axle hangs freely.
2. Remove the muffler clamps.
3. Remove the tailpipe hanger clamp.
4. Spray the joint liberally with a penetrant/rust dissolver compound such as Liquid Wrench®, WD-40®, or equivalent.
5. If the tailpipe cannot be pulled or twisted free from the muffler, drive a chisel between the muffler and tailpipe at several places to free it.
6. Disconnect the muffler hanger.
7. If the pipe leading into the muffler is not to be replaced, and cannot be pulled free of the muffler, heat the joint with an oxyacetylene torch until it is cherry red. Place a block of wood

against the front of the muffler and drive it rearward to disengage it from the pipe.
If the pipe is being replaced, use a chisel to free it.

— CAUTION —
When using a torch, make certain that no combustibles or brake or fuel lines are in the immediate area of the torch.

8. When installing the new muffler, make sure that the locator slot and tab at the tailpipe joint index each other.
9. Drive the muffler onto the front pipe.
10. Position the system, without muffler clamps, under the Jeep and install the hangers. Make certain that there is sufficient clearance between the system components and the floor pan and axle. Then, install the muffler clamps and tighten the hangers.

NOTE: Install the muffler clamps so that the shafts of the U-bolts covers the slots in the joint flanges.

Front Exhaust Pipe (Head Pipe)

REMOVAL AND INSTALLATION

1. Raise and support the front end on jackstands.
2. Disconnect any oxygen sensor wires or air injection pipes.
3. Disconnect the front pipe at the manifold(s).

1. Head pipe
2. Intermediate pipe
3. Muffler
4. Tail pipe
5. Y-pipe
6. Head pipe
7. Head pipe clamp
8. Muffler clamp
9. Tail pipe hanger
A. 4-cylinder exhaust system
B. V6 exhaust system

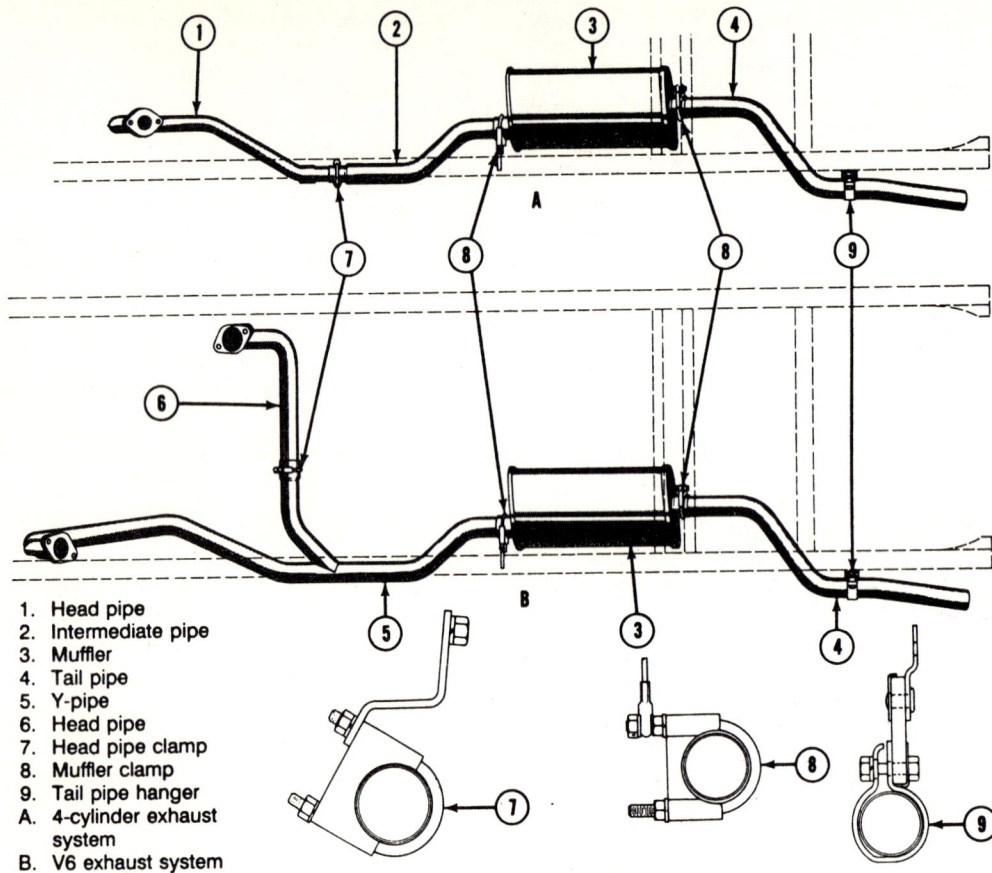

Exhause system for late model F4–134 and 6–225 CJs

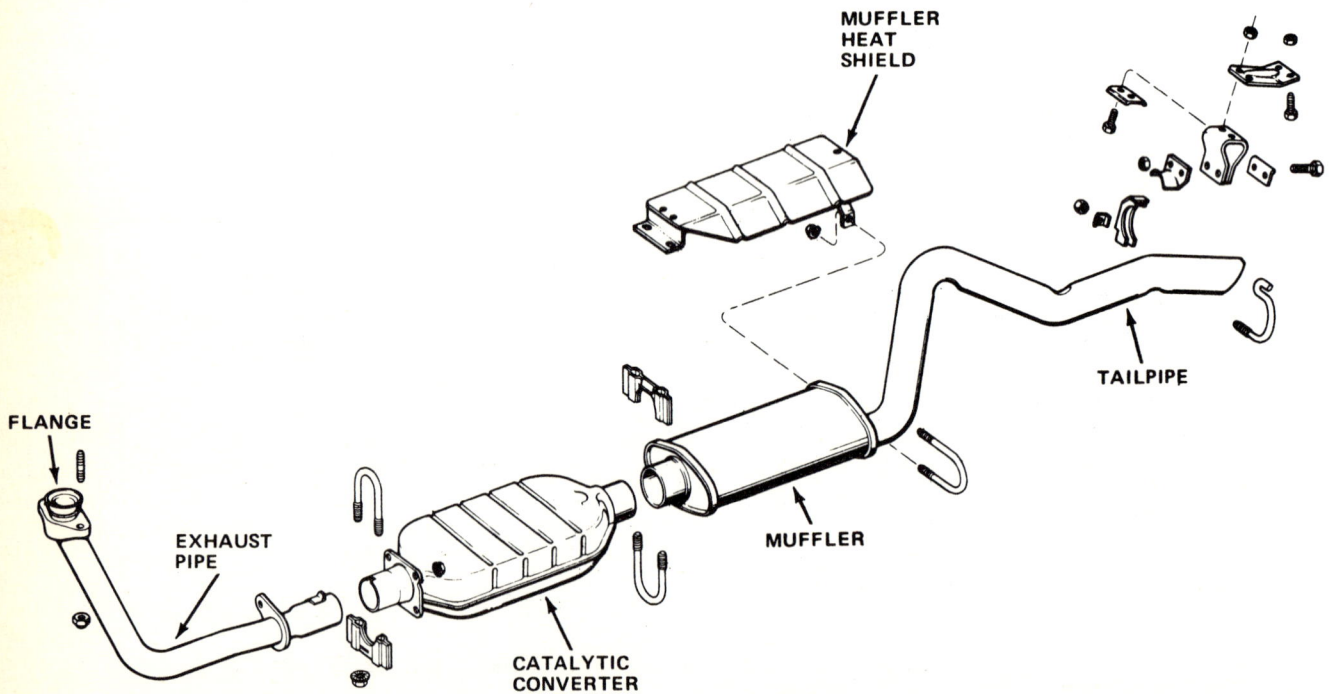

MUFFLER HEAT SHIELD

TAILPIPE

FLANGE

EXHAUST PIPE

MUFFLER

CATALYTIC CONVERTER

Exhaust system used with the 4–151 engine

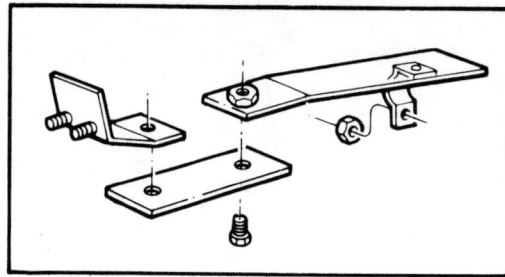

HEAT SHIELD
3 & 4-SPEED TRANSMISSION

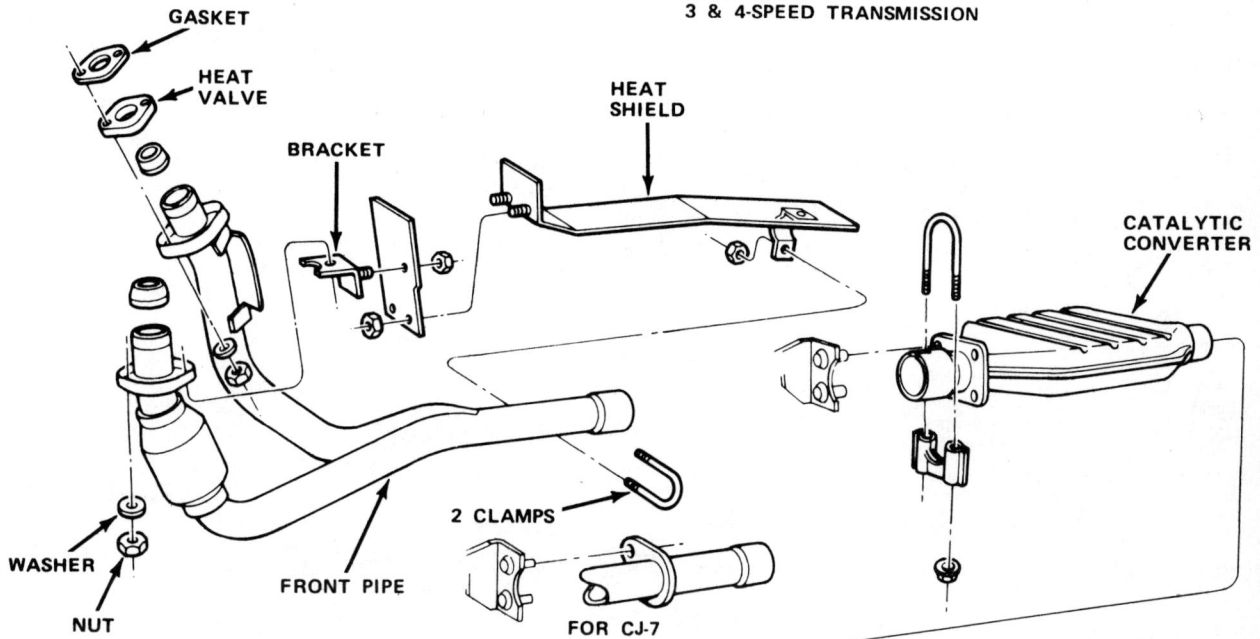

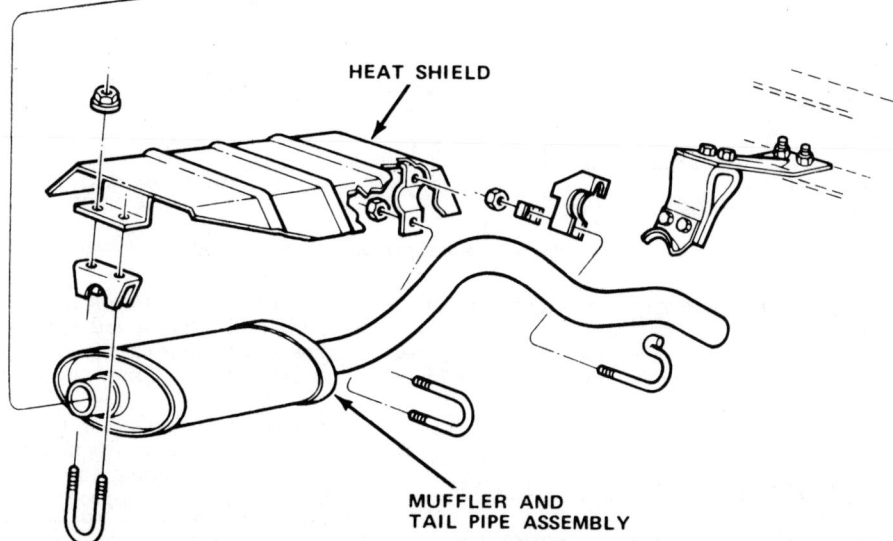

GASKET

HEAT VALVE

BRACKET

HEAT SHIELD

CATALYTIC CONVERTER

WASHER

FRONT PIPE

2 CLAMPS

NUT

FOR CJ-7

HEAT SHIELD

MUFFLER AND TAIL PIPE ASSEMBLY

Exhaust system used with the 8–304 engine

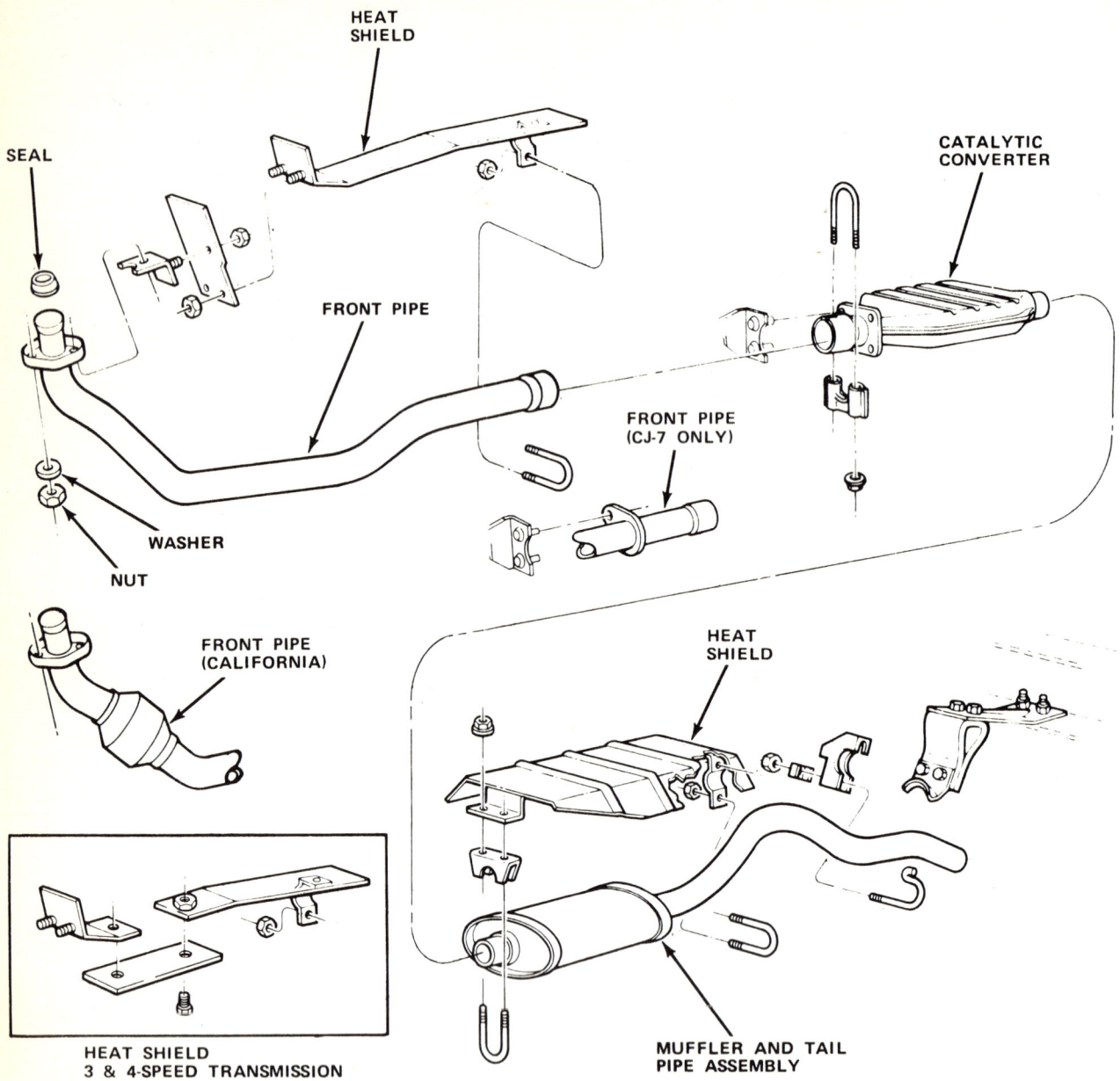

1975–79 6-258 exhaust system

4. Disconnect the rear end of the pipe from the muffler or catalytic converter.

5. Installation is the reverse of removal. Torque the pipe-to-manifold nuts to 17 ft. lbs. on 4-cylinder engines, or 20 ft. lbs. on the 6- and 8-cylinder engines. Make sure the pipe is properly aligned.

Rear Exhaust Pipe or Tailpipe

REMOVAL AND INSTALLATION

NOTE: Some vehicle use an intermediate pipe, also called a rear exhaust pipe. This pipe connects the front pipe with the muffler, or runs between the converter and muffler.

1. Raise and support the rear end on jackstands.

2. If just the intermediate pipe is being replaced, cut it at the joints and collapse and remove the remainder from the front pipe and muffler or converter. If adjoining parts are also being replaced, the pipe may be chiseled off.

3. If just the tailpipe is being replaced, cut it just behind the muffler and collapse and remove the remainder from the muffler flange. Remove the tailpipe hanger.

4. When installing any pipe, position it in the system and make sure that it is properly aligned and has sufficient clearance at the floor pan. Position U-bolts so that the bolt shafts cover any slots in the pipe flanges. When the system is correctly aligned, tighten all U-bolts and hangers.

Catalytic Converter

REMOVAL AND INSTALLATION

1. Raise and support the rear end on jackstands.

CATALYTIC
CONVERTER
HEAT SHIELD (CALIFORNIA ONLY)

SEAL

CATALYTIC
CONVERTER

CONVERTER
AIR INJECTION
TUBE (CALIFORNIA ONLY)

FRONT
PIPE CJ-7

FRONT PIPE
CJ-5

WASHER

NUT

MUFFLER
HEAT
SHIELD

MUFFLER AND TAIL
PIPE ASSEMBLY

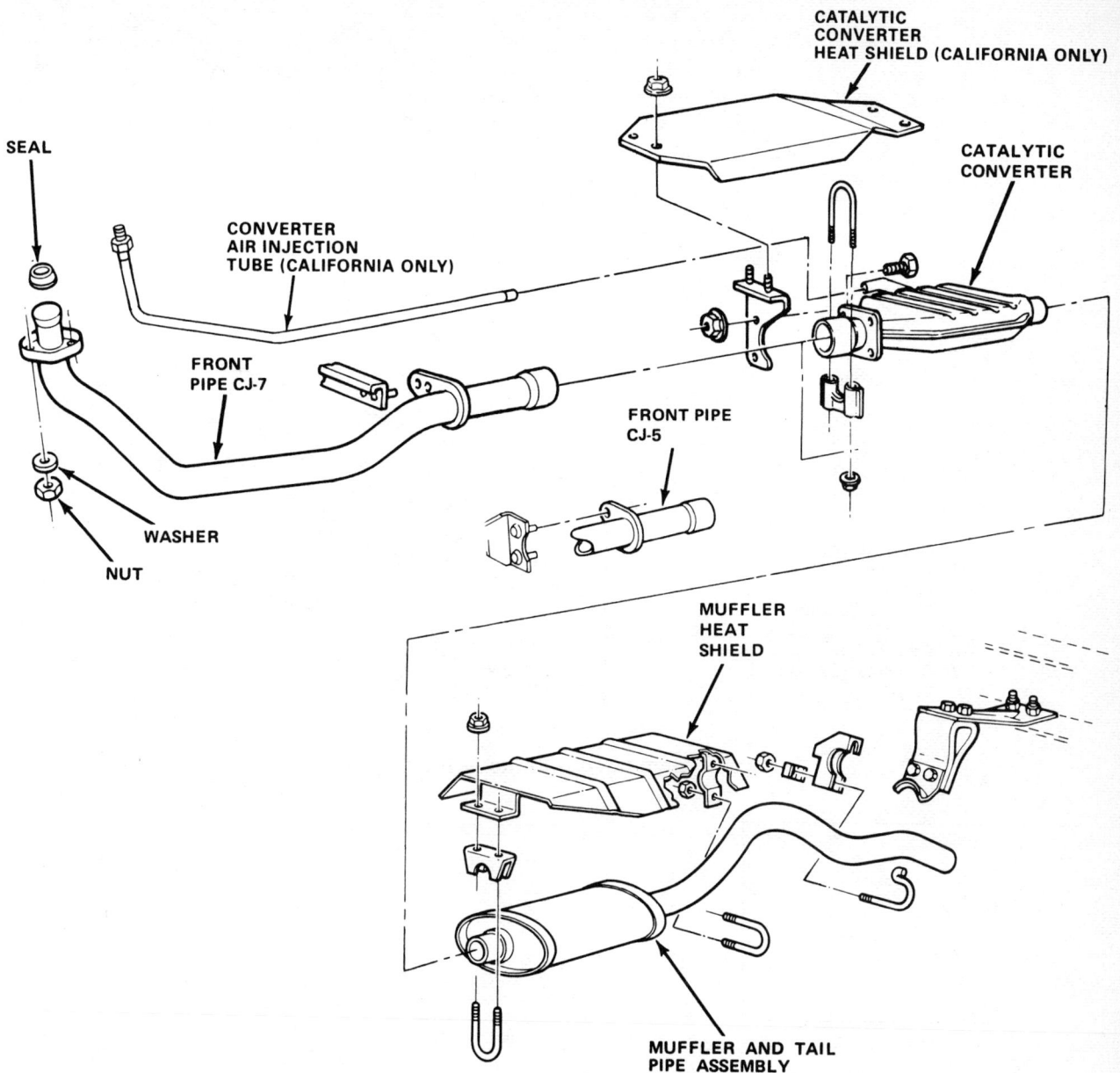

1980–86 6–258 exhaust system

2. Disconnect the downstream air injection tube at the converter.

3. On the 4-cylinder engine, the front pipe is bolted to the converter at a facing flange. On the 6- and 8-cylinder engines, the front pipe and converter are clamped together at a slip-fit joint. On all engines, the rear joint of the converter is a slip-fit.

4. To avoid damaging any components, it will probably be necessary to heat any slip-fit joint with an oxyacetylene torch, until the joint is cherry red. Then, place a block of wood against the converter and drive it off of the pipe.

────────── **CAUTION** ──────────

When using a torch, make certain that no combustibles or brake or fuel lines are in the immediate area of the torch.

5. Position the replacement converter in the system and install the rear clamp. Hand tighten the nuts.

6. On 4-cylinder engines: bolt the flanges together at the front end. Tighten the bolts to 25 ft. lbs. Tighten the rear clamp nuts to 45 ft. lbs.

7. On 6- and 8-cylinder engines, install the front clamp, make sure that the converter is properly positioned and tighten the front and rear clamp nuts to 45 ft. lbs.

8. Install the downstream air injection tube and tighten the clamps to 36–48 inch lbs.

9. Lower the Jeep.

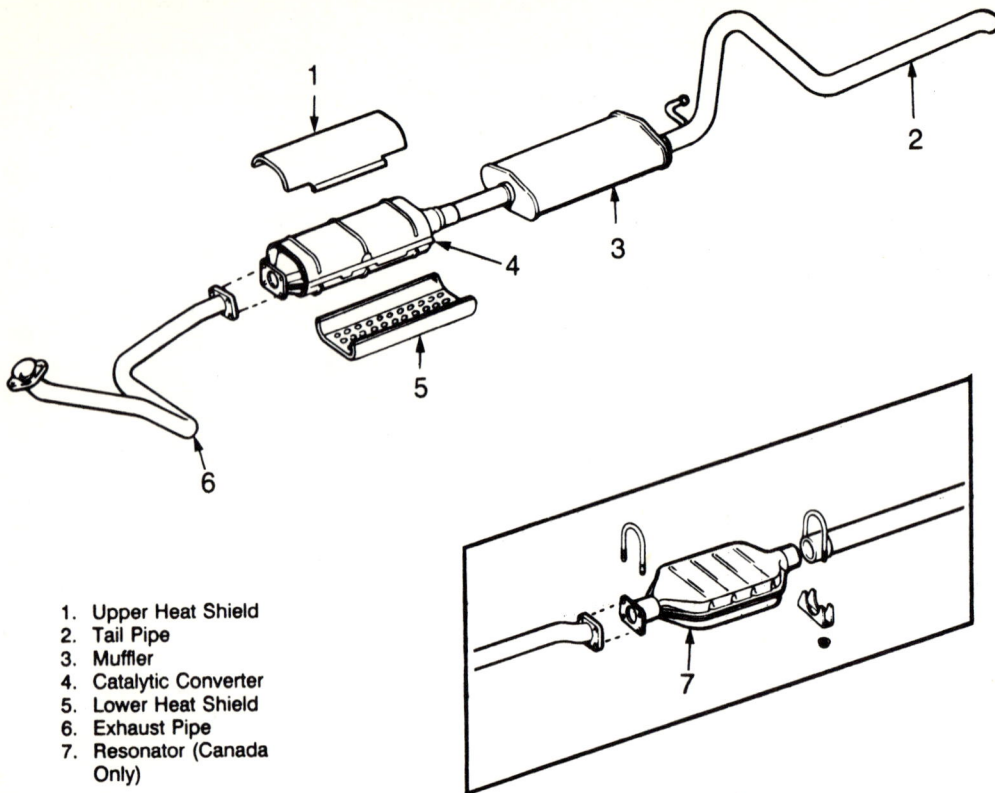

1. Upper Heat Shield
2. Tail Pipe
3. Muffler
4. Catalytic Converter
5. Lower Heat Shield
6. Exhaust Pipe
7. Resonator (Canada Only)

Exhaust system used with the 4–150 engine

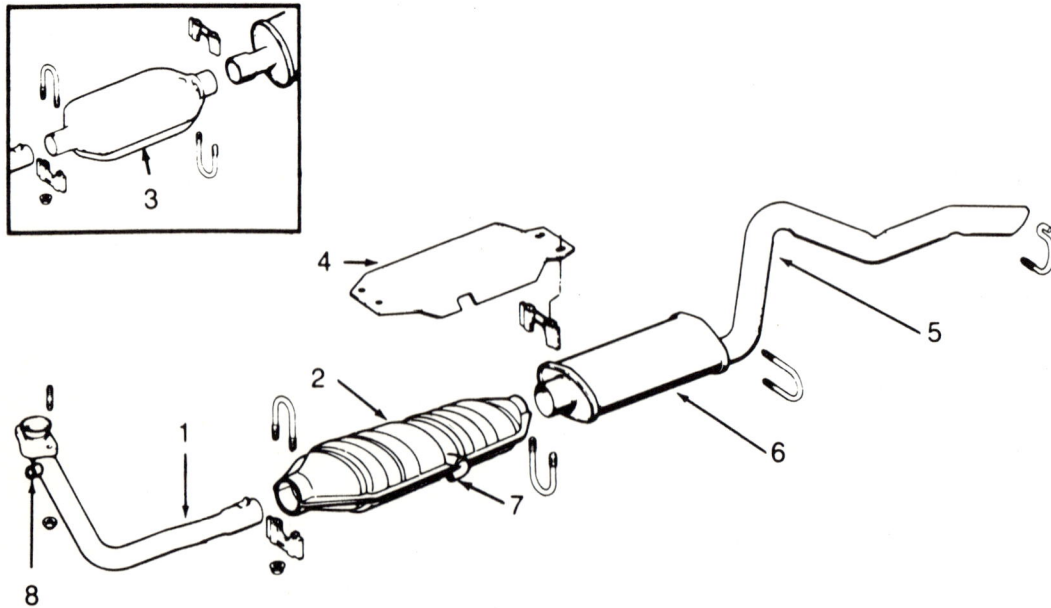

1987–90 6–258 exhaust system

Emission Controls

4

EMISSION CONTROLS

There are three types of automotive pollutants: crankcase fumes, exhaust gases and gasoline evaporation. The equipment that is used to limit these pollutants is commonly called emission control equipment.

Crankcase Emission Controls

The crankcase emission control equipment consists of a positive crankcase ventilation valve (PCV), a closed or open oil filler cap and hoses to connect this equipment.

When the engine is running, a small portion of the gases which are formed in the combustion chamber during combustion, leak by the piston rings and enter the crankcase. Since these gases are under pressure, they tend to escape from the crankcase and enter the atmosphere. If these gases were allowed to remain in the the crankcase for any length of time, they would contaminate the engine oil and cause sludge to build up. If the gases were allowed to escape into the atmosphere, they would pollute the air, as they contain unburned hydrocarbons. The crankcase emission control equipment recycles these gases back into the engine combustion chamber where they are burned.

Crankcase gases are recycled in the following manner: while the engine is running, clean filtered air is drawn into the crankcase either directly through the oil filler cap, or through the carburetor air filter and then through a hose leading to the oil filler cap. As the air passes through the crankcase, it picks up the combustion gases and carries them out of the crankcase, up through the PCV valve and into the intake manifold. After they enter the intake manifold, they are drawn into the combustion chamber and burned.

The most critical component in the system is the PCV valve. This vacuum controlled valve regulates the amount of gases which are recycled into the combustion chamber. At low engine speeds, the valve is partially closed, limiting the flow of gases into the intake manifold. As engine speed increases, the valve opens to admit greater quantities of the gases into the intake manifold. If the valve should become blocked or plugged, the gases will be prevented from escaping from the crankcases by the normal route. Since these gases are under pressure, they will find their own way out of the crankcase. This alternate route is usually a weak oil seal or gasket in the engine. As the gas escapes by the gasket, it also creates an oil leak. Besides causing oil leaks, a clogged PCV valve also allows these gases to remain in the crankcase for an extended period of time, promoting the formation of sludge in the engine.

The above explanation and the troubleshooting procedure which follows applies to all engines with PCV systems.

TROUBLESHOOTING

With the engine running, pull the PCV valve and hose from the engine. Block off the end of the valve with your finger. The engine speed should drop at least 50 rpm when the end of the valve is blocked. If the engine speed does not drop at least 50 rpm, then the valve is defective and should be replaced.

REMOVAL AND INSTALLATION

1. Pull the PCV valve and hose from the engine.
2. Remove the PCV valve from the hose. Inspect the inside of the PCV valve from the hose. If it is dirty, disconnect if from the intake manifold and clean it.
3. If the PCV valve hose was removed, connect it to the intake manifold.
4. Connect the PCV valve to its hose.
5. Install the PCV valve on the engine.

Exhaust Emission Controls

All of the gasoline engines used in these Jeep vehicles have, at one time or another, incorporated the air injection system for controlling the emission of exhaust gases into the atmosphere. Since this type of emission control system is common to most of the engines, it will be explained here.

The exhaust emission air injection system consists of a belt driven air pump which directs compressed air through connecting hoses to a steel distribution manifold into stainless steel injection tubes in the exhaust port adjacent to each exhaust valve. The air, with its normal oxygen content, reacts with the hot, but incompletely burned exhaust gases and permits further combustion in the exhaust port or manifold.

AIR PUMP

The air injection pump is a positive displacement vane type which is permanently lubricated and requires little periodic maintenance. The only serviceable parts on the air pump are the filter, exhaust tube, and relief valve. The relief valve relieves the air flow when the pump pressure reaches a preset level. This occurs at high engine rpm. This serves to prevent damage to the pump and to limit maximum exhaust manifold temperatures.

Pump Air Filter

The air filter attached to the pump is a replaceable element type. The filter should be replaced every 12,000 miles under normal conditions and sooner under off-road use. Some models draw their air supply through the carburetor air filter.

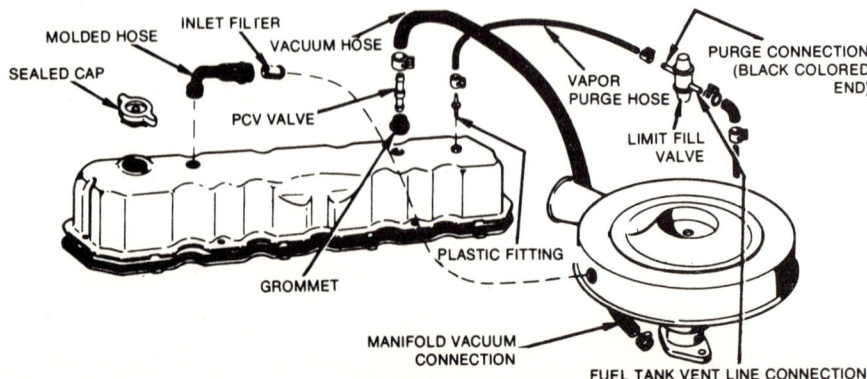

Typical PCV system

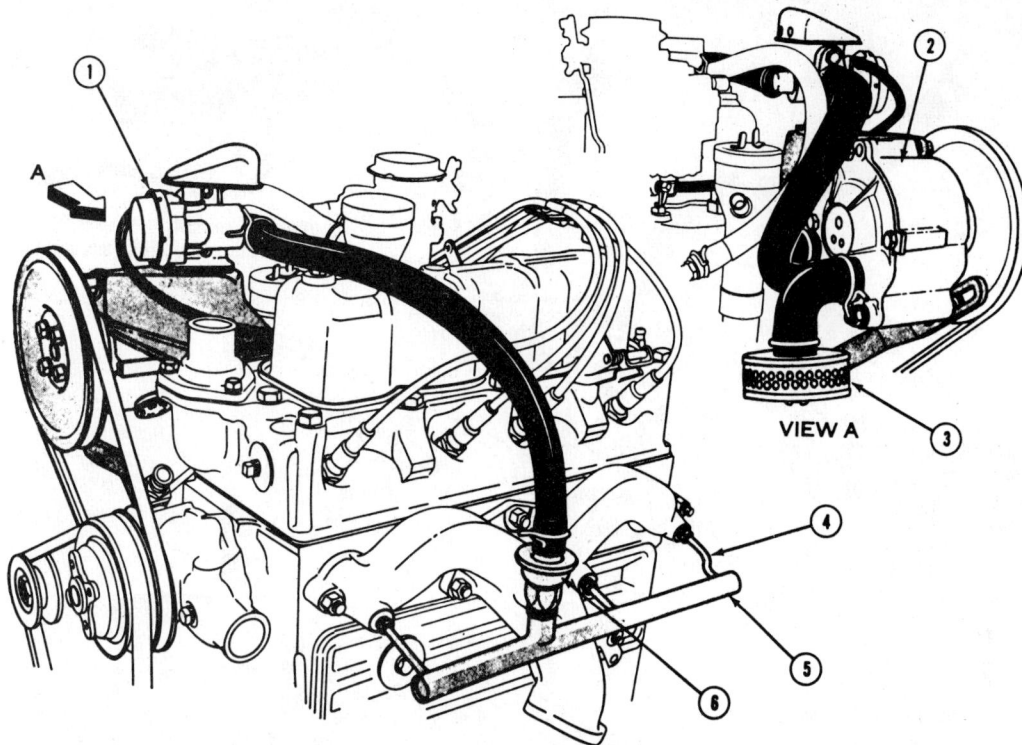

1. Anti-backfire diverter valve
2. Air pump
3. Pump air filter
4. Air injection tubes
5. Air delivery manifold
6. Check valve

F4–134 air pump system

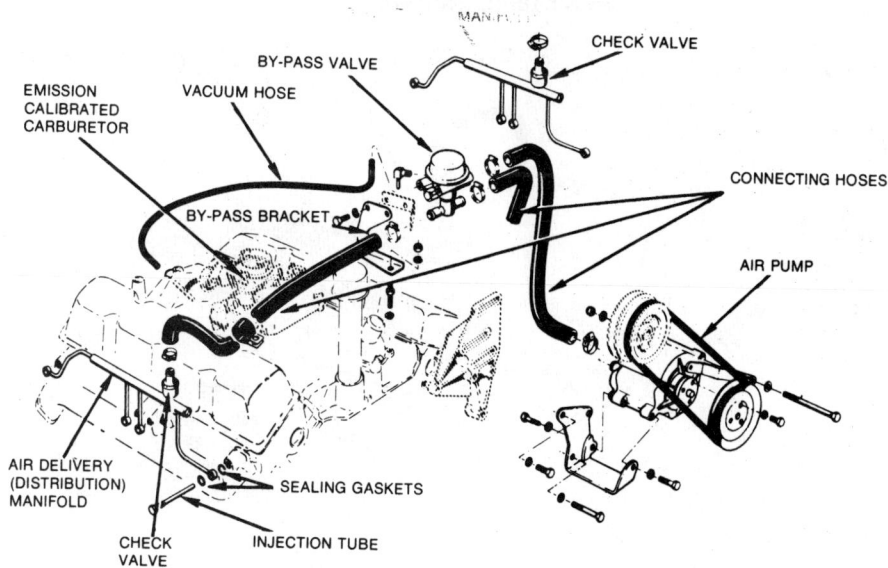

Typical V8 air pump system

Air Delivery Manifold

The air delivery manifold distributes the air from the pump to each of the air delivery tubes in a uniform manner. A check valve is integral with the air delivery manifold. Its function is to prevent the reverse flow of exhaust gases to the pump should the pump fail. This reverse flow would damage the air pump and connecting hose.

Air Injection Tubes

The air injection tubes are inserted into the exhaust ports. The tubes project into the exhaust ports, directing air into the vicinity of the exhaust valve.

Anti-Backfire Valve

The anti-backfire diverter valve prevents engine backfire by

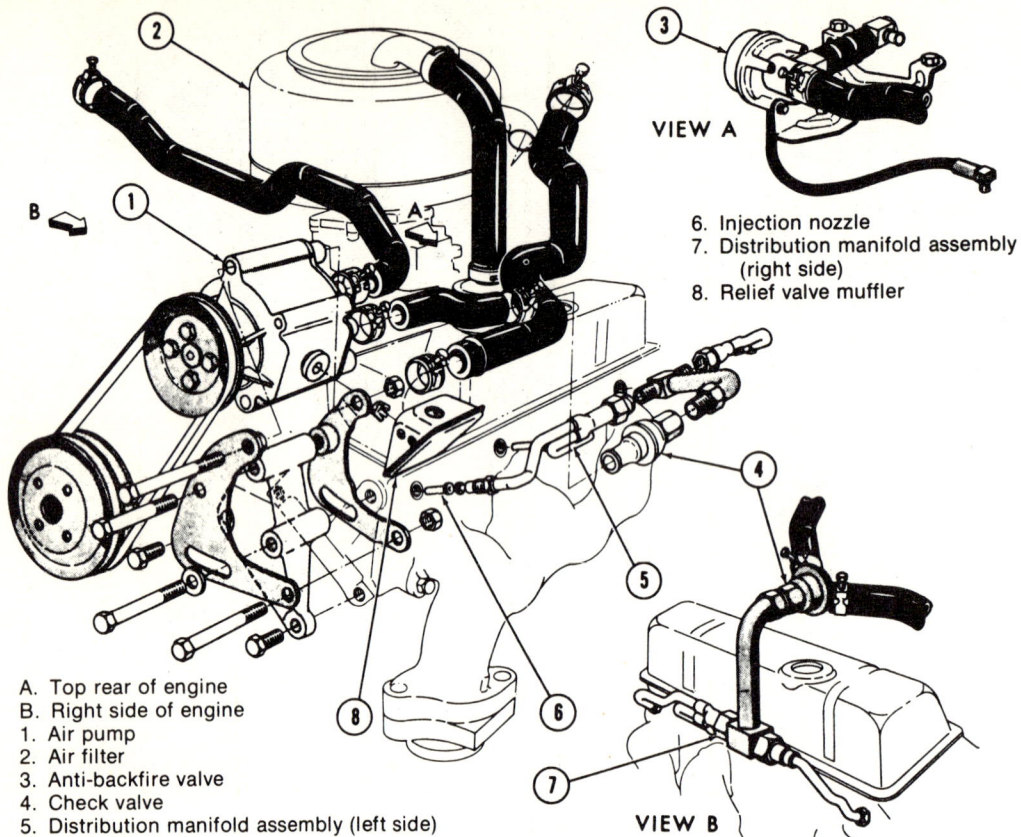

A. Top rear of engine
B. Right side of engine
1. Air pump
2. Air filter
3. Anti-backfire valve
4. Check valve
5. Distribution manifold assembly (left side)

VIEW A

6. Injection nozzle
7. Distribution manifold assembly (right side)
8. Relief valve muffler

VIEW B

V6 air pump system

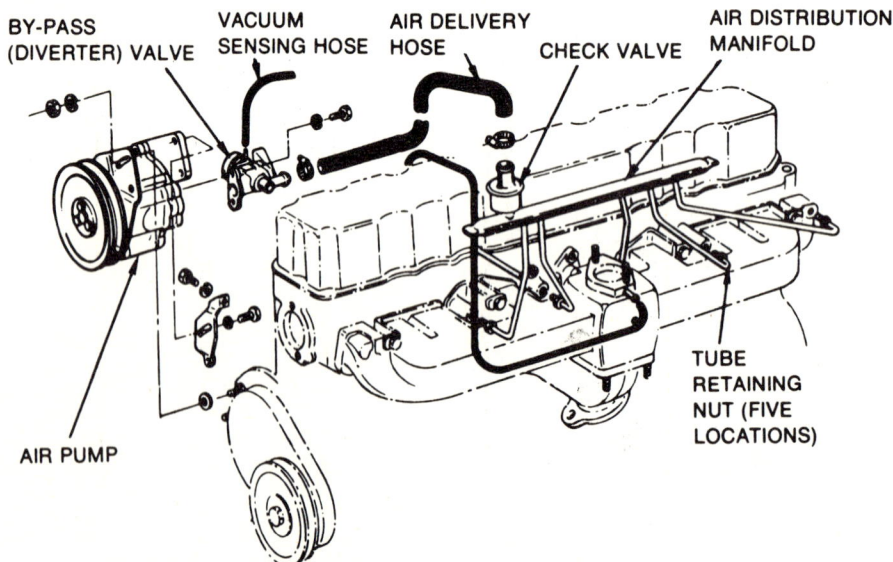

BY-PASS (DIVERTER) VALVE
VACUUM SENSING HOSE
AIR DELIVERY HOSE
CHECK VALVE
AIR DISTRIBUTION MANIFOLD
TUBE RETAINING NUT (FIVE LOCATIONS)
AIR PUMP

Air injection system, 1971 and later inline six

briefly interrupting the air being injected into the exhaust manifold during periods of deceleration or rapid throttle closure. On the 4–134 and all of the 1971 and later American Motors engines, the valve opens when a sudden increase in manifold vacuum overcomes the diaphragm spring tension. With the valve in the open position, the air flow from the air pump is directed to the atmosphere.

On the 6–225 and the 1972 6–232, the anti-backfire valve is what is commonly called a gulp valve. During rapid deceleration the valve is opened by the sudden high vacuum condition in the intake manifold and gulps air into the intake manifold.

Both of these valves prevent backfiring in the exhaust manifold. Both valves also prevent an over right fuel mixture from being burned in the exhaust manifold, which would cause backfiring and possible damage to the engine.

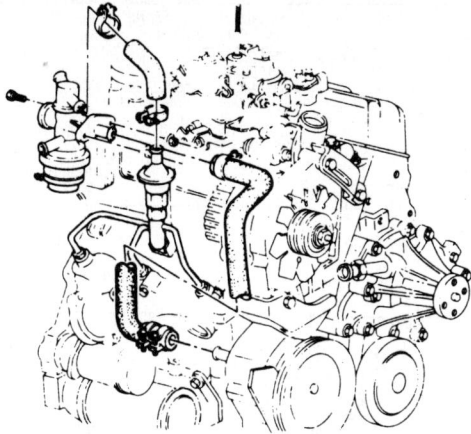

4-150 AIR diverter valve and minifold

4-150 AIR pump mounting

Carburetor

The carburetors used on engines equipped with emission controls have specific flow characteristics that differ from the carburetors used on vehicles not equipped with emission control devices. The carburetors are identified by number. The correct carburetor should be used when replacement is necessary.

A carburetor dashpot is used on the 4-134 to control throttle closing speed.

Thermostatically Controlled Air Cleaner System (TAC)

This system consists of a heat shroud which is integral with the right side exhaust manifold, a hot air hose and a special air cleaner assembly equipped with a thermal sensor and a vacuum motor and air valve assembly.

The thermal sensor incorporates an air bleed valve which regulates the amount of vacuum applied to the vacuum motor, controlling the air valve position to supply either heated air from the exhaust manifold or air from the engine compartment.

During the warm-up period when underhood temperatures are low, the air bleed valve is closed and sufficient vacuum is applied to the vacuum motor to hold the air valve in the closed (heat on) position.

As the temperature of the air entering the air cleaner approaches approximately 115°F (46°C), the air bleed valve opens

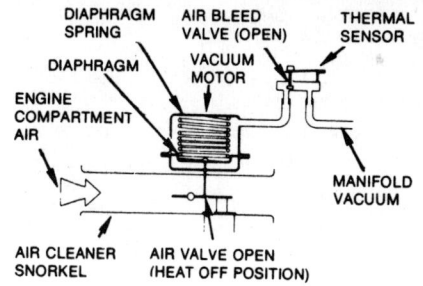

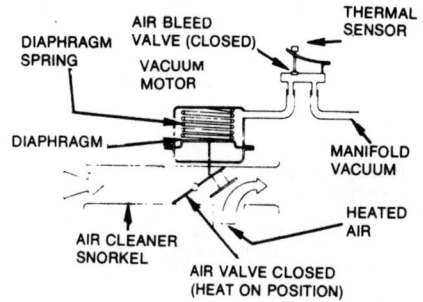

Vacuum controlled thermostatic air cleaner

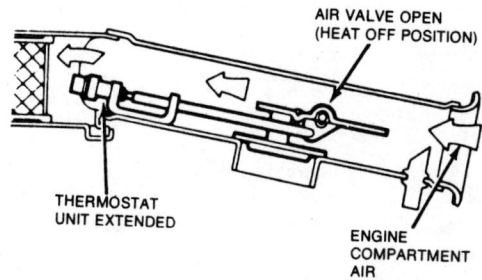

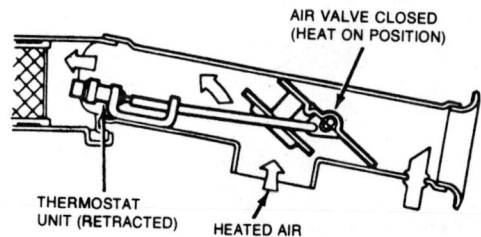

Non-vacuum thermostatically controlled air cleaner

to decrease the amount of vacuum applied to the vacuum motor. The diaphragm spring in the vacuum motor then moves the air valve into the open (heat off) position, allowing only underhood air to enter the air cleaner.

The air valve in the air cleaner will also open, regardless of air temperature, during heavy acceleration to obtain maximum air flow through the air cleaner.

Transmission Controlled Spark System

The purpose of this system is to reduce the emission of oxides of nitrogen by lowering the peak combustion pressure and temperature during the power stroke.

The system incorporates the following components:

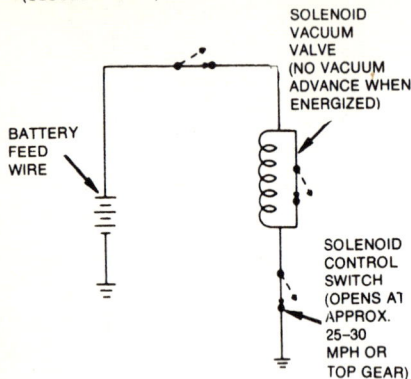

TCS electrical diagram

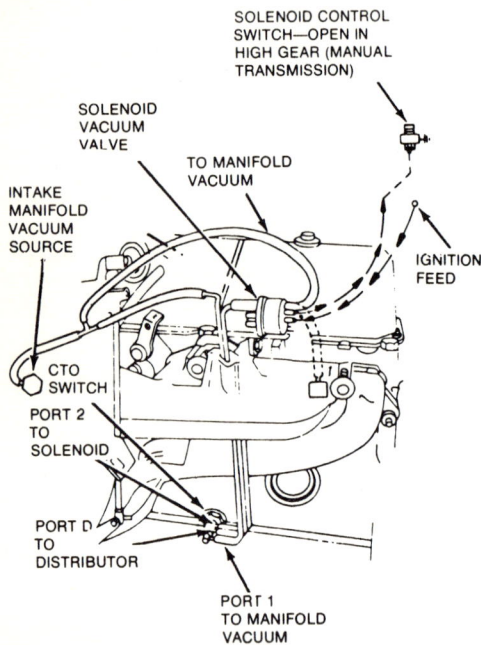

Inline 6-cylinder TCS system

V8 TCS system

SOLENOID CONTROL SWITCH

This switch is located in the transmission valve body. It opens or closes in relation to car speed and gear range. When the transmission is in high gear, the switch opens and breaks the ground circuit to the solenoid vacuum valve. In lower gear ranges the switch closes and completes the ground circuit to the solenoid vacuum valve. With a manual transmission, the switch is operated by the transmission shifter shaft. With automatic transmissions, the switch is controlled by the speedometer gear speed. Under speeds of 25 mph, the switch is activated.

COOLANT TEMPERATURE OVERRIDE SWITCH

This switch is used only on the 8–304. It is threaded into the thermostat housing. The switch reacts to coolant temperatures to route either intake manifold or carburetor vacuum to the distributor advance diaphragm.

When the coolant temperature is below 160°F (71°C), intake manifold vacuum is applied through a hose connection to the distributor advance diaphragm, resulting in full vacuum advance.

When the coolant temperature is above 160°F (71°C), intake manifold vacuum is blocked off and carburetor vacuum is then applied through the solenoid vacuum valve to the distributor advance diaphragm, resulting in decreased vacuum advance.

The relationship between distributor vacuum advance and the operation of the TCS system and coolant temperature override switch can be determined by referring to the Emission Control Distributor Vacuum Application Charts.

AMBIENT TEMPERATURE OVERRIDE SWITCH

This switch, located at the firewall, senses ambient temperatures and completes the electrical circuit from the battery to the solenoid vacuum valve when the ambient temperatures are above 63°F (17°C).

SOLENOID VACUUM VALVE

This valve is attached to the ignition coil bracket at the right side of the engine (V8 engines) or to a bracket at the rear of the intake manifold (Sixes). When the valve is energized, carburetor vacuum is blocked off and the distributor vacuum line is vented to the atmosphere through a port in the valve, resulting in no vacuum advance. When the valve is de-energized, vacuum is applied to the distributor resulting in normal vacuum advance.

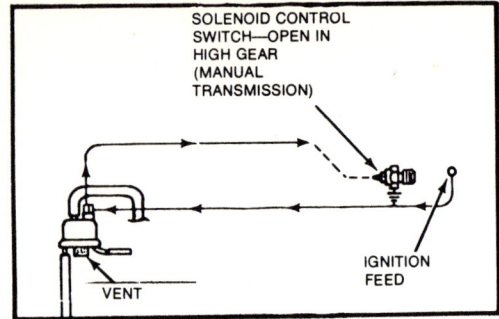

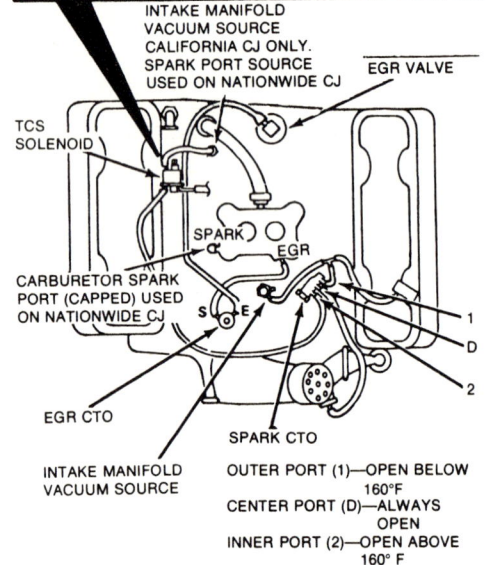

EMISSION CONTROLLED DISTRIBUTOR VACUUM APPLICATION
Vehicles Equipped with TCS

Manual Transmission Gear		Automatic Transmission Vehicle Speed (mph)	Ambient Temperature (Deg. F)	Coolant Temperature (Deg. F)	Vacuum Applied to Distributor
3 Speed	4 Speed				
1-2	1-2-3	Under 25	Below 63	Below 163	Manifold
1-2	1-2-3	Under 25	Below 63	Above 160	Ported
1-2	1-2-3	Under 25	Above 63	Above 160	None
1-2	1-2-3	Under 25	Above 63	Below 160	Manifold
3	4	25–30	Below 63	Below 160	Manifold
3	4	25–30	Below 63	Above 160	Ported
3	4	25–30	Above 63	Above 160	Ported
3	4	25–30	Above 63	Below 160	Manifold

MAINTENANCE AND SERVICE

Efficient performance of the exhaust emission control system is dependent upon precise maintenance.

Carburetor

Check the carburetor for the proper application. Check the dashpot for proper operation and adjust as required. When the throttle is released quickly, the arm of the dashpot should fully extend itself and should catch the throttle lever, letting it back to idle position gradually.

Proper idle mixture adjustment is imperative for best exhaust emission control. The idle adjustment should be made with the engine at normal operating temperature and the air cleaner in place. All lights and accessories must be turned off and the transmission must be in neutral. See Section 2 for adjustment procedures.

Distributor

Check the distributor number for proper application. Check the distributor cam dwell angle and point condition and adjust to specifications or replace as required. See Section 2 for procedures.

Anti-Backfire Diverter Valve

On the 4-134, the anti-backfire valve remains open except when the throttle is closed rapidly from an open position.

To check the valve for proper operation, accelerate the engine in neutral, allowing the throttle to close rapidly. The valve is operating satisfactorily when no exhaust system backfire occurs. A further check can be made by removing the large hose that runs from the anti-backfire valve to the check valve and accelerating the engine and allowing the throttle to close rapidly. If there is an audible momentary interruption of the flow of air, then it can be assumed that the valve is working correctly.

To check the valve on a 6-225 or 1972 6-232, listen for backfire when the throttle is released quickly. If none exists, the valve is doing its job. To check further, remove the large hose that connects the valve with the air pump. Place a finger over the open end of the hose, not the valve, and accelerate the engine, allowing the throttle to close rapidly. The valve is operating satisfactorily if there is a momentary audible rush of air.

Check Valve

The check valve in the air distribution manifold prevents the reverse flow of exhaust gases to the pump in the event the pump should become inoperative or should exhaust pressure ever exceed the pump pressure.

To check this valve for proper operation, remove the air supply hose from the pump at the distribution manifold. With the engine running, listen for exhaust leakage where the check valve is connected to the distribution manifold. If leakage is audible, the valve is not operating correctly.

Air Pump

Check for the proper drive belt tension and adjust as necessary. Do not pry on the die cast pump housing. Check to see if the pump is discharging air. Remove the air outlet hose at the pump. With the engine running, air should be felt at the pump outlet opening.

REMOVAL AND INSTALLATION

Air Pump

1. Loosen the air pump adjusting bracket bolts.
2. Remove the drive belt.
3. Remove the air pump intake and discharge hoses.
4. Remove the air pump from the engine.
5. To install, reverse the above procedure.

Air Distribution Manifold and Air Injection Tubes

It is necessary to remove the exhaust manifold only on the 4-134 prior to removing the air distribution manifold and the air injection tubes. On all the other engines, these components can be removed with the manifolds on the engine.

1. Disconnect the air delivery hose from the air injection manifold. Remove the exhaust manifold on the 4-134.
2. Remove the air distribution manifold from the air injection tubes on the 4-134 only.
3. Unscrew the air injection tube from the exhaust manifold or the head. Some resistance may be encountered because of the normal buildup of carbon. The application of heat may be helpful in removing the air injection tubes.
4. Install in the reverse order of removal.

NOTE: There are two lengths of tubes used with the 4-134. The shorter tubes are installed in number 1 and 4 cylinders. The air injection tubes must be installed on the exhaust manifold prior to installing the exhaust manifold on the engine.

Exhaust Gas Recirculation (EGR) System

The EGR system consists of a diaphragm actuated flow control valve, coolant temperature override switch, low temperature vacuum signal modulator, high temperature vacuum signal modulator.

All 1977 and later California units have a back pressure sensor which modulates EGR signal vacuum according to the rise or fall of exhaust pressure in the manifold. A restrictor plate is not used in these applications.

The purpose of the EGR system is to limit the formation of nitrogen oxides by diluting the fresh air intake charge with a metered amount of exhaust gas, thereby reducing the peak temperatures of the burning gases in the combustion chambers.

EGR VALVE

The EGR valve is mounted on a machined surface at the rear of the intake manifold on the V8s and on the side of the intake manifold on the sixes.

The valve is held in a normally closed position by a coil spring located above the diaphragm. A special fitting is provided at the carburetor to route ported (above the throttle plates) vacuum through hose connections to a fitting located above the diaphragm on the valve. A passage in the intake manifold directs exhaust gas from the exhaust crossover passage (V8) or from below the riser area (Sixes) to the EGR valve. When the diaphragm is actuated by vacuum, the valve opens and meters exhaust gas through another passage in the intake manifold to the floor of the intake manifold below the carburetor.

COOLANT TEMPERATURE OVERRIDE SWITCH

This switch is located in the intake manifold at the coolant passage adjacent to the oil filler tube on the V8s or at the left side of the engine block (formerly the drain plug) on the Sixes. The outer port of the switch is open and not used. The inner port is connected by a host to the EGR fitting at the carburetor. The center port is connected to the EGR valve. When coolant temperature is below 115°F (46°C) (160°F [71°C] on the 8–304 with manual transmission), the center port of the switch is closed and no vacuum signal is applied to the EGR valve. Therefore, no exhaust gas will flow through the valve. When the cool-

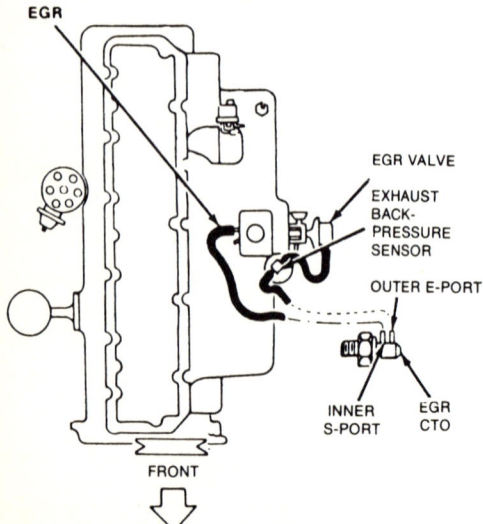

1974–79 8–304 EGR system

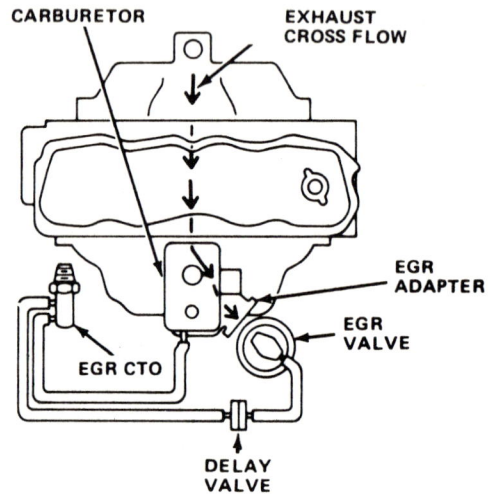

4–151 EGR system

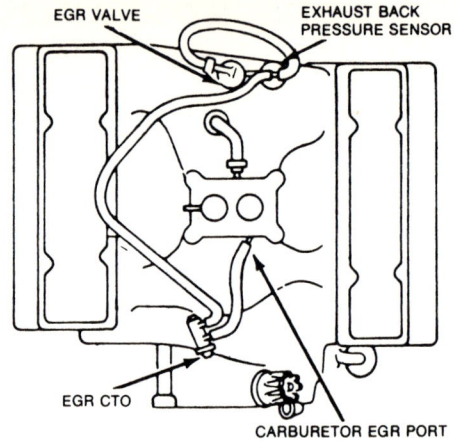

1974–79 6-cylinder EGR system

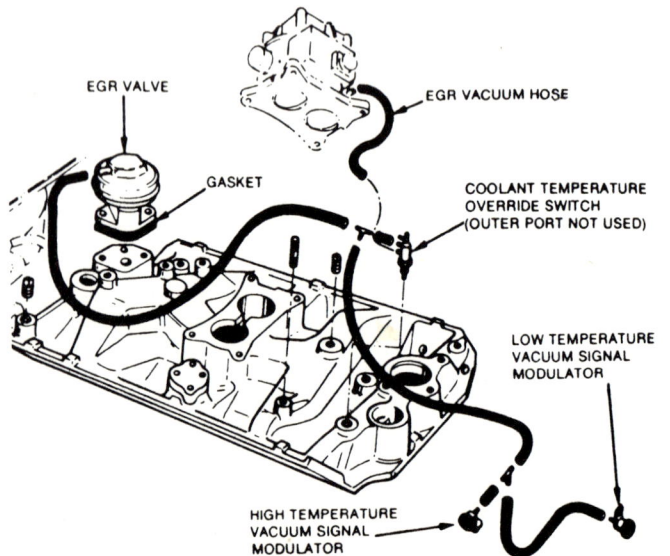

1973 V8 EGR system. The modulator isn't used on later models

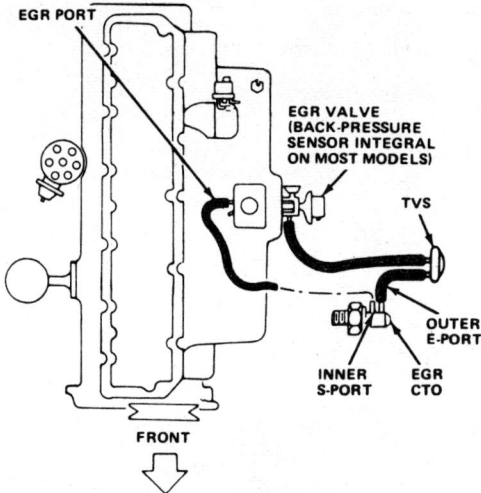

1980 and later 6-cylinder EGR system

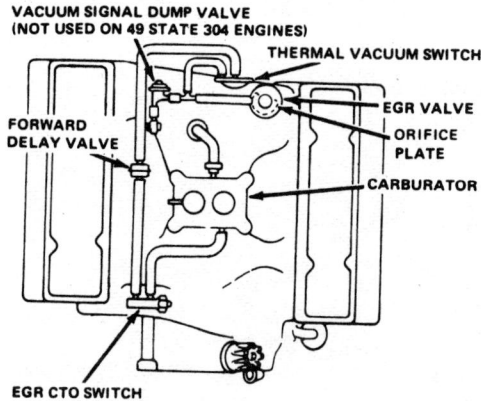

1980–81 8–304 EGR system

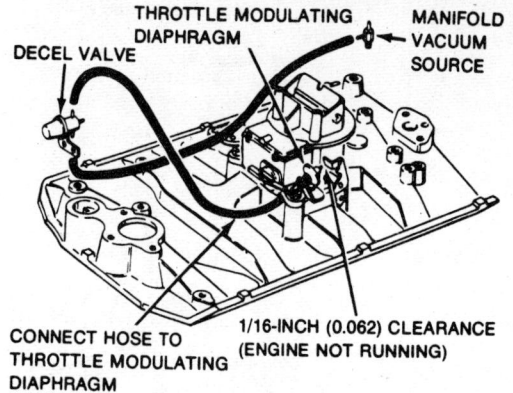

Typical vacuum throttle modulating system

ant temperature reaches 115°F (46°C), both the center port and the inner port of the switch are open and a vacuum signal is applied to the EGR valve. This vacuum signal is, however, subject to regulation by the low and high temperature signal modulators.

LOW TEMPERATURE VACUUM SIGNAL MODULATOR

This unit is located just to the right of the radiator behind the grill opening. The low temperature vacuum signal modulator vacuum hose is connected by a plastic T-fitting to the EGR vacuum signal hose. The modulator is open when ambient temperatures are below 60°F (16°C). This causes a weakened vacuum signal to the EGR valve and a resultant decrease in the amount of exhaust gas being recirculated.

HIGH TEMPERATURE VACUUM SIGNAL MODULATOR

This unit is located at the right front fender inner panel. The high temperature vacuum signal modulator is connected to the EGR vacuum signal hose by a plastic T-fitting. The modulator opens when the underhood air temperatures reach 115°F (46°C) and it causes a weakened vacuum signal to the EGR valve, thus reducing the amount of exhaust gases being recirculated.

Electric Assist Choke

An electric assist choke is used to more accurately match the choke operation to engine requirements. It provides extra heat to the choke bimetal spring to speed up the choke valve opening after the underhood air temperature reaches 95°F ± 15°F (35°C). Its purpose is to reduce the emission of carbon monoxide (CO) during the engine's warmup period.

A special AC terminal is provided at the alternator to supply a 7 volt power source for the electric choke. A thermostatic switch within the choke cover closes when the underhood air temperature reaches 95°F ± 15°F (35°C) and allows current to flow to a ceramic heating element. The circuit is completed through the choke cover ground strap and choke housing to the engine. As the heating element warms up, heat is absorbed by an attached metal plate which in turn heats the coke bimetal spring.

After the engine is turned off, the thermostatic switch remains closed until the underhood temperature drops below approximately 65°F (18°C). Therefore, the heating element will immediately begin warming up when the engine is restarted, if the underhood temperature is above 65°F (18°C).

Fuel Tank Vapor Emission Control System

A closed fuel tank system is used on some models through 1974, some models 1975–78, and all 1979 and later models, to route raw fuel vapor from the fuel tank into the PCV system (sixes) or air cleaner snorkle (V8s), where it is burned along with the fuel-air mixture. The system prevents raw fuel vapors from entering the atmosphere.

The fuel vapor system consists of internal fuel tank venting, a vacuum-pressure fuel tank filler cap, an expansion tank or charcoal filled canister, liquid limit fill valve, and internal carburetor venting.

Fuel vapor pressure in the fuel tank forces the vapor through vent lines to the expansion tank or charcoal filled storage canister. The vapor then travels through a single vent line to the limit fill valve, which regulates the vapor flow to the valve cover or air cleaner.

The fuel tank vent line is routed through the limit fill valve to the valve cover on the left side on the 1972 V8s. On the 1973 Sixes, it travels to the intake manifold and on the V8s it is routed to the carburetor air cleaner.

LIMIT FILL VALVE

This valve is essentially a combination vapor flow regulator and pressure relief valve. It regulates vapor flow from the fuel tank vent line into the valve cover. The valve consists of a hous-

1972–73 fuel tank and vant lines

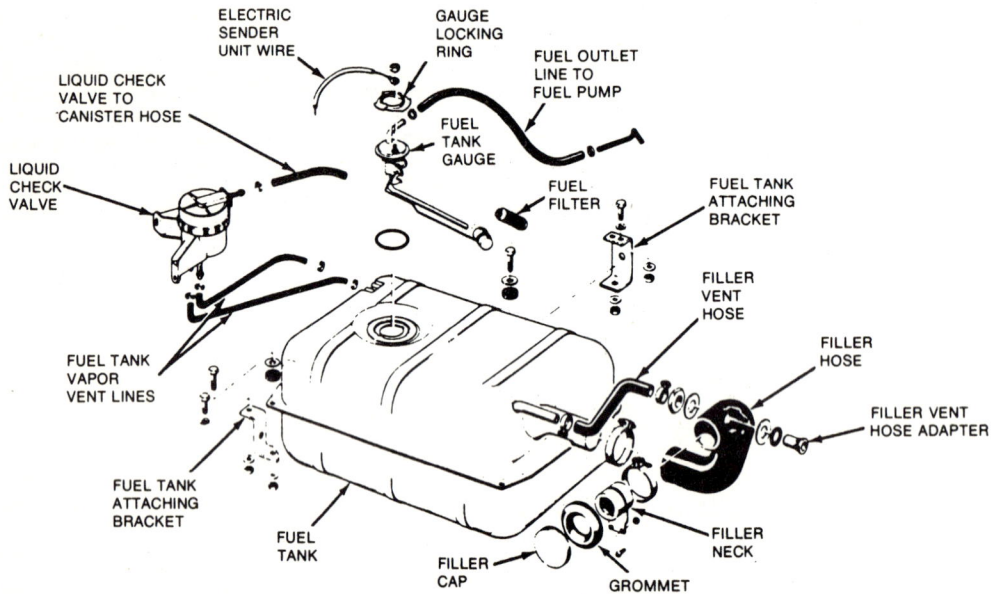

1974–78 fuel tank and vapor emission control system

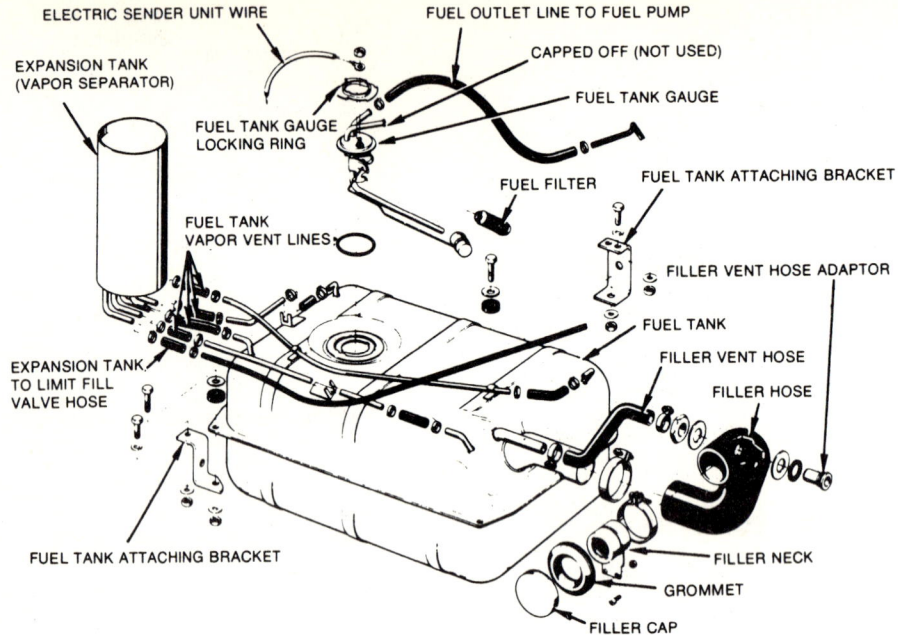

ing, a spring loaded diaphragm and a diaphragm cover. As tank vent pressure increases, the diaphragm lifts, permitting vapor to flow through. The pressure at which this occurs is 4–6 in. H_2O column. This action regulates the flow of vapors under severe conditions, but generally prohibits the flow of vapor during normal temperature operation, thus minimizing driveability problems.

LIQUID CHECK VALVE

The liquid check valve prevents liquid fuel from entering the vapor lines leading to the storage canister. The check valve incorporates a float and needle valve assembly. If liquid fuel should enter the check valve, the float will rise and force the needle upward to close the vent passage. With no liquid fuel present in the check valve, fuel vapors pass freely from the tank, through the check valve, and on to the storage canister.

Feedback Systems

Two different feedback systems are used with 1981 and later Jeep vehicles. One, the C4 system is used with 4–151 engines built for sale in California. The other, the Computerized Emission Control (CEC) System is used on 6–258 engines built for sale in California. Each system is designed for the same purpose, to reduce exhaust emission using a Three-Way Catalytic Converter (TWC).

Each system is computerized, utilizing microprocessors, and

ROLLOVER
CHECK VALVE

LIQUID
CHECK
VALVE

FUEL GAUGE
SENDING
UNIT WIRE

LOCKING
RING

SENDING
UNIT

FUEL OUTLET
LINE TO
FUEL PUMP

FUEL RETURN
LINE FROM
FUEL FILTER
(SIX- EIGHT-
CYLINDER
ENGINES
ONLY)

GROUND WIRE

TO CHARCOAL
CANISTER

FUEL PICKUP
TUBE FILTER

VENT
HOSE

FILLER
HOSE

FILLER
NECK

FUEL TANK
VAPOR
VENT HOSES

FUEL
TANK

FILLER CAP

FILLER CAP PROTECTOR

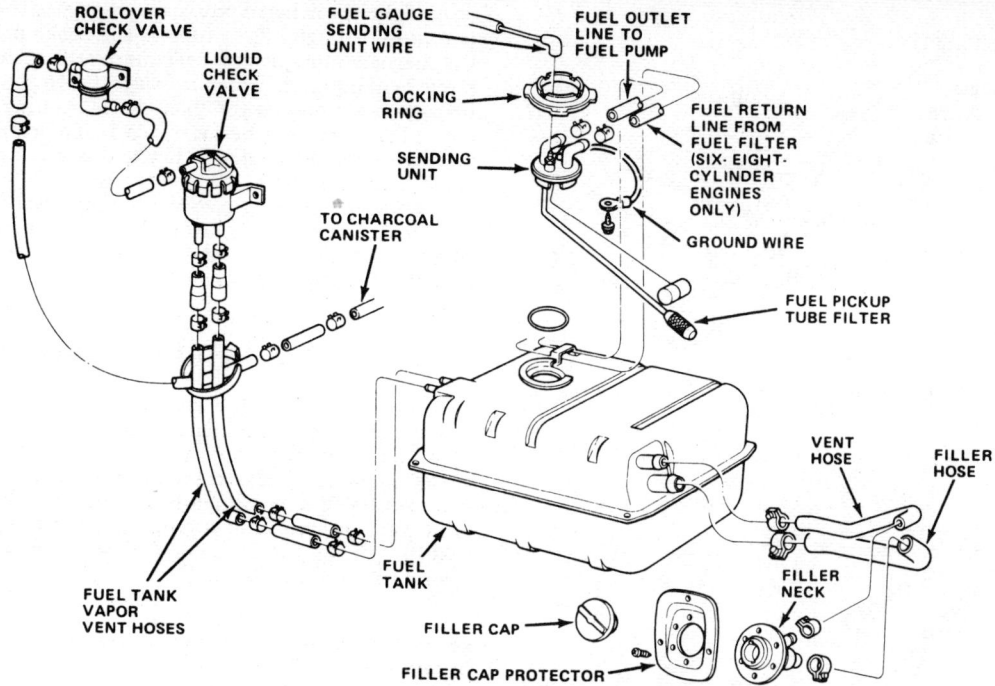

1979–81 fuel tank and related components

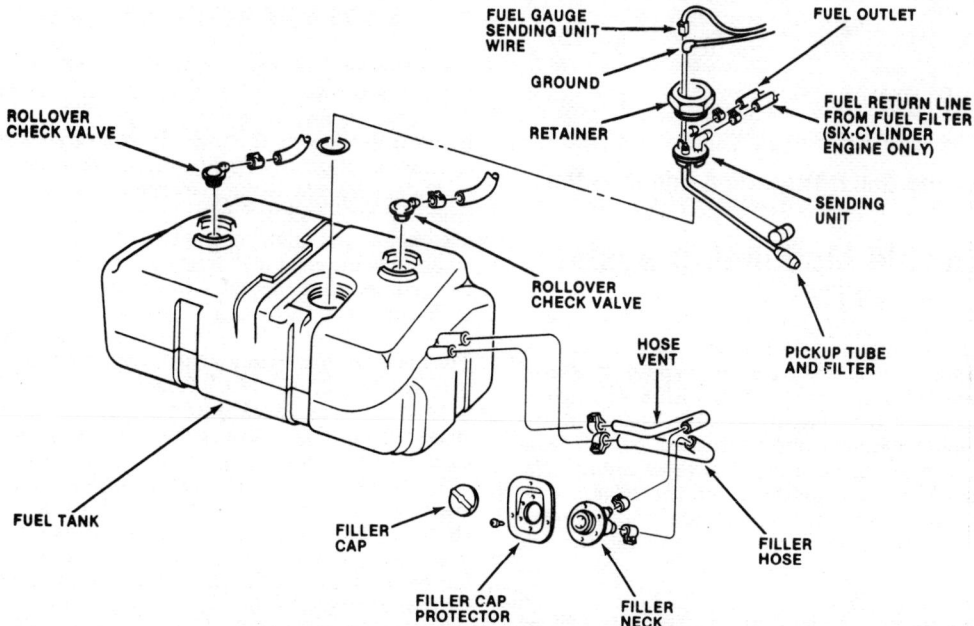

FUEL GAUGE
SENDING UNIT
WIRE

GROUND

RETAINER

FUEL OUTLET

FUEL RETURN LINE
FROM FUEL FILTER
(SIX-CYLINDER
ENGINE ONLY)

SENDING
UNIT

ROLLOVER
CHECK VALVE

ROLLOVER
CHECK VALVE

PICKUP TUBE
AND FILTER

HOSE
VENT

FUEL TANK

FILLER
CAP

FILLER CAP
PROTECTOR

FILLER
NECK

FILLER
HOSE

1982–86 fuel tank and related components

each is highly complex, requiring professional service. Therefore, no service procedures are given in this book for the diagnosis or repair of these systems.

Catalytic Converter

The catalytic converter is a muffler like device inserted in the exhaust system. Exhaust gases flow through the converter where a chemical change takes place, reducing carbon monoxide and hydrocarbons to carbon dioxide and water; the latter two elements being harmless. The catalysts promoting this reaction are platinum and palladium coated beads of alumina. Because of the chemical reaction which does take place in the converter, the temperature of the converter during operation is higher than the exhaust gases when they leave the engine. However, insulation keeps the outside skin of the converter about the same temperature as the muffler. An improperly adjusted carburetor or ignition problem which would permit unburned fuel to enter the converter could produce excessive heat. Excessive heat in the converter could result in bulging or other distortion of the converter's shape. If the converter is heat damaged and must be replaced, the ignition or carburetor problem must be corrected for.

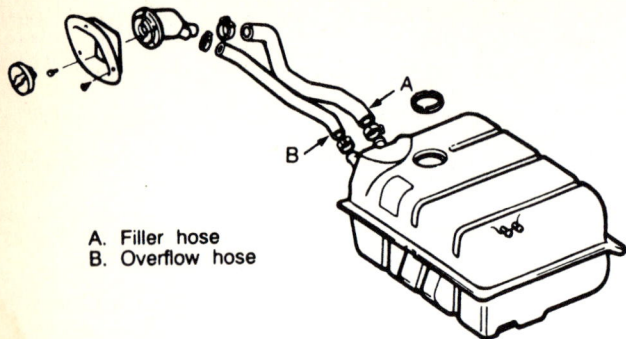

A. Filler hose
B. Overflow hose

1987–90 fuel tank and lines

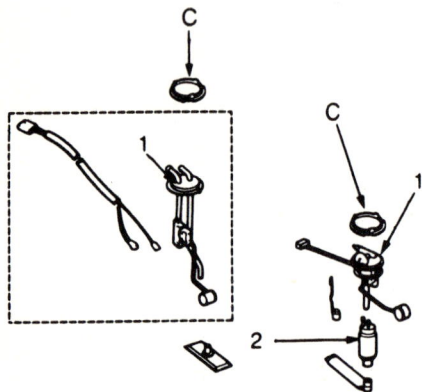

C. Lockring
1. Pick-up unit and float
2. Fuel pump

1987–90 fuel sending units. The 6–258 unit is in the dotted box

Vacuum Throttle Modulating System (VTM)

This system is designed to reduce the level of hydrocarbon emission during rapid throttle closure at high speed. It is used on some 49 state and all California models, with a V8 engine.

The system consists of a deceleration valve located at the right front of the intake manifold, and a throttle modulating diaphragm located at the carburetor base. The valve and the diaphragm are connected by a vacuum hose and the valve is connected to direct manifold vacuum. During deceleration, manifold vacuum acts to delay, slightly, the closing of the throttle plate.

To adjust
1. Run the engine to normal operating temperature and set the idle speed to specification. Shut off the engine.
2. Position the throttle lever against the curb idle adjusting screw.
3. Measure the clearance between the throttle modulating diaphragm plunger and the throttle lever. A clearance of $\frac{1}{16}$ in. (1.6mm) should exist.
4. Adjust the clearance, if necessary, by loosening the jam nut and turning the diaphragm assembly.

V8 Choke Heat By-Pass Valve (CHBPV)

When the engine is first started and begins to warm up, heat-ed air from the exhaust crossover passage in the intake manifold is routed through a heat tube to the choke housing containing the thermostatic spring for regulating the choke flap. A thermostatic by-pass valve, which is integral with the choke heat tube, helps prevent premature choke valve opening during the early part of the warmup period. This is important when ambient temperatures are relatively low and adverse driveability could occur if the choke was opened too soon.

The thermostatic by-pass valve regulates the temperature of the hot airflow to the choke housing by allowing outside unheated air to enter the heat tube. A thermostatic disc in the valve is calibrated to close the valve at 75°F (24°C) and open it at 55°F (13°C).

Fuel Return System

The purpose of the fuel return system is to reduce high temperature fuel vapor problems. The system consists of a fuel return line to the fuel tank and special fuel filter with an extra outlet nipple to which the return line is connected. During normal operation, a small amount of fuel is returned to the fuel tank. During periods of high underhood temperatures, vaporized fuel in the fuel line is returned to the fuel tank and not passed through the carburetor.

NOTE: The extra nipple on the special fuel filter should be positioned upward to ensure proper operation of the system.

Emission Control Checks

ANTI-BACKFIRE DIVERTER VALVE

On the F4–134, the anti-backfire valve remains open except when the throttle is closed rapidly from an open position.

To check the valve for proper operation, accelerate the engine in neutral, allowing the throttle to close rapidly. The valve is operating satisfactorily when no exhaust system backfire occurs. A further check can be made by removing the large hose that runs from the anti-backfire valve to the check valve and accelerating the engine and allowing the throttle to close rapidly. If there is an audible momentary interruption of the flow of air then it can be assumed that the valve is working correctly.

To check the valve on a V6, listen for backfire when the throttle is released quickly. If none exists, the valve is doing its job. To check further, remove the large hose that connects the valve with the air pump. Place a finger over the open end of the hose, not the valve, and accelerate the engine, allowing the throttle to close rapidly. The valve is operating satisfactorily if there is a momentary audible rush of air.

To check the diverter valve on American Motors engines, start the engine and let it idle. With the engine idling, there should be little or no air coming out the vents. When the engine is accelerated to 2,000–3,000 rpm, a strong flow of air should be felt at the vents. If the flow of air from the air pump is not diverted through the diverter valve vents when the engine is accelerated to the above mentioned rpm, check and make sure that the vacuum sensing line leading to the valve has vacuum and is not leaking or disconnected. The diverter valve should bleed air when 20 in.Hg or more vacuum is applied to the vacuum sensing line or when the output of the air pump exceeds 5 psi. When the engine is slowly accelerated, the diverter valve should begin to bleed off air between 2,500 and 3,500 rpm.

CHECK VALVE

The check valve in the air distribution manifold prevents the reverse flow of exhaust gases to the pump in the event the pump should become inoperative or should exhaust pressure ever exceed the pump pressure.

To check this valve for proper operation, remove the air supply hose from the pump at the distribution manifold. With the engine running, listen for exhaust leakage where the check valve is connected to the distribution manifold. If leakage is audible, the valve is not operating correctly. A small amount of leakage is normal.

AIR PUMP

Check for the proper drive belt tension and adjust as necessary. Do not pry on the die cast pump housing. Check to see if the pump is discharging air. Remove the air outlet hose at the pump. With the engine running, air should be felt at the pump outlet opening.

EGR VALVE

With the engine idling and at normal operating temperature, manually depress the EGR valve diaphragm. This should cause engine speed to drop about 200 rpm. This indicates that the EGR valve had been properly cutting off the flow of exhaust gas at idle and is operating properly.

If the engine speed did not change and the idle is smooth, exhaust gases are not reaching the combustion chambers. The probable cause of this is a plugged passage between the EGR valve and the intake manifold.

If the engine idle is rough and rpm is not affected by depressing the EGR valve diaphragm, the EGR valve is not closing off the flow of exhaust at idle like it's supposed to and there is most likely a fault in the hoses, hose routing, or the EGR valve itself.

NOTE: The EGR valve can be removed and cleaned with a wire brush and a 9/16 in. drill bit coated with grease (to hold dirt particles) inserted in discharge passage. The drill should be held with a pair of pliers only.

EGR CTO SWITCH

Before checking the operating of the EGR CTO switch, make sure that the engine coolant is below 100°F (38°C).
1. Check the vacuum lines for leaks and proper routing.
2. Disconnect the vacuum lines at the backpressure sensor, if so equipped, or at the EGR valve, and connect the line to a vacuum gauge.
3. Operate the engine at 1,500 rpm. No vacuum should be indicated at the gauge. If vacuum is indicated, replace the EGR CTO switch.
4. Allow the engine to idle until the coolant temperature exceeds 115°F (46°C).
5. Accelerate the engine to 1,500 rpm. Vacuum should be present at the gauge. If not, replace the EGR CTO switch.

EXHAUST BPS UNIT

1. Make sure that all the EGR vacuum lines are routed correctly and are not leaking.
2. Install a tee in the vacuum line between the EGR valve and BPS, and attach a vacuum gauge to the tee.
3. Start the engine and allow it to idle. No vacuum should be present. If vacuum is indicated at idle speed, make sure of correct line connections. Also, be sure that manifold vacuum is not the source. If the carburetor is providing the vacuum, look for a partially open throttle plate which could cause premature ported vacuum to the BPS unit.

4. Accelerate the engine to 2,000 rpm and observe the vacuum gauge for the following:
    a. If the coolant is below 115°F (46°C), no vacuum should be present.
    b. With coolant temperature above 115°F (46°C), ported vacuum should be indicated.
    c. If no vacuum is indicated at any time, make sure that vacuum is being applied to the inlet side of the BPS. If correct, remove the BPS and either clean it with a wire brush (if blocked) or replace it.

SPARK CTO SWITCH

Before testing the spark CTO switch, make sure that the engine coolant temperature is below 160°F (71°C).
1. Remove all the hoses from the CTO switch and plug those which will create a vacuum leak.
2. Connect a vacuum line from a manifold vacuum source to the top port of the CTO switch.
3. Connect a vacuum gauge to the center port.
4. Start the engine. Manifold vacuum should be indicated on the gauge. If not, replace the switch.
5. With the engine still running and the coolant temperature still below 160°F (71°C), disconnect the vacuum line from the top port and connect it to the bottom port.
6. No vacuum should be indicated. Replace the switch if there is vacuum.
7. Allow the engine to run until the coolant temperature exceeds 160°F (71°C). Manifold vacuum should be indicated. If not, replace the CTO switch.
8. Disconnect the hose from the bottom port and connect it to the top port again. With the coolant temperature above 160°F (71°C), no vacuum should be indicated. If there is, replace the CTO switch.

TVS FUNCTIONAL TEST

1. Allow the air cleaner to cool to between 40 and 50°F (4–10°C).
2. Disconnect the vacuum hoses from the TVS and connect an external vacuum source to one nipple and a vacuum gauge to the other.
3. Apply vacuum to the TVS. Vacuum should not be present when the air temperature is 40–50°F (4–10°C). If vacuum is present, replace the switch.
4. Start the engine and allow the air cleaner to warm above 50°F (10°C). Vacuum should be present.

Oxygen Sensor

REMOVAL AND INSTALLATION

1. Raise the vehicle and support it safely. Allow the exhaust system to cool sufficiently to permit servicing.
2. Disconnect the wire connector from the oxygen sensor.
3. Remove the oxygen sensor from the exhaust manifold.
4. If not already done, coat the threads of the replacement sensor with anti-seize compound. Be careful not to contaminate the oxygen sensor probe with the anti-seize.
5. Install the oxygen sensor into the exhaust manifold and tighten to 35 ft. lbs. (48 Nm). Reconnect the wire connector.

COMPUTERIZED EMISSION CONTROL (CEC) SYSTEM

General Information

The Computerized Emission Control System (CEC) is used on all gasoline engines. There are two primary modes of operation for the CEC feedback system, open loop and closed loop. The system will be in the open loop mode of operation (or a variation of it) whenever the engine operating conditions do not meet the programmed criteria for closed loop operation. During open loop operation, the air/fuel mixture is maintained at a programmed ratio that is dependent on the type of engine operation involved. The oxygen sensor data is not accepted by the system during this mode of operation. The following conditions involve open loop operation.

- Engine start-up
- Coolant temperature too low
- Oxygen sensor temperature too low
- Engine idling
- Wide open throttle (WOT)
- Battery voltage too low

When all input data meets the programmed criteria for closed loop operation, the exhaust gas oxygen content signal from the oxygen sensor is accepted by the computer. This results in an air/fuel mixture that will be optimum for the engine operating condition and also will correct any pre-existing mixture condition which is too lean or too rich.

NOTE: A high oxygen content in the exhaust gas indicates a lean air/fuel mixture. A low oxygen content indicates a rich air/fuel mixture. The optimum air/fuel mixture ratio is 14.7:1.

Micro Computer Unit (MCU)

The micro computer unit, or MCU, is the heart of the CEC system. The MCU receives signals from various engine sensors to constantly monitor the engine operating conditions, then it uses this information to make adjustments in order to achieve the optimum performance and economy with a minimum of engine emissions. The MCU monitors the oxygen sensor voltage and, based upon the mode of operation, generates an output control signal for the carburetor stepper motor or mixture control solenoid. If the system is in the closed loop mode of operation, the air/fuel mixture will vary according to the oxygen content in the exhaust gas and engine operating conditions. If the system is in the open loop mode of operation, the air/fuel mixture will be based on a predetermined ratio that is dependent on engine rpm. In addition, the MCU generates output signals to control ignition timing and engine idle speed, PCV flow and Pulse Air System operation.

Mixture Control Solenoid

On engines with the Carter YFA or Rochester E2SE carburetors, a mixture control (MC) solenoid is used to regulate the air/fuel mixture. During open loop operation, the MC solenoid supplies a preprogrammed amount of air to the carburetor idle circuit and main metering circuit where it mixes with the fuel. During closed loop operation, the MCU operates the MC solenoid to provide additional or less air to the fuel mixture, depending on the engine operating conditions as monitored by the various engine sensors.

Idle Relay and Solenoid

The idle relay is energized by the MCU to control the vacuum actuator portion of the Sole-Vac throttle positioner by providing a ground for the idle relay. The relay energizes the idle solenoid, which allows vacuum to operate the Sole-Vac vacuum actuator.

This, in turn, opens the throttle and increases engine speed. The idle solenoid is located on a bracket on the left front inner fender panel and can be identified by the red connecting wires.

Sole-Vac Throttle Positioner

The Sole-Vac throttle positioner is attached to the carburetor. The unit consists of a closed throttle switch, a holding solenoid and a vacuum actuator. The holding solenoid maintains the throttle position, while the vacuum actuator provides additional engine idle speed when accessories such as the air conditioner or rear window defogger are in use. The vacuum actuator is also activated during deceleration and if the steering wheel is turned to the full stop position on vehicles equipped with power steering.

Upstream and Downstream Air Switch Solenoids

The upstream and downstream solenoids of the pulse air system distribute air to the exhaust pipe and catalytic converter. Both solenoids are energized by the MCU to route air into the the exhaust pipe at a point after the oxygen sensor. When energized, the downstream solenoid routes air into the second bed of the dual-bed catalytic converter. This additional air reacts with the exhaust gases to reduce engine emissions.

The solenoids are located on a bracket attached to the left inner front fender panel. The idle solenoid is also located on this same bracket.

PCV Shutoff Solenoid

The positive crankcse ventilation shutoff solenoid is installed in the PCV valve hose and is energized by the MCU to turn off the crankcase ventilation system when the engine is at idle speed. An anti-diesel relay system on 4 cylinder engines, consisting of an anti-diesel relay and a delay relay, prevents engine run-on when the ignition is switched off by momentarily energizing the PCV valve solenoid when the ignition is switched off to prevent air entering below the throttle plate.

Bowl Vent Solenoid

The bowl vent solenoid is located in the hose between the carburetor bowl vent and the canister. The bowl vent solenoid is closed and allows no fuel vapor to flow when the engine is operating. When the engine is not operating, the solenoid is open and allows vapor to flow to the charcoal canister to control hydrocarbon emissions from the carburetor float bowl. The bowl vent solenoid is electrically energized when the ignition is switched ON and is not controlled by the MCU.

Intake Manifold Heater Switch

The intake manifold heater switch is located in the intake manifold and is controlled by the temperature of the engine coolant. Below 160°F (71°C) the manifold heater switch activates the intake manifold heater to improve fuel vaporization. The switch is not controlled by the MCU and does not provide input information to it.

Oxygen Sensor

This component of the system provides a variable voltage (millivolts) for the micro computer unit (MCU) that is proportional to the oxygen content in the exhaust gas. In addition to the oxygen sensor, the following data senders are used to supply the MCU with engine operation data.

Knock Sensor

The knock sensor is a tuned piezoelectric crystal transducer that is located in the cylinder head. The knock sensor provides the MCU with an electrical signal that is created by vibrations that correspond to its center frequency (5550 Hz). Vibrations from engine knock (detonation) cause the crystal inside the sensor to vibrate and produce an electrical signal that is used by the MCU to selectively retard the ignition timing of any single cylinder or combination of cylinders to eliminate the knock condition.

Vacuum Switches

Two vacuum-operated electrical switches (ported and manifold) are used to detect and send throttle position data to the MCU for idle (closed), partial and wide open throttle (WOT). These switches are located together in a bracket attached to the dash panel in the engine compartment. The 4 in. Hg vacuum switch can be identified by its natural (beige) color, while the 10 in. Hg vacuum switch is green in color. The 4 in. Hg switch is controlled by ported vacuum and its electrical contact is normally in the open position when the vacuum level is less than 4 in. Hg. When the vacuum exceeds 4 in. Hg, the switch closes. The 4 in. Hg vacuum switch tells the MCU when either a closed or deep throttle condition exists.

The 10 in. Hg vacuum switch is controlled by manifold vacuum. Its electrical contact is normally closed when the vacuum level is less than 10 in. Hg; if the vacuum level exceeds 10 in. Hg, the switch opens. This switch tells the MCU that either a partial or medium throttle condition exists.

Engine RPM Voltage

This voltage is supplied from a terminal on the distributor. Until a voltage equivalent to a predetermined rpm is received by the MCU, the system remains in the open loop mode of operation. The result is a fixed rich air/fuel mixture for starting purposes.

Coolant Temperature Switch

The temperature switch supplies engine coolant temperature data to the MCU. Until the engine is sufficiently warmed (above 135°F/57°C), the system remains in the open loop mode of operation (i.e., a fixed air/fuel mixture based upon engine rpm).

Thermal Electric Switch

The thermal electric switch is located inside the air cleaner to sense the incoming air temperature and indicate a cold weather start-up condition to the MCU when the air temperature is below 50°F (10°C). Above 65°F (18°C), the switch opens to indicate a normal engine start-up condition to the MCU.

Wide Open Throttle (WOT) Switch

The wide open throttle switch is attached to the base of the carburetor by a mounting bracket. It is a mechanically operated electrical switch that is controlled by the position of the throttle. When the throttle is placed in the wide open position, a cam on the throttle shaft actuates the switch about 15° before wide open position to indicate a full-throttle demand to the MCU.

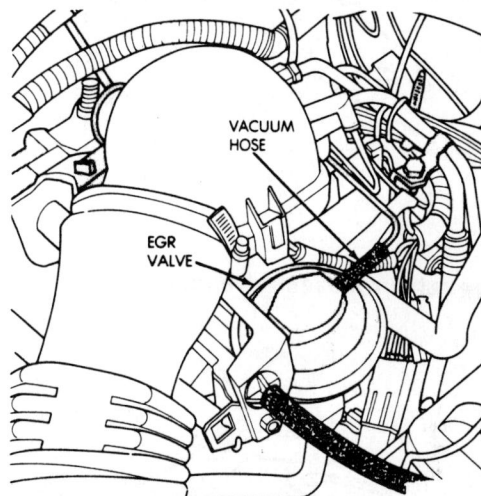

EGR valve location on the 4–150

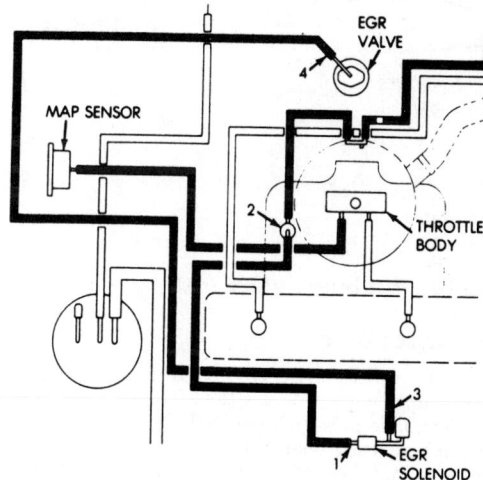

EGR system test points on a 4–150 with TBI

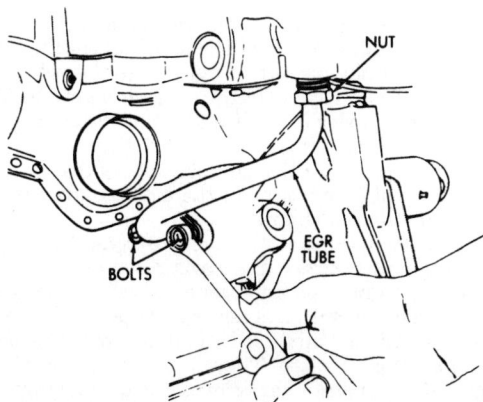

4–150 EGR tube removal

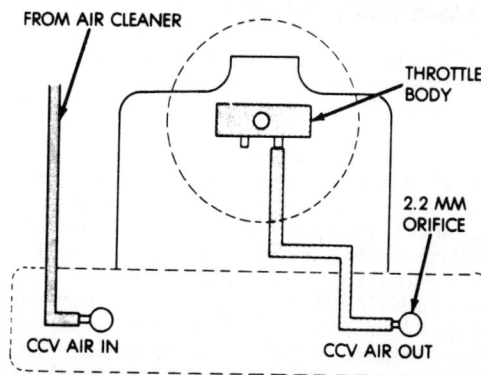

CCV system on the 4–150

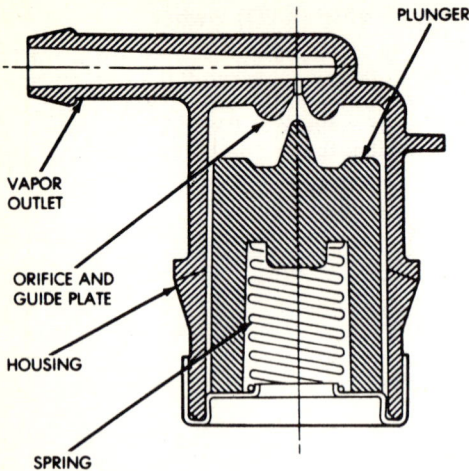

Pressure relief/rollover valve operation

Labels: PLUNGER, VAPOR OUTLET, ORIFICE AND GUIDE PLATE, HOUSING, SPRING

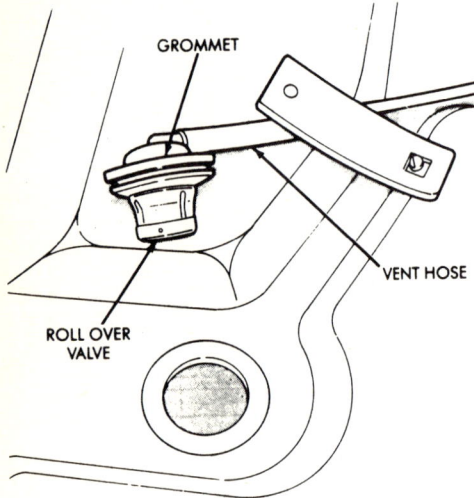

Pressure relief/rollover valve installation

Labels: GROMMET, VENT HOSE, ROLL OVER VALVE

Altitude Jumper Wire

The altitude jumper wire connector is located next to the MCU. The jumper wire provides the MCU with an indication of whether the vehicle is being operated above or below a 4000 ft. elevation (high altitude operation). The connector normally has no jumper wire installed. If a vehicle is to be operated in a designated high altitude area, a jumper wire must be installed.

CEC SYSTEM OPERATION - 4-150

The open loop mode of operation occurs when:
1. Starting engine, engine is cold or air cleaner air is cold.
2. Engine is at idle speed, accelerating to partial throttle or decelerating from partial throttle to idle speed.
3. Carburetor is either at or near wide open throttle (WOT).
When any of these conditions occur, the mixture control (MC) solenoid provides a predetermined air/fuel mixture ratio for each condition. Because the air/fuel ratios are predetermined and no feedback relative to the results is accepted, this type of operation is referred to as open loop operation. All open loop operations are characterized by predetermined air/fuel mixture ratios. Each operation (except closed loop) has a specific air/fuel ratio and because more than one of the engine operational selection conditions can be present at one time, the MCU is programmed with a priority ranking for the operations. It complies with the conditions that pertain to the operation having the highest priority. The priorities are as described below.

Cold Weather Engine Start-Up and Operation

If the air cleaner air temperature is below the calibrated value (55°F or 13°C) of the thermal electric switch (TES), the air/fuel mixture is at a "rich" ratio. Lean air/fuel mixtures are not permitted for a preset period following a cold weather start-up.

At or Near Wide Open Throttle (WOT) Operation (Cold Engine)

This open loop operation occurs whenever the coolant temperature is below the calibrated switching value (95°F or 35°C) of the open loop coolant temperature switch and the WOT vacuum switch (cold) has been closed because of the decrease in manifold vacuum (less than 5 in. Hg or 17 kPa). When this open loop condition occurs the MC solenoid provides a rich air/fuel mixture for cold engine operation at wide open throttle.

NOTE: Temperature and switching vacuum levels are nominal values. The actual switching temperature or vacuum level will vary slightly from switch to switch.

At or Near Wide Open Throttle (WOT) Operation (Warm Engine)

This open loop operation occurs whenever the coolant temperature is above the calibrated switching temperature (135°F or 57°C) of the enrichment coolant temperature switch and the WOT vacuum switch (warm) has been opened because of the decrease in manifold vacuum (less than 3 in. Hg or 10 kPa). When this open loop condition occurs the MC solenoid provides a rich air/fuel mixture for warm engine operation at wide open throttle.

Adaptive Mode of Operation

This open loop operation occurs when the engine is either at idle speed, accelerating from idle speed or decelerating to idle speed. If the engine rpm (tach) voltage is less than the calibrated value and manifold vacuum is above the calibrated switching level for the adaptive vacuum switch (i.e., switch closed), an engine idle condition is assumed to exist. If the engine rpm (tach) voltage is greater than the calibrated value and manifold vacuum is above the calibrated switching level of the adaptive vacuum switch (i.e., switch closed), an engine-deceleration-to-idle speed condition is assumed to exist. During the adaptive mode of operation the MC solenoid provides a predetermined air/fuel mixture.

Closed Loop Operation

Closed loop operation occurs whenever none of the open loop engine operating conditions exist. The MCU causes the MC solenoid to vary the air/fuel mixture in reaction to the voltage input from the oxygen sensor located in the exhaust manifold. The oxygen sensor voltage varies in reaction to changes in oxygen content present in the exhaust gas. Because the content of oxygen in the exhaust gas indicates the completeness of the combustion process, it is a reliable indicator of the air/fuel mixture that is entering the combustion chamber.

Because the oxygen sensor only reacts to oxygen, manifold air leak or malfunction between the carburetor and sensor may cause the sensor to provide an erroneous voltage output. The engine operation characteristics never quite permit the MCU to compute a single air/fuel mixture ratio that constantly provides the optimum air/fuel mixture. Therefore, closed loop operation is characterized by constant variation of the air/fuel mixture because the MCU is forced constantly to make small corrections in an attempt to create an optimum air/fuel mixture ratio.

DIAGNOSIS AND TESTING

The CEC System should be considered as a possible source of trouble for engine performance, fuel economy and exhaust emission complaints only after normal tests and inspections that would apply to an automobile without the system have been performed. The steps in each test will provide a systematic evaluation of each component that could cause an operational malfunction.

To determine if fault exists with the system, a system operational test is necessary. This test should be performed when the CEC System is suspected because no other reason can be determined for a specific complaint. A dwell meter, digital volt-ohmmeter, tachometer, vacuum gauge and jumper wires are required to diagnose system problems. Although most dwell meters should be acceptable, if one causes a change in engine operation when it is connected to the mixture control (MC) solenoid dwell pigtail wire test connector, it should not be used.

The dwell meter, set for the six-cylinder engine scale and connected to a pigtail wire test connector leading from the mixture control (MC) solenoid, is used to determine the air/fuel mixture dwell. When the dwell meter is connected, do not allow the connector terminal to contact any engine component that is connected to engine ground. This includes hoses because they may be electrically conductive. With a normally operating engine, the dwell at both idle speed and partial throttle will be between 10 degrees and 50 degrees and will be varying. Varying means the pointer continually moves back and forth across the scale. The amount it varies is not important, only the fact that is does vary. This indicates closed loop operation, indicating the mixture is being varied according to the input voltage to the MCU from the oxygen sensor. With wide open throttle (WOT) and/or cold engine operation, the air/fuel mixture ratio will be predetermined and the pointer will only vary slightly. This is open loop operation, indicating the oxygen sensor output has no effect on the air/fuel mixture. If there is a question whether or not the system is in closed loop operation, richening or leaning the air/fuel mixture will cause the dwell to vary more if the system is in closed loop operation.

Test Equipment

The equipment required to perform the checks and tests includes a tachometer, a hand vacuum pump and a digital volt-ohmmeter (DVOM) with a minimum ohms per volt of 10 megaohms.

――――――――――― CAUTION ―――――――――――
The use of a voltmeter with less than 10 mega-ohms per volt input impedance can destroy the oxygen sensor. Since it is necessary to look inside the carburetor with the engine running, observe the following precautions.

1. Shape a sheet of clear acrylic plastic at least 0.250 in. thick and 15 in. × 15 in.

2. Secure the acrylic sheet with an air cleaner wing nut after the top of the air cleaner has been removed.

3. Wear eye protection whenever performing checks and tests.

4. When engine is operating, keep hands and arms clear of fan, drive pulleys and belts. Do not wear loose clothing. Do not stand in line with fan blades.

5. Do not stand in front of running car.

CEC SYSTEM OPERATION 6-CYLINDER ENGINE

The open loop mode of operation occurs when:
1. Starting the engine, engine is cold or air cleaner air is cold.
2. Engine is at idle speed.
3. Carburetor is either at or near wide open throttle (WOT).
When any of these conditions occur, the metering pins are driven to a predetermined (programmed) position for each condition. Because the positions are predetermined and no feedback relative to the results is accepted, this type of operation is referred to as open loop operation. The five open loop operations are characterized by the metering pins being driven to a position where they are stopped and remain stationary.

Each operation (except closed loop) has a specific metering pin position and because more than one of the operation selection conditions can be present at one time, the MCU is programmed with a priority ranking for the operations. It complies with conditions that pertain to the operation having the highest priority. The priorities are as described below.

Cold Weather Engine Start-Up and Operation

If the air cleaner air temperature is below the calibrated value of the thermal electric switch (TES), the stepper motor is positioned a predetermined number of steps rich of the initialization position and air injection is diverted upstream. Lean air/fuel mixtures are not permitted for a preset period following a cold weather start-up.

Open Loop 1

Open Loop 1 will be selected if the air cleaner air temperature is above a calibrated value and open loop 2, 3, or 4 is not selected, and if the engine coolant temperature is below the calibrated value. The OL1 mode operates in lieu of normal closed loop operation during a cold engine operating condition. If OL1 operation is selected, one of two predetermined stepper motor positions are chosen, dependent if the altitude circuit (lean limit) jumper wire is installed. With each engine start-up, a start-up timer is activated. During this interval, if the engine operating condition would otherwise trigger normal closed loop operation, OL1 operation is selected.

Open Loop 2, Wide Open Throttle (WOT)

Open Loop 2 is selected whenever the air cleaner air temperature is above the calibrated value of the thermal electric switch (TES) and the WOT switch has been engaged. When the Open Loop 2 mode is selected, the stepper motor is driven to a calibrated number of steps rich of initialization and the air control valve switches air "downstream". However, if the "lean limit" circuit (with altitude jumper wire) is being used, the air is instead directed "upstream". The WOT timer is activated whenever OL2 is selected and remains active for a preset period of time. The WOT timer remains inoperative if the "lean limit" circuit is being used.

Open Loop 3

Open Loop 3 is selected when the ignition advance vacuum level falls below a predetermined level. When the OL3 mode is selected, the engine rpm is also determined. If the rpm (tach) voltage is greater than the calibrated value, an engine deceleration condition is assumed to exist. If the rpm (tach) voltage is less than the calibrated value, an engine idle speed condition is assumed to exist.

Open Loop 4

Open Loop 4 is selected whenever manifold vacuum falls below a predetermined level. During OL4 operation, the stepper motor is positioned at the initialization position. Air injection is switched "upstream" during OL4 operation. However, air is switch "downstream" if the extended OL4 timer is activated and if the "lean limit" circuit is not being used (without altitude jumper wire). Air is also switch "downstream" if the WOT timer is activated.

Closed Loop

Closed loop operation is selected after either OL1, OL2, OL3 or OL4 modes have been selected and the start-up timer has timed out. Air injection is routed "downstream" during closed loop operation. The predetermined "lean" air/fuel mixture ceiling is selected for a preset length of time at the onset of closed loop operation.

High Altitude Adjustment

An additional function of the MCU is to correct for a change in ambient conditions (e.g., high altitude). During closed loop operation the MCU stores the number of steps and direction that the metering pins are driven to correct the oxygen content of the exhaust. If the movements are consistently to the same position, the MCU will vary all open loop operation predetermined metering pin positions a corresponding amount. This function allows the open loop air/fuel mixture ratios to be "tailored" to the existing ambient conditions during each uninterrupted use of the system. This optimizes emission control and engine performance.

Closed Loop Operation

The CEC system controls the air/fuel ratio with movable air metering pins, visible from the top of the carburetor air horn, that are driven by the stepper motor. The stepper motor moves the metering pins in increments or small steps via electrical impulses generated by the MCU. The MCU causes the stepper motor to drive the metering pins to a "richer" or "leaner" position in reaction to the voltage input from the oxygen content present in the exhaust gas. Because the content of oxygen in the exhaust gas indicates the completeness of the combustion process, it is a reliable indicator of the air/fuel mixture that is entering the combustion chamber.

Because the oxygen sensor only reacts to oxygen, any air leak or malfunction between the carburetor and sensor may cause the sensor to provide an erroneous voltage output. This could be caused by a manifold air leak or malfunctioning secondary air check value. The engine operation characteristics never quite permit the MCU to compute a single metering pin position that constantly provides the optimum air/fuel mixture. Therefore, closed loop operation is characterized by constant movement of the metering pins because the MCU is forced constantly to make small corrections in the air/fuel mixture in an attempt to create an optimum air/fuel mixture ratio.

DIAGNOSIS AND TESTING

The idle speed control system is interrelated with the CEC system and must be diagnosed in conjunction with the CEC System. Refer to Diagnostic Tests 9, 10 and 11, if a malfunction occurs.

The electronic ignition retard function of the ignition control module is interrelated with CEC System and must be diagnosed in conjunction with CEC System. Refer to Diagnostic Test 4 if a malfunction occurs.

The air injection system is interrelated with the CEC System and must be diagnosed in conjunction with the CEC System. Refer to Diagnostic Test 6, 7 and 8 if a malfunction occurs.

Preliminary Tests

Before performing the Diagnostic Tests, other engine associated systems that can affect air/fuel mixture, combustion efficiency or exhaust gas composition should be tested for faults. These systems include:
1. Basic carburetor adjustments.
2. Mechanical engine operation (spark plugs, valves, rings, etc.).
3. Ignition system components and operation.
4. Gaskets (intake manifold, carburetor or base plate); loose vacuum hoses or fittings, or loose electrical connections.

Initialization

When the ignition system is turned off, the MCU is also turned off. It has no long term memory circuit for prior operation. As a result, it has an initialization function that is activated when the ignition switch is turned ON.

The MCU initialization function moves the metering pins to the predetermined starting position by first driving them all the way to the rich end stop and then driving them in the lean direction by a predetermined number of steps. No matter where they were before initialization, they will be at the correct position at the end of every initialization period. Because each open loop operation metering pin position is dependent on the initialization function, this function is the first test in the diagnostic procedure.

NOTE: The CEC System should be considered as a possible source of trouble for engine performance, fuel economy and exhaust emission complaints only after normal tests that would apply to an automobile without the system have been performed.

Jeep Self-Diagnostic System

Late model Jeep vehicles equipped with a six cylinder engine and California emissions package have a self-diagnostic system with a CHECK ENGINE light mounted in the instrument panel. The self-diagnostic system is designed to detect problems most likely to occur within the various system components.

When a jumper wire is connected between the trouble code test terminals 6 and 7 of the 15-terminal diagnostic connector (D2), the CHECK ENGINE light will flash a trouble code or codes that indicate a problem area. For a bulb and system check, the CHECK ENGINE light will illuminate when the ignition switch is ON and the engine not started. If the test terminals are then grounded, the light will flash a code 12 that indicates the self-diagnostic system is operational. A code 12 consists of one flash, followed by a short pause, then two more flashes in quick succession. After a longer pause, the code will repeat two more times.

When the engine is started, the CHECK ENGINE light will remain ON momentarily and then be turned off. If the CHECK ENGINE light remains on, the self-diagnostic system has detected a problem. If the trouble code test terminals are then grounded, the trouble code will be flashed three times; if more than one trouble code is stored, each will be flashed three more times in numerical order, from the lowest to the highest numbered code. This trouble code series will repeat as long as the test terminals are grounded.

A trouble code indicates a problem in a particular circuit or component. Trouble code 14, for example, indicates a problem in the coolant sensor circuit. It should be noted that the self-diagnostic system doesn't pinpoint where the problem is in the coolant sensor circuit, which includes the coolant sensor, connector, wire harness and the electronic control unit itself. The following diagnostic charts contain procedures for isolating the problem to a particular point in the circuit to avoid the unnecessary replacement of working components. A diagnosis chart is provided for each trouble code.

Because the self-diagnostic system does not detect all possible problems, the absence of a trouble code does not necessarily mean the system is functioning normally. The System Performance Test should be performed when the self-diagnostic system does not indicate a problem, but the system operation is suspect.

All system connectors in the engine compartment are sealed against debris and moisture. Because the system operates on low voltage and low current, corrosion on the connectors can cause problems. Before repairing or replacing any component, disconnect the appropriate connector(s) and check for proper installation, bent, broken or dirty terminals or mating tabs.

Clean, straighten or replace the connectors as required, then reconnect everything and recheck the system operation to see if the problem has been corrected. The system should be considered as a possible cause of trouble only after normal engine diagnosis for ignition timing, carburetor and idle speed adjustments has been performed. The electronic control module (ECM) is located in the passenger compartment at the right side of the steering column below the instrument panel.

TROUBLE CODE MEMORY

When a problem develops in the feedback system, the CHECK ENGINE light will illuminate and a trouble code will be stored in the on-board computer memory. If the fault is intermittent, the CHECK ENGINE light will be turned off 10 seconds after the problem disappears. The trouble code will be retained in the memory until the battery voltage to the control unit is removed. Disconnecting the battery for 10 seconds will erase all stored trouble codes.

The CHECK ENIGNE light will be illuminated only if a problem exists that pertains to the conditions listed below. It takes up to 5 seconds minimum for the light to come on when a problem occurs. Code 12 is not stored in memory and any codes stored will be cleared if the problem does not reoccur within 50 engine starts. The trouble codes indicate problems as follows:

CODE 12: No distributor reference pulses to the ECM. This code is not stored in memory and will only flash while the trouble exists. This code is normal when the igition is switched ON with the engine not running.

CODE 13: Oxygen sensor circuit. The engine must operate for up to 5 minutes at part throttle, under road load, before this code will be set.

CODE 14: Shorted coolant sensor circuit. The engine must operate for up to 5 minutes before this code will be set.

CODE 15: Open coolant sensor circuit. The engine must operate for up to 5 minutes before this code will be set.

CODE 21: Throttle position sensor circuit. The engine must operate for at least 25 seconds at curb idle speed before this code will be set.

CODE 23: Mixture control solenoid circuit is shorted or open.

CODE 34: Vacuum sensor circuit. The engine must operate for up to 5 minutes at curb idle speed before this code will be set.

CODE 41: No distributor reference pulses to the ECM at the specified engine manifold vacuum. This code will be stored in memory.

CODE 42: Electronic spark timing (EST) bypass circuit or EST circuit has short circuit to ground or an open circuit.

CODE 44: Lean exhaust indication. The engine must operate for up to 5 minutes, be in closed loop operation and at part throttle before this code will be set.

CODE 44 & 45: If these two codes appear at the same time, it indicates a problem in the oxygen sensor circuit.

CODE 45: Rich exhaust indication. The engine must operate for up to 5 minutes, be in closed loop and at part throttle before this code will be set.

CODE 51: Faulty calibration unit (PROM) or installation. It requires up to 30 seconds for this code to be set.

CODE 54: Mixture control (MC) solenoid circuit is shorted or the ECM is faulty.

CODE 55: Voltage reference has short circuit to ground (terminal 21), faulty oxygen sensor or faulty ECM.

VACUUM DIAGRAMS

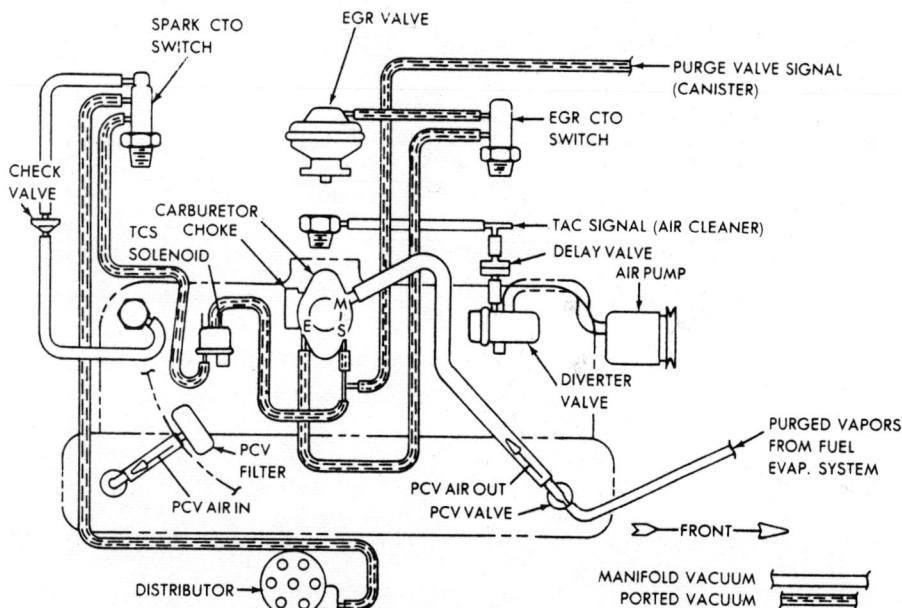

1978–79 California 6–258 with automatic transmission

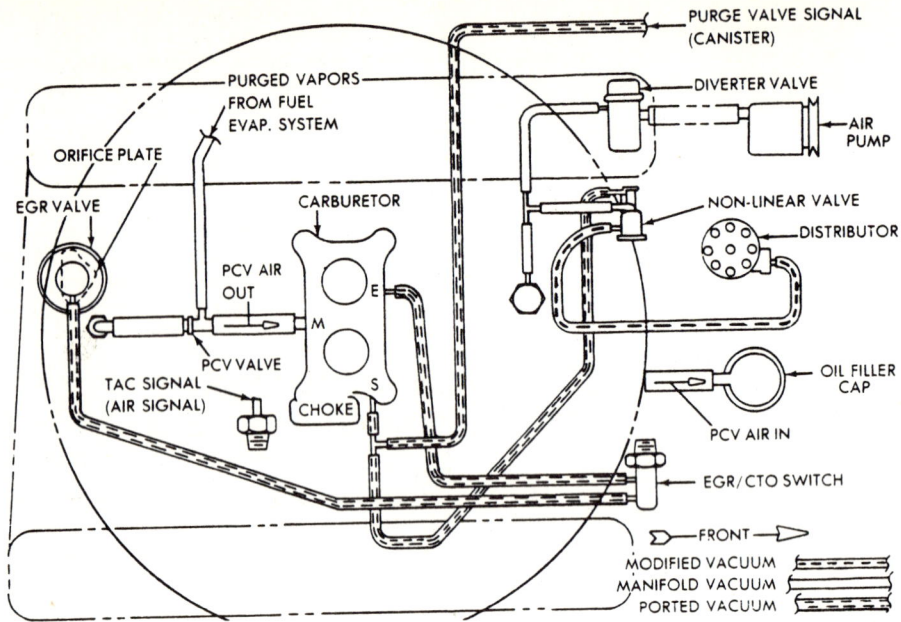

1978–79 8–304 with automatic transmission, except California

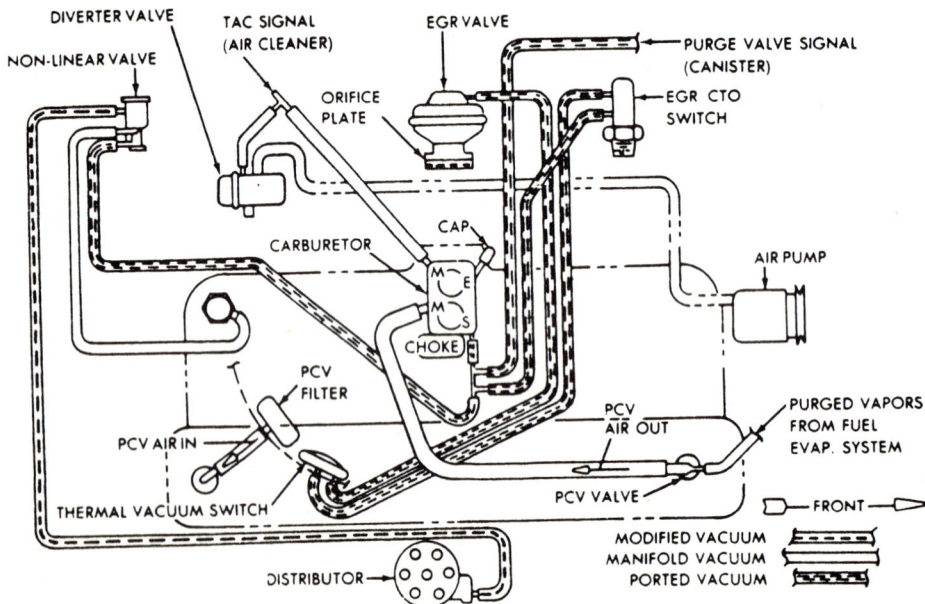

1978–79 6–258 with manual transmission, except California

1980–86 6–258 with manual transmission, except California

1980–86 California 6–258 with automatic transmission

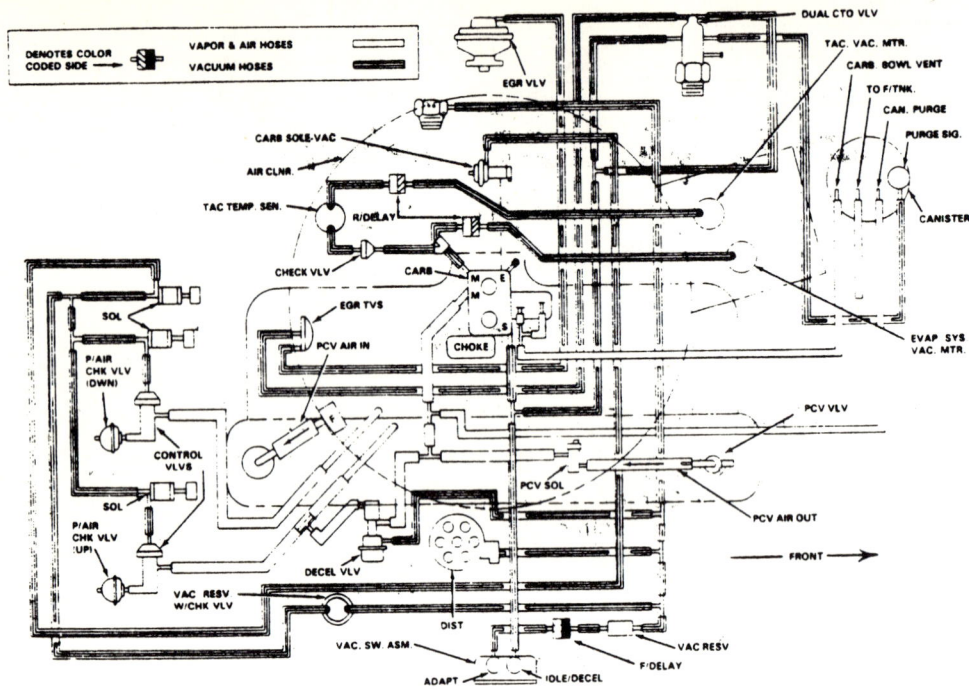

1980–86 California 6–258 with manual transmission

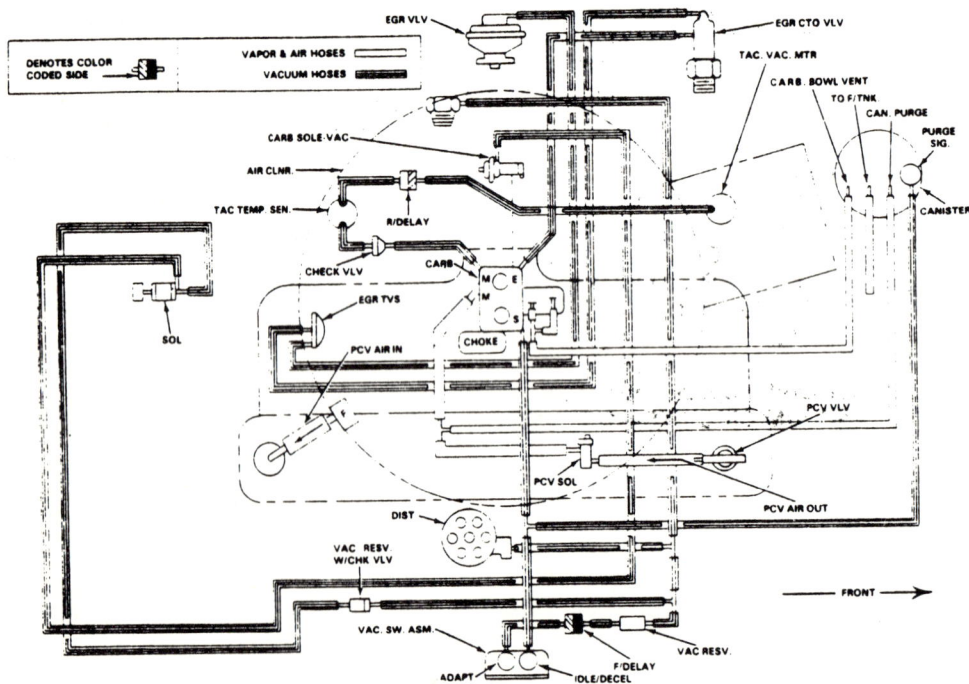

1980–86 Canadian 6–258 with manual or automatic transmission

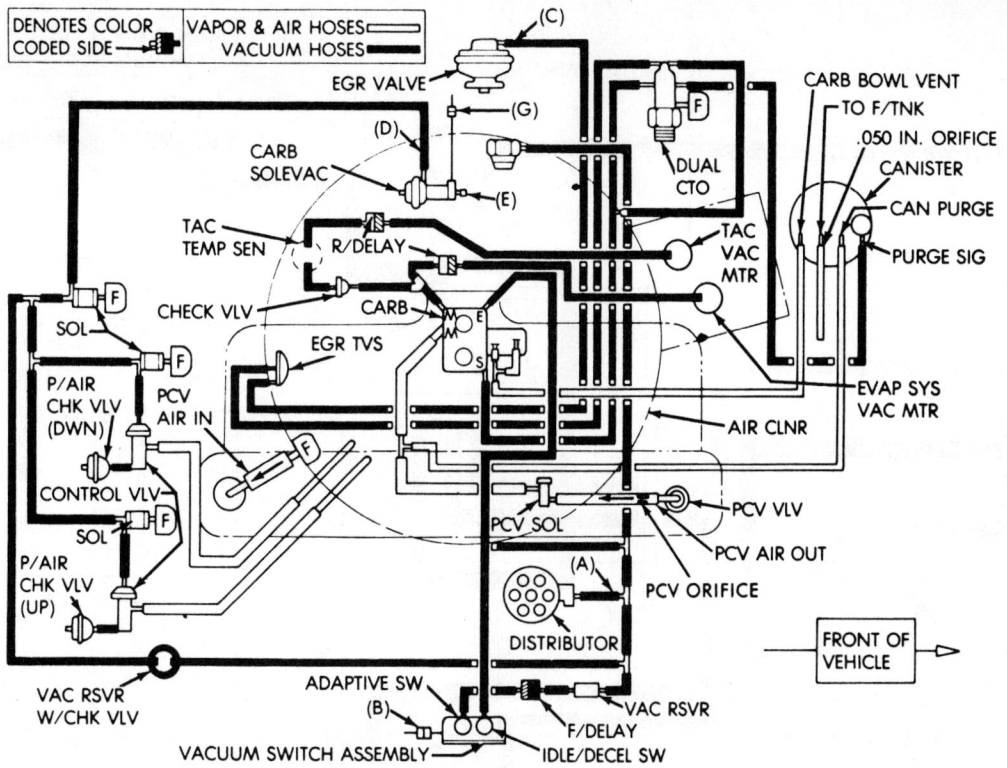

1987–90 6–258 with automatic transmission

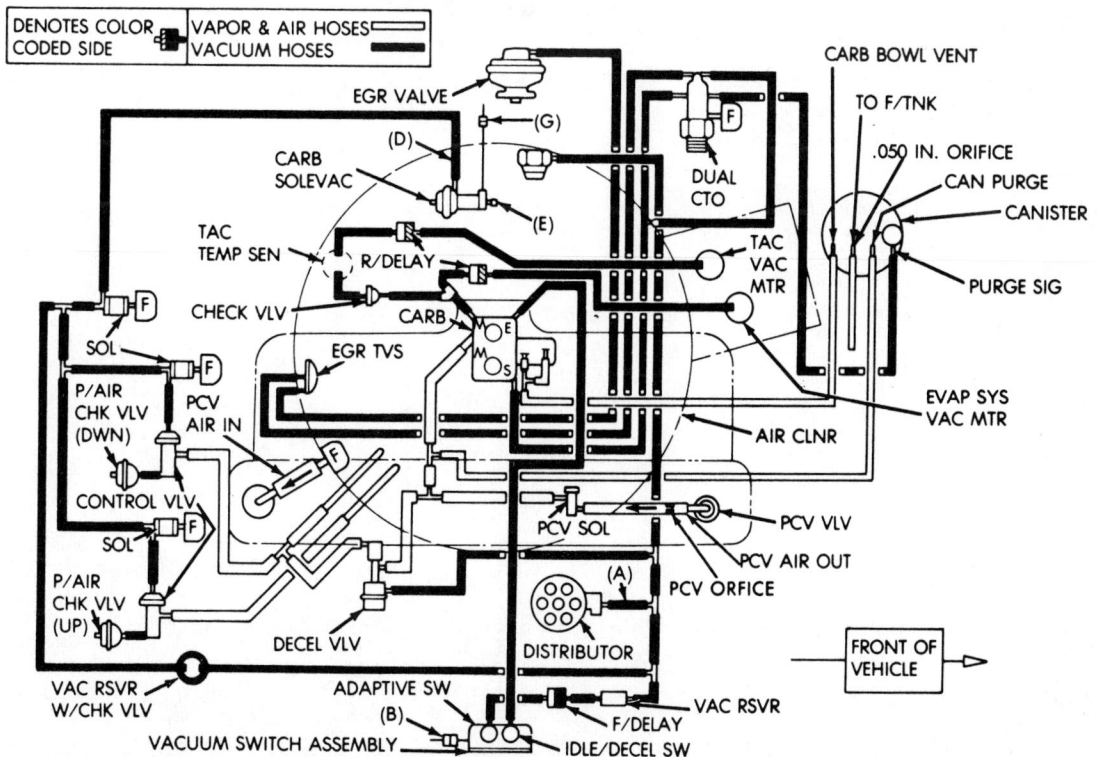

1987–90 6–258 with manual transmission

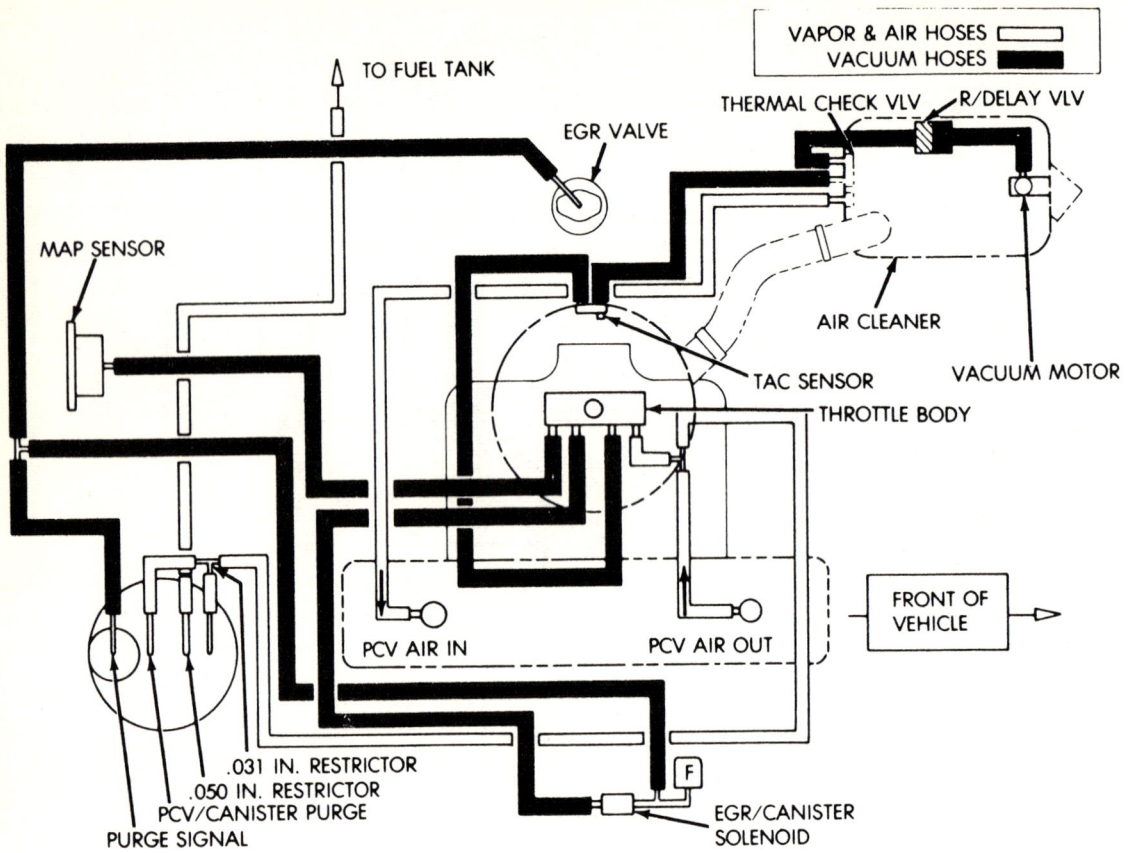

VAPOR & AIR HOSES ☐
VACUUM HOSES ■

TO FUEL TANK

EGR VALVE

THERMAL CHECK VLV   R/DELAY VLV

MAP SENSOR

AIR CLEANER

TAC SENSOR

VACUUM MOTOR

THROTTLE BODY

PCV AIR IN   PCV AIR OUT

FRONT OF VEHICLE

.031 IN. RESTRICTOR
.050 IN. RESTRICTOR
PCV/CANISTER PURGE
PURGE SIGNAL

EGR/CANISTER SOLENOID

1987-90 4-150

Fuel System

5

QUICK REFERENCE INDEX

GENERAL INDEX

CARBURETED FUEL SYSTEM

CARBURETOR SPECIFICATIONS
Carter BBD

Engine	Years	Float Level Dry (in.)	Step-Up Piston Gap (in.)	Fast Idle Cam Setting (in.)	Fast Idle rpm	Choke Unloader (in.)	Initial Choke Valve Clearance (in.)	Automatic Choke Setting
6-258	1977–78	1/4	3/64	3/32	1,700	9/64	1/8	2 rich
	1979	1/4	1/32	①	②	9/32	③	1 rich
	1980	1/4	1/32	④	⑤	9/32	⑥	⑦
	1981	1/4	1/32	3/32	⑤	9/32	9/64	1 rich
	1982	1/4	1/32	⑧	1,700	9/32	5/32	1 rich
	1983–84	1/4	1/32	3/32	⑤	9/32	5/32	1 rich
	1985–90	1/4	1/32	3/32	⑤	9/32	9/64	TR

TR: Tamper Resistant
① Except High Altitude: 7/64
   High Altitude: 3/64
② Man. Trans: 1500
   Auto Trans: 1600
③ Man. Trans., except High Altitude: 0.140
   Man. Trans., High Altitude: 0.128
   Auto. Trans.: 0.150
④ Except High Altitude: 3/32
   High Altitude: 5/64

⑤ Man. Trans.: 1,700
   Auto. Trans.: 1,850
⑥ Except High Altitude: 0.140
   High Altitude: 0.116
⑦ Except High Altitude: 2 rich
   High Altitude: 1 rich
⑧ Auto. Trans.: 7/64
   Man. Trans. 5/32

CARBURETOR SPECIFICATIONS
Carter YF

Engine	Years	Float Level (in.)	Float Drop (in.)	Dashpot Setting (in.)	Fast Idle rpm	Choke Unloader (in.)	Initial Choke Valve Clearance (in.)	Automatic Choke Setting
4-134	1971	29/64	1 1/4	3/32	1,600	19/64	15/64	Notch
6-232, 6-258	1972–73	29/64	1 1/4	3/32	1,800	9/32	7/32	1 rich
6-258	1974	31/64	1 25/64	3/32	1,600	9/32	7/32	1 rich
	1975	31/64	1 25/64	5/64	1,600	9/32	7/32	1 rich
	1976	31/64	1 3/8	5/64	1,600	9/32	7/32	1 rich

CARBURETOR SPECIFICATIONS
Rochester 2G

Engine	Years	Float Level (in.)	Fast Float Drop (in.)	Idle Speed rpm	Choke Unloader (in.)	Initial Choke Valve Clearance (in.)
6-225	1971	15/32	17/8	1,800	3/64	1/4

CARBURETOR SPECIFICATIONS
Carter YFA

Engine	Years	Float Level (in.)	Initial Choke Valve Clearance (in.)	Fast Idle Cam Setting (in.)	Choke Unloader (in.)	Fast Idle Speed rpm	Automatic Choke Setting
4-150	1984	39/64	15/64	11/64	15/64	2,000 MT 2,300 AT	TR
	1985–86	39/64	9/32	11/64	15/64	2,000 MT 2,300 AT	TR

MT: Manual Transmission
AT: Automatic Transmission
TR: Tamper Resistant

CARBURETOR SPECIFICATIONS
Rochester 2GV

Engine	Years	Float Level (in.)	Fast Float Drop (in.)	Idle Speed (rpm)	Choke Unloader (in.)	Initial Choke Valve Clearance (in.)
6-225	1971	15/32	13/4	1,800	9/64	1/4

CARBURETOR SPECIFICATIONS
Rochester 2SE/E2SE

Engine	Years	Float Level (in.)	Choke Coil Lever (in.)	Fast Idle Cam Clearance (in.)	Choke Vacuum Break (in.) Primary	Choke Vacuum Break (in.) Secondary	Choke Unloader (in.)	Air Valve Rod (deg.)
4-151	1980–83	①	5/64	1/8	9/64	②	17/64	2

① 2SE/E2SE Man. Trans. and Calif.: 13/64
　2SE Auto. Trans.: 1/4
　E2SE Auto. Trans.: 13/64
② Man. Trans. 3/64–5/64
　Auto. Trans.: 5/64

CARBURETOR SPECIFICATIONS
Autolite/Motorcraft 2100

Engine	Float Level Dry (in.)	Float Level Wet (in.)	Dashpot Setting (in.)	Fast Idle rpm	Choke Unloader (in.)	Initial Choke Valve Clearance (in.)	Automatic Choke Setting
8-304	2/5	25/32	9/64	2,200	①	0.130	2 rich

① 1972–73: 1/4
　1974 and later: 9/32

CARBURETOR SPECIFICATIONS
Motorcraft 2150

Engine	Years	Float Level Dry (in.)	Float Level Wet (in.)	Fast Idle Cam Setting (in.)	Fast Idle rpm	Initial Choke Unloader (in.)	Choke Valve Clearance (in.)	Automatic Choke Setting
8-304	1976–79	$17/32$	$59/64$	①	②	$9/32$	$7/64$	②
	1980–81	$11/32$	$59/64$	$5/64$	②	$1/3$	$7/64$	②

① Man. Trans.: 0.086
  Auto. Trans.: 0.093
② Man. Trans., except Calif.: 2 rich
  Calif. and Auto Trans.: 1 rich

Troubleshooting Basic Fuel System Problems

Problem	Cause	Solution
Engine cranks, but won't start (or is hard to start) when cold	· Empty fuel tank · Incorrect starting procedure · Defective fuel pump · No fuel in carburetor · Clogged fuel filter · Engine flooded · Defective choke	· Check for fuel in tank · Follow correct procedure · Check pump output · Check for fuel in the carburetor · Replace fuel filter · Wait 15 minutes; try again · Check choke plate
Engine cranks, but is hard to start (or does not start) when hot— (presence of fuel is assumed)	· Defective choke	· Check choke plate
Rough idle or engine runs rough	· Dirt or moisture in fuel · Clogged air filter · Faulty fuel pump	· Replace fuel filter · Replace air filter · Check fuel pump output
Engine stalls or hesitates on acceleration	· Dirt or moisture in the fuel · Dirty carburetor · Defective fuel pump · Incorrect float level, defective accelerator pump	· Replace fuel filter · Clean the carburetor · Check fuel pump output · Check carburetor
Poor gas mileage	· Clogged air filter · Dirty carburetor · Defective choke, faulty carburetor adjustment	· Replace air filter · Clean carburetor · Check carburetor
Engine is flooded (won't start accompanied by smell of raw fuel)	· Improperly adjusted choke or carburetor	· Wait 15 minutes and try again, without pumping gas pedal · If it won't start, check carburetor

Fuel Pump

REMOVAL AND INSTALLATION

All Engines

1. Disconnect the inlet and outlet fuel lines, and any vacuum lines.
2. Remove the two fuel pump body attaching nuts and lockwashers.
3. Pull the pump and gasket, or O-ring, free of the engine. Make sure that the mating surfaces of the fuel pump and the engine are clean.
4. Cement a new gasket to the mounting flange of the fuel pump.
5. Position the fuel pump on the engine block so that the lever of the fuel pump rests on the fuel pump cam of the camshaft.
6. Secure the fuel pump to the block with the two cap screws and lock washers.
7. Connect the intake and outlet fuel lines to the fuel pump, and any vacuum lines.

FUEL PUMP TESTING

Volume Check

Disconnect the fuel line from the carburetor. Place the open

1. Fuel outlet 2. Vapor return 3. Fuel inlet

Non-serviceable V6 fuel pump

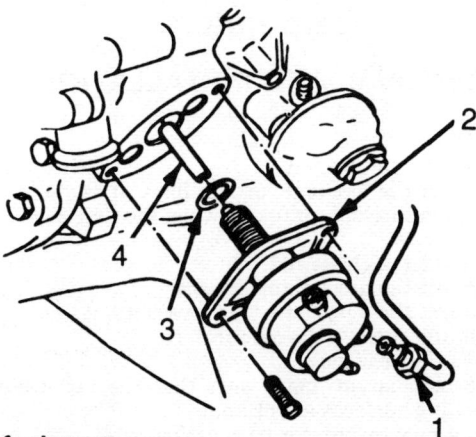

4-150 fuel pump

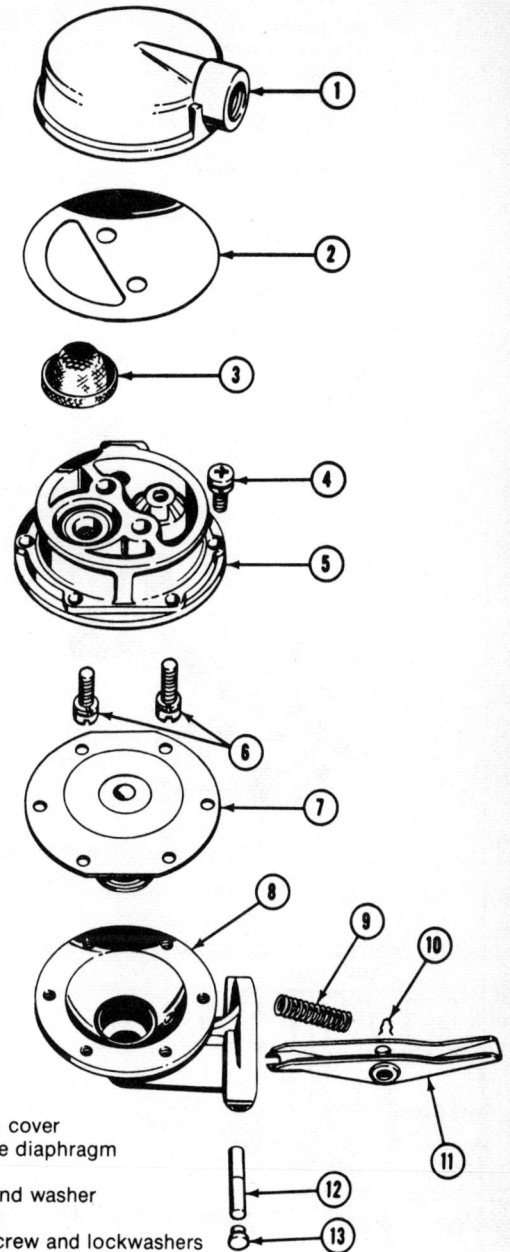

1. Housing cover
2. Air dome diaphragm
3. Strainer
4. Screw and washer
5. Housing
6. Cover screw and lockwashers
7. Main diaphragm
8. Pump body
9. Cam lever return spring
10. Pin retainer
11. Cam lever
12. Cam lever pin
13. Lever seal shaft plug

F4-134 fuel pump used with electric wipers

end in a suitable container. Start the engine and operate it at normal idle speed. The pump should deliver at least one pint in 30 seconds.

Pressure Check

Disconnect the fuel line at the carburetor. Disconnect the fuel return line from the fuel filter if so equipped, and plug the nipple on the filter. Install a T-fitting on the open end of the fuel line and refit the line to the carburetor. Plug a pressure gauge

1. Cover screw
2. Lockwasher
3. Diaphragm spring
4. Spring seat
5. Diaphragm and rod
6. Oil seal
7. Valve assembly
8. Body
9. Rocker arm pin spring
10. Fuel diaphragm
11. Oil seal retainer
12. Diaphragm and rod
13. Valve retainer
14. Cover
15. Gasket
16. Screen
17. Bowl
18. Bail
19. Gasket
20. Screw
21. Rocker arm spring
22. Link spacer
23. Rocker arm
24. Washer
25. Body

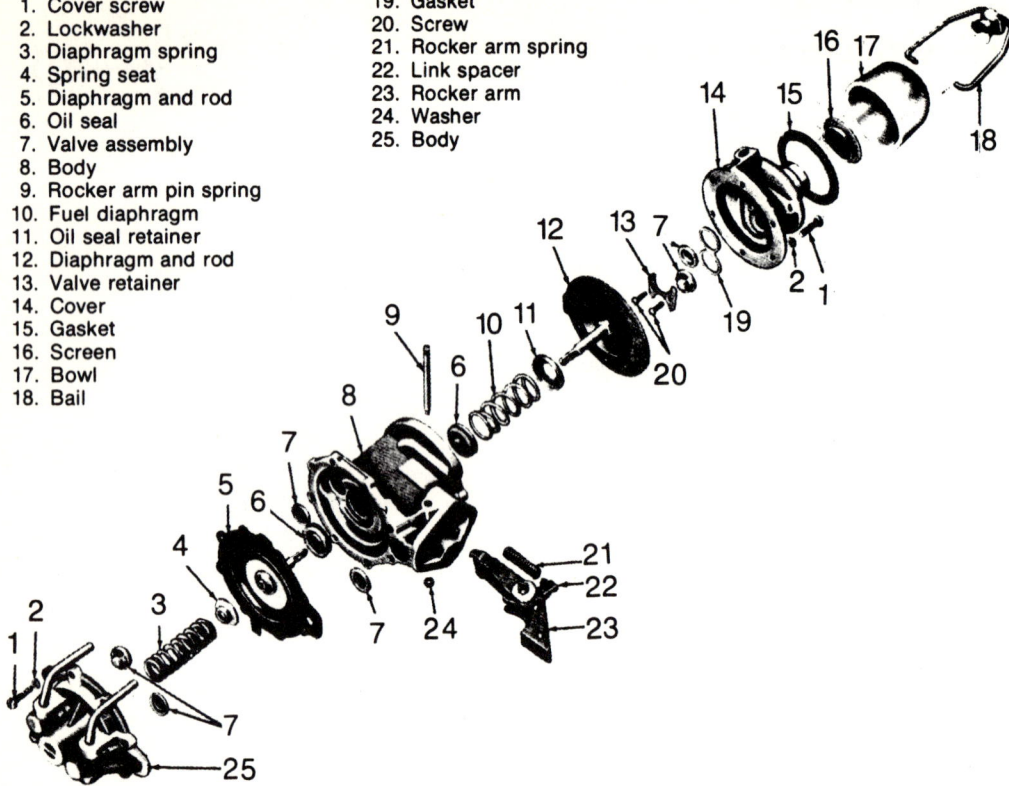

4–134 combined fuel and vacuum pump

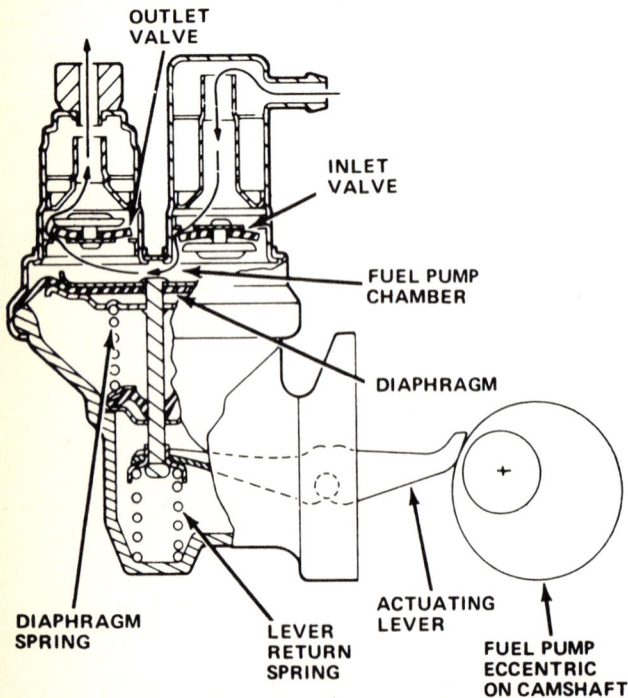

Fuel pump used on late model 6–232 and all 6–258, 8–304, 4–151 and 1984–86 4–150 engines

into the remaining opening of the T-fitting. The hose leading to the pressure gauge should not be any longer than 6 inches. Start the engine and let it run at idle speed. Bleed any air out of the hose between the gauge and the T-fitting. On pumps with a fuel return line, the line must be plugged. Start the engine. Fuel pressures are as follows:

- 4–134: 2.50–3.75psi @ 1,800 rpm
- 4–150: 4.00–5.00psi @ idle
- 4–151: 6.50–8.00psi @ idle
- 6–225: 3.75psi minimum @ 600 rpm
- 6–232: 3.00–5.00psi @ idle
- 6–258: 3.00–5.00psi @ idle
- 8–304: 4.00–6.00psi @ idle

Carburetors

REMOVAL AND INSTALLATION

Carter BBD

1. Remove the air cleaner.
2. Disconnect the fuel and vacuum lines. It might be a good idea to tag them to avoid confusion when the time comes to put them back.
3. Disconnect the choke rod.
4. Disconnect the accelerator linkage.
5. Disconnect the automatic transmission linkage.
6. Unbolt and remove the carburetor.
7. Remove the base gasket.
8. Before installation, make sure that the carburetor and manifold sealing surfaces are clean.
9. Install a new carburetor base gasket.
10. Install the carburetor and start the fuel and vacuum lines.

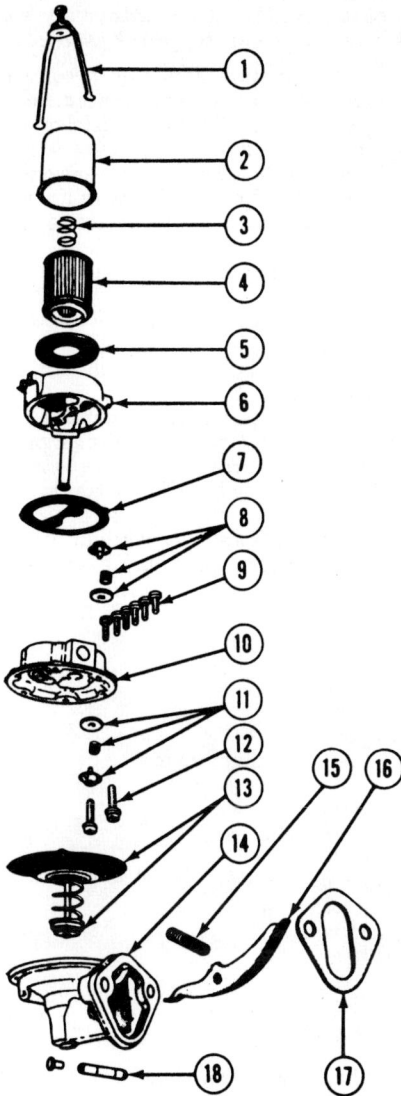

1. Ball
2. Bowl
3. Spring
4. Filter
5. Gasket
6. Pump body
7. Gasket
8. Valve assembly
9. Screws
10. Valve housing
11. Valve assembly
12. Screws
13. Diaphragm and oil seal
14. Pump body
15. Cam lever spring
16. Cam lever
17. Gasket
18. Cam lever pin and plug

Early 1971 6–232 fuel pump. This pump was also used on some 4–134 engines

11. Bolt down the carburetor evenly, tightening the bolts in a criss-cross pattern to 7 ft. lbs. (9 Nm), then to 14 ft. lbs. (19 Nm).

12. Tighten the fuel and vacuum lines.

13. Connect the accelerator and automatic transmission linkage. If the transmission linkage was disturbed, it will have to be adjusted.

14. Connect the choke rod.

15. Install the air cleaner. Adjust the idle speed and mixture.

Rochester 2SE/E2SE

1. Remove air cleaner and gasket.
2. Disconnect fuel pipe and vacuum lines.

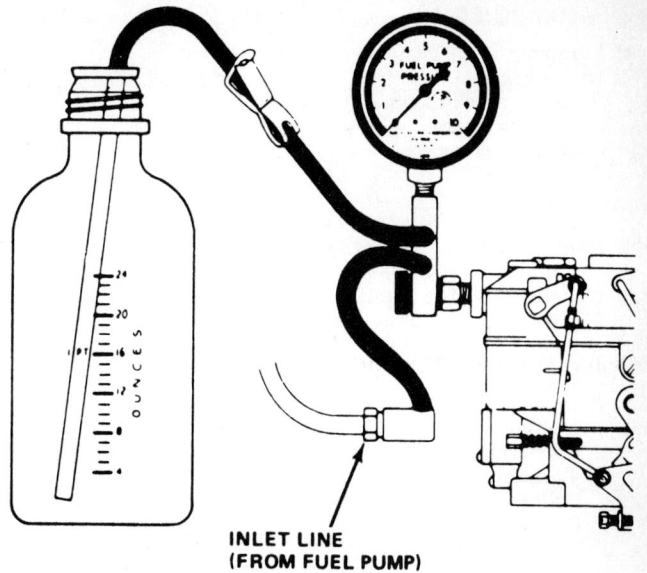

Fuel pump pressure and volume test

3. Disconnect electrical connectors.
4. Disconnect accelerator linkage.
5. If equipped with automatic transmission, disconnect downshift cable.
6. If equipped with cruise control, disconnect linkage.
7. Remove carburetor attaching bolts.
8. Remove carburetor and EFE heater/insulator (if used).
9. Fill carburetor bowl before installing carburetor. A small supply of no-lead fuel will enable the carburetor to be filled and the operation of the float and inlet needle and seat to be checked. Operate throttle lever several times and check discharge from pump jets before installing carburetor.
10. Inspect EFE heater/insulator for damage. Be certain throttle body and EFE heater/insulator surfaces are clean.
11. Install EFE heater; insulator.
12. Install carburetor and tighten nuts alternately to the correct torque.
13. Connect downshift cable as required.
14. Connect cruise control cable as required.
15. Connect accelerator linkage.
16. Connect electrical connections.
17. Connect fuel pipe sand vacuum hoses.
18. Check base (slow) and fast idle.
19. Install air cleaner.

Carter YF/YFA

1. Remove the air cleaner.
2. Tag and disconnect all hoses leading to the carburetor.
3. Disconnect the control shaft from the throttle lever.
4. Disconnect the in-line fuel filter, pullback spring and all electrical connectors.
5. Remove the carburetor mounting nuts and lift off the carburetor.
6. Remove the carburetor mounting gasket from the spacer.
7. Clean all gasket mating surfaces on the spacer and carburetor.
8. Install a new gasket on the spacer, then install the carburetor and secure it with the mounting nuts.
9. Reconnect all vacuum hoses, fuel lines and electrical connectors. Connect the control shaft and pullback spring.
10. Install the air cleaner and adjust the curb and fast idle speed as previously described.

Rochester 2G/2GV

1. Remove the air cleaner.
2. Disconnect:
   a. Accelerator linkage.
   b. Vacuum hoses.
   c. Return spring.
   f. PCV hose.
   g. Fuel line.
3. Remove the carburetor holddown nuts.
4. Remove the carburetor and mounting gasket.
5. Clean all gasket surfaces thoroughly.
6. Installation is the reverse of removal. Always use new gaskets. Tighten the nuts in a criss-cross pattern.

Autolite/Motorcraft 2100/2150

1. Remove the air cleaner.
2. Disconnect:
   a. Accelerator cable.
   b. Vacuum hoses.
   c. Pullback spring.
   d. Choke clean air tube.
   e. Solenoid wire.
   f. PCV hose.
   g. Fuel filter.
   h. Choke heat tube.
3. Remove the carburetor holddown nuts.
4. Remove the carburetor and mounting gasket. If an EGR spacer is used, remove this also and its gasket.
5. Clean all gasket surfaces thoroughly.
6. Installation is the reverse of removal. Always use new gaskets. Tighten the nuts in a criss-cross pattern.

OVERHAUL

Efficient carburetion depends greatly on careful cleaning and inspection during overhaul, since dirt, gum, water, or varnish in or on the carburetor parts are often responsible for poor performance.

Overhaul your carburetor in a clean, dust-free area. Carefully disassemble the carburetor, referring often to the exploded views. Keep all similar and look-alike parts segregated during disassembly and cleaning to avoid accidental interchange during assembly. Make a note of all jet sizes.

When the carburetor is disassemble, wash all parts (except diaphragms, electric choke units, pump plunger, and any other plastic, leather, fiber or rubber parts) in clean carburetor solvent. Do not leave parts in the solvent any longer than is necessary to sufficiently loosen the deposits. Excessive cleaning may remove the special finish from the float bowl and choke valve bodies, leaving these parts unfit for service. Rinse all parts in clean solvent and blow them dry with compressed air or allow them to air dry. Wipe clean all cork, plastic, leather and fiber parts with a clean, lint-free cloth.

Blow out all passages and jets with compressed air and be sure that there are not restrictions or blockages. Never use wire or similar tools to clean jets, fuel passages, or air bleeds. Clean all jets and valves separately to avoid accidental interchange.

Check all parts for wear or damage. If wear or damage is found, replace the defective parts. Especially check the following:
1. Check the float needle and seat for wear. If wear is found, replace the complete assembly.
2. Check the float hinge pin for wear and the float(s) for dents or distortion. Replace the float if fuel has leaked into it.
3. Check the throttle and choke shaft bores for wear or an out-of-round condition. Damage or wear to the throttle arm, shaft, or shaft bore will often require replacement of the throttle body. These parts require a close tolerance of fit. Wear may allow air leakage, which could affect starting and idling.

NOTE: Throttle shafts and bushings are not included in overhaul kits. They can be purchased separately.

4. Inspect the idle mixture adjusting needles for burrs or grooves. Any such condition requires replacement of the needle, since you will not be able to obtain a satisfactory idle.
5. Test the accelerator pump check valves. They should pass air one way but not the other. Test for proper seating by blowing and sucking on the valve. Replace the valve if necessary. If the valve is satisfactory, wash the valve again to remove breath moisture.
6. Check the bowl cover for warped surfaces with a straightedge.
7. Closely inspect the valves and seats for wear and damage, replacing as necessary.
8. After the carburetor is assembled, check the choke valve for freedom of operation.

Carburetor overhaul kits are recommended for each overhaul. These kits contain all gaskets and new parts to replace those that deteriorate most rapidly. Failure to replace all parts supplied with the kit (especially gaskets) can result in poor performance later.

Carburetor manufacturers supply overhaul kits of three basic types: minor repair, major repair, and gasket kits. Basically, they contain the following:

Minor Repair Kits
- All gaskets
- Float needle valve
- Volume control screw
- All diaphragms
- Spring for the pump diaphragm

Major Repair Kits
- All jets and gaskets
- All diaphragms
- Float needle valve
- Volume control screw
- Pump ball valve
- Main jet carrier
- Float
- Complete intermediate rod
- Intermediate pump lever
- Complete injector tube
- Some cover holddown screws and washers

Gasket Kits
- All gaskets

After cleaning and checking all components, reassemble the carburetor, using new parts and referring to the exploded view. When reassembling, make sure that all screws and jets are tight in their seats, but do not overtighten, as the tips will be distorted. Tighten all screws gradually, in rotation. Do not tighten needle valves into their seats; uneven jetting will result. Always use new gaskets. Be sure to adjust the float level when reassembling.

Carter BBD

1. Place carburetor on a repair stand to protect throttle valves from damage and to provide a stable base for working.
2. Remove retaining clip from accelerator pump arm link and remove line. Remove the stepper motor.
3. Remove step-up piston cover plate and gasket from top of air horn.
4. Remove the screws and locks from the accelerator pump arm and the vacuum piston rod lifter. Then slide the pump lever out to the air horn. The vacuum piston and step-up rods can now be lifted straight up and out of the air horn as an assembly.
5. Remove vacuum hose from between carburetor main body and choke vacuum diaphragm at main body tap.

1. Diaphragm connector link
2. Screw
3. Choke vacuum diaphragm
4. Hose
5. Valve
6. Metering rod
7. S-Link
8. Pump arm
9. Gasket
10. Rollover check valve
11. Screw
12. Lock
13. Rod lifter
14. Bracket
15. Nut
16. Solenoid
17. Screw
18. Air horn retaining screw (short)
19. Air horn retaining screw (long)
20. Pump lever
21. Venturi cluster screw
22. Idle fuel pick-up tube
23. Gasket
24. Venturi cluster
25. Gasket
26. Check ball (small)
27. Float
28. Fulcrum pin
29. Baffle
30. Clip
31. Choke link
32. Screw
33. Fast idle cam
34. Gasket
35. Thermostatic choke shaft
36. Spring
37. Screw
38. Pump link
39. Clip
40. Gasket
41. Limiter cap
42. Screw
43. Throttle body
44. Choke housing
45. Baffle
46. Gasket
47. Retainer
48. Choke coil
49. Lever
50. Choke rod
51. Clip
52. Needle and seat assembly
53. Main body
54. Main metering jet
55. Check ball (large)
56. Accelerator pump plunger
57. Fulcrum pin retainer
58. Gasket
59. Spring
60. Air horn
61. Lever

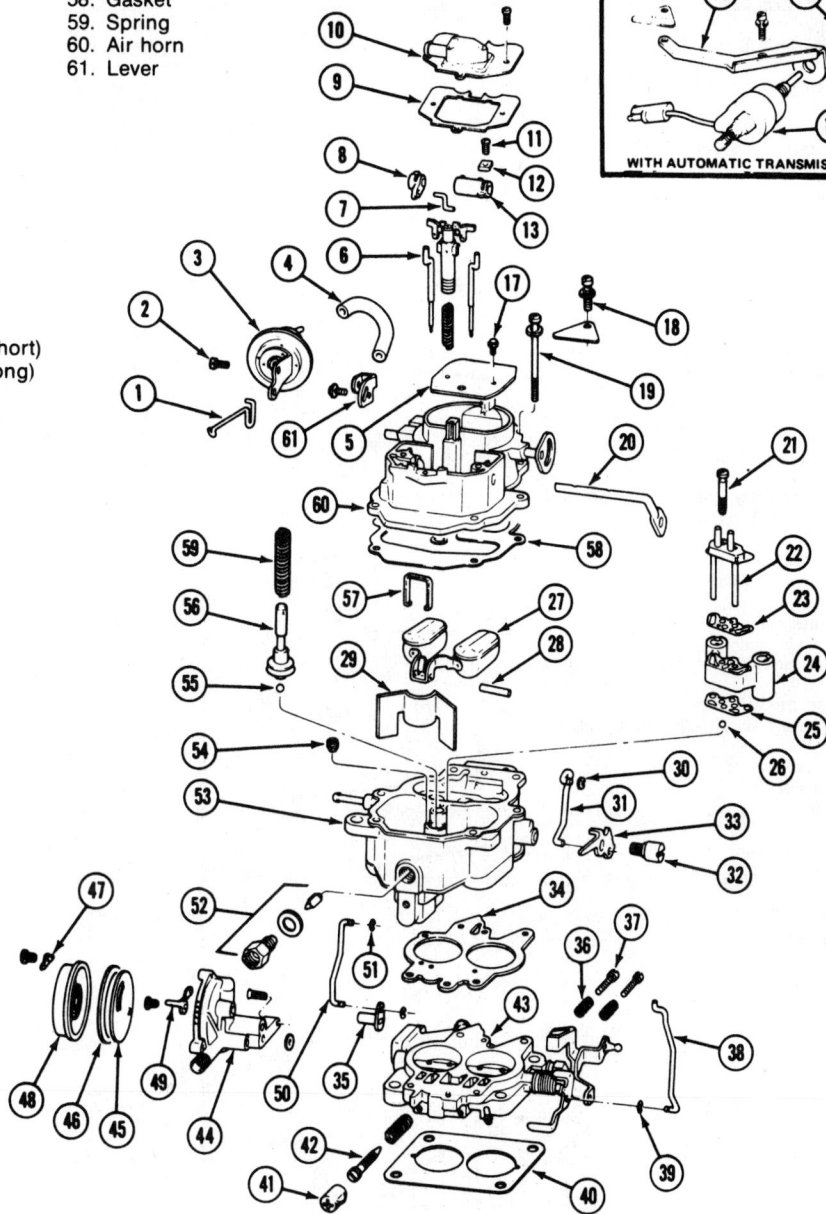

WITH AUTOMATIC TRANSMISSION

Carter BBD used on the 6–258 through 1983

6. Remove choke diaphragm, linkage and bracket assembly and place to one side to be cleaned as a separate item.

7. Remove retaining clip from fast idle cam link and remove link from choke shaft lever.

8. Remove fast idle cam retaining screw and remove fast idle cam and linkage.

9. Remove air horn retaining screws and lift air horn straight up and away from main body. Discard gasket.

10. Invert the air horn and compress accelerator pump drive spring and remove "S" link from pump shaft. Pump assembly can now be removed.

11. Remove fuel inlet needle valve, seat and gasket from main body.

12. Lift out float fulcrum pin retainer and baffle. Then lift out floats and fulcrum pin.

13. Remove main metering jets.

14. Remove venturi cluster screws, then lift venturi cluster and gaskets up and away from main body. Discard gaskets. Do not remove the idle orifice tubes or main vent tubes from the cluster. They can be cleaned in a solvent and dried with compressed air.

15. Invert carburetor and drop out accelerator pump discharge and intake check balls.

16. Remove idle mixture screws from throttle body, after lightly bottoming and noting the number of turns. Record the turns in for reference when installing the mixture screws.

1. Rollover check valve and bowl vent
2. Lock
3. Dashpot
4. Solenoid and bracket
5. Cluster screw
6. Idle fuel pickup tube
7. Gasket
8. Venturi cluster
9. Gasket
10. Check ball (small)
11. Stepper motor (actuator)
12. Clip
13. Screw
14. Fast idle cam
15. Choke link
16. Gasket
17. Screw
18. Pump link
19. Throttle body
20. Flange gasket
21. Idle mixture screw
22. Choke housing
23. Baffle
24. Gasket
25. Choke coil
26. Retainer
27. Lever
28. Wide open throttle switch and bracket
29. Needle and seat assembly
30. Main body
31. Main metering jet
32. Pin
33. Baffle
34. Fulcrum retainer
35. Float
36. Spring and accelerator pump plunger
37. Air horn
38. Accelerator pump lever
39. Choke vacuum diaphragm and housing
40. Hose
41. Metering rod
42. Vacuum piston
43. Pump arm
44. Rod lifter
45. Gasket
46. Spring
47. S-Link
48. Choke rod and shaft

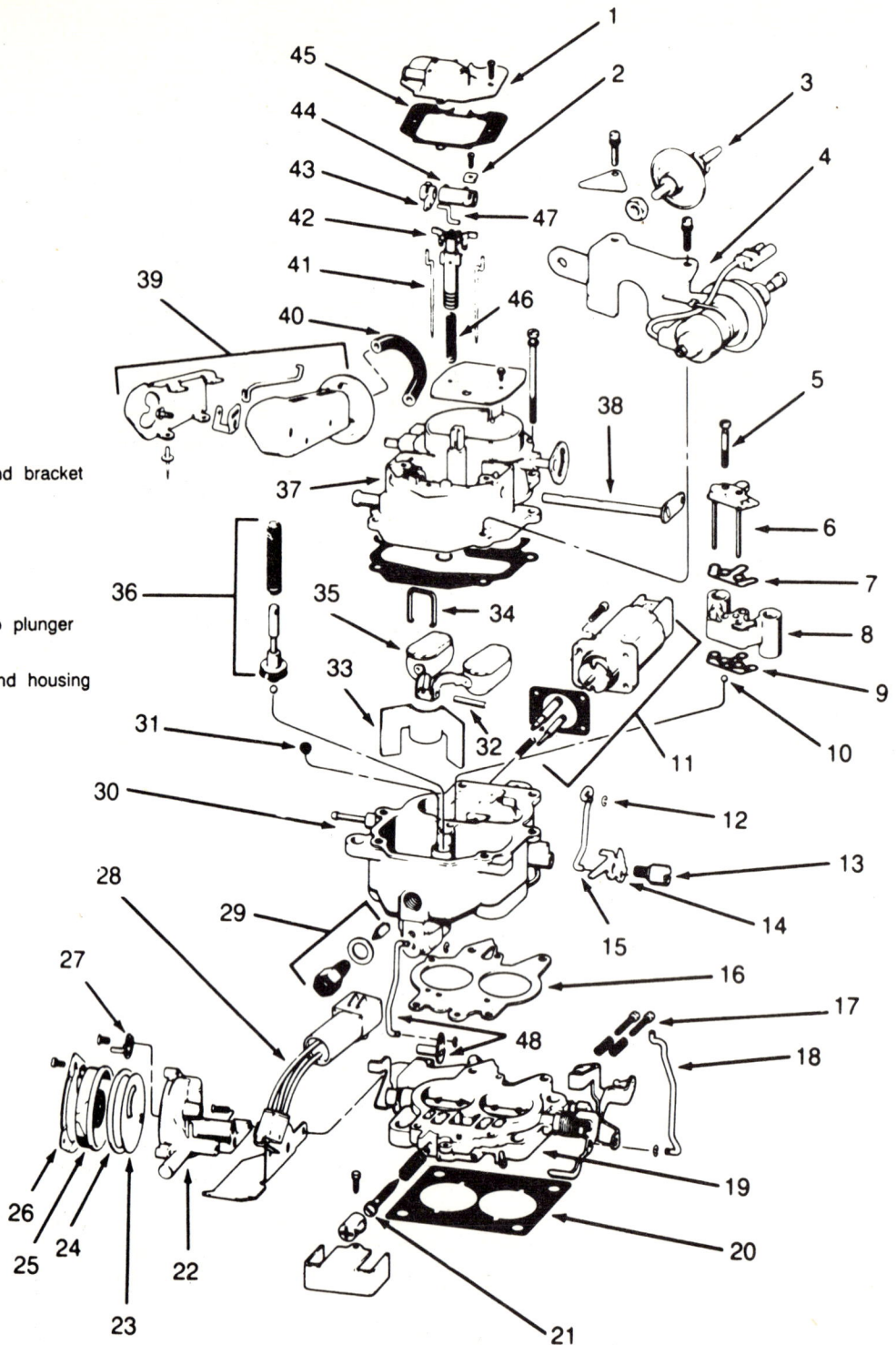

Carter BBD used on 1984–90 6–258 engines

17. Remove screws that attach throttle body to main body. Separate bodies.

18. Test freeness of choke mechanism in air horn. The choke shaft must float free to operate correctly. If choke shaft sticks, or appears to be gummed from deposits in air horn, a thorough cleaning will be required.

19. The carburetor now has been disassembled into three main units: the air horn, main body, and throttle body. Compo-

nents are disassembled as far as necessary for cleaning and inspection.

THROTTLE BODY

Check throttle shaft for excessive wear in throttle body. If wear is extreme, it is recommended that throttle body assembly be replaced rather than installing a new shaft in old body. Install idle mixture screws in body. The tapered portion must be straight and smooth. If tapered portion is grooved or ridged, a new idle mixture screw should be installed to insure having correct idle mixture control.

MAIN BODY

1. Invert main body and place insulator in position, then place throttle body on main body and align. Install screws and tighten securely.
2. Install accelerator pump discharge check ball ($^5/_{32}$ inch diameter) in discharge passage. Drop accelerator pump intake check ball ($^3/_{16}$ inch diameter) into bottom of the pump cylinder.
3. To check the accelerator pump system; fuel inlet and discharge check balls, proceed as follows:
4. Pour clean gasoline into carburetor bowl, approximately ½ inch deep. Remove pump plunger from container of mineral spirits and slide down into pump cylinder. Raise plunger and press lightly on plunger shaft to expel air from pump passage.
5. Using a small clean brass rod, hold discharge check ball down firmly on its seat. Again raise plunger and press downward. No fuel should be emitted form either intake or discharge passage. If any fuel does emit from either passage, it indicates the presence of dirt or a damaged check ball or seat. Clean passage again and repeat test. If leakage is still evident, stake check ball seats (place a piece of drill rod on top of check ball and lightly tap drill rod with a hammer to form a new seat), remove and discard old balls, and install new check balls. The fuel inlet check ball is located at bottom of the plunger well. Remove fuel from bowl.
6. Install discharge check ball. Install new gaskets on venturi cluster, then install in position in main body. Install cluster screws and tighten securely.
7. Install main metering jets.

FLOAT

The carburetors are equipped with a synthetic rubber tipped fuel inlet needle. The needle tip is viton, which is not affected by gasoline and is stable over a wide range of temperatures. The tip is flexible enough to make a good seal on the needle seat, and to give increased resistance to flooding.

NOTE: The use of the synthetic rubber tipped inlet needle requires that care be used in adjusting the float setting. Care should be taken to perform this accurately in order to secure the best performance and fuel economy.

When replacing the needle and seat assembly (inlet fitting), the float level must also be checked, as dimensional difference between the old and new assemblies may change the float level. Refer to carburetor adjustments for adjusting procedure.
1. Install floats and fulcrum pin.
2. Install fulcrum pin retainer and baffle.
3. Install fuel inlet needle valve, seat and gasket in main body.

AIR HORN

1. Place the accelerator pump drive spring on pump plunger shaft then insert shaft into air horn. Compress spring far enough to insert "S" link.
2. Install the pump lever, vacuum piston rod lifter and accelerator pump arm in the air horn.
3. Drop intake check ball into pump bore. Install baffle into main body. Place step-up piston spring in piston vacuum bore. Position a new gasket on the main body and install air horn.

4. Install air horn retaining screws and tighten alternately, a little at a time, to compress the gasket securely.
5. To qualify the step-up piston, adjust gap by turning the allen head calibration screws on top of the piston. See the adjustment section for proper measurement. Record number of turns and direction to obtain this dimension for this must be reset to its original position after the vacuum step-up piston adjustment has been made.
6. Carefully position vacuum piston metering rod assembly into bore in air horn making sure metering rods are in main metering jets. Then place the two lifting tangs of the plastic rod lifter under piston yoke. Slide shaft of the accelerator pump lever through rod lifter and pump arm. Install two locks and adjusting screws, but do not tighten until after adjustment is made.
7. Install fast idle cam and linkage. Tighten retaining screw securely.
8. Connect accelerator pump linkage to pump lever and throttle lever. Install retaining clip.

CHOKE VACUUM DIAPHRAGM

Inspect the diaphragm vacuum fitting to insure that the passage is not plugged with foreign material. Leak check the diaphragm to determine if it has internal leaks. To do this, first depress the diaphragm stem, then place a finger over the vacuum fitting to seal the opening. Release the diaphragm stem. If the stem moves more than $^1/_{16}$ inch in 10 seconds, the leakage is excessive and the assembly must be replaced.

Install the diaphragm assembly on the air horn as follows:
1. Engage choke link in slot in choke lever.
2. Install diaphragm assembly, secure with attaching screws.
3. Inspect rubber hose for cracks before placing it on correct carburetor fitting. Do not connect vacuum hose to diaphragm fitting until after vacuum kick adjustment has been made.
4. Loosen choke valve attaching screws slightly. Hold valve closed, with fingers pressing on high side of valve. Tap valve lightly with a screwdriver to seat in air horn. Tighten attaching screws securely and stake by squeezing with pliers.

CHOKE VACUUM KICK

The choke diaphragm adjustment controls the fuel delivery while the engine is running. It positions the choke valve within the air horn by action of the linkage between the choke shaft and the diaphragm. The diaphragm must be energized to measure the vacuum kick adjustment. Vacuum can be supplied by an auxiliary vacuum source.

Carter YF/YFA

1. Drill out the choke retainer rivet heads using a No. 30 (⅛ in.) drill bit. After the rivet heads are removed, drive out the remaining portion of the rivets with a ⅛ in. punch. This procedure must be followed exactly to retain the hole sizes.
2. Remove the screw holding the retainer.
3. Remove the retainer, thermostatic spring housing assembly, spring housing gasket and the locking/indexing plate.
4. Remove the vacuum break.
5. Disengage and remove the vacuum break connector link from the choke shaft lever.
6. Remove the sole-vac and the mounting bracket.
7. Remove the duty cycle solenoid from the air horn.
8. Remove the air horn attaching screws.
9. Remove the fast idle cam link.
10. Remove the air horn and gasket from the carburetor main body.
11. To remove the float from the air horn, hold the air horn bottom side up and remove the float pin and float.
12. Invert the air horn and catch the needle pin, spring and needle.
13. Remove the needle seat and gasket.
14. To remove the pump check ball and weight, turn the main body casting upside down and catch the accelerator pump check ball and weight.

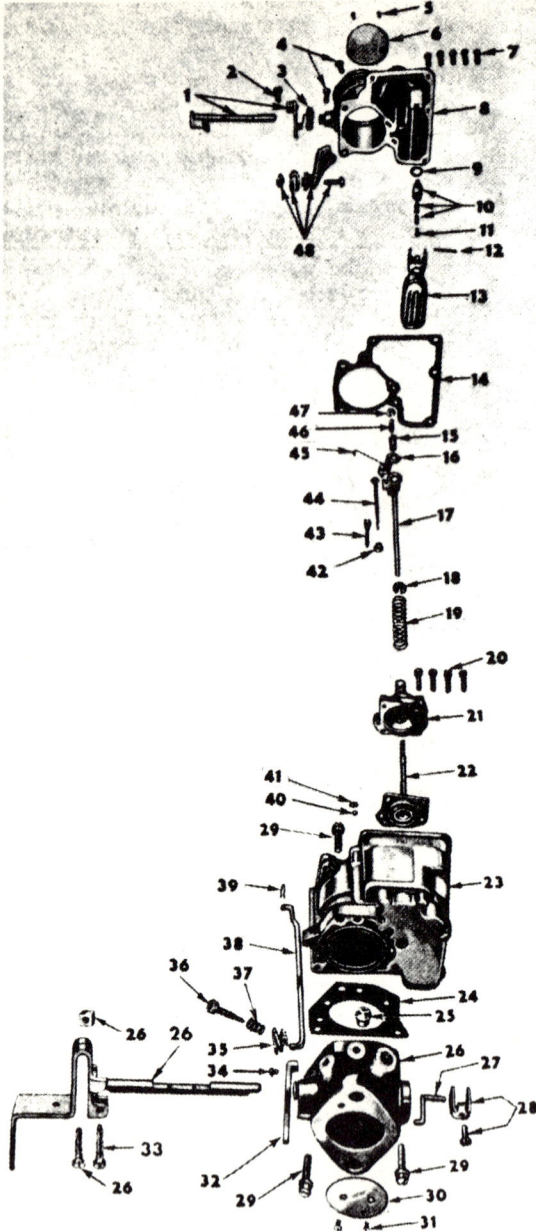

1. Choke shaft and lever
2. Screw
3. Choke lever spring
4. Screw and washer
5. Choke valve screw
6. Choke valve
7. Screw and washer
8. Air horn
9. Needle seat gasket
10. Needle spring and seat
11. Needle pin
12. Float pin
13. Float
14. Gasket
15. Pump spring
16. Metering rod arm
17. Pump link
18. Pump spring retainer
19. Vacuum diaphragm spring
20. Screw and washer
21. Diaphragm housing
22. Diaphragm
23. Body
24. Gasket
25. Idle port plug
26. Throttle body lever and shaft assembly
27. Pump link connector
28. Throttle shaft arm
29. Screw and washer
30. Throttle valve
31. Throttle valve screw
32. Fast idle arm
33. Adjusting screw
34. Body flange plug
35. Clevis clip
36. Idle adjusting screw
37. Idle screw spring
38. Fast idle connector rod
39. Pin spring
40. Ball check valve
41. Ball check valve retainer ring
42. Metering rod jet
43. Low speed jet
44. Metering rod
45. Metering rod spring
46. Inner pump spring
47. Pump spring retainer
48. Bracket and clamp assembly (choke and throttle)

Carter YF carburetor used on the 4–134 engine

15. Loosen the throttle shaft arm screw and remove the arm and pump connector link.

16. Remove the retaining screws and separate the throttle body from the main body.

17. Remove the wide open throttle switch and mounting bracket.

18. Remove the accelerator pump housing screws from the main body.

19. Lift out the pump assembly, pump lifter link and metering rod as a unit.

20. Disassemble the pump as follows:

   a. Disengage the metering rod arm spring from the metering rod.

   b. Remove the metering rod from the metering rod assembly.

   c. Compress the upper pump spring and remove the spring retainer cup.

   d. Remove the upper spring, metering rod arm assembly and pump lifter link from the pump diaphragm shaft.

   e. Compress the pump diaphragm spring and remove the pump diaphragm spring retainer, spring and pump diaphragm assembly from the pump diaphragm housing.

21. Remove the low speed jet and the main metering jet.

22. Using a sharp punch, remove the accelerator pump bleed valve plug from outside the main body casting. Loosen the bleed valve screw and remove the valve.

23. Drill out and remove the tamperproof plug. After removing the plug, count the number of turns required to seat the idle mixture screw lightly. Remove the idle mixture screw and O-ring.

24. Thoroughly clean and inspect all components and replace any that are damaged, worn or malfunctioning. Check the idle mixture screw needle for scoring or damage and replace it if any grooves are noted.

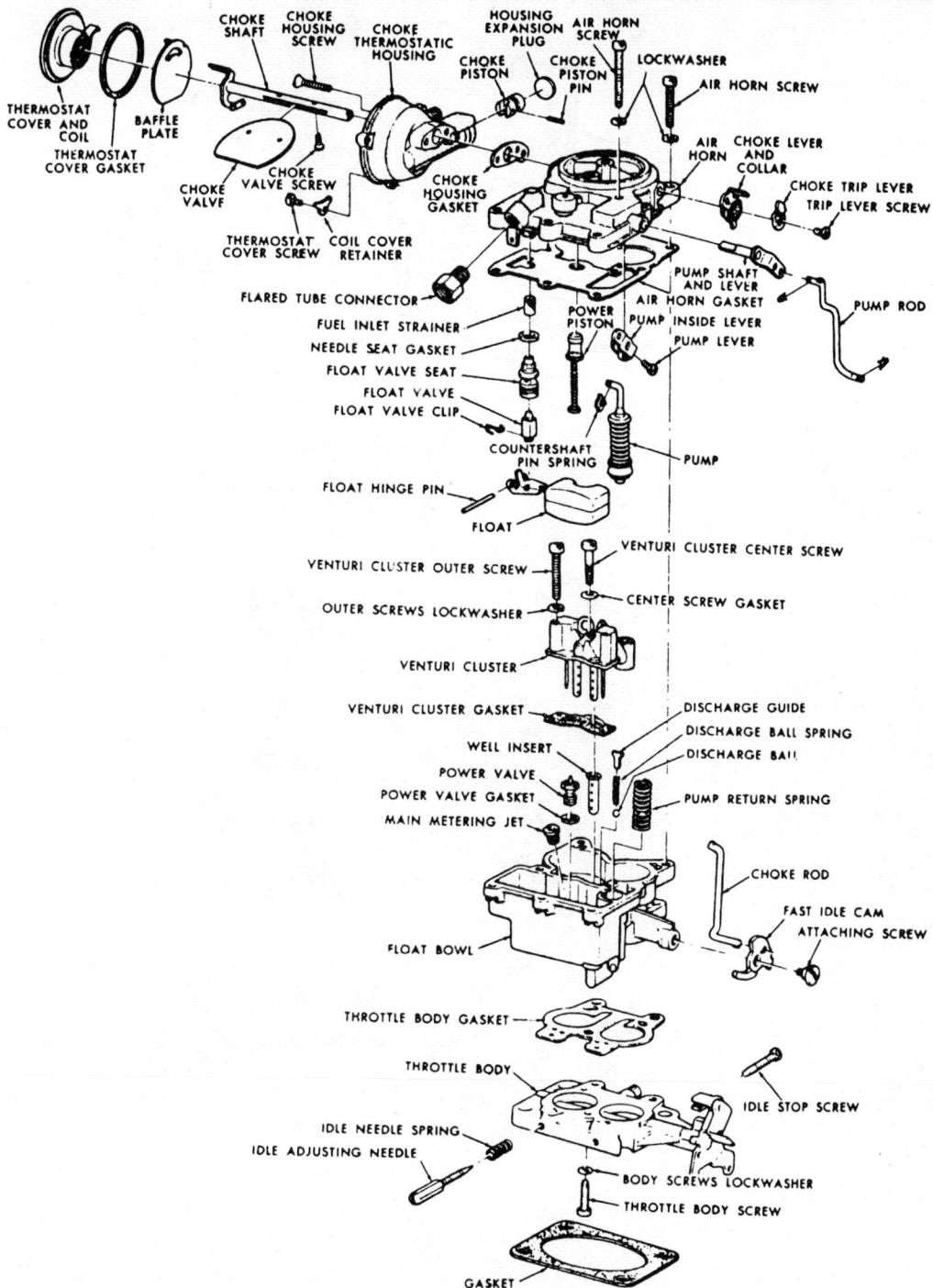

Rochester 2GC carburetor used on V6–225 engines

25. Install the throttle body to the main body with the retaining srews, then install the idle mixture screw.
26. Install the low speed jet and main metering jet.
27. Install the pump bleed valve and spring.
28. Install the accelerator pump assembly.
29. Install the pump passage tube.
30. Install the throttle shaft arm and pump connector link and retaining screw.
31. Install the wide open throttle switch actuator.
32. Install the throttle shaft retaining bolt.

33. Install the wide open throttle switch and bracket.
34. To adjust the metering rod:
   a. Make sure the idle speed adjusting screw allows the throttle plate to close tightly in the throttle bore.
   b. Press down on top of the pump diaphragm shaft until the assembly bottoms.
   c. While holding the pump diaphragm down, adjust the metering rod by turning the metering rod adjusting screw counterclockwise until the metering rod just bottoms in the main metering jet.

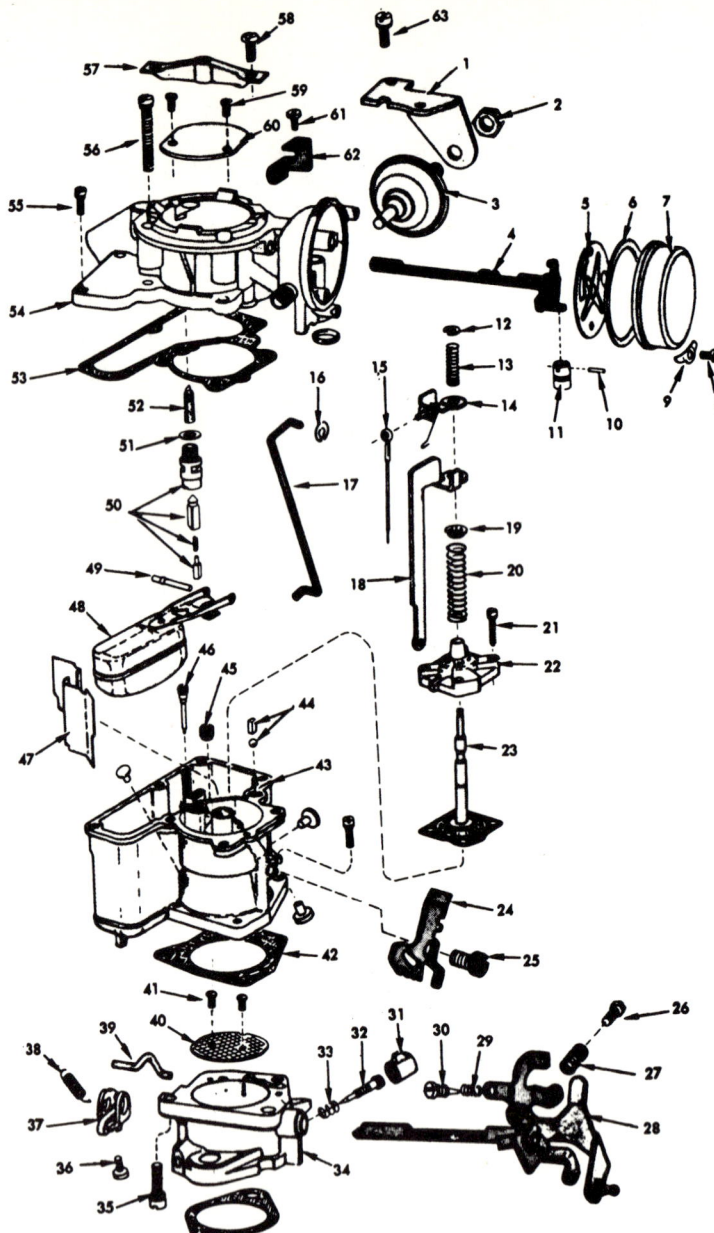

1. Dashpot bracket
2. Dashpot lock nut
3. Dashpot
4. Choke shaft and lever assembly
5. Baffle plate
6. Choke cover gasket
7. Choke cover
8. Choke cover retaining screw (3)
9. Choke cover retainer (3)
10. Choke piston pin
11. Choke piston
12. Upper pump spring retainer
13. Upper pump spring
14. Metering rod arm and spring
15. Metering rod
16. Choke rod retaining clip
17. Choke rod
18. Pump lifter link
19. Lower pump spring retainer
20. Lower pump spring
21. Pump housing retaining screw (4)
22. Pump housing
23. Pump diaphragm assembly
24. Fast idle cam
25. Fast idle cam retaining screw
26. Curb idle speed adjusting screw
27. Curb idle screw spring
28. Throttle shaft and lever assembly
29. Fast idle screw spring
30. Fast idle speed adjusting screw
31. Idle limiter cap
32. Idle mixture screw
33. Idle mixture screw spring
34. Throttle body
35. Throttle body retaining screw (3)
36. Throttle shaft arm set screw
37. Throttle shaft arm
38. Throttle shaft return spring
39. Pump connector link
40. Throttle valve
41. Throttle valve retaining screw (2)
42. Throttle body gasket
43. Main body

44. Pump discharge check ball and weight
45. Metering rod jet
46. Low speed jet
47. Fuel bowl baffle
48. Float and lever assembly
49. Float pin
50. Needle and seat assembly
51. Needle seat gasket
52. Screen
53. Air horn gasket
54. Air horn
55. Short air horn retaining screw (3)

56. Long air horn retaining screw (3)
57. Air cleaner bracket
58. Air cleaner bracket retaining screw (2)
59. Choke valve retaining screw (2)
60. Choke valve
61. Choke lever retaining screw
62. Choke lever
63. Dashpot bracket retaining screw

Carter YF carburetor used on 6–232, 6–258 engines

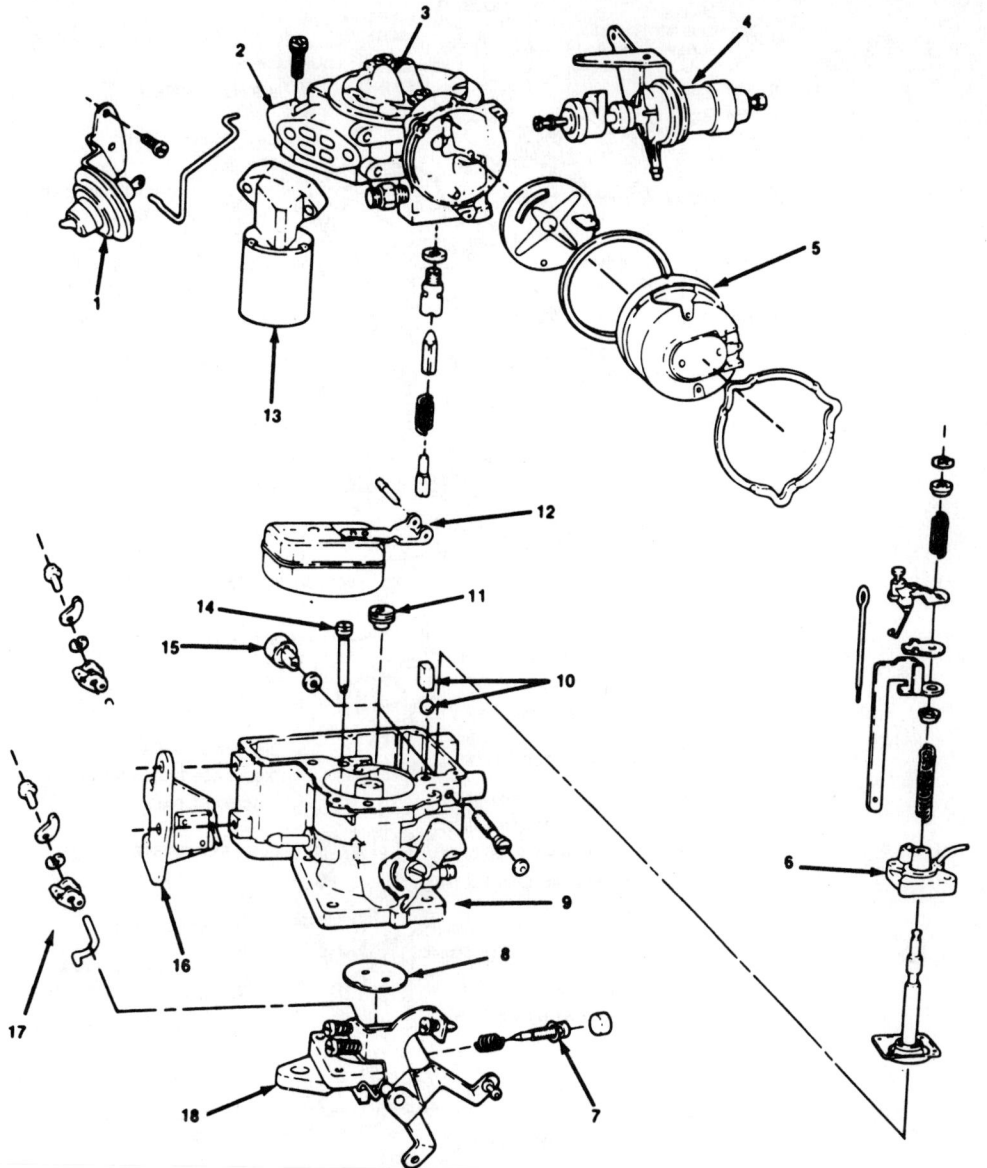

1. Vacuum break
2. Air horn
3. Choke plate
4. Sole-vac throttle positioner
5. Choke assembly
6. Accelerator pump assembly
7. Idle mixture screw with O-ring
8. Throttle plate
9. Main body
10. Accelerator pump check ball and weight
11. Main metering jet
12. Float assembly
13. Mixture control solenoid
14. Low speed jet
15. Accelerator pump vent valve
16. Wide open throttle (WOT) switch
17. Throttle shaft and lever
18. Throttle body

Carter YFA

d. Turn the metering rod adjusting screw clockwise one turn for final adjustment.

35. Install the needle pin, spring, needle, seat, gasket and strainer.

36. Install the float and pin.

37. Invert the air horn assembly and check the clearance from the top of the float to the bottom of the air horn with the gauge. The float arm should be resting on the needle pin. Bend the float arm as necessary to adjust the float level.

38. Install the accelerator pump check ball and weight, then install the air horn and gasket to the main body.

39. Install the fast idle cam link.

40. Install the sole-vac and mounting bracket, then install the duty cycle solenoid and gasket.

41. Install the locking and indexing plate, spring housing gasket, thermostatic spring housing assembly, choke cover retainer and attaching screws.

42. Position the fast idle cam screw on the second step of the fast idle cam and against the shoulder of the high step. Adjust by bending the fast idle cam link to obtain the specified clearance between the lower edge of the choke plate and the carburetor air horn.

43. Position the fast idle screw on the top step of the fast idle cam.

THERMOSTAT COVER AND COIL

THERMOSTAT COVER GASKET

BAFFLE PLATE

CHOKE SHAFT

CHOKE HOUSING SCREW

CHOKE THERMOSTATIC HOUSING

HOUSING EXPANSION PLUG

CHOKE PISTON

CHOKE PISTON PIN

AIR HORN SCREW

LOCKWASHER

AIR HORN SCREW

CHOKE VALVE SCREW

CHOKE VALVE

THERMOSTAT COVER SCREW

COIL COVER RETAINER

CHOKE HOUSING GASKET

AIR HORN

CHOKE LEVER AND COLLAR

CHOKE TRIP LEVER

TRIP LEVER SCREW

PUMP SHAFT AND LEVER

PUMP ROD

FLARED TUBE CONNECTOR

FUEL INLET STRAINER

NEEDLE SEAT GASKET

FLOAT VALVE SEAT

FLOAT VALVE

FLOAT VALVE CLIP

COUNTERSHAFT PIN SPRING

FLOAT HINGE PIN

FLOAT

POWER PISTON

AIR HORN GASKET

PUMP INSIDE LEVER

PUMP LEVER

PUMP

VENTURI CLUSTER CENTER SCREW

VENTURI CLUSTER OUTER SCREW

CENTER SCREW GASKET

OUTER SCREWS LOCKWASHER

VENTURI CLUSTER

VENTURI CLUSTER GASKET

WELL INSERT

POWER VALVE

POWER VALVE GASKET

MAIN METERING JET

DISCHARGE GUIDE

DISCHARGE BALL SPRING

DISCHARGE BALL

PUMP RETURN SPRING

CHOKE ROD

FAST IDLE CAM

ATTACHING SCREW

FLOAT BOWL

THROTTLE BODY GASKET

THROTTLE BODY

IDLE STOP SCREW

IDLE NEEDLE SPRING

IDLE ADJUSTING NEEDLE

BODY SCREWS LOCKWASHER

THROTTLE BODY SCREW

GASKET

Rochester 2GV exploded view

44. Seat the vacuum break using a hand vacuum pump.

45. Apply a light closing pressure to the choke plate to position the plate as far closed as possible without forcing it. Measure the distance between the choke plate and the air horn. To adjust, bend the vacuum break connector link.

46. Adjust the sole-vac for curb idle speed. Refer to the Sole-Vac Adjustment procedure.

47. Set the idle mixture screw to the same number of turns as noted during disassembly. The idle mixture screw must be set to the **exact** number of turns as noted during disassembly. Install a new tamperproof plug.

48. Install the carburetor. To adjust the fast idle, turn the fast idle adjusting screw to contact the fast idle cam until the desired engine rpm is achieved. See the underhood emission control sticker for rpm specifications.

NOTE: Make sure the curb idle speed and mixture are adjusted to specifications before attempting the fast idle adjustment.

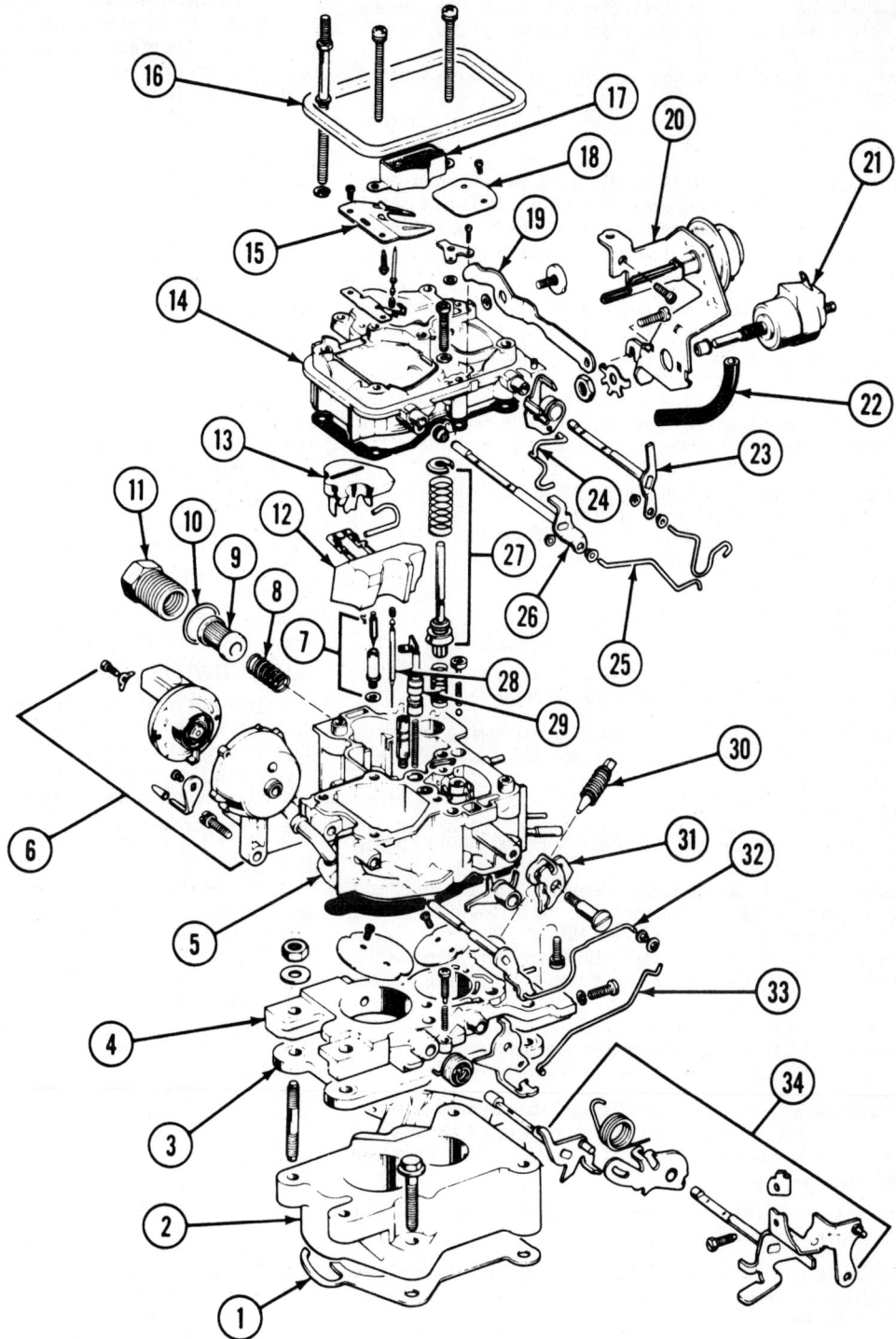

1. Gasket
2. Intake adapter
3. Insulator
4. Throttle body
5. Main body
6. Electric stat cover and coil
7. Needle seat assembly
8. Spring
9. Fuel inlet filter
10. Gasket
11. Fuel inlet fitting
12. Float assembly
13. Float baffle
14. Air horn
15. Air valve
16. Air horn gasket
17. Vent screen
18. Choke valve
19. Pump lever
20. Vacuum break and bracket
21. Idle stop solenoid
22. Vacuum hose
23. Vacuum break lever
24. Choke link
25. Air valve rod
26. Air valve lever
27. Accelerator pump
28. Metering rod
29. Power piston
30. Idle needle and spring
31. Fast idle cam
32. Intermediate choke rod
33. Pump rod
34. Throttle lever assembly

Rochester 2SE carburetor used on 49 states 4–151 engines; the E2SE used on California engines is similar

Rochester 2SE/E2SE

AIR HORN

1. Invert the carburetor, then remove the plug covering the idle mixture needle as previously described.

2. Install the carburetor in a suitable holding stand.

3. Remove the primary and secondary vacuum break assemblies. Be sure to take note of the linkage positions for installation.

4. Remove the three screws from the mixture control solenoid. Remove the mixture control solenoid with gasket and discard the gasket.

5. Remove the two screws from the vent stack and remove the vent stack. Remove the intermidiate choke shaft link retainer at the choke lever and discard it.

6. Remove the choke link and bushing from choke lever and save the bushing.

7. Remove the retainer and bushing from the fast idle cam link and discard the retainer.

NOTE: Do not remove fast idle cam screw and cam from the float bowl. If removed, the cam might not operate properly when reassembled. If needed, a replacement float bowl will include a secondary locknut lever, fast idle cam, and cam screw.

8. Remove the retainer from the pump link. Do not remove the screw attaching the pump lever to the air horn assembly. When reassembled, the screw might not hold properly.

9. Remove the seven screw assemblies of various length that retain the air horn to the carburetor and remove the air horn assembly. Tilt the air horn to disconnect fast idle cam link from the slot in fast idle cam and the pump link from the hole in the pump lever.

10. Remove the cam link from the choke lever. Be sure to line up the "squirt" on link with slot in lever.

11. Invert the air horn and remove the TPS actuator plunger. The TPS adjusting screw and plug should not be removed.

12. Remove the stakings that holds the TPS plunger seal retainer and pump stem seal retainer.

13. Remove the retainers and seals and discard them.

14. Further disassembly of the air horn is not required for cleaning purposes. The choke valve and choke valve screws, the air valve and air valve shaft should not be removed.

NOTE: Do not turn the secondary metering rod adjusting screw. The rod could come out of jet and possibly cause damage.

FLOAT BOWL

1. Remove the accelerator pump, air horn gasket and pump return spring.

2. Remove the Throttle Position Sensor (TPS) assembly and spring. Inspect the TPS connector wires for broken insulation, which could cause grounding of the TPS.

3. Remove the upper insert and the hinge pin. Remove the float and lever assembly with the float stabilizing spring if used. Remove the float needle and pull clip.

4. Remove the lower insert, if used.

5. Remove the float needle seat and seat gasket.

6. Remove the jet and lean mixture needle assembly.

NOTE: Do not remove or change the preset adjustment of calibration needle in the metering jet unless the Computer Command Control system performance check requires it.

7. Remove the pump discharge spring guide, using a suitable slide hammer puller only.

NOTE: Do not pry the guide. Damage could occur to the sealing surfaces, and could require replacement of the float bowl.

8. Remove the spring and check ball, by inverting the bowl and catching them as they fall out.

9. Remove the fuel inlet nut and the fuel filter spring.

10. Remove the fuel filter assembly and discard it. Remove the filter gasket and discard it.

CHOKE ASSEMBLY AND THROTTLE BODY

1. Remove the choke cover as follows:
   a. Use a $\frac{5}{32}$ in. (4mm) drill bit to remove the heads (only) from the rivets.
   b. Remove the choke cover retainers. Remove the remaining pieces of rivets, using drift and small hammer.

c. Remove the electric choke cover and stat assembly.

2. Remove the stat lever screw, Stat lever, intermediate choke shaft, lever and link assembly.

3. Remove the two screws and the choke housing.

4. Remove the four screws, and the throttle body assembly from the inverted float bowl.

5. Remove the gasket, pump link and line up the "squirt" on link with the slot in the lever.

6. Count and make a record of the number of turns needed to lightly bottom the idle mixture needle (69), then back out and remove needle and spring assembly using Idle mixture socket tool J-29030-B or BT-7610-B or equivalent.

7. Do not disassemble throttle body further.

INSPECTION AND CLEANING

1. Place the metal parts in immersion carburetor cleaner.

NOTE: Do not immerse idle stop solenoid, mixture control solenoid, throttle lever actuator, TPS, electric choke, rubber and plastic parts, diaphragms, and pump in the cleaner, as they may be damaged. Plastic bushing in throttle lever will withstand normal cleaning.

2. Blow dry the parts with shop air. Be sure all fuel and air passages are free of burrs and dirt. Do not pass drill bits or wires through jets and passages.

3. Be sure to check the mating surfaces of casting for damage. Replace if necessary. Check for holes in levers for wear or out-of-round conditions. Check the bushings for damage and excessive wear. Replace if necessary.

CARBURETOR REASSEMBLY

1. Install the mixture needle and spring assembly using Idle mixture socket tool J-29030-B or BT-7610-B or equivalent. Lightly bottom the needle and back it out the number of turns recorded during removal, as a preliminary adjustment. Refer to idle mixture adjustment procedure in this section for the final idle mixture adjustment.

2. Install the pump link and a new gasket on the inverted float bowl.

3. Install the throttle body to the float bowl assembly and finger tighten the four retaining screws. If the secondary actuating lever engages the lockout lever, and linkage moves without binding, tighten retaining screws.

4. If the float bowl assembly was replaced, stamp or engrave the model number on the new float bowl in same location as on old bowl.

5. Place the throttle body and float bowl together on a suitable carburetor holding stand.

6. Install the choke housing on the throttle body, with the retaining screws.

7. Install the intermediate choke shaft, lever and link assembly.

8. Install the choke stat lever on the intermediate choke shaft. The intermediate choke lever must be upright.

9. Install the choke lever attaching screw in the shaft.

10. Install the gasket on the fuel inlet nut and install the new filter assembly in the nut.

11. Install the filter spring and then install the fuel inlet nut. Tighten the fuel inlet nut to 18 ft. lbs. (24 Nm).

CAUTION

Tightening beyond this limit may damage gasket and could cause a fuel leak, which might result in personal injury.

12. Install the pump discharge ball and spring.

13. Install a new spring guide and tap it until the top is flush with the bowl casting.

14. Install the needle seat with gasket. If used, lower the insert.

15. Install the jet and lean mixture needle assembly. Using lean mixture adjusting tool J-28696-10 or BT-7928 or equiva-

lent, lightly bottom the lean mixture needle. Back it out 2½ turns, as a preliminary adjustment.

NOTE: Only adjust the lean mixture needle screw if it was removed or touched during disassembly or if the Computer Command Control system performance check, made before disassembly, indicated an incorrect lean mixture needle setting:

16. Bend the float lever upward slightly at the notch.

17. If used, install the float stabilizing spring on float. Install the hinge pin in float lever, with ends toward the pump well.

18. Install the needle with pull clip assembly on the edge of the float lever. Install the float and lever assembly in float bowl.

NOTE: Adjust the float level using Float Level T-Scale J-9789-90 or BT-8037 or equivalent.

19. Install the upper insert over the hinge pin, with the top flush with the bowl. Install the TPS spring and the TPS assembly. The parts must be below the surface of the bowl.

20. Install the gasket over the dowels. Install the spring and pump assembly.

21. Install a new pump stem seal with the lip facing outside of carburetor and install the retainer. Be sure to stake it at new locations.

22. Install a new TPS actuator plunger seal with the lip facing outside of carburetor and install the retainer. Be sure to stake it at new locations.

23. Install the TPS plunger through the seal in the air horn. Use lithium base grease, liberally to pin, if used, where contacted by spring.

24. Install the fast idle cam link in the choke lever. Be sure to line-up the "squirt" on link with slot in lever.

25. Rotate the cam to the highest position. The lower end of the fast idle cam link goes in cam slot, and the pump link end goes into the hole in the lever.

26. Hold the pump down, and than lower the air horn assembly onto the float bowl. Be sure to guide the pump stem through the seal.

27. Install the one of the air horn retaining screws, finger tight to hold the air horn in place.

28. Install the cam link in the slot of the cam. Install a new bushing and retainer to the link, with the large end of bushing facing the retainer. Check for freedom of movement.

29. Install the rest of the air horn retaining screws.

30. Install the spacer and a new seal, lightly coat the seal with automatic transmission fluid. Assemble the seal on the solenoid stem, touching the spacer.

31. Install a new retainer and a new gasket on the air horn. Install the mixture control solenoid lining up the stem with the recess in the bowl.

32. Install the solenoid retaining screws. Install the vent stack, with two retaining screws, (unless lean mixture needle requires on-vehicle adjustment).

33. Install a new retainer on the pump link. Adjust the air valve spring, if adjustable.

34. Install the bushing on the choke link. With the intermediate choke lever upright, install the link in the choke lever hole. Install the new link retainer.

35. The following procedures is for reassembly of any small components that have been removed from the carburetor, if part replacement is necessary or for any other reason.

    a. Install the idle stop solenoid, retainer and nut to the secondary side vacuum break bracket. Bend the retainer tab to secure nut.

    b. Install the bushing to the link and the link to the vacuum break plunger. Install the retainer to the link.

    c. Rotate the assembly, insert the end of the link in the upper slot of the choke lever. Install the bracket screws.

    d. Install the idle speed device, retainer and nut to the primary side vacuum break bracket. Bend the retainer tab to secure the nut.

    e. Install the bushing to the vacuum break link. Install the link to vacuum break plunger. Install the retainer to the link and the bushing to the air valve link. Install the link to the plunger and the retainer to the link.

36. Rotate the vacuum break assembly (primary side) and insert the end of the air valve link into the air valve lever and the vacuum break link into the lower slot of the choke lever.

37. Install the bracket screws, vacuum hose between the throttle body tube and the vacuum break assembly.

38. Install the choke thermostat lever. Install the choke cover and thermostat assembly in the choke housing.

39. If the thermostat has a "trap" (box-shaped pick-up tang), the trap surrounds the lever.

40. Line up the notch in the cover with projection on the housing flange. Install the retainers and rivets with rivet tool. If necessary, use an adapter. Adjust the choke as previously described.

Autolite/Motorcraft 2100/2150

DISASSEMBLY

1. Remove the air cleaner anchor screw.

2. Remove the automatic choke rod retainer from the thermostatic choke shaft lever.

3. Remove the air horn attaching screws, lockwashers and carburetor identification tag. Remove the air horn and gasket.

4. Remove the choke rod by loosening the screw that secures the choke shaft lever to the choke shaft. Remove the rod and plastic dust seal from the air horn.

5. Remove the choke modulator assembly.

6. Remove the fast idle cam retainer.

7. Remove the choke shield.

8. Remove the thermostatic choke spring housing retaining screws and clamp, and the housing and gasket.

9. Remove the cam rod from the fast idle cam lever.

10. Remove the retaining screws and remove the choke housing and gasket.

11. Remove the fast idle cam.

12. Remove the thermostat lever retaining screw and washer. Remove the thermostatic choke shaft and fast idle cam lever. from the choke housing.

13. Pry the float shaft retainer from the fuel inlet seat. Remove the float, float shaft retainer and fuel inlet needle. Remove the retainer and float shaft from the float lever.

14. Remove the fuel inlet needle seat and gasket. Remove the main jets.

15. Remove the accelerator pump discharge screw, air distribution plate, booster venturi and gasket.

NOTE: Do not attempt to remove the tubes from the venturi assembly.

16. Invert the main body and catch the accelerator pump discharge weight and ball.

17. Disconnect the accelerator pump operating rod from the overtravel lever. Remove the rod and retainer.

18. Remove the accelerator pump cover attaching screws. Remove the bowl vent bellcrank and bracket assembly, accelerator pump cover, diaphragm assembly and spring.

19. Remove the elastomer valve by grasping it firmly and pulling it out.

NOTE: If the valve tip breaks off during removal, be sure to remove the tip from the fuel bowl.
Whenever the elastomer valve is removed, it must be replaced. It cannot be reused!

20. Invert the main body and remove the power valve cover, gasket and screws. Remove the power valve. Remove and discard the gasket.

21. Remove the limiter caps from the idle mixture adjusting

1. Pivot pin
2. Modulator arm
3. Choke valve retaining screw (2)
4. Choke valve
5. Choke shaft
6. Air horn
7. Air horn retaining screw (4)
8. Air horn gasket
9. Float shaft retainer
10. Float and lever assembly
11. Needle retaining clip
12. Deflector
13. Needle and seat assembly
14. Needle seat gasket
15. Fuel bowl baffle
16. Float shaft
17. Curb idle adjusting screw
18. Curb idle adjusting screw spring
19. Throttle shaft and lever assembly
20. Dashpot
21. Dashpot locknut
22. Dashpot bracket
23. Dashpot bracket retaining screw
24. Throttle valve retaining screw (4)
25. Throttle valve (2)
26. Main jet (2)
27. Main body
28. Pump rod retainer
29. Pump rod
30. Elastomer valve
31. Pump return spring
32. Pump diaphragm
33. Pump lever pin
34. Pump cover
35. Pump lever
36. Pump cover retaining screw (4)
37. Fuel inlet fitting
38. Power valve gasket
39. Power valve
40. Power valve cover gasket
41. Power valve cover

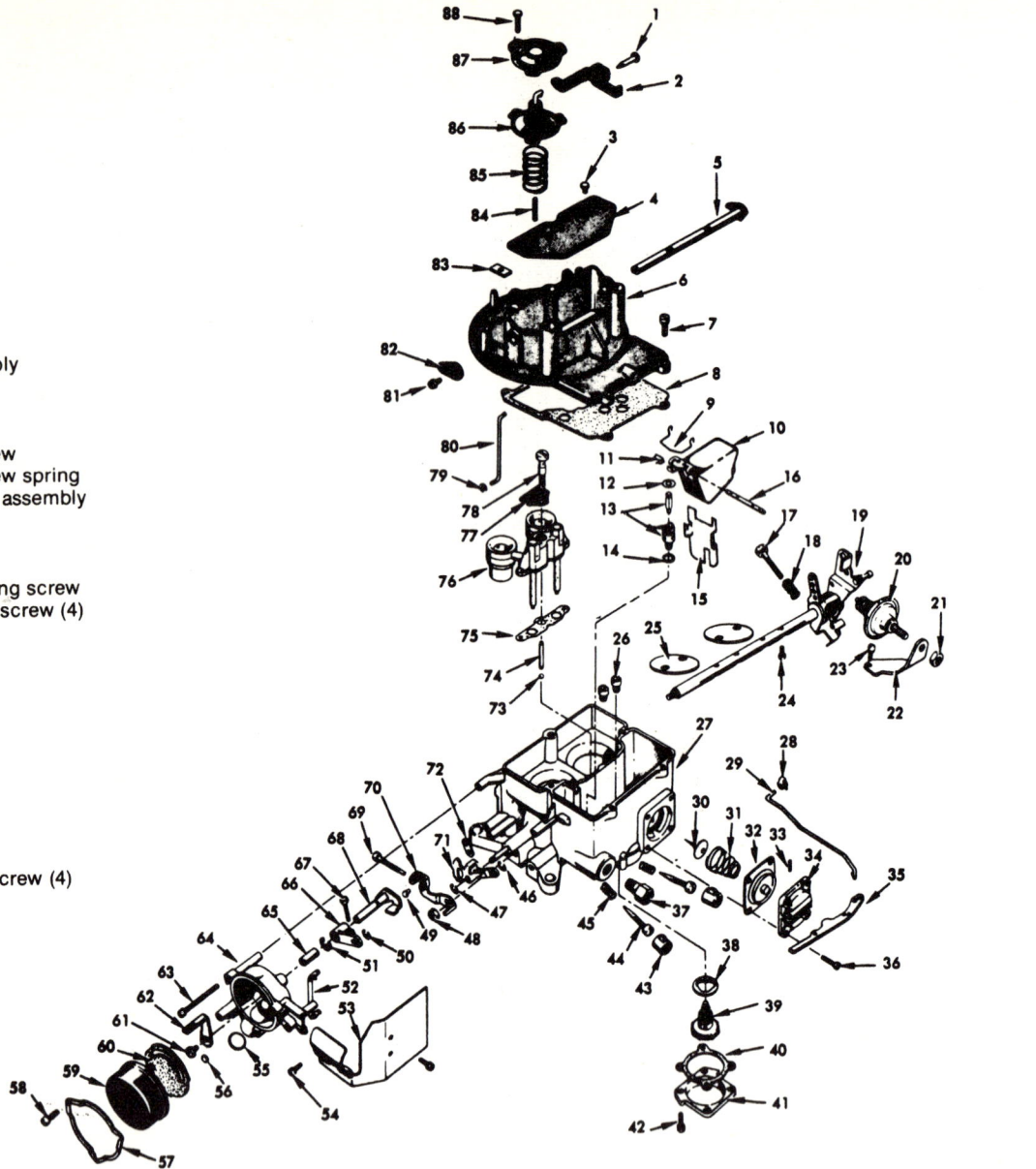

42. Power valve cover retaining screw (4)
43. Idle limiter cap (2)
44. Idle mixture screw (2)
45. Idle mixture screw spring (2)
46. Retainer
47. Retainer
48. Fast idle lever retaining nut
49. Fast idle lever pin
50. Retainer
51. Retainer
52. Fast idle cam rod
53. Choke shield
54. Choke shield retaining screw (2)

55. Piston passage plug
56. Heat passage plug
57. Choke cover retaining clamp
58. Choke cover retaining screw (3)
59. Choke cover
60. Choke cover gasket
61. Thermostat lever retaining screw
62. Thermostat lever
63. Choke housing retaining screw (3)
64. Choke housing
65. Choke shaft bushing
66. Fast idle cam lever
67. Fast idle cam lever adjusting screw
68. Thermostatic choke shaft
69. Fast idle speed adjusting screw
70. Fast idle lever
71. Fast idle cam

72. Choke housing gasket
73. Pump discharge check ball
74. Pump discharge weight
75. Booster venturi gasket
76. Booster venturi assembly
77. Air distribution plate
78. Pump discharge screw
79. Retainer
80. Choke rod
81. Choke lever retaining screw
82. Choke lever
83. Choke rod seal
84. Stop screw
85. Modulator return spring
86. Modulator diaphragm assembly
87. Modulator cover
88. Modulator retaining screw (3)

Autolite 2100 carburetor used on 8–304 engines

1. Modulator cover (if equipped)
2. Modulator retaining screw (3) (if equipped)
3. Pivot pin
4. Modulator arm
5. Choke valve retaining screw (2)
6. Choke valve
7. Choke shaft
8. Air horn
9. Air horn retaining screw (4)
10. Air horn gasket
11. Float and lever assembly
12. Float shaft retainer
13. Float shaft
14. Needle retaining clip
15. Curb idle adjusting screw
16. Curb idle adjusting screw spring
17. Throttle shaft and lever assembly
18. Dashpot
19. Dashpot locknut
20. Dashpot bracket
21. Dashpot bracket retaining screw
22. Adjusting screw
23. Carriage
24. Electric solenoid
25. Mounting bracket
26. Throttle valve retaining screw (4)
27. Throttle valve (2)
28. Needle and seat assembly
29. Needle seat gasket
30. Main jet (2)
31. Main body
32. Elastomer valve
33. Pump return spring
34. Pump diaphragm
35. Pump lever pin
36. Pump cover
37. Pump rod
38. Pump rod retainer
39. Pump lever
40. Bowl vent bellcrank
41. Fuel inlet fitting
42. Power valve gasket
43. Power valve
44. Power valve cover gasket
45. Power valve cover

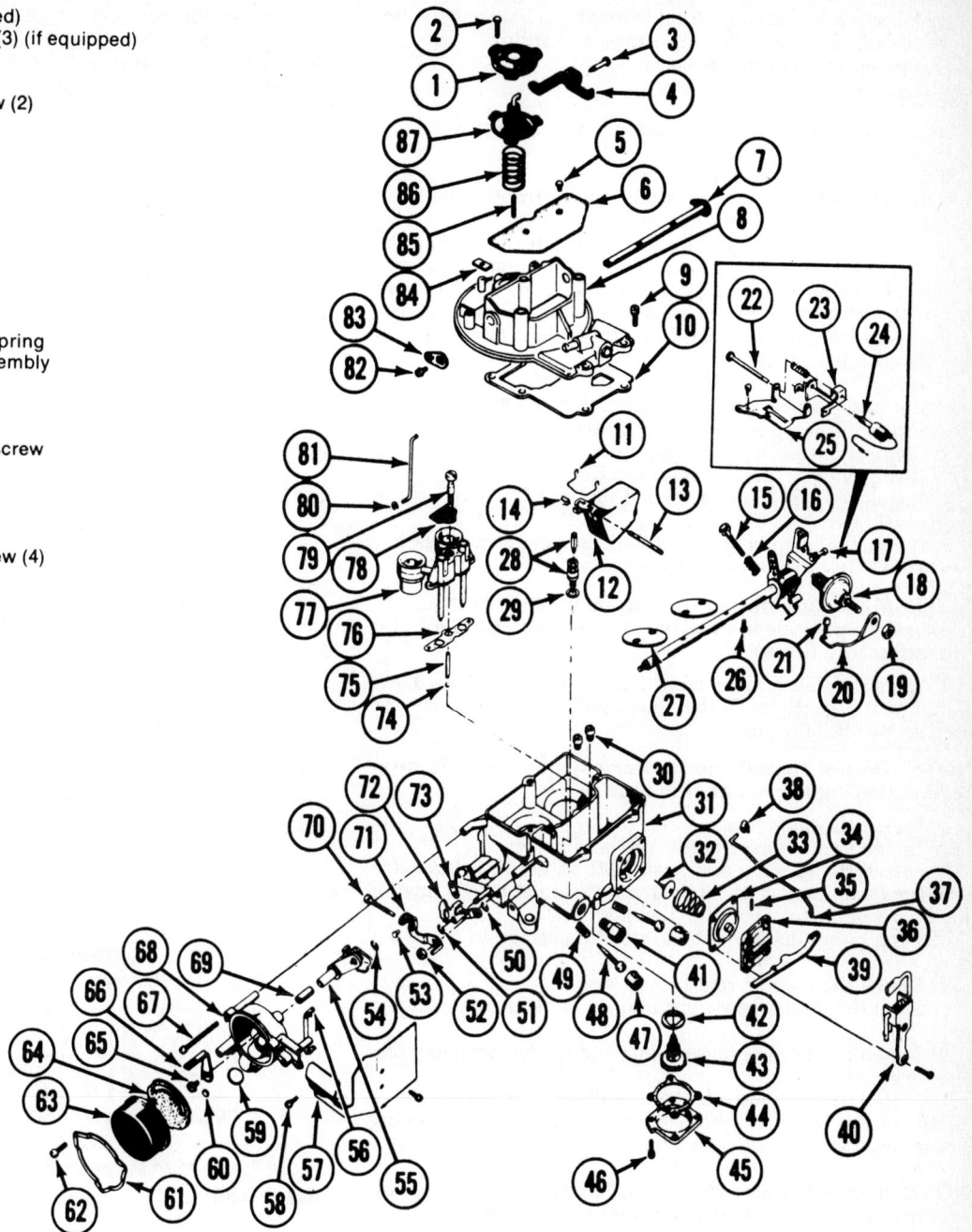

46. Power valve cover retaining screw (4)
47. Idle limiter cap (2)
48. Idle mixture screw (2)
49. Idle mixture screw spring (2)
50. Retainer
52. Fast idle lever retaining nut
53. Fast idle lever pin
54. Retainer
55. Lever and shaft
56. Fast idle cam rod
57. Choke shield
58. Choke shield retaining screw (2)
59. *Piston* passage plug

60. Heat passage plug
61. Choke cover retaining clamp
62. Choke cover retaining screw (3)
63. Choke cover and coil
64. Choke cover gasket
65. Coil lever retaining screw
66. Coil lever
67. Choke housing retaining screw (3)
68. Choke housing
69. Choke shaft bushing
70. Fast idle speed adjusting screw
71. Fast idle lever
72. Fast idle cam
73. Choke housing gasket

74. Pump discharge check ball
75. Pump discharge weight
76. Booster venturi gasket
77. Booster venturi assembly
78. Air distribution plate
79. Pump discharge screw
80. Retainer
81. Choke rod
82. Choke lever retaining screw
83. Choke plate lever
84. Choke rod seal
85. Stop screw
86. Modulator return spring (if equipped)
87. Modulator diaphragm assembly (if equipped)

Motorcraft 2150 carburetor used on 8–304 engines

screws. Use a soldering gun to cut through the limiter caps. Remove the idle mixture adjusting screws and springs.

22. If equipped, remove the solenoid.

ASSEMBLY

1. Install the fast idle adjusting screw and spring on the fast idle lever.

2. If equipped, install the solenoid.

3. Install the fast idle lever assembly on the throttle shaft and install the retaining washer and nut.

4. Lubricate the tip of the new elastomer valve and install the tip of the elastomer valve into the accelerator pump cavity center hole. Using needle-nosed pliers, reach into the fuel bowl and grasp the valve tip. Pull the valve tip in until it seats in the pump cavity wall. Cut off the tip forward of the retaining shoulder. Remove the cut-off tip.

5. Install the bowl vent bellcrank and bracket assembly, accelerator pump cover, diaphragm assembly and spring. Install the accelerator pump cover attaching screws.

6. Connect the accelerator pump operating rod to the inboard hole of the pump actuating lever. Install the retainer.

7. Install the other end of the rod into the No.3 hole of the overtravel lever.

8. Invert the main body and install the power valve, gasket and screws. Install the power valve cover.

NOTE: Make sure the limiter stops in the cover are in position to provide a positive stop for the tabs on the idle adjusting limiters.

9. Install the idle mixture adjusting screws and springs. Turn the screws in gently until they just touch the seat, then back each screw off 2 full turns.

NOTE: Do not install the limiter caps on the idle mixture adjusting screws at this time.

10. Install the main jets.

11. Install the fuel inlet needle seat and gasket.

12. Install the retainer and float shaft on the float lever. Install the float, float shaft retainer and fuel inlet needle. Install the float shaft retainer on the fuel inlet seat.

13. Drop pump discharge ball into the passage in the main body.

14. Install the booster venturi and gasket.

15. Drop the discharge weight into the booster on top of the ball.

16. Install the air distribution plate and accelerator pump discharge screw.

17. Install the thermostatic choke shaft and fast idle cam lever on the choke housing. Install the thermostat lever retaining screw and washer.

NOTE: The bottom of the fast idle cam lever adjusting screw must rest against the tang of the choke shaft.

18. Insert the choke shaft into the rear of the choke housing. Position the choke shaft so that the choke hole in the shaft is to the left side of the choke housing.

19. Install the cam rod on the fast idle cam lever.

20. Install the fast idle cam.

21. Install the fast idle cam retainer.

22. Install the choke housing and gasket. Install the retaining screws.

23. Install the thermostatic choke spring housing retaining screws and clamp, and the housing and gasket. Turn the cover ¼ turn rich and tighten one screw.

24. Install the choke shield.

25. Install the choke rod. The lower end of the rod must protrude through the air horn. Install the rod and plastic dust seal from the air horn.

26. Install the air horn and new gasket. Install the air horn attaching screws, lockwashers and carburetor identification tag.

27. Install the automatic choke rod retainer on the thermostatic choke shaft lever.

28. Install the choke modulator assembly.

29. Make all necessary adjustments.

30. Install the air cleaner anchor screw.

Adjustments

FLOAT AND FUEL LEVEL ADJUSTMENT

4–134 Carter YF

1. Remove and invert the bowl cover.

2. Remove the bowl cover gasket.

3. Allow the weight of the float to rest on the needle and spring. Be sure that there is no compression of the spring other than by the weight of the float.

4. Adjust the level by bending the float arm lip that contacts the needle (not the arm) to provide $^{17}/_{64}$ in. (6.75mm) between the gasket surface and the tip of the float.

4–150 Carter YFA

1. Remove the top of the carburetor and the gasket.

2. Invert the carburetor top and check the clearance from the top of the float to the bottom edge of the air horn with a float level gauge. Hold the carburetor top at eye level when making the check. The float arm should be resting on the inlet needle pin. To adjust, bend the float arm. DO NOT BEND THE TAB AT THE END OF THE ARM! See the Carburetor Specifications chart for the correct clearance.

4–151 2SE, E2SE

NOTE: Special tools are needed for this job.

1. Start the engine and run it to normal operating temperature.

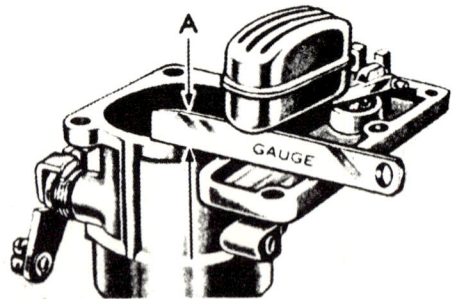

Float level measurement for the Carter YF used on the 4–134

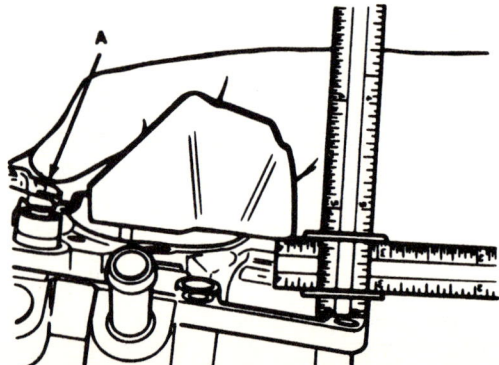

Measuring float clearance on the 4-150

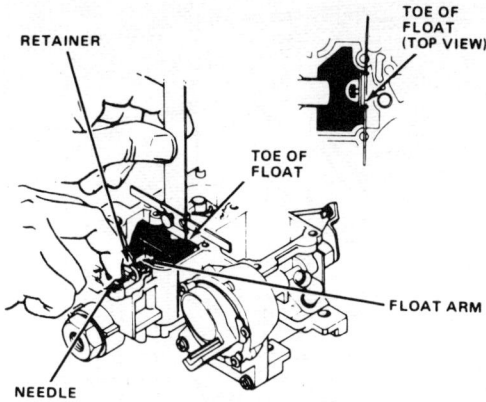

2SE and E2SE float adjustment

2. Remove the vent stack screws and the vent stack.

3. Remove the air horn screw adjacent to the vent stack.

4. With the engine idling and the choke fully opened, carefully insert float gauge J-9789-136 for E2SE carburetors and tool J-9789-138 for 2SE carburetors, into the air horn screw hole and vent hole. Allow the gauge to rest freely on the float. DO NOT PRESS DOWN ON THE FLOAT!

5. With the gauge at eye level, observe the mark that aligns with the top of the casting at the vent hole. The float level should be within $\frac{1}{16}$ in. (1.6mm) of the figure in the Carburetor Specifications Charts. If not, remove the air horn and adjust the float as follows:

    a. Hold the retainer pin firmly in place and push the float down, lightly, against the inlet needle.

    b. Using an adjustable T-scale, at a point $\frac{3}{16}$ in. (5mm) from the end of the float, at the toe, measure the distance from the float bowl top surface (gasket removed) to the top of the float at the toe. The distance should be that shown in the Carburetor Specifications Charts. If not, remove the float and bend the arm.

6–225 Rochester 2G

The procedure for adjusting the float level of the two barrel carburetor installed on the V6 is the same as the procedure for the 4–134 up to step 4.

The actual measurement is taken from the air horn gasket to the lip at the toe of the float. This distance should be $\frac{5}{32}$ in. (4mm). To adjust the float level, bend the float arm as required.

The float drop adjustment is accomplished in the following manner: With the bowl cover turned in the upright position, measure the distance from the gasket to the notch at the toe of

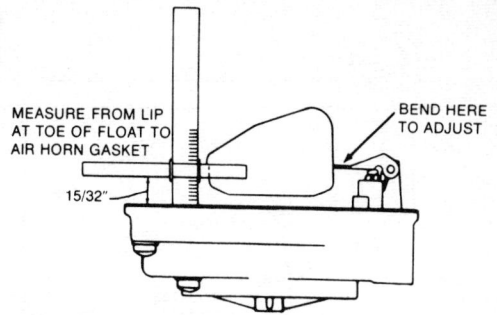

V6 float level adjustment

the float. Bend the tang as required to obtain a measurement of $1\frac{7}{32}$ in. (31mm).

6–232, 6–258 Carter YF

Remove and invert the air horn assembly and remove the gasket. Measure the distance between the top of the float at the free end, and the air horn casting. The measurement should be $\frac{29}{64}$–$\frac{31}{64}$ in. (11.5–12.3mm). Adjust by bending the float lever.

NOTE: The fuel inlet needle must be held off its seat while bending the float lever in order to prevent damage to the needle and seat.

To adjust the float drop, hold the air horn in the upright position and measure the distance between the top of the float, at the extreme outer end, and the air horn casting. The measure-

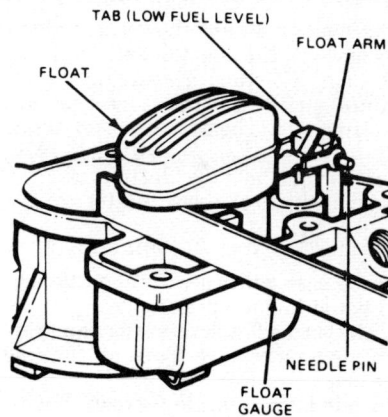

6–232, 6–258 Carter YF float adjustment

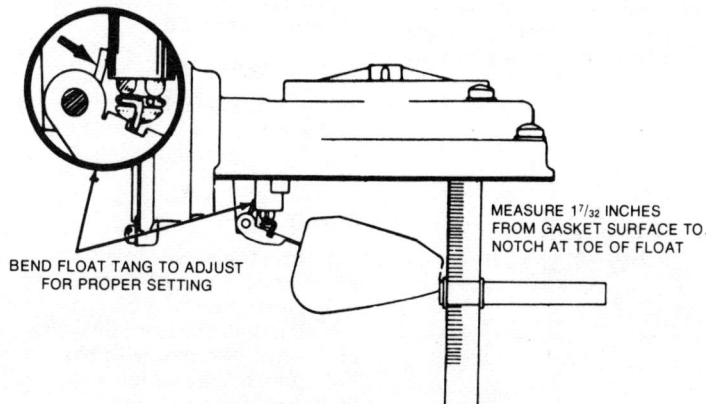

V6 float drop adjustment

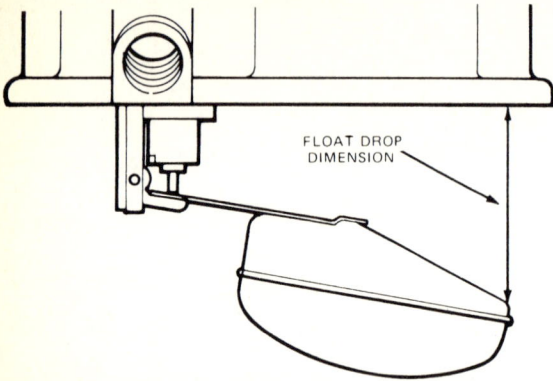

6–232, 6–258 Carter YF float drop adjustment

ment should be 1¼ in. (31.75mm) to 1973, 1⅜ in. (35mm) 1974–78. Adjust by bending the tab at the rear of the float lever.

6–258 Carter BBD 2–bbl

1. Remove the air horn.
2. Apply light finger pressure to the vertical float tab to exert GENTLE pressure against the inlet needle.
3. Lay a straight edge across the float bowl and measure the gap between the straight edge and the top of the float at its highest point. The gap should be ¼ in. (6mm).
4. To adjust, remove the float and bend the lower tab. Replace the float and check the gap.

8–304 Autolite 2100, Motorcraft 2150 Dry Adjustment

With the air horn assembly and the gasket removed raise the float by pressing down on the float tab until the fuel inlet needle is lightly seated. Using a T-scale, measure the distance from the fuel bowl machined surface to either corner of the float ⅛ in. (3mm) from the free end. The measurement should be ¾ in. (19mm) through 1975, ³⁄₁₆–¹⁵⁄₃₂ in. (5-12mm) 1976–81. To adjust bend the float tab and hold the fuel inlet needle off its eat in order to prevent damage to the seat and the tip of the needle.

8–304 Motorcraft 2150 Wet Adjustment

Exercise extreme care when performing this adjustment as fuel vapors and liquid fuel are present!
1. Place the vehicle on a flat, level surface and run the engine to normal operating temperature. Turn off the engine and remove the air cleaner.
2. Remove the air horn attaching screws, but leave the air horn in place.
3. Start the engine and let it idle for one minute. Shut off the engine and remove the air horn and gasket.
4. Use a T-scale to measure the vertical distance between the machined surface of the carburetor body and the fuel level in the bowl. Make this measurement as near the center of the bowl as

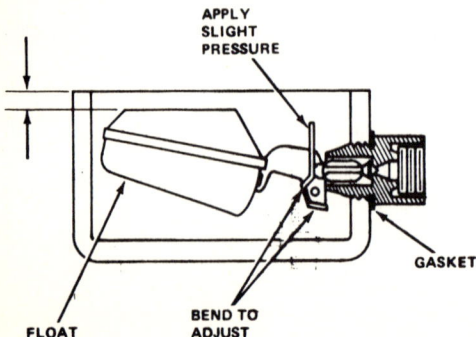

BBD float adjustment

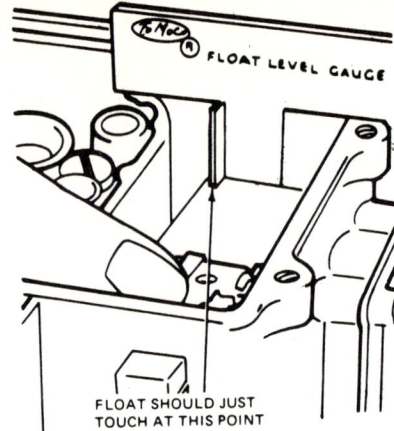

Autolite 2100 and Motorcraft 2150 dry float adjustment

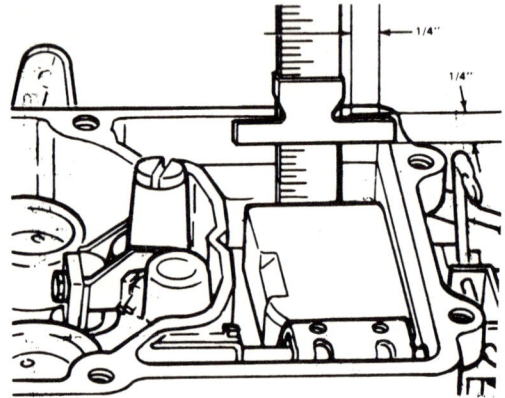

Motorcraft 2150 wet float adjustment

possible. The proper distance is ⁵⁹⁄₆₄ in. (23.4mm). To adjust, bend the float tab.

NOTE: Every time an adjustment is made, the air horn must be replaced, and the engine started and idled for one minute to stabilize the fuel level.

5. Install the air horn and gasket when adjustment is completed.

FAST IDLE LINKAGE ADJUSTMENT

NOTE: This adjustment is performed with the air cleaner removed.

4–134 Carter YF

With the choke held in the wide open position, the lip on the fast idle rod should contact the boss on the body casting. Adjust it by bending the fast idle link at the offset in the link.

4–150

1. Run the engine to normal operating temperature. Connect a tachometer according to the maker's instructions.
2. Disconnect and plug the EGR valve vacuum hose.
3. Position the fast idle adjustment screw on the second stop of the fast idle cam with the transmission in neutral.
4. Adjust the fast idle speed to 2300 rpm for auto. trans. and 2000 rpm for man. trans.
5. Idle the engine and reconnect the EGR hose.

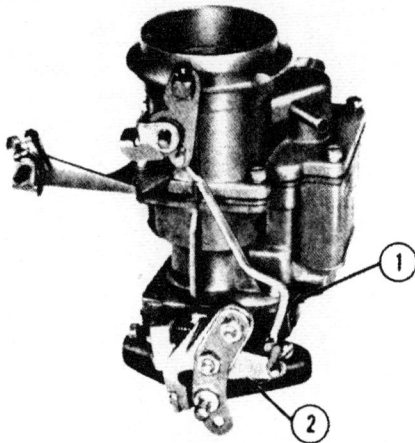

1. Fast idle connector rod
2. Fast idle link

4–134 YF fast idle adjustment

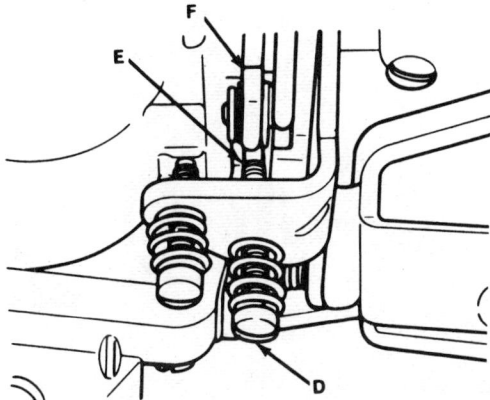

Carter YF fast idle adjustment on the 6–232, 6–258

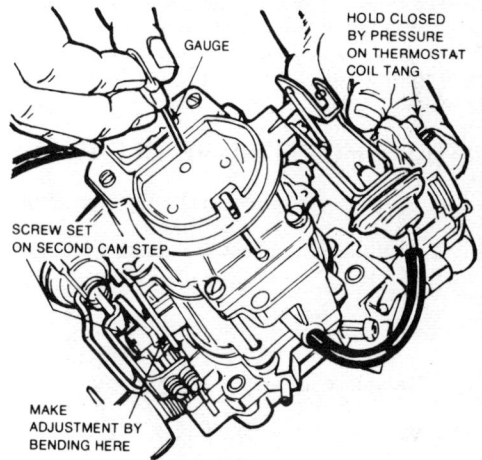

Carter YFA fast idle adjustment. (D) is the fast idle cam screw; (E) is the second step; (F) is the cam

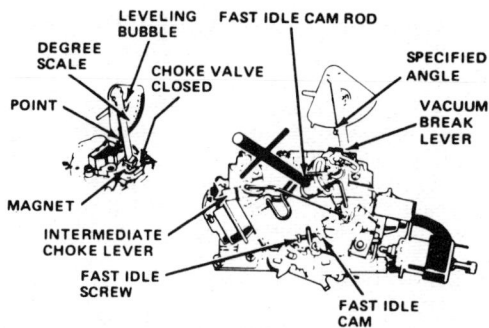

BBD fast idle cam adjustment

LEVELING
BUBBLE
FAST IDLE CAM ROD
DEGREE
SCALE
CHOKE VALVE
CLOSED
SPECIFIED
ANGLE
POINT
VACUUM
BREAK
LEVER
MAGNET
INTERMEDIATE
CHOKE LEVER
FAST IDLE
SCREW
FAST IDLE
CAM

2SE, E2SE fast idle cam position adjustment

4. Rotate the degree scale until the 25 degree mark is opposite the pointer. On carburetors with choke cover sticker number 70172, the angle is 18 degrees.

5. Place the fast idle screw on the second step of the cam.

6. Close the choke plate by pushing on the intermediate choke lever.

7. Push the vacuum brake lever toward the open choke position until the lever is against the rear tank on the choke lever.

8. Adjust by bending the fast idle cam rod until the bubble is centered.

6–225 Rochester 2G

No fast idle speed adjustment is required. Fast idle is controlled by the curb adjustment screw. If the curb idle speed is set correctly and the choke rod is properly adjusted, fast idle speed will be correct.

6–232, 6–258 through 1973 Carter YF

Partially open the throttle and close the choke valve to rotate the fast idle cam into the cold start position. While holding the choke valve closed, release the throttle. With the fast idle cam in this position, the fast idle adjusting screw must be aligned with the index mark at the back side of the cam. Adjust by bending the choke rod at its upper angle.

4–151 Rochester 2SE, E2SE

1. Make sure the choke coil adjustment is correct and that the fast idle speed is correct.

2. Obtain a Choke Angle Gauge, tool #J-26701-A. Rotate the degree scale to the zero degree mark opposite the pointer.

3. With the choke valve completely closed, place the magnet on the tool squarely on the choke plate. Rotate the bubble unit until it is centered.

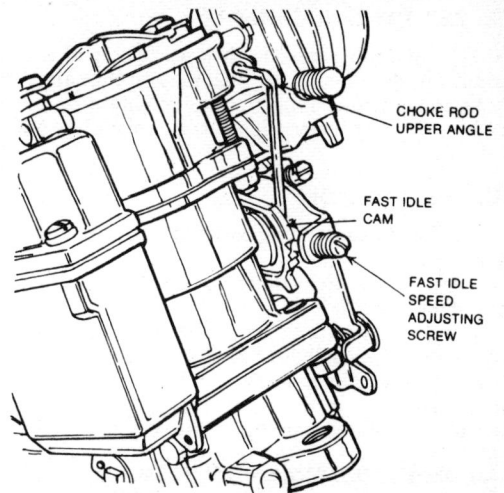

6–232, 6–258 1974–78 Carter YF

Position the fast idle screw on the second step of the fast idle cam, against the shoulder of the high step on the cam. Adjust by bending the choke plate connecting rod to obtain $\frac{13}{64}$ in. (5mm) clearance between the lower edge of the choke plate and the air horn wall.

6–258 Carter BBD 2-bbl

1. Loosen the choke housing cover and turn it ¼ turn right. Tighten one screw.
2. Slightly open the throttle and place the fast idle screw on the second cam step.
3. Measure the distance between the choke plate and the air horn wall. The distance should be $\frac{7}{64}$ in. (3mm).
4. If adjustment is necessary, bend the fast idle cam link down to increase and up to decrease the gap.
5. Return the choke cover cap to the original setting.

8–304 Autolite 2100, Motorcraft 2150

Push down on the fast idle cam lever until the fast idle speed adjusting screw is contacting the second step (index), and against the shoulder of the high step. Measure the clearance between the lower edge of the choke valve and air horn wall. Adjust by turning the fast idle cam lever screw to obtain $\frac{19}{64}$ in. (7.5mm) through 1975 and ⅛ in. (3mm) 1976–79. Adjust the automatic choke.

INITIAL CHOKE VALVE CLEARANCE

1. Position the fast idle screw on the top step of the fast idle cam.
2. Using a vacuum pump, seat the choke vacuum break.

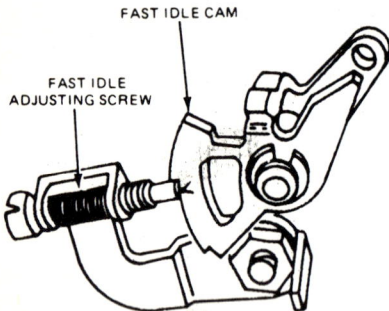

CONVENTIONAL ONE-PIECE FAST IDLE LEVER

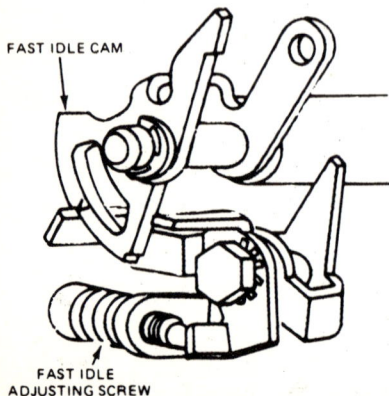

Fast idle cam index setting for the Autolite (Motorcraft) 2100

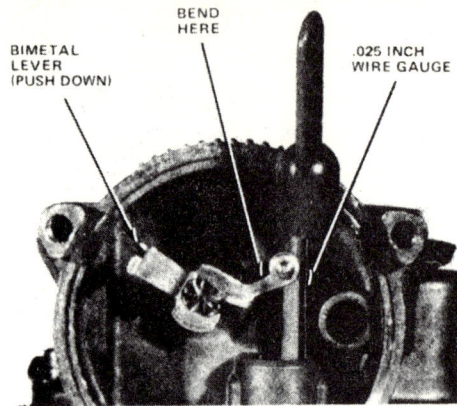

YF Initial choke valve clearance adjustment

BBD initial choke valve clearance adjustment

3. Apply light closing pressure in the choke plate to position the plate as far closed as possible without forcing it.
4. Measure the distance between the air horn wall and the choke plate. If it is not that specified in the Carburetor Specifications Chart, bend the choke vacuum break link until it is.

CHOKE SETTING ADJUSTMENT

4–134 and V6–225

The choke is manually operated by a cable that runs from the dash mounted control pull knob to the set screw on the choke actuating arm. To adjust the choke, loosen the set screw at the choke actuating lever and push in the dash knob as far as it will go. Open the choke plate as far as it will go and hold it with your finger while the set screw is tightened.

6–232, 258 and 8–304

The automatic choke setting is made by loosening the choke cover in the desired direction as indicated by an arrow on the face of the cover. the original setting will be satisfactory for most driving conditions. However, if the engine stumbles or stalls on acceleration during warmup, the choke may be set richer or leaner no more than two graduations from the original setting.

4–150, 4–151

NOTE: Once the rivets and choke cover are removed, a choke cover retainer kit is necessary for assembly.

1. Remove the rivets, retainers, choke cover and coil following the instructions found in the cover retainer kit.

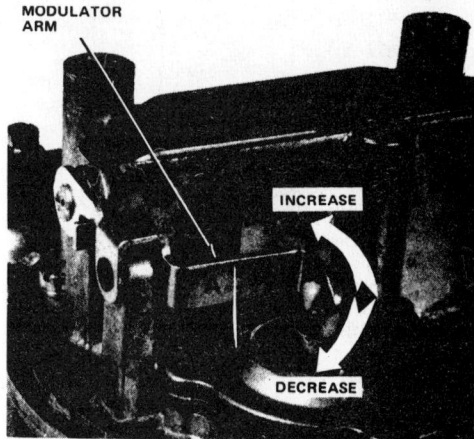

2100 Initial choke valve clearance adjustment

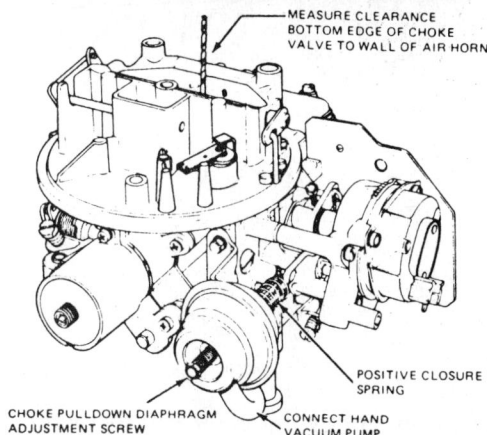

2150 Initial choke valve clearance adjustment

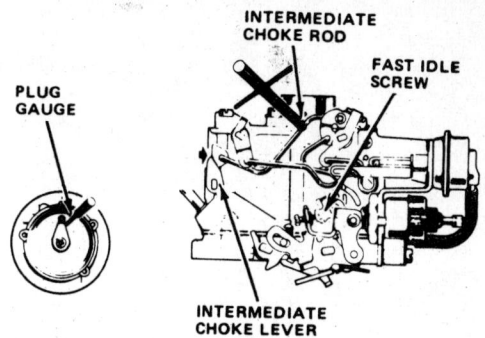

2SE, E2SE choke coil lever adjustment

2. Position the fast idle adjustment screw on the highest stop of the fast idle cam.

3. Push on the intermediate choke lever and close the choke plate.

4. Insert the proper plug gauge, 0.050–0.080 in. (1.27–2.032mm) for manual trans. and 0.85 in. (21.6mm) for automatic trans., in the hole adjacent to the coil lever. The edge of the lever should barely contact the plug gauge.

5. Bend the intermediate choke rod to adjust.

UNLOADER ADJUSTMENT

6–232, 258

With the throttle held fully open, apply pressure on the choke valve toward the closed position and measure the clearance between the lower edge of the choke valve and the air horn wall. The measurement should be ¼ in. (6mm) 1972–73, $\frac{9}{32}$ in. (7mm) 1974–87. Adjust by bending the tang on the throttle lever which contacts the fast idle cam. Bend toward the cam to increase the clearance.

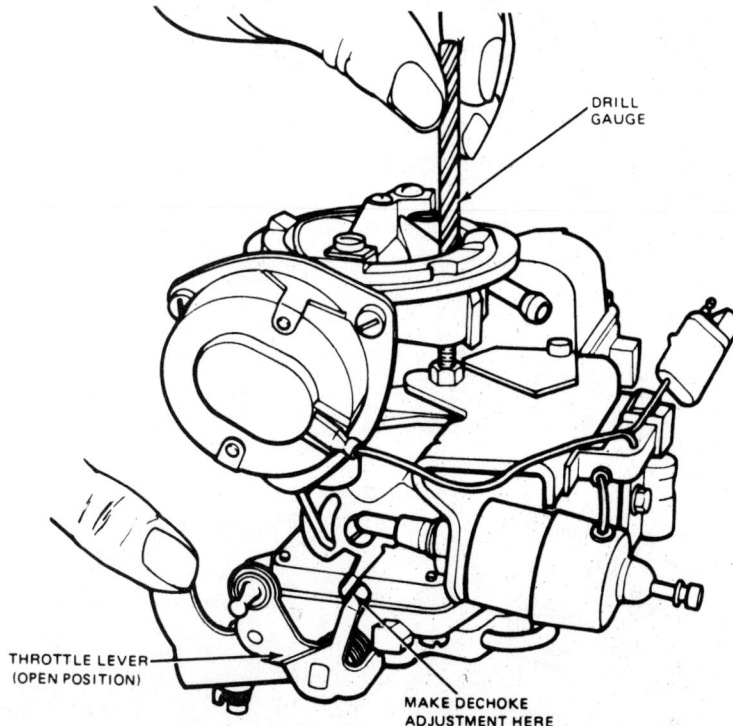

Carter YF choke unloader adjustment

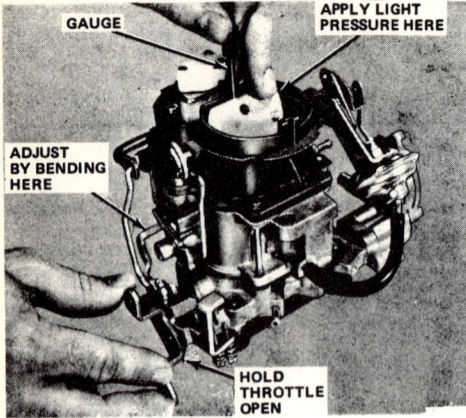

BBD choke unloader adjustment

NOTE: Do not bend the unloader down from a horizontal plane. After making the adjustment, make sure the unloader tang does not contact the main body flange when the throttle is fully open. A clearance of 0.070 in. (1.8mm) must be present. Final unloader adjustment must always be done on the vehicle. The throttle should be fully opened by depressing the accelerator pedal to the floor. This is to assure that full throttle is obtained.

4-150, 4-151

1. Obtain a Carburetor Choke Angle Gauge, tool #J-26701-A. Rotate the scale on the gauge until the 0 mark is opposite the pointer.
2. Close the choke plate completely and set the magnet squarely on top of it.
3. Rotate the bubble until it is centered.
4. Rotate the degree scale until the 32° mark is opposite the pointer. On carburetors with choke cover sticker number 70172 the setting is 19°.
5. Hold the primary throttle valve wide open.
6. Bend the throttle lever tang until the bubble is centered.

6-225, 8-304

With the throttle held fully open, apply pressure on the choke valve toward the closed position and measure the clearance between the lower edge of the choke valve and the air horn wall. The setting should be 1/4 in. (6mm); 5/16 in. (8mm) for 1979. Adjust by bending the tang on the fast idle lever, which is located on the throttle linkage. Refer to the "Note" under the procedure for adjusting the unloader on the Sixes.

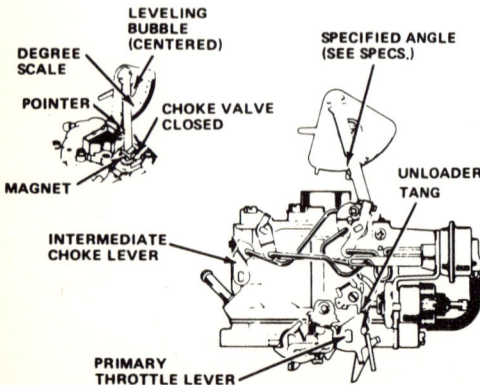

Motorcraft 2150 choke unloader adjustment

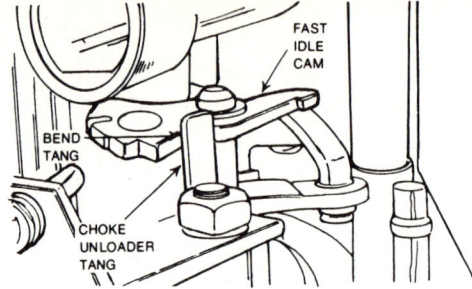

Motorcraft 2150 choke unloader/fast idle cam clearance

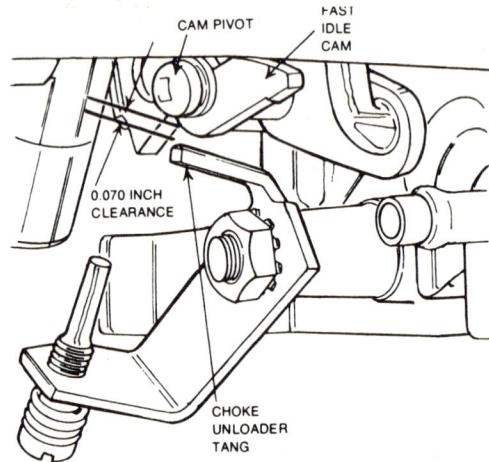

2SE, E2SE choke unloader adjustment

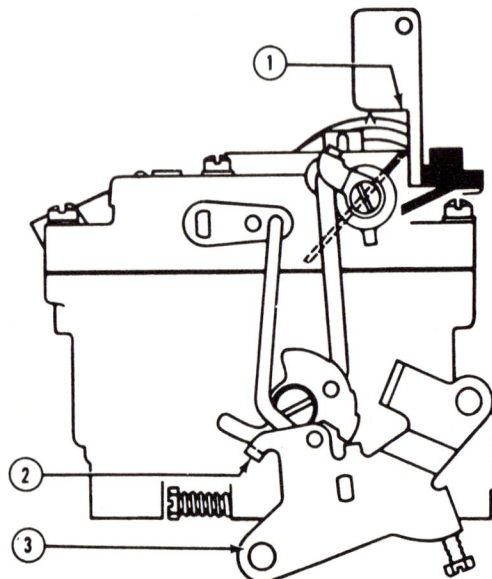

1. Gauge
2. Unloader tang
3. Throttle lever

2GV choke unloader adjustment

DASHPOT ADJUSTMENT

6-232
6-258
8-304

With the throttle set at curb idle position fully depress the dashpot stem and measure the clearance between the stem and the throttle lever. Adjust by loosening the lock nut and turning the dashpot.

4–134
6-225

The adjustment is made with the engine idling. Loosen the dashpot locknut and turn the assembly until the plunger contacts the throttle lever without being depressed. Then, turn the assembly 2½ turns against the lever, depressing the plunger. Tighten the locknut.

VACUUM (STEP UP) PISTON GAP

Carter BBD

1. Turn the adjusting screw, mounted on top of the unit, so that the gap between the metering rod lifter lower edge, and the

1. Throttle lever   3. Dashpot
2. Plunger   4. Locknut

Dashpot adjustment for the Carter YF, 4–134

1. Throttle lever   3. Dashpot
2. Plunger   4. Locknut

Dashpot adjustment for the Rochester 2GV

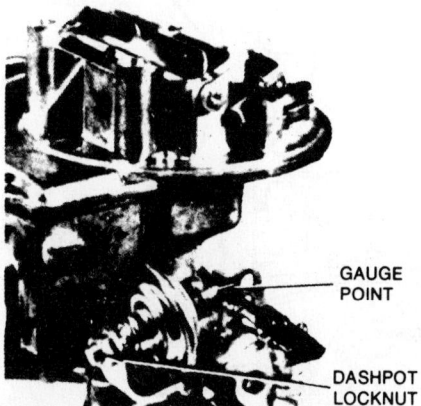

GAUGE POINT

DASHPOT LOCKNUT

Dashpot adjustment for the Carter YF

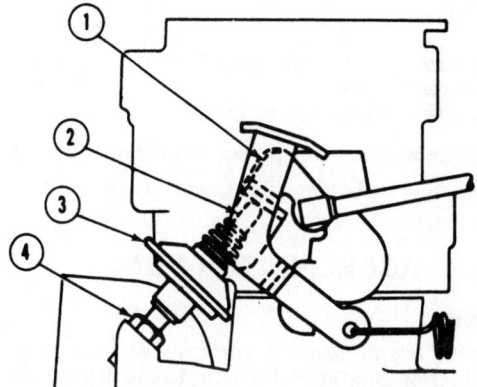

GAUGE POINT

DASHPOT LOCKNUT

Dashpot adjustment for the Autolite 2100

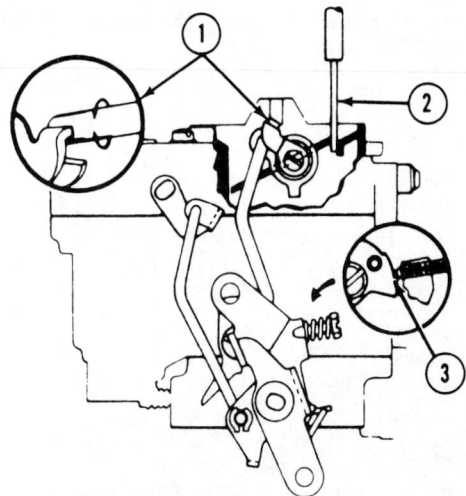

1. Choke lever tang
2. Wire gauge
3. Idle speed adjustment screw

2GV choke rod adjustment

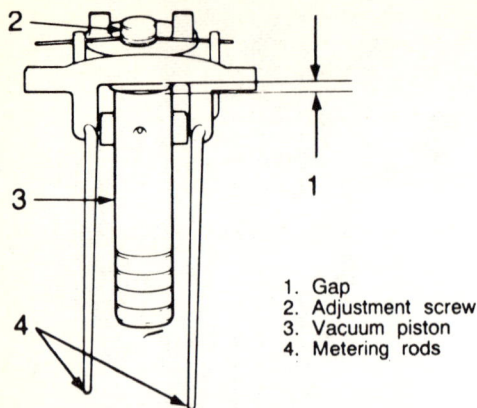

1. Gap
2. Adjustment screw
3. Vacuum piston
4. Metering rods

Carter BBD vacuum piston gap adjustment

top of the vacuum piston, is as specified in the Carburetor Specifications Charts.

2. Counting the number of turns involved, turn the curb idle adjustment screw counterclockwise, until the throttle valves are completely closed.

3. Fully depress the vacuum piston, while exerting moderate pressure on the metering rod lifter tab. In this position, tighten the rod lifter lock screw.

4. Release the piston and rod lifter.

NOTE: The accelerator pump should now be adjusted.

5. Return the curb idle adjustment screw to its original position.

ACCELERATOR PUMP

Carter BBD

1. Counting the number of turns involved, turn the curb idle adjustment screw counterclockwise, until the throttle valves are completely closed.

2. Open the choke valve so that the fast idle cam will allow the throttle valves to seat in their bores.

3. Turn the curb idle adjustment screw clockwise, so that it just barely touches the stop, then, turn it 2 full turns further.

4. Measure the distance between the surface of the air horn and the top of the accelerator pump shaft with a T-scale. The distance should be $\frac{33}{64}$ in. (13mm).

5. If the dimension is not correct, loosen the pump arm ad-

justing screw and rotate the sleeve to adjust the pump travel. Tighten the lock screw.

6. Return the curb idle screw to its original position.

Autolite/Motorcraft 2100 and 2150

Under normal driving conditions, the pump rod should be in the inboard hole of the pump actuating lever and the third hole (counting from the bottom up) of the overtravel lever.

Under extremely hot conditions, the rod should be placed in the second hole of the overtravel lever.

Under extremely cold conditions, the rod should be placed in the fourth hole of the overtravel lever.

METERING ROD

Carter YFA

1. Remove the air horn and gasket.

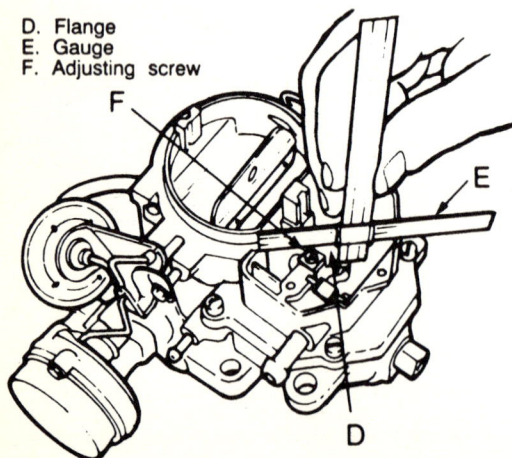

2100, 2150 accelerator pump adjustment points

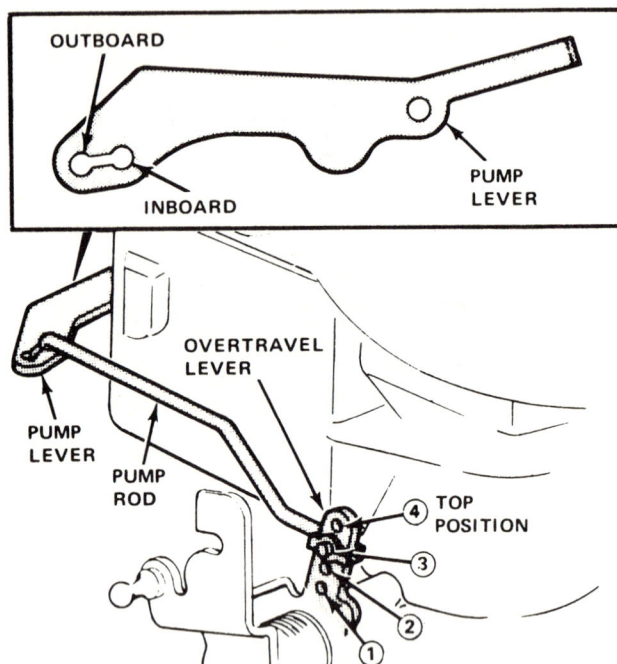

D. Flange
E. Gauge
F. Adjusting screw

Carter BBD accelerator pump adjustment

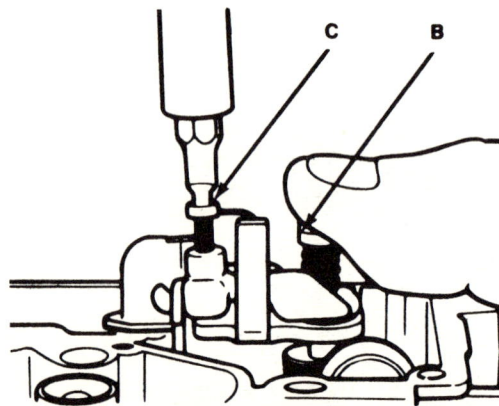

B. Diaphragm shaft
C. Adjustment screw

Metering rod adjustment

2. Make sure that the idle speed adjustment screw allows the throttle plate to close tightly in the bore.

3. Press down on the top of the pump diaphragm shaft until it bottoms.

4. In this position, adjust the metering rod by turning the adjusting screw counterclockwise, until the metering rod lightly bottoms in the main metering jet.

5. Turn the adjusting screw 1 full turn more.

6. Install the air horn and gasket, and adjust the curb idle speed.

THROTTLE BODY INJECTION FUEL SYSTEM

Components and Operation

The Renix throttle body fuel injection is a "pulse time" system that uses a single solenoid-type injector to meter fuel into the throttle body above the throttle blade. Fuel is metered to the engine by an electronic control unit (ECU), which controls the amount of fuel delivery according to input from various engine sensors that monitor exhaust gas oxygen content, coolant temperature, manifold absolute pressure, crankshaft position and throttle position. These sensors provide an electronic signal by varying resistance within the sensor itself. By reading the difference in resistance, the ECU can determine engine operating conditions and calculate the correct air/fuel mixture, and ignition timing under varying engine loads and temperatures. In addition, the ECU controls idle speed, emission control and fuel pump operation, the upshift indicator lamp and the A/C compressor clutch.

Renix TBI fuel injection has two main subsystems; a fuel subsystem and a control subsystem. The fuel subsystem consists of an electric fuel pump (mounted in the fuel tank), a fuel filter, a pressure regulator and the fuel injector. The control subsystem consists of a manifold air/fuel mixture temperature sensor (MAT), a coolant temperature sensor (CTS), a manifold absolute pressure sensor (MAP), a knock sensor, an exhaust gas oxygen (O_2) sensor, an electronic control unit (ECU), a gear position indicator (automatic transmission only), a throttle position sensor and power steering pressure switch with a load swap relay. In addition to these sensors which send signals to the ECU, there are various devices which receive signals from the ECU to control different functions such as exhaust gas recirculation, idle speed control, air conditioner operation, etc.

Electronic Control Unit (ECU)

The electronic control unit (ECU) is a sealed microprocessor unit located above the accelerator pedal under the instrument panel or below the glove box, next to the fuse panel, and is the heart of the electronic engine control system. The throttle position sensor (or wide open throttle switch) is mounted on the throttle body assembly and provides the ECU with an input signal of up to 5 volts to indicate throttle position. At minimum throttle opening (idle speed), a signal input of approximately one volt is transmitted to the ECU. As the throttle opening increases, the signal voltage to the ECU increases.

Manifold Absolute Pressure (MAP) Sensor

The manifold absolute pressure (MAP) sensor is attached to the plenum chamber near the hood latch. It reacts to absolute pressure in the intake manifold and provides an input voltage to the ECU. Manifold pressure is used to supply mixture density

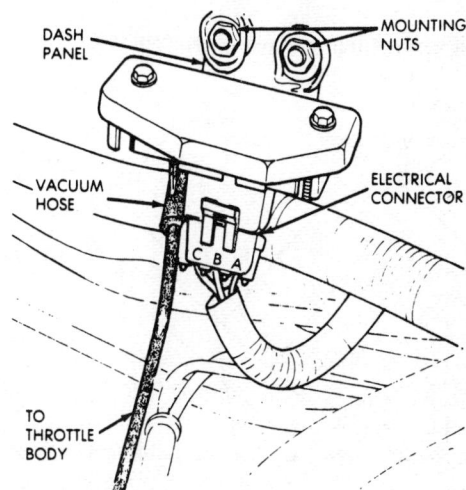

MAP sensor with TBI

information and ambient barometric pressure information that is necessary for computing the air/fuel mixture. A vacuum line from the throttle body attaches to the MAP sensor to provide its input pressure. The manifold air temperature (MAT) sensor is located in the intake manifold and measures the air/fuel mixture temperature to allow the ECU to compensate for air density changes during high temperature operation.

Coolant Temperature Sensor

The coolant temperature sensor (CTS) is located in the intake manifold coolant jacket and provides an engine coolant temperature signal to the ECU. The ECU uses the coolant temperature signal to enrich the air fuel mixture when the engine is cold, compensate for fuel condensation in the intake manifold, control engine warm-up speed, increase the ignition advance when the engine is cold and to cut off the EGR system when the engine is cold.

Knock Sensor

The knock sensor is located in the cylinder head and provides a signal to the ECU to detect detonation (spark knock) during engine operation. When detonation occurs, the ECU retards the ignition timing to elimintate it. On automatic transmission models, a transmission gear position indicator provides an input to the ECU to determine whether the transaxle is in a driving gear and not in Park or Neutral.

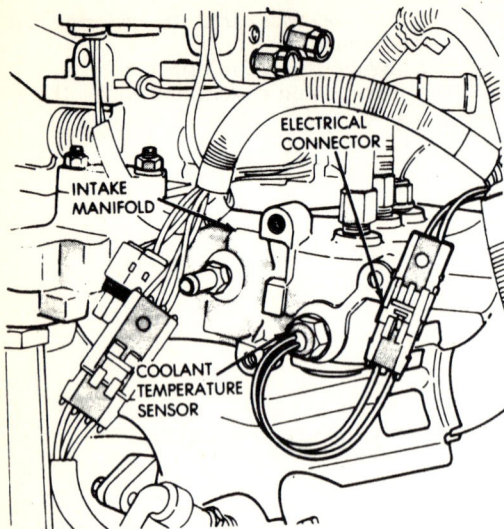

Coolant temperature sensor with TBI

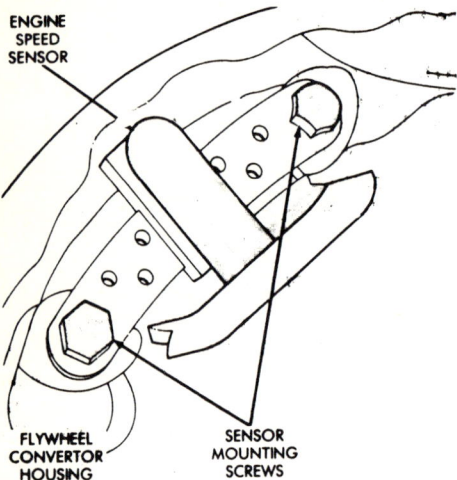

Engine speed sensor attaching bolts with TBI

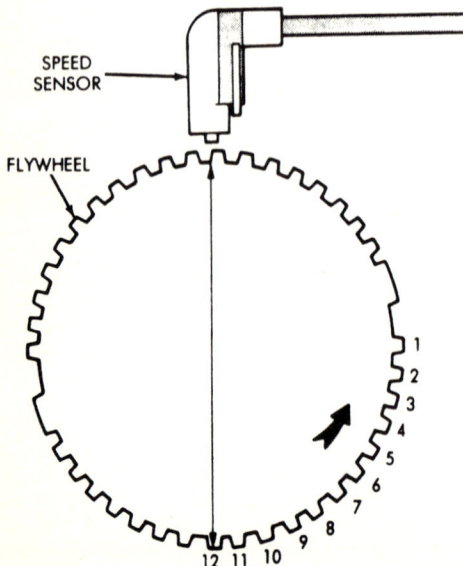

Engine speed sensor-to-flywheel position with TBI

Pressure Sensing Switch

A pressure sensing switch is included in the power steering system to increase the idle speed during periods of high pump load and low engine rpm. Input signals from the pressure switch to the ECU are routed through the A/C request and A/C select input circuits. When pump pressure exceeds 250–300 psi, the switch contacts close and transmit an input signal to the ECU. The ECU raises engine idle speed immediately after receiving the pressure switch input signal.

Load Swap Relay

The load swap relay is used on models with air conditioning and power steering. The relay works in conjunction with the power steering pressure switch to disengage the A/C compressor clutch. If the A/C compressor clutch is engaged when the power steering pressure switch contacts close, the input signal from the switch to the ECU also activates the load swap relay. The relay contacts then open, cutting off electrical feed to the compressor clutch. The clutch remains disengaged until the pressure switch contacts open and the engine returns to normal idle speed. The load swap relay does not reengage the compressor clutch immediately. The relay has a timer that delays energizing the clutch for 0.5 seconds to permit smooth engagement.

System Power Relay

The system power relay is located on the right strut tower and is initially energized when starting the engine. The relay remains energized for 3–5 seconds after the engine stops to enable the ECU to extend the idle speed actuator (ISA) for the next start-up. The fuel pump control relay is also located on the front of the right strut tower. Battery voltage is applied to the fuel pump control relay through the ignition switch and is energized when a ground is provided by the ECU. In this manner, the ECU controls fuel pump operation.

EGR/Canister Purge Solenoid

The vacuum for both the EGR valve and the vapor canister purge function is controlled by the EGR/Canister Purge Solenoid. When energized by the ECU, it cuts off vacuum to the EGR valve and canister. The solenoid is energized during engine warm-up, closed throttle (idle), wide open throttle (WOT) and rapid acceleration/deceleraton. If the solenoid wire connector is disconnected, the EGR valve and canister purge function will be operational at all times.

Idle Speed Actuator

The idle speed actuator (ISA) is mounted on the throttle body and controls idle speed and engine deceleration throttle stop angle. The actuator changes the throttle stop angle by being a movable throttle stop. The ECU controls the ISA motor by providing the appropriate voltage outputs to produce the idle speed or throttle stop angle required for the particular engine operating condition. There is no idle speed adjustment.

Speed Sensor

The speed sensor is attached to the flywheel drive plate housing. This sensor detects the flywheel/drive plate teeth as they pass during engine operation and provides engine speed and crankshaft angle information to the ECU. The flywheel/drive plate has a large trigger tooth and notch located every 90 degrees and 12 smaller teeth before each top dead center (TDC) position. When a small tooth and notch pass the magnet core in the sensor, the concentration and then collapse of the magnetic flux induces a small voltage spike into the sensor pickup coil winding. The higher voltage spike indicates to the ECU that a piston will be at TDC position 12 teeth later. The ignition timing for the cylinder is either advanced or retarded as necessary by the ECU according to the sensor inputs.

DIAGNOSIS AND TESTING

Before performing any system tests, first determine that the problem is not being caused by a component other than the fuel injection system, such as spark plugs, distributor, ignition timing, etc. Also make sure that no air is entering the intake and exhaust system above the catalytic converter and that fuel is reaching the injector under normal pressure.

NOTE: The diagnostic connectors D1 and D2 are located on the dash panel in the engine compartment.

Oxygen Sensor Heating Element

The oxygen sensor heating element can be tested by connecting an ohmmeter test leads to terminals **A** and **B** of the sensor connector. Resistance should be between 5–7 ohms. Replace the sensor if an infinite reading is obtained.

Fuel Pump Pressure Test

Fuel pump operating pressure is 14.5 psi. The fuel pressure regulator is adjustable by means of a torx head screw on the bottom of the pressure regulator.

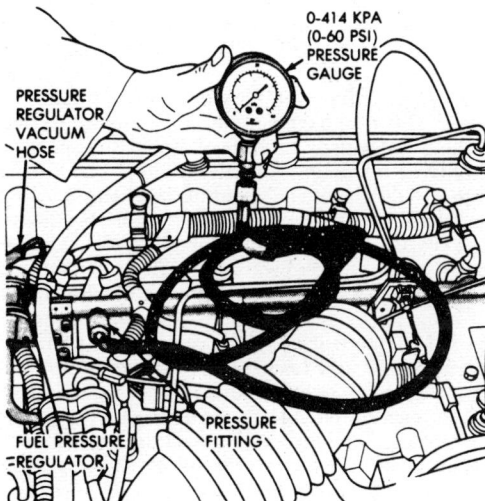

Fuel pressure test

1. Remove the air cleaner assembly. Remove the screw plug on the throttle body and install a fuel pressure test fitting (No. 8983 501 572).
2. Connect an accurate pressure test gauge to the test fitting.
3. Connect a tachometer to diagnostic connector terminals D1-1 and D1-3, then start the engine and accelerate it to 2000 rpm.
4. Read the fuel pressure on the gauge. If necessary, turn the adjustment screw on the bottom of the fuel pressure regulator to obtain 14.5 psi (1 bar) of fuel pressure. Turning the screw inward increases the pressure and turning the screw outward decreases the pressure.
5. Once all adjustments are complete, install a lead seal ball to cover the regulator adjusting screw. Turn the ignition OFF, then disconnect the tachometer and remove the fuel pressure gauge. Install the original plug screw into the throttle body, then install the air cleaner assembly.

Throttle Position Sensor Adjustment

1. Turn the ignition key ON.
2. Check the sensor input voltage. Connect the negative lead of a voltmeter to sensor terminal **B**, then connect the voltmeter positive lead to sensor terminal **C**.

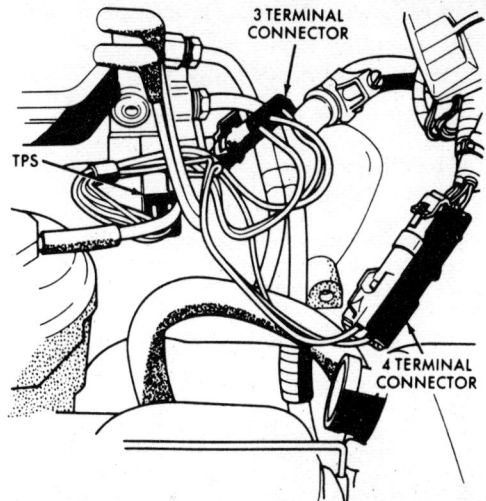

Throttle position sensor on TBI with automatic transmission

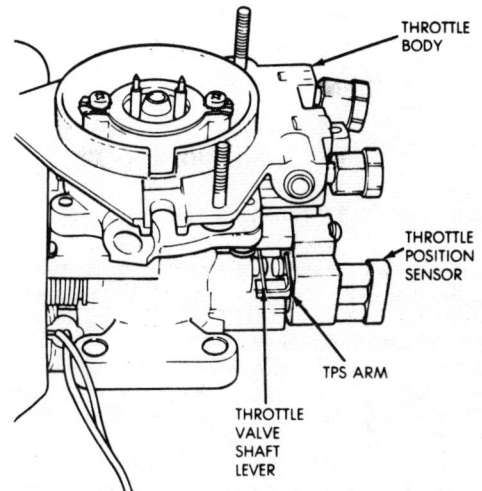

Throttle position sensor on TBI with manual transmission

NOTE: Do not disconnect the sensor wire harness connector. Insert the voltmeter test leads through the back of the wire harness connector to make contact with the sensor terminals during testing. It may be necessary to remove the throttle body from the intake manifold to gain access to the wire harness connector.

3. Move and hold the throttle plate in the wide open position. Make sure the throttle linkage contacts the stop.
4. Note the voltmeter reading. Input voltage at terminals **B** and **C** should be 5.0 volts at wide open throttle.
5. Return the throttle plate to the closed position.
6. Check the sensor output voltage. Disconnect the voltmeter positive lead from sensor terminal **C** and connect it to terminal **A**.
7. Move and hold the throttle plate in the wide open position. Make sure the throttle linkage contacts the stop.
8. Note the voltmeter reading. Output voltage should be 4.6–4.7 volts. Adjust the output voltage by loosening the lower sensor retaining screw and pivoting the sensor in the adjustment slot for coarse adjustment. Loosen the other retaining screw and pivot the sensor for fine adjustment.
9. Remove the voltmeter and return the throttle plate to the

closed position. Make sure the sensor retaining screws are tightened securely.

Wide-Open Throttle (WOT) Switch Test

1. Disconnect the harness terminal connector from the WOT switch.

2. Test the on-off operation of the switch with a digital volt-ohmmeter while operating the switch manually.

3. The resistance should be infinite when the throttle is closed and a low resistance should be indicated when the throttle is wide open. Test the switch operation several times and replace the WOT switch is defective.

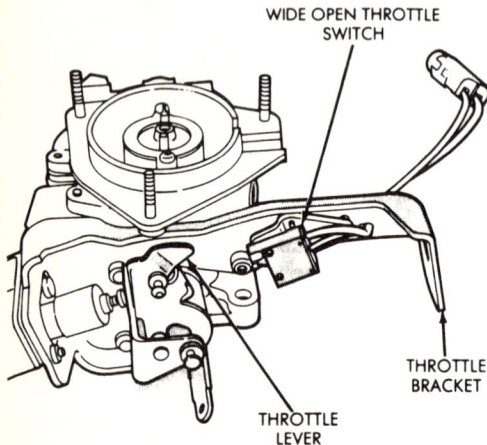

Wide open throttle switch with TBI

4. Connect the wire harness connector. With the ignition switch ON, test the WOT switch voltage at the diagnostic connector terminals D2-6 (+) and D2-7 (−). The voltage should be zero at the WOT position and greater than 2 volts if not at the WOT position.

5. If the voltage is always zero, test for a short circuit to ground in the wire harness or switch, or an open circuit between terminal 8 of the ECU connector and the switch connector. Repair or replace the wire harness as necessary.

6. If the voltage is always greater than 2 volts, test for an open circuit in the wire or connector between the switch and ground. Repair as necessary.

Closed Throttle (Idle) Switch Test

NOTE: It is important that all testing be done with the idle speed actuator (ISA) motor plunger in the fully extended position, as it would be after normal engine shutdown. If it is necessary to extend the ISA motor plunger to test the switch, an ISA motor failure can be suspected.

1. With the ignition switch ON, test the switch voltage at the diagnostic connector terminals D2-13 (+) and D2-7 (−). The voltage should be close to zero at closed throttle and greater than 2 volts when off the closed throttle position.

2. If the voltage is always zero, test for a short circuit to ground in the wire harness or switch, or for an open circuit between ECU connector terminal 25 and the switch.

3. If the voltage is always more than 2 volts, test for an open circuit in the wire harness between the ECU and the switch connector, and between the switch connector and ground. Repair or replace the wire harness as necessary.

Manifold Absolute Pressure (MAP) Sensor Test

1. Inspect the MAP sensor vacuum hose connections at the throttle body and sensor and repair as necessary.

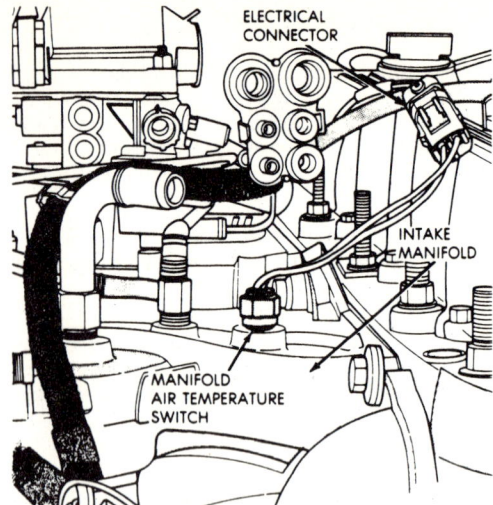

Manifold air temperature sensor with TBI

2. Test the MAP sensor output voltage at the MAP sensor connector terminal **B** as marked on the sensor body, with the ignition switch ON (engine OFF). The output voltage should be 4–5 volts.

3. Test ECU terminal 33 for the same voltage described above to verify the wire harness condition. Repair as necessary.

4. Test the MAP sensor supply voltage at the sensor connector terminal **C** with the ignition ON. It should be 4.5–5.5 volts. This voltage should also be at terminal 16 of the ECU wire harness connector. Repair or replace the wire harness as necessary. Test the ECU with Diagnostic Tester MS 1700, if necessary.

5. Test the MAP sensor ground circuit at sensor connector terminal **A** and ECU connector terminal 17. Repair the wire harness, if necessary.

6. Test the MAP sensor ground circuit at the ECU connector between terminal 17 and terminal 2 with an ohmmeter. If the ohmmeter indicates an open circuit, inspect for a defective sensor ground connection on the flywheel/drive plate housing near the starter motor. If the ground connection is good, replace the ECU. If terminal 17 has a short circuit to 12 volts, correct this condition before replacing the ECU.

Manifold Air Temperature (MAT) Sensor and Coolant Temperature Sensor (CTS) Test

These two sensors are tested in the same manner and should yield the same results. The only difference is the pin number of the test points. Disconnect the wire harness connector from the sensor, then test the resistance with a digital volt-ohmmeter. The resistance should be less than 1000 ohms with the engine warm. Refer to the chart to check the temperature-to-resistance values and replace the sensor if the resistance is not within the specified range. If the MAT sensor is being tested, check the resistance between ECU harness connector terminals 14 and 32 and the sensor connector terminals. If the CTS sensor is being tested, check the resistance between ECU harness connector terminals 15 and 32 and the sensor terminals. In either case, repair the wire harness if an open circuit is indicated.

Fuel Pump

REMOVAL AND INSTALLATION

1. Disconnect the negative battery cable. Remove all necessary components in order to gain access to the fuel tank sending unit.

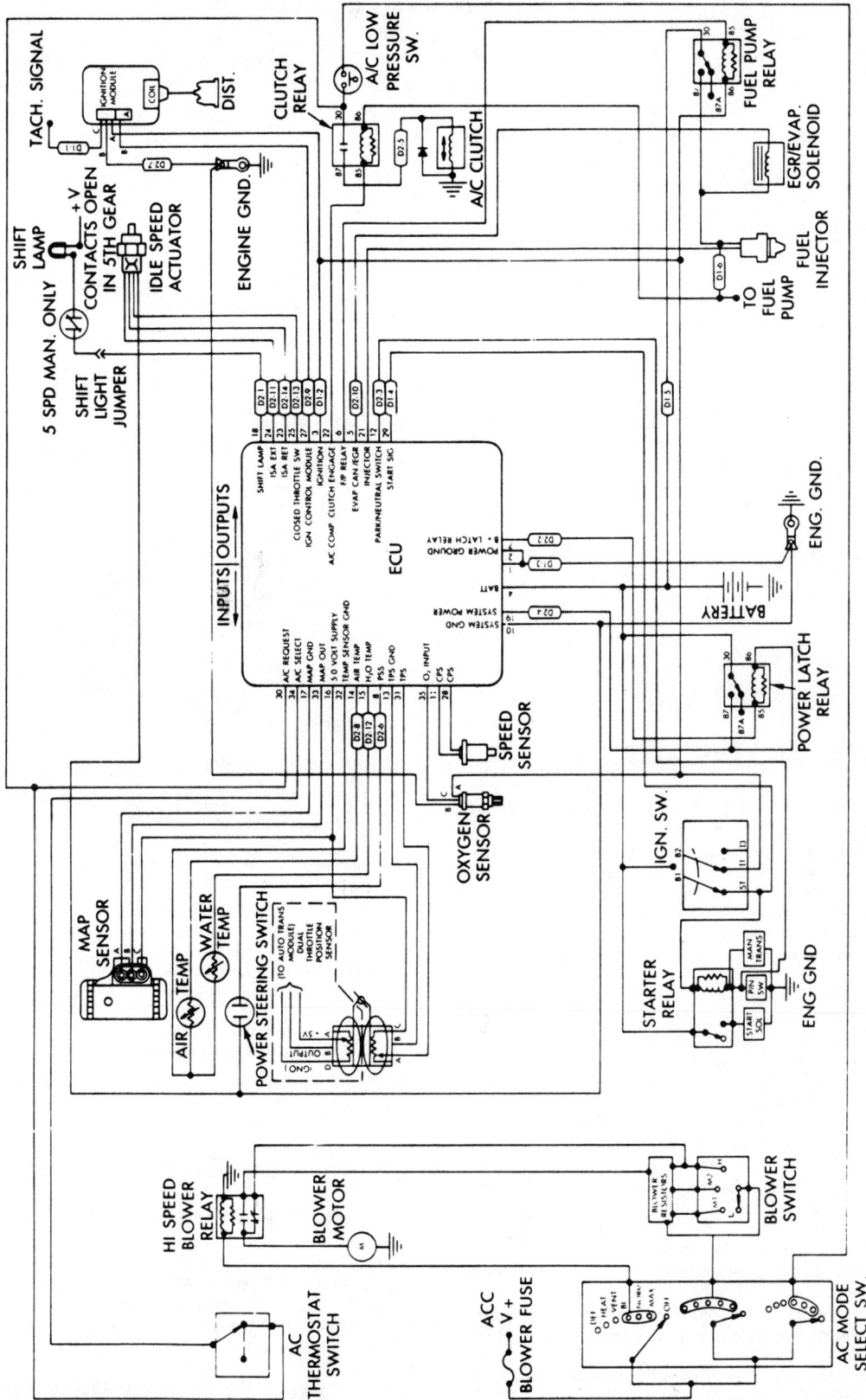

TBI engine control schematic

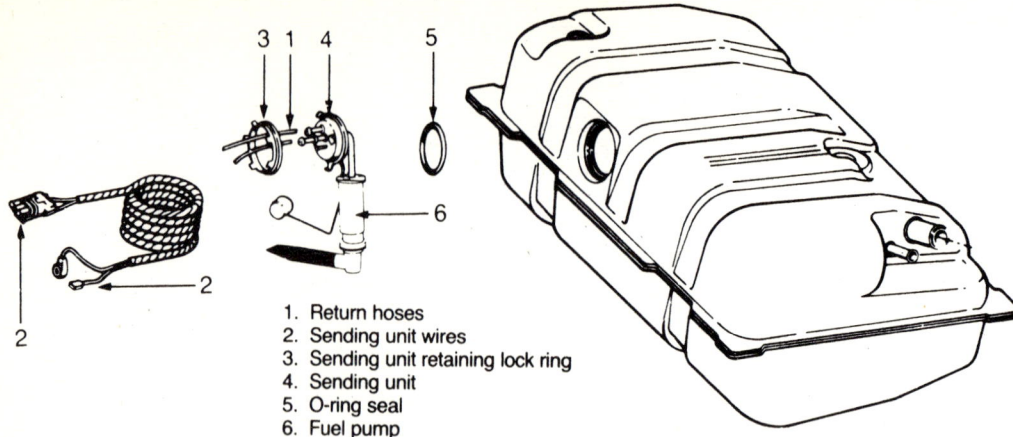

1. Return hoses
2. Sending unit wires
3. Sending unit retaining lock ring
4. Sending unit
5. O-ring seal
6. Fuel pump

Fuel pump used on the fuel Injected 4–150

2. Drain the fuel from the fuel tank. Raise and support the vehicle safely.

3. Remove the fuel inlet and outlet hoses from the sending unit. Remove the sending unit wires.

4. Remove the sending unit retaining lock ring. Remove the sending unit, which incorporates the electric fuel pump, along with the O-ring seal from the fuel tank.

5. Installation is the reverse of removal. Be sure to use a new O-ring seal.

Throttle Body

REMOVAL AND INSTALLATION

1. Disconnect the negative battery cable. Remove the upper air cleaner assembly.

2. Remove the lower air cleaner assembly retaining bolts. Remove the lower air cleaner assembly.

3. Remove the throttle cable and the return spring. Disconnect the wire harness connector from the injector.

4. Disconnect the wire harness connector from the wide open throttle switch. Disconnect the wire harness connector from the ISC motor.

5. Disconnect the fuel supply pipe from the throttle body. Disconnect the fuel return pipe from the throttle body.

6. Disconnect the vacuum hoses from the throttle body assembly. Disconnect the potentiometer wire connector.

7. Remove the throttle body to manifold retaining bolts. Remove the throttle body assembly from the intake manifold.

8. Clean the manifold and throttle body mating surfaces. Be sure to use a new gasket between the throttle body assembly and the intake manifold.

9. Install the throttle body assembly from the intake manifold. Torque the nuts to 15 ft. lbs.

10. Connect the vacuum hoses to the throttle body assembly. Connect the potentiometer wire connector.

11. Connect the fuel supply pipe to the throttle body. Connect the fuel return pipe to the throttle body.

12. Connect the wire harness connector to the wide open throttle switch. Connect the wire harness connector to the ISC motor.

13. Install the throttle cable and the return spring. Connect the wire harness connector to the injector.

14. Install the lower air cleaner assembly. Install the lower air cleaner assembly retaining bolts.

15. Install the upper air cleaner assembly.

16. Connect the negative battery cable.

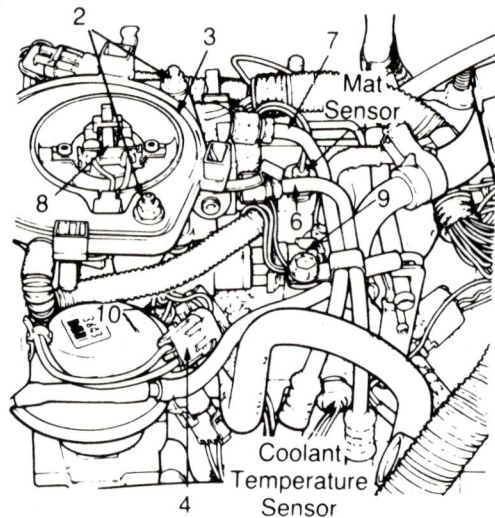

2. Lower air cleaner retaining bolts
3. Lower air cleaner assembly
4. Wide open throttle switch
6. Fuel inlet line
7. Fuel return line
8. Injector wiring connector
9. Potentiometer wire
10. Throttle body retaining bolts

Points of disconnection when removing the throttle body

Fuel Body Assembly

REMOVAL AND INSTALLATION

1. Remove the throttle body assembly from the vehicle.

2. Remove the Torx® head screws that retain the fuel body to the throttle body. Remove and discard the gasket.

3. Installation is the reverse of removal. Be sure to use a new gasket.

Fuel Pressure Regulator

REMOVAL AND INSTALLATION

1. Remove the throttle body assembly from the vehicle.

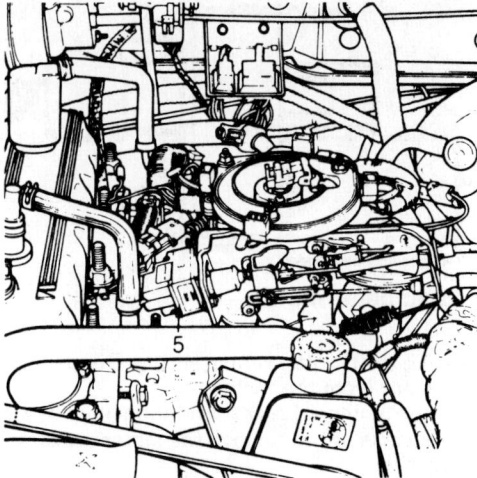

5. ISC motor

ISC motor location

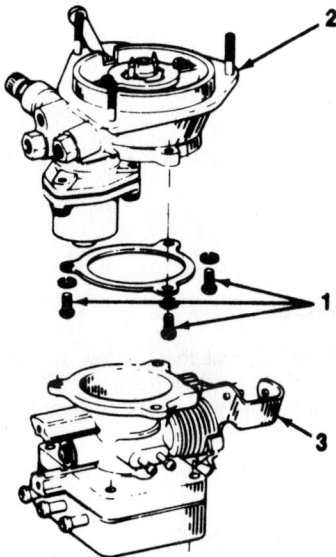

1. Relief valve 2. Diaphragm

Fuel pressure regulator

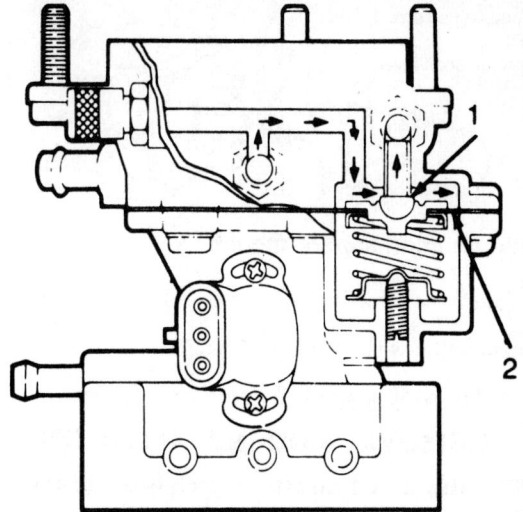

1. Fuel body retaining screws
2. Fuel body
3. Throttly body

Fuel body assembly removal

2. Remove the three retaining screws that hold the pressure regulator to the fuel body.

3. Remove the pressure regulator assembly. Note the location of the components for reassembly. Discard the gasket.

4. Installation is the reverse of the removal procedure. Be sure to use a new gasket.

Fuel Injector

REMOVAL AND INSTALLATION

1. Remove the air cleaner and hose assembly.

2. Remove the fuel injector wire. Remove the fuel injector retainer clip screws. Remove the fuel injector retainer clip.

3. Using a small pair of pliers, gently grasp the center collar of the injector, between the electrical terminals, and carefully remove the injector using a lifting-twisting motion.

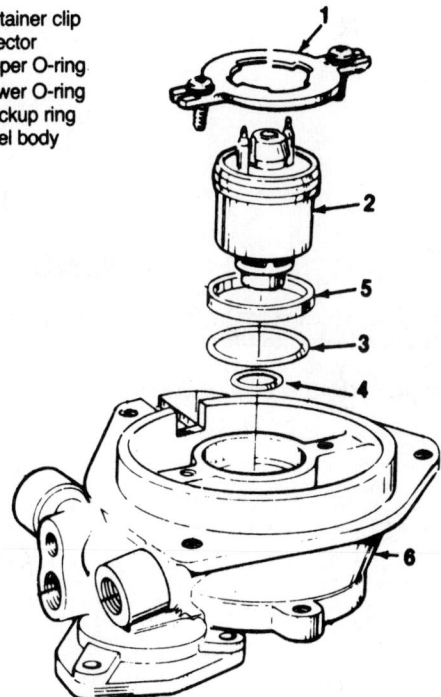

1. Retainer clip
2. Injector
3. Upper O-ring
4. Lower O-ring
5. Backup ring
6. Fuel body

Injector removal

4. Discard the upper and lower O-rings. Note that the back up ring fits over the upper O-ring.

5. Installation is the reverse of the removal procedure. Lubricate both O-rings with light oil before installation.

Throttle Position Sensor

REMOVAL AND INSTALLATION

1. Remove the upper and lower air cleaner assemblies.

2. Remove the throttle body assembly from the vehicle.

3. Remove the two Torx® head retaining screws holding the TPS assembly to the throttle body.

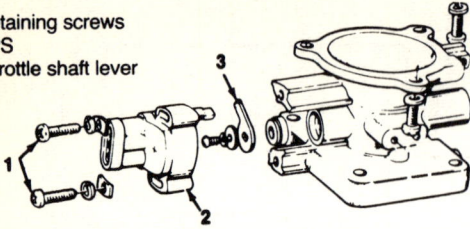

1. Retaining screws
2. TPS
3. Throttle shaft lever

Throttle position sensor mounting

4. Remove the throttle position sensor from the throttle shaft lever.
5. Installation is the reverse of removal.

Idle Speed Actuator Motor
REMOVAL AND INSTALLATION

NOTE: The closed throttle switch is integral with the motor.

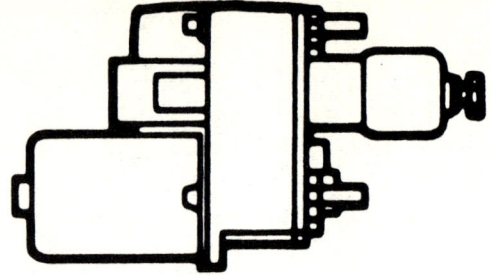

Idle speed actuator motor

1. Disconnect the throttle return spring. Disconnect the wire harness connector from the motor.
2. Remove the motor to bracket retaining nuts. Be sure to use a back up wrench as not to remove the motor studs which hold the motor together.
3. Remove the motor from the bracket.
4. Installation is the reverse of removal.

FUEL TANK

REMOVAL AND INSTALLATION

The fuel tank is attached to the frame by brackets and bolts. The brackets are attached to the tank at the seam flange or the skid plate.

Before removing the fuel tank, make sure that the level of the fuel inside the tank is at least below any of the various hoses connected. It is best to either drain or siphon the majority of fuel out of the tank to make it easier to handle while removing it.

SPECIAL TOOLS

Mot. LM

J-9789-136
J-9789-138
FLOAT LEVEL GAUGES

 Mot. 854

B.Vi. 28-01

B.Vi. 859

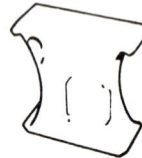 Mot. 856

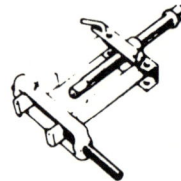 Mot. 861

6 Chassis Electrical

QUICK REFERENCE INDEX

GENERAL INDEX

UNDERSTANDING AND TROUBLESHOOTING ELECTRICAL SYSTEMS

With the rate at which automobile manufacturers are incorporating electronic control systems into their production lines, it won't be long before every new vehicle is equipped with one or more on-board computer. These electronic components (with no moving parts) should theoretically last the life of the vehicle, provided nothing external happens to damage the circuits or memory chips.

While it is true that electronic components should never wear out, in the real world malfunctions do occur. It is also true that any computer-based system is extremely sensitive to electrical voltages and cannot tolerate careless or haphazard testing or service procedures. An inexperienced individual can literally do major damage looking for a minor problem by using the wrong kind of test equipment or connecting test leads or connectors with the ignition switch ON. When selecting test equipment, make sure the manufacturers instructions state that the tester is compatible with whatever type of electronic control system is being serviced. Read all instructions carefully and double check all test points before installing probes or making any test connections.

The following section outlines basic diagnosis techniques for dealing with computerized automotive control systems. Along with a general explanation of the various types of test equipment available to aid in servicing modern electronic automotive systems, basic repair techniques for wiring harnesses and connectors is given. Read the basic information before attempting any repairs or testing on any computerized system, to provide the background of information necessary to avoid the most common and obvious mistakes that can cost both time and money. Although the replacement and testing procedures are simple in themselves, the systems are not, and unless one has a thorough understanding of all components and their function within a particular computerized control system, the logical test sequence these systems demand cannot be followed. Minor malfunctions can make a big difference, so it is important to know how each component affects the operation of the overall electronic system to find the ultimate cause of a problem without replacing good components unnecessarily. It is not enough to use the correct test equipment; the test equipment must be used correctly.

Safety Precautions

CAUTION

Whenever working on or around any computer based microprocessor control system, always observe these general precautions to prevent the possibility of personal injury or damage to electronic components.

• Never install or remove battery cables with the key ON or the engine running. Jumper cables should be connected with the key OFF to avoid power surges that can damage electronic control units. Engines equipped with computer controlled systems should avoid both giving and getting jump starts due to the possibility of serious damage to components from arcing in the engine compartment when connections are made with the ignition ON.

• Always remove the battery cables before charging the battery. Never use a high output charger on an installed battery or attempt to use any type of "hot shot" (24 volt) starting aid.

• Exercise care when inserting test probes into connectors to insure good connections without damaging the connector or spreading the pins. Always probe connectors from the rear (wire) side, NOT the pin side, to avoid accidental shorting of terminals during test procedures.

• Never remove or attach wiring harness connectors with the ignition switch ON, especially to an electronic control unit.

• Do not drop any components during service procedures and never apply 12 volts directly to any component (like a solenoid or relay) unless instructed specifically to do so. Some component electrical windings are designed to safely handle only 4 or 5 volts and can be destroyed in seconds if 12 volts are applied directly to the connector.

• Remove the electronic control unit if the vehicle is to be placed in an environment where temperatures exceed approximately 176°F (80°C), such as a paint spray booth or when arc or gas welding near the control unit location in the car.

ORGANIZED TROUBLESHOOTING

When diagnosing a specific problem, organized troubleshooting is a must. The complexity of a modern automobile demands that you approach any problem in a logical, organized manner. There are certain troubleshooting techniques that are standard:

1. Establish when the problem occurs. Does the problem appear only under certain conditions? Were there any noises, odors, or other unusual symptoms?

2. Isolate the problem area. To do this, make some simple tests and observations; then eliminate the systems that are working properly. Check for obvious problems such as broken wires, dirty connections or split or disconnected vacuum hoses. Always check the obvious before assuming something complicated is the cause.

3. Test for problems systematically to determine the cause once the problem area is isolated. Are all the components functioning properly? Is there power going to electrical switches and motors? Is there vacuum at vacuum switches and/or actuators? Is there a mechanical problem such as bent linkage or loose mounting screws? Doing careful, systematic checks will often turn up most causes on the first inspection without wasting time checking components that have little or no relationship to the problem.

4. Test all repairs after the work is done to make sure that the problem is fixed. Some causes can be traced to more than one component, so a careful verification of repair work is important to pick up additional malfunctions that may cause a problem to reappear or a different problem to arise. A blown fuse, for example, is a simple problem that may require more than another fuse to repair. If you don't look for a problem that caused a fuse to blow, for example, a shorted wire may go undetected.

Experience has shown that most problems tend to be the result of a fairly simple and obvious cause, such as loose or corroded connectors or air leaks in the intake system; making careful inspection of components during testing essential to quick and accurate troubleshooting. Special, hand held computerized testers designed specifically for diagnosing the system are available from a variety of aftermarket sources, as well as from the vehicle manufacturer, but care should be taken that any test equipment being used is designed to diagnose that particular computer controlled system accurately without damaging the control unit (ECU) or components being tested.

NOTE: Pinpointing the exact cause of trouble in an electrical system can sometimes only be accomplished by the use of special test equipment. The following describes commonly used test equipment and explains how to put it to best use in diagnosis. In addition to the information covered below, the manufacturer's instructions booklet provided with the tester should be read and clearly understood before attempting any test procedures.

TEST EQUIPMENT

Jumper Wires

Jumper wires are simple, yet extremely valuable, pieces of test equipment. Jumper wires are merely wires that are used to bypass sections of a circuit. The simplest type of jumper wire is merely a length of multistrand wire with an alligator clip at each end. Jumper wires are usually fabricated from lengths of standard automotive wire and whatever type of connector (alligator clip, spade connector or pin connector) that is required for the particular vehicle being tested. The well equipped tool box will have several different styles of jumper wires in several different lengths. Some jumper wires are made with three or more terminals coming from a common splice for special purpose testing. In cramped, hard-to-reach areas it is advisable to have insulated boots over the jumper wire terminals in order to prevent accidental grounding, sparks, and possible fire, especially when testing fuel system components.

Jumper wires are used primarily to locate open electrical circuits, on either the ground (−) side of the circuit or on the hot (+) side. If an electrical component fails to operate, connect the jumper wire between the component and a good ground. If the component operates only with the jumper installed, the ground circuit is open. If the ground circuit is good, but the component does not operate, the circuit between the power feed and component is open. You can sometimes connect the jumper wire directly from the battery to the hot terminal of the component, but first make sure the component uses 12 volts in operation. Some electrical components, such as fuel injectors, are designed to operate on about 4 volts and running 12 volts directly to the injector terminals can burn out the wiring. By inserting an inline fuseholder between a set of test leads, a fused jumper wire can be used for bypassing open circuits. Use a 5 amp fuse to provide protection against voltage spikes. When in doubt, use a voltmeter to check the voltage input to the component and measure how much voltage is being applied normally. By moving the jumper wire successively back from the lamp toward the power source, you can isolate the area of the circuit where the open is located. When the component stops functioning, or the power is cut off, the open is in the segment of wire between the jumper and the point previously tested.

--- CAUTION ---

Never use jumpers made from wire that is of lighter gauge than used in the circuit under test. If the jumper wire is of too small gauge, it may overheat and possibly melt. Never use jumpers to bypass high resistance loads (such as motors) in a circuit. Bypassing resistances, in effect, creates a short circuit which may, in turn, cause damage and fire. Never use a jumper for anything other than temporary bypassing of components in a circuit.

12 Volt Test Light

The 12 volt test light is used to check circuits and components while electrical current is flowing through them. It is used for voltage and ground tests. Twelve volt test lights come in different styles but all have three main parts; a ground clip, a probe, and a light. The most commonly used 12 volt test lights have pick-type probes. To use a 12 volt test light, connect the ground clip to a good ground and probe wherever necessary with the pick. The pick should be sharp so that it can penetrate wire insulation to make contact with the wire, without making a large hole in the insulation. The wrap-around light is handy in hard to reach areas or where it is difficult to support a wire to push a probe pick into it. To use the wrap around light, hook the wire to probed with the hook and pull the trigger. A small pick will be forced through the wire insulation into the wire core.

--- CAUTION ---

Do not use a test light to probe electronic ignition spark plug or coil

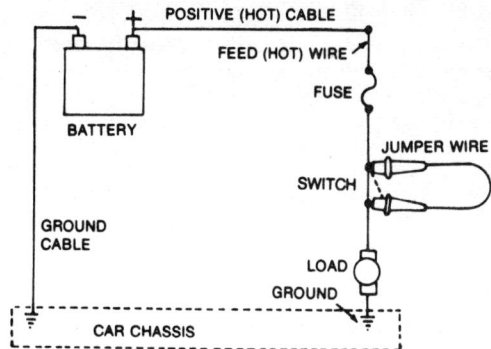

Bypassing a switch with a jumper wire

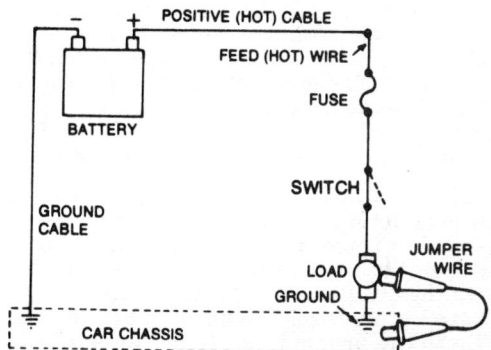

Checking for a bad ground with a jumper wire

wires. Never use a pick-type test light to probe wiring on computer controlled systems unless specifically instructed to do so. Any wire insulation that is pierced by the test light probe should be taped and sealed with silicone after testing.

Like the jumper wire, the 12 volt test light is used to isolate opens in circuits. But, whereas the jumper wire is used to bypass the open to operate the load, the 12 volt test light is used to locate the presence of voltage in a circuit. If the test light glows, you know that there is power up to that point; if the 12 volt test light does not glow when its probe is inserted into the wire or connector, you know that there is an open circuit (no power). Move the test light in successive steps back toward the power source until the light in the handle does glow. When it does glow, the open is between the probe and point previously probed.

NOTE: The test light does not detect that 12 volts (or any particular amount of voltage) is present; it only detects that some voltage is present. It is advisable before using the test light to touch its terminals across the battery posts to make sure the light is operating properly.

Self-Powered Test Light

The self-powered test light usually contains a 1.5 volt penlight battery. One type of self-powered test light is similar in design to the 12 volt test light. This type has both the battery and the light in the handle and pick-type probe tip. The second type has the light toward the open tip, so that the light illuminates the contact point. The self-powered test light is dual purpose piece of test equipment. It can be used to test for either open or short circuits when power is isolated from the circuit (continuity test). A powered test light should not be used on any computer controlled system or component unless specifically instructed to do so. Many engine sensors can be destroyed by even this small amount of voltage applied directly to the terminals.

Open Circuit Testing

To use the self-powered test light to check for open circuits, first isolate the circuit from the vehicle's 12 volt power source by disconnecting the battery or wiring harness connector. Connect the test light ground clip to a good ground and probe sections of the circuit sequentially with the test light. (start from either end of the circuit). If the light is out, the open is between the probe and the circuit ground. If the light is on, the open is between the probe and end of the circuit toward the power source.

Short Circuit Testing

By isolating the circuit both from power and from ground, and using a self-powered test light, you can check for shorts to ground in the circuit. Isolate the circuit from power and ground. Connect the test light ground clip to a good ground and probe any easy-to-reach test point in the circuit. If the light comes on, there is a short somewhere in the circuit. To isolate the short, probe a test point at either end of the isolated circuit (the light should be on). Leave the test light probe connected and open connectors, switches, remove parts, etc., sequentially, until the light goes out. When the light goes out, the short is between the last circuit component opened and the previous circuit opened.

NOTE: The 1.5 volt battery in the test light does not provide much current. A weak battery may not provide enough power to illuminate the test light even when a complete circuit is made (especially if there are high resistances in the circuit). Always make sure that the test battery is strong. To check the battery, briefly touch the ground clip to the probe; if the light glows brightly the battery is strong enough for testing. Never use a self-powered test light to perform checks for opens or shorts when power is applied to the electrical system under test. The 12 volt vehicle power will quickly burn out the 1.5 volt light bulb in the test light.

Voltmeter

A voltmeter is used to measure voltage at any point in a circuit, or to measure the voltage drop across any part of a circuit. It can also be used to check continuity in a wire or circuit by indicating current flow from one end to the other. Voltmeters usually have various scales on the meter dial and a selector switch to allow the selection of different voltages. The voltmeter has a positive and a negative lead. To avoid damage to the meter, always connect the negative lead to the negative (−) side of circuit (to ground or nearest the ground side of the circuit) and connect the positive lead to the positive (+) side of the circuit (to the power source or the nearest power source). Note that the negative voltmeter lead will always be black and that the positive voltmeter will always be some color other than black (usually red). Depending on how the voltmeter is connected into the circuit, it has several uses.

A voltmeter can be connected either in parallel or in series with a circuit and it has a very high resistance to current flow. When connected in parallel, only a small amount of current will flow through the voltmeter current path; the rest will flow through the normal circuit current path and the circuit will work normally. When the voltmeter is connected in series with a circuit, only a small amount of current can flow through the circuit. The circuit will not work properly, but the voltmeter reading will show if the circuit is complete or not.

Available Voltage Measurement

Set the voltmeter selector switch to the 20V position and connect the meter negative lead to the negative post of the battery. Connect the positive meter lead to the positive post of the battery and turn the ignition switch ON to provide a load. Read the voltage on the meter or digital display. A well charged battery should register over 12 volts. If the meter reads below 11.5 volts, the battery power may be insufficient to operate the electrical system properly. This test determines voltage available from the battery and should be the first step in any electrical trouble diagnosis procedure. Many electrical problems, especially on computer controlled systems, can be caused by a low state of charge in the battery. Excessive corrosion at the battery cable terminals can cause a poor contact that will prevent proper charging and full battery current flow.

Normal battery voltage is 12 volts when fully charged. When the battery is supplying current to one or more circuits it is said to be "under load". When everything is off the electrical system is under a "no-load" condition. A fully charged battery may show about 12.5 volts at no load; will drop to 12 volts under medium load; and will drop even lower under heavy load. If the battery is partially discharged the voltage decrease under heavy load may be excessive, even though the battery shows 12 volts or more at no load. When allowed to discharge further, the battery's available voltage under load will decrease more severely. For this reason, it is important that the battery be fully charged during all testing procedures to avoid errors in diagnosis and incorrect test results.

Voltage Drop

When current flows through a resistance, the voltage beyond the resistance is reduced (the larger the current, the greater the reduction in voltage). When no current is flowing, there is no voltage drop because there is no current flow. All points in the circuit which are connected to the power source are at the same voltage as the power source. The total voltage drop always equals the total source voltage. In a long circuit with many connectors, a series of small, unwanted voltage drops due to corrosion at the connectors can add up to a total loss of voltage which impairs the operation of the normal loads in the circuit.

INDIRECT COMPUTATION OF VOLTAGE DROPS

1. Set the voltmeter selector switch to the 20 volt position.
2. Connect the meter negative lead to a good ground.
3. Probe all resistances in the circuit with the positive meter lead.
4. Operate the circuit in all modes and observe the voltage readings.

DIRECT MEASUREMENT OF VOLTAGE DROPS

1. Set the voltmeter switch to the 20 volt position.
2. Connect the voltmeter negative lead to the ground side of the resistance load to be measured.
3. Connect the positive lead to the positive side of the resistance or load to be measured.
4. Read the voltage drop directly on the 20 volt scale.

Too high a voltage indicates too high a resistance. If, for example, a blower motor runs too slowly, you can determine if there is too high a resistance in the resistor pack. By taking voltage drop readings in all parts of the circuit, you can isolate the problem. Too low a voltage drop indicates too low a resistance. If, for example, a blower motor runs too fast in the MED and/or LOW position, the problem can be isolated in the resistor pack by taking voltage drop readings in all parts of the circuit to locate a possibly shorted resistor. The maximum allowable voltage drop under load is critical, especially if there is more than one high resistance problem in a circuit because all voltage drops are cumulative. A small drop is normal due to the resistance of the conductors.

HIGH RESISTANCE TESTING

1. Set the voltmeter selector switch to the 4 volt position.
2. Connect the voltmeter positive lead to the positive post of the battery.
3. Turn on the headlights and heater blower to provide a load.

4. Probe various points in the circuit with the negative voltmeter lead.

5. Read the voltage drop on the 4 volt scale. Some average maximum allowable voltage drops are:

FUSE PANEL — 7 volts
IGNITION SWITCH — 5volts
HEADLIGHT SWITCH — 7 volts
IGNITION COIL (+) — 5 volts
ANY OTHER LOAD — 1.3 volts

NOTE: Voltage drops are all measured while a load is operating; without current flow, there will be no voltage drop.

Ohmmeter

The ohmmeter is designed to read resistance (ohms) in a circuit or component. Although there are several different styles of ohmmeters, all will usually have a selector switch which permits the measurement of different ranges of resistance (usually the selector switch allows the multiplication of the meter reading by 10, 100, 1000, and 10,000). A calibration knob allows the meter to be set at zero for accurate measurement. Since all ohmmeters are powered by an internal battery (usually 9 volts), the ohmmeter can be used as a self-powered test light. When the ohmmeter is connected, current from the ohmmeter flows through the circuit or component being tested. Since the ohmmeter's internal resistance and voltage are known values, the amount of current flow through the meter depends on the resistance of the circuit or component being tested.

The ohmmeter can be used to perform continuity test for opens or shorts (either by observation of the meter needle or as a self-powered test light), and to read actual resistance in a circuit. It should be noted that the ohmmeter is used to check the resistance of a component or wire while there is no voltage applied to the circuit. Current flow from an outside voltage source (such as the vehicle battery) can damage the ohmmeter, so the circuit or component should be isolated from the vehicle electrical system before any testing is done. Since the ohmmeter uses its own voltage source, either lead can be connected to any test point.

NOTE: When checking diodes or other solid state components, the ohmmeter leads can only be connected one way in order to measure current flow in a single direction. Make sure the positive (+) and negative (−) terminal connections are as described in the test procedures to verify the one-way diode operation.

In using the meter for making continuity checks, do not be concerned with the actual resistance readings. Zero resistance, or any resistance readings, indicate continuity in the circuit. Infinite resistance indicates an open in the circuit. A high resistance reading where there should be none indicates a problem in the circuit. Checks for short circuits are made in the same manner as checks for open circuits except that the circuit must be isolated from both power and normal ground. Infinite resistance indicates no continuity to ground, while zero resistance indicates a dead short to ground.

RESISTANCE MEASUREMENT

The batteries in an ohmmeter will weaken with age and temperature, so the ohmmeter must be calibrated or "zeroed" before taking measurements. To zero the meter, place the selector switch in its lowest range and touch the two ohmmeter leads together. Turn the calibration knob until the meter needle is exactly on zero.

NOTE: All analog (needle) type ohmmeters must be zeroed before use, but some digital ohmmeter models are automatically calibrated when the switch is turned on. Self-calibrating digital ohmmeters do not have an

adjusting knob, but its a good idea to check for a zero readout before use by touching the leads together. All computer controlled systems require the use of a digital ohmmeter with at least 10 meagohms impedance for testing. Before any test procedures are attempted, make sure the ohmmeter used is compatible with the electrical system or damage to the on-board computer could result.

To measure resistance, first isolate the circuit from the vehicle power source by disconnecting the battery cables or the harness connector. Make sure the key is OFF when disconnecting any components or the battery. Where necessary, also isolate at least one side of the circuit to be checked to avoid reading parallel resistances. Parallel circuit resistances will always give a lower reading than the actual resistance of either of the branches. When measuring the resistance of parallel circuits, the total resistance will always be lower than the smallest resistance in the circuit. Connect the meter leads to both sides of the circuit (wire or component) and read the actual measured ohms on the meter scale. Make sure the selector switch is set to the proper ohm scale for the circuit being tested to avoid misreading the ohmmeter test value.

—————— **CAUTION** ——————
Never use an ohmmeter with power applied to the circuit. Like the self-powered test light, the ohmmeter is designed to operate on its own power supply. The normal 12 volt automotive electrical system current could damage the meter.

Ammeters

An ammeter measures the amount of current flowing through a circuit in units called amperes or amps. Amperes are units of electron flow which indicate how fast the electrons are flowing through the circuit. Since Ohms Law dictates that current flow in a circuit is equal to the circuit voltage divided by the total circuit resistance, increasing voltage also increases the current level (amps). Likewise, any decrease in resistance will increase the amount of amps in a circuit. At normal operating voltage, most circuits have a characteristic amount of amperes, called "current draw" which can be measured using an ammeter. By referring to a specified current draw rating, measuring the amperes, and comparing the two values, one can determine what is happening within the circuit to aid in diagnosis. An open circuit, for example, will not allow any current to flow so the ammeter reading will be zero. More current flows through a heavily loaded circuit or when the charging system is operating.

An ammeter is always connected in series with the circuit being tested. All of the current that normally flows through the circuit must also flow through the ammeter; if there is any other path for the current to follow, the ammeter reading will not be accurate. The ammeter itself has very little resistance to current flow and therefore will not affect the circuit, but it will measure current draw only when the circuit is closed and electricity is flowing. Excessive current draw can blow fuses and drain the battery, while a reduced current draw can cause motors to run slowly, lights to dim and other components to not operate properly. The ammeter can help diagnose these conditions by locating the cause of the high or low reading.

Multimeters

Different combinations of test meters can be built into a single unit designed for specific tests. Some of the more common combination test devices are known as Volt/Amp testers, Tach/Dwell meters, or Digital Multimeters. The Volt/Amp tester is used for charging system, starting system or battery tests and consists of a voltmeter, an ammeter and a variable resistance carbon pile. The voltmeter will usually have at least two ranges for use with 6, 12 and 24 volt systems. The ammeter also has more than one range for testing various levels of battery loads and starter current draw and the carbon pile can be adjusted to offer different amounts of resistance. The Volt/Amp tester has heavy leads to carry large amounts of current and many later models have an inductive ammeter pickup that clamps around the wire to simplify test connections. On some models, the ammeter also has a zero-center scale to allow testing of charging and starting systems without switching leads or polarity. A digital multimeter is a voltmeter, ammeter and ohmmeter combined in an instrument which gives a digital readout. These are often used when testing solid state circuits because of their high input impedance (usually 10 megohms or more).

The tach/dwell meter combines a tachometer and a dwell (cam angle) meter and is a specialized kind of voltmeter. The tachometer scale is marked to show engine speed in rpm and the dwell scale is marked to show degrees of distributor shaft rotation. In most electronic ignition systems, dwell is determined by the control unit, but the dwell meter can also be used to check the duty cycle (operation) of some electronic engine control systems. Some tach/dwell meters are powered by an internal battery, while others take their power from the car battery in use. The battery powered testers usually require calibration much like an ohmmeter before testing.

Special Test Equipment

A variety of diagnostic tools are available to help troubleshoot and repair computerized engine control systems. The most sophisticated of these devices are the console type engine analyzers that usually occupy a garage service bay, but there are several types of aftermarket electronic testers available that will allow quick circuit tests of the engine control system by plugging directly into a special connector located in the engine compartment or under the dashboard. Several tool and equipment manufacturers offer simple, hand held testers that measure various circuit voltage levels on command to check all system components for proper operation. Although these testers usually cost about $300-$500, consider that the average computer control unit (or ECM) can cost just as much and the money saved by not replacing perfectly good sensors or components in an attempt to correct a problem could justify the purchase price of a special diagnostic tester the first time it's used.

These computerized testers can allow quick and easy test measurements while the engine is operating or while the car is being driven. In addition, the on-board computer memory can be read to access any stored trouble codes; in effect allowing the computer to tell you where it hurts and aid trouble diagnosis by pinpointing exactly which circuit or component is malfunctioning. In the same manner, repairs can be tested to make sure the problem has been corrected. The biggest advantage these special testers have is their relatively easy hookups that minimize or eliminate the chances of making the wrong connections and getting false voltage readings or damaging the computer accidentally.

NOTE: It should be remembered that these testers check voltage levels in circuits; they don't detect mechanical problems or failed components if the circuit voltage falls within the preprogrammed limits stored in the tester PROM unit. Also, most of the hand held testes are designed to work only on one or two systems made by a specific manufacturer.

A variety of aftermarket testers are available to help diagnose different computerized control systems. Owatonna Tool Company (OTC), for example, markets a device called the OTC Monitor which plugs directly into the assembly line diagnostic link (ALDL). The OTC tester makes diagnosis a simple matter of pressing the correct buttons and, by changing the internal PROM or inserting a different diagnosis cartridge, it will work on any model from full size to subcompact, over a wide range of years. An adapter is supplied with the tester to allow connection to all types of ALDL links, regardless of the number of pin ter-

minals used. By inserting an updated PROM into the OTC tester, it can be easily updated to diagnose any new modifications of computerized control systems.

Wiring Harnesses

The average automobile contains about ½ mile of wiring, with hundreds of individual connections. To protect the many wires from damage and to keep them from becoming a confusing tangle, they are organized into bundles, enclosed in plastic or taped together and called wire harnesses. Different wiring harnesses serve different parts of the vehicle. Individual wires are color coded to help trace them through a harness where sections are hidden from view.

A loose or corroded connection or a replacement wire that is too small for the circuit will add extra resistance and an additional voltage drop to the circuit. A ten percent voltage drop can result in slow or erratic motor operation, for example, even though the circuit is complete. Automotive wiring or circuit conductors can be in any one of three forms:

1. Single strand wire
2. Multistrand wire
3. Printed circuitry

Single strand wire has a solid metal core and is usually used inside such components as alternators, motors, relays and other devices. Multistrand wire has a core made of many small strands of wire twisted together into a single conductor. Most of the wiring in an automotive electrical system is made up of multistrand wire, either as a single conductor or grouped together in a harness. All wiring is color coded on the insulator, either as a solid color or as a colored wire with an identification stripe. A printed circuit is a thin film of copper or other conductor that is printed on an insulator backing. Occasionally, a printed circuit is sandwiched between two sheets of plastic for more protection and flexibility. A complete printed circuit, consisting of conductors, insulating material and connectors for lamps or other components is called a printed circuit board. Printed circuitry is used in place of individual wires or harnesses in places where space is limited, such as behind instrument panels.

Wire Gauge

Since computer controlled automotive electrical systems are very sensitive to changes in resistance, the selection of properly sized wires is critical when systems are repaired. The wire gauge number is an expression of the cross section area of the conductor. The most common system for expressing wire size is the American Wire Gauge (AWG) system.

Wire cross section area is measured in circular mils. A mil is $\frac{1}{1000}$ in. (0.001 in.); a circular mil is the area of a circle one mil in diameter. For example, a conductor ¼ in. in diameter is 0.250 in. or 250 mils. The circular mil cross section area of the wire is 250 squared (250^2) or 62,500 circular mils. Imported car models usually use metric wire gauge designations, which is simply the cross section area of the conductor in square millimeters (mm^2).

Gauge numbers are assigned to conductors of various cross section areas. As gauge number increases, area decreases and the conductor becomes smaller. A 5 gauge conductor is smaller than a 1 gauge conductor and a 10 gauge is smaller than a 5 gauge. As the cross section area of a conductor decreases, resistance increases and so does the gauge number. A conductor with a higher gauge number will carry less current than a conductor with a lower gauge number.

NOTE: Gauge wire size refers to the size of the conductor, not the size of the complete wire. It is possible to have two wires of the same gauge with different diameters because one may have thicker insulation than the other.

12 volt automotive electrical systems generally use 10, 12, 14, 16 and 18 gauge wire. Main power distribution circuits and larg-

er accessories usually use 10 and 12 gauge wire. Battery cables are usually 4 or 6 gauge, although 1 and 2 gauge wires are occasionally used. Wire length must also be considered when making repairs to a circuit. As conductor length increases, so does resistance. An 18 gauge wire, for example, can carry a 10 amp load for 10 feet without excessive voltage drop; however if a 15 foot wire is required for the same 10 amp load, it must be a 16 gauge wire.

An electrical schematic shows the electrical current paths when a circuit is operating properly. It is essential to understand how a circuit works before trying to figure out why it doesn't. Schematics break the entire electrical system down into individual circuits and show only one particular circuit. In a schematic, no attempt is made to represent wiring and components as they physically appear on the vehicle; switches and other components are shown as simply as possible. Face views of harness connectors show the cavity or terminal locations in all multi-pin connectors to help locate test points.

If you need to backprobe a connector while it is on the component, the order of the terminals must be mentally reversed. The wire color code can help in this situation, as well as a keyway, lock tab or other reference mark.

NOTE: Wiring diagrams are not included in this book. As trucks have become more complex and available with longer option lists, wiring diagrams have grown in size and complexity. It has become almost impossible to provide a readable reproduction of a wiring diagram in a book this size. Information on ordering wiring diagrams from the vehicle manufacturer can be found in the owner's manual.

WIRING REPAIR

Soldering is a quick, efficient method of joining metals permanently. Everyone who has the occasion to make wiring repairs should know how to solder. Electrical connections that are soldered are far less likely to come apart and will conduct electricity much better than connections that are only "pig-tailed" together. The most popular (and preferred) method of soldering is with an electrical soldering gun. Soldering irons are available in many sizes and wattage ratings. Irons with higher wattage ratings deliver higher temperatures and recover lost heat faster. A small soldering iron rated for no more than 50 watts is recommended, especially on electrical systems where excess heat can damage the components being soldered.

There are three ingredients necessary for successful soldering; proper flux, good solder and sufficient heat. A soldering flux is necessary to clean the metal of tarnish, prepare it for soldering and to enable the solder to spread into tiny crevices. When soldering, always use a resin flux or resin core solder which is non-corrosive and will not attract moisture once the job is finished. Other types of flux (acid core) will leave a residue that will attract moisture and cause the wires to corrode. Tin is a unique metal with a low melting point. In a molten state, it dissolves and alloys easily with many metals. Solder is made by mixing tin with lead. The most common proportions are 40/60, 50/50 and 60/40, with the percentage of tin listed first. Low priced solders usually contain less tin, making them very difficult for a beginner to use because more heat is required to melt the solder. A common solder is 40/60 which is well suited for all-around general use, but 60/40 melts easier, has more tin for a better joint and is preferred for electrical work.

Soldering Techniques

Successful soldering requires that the metals to be joined be heated to a temperature that will melt the solder—usually 360-460°F (182-238°C). Contrary to popular belief, the purpose of the soldering iron is not to melt the solder itself, but to heat the parts being soldered to a temperature high enough to melt the solder when it is touched to the work. Melting flux-cored solder on the soldering iron will usually destroy the effectiveness of the flux.

NOTE: Soldering tips are made of copper for good heat conductivity, but must be "tinned" regularly for quick transference of heat to the project and to prevent the solder from sticking to the iron. To "tin" the iron, simply heat it and touch the flux-cored solder to the tip; the solder will flow over the hot tip. Wipe the excess off with a clean rag, but be careful as the iron will be hot.

After some use, the tip may become pitted. If so, simply dress the tip smooth with a smooth file and "tin" the tip again. An old saying holds that "metals well cleaned are half soldered." Flux-cored solder will remove oxides but rust, bits of insulation and oil or grease must be removed with a wire brush or emery cloth. For maximum strength in soldered parts, the joint must start off clean and tight. Weak joints will result in gaps too wide for the solder to bridge.

If a separate soldering flux is used, it should be brushed or swabbed on only those areas that are to be soldered. Most solders contain a core of flux and separate fluxing is unnecessary. Hold the work to be soldered firmly. It is best to solder on a wooden board, because a metal vise will only rob the piece to be soldered of heat and make it difficult to melt the solder. Hold the soldering tip with the broadest face against the work to be soldered. Apply solder under the tip close to the work, using enough solder to give a heavy film between the iron and the piece being soldered, while moving slowly and making sure the solder melts properly. Keep the work level or the solder will run to the lowest part and favor the thicker parts, because these require more heat to melt the solder. If the soldering tip overheats (the solder coating on the face of the tip burns up), it should be retinned. Once the soldering is completed, let the soldered joint stand until cool. Tape and seal all soldered wire splices after the repair has cooled.

Wire Harness and Connectors

The on-board computer (ECM) wire harness electrically connects the control unit to the various solenoids, switches and sensors used by the control system. Most connectors in the engine compartment or otherwise exposed to the elements are protected against moisture and dirt which could create oxidation and deposits on the terminals. This protection is important because of the very low voltage and current levels used by the computer and sensors. All connectors have a lock which secures the male and female terminals together, with a secondary lock holding the seal and terminal into the connector. Both terminal locks must be released when disconnecting ECM connectors.

These special connectors are weather-proof and all repairs require the use of a special terminal and the tool required to service it. This tool is used to remove the pin and sleeve terminals. If removal is attempted with an ordinary pick, there is a good chance that the terminal will be bent or deformed. Unlike standard blade type terminals, these terminals cannot be straightened once they are bent. Make certain that the connectors are properly seated and all of the sealing rings in place when connecting leads. On some models, a hinge-type flap provides a backup or secondary locking feature for the terminals. Most secondary locks are used to improve the connector reliability by retaining the terminals if the small terminal lock tangs are not positioned properly.

Molded-on connectors require complete replacement of the connection. This means splicing a new connector assembly into the harness. All splices in on-board computer systems should be soldered to insure proper contact. Use care when probing the connections or replacing terminals in them as it is possible to short between opposite terminals. If this happens to the wrong terminal pair, it is possible to damage certain components. Always use jumper wires between connectors for circuit checking and never probe through weatherproof seals.

Open circuits are often difficult to locate by sight because corrosion or terminal misalignment are hidden by the connectors. Merely wiggling a connector on a sensor or in the wiring harness may correct the open circuit condition. This should always be considered when an open circuit or a failed sensor is indicated. Intermittent problems may also be caused by oxidized or loose connections. When using a circuit tester for diagnosis, always probe connections from the wire side. Be careful not to damage sealed connectors with test probes.

All wiring harnesses should be replaced with identical parts, using the same gauge wire and connectors. When signal wires are spliced into a harness, use wire with high temperature insulation only. With the low voltage and current levels found in the system, it is important that the best possible connection at all wire splices be made by soldering the splices together. It is seldom necessary to replace a complete harness. If replacement is necessary, pay close attention to insure proper harness routing. Secure the harness with suitable plastic wire clamps to prevent vibrations from causing the harness to wear in spots or contact any hot components.

NOTE: Weatherproof connectors cannot be replaced with standard connectors. Instructions are provided with replacement connector and terminal packages. Some wire harnesses have mounting indicators (usually pieces of colored tape) to mark where the harness is to be secured.

In making wiring repairs, it's important that you always replace damaged wires with wires that are the same gauge as the wire being replaced. The heavier the wire, the smaller the gauge number. Wires are color-coded to aid in identification and whenever possible the same color coded wire should be used for replacement. A wire stripping and crimping tool is necessary to install solderless terminal connectors. Test all crimps by pulling on the wires; it should not be possible to pull the wires out of a good crimp.

Wires which are open, exposed or otherwise damaged are repaired by simple splicing. Where possible, if the wiring harness is accessible and the damaged place in the wire can be located, it is best to open the harness and check for all possible damage. In an inaccessible harness, the wire must be bypassed with a new insert, usually taped to the outside of the old harness.

When replacing fusible links, be sure to use fusible link wire, NOT ordinary automotive wire. Make sure the fusible segment is of the same gauge and construction as the one being replaced and double the stripped end when crimping the terminal connector for a good contact. The melted (open) fusible link segment of the wiring harness should be cut off as close to the harness as possible, then a new segment spliced in as described. In the case of a damaged fusible link that feeds two harness wires, the harness connections should be replaced with two fusible link wires so that each circuit will have its own separate protection.

NOTE: Most of the problems caused in the wiring harness are due to bad ground connections. Always check all vehicle ground connections for corrosion or looseness before performing any power feed checks to eliminate the chance of a bad ground affecting the circuit.

Repairing Hard Shell Connectors

Unlike molded connectors, the terminal contacts in hard shell connectors can be replaced. Weatherproof hard-shell connectors with the leads molded into the shell have non-replaceable terminal ends. Replacement usually involves the use of a special terminal removal tool that depress the locking tangs (barbs) on the connector terminal and allow the connector to be removed from the rear of the shell. The connector shell should be replaced if it shows any evidence of burning, melting, cracks, or breaks. Replace individual terminals that are burnt, corroded, distorted or loose.

NOTE: The insulation crimp must be tight to prevent the insulation from sliding back on the wire when the wire is pulled. The insulation must be visibly compressed under the crimp tabs, and the ends of the crimp should be turned in for a firm grip on the insulation.

The wire crimp must be made with all wire strands inside the crimp. The terminal must be fully compressed on the wire strands with the ends of the crimp tabs turned in to make a firm grip on the wire. Check all connections with an ohmmeter to insure a good contact. There should be no measurable resistance between the wire and the terminal when connected.

Mechanical Test Equipment

Vacuum Gauge

Most gauges are graduated in inches of mercury (in.Hg), although a device called a manometer reads vacuum in inches of water (in. H_2O). The normal vacuum reading usually varies between 18 and 22 in.Hg at sea level. To test engine vacuum, the vacuum gauge must be connected to a source of manifold vacuum. Many engines have a plug in the intake manifold which can be removed and replaced with an adapter fitting. Connect the vacuum gauge to the fitting with a suitable rubber hose or, if no manifold plug is available, connect the vacuum gauge to any device using manifold vacuum, such as EGR valves, etc. The vacuum gauge can be used to determine if enough vacuum is reaching a component to allow its actuation.

Hand Vacuum Pump

Small, hand-held vacuum pumps come in a variety of designs. Most have a built-in vacuum gauge and allow the component to be tested without removing it from the vehicle. Operate the pump lever or plunger to apply the correct amount of vacuum required for the test specified in the diagnosis routines. The level of vacuum in inches of Mercury (in.Hg) is indicated on the pump gauge. For some testing, an additional vacuum gauge may be necessary.

Intake manifold vacuum is used to operate various systems and devices on late model vehicles. To correctly diagnose and solve problems in vacuum control systems, a vacuum source is necessary for testing. In some cases, vacuum can be taken from the intake manifold when the engine is running, but vacuum is normally provided by a hand vacuum pump. These hand vacuum pumps have a built-in vacuum gauge that allow testing while the device is still attached to the component. For some tests, an additional vacuum gauge may be necessary.

HEATING AND AIR CONDITIONING

Blower Motor

REMOVAL AND INSTALLATION

1971-77

1. Disconnect the battery ground cable. Detach any interfering control cables.
2. Disconnect the electrical connections:
   a. Heater switch
   b. Ground wire
   c. Battery connector
3. Remove the screws that hold the motor to the heater assembly and remove the blower motor housing and motor.
4. Remove fan and blower motor from blower motor housing.

1978-90

WITHOUT AIR CONDITIONING

The heater housing assembly has to be removed to get out the blower motor.

1. Drain about two quarts of coolant.

CAUTION

When draining the coolant, keep in mind that cats and dogs are attracted by the ethylene glycol antifreeze, and are quite likely to drink any that is left in an uncovered container or in puddles on the ground. This will prove fatal in sufficient quantity. Always drain the coolant into a sealable container. Coolant should be reused unless it is contaminated or several years old.

2. Disconnect the heater hoses at the engine side of the firewall.
3. Detach the heater control cables.
4. Disconnect the motor wiring.
5. Detach the water drain hose and the defroster hose.
6. Remove the nuts from the studs in the engine compartment.
7. Tilt the heater housing assembly down and pull it back toward the inside of the vehicle.
8. Remove the attaching screws and the blower motor.
9. On installation, make sure that the seals around the core tubes and blower motor are in place.

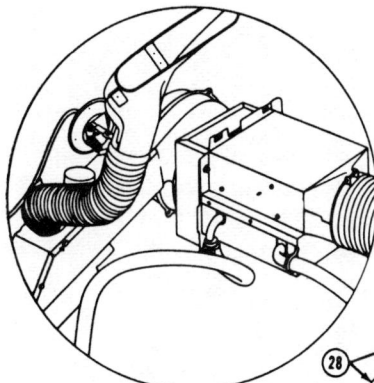

VIEW INDICATED BY ARROW

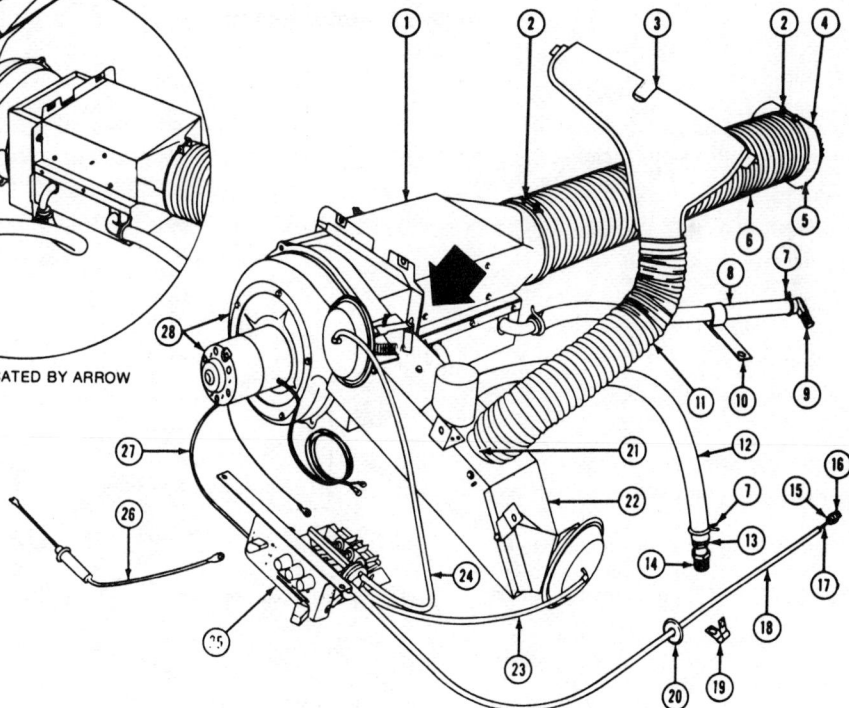

1. Heater assembly	11. Defroster hose	21. Defroster bushing
2. Hose clamp	12. Hot water hose	22. Heat distributor assembly
3. Defroster nozzle	13. Heater nipple	23. Heater control tube
4. Air duct screen	14. Reducing bushing	24. Heater control tube
5. Air duct and heater collar	15. Inverted flared tube nut	25. Heater control assembly
6. Air duct intake tube	16. Inverted flared tube connector	26. Fuse holder assembly
7. Hose clamp	17. Heater vacuum to engine tube	27. Bowden wire (control panel to heater)
8. Straight hot water hose	18. Heater control tube	
9. Heater tube elbow	19. Clip	28. Blower and air inlet assembly
10. Heater hose support bracket	20. Grommet	

1971 heater assembly

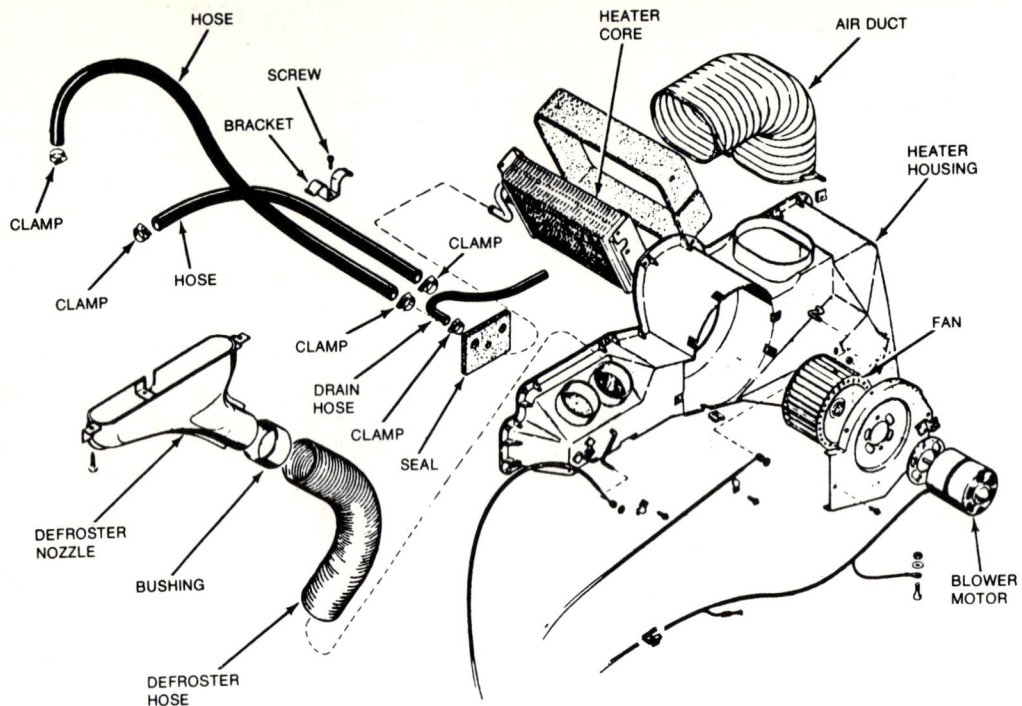

1972–77 heater assembly

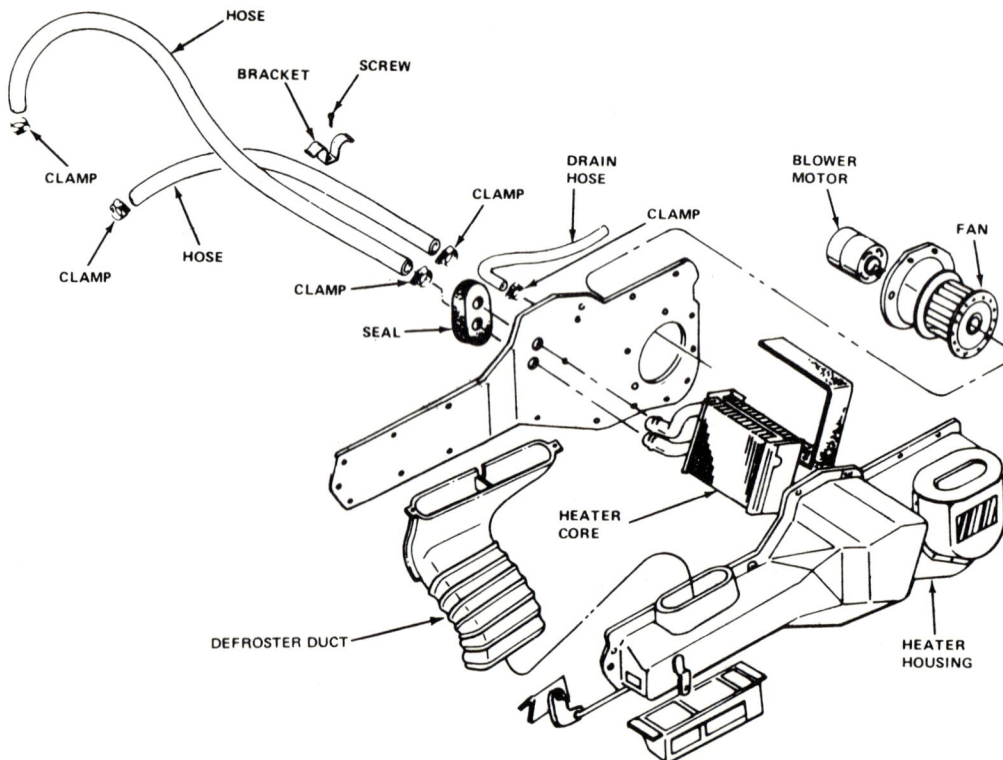

1979–83 heater/defroster components

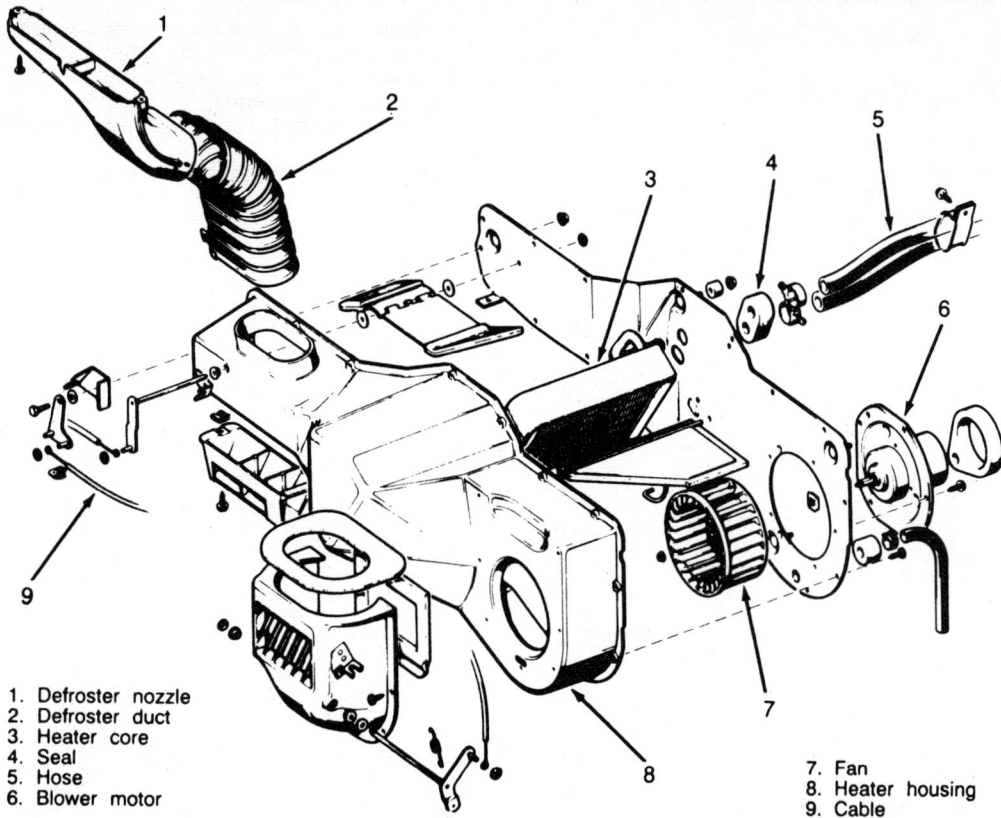

1984–90 heater assembly

1. Defroster nozzle
2. Defroster duct
3. Heater core
4. Seal
5. Hose
6. Blower motor

7. Fan
8. Heater housing
9. Cable

WITH AIR CONDITIONING

NOTE: It is not necessary to discharge the refrigerant system.

1. Remove the hose clamps and dash grommet retaining screws.
2. Remove the evaporator housing-to-instrument panel retaining screws and the housing mounting bracket screw.
3. Remove the blower mounting screws and remove the blower. Disconnect the wiring.
4. Installation is the reverse of removal.

Heater Core

REMOVAL AND INSTALLATION

1971-74

1. Drain the cooling system.

─────────── CAUTION ───────────

When draining the coolant, keep in mind that cats and dogs are attracted by the ethylene glycol antifreeze, and are quite likely to drink any that is left in an uncovered container or in puddles on the ground. This will prove fatal in sufficient quantity. Always drain the coolant into a sealable container. Coolant should be reused unless it is contaminated or several years old.

2. Mark the duct halves to be sure they are reassembled properly.
3. Remove the screws that fasten the two halves of the duct together.
4. Remove the screws that secure the heater core to the duct.
5. Remove the heater core from the vehicle.

6. Install in reverse order of the above procedure.

1975-76

1. Drain about two quarts of coolant from the radiator.

─────────── CAUTION ───────────

When draining the coolant, keep in mind that cats and dogs are attracted by the ethylene glycol antifreeze, and are quite likely to drink any that is left in an uncovered container or in puddles on the ground. This will prove fatal in sufficient quantity. Always drain the coolant into a sealable container. Coolant should be reused unless it is contaminated or several years old.

2. Disconnect the battery cables, remove the battery and battery box.
3. Disconnect the heater hoses.
4. Disconnect the damper door control cables.
5. Disconnect the blower motor wiring harness at the switch and ground wire at the instrument panel.
6. Remove the glove box.
7. Disconnect the water drain hose and defroster hose.
8. Disconnect the heater-to-air deflector duct at the heater housing.
9. Remove the nuts from the heater housing studs in the engine compartment and remove the heater housing assembly.
10. Remove the heater core from the heater housing.
11. Install the heater core in the reverse order of removal, refill the radiator, run the engine and check for leaks.

1978-90

The heater housing assembly has to be removed to get out the heater core. The procedure is the same as for blower motor removal and installation.

Evaporator

REMOVAL AND INSTALLATION

1. Discharge the system. See Section 1.
2. Disconnect the inlet (suction) and outlet hoses at the evaporator. Cap the openings at once.
3. Remove the hose clamps and dash grommet screws.
4. Remove the evaporator housing-to-dash panel screws and the housing mounting bracket screw.
5. Lower the housing and pull the hoses and hose grommets through the dash opening.
6. The blower motor and evaporator core can now be removed for service.
7. After installing all the parts, evacuate, charge and leak test the system.

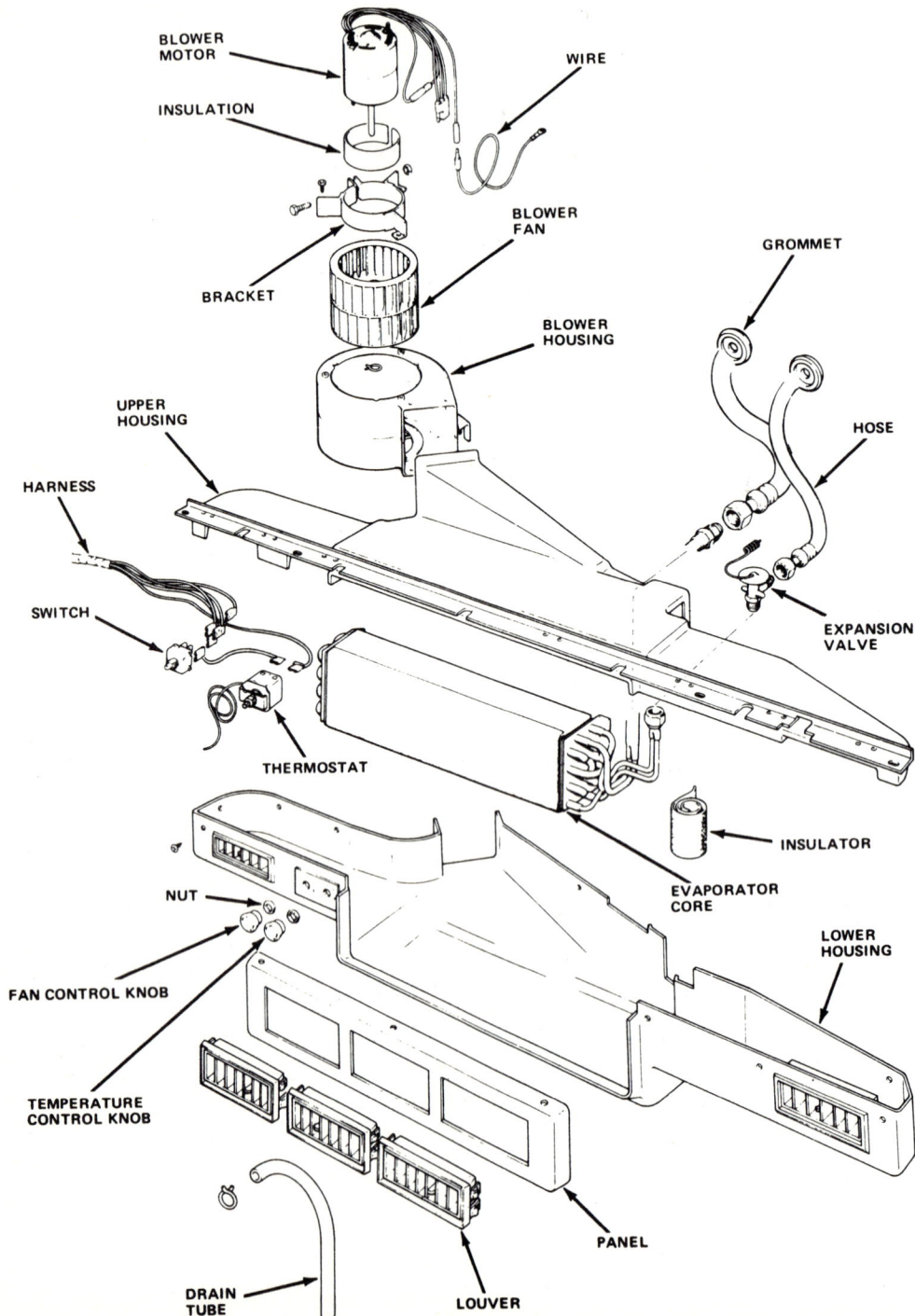

Evaporator assembly used on all models with factory installed air conditioning

5. Bezel screws
6. Panel

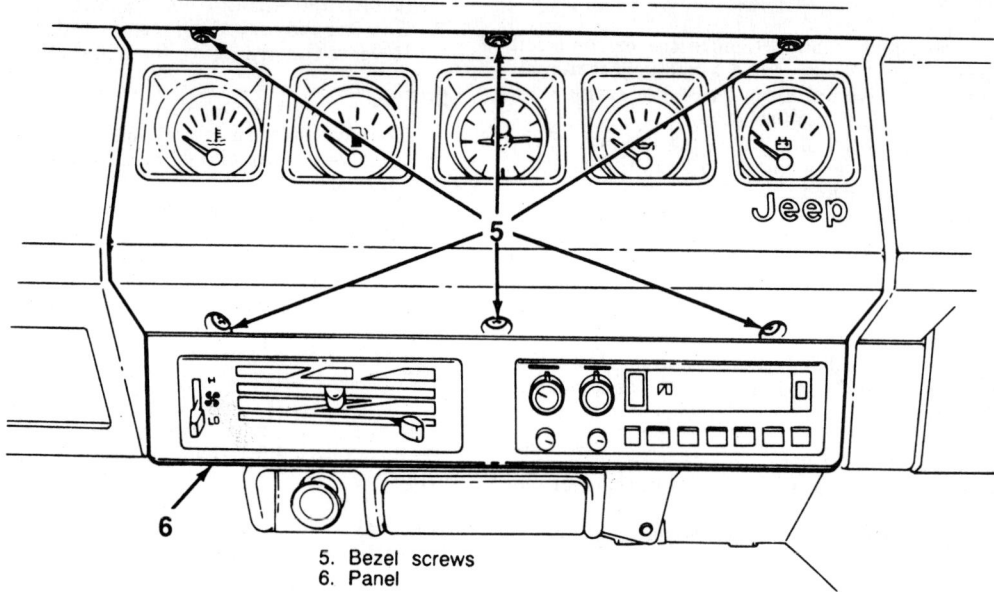

1987–90 heater/air conditioning control panel

Control Panel

REMOVAL AND INSTALLATION

NOTE: Through 1986, there was no control panel as such. The fan switch was removable. The following applies, therefore, to 1987-90 models only.

1. Remove the instrument cluster bezel screws.
2. Remove the bezel.
3. Remove the control panel attaching screws.
4. Pull the panel toward you and disconnect the cables, hoses and wires.
5. Installation is the reverse of removal.

WINDSHIELD WIPERS

Wiper Blades and Arms

REMOVAL AND INSTALLATION

1. Pull the blade away from the windshield.
2. Push against the tip of the wiper arm to compress the locking spring and disengage the retaining pin.
3. Pivot the blade clockwise to unhook it from the arm.
4. To install the blade, just snap it into position.
5. To remove the arm, simply pry it straight off carefully. When you reinstall it, make sure that it doesn't hit the rubber molding at either edge of the windshield while running.

Motor

REMOVAL AND INSTALLATION

1971

1. Remove the windshield wiper assembly from the pivot shaft.
2. Remove the vacuum hose or wire from the motor.
3. Remove all attaching screws that hold the motor to the windshield assembly and remove the motor from the vehicle.
4. Install in the reverse order.

1972-75

1. Remove the crash pad, if any. Remove the extreme left plastic hole plug from the bottom of the windshield frame air duct and disconnect the drive link from the motor crank.
2. Loosen the wiper control knob setscrew.
3. Remove the control switch and mark the location of the wires on the switch prior to removing them from the switch.
4. Remove the motor cover and the motor.
5. Install in the reverse order of the above procedure.

NOTE: The motor cover must be sealed when installing.

1976-86

1. If your Jeep has crash padding, you have to fold the windshield down for access. Even if you don't have the padding, you can't get the wires out to remove the motor from the vehicle, unless the windshield is down.
2. Remove the wiper motor cover.
3. Remove the left access plug from the bottom of the windshield.
4. Disconnect the drive link from the left wiper pivot by sliding the clip off.

5. Detach the wiring from the switch.
6. Remove the mounting screws and the wiper motor.
7. Reverse the procedure for installation.

1987-90

1. Remove the hard or soft top at the windshield frame.
2. Remove the windshield holddown bolts and fold the windshield forward.
3. Remove the wiper motor mounting screws.
4. Remove the harness clips.
5. Disconnect the drive link from the left wiper pivot by prying the clip off.
6. Grasp the motor and pull the drive arm out of the access hole.
7. Pry the drive arm off of the motor pivot. DO NOT REMOVE THE PIVOT ATTACHING NUT!
8. Remove the 2 screws holding the intermittant wiper module bracket to the bottom of the instrument panel.
9. Reach up behind the instrument panel and disconnect the wiring harness.
10. Remove the motor.

To install:

11. Install and connect the wiring harness.

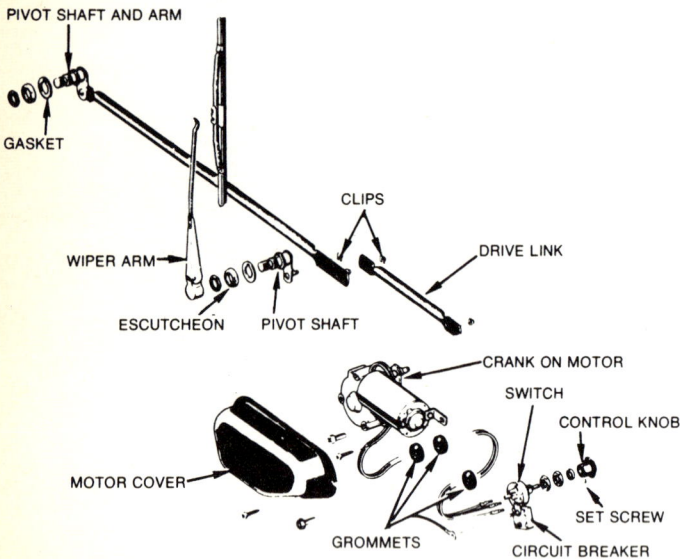

1972–75 wiper assembly

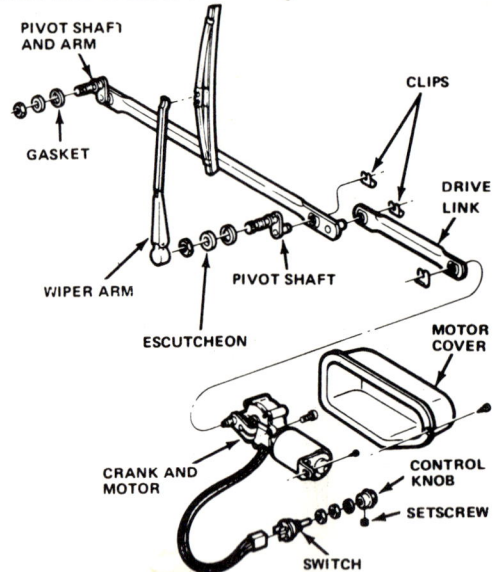

1976–78 windshield wiper components

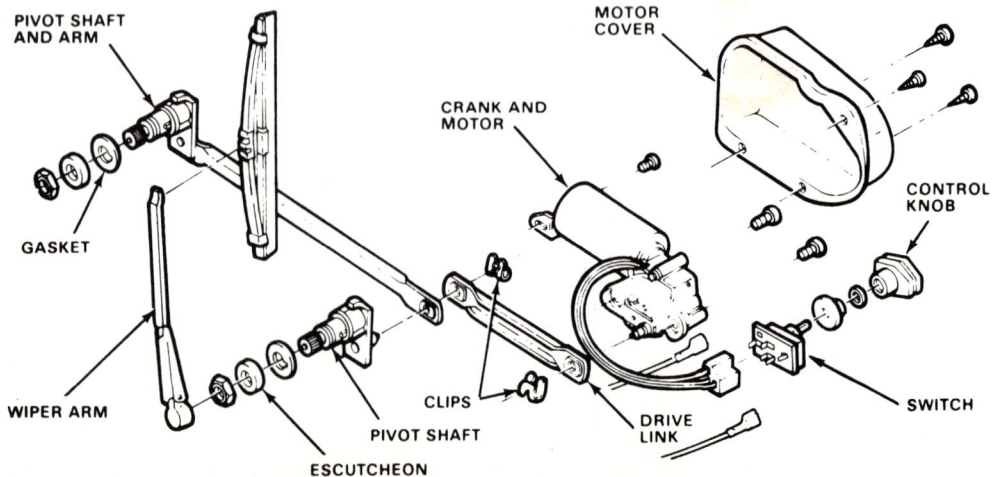

1979 and later windshield wiper components

12. Remove the 2 screws holding the intermittant wiper module bracket to the bottom of the instrument panel.
13. Turn the wipers on to make sure the motor cycles to the PARK position.
14. Install the drive arm on the motor pivot.
15. Install the motor.
16. Connect the drive link at the left wiper pivot.
17. Install the harness clips.
18. Install the wiper motor mounting screws. Torque the screws to 96 inch lbs.
19. Install the windshield.
20. Install the hard or soft top at the windshield frame.

Linkage
REMOVAL AND INSTALLATION

1971
1971 models have no windshield wiper linkage.

1972-75
1. Remove the wiper arms and pivot shaft nuts, washers, escutcheons and gaskets.
2. Disconnect the drive arm from the motor crank.
3. Remove the individual links where necessary, to remove the pivot shaft bodies without excessive interference.
4. Reverse the procedure for installation.

1976-86
1. Remove the wiper arms.
2. Remove the nuts attaching the pivots to the windshield frame.
3. Remove the necessary components from the top of the windshield frame.
4. Remove the windshield holddown knobs and fold the windshield forward.

5. Remove the access hole covers on both sides of the windshield.
6. Disconnect the wiper motor drive link from the left wiper pivot.
7. Remove the wiper pivot shafts and linkage from the access hole.
8. Install the linkage in the reverse order.

1987-90
1. Remove the wiper arms.
2. Remove the nuts attaching the pivots to the windshield frame.
3. Remove the hardtop or soft top components from the top of the windshield frame.
4. Remove the windshield holddown bolts and fold the windshield forward.
5. Remove the motor attaching screws.
6. Disconnect the wiper motor drive link from the left wiper pivot.
7. Grasp the motor and pull the motor and drive arm from the access hole.
8. Remove the pivot shaft assembly through the access hole.
9. Pry the drive arm off of the motor pivot. DO NOT REMOVE THE PIVOT ATTACHING NUT!
To install:
10. Install the drive arm on the motor pivot.
11. Install the pivot shaft assembly through the access hole.
12. Install the motor and drive arm through the access hole. Torque the motor attaching screws to 96 inch lbs.
13. Disconnect the wiper motor drive link from the left wiper pivot.
14. Raise the windshield and install the holddown bolts.
15. Install the hardtop or soft top components at the top of the windshield frame.
16. Turn the wiper motor on and allow it to cycle to the PARK position.
17. Install the wiper arms.

INSTRUMENT AND SWITCHES

Instrument Cluster
REMOVAL AND INSTALLATION

1971-75
1. Disconnect one battery cable.
2. Separate the speedometer cable from the speedometer head.
3. Remove the screws that hold up the heater control bracket (1972 and later only).
4. Remove the attaching nuts that hold the cluster to the dash.

5. Remove the gauge wires and remove the cluster assembly.
6. Install in the reverse order. After installing the cluster, connect the battery and check all of the lights and gauges for proper operation.

1976-86
1. Disconnect the negative battery cable.
2. Disconnect the speedometer cable from the back of the speedometer.
3. Remove the instrument cluster attaching nuts and remove the cluster.
4. Disconnect the instrument cluster electrical connectors and remove the cluster from the vehicle.
5. Install in the reverse order.

1987-90

GAUGE CLUSTER

1. Disconnect the battery ground.
2. Remove the six bezel screws.
3. Remove the six cluster attaching screws.
4. Pull the cluster, gently, towards you and disconnect the wiring connector.
5. Lift out the cluster.
6. Release the seven locking tabs and remove the plastic lens shield.
7. Remove the metal shield.
8. Remove the gauge nuts to free the gauges.
9. When installing the gauges, the gauge contact pins will be driven into place as you tighten the gauge nuts, so be careful to align them properly. When installing the cluster, be careful to avoid overtorquing the fasteners!

SPEEDOMETER

1. Remove the five shroud attaching screws.
2. Exert downward pressure on the top of the shroud and up- ward pressure on the bottom of the shroud while pulling the shroud towards you. This will release the retaining tabs.
3. Remove the two speedometer attaching screws and pull the speedometer towards you. Disconnect the cable.
4. Installation is the reverse of removal.

TACHOMETER

1. Remove the five shroud attaching screws.
2. Exert downward pressure on the top of the shroud and up- ward pressure on the bottom of the shroud while pulling the shroud towards you. This will release the retaining tabs.
3. Remove the two tachometer attaching screws and pull the tachometer towards you. Disconnect the wiring.
4. Installation is the reverse of removal.

Headlight Switch

REMOVAL AND INSTALLATION

NOTE: For column-mounted switches, see Section 8.

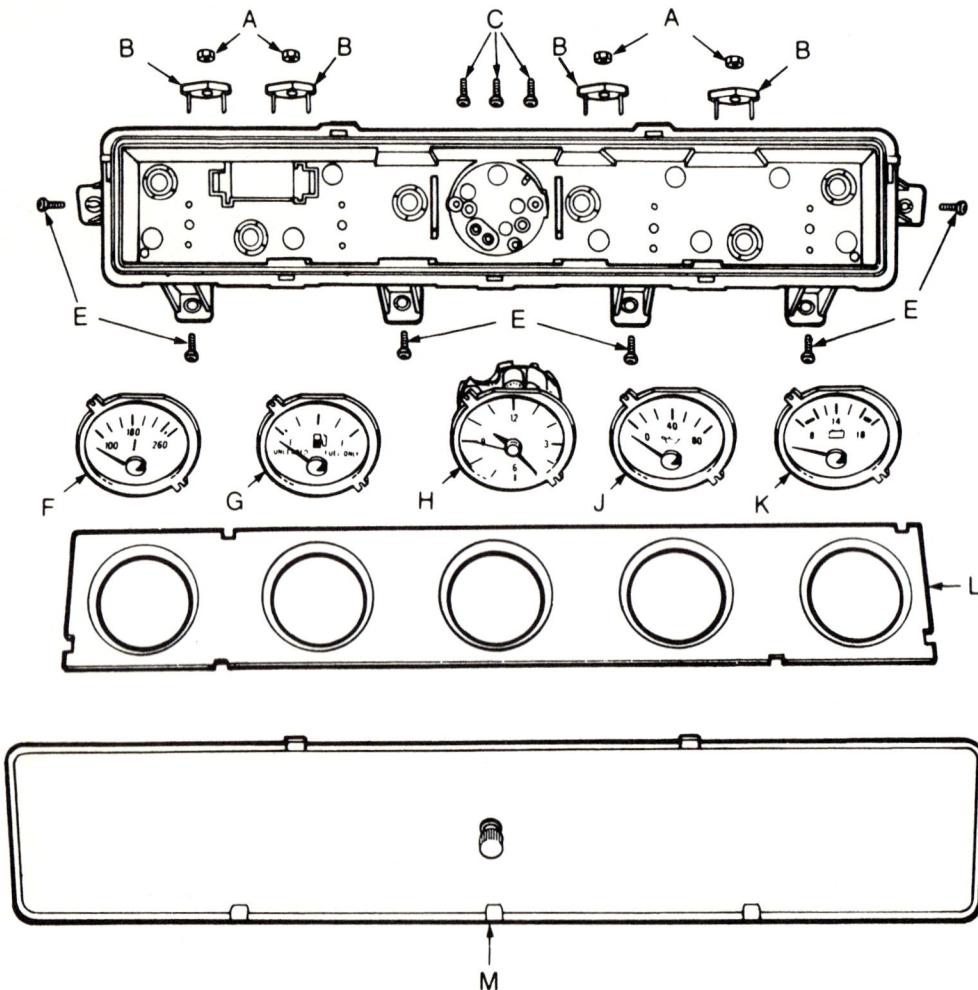

A. Gauge nuts(s)	G. Fuel gauge
B. Gauge contact pins	H. Clock
C. Clock screw(s)	J. Oil pressure
D. Contact spring(s)	K. Voltmeter
E. Gauge housing mounting screws	L. Metal shield
F. Coolant temperature gauge	M. Plastic shield

1987–90 gauge cluster

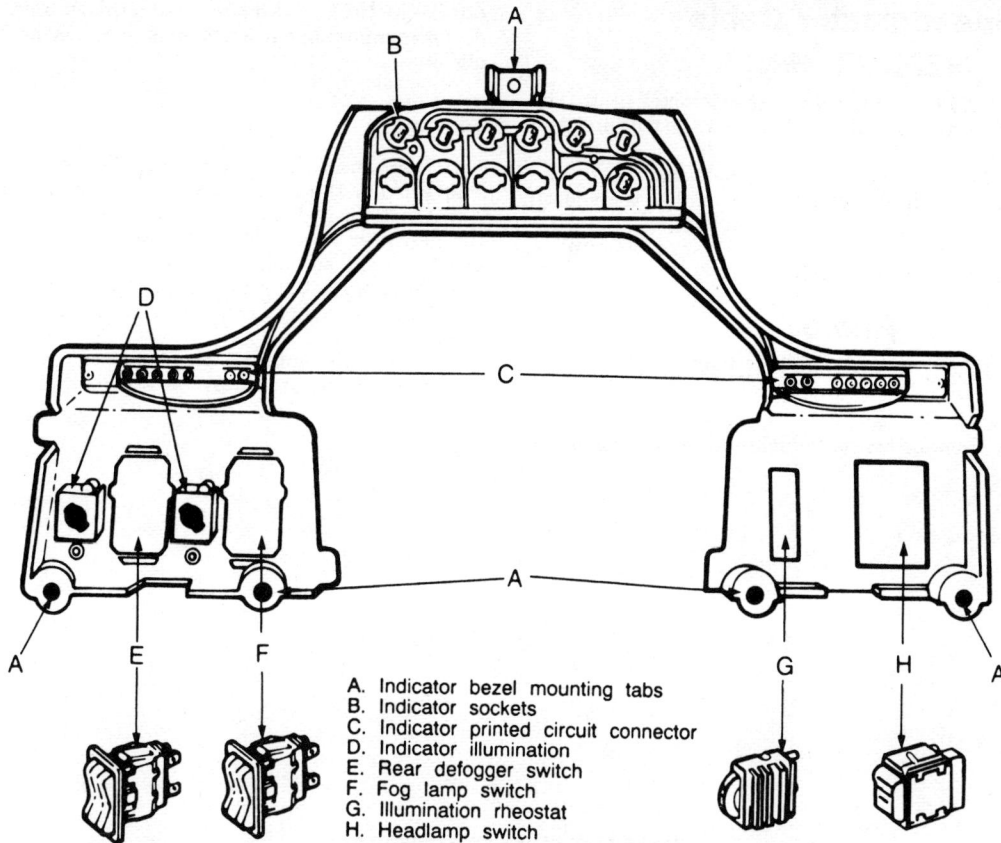

A. Indicator bezel mounting tabs
B. Indicator sockets
C. Indicator printed circuit connector
D. Indicator illumination
E. Rear defogger switch
F. Fog lamp switch
G. Illumination rheostat
H. Headlamp switch

1987–90 indicator bezel

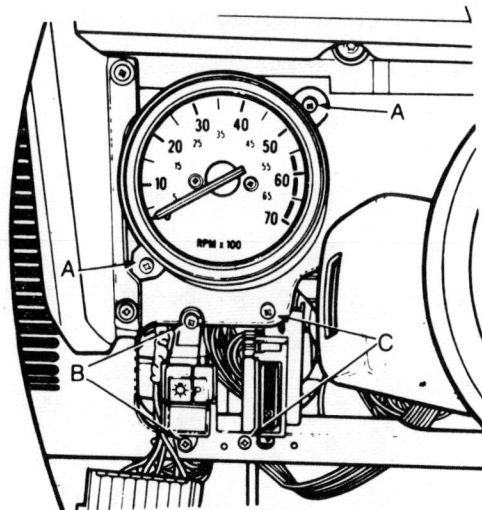

1987–90 tachometer and headlight switch removal. A, B and C are attaching screws

Early Dash-Mounted Switches

1. Loosen the set-screw and pull the knob from the switch.
2. Unscrew the switch retaining nut and pull the switch from the dash.
3. Disconnect the wiring.
4. Installation is the reverse of removal.

1987-90

1. Disconnect the battery ground.
2. Remove the 5 instrument shroud screws and slide the shroud towards the steering wheel.
3. Apply upward pressure on the shroud and downward pressure on the indicator panel. This will release the holding tabs.
4. Remove the shroud.
5. Remove the 2 switch attaching screws, pull out the switch and unplug the connector.
6. Installation is the reverse of removal.

Ignition Switch

REMOVAL AND INSTALLATION

NOTE: For column-mounted switches, see Section 8.

Dash-Mounted Switches

1. Remove the switch retaining bezel.
2. Pull the switch out of the dash.
3. Disconnect the wiring.
4. To remove the lock cylinder:
   a. Insert the key into the lock and turn it to the left.
   b. Unbend a paper clip and insert the end into the lock release hole in the switch body. Press in on the clip to compress the lock cylinder retainer and pull the lock cylinder from the switch.
5. Installation is the reverse of removal.

Speedometer Cable
REPLACEMENT

1. Reach up behind the center of the speedometer head. The cable is connected by a threaded ring. Unscrew the ring and pull the cable sheath from the head.
2. The cable core can be pulled from the sheath.
3. If the core is broken, detach the other end of the sheath from the transmission. Pull out the broken end.
4. When installing the cable, apply a very small amount of speedometer cable graphite lubricant.

Radio
REMOVAL AND INSTALLATION

Through 1975

The only factory installed radio available on these models was offered in 1975. It is a simple underdash unit, similar to those dealer installed in earlier models. Removal and installation are obvious.

1976-90

1. Disconnect the battery ground cable.
2. Remove the control knobs, nuts, and bezel.
3. On 1976 and early 1977 models, you may have to detach the defroster hose. With air conditioning, remove the screws and lower the assembly.
4. Disconnect the radio bracket from the instrument panel.
5. Tilt the radio down and remove it toward the steering wheel.
6. Detach the antenna, speaker, and power wires.
7. Reverse the procedure for installation.

CRUISE CONTROL

Control Module

REPLACEMENT

The module is located on the firewall by the fuse panel.
1. Remove the mounting screws or tape that holds the module in place.
2. Unplug the wiring connector.

CHILTON TIP: A screwdriver can be used to pry apart the connector. A slot is provided for that purpose.

3. Installation is the reverse of removal.

Servo

REMOVAL AND INSTALLATION

The servo is mounted on a bracket in the engine compartment
1. Remove the locknut holding the servo to the bracket.
2. Remove the 2 vacuum hoses from the servo.
3. Unplug the electrical connector.
4. Remove the 2 nuts and cable housing from the servo.
5. Release the cable clip.
6. Installation is the reverse of removal. Tighten the locknut to 60 inch lbs.

Speed Sensor

REMOVAL AND INSTALLATION

The speed sensor is located above the front driveshaft and is inline with the speedometer cable.

1. Unscrew the 2 attaching nuts from the sensor.
2. Disconnect the wire lead from the cruise harness.
3. Installation is the reverse of removal.

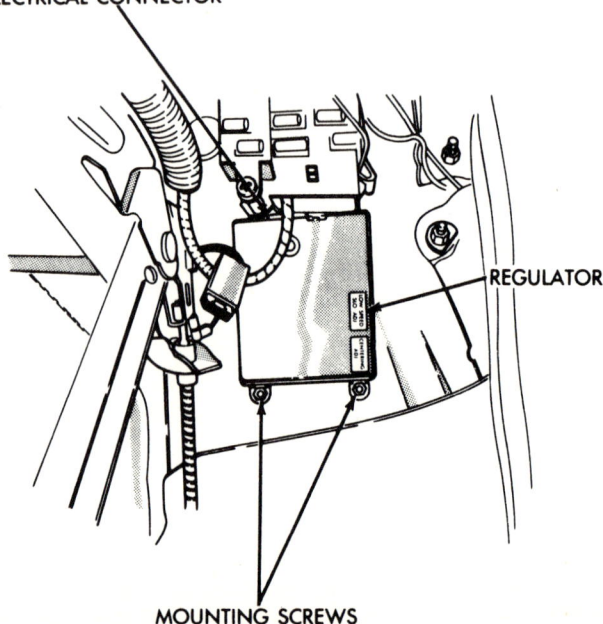

ELECTRICAL CONNECTOR

REGULATOR

MOUNTING SCREWS

Cruise control module location

Control Switch

REMOVAL AND INSTALLATION

1. Disconnect the battery ground.
2. Cover the painted areas of the column.
3. Remove the column-to-dash bezel.
4. Loosen the toe plate screws.
5. With tilt columns, place the column in the non-tilt position.
6. Remove the steering wheel.
7. Remove the lock plate cover.
8. Compress the lock plate and unseat the steering shaft snapring as follows:
   a. Check the steering shaft nut threads. Metric threads have an identifying groove in the steering wheel splines. SAE threads do not.
   b. With SAE threads use a compressor tool such as tool J-23653 to compress the lock plate and remove the snapring.
   c. If the shaft has metric threads, replace the forcing screw in the compressor with metric forcing screw J-23653-4 before using.
9. Remove the compressor and snapring.
10. Remove the lock plate, canceling cam and upper bearing preload spring.
11. Place the turn signal lever in the right turn position and remove the lever.
12. Remove the hazard warning knob. Press the knob inward and turn counterclockwise to remove it.
13. Remove the wiring harness protectors.
14. Disconnect the wiring harness connectors.
15. Remove the turn signal switch attaching screws and lift out the switch.
16. Unplug the cruise control switch connector.
17. Pull the control harness from the column.
18. Insert the key. Position the key as follows.
* 1976-83: With manual transmission, put it in the ON position; with automatic, put it in OFF/LOCK.
* 1984-86: Two detent positions, clockwise, beyond OFF/LOCK.
* 1987-90: ON.
19. Remove the key warning buzzer switch and retaining clip with a paper clip inserted below the retainer to flatten the retainer.

CHILTON TIP: Do not attempt to remove the buzzer switch and clip separately. The clip will probably fall into the column jacket.

20. Working through the slot next to the turn signal switch mounting boss, use a thin screwdriver to release the lock cylinder.
21. Remove the screws that attach the housing and sgroud assembly and remove them.

NOTE: Be carefulf to avoid dropping the dimmer switch rod, lock pin or lock rack.

22. Remove the turn signal/wiper lever.
23. Remove the wiper switch cover from the back of the housing.
24. Remove the pivot screw from the housing and remove the wiper switch.
To install:
25. Install a new switch and cover.
26. Push on the dimmer switch rod to make sure it's connected, then carefully position the housing on the column.

WARNING: Make sure that the nylon spring retainer on the lock pin is positioned forward of the retaining slot in the lock rack.

Position the first tooth of the gear (farthest from the

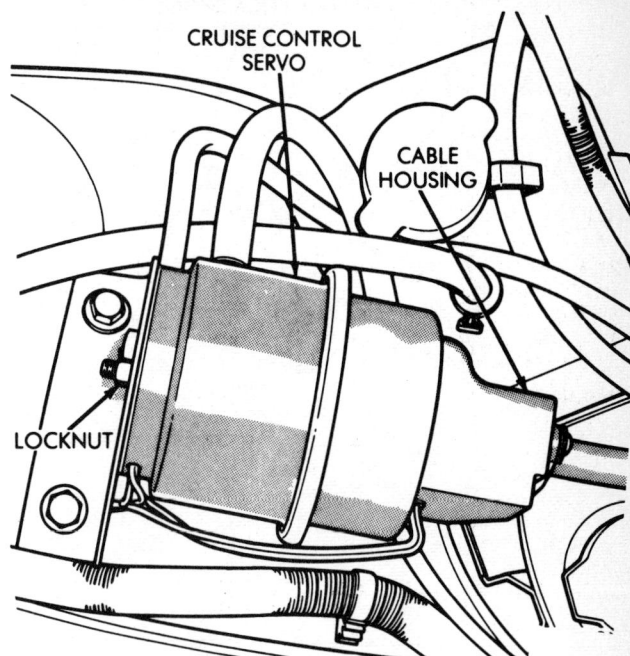

Cruise control servo

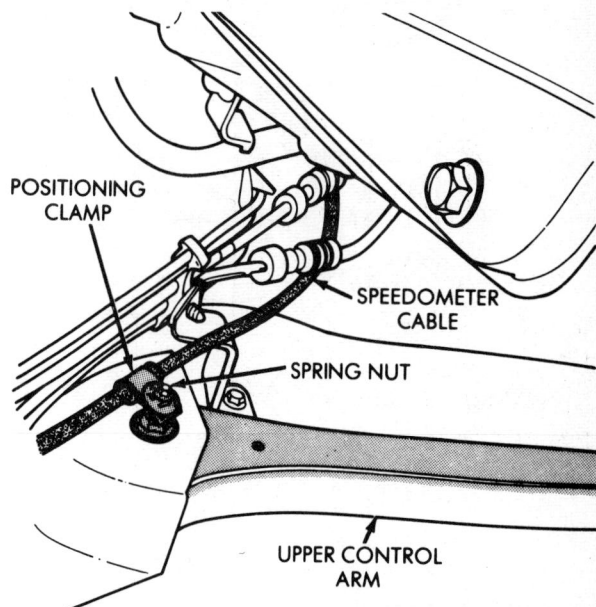

Speedometer cable positioning clamp

block tooth) with the most forward tooth of the lock rack.

27. Install the housing and shroud screws.
28. Insert the key in the new lock cylinder. Hold the sleeve and turn the key clockwise until it stops. Align the cylinder retaining tab with the housing slot and insert the cylinder. Push the cylinder in, rotate to engage, then push in until the retaining tab engages the housing groove.
29. Install the turn signal switch and attaching screws. Torque the screws to 35 inch lbs.
30. Connect the wiring harness connectors.
31. Install the wiring harness protectors.

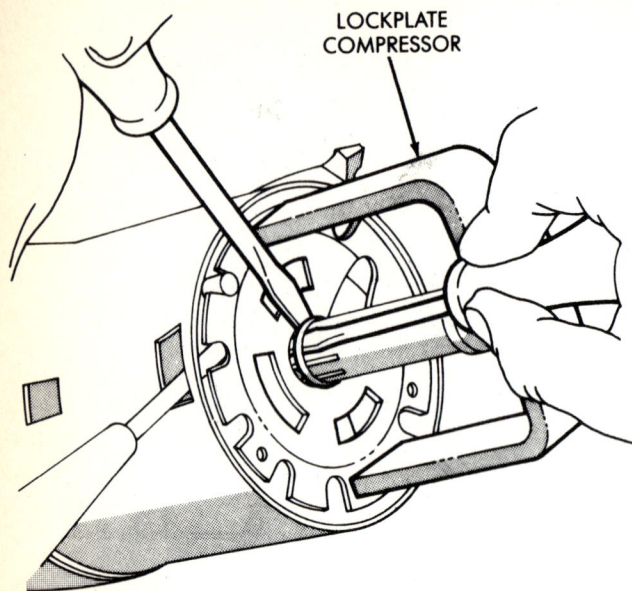

LOCKPLATE
COMPRESSOR

Lockplate removal

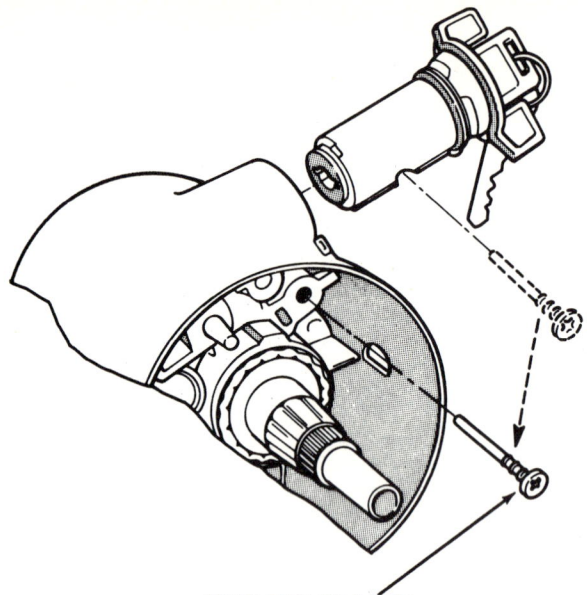

KEY/LOCK CYLINDER
RETAINING SCREW

Lock cylinder removal

WARNING BUZZER
SWITCH

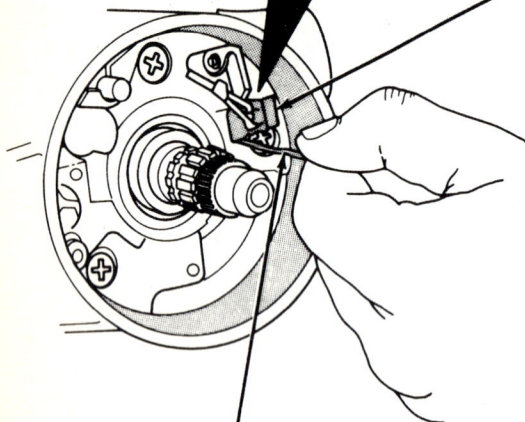

PAPER CLIP

Buzzer switch removal

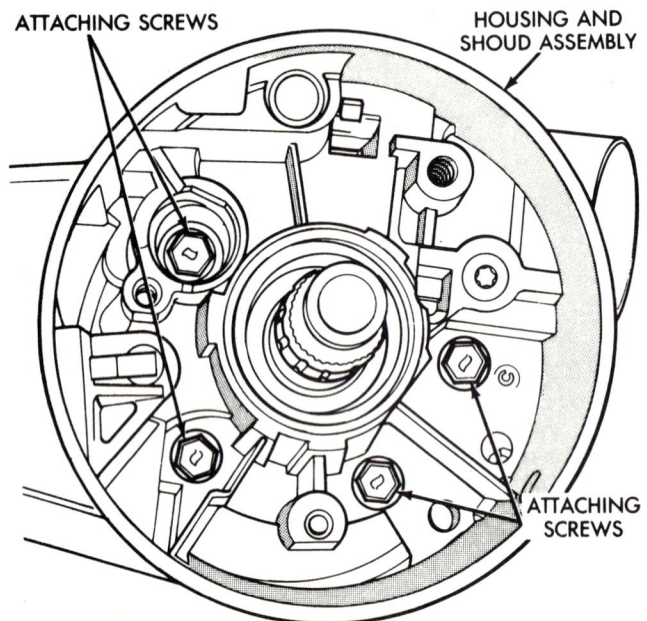

ATTACHING SCREWS

HOUSING AND
SHOUD ASSEMBLY

ATTACHING
SCREWS

Steering column housing removal

32. With the turn signal switch in the neutral position, install the hazard warning knob. Press the knob inward and turn clockwise to install it.

33. Install the lever. Torque the attaching screws to 35 inch lbs.

34. Install the upper bearing preload spring.
35. Install the canceling cam.
36. Install the lock plate.
37. Using the compressor, install the snapring.
38. Install the lock plate cover.
39. Install the steering wheel.
40. Loosen the toe plate screws.
41. Install the column-to-dash bezel.
42. Connect the battery ground.

Servo Cable

REMOVAL AND INSTALLATION

1. Using your fingers only – NO TOOLS – remove the cruise control cable connector at the bell crank by pushing the connector off the bell crank. DO NOT pull the connector off perpendicular to the bell crank!

2. Squeeze the tabs on the cable and lift the cable out of the locking plate

3. Installation is the reverse of removal.

WIPER SWITCH COVER

PIVOT SCREW

REMOVE SCREW, IF EQUIPPED

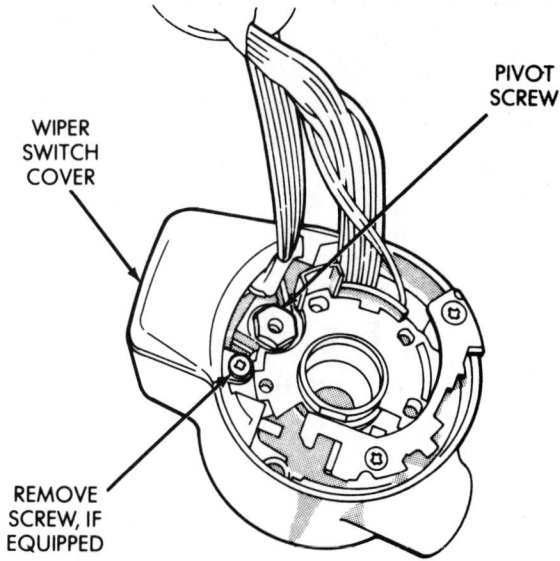

Pivot screw removal

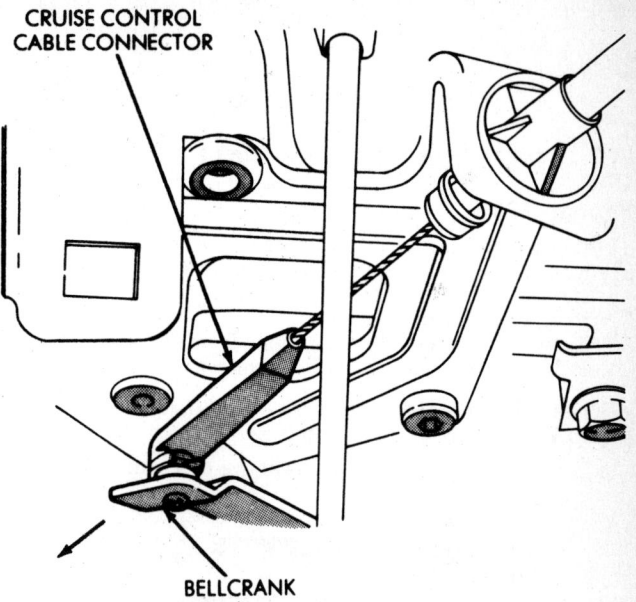

CRUISE CONTROL CABLE CONNECTOR

BELLCRANK

Removing the bellcrank connector

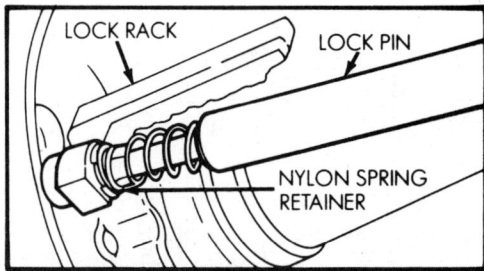

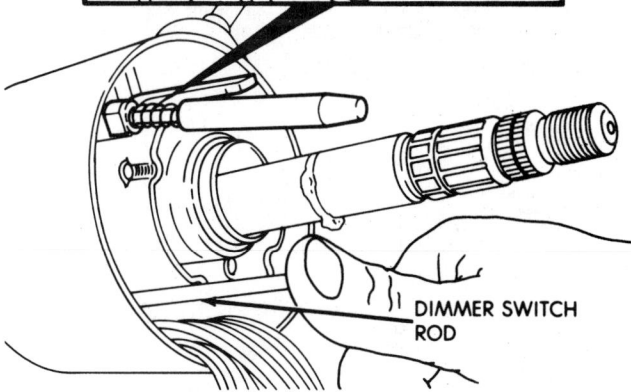

LOCK RACK

LOCK PIN

NYLON SPRING RETAINER

DIMMER SWITCH ROD

Dimmer switch rod and lock pin

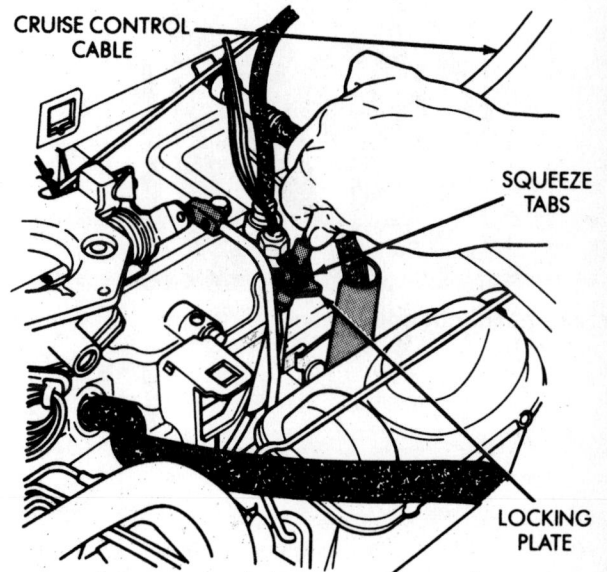

CRUISE CONTROL CABLE

SQUEEZE TABS

LOCKING PLATE

Removing the cruise control cable from the locking plate

LIGHTING

Headlights

REMOVAL AND INSTALLATION

Through 1986

1. Remove the one lower attaching screw from the headlight trim ring. Pull out slightly at the bottom and push up to disengage the upper retaining tab.
2. Remove the trim ring.
3. Remove the three retaining screws from the retaining ring.
4. Pull the headlamp out and disconnect the wire harness.

NOTE: When installing the headlamp, the number 2 is placed at the top of the lamp.

5. Install in reverse order of the above procedure. Check for proper seating of the lamp in its mounting ring and check for proper alignment.

1987-90

1. Remove the four trim ring screws and the trim ring.
2. Remove the four retaining ring screws.
3. Pull out on the lamp and disconnect the wiring plug.
4. Installation is the reverse of removal.

Fog Lights

AIMING

1. Park the Jeep on level ground, facing, perpendicular to, and about 25 ft. from a flat wall.

1972–86 headlight adjustment

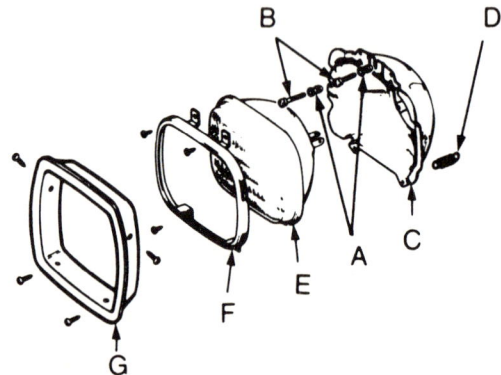

A. Plastic adjuster nut(s)
B. Adjusting screw(s)
C. Headlamp bucket
D. Adjusting spring
E. Headlamp
F. Retaining ring
G. Trim ring

1987–90 headlight assembly

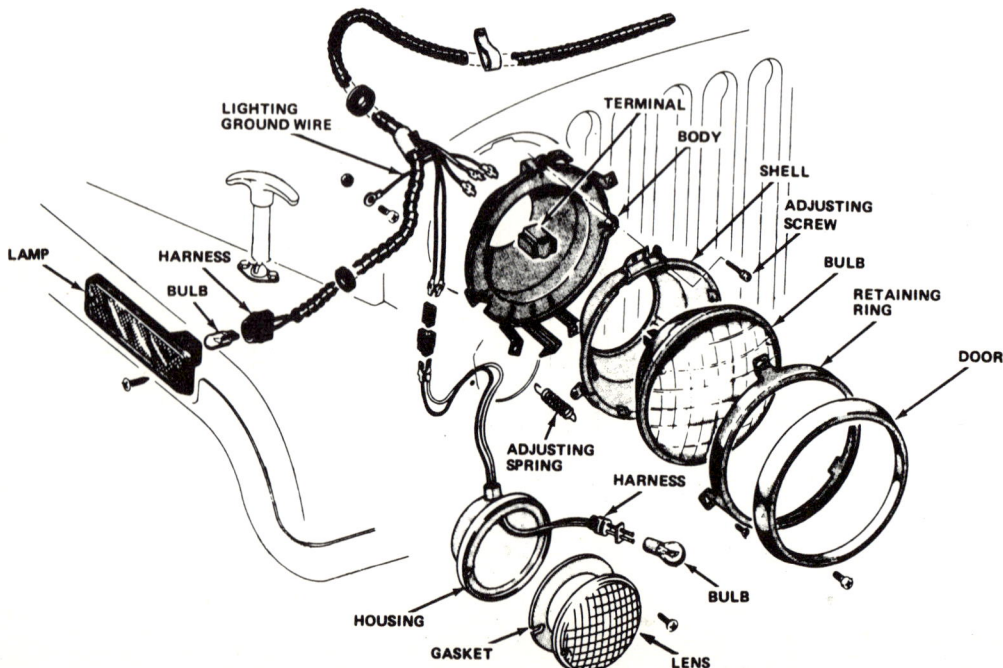

Typical front end lighting through 1986

1987–90 headlight retaining ring and headlight removal

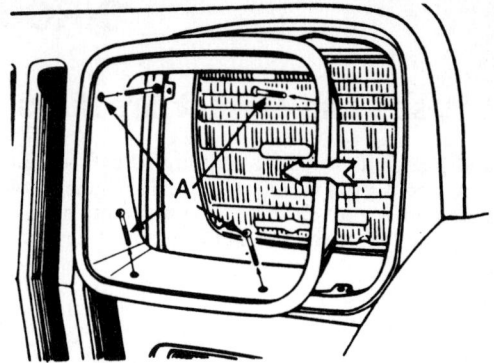

1987–90 headlight bezel

Signal, Parking, Brake and Marker Lights

REMOVAL AND INSTALLATION

These lights, depending on year and model, can be removed in one of two ways:

a. Reach up behind the light, diconnect the wire and turn the bulb socket counterclockwise to remove the socket from the lamp, or

b. Remove the lamp lens screws, remove the lens, turn the bulb and remove it from the socket.

2. Remove any stone shields and switch on the fog lights.
3. Loosen the mounting hardware of the lights so you can aim them as follows:

    a. The horizontal distance between the light beams on the wall should be the same as between the lights themselves.

    b. The vertical height of the light beams above the ground should be 4 inches less than the distance between the ground and the center of the lamp lenses.

4. Tighten the mounting hardware.

TRAILER WIRING

Wiring the Jeep for towing is fairly easy. There are a number of good wiring kits available and these should be used, rather than trying to design your own. All trailers will need brake lights and turn signals as well as tail lights and side marker lights. Most states require extra marker lights for overwide trailers. Also, most states have recently required back-up lights for trailers, and most trailer manufacturers have been building trailers with back-up lights for several years.

Additionally, some Class I, most Class II and just about all Class III trailers will have electric brakes.

Add to this number an accessories wire, to operate trailer internal equipment or to charge the trailer's battery, and you can have as many as seven wires in the harness.

Determine the equipment on your trailer and buy the wiring kit necessary. The kit will contain all the wires needed, plus a plug adapter set which included the female plug, mounted on the bumper or hitch, and the male plug, wired into, or plugged into the trailer harness.

When installing the kit, follow the manufacturer's instructions. The color coding of the wires is standard throughout the industry.

One, final point, the best kits are those with a spring loaded cover on the vehicle mounted socket. This cover prevent dirt and moisture from corroding the terminals. Never let the vehicle socket hang loosely. Always mount it securely to the bumper or hitch.

CIRCUIT PROTECTION

Fuses and Circuit Breakers

An under-dash fuse panel contains the fuses and circuit breakers protecting the various electrical systems, as well as the turn signal and hazard flashers. Fuses are replaced by simply unplugging the defective fuse and inserting a new one.

Fusible Links

REPLACEMENT

The fuse link is a short length of special, Hypalon (high temperature) insulated wire, integral with the engine compartment

REMOVE EXISTING VINYL TUBE SHIELDING
REINSTALL OVER FUSE LINK BEFORE CRIMPING
FUSE LINK TO WIRE ENDS

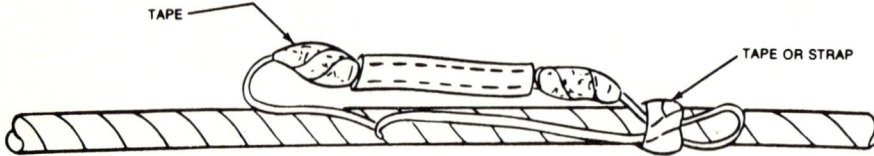

TAPE

TAPE OR STRAP

TYPICAL REPAIR USING THE SPECIAL #17 GA. (9.00" LONG-YELLOW) FUSE LINK REQUIRED FOR THE AIR/COND. CIRCUITS (2) #687E and #261A LOCATED IN THE ENGINE COMPARTMENT

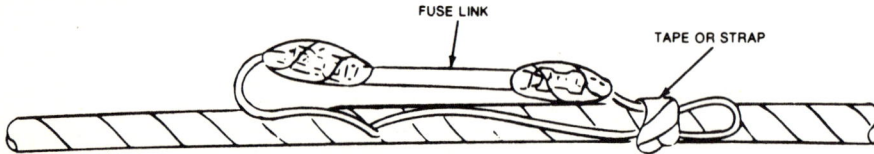

FUSE LINK

TAPE OR STRAP

TYPICAL REPAIR FOR ANY IN-LINE FUSE LINK USING THE SPECIFIED GAUGE FUSE LINK FOR THE SPECIFIC CIRCUIT

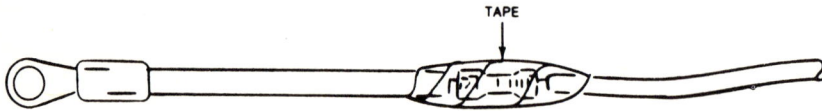

TAPE

TYPICAL REPAIR USING THE EYELET TERMINAL FUSE LINK OF THE SPECIFIED GAUGE FOR ATTACHMENT TO A CIRCUIT WIRE END

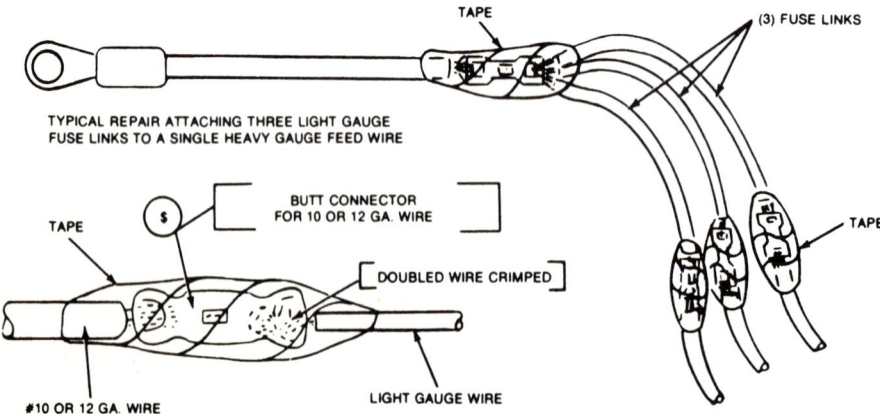

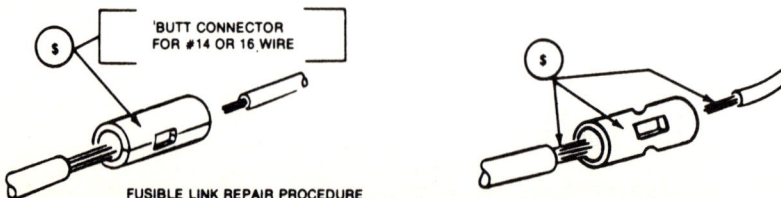

TAPE

(3) FUSE LINKS

TAPE

TYPICAL REPAIR ATTACHING THREE LIGHT GAUGE FUSE LINKS TO A SINGLE HEAVY GAUGE FEED WIRE

BUTT CONNECTOR FOR 10 OR 12 GA. WIRE

DOUBLED WIRE CRIMPED

TAPE

#10 OR 12 GA. WIRE

LIGHT GAUGE WIRE

BUTT CONNECTOR FOR #14 OR 16 WIRE

FUSIBLE LINK REPAIR PROCEDURE

General fuse link repair procedure

wiring harness and should not be confused with standard wire. It is several wire gauges smaller than the circuit which it protects. Under no circumstances should a fuse link replacement repair be made using a length of standard wire cut from bulk stock or from another wiring harness.

To repair any blown fuse link use the following procedure:

1. Determine which circuit is damaged, its location and the cause of the open fuse link. If the damaged fuse link is one of three fed by a common No. 10 or 12 gauge feed wire, determine the specific affected circuit.

2. Disconnect the negative battery cable.

3. Cut the damaged fuse link from the wiring harness and discard it. If the fuse link is one of three circuits fed by a single feed wire, cut it out of the harness at each splice end and discard it.

4. Identify and procure the proper fuse link and butt connectors for attaching the fuse link to the harness.

5. To repair any fuse link in a 3-link group with one feed:

a. After cutting the open link out of the harness, cut each of the remaining undamaged fuse links close to the feed wire weld.

b. Strip approximately ½ in. of insulation from the detached ends of the two good fuse links. Then insert two wire ends into one end of a butt connector and carefully push one stripped end of the replacement fuse link into the same end of the butt connector and crimp all three firmly together.

NOTE: Care must be taken when fitting the three fuse links into the butt connector as the internal diameter is a snug it for three wires. Make sure to use a proper crimping tool. Pliers, side cutters, etc. will not apply the proper crimp to retain the wires and withstand a pull test.

c. After crimping the butt connector to the three fuse links, cut the weld portion from the feed wire and strip approximately ½ in. of insulation from the cut end. Insert the stripped end into the open end of the butt connector and crimp very firmly.

d. To attach the remaining end of the replacement fuse link, strip approximately ½ in. of insulation from the wire end of the circuit from which the blown fuse link was removed, and firmly crimp a butt connector or equivalent to the stripped wire. Then, insert the end of the replacement link into the other end of the butt connector and crimp firmly.

e. Using rosin core solder with a consistency of 60 percent tin and 40 percent lead, solder the connectors and the wires at the repairs and insulate with electrical tape.

6. To replace any fuse link on a single circuit in a harness, cut out the damaged portion, strip approximately ½ in. of insulation from the two wire ends and attach the appropriate replacement fuse link to the stripped wire ends with two proper size butt connectors. Solder the connectors and wires and insulate the tape.

7. To repair any fuse link which has an eyelet terminal on one end such as the charging circuit, cut off the open fuse link behind the weld, strip approximately ½ in. of insulation from the cut end and attach the appropriate new eyelet fuse link to the cut stripped wire with an appropriate size butt connector. Solder the connectors and wires at the repair and insulate with tape.

8. Connect the negative battery cable to the battery and test the system for proper operation.

NOTE: Do not mistake a resistor wire for a fuse link. The resistor wire is generally longer and has print stating, "Resistor: don't cut or splice."

Troubleshooting Basic Turn Signal and Flasher Problems

Most problems in the turn signals or flasher system can be reduced to defective flashers or bulbs, which are easily replaced. Occasionally, problems in the turn signals are traced to the switch in the steering column, which will require professional service.

F = Front   R = Rear   ● = Lights off   ○ = Lights on

Problem		Solution
Turn signals light, but do not flash	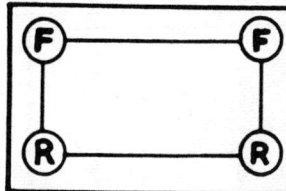	• Replace the flasher
No turn signals light on either side	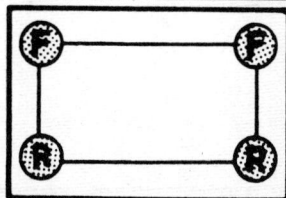	• Check the fuse. Replace if defective. • Check the flasher by substitution • Check for open circuit, short circuit or poor ground

Troubleshooting Basic Turn Signal and Flasher Problems

Most problems in the turn signals or flasher system can be reduced to defective flashers or bulbs, which are easily replaced. Occasionally, problems in the turn signals are traced to the switch in the steering column, which will require professional service.

F = Front   R = Rear   • = Lights off   o = Lights on

Problem		Solution
Both turn signals on one side don't work		• Check for bad bulbs • Check for bad ground in both housings
One turn signal light on one side doesn't work		• Check and/or replace bulb • Check for corrosion in socket. Clean contacts. • Check for poor ground at socket
Turn signal flashes too fast or too slow		• Check any bulb on the side flashing too fast. A heavy-duty bulb is probably installed in place of a regular bulb. • Check the bulb flashing too slow. A standard bulb was probably installed in place of a heavy-duty bulb. • Check for loose connections or corrosion at the bulb socket
Indicator lights don't work in either direction		• Check if the turn signals are working • Check the dash indicator lights • Check the flasher by substitution
One indicator light doesn't light		• On systems with 1 dash indicator: See if the lights work on the same side. Often the filaments have been reversed in systems combining stoplights with taillights and turn signals. Check the flasher by substitution • On systems with 2 indicators: Check the bulbs on the same side Check the indicator light bulb Check the flasher by substitution

Troubleshooting Basic Lighting Problems

Problem	Cause	Solution
Lights		
One or more lights don't work, but others do	• Defective bulb(s) • Blown fuse(s) • Dirty fuse clips or light sockets • Poor ground circuit	• Replace bulb(s) • Replace fuse(s) • Clean connections • Run ground wire from light socket housing to car frame
Lights burn out quickly	• Incorrect voltage regulator setting or defective regulator • Poor battery/alternator connections	• Replace voltage regulator • Check battery/alternator connections
Lights go dim	• Low/discharged battery • Alternator not charging • Corroded sockets or connections • Low voltage output	• Check battery • Check drive belt tension; repair or replace alternator • Clean bulb and socket contacts and connections • Replace voltage regulator
Lights flicker	• Loose connection • Poor ground • Circuit breaker operating (short circuit)	• Tighten all connections • Run ground wire from light housing to car frame • Check connections and look for bare wires
Lights "flare"—Some flare is normal on acceleration—if excessive, see "Lights Burn Out Quickly"	• High voltage setting	• Replace voltage regulator
Lights glare—approaching drivers are blinded	• Lights adjusted too high • Rear springs or shocks sagging • Rear tires soft	• Have headlights aimed • Check rear springs/shocks • Check/correct rear tire pressure
Turn Signals		
Turn signals don't work in either direction	• Blown fuse • Defective flasher • Loose connection	• Replace fuse • Replace flasher • Check/tighten all connections
Right (or left) turn signal only won't work	• Bulb burned out • Right (or left) indicator bulb burned out • Short circuit	• Replace bulb • Check/replace indicator bulb • Check/repair wiring
Flasher rate too slow or too fast	• Incorrect wattage bulb • Incorrect flasher	• Flasher bulb • Replace flasher (use a variable load flasher if you pull a trailer)
Indicator lights do not flash (burn steadily)	• Burned out bulb • Defective flasher	• Replace bulb • Replace flasher
Indicator lights do not light at all	• Burned out indicator bulb • Defective flasher	• Replace indicator bulb • Replace flasher

Troubleshooting Basic Dash Gauge Problems

Problem	Cause	Solution
Coolant Temperature Gauge		
Gauge reads erratically or not at all	• Loose or dirty connections • Defective sending unit	• Clean/tighten connections • Bi-metal gauge: remove the wire from the sending unit. Ground the wire for an instant. If the gauge registers, replace the sending unit.
	• Defective gauge	• Magnetic gauge: disconnect the wire at the sending unit. With ignition ON gauge should register COLD. Ground the wire; gauge should register HOT.
Ammeter Gauge—Turn Headlights ON (do not start engine). Note reaction		
Ammeter shows charge Ammeter shows discharge Ammeter does not move	• Connections reversed on gauge • Ammeter is OK • Loose connections or faulty wiring • Defective gauge	• Reinstall connections • Nothing • Check/correct wiring • Replace gauge
Oil Pressure Gauge		
Gauge does not register or is inaccurate	• On mechanical gauge, Bourdon tube may be bent or kinked	• Check tube for kinks or bends preventing oil from reaching the gauge
	• Low oil pressure	• Remove sending unit. Idle the engine briefly. If no oil flows from sending unit hole, problem is in engine.
	• Defective gauge	• Remove the wire from the sending unit and ground it for an instant with the ignition ON. A good gauge will go to the top of the scale.
	• Defective wiring	• Check the wiring to the gauge. If it's OK and the gauge doesn't register when grounded, replace the gauge.
	• Defective sending unit	• If the wiring is OK and the gauge functions when grounded, replace the sending unit
All Gauges		
All gauges do not operate	• Blown fuse • Defective instrument regulator	• Replace fuse • Replace instrument voltage regulator
All gauges read low or erratically	• Defective or dirty instrument voltage regulator	• Clean contacts or replace
All gauges pegged	• Loss of ground between instrument voltage regulator and car • Defective instrument regulator	• Check ground • Replace regulator

Troubleshooting Basic Dash Gauge Problems

Problem	Cause	Solution
Warning Lights		
Light(s) do not come on when ignition is ON, but engine is not started	• Defective bulb • Defective wire • Defective sending unit	• Replace bulb • Check wire from light to sending unit • Disconnect the wire from the sending unit and ground it. Replace the sending unit if the light comes on with the ignition ON.
Light comes on with engine running	• Problem in individual system • Defective sending unit	• Check system • Check sending unit (see above)

Troubleshooting the Heater

Problem	Cause	Solution
Blower motor will not turn at any speed	• Blown fuse • Loose connection • Defective ground • Faulty switch • Faulty motor • Faulty resistor	• Replace fuse • Inspect and tighten • Clean and tighten • Replace switch • Replace motor • Replace resistor
Blower motor turns at one speed only	• Faulty switch • Faulty resistor	• Replace switch • Replace resistor
Blower motor turns but does not circulate air	• Intake blocked • Fan not secured to the motor shaft	• Clean intake • Tighten security
Heater will not heat	• Coolant does not reach proper temperature • Heater core blocked internally • Heater core air-bound • Blend-air door not in proper position	• Check and replace thermostat if necessary • Flush or replace core if necessary • Purge air from core • Adjust cable
Heater will not defrost	• Control cable adjustment incorrect • Defroster hose damaged	• Adjust control cable • Replace defroster hose

Troubleshooting Basic Windshield Wiper Problems

Problem	Cause	Solution
Electric Wipers		
Wipers do not operate— Wiper motor heats up or hums	• Internal motor defect • Bent or damaged linkage • Arms improperly installed on linking pivots	• Replace motor • Repair or replace linkage • Position linkage in park and reinstall wiper arms

6 CHASSIS ELECTRICAL

Troubleshooting Basic Windshield Wiper Problems

Problem	Cause	Solution
Electric Wipers		
Wipers do not operate— No current to motor	• Fuse or circuit breaker blown • Loose, open or broken wiring • Defective switch • Defective or corroded terminals • No ground circuit for motor or switch	• Replace fuse or circuit breaker • Repair wiring and connections • Replace switch • Replace or clean terminals • Repair ground circuits
Wipers do not operate— Motor runs	• Linkage disconnected or broken	• Connect wiper linkage or replace broken linkage
Vacuum Wipers		
Wipers do not operate	• Control switch or cable inoperative • Loss of engine vacuum to wiper motor (broken hoses, low engine vacuum, defective vacuum/fuel pump) • Linkage broken or disconnected • Defective wiper motor	• Repair or replace switch or cable • Check vacuum lines, engine vacuum and fuel pump • Repair linkage • Replace wiper motor
Wipers stop on engine acceleration	• Leaking vacuum hoses • Dry windshield • Oversize wiper blades • Defective vacuum/fuel pump	• Repair or replace hoses • Wet windshield with washers • Replace with proper size wiper blades • Replace pump

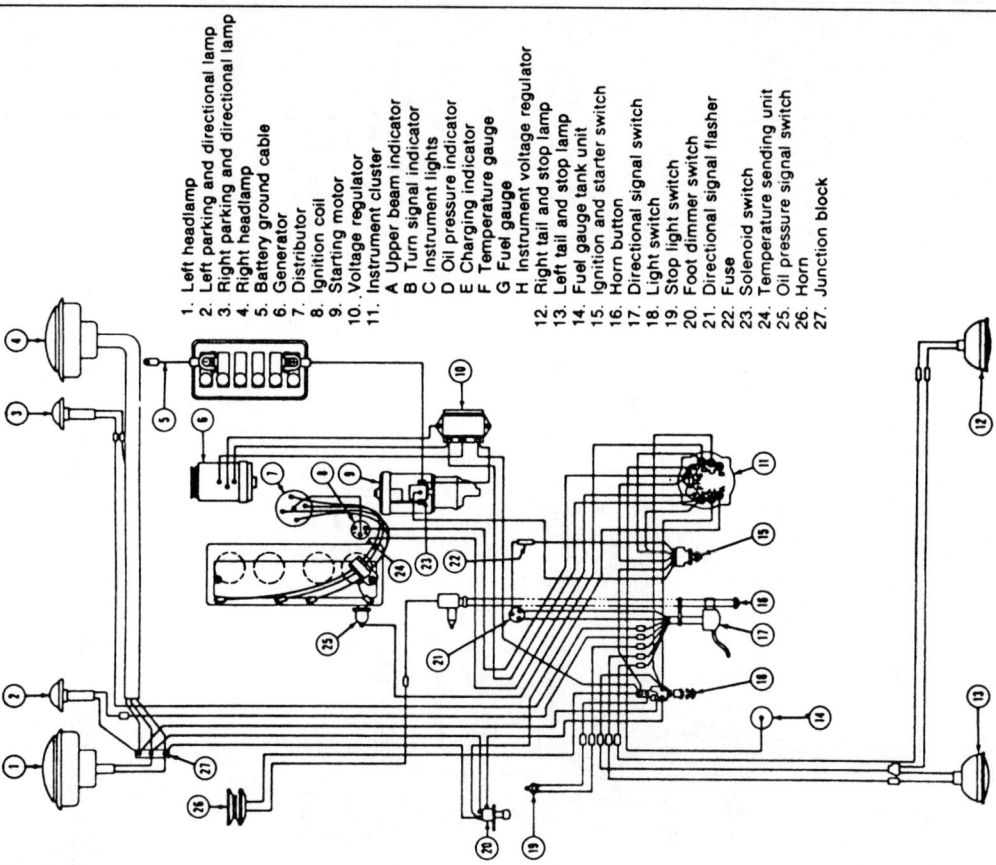

1. Left headlamp
2. Left parking and directional lamp
3. Right parking and directional lamp
4. Right headlamp
5. Battery ground cable
6. Generator
7. Distributor
8. Ignition coil
9. Starting motor
10. Voltage regulator
11. Instrument cluster
   A Upper beam indicator
   B Turn signal indicator
   C Instrument lights
   D Oil pressure indicator
   E Charging indicator
   F Temperature gauge
   G Fuel gauge
   H Instrument voltage regulator
12. Right tail and stop lamp
13. Left tail and stop lamp
14. Fuel gauge tank unit
15. Ignition and starter switch
16. Horn button
17. Directional signal switch
18. Light switch
19. Stop light switch
20. Foot dimmer switch
21. Directional signal flasher
22. Fuse
23. Solenoid switch
24. Temperature sending unit
25. Oil pressure signal switch
26. Horn
27. Junction block

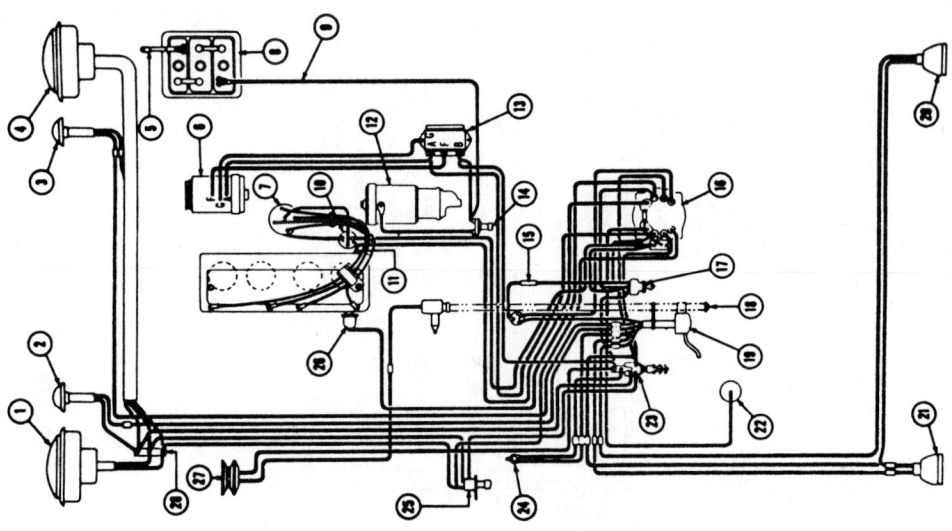

1. Left headlamp
2. Left parking lamp
3. Right parking lamp
4. Right headlamp
5. Negative ground cable
6. Generator
7. Distributor
8. Battery
9. Positive cable
10. Ignition coil
11. Temperature sending unit
12. Starting motor
13. Voltage regulator
14. Starting switch
15. Fuse
16. Instrument switch
17. Ignition switch
18. Horn button
19. Directional signal switch
20. Right tail and stop lamp
21. Left tail and stop lamp
22. Fuel gauge sending unit
23. Light switch
24. Stop light switch
25. Dimmer switch
26. Oil pressure sending unit
27. Horn
28. Junction block

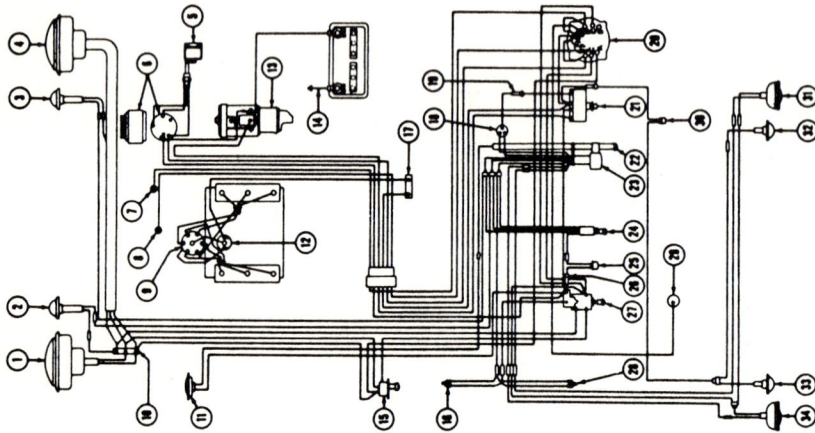

CJ-5 and CJ-6 with V6-225 engine

1. Left headlamp
2. Left parking and signal lamp
3. Right parking and signal lamp
4. Right headlamp
5. Voltage regulator
6. Alternator
7. Oil pressure sender
8. Temperature sender
9. Ignition distributor
10. Junction block
11. Horn
12. Ignition coil
13. Starting motor
14. Battery ground cable
15. Foot dimmer switch
16. Stop light switch—front
17. Ballast
18. Flasher (directional signal)
19. Fuse
20. Instrument cluster
A. Hi-beam indicator
B. Auxiliary
C. Instrument lights
D. Oil pressure indicator
E. Charging indicator
F. Temperature indicator
G. Fuel gauge
H. Instrument voltage regulator
21. Ignition and starter switch
22. Horn button
23. Directional signal switch
24. 4-Way flasher switch
25. Flasher (4-way)
26. Fuse
27. Main light switch
28. Stop light switch—rear
29. Fuel gauge tank unit
30. Back-up light switch
31. Right tail and stop lamp
32. Right back-up lamp
33. Left back-up lamp
34. Left tail and stop lamp

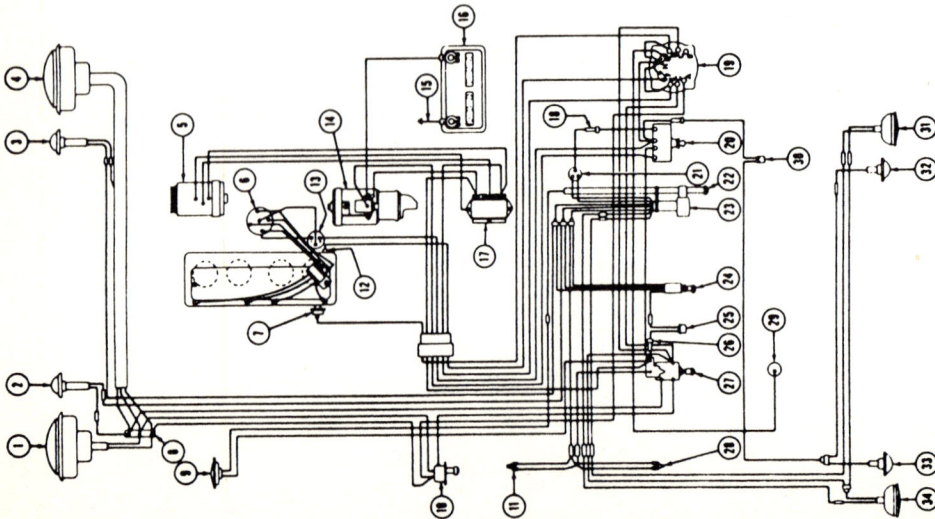

CJ-5 and CJ-6 with later F4-134 engine

1. Left headlamp
2. Left parking and signal lamp
3. Right parking and signal lamp
4. Right headlamp
5. Generator
6. Ignition distributor
7. Oil pressure sending unit
8. Junction block
9. Horn
10. Foot dimmer switch
11. Stop light switch—front
12. Temperature sending unit
13. Ignition coil
14. Starting motor
15. Battery ground cable
16. Battery
17. Voltage regulator
18. Fuse
19. Instrument cluster
A. Hi-beam indicator
B. Auxiliary
C. Instrument lights
D. Oil pressure indicator
E. Charging indicator
F. Temperature indicator
G. Fuel gauge
H. Instrument voltage regulator
20. Ignition and starter switch
21. Flasher (directional signal)
22. Horn button
23. Directional signal switch
24. 4-Way flasher switch
25. Flasher (4-way)
26. Fuse
27. Main light switch
28. Stop light switch—rear
29. Fuel gauge tank unit
30. Back-up light switch
31. Right tail and stop lamp
32. Right back-up lamp
33. Left back-up lamp
34. Left tail and stop lamp

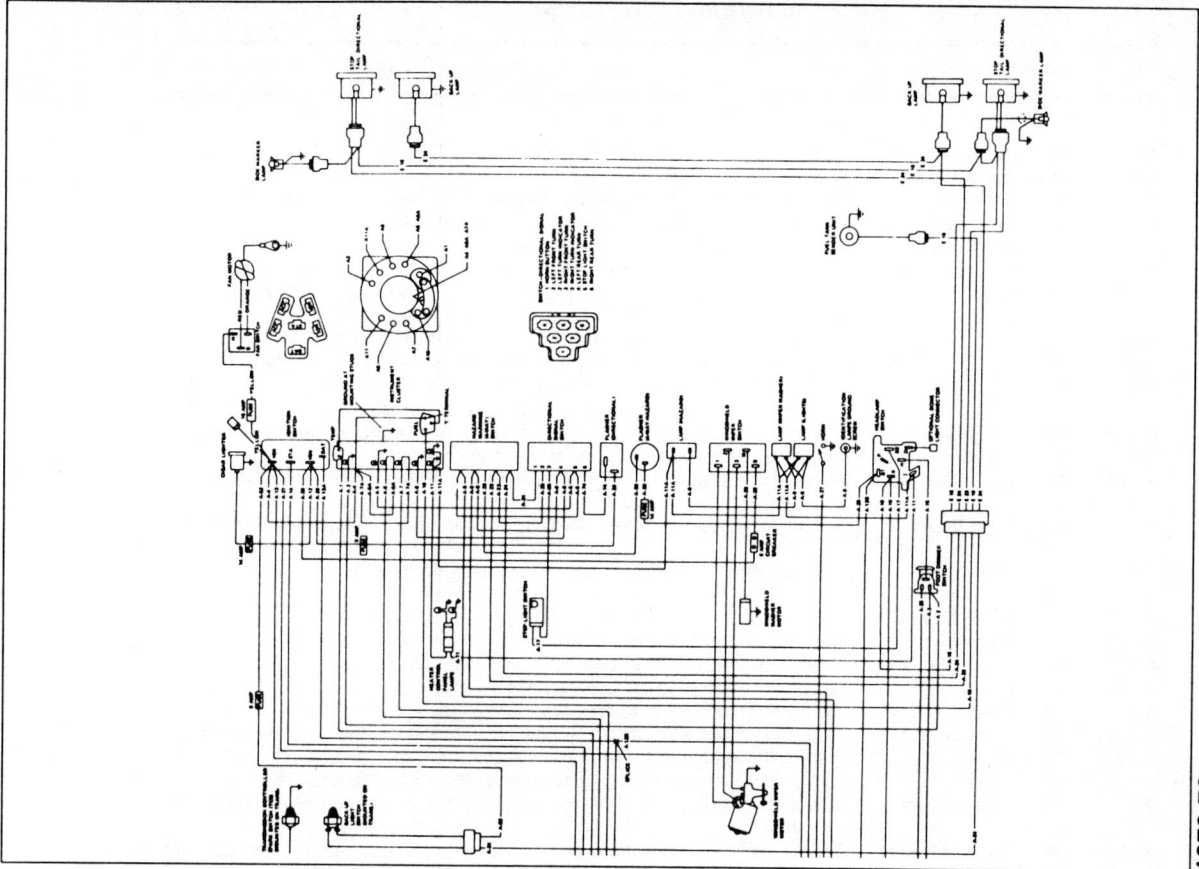

1972-73

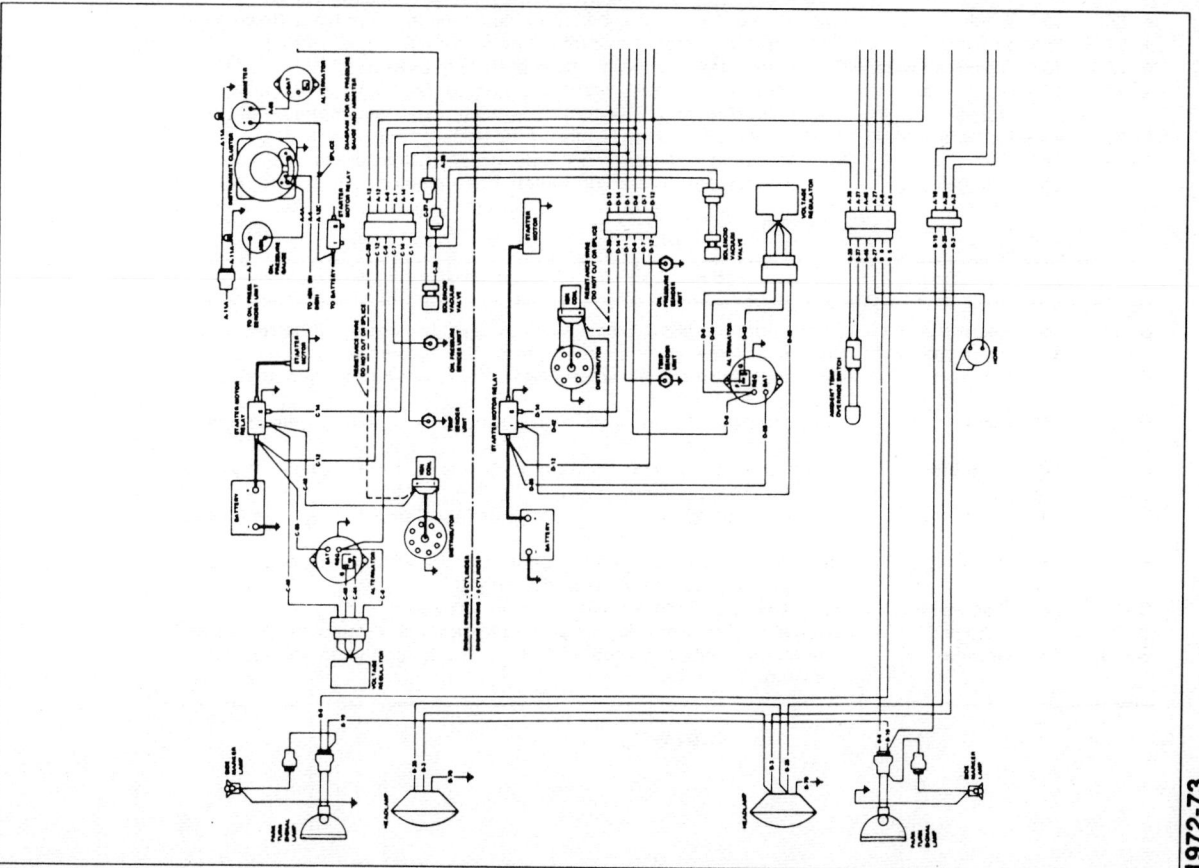

1972-73

Chassis Electrical

No.	GA.	Color	Instrument and Control Harness
A–1	18	Blue-yellow tr.	Cluster "A" temp. to connector (temp. sender)
A–2	18	White-red tr.	Cluster "L" Hi-beam indicator to dimmer switch (hi-beam)
A–3	14	Red-white tr.	Dimmer switch (hi-beam) to headlamp junction block (hi-beam)
	18	Green-white tr.	Cluster "K" ignition to ignition switch (ignition term.)
	18	Black-white tr.	Cluster "J" right turn indicator to directional signal switch (right turn)
A–4	16	Black	Directional signal switch (right turn) to directional signal lamp (right turn)
A–5	18	Gray	Cluster "H" (charge indicator) to connector (alternator auxiliary term.)
	18	Purple	Cluster "G" (oil indicator) to connector (oil pressure sender)
A–6	18	Yellow-black tr.	Cluster "E" left turn indicator to directional signal switch (left turn)
A–7	16	Yellow	Directional signal switch (left turn) to directional signal light (left turn)
A–8	18	Black-yellow tr.	Cluster "D" ground to instrument panel ground (mounting)
A–9	18	White	Cluster "C" gas gauge to connector (frame harness)
A–10	18	Red-blue tr.	Cluster "B" panel lights to dimmer switch to connector (gear selector)
A–11	14	Green	Ignition switch (ignition term.) to resistance wire
A–14	16	Light blue	Connector (ignition switch starter term.) to connector (neutral safety switch)
A–14A	16	Light blue	Connector (neutral safety switch) to connector (starter motor starter term.)
A–15	10	Red	Auxiliary circuit breaker to connector (solenoid "B" term.)
A–16	14	Red-white tr.	Light switch circuit breaker to auxiliary circuit breaker
A–17	14	Red-white tr.	Ignition switch "B" term. to auxiliary circuit breaker
A–18	14	Red-white tr.	Auxiliary circuit breaker to horn relay
A–19	14	Red-white tr.	Cigar lighter to auxiliary circuit
A–20	16	Brown	Auxiliary circuit breaker to stop light switch
A–21	14	Green	Light switch "H" term. to foot dimmer switch "B" term.
A–22	16	Light blue	Light switch (parking term.) to headlamp junction block (parking term.)
A–23	16	Yellow	Light switch (tail light term.) to instrument dimmer switch
A–23	16	Yellow	Instrument dimmer switch to connector (frame harness tail lamps)
A–24	18	Green-white tr.	Connector (ignition switch accessory term.) to connector (backup light switch)
A–25	18	Green-white tr.	Connector (backup light switch—standard) to connector (backup light switch—auto) to connector (frame harness—backup lights)
A–26	16	Black	Foot dimmer switch (low-beam) to headlamp junction block (low-beam)
A–27	16	Brown	Connector (turn signal switch) to connector (stop light switch)
A–28	18	Black-yellow tr.	Connector (steering column horn button) to horn relay
A–29	16	Light blue	Connector (turn signal switch) to connector (signal lamp rear left)
A–30	16	Orange	Connector (turn signal switch) to connector (signal lamp rear right)
A–33	14	Red-white tr.	Auxiliary circuit breaker to flasher (hazard warning)
A–34	18	Red	Ignition switch (accessory term.) (fused) to flasher (directional signal)
A–35	16	Orange	Connector (vacuum solenoid switch) to connector (temp. override switch)
A–36	16	Orange	connector (ignition switch—ignition term.) to connector (temp. override switch)

No.	GA.	Color	Harness Assembly —Headlights, Parking and Signal Lamps
B–1	14	Red-white tr.	Connector (circuit breaker—auxiliary feed) to horn relay (battery)
B–2	18	Black-yellow tr.	Connector (steering column—horn button) to horn relay (horn button)
B–3	16	Yellow	Connector (turn signal switch—left turn) to connector (directional signal lamp—left turn)
B–4	16	Black	Connector (turn signal switch—right turn) to connector (directional signal lamp—right turn)
B–5	16	Light blue	Junction block (parking term.) to connector (parking lamp—left) to connector (parking lamp—right)
B–6	14	Red-white tr.	Junction block (hi-beam) to connector (headlamp hi-beam—left) to connector (headlamp hi-beam—right)
B–7	16	Black	Junction block (low-beam) to connector (headlamp low-beam—left) to connector (headlamp low-beam—right)
B–8	18	Black-white tr.	Headlamp ground to ground mounting (2 cables)
B–9	16	Orange	Connector (vacuum solenoid switch) to connector (temp. override switch)
B–10	16	Orange	Connector ignition switch ignition term. to connector temp. override switch

1972-73

Chassis Electrical (cont.)

No.	GA.	Color	Engine Harness (V-8)
C-1	14	Green	Resistance wire (ignition) to coil (+) term.
C-2	14	Green	Coil (+) term. to starting solenoid (ignition)
C-3	16	Yellow	Starter solenoid (ignition) to alternator regulator (ignition)
C-4	16	Light blue	Connector (ignition switch—starter term.) to starter solenoid (starter term.)
C-5	18	Purple	Connector (oil pressure indicator) to oil pressure sender
C-6	18	Blue-yellow tr.	Connector (temp. indicator) to temp. sender
C-7	18	Gray	Connector (cluster "H" term.) to alternator (auxiliary term.) to alternator, regulator (auxilary term.)
C-8	18	Green-white tr.	Alternator regulator (field term.) to alternator (fiel term.)
C-9	16	Black	Alternator regulator (ground term.) to alternator (ground)
C-10	10	Red	Starter solenoid ("B" term.) to connector (circuit breaker—feed)
C-11	10	Yellow	Starter solenoid ("B" term.) to alternator (output term.)
C-12	16	Brown	Connector (transistor solenoid) (T.C.S.) vacuum switch
C-13	16	Black	Connector (sensor switch) (T.C.S.) to vacuum switch

No.	GA.	Color	Engine Harness (6 Cyl.)
D-1	14	Green	Resistance wire (ignition) to coil (+) term.
D-2	14	Green	Coil (+) term. to starting solenoid (ignition)
D-3	16	Yellow	Starter solenoid (ignition) to alternator regulator (ignition)
D-4	16	Light blue	Connector (ignition switch—starter term.) to starter solenoid (starter term.)
D-5	18	Purple	Connector (oil pressure indicator) to oil pressure sender
D-6	18	Blue-yellow tr.	Connector (temperature indicator) to temperature sender
D-7	18	Gray	Connector (cluster "H" term.) to alternator (auxiliary term.) to alternator regulator (auxiliary term.)
D-8	18	Green-white tr.	Alternator regulator (field term.) to alternator (field term.)
D-9	16	Black	Alternator regulator (ground term.) to alternator (ground term.)
D-10	10	Red	Starter solenoid ("B" term.) to connector (circuit breaker feed)
D-11	10	Yellow	Starter solenoid ("B" term.) to alternator (output term.)

1972-73

Chassis Electrical

No.	GA.	Color	Instrument and Control Harness
1	18	Purple w/tr.	Bulkhead connector (temperature sender) to temperature gauge
2	18	Gray w/tr.	Foot dimmer switch (hi-beam) to instrument cluster (hi-beam indicator)
3	14	Gray w/tr.	Foot dimmer switch (hi-beam) to bulkhead connector (hi-beam)
4	18	Red	Fuse panel (cluster feed) to instrument constant voltage regulator
4A	18	Red	Instrument constant voltage regulator (ignition terminal) to oil pressure gauge (ignition terminal)
5A	18	Green	Bulkhead connector (right turn & hazard front) to hazard switch
5B	18	Green	Hazard switch to steering column connector (right turn & hazard front)
5C	18	Green	Steering column connector to instrument cluster lamp (right turn)
7	18	Purple	Bulkhead connector (oil pressure sender) to oil pressure gauge
8A	18	Green w/tr.	Bulkhead connector (left turn & hazard front) to hazard switch
8B	18	Green w/tr.	Hazard switch to steering column connector (left turn & hazard front)
8C	18	Green w/tr.	Steering column connector to instrument cluster lamp (left turn)
9A	18	Black	Windshield wiper & washer switch light to instrument panel lights ground
9B	18	Black	Windshield wiper & washer light to light switch light
9C	18	Black	Instrument panel lights ground to hazard light
9D	18	Black	Hazard light to voltmeter (−) terminal
10	18	Pink	Bulkhead connector (frame harness-fuel sender unit) to instrument cluster fuel gauge (S-terminal)
11A	18	Orange	Fuse panel (lights—accessories) to splice "D"
11B	18	Orange	Light switch light to windshield wiper & washer switch light
11C	18	Orange	Splice "D" to windshield wiper & washer switch light
11D	18	Orange	Splice "D" to hazard light
11E	18	Orange	Splice "D" to splice "C"
11F	18	Orange	Splice "C" to right instrument panel light
11G	18	Orange	Splice "C" to left instrument panel light connector
11H	18	Orange	Left instrument panel light connector to left instrument panel light

1974-75

Chassis Electrical (cont.)

No.	GA.	Color	Instrument and Control Harness
11J	18	Orange	Splice "C" to instrument cluster oil pressure gauge light
11K	18	Orange	Splice "C" to instrument cluster voltmeter light
12A	10	Red	Bulkhead connector (alternator & voltage regulator) to splice "A"
12B	12	Red	Bulkhead connector (horn) to splice "A"
12C	12	Red	Fuse panel (traffic hazard) to splice "A"
12D	12	Red	Splice "A" to light switch (battery feed)
12E	12	Red	Splice "A" to ignition switch
12F	12	Red	Fuse panel (cigar lighter) to splice "A"
13A	14	Red w/tr.	Bulkhead connector (coil) to tachometer connector
13B	14	Red w/tr.	Tachometer connector to ignition switch
14	16	Lt blue	Bulkhead connector (starting motor solenoid) to ignition switch
15	14	Red w/tr.	Light switch (foot dimmer switch feed) to foot dimmer switch
17	14	Red w/tr.	Fuse panel (tail—stop) to stop light switch
18	16	White	Bulkhead connector (chassis harness—tail lamps) to (bulkhead connector—headlamp harness marker lights
19	16	White	Bulkhead connector (marker lamps) to light switch (parking lamps)
23	16	Lt green w/tr	Bulkhead connector (chassis harness left turn & hazard) to hazard switch (left turn & hazard—rear)
23A	16	Lt green w/tr.	Hazard switch to steering column connector (left turn & hazard—rear)
24	16	Lt green	Bulkhead connector (chassis harness right turn & hazard) to hazard switch (right turn & hazard—rear)
24A	16	Lt green	Hazard switch to steering column connector (right turn & hazard—rear)
25	16	Gray	Bulkhead connector (headlamps) to foot dimmer switch (lo-beam)
26	14	Red w/tr.	Fuse panel (heater—battery) to heater blower switch
27	18	Black w/tr.	Bulkhead connector (horn) to horn button
30	16	Yellow	Bulkhead connector (windshield wiper & washer) to windshield wiper & washer switch
33	14	Red w/tr.	Fuse panel (radio) to windshield wiper & washer switch
34	18	White w/tr.	Bulkhead connector (back-up light switch) to bulkhead connector (chassis harness—back-up lights)
39	16	Pink	Fuse panel (traffic hazard flash) to hazard flasher
52	18	Red	Fuse panel (back-up lamps) to bulkhead connector (back-up light switch)
57	18	Black	Bulkhead connector (brake failure switch) to brake warning light (ground)
60	14	Red	Cigar lighter connector to cigar lighter
65	16	Red w/tr.	Stop light switch to steering column connector (brake switch & hazard feed)
66	18	Red w/tr.	Fuse panel (panel lamps) to light switch (panel lights feed)
67A	12	Yellow	Fuse panel lamps to ignition switch

1974-75

Chassis Electrical (cont.)

No.	GA.	Color	Instrument and Control Harness
67B	18	Yellow	Ignition switch to instrument cluster voltmeter (+) terminal
74	18	Red w/tr.	Fuse panel (flash—directional signal) to steering column connector (flasher & directional signal feed)
75	12	Red w/tr.	Ignition switch to splice "B"
75A	12	Red w/tr	Fuse panel (heater—battery) to splice "B"
75B	12	Red w/tr.	Fuse panel lamps to splice "B"
77	16	Black	Bulkhead connector (brake failure switch) to brake warning light connection
78	16	Red w/tr.	Bulkhead connector (alternator—voltage regulator) to splice "B" in circuit 75
77A	16	Black	Fuse panel (warning light) to brake warning light

No.	GA.	Color	Harness Assembly—Headlamp, Parking and Signal Lamps
3A	14	Gray w/tr.	Bulkhead connector (hi-beam) to left headlamp connector (hi-beam)
3B	14	Gray w/tr.	Left headlamp connector (hi-beam) to right headlamp connector (hi-beam)
5A	16	Green	Bulkhead connector (right turn signal) to right turn signal splice "K"
5B	16	Green	Right turn splice "K" to right side marker lamp assembly
5C	16	Green	Right turn splice"K" to right front park & turn signal lamp assembly
8A	16	Green w/tr.	Bulkhead connector (left turn sgnal) to left turn signal splice "H"
8B	16	Green w/tr.	Left turn splice "H" to left side marker lamp assembly
8C	16	Green w/tr.	Left turn splice "H" to left front park & turn signal lamp assembly
19A	16	White	Bulkhead connector (parking lights) to splice "J"
19B	16	White	Parking lights splice "J" to left side marker lamp assembly
19C	16	White	Parking lights splice "L" to right front park & turn signal lamp assembly
19D	16	White	Parking lights splice "M" to left front park & turn signal lamp assembly
19E	16	White	Parking lights splice "L" to right side marker lamp assembly
19F	16	White	Left parking lamps splice "M" to right parking lamps splice "L"
19G	16	White	Splice "J" to splice "M"
25A	16	Gray	Bulkhead connector (lo-beam) to left headlamp connector (lo-beam)
25B	16	Gray	Left headlamp connector (lo-beam) to right headlamp connector (lo-beam)
27	14	Black w/tr.	Bulkhead connector (horn) to horn assembly
30	16	Yellow	Bulkhead connector (windshield wiper & washer switch) to windshield washer motor
45	14	Red w/tr.	Bulkhead connector (horn) to horn assembly
57	16	Black	Bulkhead connector (brake failure switch) to brake failure switch connector to brake failure switch
70	16	Black	Left and right headlamp ground terminals to ground mounting
77	16	Black w/tr.	Bulkhead connector (brake failure switch) to brake failure swich connector to brake failure switch

W/tr. = With tracer

No.	GA.	Color	Harness Assembly—Engine (Six cylinder)
1	18	Purple w/tr.	Bulkhead connector (temperature gauge) to temperature sender
7	18	Purple	Bulkhead connector (oil pressure gauge) to oil pressure sender
12A	14	Red	$5/16$ stud to splice "E" (fusible link in alternator/regulator circuit)
12B	10	Red	Splice "E" to ¼ stud (alternator/regulator circuit)
12C	10	Red	Bulkhead connector (alternator/regulator) to splice "F" at fusible link
12D	14	Red	Splice "F" to $5/16$ stud (fusible link in alternator/regulator circuit)
12E	14	Red	¼ stud to alternator/regulator assembly
13	14	Red w/tr.	Bulkhead connector (ignition switch) to coil (+) terminal
14	16	Lt blue	Bulkhead connector (ignition switch) to starting motor solenoid (starting terminal)
34	18	White w/tr.	Bulkhead connector (back-up lamps) to back-up light switch connector
34A	18	White w/tr.	Back-up light switch connector to back-up light switch
35	16	Red w/tr.	Back-up light switch connector to vacuum solenoid switch
37	16	Orange	Back-up light switch connector to vacuum solenoid switch
37A	16	Orange	Back-up light switch connector to transmission controlled spark switch (T.C.S.)
52	18	White w/tr.	Bulkhead connector (back-up lamps) to back-up light switch connector

1974-75

Chassis Electrical (cont.)

No.	GA.	Color	Instrument and Control Harness
52A	18	White w/tr.	Back-up light switch connector to back-up light switch
78	24	Black w/tr.	Bulkhead connector alternator/regulator) to alternator & voltage regulator
79	16	Green	Coil (−) terminal to electronic ignition pack
80	16	Blue	Distributor to electronic ignition pack
81	16	Yellow	Distributor to electronic ignition pack
82	16	Red w/tr.	Coil (+) terminal to electronic ignition pack

No.	GA.	Color	Harness Assembly—Engine (V-8)
1*	18	Purple w/tr.	Bulkhead connector (temperature gauge) to temperature sender
7	18	Purple	Bulkhead connector (oil pressure gauge) to oil pressure sender
12A*	14	Red	5/16 stud to splice "E" (fusible link in alternator circuit)
12B	10	Red	Splice "E" to 1/4 stud (alternator circuit)
12C*	10	Red	Bulkhead connector (alternator) to splice "F" at fusible link
12D*	14	Red	Splice "F" to 5/16 stud (fusible link in alternator circuit)
13	14	Red w/tr.	Bulkhead connector (ignition switch) to splice "G"
13A	14	Red W/TR	Splice "G" to coil (+) terminal
14*	16	Lt blue	Bulkhead connector (ignition switch) to starting motor solenoid (starting terminal)
34*	18	White w/tr.	Bulkhead connector (back-up lights) to back-up light switch connector
34A*	18	White w/tr.	Back-up light switch connector to back-up light switch
40	16	Yellow	Splice "G" (circuit no. 13) to voltage regulator
41	18	Gray	Voltage regulator to alternator
43	16	Black	Voltage regulator to alternator
44	18	Green	Voltage regulator to alternator
52	18	White w/tr.	Bulkhead connector (back-up lights) to back-up light switch connector
52A	18	White w/tr.	Back-up light switch connector to back-up light switch
79	16	Green	Coil (−) terminal to electronic ignition pack
80	16	Blue	Distributor to electronic ignition pack
81	16	Yellow	Distributor to electronic ignition pack
82	16	Red w/tr.	Splice "G" (circuit no. 13) to electronic ignition pack

No.	GA.	Color	Harness Assembly—Chassis
10	16	Pink	Bulkhead connector (fuel gauge—instrument unit) to connector (fuel tank sending unit)
18	16	White	Bulkhead connector (tail lamps) to connector (left tail, stop & license lamp)
18A	16	White	Connector (left tail, stop & license lamp) to connector (left rear marker lamp)
18B	16	White	Connector (left rear marker lamp) to connector (right tail & stop lamp)
18C	16	White	Connector (right tail & stop lamp) to connector (right rear marker lamp)
23	16	Lt green w/tr.	Bulkhead connector (left turn & hazard) to connector (left tail, stop & license lamp)
24	16	Lt green	Bulkhead connector (right turn & hazard) to connector (right tail & stop lamp)
34	18	White w/tr.	Bulkhead connector (back-up lights) to connector (left back-up lamp)
34A	18	White w/tr.	Connector (left back-up lamp) to connector (right back-up lamp)

No.	GA.	Color	Harness Assembly—Directional signal Switch
5B	18	Green	Connector (steering column) to right front position contact
8B	18	Green w/tr.	Connector (steering column) to left front position contact
23A	16	Lt green w/tr.	Connector (steering column) to left rear position contact
24A	16	Lt green	Connector (steering column) to right rear position contact
65	16	Red w/tr.	Connector (steering column) to brake switch position contact
74	18	Red w/tr.	Connector (steering column) to flasher position contact

(*)Combined with 6 cylinder wiring
W/tr. = With tracer

1974-75

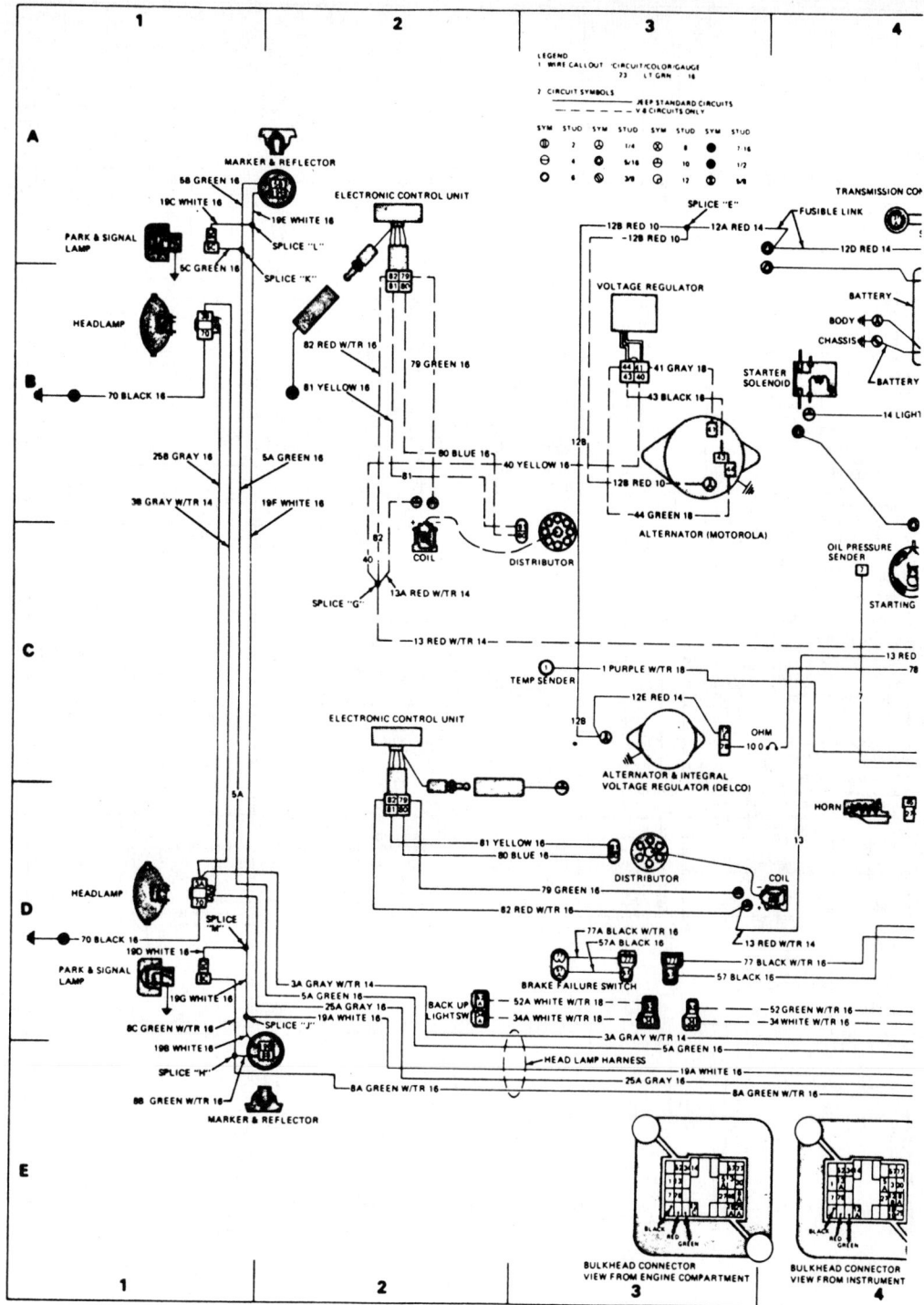

1974-75

1974-75

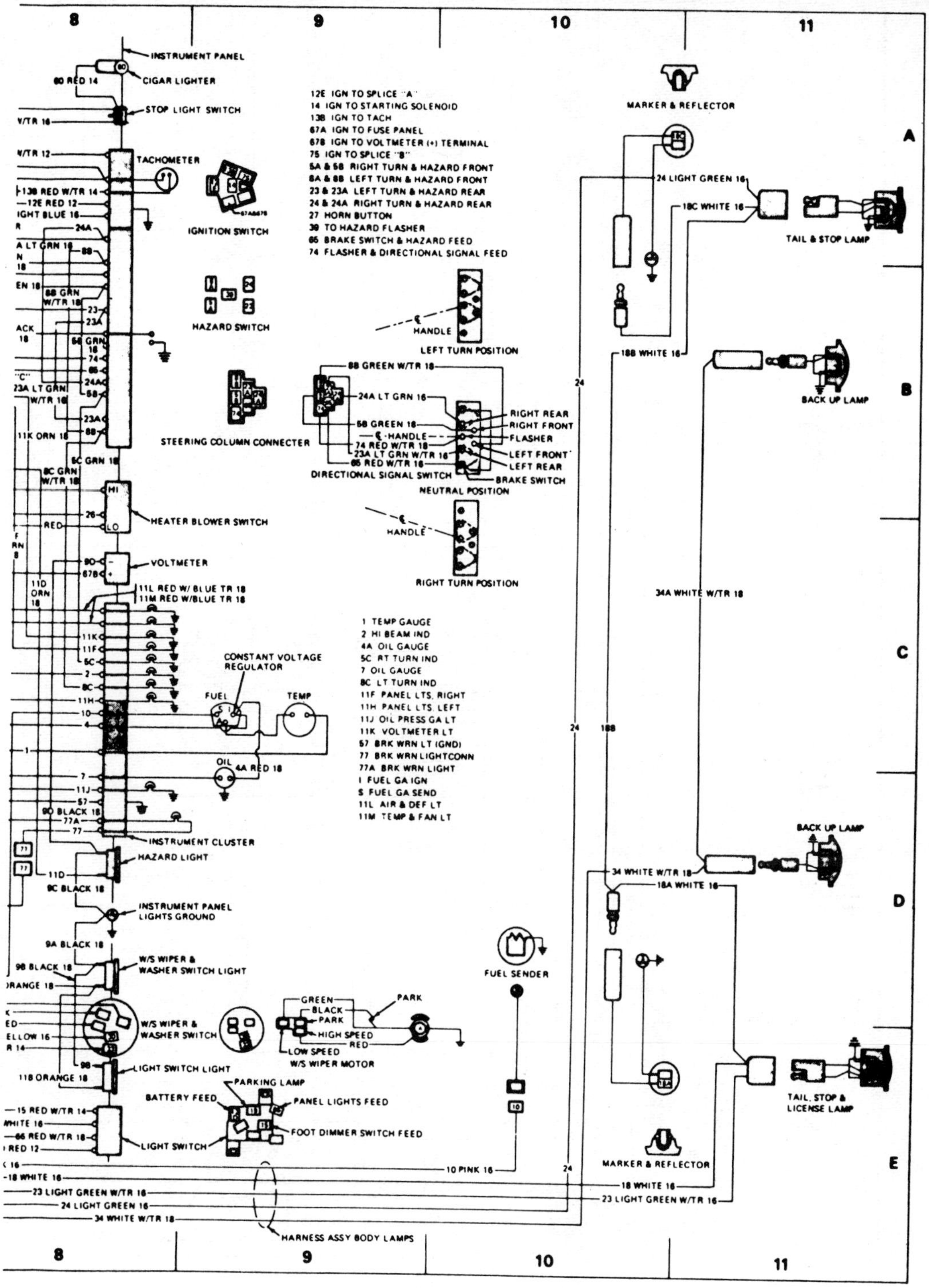

12E IGN TO SPLICE "A"
14 IGN TO STARTING SOLENOID
13B IGN TO TACH
67A IGN TO FUSE PANEL
67B IGN TO VOLTMETER (+) TERMINAL
75 IGN TO SPLICE "B"
5A & 5B RIGHT TURN & HAZARD FRONT
8A & 8B LEFT TURN & HAZARD FRONT
23 & 23A LEFT TURN & HAZARD REAR
24 & 24A RIGHT TURN & HAZARD REAR
27 HORN BUTTON
30 TO HAZARD FLASHER
65 BRAKE SWITCH & HAZARD FEED
74 FLASHER & DIRECTIONAL SIGNAL FEED

1 TEMP GAUGE
2 HI BEAM IND
4A OIL GAUGE
5C RT TURN IND
7 OIL GAUGE
8C LT TURN IND
11F PANEL LTS. RIGHT
11H PANEL LTS. LEFT
11J OIL PRESS GA LT
11K VOLTMETER LT
57 BRK WRN LT (GND)
77 BRK WRN LIGHTCONN
77A BRK WRN LIGHT
I FUEL GA IGN
S FUEL GA SEND
11L AIR & DEF LT
11M TEMP & FAN LT

1974-75

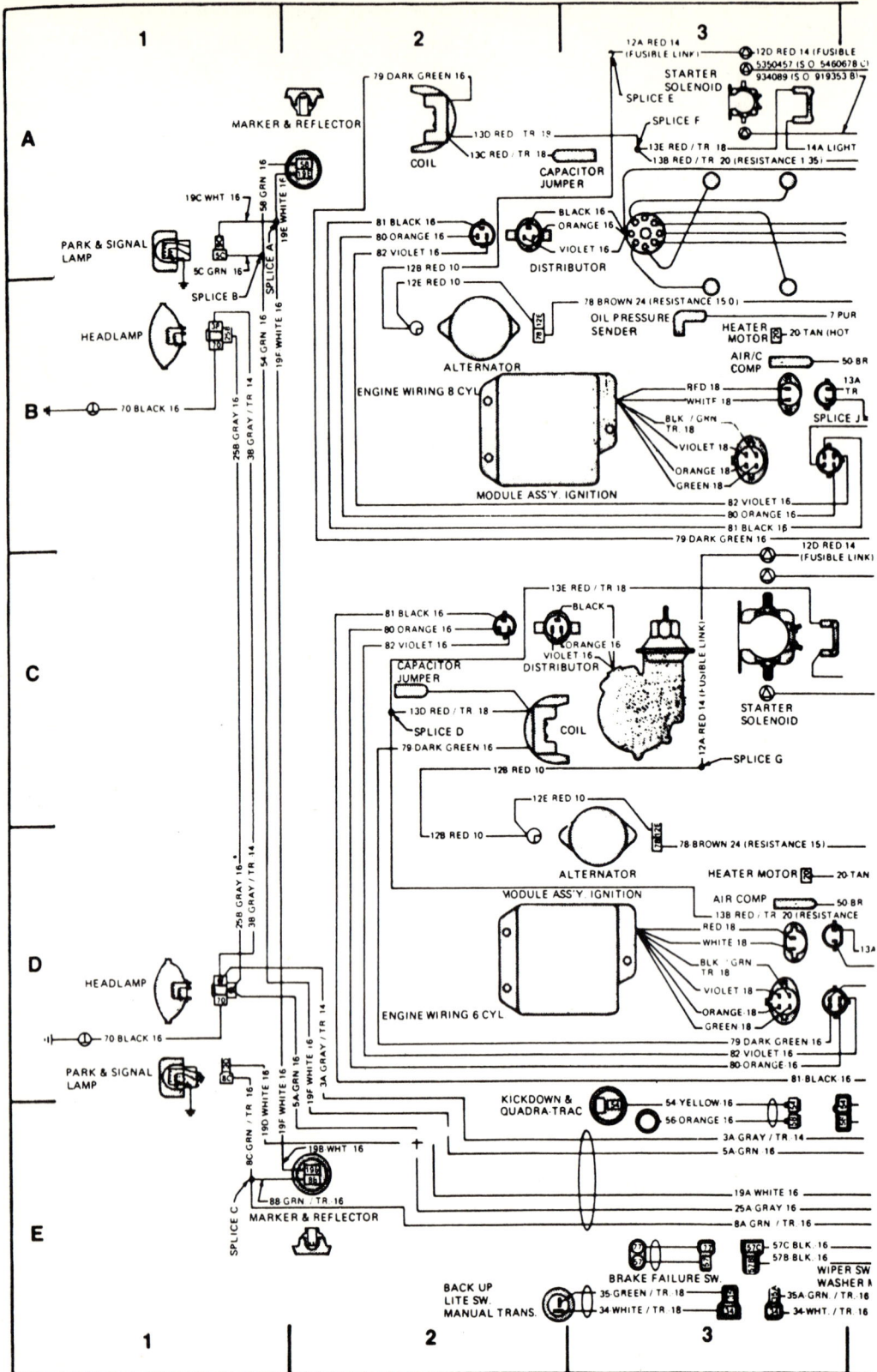

1976-79

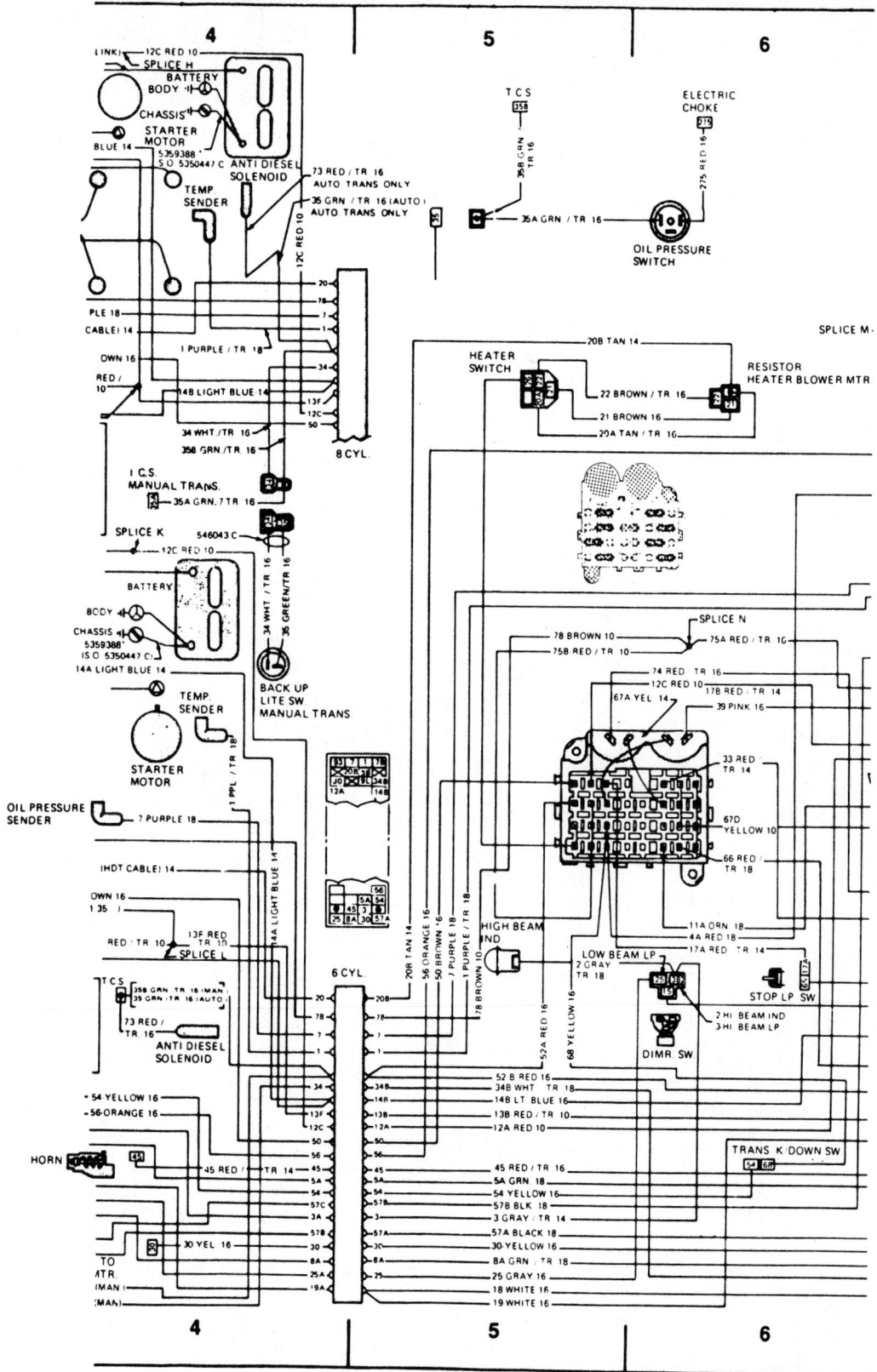

1976-79

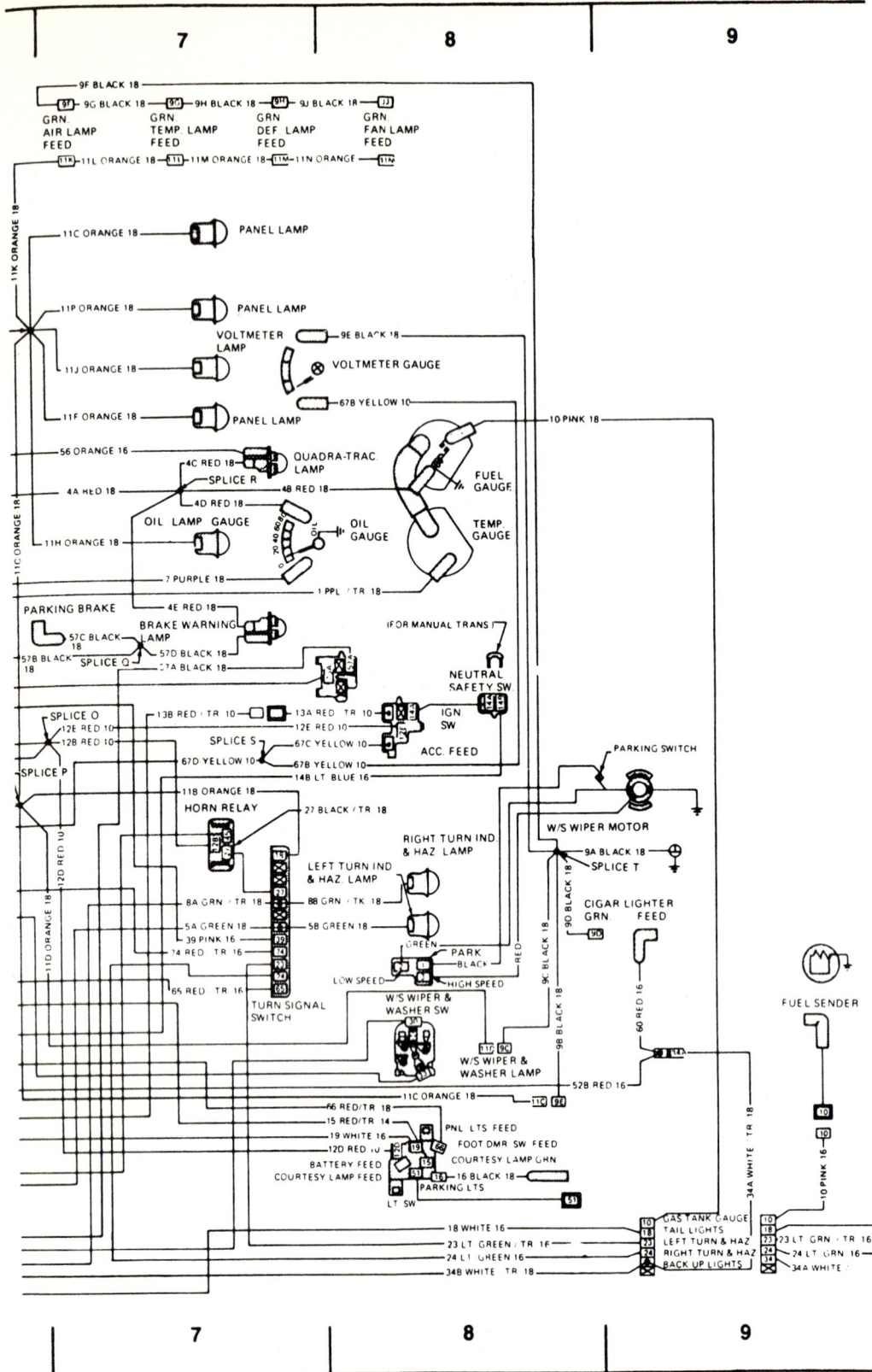

9F BLACK 18

9F · 9G BLACK 18 · 9G · 9H BLACK 18 · 9H · 9J BLACK 18 · 9J

GRN. AIR LAMP FEED GRN. TEMP. LAMP FEED GRN. DEF. LAMP FEED GRN. FAN LAMP FEED

11K · 11L ORANGE 18 · 11L · 11M ORANGE 18 · 11M · 11N ORANGE · 11N

11K ORANGE 18

11C ORANGE 18 — PANEL LAMP

11P ORANGE 18 — PANEL LAMP

VOLTMETER LAMP   9E BLACK 18

11J ORANGE 18 — VOLTMETER GAUGE

67B YELLOW 10

11F ORANGE 18 — PANEL LAMP

10 PINK 18

11C ORANGE 18

56 ORANGE 16 — QUADRA-TRAC. LAMP

4C RED 18 — SPLICE R

4A RED 18   4B RED 18   FUEL GAUGE

4D RED 18

OIL LAMP GAUGE   OIL GAUGE   TEMP. GAUGE

11H ORANGE 18

7 PURPLE 18

1 PPL / TR 18

PARKING BRAKE   4E RED 18

57C BLACK 18   BRAKE WARNING LAMP   (FOR MANUAL TRANS)

57B BLACK 18   SPLICE Q   57D BLACK 18

57A BLACK 18   NEUTRAL SAFETY SW

SPLICE O   13B RED 10   13A RED / TR 10   IGN SW

12E RED 10

12B RED 10   SPLICE S   67C YELLOW 10

67D YELLOW 10   67B YELLOW 10   ACC. FEED

SPLICE P   14B LT BLUE 16   PARKING SWITCH

11B ORANGE 18   W/S WIPER MOTOR

HORN RELAY   27 BLACK / TR 18

9A BLACK 18   SPLICE T

RIGHT TURN IND. & HAZ. LAMP

9D BLACK 18   CIGAR LIGHTER GRN FEED

LEFT TURN IND. & HAZ. LAMP   9C

8A GRN / TR 18   8B GRN / TK 18

5A GREEN 18   5B GREEN 18

39 PINK 16   GREEN   PARK BLACK   FUEL SENDER

74 RED / TR 16   LOW SPEED   HIGH SPEED

65 RED 16   W/S WIPER & WASHER SW

TURN SIGNAL SWITCH   11C 9C

W/S WIPER & WASHER LAMP

11D ORANGE 18   12D RED 10   9B BLACK 18   60 RED 16

11C ORANGE 18   11C 9C

66 RED/TR 18

15 RED/TR 14   PNL LTS FEED

19 WHITE 16   FOOT DMR. SW FEED

12D RED 10   COURTESY LAMP GRN

BATTERY FEED   52B RED 18

COURTESY LAMP FEED   16 BLACK 18

PARKING LTS   34A WHITE / TR 18   10 PINK 16

LT SW

18 WHITE 16   GAS TANK GAUGE   10

TAIL LIGHTS   18

23 LT GREEN / TR 16   LEFT TURN & HAZ   23 LT GRN / TR 16

24 LT GREEN 16   RIGHT TURN & HAZ   24 LT GRN 16

34B WHITE / TR 18   BACK UP LIGHTS   34A WHITE

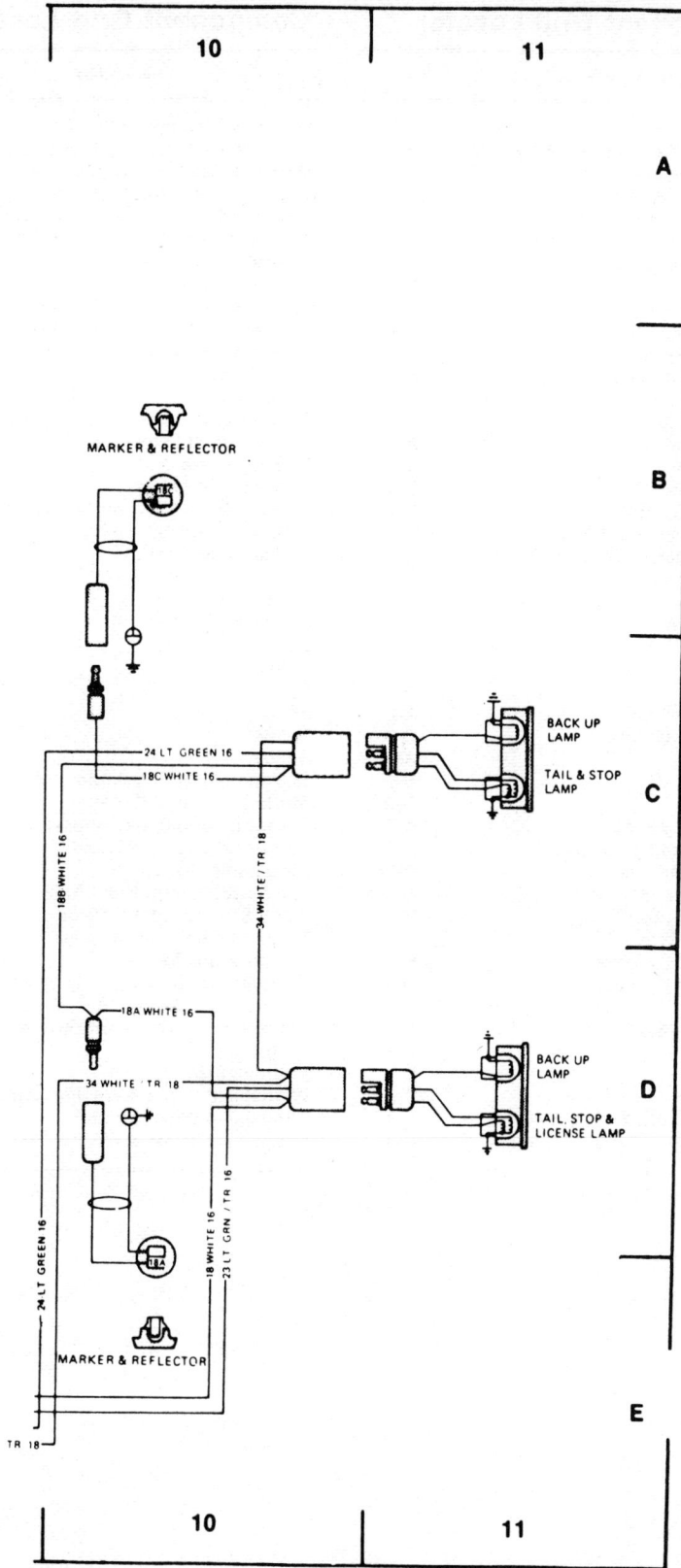

10   11

A

B

MARKER & REFLECTOR

BACK UP
LAMP

24 LT GREEN 16

18C WHITE 16

TAIL & STOP
LAMP

C

18B WHITE 16

34 WHITE / TR 18

18A WHITE 16

34 WHITE / TR 18

BACK UP
LAMP

D

TAIL, STOP &
LICENSE LAMP

24 LT GREEN 16

18 WHITE 16

23 LT GRN / TR 16

MARKER & REFLECTOR

E

TR 18

10   11

1976-79

Component Grid Locator

Nomenclature	Location
Accessory Feed	C-8
Air Conditioner Compressor, 8-Cylinder	B-3
Alternator, 8-Cylinder	B-2
Alternator, 6-Cylinder	D-3
Anti-Diesel Solenoid, 8-Cylinder	A-4
Anti-Diesel Solenoid, 6-Cylinder	D-4
Back-up Lamp, Right Side	C-11
Back-up Lamp, Left Side	D-11
Back-up Light Switch Manual Transmission	C-4
Back-up Light Switch Manual Transmission	E-2
Battery, 8-Cylinder	A-4
Battery, 6-Cylinder	C-4
Body, 8-Cylinder	A-4
Body, 6-Cylinder	C-4
Brake Failure Switch	E-3
Brake Warning Lamp	C-7
Capacitor Jumper, 8-Cylinder	A-3
Capacitor Jumper, 6-Cylinder	C-2
Chassis, 8-Cylinder	A-4
Chassis, 6-Cylinder	C-4
Cigar Lighter Feed	D-9
Coil, 8-Cylinder	A-2
Coil, 6-Cylinder	C-2
Dimmer Switch	D-6
Distributor, 8-Cylinder	A-2
Distributor, 6-Cylinder	C-3
Electric Choke	A-6
Fuel Gauge	B-6
Fuel Sender	D-9
Ground Air Lamp Feed	A-7
Ground Defogger Lamp Feed	A-7
Ground Fan Lamp Feed	A-8
Ground Temperature Lamp Feed	A-7
Headlamp, Right Side	B-1
Headlamp, Left Side	D-1
Heater Motor, 8-Cylinder	B-3
Heater Motor, 6-Cylinder	D-3
Heater Switch	B-5
High Beam Indicator	D-5
Horn	E-4
Horn Relay	D-7
Ignition Switch	C-8
Kickdown & Quadra-Trac, 6-Cylinder	E-3
Left Turn Indicator & Hazard Lamp	D-8
Low Beam L.P.	D-6

Component Grid Locator (cont.)

Nomenclature	Location
Marker & Reflector, Right Side	A-2
Marker & Reflector, Left Side	E-2
Marker & Reflector, Right Side	B-10
Marker & Reflector, Left Side	E-10
Module Assembly Ignition, 8-Cylinder	B-2
Module Assembly Ignition, 6-Cylinder	D-3
Neutral Safety Switch	C-8
Oil Lamp Gauge	B-7
Oil Pressure Sender, 8-Cylinder	B-3
Oil Pressure Sender, 6-Cylinder	C-4
Oil Pressure Switch	A-6
Panel Lamp	A-7
Panel Lamp	B-7
Park & Signal Lamp, Right Side	A-1
Park & Signal Lamp, Left Side	D-1
Parking Brake	C-7
Quadra-Trac Lamp	B-7
Resistor Heater Blower Motor	B-6
Splice, 8-Cylinder	B-4
Splice, 6-Cylinder	A-1
Splice, 6-Cylinder	C-4
Splice, 6-Cylinder	A-1
Splice, 6-Cylinder	C-2
Splice, 6-Cylinder	C-3
Splice	B-6
Splice	C-6
Splice	C-6
Splice	C-7
Splice	C-7
Starting Motor, 8-Cylinder	A-4
Starting Motor, 6-Cylinder	C-4
Starter Solenoid, 8-Cylinder	A-3
Starter Solenoid, 6-Cylinder	C-3
Stop Lamp Switch	D-6
Tail & Stop Lamp, Right Side	C-11
Tail & Stop Lamp, Left Side	D-11
T.C.S. Manual Transmission	B-4
Temperature Gauge	C-8
Temperature Sender, 8-Cylinder	A-4
Temperature Sender, 6-Cylinder	C-4
Transmission Kickdown Switch	E-6
Turn Signal Switch	D-7
Voltmeter Gauge	B-7
Windshield Wiper & Washer Lamp	D-8
Windshield Wiper Motor	C-9
Windshield Wiper & Washer Switch	D-8

1976-79

1980

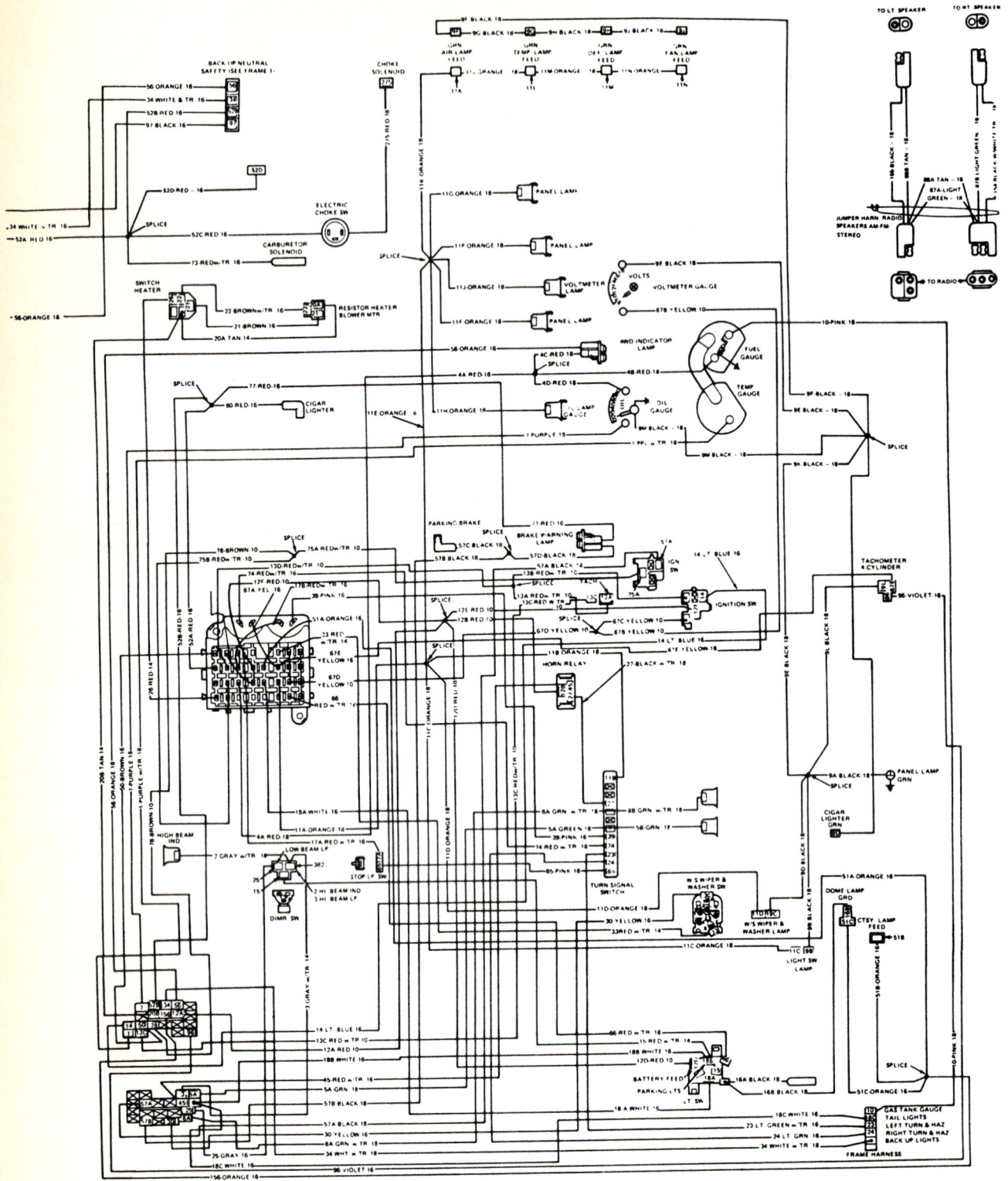

1980

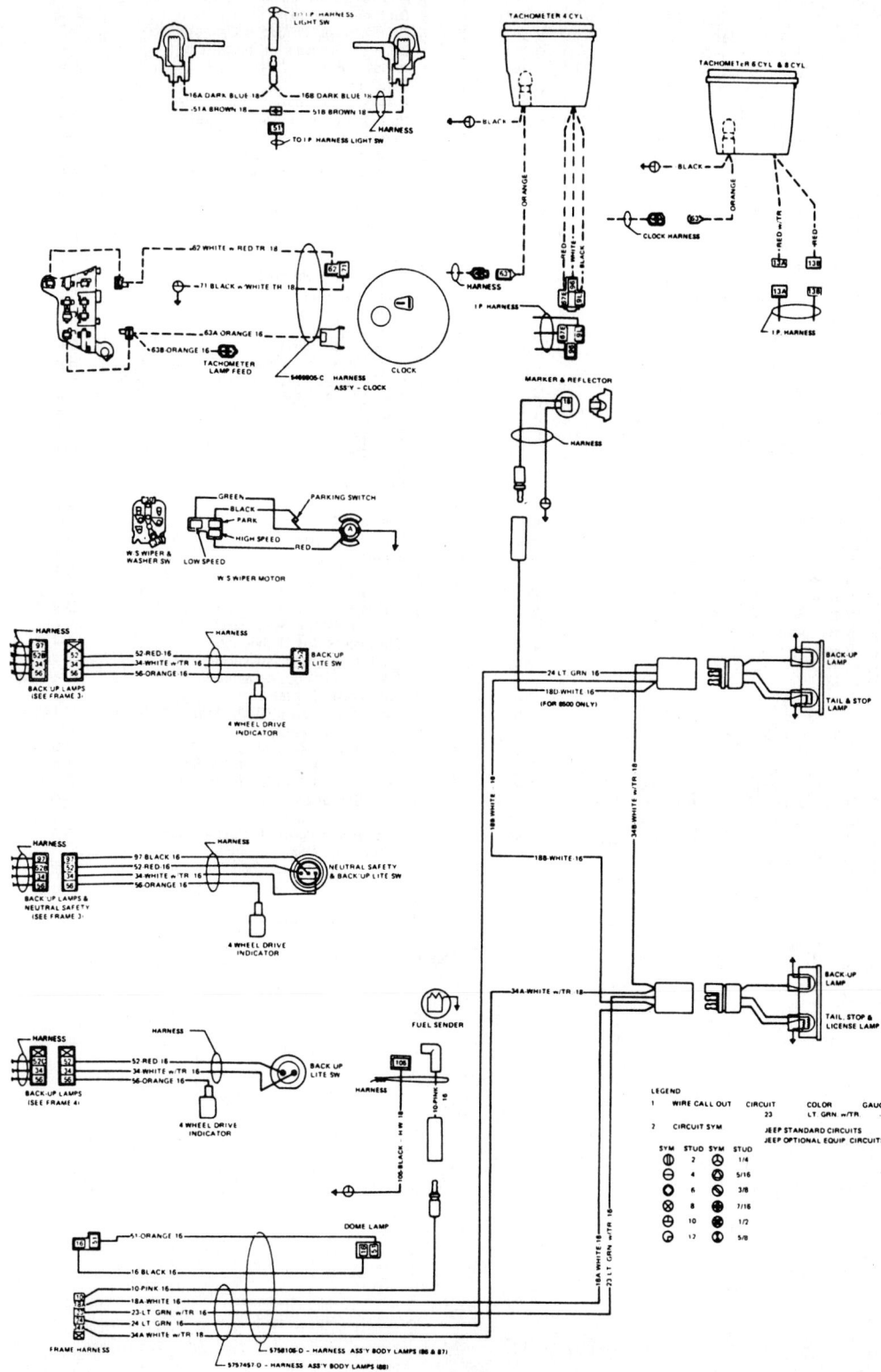

Wiring Diagram CJ Models

1980

NOMENCLATURE	LOCATION	NOMENCLATURE	LOCATION
Accessory Feed	C-8	Neutral Safety Switch	D-12
Air Conditioner Compressor, 8-Cylinder	B-4	Oil Lamp Gauge	B-8
Alternator, 8-Cylinder	B-3	Oil Pressure Sender, 8-Cylinder	B-3
Alternator, 6-Cylinder	D-3	Oil Pressure Sender, 6-Cylinder	C-4
Back-up Lamp, Right Side	C-14	Panel Lamp	A-8
Back-up Lamp, Left Side	D-14	Panel Lamp	B-8
Back-up Light Switch Manual Transmission	C-12	Park & Signal Lamp, Right Side	A-1
Back-up Light Switch Automatic Transmission	D-12	Park & Signal Lamp, Left Side	D-1
Battery, 8-Cylinder	A-4	Parking Brake	C-7
Battery, 6-Cylinder	C-4	Four Wheel Drive Indicator	B-8
Body Ground, 8-Cylinder	A-4	Resistor, Heater Blower Motor	B-6
Body Ground, 6-Cylinder	C-4	Splice A	A-2
Brake Failure Switch	E-3	Splice B	A-2
Brake Warning Lamp	C-8	Splice C	E-2
Capacitor Jumper, 8-Cylinder	A-3	Splice D	E-2
Capacitor Jumper, 6-Cylinder	C-2	Splice E	E-2
Chassis Ground, 8-Cylinder	A-4	Splice F, 8-Cylinder	A-3
Chassis Ground, 6-Cylinder	C-4	Splice G, 8-Cylinder	A-3
Cigar Lighter Feed	D-9	Splice H, 6-Cylinder	C-2
Coil, 8-Cylinder	A-2	Splice I, 6-Cylinder	C-3
Coil, 6-Cylinder	C-3	Splice J, 6-Cylinder	D-4
Dimmer Switch	D-6	Splice K, 6-Cylinder	B-4
Distributor, 8-Cylinder	A-3	Splice L, 8-Cylinder	B-4
Distributor, 6-Cylinder	C-3	Splice M, 8-Cylinder	A-4
Electric Choke	A-7	Splice N	A-6
Fuel Gauge	B-9	Splice O	B-6
Fuel Sender	D-12	Splice P	C-6
Ground Air Lamp Feed	A-7	Splice Q	B-7
Ground Defogger Lamp Feed	A-8	Splice R	C-7
Ground Fan Lamp Feed	A-8	Splice S	C-7
Ground Temperature Lamp Feed	A-8	Splice T	C-7
Headlamp, Right Side	B-1	Splice U	B-8
Headlamp, Left Side	D-1	Splice V	C-8
Heater Motor, 8-Cylinder	B-4	Splice W	C-8
Heater Motor, 6-Cylinder	D-4	Splice X	E-10
Heater Switch	B-6	Splice Y	D-9
High Beam Indicator	D-6	Starting Motor, 8-Cylinder	A-4
Horn	E-4	Starting Motor, 6-Cylinder	C-4
Horn Relay	C-8	Starter Solenoid, 8-Cylinder	A-4
Ignition Switch	C-9	Starter Solenoid, 6-Cylinder	C-4
Left Turn Indicator & Hazard Lamp	D-9	Stop Lamp Switch	D-7
Low Beam L.P.	D-6	Tail & Stop Lamp, Right Side	C-14
Marker & Reflector, Right Side — Front	A-2	Tail & Stop Lamp, Left Side	D-14
Marker & Reflector, Left Side — Front	E-2	Temperature Gauge	B-9
Marker & Reflector, Right Side — Rear	B-13	Temperature Sender, 8-Cylinder	A-4
Marker & Reflector, Left Side — Rear	E-13	Temperature Sender, 6-Cylinder	C-4
Module Assembly Ignition, 8-Cylinder	B-3	Turn Signal Switch	D-8
Module Assembly Ignition, 6-Cylinder	D-3	Voltmeter Gauge	B-8
		Windshield Wiper & Washer Lamp	D-9
		Windshield Wiper Motor	C-11
		Windshield Wiper & Washer Switch	D-8

1980

1981

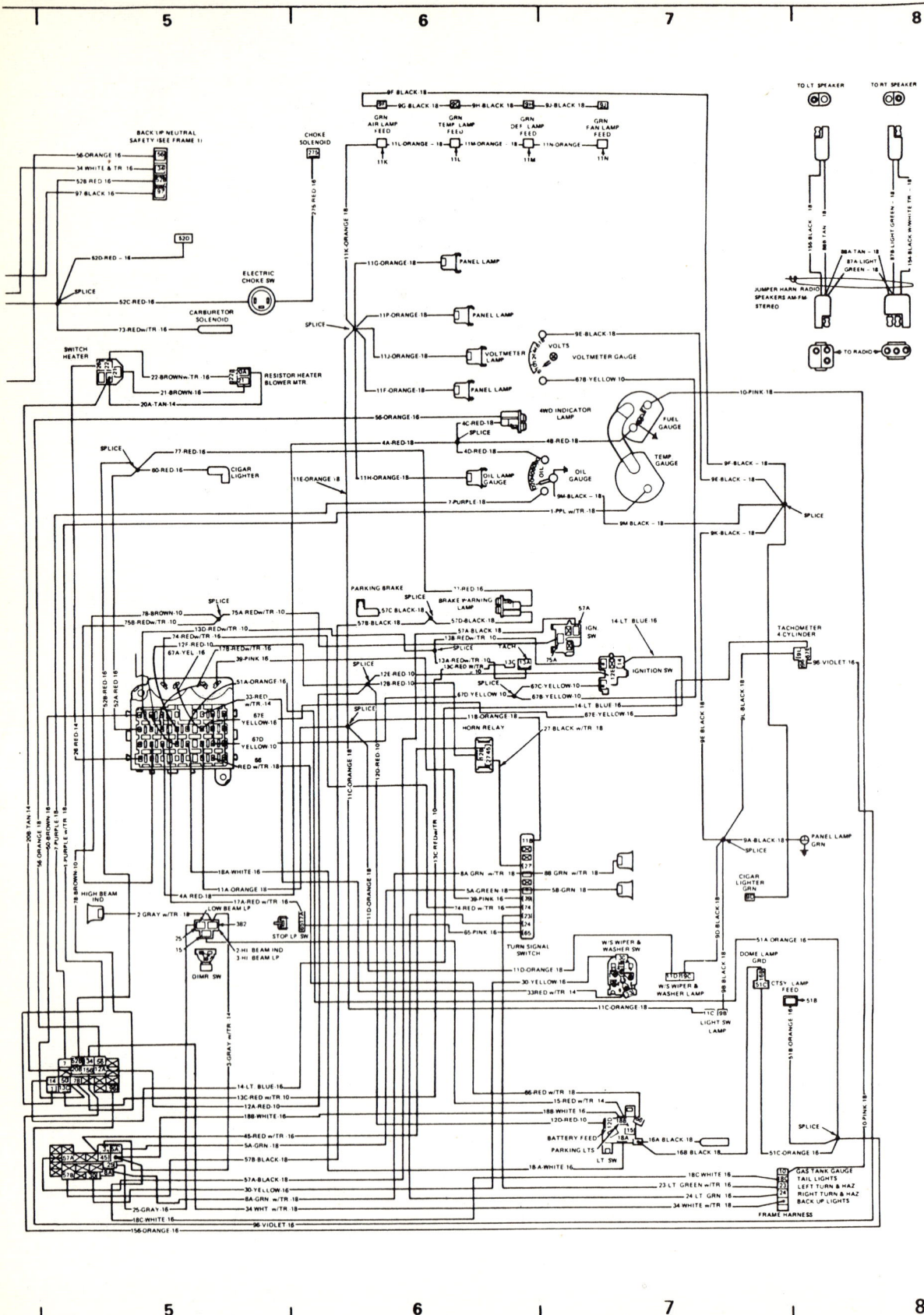

1981

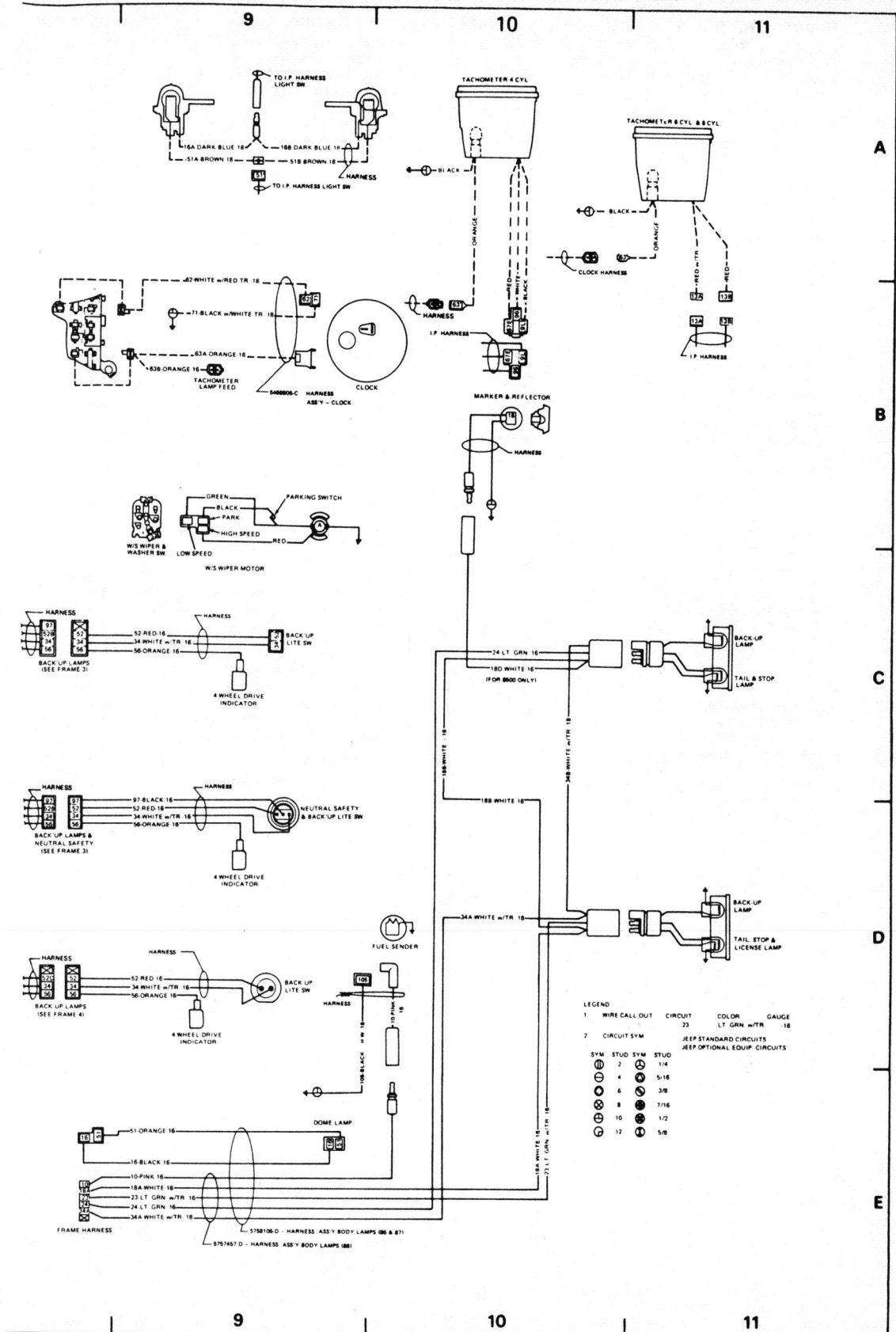

1981

NOMENCLATURE	LOCATION
Accessory Feed	C-7
Air Conditioner Compressor, 8-Cylinder	B-3
Alternator, 8-Cylinder	B-2
Alternator, 6-Cylinder	D-3
Back-up Lamp, Right Side	C-11
Back-up Lamp, Left Side	D-11
Back-up Light Switch Manual Transmission	C-9, D-9
Back-up Light Switch Automatic Transmission	D-9
Battery, 8-Cylinder	A-4
Battery, 6-Cylinder	C-4
Body Ground, 8-Cylinder	A-4
Body Ground, 6-Cylinder	C-4
Brake Failure Switch	E-3
Brake Warning Lamp	C-6
Capacitor Jumper, 8-Cylinder	A-3
Capacitor Jumper, 6-Cylinder	C-2
Chassis Ground, 8-Cylinder	A-4
Chassis Ground, 6-Cylinder	C-4
Cigar Lighter Feed	D-7
Coil, 8-Cylinder	A-2
Coil, 6-Cylinder	C-3
Dimmer Switch	D-5
Distributor, 8-Cylinder	A-3
Distributor, 6-Cylinder	C-3
Electric Choke	B-5, E-1
Fuel Gauge	B-7
Fuel Sender	D-10
Ground Air Lamp Feed	A-6
Ground Defogger Lamp Feed	A-6
Ground Fan Lamp Feed	A-7
Ground Temperature Lamp Feed	A-6
Headlamp, Right Side	B-1
Headlamp, Left Side	D-1
Heater Motor, 8-Cylinder	B-3
Heater Motor, 6-Cylinder	D-3
Heater Switch	B-5
High Beam Indicator	D-5
Horn	E-3
Horn Relay	C-6
Ignition Switch	C-7
Left Turn Indicator & Hazard Lamp	E-7
Low Beam L.P.	D-5
Marker & Reflector, Right Side — Front	A-2
Marker & Reflector, Left Side — Front	E-2
Marker & Reflector, Right Side — Rear	B-10
Marker & Reflector, Left Side — Rear	E-10
Module Assembly Ignition, 8-Cylinder	B-3
Module Assembly Ignition, 6-Cylinder	D-3

NOMENCLATURE	LOCATION
Neutral Safety Switch	D-9
Oil Lamp Gauge	B-6
Oil Pressure Sender, 8-Cylinder	B-3
Oil Pressure Sender, 6-Cylinder	D-3
Panel Lamp	B-6
Park & Signal Lamp, Right Side	A-1
Park & Signal Lamp, Left Side	D-1
Parking Brake	C-6
Four Wheel Drive Indicator	C-9, D-9
Resistor, Heater Blower Motor	B-5
Splice A	A-2
Splice B	B-2
Splice C	C-2
Splice D	E-2
Splice E	E-2
Splice F	E-2
Splice G	A-3
Splice H	A-3
Splice I	A-3
Splice J	B-3
Splice K	C-3
Splice L	C-3
Splice M	D-4
Splice N	B-5
Splice O	B-5
Splice P	C-5
Splice Q	B-6
Splice R	B-6
Splice S	C-6
Splice T	C-6
Splice U	C-6
Splice V	C-6
Splice W	C-6
Splice X	D-7
Splice Y	E-8
Starting Motor, 8-Cylinder	A-3
Starting Motor, 6-Cylinder	C-4
Starter Solenoid, 8-Cylinder	A-3
Starter Solenoid, 6-Cylinder	C-3
Stop Lamp Switch	D-5
Tail & Stop Lamp, Right Side	C-11
Tail & Stop Lamp, Left Side	D-11
Temperature Gauge	B-7
Temperature Sender, 8-Cylinder	A-4
Temperature Sender, 6-Cylinder	C-4
Turn Signal Switch	D-6
Voltmeter Gauge	B-7
Windshield Wiper & Washer Lamp	D-7
Windshield Wiper Motor	C-9
Windshield Wiper & Washer Switch	D-7

1981

1982-86 6-cylinder feedback system

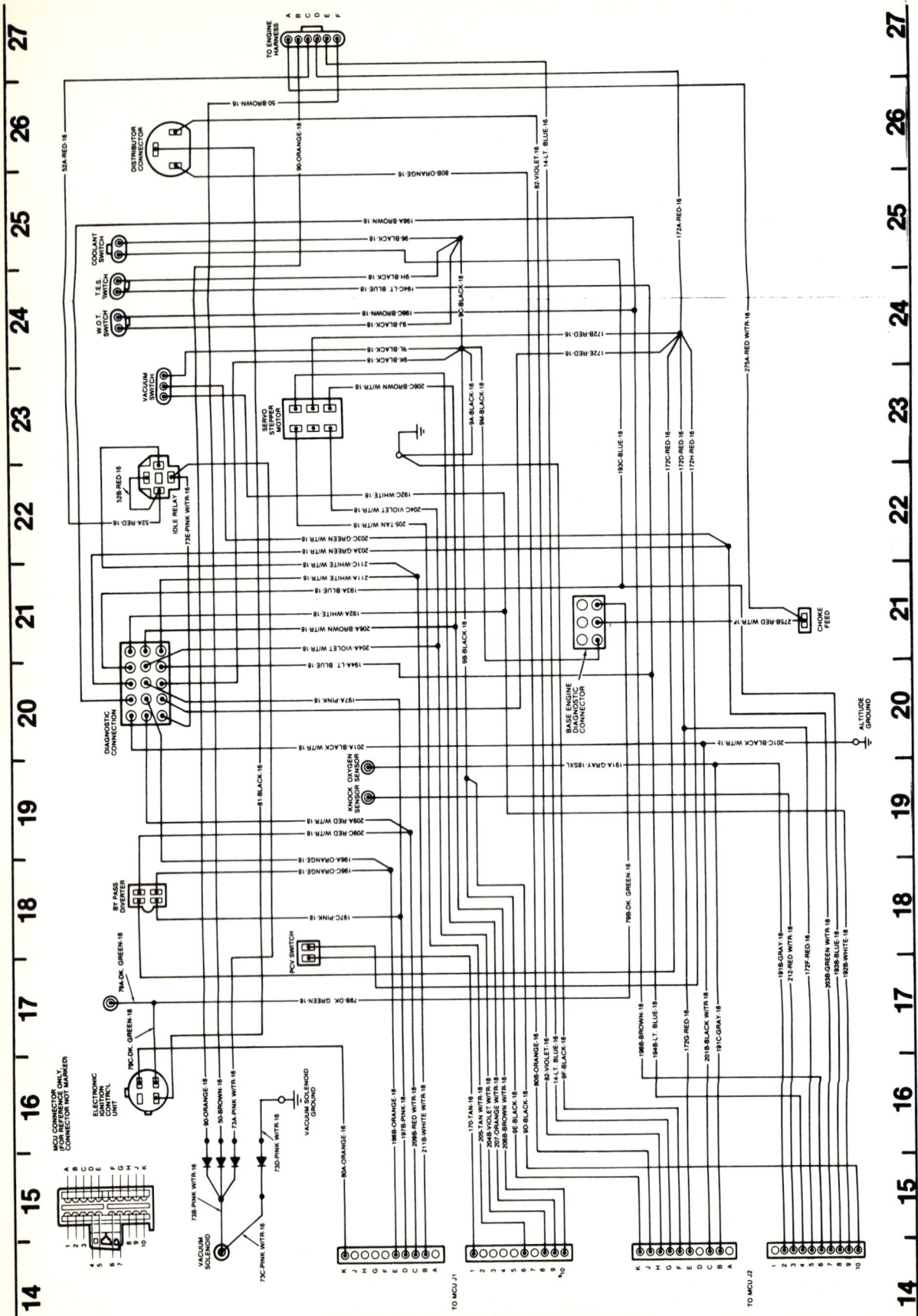

1982-86 with 6-cylinder engine

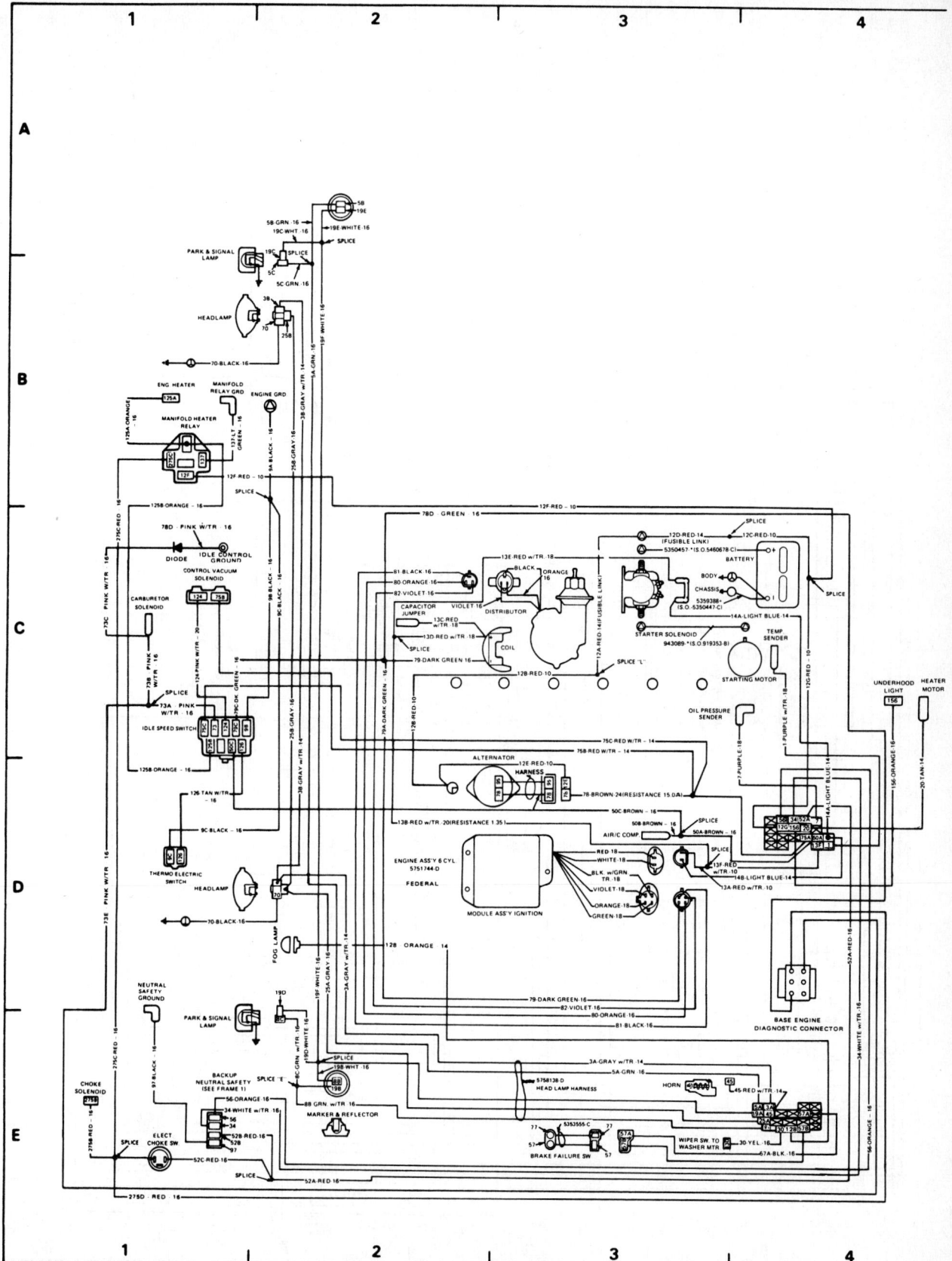

1982-86 with 6-cylinder engine

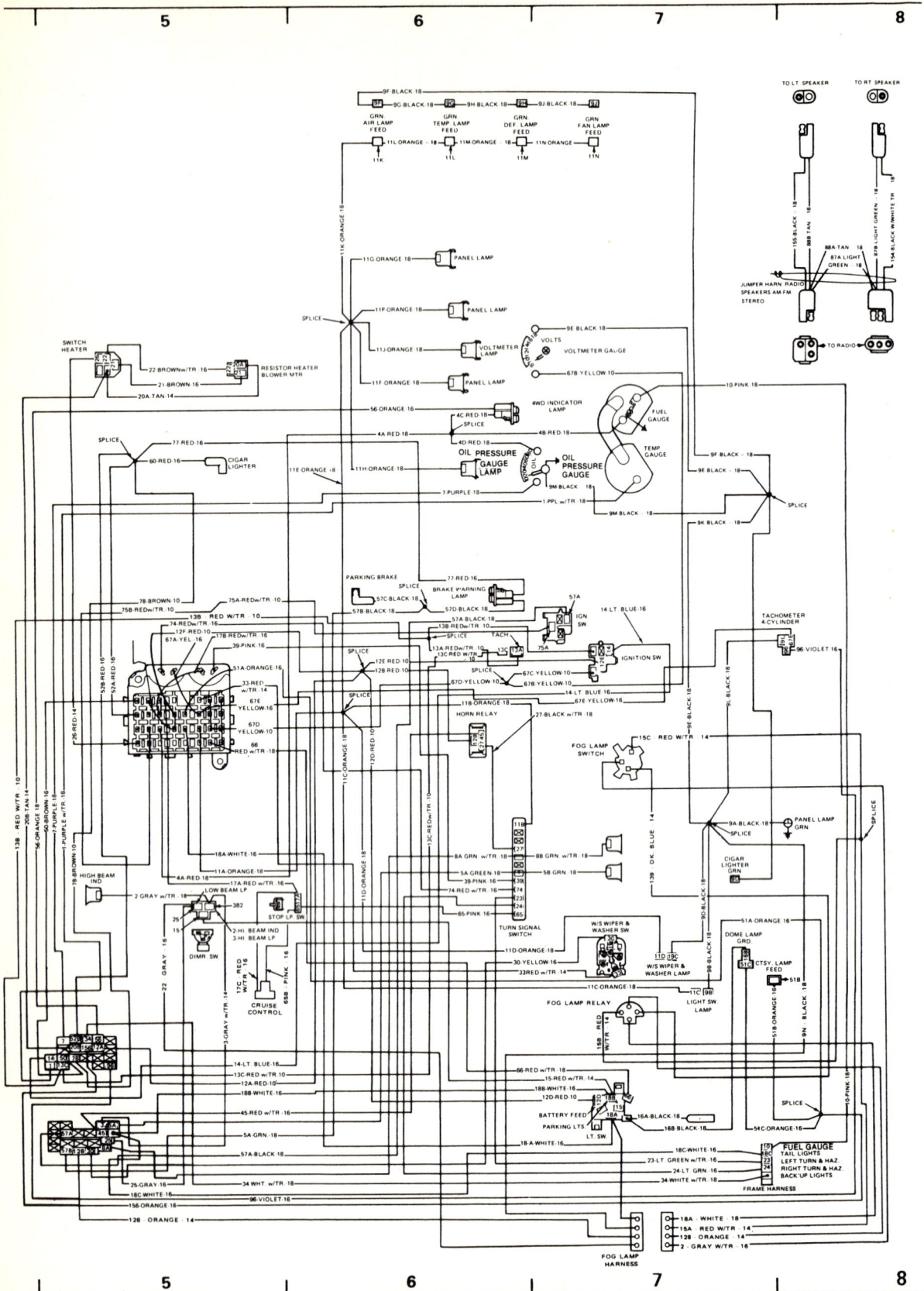

1982-86 with 6-cylinder engine

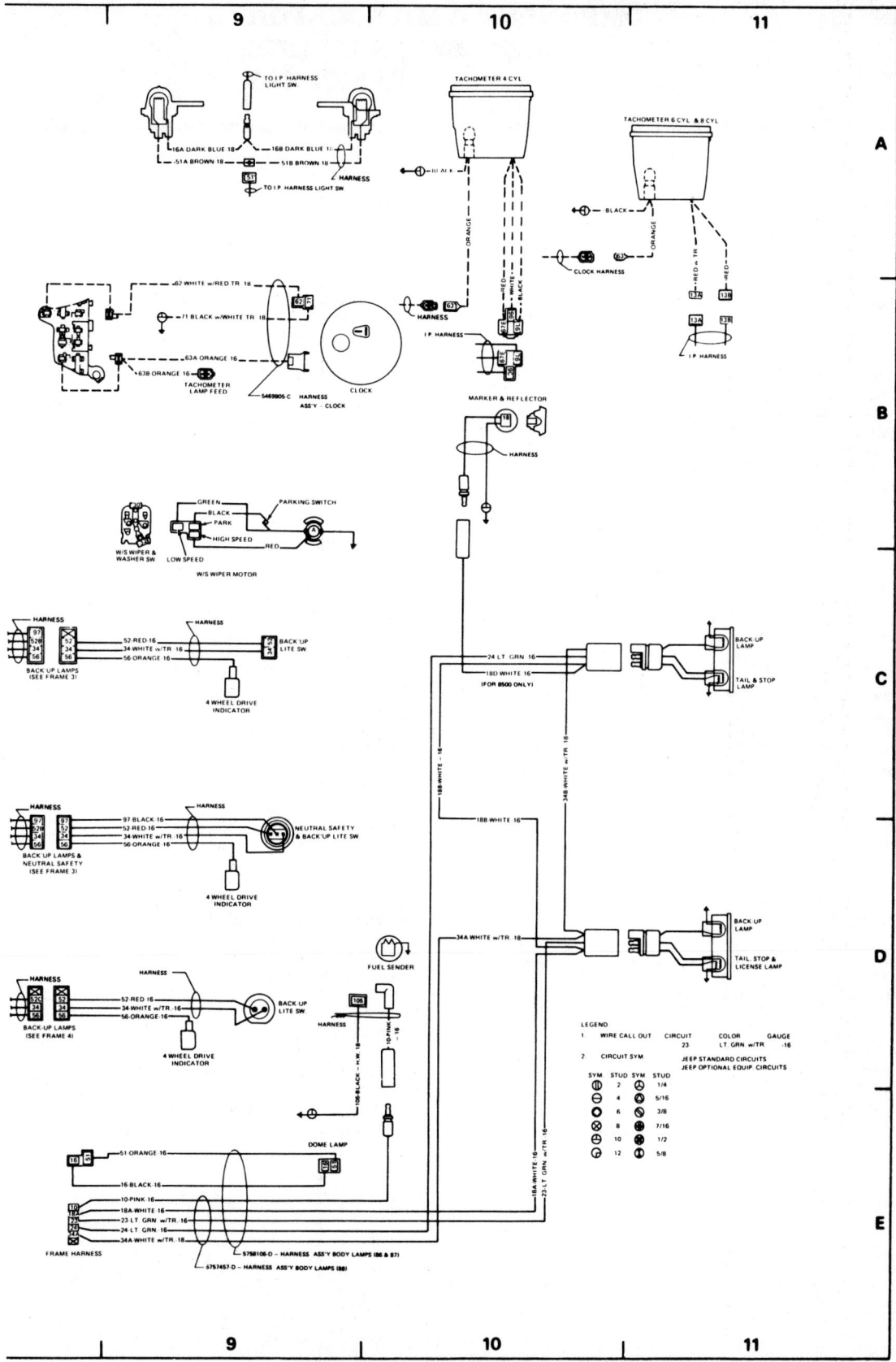

1982-86 with 6-cylinder engine

COMPONENT GRID LOCATOR
6-CYLINDER ENGINE
CJ MODELS

NOMENCLATURE	LOCATION
Air Conditioning Compressor	D-3
Alternator	D-3
Back-Up Lamp, Right Side	C-11
Back-Up Lamp, Left Side	D-11
Back-Up Light Switch Automatic Transmission	D-9
Back-Up Light Switch Manual Transmission	C-9, D-9
Back-Up Neutral Safety Switch	E-1
Battery	C-4
Body Ground	C-3
Brake Failure Switch	E-3
Brake Warning Lamp	C-6
Capacitor Jumper	C-2
Chassis Ground	C-3
Choke Solenoid	E-1
Cigar Lighter	B-5
Clock	B-9
Coil	C-3
Control Vacuum Solenoid	C-1
Cruise Control	D-5
Dimmer Switch	D-5
Distributor	C-3
Electric Choke Switch	E-1
Engine Heater	B-1
Fog Lamp	D-2
Fog Lamp Relay	D-7
Fog Lamp Switch	C-7
Four Wheel Drive Indicator	D-9
Four Wheel Drive Indicator Lamp	B-6
Fuel Gage	B-7
Fuel Sender	D-10
Headlamp, Left Side	D-1
Headlamp, Right Side	B-1
Heater Motor	C-4
Heater Switch	B-5
High Beam Indicator	D-5
Horn	E-3
Horn Relay	C-6
Ignition Switch	C-7
Light Switch Lamp	D-7
Low Beam L/P	D-5
Manifold Heater Relay	B-1
Marker and Reflector, Left Side, Front	E-2
Marker and Reflector, Right Side, Front	A-2
Marker and Reflector, Left Side, Rear	E-10
Marker and Reflector, Right Side, Rear	B-10
Module Assembly Ignition	D-3
Neutral Safety Ground	E-1
Neutral Safety Switch	D-9

NOMENCLATURE	LOCATION
Oil Gauge	B-7
Oil Lamp Gauge	B-6
Oil Pressure Sender	D-3
Panel Lamp	B-6
Park and Signal Lamp, Left Side	D-1
Park and Signal Lamp, Right Side	A-1
Parking Brake	C-6
Resistor, Heater Blower Motor	B-5
Splice A	A-2
Splice B	B-2
Splice C	C-2
Splice D	E-2
Splice E	E-2
Splice F	E-2
Splice G	A-3
Splice H	A-3
Splice I	A-3
Splice J	B-3
Splice K	C-3
Splice L	C-3
Splice M	D-4
Splice N	B-5
Splice O	B-5
Splice P	C-5
Splice Q	B-6
Splice R	B-6
Splice S	C-6
Splice T	C-6
Splice U	C-6
Splice V	C-6
Splice W	C-6
Splice X	D-7
Splice Y	E-8
Starting Motor	C-4
Starter Solenoid	C-3
Stoplamp Switch	D-5
Tachometer	A-10
Tail and Stop Lamp, Left Side	D-11
Tail and Stop Lamp, Right Side	C-11
Temperature Gauge	B-7
Temperature Sender	C-4
Thermo Electric Switch	D-1
Turn Signal Switch	D-6
Underhood Light	C-4
Voltmeter Gauge	B-7
Voltmeter Lamp	B-6
Windshield Wiper and Washer Lamp	D-7
Windshield Wiper Motor	C-9
Windhsield Wiper and Washer Switch	D-7

1982-86 4-cylinder engine

WIRE ROUTING AND COMPONENT LOCATION
ELECTRICAL WIRING DIAGRAMS 80 SERIES
ALPHABETICAL INDEX

WIRE ROUTING AND COMPONENT LOCATION
ELECTRICAL WIRING DIAGRAMS 80 SERIES
ALPHABETICAL INDEX

1987-90

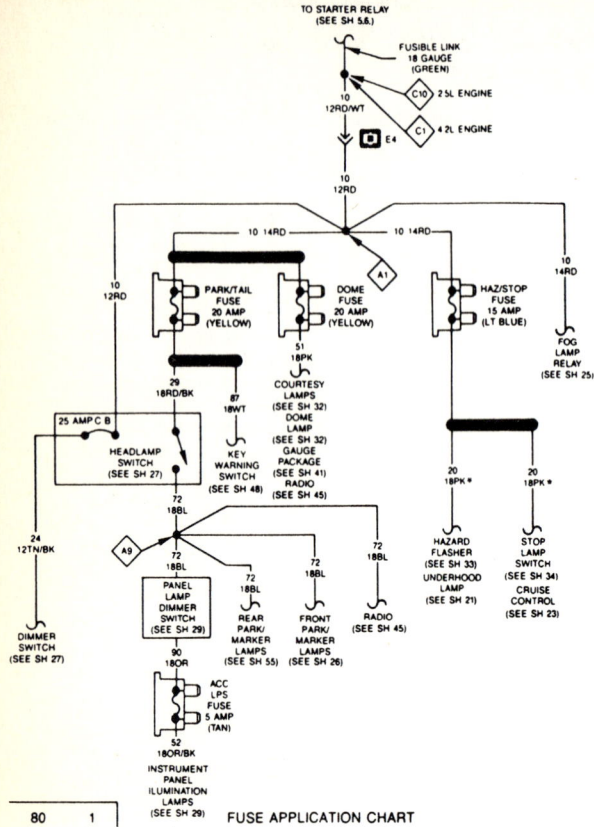

FUSE APPLICATION CHART

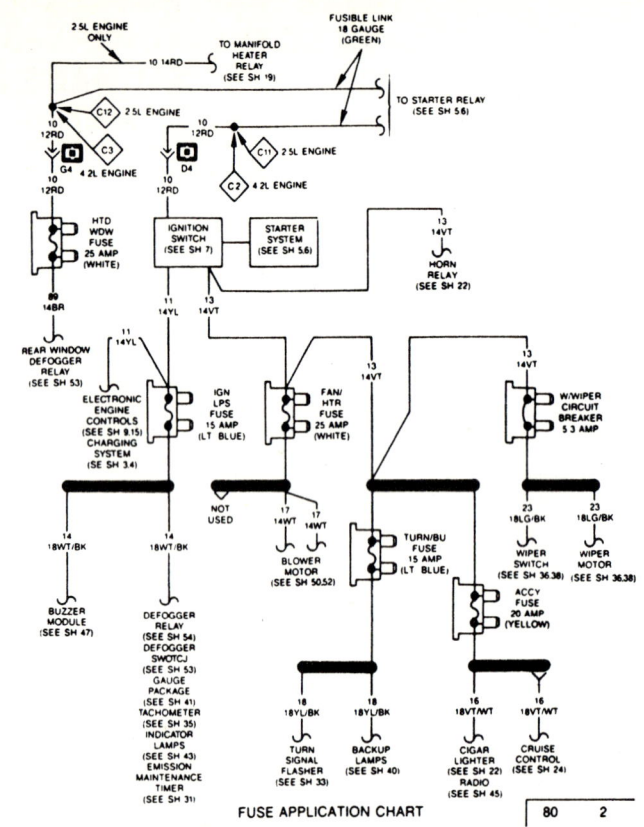

FUSE APPLICATION CHART

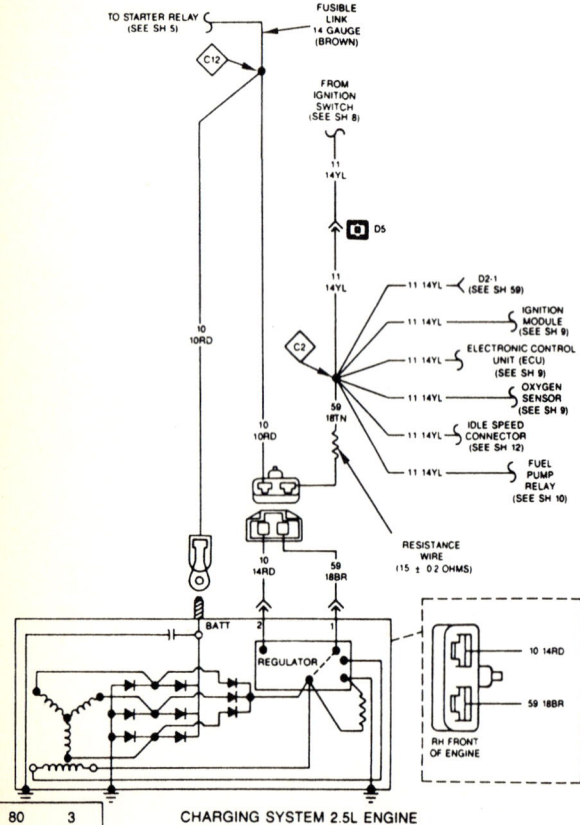

CHARGING SYSTEM 2.5L ENGINE

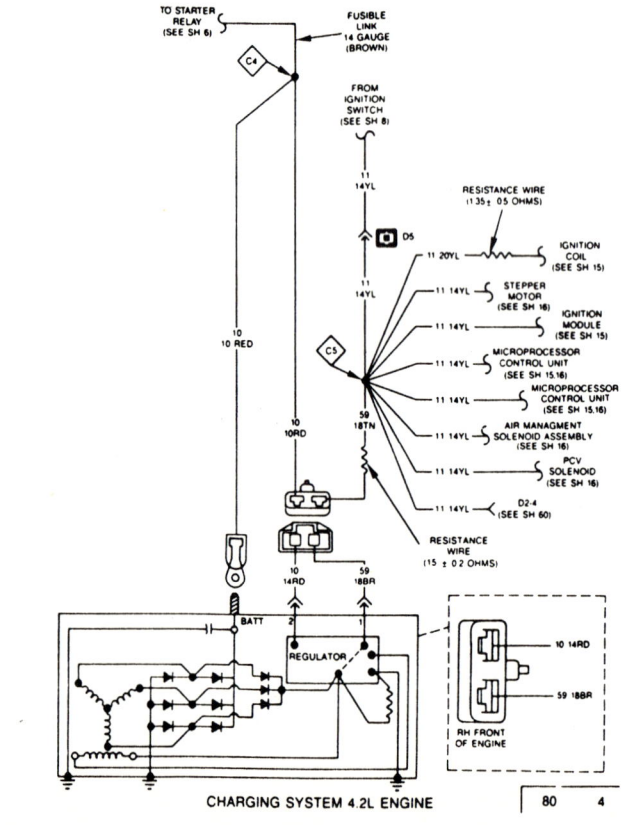

CHARGING SYSTEM 4.2L ENGINE

1987-90

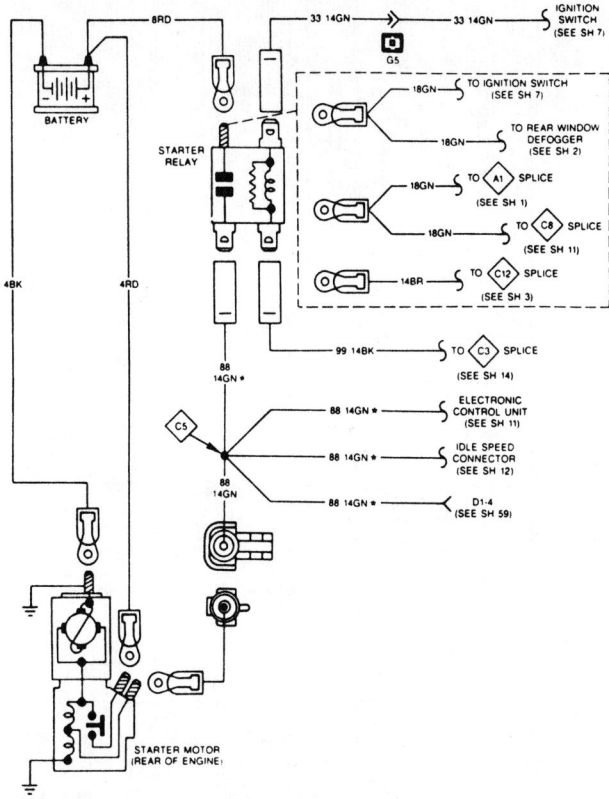

STARTER SYSTEM 2.5L ENGINE

| 80 | 5 |

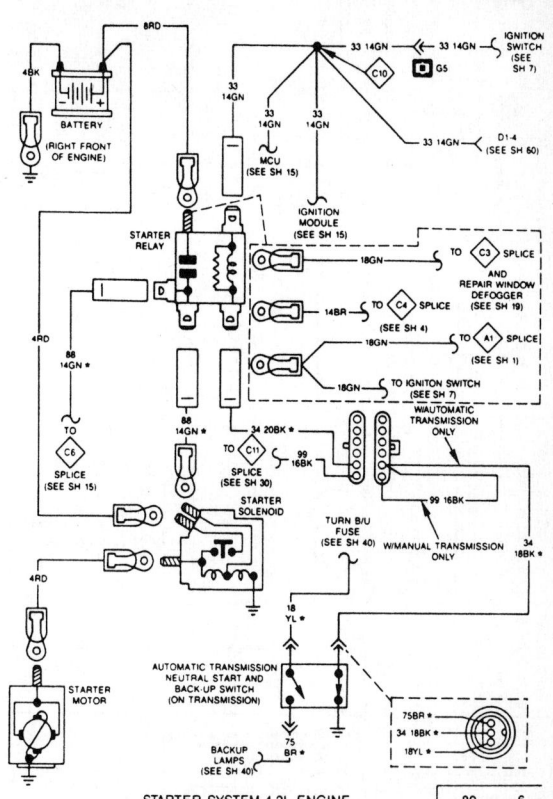

STARTER SYSTEM 4.2L ENGINE

| 80 | 6 |

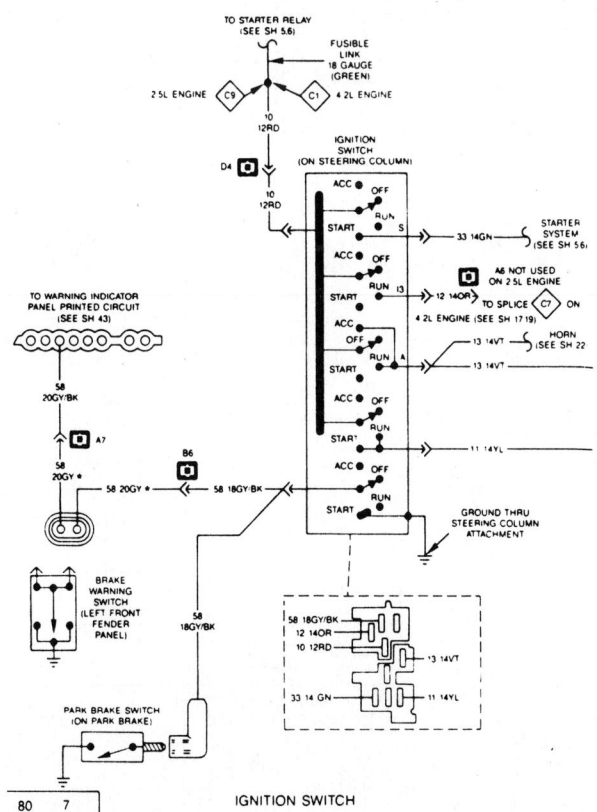

IGNITION SWITCH

| 80 | 7 |

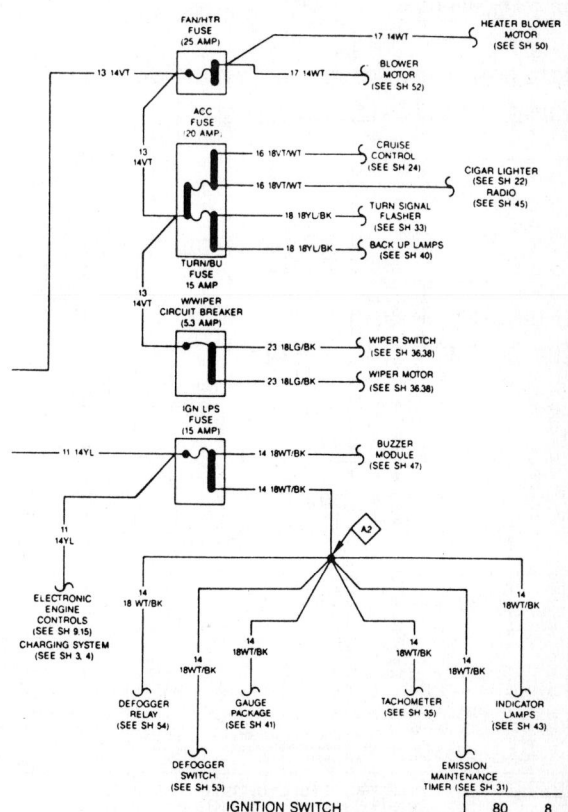

IGNITION SWITCH

| 80 | 8 |

1987-90

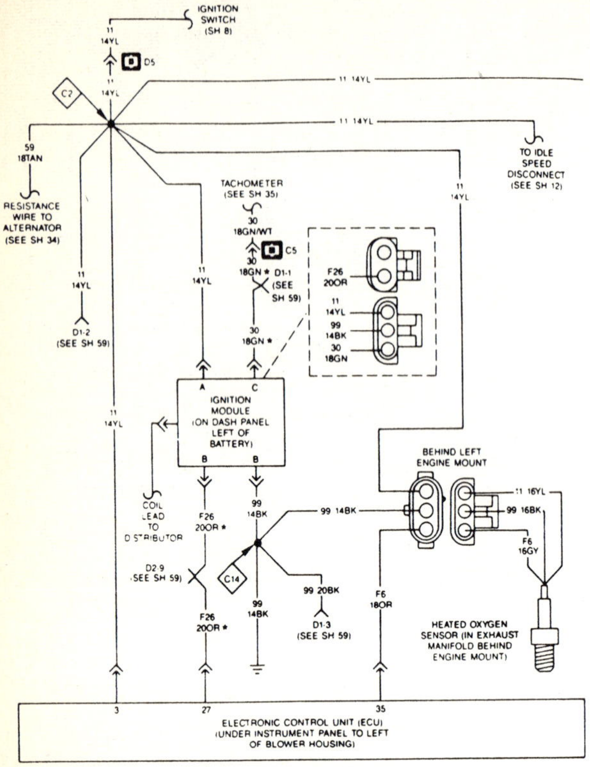

Top-left diagram:

IGNITION SWITCH (SH 8)

D5

C2

59 18TAN

RESISTANCE WIRE TO ALTERNATOR (SEE SH 34)

11 14YL

D1-2 (SEE SH 59)

11 14YL 14YL

11 14YL

11 14YL

11 14YL

11 14YL

TO IDLE SPEED DISCONNECT (SEE SH 12)

TACHOMETER (SEE SH 35)

30 18GN/WT

C5 D1-1

30 18GN * (SEE SH 59)

11 14YL

99 14BK

30 18GN *

F26 20OR

IGNITION MODULE (ON DASH PANEL LEFT OF BATTERY)

A C

B B

COIL LEAD TO DISTRIBUTOR

F26 20OR *

D2-9 (SEE SH 59)

C14

F26 20OR *

99 14BK

99 20BK *

D1-3 (SEE SH 59)

F6 18OR

99 14BK

99 14BK

BEHIND LEFT ENGINE MOUNT

99 16YL

99 16BK

F6 16GY

HEATED OXYGEN SENSOR (IN EXHAUST MANIFOLD BEHIND ENGINE MOUNT)

3 27 35

ELECTRONIC CONTROL UNIT (ECU) (UNDER INSTRUMENT PANEL TO LEFT OF BLOWER HOUSING)

ELECTRONIC FUEL INJECTION IGNITION SYSTEM 2.5L ENGINE

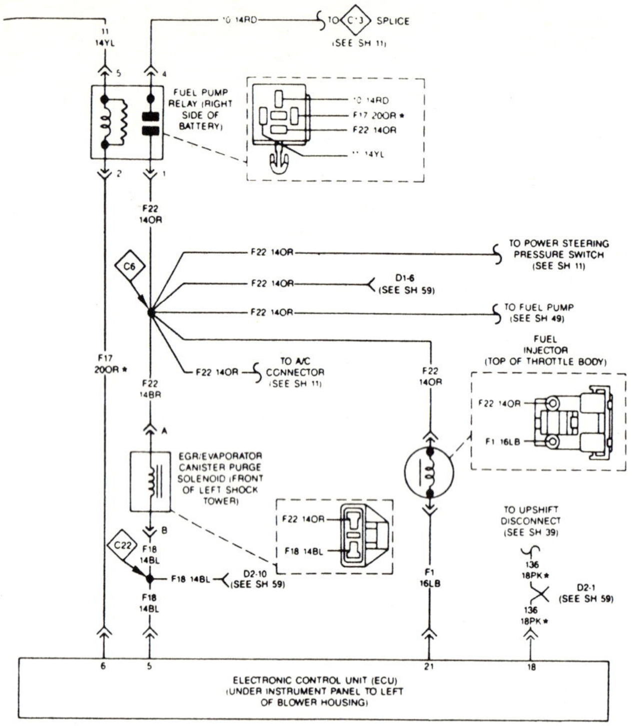

80 9

Top-right diagram:

TO C-13 SPLICE (SEE SH 11)

10 14RD

11 14YL

5 1

FUEL PUMP RELAY (RIGHT SIDE OF BATTERY)

10 14RD
F17 20OR *
F22 14OR
11 14YL

2 3

F22 14OR

C6

F22 14OR

F22 14OR

F22 14OR

TO POWER STEERING PRESSURE SWITCH (SEE SH 11)

D1-6 (SEE SH 59)

TO FUEL PUMP (SEE SH 49)

F17 20OR *

F22 14OR

TO A/C CONNECTOR (SEE SH 11)

F22 14BR

A

F22 14OR

FUEL INJECTOR (TOP OF THROTTLE BODY)

F22 14OR

F1 16LB

EGR/EVAPORATOR CANISTER PURGE SOLENOID (FRONT OF LEFT SHOCK TOWER)

A B

F22 14OR

F18 14BL

C22

F18 14BL

D2-10 (SEE SH 59)

F18 14BL

F18 14BL

F1 16LB

TO UPSHIFT DISCONNECT (SEE SH 39)

136 18PK *

D2-1 (SEE SH 59)

136 18PK *

6 5 21 18

ELECTRONIC CONTROL UNIT (ECU) (UNDER INSTRUMENT PANEL TO LEFT OF BLOWER HOUSING)

ELECTRONIC FUEL INJECTION IGNITION SYSTEM 2.5L ENGINE

80 10

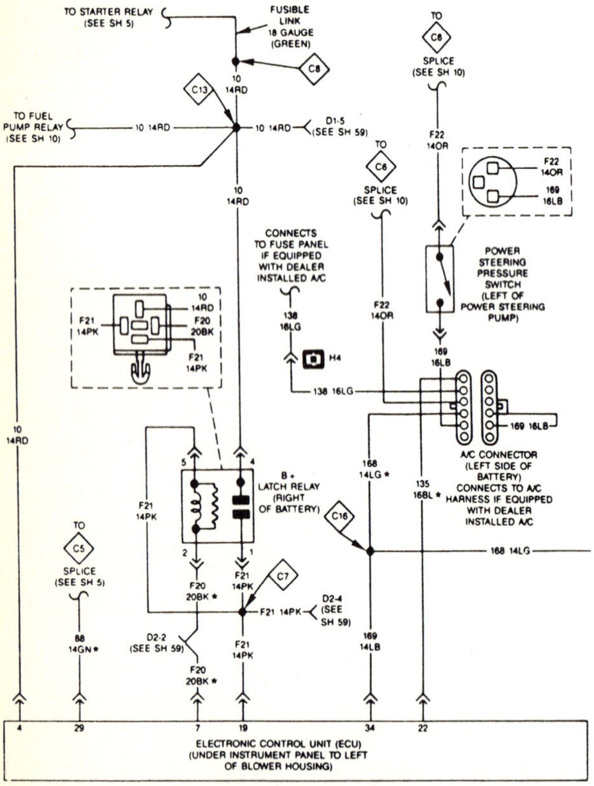

Bottom-left diagram:

TO STARTER RELAY (SEE SH 5)

FUSIBLE LINK 18 GAUGE (GREEN)

TO C8 SPLICE (SEE SH 10)

C13

10 14RD

C8

10 14RD

D1-5 (SEE SH 59)

TO FUEL PUMP RELAY (SEE SH 10)

10 14RD

10 14RD

F22 14OR

C6 SPLICE (SEE SH 10)

F22 14OR

169 16LB

POWER STEERING PRESSURE SWITCH (LEFT SIDE OF POWER STEERING PUMP)

CONNECTS TO FUSE PANEL IF EQUIPPED WITH DEALER INSTALLED A/C

138 18LG

H4

F22 14OR

169 16LB

F21 14PK
F20 20BK
F21 14PK

10 14RD

138 16LG

169 16LB

A/C CONNECTOR (LEFT SIDE OF BATTERY) CONNECTS TO A/C HARNESS IF EQUIPPED WITH DEALER INSTALLED A/C

10 14RD

5 4

F21 14PK

B+ LATCH RELAY (RIGHT OF BATTERY)

168 14LG *

135 16BL *

C16

TO C5 SPLICE (SEE SH 5)

2 1

F20 20BK *

F21 14PK

C7

D2-4 (SEE SH 59)

88 14GN *

D2-2 (SEE SH 59)

F21 14PK

F20 20BK *

168 14LG

169 14LB

4 29 7 19 34 22

ELECTRONIC CONTROL UNIT (ECU) (UNDER INSTRUMENT PANEL TO LEFT OF BLOWER HOUSING)

ELECTRONIC FUEL INJECTION IGNITION SYSTEM 2.5L ENGINE

80 11

Bottom-right diagram:

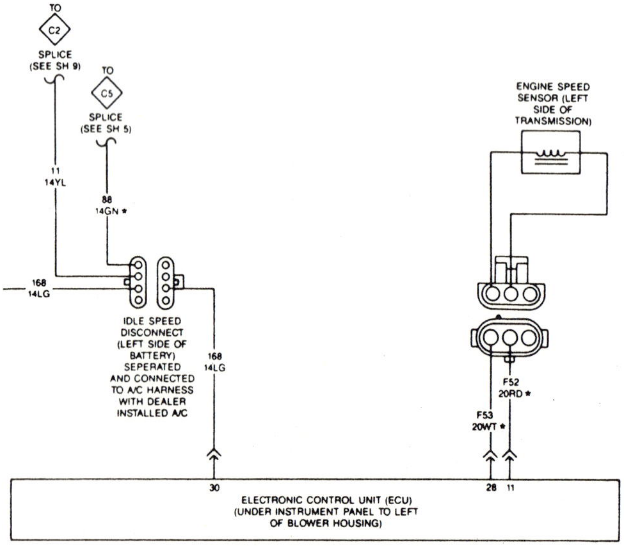

TO C2 SPLICE (SEE SH 9)

TO C5 SPLICE (SEE SH 5)

11 14YL

88 14GN *

ENGINE SPEED SENSOR (LEFT SIDE OF TRANSMISSION)

168 14LG

IDLE SPEED DISCONNECT (LEFT SIDE OF BATTERY SEPERATED AND CONNECTED TO A/C HARNESS WITH DEALER INSTALLED A/C)

168 14LG

F52 20RD *

F53 20WT *

30 28 11

ELECTRONIC CONTROL UNIT (ECU) (UNDER INSTRUMENT PANEL TO LEFT OF BLOWER HOUSING)

ELECTRONIC FUEL INJECTRION IGNITION SYSTEM 2.5L ENGINE

80 12

1987-90

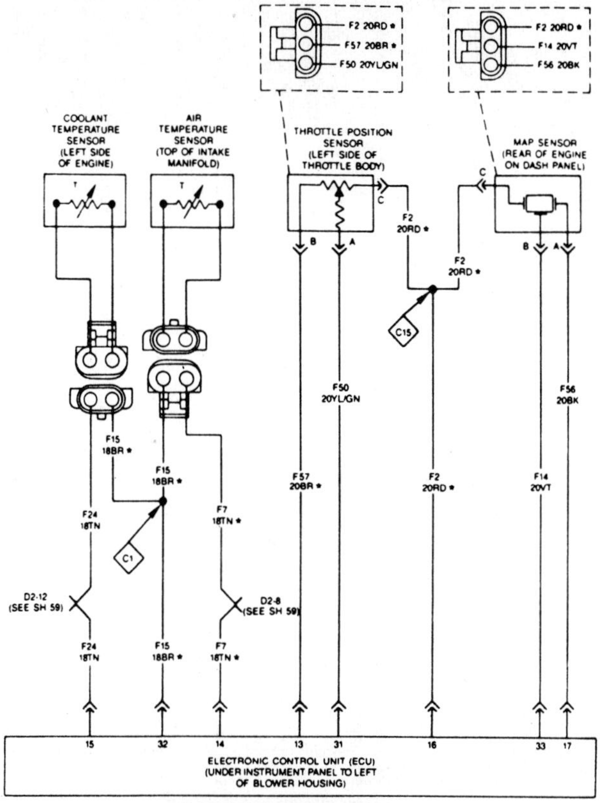

Top-left diagram

COOLANT TEMPERATURE SENSOR (LEFT SIDE OF ENGINE)

AIR TEMPERATURE SENSOR (TOP OF INTAKE MANIFOLD)

THROTTLE POSITION SENSOR (LEFT SIDE OF THROTTLE BODY)

MAP SENSOR (REAR OF ENGINE ON DASH PANEL)

F2 20RD *
F57 20BR *
F50 20YL/GN

F2 20RD *
F14 20VT
F56 20BK

ELECTRONIC CONTROL UNIT (ECU) (UNDER INSTRUMENT PANEL TO LEFT OF BLOWER HOUSING)

ELECTRONIC FUEL INJECTION IGNITION SYSTEM 2.5L ENGINE

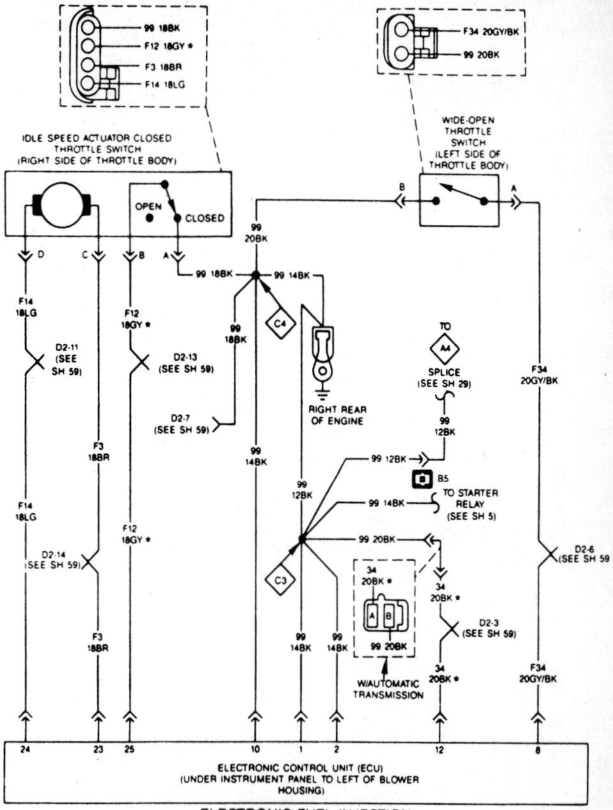

80 13

Top-right diagram

IDLE SPEED ACTUATOR CLOSED THROTTLE SWITCH (RIGHT SIDE OF THROTTLE BODY)

WIDE-OPEN THROTTLE SWITCH (LEFT SIDE OF THROTTLE BODY)

F9 18BK
F12 18GY *
F3 18BR
F14 18LG

F34 20GY/BK
99 20BK

OPEN   CLOSED

RIGHT REAR OF ENGINE

TO SPLICE (SEE SH 29)

TO STARTER RELAY (SEE SH 5)

W/AUTOMATIC TRANSMISSION

ELECTRONIC CONTROL UNIT (ECU) (UNDER INSTRUMENT PANEL TO LEFT OF BLOWER HOUSING)

ELECTRONIC FUEL INJECTION IGNITION SYSTEM 2.5L ENGINE

80 14

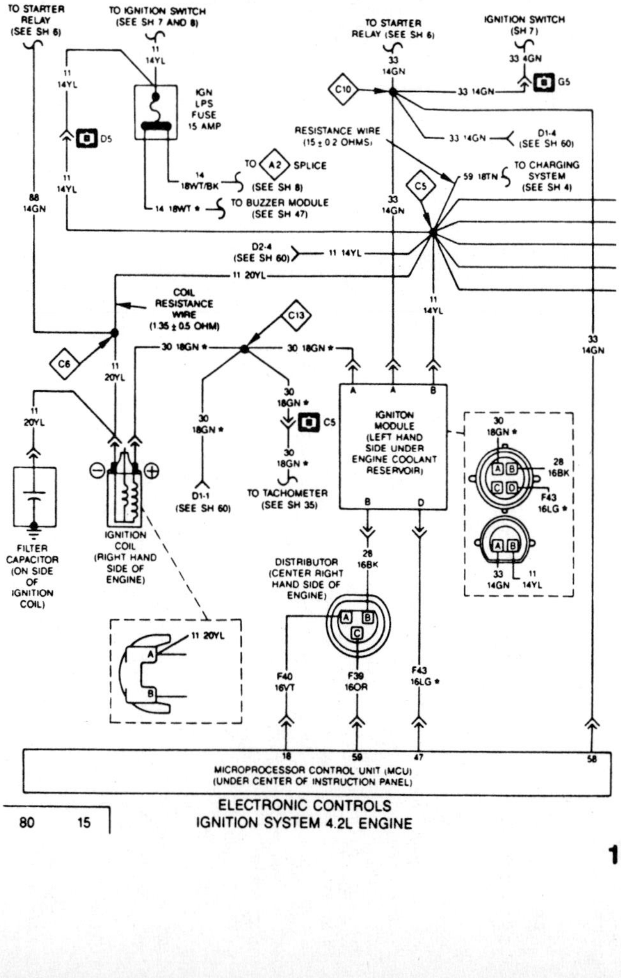

Bottom-left diagram

TO STARTER RELAY (SEE SH 6)

TO IGNITION SWITCH (SEE SH 7 AND 8)

TO STARTER RELAY (SEE SH 6)

IGNITION SWITCH (SH 7)

IGN LPS FUSE 15 AMP

RESISTANCE WIRE (15 ± 0.2 OHMS)

TO A2 SPLICE (SEE SH 8)

TO BUZZER MODULE (SEE SH 47)

TO CHARGING SYSTEM (SEE SH 4)

COIL RESISTANCE WIRE (1.35 ± 0.5 OHM)

FILTER CAPACITOR (ON SIDE OF IGNITION COIL)

IGNITION COIL (RIGHT HAND SIDE OF ENGINE)

IGNITION MODULE (LEFT HAND SIDE UNDER ENGINE COOLANT RESERVOIR)

TO TACHOMETER (SEE SH 35)

DISTRIBUTOR (CENTER RIGHT HAND SIDE OF ENGINE)

MICROPROCESSOR CONTROL UNIT (MCU) (UNDER CENTER OF INSTRUCTION PANEL)

ELECTRONIC CONTROLS IGNITION SYSTEM 4.2L ENGINE

80 15

Bottom-right diagram

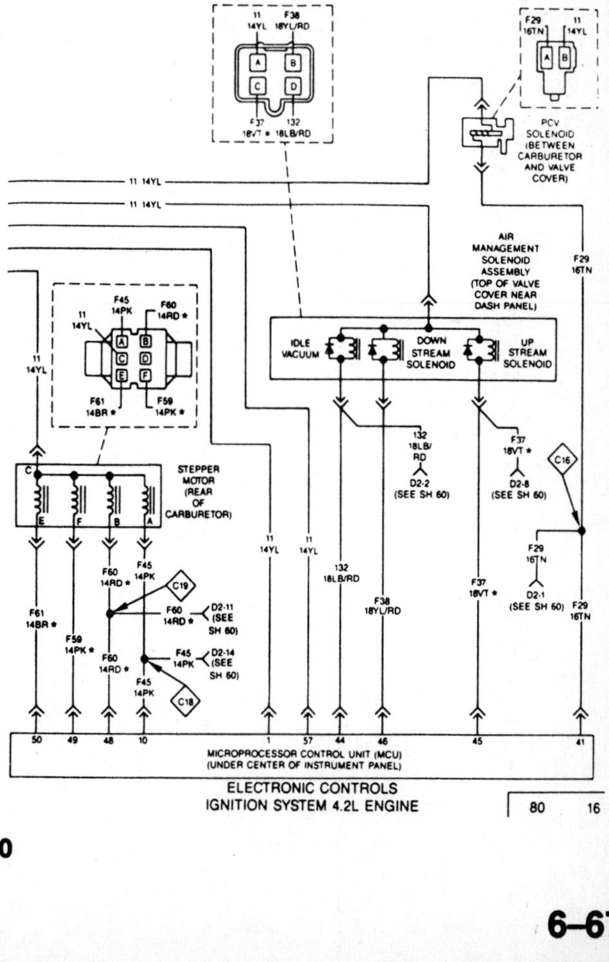

PCV SOLENOID (BETWEEN CARBURETOR AND VALVE COVER)

AIR MANAGEMENT SOLENOID ASSEMBLY (TOP OF VALVE COVER NEAR DASH PANEL)

IDLE VACUUM   DOWN STREAM SOLENOID   UP STREAM SOLENOID

STEPPER MOTOR (REAR OF CARBURETOR)

MICROPROCESSOR CONTROL UNIT (MCU) (UNDER CENTER OF INSTRUMENT PANEL)

ELECTRONIC CONTROLS IGNITION SYSTEM 4.2L ENGINE

80 16

1987-90

TO IGNITION SWITCH (SEE SH 7)

12 14OR

A6

12 14OR

C7

12 14OR

12 14OR

TO ELECTRIC CHOKE SWITCH (SEE SH 19)

86 85 87 30

IDLE SPEED-UP RELAY (RIGHT OF BATTERY)

12 14OR 14OR 1 2 3 4 5

F3 20BR F3 20BR 102 16GY

F3 20BR

D2-13 (SEE SH 60)

F2 1 AMP

F1 1 AMP

F3 20BR

F36 18YL/BL F35 YL/WT 99 20BK A B C

DIODE/FUSE ASSEMBLY (RIGHT REAR CORNER OF ENGINE COMPARTMENT BEHIND DIAGNOSTIC CONNECTION)

D2-15 (SEE SH 60)

F36 18YL/BL

F36 18YL/BL F35 18YL/WT D2-9 (SEE SH 60)

C25

F35 18YL/WT

F35 18YL/WT

C23

F36 18YL/BL F35 18YL/WT

A B C

PORTED VACUUM SWITCH (CLOSED WITH 4 IN OF HG)

ADAPTIVE VACUUM SWITCH (OPEN WITH 10 IN OF HG)

102 16GY 62 14BR * 99 20BK 32 16OR 139 16VT/RD A B C D E

62 14BR * TO REAR WINDOW DEFOGGER (SEE SH 53)

NOT USED

32 16OR TO A/C (SEE SH 51)

139 16VT/RD

*D1-5 (SEE SH 60)

139 16VT/RD

VACUUM SWITCH ASSEMBLY (BEHIND VALVE COVER ON DASH PANEL)

BLACK

99 14BK

(RIGHT REAR OF ENGINE)

IDLE SOLENOID (LEFT SIDE OF CARBURETOR)

99 20BK

99 14BK 99 18BK

43 16 55 17

MICROPROCESSOR CONTROL UNIT (MCU) (UNDER CENTER OF INSTRUMENT PANEL)

ELECTRONIC CONTROLS IGNITION SYSTEM 4.2L ENGINE

80 17

---

99 18BK F34 18GY/BK B A

F24 18TN F34 18GY/BK A B

99 20BK F33 18BL/OR A B

WIDE OPEN THROTTLE SWITCH (CLOSES AT WOT) (RIGHT BOTTOM OF CARBURETOR)

ENGINE COOLANT TEMPERATURE SWITCH (CLOSES WITH WARM ENGINE AT 135°F) (LEFT REAR CORNER OF ENGINE)

THERMO-ELECTRIC SWITCH OPEN ABOVE 15°F CLOSED BELOW 0°F (RIGHT SIDE OF AIR CLEANER)

99 18BK

F34 18GY/BK

99 20BK

F24 18TN

99 20BK

C12

99 18BK TO CRUISE CONTROL (SEE SH 24)

99 20BK (SEE SH 60) D2-7

99 20BK A B C F26 18OR *

F26 18OR *

HIGH ALTITUDE JUMPER (NEAR MCU)

F33 18BL/OR

D2-10 (SEE SH 60)

F33 18BL/OR C24

KNOCK SENSOR

(INTAKE MANIFOLD IN FRONT OF CARBURETOR)

OXYGEN SENSOR (BELOW EGR VALVE)

99 14BK

99 14BK

C26

F24 18TN

D2-12 (SEE SH 60)

C22

F34 18GY/BK

D2-6 (SEE SH 60)

F24 18TN

F33 18BL/OR

F54 18VT *

C14 F6 18GY

F26 18OR *

F26 18OR * D2-3 (SEE SH 60)

F34 18GY/BK

F24 18TN

F6 18GY F6 18GY

60 20 11 54 56 53 51 8 9

MICROPROCESSOR CONTROL UNIT (MCU) (UNDER CENTER OF INSTRUMENT PANEL)

ELECTRIC CONTROLS IGNITION SYSTEM 4.2L ENGINE

80 18

TO IGNITION SWITCH (SEE SH 7)

12 14OR

A6

12 14OR

C7

12 14OR

TO IDLE SPEED-UP RELAY (SEE SH 17)

12 14OR

ELECTRIC CHOKE SWITCH CLOSED WITH OIL PRESSURE ABOVE 4 PSI

A B C

27 16BL *

27 16BL * D1-2 (SEE SH 60)

27 16BL *

C8

27 16BL *

BI-METALLIC ELECTRIC CHOKE (RIGHT SIDE OF CARBURETOR)

TO STARTER RELAY (SEE SH 6)

FUSIBLE LINK

18GN

10 12RD TO HTD WDW FUSE (SEE SH 53)

C3

10 14RD

27 16BL * 10 14RD 104 20LG * 103 14PK 1 2 3 4

10 14RD

MANIFOLD HEATER RELAY (RIGHT OF BATTERY)

86 85 30 87

104 20LG * 103 14PK

MANIFOLD HEATER SWITCH OPENS ABOVE 160°F (70°C) (INTAKE MANIFOLD BEHIND CARBURETOR THROTTLE LINKAGE)

MANIFOLD HEATER (UNDER CARBURETOR)

CARBURETOR CIRCUITS 4.2L ENGINE

80 19

TO GAUGE PACKAGE (SEE SH 41)

1 2 3 4 5 6 7 8 9 10 11

55 18VT

54 18LB

F5 C4

54 18LB

54 18LB

55 18VT

OIL PRESSURE SENDER
0 PSI = 1 OHM
40 PSI = 46 OHMS
80 PSI = 90 OHMS
(RIGHT REAR CORNER OF CYLINDER BLOCK)

COOLANT TEMPERATURE SENDER
260°F = 55.1 OHMS
220°F = 93.5 OHMS
100°F = 1365 OHMS
(LEFT REAR CORNER OF CYLINDER HEAD)

OIL PRESSURE AND TEMPERATURE SYSTEMS

80 20

1987-90

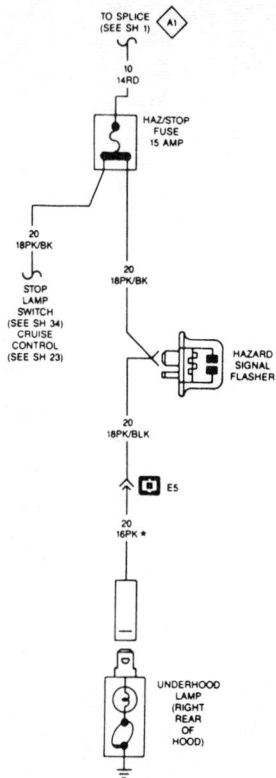

UNDERHOOD LAMP

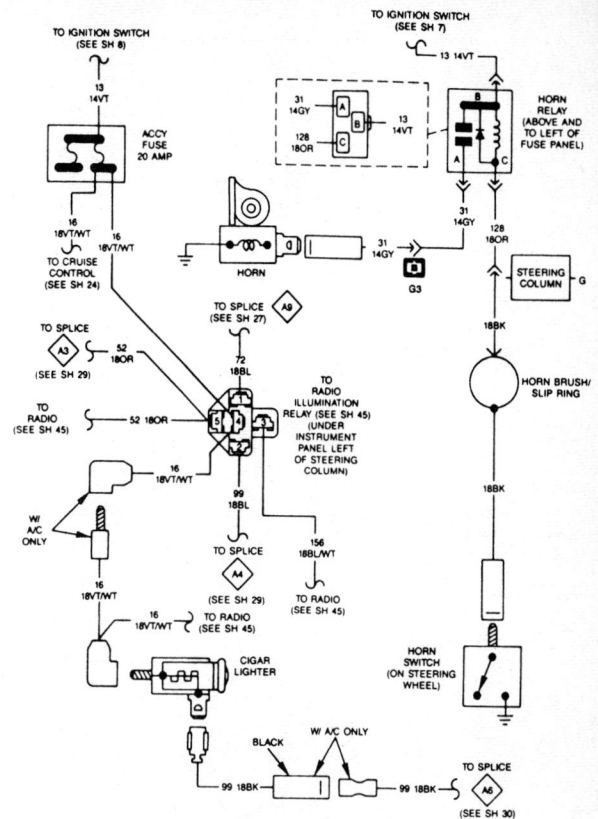

| 80 | 21 |

HORN SYSTEM AND CIGAR LIGHTER

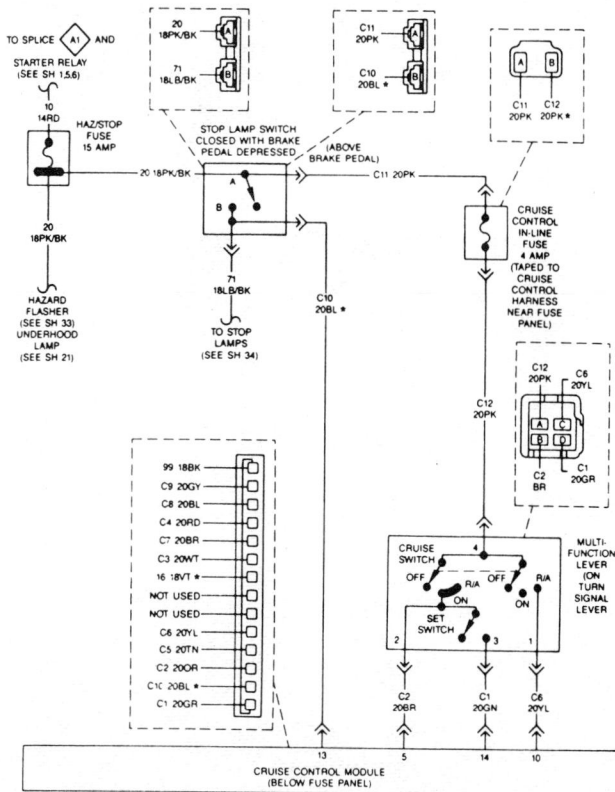

| 80 | 22 |

CRUISE CONTROL SYSTEM

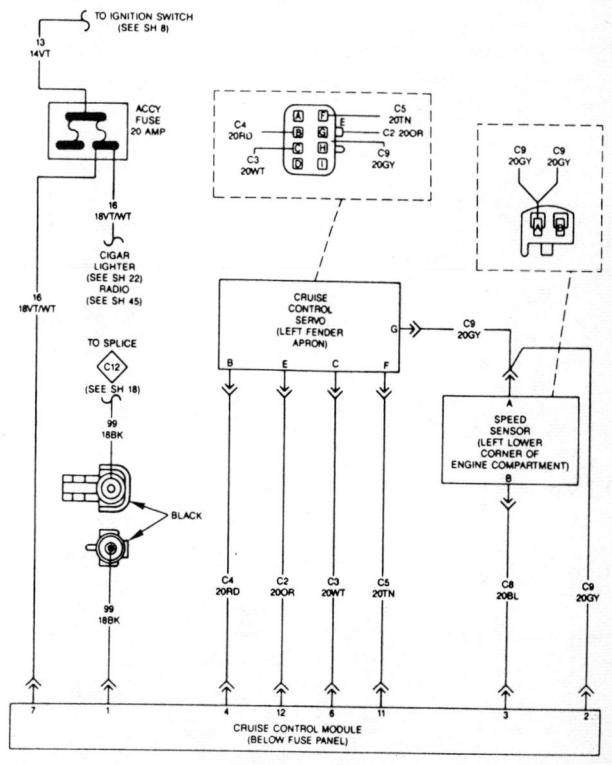

| 80 | 23 |

CRUISE CONTROL SYSTEM

| 80 | 24 |

1987-90

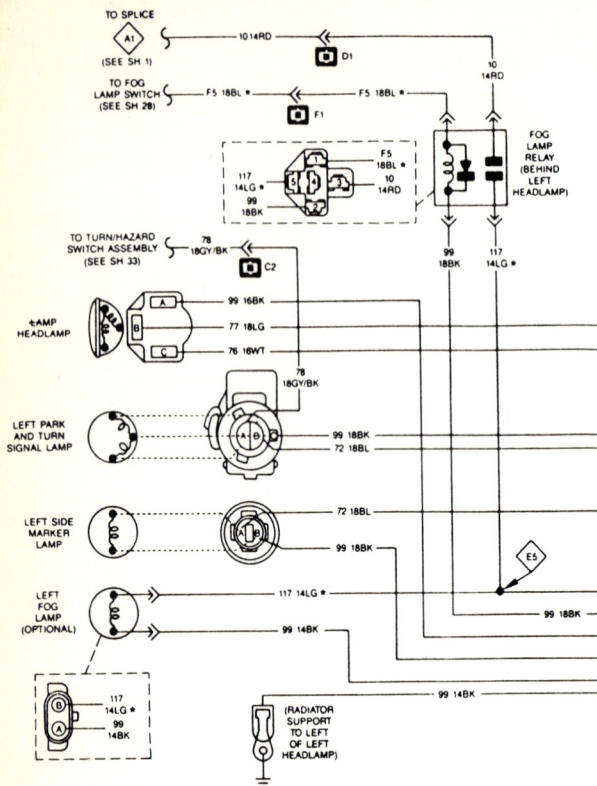

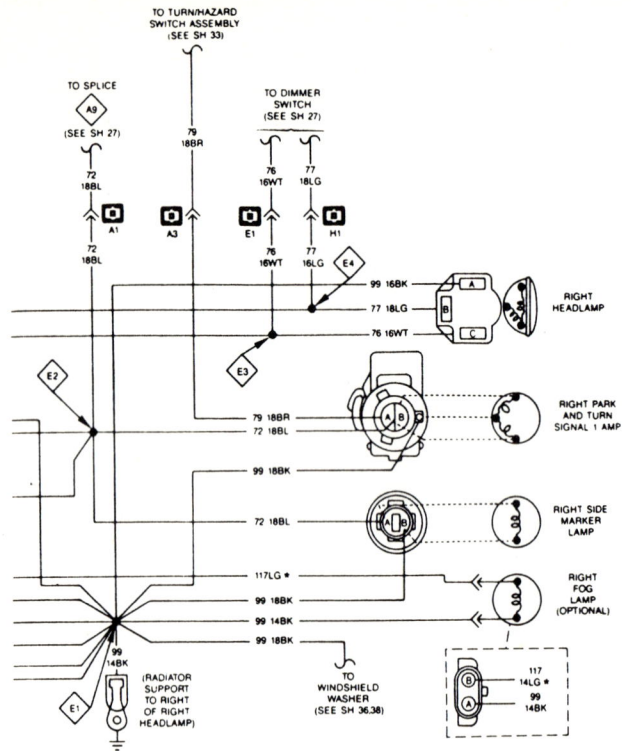

FRONT END LIGHTING

80 25

FRONT END LIGHTING

80 26

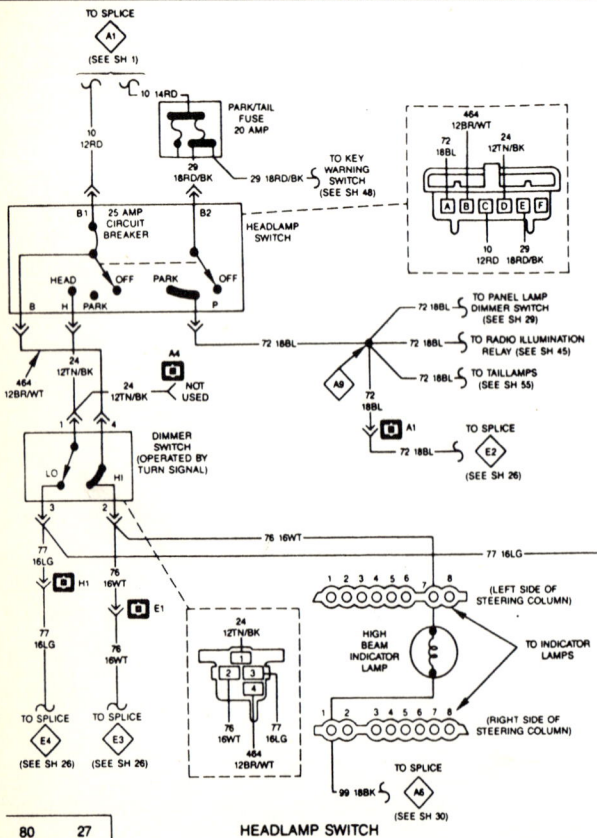

HEADLAMP SWITCH

80 27

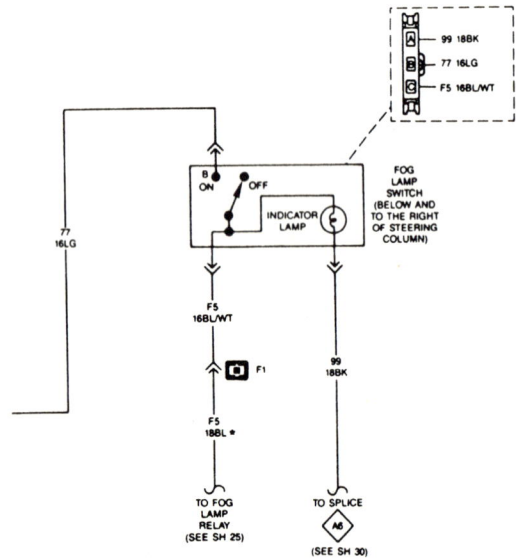

FOG LAMP SWITCH

80 28

1987-90

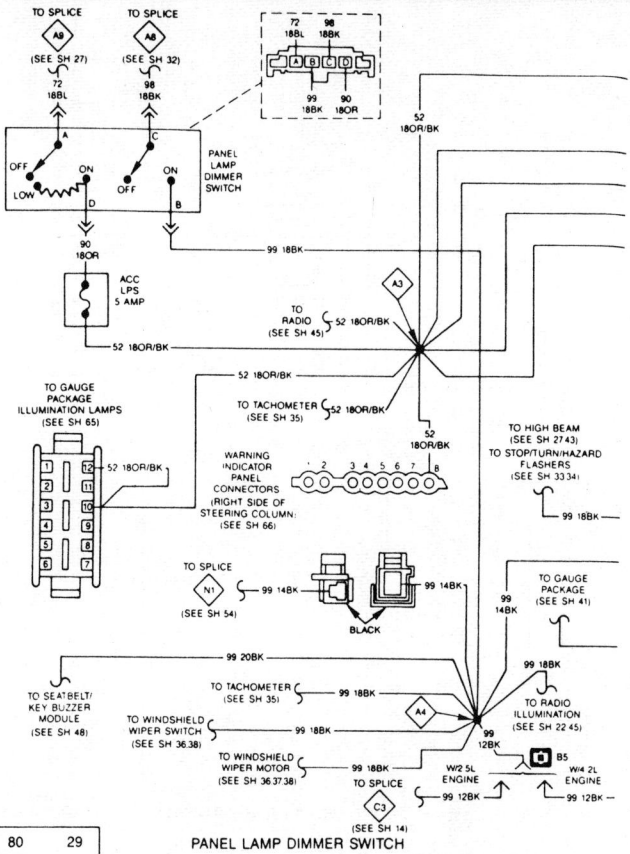

PANEL LAMP DIMMER SWITCH

| 80 | 29 |

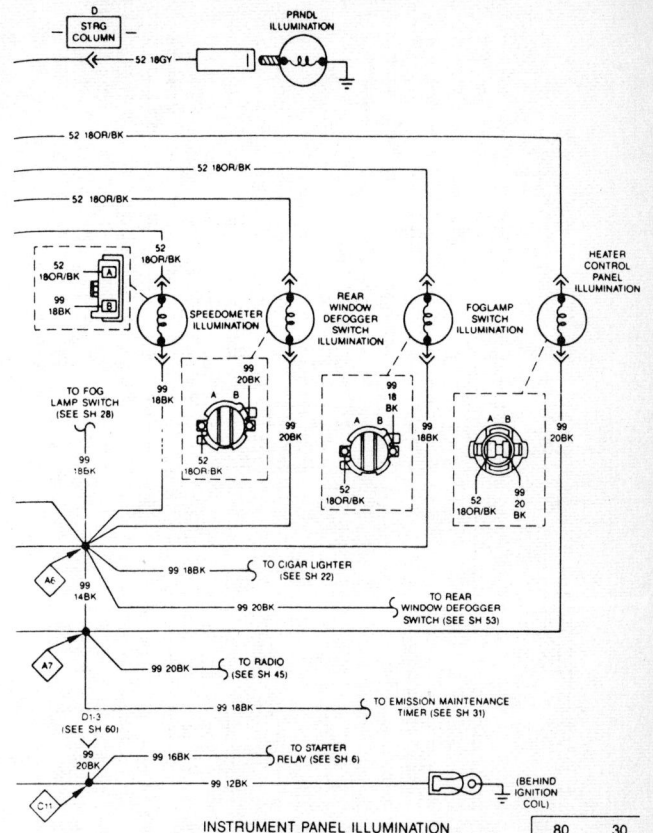

INSTRUMENT PANEL ILLUMINATION

| 80 | 30 |

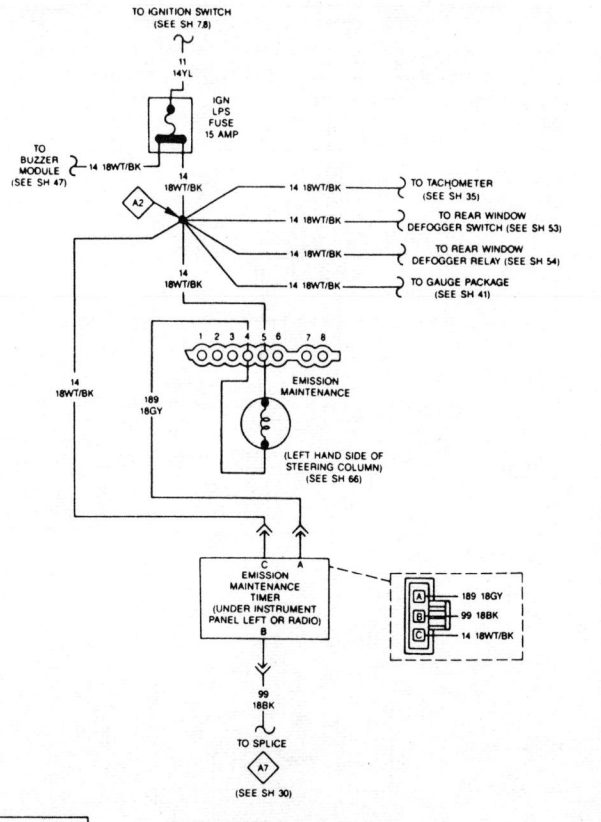

EMISSION MAINTENANCE TIMER

| 80 | 31 |

1987-90

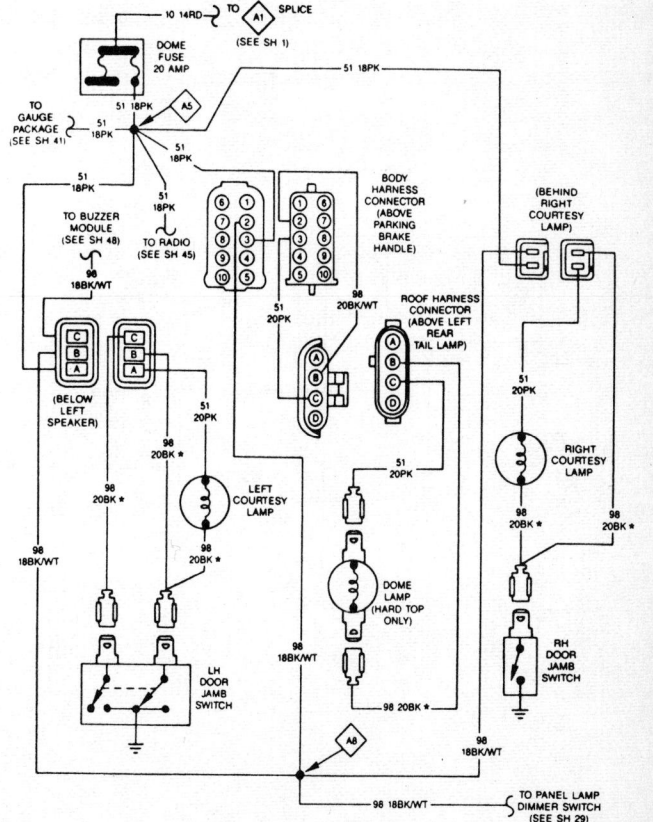

DOME AND COURTESY LAMPS

| 80 | 32 |

TO IGNITION SWITCH (SEE SH 8)

13 14VT

TURN/BU FUSE 15 AMP

TO BACKUP LAMPS SWITCH (SEE SH 40)

18 18YL/BK

18 18YL/BK

TURN SIGNAL FLASHER (ON FUSE PANEL)

TO UNDERHOOD LAMP (SEE SH 21)

TO SPLICE

A1 (SEE SH 1)

10 14RD

HAZ/STOP FUSE 15 AMP

20

20 18PK/BK

20 18PK/BK

20 18PK/BK

80 18VT/WT

HAZARD SIGNAL FLASHER (ON FUSE PANEL)

24 18GY

STRG COLUMN L

18VT

STRG COLUMN K

18BR

TURN/HAZARD SWITCH ASSEMBLY

18LB G

NORMAL

HAZARD

1 2 3

18DB F

A 18WT

TURN RIGHT

TURN LEFT

STRG COLUMN H

D

C B

STRG COLUMN J

18YL

STRG COLUMN M

18DG

STRG COLUMN N

78 18GY/BK

78 18GY/BK

79 18BR

78 18GY/BK

WARNING INDICATOR PANEL CONNECTORS

(LEFT OF STEERING COLUMN)

LEFT TURN INDICATOR

(RIGHT OF STEERING COLUMN)

RIGHT TURN INDICATOR

78 18GY/BK C2

79 18BR A3

99 18BK

TO SPLICE A6 (SEE SH 30)

78 18GY/BK

TO LEFT FRONT PARK/TURN LAMP (SEE SH 25)

79 18BR

TO RIGHT FRONT PARK/TURN LAMP (SEE SH 26)

80 33 STOP/TURN AND HAZARD FLASHER SYSTEMS

C11 20PK

C10 20BL *

A
B

20 18PK/BK

STOP LAMP SWITCH (CLOSED WITH BRAKE PEDAL DEPRESSED) (ABOVE BRAKE PEDAL)

20 18PK/BK

71 18LB/BK

A
B

A

B

C11 20PK

TO CRUISE CONTROL IN-LINE FUSE (SEE SH 23)

C10 20BL *

TO CRUISE CONTROL MODULE (SEE SH 23)

71 18LB/BK

STRG COLUMN P

BODY HARNESS CONNECTOR (ABOVE PARKING BRAKE HANDLE)

6 7 8 9 10
1 2 3 4 5

6 7 8 9 10
1 2 3 4 5

78 18GY/BK

TO LEFT REAR STOP/TURN LAMP (SEE SH 55)

79 18BR

79 18BR

TO RIGHT REAR STOP/TURN LAMP (SEE SH 55)

STOP/TURN AND HAZARD FLASHER SYSTEMS   80 34

---

14 18WT/BK
52 18OR/BK

30 18GN/WT
99 18BK

D F
B E
C A

TACHOMETER

TACHOMETER

TO IGNITION SWITCH (SEE SH 8)

11 14YL

IGN LPS FUSE 15 AMP

14 18WT/BK

14 18WT/BK

TO SPLICE A2 (SEE SH 8)

14 18WT/BK

TO BUZZER MODULE (SEE SH 47)

30 18GN/WT

30 18GN * C5

TO SPLICE C13 4.2L ENGINE (SEE SH 15)

TO IGNITION MODULE 2.5L ENGINE (SEE SH 9)

99 18BK

52 18OR/BK

TO SPLICE A4 (SEE SH 29)

TO SPLICE A3 (SEE SH 29)

80 35   TACHOMETER

WIPER SWITCH
OFF LO HI
OFF LO HI

ALSO MOVES WIPER SWITCH TO LO

WASHER SWITCH
OFF ON

WIPER/WASHER SWITCH

TO IGNITION SWITCH (SEE SH 8)

13 14VT

W/WIPER CIRCUIT BREAKER 5.3 AMP

GY VT WT

YL

PK

A B C D E F G

NOT USED

(UNDER INSTRUMENT PANEL NEXT TO STEERING COLUMN)

A B C D E F G

23 18LG/BK

23 18LG/BK

44 18VT/WT

TO SPLICE 99 18BK A4 (SEE SH 29)

48 18TN/BK

46WT/BK

47 18BL/WT

A B C D
H G F E

(LOWER LEFT CORNER OF WINDSHIELD HOUSING)

99 18BK

F3

44 18VT/WT

A B C D
H G F E

(LOWER LEFT CORNER OF WINDSHIELD HOUSING)

BL

WT

GY

RD

BK

WASHER PUMP

99 18BK

TO SPLICE E1 (SEE SH 26)

PARK SWITCH

WIPERS UP

WIPERS DOWN

LO HI

WIPER MOTOR (LOWER LEFT CORNER OF WINDSHIELD HOUSING)

WINDSHIELD WIPER SYSTEM   80 36

1987-90

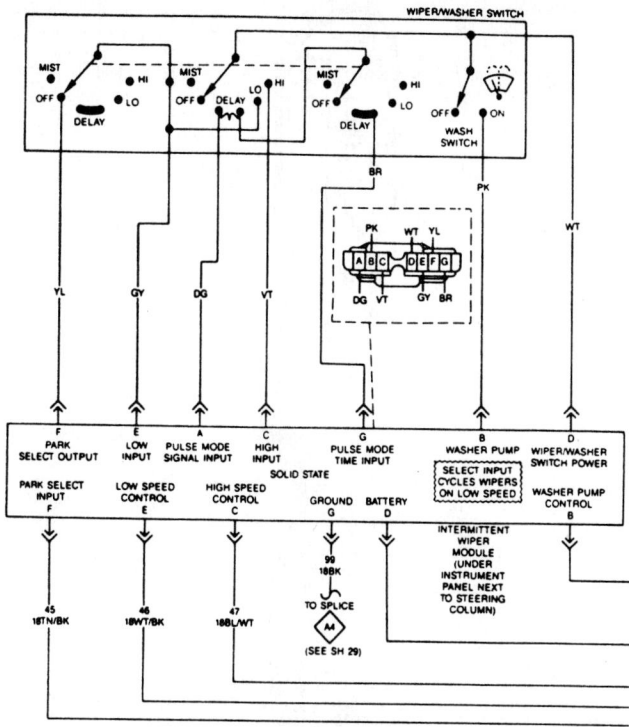

WIPER/WASHER SWITCH

MIST OFF HI LO DELAY

MIST OFF DELAY LO HI

MIST OFF DELAY LO HI

WASH SWITCH OFF ON

BR PK WT

PK WT YL
A B C D E F G
DG VT GY BR

YL GY DG VT G B WT

F	E	A	C	G	B	D
PARK SELECT OUTPUT	LOW INPUT	PULSE MODE SIGNAL INPUT	HIGH INPUT	PULSE MODE TIME INPUT	WASHER PUMP	WIPER/WASHER SWITCH POWER
PARK SELECT INPUT F	LOW SPEED CONTROL E	HIGH SPEED CONTROL C		GROUND G	BATTERY D	WASHER PUMP CONTROL B

SOLID STATE

SELECT INPUT CYCLES WIPERS ON LOW SPEED

INTERMITTENT WIPER MODULE (UNDER INSTRUMENT PANEL NEXT TO STEERING COLUMN)

45 18TN/BK   46 18WT/BK   47 18BL/WT

99 18BK
TO SPLICE
A4
(SEE SH 29)

80 37   **INTERMITTENT WIPER SYSTEM**

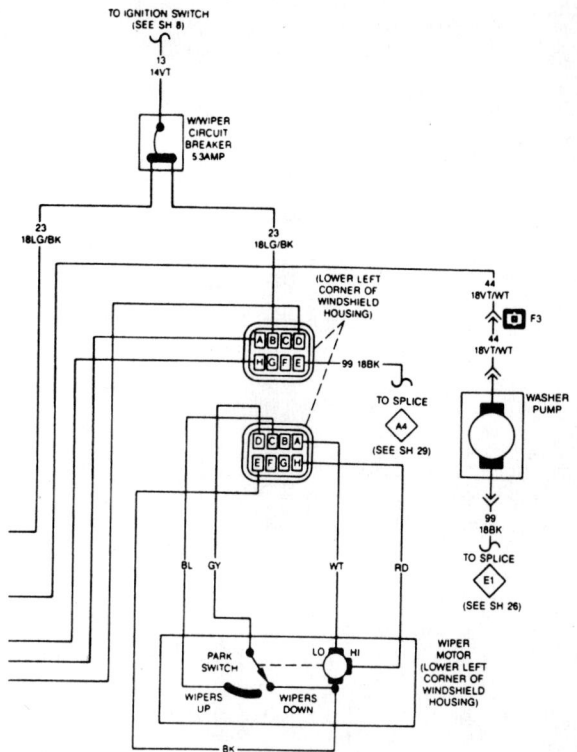

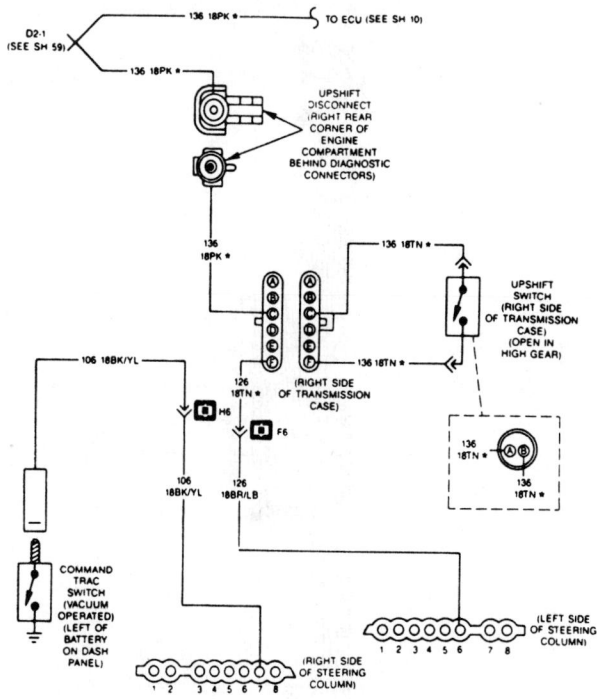

TO IGNITION SWITCH (SEE SH 8)

13 14VT

W/WIPER CIRCUIT BREAKER 53AMP

23 18LG/BK   23 18LG/BK   44 18VT/WT

44 18VT/WT F3

(LOWER LEFT CORNER OF WINDSHIELD HOUSING)
A B C D
H G F E

99 18BK
TO SPLICE
A4
(SEE SH 29)

D C B A
E F G H

WASHER PUMP

99 18BK
TO SPLICE
E1
(SEE SH 26)

BL GY WT RD

PARK SWITCH
WIPERS UP   WIPERS DOWN
LO HI

WIPER MOTOR (LOWER LEFT CORNER OF WINDSHIELD HOUSING)

BK

INTERMITTENT WIPER SYSTEM   80 38

D2-1 (SEE SH 59)   136 18PK *   TO ECU (SEE SH 10)

136 18PK *

UPSHIFT DISCONNECT (RIGHT REAR CORNER OF ENGINE COMPARTMENT BEHIND DIAGNOSTIC CONNECTORS)

136 18PK *

136 18TN *

UPSHIFT SWITCH (RIGHT SIDE OF TRANSMISSION CASE) (OPEN IN HIGH GEAR)

136 18TN *

106 18BK/YL   126 18TN *   (RIGHT SIDE OF TRANSMISSION CASE)

H6   F6

136 18TN *
A B
136 18TN *

106 18BK/YL   126 18BR/LB

COMMAND TRAC SWITCH (VACUUM OPERATED) (LEFT OF BATTERY ON DASH PANEL)

(LEFT SIDE OF STEERING COLUMN)
1 2 3 4 5 6 7 8

(RIGHT SIDE OF STEERING COLUMN)
1 2 3 4 5 6 7 8

4 WD COMMAND TRAC SWITCH - ALL 4 WD MODELS AND UPSHIFT SWITCH - W/2.5L ENGINE AND 5-SPEED TRANSMISSION

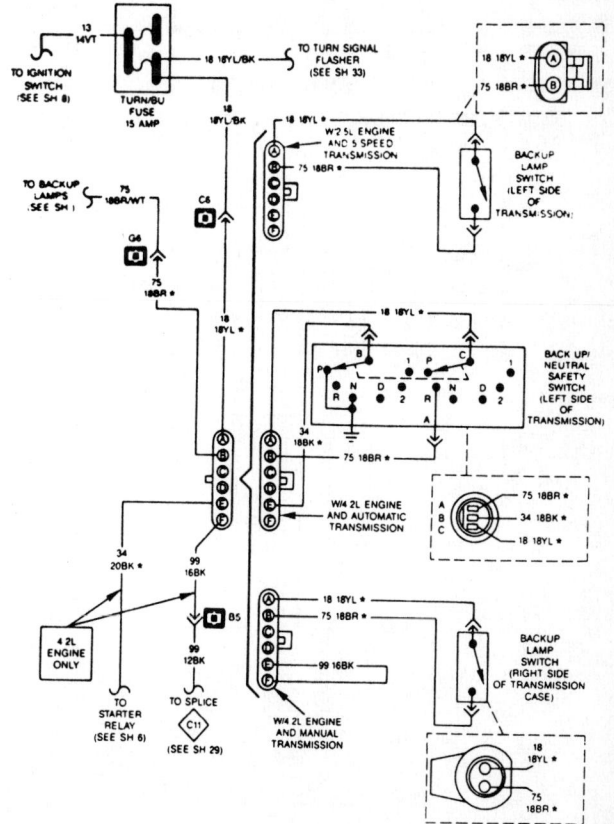

80 39

---

13 14VT

TO IGNITION SWITCH (SEE SH 8)

TURN/BU FUSE 15 AMP

18 18YL/BK   TO TURN SIGNAL FLASHER (SEE SH 33)

18 18YL/BK   18 18YL *

18 18YL *   A
75 18BR *   B

W/2.5L ENGINE AND 5 SPEED TRANSMISSION

75 18BR *   BACKUP LAMP SWITCH (LEFT SIDE OF TRANSMISSION)

TO BACKUP LAMPS (SEE SH)   75 18BR/WT

C6   G6   75 18BR *

18 18YL *

18 18YL *

BACK UP/ NEUTRAL SAFETY SWITCH (LEFT SIDE OF TRANSMISSION)
P R N D 2 R N D 2
B P C
R N A

34 18BK *
75 18BR *

W/4.2L ENGINE AND AUTOMATIC TRANSMISSION

75 18BR *
A
34 18BK *
B
18 18YL *
C

34 20BK *   99 16BK

4.2L ENGINE ONLY

B5

18 18YL *
75 18BR *
99 16BK

99 12BK

BACKUP LAMP SWITCH (RIGHT SIDE OF TRANSMISSION CASE)

W/4.2L ENGINE AND MANUAL TRANSMISSION

TO STARTER RELAY (SEE SH 6)   TO SPLICE C11 (SEE SH 29)

18 18YL *
75 18BR *

1987-90   **BACKUP LAMP SWITCH**   80 40

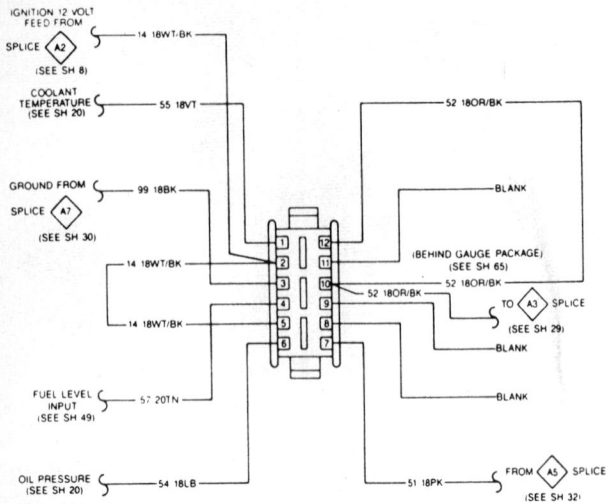

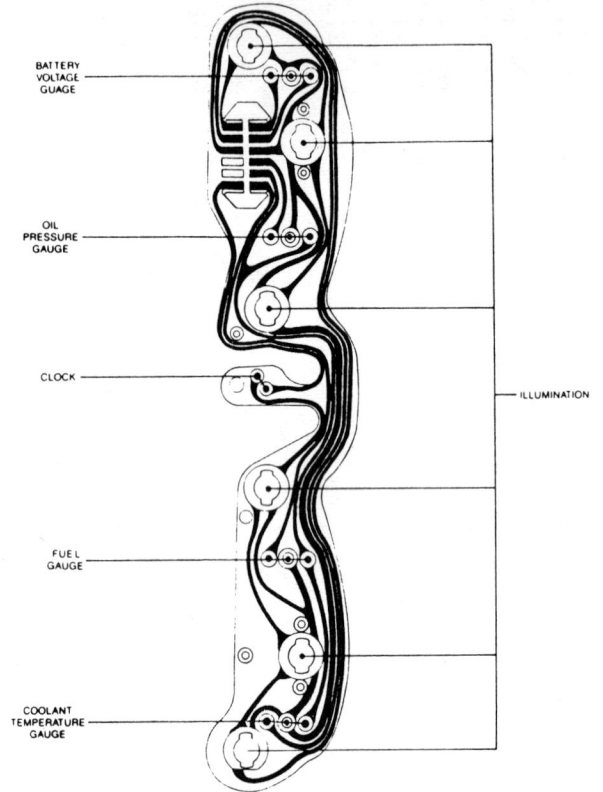

Gauge Package Printed Circuit Board (Connectors)

IGNITION 12 VOLT FEED FROM
SPLICE A2 (SEE SH 8)

14 18WT-BK

COOLANT TEMPERATURE (SEE SH 20)

55 18VT

52 18OR/BK

BLANK

GROUND FROM SPLICE A7 (SEE SH 30)

99 18BK

14 18WT/BK

(BEHIND GAUGE PACKAGE) (SEE SH 65)

52 18OR/BK

52 18OR/BK TO A3 SPLICE (SEE SH 29)

14 18WT/BK

BLANK

FUEL LEVEL INPUT (SEE SH 49)

57 20TN

BLANK

OIL PRESSURE (SEE SH 20)

54 18LB

51 18PK FROM A5 SPLICE (SEE SH 32)

GAUGE PACKAGE PRINTED CIRCUIT BOARD

| 80 | 41 |

Gauge Package Printed Circuit Board

BATTERY VOLTAGE GUAGE

OIL PRESSURE GAUGE

CLOCK

FUEL GAUGE

COOLANT TEMPERATURE GAUGE

ILLUMINATION

GAUGE PACKAGE PRINTED CIRCUIT BOARD

| 80 | 42 |

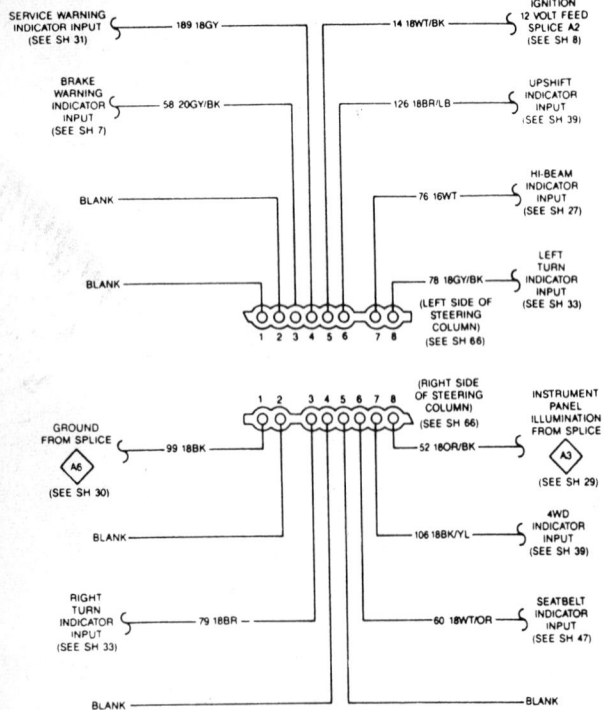

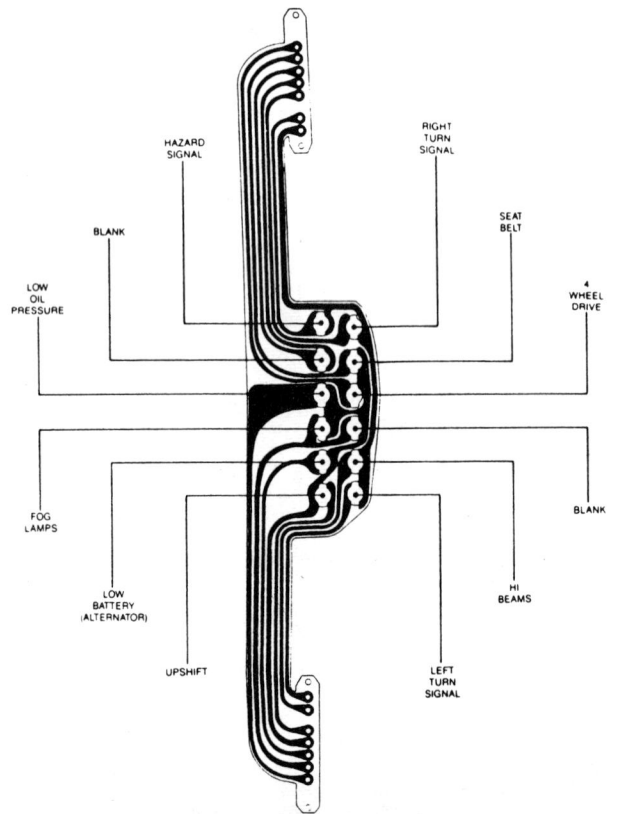

Warning Indicator Panel Printed Circuit Board Connectors

SERVICE WARNING INDICATOR INPUT (SEE SH 31)

189 18GY

14 18WT/BK IGNITION 12 VOLT FEED SPLICE A2 (SEE SH 8)

BRAKE WARNING INDICATOR INPUT (SEE SH 7)

58 20GY/BK

126 18BR/LB UPSHIFT INDICATOR INPUT (SEE SH 39)

BLANK

76 16WT HI-BEAM INDICATOR INPUT (SEE SH 27)

BLANK

78 18GY/BK LEFT TURN INDICATOR INPUT (SEE SH 33)

(LEFT SIDE OF STEERING COLUMN) (SEE SH 66)

1 2 3 4 5 6 7 8

1 2 3 4 5 6 7 8

(RIGHT SIDE OF STEERING COLUMN) (SEE SH 66)

GROUND FROM SPLICE A6 (SEE SH 30)

99 18BK

52 18OR/BK INSTRUMENT PANEL ILLUMINATION FROM SPLICE A3 (SEE SH 29)

BLANK

106 18BK/YL 4WD INDICATOR INPUT (SEE SH 39)

RIGHT TURN INDICATOR INPUT (SEE SH 33)

79 18BR

60 18WT/OR SEATBELT INDICATOR INPUT (SEE SH 47)

BLANK

BLANK

WARNING INDICATOR PANEL PRINTED CIRCUIT BOARD CONNECTORS

1987-90

| 80 | 43 |

Warning Indicator Panel Printed Circuit Board

HAZARD SIGNAL

RIGHT TURN SIGNAL

BLANK

SEAT BELT

LOW OIL PRESSURE

4 WHEEL DRIVE

FOG LAMPS

BLANK

LOW BATTERY (ALTERNATOR)

HI BEAMS

UPSHIFT

LEFT TURN SIGNAL

WARNING INDICATOR PANEL PRINTED CIRCUIT BOARD

| 80 | 44 |

TO IGNITION SWITCH (SEE SH 8)

ACCY FUSE 20 AMP

13 14VT

16 18VT/WT

TO CRUISE CONTROL (SEE SH 24)

72 18BL
18BL
52 18OR/BK
52 18OR/BK
16 18VT/WT
156 18BL/WT
99 18BK

TO A9 (SEE SH 27) 72 18BL

16 18VT/WT

TO SPLICE A4 (SEE SH 29)

RADIO ILLUMINATION RELAY (UNDER INSTRUMENT PANEL LEFT OF STEERING COLUMN)

99 18BK

52 18OR/BK

16 18VT/WT

TO CIGAR LIGHTER (SEE SH 22)

TO SPLICE A3 (SEE SH 29)

52 18OR/BK

156 18BL/WT

A2 20WT
A3 20WT/BK
A4 20GN
A5 20GN/WT
51 18PK
A1 20BK
A10 20BR/WT
A11 20BR

TO SPLICE A7 (SEE SH 30)

99 20BK

TO SPLICE A5 (SEE SH 32)

GROUND TERMINAL

CONNECTOR REAR OR RADIO

IN-LINE 5-AMP FUSE

ELECTRONICAL TUNED RADIO

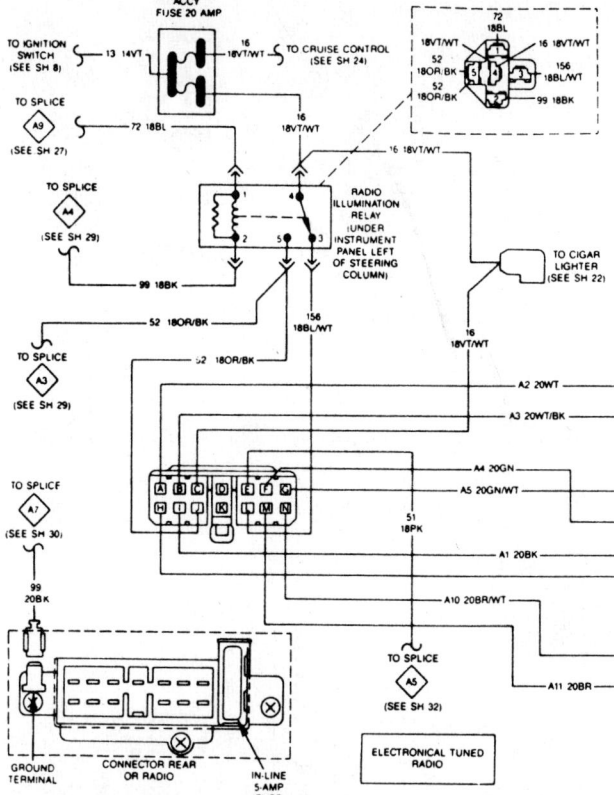

80 45 | RADIO SYSTEM

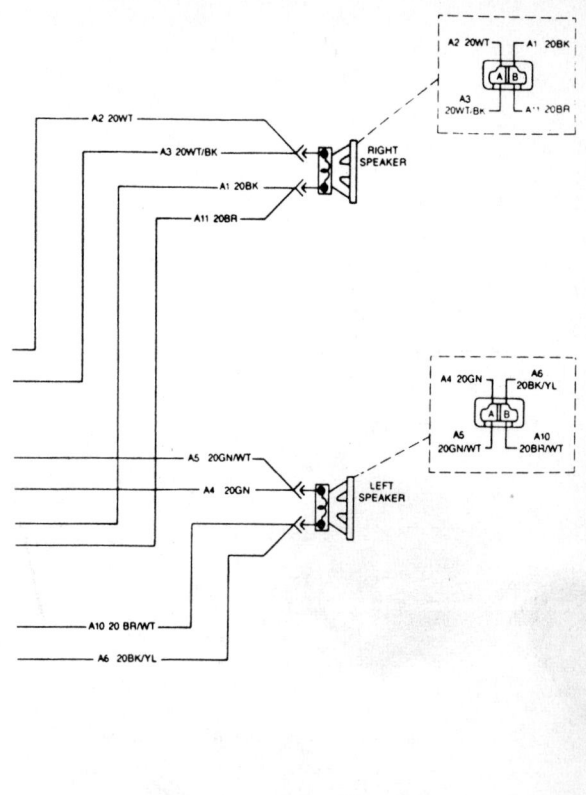

A2 20WT
A2 20WT
A3 20WT/BK
A1 20BK
A11 20BR
A1 20BK
A11 20BR

RIGHT SPEAKER

A2 20WT | A1 20BK
A3 20WT·BK | A11 20BR

A5 20GN/WT
A4 20GN
A4 20GN

LEFT SPEAKER

A10 20 BR/WT
A6 20BK/YL

A4 20GN | A6 20BK/YL
A5 20GN/WT | A10 20BR/WT

80 46 | RADIO SYSTEM

TO IGNITION SWITCH (SEE SH 8)

11 14YL

IGN LPS FUSE 15 AMP

14 18WT/BK TO GAUGE PACKAGE (SEE SH 41)

14 18WT/BK TO TACHOMETER (SEE SH 35)

14 18WT/BK

14 18WT/BK TO EMISSION MAINTENANCE TIMER (SEE SH 31)

14 18WT/BK

TO DEFOGGER SWITCH (SEE SH 54)

14 18WT/BK

A2

14 18WT/BK

BODY HARNESS CONNECTOR (ABOVE PARKING BRAKE HANDLE)

1 2 3 4 5 6 7 8 (LEFT OF STEERING COLUMN)

FASTEN SEATBELT WARNING INDICATOR

WARNING INDICATOR PANEL CONNECTORS (SEE SH 66)

1 2 3 4 5 6 7 8 (RIGHT OF STEERING COLUMN)

85 18WT

85 18WT

SEATBELT SWITCH (OPEN) WITH DRIVERS SEATBELT BUCKLED (IN DRIVERS SEATBELT BUCKLE)

REAR WINDOW DEFOGGER RELAY CONNECTOR (BEHIND PARK BRAKE PEDAL ASSEMBLY)

14 18WT/BK

14 18WT/BK

60 18WT/OR

14 18WT/BK

14 18WT/BK
85 14WT

1 4 2

BUZZER MODULE (PLUGS INTO FUSE PANEL)

PARK/TAIL FUSE 20 AMP

29 18RD/BK TO HEADLAMP SWITCH (SEE SH 27)

98 18BK/WT

TO SPLICE A8 (SEE SH 32)

10 14RD

87 18WT

C B A (BELOW LEFT SPEAKER)

C B A

51 20BK

LEFT COURTESY LAMP

TO SPLICE A1 (SEE SH 1)

STRG COLUMN — E

18BK

51 18PK

TO SPLICE A5 (SEE SH 32)

98 18BK/WT

98 20BK *

98 20BK *

98 20BK *

IGNITION KEY WARNING SWITCH (CLOSED WITH KEY IN IGNITION) (IN STEERING COLUMN BEHIND LOCK CYLINDER)

18PK

STRG COLUMN — F

86 18WT

TO SPLICE A4 (SEE SH 29)

99 20BK

LH DOOR JAMB SWITCH

7 6 3

BUZZER MODULE (PLUGS INTO FUSE PANEL)

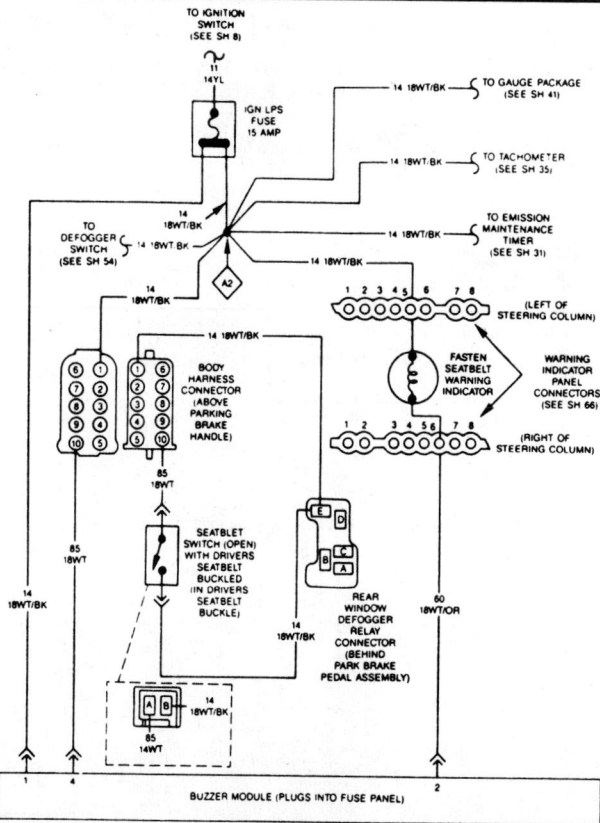

80 47 | SEATBELT/IGNITION KEY WARNING BUZZER | **1987-90**

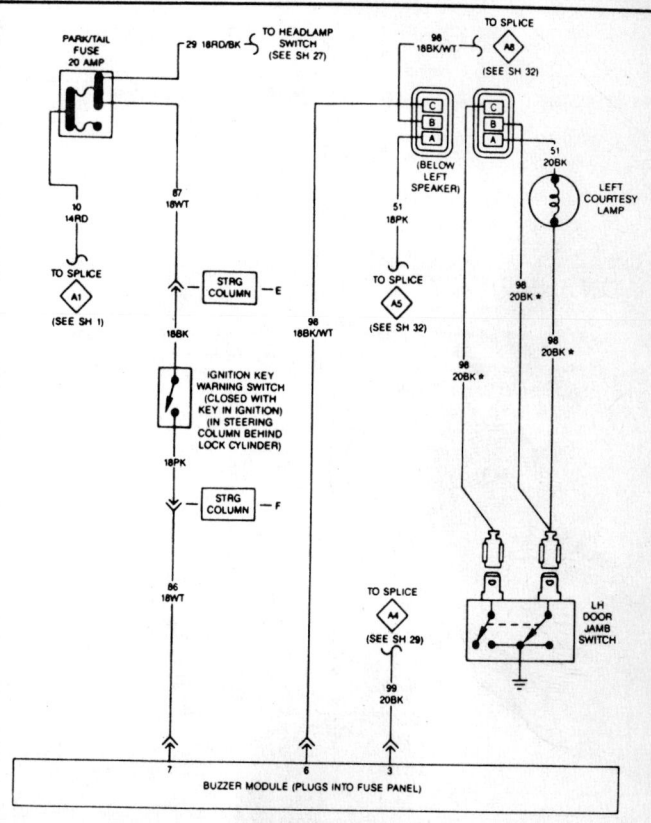

80 48

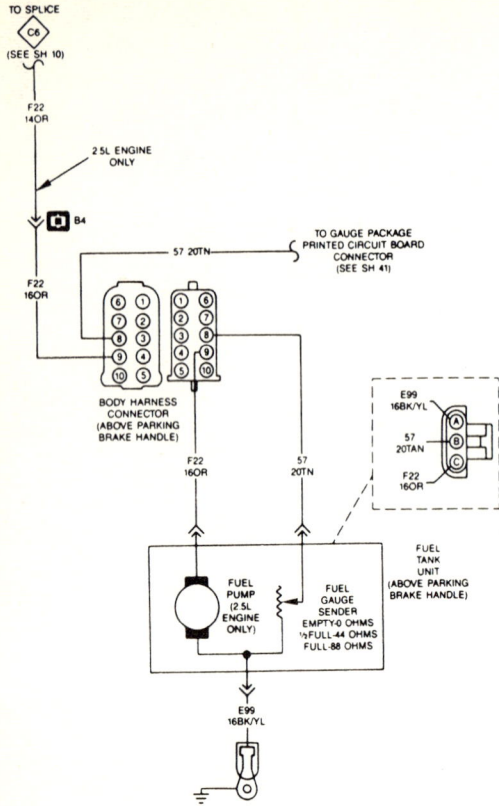

Fuel Tank Systems

TO SPLICE C6 (SEE SH 10)

F22 14OR

2.5L ENGINE ONLY

B4

F22 16OR

57 20TN

TO GAUGE PACKAGE PRINTED CIRCUIT BOARD CONNECTOR (SEE SH 41)

BODY HARNESS CONNECTOR (ABOVE PARKING BRAKE HANDLE)

F22 16OR

57 20TN

E99 16BK/YL
57 20TAN
F22 16OR

FUEL TANK UNIT (ABOVE PARKING BRAKE HANDLE)

FUEL PUMP (2.5L ENGINE ONLY)

FUEL GAUGE SENDER
EMPTY-0 OHMS
½ FULL-44 OHMS
FULL-88 OHMS

E99 16BK/YL

80 49 — FUEL TANK SYSTEMS

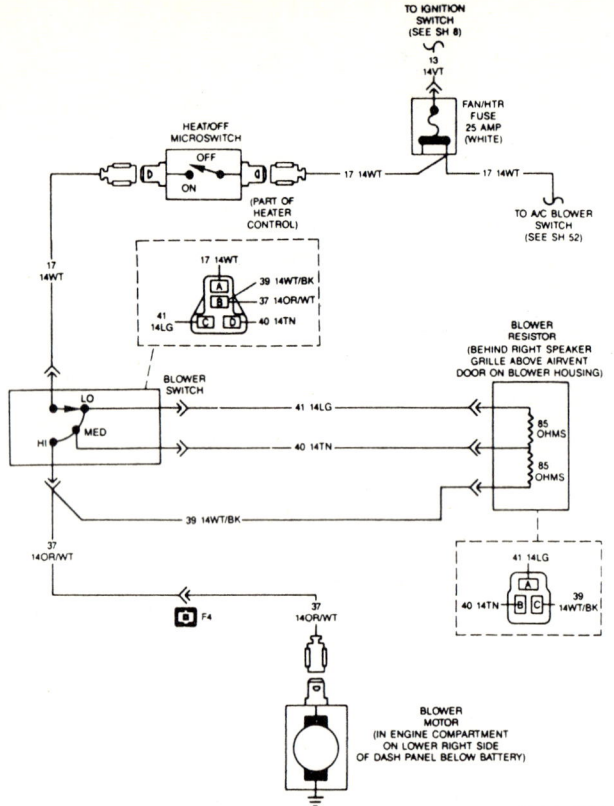

Heater System

TO IGNITION SWITCH (SEE SH 8)

13 14VT

FAN/HTR FUSE 25 AMP (WHITE)

17 14WT

17 14WT

HEAT/OFF MICROSWITCH OFF ON (PART OF HEATER CONTROL)

TO A/C BLOWER SWITCH (SEE SH 52)

17 14WT

39 14WT/BK
37 14OR/WT
41 14LG
40 14TN

BLOWER SWITCH LO MED HI

41 14LG

40 14TN

39 14WT/BK

BLOWER RESISTOR (BEHIND RIGHT SPEAKER GRILLE ABOVE AIRVENT DOOR ON BLOWER HOUSING)

85 OHMS
85 OHMS

41 14LG
40 14TN
39 14WT/BK

37 14OR/WT

F4

37 14OR/WT

BLOWER MOTOR (IN ENGINE COMPARTMENT ON LOWER RIGHT SIDE OF DASH PANEL BELOW BATTERY)

80 50 — HEATER SYSTEM

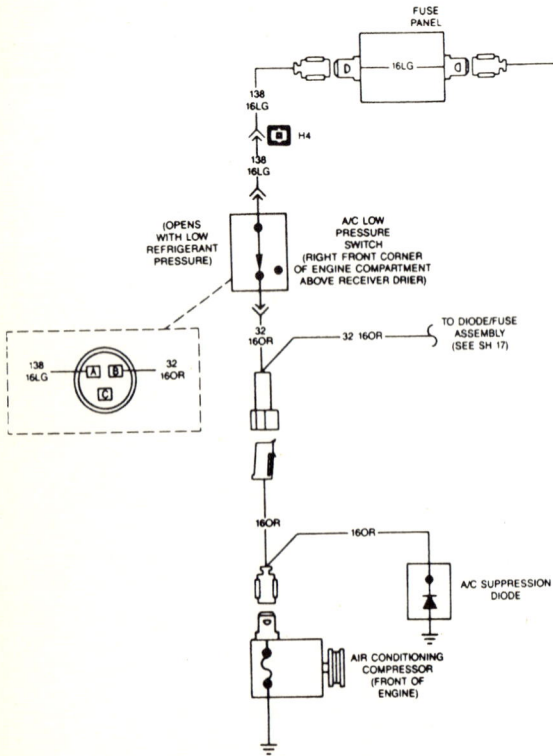

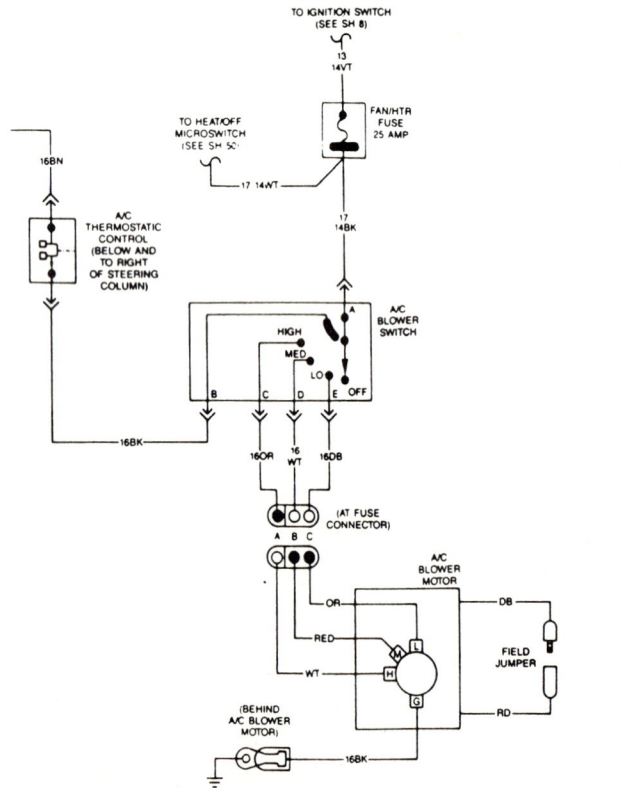

Air Conditioning System (left)

FUSE PANEL 16LG

138 16LG

H4

138 16LG

(OPENS WITH LOW REFRIGERANT PRESSURE)

A/C LOW PRESSURE SWITCH (RIGHT FRONT CORNER OF ENGINE COMPARTMENT ABOVE RECEIVER DRIER)

138 16LG
32 16OR

32 16OR

TO DIODE/FUSE ASSEMBLY (SEE SH 17)

16OR
16OR

A/C SUPPRESSION DIODE

AIR CONDITIONING COMPRESSOR (FRONT OF ENGINE)

80 51 — AIR CONDITIONING SYSTEM

Air Conditioning System (right)

TO IGNITION SWITCH (SEE SH 8)

13 14VT

FAN/HTR FUSE 25 AMP

TO HEAT/OFF MICROSWITCH (SEE SH 50)

17 14WT

17 14BK

16BN

A/C THERMOSTATIC CONTROL (BELOW AND TO RIGHT OF STEERING COLUMN)

A/C BLOWER SWITCH HIGH MED LO OFF

16BK

16OR 16WT 16DB

(AT FUSE CONNECTOR) A B C

A/C BLOWER MOTOR

OR RED WT DB RD

FIELD JUMPER

(BEHIND A/C BLOWER MOTOR)

16BK

80 52 — AIR CONDITIONING SYSTEM

1987-90

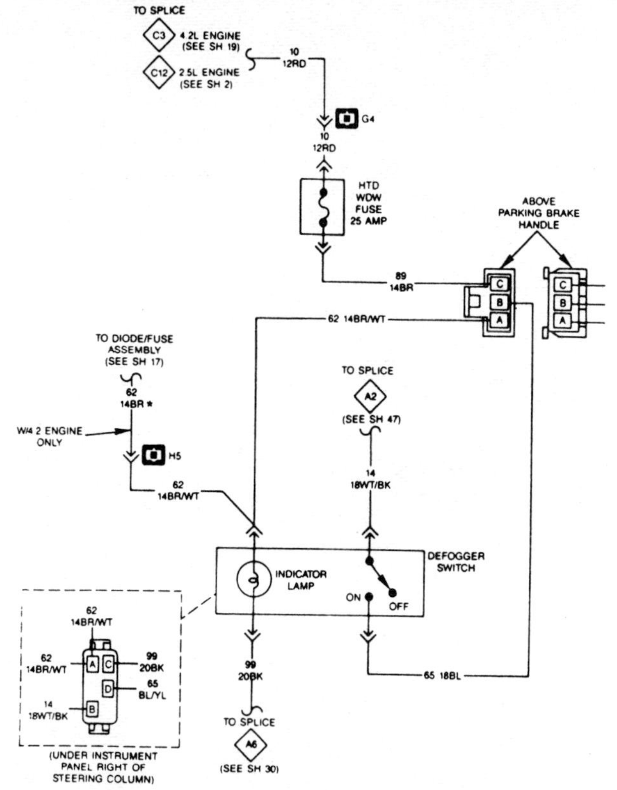

TO SPLICE
C3 4.2L ENGINE (SEE SH 19)
C12 2.5L ENGINE (SEE SH 2)
10 12RD
G4
10 12RD
HTD WDW FUSE 25 AMP
ABOVE PARKING BRAKE HANDLE
89 14BR
62 14BR/WT
TO DIODE/FUSE ASSEMBLY (SEE SH 17)
62 14BR *
W/4.2 ENGINE ONLY
H5
62 14BR/WT
TO SPLICE
A2 (SEE SH 47)
14 18WT/BK
INDICATOR LAMP
DEFOGGER SWITCH
ON OFF
99 20BK
65 18BL
TO SPLICE
A6 (SEE SH 30)
62 14BR/WT
62 14BR/WT 99 20BK
65 BL/YL
14 18WT/BK
(UNDER INSTRUMENT PANEL RIGHT OF STEERING COLUMN)
80 53 REAR WINDOW DEFOGGER SYSTEM

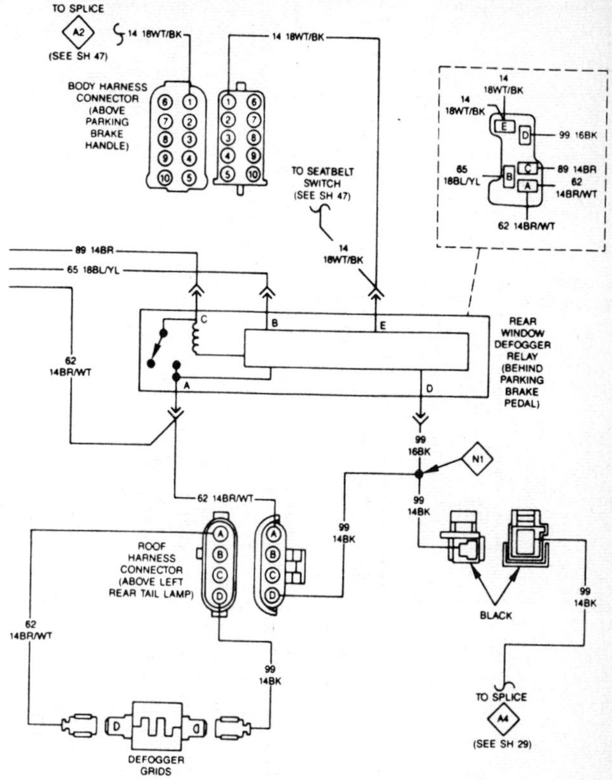

TO SPLICE
A2 (SEE SH 47)
14 18WT/BK
14 18WT/BK
BODY HARNESS CONNECTOR (ABOVE PARKING BRAKE HANDLE)
TO SEATBELT SWITCH (SEE SH 47)
14 18WT/BK
14 18WT/BK
65 18BL/YL
85 18BL/YL 89 14BR
62 14BR/WT
89 14BR
65 18BL/YL
62 14BR/WT
REAR WINDOW DEFOGGER RELAY (BEHIND PARKING BRAKE PEDAL)
99 16BK
N1
99 14BK
ROOF HARNESS CONNECTOR (ABOVE LEFT REAR TAIL LAMP)
62 14BR/WT
99 14BK
BLACK
99 14BK
TO SPLICE
A4 (SEE SH 29)
DEFOGGER GRIDS
REAR WINDOW DEFOGGER SYSTEM 80 54

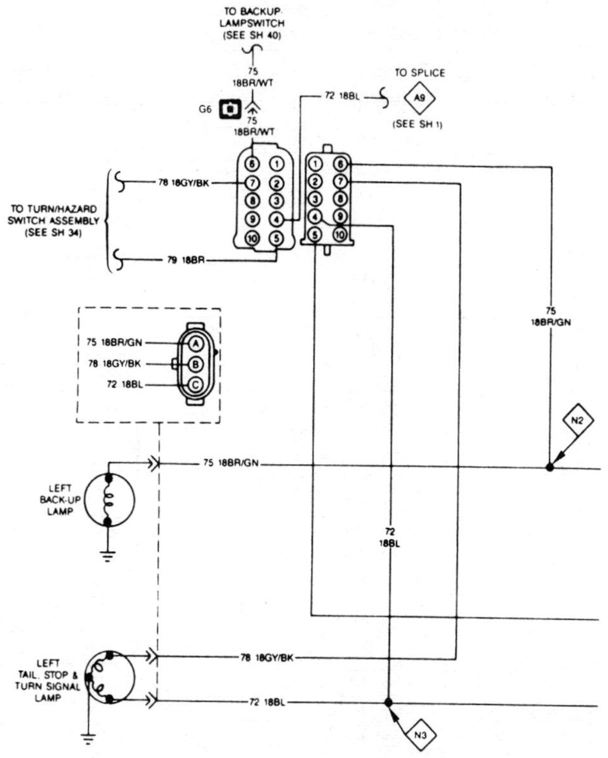

TO BACKUP LAMP SWITCH (SEE SH 40)
75 18BR/WT
G6
75 18BR/WT
72 18BL
TO SPLICE A9 (SEE SH 1)
TO TURN/HAZARD SWITCH ASSEMBLY (SEE SH 34)
78 18GY/BK
79 18BR
75 18BR/GN
75 18BR/GN
78 18GY/BK
72 18BL
N2
72 18BL
LEFT BACK-UP LAMP
75 18BR/GN
LEFT TAIL, STOP & TURN SIGNAL LAMP
78 18GY/BK
72 18BL
N3
80 55 REAR LIGHTING

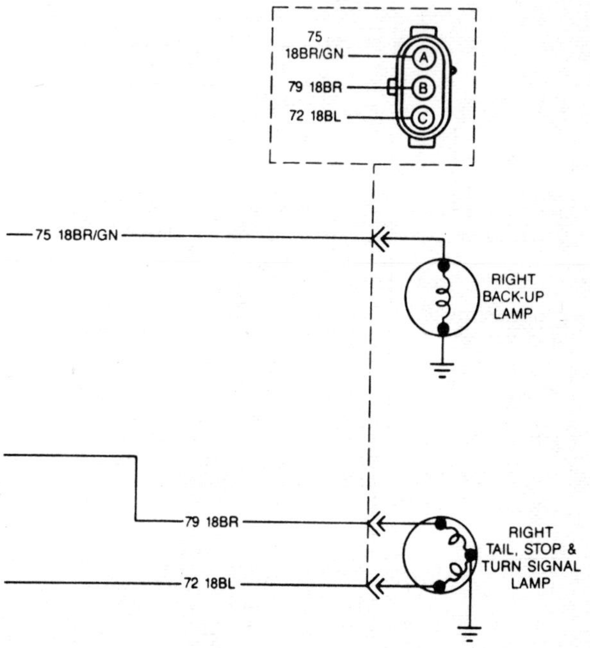

75 18BR/GN
79 18BR
72 18BL
75 18BR/GN
RIGHT BACK-UP LAMP
79 18BR
72 18BL
RIGHT TAIL, STOP & TURN SIGNAL LAMP
REAR LIGHTING 80 56

1987-90

6-77

FUSE NUMBER	AMPS	COLOR	SHEET
1	—	—	—
2	—	—	—
3	15 HAZ STOP	LIGHT BLUE	1,21,23,33
4	15 TURN/BU	LIGHT BLUE	2,8,33,40
5	20 DOME	YELLOW	1,32
6	25 HTD WDW	WHITE	2,53
7	20 ACC	YELLOW	2,8,22,24,45
8	20 PARK/TAIL	YELLOW	1,27,48
9	15 IGN LPS	LIGHT BLUE	2,8,15,31,35,47
10	5 ACC LPS	TAN	1,29
11	5.3 CIRCUIT BREAKER	SILVER CAN	2,8,36,38
12	25 FAN HTR	WHITE	2,8,50,52

80 57

80 58

CAV	DIAGNOSTIC CONNECTOR — #1	
1	30 18GN*	TACHOMETER
1	30 18GN*	TACHOMETER
2	11 14YL	I-1 IGNITION SWITCH
3	99 20BK	ECU GROUND
4	88 14GN*	START SIGNAL (—)
5	10 14RD	BATTERY
6	F22 14OR	FUEL PUMP RELAY

CAV	DIAGNOSTIC CONNECTOR — #1	
1	30 18GN*	TACHOMETER
2	27 16BL*	ELECTRIC CHOKE SWITCH
3	99 20BK	GROUND
4	33 14GN	START SIGNAL
5	139 16VT/RD	IDLE SOLENOID
6	—	—

CAV	DIAGNOSTIC CONNECTOR — #2	
1	136 18PK*	UNSHIFT LAMP - MANUAL
1	136 18PK*	ECU SERIAL DATA - AUTO
2	F20 20BK*	B+ LATCH - COIL GROUND
2	F20 20BK*	B+ LATCH - COIL GROUND
3	34 20BK*	PARK/NEUTRAL - AUTO
3	34 20BK*	ECU SERAIL DATA - MANUAL
4	F21 14PK	B+ LATCH RELAY - COIL FEED
5	—	—
6	F34 20GY*	WIDE OPEN THROTTLE SWITCH
6	F34 20GY*	WIDE OPEN THROTTLE SWITCH
7	99 18BK	SYSTEM GROUND
8	F7 18TN*	AIR TEMPERATURE SENSOR
8	F7 18TN*	AIR TEMPERATURE SENSOR
9	F26 20OR*	IGNITION TIMING
9	F26 20OR*	IGNITION TIMING
10	F18 14BL	EGR PURGE SOLENOID
11	F4 18LG	ISA - EXTENDED
11	F4 18LG	ISA - EXTENDED
12	F24 18TN	COOLANT TEMPERATURE SENSOR
12	F24 18TN	COOLANT TEMPERATURE SENSOR
13	F12 18GY*	ISA - CLOSED THROTTLE SWITCH
13	F12 18GY*	ISA - CLOSED THROTTLE SWITCH
14	F3 18BR	ISA - RETRACT
14	F3 18BR	ISA - RETRACT
15	—	—

CAV	DIAGNOSTIC CONNECTOR — #2	
1	F29 16TN	PCV SOLENOID
2	132 18LB/RD	IDLE VACUUM SOLENOID
3	F26 18OR*	HIGH ALTITUDE JUMPER
4	11 14YL	I-1 IGNITION SWITCH
5	F38 18YL/RD	DOWN STREAM SOLENOID
6	F34 18GY*	WIDE OPEN THROTTLE SWITCH
7	99 20BK	ECU GROUND
8	F37 18VT*	DOWN STREAM SOLENOID
9	F35 18YL/WT	VACUUM SWITCH
10	F33 18BL/OR	THERMO - ELECTRIC SWITCH
11	F60 14RD*	STEPPER MOTOR
12	F24 18TN	COOLANT TEMPERATURE SWITCH
13	F3 20BR	IDLE RELAY
14	F45 14PK	STEPPER MOTOR
15	F36 18YL/BL	VACUUM SWITCH

(2.5L) 6 WAY AND 15WAY
DIAGNOSTIC CONNECTORS

80 59

(4.2L) 6 WAY AND 15 WAY
DIAGNOSTIC CONNECTORS

80 60

1987-90

CAV	ECU SYSTEMS CIRCUITS	
1	99 14BK	POWER GROUND
2	99 14BK	POWER GROUND
3	11 14YL	IGNITION
4	10 14RD	BATTERY
5	F18 14BL	EGR CONTROL
6	F17 20OR*	FUEL PUMP RELAY
7	F20 20BK*	LATCH RELAY
8	F34 20GY/BK	WIDE OPEN THROTTLE
9	—	
10	99 14BK	SYSTEM GROUND
11	F52 20RD*	ENGINE SPEED INPUT
12	34 20BK*	START SIGNAL (—)
13	F57 20BR*	TPS GROUND
14	F7 18TN*	AIR TEMPERATURE INPUT
15	F24 18TN	COOLANT TEMPERATURE INPUT
16	F2 20RD*	5VOLT SUPPLY (TPS/MAP)
17	F56 20BK	MAP GROUND
18	136 18PK*	UPSHIFT INDICATOR
19	F21 14PK	B+ LATCH
20	—	
21	F1 16LB	INJECTOR CONTROL
22	135 16BL*	A/C CLUTCH CONTROL
23	F3 18BR	ISCA (RETRACT)
24	F14 18LG	ISCA (EXTENDED)
25	F12 18GY*	CLOSED THROTTLE INPUT
26	—	
27	F26 20OR*	TIMING OUTPUT
28	F53 20WT*	ENGINE SPEED INPUT
29	88 14GN*	START SIGNAL (+)
30	168 14LG	LOAD SWAP - A/C SELECT
31	F50 20YL/GN	TPS INPUT
32	F15 18BR*	TEMPERATURE SENSORS GROUND
33	F14 20VT	MAP INPUT
34	169 14LB	LOAD SWAP - A/C REQUEST
35	F6 18OR	OXYGEN SENSOR INPUT

(2.4L) 35 WAY
ENGINE CONTROLLER CONNECTOR

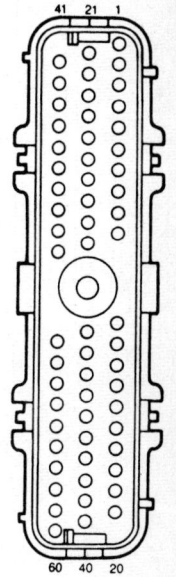

80	61

CAV	MCU SYSTEM CIRCUITS	
1	11 14YL	I-1 IGNITION SWITCH
8	F6 18GY	OXYGEN SENSOR
9	F6 18GY	OXYGEN SENSOR
10	F45 14PK	STEPPER MOTOR
11	F26 18OR*	HIGH ALTITUDE
16	F36 18YL/BL	4INCH VACUUM SWITCH
17	99 18BK	OXYGEN SELECT GROUND
18	F40 16VT	MAGNETIC PICKUP (—)
20	99 14BK	MCU GROUND
41	F29 18TN	PCV SOLENOID
43	F3 20BR	IDLE SPEEDUP RELAY
44	132 18LB/RD	IDLE VACUUM SOLENOID
45	F37 18VT*	UP STREAM SOLENOID
46	F38 18YL/RD	DOWN STREAM SOLENOID
47	F43 16LG*	TIMING OUTPUT
48	F60 14RD*	STEPPER MOTOR
49	F59 14PK*	STEPPER MOTOR
50	F61 14BR*	STEPPER MOTOR
51	F54 18VT*	KNOCK SENSOR
53	F33 18BL/OR	THERMO - ELECTRIC SWITCH
54	F34 18GY/BK	WIDE OPEN THROTTLE
55	F35 18YL/WT	10 INCH VACUUM SWITCH
56	F24 18TN	COOLANT TEMPERATURE SWITCH
57	11 14YL	I-1 IGNITION SWITCH
58	33 14GN	START SIGNAL
59	F39 16OR	MAGNETIC PICKUP (+)
60	99 14BK	MCU GROUND

(4.2L) 60 WAY
ENGINE CONTROLLER CONNECTOR

80	62

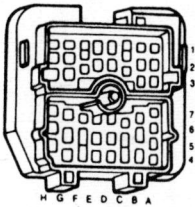

ENGINE COMPARTMENT CIRCUITS		CAV
72 18BL	PARKING LAMPS	A1
—	—	A2
79 18BR	RIGHT TURN SIGNAL LAMPS	A3
—	—	A4
—	—	A5
12 14OR	I-3 IGNITION (4.2L ONLY)	A6
58 20GY*	BRAKE WARNING SWITCH	A7
—	—	B1
—	—	B2
—	—	B3
F22 14OR	FUEL PUMP RELAY OUTPUT (2.5L ONLY)	B4
99 12BK	GROUND	B5
58 20GY*	BRAKE BULB CHECK	B6
—	—	C1
78 18GY*	LEFT TURN SIGNAL LAMPS	C2
—	—	C3
55 18VT	COOLANT TEMPERATURE SENSOR	C4
30 18GN*	TACHOMETER	C5
18 18YL*	BACKUP SWITCH - TRANSMISSION	C6
10 14RD	FOG LAMP RELAY - FEED	D1
—	—	D2
10 12RD	BATTERY	D4
11 14YL	I-1 IGNITION SWITCH	D5
76 16WT	HI BEAM HEADLAMPS	E1
—	—	E2
10 12RD/WT	BATTERY	E4
20 16PK*	UNDERHOOD LAMP	E5
F5 18BL*	FOG LAMP RELAY - COIL FEED	F1
—	—	F2
44 18VT*	WINDSHIELD WASHER PUMP FEED	F3
32 14OR/WT	BLOWER MOTOR FEED	F4
54 18LG	OIL PRESSURE SENDER	F5
126 18BR	UPSHIFT SWITCH (2.5L ONLY)	F6
—	—	G1
—	—	G2
31 14GY	HORN	G3
10 12RD	BATTERY	G4
33 14GN	STARTER RELAY - CRANK	G5
75 18BR*	BACKUP LAMPS	G6
77 16LG	LO BEAM HEADLAMPS	H1
—	—	H2
—	—	H3
138 16LG	A/C CLUTCH	H4
62 14BR*	DIODE FUSE ASSY (4.2L ONLY)	H5
106 18BK/YL	4 WHEEL DRIVE VACUUM SWITCH	H6

80	63

CAV	INSTRUMENT PANEL CIRCUITS	
A1	72 18BL	PARKING LAMPS FEED
A2	—	—
A3	79 18BR	RIGHT TURN SIGNAL LAMPS
A4	24 14TN/BL	NOT USED (EXPORT)
A5	—	—
A6	12 14OR	I-3 IGNITION SWITCH
A7	58 18GY/BK	BRAKE INDICATOR
B1	—	—
B2	—	—
B3	—	—
B4	F22 14OR	FUEL PUMP
B5	99 12BK	GROUND
B6	58 18GY/BK	BRAKE BULB CHECK
C1	—	—
C2	78 18GY/BK	LEFT TURN SIGNAL LAMPS
C3	—	—
C4	55 18VT	COOLANT TEMPERATURE
C5	30 18GN/WT	TACHOMETER
C6	18 18YL/BK	BACKUP LAMPS
D1	10 14RD	BATTERY
D2	—	—
D4	10 12RD	BATTERY
D5	11 14YL	I-1 IGNITION SWITCH
E1	76 16WT	DIMMER SWITCH - HI BEAM
E2	—	—
E4	10 12RD	BATTERY
E5	20 16PK/BK	HAZARD/STOP FUSE-UNDERHOOD LAMP
F1	F5 16BL/WT	FOG LAMP SWITCH
F2	—	—
F3	44 18VT/WT	WINDSHIELD WIPER WASHER PUMP FEED
F4	37 14OR/WT	BLOWER MOTOR FEED
F5	54 20LB	OIL PRESSURE
F6	126 18BR/LB	UPSHIFT INDICATOR
G1	—	—
G2	—	—
G3	31 14GY	HORN RELAY
G4	10 12RD	BATTERY
G5	33 14GN	IGNITION SWITCH - START
G6	75 18BR/WT	REVERSE SWITCH
H1	77 16LG	DIMMER SWITCH - LO BEAM
H2	—	—
H3	—	—
H4	138 16LG	A/C REQUEST
H5	62 14BR/WT	REAR DEFOGGER FEED
H6	106 18BK/YL	4 WHEEL DRIVE LAMP

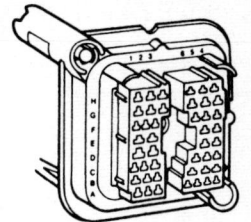

80	64

1987-90

CAV	GUAGE PACKAGE CIRCUITS	
1	55 18VT	COOLANT TEMPERATURE
2	14 18WT/BK	GAUGES FUSE
2	14 18WT/BK	GAUGES FUSE
3	99 18BK	GROUND
4	57 20TN	FUEL LEVEL
5	14 18WT/BK	GAUGES FUSE
6	54 18LB	OIL PRESSURE
7	51 18PK	CLOCK FEED
8	—	—
9	—	—
10	52 18OR/BK	ILLUMINATION
10	52 18OR/BK	ILLUMINATION
11	—	—
12	52 18OR/BK	ILLUMINATION

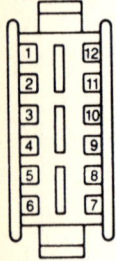

(BEHIND GAUGE PACKAGE)

CAV	INDICATOR CIRCUITS	
1	—	—
2	—	—
3	58 20GY/BK	PARKING BRAKE
4	189 18GY	EMISSION MAINTENANCE
	189 18GY	WARNING INDICATOR
5	14 18WT/BK	GAUGES FUSE
6	126 18BR/LB	UPSHIFT (2.5L/5 SPD ONLY)
7	76 16WT	HI BEAM
8	78 18GY/BK	LEFT TURN SIGNAL

(LEFT OF STEERING COLUMN)

CAV	INDICATOR CIRCUITS	
1	99 18BK	GROUND
2	—	—
3	79 18BR	RIGHT TURN SIGNAL
4	—	—
5	—	—
6	60 18WT/OR	SEAT BELT INDICATOR
7	106 18BK/YL	4 WHEEL DRIVE
8	52 18OR/BK	INSTRUMENT PANEL ILLUMINATION

(RIGHT OF STEERING COLUMN)

80	65	12-WAY GAUGE PACKAGE CONNECTOR

8-WAY INDICATOR CONNECTORS	80	66

CAV	TURN SIGNAL SWITCH CIRCUITS	
P	71 18LB/BK	STOP LAMPS
N	79 18BR	RIGHT TURN-REAR
M	78 18GY/BK	LEFT TURN-REAR
L	74 18GY	TURN SIGNAL FLASHER
K	80 18VT/WT	HAZARD FLASHER
J	79 18GY/BK	RIGHT TURN-FRONT/INDICATOR
H	78 18GY/BK	LEFT TURN-FRONT/INDIATOR
G	128 18OR	HORN SWITCH
F	86 18WT	IGNITION KEY WARNING SWITCH
E	87 18WT	IGNITION KEY WARNING SWITCH
D	52 18OR/BK	PRNDL ILLUMINATION

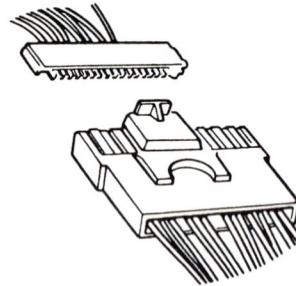

80	67	11 WAY STEERING COLUMN CONNECTOR

1987-90

7 Drive Train

QUICK REFERENCE INDEX

GENERAL INDEX

UNDERSTANDING THE MANUAL TRANSMISSION

Because of the way an internal combustion engine breathes, it can produce torque, or twisting force, only within a narrow speed range. Most modern, overhead valve engines must turn at about 2,500 rpm to produce their peak torque. By 4,500 rpm they are producing so little torque that continued increases in engine speed produce no power increases.

The manual transmission and clutch are employed to vary the relationship between engine speed and the speed of the wheels so that adequate engine power can be produced under all circumstances. The clutch allows engine torque to be applied to the transmission input shaft gradually, due to mechanical slippage. The car can, consequently, be started smoothly from a full stop.

The transmission changes the ratio between the rotating speeds of the engine and the wheels by the use of gears. On Jeep vehicles, 3-speed or 4-speed transmissions are most common. The lower gears allow full engine power to be applied to the rear wheels during acceleration at low speeds.

The clutch drive plate is a thin disc, the center of which is splined to the transmission input shaft. Both sides of the disc are covered with a layer of material which is similar to brake lining and which is capable of allowing slippage without roughness or excessive noise.

The clutch cover is bolted to the engine flywheel and incorporates a diaphragm spring which provides the pressure to engage the clutch. The cover also houses the pressure plate. The driven disc is sandwiched between the pressure plate and the smooth surface of the flywheel when the clutch pedal is released, thus forcing it to turn at the same speed as the engine crankshaft.

The transmission contains a mainshaft which passes all the way through the transmission, from the clutch to the driveshaft. This shaft is separated at one point, so that front and rear portions can turn at different speeds.

Power is transmitted by a countershaft in the lower gears and reverse. The gears of the countershaft mesh with gears on the mainshaft, allowing power to be carried from one to the other. All the countershaft gears are integral with that shaft, while several of the mainshaft gears can either rotate independently of the shaft or be locked to it. Shifting from one gear to the next causes one of the gears to be freed from rotating with the shaft and locks another to it. Gears are locked and unlocked by internal dog clutches which slide between the center of the gear and the shaft. The forward gears usually employ synchronizers; friction members which smoothly bring gear and shaft to the same speed before the toothed dog clutches are engaged.

The clutch is operating properly if:
1. It will stall the engine when released with the vehicle held stationary.
2. The shift lever can be moved freely between first and reverse gears when the vehicle is stationary and the clutch disengaged.

A clutch pedal free-play adjustment is incorporated in the linkage. If there is about 1-2 in. (25-51mm) of motion before the pedal begins to release the clutch, it is adjusted properly. Inadequate free-play wears all parts of the clutch releasing mechanisms and may cause slippage. Excessive free-play may cause inadequate release and hard shifting of gears.

Some clutches use a hydraulic system in place of mechanical linkage. If the clutch fails to release, fill the clutch master cylinder with fluid to the proper level and pump the clutch pedal to fill the system with fluid. Bleed the system in the same way as a brake system. If leaks are located, tighten loose connections or overhaul the master or slave cylinder as necessary.

Troubleshooting the Manual Transmission and Transfer Case

Problem	Cause	Solution
Transmission shifts hard	• Clutch adjustment incorrect • Clutch linkage or cable binding • Shift rail binding	• Adjust clutch • Lubricate or repair as necessary • Check for mispositioned selector arm roll pin, loose cover bolts, worn shift rail bores, worn shift rail, distorted oil seal, or extension housing not aligned with case. Repair as necessary.
	• Internal bind in transmission caused by shift forks, selector plates, or synchronizer assemblies • Clutch housing misalignment	• Remove, dissemble and inspect transmission. Replace worn or damaged components as necessary. • Check runout at rear face of clutch housing
	• Incorrect lubricant • Block rings and/or cone seats worn	• Drain and refill transmission • Blocking ring to gear clutch tooth face clearance must be 0.030 inch or greater. If clearance is correct it may still be necessary to inspect blocking rings and cone seats for excessive wear. Repair as necessary.

Troubleshooting the Manual Transmission and Transfer Case (cont.)

Problem	Cause	Solution
Gear clash when shifting from one gear to another	• Clutch adjustment incorrect • Clutch linkage or cable binding • Clutch housing misalignment • Lubricant level low or incorrect lubricant • Gearshift components, or synchronizer assemblies worn or damaged	• Adjust clutch • Lubricate or repair as necessary • Check runout at rear of clutch housing • Drain and refill transmission and check for lubricant leaks if level was low. Repair as necessary. • Remove, disassemble and inspect transmission. Replace worn or damaged components as necessary.
Transmission noisy	• Lubricant level low or incorrect lubricant • Clutch housing-to-engine, or transmission-to-clutch housing bolts loose • Dirt, chips, foreign material in transmission • Gearshift mechanism, transmission gears, or bearing components worn or damaged • Clutch housing misalignment	• Drain and refill transmission. If lubricant level was low, check for leaks and repair as necessary. • Check and correct bolt torque as necessary • Drain, flush, and refill transmission • Remove, disassemble and inspect transmission. Replace worn or damaged components as necessary. • Check runout at rear face of clutch housing
Jumps out of gear	• Clutch housing misalignment • Gearshift lever loose • Offset lever nylon insert worn or lever attaching nut loose • Gearshift mechanism, shift forks, selector plates, interlock plate, selector arm, shift rail, detent plugs, springs or shift cover worn or damaged • Clutch shaft or roller bearings worn or damaged	• Check runout at rear face of clutch housing • Check lever for worn fork. Tighten loose attaching bolts. • Remove gearshift lever and check for loose offset lever nut or worn insert. Repair or replace as necessary. • Remove, disassemble and inspect transmission cover assembly. Replace worn or damaged components as necessary. • Replace clutch shaft or roller bearings as necessary
Jumps out of gear (cont.)	• Gear teeth worn or tapered, synchronizer assemblies worn or damaged, excessive end play caused by worn thrust washers or output shaft gears • Pilot bushing worn	• Remove, disassemble, and inspect transmission. Replace worn or damaged components as necessary. • Replace pilot bushing

Troubleshooting the Manual Transmission and Transfer Case (cont.)

Problem	Cause	Solution
Will not shift into one gear	• Gearshift selector plates, interlock plate, or selector arm, worn, damaged, or incorrectly assembled • Shift rail detent plunger worn, spring broken, or plug loose • Gearshift lever worn or damaged • Synchronizer sleeves or hubs, damaged or worn	• Remove, disassemble, and inspect transmission cover assembly. Repair or replace components as necessary. • Tighten plug or replace worn or damaged components as necessary • Replace gearshift lever • Remove, disassemble and inspect transmission. Replace worn or damaged components.
Locked in one gear—cannot be shifted out	• Shift rail(s) worn or broken, shifter fork bent, setscrew loose, center detent plug missing or worn • Broken gear teeth on countershaft gear, clutch shaft, or reverse idler gear Gearshift lever broken or worn, shift mechanism in cover incorrectly assembled or broken, worn damaged gear train components	• Inspect and replace worn or damaged parts • Inspect and replace damaged part • Disassemble transmission. Replace damaged parts or assemble correctly.
Transfer case difficult to shift or will not shift into desired range	• Vehicle speed too great to permit shifting • If vehicle was operated for extended period in 4H mode on dry paved surface, driveline torque load may cause difficult shifting • Transfer case external shift linkage binding • Insufficient or incorrect lubricant • Internal components binding, worn, or damaged	• Stop vehicle and shift into desired range. Or reduce speed to 3–4 km/h (2–3 mph) before attempting to shift. • Stop vehicle, shift transmission to neutral, shift transfer case to 2H mode and operate vehicle in 2H on dry paved surfaces • Lubricate or repair or replace linkage, or tighten loose components as necessary • Drain and refill to edge of fill hole with SAE 85W-90 gear lubricant only • Disassemble unit and replace worn or damaged components as necessary
Transfer case noisy in all drive modes	• Insufficient or incorrect lubricant	• Drain and refill to edge of fill hole with SAE 85W-90 gear lubricant only. Check for leaks and repair if necessary. Note: If unit is still noisy after drain and refill, disassembly and inspection may be required to locate source of noise.

Troubleshooting the Manual Transmission and Transfer Case (cont.)

Problem	Cause	Solution
Noisy in—or jumps out of four wheel drive low range	• Transfer case not completely engaged in 4L position	• Stop vehicle, shift transfer case in Neutral, then shift back into 4L position
	• Shift linkage loose or binding	• Tighten, lubricate, or repair linkage as necessary
	• Shift fork cracked, inserts worn, or fork is binding on shift rail	• Disassemble unit and repair as necessary
Lubricant leaking from output shaft seals or from vent	• Transfer case overfilled • Vent closed or restricted	• Drain to correct level • Clear or replace vent if necessary
Lubricant leaking from output shaft seals or from vent (cont.)	• Output shaft seals damaged or installed incorrectly	• Replace seals. Be sure seal lip faces interior of case when installed. Also be sure yoke seal surfaces are not scored or nicked. Remove scores, nicks with fine sandpaper or replace yoke(s) if necessary.
Abnormal tire wear	• Extended operation on dry hard surface (paved) roads in 4H range	• Operate in 2H on hard surface (paved) roads

Manual Transmission

REMOVAL AND INSTALLATION

1971

—————— CAUTION ——————

The clutch driven disc contains asbestos, which has been determined to be a cancer causing agent. Never clean clutch surfaces with compressed air! Avoid inhaling any dust from any clutch surface! When cleaning clutch surfaces, use a commercially available brake cleaning fluid.

1. Raise and support the vehicle on jackstands.
2. Drain the transmission and transfer case.
3. Remove the shift lever and shift housing.
4. Remove the set screw from the transfer case shift lever pivot pin. Remove the pivot pin, shift levers, and shift lever springs.
5. If the vehicle is equipped with a power take-off, remove the shift lever plate screws and lift out the lever.
6. Disconnect the front and rear driveshafts from the transfer case. If the vehicle is equipped with a power take-off, disconnect the transfer case end of the PTO driveshaft.
7. Disconnect the speedometer cable at the transfer case.
8. Disconnect the hand brake cable.
9. Disconnect the clutch cable at the bellcrank.

10. Place jacks under the transmission and engine. Protect the oil pan with a wood block.
11. Remove the nuts holding the rear mount to the crossmember.
12. Remove the transfer case-to-crossmember bolt.
13. Remove the frame center crossmember-to-frame side rail bolts and remove the crossmember. Remove the transmission-to-bellhousing bolts.
14. Force the transmission to the right to disengage the clutch control lever tube ball joint.
15. Lower the engine and transmission. Slide the transmission and transfer case assemblies toward the rear until they clear the clutch.
16. Remove the six screws and lockwashers attaching the transfer case rear cover and remove the cover. If the vehicle is equipped with a power take-off, remove the PTO shift unit.
17. Remove the cotter pin, nut, and washer holding the transfer case main drive gear on the rear end of the transmission mainshaft. If possible, remove the main drive gear. If that's not possible, see step 19.
18. Remove the transmission-to-transfer case bolts.
19. Install a transmission mainshaft retaining plate tool #W-194 to prevent the mainshaft from pulling out of the case. If this tool is not available, loop a piece of wire around the mainshaft directly in back of the second speed gear. Install the shift housing right and left bolts part way into the case. Attach each end of

the wire to the bolts. Support the transfer case, and with a soft mallet, tap lightly on the end of the mainshaft to loosen the gear and separate the two units.

20. Join the transfer case and transmission. Remove the wire or special holding tool. Remove the shift housing right and left bolts.

21. Install the transmission-to-transfer case bolts. Torque them to 35 ft. lbs.

22. Install the main drive gear. Install the cotter pin, nut, and washer holding the transfer case main drive gear on the rear end of the transmission mainshaft.

23. Install the transfer case rear cover and Install the cover. If the vehicle is equipped with a power take-off, Install the PTO shift unit.

24. Slide the transmission and transfer case assemblies forward until they engage the clutch.

25. Force the transmission to the right to engage the clutch control lever tube ball joint.

26. Install the transmission-to-bellhousing bolts. Torque them to 30 ft. lbs.

27. Install the frame center crossmember. Torque the bolts to 50 ft. lbs.

28. Install the transfer case-to-crossmember bolt. Torque it to 45 ft. lbs.

29. Install the nuts holding the rear mount to the crossmember. Torque them to 40 ft. lbs.

30. Connect the clutch cable at the bellcrank.

31. Connect the hand brake cable.

32. Connect the speedometer cable at the transfer case.

33. Connect the front and rear driveshafts to the transfer case. If the vehicle is equipped with a power take-off, connect the transfer case end of the pto driveshaft.

34. If the vehicle is equipped with a power take-off, install the shift lever.

35. Install the set screw in the transfer case shift lever pivot pin. Install the pivot pin, shift levers, and shift lever springs.

36. Install the shift lever and shift housing.

37. Fill the transmission and transfer case.

1972-86

CAUTION

The clutch driven disc contains asbestos, which has been determined to be a cancer causing agent. Never clean clutch surfaces with compressed air! Avoid inhaling any dust from any clutch surface! When cleaning clutch surfaces, use a commercially available brake cleaning fluid.

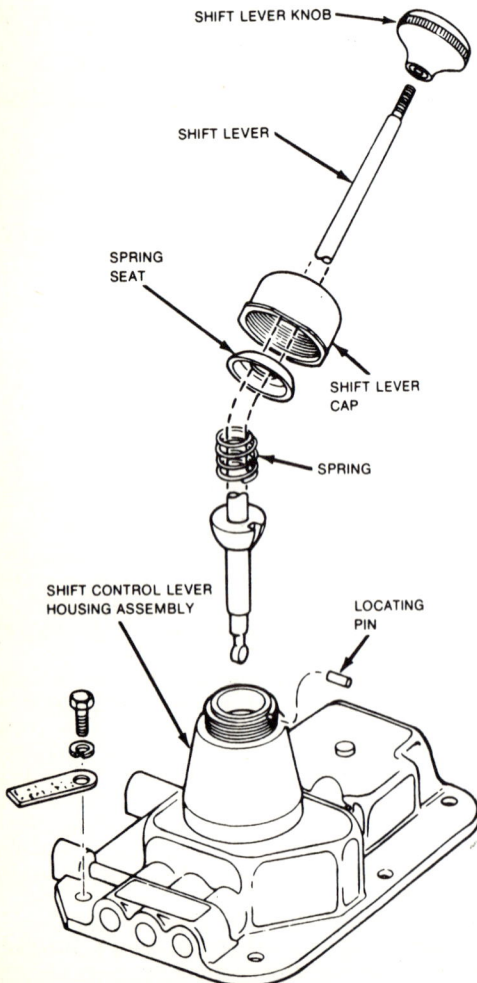

SR-4 shift lever removal

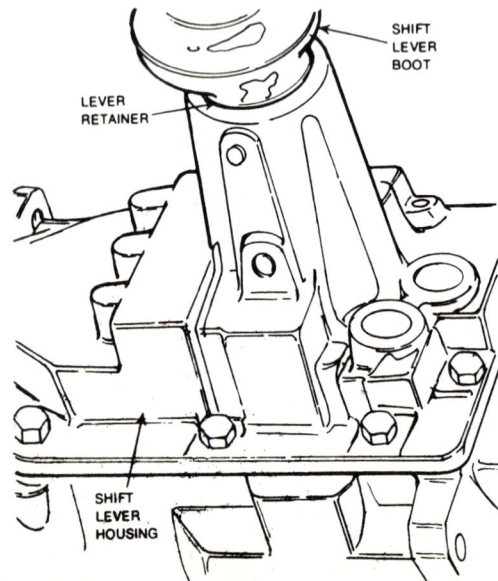

Shift lever removal for T-18A and all 3-speed units starting in 1974

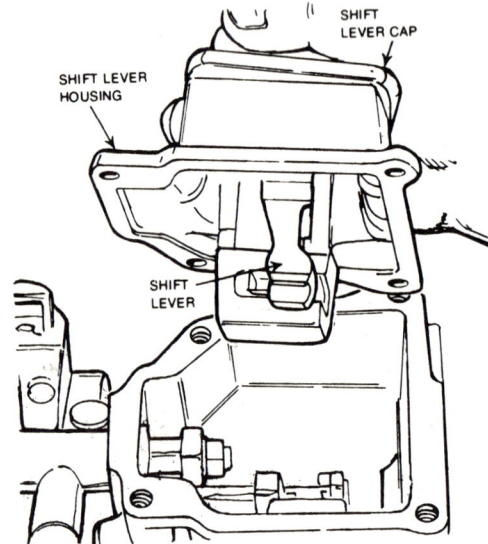

T-176 shift lever removal

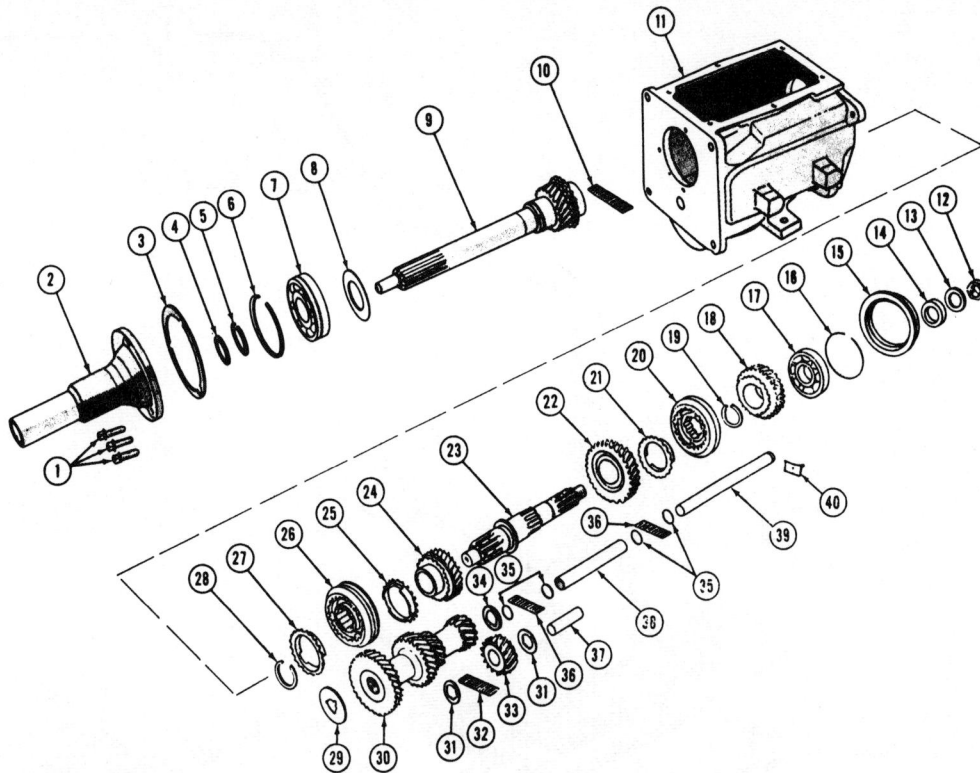

1. Retainer screws
2. Main drive gear bearing retainer
3. Retainer gasket
4. Oil seal
5. Snapring (small)
6. Snapring (large)
7. Main drive gear bearing
8. Oil retaining washer (slinger)
9. Main drive gear
10. Mainshaft plot bearing rollers
11. Case
12. Nut
13. Flatwasher
14. Spacer
15. Bearing adapter
16. Snapring
17. Mainshaft bearing
18. Reverse gear
19. Snapring
20. Low synchronizer assembly
21. Synchronizer blocking ring
22. Low gear
23. Mainshaft
24. Second gear
25. Synchronizer blocking ring
26. Second-third synchronizer assembly
27. Synchronizer blocking ring
28. Snapring
29. Countershaft front thrust washer (large)
30. Countershaft gear
31. Reverse idler gear bearing washer
32. Reverse idler gear roller bearings
33. Reverse idler gear
34. Countershaft rear thrust washer (small)
35. Countershaft bearing spacer washer
36. Reverse idler shaft
37. Countershaft roller bearings
38. Spacer washer
39. Countershaft
40. Lockplate

T-14A exploded view

1. Remove all floor lever knobs, trim rings and boots.

2. Remove the floor pan section from above the transmission shift control and unbolt the lever assembly from the transmission (early 3-speed models). On the 4-speed models except SR-4 and T-176, and 3-speeds starting 1976, unscrew the shift control housing cap, remove the washer, spring, shift lever and pin. On SR-4 models, remove the shift lever housing bolts and remove the shift lever and housing assembly.

3. On T-18A, remove the transfer case shift lever and bracket assembly. On T-176 models, press and turn the lever retainer and remove the shift lever assembly.

4. Raise the vehicle.

5. Index mark the driveshafts for proper alignment at installation.

6. Remove the front driveshaft.

7. Disconnect the front end of the rear driveshaft from the transfer case.

8. Disconnect the clutch cable and remove the cable mounting bracket from the transfer case on 1972 models only.

9. Disconnect the speedometer cable, back-up light switch wires, transmission controlled spark advance, and parking brake cable if connected to the crossmember.

10. If equipped with a V8 engine, disconnect the exhaust pipe at the manifolds and lower them. Support the engine with a jack. Disconnect the support support crossmember from the frame side rail.

11. Remove the bolts that attach the transmission to the clutch housing.

12. Lower the transmission slightly.

13. Move the transmission and transfer case assembly and crossmember backward far enough for the transmission clutch shaft to clear the clutch housing.

14. Remove the assembly from under the vehicle.

· 15. If the transmission and transfer case were separated, join them and torque the bolts to 30 ft. lbs.

16. Position the wave washer and the throwout bearing and sleeve assembly in the throwout fork. Center the bearing over the pressure plate release levers.

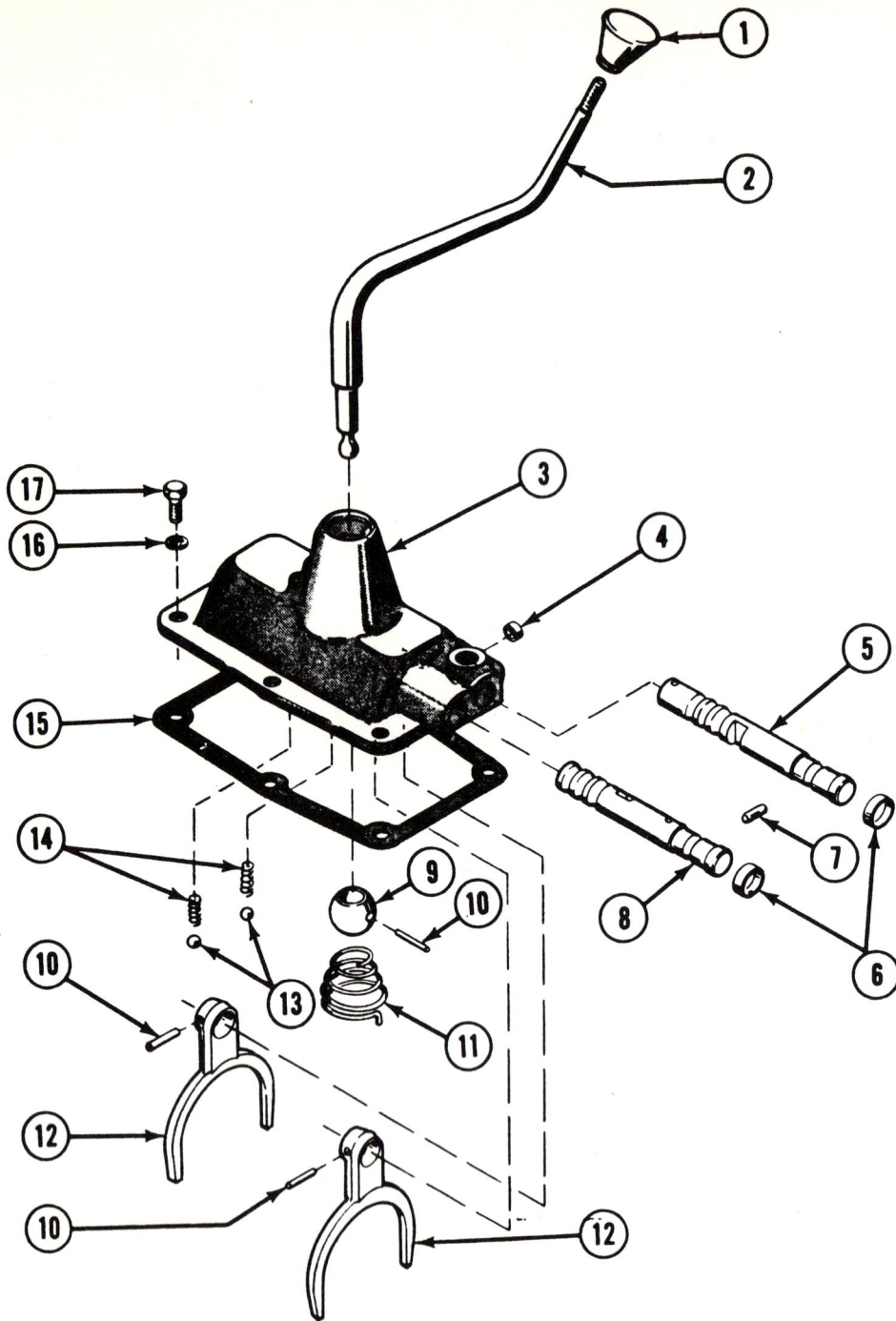

1. Shift lever knob
2. Shift lever
3. Control housing
4. Plug
5. Shift rail (second-third)
6. Shift rail cap
7. Interlock plunger
8. Shift rail (low-reverse)
9. Shift lever fulcrum ball
10. Pin
11. Shift lever support spring
12. Shift fork
13. Poppet ball
14. Poppet spring
15. Gasket
16. Lockwasher
17. Bolt

T-14A control tower

17. Protect the splines and throwout bearing alignment and slowly slide the transmission into position. Some maneuvering may be necessary in order to match the transmission input shaft splines and the clutch driven plate splines.

18. Install the bolts that attach the transmission to the clutch housing. Torque them to 54 ft. lbs.

19. If equipped with a V8 engine, Connect the exhaust pipe at the manifolds and lower them. Support the engine with a jack. Connect the support crossmember to the frame side rail. Torque the bolts to 30 ft. lbs.

20. Connect the speedometer cable, back-up light switch wires, transmission controlled spark advance, and parking brake cable if connected to the crossmember.

21. Install the clutch cable mounting bracket on the transfer case on 1972 models only. Connect the clutch cable.

22. Connect the front end of the rear driveshaft to the transfer case.

23. Install the front driveshaft.

24. Lower the vehicle.

25. On T-18A, install the transfer case shift lever and bracket assembly. On T-176 models, press and turn the lever retainer and install the shift lever assembly.

26. Install the floor pan section from above the transmission shift control and install the shift lever assembly from the transmission (early 3-speed models). On the 4-speed models exc. SR-4 and T-176, and 3-speeds starting 1976, unscrew the shift control housing cap, install the washer, spring, shift lever and pin. On SR-4 models, install the shift lever housing bolts and install the shift lever and housing assembly.

27. Install all floor lever knobs, trim rings and boots.

1987-90

AX-5

—————————— CAUTION ——————————

The clutch driven disc contains asbestos, which has been determined to be a cancer causing agent. Never clean clutch surfaces with compressed air! Avoid inhaling any dust from any clutch surface! When cleaning clutch surfaces, use a commercially available brake cleaning fluid.

1. Raise the outer gearshift lever boot and remove the upper part of the console.

2. Remove the lower part of the console.

3. Remove the inner boot.

4. Remove the gearshift lever and stub shaft by pressing down on the stub shaft retainer and rotating the retainer counterclockwise to release it from the lugs in the shift tower. Then, lift the retainer, stub shaft and shift lever up and out of the tower Don't remove the shift lever from the stub shaft.

5. Raise and support the truck on jackstands.

6. Drain the transmission and transfer case.

7. Matchmark the rear driveshaft and yoke for installation alignment.

8. Unbolt and remove the rear driveshaft.

9. Position a floor jack under the transmission and take up the weight slightly.

10. Unbolt and remove the rear crossmember.

11. Disconnect the hydraulic line from the clutch slave cylinder. Disconnect the speedometer cable.

12. Disconnect the back-up light switch.

13. Disconnect the transfer case vent hose at the case.

14. Disconnect all linkage and hoses from the transfer case and transmission.

15. Matchmark the front driveshaft and yoke.

16. Remove the front driveshaft.

17. Chain the transmission to the jack.

18. Unbolt the transmission from the engine and lower the jack while pulling back.

19. Install the transmission to the engine. Transmission-to-transfer case adapter nut: 26 ft. lbs.

20. Install the front driveshaft. U-joint flange nut-to-transfer case: 35 ft. lbs.

21. Connect all linkage and hoses at the transfer case and transmission.

22. Connect the transfer case vent hose at the case.

23. Connect the back-up light switch.

24. Connect the hydraulic line from the clutch slave cylinder.

25. Connect the speedometer cable.

26. Install the rear crossmember. Rear crossmember-to-side sill: 30 ft. lbs. Rear support isolator-to-transmission: 33 ft. lbs.

27. Install the rear driveshaft. U-joint flange nut-to-transfer case: 35 ft. lbs.

28. Fill the transmission and transfer case.

29. Lower the truck.

30. Install the gearshift lever and stub shaft by pressing down on the stub shaft retainer and rotating the retainer clockwise.

31. Install the inner boot.

32. Install the lower part of the console.

33. Raise the outer gearshift lever boot and install the upper part of the console.

BA10/5

—————————— CAUTION ——————————

The clutch driven disc contains asbestos, which has been determined to be a cancer causing agent. Never clean clutch surfaces with compressed air! Avoid inhaling any dust from any clutch surface! When cleaning clutch surfaces, use a commercially available brake cleaning fluid.

1. Remove the shift lever knob.

2. Remove the shift lever outer boots and the transmission tower dust boot.

3. Remove the shift lever stub shaft retaining plate (snapring on some models) and remove the shift lever and stub shaft as an assembly.

4. Raise and support the truck on jackstands.

5. Drain the transmission and transfer case.

6. Matchmark the rear driveshaft and yoke for installation alignment.

7. Unbolt and remove the rear driveshaft.

8. Position a floor jack under the transmission and take up the weight slightly.

9. Unbolt and remove the rear crossmember.

10. Disconnect the hydraulic line from the clutch slave cylinder. Disconnect the speedometer cable.

11. Disconnect the back-up light switch.

12. Disconnect the transfer case vent hose at the case.

13. Disconnect all linkage and hoses from the transfer case and transmission.

14. Matchmark the front driveshaft and yoke.

15. Remove the front driveshaft.

16. Chain the transmission to the jack.

17. Unbolt the transmission from the engine and lower the jack while pulling back.

18. Slide the transmission and transfer case assembly into place, engaging the clutch assembly. Transmission-to-transfer case adapter nut: 26 ft. lbs. Transmission case-to-engine: 28 ft. lbs.

19. Install the front driveshaft. U-joint strap bolts: 15 ft. lbs.

20. Connect all linkage and hoses at the transfer case and transmission.

21. Connect the transfer case vent hose at the case.

22. Connect the back-up light switch.

23. Connect the hydraulic line to the clutch slave cylinder.

24. Connect the speedometer cable.

25. Install the rear crossmember. Rear crossmember-to-side sill: 30 ft. lbs. Rear support isolator-to-transmission: 33 ft. lbs.

26. Position a floor jack under the transmission and take up the weight slightly.

27. Install the rear driveshaft. U-joint strap bolts: 15 ft. lbs.

28. Matchmark the rear driveshaft and yoke for installation alignment.

29. Drain the transmission and transfer case.

30. Raise and support the truck on jackstands.

31. Install the shift lever stub shaft retaining plate (snapring on some models) and install the shift lever and stub shaft as an assembly.

32. Install the shift lever outer boots and the transmission tower dust boot.

33. Install the shift lever knob.

Back-Up Light Switch

REMOVAL AND INSTALLATION

The switch is threaded into the transmission and is replaced by unscrewing. No adjustments are possible. Switch locations are as follows:

- AX-5 and BA 10/5: right side of the case
- SR-4, T4 and T5: left side of the case
- T-176: on the top cover behind the shifter
- T-18A, T-98A and T-150: in the top cover

T-86AA 3-Speed Overhaul

DISASSEMBLY

1. Drain the lubricant.

2. Remove the transfer case rear cover.

3. If so equipped, remove the pto shift unit.

4. Remove the cotter pin, nut and washer and remove the transfer case main drive gear.

5. Remove the transmission shift cover.

6. Loop a piece of wire around the mainshaft just behind 2nd gear. Twist the wire and attach one end to the right front cover screw and the other end to the left front cover screw. Tighten the wire to prevent the mainshaft from pullinf out of the case when the transfer case is removed. If the mainshaft pulls out, the synchronizer parts will drop to the bottom of the case.

7. Remove the transfer case screws, then tap lightly on the end of the mainshaft to separate the two units. The trnamsission mainshaft bearing should slide out of the transfer case and stay with the mainshaft.

8. Remove the front main drive gear bearing retainer and gasket.

9. Remove the oil collector screws.

10. Remove the lockplate from the reverse idler shaft and countershaft at the rear of the case.

11. Using a dummy shaft and hammer, drive the countershaft out through the rear of the case.

12. Remove the mainshaft assembly through the rear opening, followed by the main drive gear.

13. Remove the countershaft gear set and 3 thrust washers from the case. Dismantle the countershaft gear assembly.

14. Drive out the reverse idler shaft and gear.

ASSEMBLY

15. Drive in the reverse idler shaft and gear.

16. Assemble the countershaft gear set. Endplay, controlled by the thickness of the rear steel thrust washer, should be 0.012-0.018 in. (0.30-0.45mm). Assemble the large bronze washer with the lip entered in the slot in the case. The bronze-faced steel washer is placed next to the gear at the rear end and the steel washer is next to the case. Use a dummy shaft to assemble the countershaft bearing rollers. Hold the rollers in place with grease. The use of a loading sleeve will make the job easier.

17. Install the countershaft gear set and 3 thrust washers from the case.

18. When assembling the mainshaft gears, low and reverse gear is installed with the shoe shoe groove towards the front.

19. Install the mainshaft assembly and main drive gear.

20. When assembling the synchronizer unit, install the 2 springs in the high and intermediate clutch hub with the tensions opposed.

21. Drive the countershaft in through the rear of the case.

22. Install the lockplate on the reverse idler shaft and countershaft.

23. Install the oil collector screws.

24. Install the front main drive gear bearing retainer and gasket.

25. Join the transfer case and transmission. Install the transfer case screws.

26. Install the transmission shift cover.

27. Install the transfer case main drive gear.

28. If so equipped, install the pto shift unit.

29. Install the transfer case rear cover.

30. Fill the case with lubricant.

T-90C 3-Speed Overhaul

DISASSEMBLY

1. Drain the lubricant.

2. Remove the transfer case rear cover.

3. If so equipped, remove the pto shift unit.

4. Remove the cotter pin, nut and washer and remove the transfer case main drive gear.

5. Remove the transmission shift cover.

6. Loop a piece of wire around the mainshaft just behind 2nd gear. Twist the wire and attach one end to the right front cover screw and the other end to the left front cover screw. Tighten the wire to prevent the mainshaft from pullinf out of the case when the transfer case is removed. If the mainshaft pulls out, the synchronizer parts will drop to the bottom of the case.

7. Remove the transfer case screws, then tap lightly on the end of the mainshaft to separate the two units. The trnamsission mainshaft bearing should slide out of the transfer case and stay with the mainshaft.

8. Remove the front main drive gear bearing retainer and gasket.

9. Remove the oil collector screws.

10. Remove the lockplate from the reverse idler shaft and countershaft at the rear of the case.

11. Using a dummy shaft and hammer, drive the countershaft out through the rear of the case.

12. Remove the mainshaft assembly through the rear opening, followed by the main drive gear.

13. Remove the countershaft gear set and 3 thrust washers from the case. Dismantle the countershaft gear assembly.

14. Drive out the reverse idler shaft and gear.

ASSEMBLY

15. Drive in the reverse idler shaft and gear.

16. Assemble the countershaft gear set. Endplay, controlled by the thickness of the rear steel thrust washer, should be 0.012-0.018 in. (0.30-0.45mm). Assemble the large bronze washer with the lip entered in the slot in the case. The bronze-faced steel washer is placed next to the gear at the rear end and the steel washer is next to the case. Use a dummy shaft to assemble the countershaft bearing rollers. Hold the rollers in place with grease. The use of a loading sleeve will make the job easier.

17. Install the countershaft gear set and 3 thrust washers from the case.

18. When assembling the mainshaft gears, low and reverse gear is installed with the shoe shoe groove towards the front.

19. Install the mainshaft assembly and main drive gear.

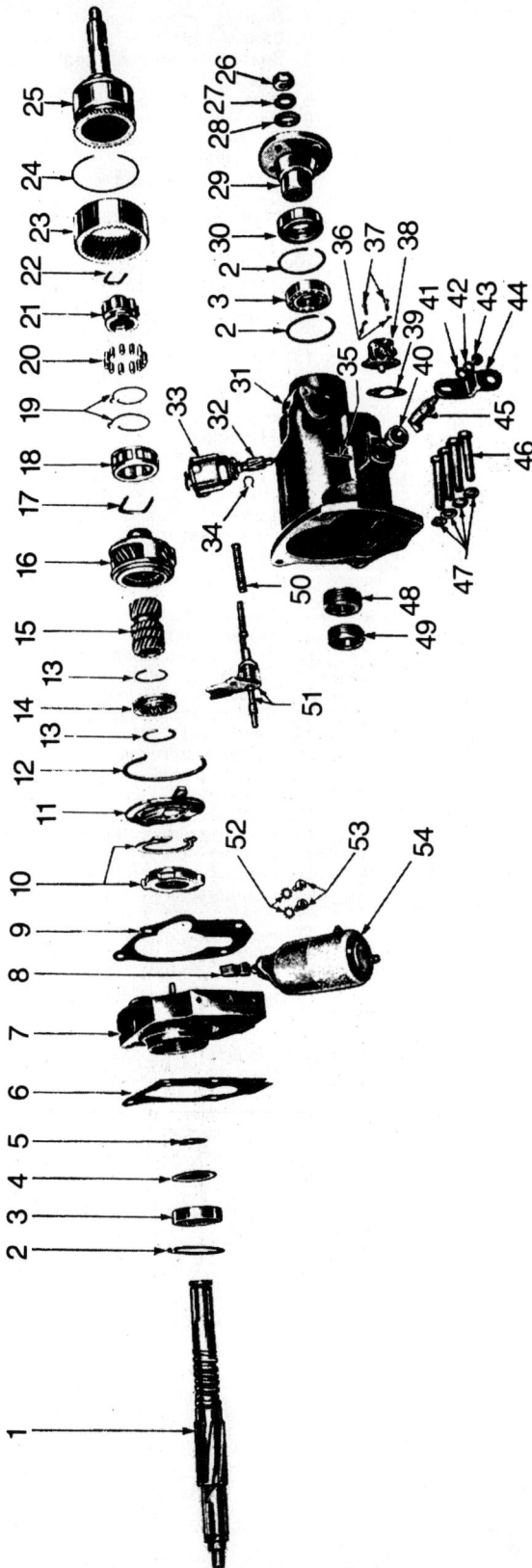

1. Transmission mainshaft
2. Mainshaft bearing snapring
3. Mainshaft bearing
4. Mainshaft oil baffle
5. Mainshaft snapring
6. Adapter to transmission gasket
7. Overdrive housing adapter
8. Sun gear pawl
9. Housing to adapter gasket
10. Balk ring and gear plate
11. Overdrive cover plate
12. Cover plate snapring
13. Sun gear snapring
14. Sun gear shifting collar
15. Sun gear
16. Planetary gear cage
17. Roller retainer clip
18. Freewheel roller retainer
19. Roller retainer spring
20. Freewheel roller
21. Freewheel cam
22. Cam retainer clip
23. Overdrive ring gear
24. Ring gear snapring
25. Mainshaft
26. Mainshaft nut
27. Mainshaft lockwasher
28. Mainshaft washer
29. Coupling flange
30. Mainshaft oil seal
31. Housing
32. Governor driven gear
33. Governor
34. Driven gear retaining ring
35. Control shaft pin
36. Rail switch lockwasher
37. Rail switch screw
38. Rail switch
39. Rail switch gasket
40. Control shaft oil seal
41. Control lever washer
42. Control lever lockwasher
43. Control lever nut
44. Control lever
45. Control shaft
46. Housing to transmission bolt
47. Housing to transmission lockwasher
48. Speedometer drive gear
49. Governor drive gear
50. Shift retractor spring
51. Shift rail and fork
52. Solenoid lockwasher
53. Solenoid bolt
54. Solenoid

T-86/T-96 with overdrive; exploded view

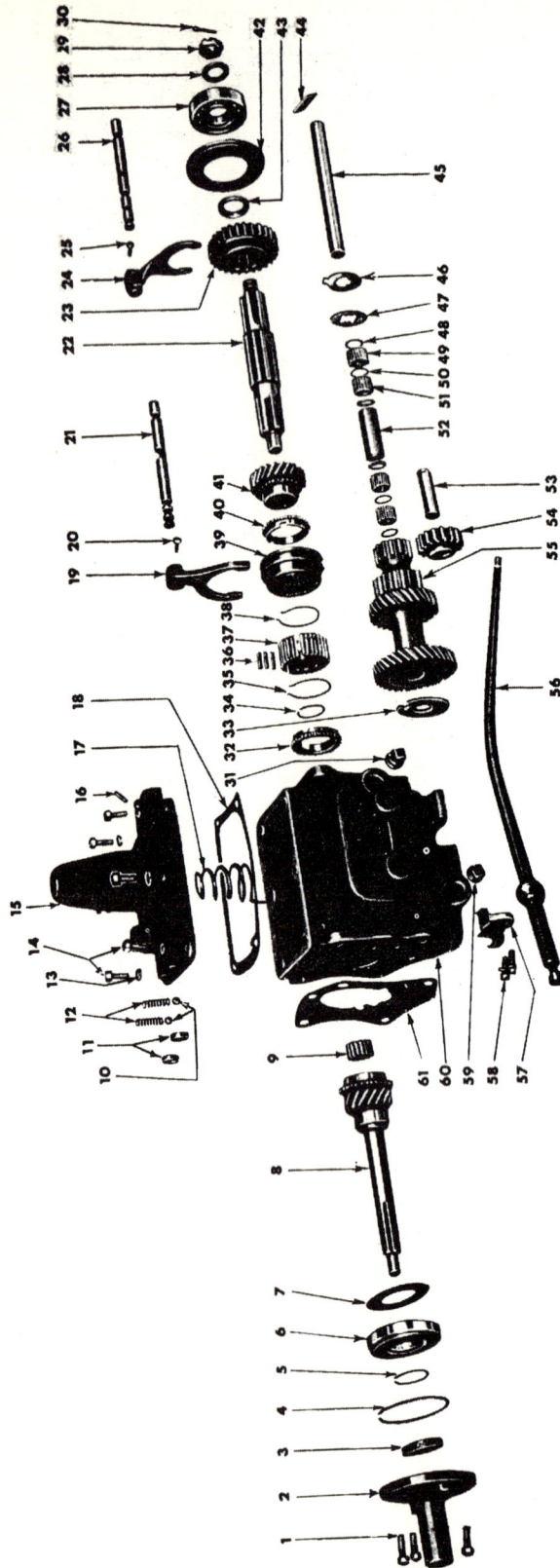

1. Bearing retainer bolt
2. Bearing retainer
3. Bearing retainer oil seal
4. Bearing snapring
5. Main drive gear snapring
6. Main drive gear bearing
7. Front bearing washer
8. Main drive gear
9. Pilot roller bearing
10. Poppet ball
11. Shift rail cap
12. Poppet spring
13. Lockwasher
14. Shift housing bolt
15. Shift housing
16. Interlock plunger
17. Shift lever spring
18. Shift housing gasket
19. High and intermediate shift fork
20. Shift fork pin
21. High and intermediate shift rail
22. Mainshaft
23. Sliding gear
24. Low and reverse shift fork
25. Shift fork pin
26. Low and reverse shift rail
27. Rear bearing
28. Mainshaft washer
29. Mainshaft nut
30. Cotter pin
31. Filler plug
32. Blocking ring
33. Front countershaft thrust washer
34. Clutch hub snapring
35. Synchronizer spring
36. Synchronizer plate
37. Clutch hub
38. Synchronizer spring
39. Clutch sleeve
40. Blocking ring
41. Second speed gear
42. Rear bearing adapter
43. Bearing spacer
44. Lockplate
45. Countershaft
46. Rear countershaft thrust washer
47. Rear countershaft thrust washer
48. Countershaft bearing washer
49. Countershaft bearing rollers
50. Countershaft bearing washer
51. Countershaft bearing rollers
52. Countershaft bearing spacer
53. Reverse gear shaft
54. Reverse idler gear
55. Countershaft gear set
56. Shift lever
57. Oil collector
58. Oil collector screw
59. Drain plug
60. Transmission case
61. Bearing retainer gasket

T-90 exploded view

20. When assembling the synchronizer unit, install the 2 springs in the high and intermediate clutch hub with the tensions opposed.

21. Drive the countershaft in through the rear of the case.

22. Install the lockplate on the reverse idler shaft and countershaft.

23. Install the oil collector screws.

24. Install the front main drive gear bearing retainer and gasket.

25. Join the transfer case and transmission. Install the transfer case screws.

26. Install the transmission shift cover.

27. Install the transfer case main drive gear.

28. If so equipped, install the pto shift unit.

29. Install the transfer case rear cover.

30. Fill the case with lubricant.

T-98A 4-Speed Overhaul
DISASSEMBLY

1. After draining the transmission and removing the parking brake drum (or shoe assembly), lock the transmission in two gears and remove the U-joint flange, oil seal, speedometer driven gear and bearing assembly. Lubricant capacity is 6½ pints.

2. Remove the output shaft bearing retainer and the speedometer drive gear and spacer.

3. Remove the output shaft bearing snapring, and remove the bearing.

4. Remove the countershaft and idler shaft retainer and the power take-off cover.

5. After removing the input shaft bearing retainer, remove the snaprings from the bearing and the shaft.

6. Remove the input shaft bearing and oil baffle.

7. Drive out the countershaft (from the front). Keep the dummy shaft in contact with the countershaft to avoid dropping any rollers.

8. After removing the input shaft and the synchronizer blocking ring, pull the idler shaft.

9. Remove the reverse gear shifter arm, the output shaft assembly, the idler gear, and the cluster gear. When removing the cluster, do not lose any of the rollers.

OUTPUT SHAFT

1. Remove the third- and high-speed synchronizer hub snapring from the output shaft, and slide the third- and high-speed synchronizer assembly and the third-speed gear off the shaft. Remove the synchronizer sleeve and the inserts from the hub. Before removing the two snaprings from the ends of the hub, check the end play of the second-speed gear. Endplay should be 0.005–0.024 in. (0.13-0.60mm).

2. Remove the second-speed synchronizer snapring. Slide the second-speed synchronizer hub gear off the hub. Do not lose any of the balls, springs, or plates. Pull the hub off the shaft, and remove the second-speed synchronizer from the second-speed gear. Remove the snapring from the rear of the second-speed gear, and remove the gear, spacer, roller bearings, and thrust washer from the output shaft. Remove the remaining snapring from the shaft.

CLUSTER GEAR

Remove the dummy shaft, pilot bearing rollers, bearing spacers, and center spacer from the cluster gear.

REVERSE IDLER GEAR

Rotate the reverse idler gear on the shaft, and if it turns freely and smoothly, disassembly of the unit is not necessary. If any roughness is noticed, disassemble the unit.

GEAR SHIFT HOUSING

1. Remove the housing cap and lever. Be sure all shafts are in neutral before disassembly.

2. Tap the shifter shafts out of the housing while holding one hand over the holes in the housing to prevent loss of the springs and balls. Remove the two shaft lock plungers from the housing.

CLUSTER GEAR ASSEMBLY

Slide the long bearing spacer into the cluster gear bore, and insert the dummy shaft in the spacer. Hold the cluster gear in a vertical position, and install one of the bearing spacers. Position the 22 pilot bearing rollers in the cluster gear bore. Place a spacer on the rollers, and install 22 more rollers and another spacer. Hold a large thrust washer against the end of cluster gear and turn the assembly over. Install the rollers and spacers in the other end of the gear.

REVERSE IDLER GEAR ASSEMBLY

1. Install a snapring in one end of the idler gear, and set the gear on end, with the snapring at the bottom.

2. Position a thrust washer in the gear on top of the snapring. Install the bushing on top of the washer, insert the 37 bearing rollers, and then a spacer followed by 37 more rollers. Place the remaining thrust washer on the rollers, and install the other snapring.

OUTPUT SHAFT ASSEMBLY

1. Install the second speed gear thrust washer and snapring on the output shaft. Hold the shaft vertically, and slide on the second speed gear. Insert the bearing rollers in the second-speed gear, and slide the spacer into the gear. Install the snapring on the output shaft at the rear of the second-speed gear. Position the blocking ring on the second-speed gear. Do not invert the shaft because the bearing rollers will slide out of the gear.

2. Press the second-speed synchronizer hub onto the shaft, and install the snapring. Position the shaft vertically in a soft-jawed vise. Position the springs and plates in the second-speed synchronizer hub, and place the hub gear on the hub.

3. Install the first speed gear and snapring on the shaft and press on the reverse gear.

4. Hold the gear above the hub spring and ball holes, and position one ball at a time in the hub, and slide the hub gear downward to hold the ball in place. Push the plate upward, and insert a small block to hold the plate in position, thereby holding the ball in the hub. Follow these procedures for the remaining balls.

5. Install the third speed gear and synchronizer blocking ring on the shaft.

6. Install the snaprings at both ends of the third and high-speed synchronizer hub. Stagger the openings of the snaprings so that they are not aligned. Place the inserts in the synchronizer sleeve, and position the sleeve on the hub.

7. Slide the synchronizer assembly onto the output shaft. The slots in the blocking ring must be in line with the synchronizer inserts. Install the snapring at the front of the synchronizer assembly.

GEAR SHIFT HOUSING

1. Place the spring on the reverse gear shifter shaft gate plunger, and install the spring and plunger in the reverse gate. Press the plunger through the gate, and fasten it with the clip. Place the spring and ball in the reverse gate poppet hole. Compress the spring and install the cotter pin.

2. Place the spring and ball in the reverse shifter shaft hole in the gear shift housing. Press down on the ball, and position the reverse shifter shaft so that the reverse shifter arm notch does not slide over the ball. Insert the shaft part way into the housing.

3. Slide the reverse gate onto the shaft, and drive the shaft into the housing until the ball snaps into the groove of the shaft. Install the lock screw lock wire to the gate.

4. Insert the two interlocking plungers in the pockets between the shifter shaft holes. Place the spring and ball in the low and second shifter shaft hole. Press down on the ball, and insert the shifter shaft part way into the housing.

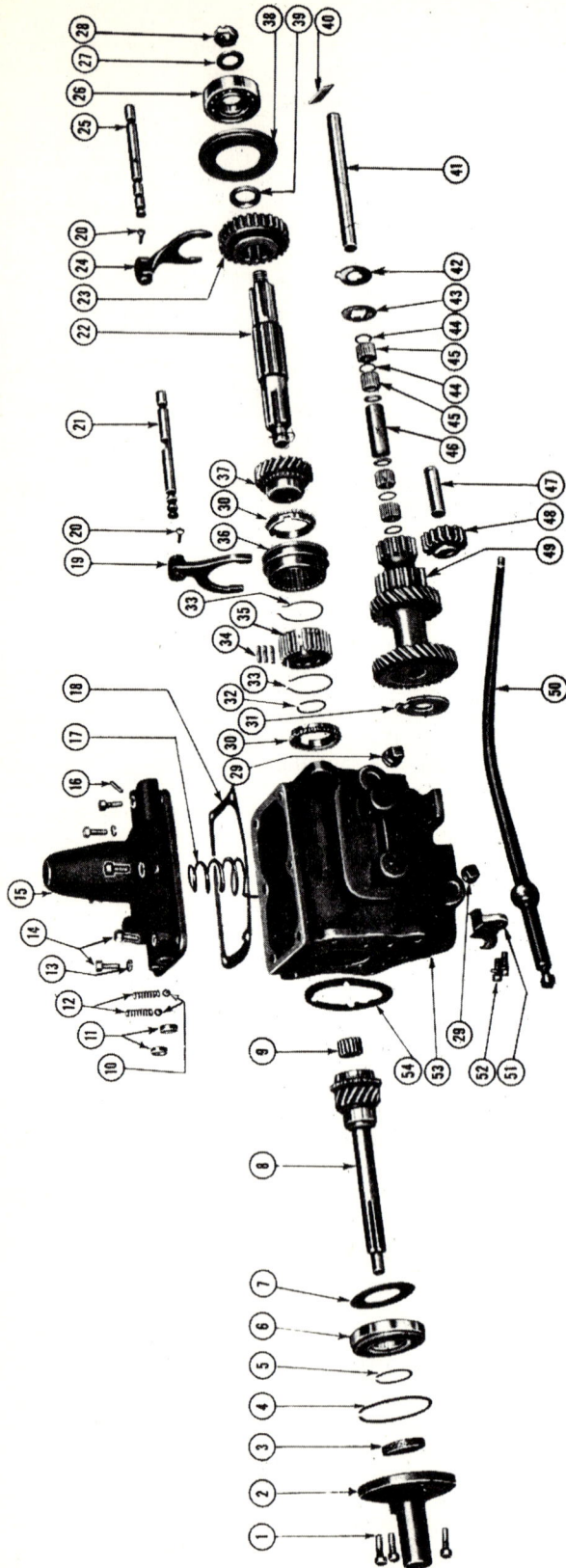

1. Bearing retainer bolt
2. Bearing retainer
3. Bearing retainer oil seal
4. Bearing snapring
5. Main drive gear snapring
6. Main drive gear bearing
7. Front bearing washer
8. Main drive gear
9. Pilot roller bearing
10. Poppet ball
11. Shift rail cap
12. Poppet spring
13. Lockwasher
14. Shift housing bolt
15. Control housing
16. Interlock plunger
17. Shift lever spring
18. Shift tower gasket
19. High and intermediate shift fork
20. Shift fork pin
21. High and intermediate shift rail
22. Mainshaft
23. Sliding gear
24. Low and reverse shift fork
25. Low and reverse shift rail
26. Rear bearing
27. Mainshaft washer
28. Mainshaft nut
29. Filler plug
30. Blocking ring
31. Front countershaft thrust washer
32. Clutch hub snapring
33. Synchronizer spring
34. Synchronizer plate
35. Clutch hub
36. Clutch sleeve
37. Second speed gear
38. Rear bearing adapter
39. Bearing spacer
40. Lockplate
41. Countershaft
42. Rear countershaft thrust washer
43. Rear countershaft thrust washer
44. Countershaft bearing washer
45. Countershaft bearing
46. Countershaft bearing spacer
47. Reverse gear shaft
48. Reverse idler gear
49. Countershaft gear set
50. Shift lever
51. Oil collector
52. Oil collector screw
53. Transmission case
54. Bearing retainer gasket

T-98A exploded view

5. Slide the low and second shifter shaft gate onto the shaft, and install the corresponding shifter fork on the shaft so that the offset of the fork is toward the rear of the housing. Push the shaft all the way into the housing until the ball engages the shaft groove. Install the lock screw and wire that fastens the fork to the shaft. Install the third and high shifter shaft in the same manner. Check the interlocking system. Install new expansion plugs in the shaft bores.

CASE ASSEMBLY

1. Coat all parts, especially the bearings, with transmission lubricant to prevent scoring during initial operation.
2. Position the cluster gear assembly in the case. Do not lose any rollers.
3. Place the idler gear assembly in the case, and install the idler shaft. Position the slot in the rear of the shaft so that it can engage the retainer. Install the reverse shifter arm.
4. Drive out the cluster gear dummy shaft by installing the countershaft from the rear. Position the slot in the rear of the shaft so that it can engage the retainer. Use thrust washers as required to get 0.006-0.020 in. (0.15-0.50mm) cluster gear end play. Install the countershaft and idler shaft retainer.
5. Position the input shaft pilot rollers and the oil baffle, so that the baffle will not rub the bearing race. Install the input shaft and the blocking ring in the case.
6. Install the output shaft assembly in the case, and use a special tool to prevent jamming the blocking ring when the input shaft bearing is installed.
7. Drive the input shaft bearing onto the shaft. Install the thickest select-fit snapring that will fit on the bearing. Install the input shaft snapring.
8. Install the output shaft bearing.
9. Install the input shaft bearing without a gasket, and tighten the bolts only enough to bottom the retainer on the bearing snapring. Measure the clearance between the retainer and the case, and select a gasket (or gaskets) that will seal in the oil and prevent end play between the retainer and the snapring. Torque the bolts to specification.
10. Position the speedometer drive gear and spacer, and install a new output shaft bearing retainer seal.
11. Install the output shaft bearing retainer. Torque the bolts to specification, and install safety wire.
12. Install the brake shoe (or drum), and torque the bolts to specification. Install the U-joint flange. Lock the transmission in two gears and torque the nut to specification.
13. Install the power take-off cover plates with new gaskets. Fill the transmission according to specifications.

T–150 3-Speed Transmission Overhaul

The Tremec T–150 transmission is used in varied vehicle applications, with or without transfer cases. The gear selection is controlled by either a top shift housing or by a remote control shift lever assembly. Although some of the gears and case applications are not interchangeable, the gear arrangement is basically the same.

DISASSEMBLY

1. Remove the bolts securing the transfer case to the transmission. Remove the transfer case.
2. Remove the transfer case drive gear locknut, flat washer, and drive gear. Remove the large fiber washer from the rear bearing adapter. Move the second-third clutch sleeve forward and the first-reverse sleeve to the rear before removing the locknut.
3. Remove the transmission oil plug and drive the counter-

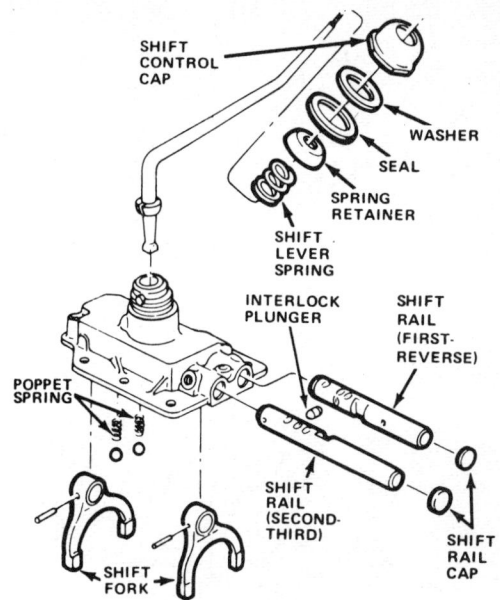

T-150 control tower

shaft out of the case with a suitable size drift. Do not lose the countershaft access plug when removing the countershaft. With the countershaft removed the countershaft gear will lie at the bottom of the case, leave it there until the mainshaft is removed.
4. Punch alignment marks in the front bearing cap and the transmission case for assembly reference.
5. Remove the front bearing cap and gasket.
6. Remove the large lock ring from the front bearing.
7. Remove the clutch shaft, front bearing, and the second-third synchronizer assembly. A special tool is required for this operation.
8. Remove the rear bearing and adapter assembly with a brass drift and hammer. Drive the adapter out the rear of the case with light blows from the hammer.
9. Remove the mainshaft assembly. Tilt the spline end of the shaft downward and lift the front end up and out of the case.
10. Remove the countershaft tool and arbor as an assembly. Remove the countershaft thrust washers, countershaft roll pin, and any pilot roller bearings that may have fallen into the case.
11. Remove the reverse idler shaft. Insert a brass drift through the clutch shaft bore in the front of the case and tap the shaft until the end with the roll pin clears the counter bore in the rear of the case. Remove the shaft.
12. Remove the reverse idler gear and thrust washers from the case.
13. Remove the retaining snapring from the front of the mainshaft. Remove the second-third synchronizer assembly and second gear. Mark the hub and sleeve for reference during assembly.

NOTE: Observe the position of the insert springs and the inserts during removal for correct assembly.

14. Remove the insert springs from the second-third synchronizer, remove the three inserts, and separate the sleeve from the synchronizer hub retaining snapring.
15. Remove the snapring and the tabbed thrust washer from the mainshaft and remove the first gear blocking ring.
16. Remove the first-reverse synchronizer hub snapring.

NOTE: Observe the position of the insert springs and the inserts during removal for correct assembly.

17. Remove the first-reverse sleeve, insert spring and the three insert from the hub. Remove the spacer from the rear of the mainshaft.

T-150 exploded view

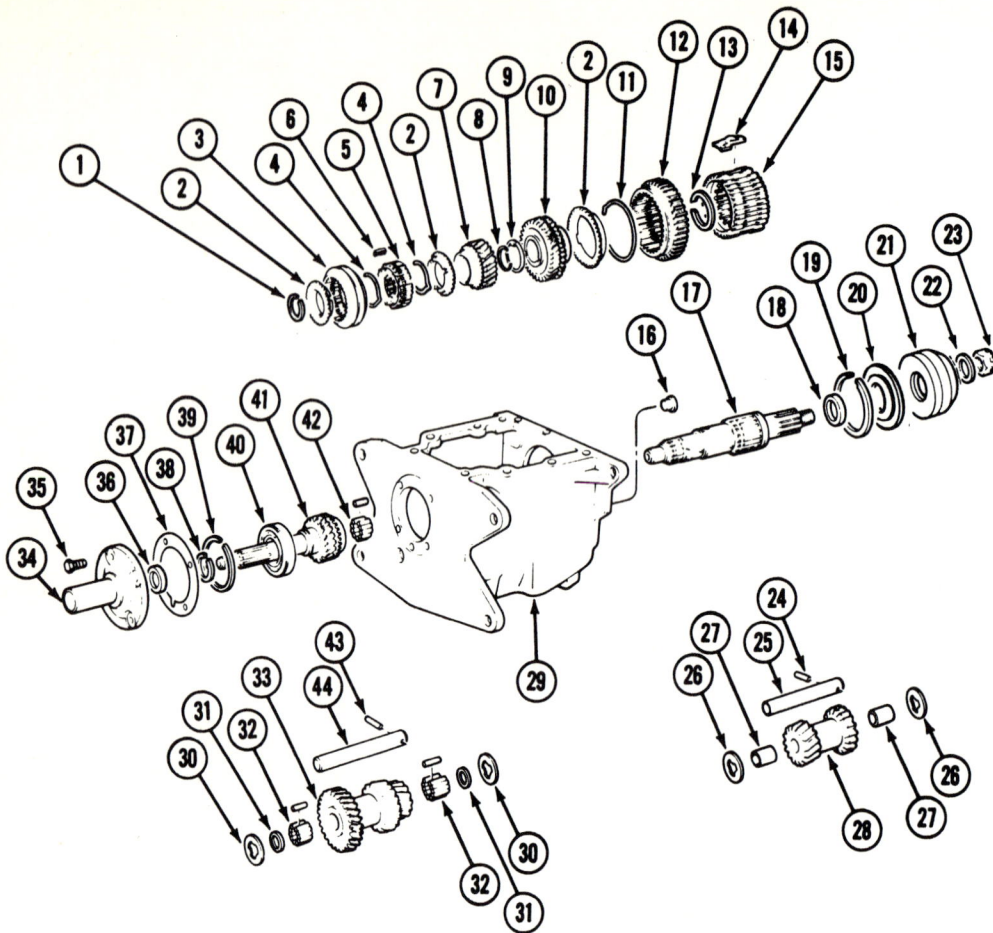

1. Mainshaft retaining snapring
2. Synchronizer blocking rings (3)
3. Second-third synchronizer sleeve
4. Second-third synchronizer insert spring (2)
5. Second-third hub
6. Second-third synchronizer insert (3)
7. Second gear
8. First gear retaining snapring
9. First gear tabbed thrust washer
10. First gear
11. First-reverse synchronizer insert spring
12. First-reverse sleeve and gear
13. First-reverse hub retaining snapring
14. First-reverse synchronizer insert (3)
15. First-reverse hub
16. Countershaft access plug
17. Mainshaft
18. Mainshaft spacer
19. Rear bearing adapter lockring
20. Oil slinger/spacer
21. Rear bearing and adapter assembly
22. Washer
23. Locknut
24. Roll pin
25. Reverse idler gear shaft
26. Thrust washer
27. Bushing (part of idler gear)
28. Reverse idler gear
29. Transmission case
30. Thrust washer (2)
31. Bearing retainer (2)
32. Countershaft needle bearings (50)
33. Countershaft gear
34. Front bearing cap
35. Bolt (4)
36. Front bearing cap oil seal
37. Gasket
38. Front bearing retainer snapring
39. Front bearing lockring
40. Front bearing
41. Clutch shaft
42. Mainshaft pilot roller bearings
43. Roll pin
44. Countershaft

CAUTION

Do not attempt to remove the press fit hub by hammering. Hammer blows will damage the hub and mainshaft.

18. Remove the front bearing retaining snapring and any remaining roller bearings from the clutch shaft.
19. Press the front bearing off the clutch shaft with an arbor press.

CAUTION

Do not attempt to remove the bearing by hammering. Hammer blows will damage the bearing and the clutch shaft.

20. Clamp the rear bearing adapter in a soft-jawed vise. Do not over-tighten.
21. Remove the rear bearing retaining snapring. Remove the bearing adapter from the vise.

22. Press the rear bearing out of the adapter with an arbor press.

23. Thoroughly wash all parts in clean solvent and dry with compressed air. Do not dry the bearings with compressed air, use a clean shop cloth. Clean the needle and clutch shaft bearings by placing them in a shallow parts cleaning tray and covering them with solvent. Allow the bearings to air dry on a clean shop cloth. Check the case for the following: Cracks in the bores, bosses, or bolt holes. Stripped threads in bolt holes. Nicks, burrs, rough surfaces in the shaft bores or on the gasket surfaces.

24. Check the gear and synchronizer assemblies for the following: Broken, chipped, or worn gear teeth. Damaged splines on the synchronizer hubs or sleeves. Bent or damaged inserts. Damaged needle bearings or bearing bores in the countershaft gear. Broken or worn teeth or excessive wear of the blocking rings. Wear of galling of the countershaft, clutch shaft, or reverse idler shaft. Worn thrust washers. Nicked, broken, or worn mainshaft or clutch shaft splines. Bent, distorted, or weak snaprings. Worn bushings in the reverse idler gear. Replace the gear if the bushings are worn. Rough, galled, or broken front or rear bearings.

ASSEMBLY

1. Lubricate the reverse idler shaft bore and bushings with transmission oil.

2. Coat the transmission case reverse idler gear thrust washer surfaces with petroleum jelly and install the thrust washers in the case.

NOTE: Make sure the locating tangs on the thrust washers are aligned in the slots in the case.

3. Install the reverse idler gear. Align the gear bore, thrust washers, and case bore. Install the reverse idler shaft from the rear of the transmission case. Be sure to align and seat the roll pin in the shaft into the counter bore in the rear of the case.

4. Measure the reverse idler gear end-play by inserting a feeler gauge between the thrust washer and the gear. End-play should be 0.004–0.018 in. (0.10-0.46mm). If end play exceeds 0.018 in. (0.46mm), remove the reverse idler gear and replace the thrust washers.

5. Coat the needle bearing bores in the countershaft gear with petroleum jelly. Insert the arbor tool in the bore of the gear and install the (25) needle bearings and the retainer washers at each end of the countershaft gear.

6. Coat the countershaft gear thrust washer surface with petroleum jelly and position the thrust washers in the case.

NOTE: Make sure the locating tangs on the thrust washers are aligned in the slots in the case.

7. Insert the countershaft into the bore at the rear of the case just far enough to hold the thrust washer in place.

8. Install the countershaft gear in the case. Do not install the roll pin at this time. Align the gear bore, thrust washers, the bores in the case, and install the countershaft.

NOTE: Do not remove the arbor tool completely.

9. Measure the countershaft gear end-play by inserting a feeler gauge between the washer and the countershaft gear. End-play should be 0.004–0.018 in. (0.10-0.46mm). If the end-play exceeds 0.018 in. (0.46mm), remove the gear and replace the thrust washer.

10. When the correct countershaft gear end-play has been obtained, install the countershaft arbor and remove the countershaft. Allow the countershaft gear to remain at the bottom of the case, leave the countershaft in the case enough to hold the thrust washer in place.

11. Coat the splines and machined surfaces on the mainshaft with transmission oil. Install the first-reverse synchronizer on

the output shaft splines by hand. The end of the hub with the slots should face the front of the shaft. Use an arbor press to complete the hub installation. Install the retaining snapring in the groove farthest to the rear.

— CAUTION —
Do not attempt to drive the hub on the shaft with a hammer.

12. Coat the splines of the first-reverse hub with transmission oil and install the first reverse sleeve and gear halfway onto the hub, with the gear end of the sleeve facing the rear of the shaft. Align the marks made during disassembly.

13. Install the insert spring in the first-reverse hub. Make sure the spring bottoms in the hub and covers all three insert slots. Position the three "T" shaped inserts in the hub with the small ends in the hub slots and the large ends inside the hub. Push the inserts fully into the hub so they seat on the insert spring, slide the first-reverse sleeve and gear over the inserts until the inserts engage in the sleeve.

14. Coat the bore and the blocking ring surface of first gear with transmission oil and place blocking ring on the tapered surface of the gear.

15. Install the first gear on the output shaft. Rotate the gear until the notches in the blocking ring engage the inserts in the first-reverse synchronizer assembly. Install the tanged thrust washer, sharp end facing out, and retaining snapring on the mainshaft.

16. Coat the bore and blocking ring surface of the second gear with transmission oil. Place the second gear blocking ring on the tapered surface of second gear.

17. Install the second gear on the output shaft with the tapered surface of the gear facing the front of the mainshaft.

18. Install one insert spring into the second-third synchronizer hub. Be sure that the spring covers all three insert slots in the hub. Align the second-third sleeve with the hub using the marks made during disassembly. Start the sleeve onto the hub.

19. Place the three inserts into the hub slots and on top of the insert spring. Push the sleeve fully onto the hub to engage the inserts in the sleeve. Install the remaining insert spring in the exact position as the first spring. The ends of both springs must cover the same slot in the hub and not be staggered.

NOTE: The inserts have a small lip on each end. When they are correctly installed, this lip will fit over the insert spring.

20. Install the second-third synchronizer assembly on the mainshaft. Rotate the second gear until the notches in the blocking ring engage the inserts in the second-third synchronizer assembly.

21. Install the retaining snapring on the mainshaft and measure the end-play between the snapring and the second-third synchronizer hub. The end-play should be 0.0040–0.014 in. (0.10-0.36mm) If the end-play exceeds the limit, replace the thrust washer and all the snaprings on the mainshaft assembly. Install the spacer on the rear of the mainshaft.

22. Install the mainshaft assembly in the case. Be sure that the first-reverse sleeve and gear is in the neutral (centered) position.

23. Press the rear bearing into the rear bearing adapter with an arbor press. Install the rear bearing retaining ring and the bearing adapter lockring.

24. Support the mainshaft assembly and install the rear bearing and adapter assembly in the case. Use a soft faced hammer to seat the adapter in the case.

25. Install the large fiber washer in the rear bearing adapter. Install the transfer drive gear, flat washer, and locknut. Tighten the locknut to 150 ft. lbs. torque.

26. Press the front bearing onto the clutch shaft. Install the bearing retaining snapring on the clutch shaft and the lockring into its groove.

27. Coat the bore of the clutch shaft assembly with petroleum jelly and install the (15) roller bearings in the clutch shaft bore.

CAUTION

Do not use chassis grease or a similar heavy grease in the clutch shaft bore. Heavy grease will plug the lubricant holes in the shaft and prevent proper lubrication of the roller bearings.

---

28. Coat the blocking ring surface of the clutch shaft with transmission oil. Position the blocking ring on the clutch shaft.

29. Support the mainshaft assembly and insert the clutch shaft through the front bearing bore in the case. Seat the mainshaft pilot in the clutch shaft roller bearings. Tap the bearings into place with a soft faced hammer.

30. Apply a thin film of sealer to the front bearing cap gasket and position the gasket on the case. Be sure the cutout in the gasket is aligned with the oil return hole in the case.

31. Remove the front bearing cap oil seal with a suitable tool. Install a new seal with a suitable driver.

32. Install the front bearing cap and tighten the bolts to 33 ft. lbs. Be sure that the marks on the cap and the transmission case are aligned and the oil return slot in the cap lines up with the oil return hole in the case.

33. Make a wire loop about 18–20 in. (457-508mm) long and pass the wire under the countershaft gear assembly. The wire loop should raise and support the countershaft gear assembly when it is pulled upward.

34. Raise the countershaft gear with the wire. Align the bore in the countershaft gear with the front thrust washer and the countershaft. Start the countershaft into the gear with a soft faced hammer.

35. Align the roll pin hole in the countershaft with the roll pin holes in the case and complete the installation of the countershaft. Install the countershaft access plug in the rear of the case and seat with a soft faced hammer.

36. Install the countershaft roll pin in the case. Use a magnet or needle nose pliers to insert and start the pin in the case. Use a ½ in. punch to seat the pin. Install the transmission filler plug.

37. Shift the synchronizer sleeves through all gear ranges and check their operation. If the clutch shaft and mainshaft appear to bind in the neutral position, check for blocking rings sticking on the first or second gear tapers.

38. Install the transfer case on the transmission. Tighten the attaching bolts to 30 ft. lbs.

SHIFT CONTROL HOUSING

1. Remove the back-up light switch and the transmission controlled spark switch (TCS) if so equipped.

2. Remove the shift control housing cap, gasket, spring retainer, and the shift lever spring as an assembly.

3. Invert the housing and mount in a soft-jawed vise.

4. Move the second-third shift rail to the rear of the housing, rotate the shift fork toward the first-reverse rail until the roll pin is accessible. Drive the roll pin out of the fork and rail with a pin punch. Remove the shift fork and the roll pin.

NOTE: The roll pin hole in the shift fork is offset. Mark the position of the shift fork for assembly reference.

5. Remove the second-third shift rail using a brass drift or hammer. Catch the shift rail plug as the rail drives it out of the housing. Cover the shift and poppet ball holes in the cover to prevent the poppet ball from flying out. Mark the location of the shift rail for assembly reference.

6. Rotate the first-reverse shift fork away from the notch in the housing until the roll pin is accessible. Drive the roll pin out of the fork and rail using a pin punch. Remove the shift fork and roll pin.

NOTE: The roll pin hole in the shift fork is offset. Mark the position of the shift fork for assembly reference.

7. Remove the first-reverse shift rail using a brass drift or hammer. Catch the shift rail plug as the rail drives it out of the housing. Cover the shift and poppet ball holes in the cover to prevent the poppet ball from flying out. Mark the location of the shift rail for assembly reference.

8. Remove the poppet balls, springs, and the interlock plunger from the housing.

9. Install the poppet springs and the detent plug in the housing.

10. Insert the first-reverse shift rail into the housing, and install the shift fork on the shift rail.

11. Install the poppet ball on the top of the spring in the first-reverse rail.

12. Using a punch or wooden dowel, push the poppet ball and spring downward into the housing bore and install the first-reverse shift rail.

13. Align the roll pin holes in the first-reverse shift fork and install the roll pin. Move the shift rail to the neutral (center) detent.

14. Insert the second-third shift rail into the housing and install the poppet ball on top of the spring in the shift rail bore.

15. Using a punch or wooden dowel, push the poppet ball and spring downward into the housing bore and install the second-third shift rail.

16. Align the roll pin holes in the second-third shift rail and the shift fork and install the roll pin. Move the shift rail to the neutral (center) position.

17. Install the shift rail plugs in the housing, and remove the shift control cover from the vise.

18. Install the shift lever, shift lever spring, spring retainer, gasket, and the shift control housing cap as an assembly. Tighten the cap securely.

11. Install the back-up light switch and the TCS switch if so equipped.

T–18A 4-Speed

The Warner T-18A transmission has four forward speeds and one reverse. The T-18A transmission is synchronized in second, third and fourth speeds only.

DISASSEMBLY

1. After draining the transmission and removing the parking brake drum (or shoe assembly), lock the transmission in two gears and remove the U-joint flange, oil seal, speedometer driven gear and bearing assembly. Lubricant capacity is 6½ pints.

2. Remove the output shaft bearing retainer and the speedometer drive gear and spacer.

3. Remove the output shaft bearing snapring, and remove the bearing.

4. Remove the countershaft and idler shaft retainer and the power take-off cover.

5. After removing the input shaft bearing retainer, remove the snaprings from the bearing and the shaft.

6. Remove the input shaft bearing and oil baffle.

7. Drive out the countershaft (from the front). Keep the dummy shaft in contact with the countershaft to avoid dropping any rollers.

8. After removing the input shaft and the synchronizer blocking ring, pull the idler shaft.

9. Remove the reverse gear shifter arm, the output shaft assembly, the idler gear, and the cluster gear. When removing the cluster, do not lose any of the rollers.

OUTPUT SHAFT

1. Remove the third- and high-speed synchronizer hub snapring from the output shaft, and slide the third- and high-speed synchronizer assembly and the third-speed gear off the shaft. Remove the synchronizer sleeve and the inserts from the

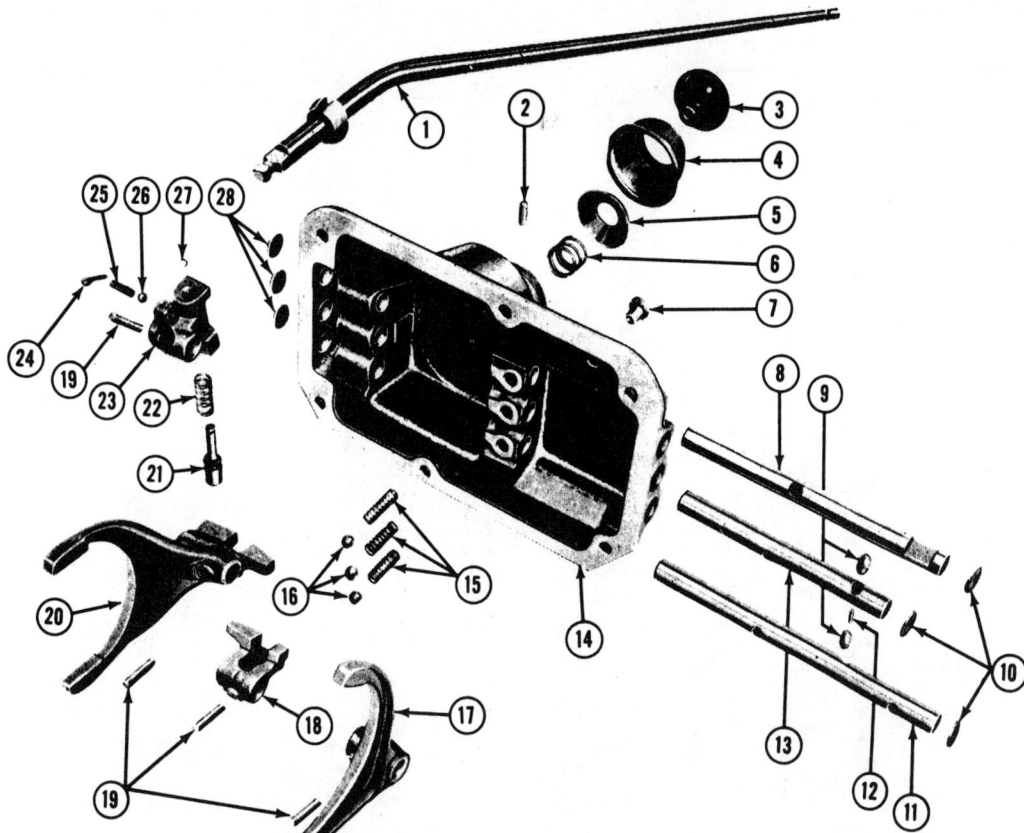

1. Shift lever
2. Control housing pin
3. Shift handle
4. Control housing cap
5. Washer
6. Control lever spring
7. Breather assembly
8. Reverse shift rail
9. Shift rail interlock plunger
10. Expansion plug
11. Low and second shift rail
12. Shift rail interlock pin
13. Direct and third shift rail
14. Control housing
15. Shift rail poppet spring
16. Shift rail poppet ball
17. Lockwire
18. Low and second shift fork
19. Lock pin
20. Shift rail end
21. Direct and third shift fork
22. Reverse plunger
23. Reverse plunger spring
24. Reverse rail end
25. Cotter pin
26. Reverse plunger poppet spring
27. Reverse plunger poppet ball
28. C-washer

T-18 control tower

hub. Before removing the two snaprings from the ends of the hub, check the end play of the second-speed gear. Endplay should be 0.005–0.024 in. (0.13-0.61mm).

2. Remove the second-speed synchronizer snapring. Slide the second-speed synchronizer hub gear off the hub. Do not lose any of the balls, springs, or plates. Pull the hub off the shaft, and remove the second-speed synchronizer from the second-speed gear. Remove the snapring from the rear of the second-speed gear, and remove the gear, spacer, roller bearings, and thrust washer from the output shaft. Remove the remaining snapring from the shaft.

CLUSTER GEAR

Remove the dummy shaft, pilot bearing rollers, bearing spacers, and center spacer from the cluster gear.

REVERSE IDLER GEAR

Rotate the reverse idler gear on the shaft, and if it turns freely and smoothly, disassembly of the unit is not necessary. If any roughness is noticed, disassemble the unit.

GEAR SHIFT HOUSING

1. Remove the housing cap and lever. Be sure all shafts are in neutral before disassembly.

2. Tap the shifter shafts out of the housing while holding one hand over the holes in the housing to prevent loss of the springs and balls. Remove the two shaft lock plungers from the housing.

CLUSTER GEAR ASSEMBLY

Slide the long bearing spacer into the cluster gear bore, and insert the dummy shaft in the spacer. Hold the cluster gear in a vertical position, and install one of the bearing spacers. Position the 22 pilot bearing rollers in the cluster gear bore. Place a spacer on the rollers, and install 22 more rollers and another spacer. Hold a large thrust washer against the end of cluster gear and turn the assembly over. Install the rollers and spacers in the other end of the gear.

REVERSE IDLER GEAR ASSEMBLY

1. Install a snapring in one end of the idler gear, and set the gear on end, with the snapring at the bottom.

2. Position a thrust washer in the gear on top of the snapring. Install the bushing on top of the washer, insert the 37 bearing rollers, and then a spacer followed by 37 more rollers. Place the remaining thrust washer on the rollers, and install the other snapring.

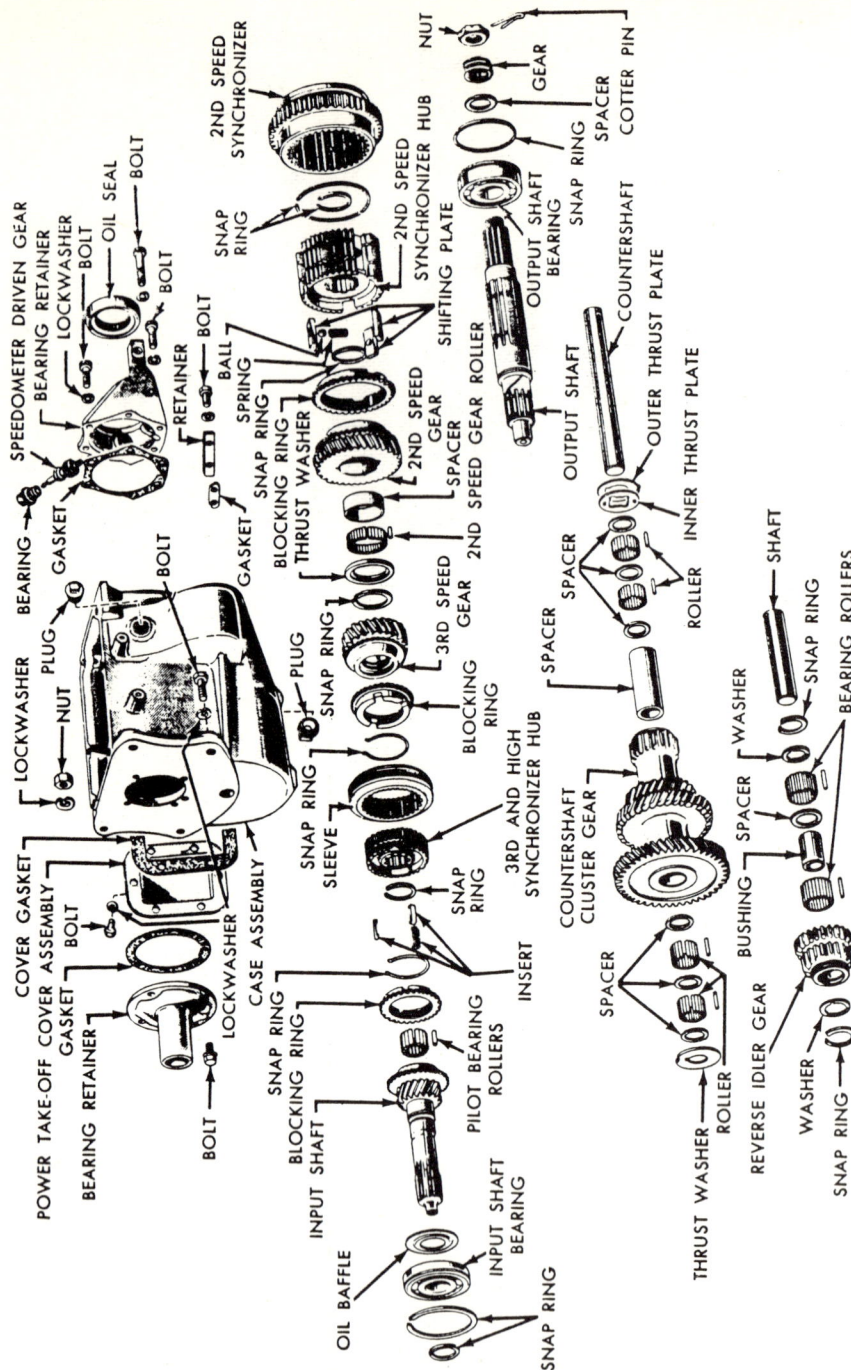

T-18 exploded view

OUTPUT SHAFT ASSEMBLY

1. Install the second speed gear thrust washer and snapring on the output shaft. Hold the shaft vertically, and slide on the second speed gear. Insert the bearing rollers in the second-speed gear, and slide the spacer into the gear. Install the snapring on the output shaft at the rear of the second-speed gear. Position the blocking ring on the second-speed gear. Do not invert the shaft because the bearing rollers will slide out of the gear.

2. Press the second-speed synchronizer hub onto the shaft, and install the snapring. Position the shaft vertically in a soft-jawed vise. Position the springs and plates in the second-speed synchronizer hub, and place the hub gear on the hub.

3. Install the first speed gear and snapring on the shaft and press on the reverse gear.

4. Hold the gear above the hub spring and ball holes, and position one ball at a time in the hub, and slide the hub gear downward to hold the ball in place. Push the plate upward, and insert a small block to hold the plate in position, thereby holding the ball in the hub. Follow these procedures for the remaining balls.

5. Install the third speed gear and synchronizer blocking ring on the shaft.

1. Mainshaft pilot bearing roller spacer
2. Third-fourth blocking ring
3. Third-fourth retaining ring
4. Third-fourth synchronizer snapring
5. Third-fourth shifting plate (3)
6. Third-fourth clutch hub
7. Third-fourth retaining ring
8. Third-fourth clutch sleeve
9. Third-fourth blocking ring
10. Third-fourth gear synchronizer assembly

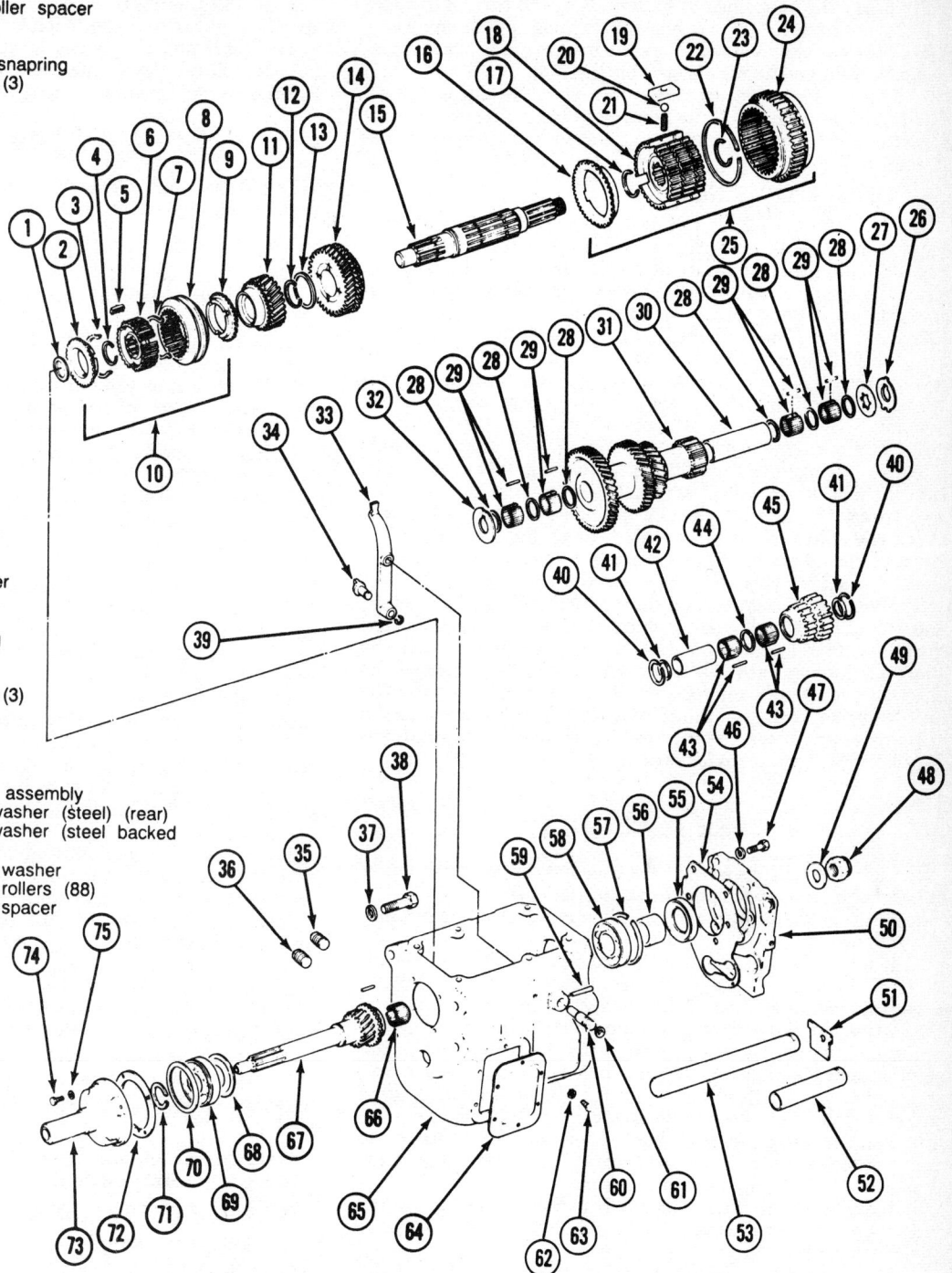

11. Third gear
12. Mainshaft snapring
13. Second gear thrust washer
14. Second gear
15. Mainshaft
16. Second gear blocking ring
17. Mainshaft snapring
18. First-second clutch hub
19. First-second shifting plate (3)
20. Poppet spring (3)
22. First-second insert spring
23. Mainshaft snapring
24. First-second clutch sleeve
25. Second gear synchronizer assembly
26. Countershaft gear thrust washer (steel) (rear)
27. Countershaft gear thrust washer (steel backed bronze) (rear)
28. Countershaft gear bearing washer
29. Countershaft gear bearing rollers (88)
30. Countershaft gear bearing spacer
31. Countershaft gear
32. Countershaft gear thrust washer (front)

33. Reverse shifting arm
34. Reverse shifting arm shoe
35. Filler plug
36. Drain plug
37. Lockwasher
38. Bolt (transmission-to-clutch Housing)
39. C-washer
40. Reverse idler gear snapring
41. Reverse idler gear thrust washer
42. Reverse idler shaft sleeve
43. Reverse idler gear bearing rollers (74)
44. Reverse idler gear bearing washer
45. Reverse idler gear
46. Lockwasher (6)
47. Adapter plate bolts (6)
48. Drive gear locknut

49. Washer
50. Adapter plate
51. Countershaft-reverse idler shaft lockplate
52. Reverse idler gear shaft
53. Countershaft
54. Adapter plate gasket
55. Adapter plate seal
56. Speedometer gear spacer
57. Rear bearing locating snapring
58. Rear bearing
59. Reverse shifting arm pivot pin
60. Reverse shifting arm pivot
61. Reverse shifting arm pivot O-ring

62. Washer (6)
63. Side cover bolt (6)
64. Side cover
65. Transmission case
66. Mainshaft pilot bearing rollers (22)
67. Clutch shaft
68. Front bearing retainer washer
69. Front bearing
70. Front bearing locating snapring
71. Front bearing lockring
72. Front bearing cap gasket
73. Front bearing cap
74. Front bearing cap bolts (4)
75. Lockwashers (4)

T-18A exploded view

6. Install the snaprings at both ends of the third and high-speed synchronizer hub. Stagger the openings of the snaprings so that they are not aligned. Place the inserts in the synchronizer sleeve, and position the sleeve on the hub.

7. Slide the synchronizer assembly onto the output shaft. The slots in the blocking ring must be in line with the synchronizer inserts. Install the snapring at the front of the synchronizer assembly.

GEAR SHIFT HOUSING

1. Place the spring on the reverse gear shifter shaft gate plunger, and install the spring and plunger in the reverse gate. Press the plunger through the gate, and fasten it with the clip. Place the spring and ball in the reverse gate poppet hole. Compress the spring and install the cotter pin.

2. Place the spring and ball in the reverse shifter shaft hole in the gear shift housing. Press down on the ball, and position the reverse shifter shaft so that the reverse shifter arm notch does not slide over the ball. Insert the shaft part way into the housing.

3. Slide the reverse gate onto the shaft, and drive the shaft into the housing until the ball snaps into the groove of the shaft. Install the lock screw lock wire to the gate.

4. Insert the two interlocking plungers in the pockets between the shifter shaft holes. Place the spring and ball in the low and second shifter shaft hole. Press down on the ball, and insert the shifter shaft part way into the housing.

5. Slide the low and second shifter shaft gate onto the shaft, and install the corresponding shifter fork on the shaft so that the offset of the fork is toward the rear of the housing. Push the shaft all the way into the housing until the ball engages the shaft groove. Install the lock screw and wire that fastens the fork to the shaft. Install the third and high shifter shaft in the same manner. Check the interlocking system. Install new expansion plugs in the shaft bores.

CASE ASSEMBLY

1. Coat all parts, especially the bearings, with transmission lubricant to prevent scoring during initial operation.

2. Position the cluster gear assembly in the case. Do not lose any rollers.

3. Place the idler gear assembly in the case, and install the idler shaft. Position the slot in the rear of the shaft so that it can engage the retainer. Install the reverse shifter arm.

4. Drive out the cluster gear dummy shaft by installing the countershaft from the rear. Position the slot in the rear of the shaft so that it can engage the retainer. Use thrust washers as required to get 0.006 to 0.020 in. (0.15-0.50mm) cluster gear end play. Install the countershaft and idler shaft retainer.

5. Position the input shaft pilot rollers and the oil baffle, so that the baffle will not rub the bearing race. Install the input shaft and the blocking ring in the case.

6. Install the output shaft assembly in the case, and use a special tool to prevent jamming the blocking ring when the input shaft bearing is installed.

7. Drive the input shaft bearing onto the shaft. Install the thickest select-fit snapring that will fit on the bearing. Install the input shaft snapring.

8. Install the output shaft bearing.

9. Install the input shaft bearing without a gasket, and tighten the bolts only enough to bottom the retainer on the bearing snapring. Measure the clearance between the retainer and the case, and select a gasket (or gaskets) that will seal in the oil and prevent end play between the retainer and the snapring. Torque the bolts to specification.

10. Position the speedometer drive gear and spacer, and install a new output shaft bearing retainer seal.

11. Install the output shaft bearing retainer. Torque the bolts to specification, and install safety wire.

12. Install the brake shoe (or drum), and torque the bolts to specification. Install the U-joint flange. Lock the transmission in two gears and torque the nut to specification.

13. Install the power take-off cover plates with new gaskets. Fill the transmission according to specifications.

SR–4 4-Speed Overhaul

The Warner SR–4 transmission is a 4-speed, constant mesh unit, providing synchromesh engagement in all forward gears.

DISASSEMBLY

1. Separate the transmission from the transfer case, if attached.

2. Drain the lubricant from the transmission by removing the lower adapter housing bolt.

3. If the shift lever housing has not been removed, place the shift lever in the neutral position, remove the retaining bolts and lift the shift lever housing from the transmission.

4. Remove the flanged nut holding the offset lever to the shift rail. Remove the offset lever.

5. Remove the adapter housing retaining bolts and the housing from the transmission case.

6. Remove the shift control housing retaining bolts and remove the cover and gasket. Mark the location of the two dowel bolts to reinstall in their original position.

7. Remove the spring clip holding the reverse lever to the reverse lever pivot bolt. Remove the reverse lever pivot bolt, allowing the removal of the reverse lever and reverse lever fork as an assembly.

8. Match mark the front bearing retainer to the transmission case and remove the bearing retainer and gasket.

9. Remove the large and small snaprings from the front and rear ball bearings on the input and output shafts.

10. With the aid of a bearing puller tool or equivalent, remove the input shaft ball bearing and remove the input shaft from the case.

11. Remove the rear (output shaft) bearing from the shaft with the aid of a bearing puller tool or equivalent.

12. Remove the output shaft assembly as a unit from the transmission case. Do not allow the synchronizer sleeves to separate from the hubs during the removal.

13. Push the reverse idler gear shaft rearward and remove the shaft and gear from the case.

14. Using a dummy countershaft, push the countershaft to the rear of the case. Remove the cluster gear assembly and dummy countershaft as a unit, from the transmission case.

15. Separate the dummy countershaft and remove the 50 needle roller bearings, spacers and thrust washers from the cluster gear.

NOTE: The cluster gear front thrust washer is of a plastic material, while the rear thrust washer is metal.

COUNTERSHAFT GEAR BEARING

1. Remove the dummy shaft, bearing retainer washers and needle bearings from the countershaft gear. Clean and inspect the parts.

2. Coat the bore at each end of the countershaft gear with grease to retain the needle bearings.

3. While holding the dummy shaft in the gear, install the needle bearings and retainer washers in each end of the gear.

4. Slide first gear off the output shaft, and remove the first speed blocker ring. Take care not to lose the sliding gear from the first and second speed synchronizer assembly.

5. Clean and inspect all parts.

6. Place a blocker ring on the cone of first gear, and slide the gear and ring assembly onto the output shaft. Make sure that the inserts in the synchronizer engage in the blocker ring notches.

DRIVE TRAIN 7

1. Third-fourth shift insert
2. Third-fourth shift fork
3. Selector interlock plate
4. Selector arm plate (2)
5. Selector arm
6. Selector arm roll pin
7. First-second shift fork insert
8. First-second shift fork

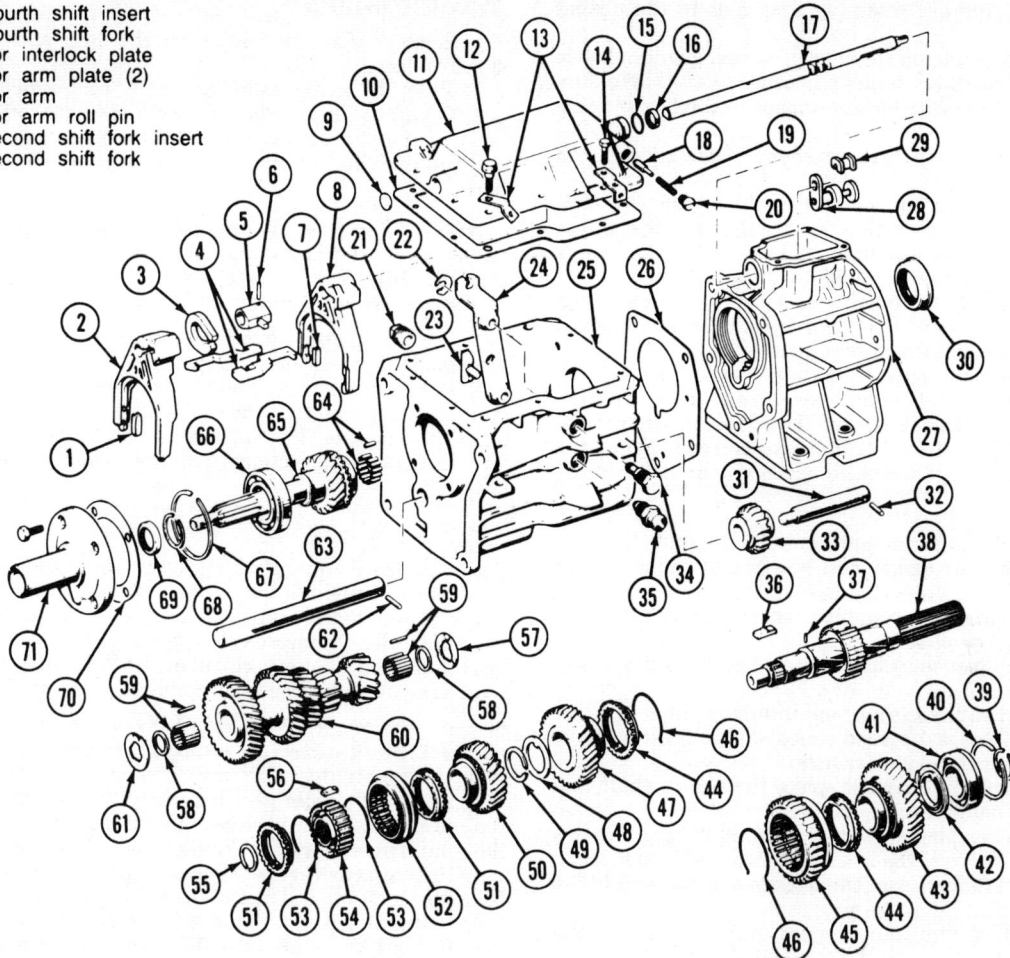

9. Shift rail plug
10. Transmission cover gasket
11. Transmission cover
12. Transmission cover dowel bolt (2)
13. Clip
14. Transmission cover bolt (8)
15. Shift rail O-ring seal
16. Shift rail Oil seal
17. Shift rail
18. Detent spring
20. Detent plug
21. Fill plug
22. Reverse lever pivot bolt C-clip
23. Reverse lever fork
24. Reverse lever
25. Transmission case
26. Gasket
27. Adapter housing
28. Offset lever
29. Offset lever insert
30. Extension housing oil seal
31. Reverse idler shaft
32. Reverse idler shaft roll pin
33. Reverse idler gear
34. Reverse lever pivot bolt
35. Backup lamp switch
36. First-second synchronizer insert (3)
37. First gear roll pin
38. Output shaft and hub assembly
39. Rear bearing retaining snapring
40. Rear bearing locating snapring

41. Rear bearing
42. First gear thrust washer
43. First gear
44. First-second synchronizer blocking ring (2)
45. First-reverse sleeve and gear
46. First-second synchronizer insert spring (2)
47. Second gear
48. Second gear thrust washer (tabbed)
49. Second gear snapring
50. Third gear
51. Third-fourth synchronizer blocking ring (2)
52. Third-fourth synchronizer sleeve
53. Third-fourth synchronizer insert spring (2)
54. Third-fourth synchronizer hub
55. Hub shaft snapring
56. Third-fourth synchronizer insert (3)
57. Countershaft gear rear thrust washer (metal)
58. Countershaft needle bearing retainer (2)
59. Countershaft needle bearing (50)
60. Countershaft gear
61. Countershaft gear front thrust washer (plastic)
62. Countershaft roll pin
63. Countershaft
64. Clutch shaft roller bearings (15)
65. Clutch shaft
66. Front bearing
67. Front bearing locating snapring
68. Front bearing retaining snapring
69. Front bearing cap oil seal
70. Front bearing cap gasket
71. Front bearing cap

SR-4 exploded view

7. Install the spring pin retaining first gear to the output shaft.

8. Install a blocker ring on the cone of second gear, and slide the gear and ring assembly onto the output shaft. Make sure that the inserts in the synchronizer engage in the blocker ring notches.

9. Install the second gear thrust washer and new snapring on the shaft.

10. Install a blocker ring on the cone of third gear, and slide the gear and ring assembly onto the output shaft. Install the third and fourth speed synchronizer. Make sure that the inserts in the synchronizer engage in the blocker ring notches.

11. Install a new third and fourth gear synchronizer snapring.

12. Place the first gear thrust washer (oil slinger) on the shaft and on the spring pin retaining first gear.

13. Assembly end play measurements are as follows: Second gear; 0.004-0.014 in. (0.10-0.36mm), measured between the second speed gear and the thrust washer. Third/fourth synchronizer hub; 0.004-0.014 in. (0.10-0.36mm), measured between the output shaft snapring and the third speed synchronizer hub.

COVER ASSEMBLY

1. Remove the detent screw, spring and plunger.

2. Pull the shifter shaft rod rearward, rotating it counterclockwise.

3. Remove the spring pin retaining the manual selector and interlock to the shifter shaft.

4. Remove the shifter shaft from the cover taking care not to damage the seal.

5. Remove the manual selector and interlock plate.

6. Remove the first and second speed shifter fork. Remove the third and fourth speed shifter fork.

7. Clean and inspect all parts. Replace the shifter shaft seal and welch plug, if damaged.

8. Assemble the two plastic inserts to each shift fork; the two projections on the inside of the inserts fit into the blind holes in the ends of the shift forks. Insert the selector arm plates into the shift forks.

9. Install the third and fourth speed shifter fork into the cover.

10. Install the first and second speed shifter fork into the cover. Lubricate the shifter shaft bore with grease.

11. Install the manual selector arm through the interlock plate, and position the two pieces into the cover, with the wide leg of the interlock plate towards the inside of the transmission case.

12. Align the shifter shaft in the cover, and insert the shaft through the shifter forks and manual selector. Coat the shifter shaft with a light coating of grease. Make sure the detent grooves face the plunger side of the cover.

13. Align the pin holes in the manual selector arm and shifter shaft. Install the spring pin flush with the surface of the selector arm.

14. Install the detent plunger, spring, and plug. Tighten the plug to 8–12 ft. lbs.

15. Check the operation of the shift forks in each gear position.

OUTPUT SHAFT

1. Scribe alignment marks on the synchronizer and blocker rings. Remove the snapring from the front of the output shaft. Slide the third and fourth speed synchronizer assembly, blocker rings and third gear off the shaft.

2. Remove the next snapring and the second gear thrust washer from the shaft. Slide second gear and the blocker ring off the shaft, taking care not to lose the sliding gear from the first and second speed synchronizer assembly. The first and second speed synchronizer hub cannot be removed from the output shaft.

3. Remove the first gear thrust washer (oil slinger) from the rear of the output shaft. Remove the spring pin retaining first gear onto the shaft.

SYNCHRONIZER

1. Scribe reference marks on the hub and sleeve of the synchronizer.

2. Push the sleeve from the hub of each synchronizer.

3. Separate the inserts and insert springs from the hubs. Do not mix the parts between the first/second speed synchronizer and the third/fourth speed synchronizer. Clean and inspect all parts.

NOTE: The first/second speed synchronizer hub is not to be removed from the shaft. They have been assembled and machined as a matched unit during manufacturing to assure concentricity.

4. To assemble, position the sleeve on the hub, aligning the previously marked reference points.

5. Position the three inserts per hub and install the insert springs, being sure that the bent end of the springs are seated in one of the inserts. The springs on each side of the hubs must face in opposite directions and the openings be 180 degrees apart.

ASSEMBLY

1. Coat the countershaft thrust washers with a vaseline type lubricant and position the plastic type washer at the front of the case and the metal washer at the rear of the case.

2. With the 50 needle roller bearings in place in the cluster gear and the dummy countershaft in place, install the countershaft/cluster gear assembly into the case.

CAUTION
Be sure the thrust washers are not displaced during the gear installation.

---

3. Align the cluster gear bore with the case bores and install the countershaft from the rear to the front of the case, pushing the dummy countershaft from the gear and case.

4. Position the reverse idler gear with the shift lever groove facing to the front and install the shaft from the rear of the case.

5. Being careful not to disturb the synchronizers, install the output shaft assembly into the transmission case. Install the fourth gear blocking ring in the third speed synchronizer sleeve, engaging the inserts on the hub with the grooves of the blocking ring.

6. Install the 15 roller bearings in the input shaft pocket and retain with a vaseline type lubricant. Install the input shaft into the case and engage the shaft in the third/fourth synchronizer, while the stub of the output shaft is installed in the pocket of the input shaft.

CAUTION
Do not jam or drop the 15 roller bearings during the input shaft installation.

---

7. Install the input shaft front bearing. Block the first speed gear against the rear of the case, align the bearing with the bearing bore in the case and drive the bearing completely onto the input shaft and into the transmission case.

NOTE: To identify the front and rear bearings, look for a notch in the front bearing race. The rear bearing has no notch.

8. Install the front bearing retaining and locating snaprings.

9. Install the front bearing cap oil seal and install the cap (bearing retainer) with a new gasket to the transmission case. Install the retaining bolts.

10. Install the first speed thrust washer on the output shaft with the oil grooves facing the first speed gear. Install the rear bearing onto the output shaft and into the case bearing bore.

NOTE: Be sure the first gear thrust washer is engaged on the first gear roll pin before installing the rear bearing.

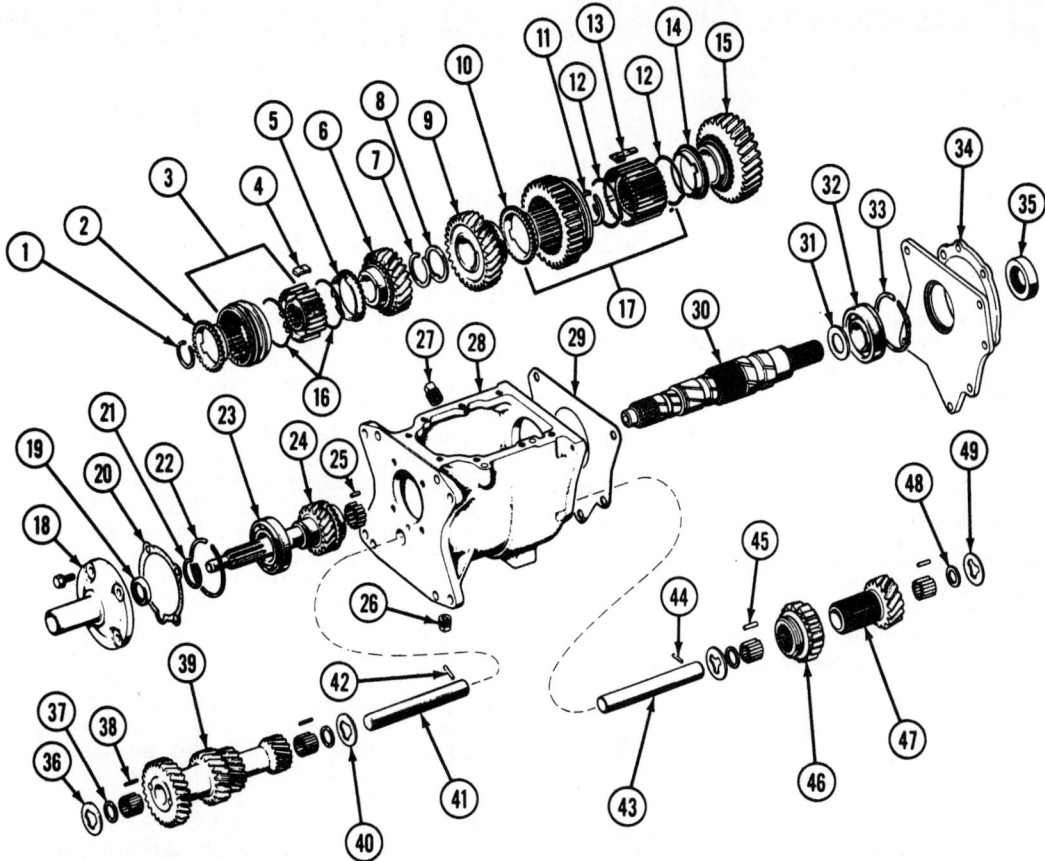

1. Third-fourth gear snapring
2. Fourth gear synchronizer ring
3. Third-fourth gear clutch assembly
4. Third-fourth gear plate
5. Third gear synchronizer ring
6. Third speed gear
7. Second gear snapring
8. Second gear thrust washer
9. Second speed gear
10. Second gear synchronizer ring
11. Mainshaft snapring
12. First-second synchronizer spring
13. Low-second plate
14. First gear synchronizer ring
15. First gear
16. Third-fourth synchronizer spring
17. First-second gear clutch assembly
18. Front bearing cap
19. Oil seal
20. Gasket
21. Snapring
22. Lockring
23. Front ball bearing
24. Clutch shaft
25. Roller bearing

26. Drain plug
27. Fill plug
28. Case
29. Gasket
30. Spline shaft
31. First gear thrust washer
32. Rear ball bearing
33. Snapring
34. Adapter plate
35. Adapter seal
36. Front countershaft gear thrust washer
37. Roller washer
38. Rear roller bearing
39. Countershaft gear
40. Rear countershaft thrust washer
41. Countershaft
42. Pin
43. Idler gear shaft
44. Pin
45. Idler gear roller bearing
46. Reverse idler sliding gear
47. Reverse idler gear
48. Idler gear washer
49. Idler gear thrust washer

T-176 exploded view

11. Install the retaining and locating snaprings on the rear bearing and output shaft.

12. Position the reverse lever in the case, on the pivot bolt and install the retaining clip. Tighten the pivot bolt. Be sure the reverse lever fork is engaged in the reverse idler gear.

13. Rotate the input shaft and output shaft gears and blocking rings to insure freeness of movement. Blocking ring to gear clutch tooth face should have a clearance of 0.030 in. (0.8mm).

14. Place the reverse lever in the neutral position and install the cover assembly on the transmission case. Place the two dow-

el bolts in their original positions and install the remaining retaining bolts.

15. Install a new oil seal in the adapter housing and, using a new gasket, install the adapter housing to the transmission case.

16. Install 3 pints of lubricant into the transmission.

17. Install the offset lever and retain with the flanged nut.

18. Depending upon the installation of the transmission into a vehicle, the shift lever housing can be installed and the transmission attached to the transfer case.

T-176 4-Speed Overhaul

The Warner T-176 transmission is a constant mesh unit, synchronized in all forward gears and with one reverse gear.

DISASSEMBLY

1. Remove the transfer case from the rear of the transmission.
2. Remove the shift control housing. Mark the location of the two dowel bolts in the housing.
3. Drain the lubricant from the transmission, if not previously done. Remove the rear adapter housing.
4. With a dummy countershaft tool, remove the countershaft from the transmission, front to rear. Allow the cluster gear to lay on the bottom of the case.
5. Remove the rear bearing locating and retaining snaprings. Remove the rear bearing with a bearing remover tool or equivalent.
6. Match mark the front bearing retainer to the case for easier installation, remove the retaining bolts and the retainer.
7. Remove the locating and retaining snaprings from the front bearing. Remove the front bearing and the input shaft using a puller tool or equivalent.
8. Remove the mainshaft pilot bearing rollers from the input shaft pocket. Engage the third speed synchronizer.
9. Remove the mainshaft assembly by lifting the front of the shaft upward and out.
10. Remove the cluster gear assembly from the case. Locate and remove any thrust washers and needle roller bearings from the case.
11. Tap the reverse idler gear shaft from the case and remove the reverse idler gear and thrust washers.
12. Separate the reverse idler gear from the sliding gear. Do not lose the needle roller bearings.

MAINSHAFT

1. Remove the third/fourth speed synchronizer snapring from the front of the mainshaft.
2. Remove the third/fourth synchronizer from the mainshaft and slide the hub from the sleeve. Remove the inserts and springs. Inspect the blocking rings for wear and damage.
3. Remove the third speed gear and the second speed gear snapring. Remove the second speed gear and the blocking ring. Remove the tabbed thrust washer.
4. Remove the snapring from the first/second synchronizer hub. Remove the hub and the reverse gear with sleeve as an assembly. Match mark the hub and sleeve for assembly references. Remove the inserts and springs as the sleeve is removed.
5. Remove the first speed gear thrust washer from the rear of the shaft and remove the first speed gear and the blocking ring.
6. Assemble the first/second synchronizer hub, inserts and springs. Install the clutch sleeve. Be sure to position the spring ends 180 degrees apart.
7. Install the assembled first/second speed synchronizer hub and the reverse gear with sleeve, on the mainshaft. Secure with a new snapring.
8. Install the first speed gear and blocking ring on the rear of the mainshaft and install the first gear thrust washer.
9. Install a new tabbed thrust washer on the mainshaft with the tab seated in the mainshaft tab bore.
10. Install the second speed gear and the blocking ring on the mainshaft and secure with a new snapring.
11. Install the third speed gear and blocking ring on the mainshaft.
12. Assemble the third/fourth speed synchronizer hub, inserts, and springs. Be sure the spring ends are 180 degrees apart.
13. Install the assembled third/fourth speed synchronizer on the mainshaft and secure with a new snapring.

14. The measured end play between the snapring and the third/fourth speed synchronizer should be 0.004-0.014 in. (0.10-0.36mm).

ASSEMBLY

1. Load the reverse idler gear with the 44 needle roller bearings and a bearing retainer on each end of the gear. Install the sliding gear on the reverse idler gear. Install lubricated thrust washers into the case.
2. Install the reverse idler assembly into the case and install the reverse idler gear shaft.
3. Be sure to engage the thrust washer locating tabs in the case locating slots.
4. Seat the reverse idler gear shaft roll pin into the counterbore in the case. The reverse idler gear end play should be 0.004-0.018 in. (0.10-0.46mm).
5. Install the 42 needle roller bearings in the cluster gear, using the dummy countershaft as a bearing holder. Use of a vaseline type lubricant is suggested to hold the bearings in place.
6. Position the lubricated thrust washers in place on the inside of the transmission case. Position the thrust washer tabs in the tab slots of the case.
7. Insert the countershaft into the rear case bore, just far enough to hold the rear thrust washer. Lower the cluster gear assembly into the case and align the gear bore with the case bore. Push the countershaft into the cluster gear, displacing the dummy countershaft out the front case bore hole. Do not completely remove the dummy countershaft.
8. Measure the cluster gear end play which should be 0.004-0.018 in. (0.10-0.46mm). Correct as required and reinstall the dummy countershaft into the cluster gear, pushing the countershaft from the gear.
9. Allow the cluster gear to remain at the bottom of the case until the input and mainshaft has been installed to provide the necessary assembly clearance.
10. With the synchronizers in the neutral position, install the mainshaft assembly into the case.
11. Install the front bearing part way on the input shaft and install the 15 roller bearing in the shaft pocket.

NOTE: Do not use a heavy grease to hold the bearings in the pocket as the grease can plug the lubrication holes. Use only a vaseline type lubricant.

12. Position the blocking ring on the third/fourth synchronizer. Support the mainshaft assembly and insert the input shaft through the front bearing bore of the case. Seat the mainshaft pilot hub into the bearing pocket of the input shaft and tap the front bearing and input shaft into the case, using a soft faced hammer.
13. When the bearing is fully seated, install the bearing retainer housing, but not the snaprings at this time.
14. Install the rear bearing on the mainshaft and the bearing bore of the case. It will be necessary to seat the rear bearing further than the locating snapring would allow, so do not install the locating snapring until after the retaining snapring is installed.
15. Remove the front bearing retainer housing and fully seat the front bearing on the input shaft. Install the retaining and locating snaprings. Install a new oil seal in the retainer housing and install on the transmission case.
16. Install the locating snapring on the rear bearing, if not previously done.
17. To install the cluster gear and countershaft, turn the transmission case on end with the input shaft down. Align the cluster gear bore and thrust washers with the case bores. Tap the countershaft into place and displace the dummy countershaft out the front of the case. Do not allow the dummy shaft to drop to the floor.
18. Level the transmission case and install the extension adapter housing with a new gasket.

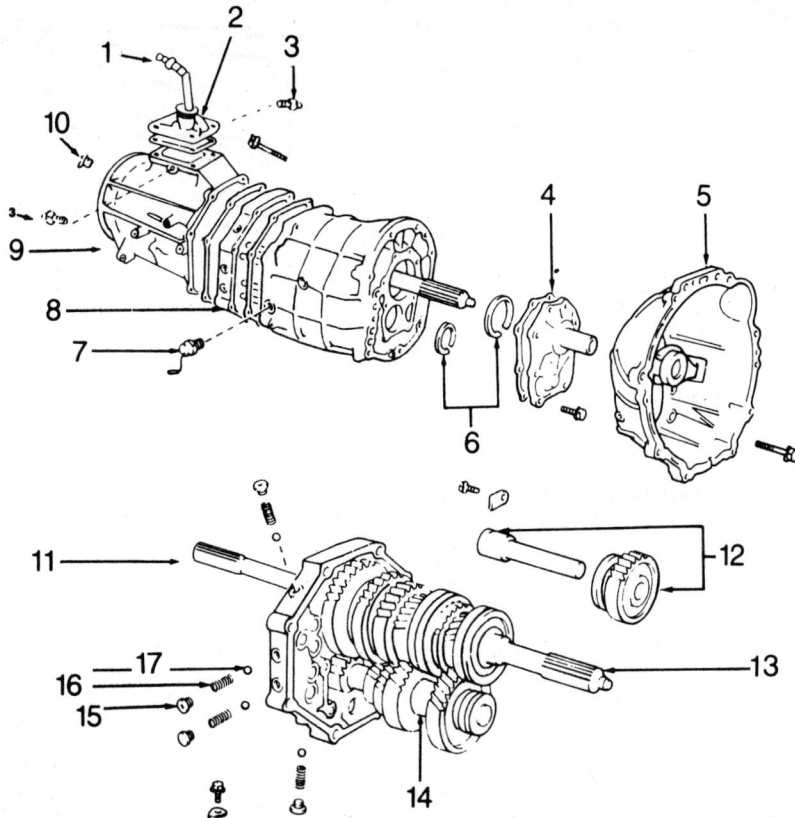

1. Shift lever
2. Shift lever retainer
3. Restrict pins
4. Front bearing retainer
5. Clutch housing
6. Snapring

7. Back-up light switch
8. Intermediate plate
9. Adapter housing
10. Adapter screw plug
11. Output shaft
12. Reverse idler gear

13. Input shaft
14. Counter gear
15. Straight screw plug
16. Spring
17. Locking ball

AX-5 exploded view

19. Shift the synchronizer sleeves by hand to insure correct operation. Install 3.5 pints of lubricant into the case and install a new gasket on the shift housing flange. With the gears in the neutral position and the shift lever forks in their neutral position, install the shift lever housing in place on the transmission case.

AX-5 Overhaul

The Aisin AX-5 is a Japanese-made 5-speed manual transmission. The transmission has synchromesh engagement in all forward gears controlled by a floor shift mechanism integrated into the transmission top cover. All fasteners and measurements are metric.

NOTE: The following components and materials must be replaced whenever the transmission is overhauled: Lip-type oil seals. Lock nuts. All roll pins. All snaprings. Loctite®Thread Lock or Loctite®242 Sealer should be used on all fasteners.

DISASSEMBLY

1. Remove the clutch housing.
2. Remove the straight screw plug, spring and ball using a Torx bit to remove the screw plug, and a magnet to remove spring and ball.
3. Remove five adapter housing bolts and one nut.
4. Remove the shift lever housing set bolt and lock plate.

5. Remove the plug at the rear of the shift fork shaft.
6. Remove the large magnet to pull the shaft out.
7. Remove the select lever from the top while rotating.
8. Remove the five adapter housing bolts two studs and one nut.
9. Using a plastic hammer, tap and remove the extension housing. Leave the gasket attached to the intermediate plate.
10. Remove the front bearing retainer and outer snaprings from the two front bearings.

A. Dust boot
B. Stub shaft

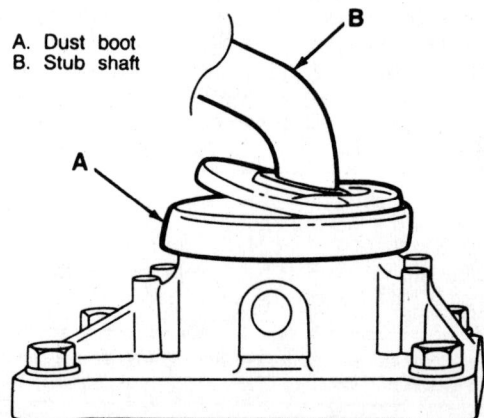

Dust boot removal from the AX-5 shifter

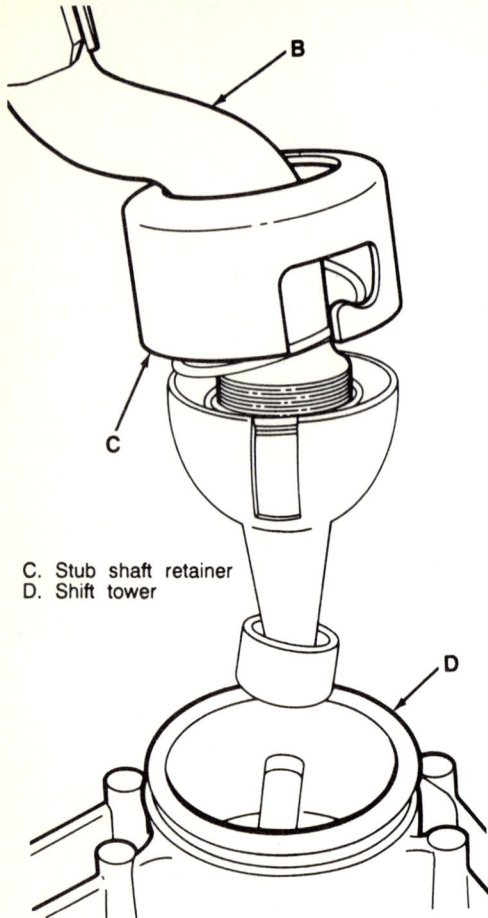

C. Stub shaft retainer
D. Shift tower

Stub shaft removal from the AX-5 shifter

11. Separate the intermediate plate from the transmission case using a small plastic hammer and remove the case.

12. Mount the intermediate plate in a vise. Be careful not to damage the plate.

NOTE: Before placing the intermediate plate in a vise, insert bolts, washers, and nuts in the open holes at the bottom of plate. Tighten vise against these bolts to prevent damage to the plate.

13. Remove the straight screw plug, locking balls and springs using a Torx bit and magnet.

14. Remove the five slotted spring pins using a hammer and punch and then remove the two E-rings from the shift rails.

─────────── CAUTION ───────────
The locking ball from the reverse shift head and locking ball and pin from the intermediate housing will fall from the holes so be sure to catch them. If they do not come out, remove them with a magnet.

15. Pull out the shift fork shaft No. 4 from the intermediate plate and catch the locking ball.

16. Remove shift fork shaft No. 4 and the 5th gear fork.

17. Pull out shift fork shaft No. 5 from the intermediate plate, and remove it with the reverse shift head.

─────────── CAUTION ───────────
The interlock pins will fall from their hole. If they do not come out, remove them with a magnet.

18. Remove the shift fork shaft No. 3 from the intermediate plate and catch the interlock pins.

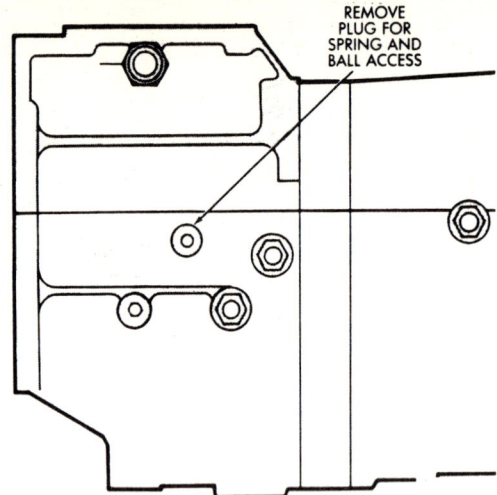

Detent ball plug location on the AX5

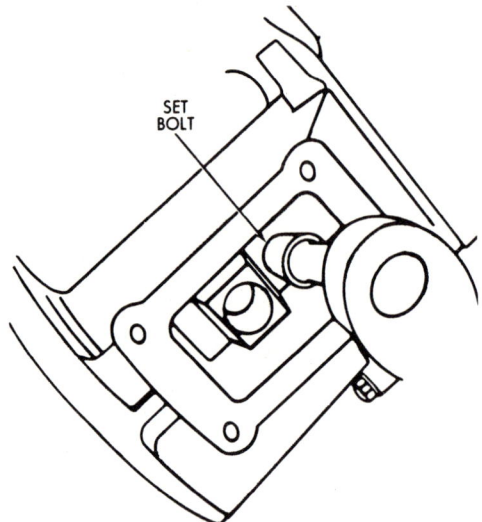

Set bolt removal from the AX5

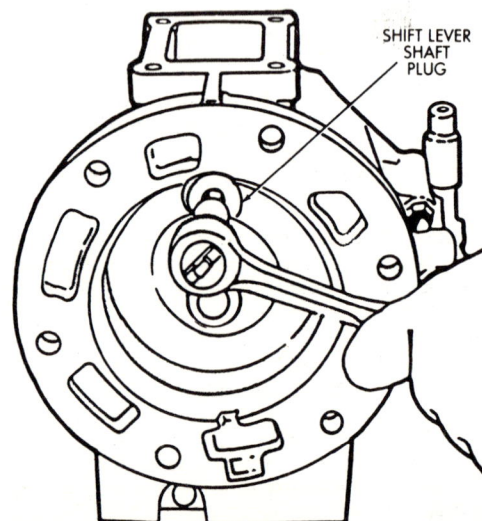

Removing the shift lever shaft plug from the AX5

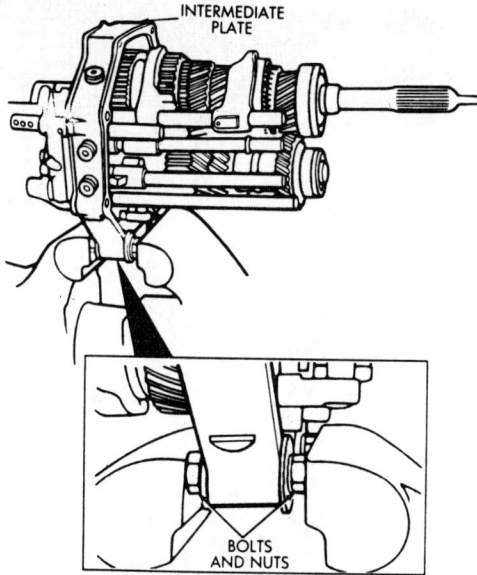

Positioning the intermediate plate in a vise

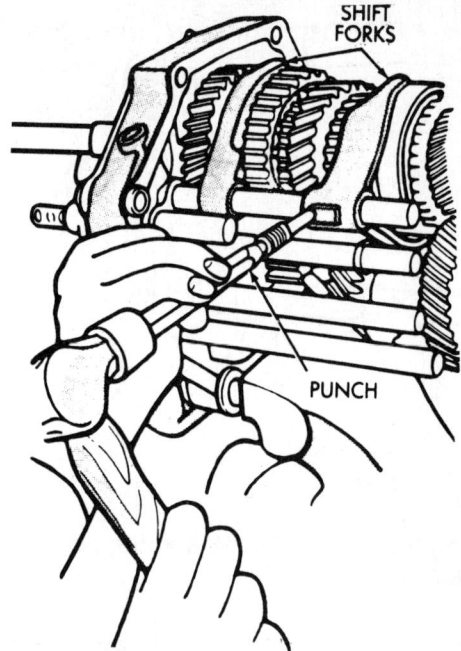

Removing the shift fork pin from the AX5

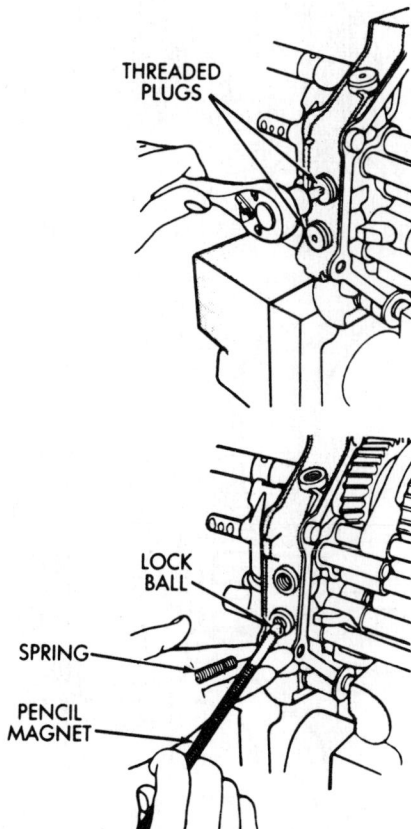

Removing the lock ball and spring from the AX5

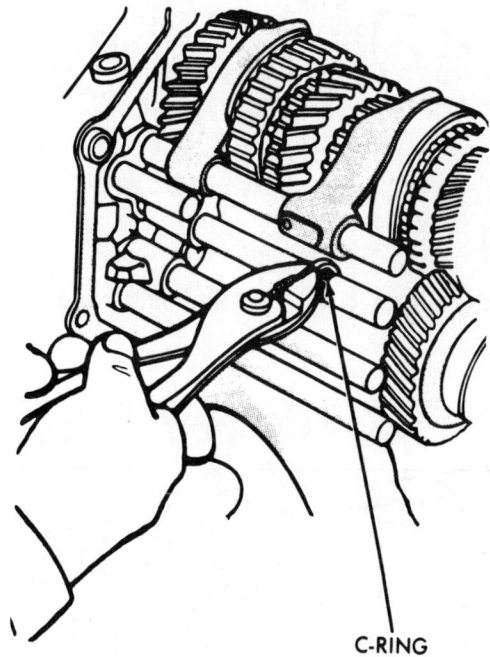

Shift rail C-ring removal from the AX5

─────────── **CAUTION** ───────────

The interlock pin will fall from the hole so be sure to catch it. If it does not come out, remove it with a magnet.

19. Remove shift fork shaft No. 1 from the intermediate plate being careful not to drop the interlock pin.

20. Remove shift fork shaft No. 2, shift fork No. 2 and shift fork No. 1.
21. Remove the reverse idle gear shaft stopper, reverse idler gear and shaft.
22. Remove the reverse shift arm from the reverse shift arm bracket.
23. Using a feeler gauge, measure the counter 5th gear thrust clearance. Standard Clearance: 0.10-0.30mm.
24. Engage two gears to lock the output shaft. Using a hammer and chisel, loosen the staked part of the nut on the countershaft.
25. Remove the lock nut. Disengage the gears.

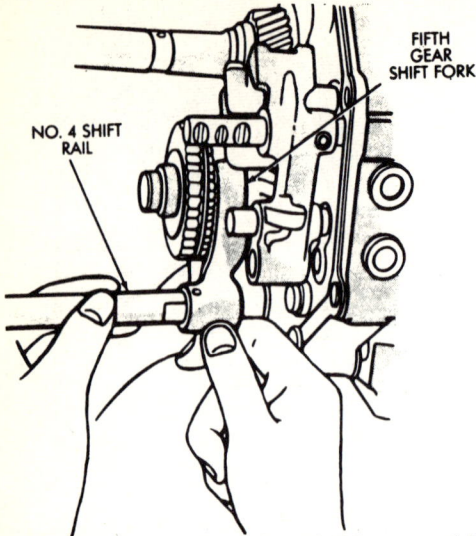

Removing the No. 4 shift rail and 5th gear shift fork from the AX5

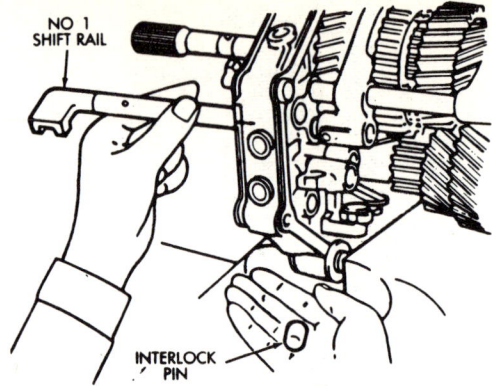

Removing the No. 1 shift rail and Interlock pin from the AX5

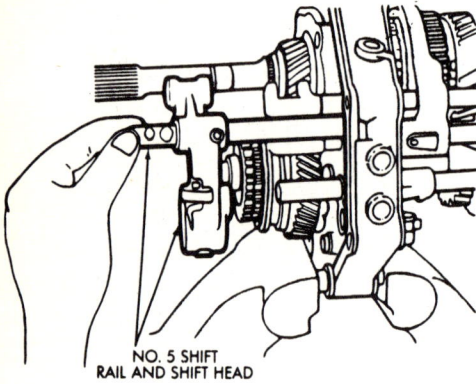

Removing the No. 5 shift rail and shift head from the AX5

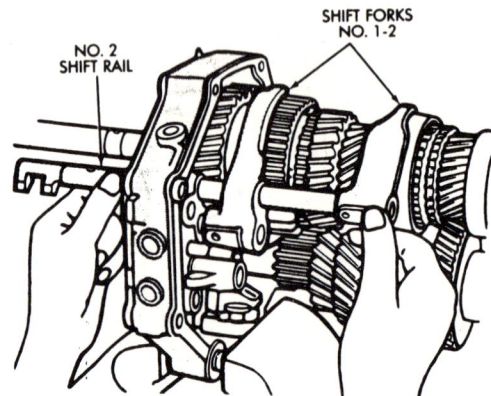

Removing the shift forks and No. 2 shift rail from the AX5

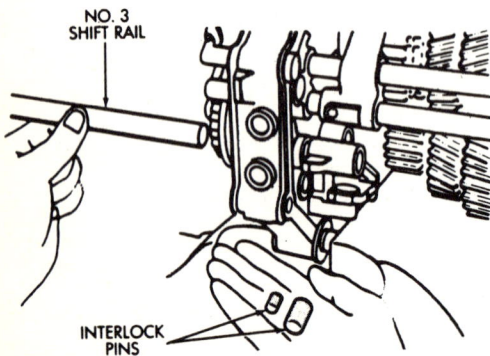

Removing the No. 3 shift rail and Interlock pin from the AX5

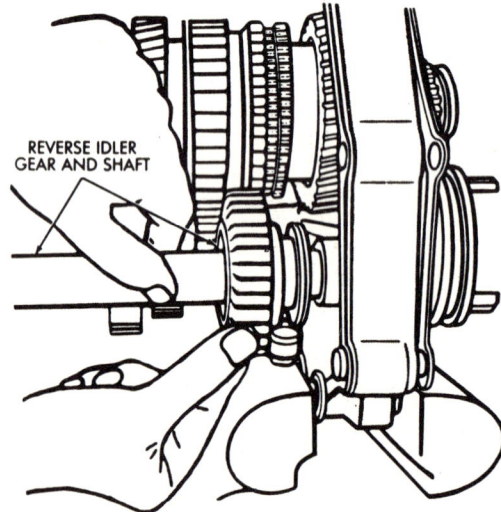

Removing the reverse Idler gear and shaft from the AX5

26. Remove the gear spline piece No. 5, synchronizer ring, needle roller bearing and counter 5th gear using tool J-22888 or equivalent.
27. Remove the spacer and use a magnet to remove the ball.
28. Remove the reverse shift arm bracket.
29. Remove the rear bearing retainer bolts with a Torx bit and the snapring using snapring pliers.
30. Remove the output shaft, counter gear and input shaft as a

unit from the intermediate plate by pulling on the counter gear and tapping on the intermediate plate with a plastic hammer.
31. Remove the input shaft with fourteen needle roller bearings from the output shaft.
32. Remove the counter rear bearing from the intermediate plate.

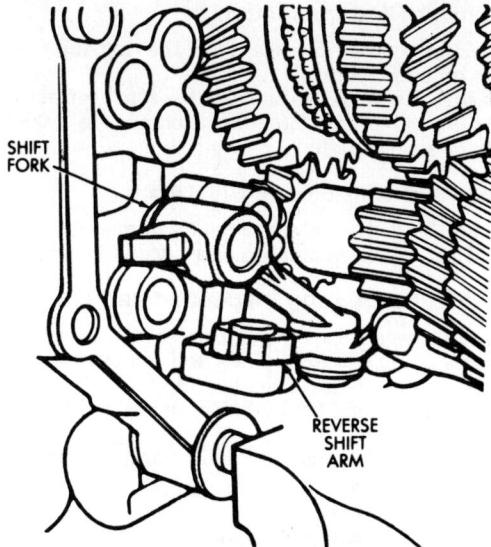

Reverse shift arm removal from the AX5

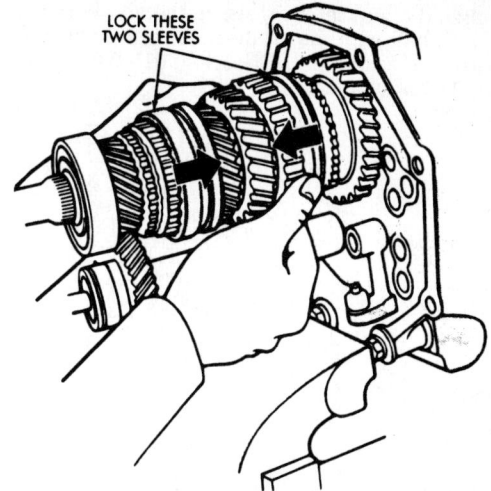

Locking the mainshaft gears on the AX5

33. Measure the thrust clearance of each gear. Standard clearance is 0.10-0.25mm.
34. Using two awls and a hammer, tap out the snapring.
35. Using a press, remove the 5th gear, rear bearing, 1st gear and the inner race.
36. Remove the needle roller bearing.
37. Remove the synchronizer ring and locking ball.
38. Using a press, remove hub sleeve No. 1 assembly, synchronizer ring, 2nd gear.
39. Remove the needle roller bearing.
40. Remove the snapring from hub sleeve No. 2.
41. Using a press, remove the hub sleeve, synchronizer ring, and 3rd gear.
42. Remove the needle roller bearing.

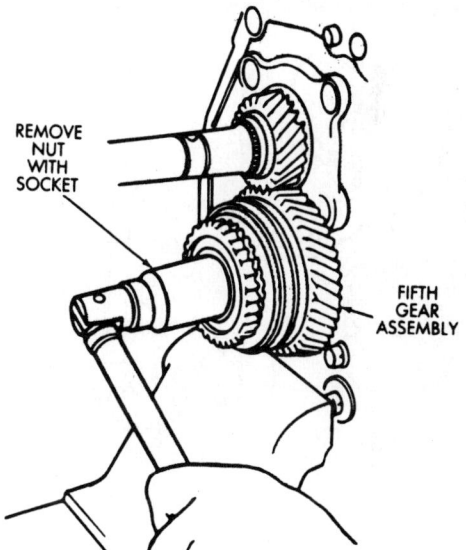

Measuring the counter 5th gear thrust clearance on the AX5

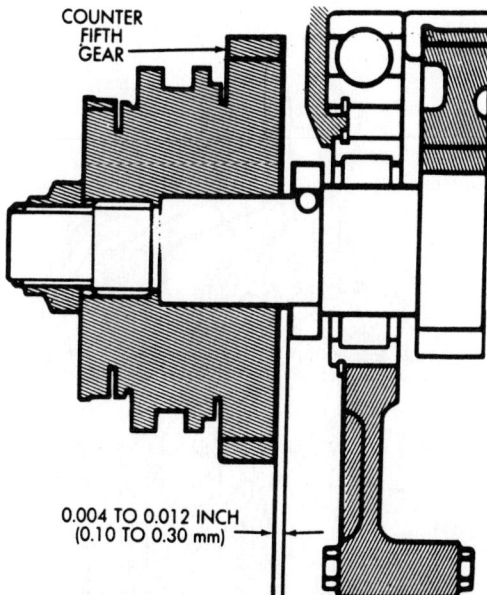

Removing the 5th gear nut from the AX5

COMPONENT INSPECTION

Output Shaft & Inner Race

1. Check the output shaft and inner race for wear or damage.
2. Using calipers, measure the output shaft flange thickness. Minimum thickness is 4.8mm.
3. Using calipers, measure the inner face flange thickness. Minimum thickness is 4.0mm.
4. Using a micrometer, measure the outer diameter of the output shaft journal surface. 2nd gear minimum is 38mm; 3rd gear minimum is 35mm.
5. Using a micrometer, measure the outer diameter of the inner race. Minimum diameter is 39mm.
6. Using a dial indicator, measure the shaft runout. Maximum Runout: 0.05mm.

1st Gear Oil Clearance

1. Using a dial indicator, measure the oil clearance between the gear and inner race with the needle roller bearing installed. Standard clearance is 0.010-0.033mm.

2. Using a dial indicator, measure the oil clearance between the gear and shaft with the needle roller bearing installed. Standard Clearance: 2nd and 3rd Gears, 0.010-0.033mm; Counter 5th Gear, 0.010-0.033mm.

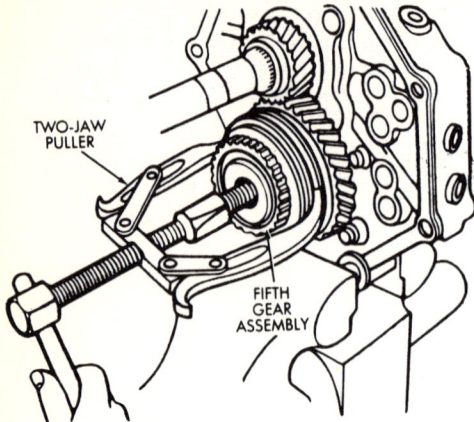

Removing the 5th gear assembly from the AX5

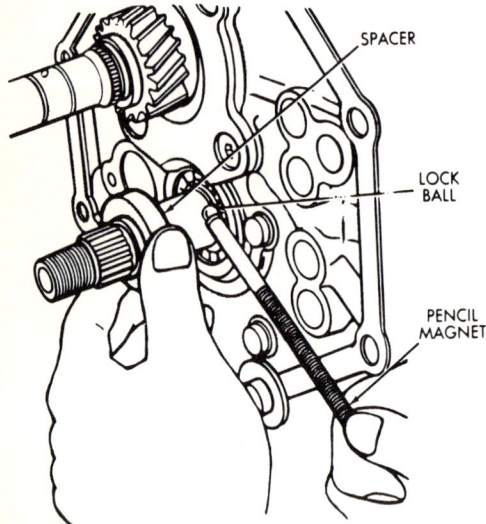

Spacer and lock ball removal from the AX5

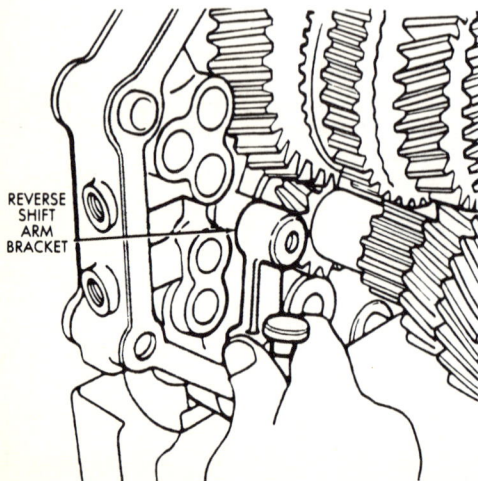

Removing the reverse shift arm bracket from the AX5

Synchronizer Ring Inspection

1. Check for wear or damage. Turn the ring and push it in to check the braking action.
2. Measure the clearance between the synchronizer ring back and the gear spline end. Standard clearance is 1.00-1.80mm; minimum clearance: 0.8mm.

Shift Fork and Hub Sleeve Clearance

Using a feeler gauge, measure the clearance between the hub sleeve and shift fork. Maximum clearance is 1.0mm.

Input Shaft and Bearing Inspection and Removal

1. Check for wear or damage. If necessary, remove the bearing snapring using snapring pliers and remove the bearing.
2. Using a press, remove the bearing.
3. Using a press and tool J-34603 or equivalent, install the new bearing.
4. Select a snapring that will allow minimum axial play and install it on the shaft.

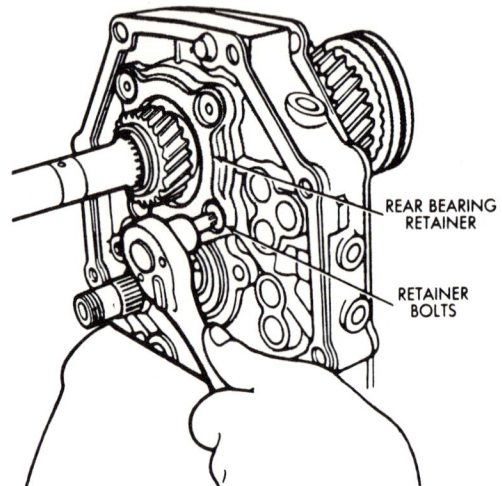

Removing the rear bearing retainer from the AX5

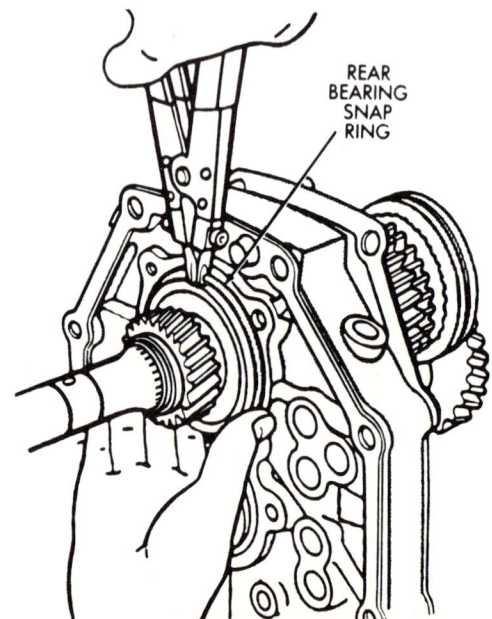

Removing the rear bearing snapring from the AX5

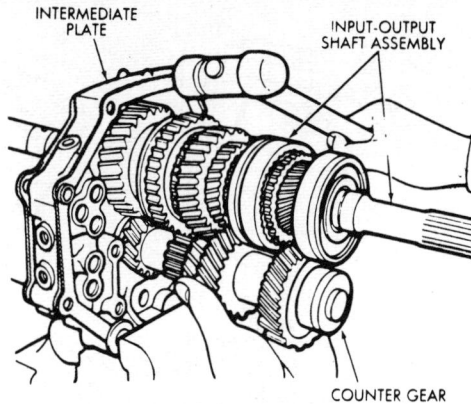

Removing the counter gear and output shaft from the AX5

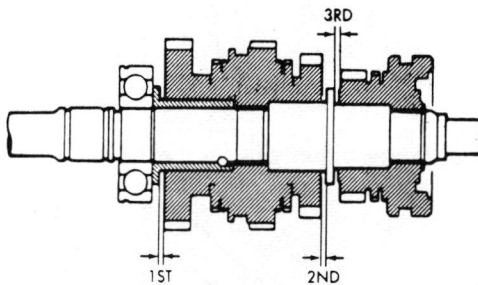

Checking the output shaft gear thrust clearance on the AX5

Counter Gear and Bearing Inspection

1. Check the gear teeth for wear or damage.
2. Check the bearing for wear or damage.

Counter Gear Front Bearing Replacement

1. Using snapring pliers, remove the snapring.
2. Press out the bearing using tool J-22912-01 or equivalent.
3. Replace the side race.
4. Using tool J-28406 or equivalent, press in the bearing and inner race.
5. Select a snapring that will allow minimum axial play and install it on the shaft.

Front Bearing Retainer Inspection

1. Check retainer for damage.
2. Check the oil seal lip for wear or damage.

OIL SEAL REPLACEMENT

1. Using a awl, pry the old seal out of the housing.
2. Press in the new oil seal using tool J-34602 or equivalent.
3. The oil seal depth is 11.20-12.20mm from the housing-to-transmission surface to the top edge of the seal.

REVERSE RESTRICT PIN REPLACEMENT

1. Check for wear or damage.
2. Using a Torx bit, remove the screw plug.

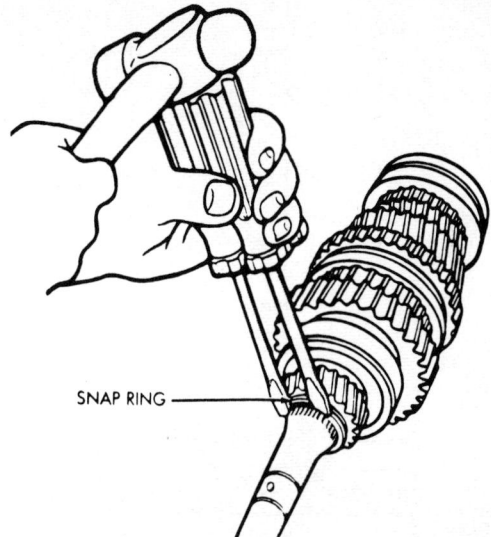

Removing the 5th gear snapring from the AX5

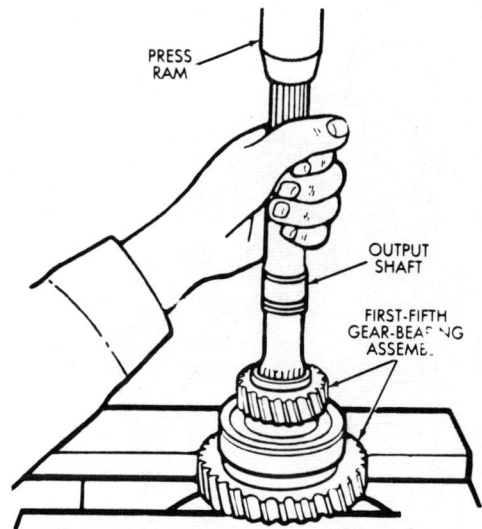

Removing the 5th gear and 1st gear bearing and race from the AX5

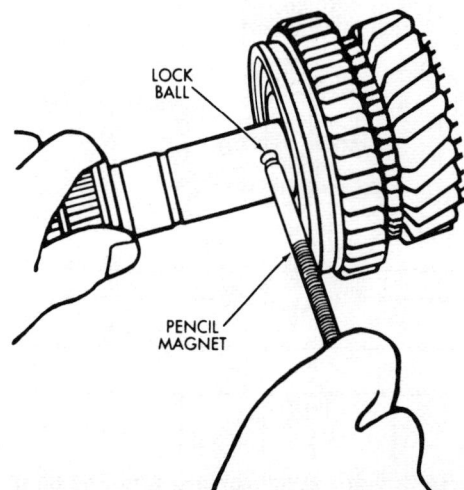

Synchronizer lock ball removal from the AX5

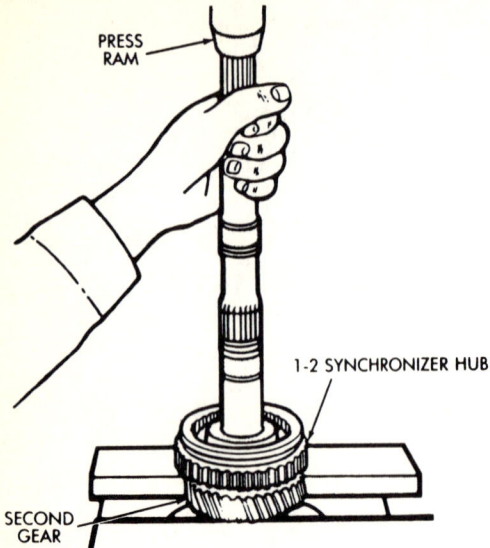

1st-2nd synchronizer/2nd gear removal from the AX5

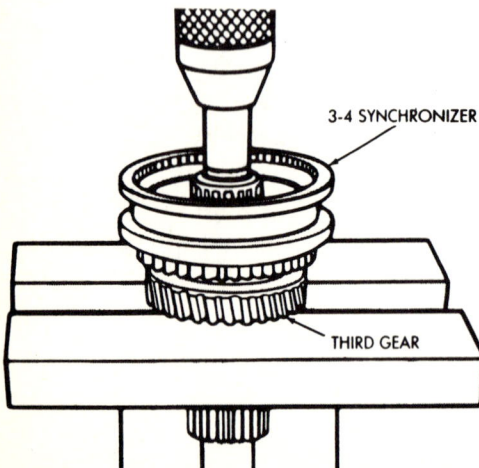

Removing the 3rd-4th synchronizer snapring from the AX5

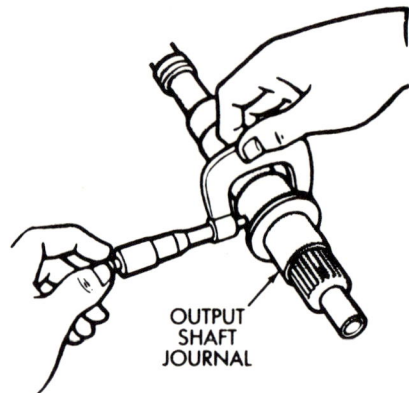

Removing the 3rd-4th synchronizer and 3rd gear from the AX5

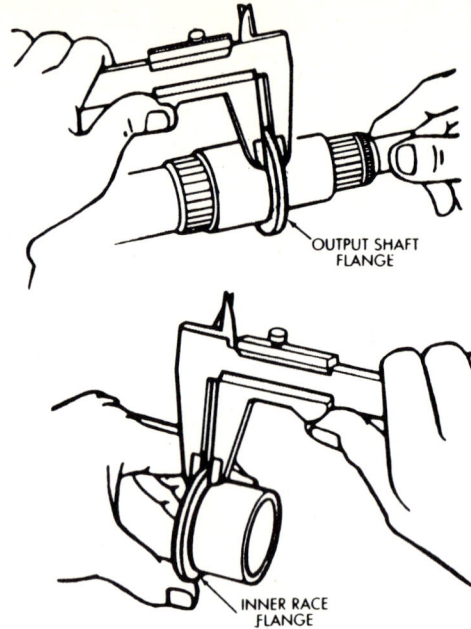

Checking flange thickness on the AX5

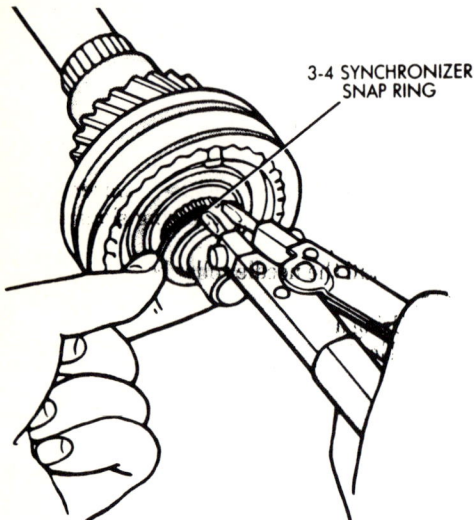

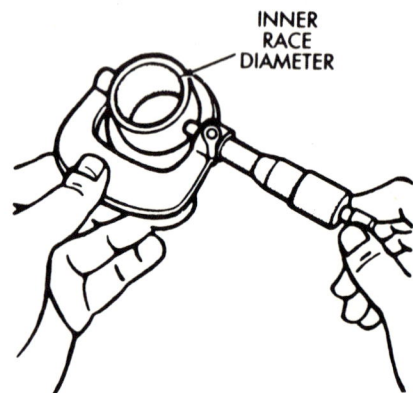

Checking shaft and race diameter on the AX5

3. Using a hammer and pin punch, drive out the slotted spring pin.

4. Pull off the lever housing and slide out the shaft.

5. Install the lever housing.

6. Using a hammer and pin punch, drive out the slotted spring pin.

7. Using a Torx bit, install and torque the screw plug to 27 ft. lbs. torque.

ADAPTER HOUSING & OIL SEAL INSPECTION & REPLACEMENT

1. Check the adapter housing for wear or damage.
2. Replace the oil seal with tool J-29184 or equivalent.

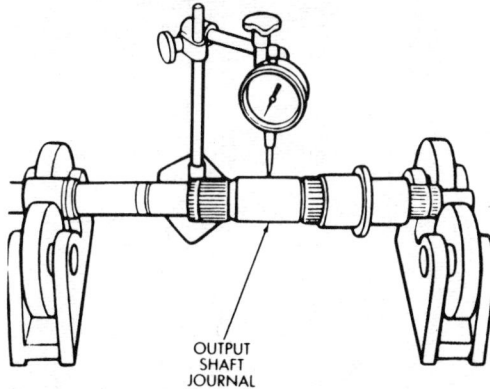

Checking output shaft runout on the AX5

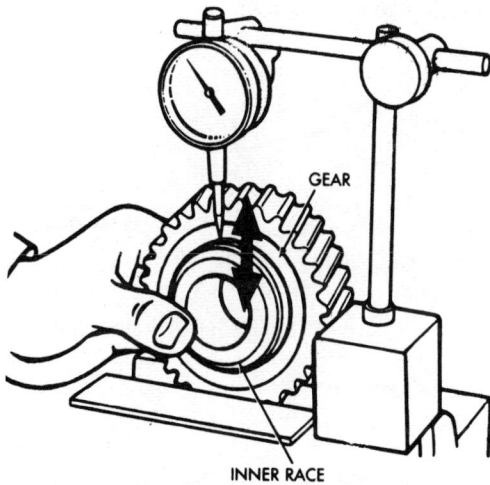

Checking gear-to-race clearance on the AX5

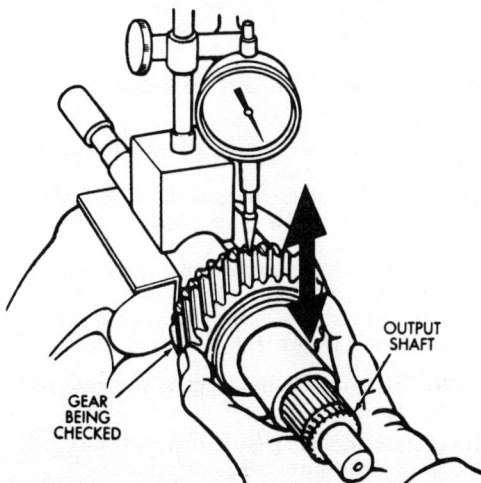

Checking gear-to-shaft clearance on the AX5

ASSEMBLY

1. Install the clutch hub No. 1 and No. 2 into hub sleeves along with the shifting keys.

----- CAUTION -----
Install the key springs so their gaps are not in line.

2. Install the shifting springs under the shifting keys.
3. Apply gear oil on the output shaft and 3rd gear needle roller bearing.
4. Place the 3rd gear synchronizer ring on the gear and align the ring slots with the shifting keys.
5. Install the needle roller bearing in the 3rd gear and hub sleeve No. 2.
6. Select a new snapring (2) that will allow minimum axial play and install it on the shaft.
7. Using a feeler gauge, measure the 3rd gear thrust clearance. Standard clearance is 0.10-0.25mm.
8. Apply gear oil on the output shaft and 2nd gear needle bearing.
9. Place the 2nd gear synchronizer ring on the 2nd gear and align the ring slots with the shifting keys.
10. Install the needle roller bearing in the 2nd gear.
11. Using a press install the 2nd gear and hub sleeve No. 1.
12. Install the 1st gear locking ball in the output shaft.
13. Apply gear oil to the needle roller bearing.

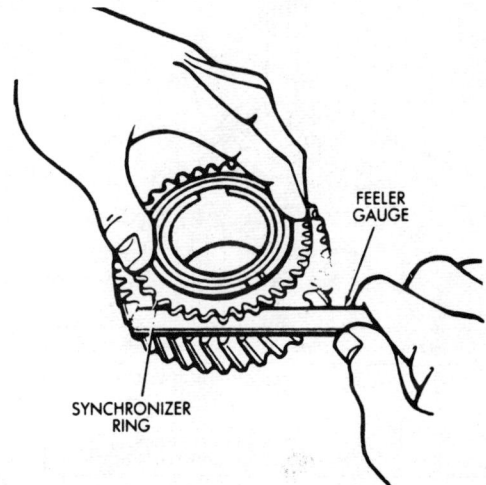

Checking synchronizer ring wear on the AX5

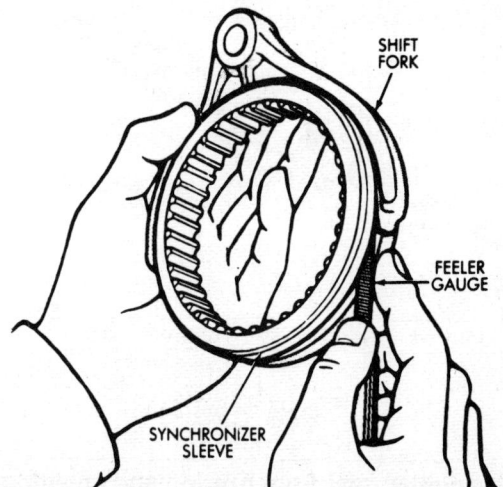

Checking fork-to-hub clearance on the AX5

7-35

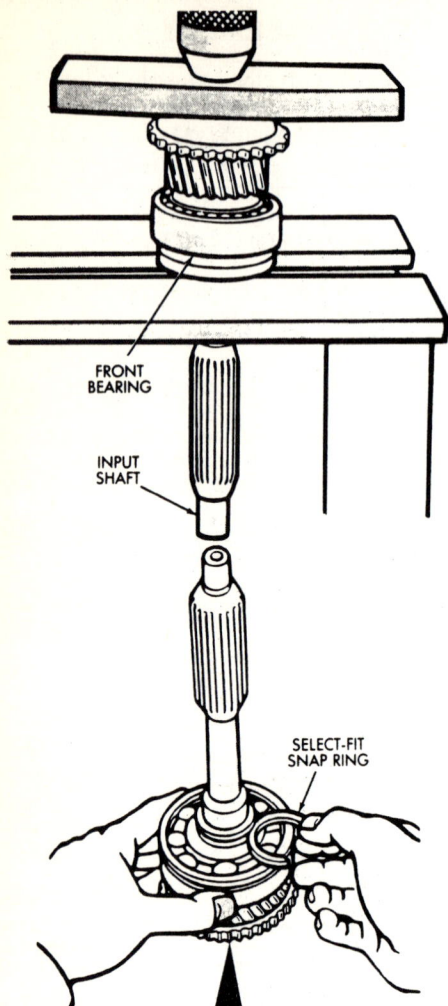

Installing front bearing and snapring on the AX5

I.D. Mark	Snap Ring Thickness mm (in.)
0	2.05-2.10 (0.0807-0.0827)
1	2.10-2.15 (0.0827-0.0846)
2	2.15-2.20 (0.0846-0.0866)
3	2.20-2.25 (0.0866-0.0886)
4	2.25-2.30 (0.0886-0.0906)
5	2.30-2.35 (0.0906-0.0925)

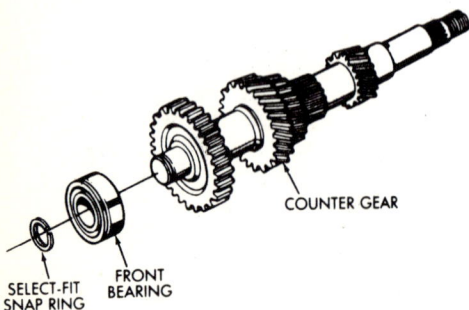

I.D. Mark	Snap Ring Thickness mm (in.)
1	2.05-2.10 (0.0807-0.0827)
2	2.10-2.15 (0.0827-0.0846)
3	2.15-2.20 (0.0846-0.0866)
4	2.20-2.25 (0.0866-0.0886)
5	2.25-2.30 (0.0886-0.0906)
6	2.30-2.35 (0.0906-0.0925)

Installing counter gear front bearing and snapring on the AX5

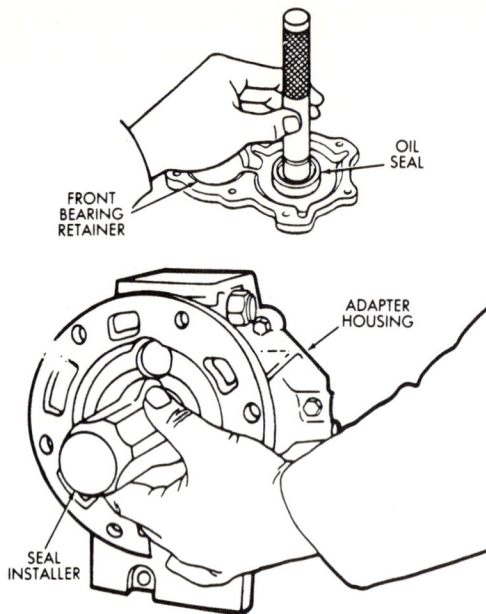

Oil seal installation on the AX5

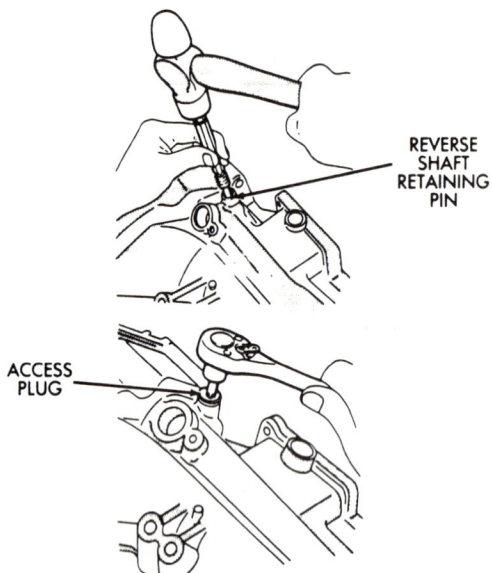

Installing the reverse shaft pin on the AX5

14. Assemble the 1st gear, synchronizer ring, needle roller bearing and bearing inner race.

15. Install the assembly on the output shaft, with the synchronizer ring slots aligned with the shifting keys.

16. Turn the inner race to align it with the locking ball.

17. Install the output shaft rear bearing using tool J-34603 or equivalent and a press.

18. Install the bearing on the output shaft with the outer race snapring groove toward the rear.

NOTE: Hold the 1st gear inner race to prevent it from falling.

19. Measure the 1st and 2nd gear thrust clearance with a feeler gauge. Standard Clearance: 0.10-0.25mm.

20. Install 5th gear on the output shaft using tool J-34603 or equivalent and a press.

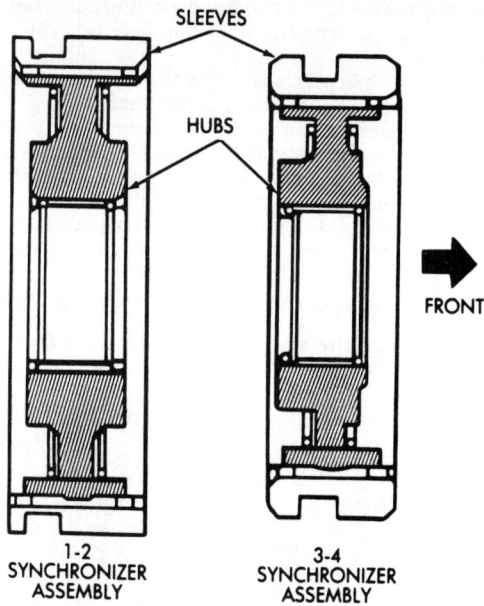

Synchronizer identification on the AX5

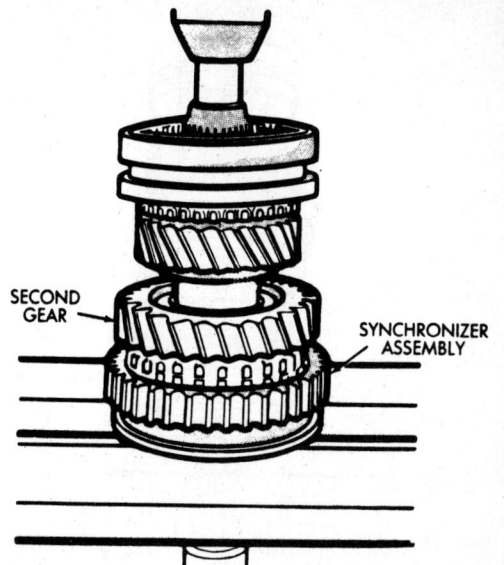

Installing 2nd gear and synchronizer on the AX5

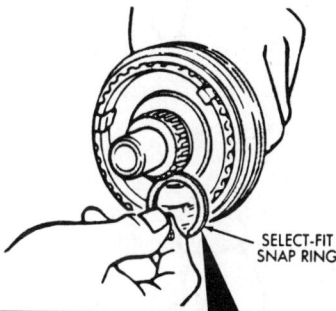

I.D. Mark	Snap Ring Thickness mm (in.)
C-1	1.75-1.80 (0.0689-0.0709)
D	1.80-1.85 (0.0709-0.0728)
D-	1.85-1.90 (0.0728-0.0748)
E	1.90-1.95 (0.0748-0.0768)
E-1	1.95-2.00 (0.0768-0.0787)
F	2.00-2.05 (0.0788-0.0807)
F-1	2.05-2.10 (0.0807-0.0827)

Installing 3rd gear and 3rd-4th synchronizer on the AX5

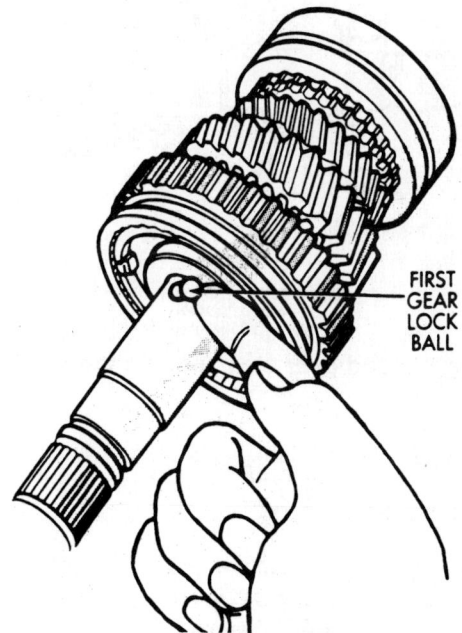

Installing 1st gear and lock ball on the AX5

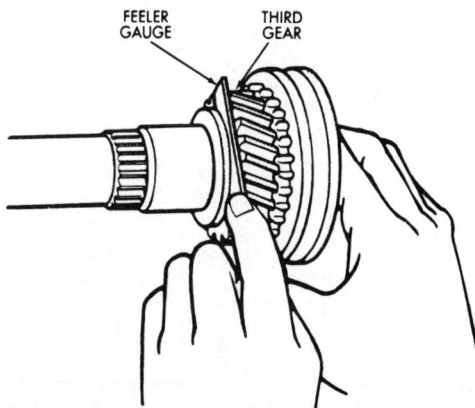

Checking 3rd gear clearance on the AX5

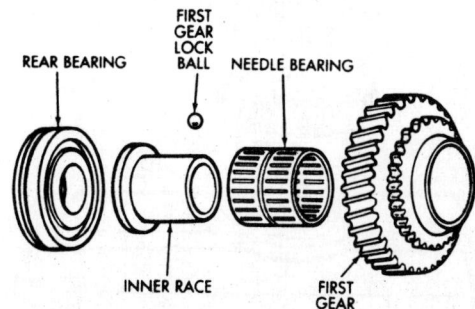

1st gear assembled on the AX5

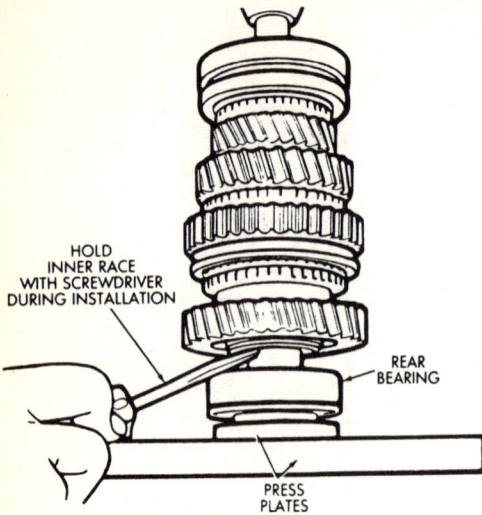

Installing the output shaft rear bearing on the AX5

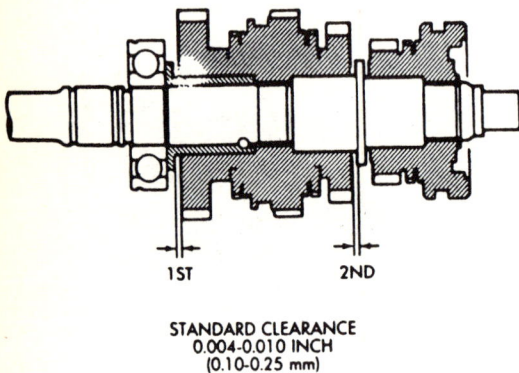

HOLD INNER RACE WITH SCREWDRIVER DURING INSTALLATION

REAR BEARING

PRESS PLATES

21. Select a snapring that will allow minimum axial play.
22. Using a screwdriver and a hammer, tap the snap into position.
23. Apply multi-purpose grease to the fourteen needle roller bearings and install them in the input shaft.
24. Install the output shaft into the intermediate plate by pulling on the output shaft and tapping on the intermediate plate.
25. Install the input shaft to the output shaft with the synchronizer ring slots aligned with the shifting keys.
26. Install the counter gear into the intermediate plate while holding the counter gear, and install the counter rear bearing with a suitable driver.
27. Install the bearing snapring using snapring pliers.

NOTE: Be sure the snapring is flush with the intermediate plate surface.

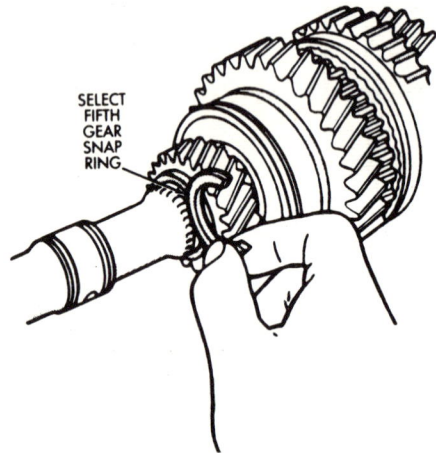

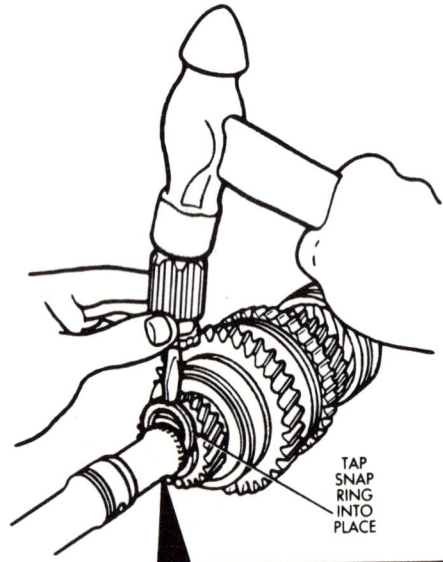

Checking 1st-2nd gear clearance on the AX5

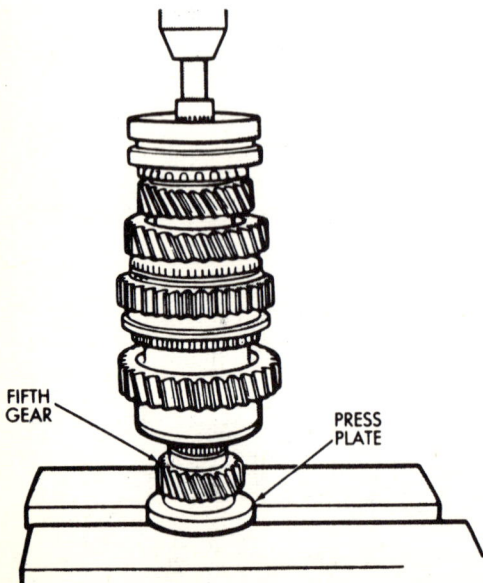

1ST   2ND

STANDARD CLEARANCE
0.004-0.010 INCH
(0.10-0.25 mm)

SELECT FIFTH GEAR SNAP RING

FIFTH GEAR

PRESS PLATE

Installing output shaft 5th gear on the AX5

TAP SNAP RING INTO PLACE

I.D. Mark	Snap Ring Thickness mm (in.)
A	2.67-2.72 (0.1051-0.1071)
B	2.73-2.78 (0.1075-0.1094)
C	2.79-2.84 (0.1098-0.1118)
D	2.85-2.90 (0.1122-0.1142)
E	2.91-2.96 (0.1146-0.1165)
F	2.97-3.02 (0.1169-0.1189)
G	3.03-3.08 (0.1193-0.1213)
H	3.09-3.14 (0.1217-0.1236)
J	3.15-3.20 (0.1240-0.1260)
K	3.21-3.26 (0.1264-0.1283)
L	3.27-3.32 (0.1287-0.1307)

Selecting and installing the 5th gear snapring on the AX5

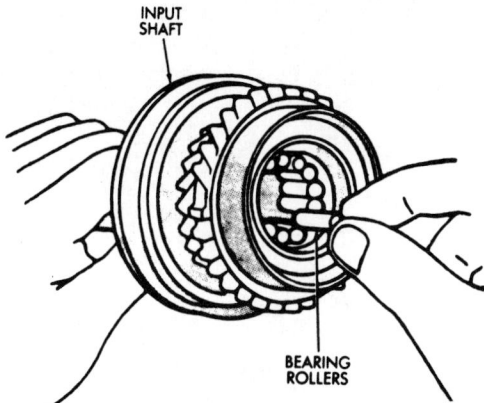

Installing the input shaft bearing rollers on the AX5

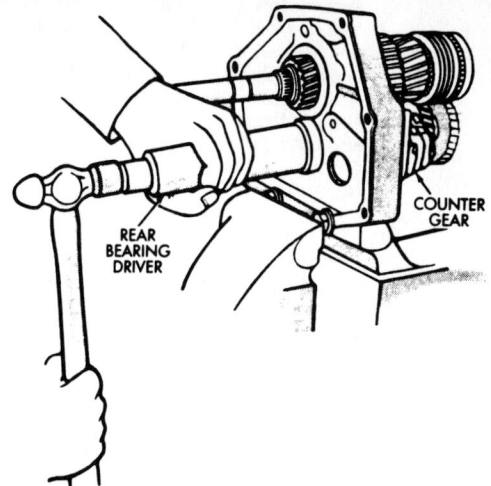

Installing the counter gear on the AX5

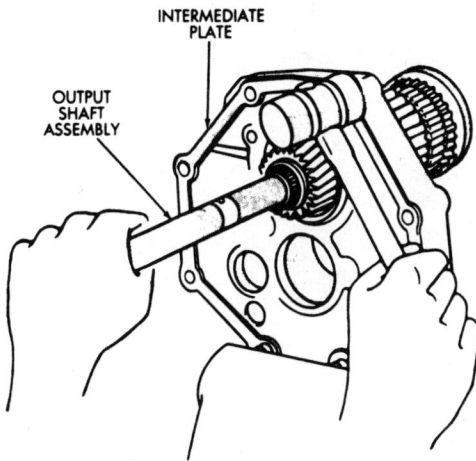

Installing the output shaft in the intermediate plate on the AX5

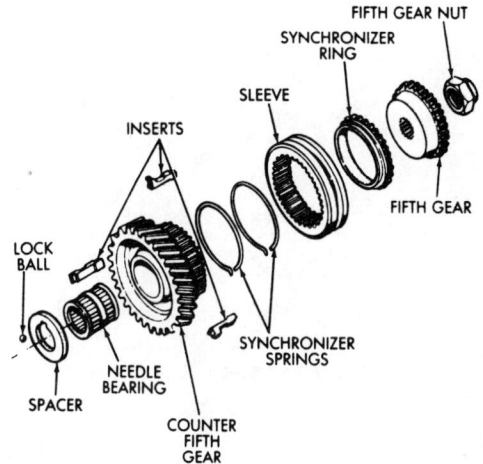

Counter 5th gear and synchronizer components on the AX5

28. Using a Torx bit, install and tighten the screws to 13 ft. lbs. torque.
29. Install the reverse shift arm bracket and tighten the bolts to 13 ft. lbs. torque.
30. Install the ball and spacer.
31. Install the shifting keys and hub sleeve No. 3 onto the counter 5th gear.

--------- **CAUTION** ---------
Install the key springs positioned so the end gaps are not in line.

32. Install shifting key springs under the shifting keys.
33. Apply gear oil to the needle roller bearing and install the counter 5th gear with hub sleeve No. 3 and needle roller bearings.
34. Install the synchronizer ring on gear spline piece.
35. Using tool J-28406 or equivalent drive in gear spline piece No. 5 with the synchronizer ring slots aligned with the shifting keys.

NOTE: When installing gear spline piece No. 5, support the counter gear in front with a 3-5 lb. hammer or equivalent.

36. Engage two gears to lock the output shaft.
37. Install and tighten the lock nut to 90 ft. lbs. torque on the counter shaft.
38. Stake the lock nut.
39. Disengage the gears.

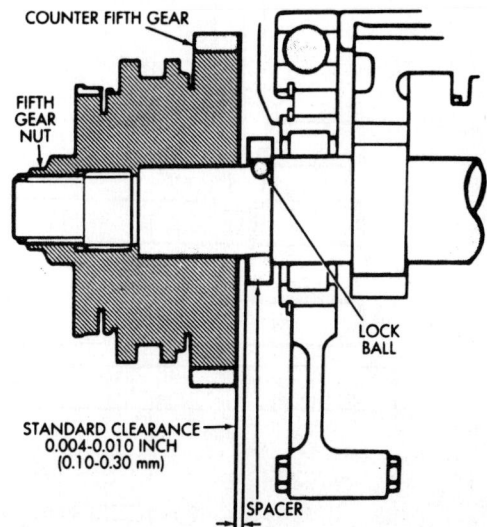

Checking 5th gear thrust clearance on the AX5

1. Reverse fork and shift arm
2. 1–2 shift fork
3. 3–4 shift fork
4. Lock ball, spring and plug
5. Bracket bolt
6. No. 3 shift rail
7. No. 1 shift rail
8. C-ring
9. No. 2 shift rail
10. C-ring
11. Lock ball, spring and plug
12. Shift arm
13. Set bolt and lock plate
14. Shift lever shaft
15. Shaft plug
16. Reverse pin
17. Retaining pin and plug
18. No. 5 shift rail
19. Interlock pin
20. Interlock pin
21. Interlock pin
22. C-ring
23. Interlock pin
24. Fifth-reverse fork
25. Reverse shift head
26. Lock balls
27. No. 4 shift rail
28. Reverse arm bracket

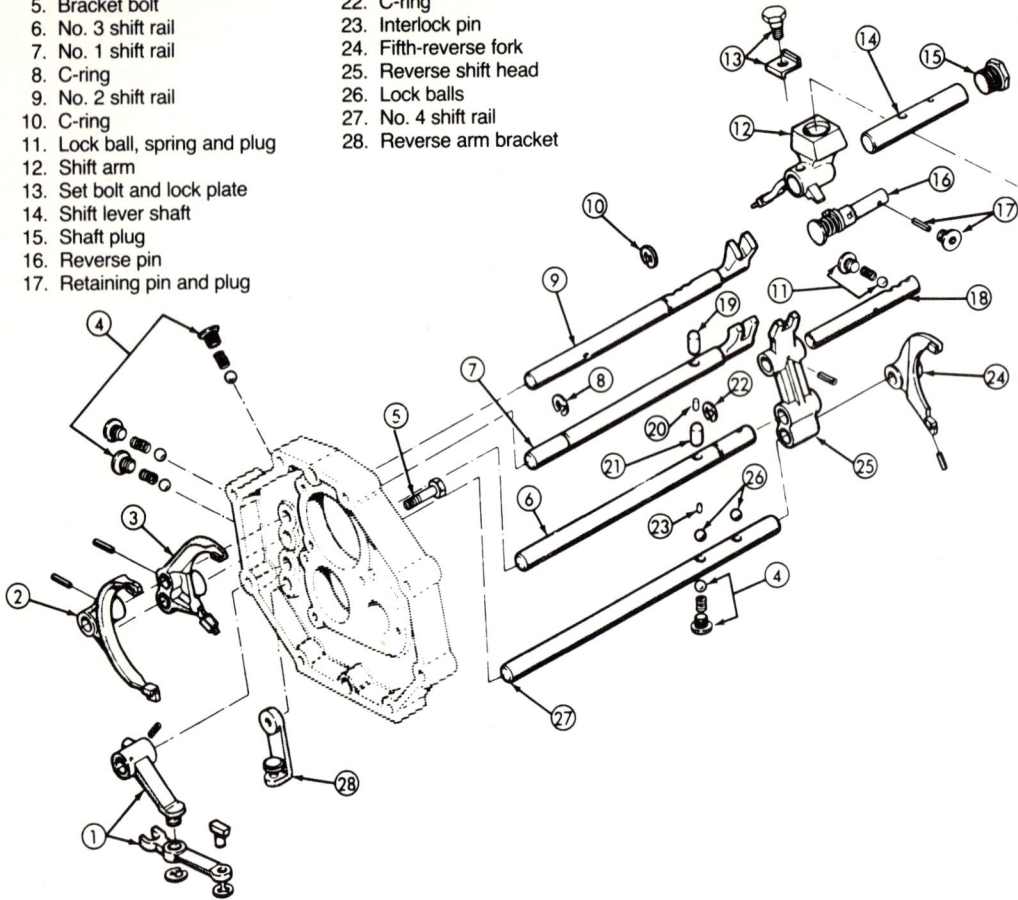

AX5 shift components

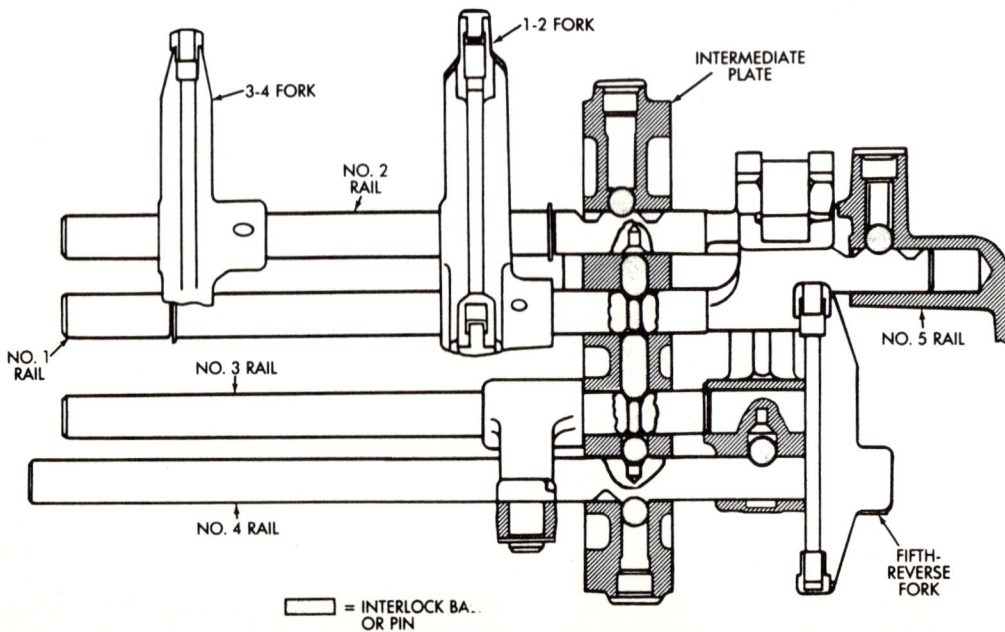

AX5 interlock ball and pin position

□ = INTERLOCK BALL OR PIN

40. Measure the counter 5th gear thrust clearance using a feeler gauge. Standard Clearance: 0.10-0.30mm.

41. Install the reverse shift arm to the pivot of the reverse shift arm bracket.

42. Install the reverse idler gear on the shaft.

43. Align the reverse shift arm shoe to the reverse idler gear groove and insert the reverse idler gear shift to the intermediate plate.

44. Install the reverse idler gear shaft stopper and tighten the bolt to 13 ft. lbs. torque.

45. Place shift forks No. 1 and No. 2 into groove of hub sleeves No. 1 and No. 2 and install fork shaft No. 2 to the shift forks No. 1 and No. 2 through the intermediate plate.

46. Apply multi-purpose grease to the interlock pins.

47. Using a magnet and screwdriver, install the interlock pin into the intermediate plate.

48. Install the interlock pin into the shaft hole.

49. Install fork shaft No. 1 to shift fork No. 1 through the intermediate plate.

50. Using a magnet and screwdriver, install the interlock pin into the intermediate plate.

51. Install the interlock pin into the shaft hole.

52. Install fork shaft No. 3 to the reverse shift arm through the intermediate plate.

53. Install the reverse shift head into fork shaft No. 5.

54. Insert fork shaft No. 5 to the intermediate plate and put in the reverse shift head to the shift fork No. 3.

55. Using a magnetic finger and screwdriver, install the locking ball into the reverse shift head hole.

56. Shift hub sleeve No. 3 to the 5th speed position.

57. Place shift fork No. 3 into the groove of hub sleeve No. 3 and install fork shaft No. 4 to shift fork No. 3 and reverse shift arm.

58. Using a magnet and screwdriver, install the locking ball into the intermediate plate and insert fork shaft No. 4 to the intermediate plate.

59. Check the interlock by positioning the shift fork shaft No. 1 to the 1st speed position.

60. Fork shafts No. 2, No. 3, No. 4 and No. 5 should not move.

61. Using a pin punch and a hammer, drive in new slotted spring pins in each shift fork, reverse shift arm and reverse shift head.

62. Install two fork shaft E-rings.

63. Apply liquid sealer to the screw plugs.

64. Install the locking balls, springs and screw plugs with a Torx bit and tighten to 14 ft. lbs. torque.

NOTE: Install the short spring into the tower of the intermediate plate.

65. Remove the intermediate plate from the vise.

66. Remove the bolts, nuts, washers and gasket.

CASE INSTALLATION

1. Align each bearing outer race, each fork shaft end and reverse idler gear with the holes in the case and install the case on the intermediate plate. If necessary, tap on the case with a plastic hammer.

2. Install two new bearing snaprings.

3. Install front bearing retainer with a new gasket.

4. Apply liquid sealer to the bolts.

5. Install and tighten the bolts to 12 ft. lbs. torque.

6. Install the new gasket to the intermediate plate.

7. Install the adapter housing.

8. Install and tighten the adapter bolts to 27 ft. lbs. torque.

9. Install the shift lever housing.

10. Insert the shift lever into the adapter and shift lever housing.

11. Install and tighten shift lever housing bolt with a lock plate to 28 ft. lbs. torque. Lock the lock plate.

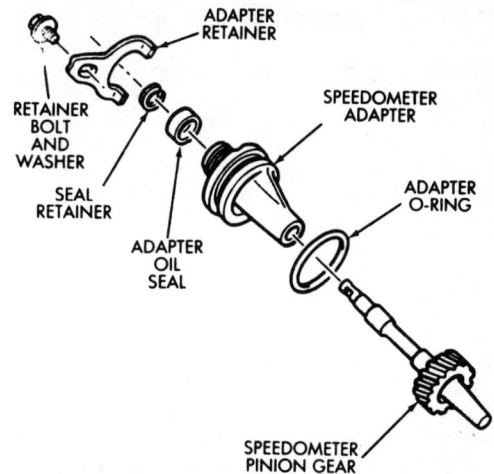

AX5 speedometer gear assembly

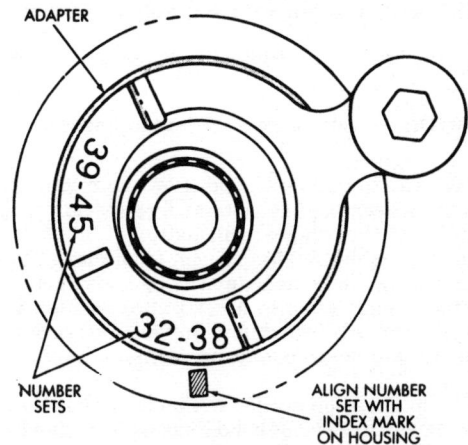

Indexing the speedometer gears on the AX5

12. Install and tighten the adapter screw plug to 13 ft. lbs. torque.

13. Apply liquid sealer to the plug.

14. Install the locking ball, spring and screw plug and tighten the plug to 14 ft. lbs. torque.

15. Check to see that the input shaft and output shafts rotate smoothly.

16. Check to see that shifting can be done smoothly to all positions.

17. Install the black restrict pin on the reverse gear/5th gear side.

18. Install the remaining pin and tighten the pins to 20 ft. lbs. torque.

19. Install the shift lever retainer with a new gasket and tighten the bolts to 13 ft. lbs. torque.

20. Install the back-up light switch and tighten to 27 ft. lbs. torque.

21. Install the clutch housing and tighten the bolts to 27 ft. lbs. torque.

BA10/5 Overhaul

The BA10/5 is a French designed Peugeot transmission in which 5th gear is overdrive. All fasteners and measurements are metric.

NOTE: The service tools necessary for overhaul can be purchased in a service tool kit at your Jeep dealer's parts department.

DISASSEMBLY

1. Mount the transmission in a holding fixture or equivalent and drain the lubricant.
2. Remove the speedometer driven gear socket set screw and remove the driven gear socket from the rear extension housing.
3. Position the transmission in the vertical position with the rear upward. Set the gear selectors in the neutral position. Remove the selection lever return spring.
4. Remove the 5th gear cover plate and its gasket. Remove the extension housing bolts.
5. Place an extractor plate tool or equivalent to the rear housing, using the three bolts of the cover plate. Remove the rear housing from the transmission.
6. Remove the 5th speed gear from the mainshaft with an appropriate puller. The rear bearing will be removed as part of the 5th speed gear.
7. Remove the 5th speed drive gear shim washer, the spacer washer, the 5th gear and its needle bearing from the 5th gear stub shaft.
8. Mark the direction of rotation of the 5th/reverse synchronizer and the position of the cage and hub, in relation to each other.
9. Engage the 5th gear and drive the 5th/reverse selector fork roll pin from the fork and rail.

WARNING: Do not damage the mating surface of the housing

10. Reset the rail to the neutral position and remove the 5th/reverse synchronizer cage and selector fork assembly, the synchronizer hub and the 5th gear subshaft.
11. Disengage the selection lever finger from the selector forks spindles. Remove the intermediate housing and retaining bolts.
12. Place the transmission in a horizontal position with the right side up. Remove the clutch fork and release bearing assembly from the front of the transmission. Remove the clutch housing and bolts.
13. Remove the six Allen headed screws from the rear bearing thrust plate. Remove the right hand housing retaining bolts and the housing.
14. Remove the countershaft from the exposed left hand housing.

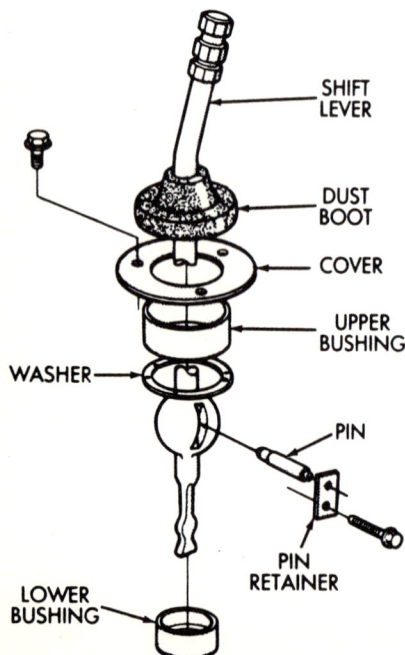

Removing the intermediate shaft bearing from the BA 10/5

NOTE: Mark and set aside the bearing races, if to be used again.

15. Lift the input and mainshaft assembly from the case as a unit. Do not separate while removing.
16. Separate the input and the mainshaft when the assembly is on a work bench. Remove the needle bearing from the bore of the input shaft. Hold the mainshaft in the 3rd gear position.
17. To disassemble the input shaft, remove the circlips, the spring washer and press the bearing from the shaft. Do not lose the adjusting shims.
18. To disassemble the mainshaft, mark the direction of rotation of the 3rd/4th synchronizer cage in relation to the synchronizer hub.

NOTE: Mark the components with a sharp piece of brass welding rod.

19. Remove the front synchronizer ring, the circlip and the spring washer from the front of the mainshaft.
20. Remove the 3rd/4th gear synchronizer hub and 3rd gear with an appropriate puller.
21. Invert the mainshaft and lock in a holding device such as a vise. Unscrew the reverse driven gear locknut from the mainshaft. Remove the reverse gear from the shaft.

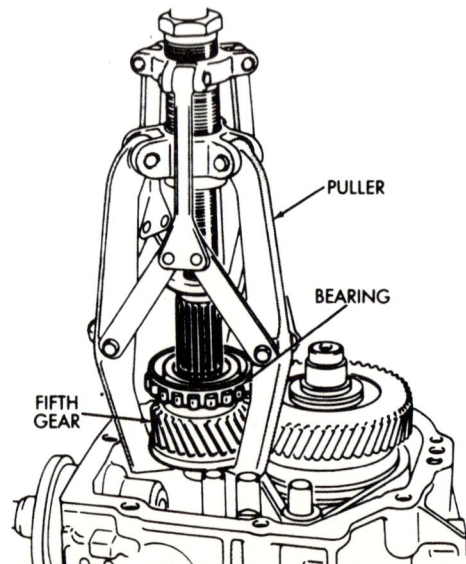

Shift lever removal from the BA 10/5

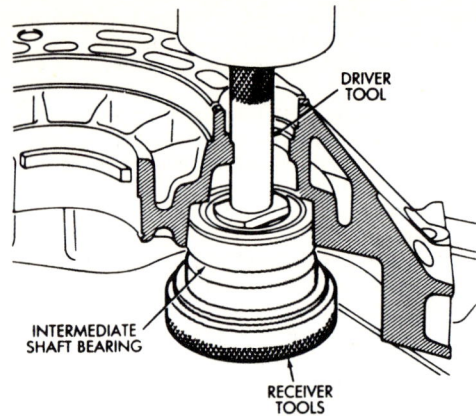

Removing the 5th gear and bearing from the BA 10/5

22. Position the mainshaft and the remaining gears on a press bench and press the mainshaft through the rear bearing. Remove the gear components from the mainshaft in the following order:

  a. Rear bearing.
  b. Bearing spacer.
  c. Adjustment shim washer.
  d. 1st speed driven gear.
  e. Needle bearing.
  f. 1st gear bushing.
  g. 1st/2nd gear synchronizer and hub.
  h. 2nd speed driven gear.
  i. Needle bearing.

23. Using a press, remove the front bearing and the shim adjusting washers, Remove the rear bearing.

24. To disassemble the rear housing, the ball bearing, the oil seal and the bearing race is pressed from the housing. Locate the shim washer in the race bore.

WARNING: The press ram should be no bigger than 24mm in diameter.

Selector Forks and Shift Rail Removal

1. Remove the spring and the 4th speed locking ball from the rear bearing bore of the half case.

2. Move the 4th gear fork and rail to the engaged position and drive the roll pins from the 1st/2nd and 3rd/4th selector forks. Return the 3rd/4th shift rail to the neutral position.

3. Remove the 1st/2nd locking ball plug, spring and ball from the passage on the center of the outer side of the case.

4. Pivot the 3rd/4th selector rail and remove the 1st/2nd rail from the case. Remove the fork.

5. Remove the reverse gear locking ball plug. spring and ball from the opposite side from the 3rd/4th locking ball passage.

6. Disengage the reverse fork and rail assembly and withdraw it from the case.

7. Remove the 3rd/4th shifting rail and remove the fork, interlock pin, the ball and the interlock plunger from the bearing bore.

8. Remove the roll pin from the reverse fork and rail. Push the shifting rail outward. Remove the fork.

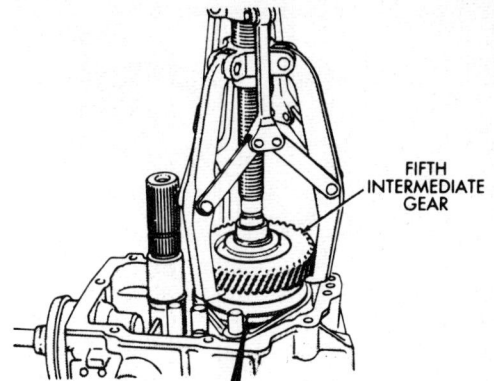

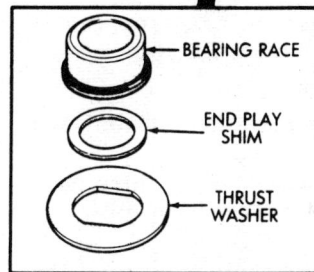

Removing the 5th intermediate gear from the BA 10/5

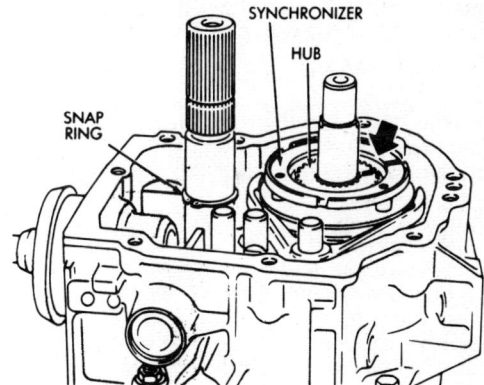

Maring the synchronizer and hub on the BA 10/5

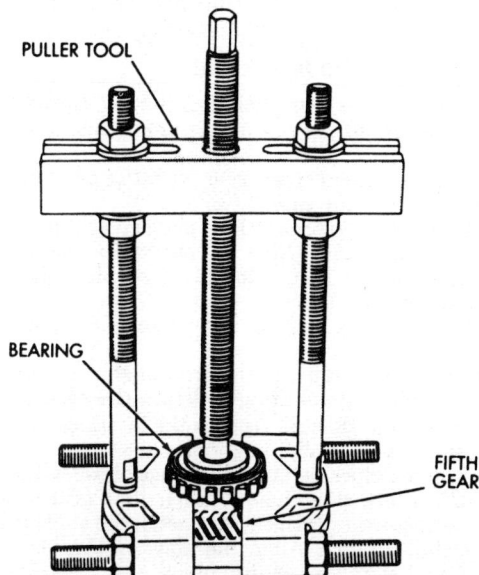

Removing the bearing from the 5th gear from the BA 10/5

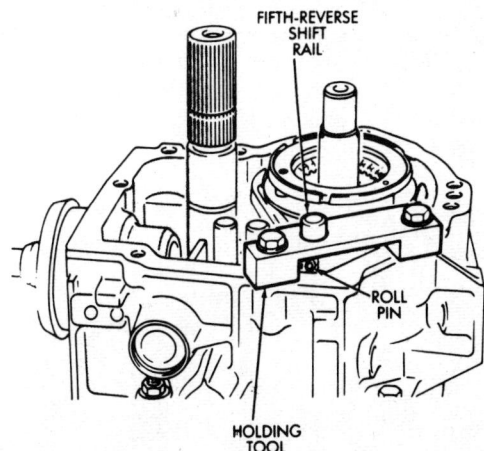

Removing the shift rail pin from the BA 10/5

7 DRIVE TRAIN

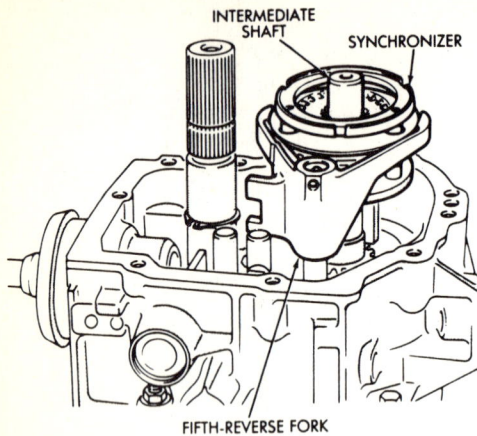

Removing the 5th-reverse fork, synchronizer and intermediate shaft from the BA 10/5

Engagement Lever in the Intermediate Cover

Remove the circlip and washer. Remove the engagement lever spindle and recover the washer, spring, thrust plate and the operating fingers from the inside of the housing.

Clutch Housing

The thrust guide sleeve can be pressed from the housing after the oil seal has been removed.

ASSEMBLY

General

Always replace the following components when overhauling the transmission:
  a. Shaft circlips.
  b. Spring washers.
  c. Roll pins.
  d. Mainshaft nut.
  e. Output and input seals.
  f. O-ring on speedometer driven bushing.
  g. Thrust washers.
  h. Necessary gaskets.
  i. Sealing compound to mating surfaces

Engagement Levers

1. Install the lever spindle while fitting the operating fingers, the spring thrust plate, the spring and the washer.
2. Install the washer and the circlip on the lever spindle, exposed on the side of the intermediate housing.

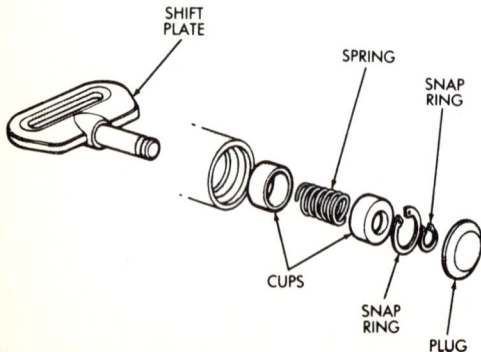

Removing the shift plate from the BA 10/5

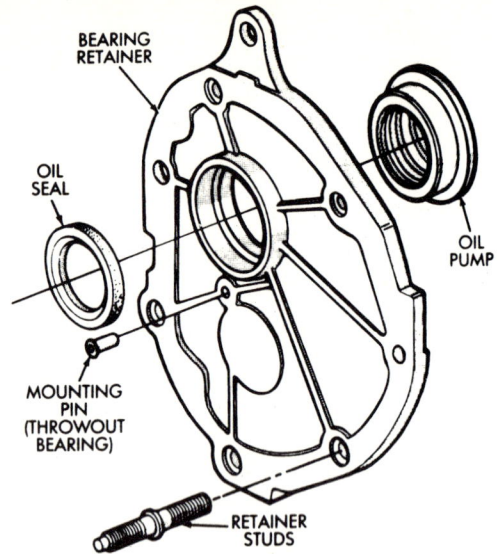

Bearing retainer and components for the BA 10/5

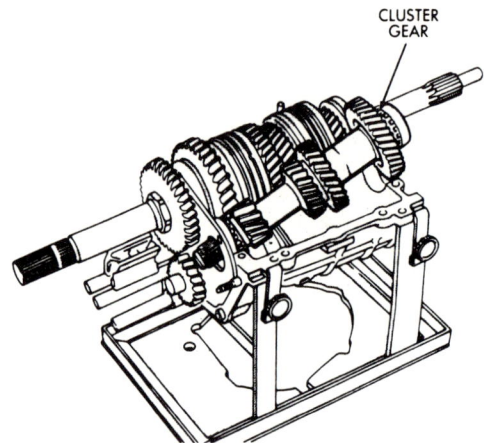

Removing the cluster gear from the BA 10/5

Forks, Shift Rails and Interlocks

1. Install the reverse gear sliding shift rail into the case and into the bracket. Align the roll pin holes and install a new roll pin.
2. Install the reverse sliding gear with the 5th/reverse fork and shifting rail into the case.
3. Install in the reverse interlock passage on the outside of the case, a ball and spring. Coat the plug with a sealer and install it into the passage. Torque to 114 in. lbs. Move the shift rail to the neutral position.
4. Install the 3rd/4th and 5th/reverse interlock plunger into its passage in the case.
5. Install an interlock pin in the 3rd/4th shift rail and retain with grease. Install the shifting rail into the case. Align the holes in the fork and the rail. Be sure the interlock pin is in the correct position and install a new roll pin.
6. Install an interlock ball in the passage between the 3rd/4th and the 1st/2nd fork rails. Install the 1st/2nd gear fork in the case with the boss towards the front. Install the shifting rail and engage the shifting fork. Align the holes in the fork and the shifting rail and install a new roll pin.
7. Install an interlock ball and spring in the 3rd/4th/2nd/1st interlock passage. Coat the plug with sealer and install it in the passage. Torque to 9.5 ft. lbs.

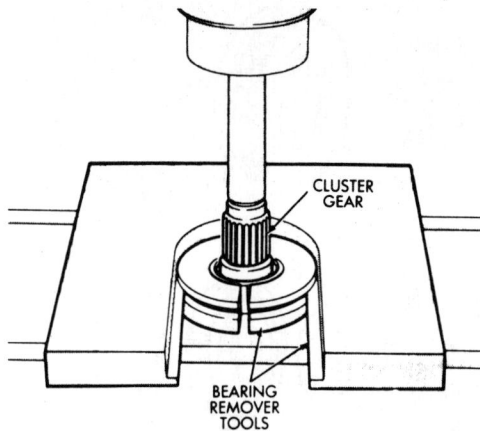

Removing the cluster gear bearings from the BA 10/5

8. Install the ball and spring in the passage of the rear bearing bore, for the 3rd/4th shifting rail.

Preparing the Input Shaft and Mainshaft for Adjustment

INPUT SHAFT

Press the front bearing on the input shaft with the snapring groove to the front. Do not install any shims.

MAINSHAFT

1. Install the 2nd speed gear and its needle bearing, the 1st/2nd synchronizer hub, the 1st speed gear spacer and washer. Install the rear bearing along with a new circlip.

NOTE: The bearing will have to be pressed on the shaft. Do not exceed 3 metric tons pressure after the bearing is seated.

2. Install the adjustment shim removed during the disassembly, the spacer and a new nut on the shaft following the rear bearing. Tighten the nut to 39 ft. lbs.

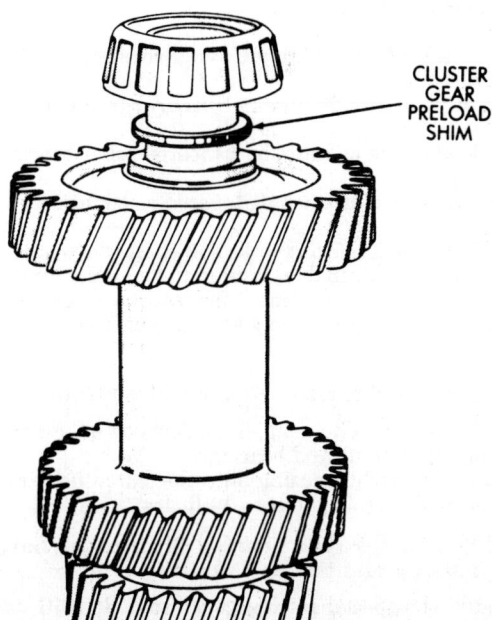

Cluster gear preload shim on the BA 10/5

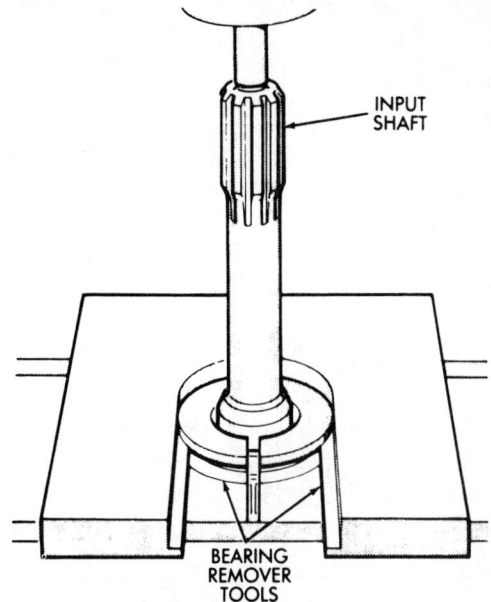

Removing the input shaft bearing from the BA 10/5

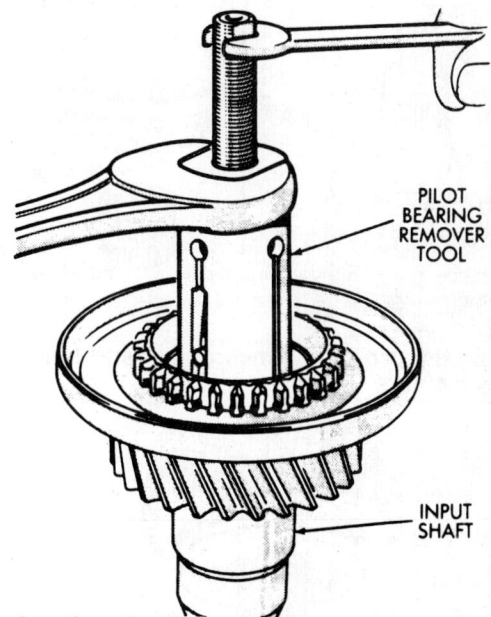

Removing the pilot bearing from the BA 10/5

COUNTERSHAFT

Press the new bearings onto the countershaft, beginning with the rear bearing and then the front

CLUTCH HOUSING

1. Install the thrust guide sleeve and new circlip in its groove.

NOTE: Do not install an oil seal in the housing until all adjustments are completed.

2. Verify the front and rear faces of the clutch housing are parallel with the use of a dial indicator. If out of parallel more than 0.10mm, replace the housing.

Adjustments Prior to Assembly

Five adjustments are necessary, prior to the complete assembly of the transmission. The adjustments are as follows:

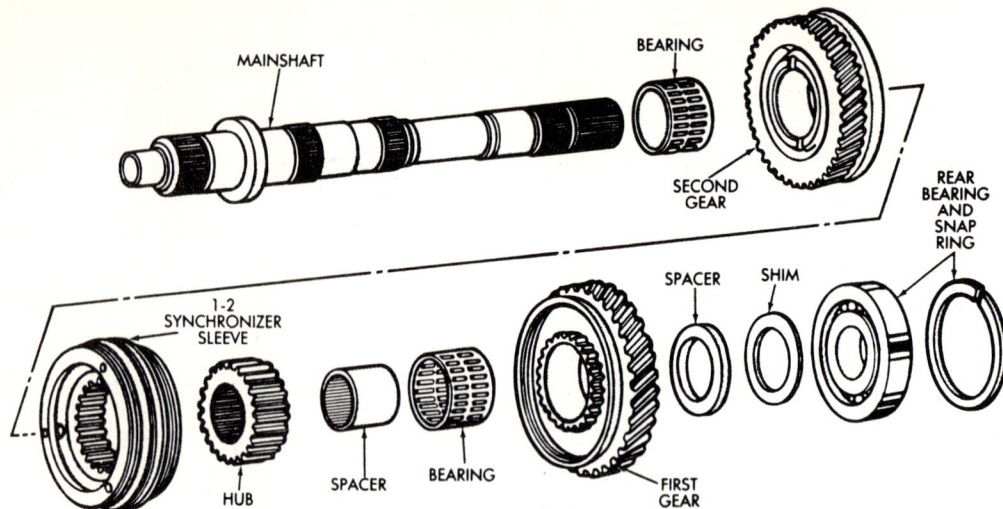

Removing the 1st-2nd gears from the BA 10/5

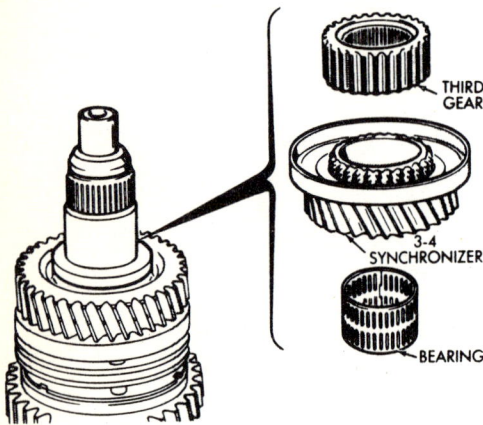

Removing the 3rd gear synchronizer and bearing from the BA 10/5

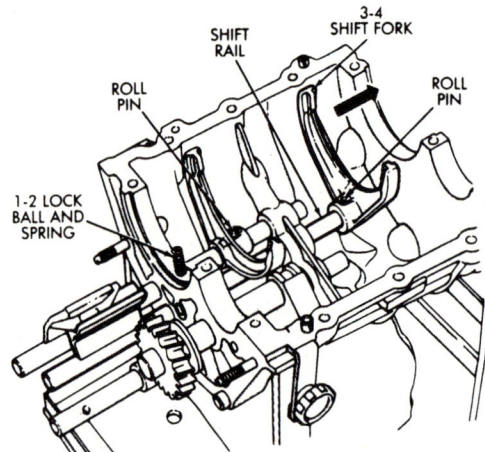

Removing 1st-2nd lock ball and shift fork roll pins from the BA 10/5

1. Position of the 4th gear synchronizer cone (prior to assembly).
2. Positon of the 2nd gear synchronizer cone (prior to assembly).
3. Preload of the countershaft tapered roller bearings (prior to assembly).
4. Preload of the mainshaft tapered roller bearings (during assembly).
5. End play of the 5th/reverse sub-shaft (during assembly).

Special tools are available through the manufacturer and other sources for the measurement procedures needed during the adjustment phases. References will be made to the special tools needed.

Position of the 4th Gear Synchronizer Cone

1. Place the clutch housing, front down on a flat surface, and install the input shaft and bearing.
2. Install the right housing onto the clutch housing and secure with two bolts. Tighten to 14 ft. lbs.

WARNING: Be sure the bearing fits into the bore of the clutch housing and the half housing.

3. Install the special setting gauge tool, 80314G or equivalent, into the clutch housing, in place of the countershaft front bearing.

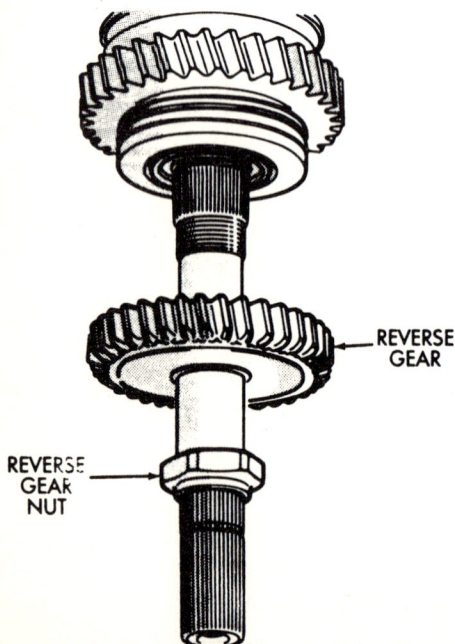

Removing reverse gear from the BA 10/5

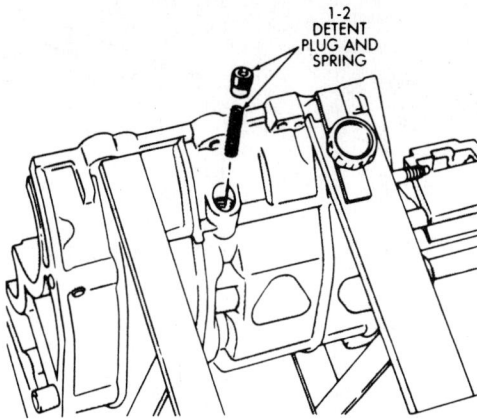

Removing 1st-2nd detent plug and spring from the BA 10/5

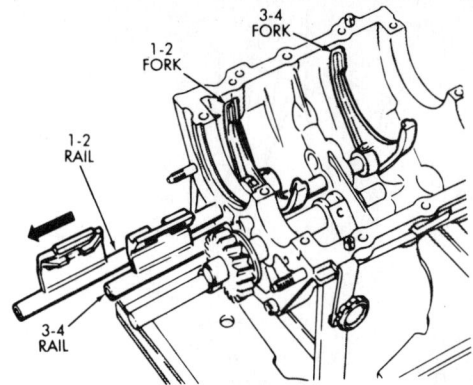

Removing 5th-reverse detent plug, spring and ball from the BA 10/5

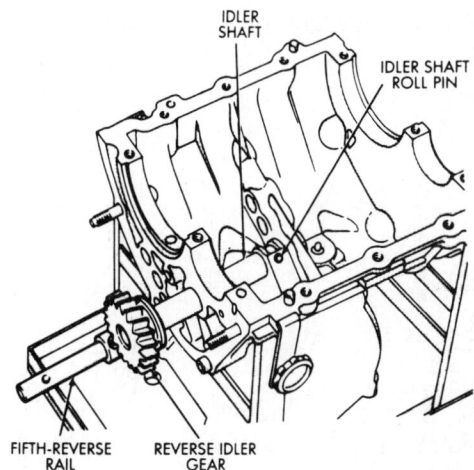

Shift component chart for the BA 10/5

Removing the shift forks and rails from the BA 10/5

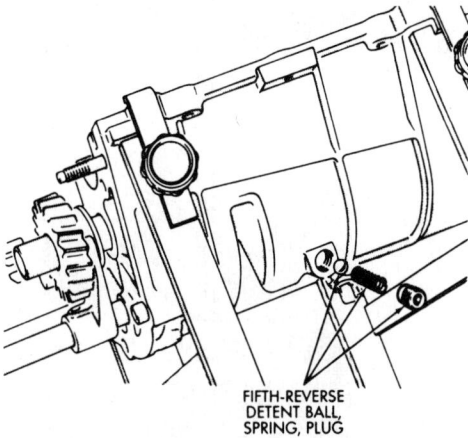

Removing the idler gear and 5th-reverse shift rail from the BA 10/5

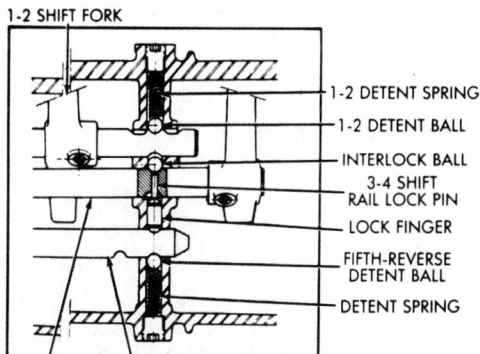

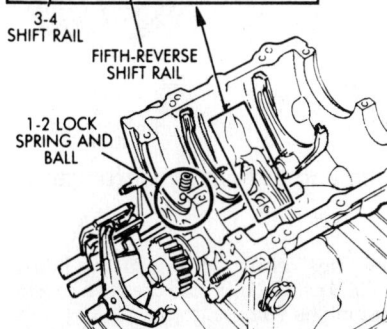

4. Place the dial indicator and the special tool base, 80310FZ or equivalent, on the top of the setting tool.

5. Align the dial indicator stem on the edge of the synchronizer cone and rotate the input shaft one complete turn and obtain an average reading. Adjust the indicator to zero at the average reading point on the cone edge.

6. Reposition the dial indicator and base on the setting gauge block and record the measurement.

7. The measurement obtained represents the thickness of shims needed between the input shaft and the front bearing, minus 0.5mm and rounded off to the nearest 0.05mm.

Example:
- Indicator reading 1.12mm
- Minus 0.50mm
- Result 0.62mm
- Rounded off to nearest 0.05mm = 0.60mm

Therefore, a shim pack of 0.60mm is needed between the input shaft and the front bearing to properly position the 4th speed synchronizer cone, in this hypothetical example.

NOTE: Shims are available in steps of 0.05mm, from 0.15mm to 0.50mm.

Position of the 2nd Gear Synchronizer Cone

1. Install the needle bearing into the bore of the input shaft. Install the mainshaft and prepared components into the input shaft, with the rear bearing seated in its bore on the right hand housing.

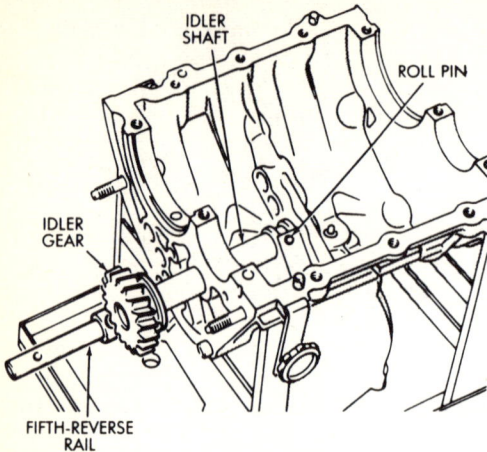

Installing the idler shaft, gear and 5th reverse rail on the BA 10/5

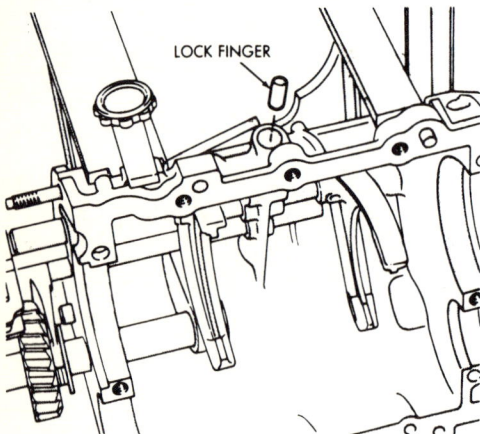

Installing the lock finger on the BA 10/5

NOTE: Be sure the bearing circlip is installed in the half housing groove.

2. Install the larger setting tool, 80314K or equivalent, into the countershaft front bearing bore of the clutch housing.

3. Install a longer stem, 80310J or equivalent, onto the indicator. Place the dial indicator and special base on the top of the right hand half housing, in a position to touch both the setting tool and the 2nd synchronizer cone.

4. With the dial indicator set to zero, position the stem on the top of the setting tool.

5. Reposition the indicator stem to the edge of the 2nd synchronizer cone and record the measurement reading. The amount of movement noted, represents the thickness of shims needed between the 1st gear spacer and the rear bearing, plus 0.50mm, rounded off to the nearest 0.05mm.

Example:
- Indicator reading 2.51mm
- Plus 0.50mm
- Result 3.01mm
- Rounded off to nearest 0.05mm = 3.00mm

6. Remove the input and mainshaft assemblies. Separate the clutch housing and the half housing.

Preload of the Countershaft Tapered Roller Bearings

1. Place the left half housing in the support stand or equivalent. Install the countershaft and its bearings into the housing.

2. Place the right housing in place on the left housing, making sure the dowel pins are in position.

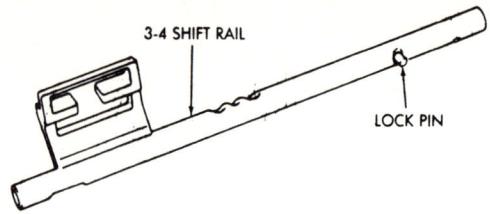

Installing the lock pin in the 3rd-4th shift rail on the BA 10/5

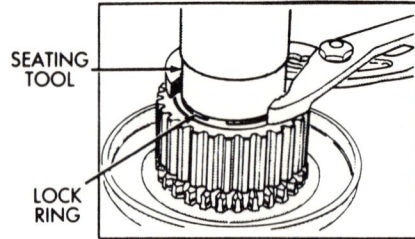

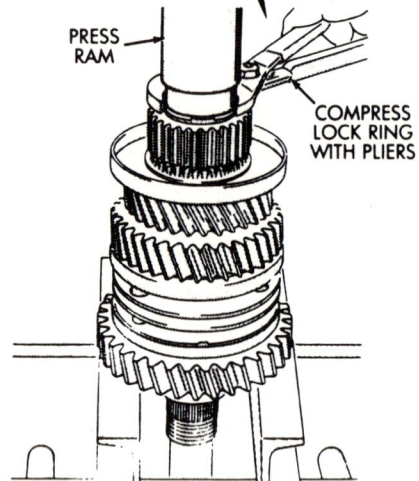

Setting the lock ring on the BA 10/5

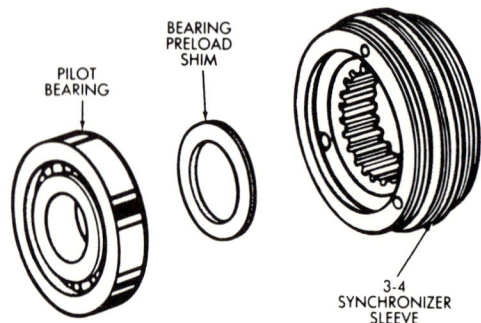

Pilot bearing, shim and 3rd-4th synchronizer sleeve on the BA 10/5

3. Install two bearings center bolts into the housings and hand tighten the bolts. Install bolts into the rear bearing thrust place and hand tighten the bolts.

4. Position the transmission with the front of the housings upward. Apply a downward pressure to the countershaft front

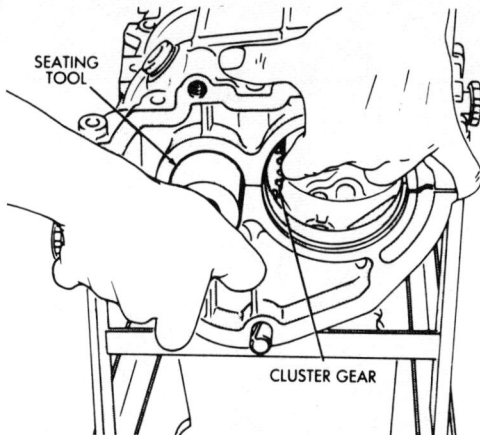

Seating the cluster bearings on the BA 10/5

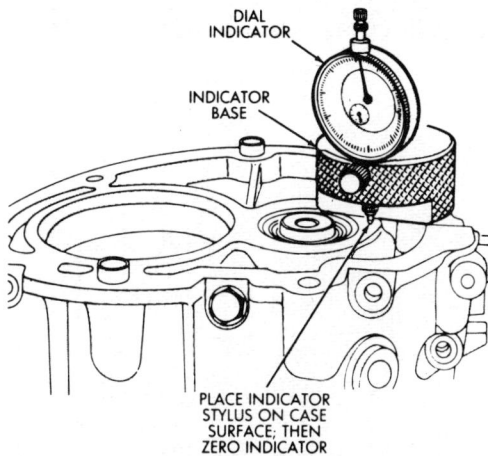

Zeroing the dial indicator on the BA 10/5

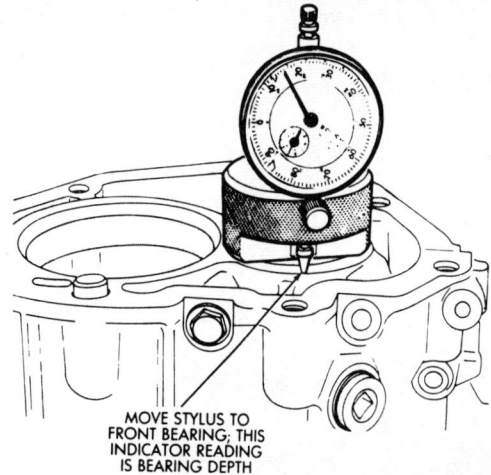

Measuring the cluster gear front bearing depth on the BA 10/5

Installing the cluster bearing preload shim on the BA 10/5

bearing while rotating the countershaft, in order to seat the bearings.

5. Tighten the bearing center bolts and the bearing thrust plate bolts to 7 ft. lb.

6. By placing the dial indicator on the end of the countershaft and rotating it one complete turn with the stem contacting the housings, the run-out between the outer race and the face of the half housings must not exceed 0.03mm.

7. Should the run-out exceed the specifications, the race must be realigned by tapping with a mallet. Should the countershaft be difficult to turn, the bearing center bolts and the bearing thrust plate bolts must be loosened and retightened and the run-out rechecked.

8. If the run-out is within specifications, set the dial indicator with a short stem, to number 2 and zero, with the stem resting on the side of the outer race. Move the indicator until the stem is contacting the face of the housing and record the measurement of the movement. Add 0.10mm to the results for the preload of the bearings and round the results to the nearest 0.05mm.

Example:
- Housing reading 4.27mm
- Bearing reading (preset) 2.00mm
- Result 2.27mm
- Plus preload 0.10mm
- Shim pack needed 2.37mm
- Rounded off to nearest 0.05mm = 2.35mm

NOTE: Shims are available from 2.15mm to 3.30mm, in increments of 0.05mm.

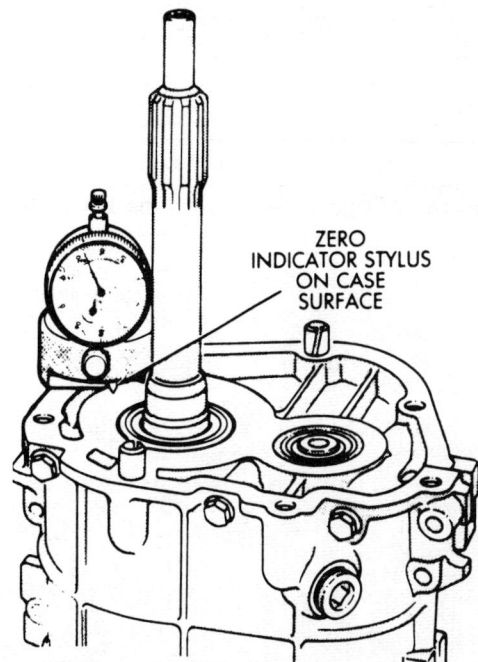

Zeroing the dial indicator on the BA 10/5

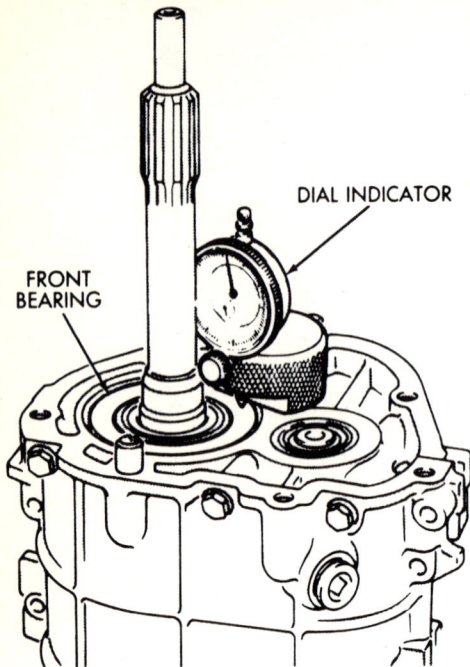

Measuring the input shaft bearing depth on the BA 10/5

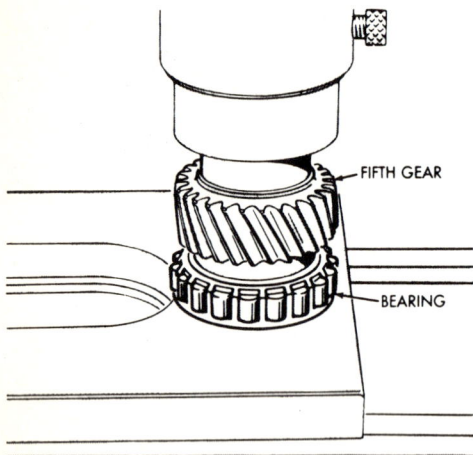

Installing the 5th gear bearing on the BA 10/5

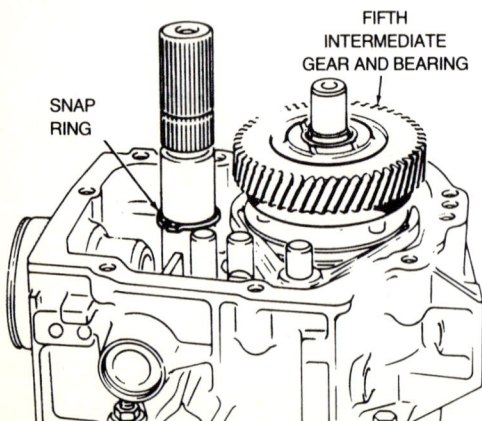

Installing the snapring and 5th intermediate gear on the BA 10/5

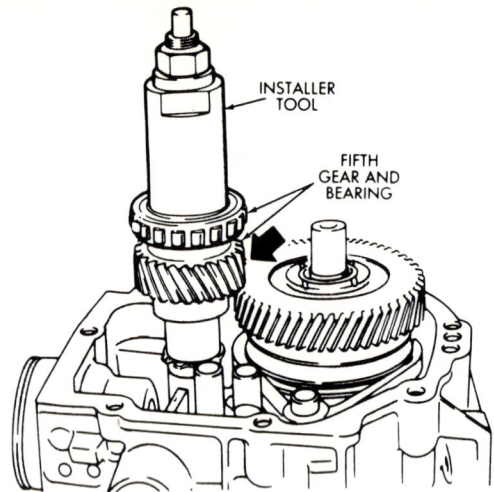

Installing the 5th gear and bearing on the BA 10/5

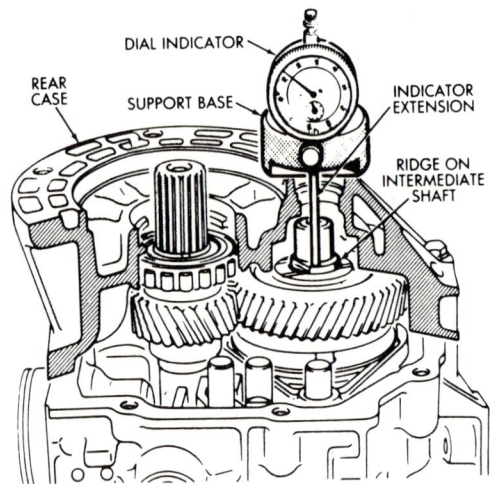

Zero and lock the dial indicator on the BA 10/5

9. Remove the countershaft and the front bearing from the countershaft.

10. Install the predeterminded shim with the chamfer facing the gear, between the gear and the bearing.

11. Press the bearing in place.

NOTE: The 4th and 5th adjustments are made during the assembly of the transmission as noted in the adjustment list.

FINAL ASSEMBLY

Input Shaft

1. Remove the bearing from the shaft and install the predetermined shim pack between the shaft and the bearing.

2. Reinstall the bearing, with the groove for the large circlip to the front. Press the bearing into position and install the spring washer and circlip on the input shaft. Be sure the circlip engages the groove completely.

Mainshaft

1. Remove the rear bearing and shims from the mainshaft, used in the measurement check.

2. Install in the following order, from the rear to the front of the shaft, the components as listed:

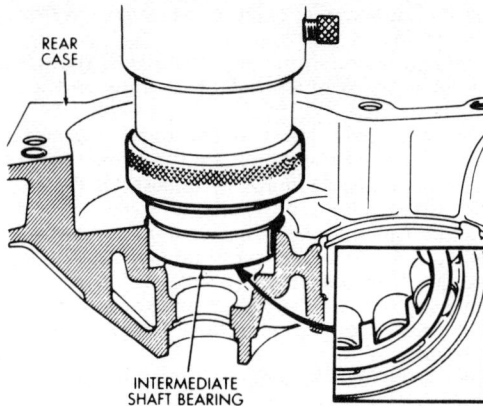

Installing the intermediate shaft bearing on the BA 10/5

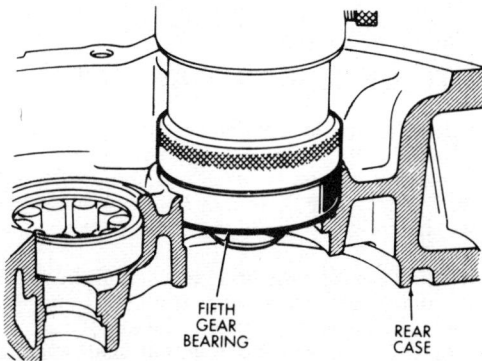

Installing the 5th gear bearing on the BA 10/5

a. The 2nd gear and its needle bearing (31mm wide).
b. The synchronizer hub and cage.
c. The 1st gear and needle bearing (29mm wide).
d. The spacer and adjusting shim (from measurement check).
e. The bearing spacer.
f. Press the bearing onto the shaft with the circlip to the rear.
g. Install the reverse driven gear with the plain face to the rear.
h. Install a new nut and tighten to 39 ft. lbs. and lock the collar to the shaft.
3. Invert the mainshaft and install the following in order.
a. 3rd gear along with its needle bearing (31mm wide).
b. 3rd-4th synchronizer hub.
4. Install a new spring washer and circlip on the end of the mainshaft. Be sure circlip is in the groove.
5. Insert the needle bearing in the bore of the input shaft, fit the 3rd-4th synchronizer cage and assemble the input and mainshaft together. Set the syncrhonizers to neutral positions.

Installing Gear Trains Into Transmission Cases

1. Be sure the reverse synchronizer is in the neutral position and the 3rd-4th gear locking ball and spring is in position in the bearing bore.
2. Install the input and mainshaft assembly into the left half housing, engaging the selector forks with the synchronizer cages.
3. Install the outer races to the countershaft and install the countershaft into the housing. Be sure the teeth of both gear trains mate properly.
4. Be sure the alignment dowel pins are in place. Coat the

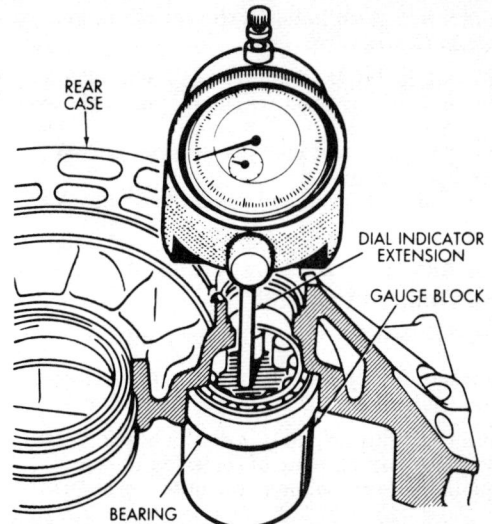

Measuring the intermediate shaft endplay on the BA 10/5

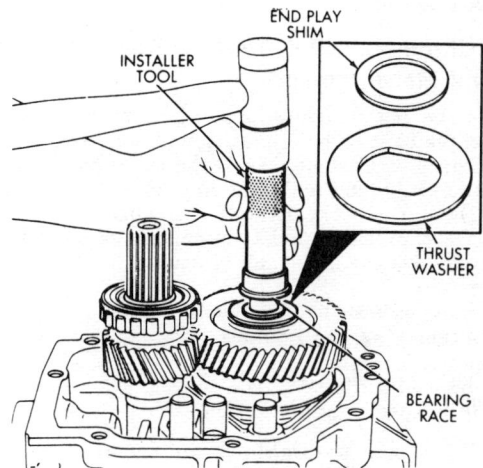

Installing the washer, shim and race on the BA 10/5

mating surfaces of the half housings with sealer and position the right hand housing onto the left. Position the rear bearing thrust plate.
5. Install the six bearing bolts and tighten to 3 ft. lbs. Install the two thrust plate bolts and tighten to 7 ft. lbs.
6. Be sure the oil seal is installed in the clutch cover and install the clutch cover in place on the transmission housing.
7. Tighten the seven securing bolts to 19 ft. lbs. Rotate the input shaft during the clutch cover retaining bolt tightening.
8. Loosen the six bearing bolts in the housings and tap the half housing with a rubber mallet while rotating the input shaft. Retighten the six bearing bolts to 11 ft. lbs.
9. Install the six assembly bolts in the housing and torque to 7 ft. lbs.
10. Invert the transmission with the rear of the mainshaft upward. Be sure the alignment dowels are in place, coat the mating surfaces of the transmission housing and the intermediate housing with sealing compound and install the intermediate housing in place, while engaging the finger in the selector fork detents.
11. Tighten the five nuts and two bolts to 13 ft. lbs. Place the 5th/reverse spindle to the 5th gear position.
12. Install the 5th/reverse stub shaft and the 5th/reverse synchronizer hub.

NOTE: If a new hub is used, the marking groove should be towards the reverse gear.

13. As a unit, install the 5th/reverse synchronizer cage and the selector fork, bringing together the marks on the synchronizer hub and the cage.

14. Align the holes and install a new roll pin. Set the unit to the neutral position.

15. Install the 5th gear and needle bearing, along with the spacer.

5th Gear Assembly and Preparation for Measurements 4th and 5th

1. Press the bearing on to the gear pinion.

NOTE: When new parts are used, match the pinion gear to the mainshaft by green or yellow color.

2. Place the 5th gear pinion with the bearing fitted, on a hot plate and place a small piece of solder on the pinion. When the solder melts, place the pinion gear onto the shaft.

NOTE: A drift may have to be used to seat the gear.

3. Remove the alignment dowels from the rear housing and set them aside for later use. Install a shim pack, 4.0mm thick and the bearing race, into the rear housing.

Preload of the Mainshaft Tapered Roller Bearing (Number 4 Measurement)

1. Place the rear housing in position on the transmission case. Fit three bolts to hold housing and hand tighten.

2. Rotate the mainshaft, loosen the three bolts of the rear housing and retighten hand tight only.

3. A gap will exist between the two housings. Measure the gap to check for parallelism and to calculate the shim thickness needed to preload the mainshaft bearing.

Example:
- Thickness of basic shim 4.00mm
- Measurement of gap 1.85mm
- Difference 2.15mm
- Plus preload 0.10mm
- Shim thickness required 2.25mm

NOTE: Shims are available in increments of 0.05mm from 1.5mm to 2.95mm.

4. Remove the rear housing and remove the rear bearing outer race and the basic 4.00mm shim pack.

Warner T4 Overhaul

CASE DISASSEMBLY

1. Drain the transmission lubricant.

2. Use a pin punch and hammer to remove the offset lever-to-shift rail roll pin.

3. Remove the extension housing adapter. Remove the housing and the offset lever as an assembly.

4. Remove the detent ball and spring from the offset lever. Remove the roll pin from the extension housing or adapter.

5. Remove the countershaft rear thrust bearing and race.

6. Remove the transmission cover and shift fork assembly. Two of the transmission cover bolts are alignment type dowel pins. Mark their location so that they may be reinstalled in their original locations.

7. Remove the reverse lever to reverse lever pivot bolt C-clip.

8. Remove the reverse lever pivot bolt. Remove the reverse lever and fork as an assembly.

9. Mark the position of the front bearing cap to case, then remove the bearing cap bolts and cap.

10. Remove the front bearing race and the shims from the bearing cap. Use a small pry bar and remove the front seal from the bearing cap.

11. Rotate the main drive gear shaft until the flat portion of the gear faces the countershaft, then remove the main drive gear shaft assembly.

12. Remove the thrust bearing and 15 roller bearings from the clutch shaft. Remove the output shaft bearing race. Tap the output shaft with a plastic hammer to loosen it if necessary.

13. Tilt the output shaft assembly upward and remove the assembly from the case.

14. Carefully pull off the countershaft rear bearing with the proper puller after marking the position for reinstallation.

15. Move the countershaft rearward and tilt it upward to remove it from the transmission case. Remove the countershaft bearing spacer.

16. Remove the reverse idler shaft roll pin, then remove the reverse idler shaft and gear.

17. Press off the countershaft front bearing. Use the appropriate pullers and remove the bearing from the main drive gear shaft.

18. Remove the extension housing or adapter oil seal and remove the back-up light switch from the case.

OUTPUT SHAFT DISASSEMBLY

1. Remove the thrust bearing washer from the front of the output shaft.

2. Scribe matchmarks on the hub and sleeve of the 3rd-4th synchronizer so that these parts may be reassembled properly.

3. Remove the 3rd-4th synchronizer blocking ring, sleeve and hub as an assembly.

4. Remove the insert springs and the inserts from the 3rd-4th synchronizer and separate the sleeve from the hub.

5. Remove the 3rd speed gear from the shaft.

6. Remove the 2nd speed gear to output shaft snapring, the tabbed thrust washer and the 2nd speed gear from the shaft.

7. Use an appropriate puller and remove the the output shaft bearing.

8. Remove the 1st gear thrust washer, the roll pin, the 1st speed gear and the blocking ring.

9. Scribe matchmarks on the 1st-2nd synchronizer sleeve and the output shaft.

10. Remove the insert spring and the inserts from the 1st-reverse sliding gear, then remove the gear from the output hub.

OUTPUT SHAFT ASSEMBLY

1. Coat the output shaft and the gear bores with transmission lubricant.

2. Align the matchmarks and install the 1st-2nd synchronizer sleeve on the output shaft hub.

3. Install the three inserts and two springs into the 1st-reverse synchronizer sleeve.

NOTE: The tanged end of each spring should be positioned on the same insert but the open face of each spring should be opposite each other.

4. Install the blocking ring and the 2nd speed gear onto the output shaft.

5. Install the tabbed thrust washer and 2nd gear snapring in the output shaft; be sure that the washer is properly seated in the notch.

6. Install the blocking ring and the 1st speed gear onto the output shaft, then install the 1st gear roll pin.

7. Press the rear bearing onto the shaft.

8. Install the remaining components onto the output shaft: The 1st gear thrust washer. The 3rd speed gear. The 3rd-4th synchronizer hub inserts and the sleeve (the hub offset must face forward). The thrust bearing washer on the rear of the countershaft.

T-4 exploded view

COVER & FORKS DISASSEMBLY

1. Place the selector arm plates and the shift rail centered in the Neutral position.
2. Rotate the shift rail counterclockwise until the selector arm disengages from the selector arm plates; the selector arm roll pin should now be accessible.
3. Pull the shift rail rearward until the selector contacts the 1st-2nd shift fork.
4. Use a 3/16 in. pin punch and remove the selector arm roll pin and the shift rail.
5. Remove the shift forks, the selector arm, the roll pin and the interlock plate.

6. Remove the shift rail oil seal and O-ring.
7. Remove the nylon inserts and the selector arm plates from the shift forks.

NOTE: Mark the position of the parts so that they may be properly installed.

COVER & FORK ASSEMBLY

1. Attach the nylon inserts to the selector arm plates and through the shift forks.
2. If removed, coat the edges of the shift rail plug with sealer and install the plug.

3. Coat the shift rail and the rail bores with petroleum jelly, then slide the shift rail into the cover until the end of the rail is flush with the inside edge of the cover.

4. Position the 1st-2nd shift fork into the cover; with the offset of the shift fork facing the rear of the cover. Push the shift rail through the fork. The 1st-2nd fork is the larger of the two forks.

5. Position the selector arm and the C-shaped interlock plate into the cover, then push the shift rail through the arm. The widest part of the interlock plate must face away from the cover and the selector arm roll pin must face downward, toward the rear of the cover.

6. Position the 3rd-4th shift fork into the cover with the fork offset facing the rear of the cover. The 3rd-4th shift selector arm plate must be positioned under the 1st-2nd shift fork selector arm plate.

7. Push the shift rail through the 3rd-4th shift fork and into the front cover rail bore.

8. Rotate the shift rail until the forward selector arm plate faces away from parallel to the cover.

9. Align the roll pin holes of the selector arm and the shift rail and install the roll pin. The roll pin must be installed flush with the surface of the selector arm to prevent selector arm plate to pin interference.

10. Install the O-ring into the groove of the shift rail oil seal, then install the oil seal carefully after lubricating it.

CASE ASSEMBLY

1. Apply a coat of Loctite® 601, or equivalent, to the outer cage of the front countershaft bearing, then press the bearing into the bore until it is flush with the case.

2. Apply petroleum jelly to the tabbed countershaft thrust washer and install the washer with the tab engaged in the corresponding case depression.

3. Tip the transmission case on end and install the countershaft into the front bearing bore.

4. Install the rear countershaft bearing spacer and coat the rear bearing with petroleum jelly. Install the rear countershaft bearing using the appropriate tools. The rear bearing is properly installed when 3mm is extended beyond the case surface.

5. Position the reverse idler into the case (the shift lever groove must face rearward) and install the reverse idler shaft into the case. Install the shaft retaining pin.

6. Install the output shaft assembly into the transmission case.

7. Install the main drive gear bearing onto the main drive shaft using the appropriate tools. Coat the roller bearings with petroleum jelly and install them in the main drive gear recess. Install the thrust bearing and race.

8. Install the 4th gear blocking ring onto the output shaft. Install the rear output shaft bearing race.

9. Install the main drive gear assembly into the case, engaging the 3rd-4th synchronizer blocking ring.

10. Install a new seal in the front bearing cap and in the rear extension or adapter.

11. Install the front bearing into the front bearing cap but do not (at this time) install the shims. Temporarily install the cap to the transmission without applying sealer.

12. Install the reverse lever, the pivot pin (coat the threads with non-hardening sealer) and the retaining C-clip. Be sure the reverse lever fork is engaged with the reverse idler gear.

13. Coat the countershaft rear bearing race and the thrust bearing with petroleum jelly, then install the parts into the extension housing or adapter.

14. Temporarily install the extension housing or adapter without sealer, tighten the retaining bolts slightly, but do not final torque them.

15. Turn the transmission case on end and mount a dial indicator in position to measure output shaft end play. To eliminate end play the bearings must be preloaded from 0.025-0.130mm. Check the endplay. Select a shim pack that measures 0.025-0.130mm thicker than the measured endplay.

16. Install the shims under the front bearing cap. Apply a 1/8 in. bead of RTV sealer to the cap. Align the reference marks and install the cap on the front of the transmission. Torque the mounting bolts to 15 ft. lbs. Recheck the output shaft end play, none should exist. Adjust if necessary.

17. Remove the extension housing or adapter. Move the shift forks and synchronizer sleeves to their neutral position. Apply a 1/8 in. bead of RTV sealer to the cover to case mounting surface. Align the forks with their sleeves and carefully lower the cover into position. Center the cover and install the alignment dowels. Install the mounting bolts and tighten to 9 ft. lbs.

NOTE: The offset lever to shift rail roll pin must be position vertically; if not, repeat Step 17.

18. Apply a 1/8 in. bead of RTV sealer to the extension housing or adapter and install over the output shaft.

NOTE: The shift rail must be positioned so that it just enters the shift cover opening.

19. Install the detent spring into the offset lever and place the steel ball into the Neutral guide plate detent. Apply pressure to the detent spring and offset lever, then slide the offset lever on the shift rail and seat the extension housing or adapter plate against the transmission case. Install and tighten the mounting bolts to 25 ft. lbs.

20. Install the roll pin into the offset lever and shift rail. Install the damper sleeve in the offset lever. Coat the back up lamp switch threads with sealer and install the switch, tighten to 15 ft. lbs.

Warner T5 Overhaul

CASE DISASSEMBLY

1. Remove drain bolt on transmission case and drain lubricant.

2. Thoroughly clean the exterior of the transmission assembly.

3. Using pin punch and hammer, remove roll pin attaching offset lever to shift rail.

4. Remove extension housing-to-transmission case bolts and remove housing and offset lever as an assembly.

NOTE: Do not attempt to remove the offset lever while the extension housing is still bolted in place. The lever has a positioning lug engaged in the housing detent plate which prevents moving the lever far enough for removal.

5. Remove detent ball and spring from offset lever and remove roll pin from extension housing or offset lever.

6. Remove plastic funnel, thrust bearing race and thrust bearing from rear of countershaft.

NOTE: The countershaft rear thrust bearing, bearing washer and plastic funnel may be found inside the extension housing.

7. Remove bolts attaching transmission cover and shift fork assembly and remove cover.

NOTE: Two of the transmission cover attaching bolts are alignment-type dowel bolts. Note the location of these bolts for assembly reference.

8. Using a punch and hammer, drive the roll pin from the 5th gearshift fork while supporting the end of the shaft with a block of wood.

9. Remove 5th synchronizer gear snapring, shift fork, 5th

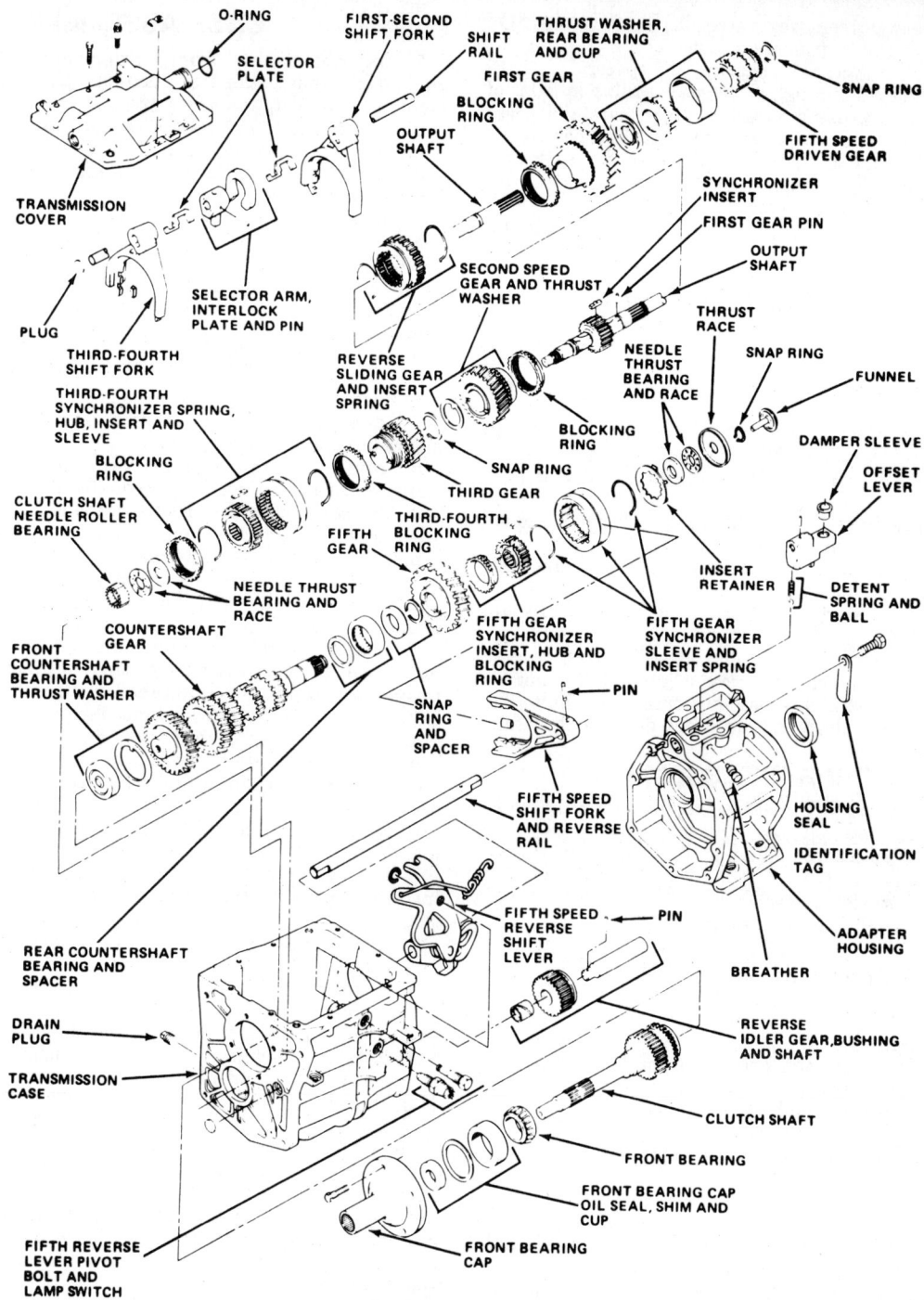

T-5 exploded view

gear synchronizer sleeve, blocking ring and 5th speed drive gear from rear of countershaft.

10. Remove snapring from 5th speed driven gear.

11. Using a hammer and punch, mark both bearing cap and case for assembly reference.

12. Remove front bearing cap bolts and remove front bearing cap. Remove front bearing race and end play shims from front bearing cap.

13. Rotate drive gear until flat surface faces countershaft and remove drive gear from transmission case.

14. Remove reverse lever C-clip and pivot bolt.

15. Remove mainshaft rear bearing race and then tilt mainshaft assembly upward and remove assembly from transmission case.

16. Unhook overcenter link spring from front of transmission case.

17. Rotate 5th gear-reverse shift rail to disengage rail from reverse lever assembly. Remove shift rail from rear of transmission case.

18. Remove reverse lever and fork assembly from transmission case.

19. Using hammer and punch, drive roll pin from forward end

7 DRIVE TRAIN

of reverse idler shaft and remove reverse idler shaft, rubber "O" ring and gear from the transmission case.

20. Remove rear countershaft snapring and spacer.
21. Insert a brass drift through drive gear opening in front of transmission case and, using an arbor press, carefully press countershaft rearward to remove rear countershaft bearing.
22. Move countershaft assembly rearward, tilt countershaft upward and remove from case. Remove countershaft front thrust washer and rear bearing spacer.
23. Remove countershaft front bearing from transmission case using an arbor press.

MAINSHAFT DISASSEMBLY

1. Remove thrust bearing washer from front end of mainshaft.
2. Scribe reference mark on 3rd-4th synchronizer hub and sleeve for reassembly.
3. Remove 3rd-4th synchronizer blocking ring, sleeve, hub and 3rd gear as an assembly from mainshaft.
4. Remove snapring, tabbed thrust washer, and 2nd gear from mainshaft.
5. Remove 5th gear with Tool J-22912-01 or its equal and arbor press. Slide rear bearing off mainshaft.
6. Remove 1st gear thrust washer, roll pin, 1st gear and synchronizer ring from mainshaft.
7. Scribe reference mark on 1st-2nd synchronizer hub and sleeve for reassembly.
8. Remove synchronizer spring and keys from 1st-reverse sliding gear and remove gear from mainshaft hub. Do not attempt to remove the 1st-2nd-reverse hub from mainshaft. The hub and shaft are assembled and machined as a matched set.

DRIVE GEAR DISASSEMBLY

1. Remove bearing race, thrust bearing, and roller bearings from cavity of drive gear.
2. Using Tool J-22912-01 or its equal and arbor press, remove bearing from drive gear.
3. Wash parts in a cleaning solvent.
4. Inspect gear teeth and drive shaft pilot for wear.

DRIVE GEAR ASSEMBLY

1. Using Tool J-22912-01 or its equal with an arbor press, install bearing on drive gear.
2. Coat roller bearings and drive gear bearing bore with grease. Install roller bearings into bore of drive gear.
3. Install thrust bearing and race in drive gear.

MAINSHAFT ASSEMBLY

1. Coat mainshaft and gear bores with transmission lubricant.
2. Install 1st-2nd synchronizer sleeve on mainshaft hub aligning marks made at disassembly.
3. Install 1st-2nd synchronizer keys and springs. Engage tang end of each spring in same synchronizer key but position open end of springs opposite of each other.
4. Install blocker ring and 2nd gear on mainshaft. Install tabbed thrust washer and 2nd gear retaining snapring on mainshaft. Be sure washer tab is properly seated in mainshaft notch.
5. Install blocker ring and 1st gear on mainshaft. Install 1st gear roll pin and then 1st gear thrust washer.
6. Slide rear bearing on mainshaft.
7. Install 5th speed gear on mainshaft using Tool J-22912-01 and arbor press. Install snapring on mainshaft.
8. Install 3rd gear, 3rd-4th synchronizer assembly and thrust bearing on mainshaft. Synchronizer hub offset must face forward.

CASE ASSEMBLY

1. Coat countershaft front bearing bore with Loctite 601 or equivalent, and install front countershaft bearing flush with facing of case using an arbor press.
2. Coat countershaft tabbed thrust washer with grease and install washer so tab engages depression in case.
3. Tip transmission case on end and install countershaft in front bearing bore.
4. Install countershaft rear bearing spacer. Coat countershaft rear bearing with grease and install bearing using Tool J-29895 and sleeve J-33032, or its equivalent. The bearing when correctly installed will extend beyond the case surface 3mm.
5. Position reverse idler gear in case with shift lever groove facing rear of case and install reverse idler shaft from rear of case. Install roll pin in idler shaft.
6. Install assembled mainshaft in transmission case. Install rear mainshaft bearing race in case.
7. Install drive gear in case, and engage in 3rd-4th synchronizer sleeve and blocker ring.
8. Install front bearing race in front bearing cap. Do not install shims in front bearing cap at this time.
9. Temporarily install front bearing cap.
10. Install 5th speed-reverse lever, pivot bolt and retaining clip. Coat pivot bolt threads with nonhardening sealer. Be sure to engage reverse lever fork in reverse idler gear.
11. Install countershaft rear bearing spacer and retaining snapring.
12. Install 5th speed gear on countershaft.
13. Insert 5th speed-reverse rail in rear of case and install in to reverse 5th speed lever. Rotate rail during installation to simplify engagement with lever. Connect spring to front of case.
14. Position 5th gear shift fork on 5th gear synchronizer assembly and install synchronizer on countershaft and shift fork on shift rail. Make sure roll pin hole in shift fork and shift rail are aligned.
15. Support 5th gear shift rail and fork on a block of wood and install roll pin.
16. Install thrust race against 5th speed synchronizer hub and install snapring. Install thrust bearing against race on countershaft. Coat both bearing and race with petroleum jelly.
17. Install lipped thrust race over needle-type thrust bearing and install plastic funnel into hole in end of countershaft gear.
18. Temporarily install extension housing and attaching bolts. Turn transmission case on end, and mount a dial indicator on extension housing with indicator on the end of mainshaft.
19. Rotate mainshaft and zero dial indicator. Pull upward on mainshaft until end play is removed and record reading. Mainshaft bearings require a preload of 0.025-0.130mm. To set preload, select a shim pack measuring 0.025-0.130mm greater than the dial indicator reading recorded.
20. Remove front bearing cap and front bearing race. Install necessary shims to obtain preload and reinstall bearing race.
21. Apply a 1/8 in. bead of RTV sealant, #732 or equivalent, on case mating surface of front bearing cap. Install bearing cap aligning marks made during disassembly and torque bolts to specification.
22. Remove extension housing.
23. Move shift forks on transmission cover and synchronizer sleeves inside transmission to the neutral position.
24. Apply a 1/8 in. bead of RTV sealant, #732 or equivalent, on cover mating surface of transmission.
25. Lower cover onto case while aligning shift forks and synchronizer sleeves. Center cover and install the 2 dowel bolts. Install remaining bolts and torque to specification. The offset lever to shift rail roll pin hole must be in the vertical position after cover installation.
26. Apply a 1/8 in. bead of RTV Sealant, #732 or equivalent, on extension housing to transmission case mating surface.
27. Install extension housing over mainshaft and shift rail to a position where shift rail just enters shift cover opening.

28. Install detent spring into offset lever and place steel ball in neutral guide plate detent. Position offset lever on steel ball and apply pressure on offset lever and at the time seat extension housing against transmission case.

29. Install extension housing bolts and torque to specification.
30. Align and install roll pin in offset lever and shift rail.
31. Fill transmission to its proper level with lubricant.

CLUTCH

Troubleshooting Basic Clutch Problems

Problem	Cause
Excessive clutch noise	Throwout bearing noises are more audible at the lower end of pedal travel. The usual causes are: • Riding the clutch • Too little pedal free-play • Lack of bearing lubrication A bad clutch shaft pilot bearing will make a high pitched squeal, when the clutch is disengaged and the transmission is in gear or within the first 2″ of pedal travel. The bearing must be replaced. Noise from the clutch linkage is a clicking or snapping that can be heard or felt as the pedal is moved completely up or down. This usually requires lubrication. Transmitted engine noises are amplified by the clutch housing and heard in the passenger compartment. They are usually the result of insufficient pedal free-play and can be changed by manipulating the clutch pedal.
Clutch slips (the car does not move as it should when the clutch is engaged)	This is usually most noticeable when pulling away from a standing start. A severe test is to start the engine, apply the brakes, shift into high gear and SLOWLY release the clutch pedal. A healthy clutch will stall the engine. If it slips it may be due to: • A worn pressure plate or clutch plate • Oil soaked clutch plate • Insufficient pedal free-play
Clutch drags or fails to release	The clutch disc and some transmission gears spin briefly after clutch disengagement. Under normal conditions in average temperatures, 3 seconds is maximum spin-time. Failure to release properly can be caused by: • Too light transmission lubricant or low lubricant level • Improperly adjusted clutch linkage
Low clutch life	Low clutch life is usually a result of poor driving habits or heavy duty use. Riding the clutch, pulling heavy loads, holding the car on a grade with the clutch instead of the brakes and rapid clutch engagement all contribute to low clutch life.

The purpose of the clutch is to disconnect and connect engine power at the transmission. A car at rest requires a lot of engine torque to get all that weight moving. An internal combustion engine does not develop a high starting torque (unlike steam engines), so it must be allowed to operate without any load until it builds up enough torque to move the car. Torque increases with engine rpm. The clutch allows the engine to build up torque by physically disconnecting the engine from the transmission, relieving the engine of any load or resistance. The transfer of engine power to the transmission (the load) must be smooth and gradual; if it weren't, drive line components would wear out or break quickly. This gradual power transfer is made possible by gradually releasing the clutch pedal. The clutch disc and pressure plate are the connecting link between the engine and transmission. When the clutch pedal is released, the disc and plate contact each other (clutch engagement), physically joining the engine and transmission. When the pedal is pushed in, the disc and plate separate (the clutch is disengaged), disconnecting the engine from the transmission.

The clutch assembly consists of the flywheel, the clutch disc, the clutch pressure plate, the throwout bearing and fork, the actuating linkage and the pedal. The flywheel and clutch pressure plate (driving members) are connected to the engine crankshaft and rotate with it. The clutch disc is located between the flywheel and pressure plate, and splined to the transmission shaft. A driving member is one that is attached to the engine and transfers engine power to a driven member (clutch disc) on the transmission shaft. A driving member (pressure plate) rotates (drives) a driven member (clutch disc) on contact and, in so doing, turns the transmission shaft. There is a circular diaphragm spring within the pressure plate cover (transmission side). In a relaxed state (when the clutch pedal is fully released), this spring is convex; that is, it is dished outward toward the transmission. Pushing in the clutch pedal actuates an attached linkage rod. Connected to the other end of this rod is the throwout bearing fork. The throwout bearing is attached to the fork. When the clutch pedal is depressed, the clutch linkage pushes the fork and bearing forward to contact the diaphragm spring of the pressure plate. The outer edges of the spring are secured to the pressure plate and are pivoted on rings so that when the center of the spring is compressed by the throwout bearing, the outer edges bow outward and, by so doing, pull the pressure plate in the same direction - away from the clutch disc. This action separates the disc from the plate, disengaging the clutch and allowing the transmission to be shifted into another gear. A coil type clutch return spring attached to the clutch pedal arm permits full release of the pedal. Releasing the pedal pulls the throwout bearing away from the diaphragm spring resulting in a reversal of spring position. As bearing pressure is gradually released from the spring center, the outer edges of the spring bow outward, pushing the pressure plate into closer contact with the clutch disc. As the disc and plate move closer together, friction between the two increases and slippage is reduced until, when full spring pressure is applied (by fully releasing the pedal), The speed of the disc and plate are the same. This stops all slipping, creating a direct connection between the plate and disc which results in the transfer of power from the engine to the transmission. The clutch disc is now rotating with the pressure plate at engine speed and, because it is splined to the transmission shaft, the shaft now turns at the same engine speed. Understanding clutch operation can be rather difficult at first; if you're still confused after reading this, consider the following analogy. The action of the diaphragm spring can be compared to that of an oil can bottom. The bottom of an oil can is shaped very much like the clutch diaphragm spring and pushing in on the can bottom and then releasing it produces a similar effect. As mentioned earlier, the clutch pedal return spring permits full release of the pedal and reduces linkage slack due to wear. As the linkage wears, clutch free-pedal travel will increase and free-travel will decrease as the clutch wears. Free-travel is actually throwout bearing lash.

The diaphragm spring type clutches used are available in two different designs: flat diaphragm springs or bent spring. The bent fingers are bent back to create a centrifugal boost ensuring quick re-engagement at higher engine speeds. This design enables pressure plate load to increase as the clutch disc wears and makes low pedal effort possible even with a heavy-duty clutch. The throwout bearing used with the bent finger design is 1¼ in. long and is shorter than the bearing used with the flat finger design. These bearings are not interchangeable. If the longer bearing is used with the bent finger clutch, free-pedal travel will not exist. This results in clutch slippage and rapid wear.

The transmission varies the gear ratio between the engine and rear wheels. It can be shifted to change engine speed as driving conditions and loads change. The transmission allows disengaging and reversing power from the engine to the wheels.

Pressure Plate and Driven Disc

REMOVAL AND INSTALLATION

4-134

----- CAUTION -----

The clutch driven disc contains asbestos, which has been determined to be a cancer causing agent. Never clean clutch surfaces with compressed air! Avoid inhaling any dust from any clutch surface! When cleaning clutch surfaces, use a commercially available brake cleaning fluid.

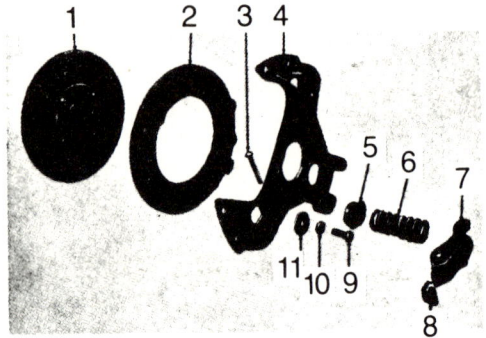

1. Driven plate and hub	7. Release lever
2. Pressure plate	8. Return spring
3. Pivot pin	9. Adjusting screw
4. Bracket	10. Jam nut
5. Spring cup	11. Washer
6. Pressure spring	

4–134 Auburn clutch

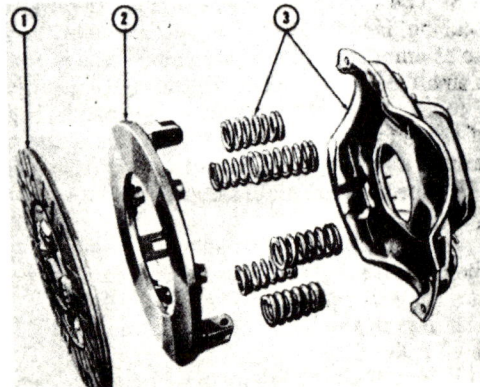

1. Driven plate and hub
2. Pressure plate
3. Backing plate and pressure spring

4–134 Rockford clutch

1. Remove the transmission and transfer case from the vehicle.
2. Remove the flywheel housing.
3. Mark the clutch pressure plate and engine flywheel with a center punch so the clutch assembly may be installed in the same position after adjustments or replacement are complete.
4. Loosen the clutch pressure plate bracket bolts equally, a little at a time, to prevent distortion and relieve the clutch springs evenly. Remove the bolts.
5. Remove the pressure plate assembly (bracket and pressure plate) and driven plate from the flywheel. The driven plate will just be resting on the pressure plate housing since it usually is mounted on the input shaft of the transmission, which has been removed. Be careful that it does not fall down and cause injury.
6. The clutch release bearing (throwout bearing) is lubricated at time of assembly and no attempt should be made to lubricate it. Put a small amount of grease in the pilot bushing.
7. Install the driven plate with the short end of the hub toward the flywheel. Use a spare transmission mainshaft or an aligning arbor to align the pressure plate assembly and the driven plate.
8. Leave the arbor in place while tightening the pressure plate screws evenly a turn or two at a time. Torque the bolts to 25 ft. lbs.
9. Install the flywheel housing. Torque the bolts to 40-50 ft. lbs.
10. Install the transmission and transfer case.

6-225

──────── CAUTION ────────

The clutch driven disc contains asbestos, which has been determined to be a cancer causing agent. Never clean clutch surfaces with compressed air! Avoid inhaling any dust from any clutch surface! When cleaning clutch surfaces, use a commercially available brake cleaning fluid.

1. Remove the transmission and transfer case.
2. Remove the clutch throwout bearing and pedal return spring from the clutch fork.
3. Remove the flywheel housing from the engine.
4. Disconnect the clutch form from the ball stud by forcing it toward the center of the vehicle.
5. Mark the clutch cover and flywheel with a center punch so that the cover an later be installed in the same position on the flywheel. This is necessary to maintain engine balance.
6. Loosen the clutch attaching bolts alternately, one turn at a time to avoid distorting the clutch cover flange, until the diaphragm spring is released.
7. Support the pressure plate and cover assembly while removing the last of the bolts; remove the pressure plate and driven plate from the flywheel.
8. If it is necessary to disassemble the pressure plate assembly, note the position of the grooves on the edge of the pressure plate and cover. These marks must be aligned during assembly

to maintain balance. The clutch diaphragm spring and two pivot rings are riveted to the clutch cover. Inspect the spring, rings and cover for excessive wear or damage. If there is a defect, replace the complete cover assembly.
9. Replace the clutch assembly in reverse order of the removal procedure, taking note of the following:
   a. Use extreme care at all times not to get the clutch driven plate dirty in any way.
   b. Lightly lubricate the inside of the clutch driven plate's spline with a coat of wheel bearing grease. Do the same to the input shaft of the transmission. Wipe off all excess grease so that none will fly off and get onto the driven plate.
   c. Lubricate the throwout bearing collar, the ball stud and the clutch fork with wheel bearing grease.
   d. Use a pilot shaft or a spare transmission main shaft to

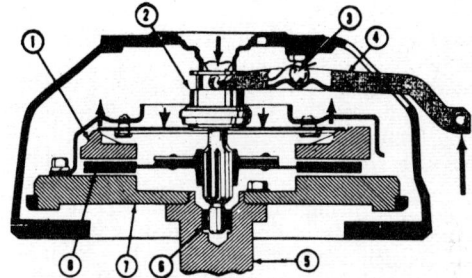

1. Pressure plate
2. Throwout bearing
3. Pivot point
4. Clutch fork
5. Engine crankshaft
6. Pilot bearing
7. Flywheel
8. Driven plate

V6 clutch cutaway

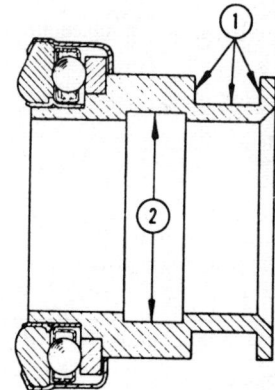

1. Coat this groove
2. Pack this recess

Lubrication points on the 6–225 throwout bearing collar

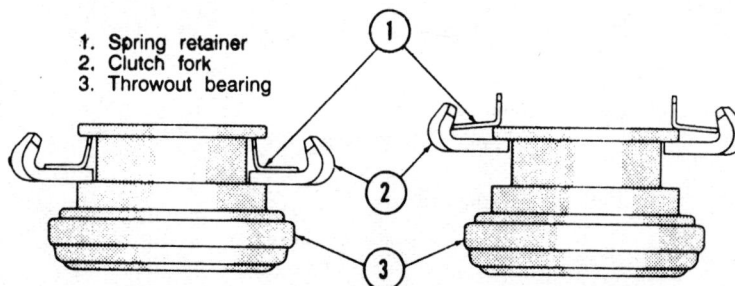

1. Spring retainer
2. Clutch fork
3. Throwout bearing

CORRECT   INCORRECT

6–225 throwout bearing installation

align the driven shaft and the clutch pressure plate when attaching the assembly to the flywheel.

e. Tighten down on the clutch-to-flywheel attaching bolts alternately so that the clutch is drawn squarely into position on the flywheel. Each bolt must be tightened one turn at a time to avoid bending the clutch cover flange. Torque the bolts to 30-40 ft. lbs.

1972-90 except 4-151

─────── **CAUTION** ───────

The clutch driven disc contains asbestos, which has been determined to be a cancer causing agent. Never clean clutch surfaces with compressed air! Avoid inhaling any dust from any clutch surface! When cleaning clutch surfaces, use a commercially available brake cleaning fluid.

1. Remove the transmission.
2. Remove the starter.
3. Remove the throwout bearing and sleeve assembly.
4. Remove the bell housing.
5. Mark the clutch cover, pressure plate and the flywheel with a center punch so that these parts can be later installed in the same position.
6. Remove the clutch cover-to-flywheel attaching bolts. When removing these bolts, loosen them in rotation, one or two turns at a time, until the spring tension is released. The clutch cover is a steel stamping which could be warped by improper removal procedures, resulting in clutch chatter when reused.
7. Remove the clutch assembly from the flywheel.
8. The clutch release bearing (throwout bearing) is lubricated at time of assembly and no attempt should be made to lubricate it. Put a small amount of grease in the pilot bushing.
9. Install the driven plate with the short end of the hub toward the flywheel. Use a spare transmission mainshaft or an aligning arbor to align the pressure plate assembly and the driven plate.
10. Leave the arbor in place while tightening the pressure

plate screws evenly a turn or two at a time. Torque the bolts to 40 ft. lbs.
11. Install the bellhousing. Torque the bolts to 40 ft. lbs.
12. Install the throwout bearing and sleeve assembly.
13. Install the starter.
14. Install the transmission.

4-151

─────── **CAUTION** ───────

The clutch driven disc contains asbestos, which has been determined to be a cancer causing agent. Never clean clutch surfaces with compressed air! Avoid inhaling any dust from any clutch surface! When cleaning clutch surfaces, use a commercially available brake cleaning fluid.

1. Remove the shift lever boot.
2. Remove the shift lever assembly.
3. Raise the vehicle and support it on jackstands.
4. Remove the transmission and transfer case.
5. Remove the slave cylinder-to-clutch housing bolts.
6. Disengage the slave cylinder pushrod from the throwout lever and move the cylinder out of the way.
7. Remove the starter.
8. Remove the throwout bearing.
9. Unbolt and remove the clutch housing.
10. Mark the position of the clutch pressure plate and remove the pressure plate bolts evenly, a little at a time in rotation.
11. Remove the pilot bushing lubricating wick from its bore in the crankshaft and soak the wick in clean engine oil.
12. The clutch release bearing (throwout bearing) is lubricated at time of assembly and no attempt should be made to lubricate it.
13. Install the driven plate with the short end of the hub toward the flywheel. Use a spare transmission mainshaft or an aligning arbor to align the pressure plate assembly and the driven plate.
14. Leave the arbor in place while tightening the pressure

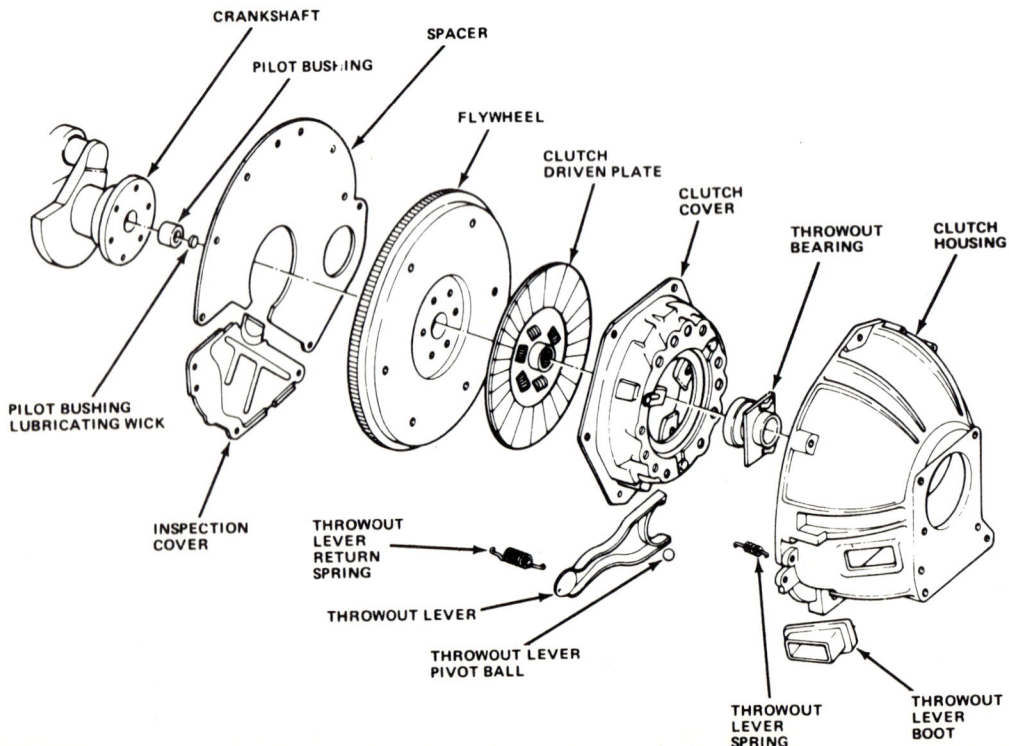

6–232, 6–258 and 8–304 clutch assembly

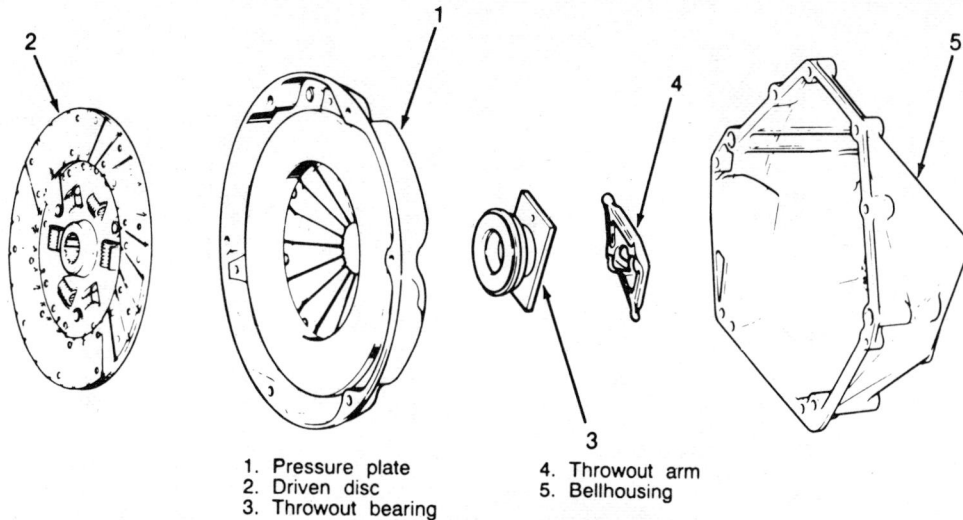

1. Pressure plate
2. Driven disc
3. Throwout bearing
4. Throwout arm
5. Bellhousing

4–150 clutch assembly

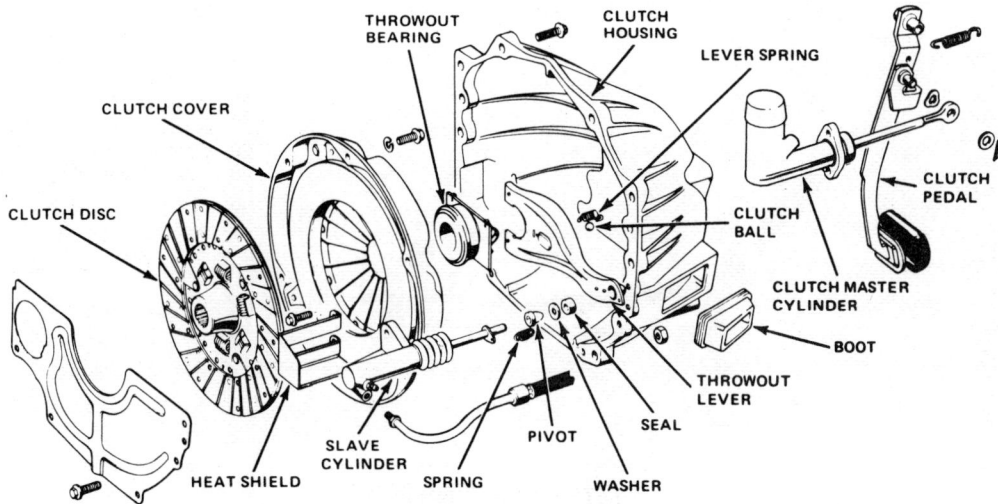

4–151 clutch assembly

plate screws evenly a turn or two at a time. Torque the pressure plate bolts to 23 ft. lbs.

15. Install the pilot bushing lubricating wick in its bore in the crankshaft.

16. Install the clutch housing. Torque the clutch housing to 54 ft. lbs.

17. Install the throwout bearing.

18. Install the starter.

19. Engage the slave cylinder pushrod in the throwout lever.

20. Install the slave cylinder-to-clutch housing bolts.

21. Install the transmission and transfer case. Torque the transmission-to-clutch housing bolts to 54 ft. lbs.; the transfer case-to-transmission bolts to 30 ft. lbs.

22. Lower the vehicle.

23. Install the shift lever assembly.

24. Install the shift lever boot.

CLUTCH LINKAGE ADJUSTMENT

1971

As the clutch facings wear out the free travel of the clutch pedal diminishes. When sufficient wear occurs, the pedal clear-ance must be adjusted to 1-1½ in. (25-38mm). The free pedal clearance is adjusted by lengthening or shortening the clutch fork cable.

To make this adjustment, loosen the jam nut on the cable clevis and lengthen or shorten the cable to obtain the proper clearance at the pedal pad, then tighten the jam nut.

PEDAL HEIGHT ADJUSTMENT

1972 Only

The clutch pedal has an adjustable stop located on the pedal support bracket directly behind the instrument cluster.

Adjust the stop to provide the specified clearance between the top of the pedal pad and the closest point on the bar floor pan. The distance must be 8 in.

CONTROL CABLE ADJUSTMENT

1972

1. Lift up the clutch pedal against the pedal support bracket stop.

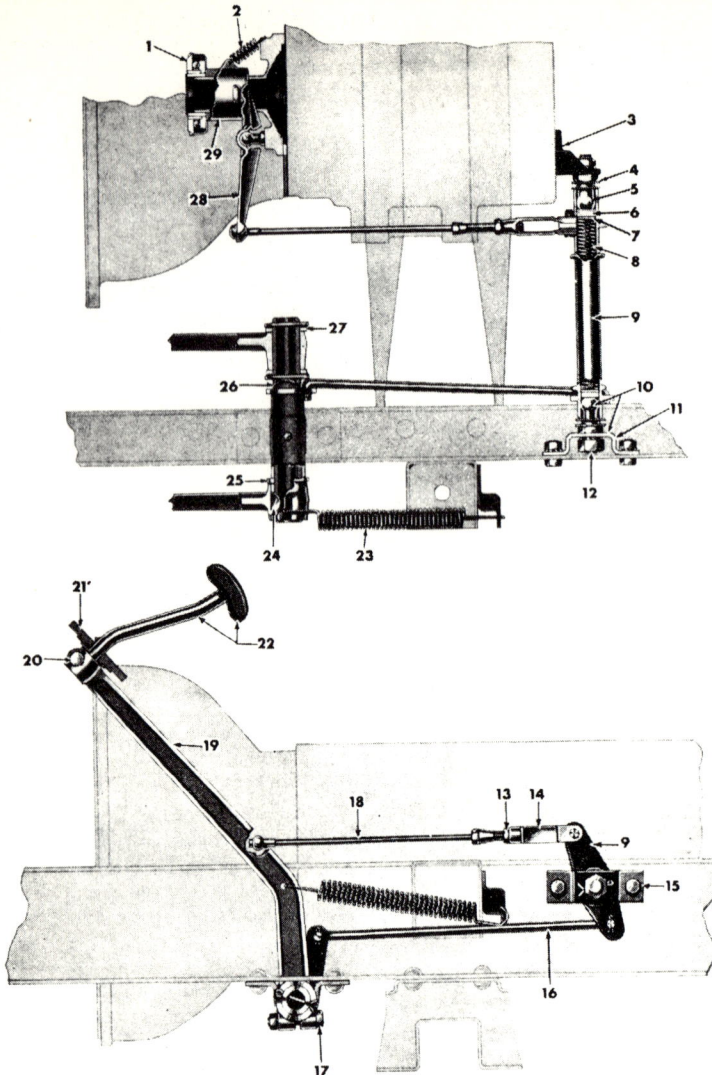

1. Clutch release bearing
2. Carrier spring
3. Bracket
4. Dust seal
5. Ball stud
6. Pad
7. Retainer
8. Control tube spring
9. Control lever and tube
10. Ball stud and bracket
11. Frame bracket
12. Ball stud nut
13. Yoke lock nut
14. Adjusting yoke
15. Bolt
16. Pedal release rod
17. Pedal clamp bolt
18. Control cable
19. Clutch pedal
20. Screw and lockwasher
21. Draft pad
22. Pedal pad and shank
23. Retracting spring
24. Pedal to shaft key
25. Washer
26. Pedal shaft
27. Master cylinder tie bar
28. Control lever
29. Bearing carrier

1971 clutch linkage

2. Unhook the clutch fork return spring.
3. Loosen the ball adjusting nut until some cable slack exists.
4. Adjust the ball adjusting nut until the slack is removed from the cable and the clutch throwout bearing contacts the pressure plate fingers.
5. Back off the ball adjusting nut ¾ of a turn to provide the proper amount of free play. Tighten the jam nut.
6. Hook the clutch fork return spring.

1973-75

1. Adjust the bellcrank outer support bracket to provide approximately ⅛ in. (3mm) of bellcrank end play.
2. Lift up the clutch pedal against the pedal stop.
3. On the clutch push rod (pedal to bellcrank) adjust the lower ball pivot assembly onto or off the rod (as required) to position the bellcrank inner lever parallel to the front face of the clutch housing (slightly forward from vertical).
4. Adjust the clutch fork release rod (bellcrank to release fork) to obtain the maximum specified clutch pedal free play of ¾ in. (19mm) on 1973-74 models and 1 in. (25mm) on 1975 models.

1976-83

NOTE: 4-151 and all 1984 and later models have a non-adjustable hydraulic clutch.

1. Lift the pedal up against the stop.
2. Loosen the release rod adjuster jam nut, under the vehicle.
3. Adjust the pedal free-play to about one inch.
4. Tighten the jam nut.

Clutch Master Cyliner

REMOVAL AND INSTALLATION

1. Raise and support the truck on jackstands.
2. Disconnect the hydraulic line at the cylinder. Cap the line.
3. Unbolt and remove the slave cylinder from the clutch housing.
4. Installation is the reverse of removal. Torque the mounting bolts to 16 ft. lbs. Refill and bleed the system.

OVERHAUL

Through 1986

1. Remove the cover and reservoir cap.
2. Remove and discard the dust boot from the pushrod.
3. Remove the pushrod seal.
4. Remove and discard the pushrod retaining snapring.
5. Remove the pushrod, washer and seal. Discard the seal.

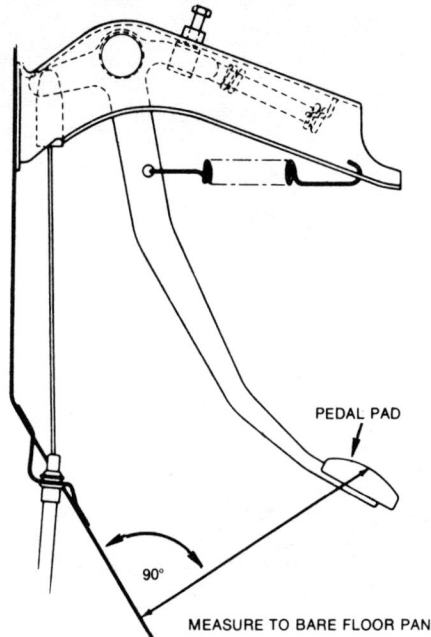

Clutch pedal height adjustment for 1972 only

6. Remove the plunger, valve spring and stem. It may be necessary to tap the assembly out of the bore with a rubber mallet.

7. Compress the valve spring enough to pry off the retainer and remove the spring and stem assembly from the plunger. The retainer tab should be pried upward with a thin screwdriver.

8. Remove the seal from the plunger and discard it.

9. Remove the spring retainer and valve stem from the spring.

10. Remove the valve stem from the retainer and remove the spring washer and stem tip seal from the end of the valve stem. Discard the stem tip seal and the spring washer.

11. Clean all parts thoroughly in a safe brake cleaning solvent. Discard any parts that show signs of wear, pitting or damage. If the core shows signs of excessive wear, deep pitting, severe corrosion or scoring, replace the entire master cylinder. Minor bore imperfections can be corrected by honing.

12. Lubricate the bore with clean brake fluid.

13. Position the new seals on the plunger and valve stem. Make sure that the lip of the plunger seal faces the stem end of the plunger. The should of the stem tip seal should fit into the undercut at the end of the valve stem.

14. Install the new spring washer on the stem.

15. Install the new plastic spring retainer on the stem and over the spring washer. The large end of the retainer faces the end of the stem.

16. Install the valve spring over the stem and the seat spring on the valve stem retainer.

17. Install the assembled valves spring, the retainer and the stem assembly on the plunger.

18. Compress the spring against the plunger. When the end of the stem passes through the valve stem retainer and seats in the small bore in the end of the plunger, bend the retainer tab on the valve stem retainer downward to lock the stem and retainer on the plunger.

19. Lubricate the spring and plunger assembly with clean brake fluid and insert the assembly, spring end first, into the bore.

20. Install the new seal and dust boot on the pushrod.

21. Lubricate the ball end of the pushrod and the lip of the seal and dust boot with clean brake fluid, or the lubricant supplied in the overhaul kit.

22. Insert the pushrod and retaining washer into the bore.

23. Install the new snapring. Properly position the seal and dust boot.

1987-90

1. Remove the cover and reservoir cap.
2. Remove and discard the dust boot from the pushrod.
3. Remove and discard the pushrod retaining snapring.
4. Remove the pushrod and washer.

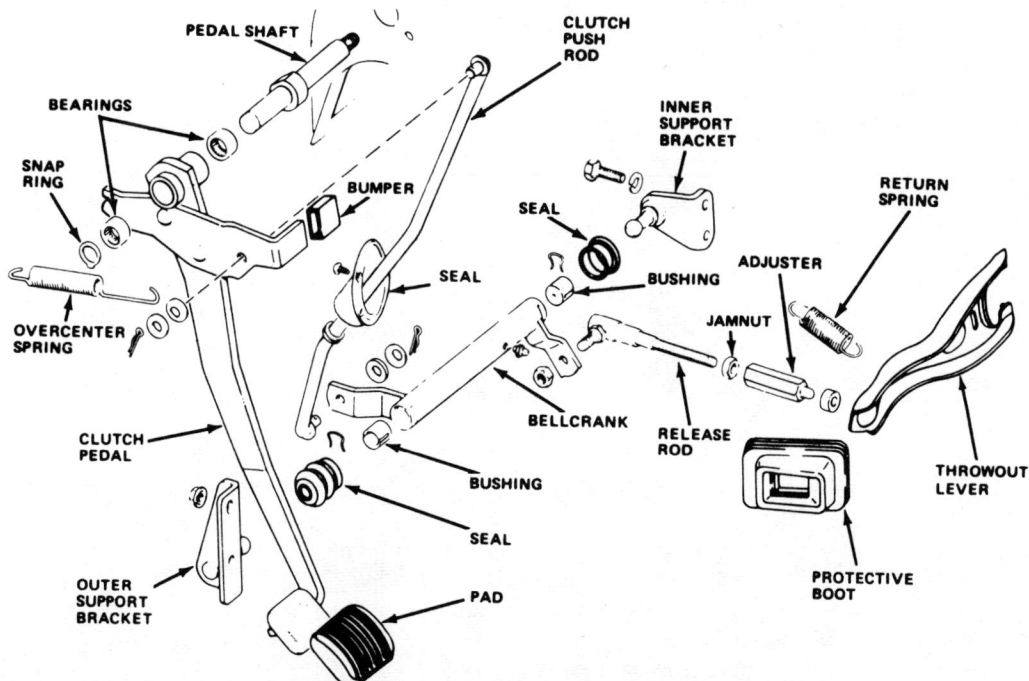

1976-83 clutch linkage for all except the 4-151

PEDAL SUPPORT

THROW OUT BEARING

STOP BRACKET

CLUTCH FORK
RETURN SPRING

CLUTCH FORK

CLUTCH PEDAL

ADJUSTER NUT JAM

BALL ADJUSTER

CLUTCH CABLE

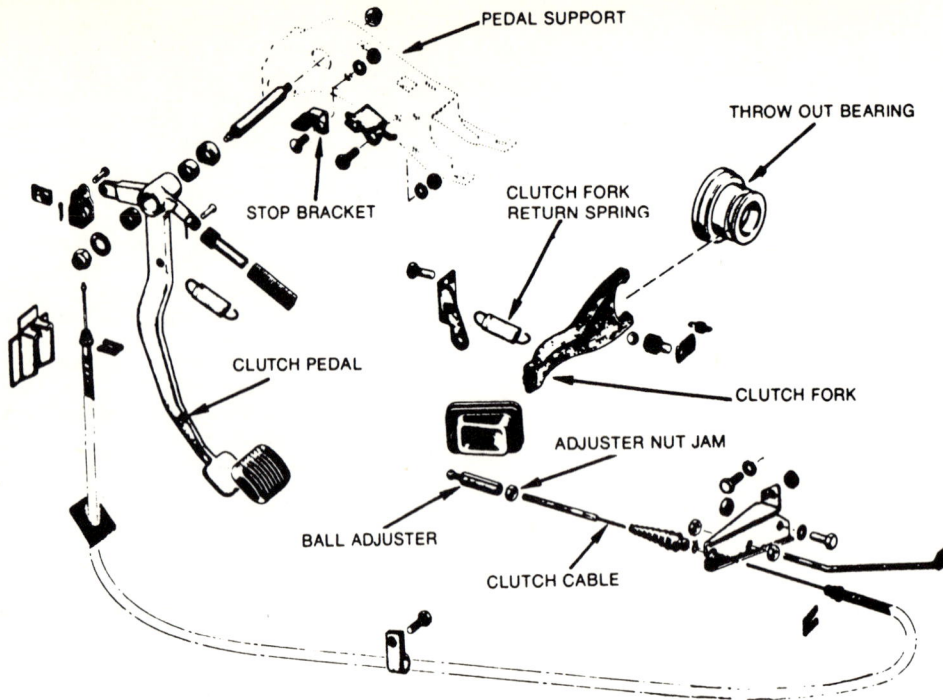

1972 clutch linkage

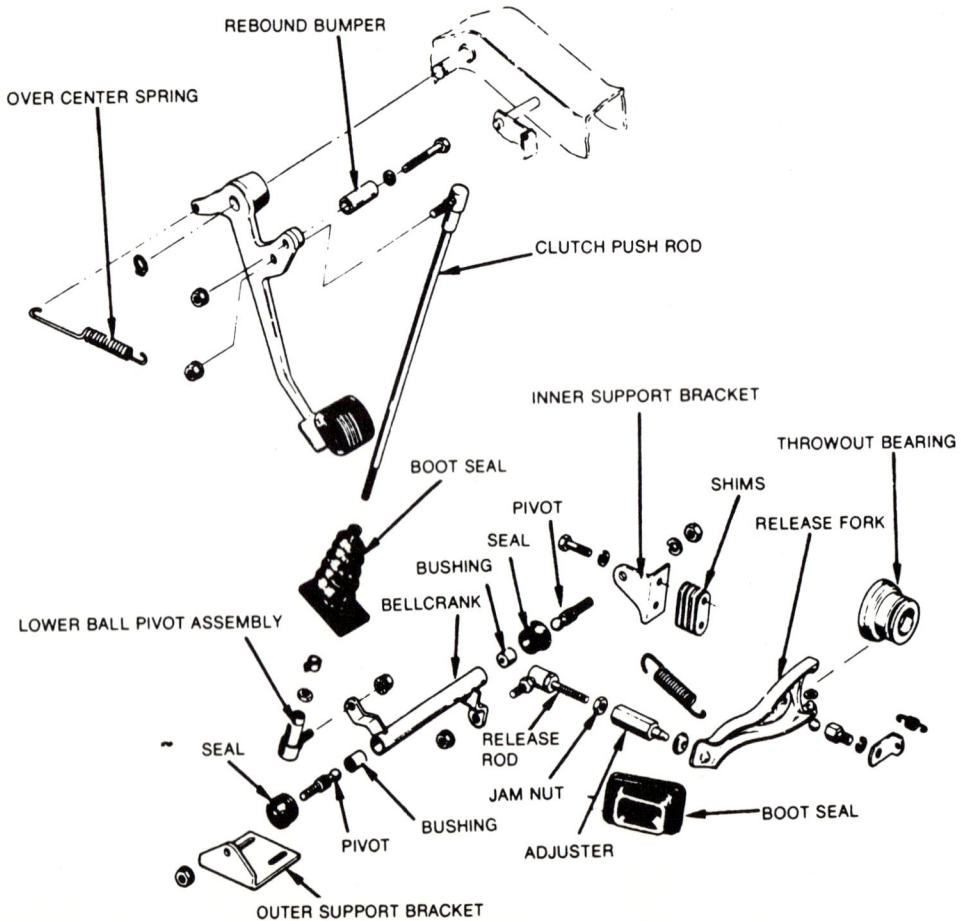

REBOUND BUMPER

OVER CENTER SPRING

CLUTCH PUSH ROD

INNER SUPPORT BRACKET

THROWOUT BEARING

SHIMS

RELEASE FORK

BOOT SEAL

PIVOT

SEAL

BUSHING

LOWER BALL PIVOT ASSEMBLY

BELLCRANK

RELEASE
ROD

SEAL

JAM NUT

BUSHING

ADJUSTER

BOOT SEAL

PIVOT

OUTER SUPPORT BRACKET

1973–75 clutch linkage

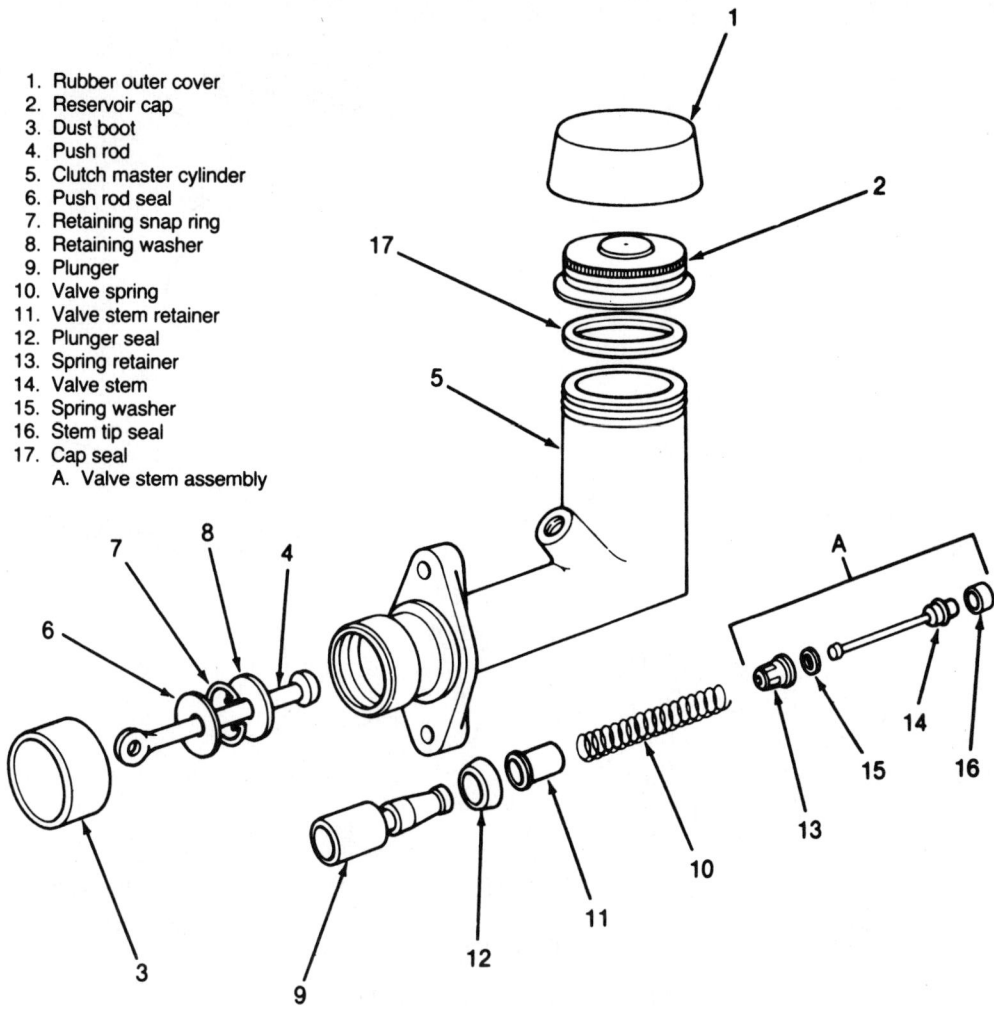

1. Rubber outer cover
2. Reservoir cap
3. Dust boot
4. Push rod
5. Clutch master cylinder
6. Push rod seal
7. Retaining snap ring
8. Retaining washer
9. Plunger
10. Valve spring
11. Valve stem retainer
12. Plunger seal
13. Spring retainer
14. Valve stem
15. Spring washer
16. Stem tip seal
17. Cap seal
   A. Valve stem assembly

1980–86 clutch master cylinder

5. Remove the plunger, valve spring and stem. It may be necessary to tap the assembly out of the bore with a rubber mallet.
6. Compress the valve spring enough to pry off the retainer and remove the spring and stem assembly from the plunger. The retainer tab should be pried upward with a thin screwdriver.
7. Remove the seals from the plunger and discard them.
8. Remove the spring retainer and valve stem from the spring.
9. Remove the valve stem from the retainer and remove the spring washer and stem tip seal from the end of the valve stem. Discard the stem tip seal and the spring washer.
10. Clean all parts thoroughly in a safe brake cleaning solvent. Discard any parts that show signs of wear, pitting or damage. If the core shows signs of excessive wear, deep pitting, severe corrosion or scoring, replace the entire master cylinder. Minor bore imperfections can be corrected by honing.
11. Lubricate the bore with clean brake fluid.
12. Position the new seal on the valve stem. The shoulder of the stem tip seal should fit into the undercut at the end of the valve stem.
13. Install the new seals on the plunger. The seal lips face the valve stem end of the plunger.
14. Install the new spring and retainer on the stem.
15. Install the plunger retainer in the spring.
16. Insert the plunger into the retainer.

17. Compress the spring against the plunger. When the end of the stem passes through the valve stem retainer and seats in the small bore in the end of the plunger, bend the retainer tab on the valve stem retainer downward to lock the stem and retainer on the plunger.
18. Lubricate the spring and plunger assembly with clean brake fluid and insert the assembly, spring end first, into the bore.
19. Install the new dust boot on the pushrod.
20. Lubricate the ball end of the pushrod and the lip of the dust boot with clean brake fluid, or the lubricant supplied in the overhaul kit.
21. Insert the pushrod and retaining washer into the bore.
22. Install the new snapring. Properly position the dust boot.

Clutch Slave Cylinder

REMOVAL AND INSTALLATION

1980-86

1. Raise and support the truck on jackstands.
2. Disconnect the hydraulic line at the cylinder. Cap the line.
3. Unbolt and remove the slave cylinder from the clutch housing.

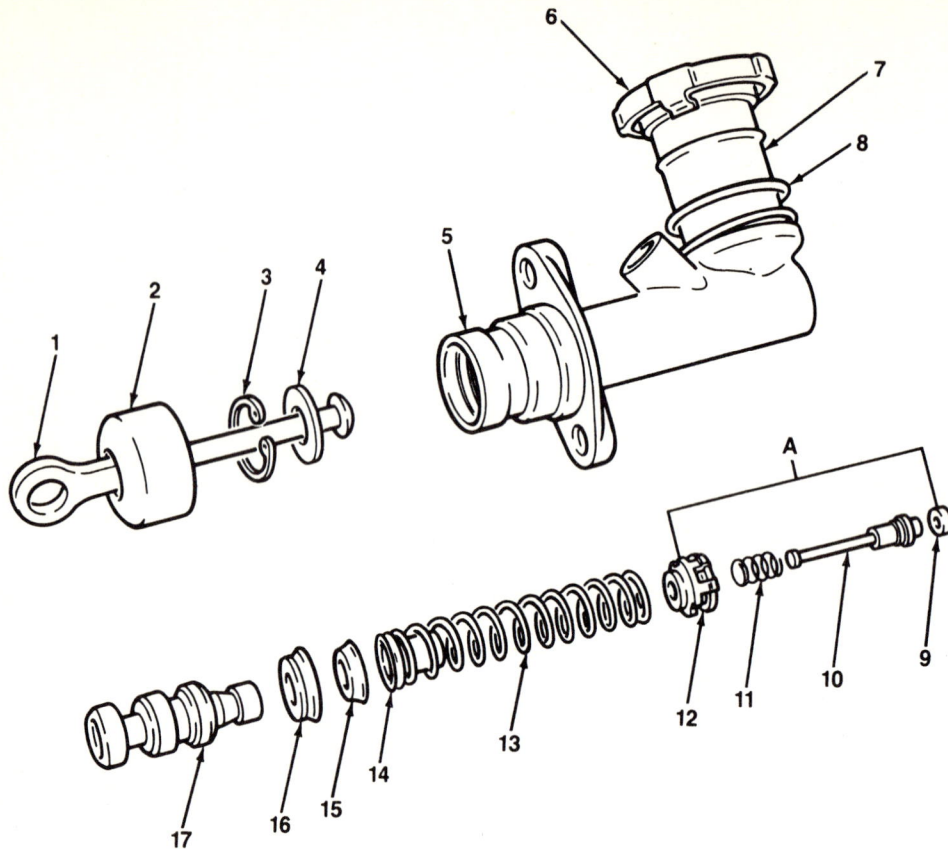

1. Push rod
2. Dust boot
3. Snapring
4. Washer
5. Master cylinder
6. Reservoir cap
7. Reservoir
8. Retaining clamp
9. Stem tip seal
10. Valve stem
11. Retainer spring
12. Spring retainer
13. Plunger spring
14. Valve stem retainer
15. Plunger rear seal
16. Plunger front seal
17. Plunger

1987–90 clutch master cylinder

4. Installation is the reverse of removal. Torque the mounting bolts to 16 ft. lbs. Refill and bleed the system.

1987-90

1. Disconnect the master cylinder line at the slave cylinder inlet line.
2. Remove the transmission and transfer case.
3. Slide the rubber insulator out of the insulator bracket and off of the hydraulic lines.
4. Unbolt the insulator bracket from the bellhousing and slide it off of the lines.
5. Remove the cylinder and bearing retaining nut. Pry the nut up and off of the mounting pin on the transmission front case.
6. Being careful to avoid kinking the lines, slide the cylinder and bearing assembly off of the input shaft.

NOTE: Don't remove the lines from the slave cylinder. They are not meant to be removed and will be damaged if removal is attempted.

Also, some replacement cylinder/bearing assemblies will come with nylon retaining straps. These straps are meant to hold the assembly together during shipment. DO NOT REMOVE THEM! They are designed to break off the first time piston movement takes place.

7. Install the assembly on the input shaft.
8. Guide the lines through the opening in the bellhousing.

9. Position the cylinder mounting boss over the pin on the transmission front face.
10. Secure the assembly to the mounting pin with the nut.
11. Install the insulator and bracket.
12. Install the transmission and transfer case.
13. Fill and bleed the system.

OVERHAUL

1980-86

1. Clean the outside of the cylinder.
2. Remove the dust boot.
3. Remove the pushrod, boot, plunger and spring as an assembly.
4. Remove the spring and seal from the plunger.
5. Remove the snapring and separate the pushrod and boot from the plunger.
6. Clean all parts and replace the cylinder if the bore is excessively worn, pitted, nicked or corroded. Minor imperfections in the bore can be removed through honing.
7. Rebuild the cylinder using a rebuilding kit. Lubricate all parts with clean brake fluid, including the cylinder bore.

1987-90

The slave cylinder is not rebuildable and must be replaced as an assembly if defective.

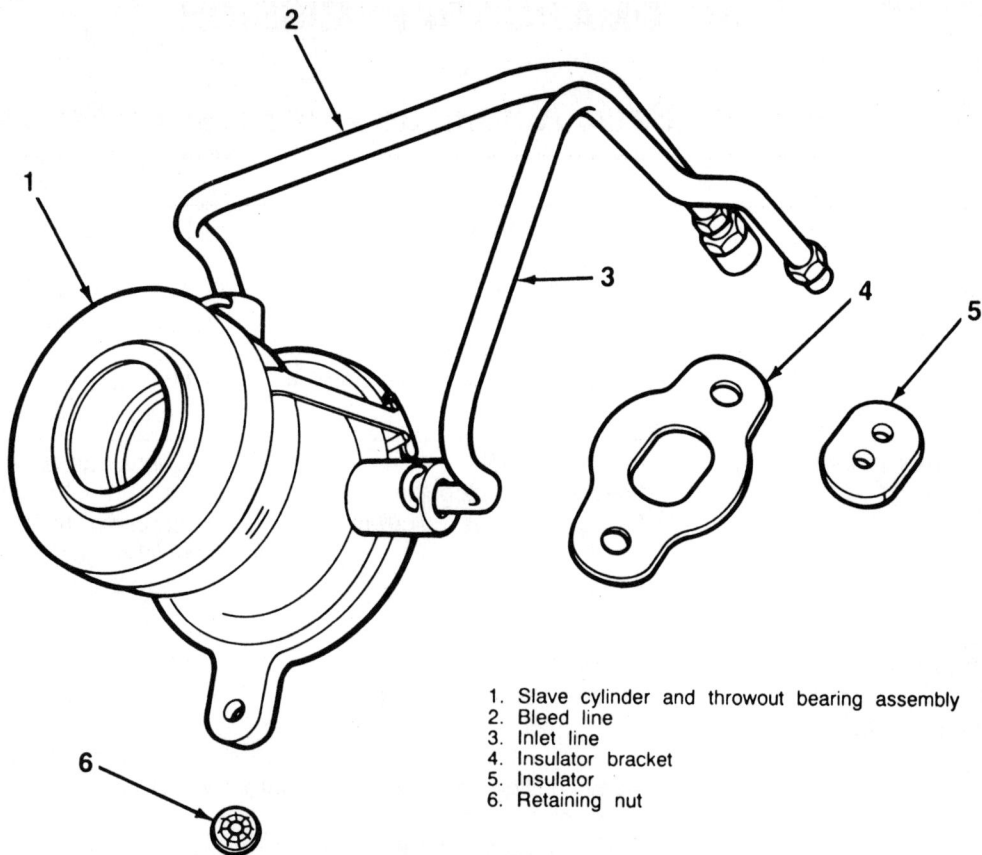

1. Slave cylinder and throwout bearing assembly
2. Bleed line
3. Inlet line
4. Insulator bracket
5. Insulator
6. Retaining nut

1987–90 clutch slave cylinder and throwout bearing

Clutch Hydraulic System

BLEEDING THE SYSTEM

1. Fill the reservoir with clean brake fluid.
2. Raise and support the truck on jackstands.
3. Remove the slave cylinder from the clutch housing, but do not disconnect the hydraulic line. There is enough play in the line to do this.
4. Remove the slave cylinder pushrod.
5. Using a wood dowel, compress the slave cylinder plunger.
6. Attach one end of a rubber hose to the slave cylinder bleeder screw and place the other end in a glass jar, filled halfway with clean brake fluid. Make sure that the hose will stay submerged.
7. Loosen the bleeder screw.
8. Have an assistant press and hold the clutch pedal to the floor. Tighten the bleeder screw with the pedal at the floor. Bubbles will have appeared in the jar when the pedal was depressed.

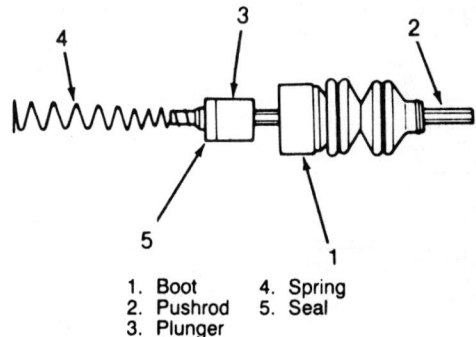

1. Boot   4. Spring
2. Pushrod   5. Seal
3. Plunger

1980–86 slave cylinder

9. Have your assistant release the pedal, then perform the sequence again, until bubbles no longer appear in the jar.
10. Install the slave cylinder and lower the truck. Test the clutch.

AUTOMATIC TRANSMISSION

Troubleshooting Basic Automatic Transmission Problems

Problem	Cause	Solution
Fluid leakage	• Defective pan gasket	• Replace gasket or tighten pan bolts
	• Loose filler tube	• Tighten tube nut
	• Loose extension housing to transmission case	• Tighten bolts
	• Converter housing area leakage	• Have transmission checked professionally
Fluid flows out the oil filler tube	• High fluid level	• Check and correct fluid level
	• Breather vent clogged	• Open breather vent
	• Clogged oil filter or screen	• Replace filter or clean screen (change fluid also)
	• Internal fluid leakage	• Have transmission checked professionally
Transmission overheats (this is usually accompanied by a strong burned odor to the fluid)	• Low fluid level	• Check and correct fluid level
	• Fluid cooler lines clogged	• Drain and refill transmission. If this doesn't cure the problem, have cooler lines cleared or replaced.
	• Heavy pulling or hauling with insufficient cooling	• Install a transmission oil cooler
	• Faulty oil pump, internal slippage	• Have transmission checked professionally
Buzzing or whining noise	• Low fluid level	• Check and correct fluid level
	• Defective torque converter, scored gears	• Have transmission checked professionally
No forward or reverse gears or slippage in one or more gears	• Low fluid level	• Check and correct fluid level
	• Defective vacuum or linkage controls, internal clutch or band failure	• Have unit checked professionally
Delayed or erratic shift	• Low fluid level	• Check and correct fluid level
	• Broken vacuum lines	• Repair or replace lines
	• Internal malfunction	• Have transmission checked professionally

Lockup Torque Converter Service Diagnosis

Problem	Cause	Solution

No lockup	• Faulty oil pump
	• Sticking governor valve
	• Valve body malfunction
	    (a) Stuck switch valve
	    (b) Stuck lockup valve
	    (c) Stuck fail-safe valve
	• Failed locking clutch
	• Leaking turbine hub seal
	• Faulty input shaft or seal ring
		• Replace oil pump
		• Repair or replace as necessary
		• Repair or replace valve body or its internal components as necessary
		
		• Replace torque converter
		• Replace torque converter
		• Repair or replace as necessary

Will not unlock	• Sticking governor valve
	• Valve body malfunction
	    (a) Stuck switch valve
	    (b) Stuck lockup valve
	    (c) Stuck fail-safe valve
		• Repair or replace as necessary
		• Repair or replace valve body or its internal components as necessary

Stays locked up at too low a speed in direct	• Sticking governor valve
	• Valve body malfunction
	    (a) Stuck switch valve
	    (b) Stuck lockup valve
	    (c) Stuck fail-safe valve
		• Repair or replace as necessary
		• Repair or replace valve body or its internal components as necessary

Locks up or drags in low or second	• Faulty oil pump
	• Valve body malfunction
	    (a) Stuck switch valve
	    (b) Stuck fail-safe valve
		• Replace oil pump
		• Repair or replace valve body or its internal components as necessary

Sluggish or stalls in reverse	• Faulty oil pump
	• Plugged cooler, cooler lines or fittings
	• Valve body malfunction
	    (a) Stuck switch valve
	    (b) Faulty input shaft or seal ring
		• Replace oil pump as necessary
		• Flush or replace cooler and flush lines and fittings
		• Repair or replace valve body or its internal components as necessary

Loud chatter during lockup engagement (cold)	• Faulty torque converter • Failed locking clutch • Leaking turbine hub seal	• Replace torque converter • Replace torque converter • Replace torque converter
Vibration or shudder during lockup engagement	• Faulty oil pump • Valve body malfunction • Faulty torque converter • Engine needs tune-up	• Repair or replace oil pump as necessary • Repair or replace valve body or its internal components as necessary • Replace torque converter • Tune engine
Vibration after lockup engagement	• Faulty torque converter • Exhaust system strikes underbody • Engine needs tune-up • Throttle linkage misadjusted	• Replace torque converter • Align exhaust system • Tune engine • Adjust throttle linkage

Lockup Torque Converter Service Diagnosis

Problem	Cause	Solution
Vibration when revved in neutral Overheating: oil blows out of dip stick tube or pump seal	• Torque converter out of balance • Plugged cooler, cooler lines or fittings • Stuck switch valve	• Replace torque converter • Flush or replace cooler and flush lines and fittings • Repair switch valve in valve body or replace valve body
Shudder after lockup engagement	• Faulty oil pump • Plugged cooler, cooler lines or fittings • Valve body malfunction • Faulty torque converter • Fail locking clutch • Exhaust system strikes underbody • Engine needs tune-up • Throttle linkage misadjusted	• Replace oil pump • Flush or replace cooler and flush lines and fittings • Repair or replace valve body or its internal components as necessary • Replace torque converter • Replace torque converter • Align exhaust system • Tune engine • Adjust throttle linkage

Transmission Fluid Indications

The appearance and odor of the transmission fluid can give valuable clues to the overall condition of the transmission. Always note the appearance of the fluid when you check the fluid level or change the fluid. Rub a small amount of fluid between your fingers to feel for grit and smell the fluid on the dipstick.

If the fluid appears:	It indicates:
Clear and red colored	• Normal operation
Discolored (extremely dark red or brownish) or smells burned	• Band or clutch pack failure, usually caused by an overheated transmission. Hauling very heavy loads with insufficient power or failure to change the fluid, often result in overheating. Do not confuse this appearance with newer fluids that have a darker red color and a strong odor (though not a burned odor).
Foamy or aerated (light in color and full of bubbles)	• The level is too high (gear train is churning oil) • An internal air leak (air is mixing with the fluid). Have the transmission checked professionally.
Solid residue in the fluid	• Defective bands, clutch pack or bearings. Bits of band material or metal abrasives are clinging to the dipstick. Have the transmission checked professionally.
Varnish coating on the dipstick	• The transmission fluid is overheating

Understanding Automatic Transmissions

The automatic transmission allows engine torque and power to be transmitted to the rear wheels within a narrow range of engine operating speeds. The transmission will allow the engine to turn fast enough to produce plenty of power and torque at very low speeds, while keeping it at a sensible rpm at high vehicle speeds. The transmission performs this job entirely without driver assistance. The transmission uses a light fluid as the medium for the transmission of power. This fluid also works in the operation of various hydraulic control circuits and as a lubricant. Because the transmission fluid performs all of these three functions, trouble within the unit can easily travel from one part to another. For this reason, and because of the complexity and unusual operating principles of the transmission, a very sound understanding of the basic principles of operation will simplify troubleshooting.

THE TORQUE CONVERTER

The torque converter replaces the conventional clutch. It has three functions:

1. It allows the engine to idle with the vehicle at a standstill, even with the transmission in gear.

2. It allows the transmission to shift from range to range smoothly, without requiring that the driver close the throttle during the shift.

3. It multiplies engine torque to an increasing extent as vehicle speed drops and throttle opening is increased. This has the effect of making the transmission more responsive and reduces the amount of shifting required.

The torque converter is a metal case which is shaped like a sphere that has been flattened on opposite sides. It is bolted to the rear end of the engine's crankshaft. Generally, the entire metal case rotates at engine speed and serves as the engine's flywheel.

The case contains three sets of blades. One set is attached directly to the case. This set forms the torus or pump. Another set is directly connected to the output shaft, and forms the turbine. The third set is mounted on a hub which, in turn, is mounted on a stationary shaft through a one-way clutch. This third set is known as the stator.

A pump, which is driven by the converter hub at engine speed, keeps the torque converter full of transmission fluid at all times. Fluid flows continuously through the unit to provide cooling. Under low-speed acceleration, the torque converter functions as follows:

The torus is turning faster than the turbine. It picks up fluid at the center of the converter and, through centrifugal force,

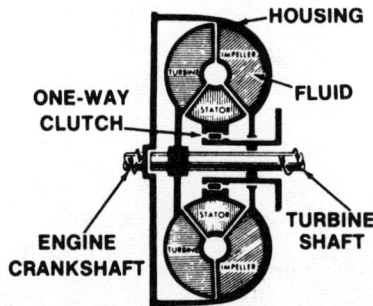

The torque converter housing is roated by the engine's crankshaft, and turns the Impeller. The Impeller spins the turbine, which gives motion to the turbine shaft, driving the gears

slings it outward. Since the outer edge of the converter moves faster than the portions at the center, the fluid picks up speed.

The fluid then enters the outer edge of the turbine blades. It then travels back toward the center of the converter case along the turbine blades. In impinging upon the turbine blades, the fluid loses the energy picked up in the torus.

If the fluid were now to immediately be returned directly into the torus, both halves of the converter would have to turn at approximately the same speed at all times, and torque input and output would both be the same.

In flowing through the torus and turbine, the fluid picks up two types of flow, or flow in two separate directions. It flows through the turbine blades, and it spins with the engine. The stator, whose blades are stationary when the vehicle is being accelerated at low speeds, converts one type of flow into another. Instead of allowing the fluid to flow straight back into the torus, the stator's curved blades turn the fluid almost 90 degrees toward the direction of rotation of the engine. Thus the fluid does not flow as fast toward the torus, but is already spinning when the torus picks it up. This has the effect of allowing the torus to turn much faster than the turbine. This difference in speed may be compared to the difference in speed between the smaller and larger gears in any gear train. The result is that engine power output is higher, and engine torque is multiplied.

As the speed of the turbine increases, the fluid spins faster and faster in the direction of engine rotation. As a result, the ability of the stator to redirect the fluid flow is reduced. Under cruising conditions, the stator is eventually forced to rotate on its one-way clutch in the direction of engine rotation. Under these conditions, the torque converter begins to behave almost like a solid shaft, with the torus and turbine speeds being almost equal.

THE PLANETARY GEARBOX

The ability of the torque converter to multiply engine torque is limited. Also, the unit tends to be more efficient when the turbine is rotating at relatively high speeds. Therefore, a planetary gearbox is used to carry the power output of the turbine to the driveshaft.

Planetary gears function very similarly to conventional transmission gears. However, their construction is different in that three elements make up one gear system, and, in that all three elements are different from one another. The three elements are: an outer gear that is shaped like a hoop, with teeth cut into the inner surface; a sun gear, mounted on a shaft and located at the very center of the outer gear; and a set of three planet gears, held by pins in a ring-like planet carrier, meshing with both the sun gear and the outer gear. Either the outer gear or the sun gear may be held stationary, providing more than one possible torque multiplication factor for each set of gears. Also, if all three gears are forced to rotate at the same speed, the gearset forms, in effect, a solid shaft.

Most modern automatics use the planetary gears to provide either a single reduction ratio of about 1.8:1, or two reduction gears: a low of about 2.5:1, and an intermediate of about 1.5:1. Bands and clutches are used to hold various portions of the gearsets to the transmission case or to the shaft on which they are mounted. Shifting is accomplished, then, by changing the portion of each planetary gearset which is held to the transmission case or to the shaft.

THE SERVOS AND ACCUMULATORS

The servos are hydraulic pistons and cylinders. They resemble the hydraulic actuators used on many familiar machines, such as bulldozers. Hydraulic fluid enters the cylinder, under pressure, and forces the piston to move to engage the band or clutches.

The accumulators are used to cushion the engagement of the

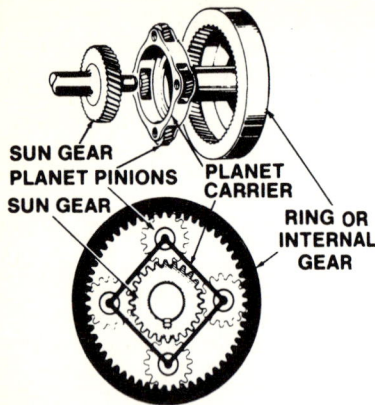

Planetary gears are similar to manual transmission gears but are composed of three parts

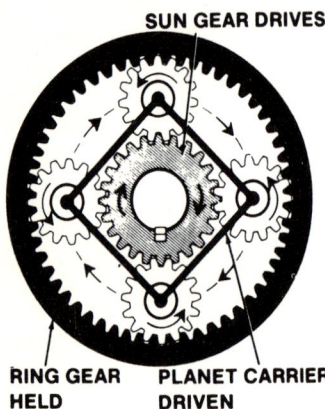

Planetary gears in the maximum reduction (low) range. The ring gear is held and a lower gear ration is obtained

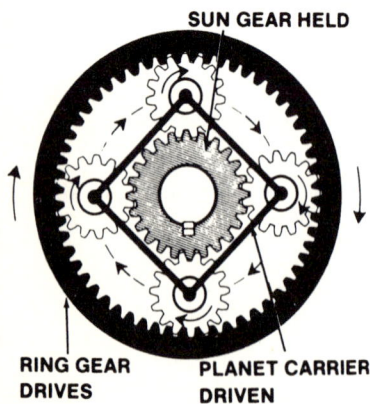

Planetary gears in the minimum reduction (drive) range. The ring gear is allowed to revolve, providing a higher gear ratio

servos. The transmission fluid must pass through the accumulator on the way to the servo. The accumulator housing contains a thin piston which is sprung away from the discharge passage of the accumulator. When fluid passes through the accumulator on the way to the servo, it must move the piston against spring pressure, and this action smooths out the action of the servo.

THE HYDRAULIC CONTROL SYSTEM

The hydraulic pressure used to operate the servos comes from the main transmission oil pump. This fluid is channeled to the various servos through the shift valves. There is generally a manual shift valve which is operated by the transmission selector lever and an automatic shift valve for each automatic upshift the transmission provides: i.e., 2-speed automatics have a low-high shift valve, while 3-speeds have a 1-2 valve, and a 2-3 valve.

There are two pressures which effect the operation of these valves. One is the governor pressure which is affected by vehicle speed. The other is the modulator pressure which is affected by intake manifold vacuum or throttle position. Governor pressure rises with an increase in vehicle speed, and modulator pressure rises as the throttle is opened wider. By responding to these two pressures, the shift valves cause the upshift points to be delayed with increased throttle opening to make the best use of the engine's power output.

Most transmissions also make use of an auxiliary circuit for downshifting. This circuit may be actuated by the throttle linkage or the vacuum line which actuates the modulator, or by a cable or solenoid. It applies pressure to a special downshift surface on the shift valve or valves.

The transmission modulator also governs the line pressure, used to actuate the servos. In this way, the clutches and bands will be actuated with a force matching the torque output of the engine.

Pan Removal and Fluid Change

Chrysler 904 and 999

1. Raise and support the truck on jackstands.
2. The pan has no drain plug, so remove the bolts at one corner and loosen the other pan bolts so that the fluid drains neatly from the one, low hanging corner.
3. Remove the remaining bolts and remove the pan. Discard the gasket.
4. Unbolt and remove the filter from the valve body.
5. Install the new filter and torque the bolts to 35 inch lbs.
6. Coat a new pan gasket with sealer and install the pan. Torque the bolts to 150 inch lbs. (12 ft. lbs.). Fill the transmission.

Turbo Hydra-Matic 400

Since the Turbo Hydra-Matic transmission doesn't have a drain plug, the fluid is drained by loosening the pan and allowing the fluid to run out over the top of the pan.

To avoid making a really big mess, place a drain pan under one corner of the transmission pan and remove the two attach-

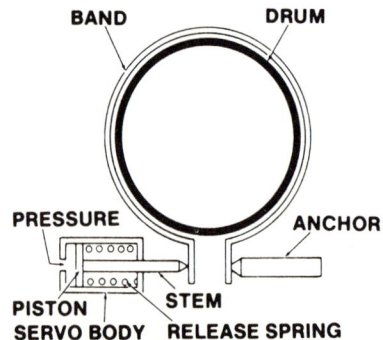

Servos, operated by pressure, are used to apply or release the bands, to either hold the ring gear or allow it to rotate

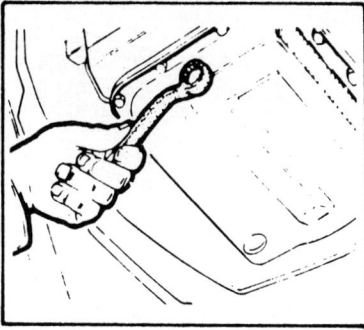

1. Position a catch pan under the transmission. If equipped, remove the drain plug. Be careful; the fluid may be hot.

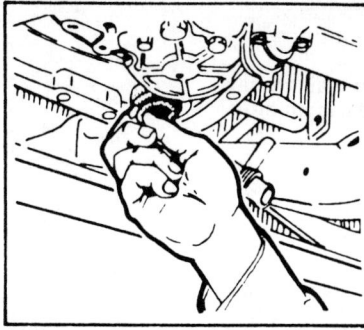

4. Remove the old O-ring from the filter neck and replace with new O-Ring supplied with filter kit.

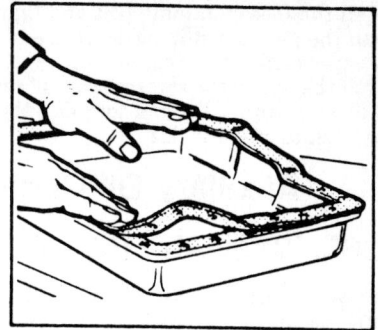

7. Install a new gasket on the pan.

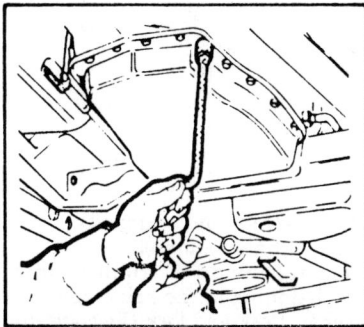

2. Many late-model vehicles have no drain plug. Loosen the pan bolts and allow one corner of the pan to tilt slightly to drain the fluid.

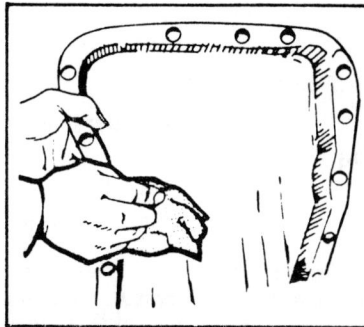

5. Clean the pan thoroughly with gasoline and allow to air dry completely.

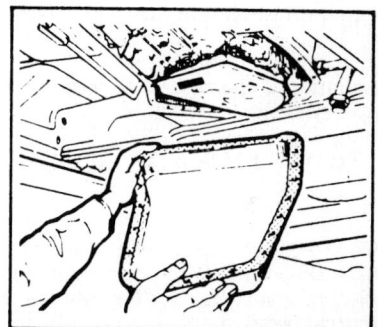

8. Install the new pan and gasket. Do not overtighten the screws.

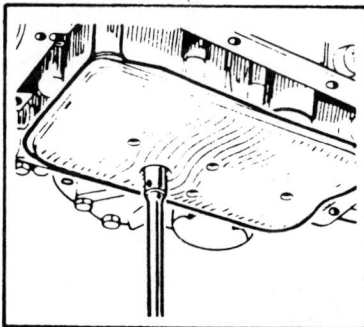

3. The filter or screen is held on by bolts or screws. Remove the filter or screen straight down.

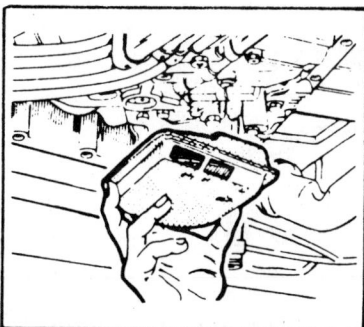

6. Install the new filter. Be sure the intake pipe is seated in the O-ring. Some transmissions use a screen which can be cleaned in gasoline and air dried.

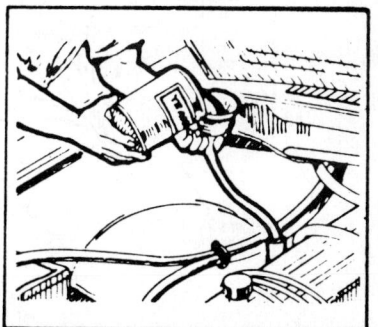

9. Fill the transmission with the required amount of fluid. Do not overfill. Start the engine and shift through all the gears. Check the fluid level and add fluid if necessary.

Follow these 9 easy steps to change your transmission fluid and service the filter

ing screws nearest to either side of that particular corner. One by one, and in a progressive manner, loosen all of the other attaching screws holding the transmission pan, leaving the ones farthest away from the drain corner tighter than the rest. When the majority of the fluid has drained, hold the pan up with one hand, remove the remaining attaching screws and carefully lower the pan. There will be some automatic transmission fluid left in the pan, so be careful not to spill any. The filter is located directly under the oil pan.

There are filter replacement kits available for changing the transmission fluid filter. The kit includes a new filter, pan gasket and, in most cases, a new rubber O-ring to seal the intake

pipe. If a new O-ring is not provided, leave the old one in place. If you can see that the old O-ring is cracked or damaged in any way, it is necessary to replace it with a new one, which can be obtained at a Jeep or GM dealer.

1. Remove the oil filter retainer bolt and remove the oil filter assembly from the transmission.

2. Remove the intake pipe from the filter and the intake pipe-to-case O-ring, if it is to be replaced.

3. Coat the new rubber O-ring with transmission fluid and position it in the groove at the inlet opening.

4. Slide the inlet pipe onto the new filter and position the filter on the transmission, guiding the inlet pipe in place.

5. Install the filter retaining bolt and tighten securely.

6. Clean the pan in a suitable solvent and wipe it dry with a clean, lint-free cloth.

7. Install the pan in the reverse order of removal using a new pan-to-transmission gasket and torquing the bolts in an alternating pattern to 10-13 ft. lbs.

Auxiliary Oil Cooler

REMOVAL AND INSTALLATION

1. Remove the attaching screws and lift off the grille panel.

2. Using masking tape, mark the cooler lines for installation.

3. Place a drain pan on the ground, under the cooler.

4. Loosen the clamps securing the hoses to the cooler and slide them out of the way.

5. Twist the hoses to free them from the cooler pipes and slide the off. Cap the hose ends and cooler outlets to prevent dirt from entering.

6. Unbolt and remove the cooler.

7. Installation is the reverse of removal. Add sufficient fluid to refill the system.

SHIFT LINKAGE ADJUSTMENT

1976-79

1. Place the column shift lever in Neutral.

2. Loosen the gearshift rod clamp adjustment locknut under the vehicle. Make sure that the lever on the transmission is fully in the Neutral position.

3. Tighten the locknut.

4. Check that the engine can be started only in Park and Neutral and that each gear position is fully engaged.

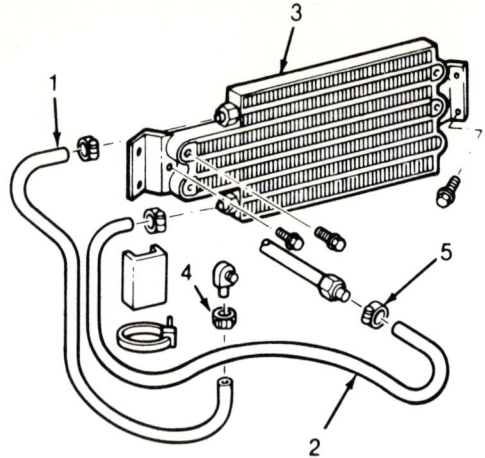

1. Hose	4. Clamp
2. Hose	5. Clamp
3. Cooler	

Auxiliary oil cooler

1980-90

1. Raise and support the vehicle on jackstands.

2. Loosen the shift rod trunnion jamnuts.

3. Remove the lockpin retaining the trunnion to the bell crank and disengage the trunnion at the bell crank.

4. Place the shift lever in Park and lock the column.

5. Move the transmission case lever as far into the Park (rearward) position as possible. Check that the driveshaft will not rotate in this position.

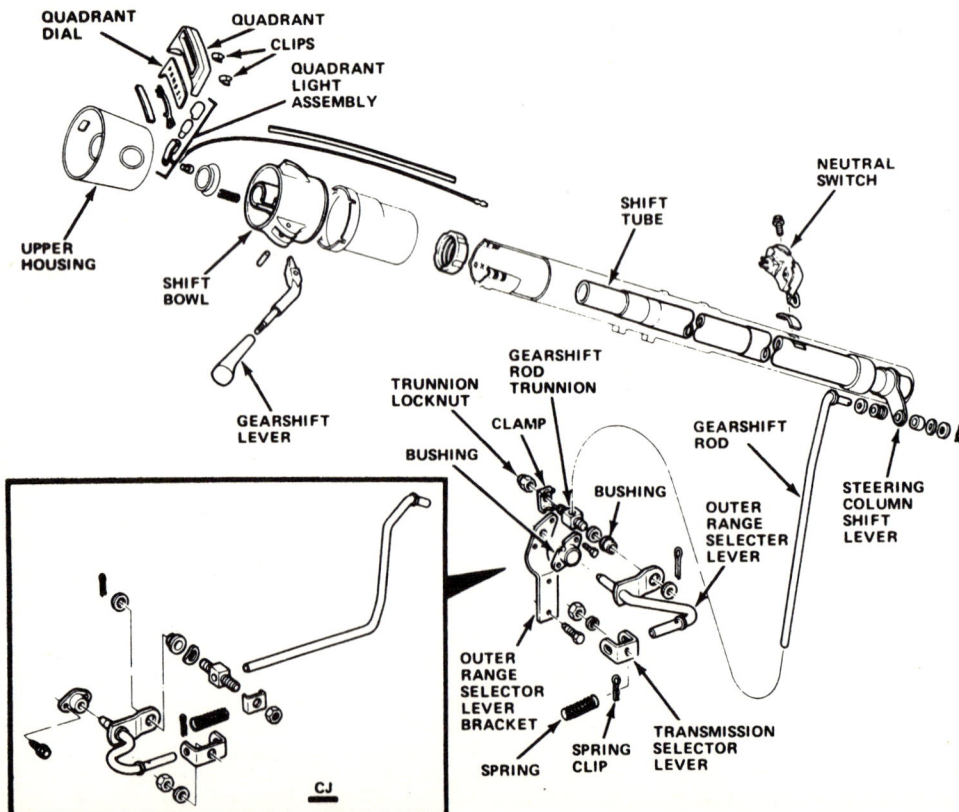

1976–79 automatic transmission shift linkage

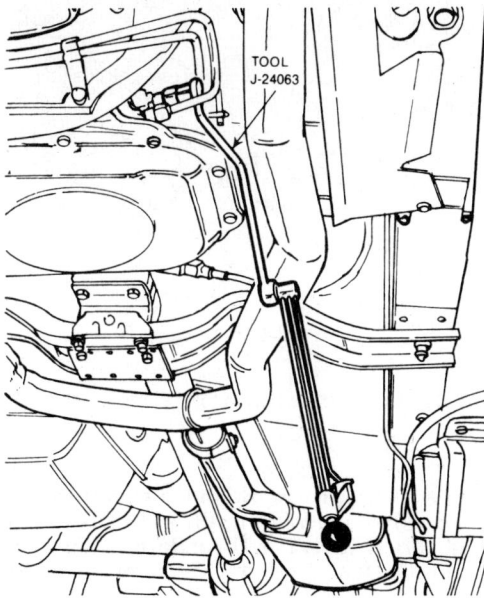

Chrysler 904 and 999 front band adjustment

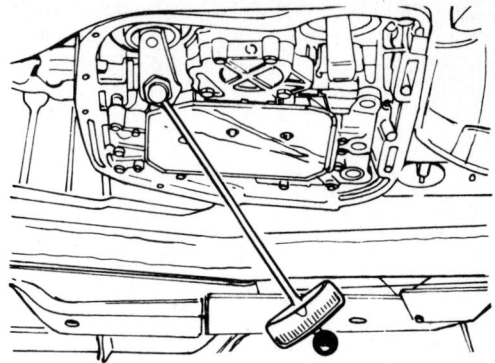

Chrysler 904 and 999 rear band adjustment

6. Adjust the shift rod trunnion to obtain free pin fit in the bell crank arm and tighten the trunnion jamnuts.

NOTE: All play must be eliminated for proper adjustment. Eliminate play by pulling downward on the shift rod and pressing on the outer bell crank.

7. Move the gearshift lever to Park and Neutral and check to see if the engine starts.
8. Road test the vehicle.

BAND ADJUSTMENTS

NOTE: The GM Turbo Hydra-Matic 400 used in 1976-79 models, does not have any band adjustments. The following apply only to the Chrysler built 904 and 999 used in 1980 and later models.

Front Band

1980-81

1. Raise and support the vehicle on jackstands.
2. Loosen the adjusting screw locknut and back the locknut off five turns.
3. Mark the adjusting screw location. Check that it turns freely. If not, squirt some Liquid Wrench®, WD-40®, or similar substance on it.
4. Tighten the adjusting screw to 36 inch. lbs. using, if necessary, an adapter such as the one pictured, and a $\frac{5}{16}$ in. square socket.

NOTE: If the adapter is not used, and the torque wrench is applied directly to the adjuster, tighten the adjuster to 72 inch lbs.

5. Back off the adjuster two full turns.
6. Hold the adjuster firmly and tighten the locknut to 35 ft. lbs.
7. Lower the vehicle.

1983-90

1. Raise and support the truck on jackstands. The front band adjusting screw is located on the left side of the case, just above the control levers.
2. Loosen the locknut and back it off about 5 full turns.

3. Make sure that the screw turns freely. Use penetrating oil if it binds.
4. Tighten the screw to 72 inch lbs.
5. Back off the screw 2½ turns.
6. Tighten the locknut to 35 ft. lbs. Hold the screw still while tightening the locknut.
7. Lower the truck.

Rear Band

1980-86

NOTE: The transmission oil pan must be removed to gain access to the adjusting screw.

1. Raise and support the vehicle on jackstands.
2. Remove the pan.
3. Remove the adjusting screw locknut.
4. Tighten the adjusting screw to 41 inch lbs. using a torque wrench and a ¼ in. hex head socket.
5. Back off the adjusting screw four full turns.
6. Hold the adjusting screw firmly, install the locknut and tighten it to 35 ft. lbs.
7. Install the oil pan and new gasket. Torque the bolts to 12 ft. lbs.
8. Lower the vehicle. Fill the transmission with Dexron®II fluid.

1987-90

1. Raise and support the truck on jackstands.
2. Drain the fluid and remove the pan.
3. Remove the adjusting screw locknut.
4. Tighten the adjusting screw to 72 inch lbs. using a torque wrench and a ¼ in. hex head socket.
5. Back off the adjusting screw 7 full turns.
6. Hold the adjusting screw still and tighten the locknut to 35 ft. lbs.
7. Replace the pan and fill the unit with Dexron®II fluid.

THROTTLE LINKAGE ADJUSTMENT

4-151

1. Remove the air cleaner.
2. Remove the spark plug wire holder from the throttle cable bracket and move the holder and wires aside.
3. Raise and support the vehicle on jackstands.
4. Hold the throttle control lever rearward against its stop. Hook one end of a spare spring to the lever and hook the opposite end to any convenient point. This will hold the lever in position.
5. Lower the vehicle.
6. Block the choke open and move the carburetor linkage completely off the fast idle cam.

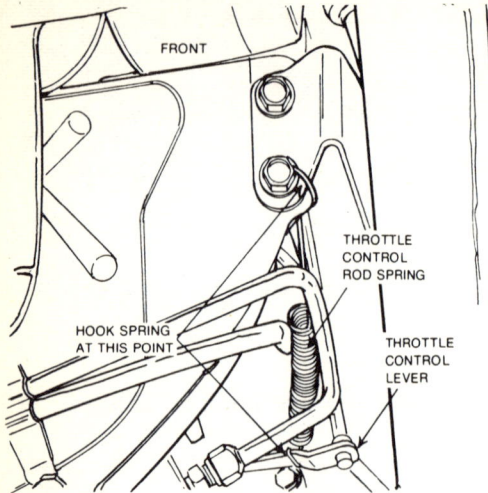

Installing the throttle control lever spring on the 4–151

7. On vehicles without air conditioning, turn the ignition to ON to energize the solenoid.

8. Unlock the throttle control cable by releasing the T shaped adjuster clamp on the cable by lifting it upward with a small screwdriver.

9. Grasp the outer sheath of the cable and move the cable and sheath forward to remove any load on the cable bell crank.

10. Adjust the cable by moving the cable and sheath rearward until there is no play at all between the plastic cable and the bell crank ball.

11. When play has been eliminated, lock the cable by pressing the T shaped clamp downward until it snaps into place.

12. Turn the ignition off. Install all parts and remove the spare spring.

6-258

1. Disconnect the throttle control rod spring at the carburetor.

2. Raise and support the vehicle on jackstands.

3. Use the throttle control rod spring to hold the throttle control lever forward against its stop, by hooking one end of the spring on the throttle control lever and the other end on the throttle linkage bell crank bracket which is attached to the transmission housing.

4. Block the choke plate open and move the throttle linkage off the fast idle cam.

5. On carburetors equipped with a throttle operated solenoid valve, turn the ignition ON to energize the solenoid, then open the throttle halfway to allow the solenoid to lock and return the carburetor to the idle position.

6. Loosen the retaining bolt on the throttle control adjusting link. DO NOT REMOVE THE SPRING CLIP AND NYLON WASHER!

7. Pull on the end of the link to eliminate play and tighten the retaining bolt.

8. Remove the throttle control rod spring and install it on the control rod from where it cam.

9. Lower the vehicle.

8-304

1. Disconnect the throttle control rod spring at the carburetor.

2. Raise and support the vehicle on jackstands.

3. Use the throttle control rod spring to hold the transmission throttle valve control lever against its stop.

4. Block the choke plate open and make sure the throttle linkage is off the fast idle cam.

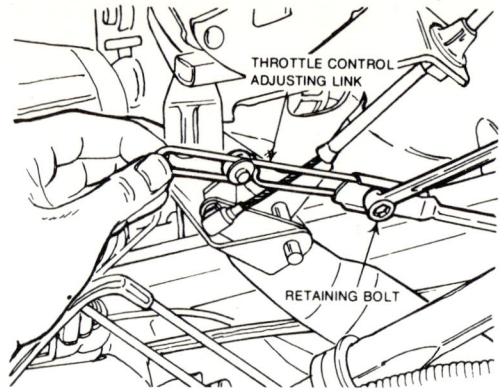

Tightening the link retaining bolt on the 1980 and later 6–258

NOTE: On carburetors equipped with a throttle operated solenoid valve, turn the ignition to ON to energize the solenoid. Then turn the throttle half way to allow the solenoid to lock and return the carburetor to idle.

5. Loosen the retaining bolt on the throttle control rod adjuster link. Remove the spring clip and move the nylon washer to the rear of the link.

6. Push on the end of the link to eliminate play and tighten the link retaining bolt.

7. Install the nylon washer and spring clip.

8. Remove the throttle control rod spring and install it in its intended position.

9. Lower the vehicle.

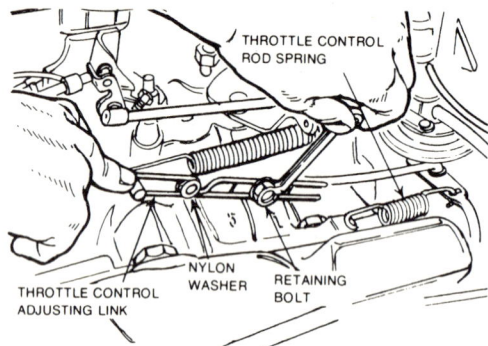

Tightening the link retaining bolt on 1980–81 8–304

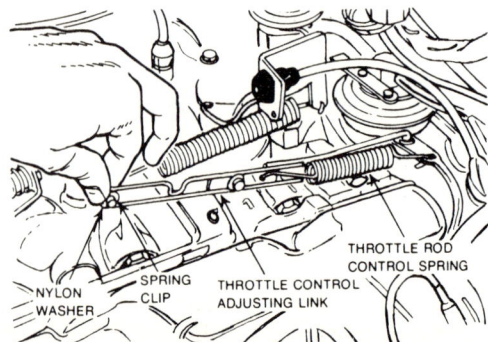

Installing the nylon washer and spring clip on the 1980 and later 8–304

Neutral Safety/Back-up Light Switch

ADJUSTMENT/REPLACEMENT

GM Turbo Hydra-Matic

This switch prevents the engine from being started in any position other than Park or Neutral. It also controls the backup lights.

1. Set the parking brake.
2. Make sure the shift linkage is adjusted correctly.
3. Remove the switch from the base of the steering column, inside the vehicle.
4. Shift into Park and lock the column by removing the key.
5. Move the switch lever until it aligns with the letter **P** on the back of the switch. Insert a $\frac{3}{32}$ in. drill bit in the hole below the letter N on the switch. Move the switch lever till it stops against the drill bit.
6. Install the switch on the column, tighten the screws, and remove the drill bit.
7. Check that the engine will start only in Park and Neutral, and that the backup lights come on only in Reverse.

Chrysler 904, 999

The switch is mounted in the transmission and has no direct adjustment. Proper operation is determined by correct shift linkage adjustment.

To replace the switch:
1. Raise and support the front end on jackstands.
2. Place a drain pan under the switch.
3. Unscrew the switch from the transmission.
4. Replace the switch, using a new seal. Tighten the switch to 24 ft. lbs.
5. Refill the transmission.

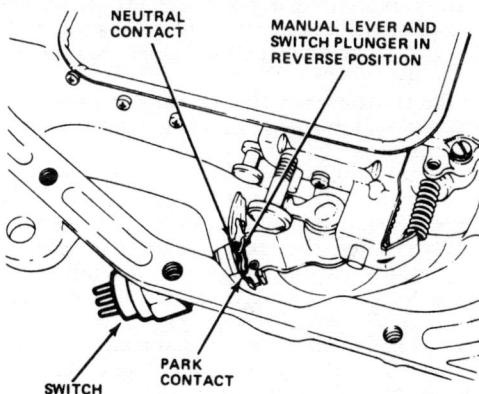

Chrysler 904 or 999 neutral start and backup light switch

Transmission

REMOVAL AND INSTALLATION

1976-79

1. Remove the dipstick.
2. Detach the radiator shroud.
3. Mark the driveshafts for reinstallation.
4. If the vehicle has low range, disconnect the shift linkage and remove the unit.
5. Detach the speedometer cable.
6. Mark and remove the Emergency Drive hoses and wire.

7. Unbolt the vacuum line bracket at the rear of the transfer case.
8. Detach the downshift solenoid wire at the transmission.
9. Remove the starter.
10. Remove the torque converter housing inspection cover.
11. Matchmark the torque converter and drive plate for reassembly. Remove the bolts.
12. Remove the rear support cushion to crossmember nuts.
13. Support the transmission with a transmission jack, using a safety chain.
14. Remove the rear crossmember.
15. Detach the transmission shift linkage. Detach the linkage bracket, bushing, and lever from the frame.
16. Detach and wire up the front driveshaft.
17. Detach the cooler lines from the transmission.
18. Disconnect the vacuum hose at the modulator.
19. Support the engine.
20. Remove the converter housing to engine bolts.
21. Remove the dipstick tube.
22. Move the transmission back till it clears the engine. Hold the converter in place and lower the transmission.
23. On installation, align the torque converter and drive plate matchmarks. Dowels on the engine must line up with holes in the converter housing. Torque the transmission-to-engine bolts 42 ft. lbs.
24. Move the transmission back till it clears the engine. Hold the converter in place and lower the transmission.
25. Install the dipstick tube.
26. Connect the vacuum hose at the modulator.
27. Attach the cooler lines from the transmission.
28. Attach the front driveshaft.
29. Attach the transmission shift linkage.
30. Attach the linkage bracket, bushing, and lever to the frame.
31. Install the rear crossmember. Torque the bolts to 30 ft. lbs.
32. Install the rear support cushion to crossmember nuts.
33. Matchmark the torque converter and drive plate for reassembly. Install the bolts. Torque them to 33 ft. lbs.
34. Install the torque converter housing inspection cover.
35. Install the starter.
36. Attach the downshift solenoid wire at the transmission.
37. Bolt the vacuum line bracket to the rear of the transfer case.
38. Install the Emergency Drive hoses and wire.
39. Attach the speedometer cable.
40. If the vehicle has low range, connect the shift linkage and install the unit.
41. Install the radiator shroud.
42. Install the dipstick.

1980-90

1. Remove the fan shroud.
2. Disconnect the transmission fill tube upper bracket.
3. Raise and support the vehicle on jackstands.
4. Remove the converter housing inspection cover.
5. Remove the fill tube.
6. Remove the starter.
7. Mark the driveshafts for installation.
8. Remove the driveshafts.
9. On 8-304 models, disconnect the exhaust pipes at the manifolds.
10. Disconnect the gearshift and throttle linkages.
11. Disconnect the neutral start switch.
12. Mark the driveplate and converter for realignment.
13. Remove the converter-to-driveplate bolts.
14. Take up the transmission weight with a floor jack. It's a good idea to chain the transmission to the jack.
15. Remove the rear crossmember-to-transmission bolts.
16. Remove the rear crossmember.

17. Lower the transmission slightly and disconnect the fluid cooler lines.
18. Remove the transmission-to-engine bolts.
19. Roll the transmission rearward to clear the crankshaft, lower the jack and remove the unit.
20. Position the transmission against the engine.
21. Install the transmission-to-engine bolts. Torque the bolts to 28 ft. lbs.
22. Connect the fluid cooler lines.
23. Install the rear crossmember. Torque the bolts to 30 ft. lbs.
24. Install the rear crossmember-to-transmission bolts. Torque the bolts to 33 ft. lbs.

25. Install the converter-to-driveplate bolts. Torque the bolts to 40 ft. lbs. on the 4-151; 26 ft. lbs. on the 6-258 and 8-304.
26. Connect the neutral start switch.
27. Connect the gearshift and throttle linkages.
28. On 8-304 models, connect the exhaust pipes at the manifolds.
29. Install the driveshafts.
30. Install the starter.
31. Install the fill tube.
32. Install the converter housing inspection cover.
33. Lower the vehicle.
34. Connect the transmission fill tube upper bracket.
35. Install the fan shroud.

MANUAL TRANSFER CASE

REMOVAL AND INSTALLATION

1971

The transfer case can be removed without removing the transmission.
1. Drain the transfer case and transmission and replace the drain plugs.
2. Disconnect the brake cable.
3. Disconnect the front and rear driveshafts at the transfer case.
4. Disconnect the speedometer cable at the transfer case.
5. Disconnect the transfer case shift levers. On vehicles equipped with two shift levers loosen the set screw and remove the pivot pin. Use a prying tool to pry the shift lever springs away from the shift levers. On models equipped with a single shift lever remove the pivot pin cotter key and the adjusting rod attaching nut to remove the shift lever.
6. Remove the cover plate on the rear face of the transfer case or power take-off shift unit. Remove the cotter key, nut and washer from the transmission main shaft.
7. If possible, remove the transfer case main drive gear from the transmission main shaft. If it is not possible, continue on.
8. Remove the transmission-to-transfer case mounting bracket bolt and nut.
9. Remove the transmission-to-transfer case attaching bolts.
10. Remove the transfer case. If the transfer case main drive gear has not been removed in step 7, proceed as follows: Brace the end of the transmission main shaft so that it cannot be moved in the transmission, then pull the transfer case to the rear to loosen the gear. Remove the gear. When separating the two housing, be careful that the transmission main shaft bearing, which bears in both housings, remains in the transmission case.

NOTE: f the transfer case is being removed from the transmission with the two units out of the vehicle, use the above procedure starting from step 6 and replacing step 10 with the following procedure:

11. Remove the transmission shift housing. Install a transmission mainshaft retaining plate, tool W-194, to prevent the mainshaft from pulling out of the transmission case. Should this tool be unavailable, loop a piece of wire around the mainshaft directly in back of the mainshaft second speed gear. Install the transmission shift housing right and left front attaching bolts part way into the transmission case. Twist the wire and attach each end to one of the screws. Tighten the wire. With the mainshaft securely in place, support the transfer case and with a rawhide mallet or brass drift and hammer, then tap lightly on the end of the mainshaft to loosen the gear and separate the two units.
12. Install the transfer case. If the transfer case main drive gear was not removed in step 7, proceed as follows:
    a. Install the gear.
    b. When separating the two housings, be careful that the transmission mainshaft bearing, which bears in both housings, remains in the transmission case. When installing the transfer case gear on the transmission rear splined driveshaft, tighten the large gear nut securely and insert the cotter pin. Sink the cotter pin well into the nut slots so it will clear the power take-off drive (if so equipped).
    c. Brace the end of the transmission main shaft so that it cannot be moved in the transmission, then pull the transfer case to the rear to loosen the gear.
13. Remove transmission mainshaft retaining plate, tool W-194, or the wire.
14. Install the transmission shift housing. When installing the rear adapter plate on a 4-speed transmission, be sure that the cap screw heads do not protrude beyond the adapter plate face and that they do not interfere with the transfer case fitting tightly against the rear adapter plate.
15. Install the transmission-to-transfer case attaching bolts.
16. Install the transmission-to-transfer case mounting bracket bolt and nut.
17. Install the transfer case main drive gear on the transmission main shaft.
18. Install the cotter key, nut and washer from the transmission main shaft.
19. Install the cover plate on the rear face of the transfer case or power take-off shift unit.
20. Connect the transfer case shift levers. On vehicles equipped with two shift levers loosen the set screw and Install

the pivot pin. On models equipped with a single shift lever install the pivot pin cotter key and the adjusting rod attaching nut to install the shift lever.

21. Connect the speedometer cable at the transfer case.
22. Connect the front and rear driveshafts at the transfer case.
23. Connect the brake cable.
24. Fill the transfer case and transmission.

1972-75

1. Remove the transfer case shift lever knob and trim ring and boot.
2. Remove the transfer case shift lever.
3. Lift and support the vehicle.
4. Drain the transfer case lubricant.
5. Mark the yokes for reference during assembly and disconnect the front and rear driveshafts from the transfer case.
6. Install the transfer case drain plug.
7. Disconnect the parking brake cable at the equalizer and mounting bracket.
8. Disconnect the speedometer cable.
9. Remove the screws which attach the transfer case to the transmission. Install two ⅜ in. x 4 in. threaded dowel pins, one on each side of the case.
10. Remove the transfer case.
11. Remove the gasket between the transmission and the transfer case.
12. Place a new gasket on the dowel pins in the transmission case before installing the transfer case back onto the transmission.
13. Shift the transfer case to 4WD Low position.
14. Position the transfer case on the dowel pins.
15. Rotate the transfer case output shaft until the gears engage with the output gear on the transmission. Slide the transfer case forward to the transmission.

NOTE: Be sure that the transfer case fits flush against the transmission. Severe damage will result if the transfer case bolts are tightened while the transfer case is binding.

16. Install one attaching screw. Remove the dowel pins and install all of the remaining attaching screws.
17. Connect the driveshafts in the same positions from which they were removed.
18. Connect the speedometer cable and parking brake cable.
19. Fill the transfer case with the proper amount of lubricant. See Section 1.
20. Lower the vehicle.
21. Install the transfer case lever, trim boot and lever knob.

1976-79

1. Remove the shift lever knob, trim ring, and boot.
2. Remove the transmission access cover from the floorpan.
3. Drain the lubricant from the transfer case and transmission.
4. Disconnect the torque reaction bracket from the frame crossmember, if so equipped.
5. Support the engine and transmission by placing a jackstand under the clutch housing.
6. Remove the rear frame crossmember.
7. mark the driveshaft yokes for reference during assembly and disconnect the front and rear driveshafts from the transfer case.
8. Disconnect the speedometer cable from the transfer case.
9. Remove the bolts attaching the transfer case to the transmission and remove the transfer case. Remove the gasket which goes between the the transmission and transmission case.

NOTE: here is one transfer case attaching bolt located at the bottom right corner of the transmission that must be removed from the front end of the case.

10. Install the transmission-to-transfer case gasket on the transmission.
11. Shift the transfer case into the 4WD low position.
12. Install a ⅜-16 x 4 in. dowel pin on each side of the transmission to assist in guiding the transfer case into place during installation.
13. Position the transfer case on the dowel pins and slide the case forward until it seats against the transmission. It may be necessary to rotate the transfer case output shaft until the mainshaft gear on the transmission engages the rear output shaft gear in the transfer case.

NOTE: Make sure that the transfer case is flush against the transmission. The case could be cracked if the attaching bolts are tightened while the transfer case is cocked or binding.

14. Install two transfer case attaching bolts, but do not tighten them completely.
15. Remove the dowel pins and install the remaining attaching bolts, retightening them all to 30 ft. lbs.
16. Fill the transfer case with SAE 80W-90 gear lubricant (API GL-4).
17. Assemble the remaining components in the reverse order of removal.

1980-86

1. On models with automatic transmission, remove the shift lever knob, trim ring, and boot from the transfer case shift lever.
2. On models with manual transmission, remove the shift lever knob, trim ring, and boot from the transmission and transfer case levers.
3. Remove the transmission access plate from the floor pan.
4. Raise and support the vehicle on jackstands.
5. Support the engine at the clutch housing and remove the rear crossmember.
6. Mark the front and rear driveshaft-to-transfer case position and disconnect them at the transfer case.
7. Disconnect the speedometer cable at the transfer case.
8. Disconnect the parking brake cable at the equalizer.
9. Disconnect the exhaust pipe bracket at the transfer case.
10. Unbolt the transfer case from the transmission and remove it.
11. Attach the transfer case to the transmission Torque the bolts to 30 ft. lbs.
12. Connect the exhaust pipe bracket at the transfer case.
13. Connect the parking brake cable at the equalizer.
14. Connect the speedometer cable at the transfer case.
15. Connect the front and rear driveshafts at the transfer case.
16. Install the rear crossmember. Torque the bolts to 30 ft. lbs.
17. Lower the vehicle.
18. Install the transmission access plate.
19. On models with manual transmission, install the shift lever knob, trim ring, and boot on the transmission and transfer case levers.
20. On models with automatic transmission, install the shift lever knob, trim ring, and boot on the transfer case shift lever.

NOTE: Some 1980-81 vehicles have experienced difficult transfer case shifting. This may be the result of the transfer case shift lever shaft being bent at the threaded end. To correct this condition, the shift lever shaft and, if necessary, the lever must be replaced. The part numbers are:

- Shaft: 5360045
- Lever w/4-cyl & SR-4, CJ-5 and CJ-7: 5360044
- Lever w/6 or 8 & T-176, CJ-7: 5360044
- Lever w/6-cyl & SR-4, CJ-7: 5360129

To correct the condition:
1. Remove the transfer case.
2. Remove the defective lever shaft, and lever.
3. Install new lever shaft and lever.
4. Install the transfer case.

1987-90

1. Shift the case into 4H.
2. Raise and support the Jeep on jackstands.
3. Drain the case.
4. Matchmark the rear driveshaft and remove it.
5. Disconnect the speedometer cable, vacuum hoses and vent hose from the case.
6. Support the transmission with a floor jack.
7. Remove the crossmember.
8. Matchmark the front driveshaft and remove it.
9. Disconnect the shift lever linkage rod at the case.
10. Remove the shift lever bracket bolts.
11. Support the transfer case with a floor jack or transmission jack and remove the attaching bolts.
12. Pull the case out of the Jeep.
13. Install the case.
14. Install the attaching bolts. Torque them to 26 ft. lbs.
15. Install the shift lever bracket bolts.
16. Connect the shift lever linkage rod at the case.
17. Install the front driveshaft.
18. Install the crossmember. Torque the frame bolts to 30 ft. lbs.; the case bolts to 33 ft. lbs.
19. Connect the speedometer cable, vacuum hoses and vent hose at the case.
20. Install the rear driveshaft.
21. Fill the case.
22. Lower the Jeep.

TRANSFER CASE SHIFT LINKAGE ADJUSTMENT

NOTE: Only the Dana/Spicer 18 and New Process 207 have adjustable linkage.

Spicer 18

This linkage should be adjusted to give ½ in. clearance between the floor pan and the lever when in four wheel drive, low range.

NP-207

1. Place the shift lever in the 2WD position.
2. Insert a ⅛ in. spacer between the gate and the forward edge of the lever.
3. Hold the lever in this position.
4. Raise and support the front end on jackstands.
5. loosen the lockbolt on the adjusting trunnion just enough to allow the linkage rod to slide freely in the trunnion.
6. Move the range lever all the way rearward, to the 2WD position.
7. Position the linkage rod so that it is a free fit in the range lever. Tighten the locknut.
8. Lower the vehicle and check the linkage operation.

Dana/Spicer 18 Overhaul

DISASSEMBLY

1. Remove the bottom cover from the case.
2. Remove the front drive shift bar poppet spring access hole plugs from the front output shaft bearing retainer. If it is necessary to remove the shift bar lock plungers, remove the 2 expansion plugs from the housing.
3. Remove the idler gear shaft lockplate bolt and lockplate.

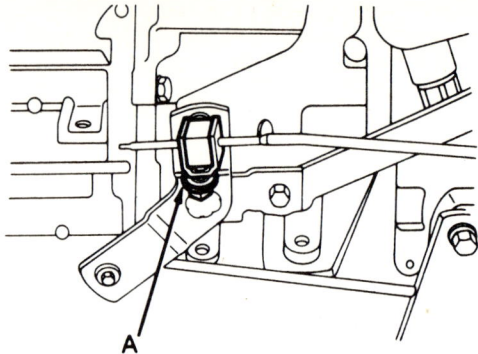

NP-207 range control adjustment; A is the adjustment point

4. Using a brass drift and hammer, drive out the idler gear shaft, driving from the front end of the case. As the shaft passes the thrust washers, remove the washers. Remove the idler gear and bearings with the spacers. Remove the bearings and spacers from the bore of the idler gear.
5. Remove the end yoke retaining lock nuts from the front and rear of the output shafts. Using a gear puller, remove the end yokes.
6. Remove the retaining springs from the shift levers. Remove the setscrew securing the shift lever pin. Remove the pin and left out each lever.
7. Remove the socket-headed screw from the underdrive shift fork.
8. Move the underdrive shift fork and rear output shaft sliding gear forward far enough for the gear to clear the splines on the rear output shaft. Swing the fork and gear toward the cover opening and lift them out.
9. Remove the rear output shaft bearing retainer and output shaft as an assembly.
10. Remove the front output shaft rear cover and shims. Tag the shims for assembly.
11. Remove the socket-headed setscrew from the front wheel drive shift fork.
12. Remove the front output shaft bearing retainer with the shift bars and gasket.
13. Pull forward on the front output shaft as far as possible to permit removal of the bearing cup. Remove the bearing cone from the front output shaft with a gear puller.
14. Remove the front output shaft bearing snapring. Using a plastic mallet, tap on the front end of the output shaft. Take out the driven gear and sliding gear along with the shift fork.
15. Remove the shift bars from the retainer.

―――――――――――― **CAUTION** ――――――――――――
Be careful to secure the poppet balls and springs. The springs are under considerable tension and the ball and springs could fly out causing personal injury!
―――――――――――――――――――――――――――――――――――

16. Remove the shift bar oil seals.
17. Remove the speedometer driven gear.
18. Place the inner face of the bearing retainer in a press and press the output shaft from the retainer.
19. Remove the oil seal from the retainer bore.
20. Remove the tapered bearing cone and cup from the rear bore of the retainer. Remove the tapered bearing cup from the front bore. Remove the O-ring.
21. Using a gear puller, remove the bearing cone from the output shaft. Remove the shims from the shaft and tag them for installation.
22. Remove the speedometer drive gear.
23. Using a gear puller, remove the bearing cone from the front output shaft.

1. Companion flange
2. Brake drum
3. Emergency brake
4. Operating lever
5. Oil seal
6. Lever stud
7. Rear cap
8. Shims
9. Screw
10. Lockwasher

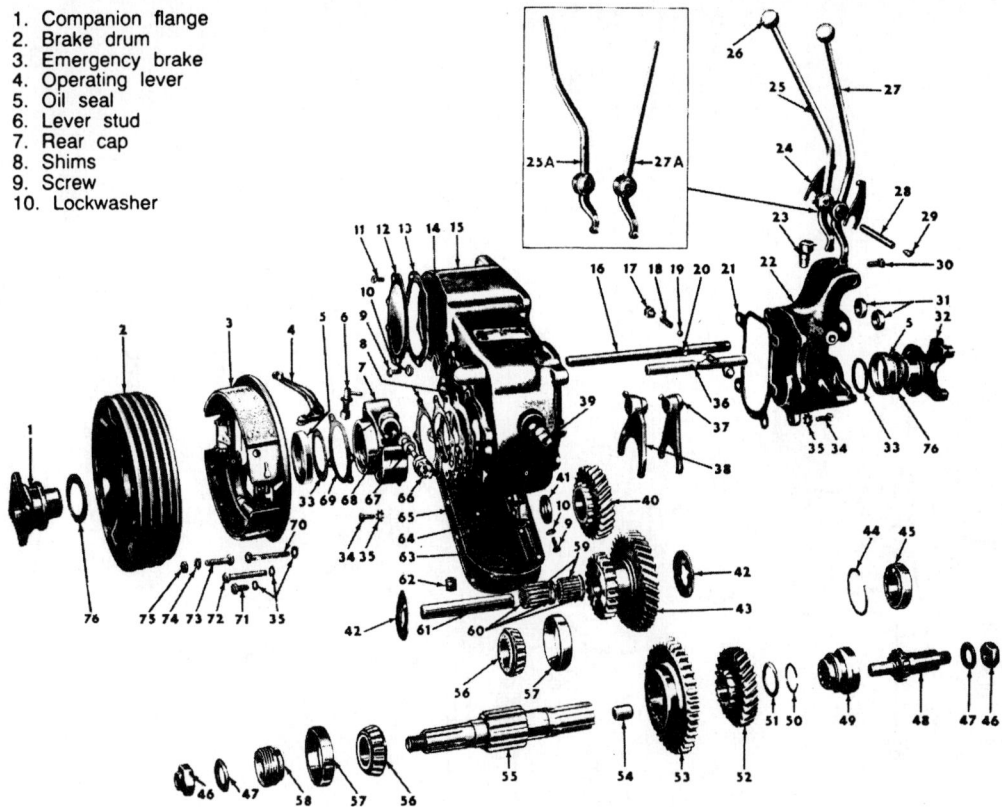

11. Bolt
12. Rear cover
13. Gasket
14. Lockplate
15. Transfer case
16. Shift rod
17. Poppet plug
18. Poppet ring
19. Poppet ball
20. Interlock
21. Gasket
22. Front cap
23. Breather
24. Shift lever spring
25. Shift lever
25a. Shift lever (used with 4-speed transmission)
26. Shift lever knob
27. Shift lever
27a. Shift lever (used with 4-speed transmission)
28. Pivot pin
29. Lubrication fitting
30. Set screw
31. Oil seal
32. Front yoke
33. Gasket
34. Bolt
35. Lockwasher
36. Shift rod
37. Shift fork
38. Shift fork
39. Filler pipe plug
40. Mainshaft gear
41. Plain washer
42. Thrust washer
43. Intermediate gear

44. Snapring
45. Bearing
46. Nut
47. Washer
48. Output clutch shaft
49. Output clutch gear
50. Snapring
51. Thrust washer
52. Output shaft gear
53. Sliding gear
54. Bushing
55. Output shaft
56. Cone and rollers
57. Bearing cup
58. Speedometer gear
59. Needle bearings
60. Bearing spacer
61. Intermediate shaft
62. Drain plug
63. Gasket
64. Nut
65. Bottom cover
66. Sleeve
67. Speedometer gear
68. Bushing
69. Gasket
70. Bolt
71. Bolt
72. Hex nut
72. Bolt
73. Bolt
74. Lockwasher
75. Nut
76. Output shaft seal

Spicer 18 dual lever transfer case

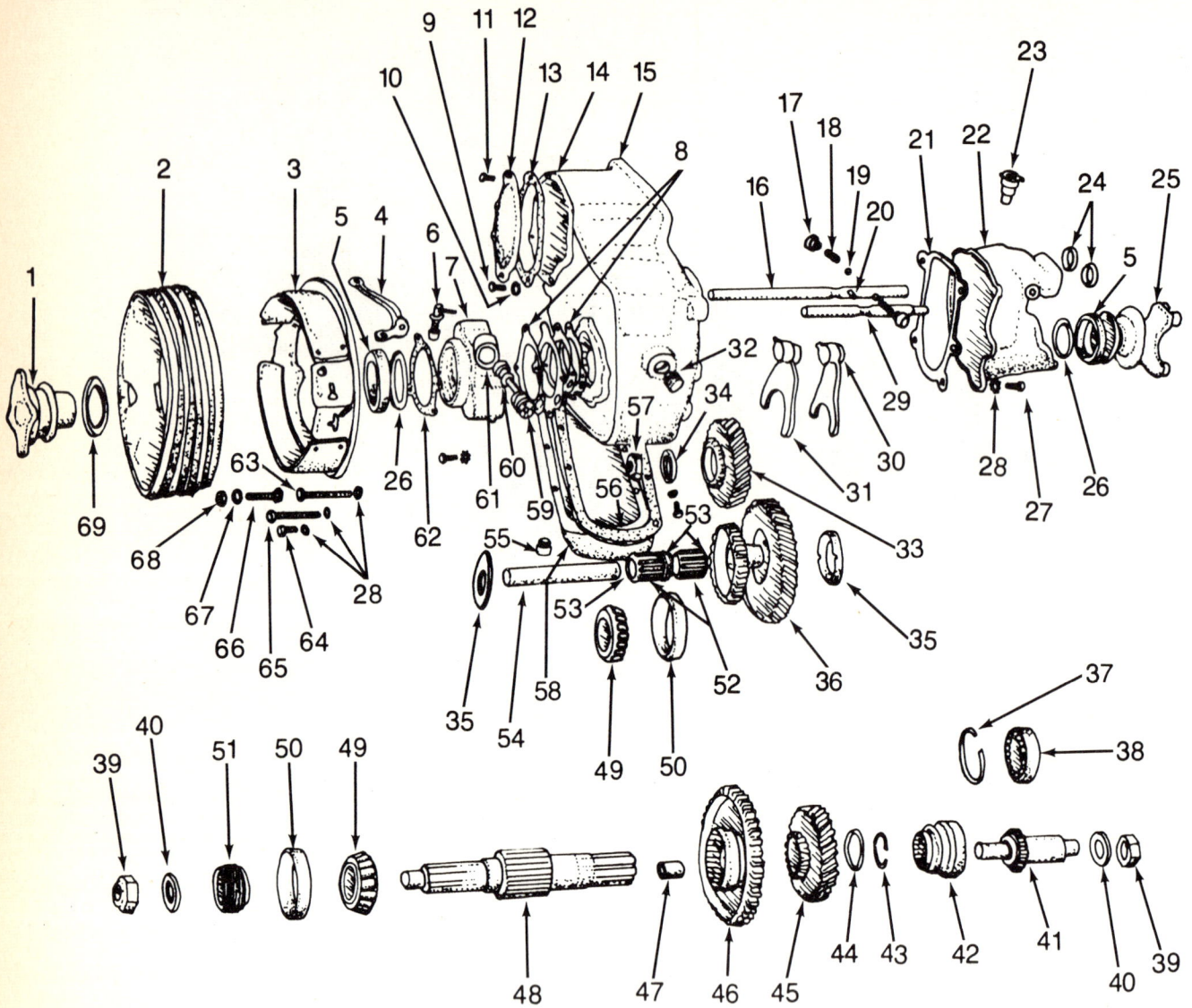

Spicer 18 single lever transfer case

1. Companion flange and oil seal guard
2. Parking brake drum
3. Parking brake
4. Brake operating lever
5. Bearing cap oil seal
6. Brake operating lever stud
7. Rear bearing cap
8. Rear bearing cap shim set
9. Bolt
10. Lockwasher
11. Bolt
12. Transfer case rear cover
13. Rear gasket cover
14. Intermediate shaft lockplate
15. Transfer case
16. Shift rod
17. Poppet plug

18. Poppet spring
19. Poppet ball
20. Shift rod interlock
21. Bearing cap gasket
22. Front bearing shaft cap
23. Breather
24. Shift rod oil seal
25. End yoke
26. Oil seal gasket
27. Bolt
28. Lockwasher
29. Shift rod
30. Front wheel drive shift fork
31. Underdrive and direct shift fork
32. Filler plug
33. Mainshaft gear
34. Mainshaft washer
35. Thrust washer
36. Intermediate gear

37. Bearing shaft snapring
38. Output clutch shaft bearing
39. Companion flange nut
40. Plain washer
41. Output clutch shaft
42. Output clutch shaft gear
43. Output shaft gear snapring
44. Thrust washer
45. Output shaft gear
46. Output shaft sliding gear
47. Pilot bushing
48. Output shaft
49. Cone and rollers
50. Output shaft bearing cup
51. Speedometer drive gear 6 teeth
52. Needle bearing
53. Bearing spacer

54. Intermediate shaft
55. Drain plug
56. Bottom cover gasket
57. Mainshaft nut
58. Transfer case bottom cover
59. Speedometer driven gear sleeve
60. Speedometer driven gear - 15 teeth
61. Speedometer driven gear bushing
62. Backing plate gasket
63. Bolt
64. Bolt
65. Bolt
66. Bolt
67. Lockwasher
68. Nut
69. Ouput shaft seal

Transmission mainshaft retaining plate installed for removing the Spicer 18 transfer case from the transmission

ASSEMBLY

24. Press the bearing cone on the front output shaft.
25. Install the speedometer drive gear.
26. Install the bearing cone and shims on the output shaft.
27. Install the O-ring and tapered bearing cup in the front bore. Install the tapered bearing cone and cup in the rear bore of the retainer.
28. Install the oil seal in the retainer bore.
29. Place the rear output shaft into the retainer and support the shaft in a press. Using an adapter slightly larger than the shaft, press the bearing cone onto the shaft. Locate the rear output shaft bearing retainer in a fixed position. Set up a dial indicator and check the shaft endplay. Endplay should be 0.004-0.008 in. (0.10-0.20mm). Add or remove shims to adjust. Endplay is critical.
30. Install the speedometer driven gear.
31. Install the shift bar oil seals.
32. Install the poppet springs and balls. The heavier spring if for the front shift bar.
33. Depress the balls and install the shift bars in the retainer. Make sure that the seal lips are not damaged.
34. Install the driven gear and sliding gear along with the shift fork, and the front output shaft bearing snapring.
35. Install the bearing cone on the front output shaft.
36. Install the front output shaft bearing retainer with the shift bars and gasket.
37. Install the socket-headed setscrew from the front wheel drive shift fork.
38. Install the front output shaft rear cover and shims. Check the endplay with a dial indicator. Endplay should be 0.004-0.008 in. (0.10-0.20mm). Add or remove shims to adjust. Endplay is critical.
39. Install the rear output shaft bearing retainer and output shaft as an assembly. Torque the bolts to 30-40 ft. lbs.
40. Install the underdrive shift fork and rear output shaft sliding gear.
41. Install the socket-headed screw from the underdrive shift fork.
42. Install the retaining springs, shift levers and setscrews. Install the pin in each lever.

43. Install the end yokes. Torque the retaining lock nuts to 200-300 ft. lbs.
44. Install the bearings and spacers in the bore of the idler gear.
45. Install the idler gear and bearings with the spacers.
46. Install the washers.
47. Install the idler gear shaft.
48. Install the idler gear shaft lockplate bolt and lockplate. Torque the bolt to 15-20 ft. lbs.
49. Install the shift bar lock plungers.
50. Install the front drive shift bar poppet spring access hole plugs.
51. Install the bottom cover from the case. Torque the bolts to 15-20 ft. lbs.

Dana/Spicer 20 Overhaul

DISASSEMBLY

1. Remove the bottom cover from the case.
2. Remove the front drive shift bar poppet spring access hole plugs from the front output shaft bearing retainer. If it is necessary to remove the shift bar lock plungers, remove the 2 expansion plugs from the housing.
3. Remove the idler gear shaft lockplate bolt and lockplate.
4. Using a brass drift and hammer, drive out the idler gear shaft, driving from the front end of the case. As the shaft passes the thrust washers, remove the washers. Remove the idler gear and bearings with the spacers. Remove the bearings and spacers from the bore of the idler gear.
5. Remove the end yoke retaining lock nuts from the front and rear of the output shafts. Using a gear puller, remove the end yokes.
6. Remove the retaining springs from the shift levers. Remove the setscrew securing the shift lever pin. Remove the pin and left out each lever.
7. Remove the socket-headed screw from the underdrive shift fork.
8. Move the underdrive shift fork and rear output shaft sliding gear forward far enough for the gear to clear the splines on the rear output shaft. Swing the fork and gear toward the cover opening and lift them out.
9. Remove the rear output shaft bearing retainer and output shaft as an assembly.
10. Remove the front output shaft rear cover and shims. Tag the shims for assembly.
11. Remove the socket-headed setscrew from the front wheel drive shift fork.
12. Remove the front output shaft bearing retainer with the shift bars and gasket.
13. Pull forward on the front output shaft as far as possible to permit removal of the bearing cup. Remove the bearing cone from the front output shaft with a gear puller.
14. Remove the front output shaft bearing snapring. Using a plastic mallet, tap on the front end of the output shaft. Take out the driven gear and sliding gear along with the shift fork.
15. Remove the shift bars from the retainer.

---------------------------- **CAUTION** ----------------------------

Be careful to secure the poppet balls and springs. The springs are under considerable tension and the ball and springs could fly out causing personal injury!

---

16. Remove the shift bar oil seals.
17. Remove the speedometer driven gear.
18. Place the inner face of the bearing retainer in a press and press the output shaft from the retainer.
19. Remove the oil seal from the retainer bore.
20. Remove the tapered bearing cone and cup from the rear bore of the retainer. Remove the tapered bearing cup from the front bore. Remove the O-ring.

7 DRIVE TRAIN

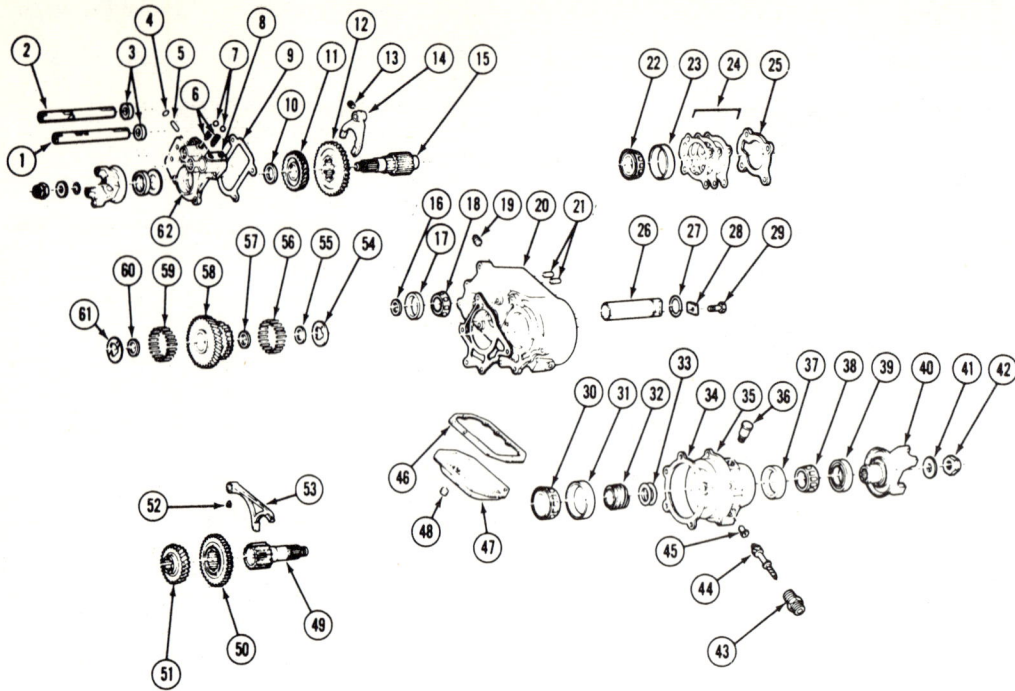

1. Shift rod—rear output shaft shift fork
2. Shift rod—front output shaft shift fork
3. Shift rod oil seal
4. Interlock plug
5. Interlock
6. Poppet ball spring
7. Poppet ball
8. Front bearing cap
9. Front bearing cap gasket
10. Front output shaft thrust washer
11. Front output shaft gear
12. Front output shaft sliding gear
13. Setscrew
14. Front output shaft shift fork
15. Front output shaft
16. Front output shaft spacer
17. Front output shaft front bearing cup
18. Front output shaft front bearing
19. Filler plug
20. Transfer case
21. Thimble cover
22. Front output shaft rear bearing
23. Front output shaft rear bearing cup
24. Front output shaft rear bearing cup shims
25. Cover plate
26. Intermediate shaft
27. Intermediate shaft O-ring
28. Lock plate
29. Lock plate bolt
30. Rear output shaft front bearing
31. Rear output shaft front bearing cup
32. Speedometer drive gear
33. Rear output shaft bearing shim
34. Rear bearing cap gasket
35. Rear bearing cap
36. Breather
37. Rear bearing cap cup
38. Rear bearing cap bearing
39. Rear bearing cap oil seal
40. Rear yoke
41. Rear yoke washer
42. Rear yoke nut
43. Speedometer sleeve
44. Speedometer driven gear
45. Speedometer bushing
46. Bottom cover gasket
47. Bottom cover
48. Drain plug
49. Rear output shaft
50. Rear output shaft sliding gear
51. Mainshaft gear
52. Setscrew
53. Rear output shaft shift fork
54. Intermediate gear thrust washer
55. Intermediate gear bearing spacer
56. Intermediate gear shaft needle bearings
57. Intermediate gear bearing spacer
58. Intermediate gear
59. Intermediate gear shaft needle bearings
60. Intermediate gear bearing spacer
61. Intermediate gear thrust washer
62. Front bearing cap

Dana 20 transfer case components

21. Using a gear puller, remove the bearing cone from the output shaft. Remove the shims from the shaft and tag them for installation.
22. Remove the speedometer drive gear.
23. Using a gear puller, remove the bearing cone from the front output shaft.

ASSEMBLY

24. Press the bearing cone on the front output shaft.
25. Install the speedometer drive gear.
26. Install the bearing cone and shims on the output shaft.
27. Install the O-ring and tapered bearing cup in the front

7-84

bore. Install the tapered bearing cone and cup in the rear bore of the retainer.

28. Install the oil seal in the retainer bore.

29. Place the rear output shaft into the retainer and support the shaft in a press. Using an adapter slightly larger than the shaft, press the bearing cone onto the shaft. Locate the rear output shaft bearing retainer in a fixed position. Set up a dial indicator and check the shaft endplay. Endplay should be 0.003-0.005 in. (0.08-0.13mm). Add or remove shims to adjust. Endplay is critical.

30. Install the speedometer driven gear.

31. Install the shift bar oil seals.

32. Install the poppet springs and balls. The heavier spring if for the front shift bar.

33. Depress the balls and install the shift bars in the retainer. Make sure that the seal lips are not damaged.

34. Install the driven gear and sliding gear along with the shift fork, and the front output shaft bearing snapring.

35. Install the bearing cone on the front output shaft.

36. Install the front output shaft bearing retainer with the shift bars and gasket.

37. Install the socket-headed setscrew from the front wheel drive shift fork.

38. Install the front output shaft rear cover and shims. Check the endplay with a dial indicator. Endplay should be 0.003-0.005 in. (0.08-0.13mm). Add or remove shims to adjust. Endplay is critical.

39. Install the rear output shaft bearing retainer and output shaft as an assembly. Torque the bolts to 30-40 ft. lbs.

40. Install the underdrive shift fork and rear output shaft sliding gear.

41. Install the socket-headed screw from the underdrive shift fork.

42. Install the retaining springs, shift levers and setscrews. Install the pin in each lever.

43. Install the end yokes. Torque the retaining lock nuts to 200-300 ft. lbs.

44. Install the bearings and spacers in the bore of the idler gear.

45. Install the idler gear and bearings with the spacers.

46. Install the washers.

47. Install the idler gear shaft.

48. Install the idler gear shaft lockplate bolt and lockplate. Torque the bolt to 15-20 ft. lbs.

49. Install the shift bar lock plungers.

50. Install the front drive shift bar poppet spring access hole plugs.

51. Install the bottom cover from the case. Torque the bolts to 15-20 ft. lbs.

Dana 300 Transfer Case Overhaul

The 300 has a cast iron case, four gear positions and employs an external floor mounted gearshift linkage for range control. It is a part time, 2 speed unit with undifferentiated high and low ranges. It is used with both manual and automatic transmission. Low range reduction is 2.6:1.

DISASSEMBLY

1. Drain the unit and remove the shift lever assembly.

2. Remove the bottom cover.

NOTE: The bottom cover has been coated with a sealant. Use a putty knife to break the seal and work the knife around the bottom of the cover to break it loose. Don't try to wedge the cover off.

3. With a puller, remove the front and rear yokes.

4. Unbolt and remove the input shaft support from the case.

The rear output shaft gear and input shaft will come with it as an assembly.

NOTE:The support has been coated with sealant. Use a putty knife to break the seal and work the knife around the bottom of the cover to break it loose. Don't try to wedge the cover off.

5. Remove the rear output shaft clutch sleeve from the case.

6. Remove and discard the snap ring retaining the rear output shaft gear on the input shaft and remove the gear.

7. Remove and discard the input bearing snapring.

8. Remove the input shaft bearing from the support. Tap the end of the shaft with a soft mallet to aid removal.

9. Remove the input shaft bearing and end-play shims from the shaft with an arbor press.

10. Remove the input shaft oil seal from the support.

11. Unbolt and remove the intermediate shaft lockplate.

12. Remove the intermediate shaft. Tap the shaft out of the case using a brass punch and plastic mallet.

13. Remove and discard the intermediate shaft O-ring seal.

14. Remove the intermediate gear assembly and thrust washers.

NOTE: The thrust washers have locating tabs which must fit into notches in the case at assembly.

15. Remove the needle bearings and spacers from the intermediate gear. There are 48 needle bearings and three spacers.

16. Remove the rear bearing cap attaching bolts and remove the cap. A plastic mallet will aid in removal.

NOTE: The rear bearing cap has been coated with sealant.

17. Remove the end play shims and speedometer drive gear from the rear output shaft.

18. Remove and discard the rear output shaft oil seal. Remove the bearings and races from the rear cap.

19. Unbolt and remove the front and rear output shaft shift forks from the shift rods.

20. Remove the shift rods. Insert a punch through the clevis pin holes in the rods and rotate the rods while pulling them out of the case.

NOTE: The shift rods are free of the case, take care to avoid losing the shift rod poppet balls and springs.

21. Remove the shift forks from the case.

22. Remove the bolts attaching the front cap to the case and remove the cap.

NOTE: The front cap has been coated with sealant.

23. Remove the front output shaft and shift rod oil seals from the front cap.

24. Remove the bearing race from the front cap.

25. Remove the cover plate bolts and remove the plate and end play shims from the case. Keep the shims together for assembly.

26. Move the front output shaft toward the front of the case.

27. Remove the front output shaft rear bearing race.

28. Remove the rear output shaft front bearing. Position the case on wood blocks. Seat the clutch gear on the case interior surface and tap the shaft out of the bearing with a soft mallet.

NOTE: If the bearing is difficult to remove, an arbor press may have to be used.

29. Remove the rear output shaft front bearing, thrust washer, clutch gear and output shaft from the case.

30. Remove the front output shaft rear bearing with an arbor press.

─────── **CAUTION** ───────

Be sure to support the case with wood blocks positioned on either side of the case bore.

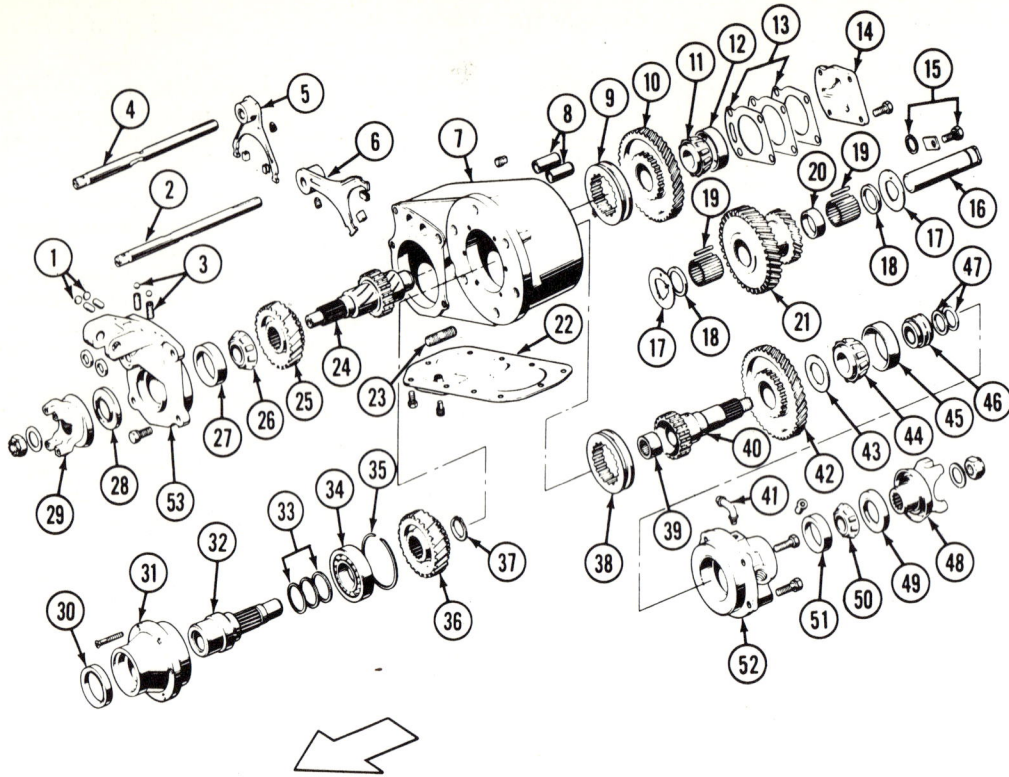

1. Interlock plugs and interlocks
2. Shift rod—rear output shaft fork
3. Poppet balls and springs
4. Shift rod—front output shaft fork
5. Front output shaft shift fork
6. Rear output shaft shift fork
7. Transfer case
8. Thimble covers
9. Clutch sleeve—front output shaft
10. Clutch gear—front output shaft
11. Bearing—front output shaft rear
12. Race—front output shaft bearing
13. End play shims—front output shaft
14. Cover plate
15. Lock plate, bolt and washer
16. Intermediate gear shaft
17. Thrust washer
18. Bearing spacer (thin)
19. Intermediate gear shaft needle bearings
20. Bearing spacer (thick)
21. Intermediate gear
22. Bottom cover
23. Stud (case-to-trans.)
24. Front output shaft
25. Front output shaft gear
26. Front ouput shaft bearing (front)
27. Front output shaft bearing race
28. Oil seal
29. Front yoke
30. Seal
31. Support—input shaft
32. Input shaft
33. Shims
34. Input shaft bearing
35. Input shaft bearing snap ring
36. Rear output shaft gear
37. Snap ring
38. Clutch sleeve—rear output shaft
39. Input shaft rear bearing (needle) (or pilot bearing)
40. Rear output shaft
41. Vent
42. Clutch gear—rear output shaft
43. Thrust washer
44. Bearing—rear output shaft front
45. Race—rear output shaft bearing
46. Speedometer drive gear
47. End play shims
48. Rear yoke
49. Rear output shaft oil seal
50. Bearing—rear output shaft rear
51. Bearing race
52. Rear bearing cap
53. Front bearing cap

Dana 300 transfer case components

31. Remove the case from the press and remove the output shaft, clutch gear and sleeve and the shaft rear bearing.

32. Remove the front output shaft front bearing with an arbor press and tool J–22912–01 or its equivalent.

33. Remove the front output shaft from the gear.

34. Remove the input shaft rear needle bearing from the rear output shaft using tool J–29369–1 or its equivalent. Support the shaft in a vise during removal.

35. Using a ⅜ in. drive and $7/16$ in. socket, remove the shift rod thimbles from the case.

ASSEMBLY

Coat all parts with SAE 85W–90 oil before assembly.

1. Apply Loctite® 220 or its equivalent to the thimbles and install them in the case.

2. Install the front output shaft gear on the front output shaft. Be sure that the clutch teeth on the gear face the shaft gear teeth.

3. Install the front bearing on the front output shaft using an arbor press. Be sure that the bearing is seated against the gear.

4. Install the front output shaft in the case and install the clutch sleeve and gear on the shaft.

5. Install the front output shaft rear bearing using an arbor press.

NOTE: Install an old yoke nut on the shaft to avoid damage to the threads.

6. Install the input shaft needle bearings in the rear output shaft with tool J–29179 or its equivalent.

7. Position the rear output shaft clutch gear in the case and insert the rear output shaft into the gear.

8. Install the thrust washer and front bearing on the rear output shaft using an arbor press.

9. Install the shims and bearing on the input shaft using an arbor press.

10. Install a new input shaft seal.

11. Using a new snapring, install the input shaft and bearing in the support.

12. Install the rear output shaft gear on the input gear and install a new gear retaining ring.

13. Measure the clearance between the input gear and the gear retaining snapring using a feeler gauge. Clearance should not exceed 0.003 in. (0.08mm). If clearance is beyond tolerance, add shims between the input shaft and bearing.

14. Install the clutch sleeve on the rear output shaft.

15. Apply Loctite® 515 or equivalent to the mating surfaces of the input shaft support and install the support assembly, shaft and gear in the case. Use two support bolts to align the support on the case and tap the support into position with a soft mallet. Torque the support bolts to 10 ft. lbs.

16. Install the rear bearing cap front bearing race.

17. Install the rear bearing cap rear bearing race.

18. Position the rear output shaft rear bearing in the rear bearing cap.

19. Install the rear output shaft yoke oil seal.

20. Install the speedometer gear and end-play shims on the rear output shaft.

21. Apply Loctite® 515 or equivalent to the mating surfaces of the cap and install the rear bearing cap. Use two cap bolts to align the cap and tap it into place with a soft mallet.

22. Tighten the cap bolts to 35 ft. lbs.

23. Install the rear output shaft yoke. Torque a new locknut to 120 ft. lbs.

24. Clamp a dial indicator on the rear output shaft bearing cap. Position the indicator stylus so that it contacts the end of the shaft.

25. Pry the shaft back and forth to check end-play. End-play should be 0.001–0.005 in. (0.025-0.13mm). If play is not correct, remove or add shims between the speedometer drive gear and the output shaft rear bearing.

26. Install the front output shaft rear bearing race.

27. Install the front output shaft end play shims and cover plate. Tighten the cover plate bolts to 35 ft. lbs.

NOTE: Apply Loctite® 220 to the bolts before installation.

28. Install the front output shaft front bearing race.

29. Install the front output shaft yoke oil seal.

30. Install the shift rod oil seals.

31. Install the front bearing cap, using Loctite® 515 on the mating surfaces. Use two bolts to align the cap and tap it into position with a soft mallet.

32. Install and tighten the bearings cap bolts to 35 ft. lbs.

33. Seat the rear bearing cup against the cover plate by tapping the end of the front output shaft with a plastic mallet.

Mount a dial indicator on the front bearing cap and position the stylus against the end of the output shaft. Pry the shaft back and forth to check end-play. End-play should be 0.001–0.005 in. (0.025-0.13mm). If the play is not correct, add or remove shims between the cover plate and case. If shims are added seat the rear bearing cup before checking.

34. Install the front output shaft yoke. Tighten the new locknut to 120 ft. lbs.

35. Install the front and rear output shaft shift forks.

36. Install the front output shaft shift rod poppet ball and spring in the front bearing cap.

37. Compress the poppet ball and spring and install the front output shaft shift rod part way in the case.

38. Insert the front output shaft shift rod through the shift fork.

39. Align the setscrew hole in the shift fork and rod. Install and tighten the setscrew to 14 ft. lbs.

40. Install the rear output shaft shift rod poppet ball and spring in the front bearing cap.

41. Compress the ball and spring and install the rear output shaft shift rail part way. The front output shaft shift rod should be in neutral and the interlocks seated in the front bearing cap bore.

42. Insert the rear output shaft shift rod through the shift fork.

43. Align the setscrew holes in the fork and rod. Torque the setscrew to 14 ft. lbs.

44. Insert tool J–25142 in the intermediate gear and install the needle bearings and spacer.

45. Install the intermediate gear thrust washers in the case. Make sure that the tangs are aligned with the grooves in the case. The thrust washers may be held in place with petroleum jelly.

46. Install a new O-ring seal on the intermediate shaft.

47. Position the intermediate gear in the case.

48. Install the intermediate shaft in the case bore. Tap the shaft into the gear until the shaft forces the tool out of the case.

49. Install the intermediate shaft lock plate and bolt. Torque the bolt to 23 ft. lbs.

50. Install the bottom cover, applying Loctite® 515 or equivalent to the mating surfaces. Install and torque the bolts to 15 ft. lbs.

51. Fill the case with 4 pints of SAE 85W–90W gear oil.

NP-207 Overhaul

CASE DISASSEMBLY

1. Remove fill and drain plugs.

2. Remove front yoke. Discard yoke seal washer and yoke nut.

3. Turn transfer case on end and position front case on wood blocks.

4. Shift transfer case to 4 Lo.

5. Remove extension housing attaching bolts. Using a hammer, tap the shoulder on the extension housing to break sealer loose.

6. Remove the snapring for the rear bearing from the main shaft and discard.

7. Remove the rear retainer attaching bolts. Using a hammer, tap the shoulder on the retainer to break sealer loose.

8. Remove the rear retainer and pump housing from the transfer case.

9. Remove the pump seal from the pump housing and discard.

10. Remove the speedometer drive gear from the main shaft.

11. Remove the pump gear from the main shaft.

12. Remove the bolts attaching the rear case to the front case and remove rear case. To separate the case, insert a prybar into the slots casted in the case ends and pry upward. DO NOT at-

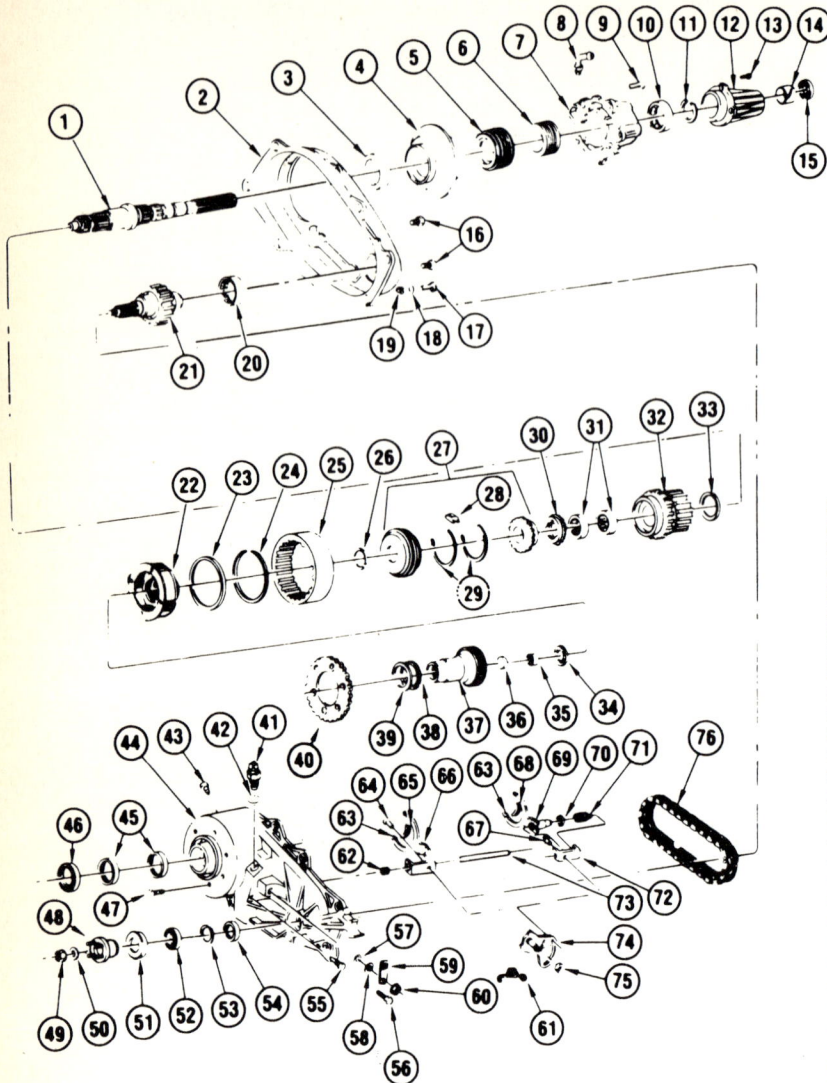

1. Main driveshaft
2. Case housing
3. Oil pump housing seal
4. Oil pump housing
5. Oil pump
6. Speed drive gear
7. Mainshaft rear bearing retaining ring
8. Case vent connector
9. Bolt
10. Mainshaft rear bearing
11. Mainshaft rear bearing retaining ring
12. Mainshaft extension
13. Hex bolt
14. Case mainshaft extension bushing
15. Main shaft extension seal
16. Case oil plug
17. Hex (m10 x 1.5mm x 35mm) (2 req'd) bolt
18. Alignment dowel washer housing; alignment

19. Housing alignment dowel
20. Front output shaft pilot bearing
21. Front output shaft
22. Planet gear assembly carrier

23. Planet gear carrier retaining ring thrust washer
24. Planet gear carrier retaining ring
25. Planet gear carrier annulus gear
26. Main driveshaft synchronizer retaining ring
27. Main driveshaft assembly synchronizer
28. Synchronizer strut
29. Synchronizer strut spring
30. Synchronizer stop ring
31. Drive chain sprocket bearing
32. Drive chain sprocket
33. Drive chain sprocket thrust washer
34. Input main drive gear thrust washer
35. Input drive gear pilot bearing
36. Cup plug
37. Input main drive assembly gear
38. Input drive gear thrust bearing
39. Input drive gear thrust bearing washer
40. Low range lockplate
41. Vacuum four wheel switch
42. Four wheel drive indicator light switch seal
43. Oil access hole plug
44. Case (front half) housing
45. Input drive bearing
46. Input drive gear seal
47. Hex bolt
48. Front output driveshaft flange yoke
49. Front output driveshaft yoke nut
50. Front output driveshaft yoke (rubber)
51. Front output driveshaft yoke deflector
52. Front output driveshaft seal
53. Front output driveshaft retaining ring
54. Front output driveshaft bearing
55. Shift sector spring screw
56. Screw
57. Shift sector & shaft oil seal
58. Shift sector & shaft retainer
59. Shifter shaft lever
60. Shift shaft lever nut
61. Shift sector assembly spring
62. Range fork bushing
63. Fork end pad
64. Range shift fork pin
65. Range shift fork center
66. Range shift assembly fork
67. Mode shft fork bracket pin
68. Mode shift fork center pad
69. Mode shift assembly fork
70. Mode shift fork spring cup
71. Mode shift fork spring
72. Mode shift fork assembly bracket
73. Shift fork shaft
74. W/shf, shift sector
75. Shift sector shaft spacer
76. Drive chain

NP-207 transfer case used in 1987 models

tempt to wedge the case halves apart at any point on the mating surfaces.

13. Remove the front output shaft and drive chain as an assembly. It may be necessary to raise the main shaft slightly for the output shaft to clear the case.

14. Pull up on the mode fork rail until rail clears range fork and rotate mode fork and rail and remove from transfer case.

15. Pull up on the main shaft until it separates from the planetary assembly. Remove the main shaft from the transfer case.

16. Remove the planetary assembly with the range fork from the transfer case.

17. Remove the planetary thrust washer, input gear thrust bearing and front thrust washer from the transfer case.

18. Remove the shift sector detent spring and retaining bolt.

19. Remove the shift sector, shaft and spacer from the transfer case.

20. Remove the locking plate retaining bolts and lock plate from the transfer case.

21. Remove the input gear pilot bearing using J-29369-1 or equivalent with a slide hammer.

22. Remove the front output shaft seal, input shaft seal and the rear extension seal using a brass drift.

23. Using J-33841 with J-8092 or equivalent, press the 2 caged roller bearings for the front input shaft gear from the transfer case.

24. Using J-29369-2 with J-33367 or a slide hammer, remove the rear bearing for the front output shaft.

25. Using a hammer and drift, remove the rear main shaft bearing from the rear retainer.

26. Using an awl, remove the snapring retaining the front output shaft bearing. Using a hammer and drift, remove the bearing from the case.

27. Remove the bushing from the extension housing using J-33839 with J-8092 or equivalent. Press bushing from the extension housing.

MAINSHAFT DISASSEMBLY

1. Remove the speedometer gear.
2. Using an awl, pry off the pump gear from the mainshaft.
3. Remove the snapring retaining the synchronizer hub from the mainshaft.
4. Using a brass hammer, tap the synchronizer hub from mainshaft.
5. Remove the drive sprocket.
6. Using J-33826 and J-8092 or equivalent, press 2 caged roller bearings from the drive sprocket.
7. Remove synchronizer keys and retaining rings from the synchronizer hub.
8. Clean and inspect all parts. Replace any parts if they show evidence of excessive wear, distortion or damage.

PLANETARY GEAR DISASSEMBLY

1. Remove the snapring retaining the planetary gear in the annulus gear.
2. Remove outer thrust ring and discard.
3. Remove planetary assembly from the annulus gear.
4. Remove inner thrust ring from the planetary assembly and discard.
5. Clean and inspect parts. Replace any parts if they show evidence of excessive wear, distortion or damage.

CLEANING & INSPECTION

Wash all parts thoroughly in clean solvent. Be sure all old lubricant, metallic particles, dirt, or foreign material are removed from the surfaces of every part. Apply compressed air to each oil feed port and channel in each case half to remove any obstructions or cleaning solvent residue.

Inspect all gear teeth for signs of excessive wear or damage and check all gear splines for burrs, nicks, wear or damage. Remove minor nicks or scratches with an oil stone. Replace any part exhibiting excessive wear or damage.

Inspect all snaprings and thrust washers for evidence of excessive wear, distortion or damage. Replace any of these parts if they exhibit these conditions.

Inspect the two case halves for cracks, porosity damaged mating surfaces, stripped bolt threads, or distortion. Replace any part that exhibits these conditions. Inspect the low range lock plate in the front case. If the lock plate teeth or the plate hub is cracked, broken, chipped, or excessively worn, replace the lock plate and the lock plate attaching bolts.

Inspect the condition of all needle, roller and thrust bearings in the front and rear case halves and the input gear. Also, check the condition of the bearing bores in both cases and in the input gear, rear output shaft and rear retainer. Replace any part that exhibits signs of excessive wear or damage.

PLANETARY GEAR ASSEMBLY

1. Install the inner thrust ring on planetary assembly.
2. Install the planetary assembly into the annulus gear.
3. Install the outer thrust ring and then the snapring.

MAINSHAFT ASSEMBLY

1. Using J-33828 and J-8092 or equivalent, install the front drive sprocket bearing. Press bearing until tool bottoms out. Bearing should be flush with front surface. Reverse tool on J-8092 or equivalent and press rear bearing into sprocket until tool bottoms out. The rear bearing should be recessed after installation.
2. Install thrust washer on the mainshaft.
3. Install drive sprocket on the mainshaft.
4. Install blocker ring and synchronizer hub on the mainshaft. Seat hub on main shaft and install a new snapring to retain.
5. Install pump gear on the mainshaft. Tap the gear with a hammer to seat on mainshaft.
6. Install speedometer gear on the mainshaft.

CASE ASSEMBLY

All of the bearings used in the transfer case must be correctly positioned to avoid covering the bearing oil feed holes. After installation of bearings, check the bearing position to be sure the feed hole is not obstructed or blocked by a bearing.

1. Install the lock plate in the transfer case. Coat case and lock plate surfaces around bolt holes with Loctite®515 or equivalent.
2. Position the lock plate to the case and align bolt holes in lock plate with case. Install attaching bolts and torque to specification.
3. Install the roller bearings for the input shaft into the transfer case using J-33830 and J-8092 or equivalent. Press bearings until tool bottoms in bore.
4. Install the front output shaft rear bearing, using J-33832 and J-8092 or equivalent. Press bearing until tool bottoms in case.
5. Install the front output shaft front bearing using J-33833 and J-8092 or equivalent. Press bearing until tool bottoms in bore.
6. Install the snapring that retains the front output shaft bearing in case.
7. Install the front output shaft seal using J-33834 or equivalent.
8. Install the input shaft seal using J-33831 or equivalent.
9. Install spacer on shift sector shaft and install sector in

transfer case. Install shift lever and retaining nut. Torque to specification.

10. Install shift sector detent spring and retaining bolt.

11. Install the pilot bearing into the input gear using J-33829 and J-8092 or equivalent. Press bearing until tool bottoms out.

12. Install the input gear front thrust bearing and input gear in transfer case.

13. Install the planetary gear thrust washer on the input gear. Position range fork on planetary assembly and install planetary assembly into the transfer case.

14. Install the mainshaft into the transfer case. Make sure the thrust washer is aligned with the input gear and planetary assembly before installing mainshaft.

15. Install mode fork on synchronizer sleeve and rotate until mode fork is aligned with range fork. Slide mode fork rail down through range fork until rail is seated in bore of transfer case.

16. Position drive chain on front output shaft and install chain on drive sprocket. Install front output shaft in the transfer case. It may be necessary to slightly raise the main shaft to seat the output shaft in the case.

17. Install the magnet into pocket of transfer case.

18. Apply ⅛ in. bead of Loctite®515 or equivalent to the mat-

ing surface of the front case. Install rear case on the front case aligning dowel pins. Install bolts and torque to 20-25 ft. lbs. Install the two bolts with washers into the dowel pin holes.

19. Install the output bearing into the rear retainer using J-33833 and J-8092 or equivalent. Press bearing until seated in bore.

20. Install pump seal in pump housing using J-33835 or equivalent. Apply petroleum jelly to pump housing tabs and install housing in rear retainer.

21. Apply ⅛ in. bead of Loctite®515 or equivalent to mating surface of rear retainer. Align retainer to case and install retaining bolts. Torque bolts to specification 15-20 ft. lbs.

22. Using a new snapring, install snapring on mainshaft. Pull up on mainshaft and seat snapring in its groove.

23. Install bushing in extension housing using J-33826 and J-8092 or equivalent. Press bushing until tool bottoms in bore.

24. Install a new seal in the extension housing using J-33843 or equivalent.

25. Apply ⅛ in. bead of Loctite®515 or equivalent to mating surface of extension housing. Align extension housing to the rear retainer and install attaching bolts. Torque bolts to specification 20-25 ft. lbs.

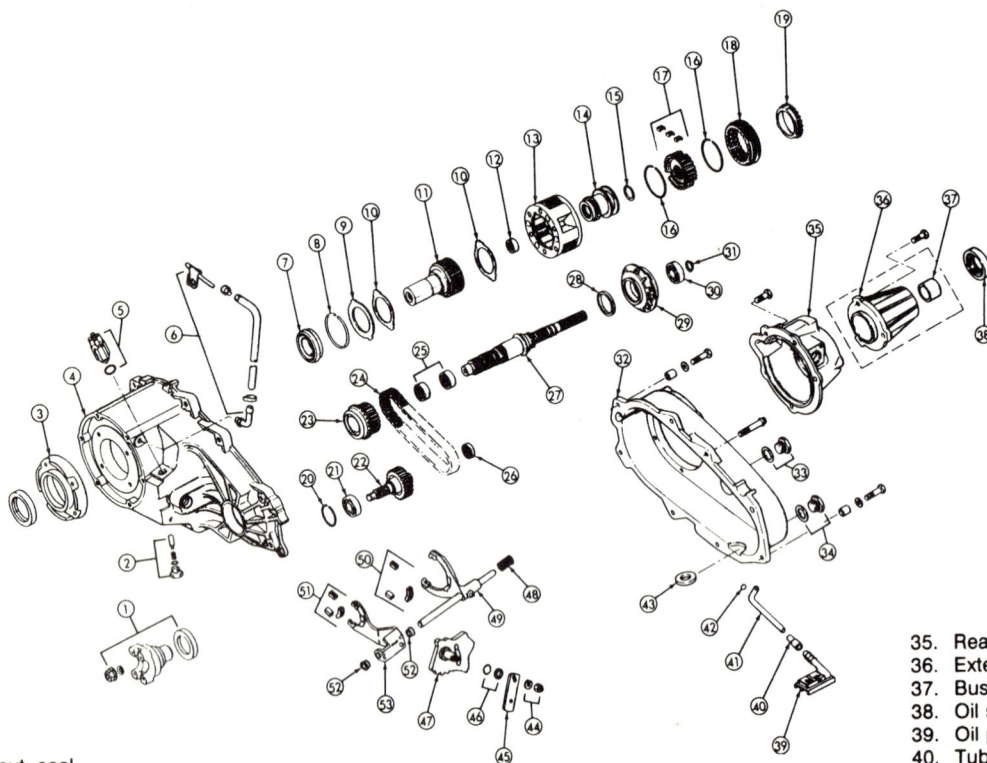

1. Front yoke, nut, seal washer, and oil seal
2. Shift detent plug, spring and pin
3. Front retainer and seal
4. Front case
5. Vacuum switch and seal
6. Vent assembly
7. Input gear bearing and snap ring
8. Low range gear snap ring
9. Input gear retainer
10. Low range gear thrust washers
11. Input gear
12. Input gear pilot bearing
13. Low range gear
14. Range fork shift hub
15. Synchro hub snap ring
16. Synchro hub springs
17. Synchro hub and inserts
18. Synchro sleeve
19. Stop ring
20. Snap ring
21. Output shaft front bearing
22. Output shaft (front)
23. Drive sprocket
24. Drive chain
25. Drive sprocket bearings
26. Output shaft rear bearing
27. Mainshaft
28. Oil seal
29. Oil pump assembly
30. Rear bearing
31. Snap ring
32. Rear case
33. Fill plug and gasket
34. Drain plug and gasket
35. Rear retainer
36. Extension housing
37. Bushing
38. Oil seal
39. Oil pickup screen
40. Tube connector
41. Oil pickup tube
42. Pickup tube O–ring
43. Magnet
44. Range lever nut and washer
45. Range lever
46. O–ring and seal
47. Sector
48. Mode spring
49. Mode fork
50. Mode fork inserts
51. Range fork inserts
52. Range fork bushings
53. Range fork

NP-231 exploded view

26. Install front yoke on output shaft. Install a new yoke seal washer with a new nut and torque to specification.
27. Install drain plug and torque to specification. Install fill plug.

NP-231 Overhaul

DISASSEMBLY

1. Remove the transfer case from the vehicle as descibed above.
2. Remove the attaching nuts from the front and rear output yokes. Remove the yokes and sealing washers.
3. Remove the bolts and tap the extension housing off of the rear retainer.
4. Remove the snapring from the rear bearing, then, remove the four bolts and separate the rear bearing retainer from the rear case half.
5. Remove the retaining bolts and separate the case halves by inserting a small pry bar in the pry slots on the case.
6. Remove the oil pump, pickup tube and pickup screen from the rear case.
7. Remove the mode spring from the shift rail.
8. Remove the drive chain by pushing the front input shaft inward and by angling the gear slightly to obtain adequate clearance to remove the chain.
9. Remove the mainshaft assembly from the front case half.
10. Remove the snapring and thrust washer from the planetary gear set assembly in the front case half.
11. Remove the annulus gear assembly and thrust washer from the front case half.
12. Lift the planetary gear assembly from the front case half.

13. Using an arbor press ram, press the input gear out of the front case assembly.
14. Remove the detent spring bolt from the front case and remove the detent spring.
15. Remove the shift selector, then the low-range lock plate from the front case assembly.

ASSEMBLY

1. Install the low-range lock plate then the shift selector in the front case assembly.
2. Install the detent spring and bolt in the front case.
3. Pess the input gear into the front case assembly.
4. Install the planetary gear assembly in the front case half.
5. Install the annulus gear assembly and thrust washer in the front case half.
6. Install the snapring and thrust washer in the planetary gear set assembly in the front case half.
7. Install the mainshaft assembly in the front case half.
8. Install the drive chain.
9. Install the mode spring on the shift rail.
10. Install the oil pump, pickup tube and pickup screen in the rear case.
11. Join the case halves and install the retaining bolts.
12. Install the snapring on the rear bearing, then install the bearing retainer.
13. Install the extension housing.
14. Install the front and rear output yokes.
15. Install the transfer case from the vehicle as descibed above.
16. Lubricate all components with Dexron®II ATF.
17. Seal the case halves and the extension housing with a ⅛ in. bead of RTV on assembly.
18. Fill the unit with DEXRON®II automatic transmission fluid.

AUTOMATIC TRANSFER CASE

Quadra-Trac®

The Warner Quadra-Trac® full-time, automatic transfer case was offered as optional equipment on CJ-7 models from 1976-79. The option was dropped after the 1979 model year.

NOTE: Complete assembly removal is normally not required except when the front output shaft, front annular bearing, transmission output shaft seals or the transfer case (front housing) require service. To service the chain, drive sprocket, differential unit, diaphragm control system, needle bearing, thrust washer or rear output shaft, the rear half of the Quadra-Trac® transfer case can be removed, giving access to these components without removing the unit from the vehicle.

REMOVAL AND INSTALLATION

1. Raise and support the vehicle.
2. Mark the front and rear output shaft yokes and universal

joints to provide alignment references to be used during assembly.
3. Disconnect the front driveshaft rear universal joint from the transfer case front yoke.
4. Disconnect the rear driveshaft front universal joint from the transfer case rear yoke.
5. Remove the bolts that attach the exhaust pipe support bracket to the transfer case. Support the transmission and remove the rear crossmember.
6. Mark and remove the diaphragm control vacuum hoses, lockout indicator switch wire and speedometer cable.
7. Disconnect the parking brake cable guide from the pivot on the right frame side.
8. Remove the two transfer case-to-transmission bolts which enter from the front side and the two that enter from the rear side.
9. Move the transfer case assembly backward until the unit is free of the transmission output shaft and lower the assembly from the vehicle.
10. Remove all gasket material from the rear of the transmission.

11. Slide the transfer case assembly forward until the unit engages the transmission output shaft.

12. Install the two transfer case-to-transmission bolts which enter from the front side and the two that enter from the rear side. Torque the bolts to 40 ft. lbs.

13. Connect the parking brake cable guide to the pivot on the right frame side.

14. Install the diaphragm control vacuum hoses, lockout indicator switch wire and speedometer cable.

15. Install the rear crossmember. Torque the bolts to 35 ft. lbs.

16. Install the bolts that attach the exhaust pipe support bracket to the transfer case.

17. Connect the rear driveshaft front universal joint to the transfer case rear yoke.

18. Connect the front driveshaft rear universal joint to the transfer case front yoke.

19. Lower the vehicle.

Quadra-Trac® Overhaul
DISASSEMBLY

1. Drain the lubricant from the case.

2. Remove the front and rear output shaft yokes and discard the yoke seal washers and yoke nuts.

3. Mark the rear retainer and rear case for an alignment reference.

4. Unbolt and remove the rear retainer. If necessary, use a soft mallet to loosen the retainer. Under no circumstances should the retainer be pried off.

5. Remove the differential shims and speedometer drive gear from the rear output shaft. Mark the shims for reference.

6. Remove the rear output bearing snapring and remove the bearing from the retainer using a soft mallet.

NOTE: The rear output bearing has one side shielded. Note this for reassembly.

7. Remove the rear output shaft seal from the retainer using a small prybar or punch.

8. Position the front case assembly on wood blocks. The blocks should have V cuts made in them for more positive support of the case.

9. Remove the case halve bolts. The case halves may be pried apart using a small prybar in the notches provided at the case ends.

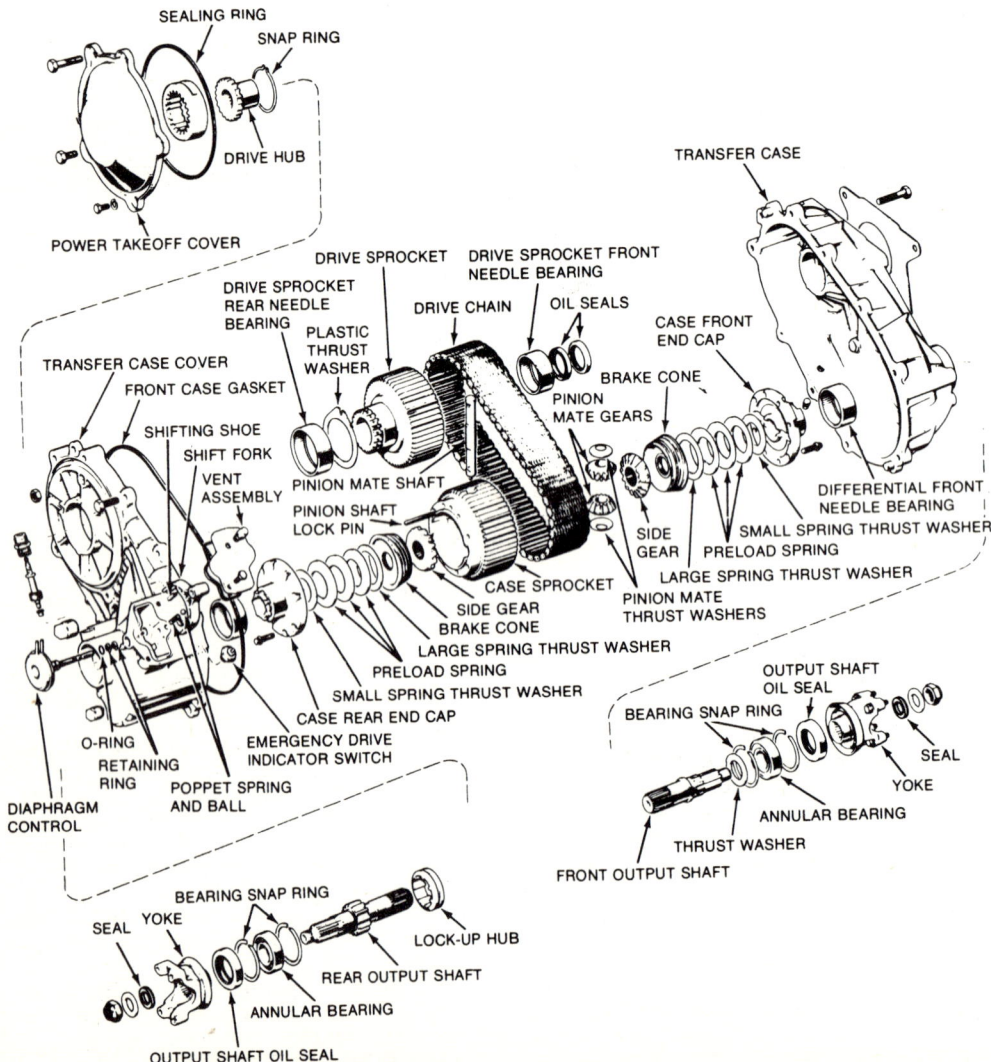

Warner Quadra-Trac® transfer case without low range

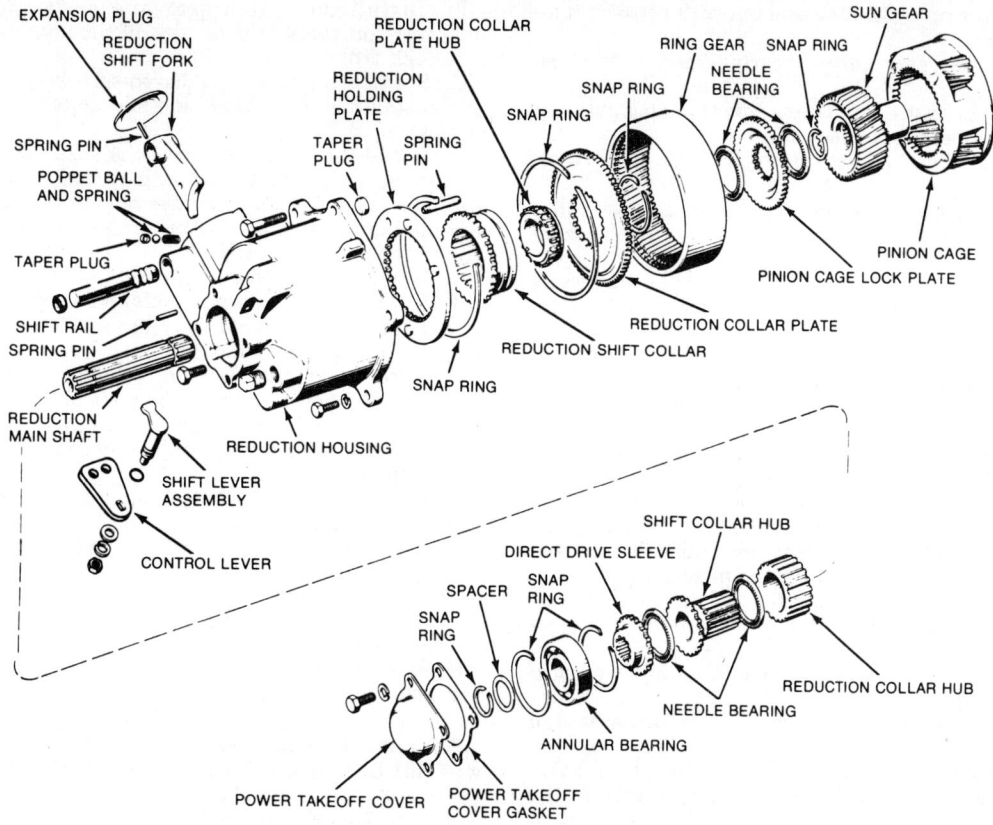

Warner Quadra-Trac® low range unit

NOTE: **The two case end bolts have flat washers and alignment dowels. Note their location for assembly.**

10. Remove the rear output shaft and viscous coupling as an assembly. Tap the shaft with a plastic mallet if necessary.
11. Remove the O-ring seal and pilot roller bearings from the main shaft.
12. Remove the rear output shaft from the viscous coupling.
13. Remove the shift rail spring from the rail.
14. Remove the plastic oil pump from the shaft bore in the rear case. Note the pump position for assembly reference. The end with the recess must face the shaft bore when installed.
15. Remove the rear output shaft bearing seal from the case. A small prybar may be used to pry it out.
16. Remove the front output shaft thrust bearing assembly. Remove the thick washer, bearing and thin washer.
17. Remove the driven sprocket retaining snapring.
18. Remove the drive sprocket, drive chain, driven sprocket, side gear clutch and clutch gear as an assembly. Place the assembly on a workbench and mark the components for assembly.
19. Remove the needle bearings and spacers from the main shaft and side gear bore. A total of 82 bearings and three spacers is used.
20. Remove the side gear/clutch gear assembly from the drive sprocket. Remove two snaprings and remove the clutch gear from the side gear.
21. Remove the side gear clutch, main shaft thrust washer and remaining main shaft needle bearing spacer.
22. Remove the front output shaft and shaft thrust bearing assembly. Note the installation sequence of the bearing assembly.
23. Remove the front output shaft seal from the front case using a small prybar or punch.
24. Remove the shift rail spring from the shift rail.

25. Remove the clutch sleeve, mode fork and spring as an assembly.
26. Remove the main shaft thrust washer and main shaft. Grasp the shaft and pull it straight up and out.
27. Move the range operating lever downward to the last detent position.
28. Disengage the range fork lug from the range sector slot.
29. Remove the annulus gear retaining snapring and thrust washer.
30. Remove the annulus gear and range fork.
31. Remove the planetary thrust washer from the hub.
32. Remove the planetary assembly.
33. Remove the main shaft thrust bearing from the input gear.
34. Remove the input gear and remove the input gear thrust bearing and race.
35. Remove the range selector detent ball and spring retaining bolt and remove the detent ball and spring.
36. Remove the range selector and operating lever attaching nut and lockwasher, and remove the lever.
37. Remove the range selector.
38. Remove the range selector O-ring and retainer.
39. Remove the input gear oil seal from the front case with a small prybar.

ASSEMBLY

Lubricate all parts before assembly with 10W–30 motor oil. Petroleum jelly will be indicated for some assemblies. Do not use chassis lube or other heavy lubricants.
1. Install new input gear and rear output shaft bearing oil seals. Seat the seals flush with the edge of the seal bore or with

the seal groove in the case. Coat the seal lips with petroleum jelly after installation.

2. Install the input gear thrust bearing race in the case counterbore.

3. Install the input gear thrust bearing on the input gear and install the gear and bearing in the case.

4. Install the main shaft thrust bearing in the bearing recess in the input gear.

5. Install the planetary assembly on the input gear. Make sure that the planetary pinion teeth mesh fully with the input gear.

6. Install the planetary thrust washer on the planetary hub.

7. Install a new sector shaft O-ring and retainer in the shaft bore in the case.

8. Install the range selector in the front case. Install the operating lever on the sector shaft and install the lever attaching washer and locknut on the shaft. Tighten the locknut to 17 ft. lbs.

9. Install the detent spring, ball and retaining bolt in the front case detent bore. Tighten the bolt to 22 ft. lbs.

10. Move the range selector to the last detent position.

11. Assemble the annulus gear and range fork. Install the assembled fork and gear over the planetary assembly. Be sure that the annulus gear is fully meshed with the planetary pinions.

12. Insert the range fork lug in the range detent slot.

13. Install the annulus thrust washer and retaining ring on the annulus gear hub.

14. Align the main shaft thrust washer in theinput gear, if necessary.

15. Install the main shaft. Be sure the shaft is fully seated in the input gear.

16. Install the main shaft thrust washer on the main shaft.

17. Install the short main shaft needle bearing spacer on the shaft.

18. Apply a liberal coating of petroleum jelly to the main shaft needle bearing surface and install 41 of the 82 needle bearings on the shaft. Be sure the bearings seat on the short spacer.

19. Install the long needle bearing spacer on the shaft. Lower the spacer onto the previously installed needle bearings carefully to avoid displacing them.

20. Align the shift rail bore in the case with the bore in the range fork and install the shift rail.

NOTE: Remove all traces of oil from the case shift rail bore before installing the rail. Oil in the case bore may prevent the rail from seating completely and prevent rear case installation.

21. Assemble the mode fork, mode fork spring and mode fork bracket.

22. Install the clutch sleeve in the mode fork. Be sure the sleeve is positioned so that the ID numbers on the sleeve face upward when the sleeve is installed.

23. Align the clutch sleeve and mode fork assembly with the shift rail and install the assembly on the shift rail and main shaft. Be sure that the clutch sleeve is meshed with the main shaft gear.

24. Lubricate the remaining 41 needle bearings and place them on the main shaft.

25. Install the side gear clutch on the main shaft with the teeth facing downward. Be sure the gear teeth mesh with the clutch sleeve.

26. Install the remaining short main shaft needle bearing spacer. Install the spacer carefully to avoid displacing previously installed bearings.

27. Install the front output shaft front thrust bearing in the front case. Correct sequence is thick race, bearing, thin race.

28. Install the front output shaft in the front case.

29. Install the clutch gear on the side gear. The tapered side of the clutch gear teeth must face the side gear teeth.

30. Install the clutch gear and drive sprocket locating snaprings on the side gear. Install the snaprings so that they face each other.

31. Position the drive and driven sprockets in the drive chain and install the assembled side and clutch gears in the drive sprocket.

32. Install the assembled drive chain, sprockets and side gear on the main shaft and front output shaft. Align the sprockets with the shaft, keeping the assembly level and carefully lower the assembly onto both shafts simultaneously. Do not displace any of the needle bearings.

33. Install the driven sprocket retaining snapring.

34. Install the front output shaft rear thrust bearing assembly on the front output shaft. Correct installation sequence is thin race, thrust bearing, thick race.

35. Install the shift rail spring on the shift rail.

36. Install a new O-ring on the main shaft pilot bearing hub.

37. Coat the main shaft pilot roller bearing hub and bearings with a liberal amount of petroleum jelly and install the rollers on the shaft.

38. Install the rear output shaft in the viscous coupling. Be sure it is fully seated.

39. Install the assembled viscous coupling and rear output shaft on the main shaft. Align the main shaft pilot hub with the pilot bearing bore in the rear output shaft and carefully lower the assembly onto the main shaft. Take care to avoid displacing the roller bearings.

40. Align the clutch gear teeth with the viscous coupling teeth and seat the coupling fully onto the clutch gear.

NOTE: When correctly installed, the clutch gear teeth will not be visible or extend out of the coupling.

41. Install the magnet in the front case, if removed.

42. Clean the mating surfaces of the case halves thoroughly.

43. Apply Loctite®515 or equivalent to the mating surfaces and all attaching bolts.

44. Join the case halves, aligning the dowels and install the bolts. Torque the bolts to 22 ft. lbs.

NOTE: The two end bolts require flat washers.

45. Install the oil pump on the rear output shaft and seat it in the case. The side with the recess should face the inside of the case.

46. Install the speedometer drive gear and differential shift, on the output shaft.

47. Install the vent chamber seal in the rear retainer.

48. Align and install the rear retainer on the case. Make the retainer finger tight only.

49. Install the yoke on the rear output shaft. Make the yoke finger tight only.

50. Mount a dial indicator on the rear retainer. Position the indicator stylus so that it contacts the top of the yoke nut.

51. Install the yoke on the front output shaft and rotate the shaft ten complete revolutions.

52. Rotate the front output shaft again and note the play indicated on the dial. End play should be 0.002–0.010 in. (0.05-0.25mm). If the end play must be adjusted, remove the rear retainer and add or subtract shims as required.

53. Remove both output shaft yokes and discard the nuts.

54. Install the front and rear yoke seals.

55. Remove the rear retainer bolts, apply Loctite®515 or equivalent to the mating surface of the retainer and to the bolts and install the bolts. Torque them to 22 ft. lbs.

56. Install new yoke seal washers on the output shafts, install yokes on the shafts and install new yoke nuts. Tighten the nuts to 110 ft. lbs.

57. Install the drain plug and tighten to 18 ft. lbs.

58. Pour 4 pints of 10W–30 motor oil into the case and install the fill plug. Tighten it to 18 ft. lbs.

DRIVELINE
Troubleshooting Basic Driveshaft and Rear Axle Problems

When abnormal vibrations or noises are detected in the driveshaft area, this chart can be used to help diagnose possible causes. Remember that other components such as wheels, tires, rear axle and suspension can also produce similar conditions.

BASIC DRIVESHAFT PROBLEMS

Problem	Cause	Solution
Shudder as car accelerates from stop or low speed	• Loose U-joint • Defective center bearing	• Replace U-joint • Replace center bearing
Loud clunk in driveshaft when shifting gears	• Worn U-joints	• Replace U-joints
Roughness or vibration at any speed	• Out-of-balance, bent or dented driveshaft • Worn U-joints • U-joint clamp bolts loose	• Balance or replace driveshaft • Replace U-joints • Tighten U-joint clamp bolts
Squeaking noise at low speeds	• Lack of U-joint lubrication	• Lubricate U-joint; if problem persists, replace U-joint
Knock or clicking noise	• U-joint or driveshaft hitting frame tunnel • Worn CV joint	• Correct overloaded condition • Replace CV joint

BASIC REAR AXLE PROBLEMS

First, determine when the noise is most noticeable.

Drive Noise—Produced under vehicle acceleration.

Coast Noise—Produced while the car coast with a closed throttle.

Float Noise—Occurs while maintaining constant car speed (just enough to keep speed constant) on a level road.

Road Noise

Brick or rough surfaced concrete roads produce noises that seem to come from the rear axle. Road noise is usually identical in Drive or Coast and driving on a different type of road will tell whether the road is the problem.

Tire Noise

Tire noises are often mistaken for rear axle problems. Snow treads or unevenly worn tires produce vibrations seeming to originate elsewhere. Temporarily inflating the tire to 40 lbs will significantly alter tire noise, but will have no effect on rear axle noises (which normally cease below about 30 mph).

Engine/Transmission Noise

Determine at what speed the noise is more pronounced, then stop the car in a quiet place. With the transmission in Neutral, run the engine through speeds corresponding to road speeds where the noise was noticed. Noises produced with the car standing still are coming from the engine or transmission.

Front Wheel Bearings

While holding the car speed steady, lightly apply the foot brake; this will often decease bearing noise, as some of the load is taken from the bearing.

Rear Axle Noises

Eliminating other possible sources can narrow the cause to the rear axle, which normally produces noise from worn gears or bearings. Gear noises tend to peak in a narrow speed range, while bearing noises will usually vary in pitch with engine speeds.

NOISE DIAGNOSIS

The Noise Is	Most Probably Produced By
• Identical under Drive or Coast	• Road surface, tires or front wheel bearings
• Different depending on road surface	• Road surface or tires
• Lower as the car speed is lowered	• Tires
• Similar with car standing or moving	• Engine or transmission
• A vibration	• Unbalanced tires, rear wheel bearing, unbalanced driveshaft or worn U-joint
• A knock or click about every 2 tire revolutions	• Rear wheel bearing
• Most pronounced on turns	• Damaged differential gears
• A steady low-pitched whirring or scraping, starting at low speeds	• Damaged or worn pinion bearing
• A chattering vibration on turns	• Wrong differential lubricant or worn clutch plates (limited slip rear axle)
• Noticed only in Drive, Coast or Float conditions	• Worn ring gear and/or pinion gear

Front and Rear Driveshafts

REMOVAL AND INSTALLATION

In order to remove the front and rear driveshafts, unscrew the holding nuts from the universal joint's U-bolts, remove the U-bolts and slide the shaft forward or backward toward the slip joint. The shaft can then be removed from the end yokes and removed from under the vehicle.

Each shaft is equipped with a splined slip joint at one end to allow for variations in length caused by vehicle spring action. Some slip joints are marked with arrows at the spline and sleeve yoke. When installing, align the arrows. If the slip joint is not marked with arrows, align the yokes at the front and rear of the shaft in the same horizontal plane. This is necessary in order to avoid vibration in the drive train. Torque the U-bolt nuts to 15 ft. lbs.

U-Joints

Most Jeep vehicles use a conventional universal joint at both ends of both driveshafts. The CJ-7 and Scrambler with automatic transmission use a double, or constant velocity, joint at the transfer case end of the front driveshaft. Universal joints on 1971 models are held together by snaprings on the outside of the bearing caps; 1972-90 models have C-type retainer rings on the inside of the bearing caps. The constant velocity joint is also assembled with snaprings on the outside.

On 1971 vehicles, three types of front axle U-joints were used; the Bendix type and the Rzeppa type, used on axle model 27, and the more familiar single cross cardan type used on axle model 27AF. All axles after 1972 use the single cross cardan type.

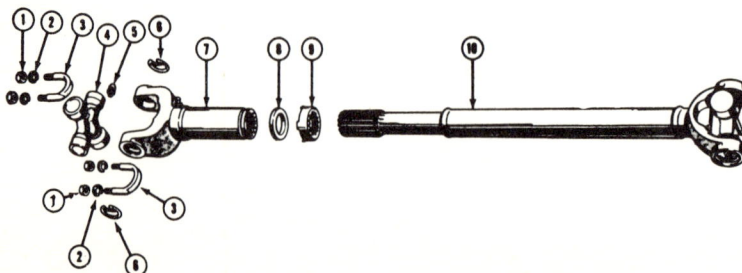

OVERHAUL

1971

1. Remove the snaprings by pinching the ends together with a pair of pliers. If the rings do not readily snap out of the groove, tap the end of the bearing lightly to relieve pressure against the rings.
2. After removing the snaprings, press on the end of one bearing until the opposite bearing is pushed from the yoke arm. Turn the joint over and press the first bearing back out of that arm by pressing on the exposed end of the journal shaft. To drive it out, use a soft drift with a flat face, about $1/32$ in. (0.8mm) smaller in diameter than the hole in the yoke; otherwise there is danger of damaging the bearing.
3. Repeat the procedure for the other two bearings, then lift out the journal assembly by sliding it to one side.
4. Wash all parts in cleaning solvent and inspect the parts after cleaning. Replace the journal assembly if it is worn extensively. Make sure that the grease channel in each journal trunnion is open.
5. Pack all of the bearing caps $1/3$ full of grease and install the rollers (bearings).

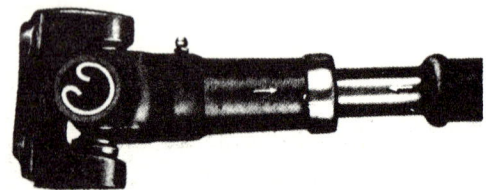

Driveshaft alignment markings

1. U-bolt nut
2. U-bolt washer
3. U-bolt
4. Universal joint journal
5. Lubrication fitting
6. Snap ring
7. Universal joint sleeve yoke
8. Rubber washer
9. Dust cap
10. Propeller shaft tube

1971 driveshaft

6. Press one of the cap/bearing assemblies into one of the yoke arms just far enough so that the cap will remain in position.

7. Place the journal in position in the installed cap, with a cap/bearing assembly placed on the opposite end.

8. Position the free cap so that when it is driven from the opposite end it will be inserted into the opening of the yoke. Repeat this operation for the other two bearings.

9. Install the retaining clips. If the U-joint binds when it is assembled, tap the arms of the yoke slightly to relieve any pressure on the bearings at the end of the journal.

Bendix Joint

With ordinary shop equipment it is nearly impossible to satisfactorily rebuild this unit. For this reason, the factory no longer supplies parts. After considerable mileage, a joint may pull apart upon removal from the vehicle. This does not mean that the joint is no longer usable. To assemble the axle shaft and universal:

1. Place the differential half of the shaft in a vise with the ground portion above the jaws.

2. Install the center ball (the one with the drilled hole) in the socket in the shaft, with the hole and groove visible.

3. Drop the center ball pin into the drilled hole in the wheel half of the shaft.

4. Place the wheel half of the shaft on the center ball. Slip the three balls into the races.

5. Turn the center ball until the groove lines up with the race for the remaining ball. Slip the ball into the race and straighten the wheel end of the shaft.

6. Turn the center ball until the pin drops into the hole in the ball.

7. Install the lock pin and center punch both ends to secure it.

Rzeppa Joint

With the joint removed, determine the method of attachment

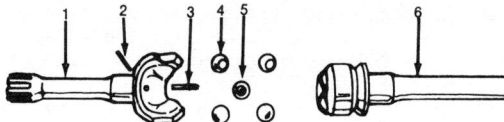

1. Outer shaft
2. Lock pin
3. Center ball pin
4. Universal joint ball
5. Center ball
6. Inner shaft

Bendix front axle U-joint

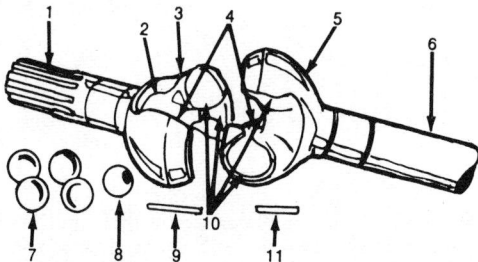

1. Outer shaft	7. Ball
2. Ground faces	8. Center ball
3. Outer yoke	9. Center ball pin
4. Flanges	10. Races
5. Inner yoke	11. Lock pin
6. Inner shaft	

Component view of the Bendix joint

of the axle to the joint. If three bolts are used, use step 1; if there are no screws, skip 1 and proceed to step 2.

1. Remove the three screws securing the front axle to the joint. Pull the shaft free of the splined inner race. Remove the retaining ring and remove the axle shaft retainer.

2. To remove the axle shaft from the joint, use a wooden pry, and exert force in the direction of the axis of the axle shaft. Use a mallet, if necessary, to exert enough force to drive the retaining ring, installed on the end of the shaft, into its groove in the spline, permitting the joint to be slipped off the shaft.

3. Push down on various points of the inner race and cage until the balls can be removed with the help of a small screwdriver.

4. There are two large rectangular holes in the cage as well as four small holes. Turn the cage so that the two bosses in the spindle shaft will drop onto the rectangular holes and lift out the cage.

5. To remove the inner race, turn it so that one of the bosses will drop into a rectangular hole in its cage and shift the race to one side. Lift it out.

6. Assembly is the reverse of disassembly. Take care to keep all parts as clean as possible.

1972-90 Single Cross Cardan Joint

1. Clamp the yoke, not the tube, in a vise.

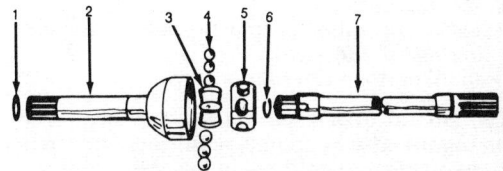

1. Outer axle shaft snap ring
2. Outer shaft
3. Universal joint inner race
4. Ball
5. Cage
6. Axle shaft retainer snap ring
7. Inner shaft

Rzeppa front axle U-joint

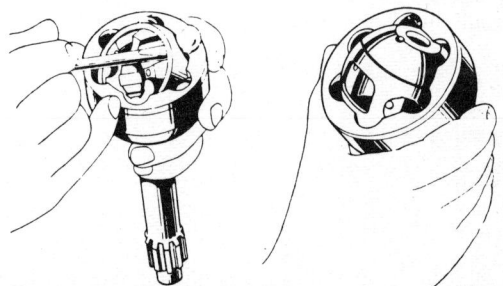

Removing the balls from the Rzeppa joint using a small screwdriver

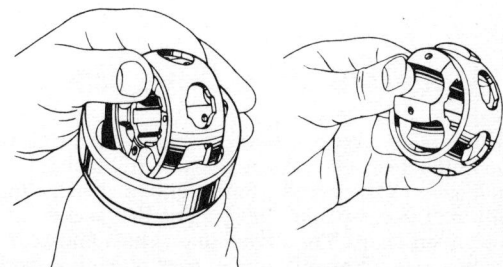

Removing the Rzeppa joint ball cage

2. Remove the bearing cap C-retainers. Tap on the bearing caps to relieve pressure as necessary.

3. Support the yoke on the vise jaws.

4. Tap one bearing cap in until the opposite one comes out.

5. Turn the yoke around and tap the exposed end of the spider to drive the remaining bearing cap out.

6. Clean all parts in solvent and dry. Use all the parts in the repair kit, even if some of the old ones seem usable.

7. Lubricate all needle bearings, bearing caps, and bearing surfaces with chassis grease.

8. Place the seals on the spider.

9. Install one cap and needle bearing assembly partway into the shaft yoke.

10. Install the spider and the opposite bearings and cap.

11. Support the yoke and seat both caps with a hammer.

12. Install the retainer C-clips. Tap the bearing caps as necessary.

13. Install the other two cap and bearing assemblies. Hold them in place with tape until the shaft is reinstalled.

Constant Velocity (Double Cardan) Joint

1. Remove the bearing cap retainer snaprings.

2. Mark all components for reassembly.

3. Use a 5/8 in. socket as a bearing cap driver and a 1 1/16 in. socket as a bearing cap receiver. Squeeze the assembly in a vise to force out the bearing caps.

4. Repeat the operation of step 3 to remove the bearing caps at the other end of the joint.

5. Clean all parts in solvent and dry.

NOTE: Do not disassemble the socket yoke, centering ball, spring, needle bearings, retainer, and thrust washers. These parts are sold as an assembly only.

6. Lubricate all bearings and contact surfaces with chassis grease.

7. Install the bearing caps on the transfer case yoke ends of the rear spider. Tape them in place.

8. Assemble the socket yoke and the rear spider.

9. Place the rear spider in the link yoke and install the bearing caps. Press them into place with the 5/8 in. socket. Install the snaprings.

10. Install the front spider, bearing caps, and snaprings in the driveshaft yoke.

11. Install the thrust washer and socket spring in the ball

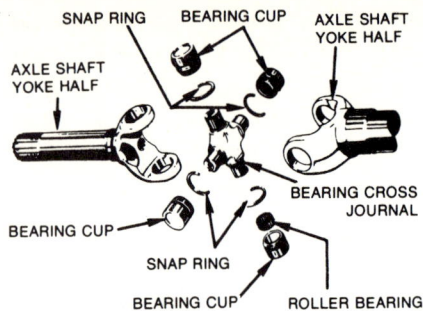

Single cross Cardan joint

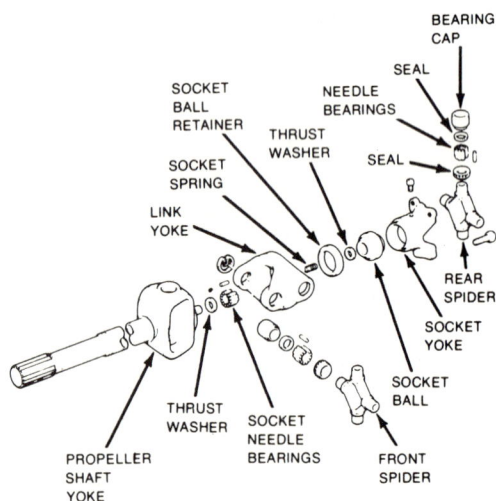

Constant velocity joint

socket bearing bore. Install the thrust washer on the ball socket bearing boss on the driveshaft yoke. Align the ball socket bearing boss with the ball socket bearing bore and insert the boss into the bore.

12. Align the front spider with the link yoke and install the bearing caps and snaprings.

DRIVE AXLE

Understanding Drive Axles

The drive axle is a special type of transmission that reduces the speed of the drive from the engine and transmission and divides the power to the wheels. Power enters the axle from the driveshaft via the companion flange. The flange is mounted on the drive pinion shaft. The drive pinion shaft and gear which carry the power into the differential turn at engine speed. The gear on the end of the pinion shaft drives a large ring gear the axis of rotation of which is 90 degrees away from the of the pinion. The pinion and gear reduce the gear ratio of the axle, and change the direction of rotation to turn the axle shafts which drive both wheels. The axle gear ratio is found by dividing the number of pinion gear teeth into the number of ring gear teeth.

The ring gear drives the differential case. The case provides the two mounting points for the ends of a pinion shaft on which are mounted two pinion gears. The pinion gears drive the two side gears, one of which is located on the inner end of each axle shaft.

By driving the axle shafts through the arrangement, the differential allows the outer drive wheel to turn faster than the inner drive wheel in a turn.

The main drive pinion and the side bearings, which bear the weight of the differential case, are shimmed to provide proper bearing preload, and to position the pinion and ring gears properly.

NOTE: The proper adjustment of the relationship of the ring and pinion gears is critical. It should be attempted only by those with extensive equipment and/or experience.

Limited-slip differentials include clutches which tend to link each axle shaft to the differential case. Clutches may be engaged either by spring action or by pressure produced by the torque on the axles during a turn. During turning on a dry pavement, the effects of the clutches are overcome, and each wheel turns at the required speed. When slippage occurs at either wheel, however, the clutches will transmit some of the power to the wheel which has the greater amount of traction. Because of the presence of clutches, limited-slip units require a special lubricant.

Determining Axle Ratio

The drive axle is said to have a certain axle ratio. This number (usually a whole number and a decimal fraction) is actually a comparison of the number of gear teeth on the ring gear and the pinion gear. For example, a 4.11 rear means that theoretically, there are 4.11 teeth on the ring gear and one tooth on the pinion gear or, put another way, the driveshaft must turn 4.11 times to turn the wheels once. Actually, on a 4.11 rear, there might be 37 teeth on the ring gear and 9 teeth on the pinion gear. By dividing the number of teeth on the pinion gear into the number of teeth on the ring gear, the numerical axle ratio (4.11) is obtained. This also provides a good method of ascertaining exactly what axle ratio one is dealing with.

Another method of determining gear ratio is to jack up and support the car so that both rear wheels are off the ground. Make a chalk mark on the rear wheel and the driveshaft. Put the transmission in neutral. Turn the rear wheel one complete turn and count the number of turns that the driveshaft makes. The number of turns that the driveshaft makes in one complete revolution of the rear wheel is an approximation of the rear axle ratio.

REAR AXLE

NOTE: Two different types of shafts have been used: the tapered shaft and the the flanged shaft. The differences are obvious. The terms, tapered and flanged, refer to the outer end of the shaft. The tapered shaft has a single retaining nut on the outer end of the axle shaft. The flanged shaft has a mounting flange for the brake drum on the outer end of the shaft and the shaft is held in place by a retaining plate. One other important point, some tapered and flanged axles have an inner oil seal fitted in the axle shaft housing, inboard of the bearing; some do not. If your axle does not have one, don't install one when replacing the bearing! Axles with an inner seal rely on chassis lube for bearing lubrication and must be prelubed prior to installation. Axles without an inner seal rely on differential oil to lubricate the bearing.

Pinion Oil Seal

REMOVAL AND INSTALLATION

Semi-Floating Axle w/Tapered Shaft

NOTE: Special tools are needed for this job.

1. Raise and support the vehicle and remove the rear wheels and brake drums.
2. Mark the driveshaft and yoke for reassembly and disconnect the driveshaft from the rear yoke.
3. With a socket on the pinion nut and an inch lb. torque wrench, rotate the drive pinion several revolutions. Check and record the torque required to turn the drive pinion.

4. Remove the pinion nut. Use a flange holding tool to hold the flange while removing the pinion nut. Discard the pinion nut.
5. Mark the yoke and the drive pinion shaft for reassembly reference.
6. Remove the rear yoke with a puller.
7. Inspect the seal surface of the yoke and replace it with a

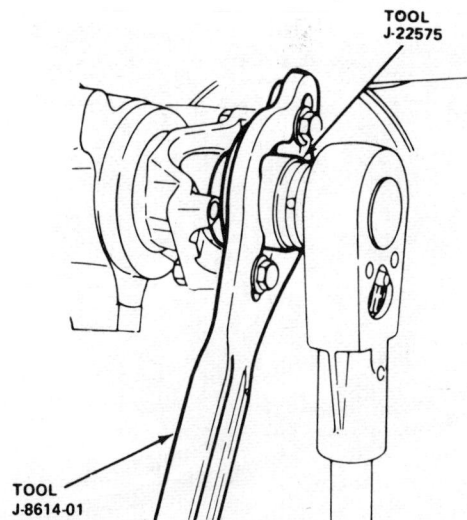

TOOL
J-22575

TOOL
J-8614-01

Pinion nut removal on all axles

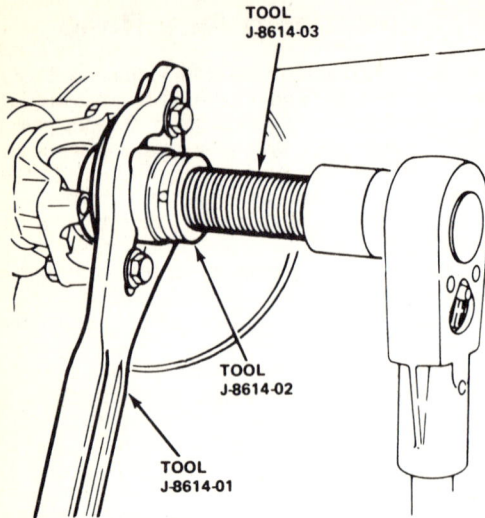

Pinion yoke removal on all axles

new one if the seal surface is pitted, grooved, or otherwise damaged.

8. Remove the pinion oil seal using tool J-9233.

9. Before installing the new seal, coat the lip of the seal with rear axle lubricant.

10. Install the seal, driving it into place with tool J-22661.

11. Install the yoke on the pinion shaft. Align the marks made on the pinion shaft and yoke during disassembly.

12. Install a new pinion nut. Tighten nut until endplay is removed from the pinion bearing. Do not overtighten.

13. Check the torque required to turn the drive pinion. The pinion must be turned several revolutions to obtain an accurate reading.

14. Tighten the pinion nut to obtain the torque reading observed during disassembly (Step 3) plus 5 inch. lbs. Tighten the nut minutely each time, to avoid overtightening. Do not loosen and then retighten the nut.

NOTE: If the desired torque is exceeded a new collapsible pinion spacer sleeve must be installed and the pinion gear preload reset.

15. Install the driveshaft, aligning the index marks made during disassembly. Install the rear brake drums and wheels.

Pulling off the wheel hub

Semi-Floating Axles w/Flanged Shaft

1. Raise and support the vehicle.
2. Mark the driveshaft and yoke for reference during assembly and disconnect the driveshaft at the yoke.
3. Remove the pinion shaft nut and washer.
4. Remove the yoke from the pinion shaft, using a puller.
5. Remove the pinion shaft oil seal with tool J-25180.
6. Install the new seal with a suitable driver.
7. Install the pinion shaft washer and nut. Tighten the nut to 210 ft. lbs.
8. Align the index marks on the driveshaft and yoke and install the driveshaft. Tighten the attaching bolts or nuts to 16 ft. lbs.
9. Remove the supports and lower the vehicle.

Axle Shaft

REMOVAL AND INSTALLATION

Tapered Shaft

1. Jack up the vehicle and remove the hub cap.
2. Remove the wheel.
3. Remove the axle nut dust cap.
4. Remove the axle shaft cotter pin, castle nut and flat washer.
5. Back off the brake adjustment.
6. Use a puller to remove the wheel hub.
7. Remove the screws attaching the brake dust protector, grease and bearing retainers, brake assembly and shim to the housing.
8. Remove the hydraulic line from the brake assembly.
9. Remove the dust shield and oil seal.

NOTE: If both shafts are being removed, keep the shims separated. Axle shaft endplay is adjusted at the left side only.

10. Use a puller to remove the axle shaft.

11. Install the axle shaft in the reverse order of removal, using a new grease seal and installing the hub assembly before the woodruff key. Tighten the axle shaft nut to 150 ft. lbs. Some axles have and inner oil seal fitted in the axle shaft housing, inboard of the bearing; some do not. If your axle does not have one, don't install one when replacing the bearing! Axles with an inner seal rely on chassis lube for bearing lubrication and must be prelubed prior to installation. Axles without an inner seal rely on differential oil to lubricate the bearing.

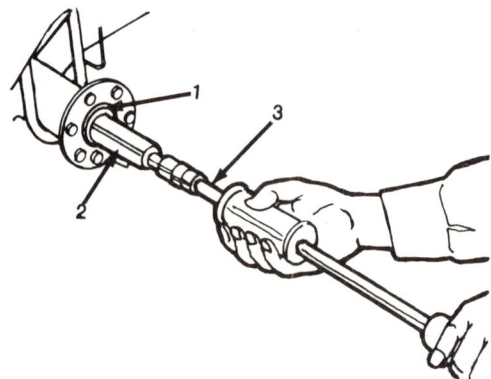

1. Cone and roller
2. Axle
3. Tool

Pulling the axle shaft

NOTE: Should the axle shaft be broken, the inner end can usually be drawn out of the housing with a wire loop after the outer oil seal is removed. However, if the broken end is less than 8 in. (203mm) long, it usually is necessary to remove the differential assembly.

Flanged Shaft

1. Jack up the vehicle and remove the wheels.
2. Remove the brake drum spring locknuts and remove the drum.
3. Remove the axle shaft flange cup plug by piercing the center with a sharp tool and prying it out.
4. Using the access hole in the axle shaft flange, remove the nuts which attach the backing plate and retainer to the axle tube flange.
5. Remove the axle shaft from the housing with an axle puller.
6. Install in reverse order of removal. Torque the bearing retainer bolts to 50 ft. lbs. in a criss-cross pattern

NOTE: Some axles have an inner oil seal fitted in the axle shaft housing, inboard of the bearing; some do not. If your axle does not have one, don't install one when replacing the bearing! Axles with an inner seal rely on chassis lube for bearing lubrication and must be prelubed prior to installation. Axles without an inner seal rely on differential oil to lubricate the bearing.

Axle Shaft Bearing

REMOVAL AND INSTALLATION

Tapered Shaft Axles

NOTE: An arbor press is necessary for this procedure.

1. With the aid of an arbor press, remove the bearing from the axle shaft.
2. The new bearing must be installed with the use of the same press used for removal, or, bearing replacing tool J-2995.

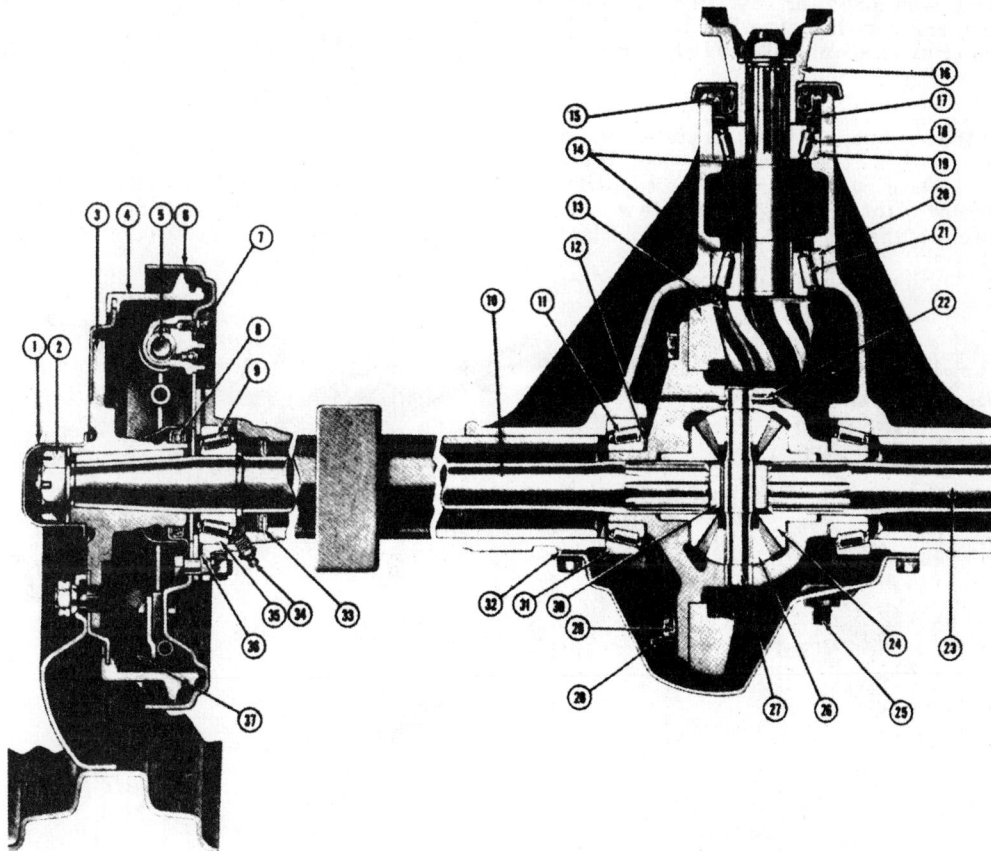

1. Hub cap	14. Pinion bearing shims	26. Pinion mate
2. Hex nut	15. Drive pinion oil seal	27. Pinion mate shaft
3. Rear wheel hub	16. Universal joint end yoke	28. Drive gear screw
4. Wheel brake drum	17. Drive pinion oil slinger	29. Drive gear screw strap
5. Brake wheel cylinder	18. Drive pinion outer bearing cone and roller	30. Axle shaft spacer (center block)
6. Backing plate	19. Drive pinion outer bearing cup	31. Differential bearing cup
7. Brake cylinder bleeder screw	20. Drive pinion inner bearing cup	32. Axle housing cover gasket
8. Axle shaft outer grease retainer	21. Drive pinion inner bearing cone and roller	33. Axle shaft oil seal (inboard)
9. Axle shaft bearing cone and roller	22. Pinion mate shaft pin and lock	34. Lubrication fitting
10. Axle shaft—left	23. Axle shaft—right	35. Axle shaft bearing cup
11. Differential bearing cone and roller	24. Side gear	36. Rear axle shaft bearing shims
12. Differential shims	25. Pipe plug (filler)	37. Brake shoe and lining
13. Axle drive gear and pinion		

1971 rear axle cutaway

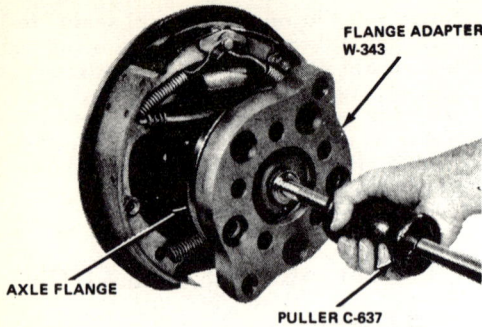

FLANGE ADAPTER
W-343

AXLE FLANGE

PULLER C-637

Removing the 1972–75 flanged axle shaft

3. If an inner seal, in the axle housing, was there when you removed the shaft, replace it with a new inner seal and pack the bearing with wheel bearing grease, making sure the grease fills the cavities between the bearing rollers.

NOTE: If there was no inner seal, don't install one. Don't pack the bearing with grease. The lack of an inner seal indicates that the bearing is lubed with axle lubricant.

4. Axle shaft endplay can be measured by installing the hub retaining nut on the shaft so that it can be pushed and pulled with relative ease. Strike the end of each axle shaft with a lead hammer to seat the bearing cups against the support plate. Mount a dial indicator on the left side support plate with the stylus resting on the end of the axle shaft. Check the endplay while pushing and pulling on the axle shaft. Endplay should be within 0.004-0.008 in. (0.1-0.2mm), with 0.006 in. (0.15mm) ideal. Add shims to increase endplay. Remove the hub retaining nut when finished checking endplay.

NOTE: When a new axle shaft is installed, a new hub must also be installed. However, a new hub can be installed on an original axle shaft if the serrations on the shaft are not worn or damaged. The procedures for installing an original hub and a new hub are different.

5. Install an original hub in the following manner:
   a. Align the keyway in the hub with the axle shaft key
   b. Slide the hub onto the axle shaft as far as possible
   c. Install the axle shaft nut and washer
   d. Install the drum, drum retaining screws, and wheel
   e. Lower the vehicle onto its wheels and tighten the axle shaft nut to 250 ft. lbs. If the cotter pin hole is not aligned, tighten the nut to the next castellation and install the pin. Do not loosen the nut to align the cotter pin hole.
6. Install a new hub in the following manner:
   a. Align the keyway in the hub with the axle shaft key.
   b. Slide the hub onto the axle shaft as far as possible.
   c. Install two well lubricated thrust washers and the axle shaft nut.
   d. Install the brake drum, drum retaining screws, and wheel.
   e. Lower the vehicle onto its wheels.
   f. Tighten the axle shaft nut until the distance from the outer face of the hub to the outer end of the axle shaft is $1\frac{5}{16}$

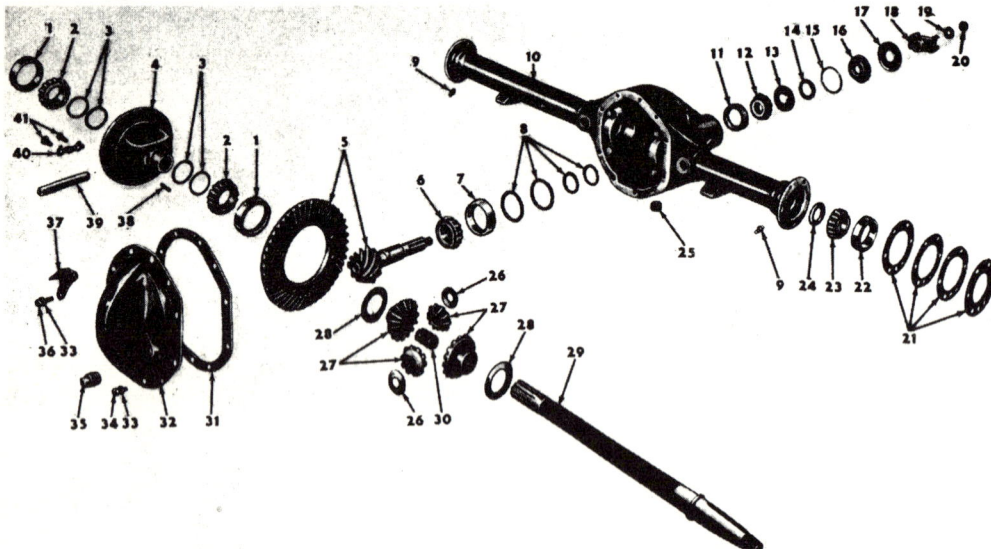

1. Bearing cup	15. Gasket	29. Axle shaft
2. Cone and rollers	16. Oil seal	30. Spacer
3. Shims	17. Dust shield	31. Gasket
4. Differential case	18. End yoke	32. Housing cover
5. Gear and pinion	19. Washer	33. Lockwasher
6. Cone and rollers	20. Pinion nut	34. Screw
7. Cup	21. Shims	35. Filler plug
8. Shims	22. Cup	36. Hex screw
9. Fitting	23. Cone and rollers	37. Tee bracket
10. Housing	24. Oil seal	38. Lock pin
11. Cup	25. Drain plug	39. Pinion shaft
12. Cone and rollers	26. Thrust washer	40. Lock strap
13. Oil slinger	27. Differential gears	41. Screw
14. Felt wick	28. Thrust washer	

1971 rear axle components

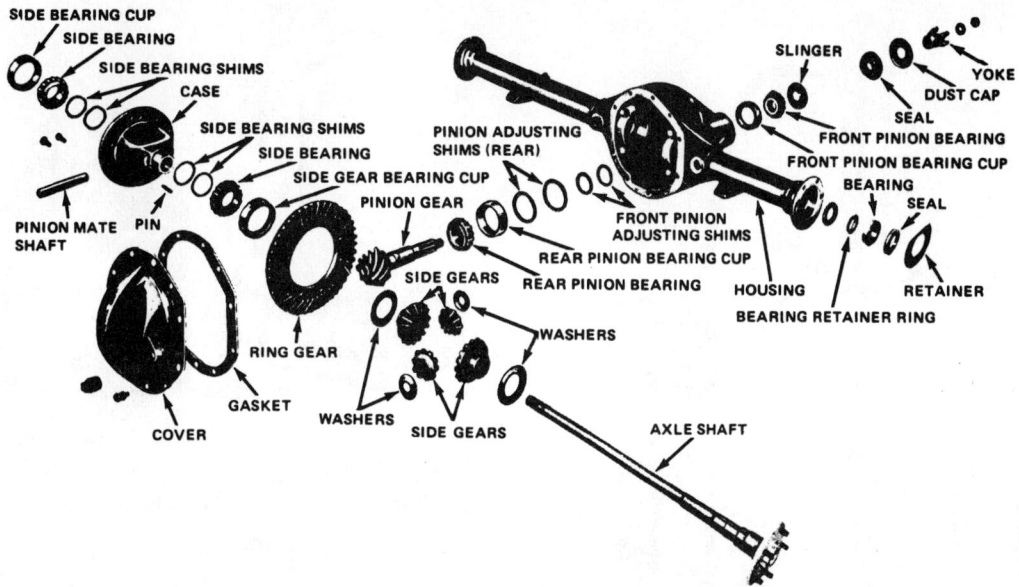

Component view of the 1972–75 flanged rear axle

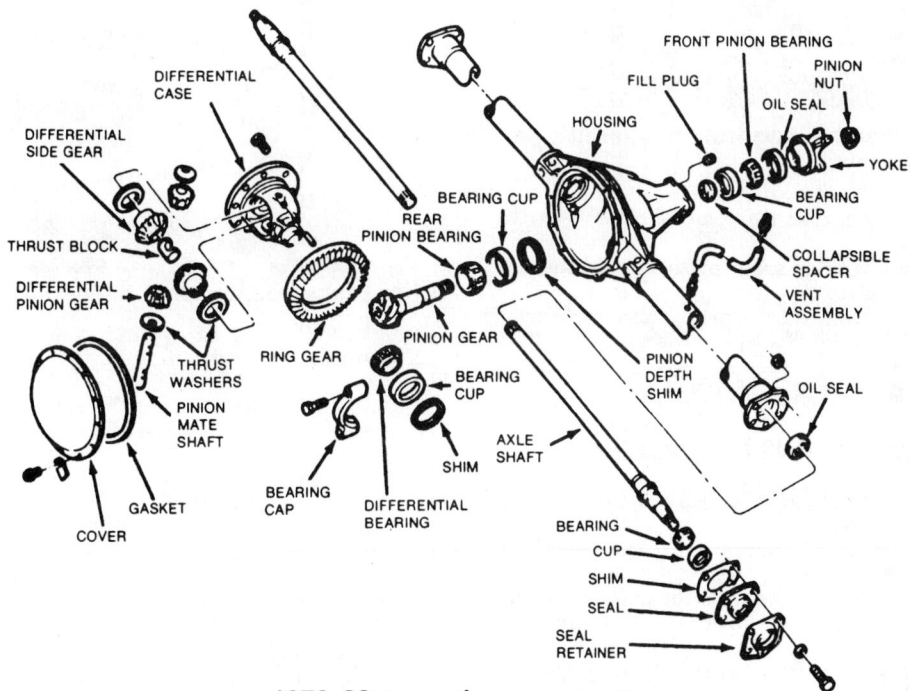

1976–90 rear axle components

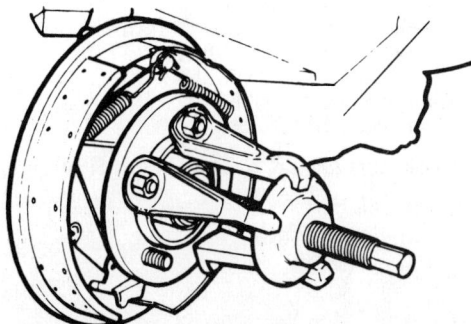

Using a puller to remove the hub on 1984–90 models

in. (33.3mm). Pressing the hub onto the axle to the specified distance is necessary to form the hub serrations properly.

g. Remove the axle shaft nut and one thrust washer.

h. Install the axle shaft nut and tighten it to 250 ft. lbs. If the cotter pin hole is not aligned, tighten the nut to the next castellation and install the pin. Do not loosen the nut to install the cotter pin.

7. Connect the brake line to the wheel cylinder and bleed the brake hydraulic system and adjust the brake shoes.

Flanged Shaft

NOTE: An arbor press is necessary for this procedure.

1. Position the axle shaft in a vise.
2. Remove the retaining ring by drilling a ¼ in. hole about ¾

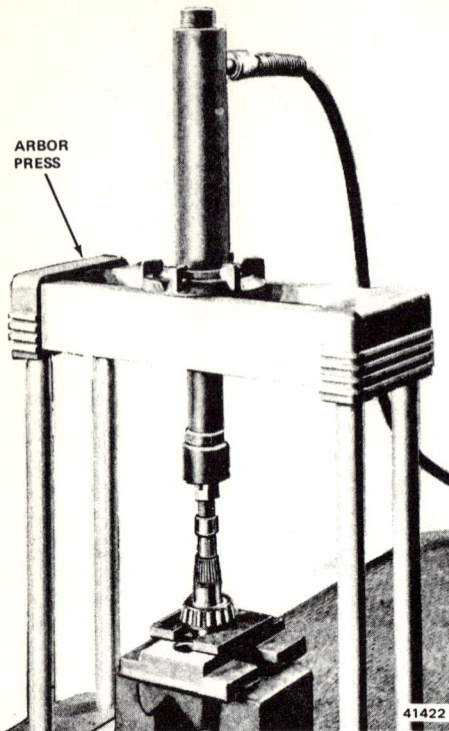

Removing the axle shaft bearing from a tapered axle shaft

of the way through the ring, then using a cold chisel over the hole, split the ring.

3. Remove the bearing with an arbor press, discard the seal and remove the retainer plate.

4. Installation is the reverse of removal. The new bearing must be pressed on. Make sure it is squarely seated.

Rear Axle Unit

REMOVAL AND INSTALLATION

1. Raise the vehicle and support it on jackstands.
2. Remove the rear wheels.
3. Place an indexing mark on the rear yoke and driveshaft, and disconnect the shaft.
4. Disconnect the shock absorbers from the axle tubes. Disconnect the track bar at the axle bracket, on vehicles so equipped.
5. Disconnect the brake hose from the tee fitting on the axle housing. Disconnect the vent tube at the axle.
6. Disconnect the parking brake cable at the frame mounting.
7. Remove the U-bolts. On vehicles with the spring mounted above the axle, disconnect the spring at the rear shackle.
8. Support the axle on a jack, remove the spring clips, and remove the axle assembly from under the vehicle.
9. Raise the axle on a jack and install the spring clips.
10. Install the U-bolts. On vehicles with the spring mounted above the axle, connect the spring at the rear shackle.
11. Connect the parking brake cable at the frame mounting.

Splitting the locking ring on the flanged shaft axle

Arbor press adapter on a flanged shaft bearing

12. Connect the brake hose at the tee fitting on the axle housing.
13. Connect the vent tube at the axle.
14. Connect the track bar at the axle bracket, on vehicles so equipped.
15. Connect the shock absorbers from the axle tubes.
16. Connect the driveshaft.
17. Install the rear wheels.
18. Lower the vehicle.

Wrangler
- Track bar bolts: 74 ft. lbs.
- Shock absorber-to-axle nut: 44 ft. lbs.
- Spring U-bolts: 90 ft. lbs.
- Spring shackle bolts: 95 ft. lbs.
- Spring-to-frame bracket bolts: 105 ft. lbs.

CJ-5, CJ-6, CJ-7 and Scrambler
- Shock absorber lower stud nut: 45 ft. lbs.
- $9/16$ in. spring U-bolt nut: 100 ft. lbs.
- $1/2$ in. spring U-bolt nut: 55 ft. lbs.
- Spring shackle nuts: 24 ft. lbs.
- Spring pivot bolts: 100 ft. lbs.

NOTE: Bleed and adjust brakes accordingly.

FRONT DRIVE AXLE

Axle Shaft, Bearing and Seal

REMOVAL AND INSTALLATION

1971

The front axle shaft and universal joint assembly is removed as an assembly.

NOTE: See the U-joint section of this section for a description of the three types used on these axles.

1. Remove the wheel.
2. Remove the hub with a puller. If there are locking hubs, remove them as detailed in Section 1.
3. Remove the axle shaft driving flange bolts.
4. Apply the foot brakes and remove the axle shaft flange with a puller.
5. Release the locking lip on the lockwasher and remove the outer nut, lockwasher, adjusting nut, and bearing lockwasher.
6. Remove the wheel hub and drum assembly with the bearings. Be careful not to damage the oil seal.

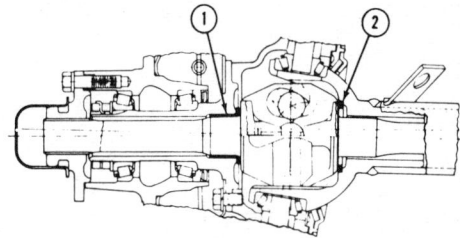

1. Bushing
2. Thrust washer

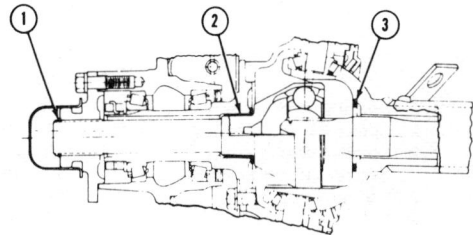

1. Snap ring
2. Bushing
3. Thrust washer

Bendex (top) and Rzeppa joint axle shafts

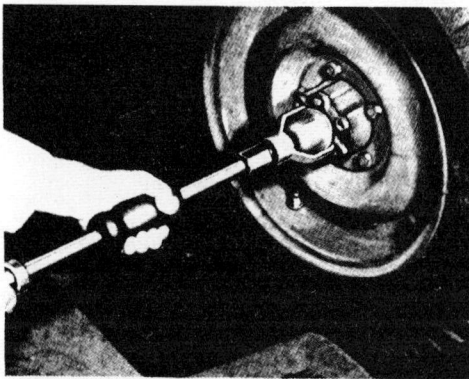

Pulling the front drive hub

Pulling off axle shaft drive flange

Wheel bearing nut wrench

7. Remove the hydraulic brake tube and the brake backing plate screws.
8. Remove the spindle.
9. Remove axle shaft and universal joint assembly.
10. Single cross cardan type installation is the reverse of the removal procedure.
11. Bendix type installation is as follows:
   a. Enter the U-joint and shaft assembly into the housing. Mesh the splined end of the shaft with the differential and push into place.
   b. Install the wheel bearing spindle.
   c. Install the brake tube and backing plate.
   d. Grease and assemble the wheel bearings and hub and drum on the spindle. Install the bearing washer and adjusting

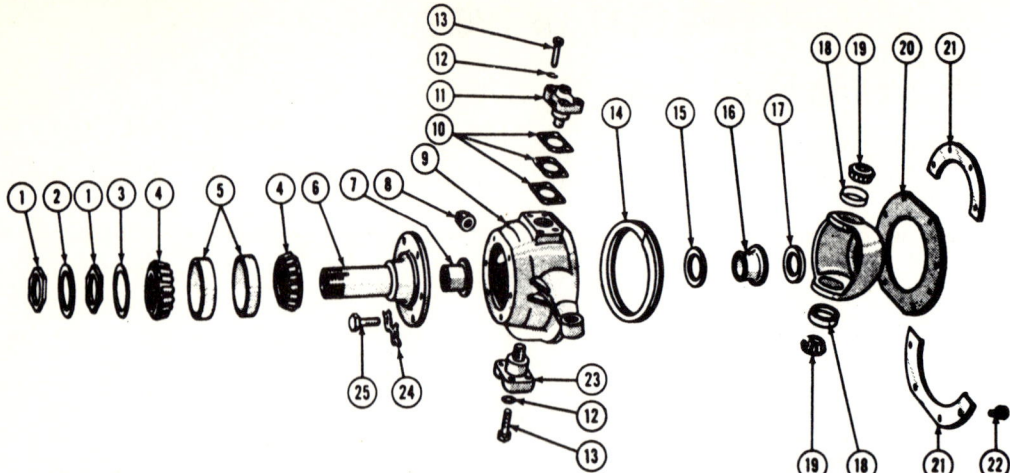

1. Bearing adjusting nut
2. Lockwasher
3. Lockwasher
4. Bearing cone and rollers
5. Bearing cup
6. Spindle
7. Bushing
8. Filler plug
9. Left knuckle
10. Shims
11. Upper bearing cap
12. Lockwasher
13. Bolt
14. Oil seal and backing ring
15. Thrust washer
16. Axle pilot
17. Oil seal
18. Bearing cup
19. Bearing cone and rollers
20. Oil seal
21. Retainer
22. Bolt
23. Lower bearing cap
24. Lockstrap
25. Bolt

Early model steering knuckle and spindle

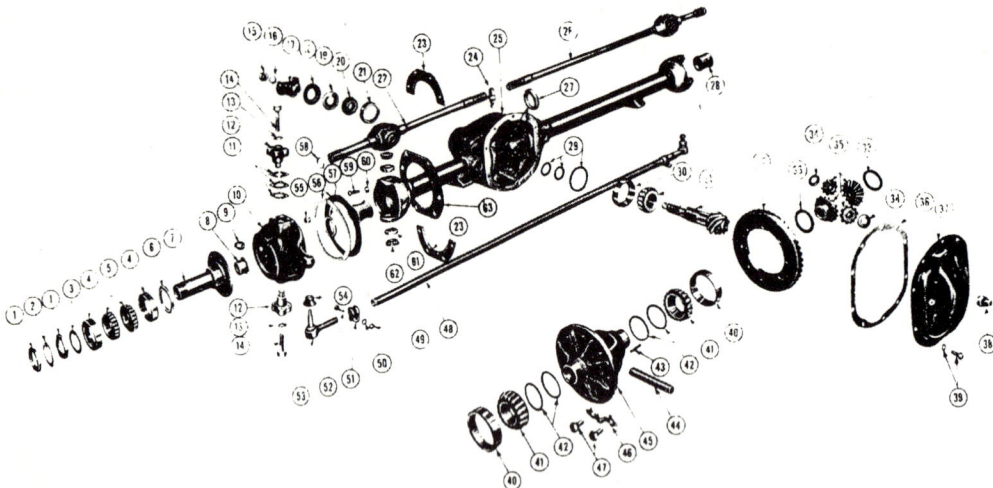

1. Nut
2. Lockwasher
3. Bearing lockwasher
4. Wheel bearing cup
5. Cone and rollers
6. Oil seal
7. Spindle
8. Spindle bushing
9. Filler plug
10. Left knuckle and arm
11. Shims
12. Pivot pin
13. Lockwasher
14. Capscrew
15. Nut
16. Washer
17. Universal joint yoke
18. Oil seal
19. Oil slinger
20. Cone and rollers
21. Bearing cup
22. Right axle shaft with universal joint
23. Knuckle oil seal retainer
24. Housing breather
25. Front axle housing
26. Left axle shaft with universal joint
27. Oil seal
28. Axle shaft guide
29. Shim pack
30. Bearing cup
31. Cone and rollers
32. Ring gear and pinion
33. Thrust washer
34. Thrust washer
35. Differential gears
36. Housing cover gasket
37. Housing cover
38. Fill plug
39. Screw and lockwasher
40. Bearing cup
41. Cone and rollers
42. Shims
43. Lock pin
44. Pinion shaft
45. Differential case
46. Lock strap
47. Bolts
48. Steering tie rod
49. Tie rod clamp nut
50. Lockwasher
51. Tie rod socket clamp
52. Screw
53. Tie rod socket
54. Dust cover
55. Nut
56. Oil seal and backing ring
57. Thrust washer
58. Snap ring
59. Stop bolt
60. Nut
61. Bearing cup
62. Cone and rollers
63. Gasket

Dana 27 front drive axle

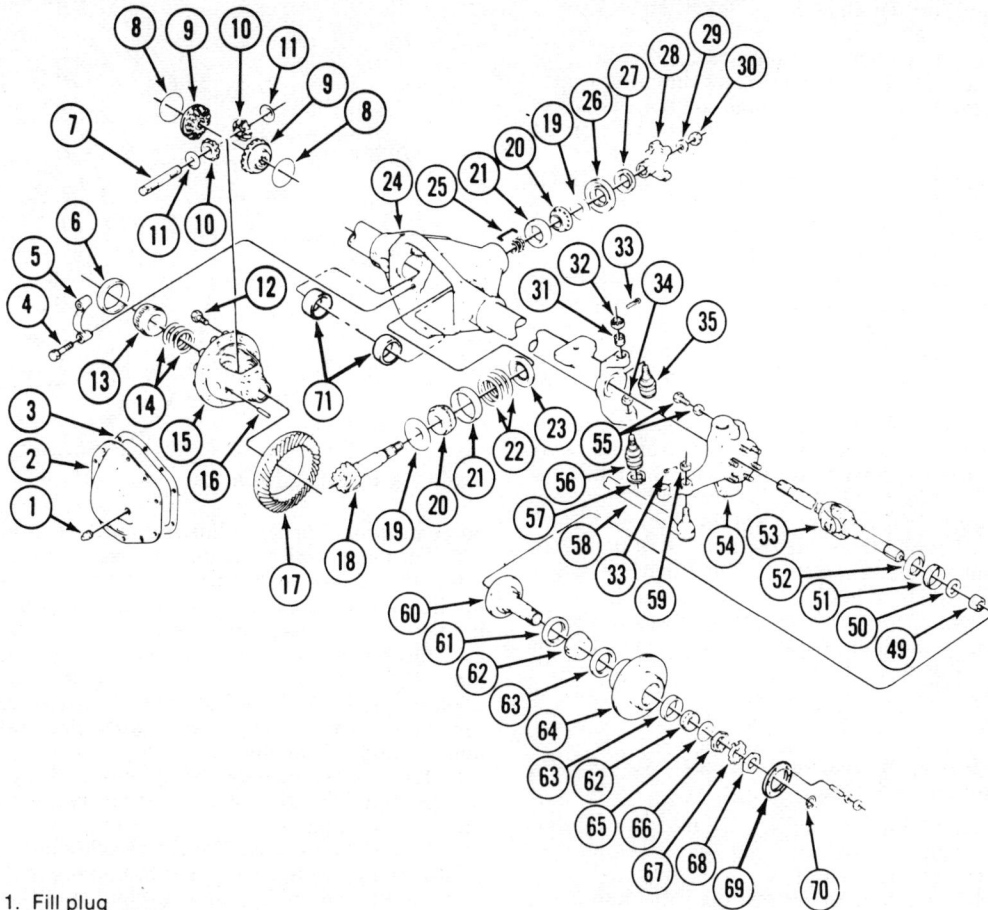

1. Fill plug
2. Axle housing cover
3. Axle housing cover gasket
4. Differential bearing cap bolt
5. Differential bearing cap
6. Differential bearing cup (2)
7. Pinion mate shaft
8. Thrust washer
9. Differential side gear
10. Differential pinion gear
11. Thrust washer
12. Ring gear mounting bolts
13. Differential bearing (2)
14. Differential bearing preload shims
15. Differential case
16. Pinion mate shaft pin
17. Ring gear
18. Pinion gear
19. Slinger
20. Pinion bearing
21. Pinion bearing cup
22. Pinion depth shims
23. Baffle
24. Axle housing
25. Pinion preload shims
26. Oil seal
27. Dust cap
28. Yoke
29. Washer
30. Pinion nut

31. Upper ball stud split ring seat
32. Upper ball stud nut
33. Cotter pin
34. Lower ball stud jamnut
35. Upper ball stud
49. Spindle bearing
50. Washer
51. Seal
52. Seal seat
53. Axle shaft
54. Steering knuckle
55. Steering stop bolt
56. Lower ball stud
57. Snap ring
58. Tie rod
59. Tie rod end nut
60. Spindle
61. Seal
62. Bearing
63. Bearing cup
64. Hub
65. Tabbed washer
66. Inner locknut
67. Lock washer
68. Outer locknut
69. Gasket
70. Snap ring
71. Inner oil seal

Dana 30 front axle assembly

nut. Tighten the nut until a slight drag is felt, then back off ⅙ turn. Install remaining parts.

12. Early Rzeppa type type requires a shimming procedure. Installation is the same as the Bendix type except that a shim pack must be installed between the driving flange and the wheel hub to determine the proper operating clearance for the U-joint. To do this:

a. Install the drive flange on the axle splines without shims.

b. Install the axle nut and tighten it snugly.

c. Install two opposite flange bolts snugly.

d. Use a feeler gauge to measure the gap between the outer end of the hub and the inner face of the driving flange. This determines the amount of shimming to be used. It is necessary to install shims of a thickness equal to the measured gap plug 0.015-0.050 in. (0.381-1.27mm). If no gap is found, install a 0.010 in. shim.

e. Install the correct amount of shims, replace the flange and install the six bolts. Install the axle shaft nut and make sure that the proper end float has been obtained. To do this, back off the shaft nut so that a 0.050 in. feeler will fit between the nut and driving flange. Tap the end of the shaft with a soft mallet which will force in the shaft the amount of end float. Measure the clearance between the nut and driving flange. Clearance should be 0.015-0.050 in. (0.381-1.27mm).

13. Late Rzeppa type installation is the same as the Bendix type except that a snapring is used to secure the outer end of the shaft controlling end float.

1972-86

NOTE: On models with locking hubs, refer to Locking Hub Removal and Installation.

1. Remove the locking hub or hub cap. On models with disc brakes, remove the caliper.

2. Remove the drive flange snapring.

3. On models with disc brakes, remove the rotor hub bolts, cover and gasket. On models with drum brakes, remove the axle flange bolts, lockwashers and flatwashers.

4. If the axle is on the vehicle, apply the foot brakes. Remove the axle flange with a puller.

5. Release the locking lip of the lockwasher, and remove the

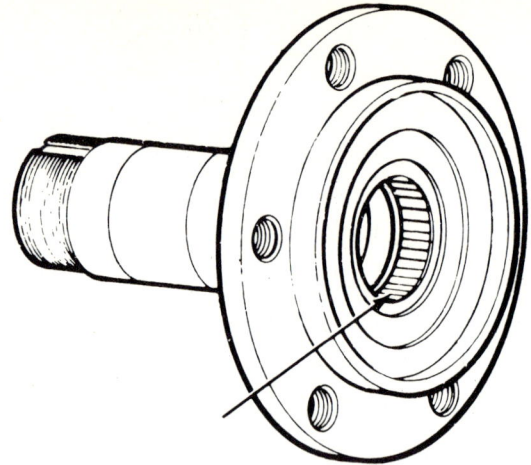

Spindle bearing location

outer nut, lockwasher, adjusting nut, and bearing lockwasher.

6. On models with the disc brakes, remove the bearing and rotor. On models with drum brakes, back off on the brake adjusting star wheel adjusters and remove the brake drum assembly with the bearings. Be careful not to damage the oil seal.

7. On models with drum brakes, remove the brake backing plate. If the axle is on the vehicle, it will first be necessary to disconnect the brake hose between the front brake line and the flexible connection. On models with disc brakes, remove the adapter and splash shield.

8. Remove the spindle and spindle bushing.

9. Remove the axle shaft and universal joint assembly.

10. Clean all parts.

11. Insert the universal joint and axle shaft assembly into the axle housing, being careful not to knock out the inner seal. Insert the splined end of the axle shaft into the differential and push into place.

12. Install the wheel bearing spindle and bushing.

13. Install the brake backing plate, or adapter and splash shield.

14. Grease and assemble the wheel bearings and oil seal.

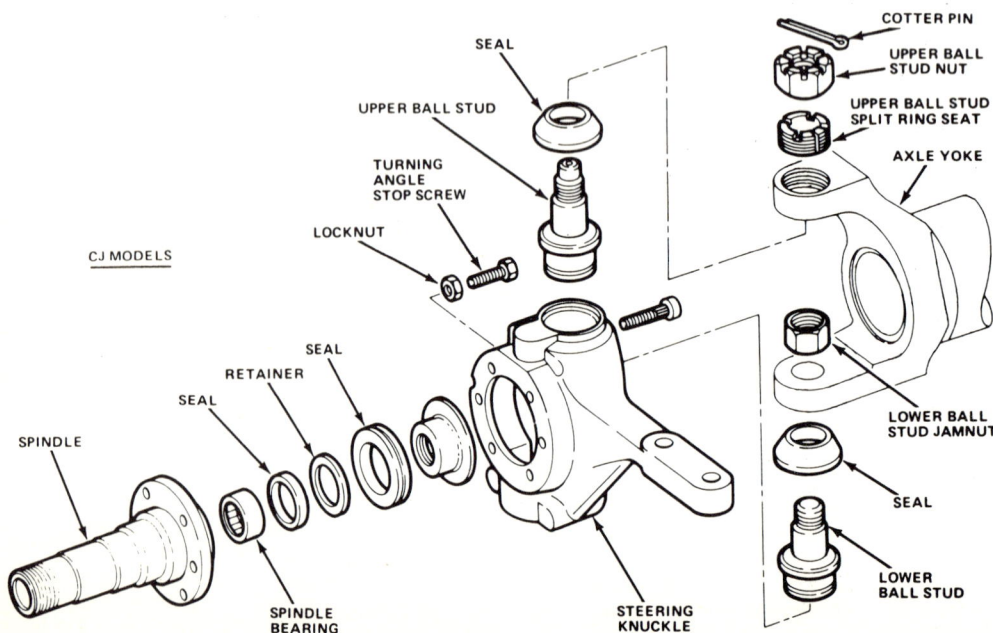

Late model spindle and knuckle assembly

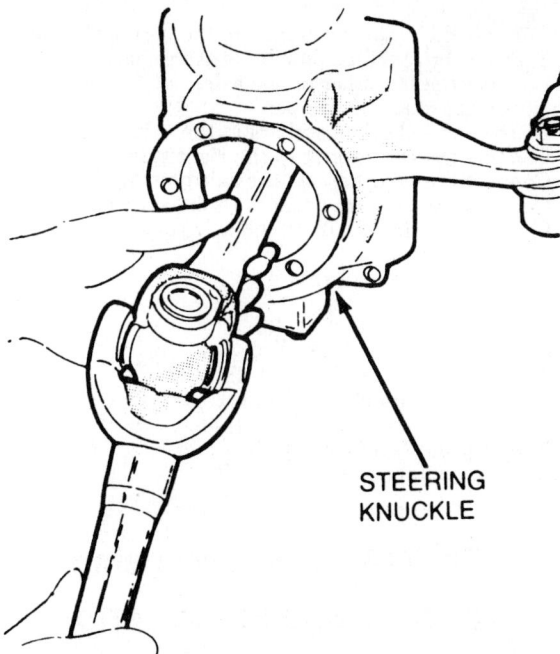

Typical front axle shaft removal on vehicles through 1979

15. Install the wheel hub and drum on the wheel bearing spindle. On disc brakes, install the rotor, hub, and caliper. Install the wheel bearing washer and adjusting nut. Tighten the nut to 50 ft. lbs., and back it off ⅙-¼ turn while rotating the hub. Install the lockwasher and nut, tighten the nut to 50 ft. lbs. and then bend the lip of the lockwasher over onto the locknut.
16. Install the drive flange and gasket onto the hub and attach with six capscrews and lockwashers. Torque the capscrews to 30 ft. lbs. in an alternate and even pattern. Install the snapring onto the outer end of the axle shaft.
17. Install the hub cap.
18. Install the wheel, lug nuts, and wheel disc.
19. If the tube was installed with the axle assembly on the vehicle, check the front wheel alignment, bleed the brakes and lubricate the front axle universal joints.

1987-90

1. Raise and support the vehicle safely.
2. Remove the wheels, calipers and rotors.
3. Remove the cotter pin, locknut and axle hub nut.
4. Remove the hub-to-knuckle attaching bolts.
5. Remove the hub and splash shield from the steering knuckle.
6. To remove the left shaft, remove the axle shaft from the housing.
7. To remove the right shaft:
   a. Disconnect the vacuum harness from the shift motor.
   b. Remove the shift motor from the housing.
   c. Remove the axle shaft from the housing.
8. To install the right axle shaft first be sure that the shift collar is in position on the intermediate shaft and that the axle shaft is fully engaged in the intermediate shaft end.
9. Install the shift motor, making sure that the fork engages with the collar. Tighten the bolts to 96 inch lbs.
10. On the left side, install the axle shaft in the housing.
11. Partially fill the hub cavity of the knuckle with chassis lube and install the hub and splash shield.
12. Tighten the hub bolts to 75 ft. lbs.

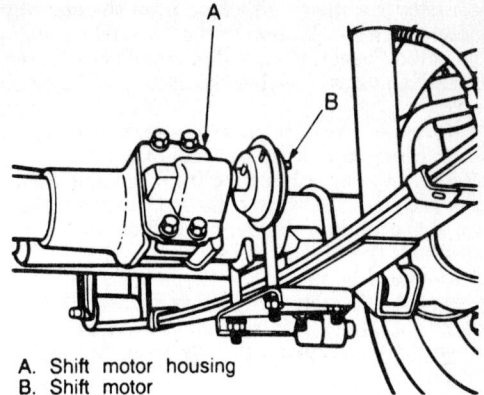

A. Shift motor housing
B. Shift motor

1987-90 front drive axle

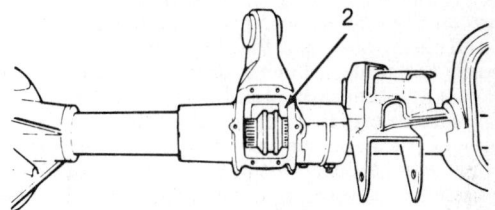

1. Shift motor
2. Shift collar

1987-90 right side front axle shaft

13. Install the hub washer and nut. Torque the nut to 175 ft. lbs. Install the locknut. Install a new cotter pin.
14. Install the rotor, caliper and wheel.

Pinion Seal and Yoke

REMOVAL AND INSTALLATION

All Models

1. Raise and support the front end on jackstands.
2. Matchmark the driveshaft and yoke and disconnect the driveshaft.
3. Using a holding tool, such as J-8614-01, on the yoke, remove the pinion nut.
4. Remove the yoke, using tools J-8614-01, -02, and -03, or their equivalents.
5. Using tool J-9233 or J-7583, or, on 1987-90 models, tool J-25180, or their equivalents, remove the seal.
6. Coat the outer rim of the new seal with sealer and intall it using a seal driver.
7. Install the yoke, pinion washer and a new pinion nut. Torque the nut to 210 ft. lbs.
8. Connect the driveshaft.

Front Axle Unit

REMOVAL AND INSTALLATION

1971-86

1. Raise and support the vehicle safely. Remove the wheels.
2. Index the driveshaft to the differential yoke for the proper alignment upon installation. Disconnect the driveshaft at the axle yoke and secure the shaft to the frame rail.

3. Disconnect the steering linkage from the steering knuckles. Disconnect the shock absorbers at the axle housing.

4. If the vehicle is equipped with a stabilizer bar, remove the nuts attaching the stabilizer bar connecting links to the spring tie plates.

5. On vehicles equipped with sway bar, remove nuts attaching sway bar connecting links to spring tie plates.

6. Disconnect the breather tube from the axle housing. Disconnect the stabilizer bar link bolts at the spring clips.

7. Remove the brake calipers, or drums, hub and rotor, or brake shoes, and the brake shield.

8. Remove the U-bolts and the tie plates.

9. Support the assembly on a jack and loosen the nuts securing the rear shackles, but do not remove the bolts.

10. Remove the front spring shackle bolts. Lower the springs to the floor.

11. Pull the jack and axle housing from underneath the vehicle.

12. Raise the axle into position.

13. Install the front spring shackle bolts.

14. Tighten the nuts securing the rear shackles.

15. Install the U-bolts and the tie plates.

16. Install the brake calipers, or drums, hub and rotor, or brake shoes, and the brake shield.

17. Connect the breather tube to the axle housing.

18. Connect the stabilizer bar link bolts at the spring clips.

19. On vehicles equipped with a sway bar, install the nuts attaching sway bar connecting links to the spring tie plates.

20. If the vehicle is equipped with a stabilizer bar, install the nuts attaching the stabilizer bar connecting links to the spring tie plates.

21. Connect the steering linkage to the steering knuckles.

22. Connect the shock absorbers at the axle housing.

23. Connect the driveshaft at the axle yoke.

24. Install the wheels.

25. Lower the vehicle.

CJ-5, CJ-6, CJ-7, Scrambler
- Connecting rod ball studs: 60 ft. lbs. minimum.
- Spring shackle bolts: 24 ft. lbs.
- Shock absorber lower mounting nut: 45 ft. lbs.
- Spring pivot bolts: 100 ft. lbs.
- $9/16$ in. spring U-bolt nuts: 100 ft. lbs.
- $1/2$ in. spring U-bolt nuts: 55 ft. lbs.

1987-90

1. Raise and support the vehicle safely. Remove the wheels.

2. Index the driveshaft to the differential yoke for the proper alignment upon installation. Disconnect the driveshaft at the axle yoke and secure the shaft to the frame rail.

3. Disconnect the vacuum harness from the shift motor.

4. Disconnect the vent hose from the axle housing.

5. Disconnect the center link at the right side of the tie rod.

6. Disconnect the shock absorbers at the axle.

7. Remove the steering damper.

8. Disconnect the track bar at the axle.

9. Loosen the stabilizer bar links at the bar and disconnect the stabilizer bar from the spring tie plates.

10. Loosen the bolts attaching the spring to the frame brackets.

11. Loosen the spring-to-shackle bolts.

12. Take up the weight of the axle with a floor jack.

13. Remove the spring U-bolts and tie plates.

14. Remove the front shackle bolts and remove the axle from the truck.

15. Raise the axle into position.

16. Install the front shackle bolts.

17. Install the spring U-bolts and tie plates.

18. Tighten the spring-to-shackle bolts.

19. Tighten the bolts attaching the spring to the frame brackets.

20. Connect the stabilizer bar from the spring tie plates.

21. Tighten the stabilizer bar links at the bar.

22. Connect the track bar at the axle.

23. Install the steering damper.

24. Connect the shock absorbers at the axle.

25. Connect the center link at the right side of the tie rod.

26. Connect the vent hose at the axle housing.

27. Connect the vacuum harness to the shift motor.

28. Connect the driveshaft at the axle yoke.

29. Install the wheels.

30. Lower the vehicle.

- Spring U-bolt nuts: 90 ft. lbs.
- Tie rod-to-knuckle: 35 ft. lbs.
- Center link-to-knuckle: 35 ft. lbs.
- Lower shock absorber bolt: 45 ft. lbs.
- Track bar-to-axle: 74 ft. lbs.
- Stabilizer bar link bolts: 45 ft. lbs.
- Steering damper-to-axle: 55 ft. lbs.

Shift Motor and Housing

REMOVAL AND INSTALLATION

Wrangler

1. Raise and support the front end on jackstands.

2. Place a drain pan under the shift motor.

3. Disconnect the vacuum harness at the motor.

4. Remove the attaching bolts and slowly lift off the motor and housing. Matchmark the shift fork and housing for installation reference.

5. Rotate the motor and remove the shift fork and motor snaprings.

6. Remove the motor from the housing and remove and discard the motor O-ring.

7. When installing the motor, always use a new O-ring. Install the motor in the housing and slide the shift fork on the shaft.

8. Position the housing and motor on the axle and add about 5 oz. of axle lubricant to the shift motor housing.

9. Engage the fork in the shift collar and install the attaching bolts. Torque the bolts to 96 inch lbs.

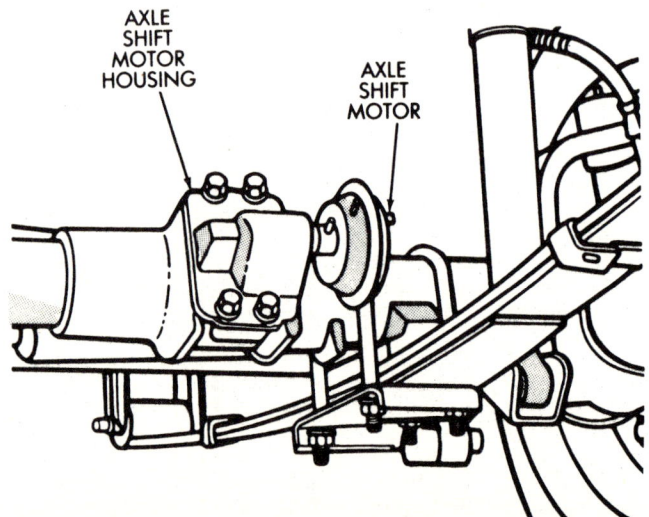

Axle vacuum shift and motor housing

Front Hub and Wheel Bearings

ADJUSTMENT

NOTE: Sodium-based grease is not compatible with lithium-based grease. Read the package labels and be careful not to mix the two types. If there is any doubt as to the type of grease used, completely clean the old grease from the bearing and hub before replacing.

Before handling the bearings, there are a few things that you should remember to do and not to do.

Remember to DO the following:
• Remove all outside dirt from the housing before exposing the bearing.
• Treat a used bearing as gently as you would a new one.
• Work with clean tools in clean surroundings.
• Use clean, dry canvas gloves, or at least clean, dry hands.
• Clean solvents and flushing fluids are a must.
• Use clean paper when laying out the bearings to dry.
• Protect disassembled bearings from rust and dirt. Cover them up.
• Use clean rags to wipe bearings.

• Keep the bearings in oil-proof paper when they are to be stored or are not in use.
• Clean the inside of the housing before replacing the bearing.

Do NOT do the following:
• Don't work in dirty surroundings.
• Don't use dirty, chipped or damaged tools.

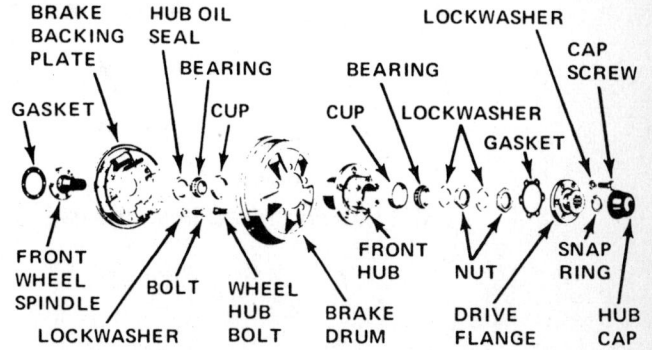

Front hub and wheel bearings with drum brakes

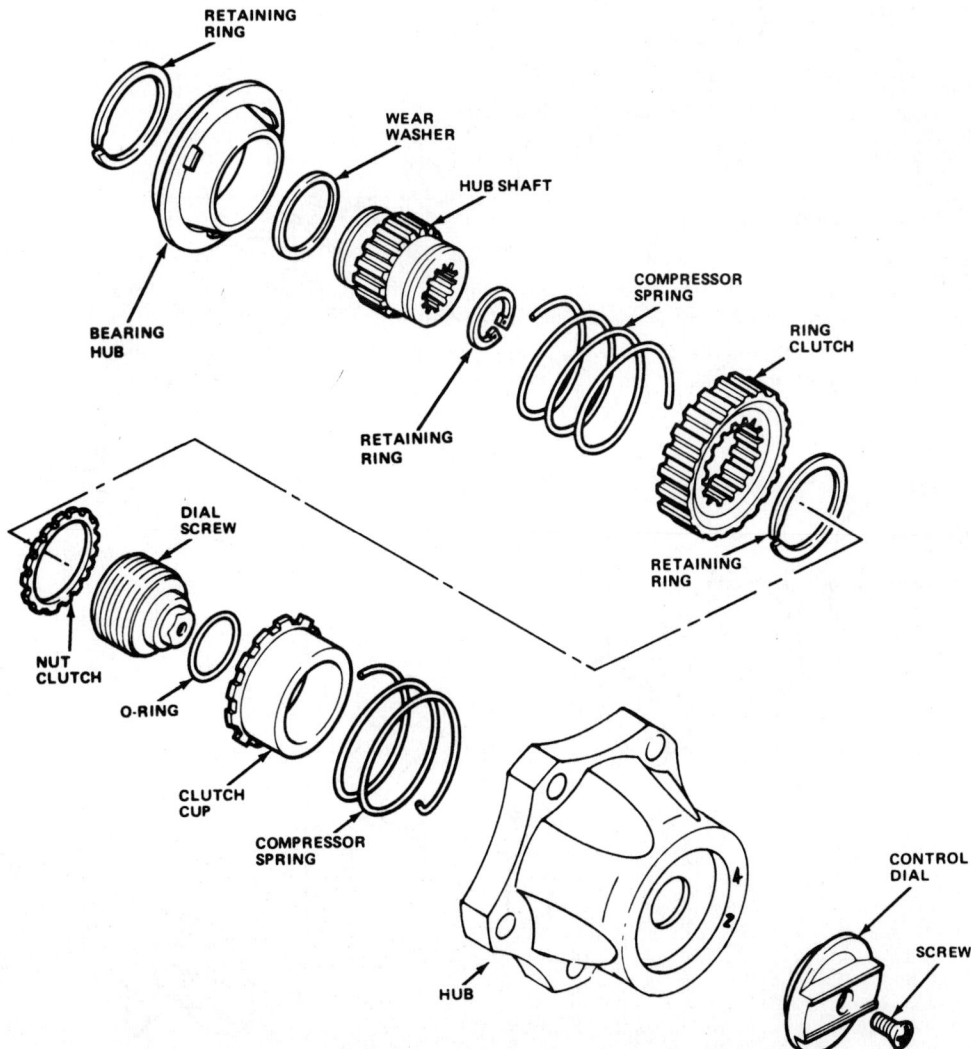

1980–83 front hub and wheel bearings with locking hubs

INNER LOCKNUT
OUTER LOCKNUT
DRIVE FLANGE
SEAL
BEARING
BEARING CUP
HUB AND ROTOR
BEARING CUP
BEARING
TABBED WASHER
GASKET
SNAP RING
HUB CAP

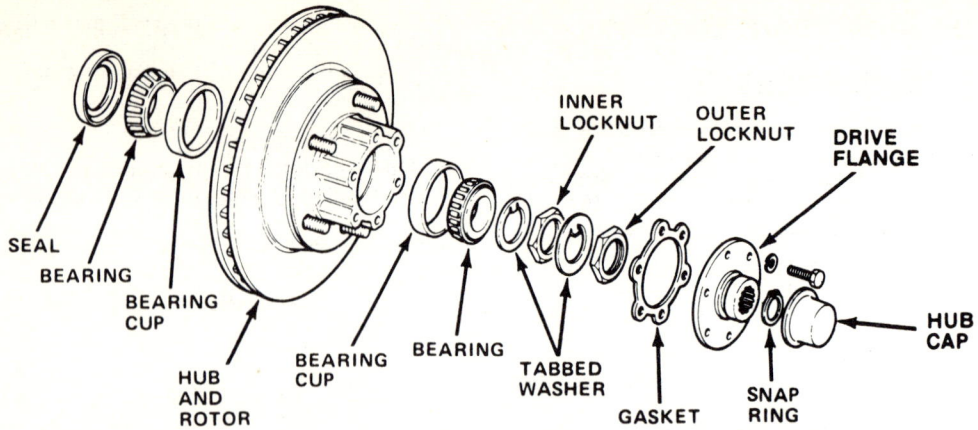

Front hub and wheel bearings with disc brakes, but without locking hubs

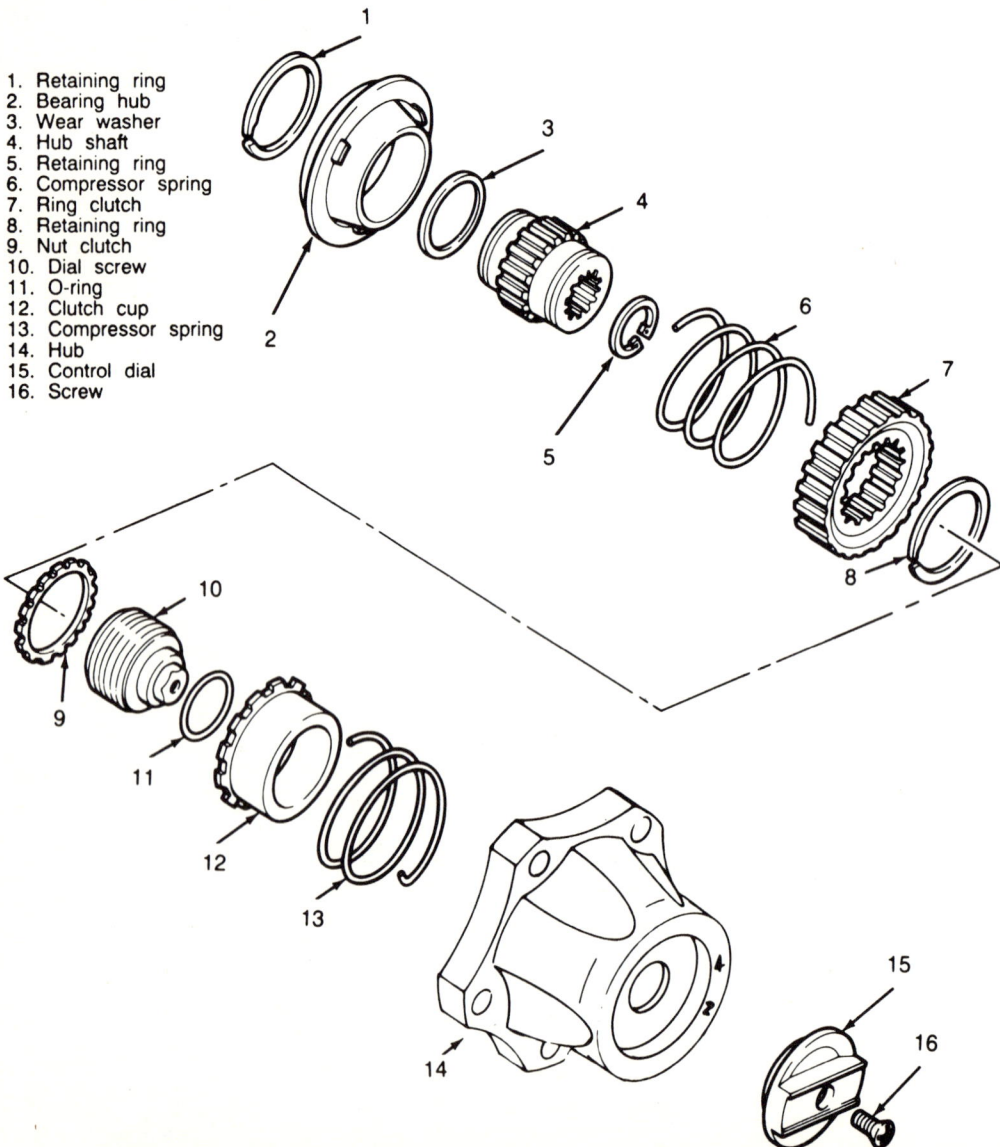

1. Retaining ring
2. Bearing hub
3. Wear washer
4. Hub shaft
5. Retaining ring
6. Compressor spring
7. Ring clutch
8. Retaining ring
9. Nut clutch
10. Dial screw
11. O-ring
12. Clutch cup
13. Compressor spring
14. Hub
15. Control dial
16. Screw

1984–86 front hub and wheel bearings with locking hubs

• Try not to work on wooden work benches or use wooden mallets.
• Don't handle bearings with dirty or moist hands.
• Do not use gasoline for cleaning; use a safe solvent.
• Do not spin-dry bearings with compressed air. They will be damaged.
• Do not spin dirty bearings.
• Avoid using cotton waste or dirty cloths to wipe bearings.
• Try not to scratch or nick bearing surfaces.
• Do not allow the bearing to come in contact with dirt or rust at any time.

1971-86

1. Raise the front of the vehicle and place jackstands under the axle.
2. Remove the wheel.
3. Remove the front hub grease cap and driving hub snapring. On models equipped with locking hubs, remove the retainer knob hub ring, agitator knob, snapring, outer clutch retaining ring and actuating cam body.
4. Remove the splined driving hub and the pressure spring. This may require slight prying with a screwdriver.
5. Remove the external snapring from the spindle shaft and remove the hub shaft drive gear.

6. Remove the wheel bearing locknut, lockring, adjusting nut and inner lockring.
7. On vehicles with drum brakes, remove the hub and drum assembly. This may require that the brake adjusting wheel be backed off a few turns. The outer wheel bearing and spring retainer will come off with the hub.
8. On vehicles with disc brakes, remove the caliper and suspend it out of the way by hanging it from a suspension or frame member with a length of wire. Do not disconnect the brake hose, and be careful to avoid stretching the hose. Remove the rotor and hub assembly. The outer wheel bearing and, on vehicles with locking hubs, the spring collar, will come off with the hub.
9. Carefully drive out the inner bearing and seal from the hub, using a wood block.
10. Inspect the bearing races for excessive wear, pitting or grooves. If they are cracked or grooved, or if pitting and excess wear is present, drive them out with a drift or punch.
11. Check the bearing for excess wear, pitting or cracks, or excess looseness.

NOTE: If it is necessary to replace either the bearing or the race, replace both. Never replace just a bearing or a race. These parts wear in a mating pattern. If just one is replaced, premature failure of the new part will result.

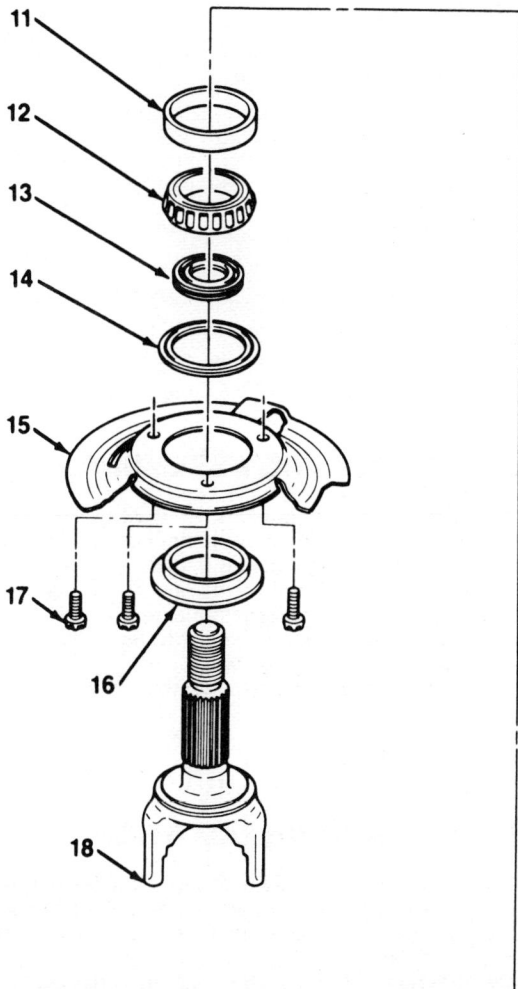

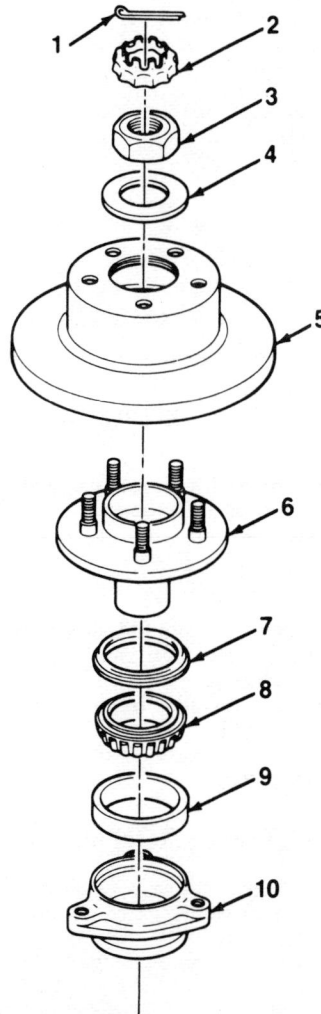

1. Cotter pin
2. Nut retainer
3. Nut
4. Washer
5. Brake rotor
6. Hub
7. Outer bearing seal
8. Outer bearing
9. Outer bearing race
10. Bearing carrier
11. Inner bearing race
12. Inner bearing
13. Inner bearing seal
14. Carrier seal
15. Rotor shield
16. Axle shaft dust slinger
17. Bearing carrier bolts
18. Axle shaft

1987–90 front hub and wheel bearings

12. If the old parts are retained, thoroughly clean them in a safe solvent and allow them to dry on a clean towel. Never spin dry them with compressed air.

13. On vehicles with drum brakes, cover the spindle with a cloth and thoroughly brush all dirt from the brakes. Never blow the dirt off the brakes, due to the presence of asbestos in the dirt, which is harmful to your health when inhaled.

14. Remove the cloth and thoroughly clean the spindle.

15. Thoroughly clean the inside of the hub.

16. Pack the inside of the hub with EP wheel bearing grease. Add grease to the hub until it is flush with the inside diameter of the bearing cup.

17. Pack the bearing with the same grease. A needle-shaped wheel bearing packer is best for this operation. If one is not available, place a large amount of grease in the palm of your hand and slide the edge of the bearing cage through the grease to pick up as much as possible, then work the grease in as best you can with your fingers.

18. If a new race is being installed, very carefully drive it into position until it bottoms all around, using a brass drift. Be careful to avoid scratching the surface.

19. Place the inner bearing in the race and install a new grease seal.

20. Place the hub assembly onto the spindle and install the inner lockring and outer bearing. Install the wheel bearing nut and torque it to 50 ft. lbs. while turning the wheel back and forth to seat the bearings. Back off the nut about ¼ turn (90°) maximum.

21. Install the lockwasher with the tab aligned with the keyway in the spindle and turn the inner wheel bearing adjusting nut until the peg on the nut engages the nearest hole in the lockwasher.

22. Install the outer locknut and torque it to 50 ft. lbs.

23. Install the spring collar, drive flange, snapring, pressure spring, and hub cap.

24. Install the caliper over the rotor.

1987-90

1. Raise and support the front end on jackstands.
2. Remove the wheel.
3. Dismount the caliper and suspend it out of the way.
4. Remove the rotor.
5. Remove the hub nut pin, cap and nut.
6. Remove the hub.
7. The hub and bearings are usually replaced as a unit. The hub and bearing carrier may, however, be disassembled and the bearings replaced as a set. Once the hub and bearing carrier have been separated, the bearings should not be reused.
8. Pack the hub cavity and bearings with wheel bearing grease and install the hub on the axle shaft. If the carrier was separated from the hub, make sure you install a new carrier seal and inner bearing seal.
9. Install the hub washer and nut. Torque the nut to 175 ft. lbs. and install the cap and new cotter pin.
10. Install the rotor, caliper and wheel.

POWER TAKE-OFF

Early 1971 Jeep vehicles were available with an optional power take-off unit. The PTO consists of four assemblies:

1. The shift unit, mounted on the transfer case.
2. The driveshaft and U-joints.
3. The shaft drive assembly.
4. The pulley drive assembly.

The shaft drive exits the rear of the vehicle and is designed to operate trailed equipment. The pulley drive is driven by the shaft drive and is designed to operate stationary equipment by a belt drive. The shaft drive assembly was installed far more frequently than was the pulley drive assembly.

Shift Assembly

REMOVAL AND INSTALLATION

Drive for the PTO is taken from the transfer case main drive gear through an internal sliding gear. The sliding gear is mounted in the shift housing.

1. Remove the bolts in the driveshaft companion flange at the PTO front U-joint.
2. Unbolt and remove the shift lever.
3. Remove the five bolts securing the shift unit to the transfer case and pull it rearward from the case.
4. Installation is the reverse of removal.

DISASSEMBLY AND ASSEMBLY

1. Carefully pry the shift rail and fork froward to clear the poppet ball and spring. Be careful to avoid damaging or losing the ball and spring. Remove the shifting sleeve.
2. Remove the attaching nut and the companion flange.
3. Drive the shaft forward out of the housing.
4. Remove the spacer and bearing from the shaft.
5. Remove the bearing from the housing.
6. Clean and inspect all parts and assemble in reverse of disassembly.

Shaft Drive Unit

REMOVAL AND INSTALLATION

The standard 6 splined 1⅜ in. (34.925mm) diameter output shaft is driven through two helical cut gears mounted in a housing attached to the vehicle at the center of the frame rear crossmember.

1. Disconnect the rear U-joint at the companion flange.
2. Remove the retaining screw and the flange.
3. Unbolt and remove the assembly from the vehicle.
4. Installation is the reverse of removal.

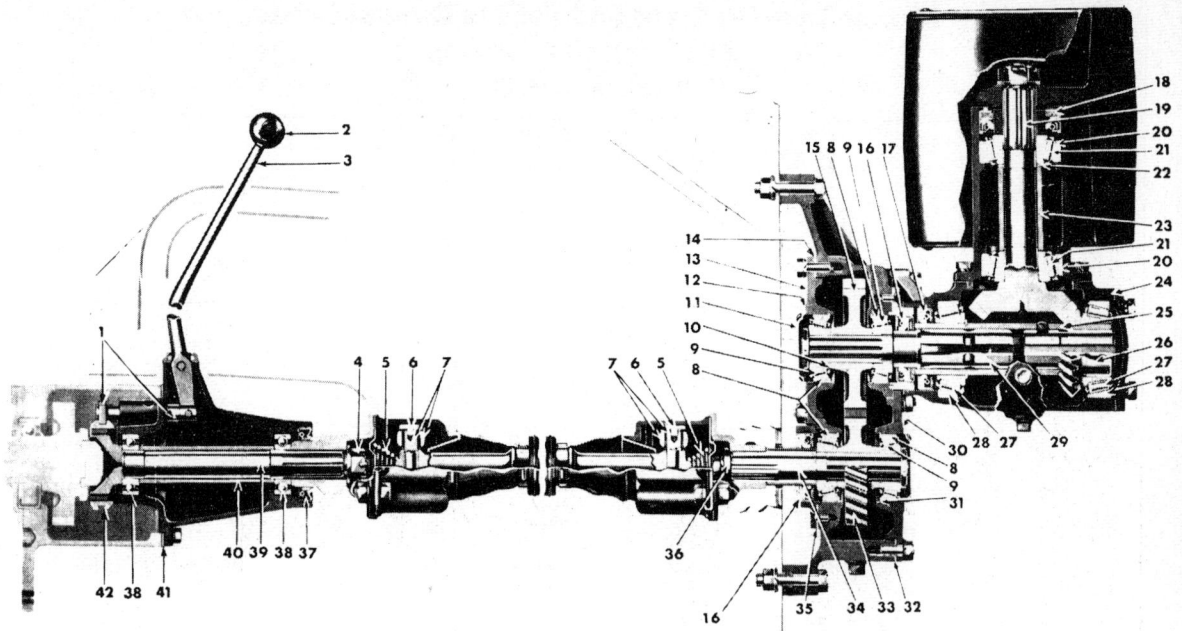

1. Fork and rod	10. Snap ring	19. Gear and shaft	28. Cup	37. Oil seal
2. Ball	11. Plate	20. Cup	29. Shaft	38. Ball bearing
3. Lever	12. Gasket	21. Cone and roller	30. Gasket	39. Gear and shaft
4. Nut	13. Retainer	22. Shims	31. Shims	40. Spacer
5. Button and spring	14. Gasket	23. Spacer	32. Gasket	41. Gasket
6. Spring	15. Gear	24. Shims	33. Gear	42. Sleeve
7. Trunnion and ball	16. Oil seal	25. Shims	34. Shaft	
8. Cup	17. Oil seal	26. Pinion	35. Gasket	
9. Bearing	18. Oil seal	27. Cone and roller	36. Washer	

Power Take-Off assembly

DISASSEMBLY AND ASSEMBLY

1. Drain the oil from the unit.
2. Remove the rear bearing cover.
3. Remove the nut and lockwasher from the input shaft.
4. Unbolt and remove the input shaft bearing retainer and remove the bearing. Take care not to lose the shims between the gear and the bearing case.
5. Remove the bearing cone, cup and snapring.
6. Remove the oil seal retainer and pilot assembly.
7. Press the shaft through the housing, removing the bearing cone, oil seal and retainer as an assembly.
8. Remove the input shaft gear through the rear opening. Push out the bearing cup and remove the snapring. Remove the bearing cone and oil seal from the shaft.
9. Remove the output shaft in the same manner as the input shaft.
10. Adjustment of the tapered roller bearings on both shafts is accomplished by shim packs placed between the gear hubs and bearing cones.
11. Assembly is the reverse of disassembly. Fill the unit with 90W gear oil.

Pulley Drive Unit

REMOVAL AND INSTALLATION

This procedure is accomplished simply by unbolting and removing the unit.

DISASSEMBLY AND ASSEMBLY

1. Remove the unit from the vehicle.
2. Drain the oil and clean the unit.
3. Remove the retaining nut and remove the pulley.
4. Unbolt and remove the pulley shaft housing from the gear housing. Don't lose the shims.
5. Press the pulley shaft through the housing, removing the inner bearing cone, spacer and shim pack.
6. Remove the oil seal and outer bearing cone.
7. Remove the bearing retaining cover from the gear housing, then remove the shim pack.
8. Using a brass drift, tap the shaft through the housing; the bearing and gear will come out with it. Be careful not to lose the shim pack.
9. Clean and inspect all parts. Assembly is the reverse of disassembly. Fill the unit with 90W gear oil.

Power Take-Off Chart and Vehicle Ground Speeds
All Gearshift Positions
Miles Per Hour

Governor Control Position	Transfer In	PTO 1 to 1 Gear Ratio						Engine Speed rpm
		Transmission Gear In						
		Low		Intermediate		High		
		PTO Shaft rpm	Jeep Speed mph	PTO Shaft rpm	Jeep Speed mph	PTO Shaft rpm	Jeep Speed mph	
1	Low	358	2.22	644	4.01	1,000	6.22	1,000
	High	358	5.40	644	9.75	1,000	15.13	
2	Low	428	2.67	773	4.81	1,200	7.47	1,200
	High	428	6.48	773	11.71	1,200	18.15	
3	Low	500	3.11	902	5.62	1,400	8.72	1,400
	High	500	7.56	902	13.66	1,400	21.17	
4	Low	571	3.56	1.301	6.42	1,600	9.96	1,600
	High	571	8.65	1,301	15.61	1,600	24.20	
5	Low	643	4.00	1,160	7.22	1,800	12.08	1,800
	High	643	9.73	1,160	17.56	1,800	27.22	
6	Low	714	4.44	1,289	8.02	2,000	12.45	2,000
	High	714	11.89	1,289	19.51	2,000	30.25	
7	Low	786	4.89	1,418	8.83	2,200	13.70	2,200
	High	786	11.89	1,418	21.46	2,200	33.27	
8	Low	857	5.34	1,547	9.63	2,400	14.84	2,400
	High	857	12.97	1,547	23.41	2,400	36.31	
9	Low	929	5.78	1,657	10.43	2,600	16.19	2,600
	High	929	14.05	1,657	25.36	2,600	39.33	

Power Take-Off Chart and Vehicle Ground Speeds All Gearshift Positions Miles Per Hour

SPECIAL TOOLS

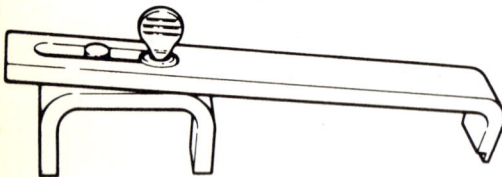

J-35514

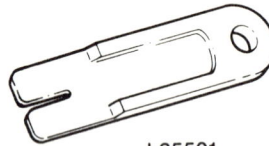

J-35591

J-25180 PULLER

J-9233

J-6221

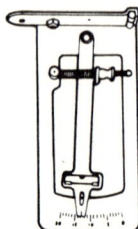

J-23498

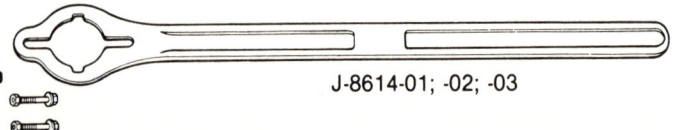

J-8614-01; -02; -03

J-22661

SPECIAL TOOLS

Suspension and Steering

8

QUICK REFERENCE INDEX

GENERAL INDEX

WHEEL ALIGNMENT SPECIFICATIONS

Year	Caster (deg.)		Camber (deg.)		Toe-in (in.)	King Pin Incl. (deg.)
	Range	Pref.	Range	Pref.		
1971–73	$2^1/_2$P to $3^1/_2$P	3P	1P to 2P	$1^1/_2$P	$3/_{64}$ to $3/_{32}$	$7^1/_2$
1974–80	$2^1/_2$P to $3^1/_2$P	3P	1P to 2P	$1^1/_2$P	$3/_{64}$ to $3/_{32}$	$8^1/_2$
1981	$5^1/_2$P to $6^1/_2$P	6P	$1^1/_4$P to $1^3/_4$P	$1^1/_2$P	$3/_{64}$ to $3/_{32}$	$8^1/_2$
1982–83	$5^1/_2$P to $6^1/_2$P	6P	$1/_2$N to $1/_2$P	0	$3/_{64}$ to $3/_{32}$	$8^1/_2$
1984–86	$5^1/_2$P to $6^1/_2$P	6P	$1/_2$N to $1/_2$P	0	0 to $3/_{32}$ in	10
1987	$7^3/_4$P to $8^1/_4$P	8P	$1/_2$N to $1/_2$P	0	$1/_{32}$ out to $1/_{32}$ in	10
1988–90	①	②	$1/_2$N to $1/_2$P	0	$1/_{32}$ in to 0	10

① Auto. trans.: 6P to 7P
  Man. trans.: $7^1/_2$P to $8^1/_2$P
② Auto. trans.: $6^1/_2$P
  Man. trans.: 8P

Troubleshooting Basic Steering and Suspension Problems

Problem	Cause	Solution
Hard steering (steering wheel is hard to turn)	• Low or uneven tire pressure • Loose power steering pump drive belt • Low or incorrect power steering fluid • Incorrect front end alignment • Defective power steering pump • Bent or poorly lubricated front end parts	• Inflate tires to correct pressure • Adjust belt • Add fluid as necessary • Have front end alignment checked/adjusted • Check pump • Lubricate and/or replace defective parts
Loose steering (too much play in the steering wheel)	• Loose wheel bearings • Loose or worn steering linkage • Faulty shocks • Worn ball joints	• Adjust wheel bearings • Replace worn parts • Replace shocks • Replace ball joints
Car veers or wanders (car pulls to one side with hands off the steering wheel)	• Incorrect tire pressure • Improper front end alignment • Loose wheel bearings • Loose or bent front end components • Faulty shocks	• Inflate tires to correct pressure • Have front end alignment checked/adjusted • Adjust wheel bearings • Replace worn components • Replace shocks
Wheel oscillation or vibration transmitted through steering wheel	• Improper tire pressures • Tires out of balance • Loose wheel bearings • Improper front end alignment • Worn or bent front end components	• Inflate tires to correct pressure • Have tires balanced • Adjust wheel bearings • Have front end alignment checked/adjusted • Replace worn parts

Troubleshooting Basic Steering and Suspension Problems

Problem	Cause	Solution
Uneven tire wear	• Incorrect tire pressure • Front end out of alignment • Tires out of balance	• Inflate tires to correct pressure • Have front end alignment checked/adjusted • Have tires balanced

FRONT SUSPENSION

All springs should be examined periodically for broken or shifted leaves, loose or missing clips, angle of the spring shackles, and position of the springs on the saddles. Springs with shifted leaves do not retain their normal strength. Missing clips may permit the spirit leaves to fan out or break on rebound. Broken leaves may make the vehicle hard to handle or permit the axle to shift out of line. Weakened springs may break causing difficulty in steering. Spring attaching clips or bolts must be tight. It is suggested that they be checked at each vehicle inspection.

Springs

REMOVAL AND INSTALLATION

1971

1. Raise the vehicle with a jack under the axle and place a jackstand under the frame side rail. Then lower the axle jack so that the load is relieved from the spring and the wheels rest on the floor.

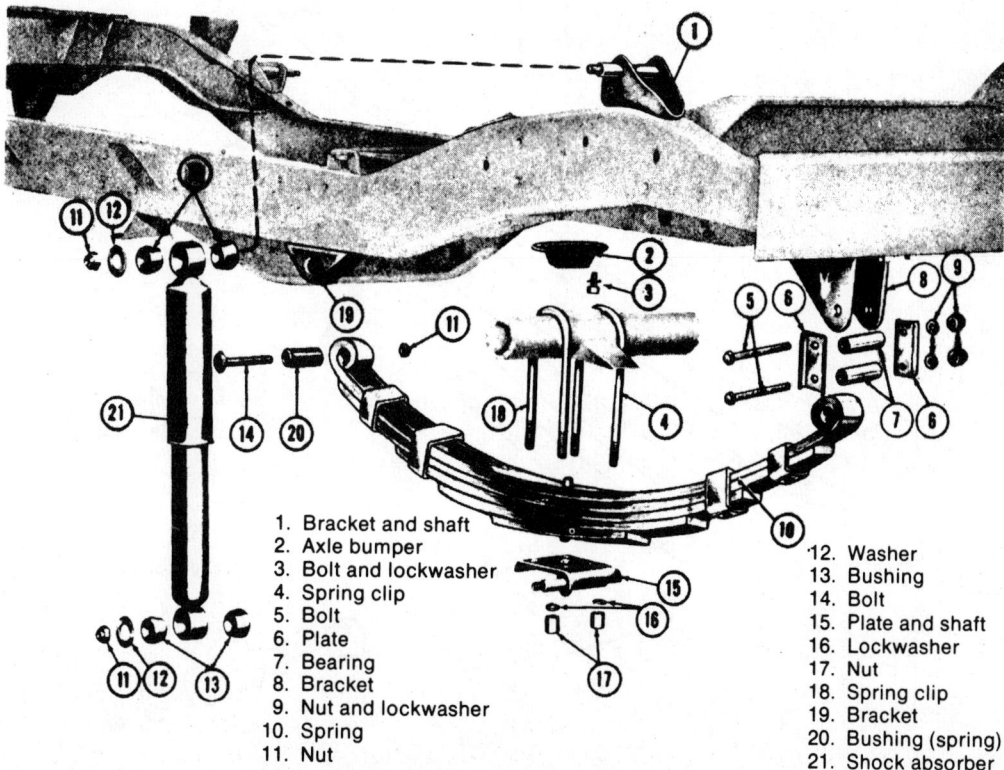

1. Bracket and shaft
2. Axle bumper
3. Bolt and lockwasher
4. Spring clip
5. Bolt
6. Plate
7. Bearing
8. Bracket
9. Nut and lockwasher
10. Spring
11. Nut
12. Washer
13. Bushing
14. Bolt
15. Plate and shaft
16. Lockwasher
17. Nut
18. Spring clip
19. Bracket
20. Bushing (spring)
21. Shock absorber

Early model front spring and shock absorber

2. Remove the nuts which secure the spring clip bolts. Remove the spring plate and clip bolts. Free the spring from the axle by raising the axle jack.

3. Remove the pivot bolt nut and drive out the pivot bolt. Disconnect the shackle either by removing the lower nuts and bolts on the rubber bushed shackles, or by removing the threaded bushings on the U-shackles.

4. To replace, first install the pivot bolt. Then, connect the shackle using the following procedures.

5. On bronze bushed pivot bolts, install the bolt and nut and tighten the nut. Then back it off two cotter pin slots and install the cotter pin. The nut must be drawn up tightly but must be sufficiently loose to allow the spring to pivot freely. Otherwise the spring might break.

6. On rubber bushed pivot bolts and locknuts (or lockwasher and nut) only tighten the bolt enough to hold the bushings in position until the vehicle is lowered from the jack.

7. Connect the shackle. On rubber bushed shackles install the bolts as in step 6 above. For U-shackles, insert the shackle through the frame bracket and eye of the spring. Holding the U-shackle tightly against the frame, start the upper bushing on the shackle, taking care that when it enters the thread in the frame it does not crossthread. Screw the bushing on the shackle tightly against the spring eye, and thread the bushing in approximately half way. Then, alternately from top bushing to lower bushing, turn them in until the head of the bushing is snug against the frame bracket and the bushing in the spring eye is $\frac{1}{32}$ in. (0.8mm) away from the spring as measured from the inside of the hexagon head in the spring. Lubricate the bushing and then try the flex of the shackle, which must be free. If a shackle is tight, rethread the bushings on the shackle.

8. Move the axle into position on the spring by lowering or raising the axle jack. Install the spring clip bolts, spring plate, lockwashers, and nuts. Torque the nuts to 50-55 ft. lbs. Avoid overtightening. Be sure the spring is free to move at both ends.

9. Remove both jacks. On rubber bushed shackles and pivot bolts, allow the weight of the vehicle to seat the bushings in their operating positions. Then torque the nuts to 27-30 ft. lbs.

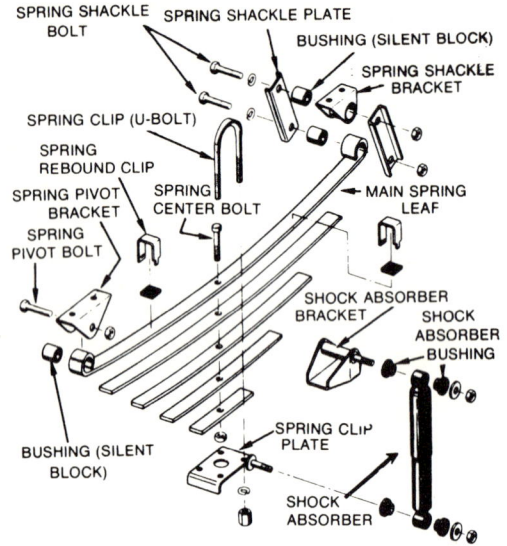

1972–75 front spring assembly

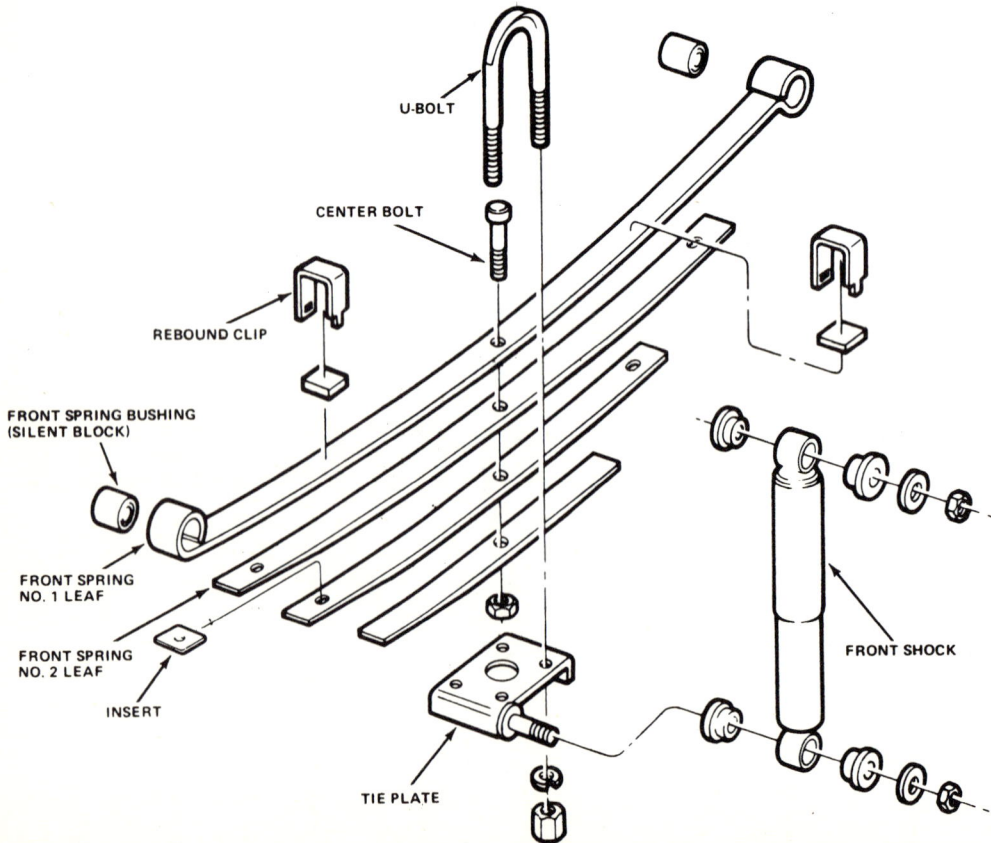

Front spring and shock absorber, 1976–78 models

SHOCK ABSORBER BUSHINGS

SPRING HANGER

SHACKLE PLATE

SPRING SHACKLE BUSHINGS

SHOCK BRACKET

SHOCK ABSORBER

SPRING SHACKLE

SPRING HANGER

CENTER BOLT

REBOUND CLIP

INSULATOR

SPRING LEAVES

SPRING EYE BUSHING

MAIN LEAF

U-BOLT

INSULATOR

SHOCK ABSORBER BUSHINGS

TIE PLATE

U-BOLT NUT

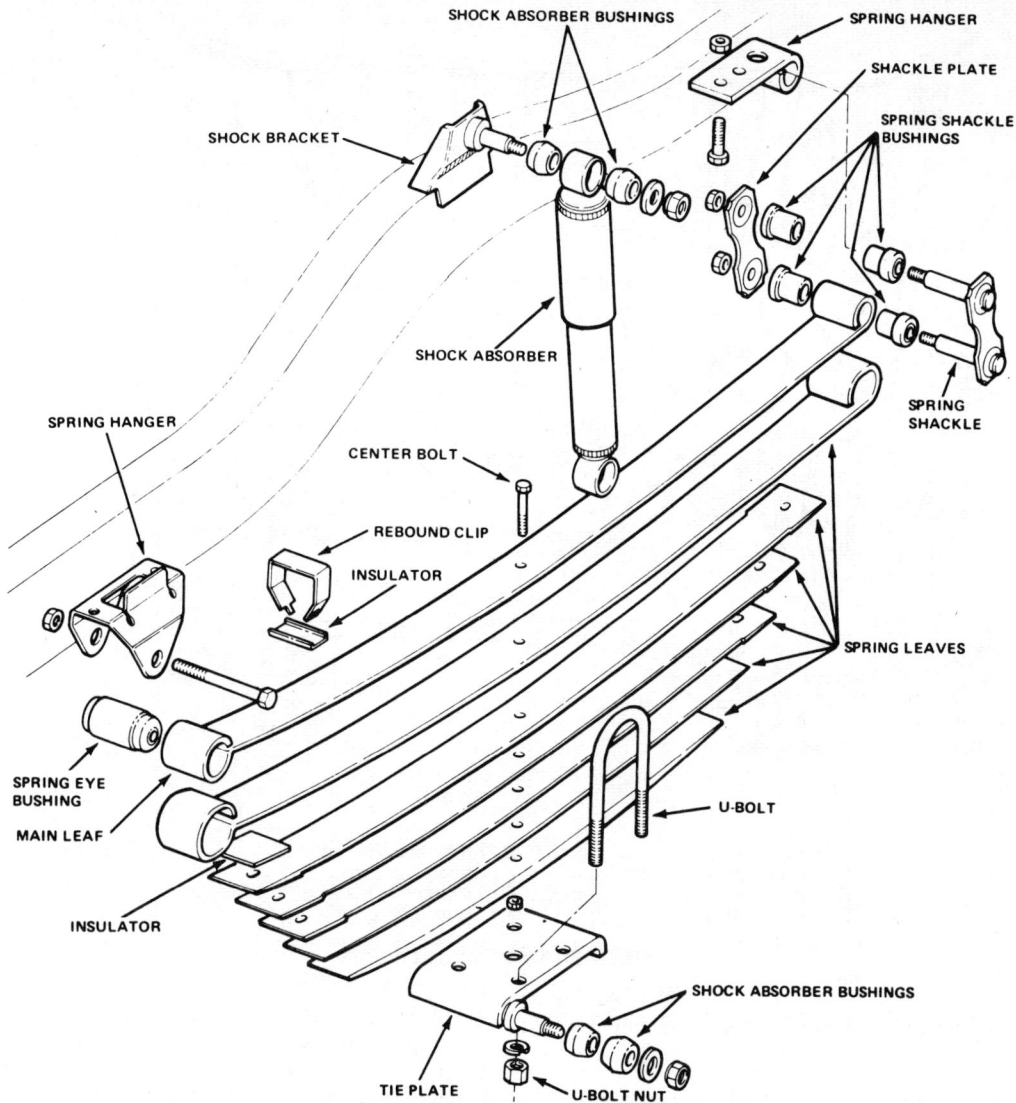

Front suspension for 1979–86 models

1972-90

1. Raise the vehicle with a jack under the axle. Place a jackstand under the frame side rail. Then lower the axle jack so the load is relieved from the spring and the wheels just touch the floor.

2. Disconnect the shock absorber from the spring clip plate. On models through 1986, disconnect the stabilizer bar. On 1987-90 models, loosen, but don't remove, the front stabilizer bar link nut.

3. Remove the nuts which secure the spring clips (U-bolts). Remove the spring plate and spring clips. Free the spring from the axle by raising the axle.

4. Remove the pivot bolt nut and drive out the pivot bolt. Disconnect the shackle.

5. With the spring removed, the spring shackle and/or shackle plate may be removed from the spring.

6. Inspect the bushings in the eye of the main spring leaf and the bushings of the spring shackle for excessive wear. Replace if necessary.

7. The spring can be disassembled for replacing an individual spring leaf, by removing the clips and the center bolts.

8. To install the spring on the vehicle, with the bushings in place and the spring shackle attached to the springs, position the spring in the pivot hanger and install the pivot bolt and lock nut. Only tighten the lock nut enough to hold the bushings in position until the vehicle is lowered from the jack.

9. Position the spring and install the shackle, shackle bolts, shackle plate if applicable, lockwasher, and nut. Only finger tighten the nuts at this time.

10. Move the axle into position on the spring by lowering the axle jack. Place the spring center bolt in the axle saddle hole. Install the spring clips, spring plate, lockwashers and nuts. Observe the following torques:

1972-86
- $^7/_{16}$ in. nuts: 36-42 ft. lbs.
- $^1/_2$ in. nuts: 45-65 ft. lbs.
- $^9/_{16}$ in. nuts: 100 ft. lbs.

1987-90
- All nuts: 90 ft. lbs.

NOTE: Be sure that the center bolt is properly centered in the axle saddle.

11. Connect the shock absorber.

12. Remove the jack and allow the weight of the vehicle to seat

8 SUSPENSION AND STEERING

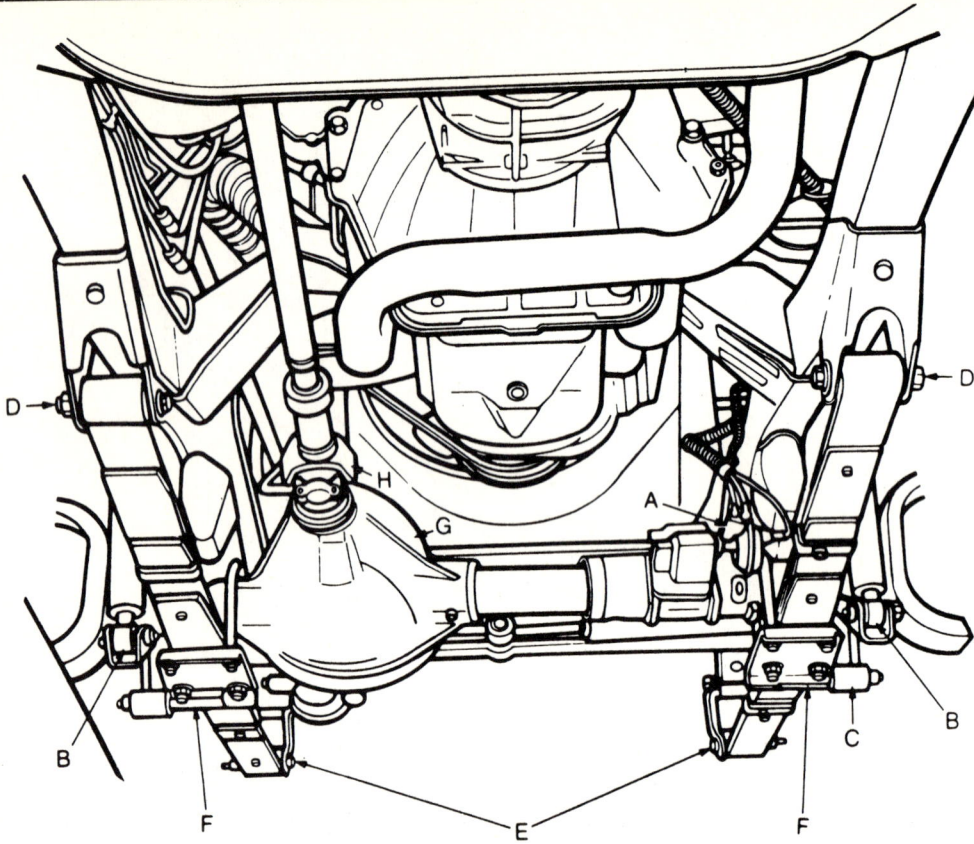

A. Shift motor
B. Shock absorber lower end
C. Link
D. Rear shackle bolt
E. Front shackle bolts
F. Spring tie plates
G. Differential housing
H. Driveshaft yoke

1987–90 front suspension

the bushings in their operating positions. Observe the following torques:

1972-76
- $7/16$ in. spring pivot bolt nuts and spring shackle nuts: 35-50 ft. lbs.
- $5/8$ in. shackle nuts: 55-75 ft. lbs.

1977-86
- Pivot bolts: 100 ft. lbs.
Shackle nuts: 24 ft. lbs.

1987-90
- Spring-to-shackle nuts: 95 ft. lbs.

Shock Absorbers

REMOVAL AND INSTALLATION

Except Wrangler

1. Remove the locknuts and washers.
2. Pull the shock absorber eyes and rubber bushings from the mounting pins.
3. Install the shocks in reverse order of the removal procedure. Torque the upper bolt to 35 ft. lbs. and the lower bolt to 45 ft. lbs.

Wrangler

1. Remove the upper end nut, washer and grommet from the shock absorber stem.
2. Raise and support the front end on jackstands.

3. Unbolt the lower end and remove the shock absorber.
4. Remove the remaining upper grommet.
5. If the shock absorbers are being replaced, make sure that you use new grommets at the upper end. If you are reusing the shocks, it's a good idea to get new grommets if they show any signs of wear or are more than a year old.

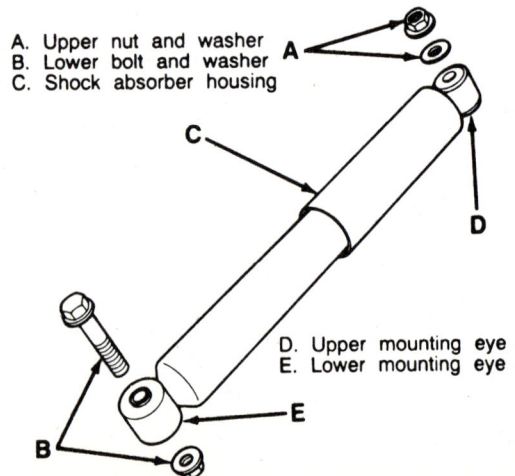

A. Upper nut and washer
B. Lower bolt and washer
C. Shock absorber housing
D. Upper mounting eye
E. Lower mounting eye

Shock absorber type used on the front and rear of all Jeep vehicles, except the front of Wrangler models

1. Upper nut
2. Washer
3. Upper end, upper grommet
4. Upper mounting stud

5. Lower mounting bolt and nut
6. Lower mounting eye
7. Shock absorber housing
8. Upper end, lower grommet

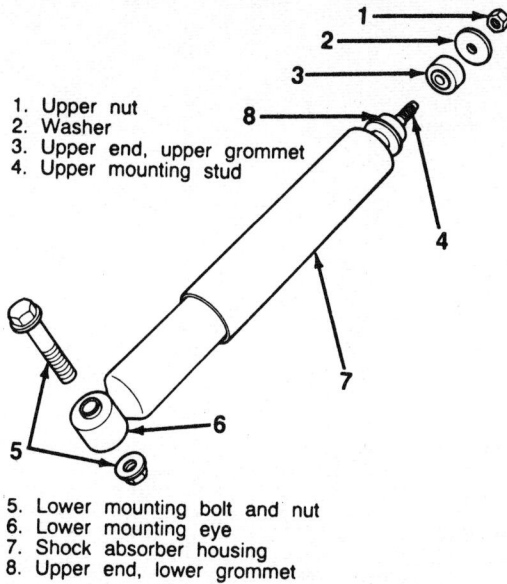

Shock absorber used on the front of the Wrangler models

6. Install the new lower grommet on the shock stem, with the shoulder facing upwards.
7. Mount the shock at the lower end, with the nut finger tight.
8. Guide the shock stem into the mounting hole in the frame bracket.
9. Torque the lower end nut to 45 ft. lbs.
10. Lower the truck.
11. Install the new upper end grommet with the shoulder facing downwards.
12. Align the shoulders of the two upper end grommets in the frame mounting hole. Install the washer and nut. Torque the nut to 96 inch lbs.

CHILTON TIP: Squeaking usually occurs when movement takes place between the rubber bushings and the metal parts. The squeaking may be eliminated by placing the bushings under greater pressure. This is accomplished either by adding additional washers where the cotter pins are used or by tightening the locknuts. Do not use mineral lubricant to stop the squeaking as it will deteriorate the rubber.

Stabilizer Bar

REMOVAL AND INSTALLATION

1. Raise and support the front end on jackstands.
2. Unbolt the stabilizer bar from the vertical links.
3. Unbolt the stabilizer bar from the frame brackets.
4. Replace any worn or damaged rubber parts.
5. Install the stabilizer bar at the links first, hand tightening the fasteners.
6. Install the frame brackets, hand tightening the fasteners.
7. Make sure everything is aligned. Observe the following torques:
- Frame bracket bolts
  Wrangler: 30 ft. lbs.
  CJ-5, CJ-6, CJ-7 and Scrambler: 35 ft. lbs.
- Link nuts
  Wrangler: 45 ft. lbs.
  CJ-5, CJ-6, CJ-7 and Scrambler: 55 ft. lbs.

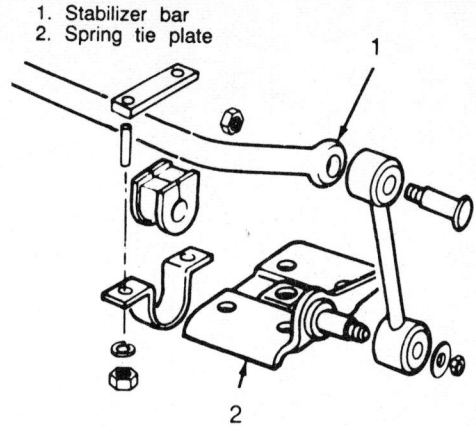

1. Stabilizer bar
2. Spring tie plate

Front stabilizer bar used on all models through 1986

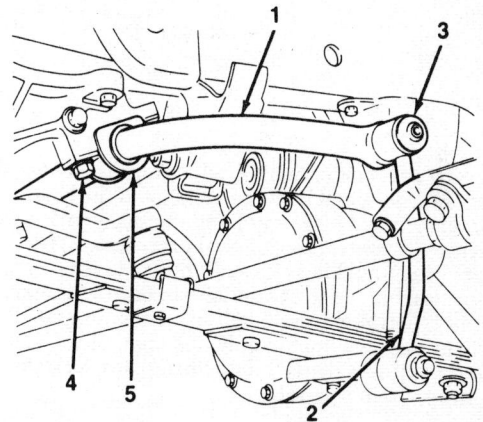

1. Stabilizer bar
2. Link
3. Link bolt
4. Mounting bracket bolt
5. Mounting bracket

1987–90 front stabilizer bar

Track Bar

REMOVAL AND INSTALLATION

Wrangler

1. Raise and support the front end on jackstands.
2. Unbolt the bar from the frame bracket and the axle bracket.
3. Installation is the reverse of removal. Torque the bolts to 74 ft. lbs.

Steering Knuckle and Pivot Pins

REMOVAL AND INSTALLATION

1971

1. Remove the eight screws that hold the oil seal retainer in place.
2. Remove the four screws which secure the lower pivot pin bearing cap.
3. Remove the four screws which hold the upper bearing cap in place.
4. Remove the bearing cap.
5. The steering knuckle can now be removed from the axle.
6. Wash all of the parts in cleaning solvent.

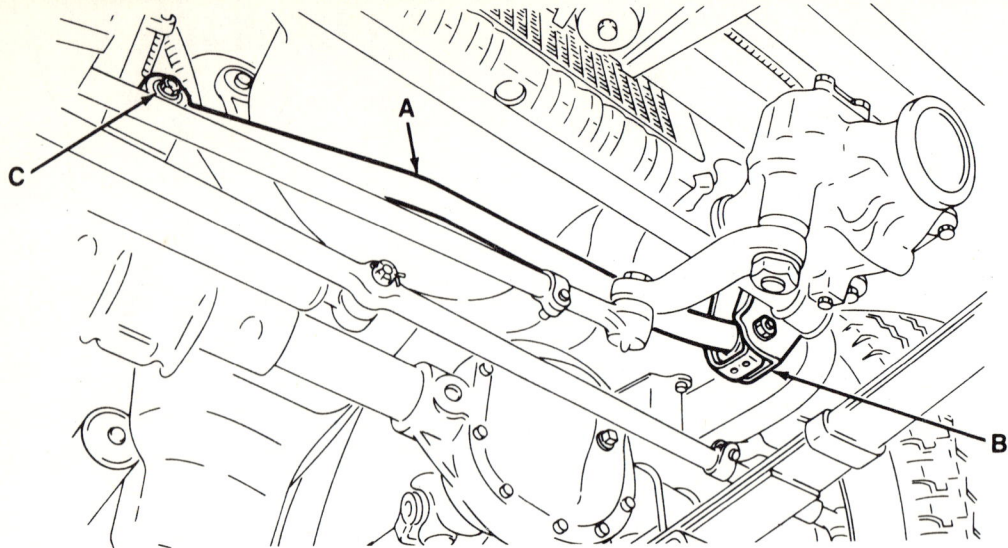

A. Bar   B. Frame bracket   C. Axle bracket

Front track bar on the Wrangler

7. Replace any worn or damaged parts. Inspect the bearings and races for scores, cracks, or chips. Should the bearing cups be damaged, they may be removed and installed with a driver.

8. To install, reverse the removal procedure. When reinstalling the steering knuckle sufficient shims must be installed under the top bearing cap to obtain the correct preload on the bearing. Shims are available in 0.003 in., 0.005 in., 0.010 in., and 0.030 in. thicknesses. Install only one shim of the above thicknesses at the top only. Install the bearing caps, lockwashers, and screws, and tighten securely.

You can check the preload on the bearings by hooking a spring scale in the hole in the knuckle arm for the tie rod sprocket. Take the scale reading when the knuckle has just started its sweep.

The pivot pin bearing preload should be 12-16 lbs. with the oil seal removed. Remove or add shims to obtain a preload within these limits. If all shims are removed and adequate preload is still not obtained, a washer may be used under the top bearing cap to increase preload. When a washer is used, shims may have to be reinstalled to obtain proper adjustment.

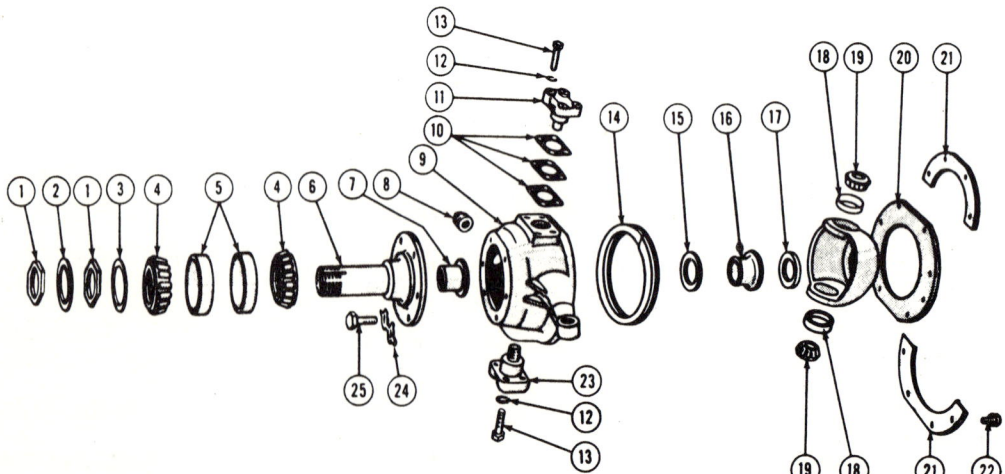

1. Bearing adjusting nut
2. Lockwasher
3. Lockwasher
4. Bearing cone and rollers
5. Bearing cup
6. Spindle
7. Bushing
8. Filler plug
9. Left knuckle and arm
10. Shims
11. Upper bearing cap
12. Lockwasher
13. Bolt
14. Oil seal and backing ring
15. Thrust washer
16. Axle pilot
17. Oil seal
18. Bearing cup
19. Bearing cone and rollers
20. Oil seal
21. Retainer
22. Bolt
23. Lower bearing cap
24. Lock strap
25. Bolt

1971 steering knuckle and wheel bearings

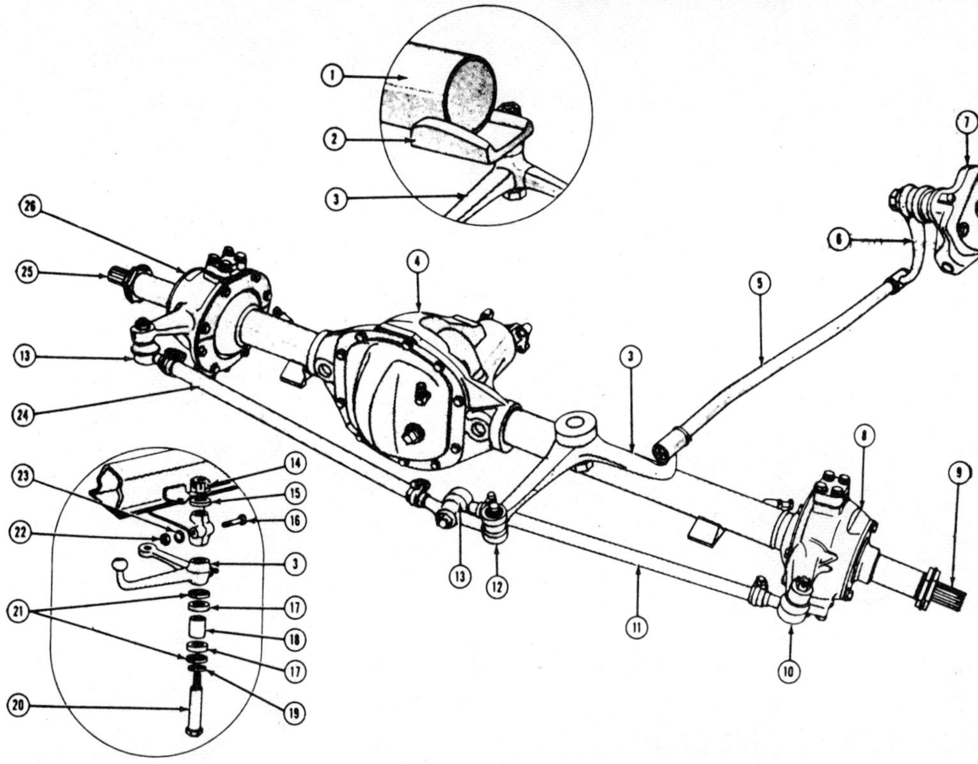

1. Frame cross tube
2. Steering bellcrank bracket
3. Steering bellcrank
4. Front axle assembly
5. Steering connecting rod
6. Steering gear arm
7. Steering gear
8. Left steering knuckle and arm
9. Left shaft and universal joint
10. Left tie rod socket
11. Left steering tie rod
12. Left tie rod socket
13. Right tie rod socket
14. Bellcrank nut
15. Washer
16. Bolt
17. Bellcrank bearing
18. Bearing spacer
19. Washer
20. Bellcrank shaft
21. Bearing seal
22. Nut
23. Lockwasher
24. Right steering tie rod
25. Right shaft and universal joint
26. Right steering knuckle and arm

1971 steering linkage

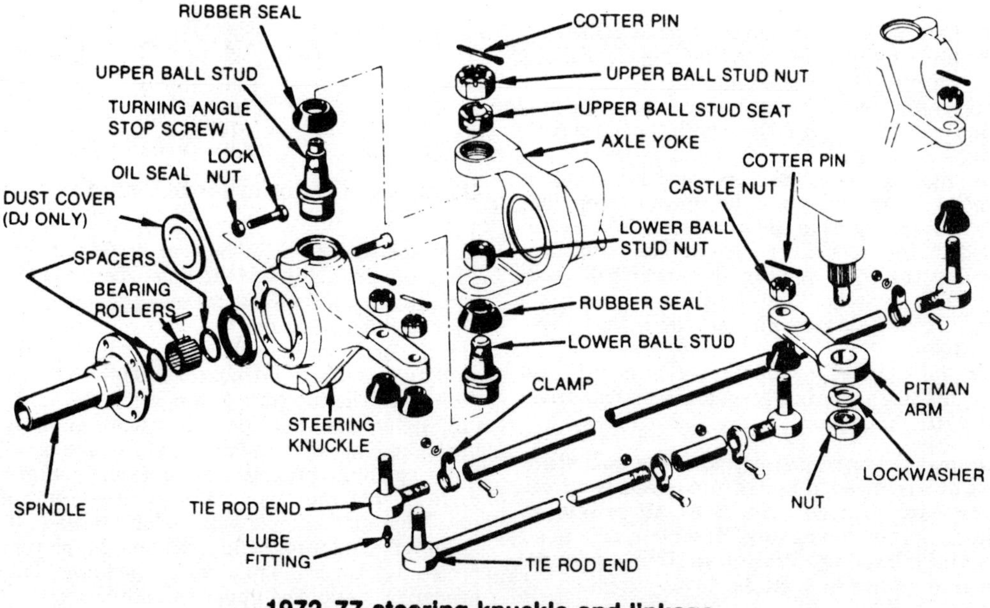

1972–77 steering knuckle and linkage

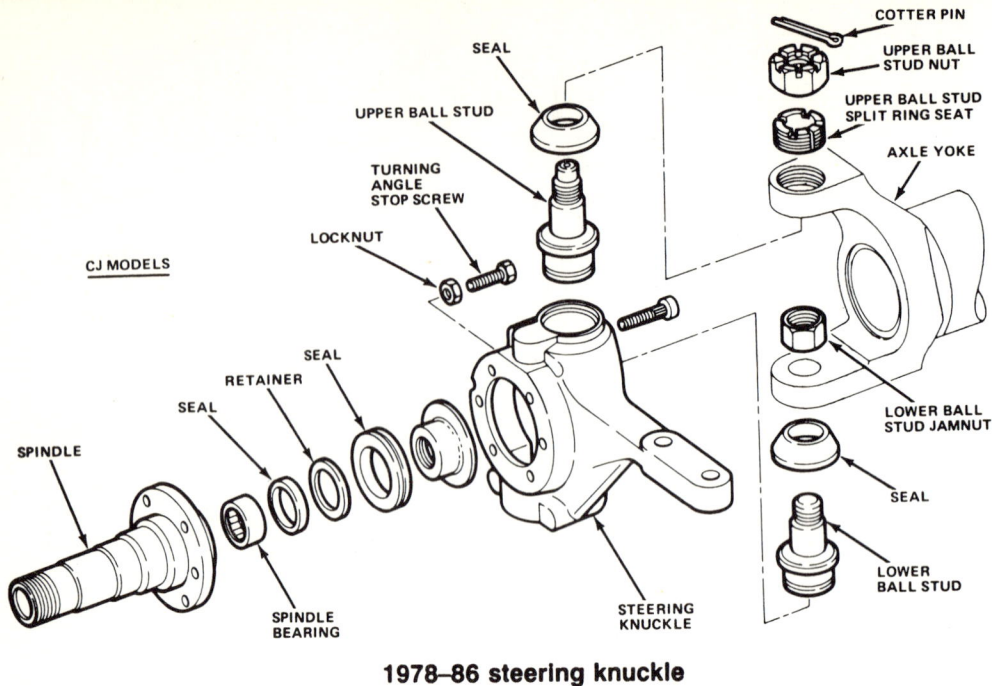

1978–86 steering knuckle

Steering Knuckle and Ball Joints

REMOVAL AND INSTALLATION

1972-86

1. Replacement of the ball joints, or ball stud, as they will be called from here on, requires the removal of the steering knuckle. To remove the steering knuckle, first remove the wheel, brake drum or disc, and hub as an assembly. Remove the brake assembly from the spindle. Position the brake assembly on the front axle in a convenient place. Remove the snapring from the axle shaft.

2. Remove the spindle and bearing assembly. It may be necessary to tap the spindle with a soft mallet to disengage it from the steering knuckle.

3. Slide the axle shaft out through the steering knuckle.

4. Disconnect the steering tie rods from the knuckle arm.

5. Remove and discard the lower ball stud nut.

6. Remove the cotter pin from the upper stud. Loosen the upper stud until the top edge of the nut is flush with the top end of the stud.

7. Use a lead hammer to unseat the upper and lower studs from the yoke. Remove the upper nut and the knuckle assembly.

8. Remove the ball stud seat from the upper hole in the axle yoke. It is threaded in the hole. There are special wrenches available for removing the seat. Remove the lower ball stud snapring.

9. Securely clamp the knuckle assembly in a vise with the upper ball stud pointed down.

10. Using a puller and adapters, or a large socket or drift, of approximately the same size as the ball stud, and a mallet, drive the lower stud out of the knuckle.

NOTE: Throughout this procedure, where a ball stud is either removed or installed, a hydraulic press or a two jawed gear puller can be used and, if at all possible, should be used to make the job easier. However, it is possible to complete the job using a mallet, drift and a large socket the same size as the ball studs.

11. Use a puller and adapters to remove the upper ball stud, or,

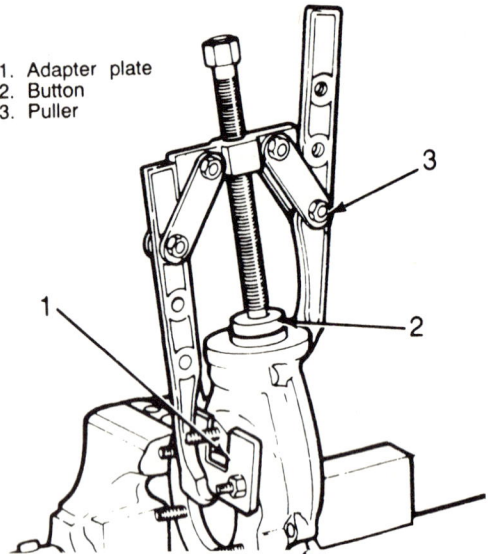

1. Adapter plate
2. Button
3. Puller

Using a puller for lower ball stud removal

place the socket on the bottom surface of the upper ball stud. Place the drift through the hole where the lower ball stud was and place in on the socket. Drive the upper ball stud out of the knuckle with a mallet.

12. Before installing the lower ball stud, run the lower ball stud nut onto the stud just far enough so the head of the stud is flush with the top edge of the nut.

13. Invert the knuckle in the vise. Position the lower ball stud in the knuckle with the nut in place. Use a puller to install the lower ball stud, or, place the same size socket over the nut and drive the ball stud into place with the drift and mallet.

14. Tighten the upper ball stud nut to 10-20 ft. lbs. to draw the lower ball stud into the tapered hold in the yoke. Install the upper stud in the same manner as the lower. The drift will not be needed to install the upper ball stud.

15. Install the upper ball stud seat into the axle yoke. Use a

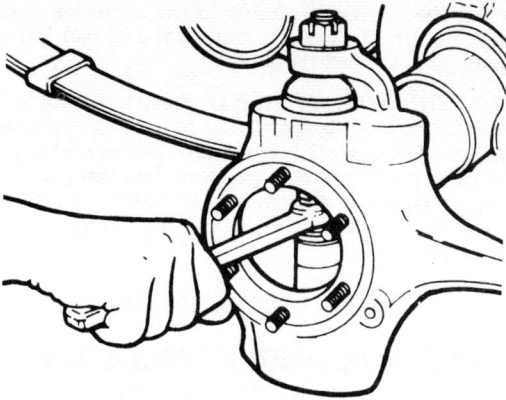

Removing lower ball and stud nut

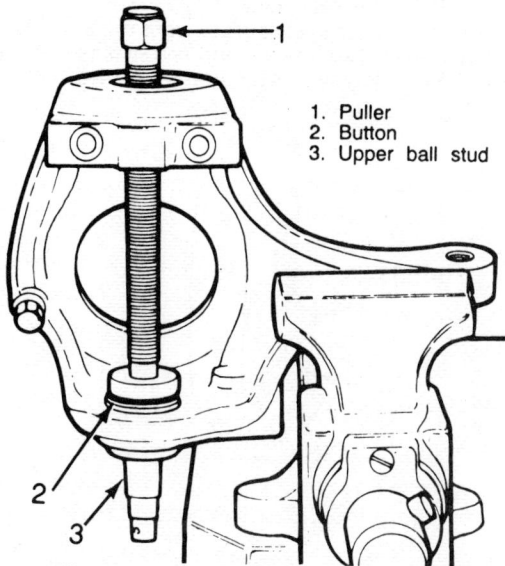

1. Puller
2. Button
3. Upper ball stud

Using a puller and adapters for upper ball stud removal

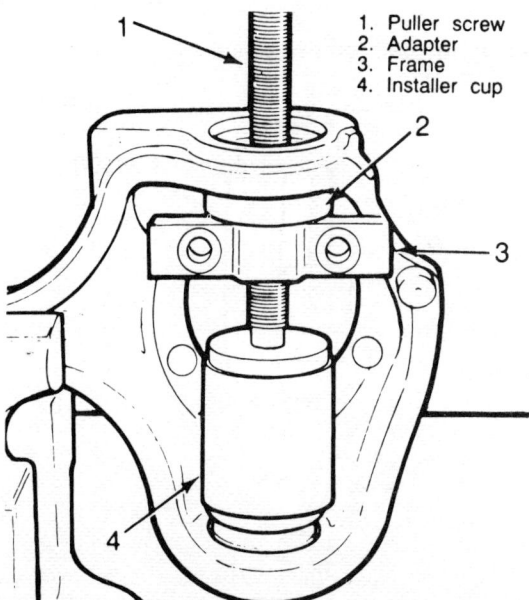

1. Puller screw
2. Adapter
3. Frame
4. Installer cup

Using a puller and adapters to install the lower ball stud

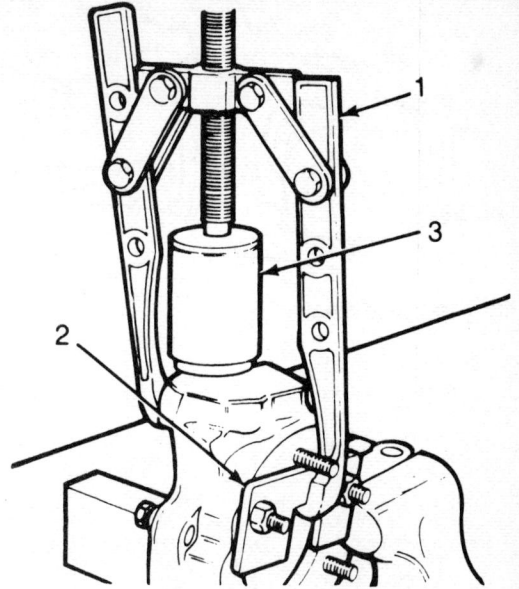

1. Puller   2. Plate   3. Installer cup

Using a puller and adapters to install the upper ball stud

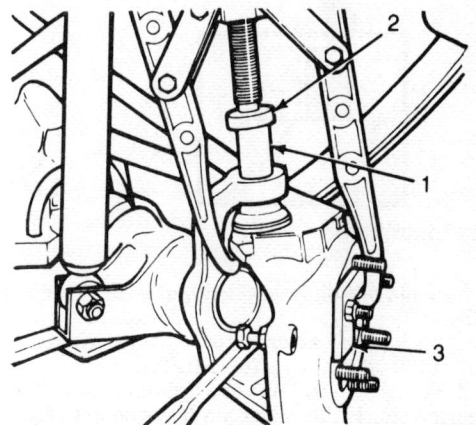

1. Nut socket   2. Button   3. Plate

Steering knuckle installation

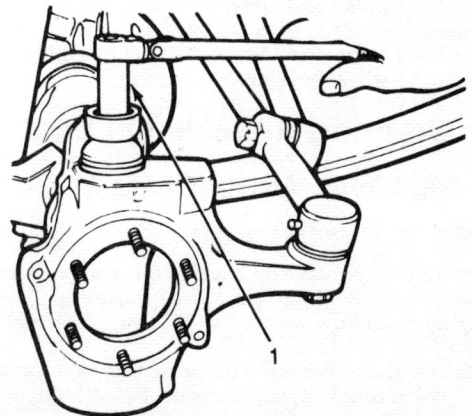

Tightening the upper ball stud seat. 1 is the nut wrench

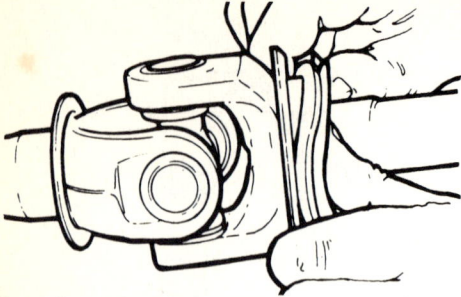

Installing the axle seal

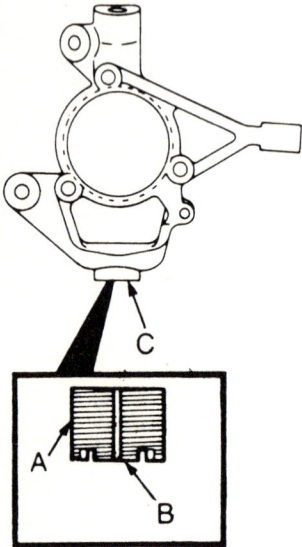

Split seat installation

axle. Install the oil seal assembly in the sequence mentioned above, making sure the backing ring (of the oil seal and backing ring assembly) is toward the wheel.

CHILTON TIP: After driving in wet, freezing weather swing the front wheels from side to side to remove moisture adhering to the oil seal and the spherical surface of the axle housing. This will prevent freezing with resultant damage to the seals. Should be vehicle be stored for any period of time, coat the surfaces with light grease to prevent rusting.

Steering Knuckle

REMOVAL AND INSTALLATION

Wrangler

1. Remove the outer axle shaft.
2. Remove the caliper anchor plate from the knuckle.
3. Remove the knuckle-to-ball joint cotter pins and nuts.
4. Drive the knuckle out with a brass hammer.

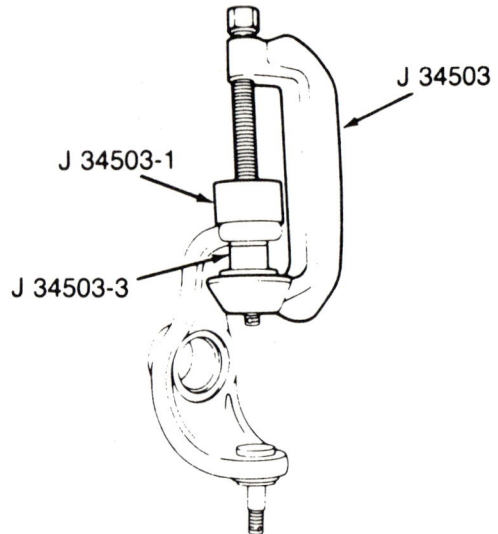

Wrangler upper ball joint removal

new one if the old one shows evidence of wear. Torque the seat to 50 ft. lbs.

16. Install the knuckle assembly onto the axle yoke. Install the lower stud nut. Tighten it to 70-90 ft. lbs.
17. Install the upper stud nut and tighten it to 100 ft. lbs. Install the cotter pin. If the cotter pin holes do not align, tighten the nut until the pin can be installed. Do not loosen the nut to align the holes.
18. Install the axle shaft.
19. Install the spindle and bearing assembly.
20. Install the brake assembly.
21. Connect the steering rods.
22. Install the drum and hub, and wheel assembly.
23. Adjust the wheel bearings.

Steering Knuckle Oil Seal

REMOVAL AND INSTALLATION

CJ-5, CJ-6, CJ-7 and Scrambler

1. Remove the old steering knuckle oil seal by removing the eight screws which hold it in place. Jeeps have a split oil seal and backing ring assembly, an oil seal felt, and two seal retainer plate halves.
2. Examine the spherical surface of the axle for scores or scratches which could damage the seal. Smooth any roughness with emery cloth.
3. Before installing the oil seal felt, make a diagonal cut across the top side of the felt so that it may be slipped over the

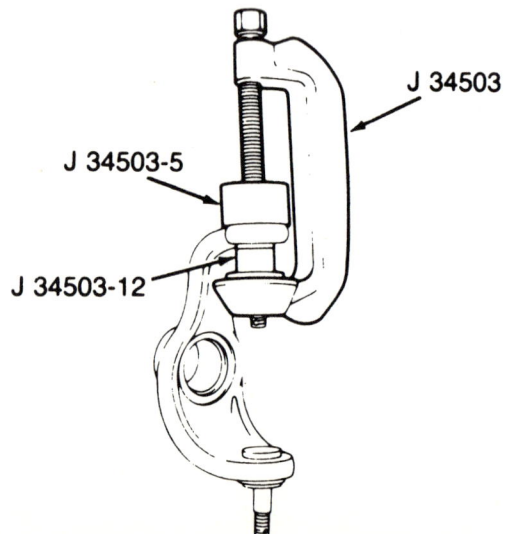

Wrangler upper ball joint installation

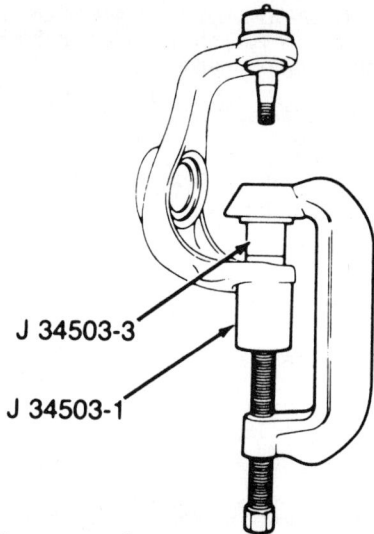

Wrangler lower ball joint removal

NOTE: A split ring seat is located in the bottom of the knuckle. During installation, this ring seat must be set to a depth of 5.23mm (0.206 in.). Measure the depth to the top of the ring seat (4).

5. Drive the knuckle in with a brass hammer.

NOTE: A split ring seat is located in the bottom of the knuckle. During installation, this ring seat must be set to a depth of 5.23mm (0.206 in.). Measure the depth to the top of the ring seat (4).

6. Install the knuckle-to-ball joint cotter pins and nuts.
7. Tighten the knuckle retaining nuts to 75 ft. lbs.
8. Install the caliper anchor plate on the knuckle. Torque the caliper anchor bolts to 77 ft. lbs.
9. Install the outer axle shaft.

Upper Ball Joint

REMOVAL AND INSTALLATION

Wrangler

NOTE: This procedure requires the use of a special tool.

1. Remove the steering knuckle.
2. Position a ball joint removal tool, J-34503-1 and 34503-3, in a C-clamp as shown, and on the upper ball joint.
3. Tighten the clamp screw to remove the joint.
4. Use tools J-34503-5 and J-34503-12, in a similar manner, as illustrated, to install the ball joint.
5. Install the knuckle.

Lower Ball Joint

REMOVAL AND INSTALLATION

Wrangler

NOTE: This procedure requires the use of a special tool.

1. Remove the steering knuckle.
2. Position a ball joint removal tool, J-34503-1 and J-34503-3, as shown, on the lower ball joint.
3. Tighten the clamp screw to remove the joint.

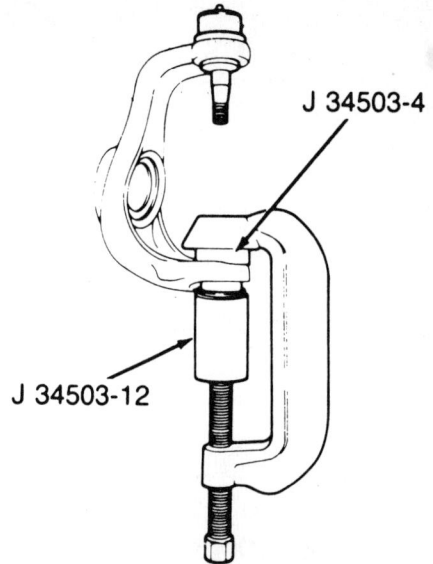

Wrangler lower ball joint installation

4. Use tool J-34503-4 and J-34503-12 to install the ball joint by reversing the removal procedure.
5. Install the knuckle.

Front End Alignment

Proper alignment of the front wheels must be maintained in order to ensure ease of steering and satisfactory tire life.

The most important factors of front wheel alignment are wheel camber, axle caster, and wheel toe-in.

Wheel toe-in is the distance by which the wheels are closer together at the front than at the rear.

Wheel camber is the amount the top of the wheels incline outward from the vertical.

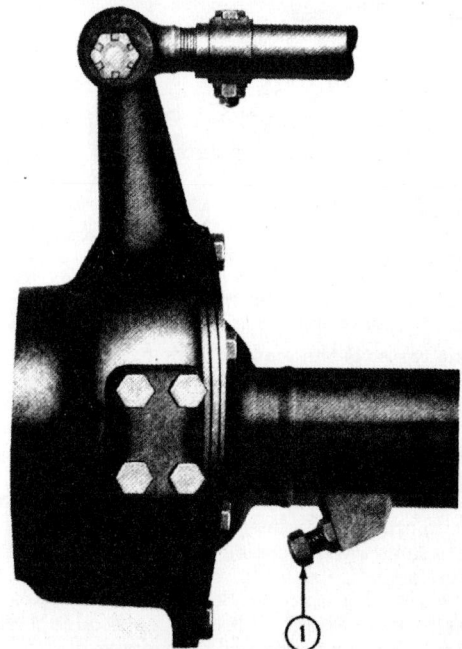

1. Stop screw

1971 turning angle adjusting screw

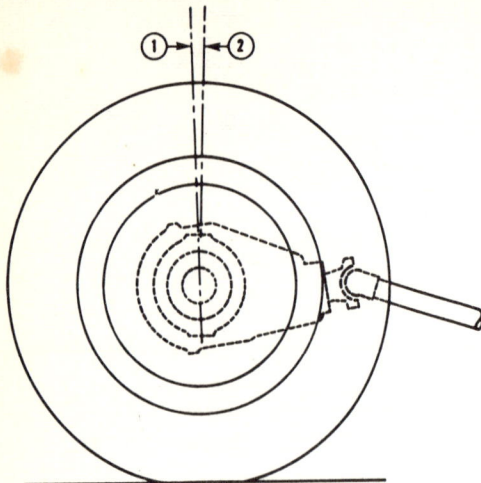

1. Vertical line   2. Caster angle

Caster

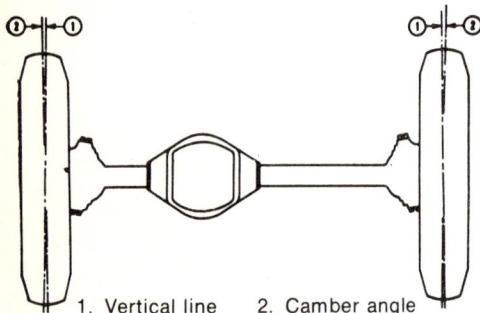

1. Vertical line   2. Camber angle

Camber

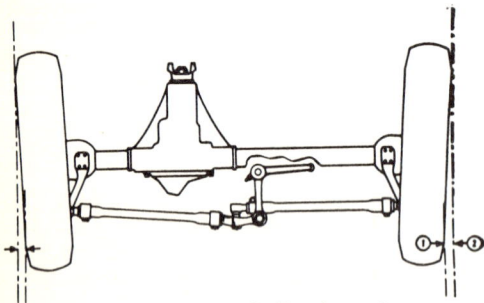

1. Vertical line   2. Toe-in angle

Toe-in

Front axle caster is the amount in degrees that the steering pivot pins are tilted toward the rear of the vehicle. Positive caster is inclination of the top of the pivot pin toward the rear of the vehicle.

These points should be checked at regular intervals, particularly when the front axle has been subjected to a heavy impact. When checking wheel alignment, it is important that wheel bearings be in proper adjustment. Loose bearings will affect instrument readings when checking the camber, pivot pin inclination, and toe-in.

Front wheel camber is preset. Some alignment shops can correct camber to some extent by installing special tapered shims between the steering knuckle and the spindle.

Caster is also preset, but can be altered by use of tapered shims between the axle pad and the springs. Wheel toe-in is adjustable.

To avoid damage to the U-joints, it is advisable to check the turning angle periodically. An adjustment turntable is advisable for properly determining the angle.

Correct turning angles are:
- All 1971: 27.5° max.
- 1972-75
  With standard (F78 x 15) tires: 34-35°
  With larger optional tires: 31°
- 1976: 31°
- 1977: 29°
- 1978-83: 31-32°
- 1984-86: 30-31°
- 1987-90: 32-33°

To adjust the turning angle, loosen the locknut (on some early models, a securing weld will have to be broken) and turn the adjusting screw. The adjusting screw is located on the axle tube near the knuckle on early models, and on the knuckle, just below the axle centerline on later models.

CASTER ADJUSTMENT

Caster angle is established in the axle design by tilting the top of the kingpins forward so that an imaginary line through the center of the kingpins would strike the ground at a point ahead of the point of the contact.

The purpose of caster is to provide steering stability which will keep the front wheels in the straight ahead position and also assist in straightening up the wheels when coming out of a turn.

If the angle of caster, when accurately measured, is found to be incorrect, correct it to the specification given in this section by either installing new parts or installing caster shims between the axle pad and the springs.

If the camber and toe-in are correct and it is known that the axle is not twisted, a satisfactory check may be made by testing the vehicle on the road. Before road testing, make sure all tires are properly inflated, being particularly careful that both front tires are inflated to exactly the same pressure.

If the vehicle turns easily to either side but is hard to straighten out, insufficient caster for easy handling of the vehicle is indicated. If correction is necessary, it can usually be accomplished by installing shims between the springs and axle pads to secure the desired result.

CAMBER ADJUSTMENT

The purpose of camber is to more nearly place the weight of the vehicle over the tire contact patch on the road to facilitate ease of steering. The result of excessive camber is irregular wear of the tires on the outside shoulders and is usually caused by bent axle parts.

The result of excessive negative or reverse camber will be hard steering and possibly a wandering condition. Tires will also wear on the inside shoulders. Negative camber is usually caused by excessive wear or looseness of the front wheel bearings, axle parts or the result of a sagging axle.

Unequal camber may cause any or a combination of the following conditions: unstable steering, wandering, kickback or road shock, shimmy or excessive tire wear. The cause of unequal camber is usually a bent steering knuckle or axle end.

Correct wheel camber is set in the axle at the time of manufacture. It is important that the camber be the same on both front wheels.

TOE-IN ADJUSTMENT

Through 1971

Toe-in may be adjusted with a line or straightedge as the vehicle tread is the same in the front and rear. To set the adjustment both tie rods must be adjusted as outlined below: Set the tie rod end of the steering bellcrank at right angles with the front axle. Place a straight edge or line against the left rear wheel and left front wheel to determine if the wheel is in a straight ahead position. If the front wheel tire does not touch the straight edge at both the front and rear, it will be necessary to adjust the left tie rod by loosening the clamps on each end and turning the rod until the tire touches the straight edge.

Check the right hand side in the same manner, adjusting the tie rod if necessary making sure that the bellcrank remains at right angles to the axle. When it is determined that the front wheels are in the straight ahead position, set the toe-in by shortening each tie rod approximately ½ turn.

1972-90

First raise the front of the vehicle to free the front wheels. Turn the wheels to the straight ahead position. Use a Steadyrest® to scribe a pencil line in the center of each tire tread as the wheel is turned by hand. A good way to do this is to first coat a strip with chalk around the circumference of the tread at the center to form a base for a fine pencil line.

Measure the distance between the scribed lines at the front and rear of the wheels using care that both measurements are made at an equal distance from the floor. The distance between the lines should be greater at the rear than at the front by 3/64 in. (1.2mm) to 3/32 in. (2.2mm). To adjust, loosen the clamp bolts and turn the tie rod with a small pipe wrench. The tie rod is threaded with right and left hand threads to provide equal adjustment at both wheels. Do not overlook retightening the clamp bolts. It is common practice to measure between the wheel rims. This is satisfactory providing the wheels run true. By scribing a line on the tire tread, measurement is taken between the road contact points reducing error caused by wheel runout.

REAR SUSPENSION

All springs should be examined periodically for broken or shifted leaves, loose or missing clips, angle of the spring shackles, and position of the springs on the saddles. Springs with shifted leaves do not retain their normal strength. Missing clips may permit the spirit leaves to fan out or break on rebound. Broken leaves may make the vehicle hard to handle or permit the axle to shift out of line. Weakened springs may break causing difficulty in steering. Spring attaching clips or bolts must be tight. It is suggested that they be checked at each vehicle inspection.

Springs

REMOVAL AND INSTALLATION

1971

1. Raise the vehicle with a jack under the axle and place a jackstand under the frame side rail. Then lower the axle jack so that the load is relieved from the spring and the wheels rest on the floor.
2. Remove the nuts which secure the spring clip bolts. Remove the spring plate and clip bolts. Free the spring from the axle by raising the axle jack.
3. Remove the pivot bolt nut and drive out the pivot bolt. Disconnect the shackle either by removing the lower nuts and bolts on the rubber bushed shackles, or by removing the threaded bushings on the U-shackles.
4. To replace, first install the pivot bolt. Then, connect the shackle using the following procedures.
5. On bronze bushed pivot bolts, install the bolt and nut and tighten the nut. Then back it off two cotter pin slots and install the cotter pin. The nut must be drawn up tightly but must be sufficiently loose to allow the spring to pivot freely. Otherwise the spring might break.
6. On rubber bushed pivot bolts and locknuts (or lockwasher and nut) only tighten the bolt enough to hold the bushings in position until the vehicle is lowered from the jack.
7. Connect the shackle. On rubber bushed shackles install the bolts as in step 6 above. For U-shackles, insert the shackle through the frame bracket and eye of the spring. Holding the U-shackle tightly against the frame, start the upper bushing on the shackle, taking care that when it enters the thread in the frame it does not crossthread. Screw the bushing on the shackle tightly against the spring eye, and thread the bushing in approximately half way. Then, alternately from top bushing to lower bushing, turn them in until the head of the bushing is snug against the frame bracket and the bushing in the spring eye is 1/32 in. (0.8mm) away from the spring as measured from the inside of the hexagon head in the spring. Lubricate the bushing and then try the flex of the shackle, which must be free. If a shackle is tight, rethread the bushings on the shackle.
8. Move the axle into position on the spring by lowering or raising the axle jack. Install the spring clip bolts, spring plate, lockwashers, and nuts. Torque the nuts to 50-55 ft. lbs. Avoid overtightening. Be sure the spring is free to move at both ends.
9. Remove both jacks. On rubber bushed shackles and pivot bolts, allow the weight of the vehicle to seat the bushings in their operating positions. Then torque the nuts to 27-30 ft. lbs.

1972-90

1. Raise the vehicle with a jack under the axle. Place a jackstand under the frame side rail. Then lower the axle jack so the load is relieved from the spring and the wheels just touch the floor.
2. Disconnect the shock absorber from the spring clip plate. On models through 1986, disconnect the stabilizer bar. On 1987-90 models, loosen, but don't remove, the front stabilizer bar link nut.
3. Remove the nuts which secure the spring clips (U-bolts). Remove the spring plate and spring clips. Free the spring from the axle by raising the axle.
4. Remove the pivot bolt nut and drive out the pivot bolt. Disconnect the shackle.
5. With the spring removed, the spring shackle and/or shackle plate may be removed from the spring.
6. Inspect the bushings in the eye of the main spring leaf and

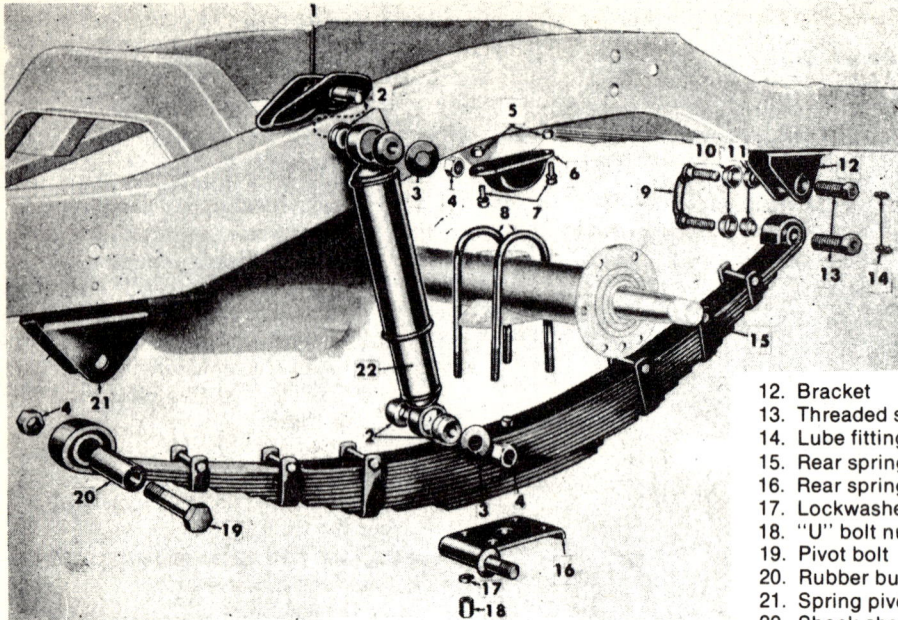

1. Shock absorber bracket
2. Mounting pin bushing
3. Washer
4. Lock nut
5. Nut
6. Rear axle bumper
7. Bolt
8. "U" bolt
9. Shackle
10. Retainer
11. Grease seal

12. Bracket
13. Threaded shackle bushing
14. Lube fitting
15. Rear spring assembly
16. Rear spring clip plate
17. Lockwasher
18. "U" bolt nut
19. Pivot bolt
20. Rubber bushing
21. Spring pivot bracket
22. Shock absorber assembly

Early model rear spring and shock absorber

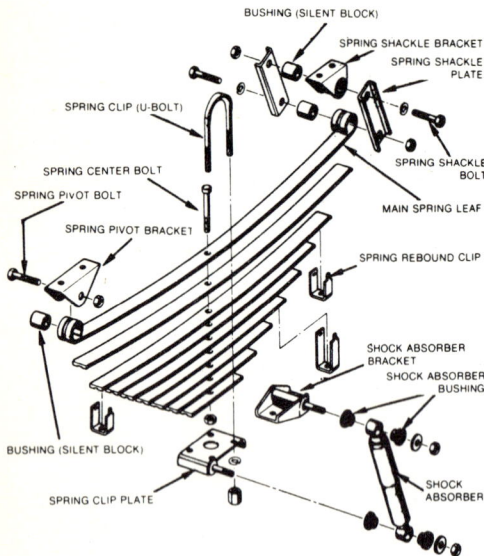

1972–75 rear spring assembly

the bushings of the spring shackle for excessive wear. Replace if necessary.

7. The spring can be disassembled for replacing an individual spring leaf, by removing the clips and the center bolts.

8. To install the spring on the vehicle, with the bushings in place and the spring shackle attached to the springs, position the spring in the pivot hanger and install the pivot bolt and lock nut. Only tighten the lock nut enough to hold the bushings in position until the vehicle is lowered from the jack.

9. Position the spring and install the shackle, shackle bolts, shackle plate if applicable, lockwasher, and nut. Only finger tighten the nuts at this time.

10. Move the axle into position on the spring by lowering the axle jack. Place the spring center bolt in the axle saddle hole. In-stall the spring clips, spring plate, lockwashers and nuts. Torque the $7/16$ in. nuts to 36-42 ft. lbs. and the $1/2$ in. nuts to 45-65 ft. lbs. and the $9/16$ in. nuts to 100 ft. lbs.

NOTE: Be sure that the center bolt is properly centered in the axle saddle.

11. Connect the shock absorber.

12. Remove the jack and allow the weight of the vehicle to seat the bushings in their operating positions. On models through 1976, torque the $7/16$ in. spring pivot bolt nuts and spring shackle nuts to 35-50 ft. lbs. Torque the $5/8$ in. shackle nuts 55-75 ft. lbs. On 1977 and later models, tighten pivot bolts to 100 ft. lbs., and shackle nuts to 24 ft. lbs.

Shock Absorbers

REMOVAL AND INSTALLATION

1. Remove the locknuts and washers.
2. Pull the shock absorber eyes and rubber bushings from the mounting pins.
3. Install the shocks in reverse order of the removal procedure. Torque the upper bolt to 35 ft. lbs. on the CJ-5, CJ-6, CJ-7 and Scrambler, 45 ft. lbs. on the Wrangler, and the lower bolt to 45 ft. lbs.

Track Bar

REMOVAL AND INSTALLATION

Wrangler

1. Raise and support the rear end on jackstands.
2. Unbolt the bar from the frame bracket.
3. Unbolt the bar from the axle bracket.
4. Remove the bar.
5. Installation is the reverse of removal. Torque the bolts to 74 ft. lbs.

U-BOLT

CENTER BOLT

REBOUND CLIP

REAR SPRING
BUSHING (SILENT
BLOCK)

INSERT

TIE PLATE

REAR LEAF SPRING
NO. 2 LEAF

REAR SHOCK

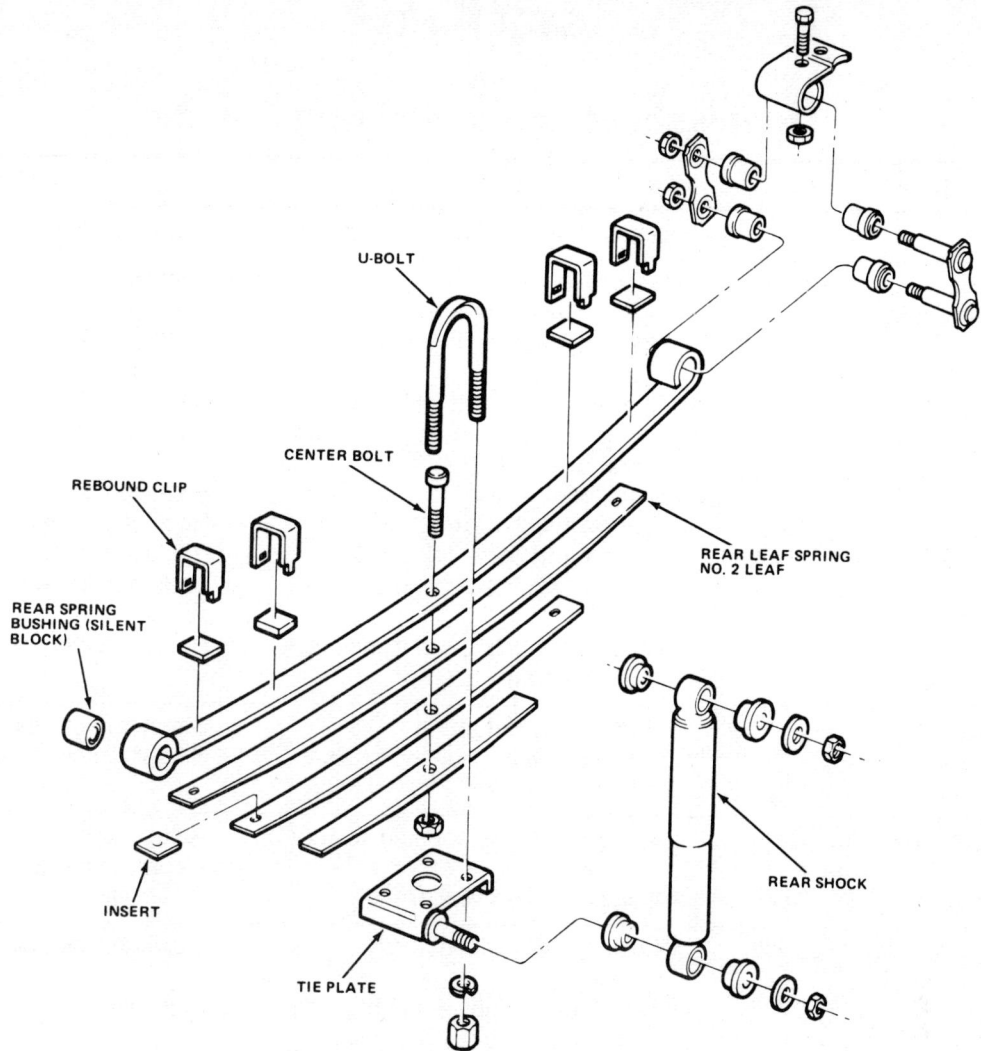

Rear suspension for 1976–86 models

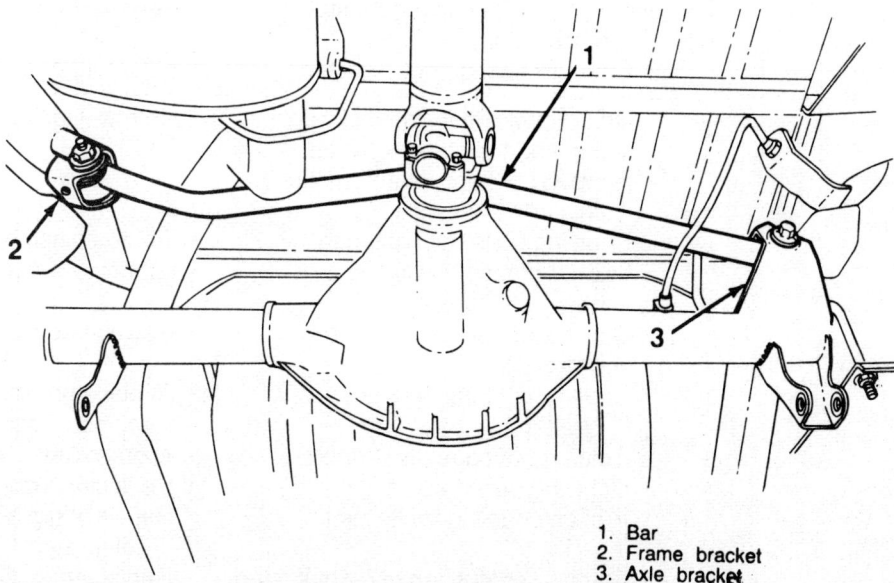

1
2
3

1. Bar
2. Frame bracket
3. Axle bracket

Rear track bar on the Wrangler

8 SUSPENSION AND STEERING

STEERING

Troubleshooting the Steering Column

Problem	Cause	Solution
Will not lock	• Lockbolt spring broken or defective	• Replace lock bolt spring
High effort (required to turn ignition key and lock cylinder)	• Lock cylinder defective • Ignition switch defective • Rack preload spring broken or deformed • Burr on lock sector, lock rack, housing, support or remote rod coupling • Bent sector shaft • Defective lock rack • Remote rod bent, deformed • Ignition switch mounting bracket bent • Distorted coupling slot in lock rack (tilt column)	• Replace lock cylinder • Replace ignition switch • Replace preload spring • Remove burr • Replace shaft • Replace lock rack • Replace rod • Straighten or replace • Replace lock rack
Will stick in "start"	• Remote rod deformed • Ignition switch mounting bracket bent	• Straighten or replace • Straighten or replace
Key cannot be removed in "off-lock"	• Ignition switch is not adjusted correctly • Defective lock cylinder	• Adjust switch • Replace lock cylinder
Lock cylinder can be removed without depressing retainer	• Lock cylinder with defective retainer • Burr over retainer slot in housing cover or on cylinder retainer	• Replace lock cylinder • Remove burr
High effort on lock cylinder between "off" and "off-lock"	• Distorted lock rack • Burr on tang of shift gate (automatic column) • Gearshift linkage not adjusted	• Replace lock rack • Remove burr • Adjust linkage
Noise in column	• One click when in "off-lock" position and the steering wheel is moved (all except automatic column) • Coupling bolts not tightened • Lack of grease on bearings or bearing surfaces • Upper shaft bearing worn or broken • Lower shaft bearing worn or broken • Column not correctly aligned • Coupling pulled apart • Broken coupling lower joint • Steering shaft snap ring not seated	• Normal—lock bolt is seating • Tighten pinch bolts • Lubricate with chassis grease • Replace bearing assembly • Replace bearing. Check shaft and replace if scored. • Align column • Replace coupling • Repair or replace joint and align column • Replace ring. Check for proper seating in groove.

Troubleshooting the Steering Column (cont.)

Problem	Cause	Solution
Noise in column	• Shroud loose on shift bowl. Housing loose on jacket—will be noticed with ignition in "off-lock" and when torque is applied to steering wheel.	• Position shroud over lugs on shift bowl. Tighten mounting screws.
High steering shaft effort	• Column misaligned • Defective upper or lower bearing • Tight steering shaft universal joint • Flash on I.D. of shift tube at plastic joint (tilt column only) • Upper or lower bearing seized	• Align column • Replace as required • Repair or replace • Replace shift tube • Replace bearings
Lash in mounted column assembly	• Column mounting bracket bolts loose • Broken weld nuts on column jacket • Column capsule bracket sheared	• Tighten bolts • Replace column jacket • Replace bracket assembly
Lash in mounted column assembly (cont.)	• Column bracket to column jacket mounting bolts loose • Loose lock shoes in housing (tilt column only) • Loose pivot pins (tilt column only) • Loose lock shoe pin (tilt column only) • Loose support screws (tilt column only)	• Tighten to specified torque • Replace shoes • Replace pivot pins and support • Replace pin and housing • Tighten screws
Housing loose (tilt column only)	• Excessive clearance between holes in support or housing and pivot pin diameters • Housing support-screws loose	• Replace pivot pins and support • Tighten screws
Steering wheel loose—every other tilt position (tilt column only)	• Loose fit between lock shoe and lock shoe pivot pin	• Replace lock shoes and pivot pin
Steering column not locking in any tilt position (tilt column only)	• Lock shoe seized on pivot pin • Lock shoe grooves have burrs or are filled with foreign material • Lock shoe springs weak or broken	• Replace lock shoes and pin • Clean or replace lock shoes • Replace springs
Noise when tilting column (tilt column only)	• Upper tilt bumpers worn • Tilt spring rubbing in housing	• Replace tilt bumper • Lubricate with chassis grease
One click when in "off-lock" position and the steering wheel is moved	• Seating of lock bolt	• None. Click is normal characteristic sound produced by lock bolt as it seats.
High shift effort (automatic and tilt column only)	• Column not correctly aligned • Lower bearing not aligned correctly • Lack of grease on seal or lower bearing areas	• Align column • Assemble correctly • Lubricate with chassis grease

Troubleshooting the Steering Column (cont.)

Problem	Cause	Solution
Improper transmission shifting—automatic and tilt column only	• Sheared shift tube joint • Improper transmission gearshift linkage adjustment • Loose lower shift lever	• Replace shift tube • Adjust linkage • Replace shift tube

Troubleshooting the Ignition Switch

Problem	Cause	Solution
Ignition switch electrically inoperative	• Loose or defective switch connector • Feed wire open (fusible link) • Defective ignition switch	• Tighten or replace connector • Repair or replace • Replace ignition switch
Engine will not crank	• Ignition switch not adjusted properly	• Adjust switch
Ignition switch wil not actuate mechanically	• Defective ignition switch • Defective lock sector • Defective remote rod	• Replace switch • Replace lock sector • Replace remote rod
Ignition switch cannot be adjusted correctly	• Remote rod deformed	• Repair, straighten or replace

Troubleshooting the Turn Signal Switch

Problem	Cause	Solution
Turn signal will not cancel	• Loose switch mounting screws • Switch or anchor bosses broken • Broken, missing or out of position detent, or cancelling spring	• Tighten screws • Replace switch • Reposition springs or replace switch as required
Turn signal difficult to operate	• Turn signal lever loose • Switch yoke broken or distorted • Loose or misplaced springs • Foreign parts and/or materials in switch • Switch mounted loosely	• Tighten mounting screws • Replace switch • Reposition springs or replace switch • Remove foreign parts and/or material • Tighten mounting screws
Turn signal will not indicate lane change	• Broken lane change pressure pad or spring hanger • Broken, missing or misplaced lane change spring • Jammed wires	• Replace switch • Replace or reposition as required • Loosen mounting screws, reposition wires and retighten screws

Troubleshooting the Turn Signal Switch (cont.)

Problem	Cause	Solution
Turn signal will not stay in turn position	• Foreign material or loose parts impeding movement of switch yoke	• Remove material and/or parts
	• Defective switch	• Replace switch
Hazard switch cannot be pulled out	• Foreign material between hazard support cancelling leg and yoke	• Remove foreign material. No foreign material impeding function of hazard switch—replace turn signal switch.
No turn signal lights	• Inoperative turn signal flasher	• Replace turn signal flasher
	• Defective or blown fuse	• Replace fuse
	• Loose chassis to column harness connector	• Connect securely
	• Disconnect column to chassis connector. Connect new switch to chassis and operate switch by hand. If vehicle lights now operate normally, signal switch is inoperative	• Replace signal switch
	• If vehicle lights do not operate, check chassis wiring for opens, grounds, etc.	• Repair chassis wiring as required
Instrument panel turn indicator lights on but not flashing	• Burned out or damaged front or rear turn signal bulb	• Replace bulb
	• If vehicle lights do not operate, check light sockets for high resistance connections, the chassis wiring for opens, grounds, etc.	• Repair chassis wiring as required
	• Inoperative flasher	• Replace flasher
	• Loose chassis to column harness connection	• Connect securely
	• Inoperative turn signal switch	• Replace turn signal switch
	• To determine if turn signal switch is defective, substitute new switch into circuit and operate switch by hand. If the vehicle's lights operate normally, signal switch is inoperative.	• Replace turn signal switch
Stop light not on when turn indicated	• Loose column to chassis connection	• Connect securely
	• Disconnect column to chassis connector. Connect new switch into system without removing old.	• Replace signal switch

Troubleshooting the Turn Signal Switch (cont.)

Problem	Cause	Solution
Stop light not on when turn indicated (cont.)	Operate switch by hand. If brake lights work with switch in the turn position, signal switch is defective. • If brake lights do not work, check connector to stop light sockets for grounds, opens, etc.	• Repair connector to stop light circuits using service manual as guide
Turn indicator panel lights not flashing	• Burned out bulbs • High resistance to ground at bulb socket • Opens, ground in wiring harness from front turn signal bulb socket to indicator lights	• Replace bulbs • Replace socket • Locate and repair as required
Turn signal lights flash very slowly	• High resistance ground at light sockets • Incorrect capacity turn signal flasher or bulb • If flashing rate is still extremely slow, check chassis wiring harness from the connector to light sockets for high resistance • Loose chassis to column harness connection • Disconnect column to chassis connector. Connect new switch into system without removing old. Operate switch by hand. If flashing occurs at normal rate, the signal switch is defective.	• Repair high resistance grounds at light sockets • Replace turn signal flasher or bulb • Locate and repair as required • Connect securely • Replace turn signal switch
Hazard signal lights will not flash—turn signal functions normally	• Blow fuse • Inoperative hazard warning flasher • Loose chassis-to-column harness connection • Disconnect column to chassis connector. Connect new switch into system without removing old. Depress the hazard warning lights. If they now work normally, turn signal switch is defective. • If lights do not flash, check wiring harness "K" lead for open between hazard flasher and connector. If open, fuse block is defective	• Replace fuse • Replace hazard warning flasher in fuse panel • Conect securely • Replace turn signal switch • Repair or replace brown wire or connector as required

Troubleshooting the Manual Steering Gear

Problem	Cause	Solution
Hard or erratic steering	• Incorrect tire pressure	• Inflate tires to recommended pressures
	• Insufficient or incorrect lubrication	• Lubricate as required (refer to Maintenance Section)
	• Suspension, or steering linkage parts damaged or misaligned	• Repair or replace parts as necessary
	• Improper front wheel alignment	• Adjust incorrect wheel alignment angles
	• Incorrect steering gear adjustment	• Adjust steering gear
	• Sagging springs	• Replace springs
Play or looseness in steering	• Steering wheel loose	• Inspect shaft spines and repair as necessary. Tighten attaching nut and stake in place.
	• Steering linkage or attaching parts loose or worn	• Tighten, adjust, or replace faulty components
	• Pitman arm loose	• Inspect shaft splines and repair as necessary. Tighten attaching nut and stake in place
	• Steering gear attaching bolts loose	• Tighten bolts
	• Loose or worn wheel bearings	• Adjust or replace bearings
	• Steering gear adjustment incorrect or parts badly worn	• Adjust gear or replace defective parts
Wheel shimmy or tramp	• Improper tire pressure	• Inflate tires to recommended pressures
	• Wheels, tires, or brake rotors out-of-balance or out-of-round	• Inspect and replace or balance parts
	• Inoperative, worn, or loose shock absorbers or mounting parts	• Repair or replace shocks or mountings
	• Loose or worn steering or suspension parts	• Tighten or replace as necessary
	• Loose or worn wheel bearings	• Adjust or replace bearings
	• Incorrect steering gear adjustments	• Adjust steering gear
	• Incorrect front wheel alignment	• Correct front wheel alignment
Tire wear	• Improper tire pressure	• Inflate tires to recommended pressures
	• Failure to rotate tires	• Rotate tires
	• Brakes grabbing	• Adjust or repair brakes
	• Incorrect front wheel alignment	• Align incorrect angles
	• Broken or damaged steering and suspension parts	• Repair or replace defective parts
	• Wheel runout	• Replace faulty wheel
	• Excessive speed on turns	• Make driver aware of conditions

8 SUSPENSION AND STEERING

Troubleshooting the Manual Steering Gear

Problem	Cause	Solution
Vehicle leads to one side	• Improper tire pressures	• Inflate tires to recommended pressures
	• Front tires with uneven tread depth, wear pattern, or different cord design (i.e., one bias ply and one belted or radial tire on front wheels)	• Install tires of same cord construction and reasonably even tread depth, design, and wear pattern
	• Incorrect front wheel alignment	• Align incorrect angles
	• Brakes dragging	• Adjust or repair brakes
	• Pulling due to uneven tire construction	• Replace faulty tire

Troubleshooting the Power Steering Gear

Problem	Cause	Solution
Hissing noise in steering gear	• There is some noise in all power steering systems. One of the most common is a hissing sound most evident at standstill parking. There is no relationship between this noise and performance of the steering. Hiss may be expected when steering wheel is at end of travel or when slowly turning at standstill.	• Slight hiss is normal and in no way affects steering. Do not replace valve unless hiss is extremely objectionable. A replacement valve will also exhibit slight noise and is not always a cure. Investigate clearance around flexible coupling rivets. Be sure steering shaft and gear are aligned so flexible coupling rotates in a flat plane and is not distorted as shaft rotates. Any metal-to-metal contacts through flexible coupling will transmit valve hiss into passenger compartment through the steering column.
Rattle or chuckle noise in steering gear	• Gear loose on frame	• Check gear-to-frame mounting screws. Tighten screws to 88 N·m (65 foot pounds) torque.
	• Steering linkage looseness	• Check linkage pivot points for wear. Replace if necessary.
	• Pressure hose touching other parts of car	• Adjust hose position. Do not bend tubing by hand.
	• Loose pitman shaft over center adjustment NOTE: A slight rattle may occur on turns because of increased clearance off the "high point." This is normal and clearance must not be reduced below specified limits to eliminate this slight rattle.	• Adjust to specifications
	• Loose pitman arm	• Tighten pitman arm nut to specifications

Troubleshooting the Power Steering Gear (cont.)

Problem	Cause	Solution
Squawk noise in steering gear when turning or recovering from a turn	• Damper O-ring on valve spool cut	• Replace damper O-ring
Poor return of steering wheel to center	• Tires not properly inflated • Lack of lubrication in linkage and ball joints • Lower coupling flange rubbing against steering gear adjuster plug • Steering gear to column misalignment • Improper front wheel alignment • Steering linkage binding • Ball joints binding • Steering wheel rubbing against housing • Tight or frozen steering shaft bearings • Sticking or plugged valve spool • Steering gear adjustments over specifications • Kink in return hose	• Inflate to specified pressure • Lube linkage and ball joints • Loosen pinch bolt and assemble properly • Align steering column • Check and adjust as necessary • Replace pivots • Replace ball joints • Align housing • Replace bearings • Remove and clean or replace valve • Check adjustment with gear out of car. Adjust as required. • Replace hose
Car leads to one side or the other (keep in mind road condition and wind. Test car in both directions on flat road)	• Front end misaligned • Unbalanced steering gear valve **NOTE:** If this is cause, steering effort will be very light in direction of lead and normal or heavier in opposite direction	• Adjust to specifications • Replace valve
Momentary increase in effort when turning wheel fast to right or left	• Low oil level • Pump belt slipping • High internal leakage	• Add power steering fluid as required • Tighten or replace belt • Check pump pressure. (See pressure test)
Steering wheel surges or jerks when turning with engine running especially during parking	• Low oil level • Loose pump belt • Steering linkage hitting engine oil pan at full turn • Insufficient pump pressure • Pump flow control valve sticking	• Fill as required • Adjust tension to specification • Correct clearance • Check pump pressure. (See pressure test). Replace relief valve if defective. • Inspect for varnish or damage, replace if necessary

Troubleshooting the Power Steering Gear (cont.)

Problem	Cause	Solution
Excessive wheel kickback or loose steering	• Air in system	• Add oil to pump reservoir and bleed by operating steering. Check hose connectors for proper torque and adjust as required.
	• Steering gear loose on frame	• Tighten attaching screws to specified torque
	• Steering linkage joints worn enough to be loose	• Replace loose pivots
	• Worn poppet valve	• Replace poppet valve
	• Loose thrust bearing preload adjustment	• Adjust to specification with gear out of vehicle
	• Excessive overcenter lash	• Adjust to specification with gear out of car
Hard steering or lack of assist	• Loose pump belt	• Adjust belt tension to specification
	• Low oil level **NOTE:** Low oil level will also result in excessive pump noise	• Fill to proper level. If excessively low, check all lines and joints for evidence of external leakage. Tighten loose connectors.
	• Steering gear to column misalignment	• Align steering column
	• Lower coupling flange rubbing against steering gear adjuster plug	• Loosen pinch bolt and assemble properly
	• Tires not properly inflated	• Inflate to recommended pressure
Foamy milky power steering fluid, low fluid level and possible low pressure	• Air in the fluid, and loss of fluid due to internal pump leakage causing overflow	• Check for leak and correct. Bleed system. Extremely cold temperatures will cause system aeration should the oil level be low. If oil level is correct and pump still foams, remove pump from vehicle and separate reservoir from housing. Check welsh plug and housing for cracks. If plug is loose or housing is cracked, replace housing.
Low pressure due to steering pump	• Flow control valve stuck or inoperative	• Remove burrs or dirt or replace. Flush system.
	• Pressure plate not flat against cam ring	• Correct
Low pressure due to steering gear	• Pressure loss in cylinder due to worn piston ring or badly worn housing bore	• Remove gear from car for disassembly and inspection of ring and housing bore
	• Leakage at valve rings, valve body-to-worm seal	• Remove gear from car for disassembly and replace seals

Troubleshooting the Power Steering Pump

Problem	Cause	Solution
Chirp noise in steering pump	• Loose belt	• Adjust belt tension to specification
Belt squeal (particularly noticeable at full wheel travel and stand still parking)	• Loose belt	• Adjust belt tension to specification
Growl noise in steering pump	• Excessive back pressure in hoses or steering gear caused by restriction	• Locate restriction and correct. Replace part if necessary.
Growl noise in steering pump (particularly noticeable at stand still parking)	• Scored pressure plates, thrust plate or rotor • Extreme wear of cam ring	• Replace parts and flush system • Replace parts
Groan noise in steering pump	• Low oil level • Air in the oil. Poor pressure hose connection.	• Fill reservoir to proper level • Tighten connector to specified torque. Bleed system by operating steering from right to left—full turn.
Rattle noise in steering pump	• Vanes not installed properly • Vanes sticking in rotor slots	• Install properly • Free up by removing burrs, varnish, or dirt
Swish noise in steering pump	• Defective flow control valve	• Replace part
Whine noise in steering pump	• Pump shaft bearing scored	• Replace housing and shaft. Flush system.
Hard steering or lack of assist	• Loose pump belt • Low oil level in reservoir **NOTE:** Low oil level will also result in excessive pump noise • Steering gear to column misalignment • Lower coupling flange rubbing against steering gear adjuster plug • Tires not properly inflated	• Adjust belt tension to specification • Fill to proper level. If excessively low, check all lines and joints for evidence of external leakage. Tighten loose connectors. • Align steering column • Loosen pinch bolt and assemble properly • Inflate to recommended pressure
Foaming milky power steering fluid, low fluid level and possible low pressure	• Air in the fluid, and loss of fluid due to internal pump leakage causing overflow	• Check for leaks and correct. Bleed system. Extremely cold temperatures will cause system aeration should the oil level be low. If oil level is correct and pump still foams, remove pump from vehicle and separate reservoir from body. Check welsh plug and body for cracks. If plug is loose or body is cracked, replace body.

Troubleshooting the Power Steering Pump (cont.)

Problem	Cause	Solution
Low pump pressure	• Flow control valve stuck or inoperative • Pressure plate not flat against cam ring	• Remove burrs or dirt or replace. Flush system. • Correct
Momentary increase in effort when turning wheel fast to right or left	• Low oil level in pump • Pump belt slipping • High internal leakage	• Add power steering fluid as required • Tighten or replace belt • Check pump pressure. (See pressure test)
Steering wheel surges or jerks when turning with engine running especially during parking	• Low oil level • Loose pump belt • Steering linkage hitting engine oil pan at full turn • Insufficient pump pressure	• Fill as required • Adjust tension to specification • Correct clearance • Check pump pressure. (See pressure test). Replace flow control valve if defective.
Steering wheel surges or jerks when turning with engine running especially during parking (cont.)	• Sticking flow control valve	• Inspect for varnish or damage, replace if necessary
Excessive wheel kickback or loose steering	• Air in system	• Add oil to pump reservoir and bleed by operating steering. Check hose connectors for proper torque and adjust as required.
Low pump pressure	• Extreme wear of cam ring • Scored pressure plate, thrust plate, or rotor • Vanes not installed properly • Vanes sticking in rotor slots • Cracked or broken thrust or pressure plate	• Replace parts. Flush system. • Replace parts. Flush system. • Install properly • Freeup by removing burrs, varnish, or dirt • Replace part

Steering Wheel

REMOVAL AND INSTALLATION

1971-75

1. Disconnect the negative battery cable.
2. Set the front tires in a straight ahead position.
3. Pull the horn button from the steering wheel.
4. Remove the steering wheel nut and horn button contact cup.
5. Scribe a line mark on the steering wheel and steering shaft if there is not one already. Release the turn signal assembly from the steering post and install a puller.
6. Remove the steering wheel and spring.

7. To install, align the scribe marks on the steering shaft with the steering wheel and secure the steering wheel spring, steering wheel, and horn button contact cup with the steering wheel nut.
8. Install the horn button.
9. Connect the battery cable and test the horn.

1976-90

NOTE: Some steering shafts have metric threads. These are identified by a groove cast into the shaft. See the accompanying illustration for an example.

1. Disconnect the negative battery cable.
2. Place the front wheels in the straight ahead position.
3. Remove the horn button from the steering wheel. Turn the

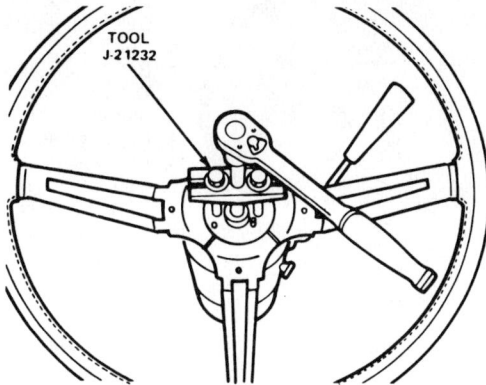

Steering wheel removal

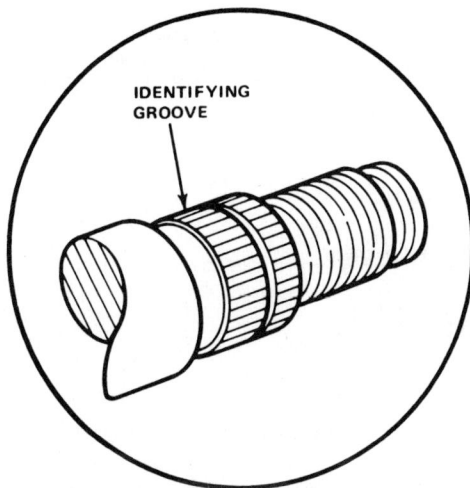

Metric steering shaft identification

Turn Signal Switch

REPLACEMENT

1971-75

The turn signal switch is attached to the steering column; the whole unit is mounted externally. To remove the switch assembly, remove the attaching screws, unfasten the wires and remove the unit from the steering column.

The most frequent causes of failure in the directional signal system are loose connections and burned out bulbs. A flashing rate of approximately twice normal usually indicates a burned out bulb in the circuit. When trouble in the signal switch is suspected, it is advisable to make a few checks to definitely locate the trouble before going tot he effort of removing the signal switch. First check the fuse. There is an inline fuse located between the ignition switch and the turn signal flasher. If the fuse checks out OK, next eliminate the flasher unit by substituting a known good flasher. If a new flasher does not cure the trouble, check the signal system wiring connections at the fuse and at the steering column connector.

NOTE: If the right front parking light and the right rear stop light are inoperative, switch failure is indicated. If the brake lights function properly, the rear signal lights are OK.

To check the switch on 1971 models, first put the control lever in the neutral position. Then disconnect the wire to the right side circuit and bridge it to the "L" terminal, thus by-passing the signal switch. If the right side circuit lights, the signal switch is inoperative and must be replaced.

To check out the switch on the 1972 and 1973 models, disconnect the switch at the six wire connector. Use a jumper wire from the white (battery feed) wire to the other wires. Circuitry is as follows:
- White to Orange: Right rear
- White to Black: Right front
- White to Yellow: Left front
- White to Blue: Left rear

If the lights in any of these circuits light then the switch is bad and must be replaced.

1976-79

1. Disconnect the negative battery cable.
2. Remove the steering wheel.
3. Loosen the anti-theft cover retaining screws on 1976 models and lift the cover from the steering column. It is not necessary to completely remove these screws.
4. Depress the lockplate and pry the round wire snapring from the steering shaft groove. A lockplate compressor tool is available for compressing the lockplate.
5. Remove the lockplate, directional signal canceling cam, upper bearing preload spring, and thrust washer from the steering shaft.
6. Move the directional signal actuating lever to the right turn position and remove the lever.
7. Depress the hazard warning light switch and remove the button by turning it counterclockwise.
8. Remove the directional signal wiring harness connector block from its mounting bracket on the right side of the lower column.
9. On vehicles equipped with an automatic transmission, use a stiff wire, such as a paper clip, to depress the lock tab which retains the shift quadrant light wire in the connector block.
10. Remove the directional signal switch retaining screws and pull the switch and wiring harness from the steering column.
11. Guide the wiring harness of the new switch into position and carefully align the switch assembly. make sure that the actuating lever pivot is correctly aligned and seated in the upper

button until the locktabs on the button align with the notches in the contact cup and pull upward to remove it. With the sport wheel, just pull the button up.
4. Remove the steering wheel nut and washer.
5. If the Jeep is equipped with a sport style steering wheel, remove the horn button, nut and washer, bottom retaining ring, and horn contact ring.
6. Remove the plastic horn contact cup retainer and remove the cup and contact plate from the steering wheel.
7. Remove the horn contact pin and bushing from the steering wheel.
8. Paint or scribe alignment marks on the steering wheel and shaft for reference during assembly.
9. Remove the steering wheel using a puller.
10. Position the steering wheel on the shaft, aligning the scribed marks.
11. Install the horn contact pin and bushing.
12. Install the plastic horn contact cup retainer and install the cup and contact plate.
13. If the Jeep is equipped with a sport style steering wheel, install the horn button, nut and washer, bottom retaining ring, and horn contact ring.
14. Install the steering wheel nut and washer. Tighten the nut to 20 ft. lbs. for 1976-77; 30 ft. lbs. for 1978-86; 25 ft. lbs. for 1987-90.
15. Install the horn button. Turn the button until the locktabs on the button align with the notches in the contact cup. With the sport wheel, just push the button on.
16. Connect the negative battery cable.

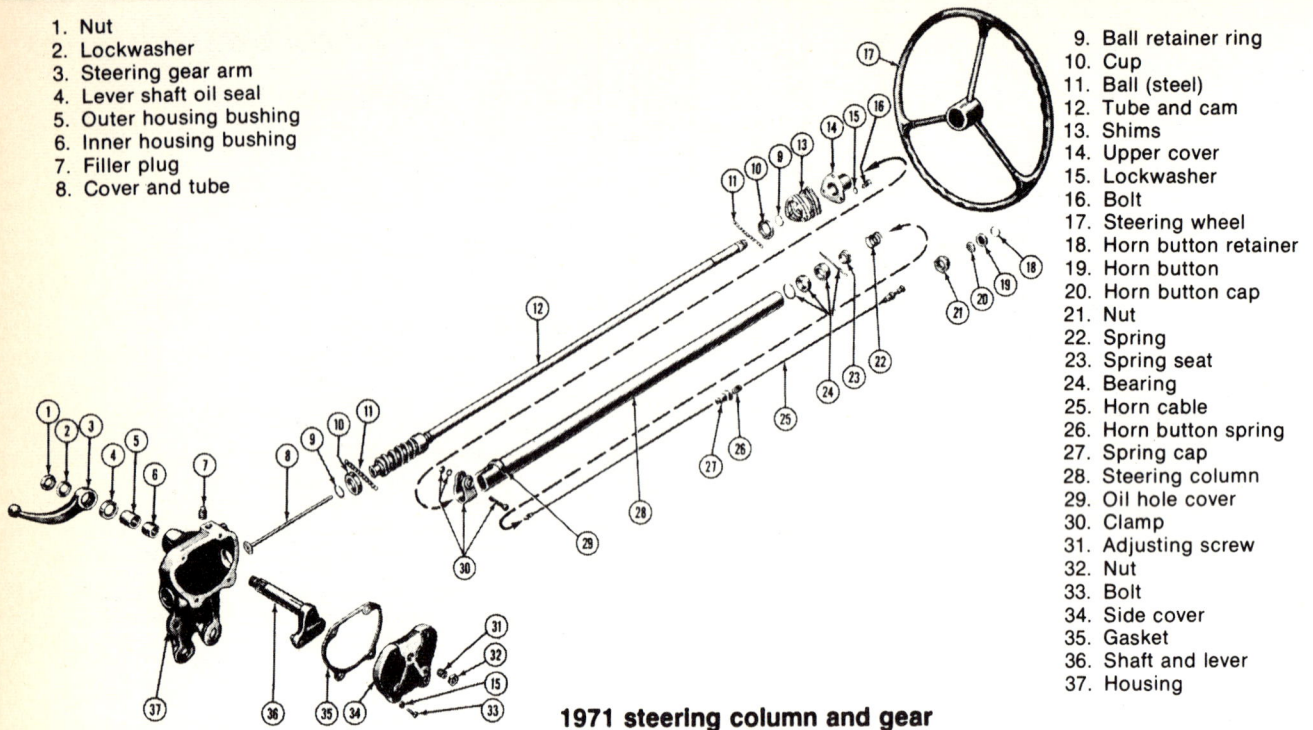

1. Nut
2. Lockwasher
3. Steering gear arm
4. Lever shaft oil seal
5. Outer housing bushing
6. Inner housing bushing
7. Filler plug
8. Cover and tube

9. Ball retainer ring
10. Cup
11. Ball (steel)
12. Tube and cam
13. Shims
14. Upper cover
15. Lockwasher
16. Bolt
17. Steering wheel
18. Horn button retainer
19. Horn button
20. Horn button cap
21. Nut
22. Spring
23. Spring seat
24. Bearing
25. Horn cable
26. Horn button spring
27. Spring cap
28. Steering column
29. Oil hole cover
30. Clamp
31. Adjusting screw
32. Nut
33. Bolt
34. Side cover
35. Gasket
36. Shaft and lever
37. Housing

1971 steering column and gear

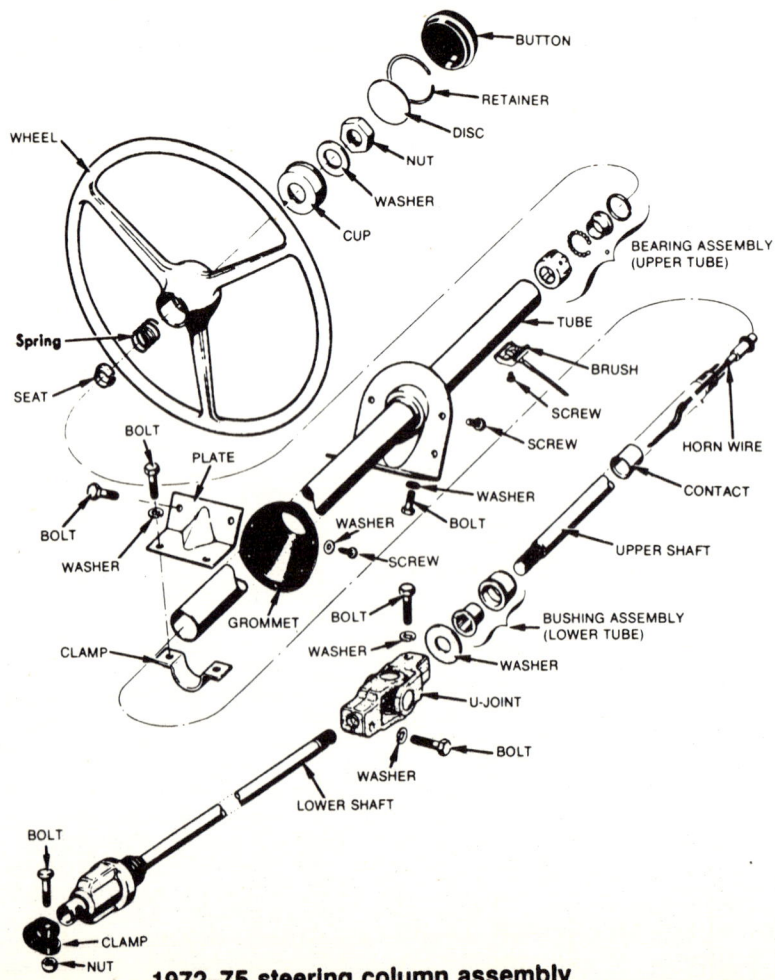

1972–75 steering column assembly

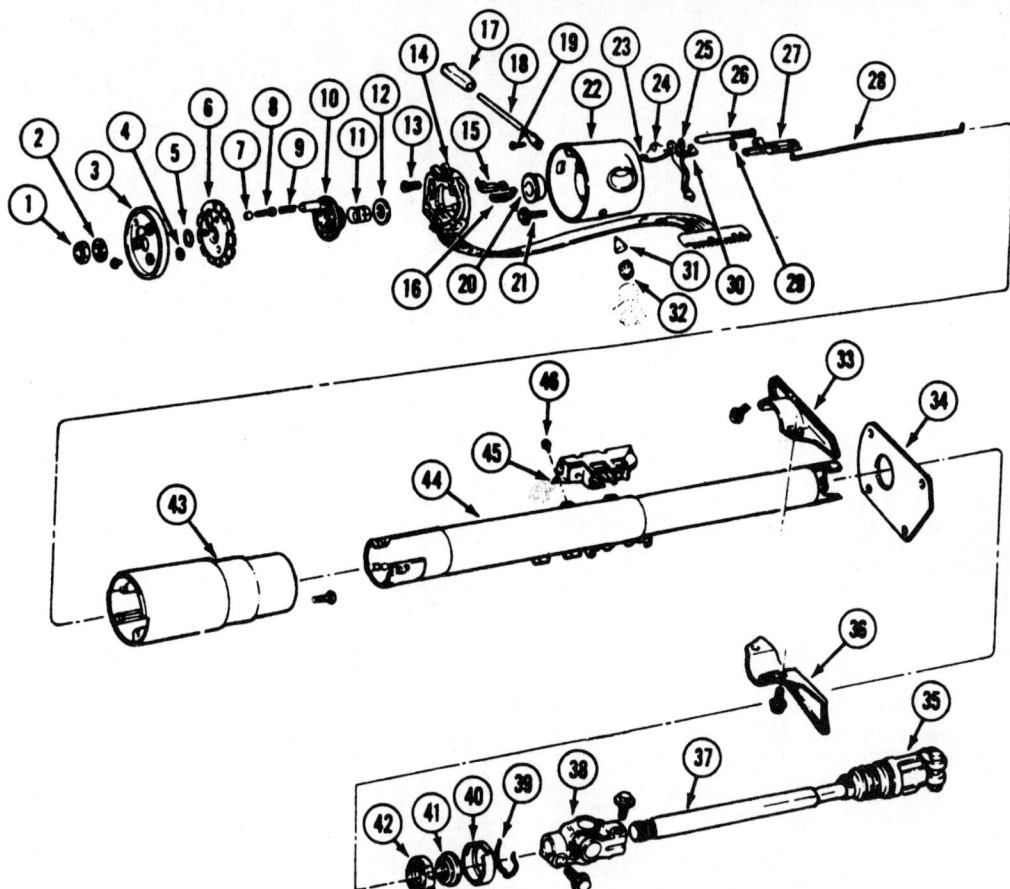

1. Steering wheel nut
2. Washer
3. Anti-theft cover
4. Anti-theft cover screw and retainer
5. Steering shaft snap-ring
6. Lockplate
7. Bushing
8. Horn contact pin
9. Spring
10. Concelling cam
11. Upper bearing preload spring
12. Thrust washer
13. Turn signal switch screw
14. Turn signal switch
15. Buzzer switch
16. Buzzer switch spring

17. Turn signal lever knob
18. Turn signal lever
19. Turn signal lever screw
20. Upper bearing
21. Housing retaining screw
22. Housing
23. Rack preload spring
24. Key release lever spring
25. Wave washer
26. Lockbolt
27. Lock rack
28. Remote rod
29. Spring washer
30. Key release lever
31. Hazard warning switch knob
32. Sector

33. Upper half of toe plate
34. Seal
35. Intermediate shaft coupling
36. Lower half of toe plate
37. Intermediate shaft
38. U-joint
39. Snap-ring
40. Retainer
41. Lower bearing
42. Lower bearing adapter
43. Shroud
44. Jacket
45. Ignition switch
46. Ignition switch screw

1976–79 steering column

housing pivot boss prior to installing the retaining screws.

12. Install the directional signal lever and actuate the directional signal switch to assure correct operation.

13. Place the thrust washer, spring, and directional signal canceling cam on the upper end of the steering shaft.

14. Align the lockplate splines with the steering shaft splines and place the lockplate in position with the directional signal canceling cam shaft protruding through the dogleg opening in the lockplate.

15. Install the snapring.

16. Install the anti-theft cover.

17. Install the steering wheel and connect the negative battery cable.

18. Check the operation of the turn signal switch.

1980-90

1. Disconnect the battery ground.
2. Cover the painted areas of the column.
3. Remove the column-to-dash bezel.
4. Loosen the toe plate screws.
5. With tilt columns, place the column in the non-tilt position.
6. Remove the steering wheel.
7. Remove the lock plate cover.
8. Compress the lock plate and unseat the steering shaft snapring as follows:

   a. Check the steering shaft nut threads. Metric threads have an identifying groove in the steering wheel splines. SAE threads do not.

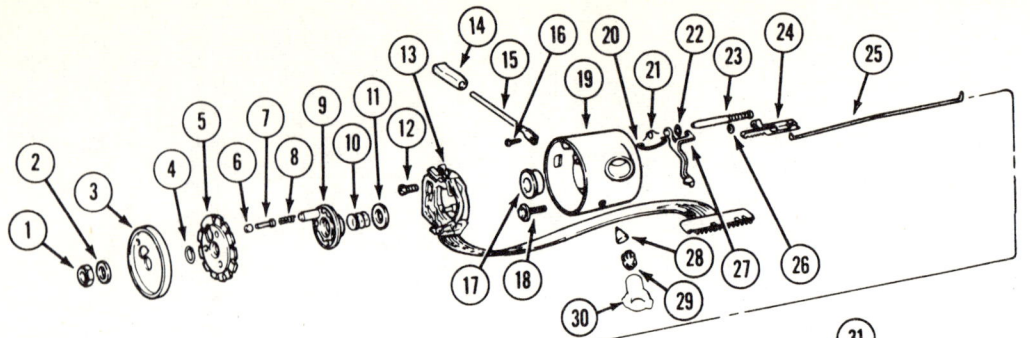

1. Steering wheel nut
2. Washer
3. Lockplate cover
4. Steering shaft snapring
5. Lockplate
6. Retainer
7. Horn contact pin
8. Spring
9. Canceling cam
10. Upper bearing preload spring
11. Thrust washer
12. Turn signal switch screw
13. Turn signal switch
14. Turn signal lever knob
15. Turn signal lever
16. Turn signal lever screw
17. Upper bearing
18. Housing retaining screw
19. Housing
20. Rack preload spring
21. Key release lever spring
22. Wave washer
23. Lock bolt
24. Lock rack
25. Remote rod
26. Spring washer
27. Key release lever
28. Hazard warning switch knob
29. Lock sector
30. Lock cylinder
31. Toe plate upper half
32. Seal
33. Intermediate shaft coupling
34. Toe plate lower half
35. Intermediate shaft

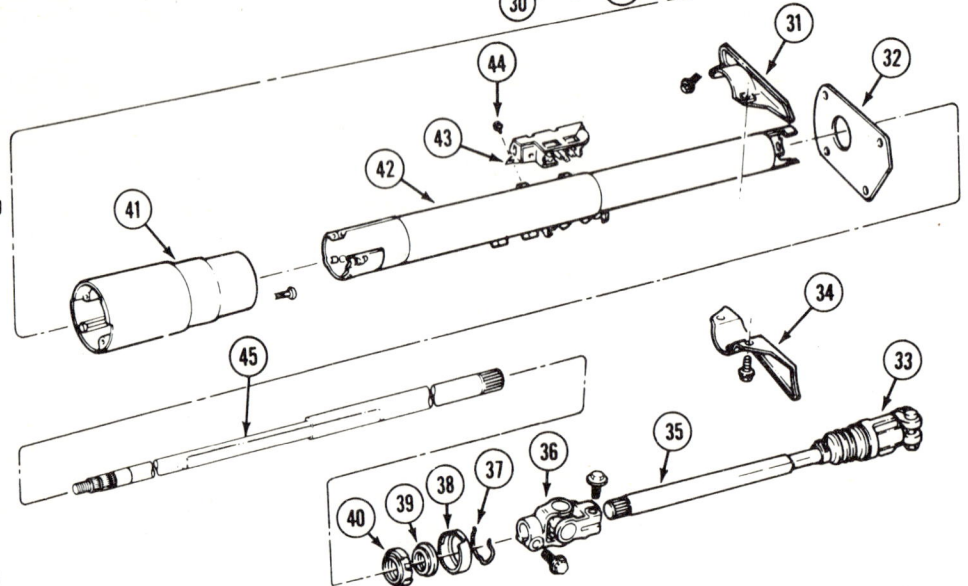

36. Intermediate shaft U-joint
37. Snapring
38. Retainer
39. Lower bearing
40. Lower bearing adapter

41. Shroud
42. Jacket
43. Ignition switch
44. Ignition switch screw
45. Steering shaft

1980–86 steering column, with manual transmission

Lockplate components

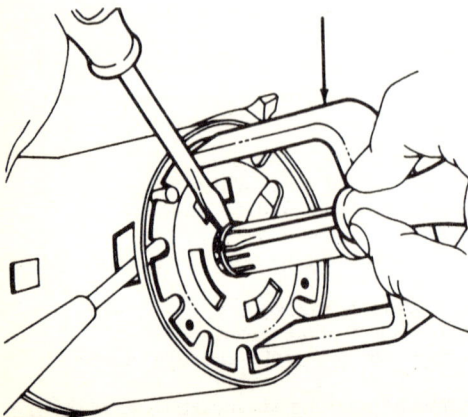

Using lockplate spring compressor

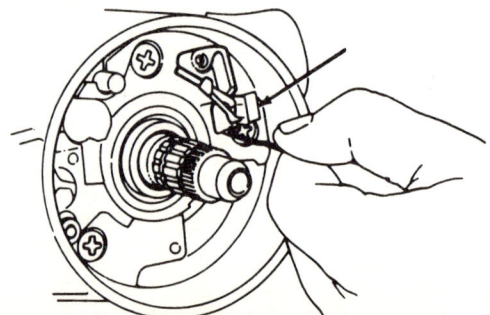

Removing the key warning switch buzzer components

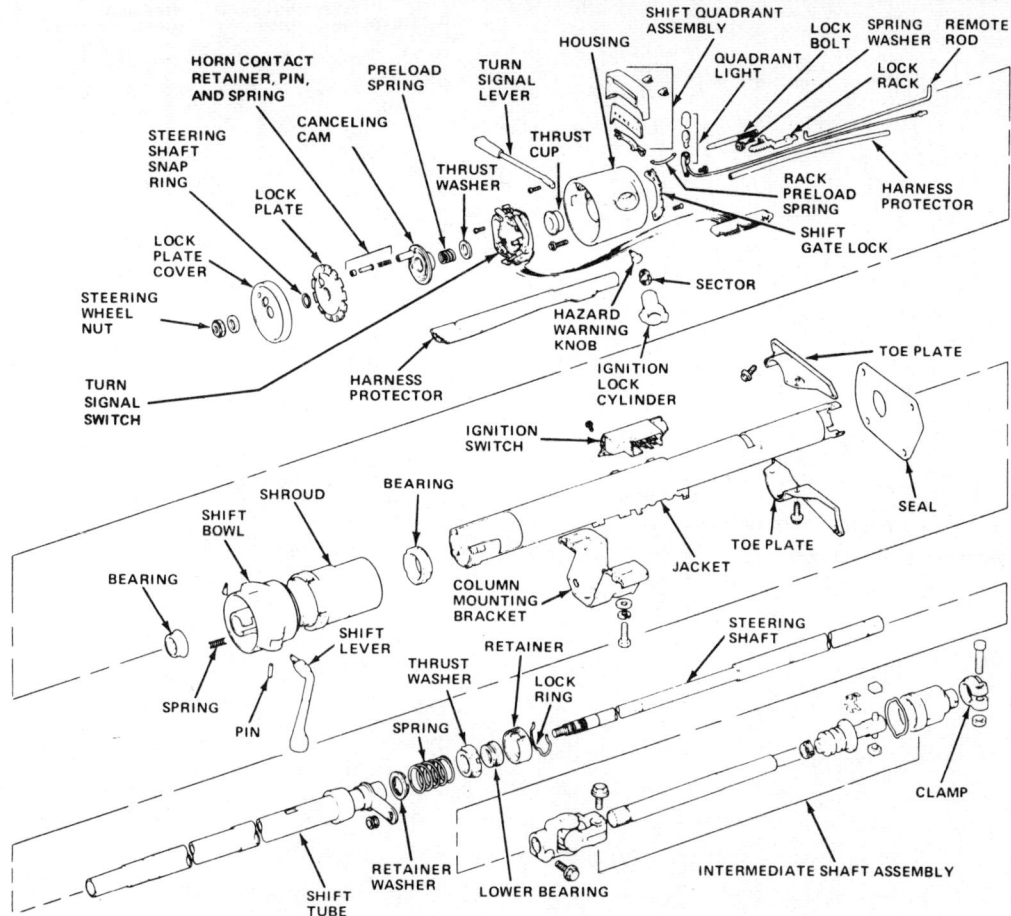

1980–86 steering column, with automatic transmission

b. With SAE threads use a compressor tool such as tool J-23653 to compress the lock plate and remove the snapring.

c. If the shaft has metric threads, replace the forcing screw in the compressor with metric forcing screw J-23653-4 before using.

9. Remove the compressor and snapring.

10. Remove the lock plate, canceling cam and upper bearing preload spring.

11. Place the turn signal lever in the right turn position and remove the lever.

12. Remove the hazard warning knob. Press the knob inward and turn counterclockwise to remove it.

13. Remove the wiring harness protectors.

14. Disconnect the wiring harness connectors.

15. Remove the turn signal switch attaching screws and lift out the switch.

To install:

16. Install the turn signal switch and attaching screws. Torque the screws to 35 inch lbs.

17. Connect the wiring harness connectors.

18. Install the wiring harness protectors.

19. With the turn signal switch in the neutral position, install the hazard warning knob. Press the knob inward and turn clockwise to install it.

20. Install the lever. Torque the attaching screws to 35 inch lbs.

21. Install the upper bearing preload spring.

22. Install the canceling cam.

23. Install the lock plate.

24. Using the compressor, install the snapring.

25. Install the lock plate cover.

26. Install the steering wheel.

27. Loosen the toe plate screws.

28. Install the column-to-dash bezel.

29. Connect the battery ground.

Ignition Switch

REPLACEMENT

1971-75

1. Disconnect the battery ground cable.

2. On models through 1972, unscrew the nut from the front of the instrument panel and remove the switch. Some early production Utility models had a switch held in place by a bezel and tension spring, rather than a threaded bezel.

3. On 1973-75 models, reach behind the panel and press the switch in against the spring. Turn the bezel counterclockwise to release.

4. Lower the switch and detach the wiring.

5. Reverse the procedure for installation.

1976-86

The ignition switch is on top of the lower part of the steering column, inside the vehicle.

1. Put the key in the lock and turn to the OFF/UNLOCKED position.

2. Disconnect the battery ground cable.

3. Detach the wire connectors at the switch.

4. Remove the switch screws.

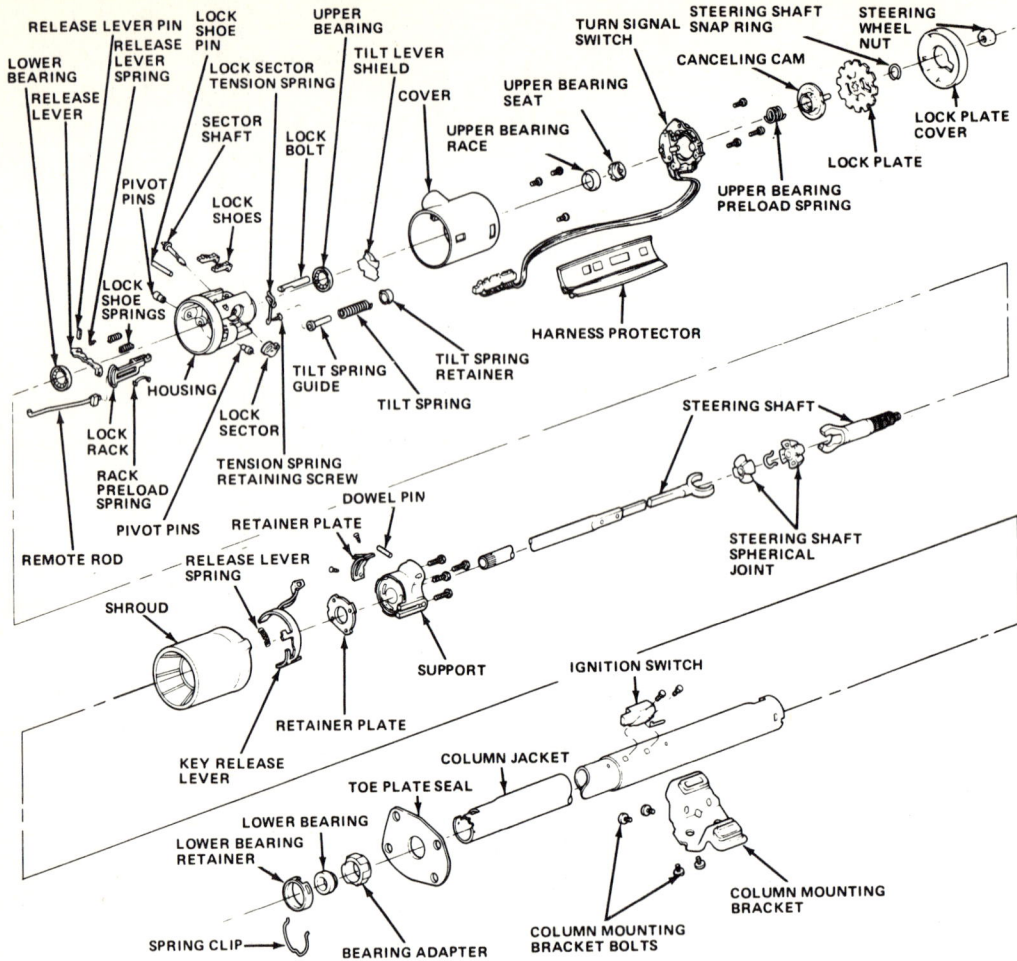

1980—86 steering column, with tilt wheel

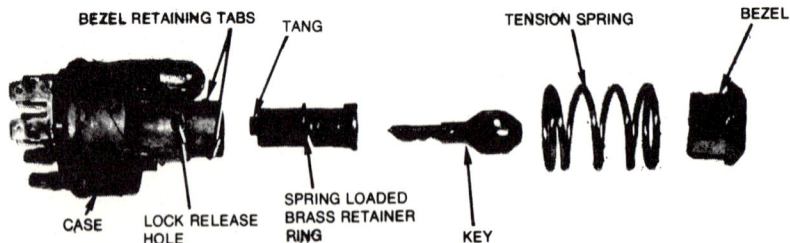

1973—75 Ignition switch and lock cylinder details

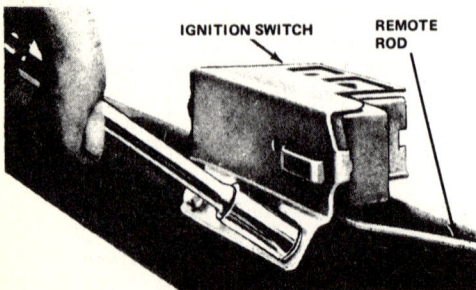

Ignition switch removal or Installation, column mounted switches

5. Disconnect the actuating rod from the switch and remove the switch.

6. Move the switch slider all the way down the column. Move it back toward the steering wheel two clicks to the center OFF/UNLOCKED position.

7. Engage the column actuating rod in the switch slider and fasten the switch down.

8. Connect the wire connectors, then the battery ground cable.

1987-90

1. Disconnect the battery ground.

2. Tape the dimmer switch actuator rod to the column to keep it from disengaging.

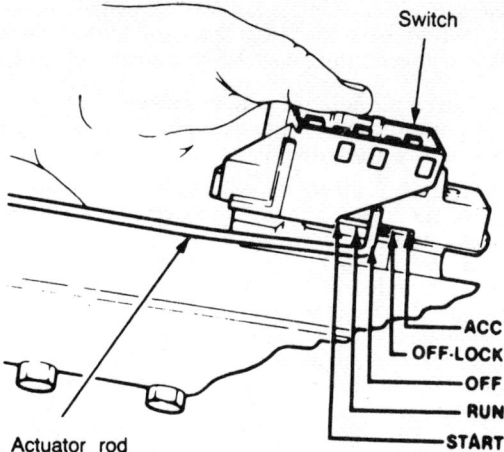

1987–90 ignition switch

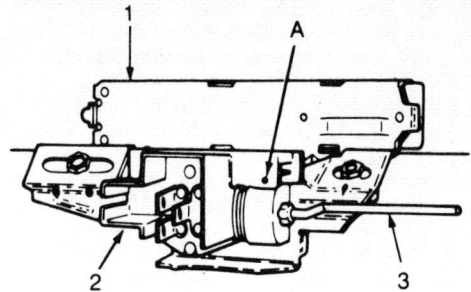

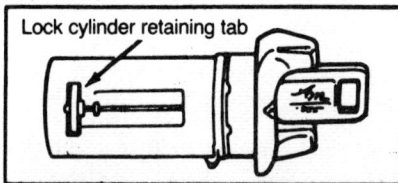

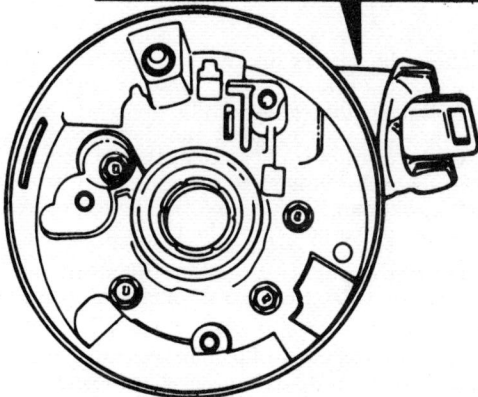

1976–86 lock cylinder

3. Remove the dimmer switch.

4. Unplug the wiring connector at the ignition switch.

5. Remove the attaching screws and lift the switch off of the column until it clears the actuator rod. Remove the switch.

6. When installing the ignition switch, engage the actuator rod in the bottom of the switch. Install the switch on the column and tighten the screws.

7. Adjust the ignition switch as follows:
   a. Insert the key and turn the lock cylinder to the OFF/UNLOCK position.
   b. Loosen the switch mounting screws.
   c. Move the switch down the column to eliminate any play and tighten the screws to 35 inch lbs.
   d. Connect the wiring to the switch.

8. Engage the actuator od in the dimmer switch and install the switch on the column.

9. Untape the rod.

10. Adjust the dimmer switch as follows:
   a. Compress the switch slightly and insert a $3/32$ in. drill bit in the switch adjusting hole.

A. $3/32''$ diameter hole   2. Dimmer switch
1. Ignition switch   3. Actuator rod

1987–90 dimmer switch

   b. Move the switch towards the steering wheel to remove any play.
   c. Tighten the attaching screws to 35 inch lbs.
   d. Connect the battery ground.
   e. Remove the drill bit.
   f. Check the operation of the switch.

NOTE: If your Jeep has a tilt column, check the switch operation in all positions.

Ignition Lock Cylinder

REPLACEMENT

1971-75

1. Remove the ignition switch.

2. Put the key in the lock and turn it to the ON position.

3. Insert a heavy paper clip wire or something similar through the release hole in the side of the switch. Push in the retaining ring until the lock cylinder can be pulled out.

4. To install the new lock cylinder, line up the tang on the cylinder with the slot in the case and push the cylinder in.

5. Replace the switch.

1976-86

1. Disconnect the battery ground cable.

2. Remove the turn signal switch as described earlier in this section. You don't have to remove the switch completely, just set it aside.

3. Insert the key. Position the key as follows.

● 1976-83: With manual transmission, put it in the ON position; with automatic, put it in OFF/LOCK.

● 1984-86: Two detent positions, clockwise, beyond OFF/LOCK.

4. Working through the slot next to the turn signal switch mounting boss, use a thin screwdriver to release the lock cylinder.

5. To install, insert the key in the new lock cylinder. Hold the sleeve and turn the key clockwise until it stops. Align the cylinder retaining tab with the housing slot and insert the cylinder. Push the cylinder in, rotate to engage, then push in until the retaining tab engages the housing groove.

6. The rest of the procedure is the reverse of removal.

1987-90

1. Disconnect the battery ground cable.

2. Remove the turn signal switch as described earlier in this section. You don't have to remove the switch completely, just set it aside.

3. Remove the wiper switch harness and any additional harnesses from the column.

4. Insert the key and turn it to the ON position.

5. Remove the key warning buzzer and clip, with needle nosed pliers, or a paper clip with a 90° bend. Don't remove the buzzer and clip separately, since the clip will fall into the column.

6. Using a thin bladed screwdriver, remove the lock cylinder retaining screw and pull the lock cylinder out of the column housing.

7. Before installation, insert the key into the cylinder. Hold the cylinder sleeve so it won't turn and rotate the key clockwise until it stops. This retracts the cylinder actuator.

8. Align the lock cylinder tab with the housing keyway.

9. Push the cylinder into the housing unti it bottoms.

10. Install the cylinder retaining screw and torque it to 40 inch lbs.

11. Turn the key to ON.

12. Install the key warning buzzer switch and clip.

13. When installing the ignition switch, engage the actuator rod in the bottom of the switch. Install the switch on the column and tighten the screws.

14. Adjust the ignition switch as follows:

    a. Insert the key and turn the lock cylinder to the OFF/UN-LOCK position.

    b. Loosen the switch mounting screws.

    c. Move the switch down the column to eliminate any play and tighten the screws to 35 inch lbs.

    d. Connect the wiring to the switch.

15. Engage the actuator rod in the dimmer switch and install the switch on the column.

16. Untape the rod.

17. Adjust the dimmer switch as follows:

    a. Compress the switch slightly and insert a $3/32$ in. drill bit in the switch adjusting hole.

    b. Move the switch towards the steering wheel to remove any play.

    c. Tighten the attaching screws to 35 inch lbs.

    d. Connect the battery ground.

    e. Remove the drill bit.

    f. Check the operation of the switch.

NOTE: If your Jeep has a tilt column, check the switch operation in all positions.

18. Install the turn signal switch and attaching screws. Torque the screws to 35 inch lbs.

19. Connect the wiring harness connectors.

20. Install the wiring harness protectors.

21. With the turn signal switch in the neutral position, install the hazard warning knob. Press the knob inward and turn clockwise to install it.

22. Install the lever. Torque the attaching screws to 35 inch lbs.

23. Install the upper bearing preload spring.

24. Install the canceling cam.

25. Install the lock plate.

26. Using the compressor, install the snapring.

27. Install the lock plate cover.

28. Install the steering wheel.

29. Loosen the toe plate screws.

30. Install the column-to-dash bezel.

31. Connect the battery ground.

Steering Column

REMOVAL AND INSTALLATION

1971-86

1. Disconnect the battery ground.

2. Remove the column cover plate at the floorboards.

3. Remove the column-to-dash lower bezel.

4. On models with automatic transmission, disconnect the shift rod at the column. On models with air conditioning, it will be necessary to remove the left side air conditioning duct.

5. Remove the column-to-dash bracket and lower the column.

NOTE: Later models have breakaway capsules in the bracket. Remove the bracket and put it in a safe place to avoid damage to the capsules.

6. Disconnect any wiring attached to column components.

7. Remove the steering column-to-gear shaft coupling and pull the column out of the Jeep.

8. On models with energy absorbing columns, it is extremely important that only specified fasteners be used. Fasteners which are not of the exact length or hardness may impair the energy absorbing action of the column. Bolts securing the column mounting bracket to the dash must be torqued exactly.

9. Connect the column to the gear shaft and tighten the coupling pinch bolt to 45 ft. lbs.

10. Connect all wiring. If you have Cruise Command, connect the white wire first, then the black one.

11. Install the toe plates, but don't fully tighten the fasteners.

12. Install the bracket on the column and torque the bolts to 20 ft. lbs.

13. Align the bracket and dash and loosely install the mounting bolts.

14. While applying a constant upward pressure on the column, torque the bracket-to-dash bolts to 20 ft. lbs.

15. Torque the toe plate bolts to 10 ft. lbs.

16. Install the bezel.

17. Connect the transmission linkage and check its operation.

1987-90

1. Disconnect the battery ground.

2. Set the parking brake.

3. Remove the steering wheel.

4. Matchmark the steering shaft and the U-joint.

5. Remove the column cover plate at the floorboards.

6. On models with automatic transmission, disconnect the shift rod at the column.

7. Remove the pinch bolt from the U-joint.

NOTE: Do not separate the steering shaft and intermediate shaft at this time. To do so would damage the components

8. Remove the speedometer/tachometer trim cover.

9. Remove the column-to-dash bracket and lower the column.

NOTE: Later models have breakaway capsules in the bracket. Remove the bracket and put it in a safe place to avoid damage to the capsules.

10. Disconnect any wiring attached to column components.

11. Pull the column from the Jeep.

12. It is extremely important that only specified fasteners be used. Fasteners which are not of the exact length or hardness may impair the energy absorbing action of the column. Bolts securing the column mounting bracket to the dash must be torqued exactly.

13. Connect the column to the gear shaft and tighten the coupling pinch bolt to 45 ft. lbs.

14. Connect all wiring.

15. Install the toe plates, but don't fully tighten the fasteners.

16. Install the bracket on the column and torque the bolts to 20 ft. lbs.

17. Align the bracket and dash and loosely install the mounting bolts.

18. While applying a constant upward pressure on the column, torque the bracket-to-dash bolts to 20 ft. lbs.

19. Torque the toe plate bolts to 10 ft. lbs.

20. Install the trim plate.

21. Connect the transmission linkage and check its operation.

Manual Steering Gear

REMOVAL AND INSTALLATION

1971

NOTE: The steering gear has to be removed down through the floor pan.

1. Remove the left front fender.
2. Remove the steering wheel.
3. Unbolt the steering column bracket from the instrument panel.
4. Disconnect the exhaust pipe at the manifold.
5. Remove the steering column cover plate from the floorboard.
6. Disconnect the horn wire.
7. Remove the drag link from the steering gear arm ball.
8. Unbolt the steering gear housing from the frame.
9. Lower the steering gear through the floor pan and over the outside of the frame rail.
10. Lift the steering gear through the floor pan and into position on the frame rail.
11. Install the bolts attaching the steering gear housing to the frame. Torque the ⅜ in. bolts to 40 ft. lbs.; the ⁷⁄₁₆ in. bolts to 55 ft. lbs.
12. Install the drag link on the steering gear arm ball.
13. Connect the horn wire.
14. Install the steering column cover plate on the floorboard.
15. Connect the exhaust pipe at the manifold.
16. Bolt the steering column bracket to the instrument panel.
17. Install the steering wheel.
18. Install the left front fender.

1972-75

1. Disconnect the steering gear from the lower steering shaft by removing the bolt and nut attaching the coupling to the worm shaft.
2. Disconnect the steering arm from the connecting rod.
3. Remove the upper steering gear-to-frame bracket bolt.
4. Remove the two lower steering gear-to-frame bracket bolts and remove the gear.
5. Installation is the reverse of removal. Observe the following torques:
- Pitman arm-to-shaft nut: 160-250 ft. lbs.
- Steering bracket-to-frame ⅜ in. bolt: 35-45 ft. lbs.
- Steering bracket-to-frame ⁷⁄₁₆ in. bolt: 60-70 ft. lbs.
- Steering gear-to-bracket bolts: 60-80 ft. lbs.

1976-86

1. Remove the intermediate shaft-to-wormshaft coupling clamp bolt and disconnect the intermediate shaft.
2. Remove the pitman arm nut and lockwasher.
3. Using a puller, remove the Pitman arm from the shaft.

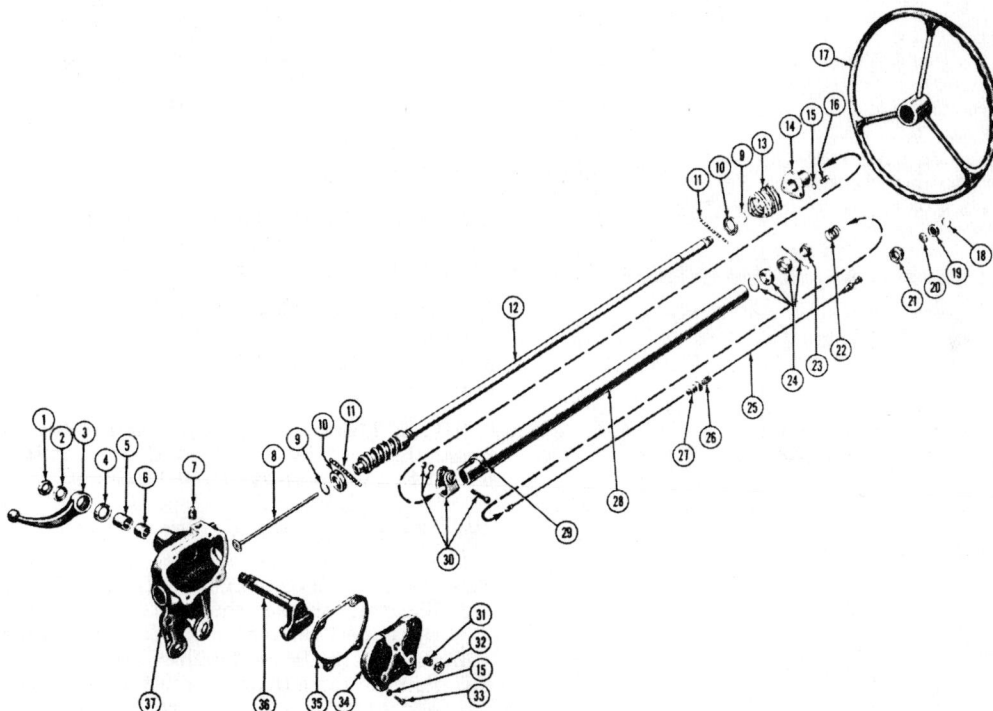

1. Nut
2. Lockwasher
3. Steering gear arm
4. Lever shaft oil seal
5. Outer housing bushing
6. Inner housing bushing
7. Filler plug
8. Cover and tube
9. Ball retaining ring
10. Cup
11. Ball (steel)
12. Tube and cam
13. Shims
14. Upper cover
15. Lockwasher
16. Bolt
17. Steering wheel
18. Horn button retainer
19. Horn button
20. Horn button cap
21. Nut
22. Spring
23. Spring seat
24. Bearing
25. Horn cable
26. Horn buton spring
27. Spring cup
28. Steering column
29. Oil hole cover
30. Clamp
31. Adjusting screw
32. Nut
33. Bolt
34. Side cover
35. Gasket
36. Shaft and lever
37. Housing

1971 steering gear and column

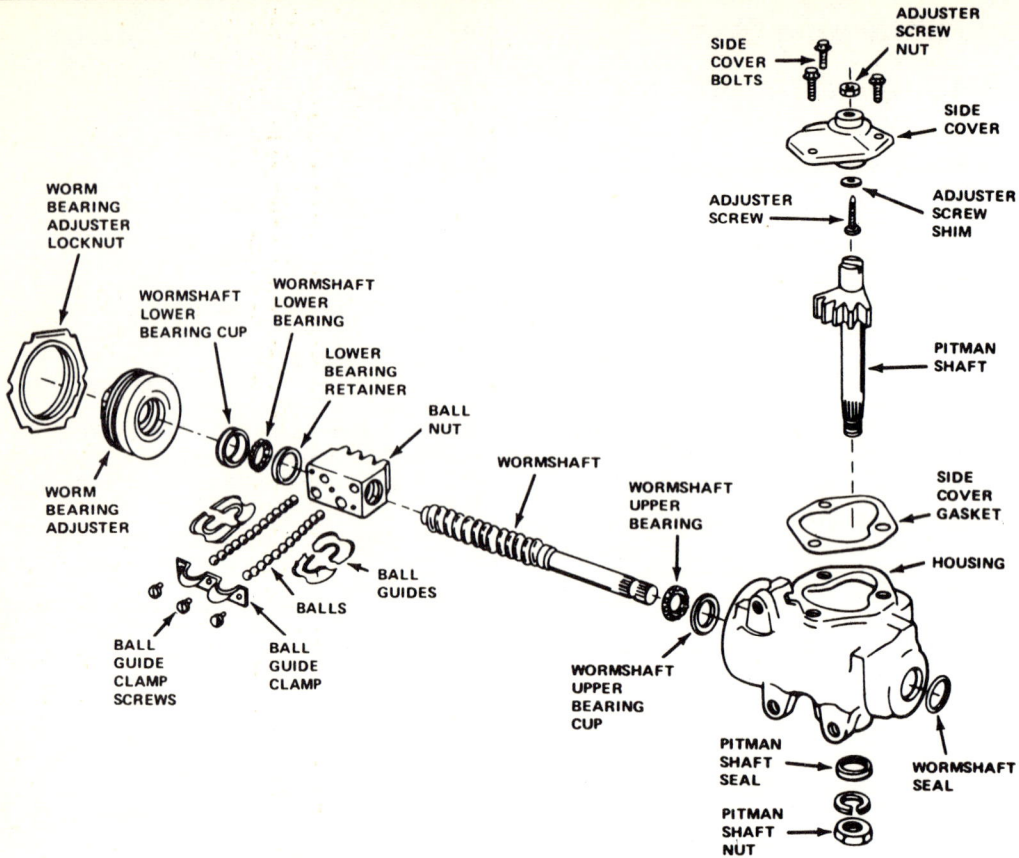

1972–83 manual steering gear

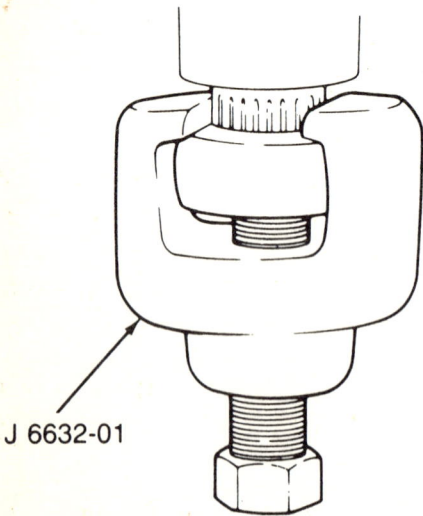

J 6632-01

Pitman arm removal

4. Raise the left side of the vehicle slightly to relieve tension on the left front spring and rest the frame on a jackstand.

5. Remove the steering gear lower bracket-to-frame bolts.

6. Remove the bolts attaching the steering gear upper bracket to the crossmember. Beginning in 1979, one of these bolts is a Torx® head bolt. This bolt, and some others may be removed with the aid of a 9 inch extension. Remove the gear.

NOTE: Loctite® 271 or similar material must be applied to all attaching bolt threads prior to installation.

7. Position the tie plate upper and lower mounting brackets on the gear and install the bolts. Torque the bracket-to-gear bolts to 70 ft. lbs. and the bracket-to-tie plate bolt to 55 ft. lbs.

8. Align and engage the intermediate shaft coupling with the steering gear wormshaft splines.

9. Position the steering gear on the frame and install the mounting bolts. Torque the bolts to 55 ft. lbs. Install the Pitman arm and torque the nut to 185 ft. lbs.

NOTE: The steering gear may produce a slight roughness, this can be eliminated by turning the steering wheel full left and right 10-15 times.

1987-90

1. Remove the intermediate shaft-to-wormshaft coupling clamp bolt and disconnect the intermediate shaft.

2. Raise and support the front end on jackstands.

3. Disconnect the center link from the pitman arm.

4. Remove the front stabilizer bar.

5. Remove the pitman arm nut and washer.

6. Matchmark the pitman arm and shaft and, with a puller, such as J-6632-01, remove the pitman arm.

7. Remove the mounting bolts and the gear.

8. Position the gear on the frame and install the mounting bolts. Torque the gear-to-frame bolts to 75 ft. lbs.

9. Install the pitman arm. Torque the pitman arm nut to 185 ft. lbs. and stake it in two places.

10. Install the front stabilizer bar. Torque the stabilizer bar-to-frame bolts to 55 ft. lbs. Torque the stabilizer bar-to-link bolts to 27 ft. lbs.

11. Connect the center link to the pitman arm. Torque the center link-to-pitman arm nut to 35 ft. lbs.

12. Install the intermediate shaft-to-wormshaft coupling clamp bolt. Torque the bolt to 33 ft. lbs.

1. Cover bolt
2. Adjusting screw locknut
3. Cover
4. Endplay shims
5. Cover gasket
6. Pitman shaft adjuster screw
7. Pitman shaft
8. Gear housing
9. Upper bearing race
10. Upper bearing
11. Wormshaft
12. Ball nut
13. Ball bearings (50)
14. Ball guide clamp and screws
15. Ball guides
16. Lower bearing retainer
17. Lower bearing
18. Lower bearing race
19. Worm bearing adjuster
20. Adjuster locknut
21. Pitman nut and washer
22. Pitman shaft seal
23. Wormshaft seal

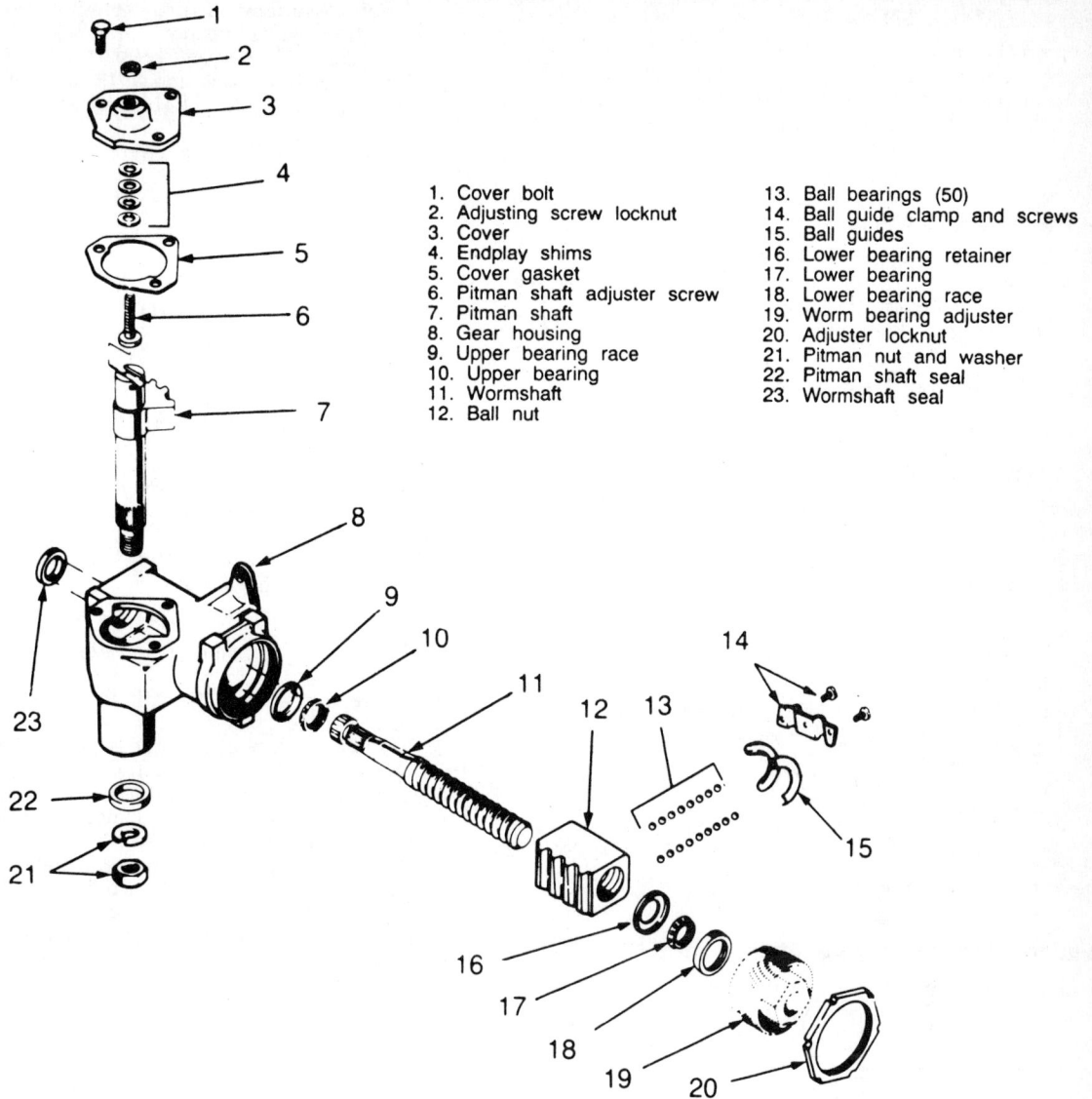

1984–90 manual steering gear

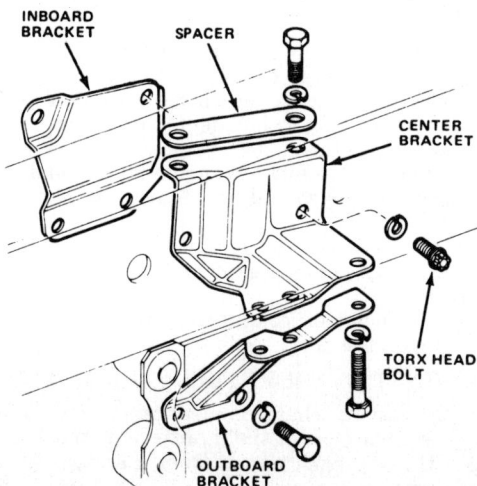

Steering gear mounting brackets on 1972 and later models

MANUAL STEERING GEAR ADJUSTMENT

NOTE: Adjustments must be made in the order given. Failure to following sequence could result in damage to the gear.

1971

Before adjusting, remove all load from the system by disconnect the drag link from the steering arm and loosening the instrument panel bracket bolts and the steering gear-to-frame bolts.

STEERING SHAFT PLAY ADJUSTMENT

1. Remove the shims installed between the steering gear housing and the upper cover.
2. Loosen the housing side cover adjusting screw.
3. Loosen the housing cover to cut and remove one or more shims as required. Proper adjustment allows a slight drag and free operation.
4. Tighten the cover.

BACKLASH ADJUSTMENT

1. Loosen the adjusting screw locknut.
2. Turn the adjusting screw in until a very slight drag is felt through the mid-point in steering wheel travel. This procedure is done with the wheels in the straight-ahead position.
3. Tighten the adjusting screw locknut.

1972-75

WORM BEARING PRELOAD ADJUSTMENT

1. Loosen the steering gear end cover.
2. Add to or subtract from the number of shims under the cover to obtain a rolling torque of 2-5 inch lbs.
3. Tighten the cover bolts alternately and evenly to 18-22 ft. lbs.

STEERING GEAR CLEARANCE ADJUSTMENT

1. Loosen the locknut and turn the adjusting screw on the side cover, counterclockwise until the worm gear shaft turns freely through its entire range of travel.

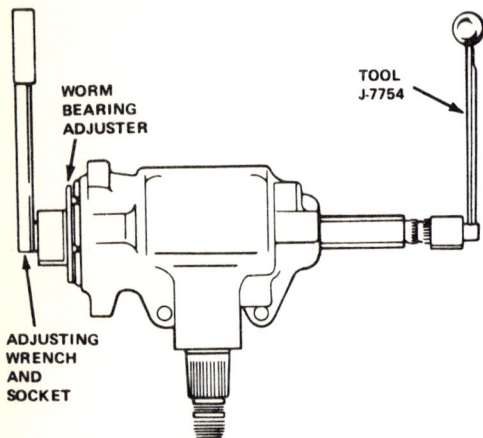

Adjusting worm bearing preload

Adjusting Pitman shaft overcenter torque drag

2. Count the number of turns necessary to rotate the worm gear shaft through its travel.
3. Turn the shaft to center point.
4. Rotate the shaft back and forth over center, and tighten the adjusting screw until the shaft binds slightly at the center point.
5. Adjust the screw to obtain a rolling torque of 7-12 inch lbs. through the center.
6. Hold the adjusting screw and tighten the locknut to 16-20 ft. lbs.

1976-79

WORM BEARING PRELOAD ADJUSTMENT

1. Check that the steering gear mounting bolts are properly torqued.
2. Matchmark and remove the pitman arm.
3. Attach an inch-pound torque wrench and socket to the pitman shaft and turn the shaft to the extreme right and left, without hitting the travel stops.
4. Loosen the adjuster locknut. Tighten the worm bearing adjuster until the torque wrench shows 8 inch lbs. within ½ turn of either extreme.
5. Tighten the adjuster locknut to 90 ft. lbs. and recheck the torque reading on the shaft.

PITMAN SHAFT OVERCENTER ADJUSTMENT

1. Rotate the wormshaft from stop-to-stop and count the number of turns.
2. Rotate the wormshaft back from the stop, ½ the total number of turns.
3. Install an inch-pound torque wrench and socket J-7754 on the splined end of the wormshaft. Check the rotating torque as the shaft passes over the center point of travel. Overcenter torque should be 4-10 inch lbs. If not, proceed with steps 4-6.
4. Loosen the adjuster locknut.
5. Rotate the shaft over center and tighten the adjuster as necessary. Do not exceed 16 inch lbs. combined total drag (overcenter + worm bearing preload.
6. Hold the adjuster screw and tighten the locknut to 23 ft. lbs. Do not allow the adjuster to turn, or the adjustment will have to be made over again!
7. Install the pitman arm, torque it to 185 ft. lbs. and stake it in two places.

1980-90

WORM BEARING PRELOAD ADJUSTMENT

1. Check that the steering gear mounting bolts are properly torqued.
2. Matchmark and remove the pitman arm.
3. Remove the horn button and cover. Attach an inch-pound torque wrench and socket to steering wheel nut and turn the shaft to the extreme right or left, gently, until you hit the stop. Then, turn it back ½ turn.
4. Turn the torque wrench through a 90° arc and check the torque reading. Torque should be 5-8 inch lbs. If not, proceed with steps 4 and 5.
5. Loosen the adjuster locknut. Tighten the worm bearing adjuster until the torque wrench shows 5-8 inch lbs. within ½ turn of either extreme.
6. Tighten the adjuster locknut to 90 ft. lbs. and recheck the torque reading on the shaft.

PITMAN SHAFT OVERCENTER ADJUSTMENT

1. With the pitman shaft removed, rotate the wormshaft from stop-to-stop and count the number of turns.
2. Rotate the wormshaft back from the stop, ½ the total number of turns.
3. Install an inch-pound torque wrench and socket on the steering wheel nut. Check the rotating torque as the shaft

passes over the center point of travel. Overcenter torque should be equal to the worm bearing preload plus 4-10 inch lbs., but not more than 18 inch lbs. If not, proceed with steps 4-6.

4. Loosen the adjuster locknut.

5. Rotate the shaft over center and tighten the adjuster as necessary.

6. Hold the adjuster screw and tighten the locknut to 25 ft. lbs. Do not allow the adjuster to turn, or the adjustment will have to be made over again!

7. Install the pitman arm, torque it to 185 ft. lbs. and stake it in two places.

Power Steering Gear

REMOVAL AND INSTALLATION

1972-75

1. Disconnect the hoses at the gear and raise them above the pump to prevent fluid loss.

2. Remove the pinch bolt from the lower flange.

3. Remove the pitman arm nut and lockwasher, and remove the pitman arm with a puller.

4. Unbolt and remove the pump.

5. Installation is the reverse of removal. Torque the pitman arm nut to 160-210 ft. lbs. and the gear-to-frame bolts to 55 ft. lbs.

1976-86

1. Disconnect the hoses at the gear and raise them above the pump to prevent fluid loss.

2. Remove the clamp bolt and nut attaching the intermediate shaft coupling to the steering gear stub shaft and disconnect the intermediate shaft.

3. Mark the pitman shaft and arm for alignment. Remove the pitman nut and lockwasher and remove the pitman arm with a puller.

4. Raise the left side of the vehicle slightly to relieve tension from the spring. Support with a jackstand under the frame.

5. Remove the three lower steering gear mounting bracket-to-frame bolts.

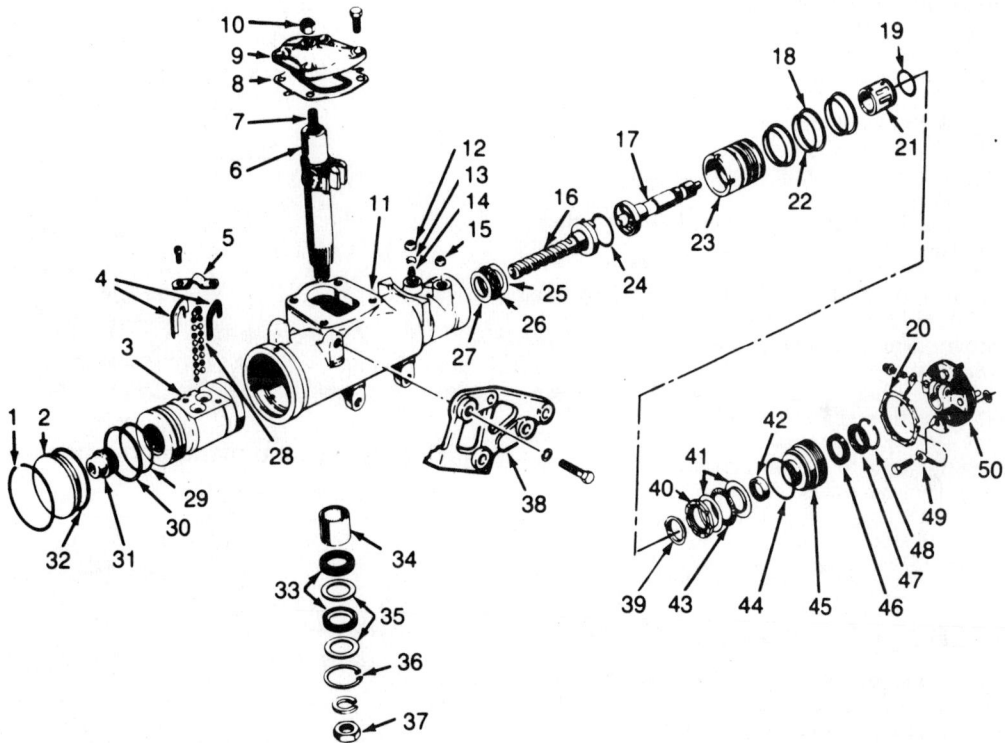

1. Retaining ring
2. Housing end plug
3. Rack piston
4. Ball return guide halves
5. Clamp
6. Pitman shaft
7. Adjusting screw
8. Gasket
9. Side cover
10. Locknut
11. Housing
12. Pressure port seat
13. Poppet valve
14. Spring
15. Return port seat
16. Worm
17. Stub shaft
18. Teflon rings
19. Damper O-ring
20. Adjuster plug locknut
21. Valve spool
22. Back-up O-rings
23. Valve body
24. O-ring
25. Race
26. Thrust bearing
27. Race
28. Ball bearings (24)
29. Back-up O-ring
30. Piston ring
31. Rack piston end plug
32. O-ring
33. Oil seals
34. Needle bearings
35. Washers
36. Retaining ring
37. Pitman arm nut
38. Spacer
39. Bearing retainer
40. Spacer
41. Races
42. Bearing
43. Thrust bearing
44. O-ring
45. Adjuster plug
46. Oil seal
47. Washer and dust seal
48. Retaining ring
49. Ground wire
50. Flexible coupling

Power steering gear

6. Remove the two steering gear-to-crossmember upper bolts. Remove the gear and brackets as an assembly.

7. Remove the brackets from the gear.

NOTE: Prior to installation, all bolts must be coated with Loctite® 271 or its equivalent.

8. Position the mounting brackets on the gear and torque the bolts to 70 ft. lbs.

9. Align and connect the intermediate shaft coupling to the steering gear stub shaft.

10. Position the steering gear on the frame and crossmember. Install and tighten the bolts to 55 ft. lbs.

11. Lower the vehicle.

12. Install the intermediate shaft coupling-to-steering gear stub shaft clamp bolt and nut. Tighten the nut to 45 ft. lbs.

13. Align and install the pitman arm, nut and lockwasher. Torque the nut to 185 ft. lbs. Stake the nut in two places.

14. Connect the hoses. Torque the hose connections to 25 ft. lbs.

1987-90

1. Place the wheels in a straight ahead position.
2. Place a drain pan under the steering gear.
3. Disconnect the hoses at the gear. Secure the hose ends in an upward position, higher than the gear. Cap the open ends.
4. Disconnect the intermediate shaft from the steering gear shaft.
5. Raise and support the front end on jackstands.
6. Matchmark the pitman arm and shaft.
7. Disconnect the center link from the pitman arm.
8. Remove the stabilizer bar.
9. Remove the pitman arm nut and washer. Using a puller, remove the pitman arm.
10. Remove the mounting bolts and remove the gear.
11. Position the gear and install the mounting bolts. Torque the steering gear-to-frame bolts to 75 ft. lbs.
12. Install the pitman arm, nut, and washer. Torque the pitman arm nut to 185 ft. lbs.
13. Install the stabilizer bar. Torque the stabilizer bar-to-frame bolts to 30 ft. lbs.; the stabilizer bar-to-link nuts to 45 ft. lbs.
14. Connect the center link to the pitman arm. Torque the center link-to-pitman arm nuts to 35 ft. lbs.
15. Connect the intermediate shaft to the steering gear shaft. Torque the intermediate shaft pinch bolt to 45 ft. lbs.
16. Connect the hoses at the gear.

POWER STEERING GEAR ADJUSTMENTS

NOTE: The gear must be adjusted off the vehicle. All adjustments must be made in the sequence described below. Worm bearing preload is always adjusted first!

Worm Bearing Preload Adjustment

1. Mount the gear assembly in a vise.
2. Torque the adjuster plug to 20 ft. lbs.
3. Mark the gear housing in line with one of the adjuster plug holes.
4. Measure counterclockwise $3/16$-$1/4$ in. (5-6mm) from the first mark on models through 1979; $1/2$ in. (13mm) from the first mark on 1980-86 models; $3/16$-$1/4$ in. (5-6mm) for 1987-90 models, and make another mark.
5. Turn the adjuster plug counterclockwise to align the hole with the second mark.
6. Hold the adjuster plug and torque the locknut to 85 ft. lbs. Do not allow the adjuster to turn.
7. Turn the stubshaft clockwise to its stop, then back $1/4$ turn.
8. Using a torque wrench of no more than 50 inch lbs. capacity and a 12 point deep socket, check the rotating torque at the

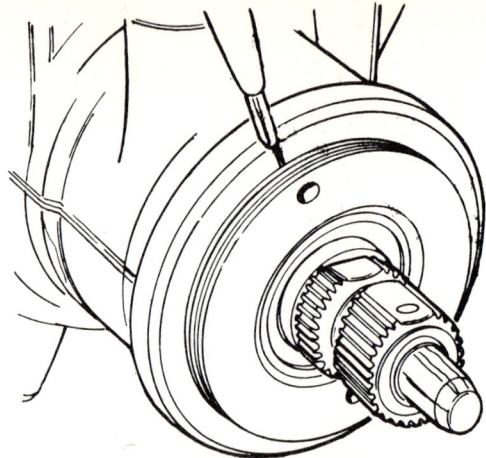
Marking power steering gear housing adjacent to the hole in the adjuster

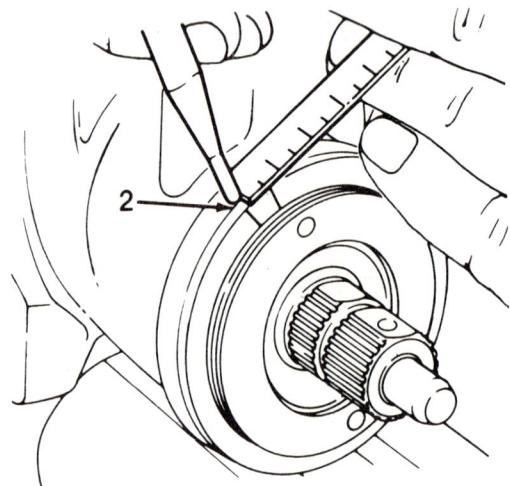
Making the second mark on the housing

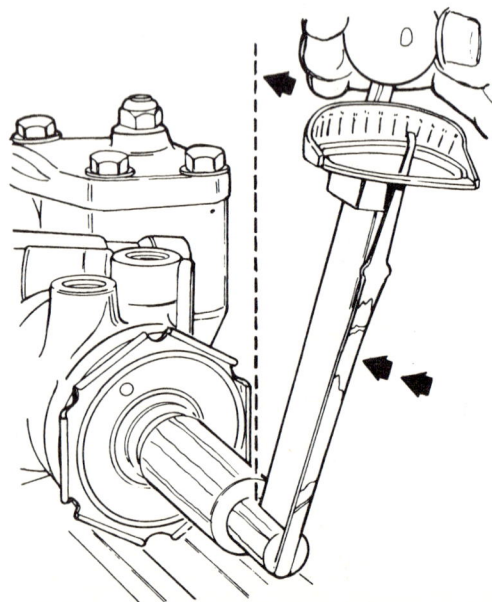
Measuring wormshaft bearing preload on power steering gears

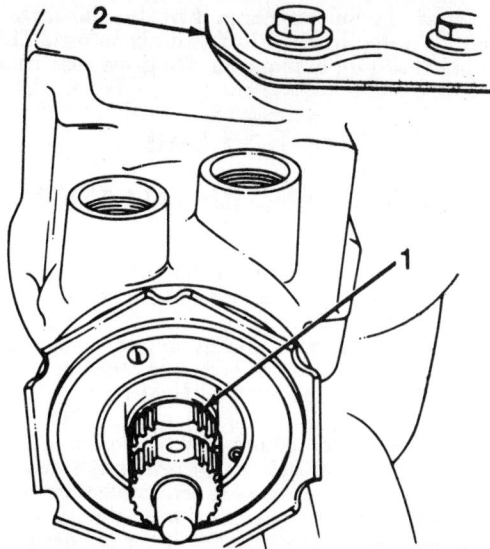

Stubshaft (1) position with gear centered. 2 is the gear housing

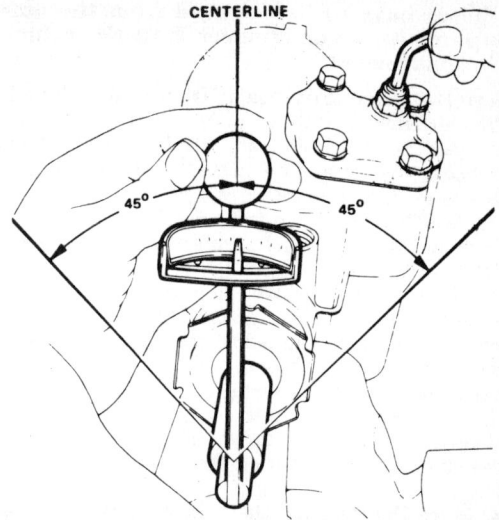

Measuring Pitman shaft overcenter torque drag on power steering gears

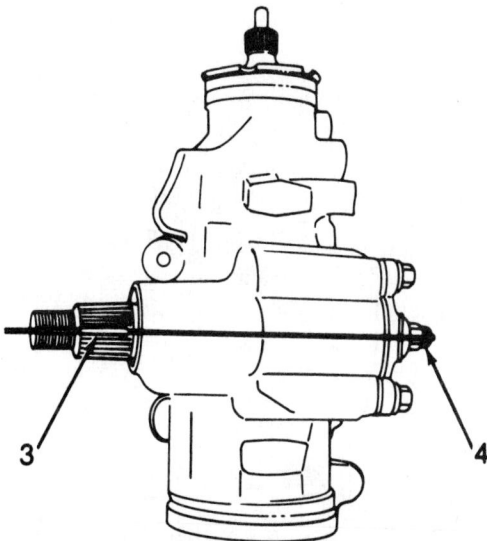

Pitman shaft master spline (3) position with the gear centered. 4 is the adjusting screw

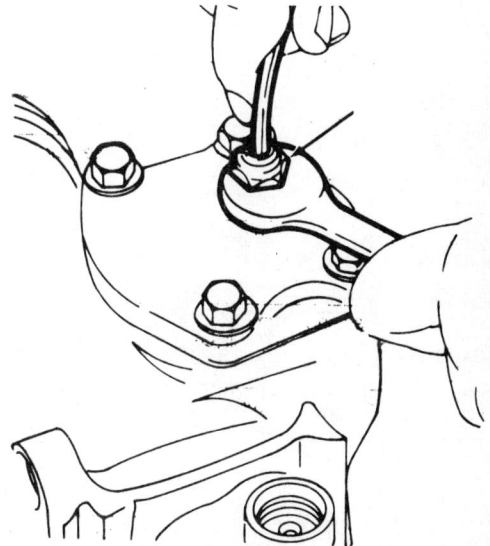

Holding the locknut while turning the adjusting nut

splined end of the stub shaft at or near a vertical position. Torque should be 4-10 inch lbs.

9. If the torque cannot be adjusted within these limits, the gear will have to be rebuilt.

Pitman Shaft Overcenter Adjustment

1. Loosen the adjuster screw locknut.
2. Turn the adjuster screw counterclockwise until the screw is fully extended. Turn the screw back in one full turn.
3. Count the number of turns to rotate the stubshaft from stop-to-stop.
4. Turn the shaft back ½ the number of turns. At this point the flat surface of the stubshaft should be upward and the master spline on the pitman shaft should be aligned with the adjuster screw.
5. Install a 50 inch lbs. torque wrench and deep 12 point socket on the splined end of the stub shaft. Place the torque wrench in a vertical position.

6. Rotate the torque wrench 45 degrees to each side and record the highest torque at or near center. Record this reading.
7. Adjust the torque by turning the adjuster screw clockwise. Adjustment is: the recorded reading plus 4-8 inch lbs. for new gears, but not exceeding 14. inch lbs. total; the previously recorded reading plus 4-5 inch lbs. for used gears, but not exceeding 14 inch lbs. combined total.
8. Tighten the adjuster screw locknut to 35 ft. lbs. while holding the adjuster screw.
9. Install the gear.

Power Steering Pump

REMOVAL AND INSTALLATION

NOTE: If the power steering pump has to be removed to service another component, it is not necessary to remove the hoses from the pump. Just disconnect the mounting fixtures and lift the pump away from the engine and lay it out of the way. The only time the power steering hoses have to be removed from the pump is when the pump has to be removed from the vehicle for service or replacement.

1. Remove the pump drive belt tension adjusting bolt. Disconnect the belt from the pump.

2. Disconnect the return and pressure hoses from the pump. Cover the hose connector and union on the pump and open ends of the hoses to avoid the entrance of dirt.

3. On the 8-304, remove the front bracket from the engine.

4. Remove the two nuts which secure the rear of the pump to the bracket, and the two bolts which secure the front of the pump to the bracket and remove the pump.

5. To install, position the pump in the bracket and install the rear attaching screws. On the 8-304, install the front bracket.

6. Connect the hydraulic hoses.

7. Adjust the drive belt tension.

8. Fill the pump reservoir to the correct level.

9. Start the engine and wait for at least three minutes before turning the steering wheel. Check the lever frequently during this time.

10. Slowly turn the steering wheel through its entire range a few times with the engine running. Recheck the level and inspect for possible leaks.

NOTE: If air becomes trapped in the fluid, the pump may become noisy until all of the air is out. This may take some time since trapped air does not bleed out rapidly.

Tie Rod End

REMOVAL AND INSTALLATION

1971

1. Raise and support the front end on jackstands.

2. Remove the cotter pin and nut and disconnect the right tie rod section from the bell crank.

3. Remove the cotter pins and nuts, and, using a puller or separator, disconnect the outer ends from the knuckles.

4. Where applicable, the left and right tie rod sections can now be spearated.

5. The tie rod ends can be removed by loosening the clamps and unscrewing the ends. Before unscrewing the ends, note the exact number of threads visible, as an installation reference.

6. All seals that show any sign of wear should be replaced. New tie rod ends should be installed if the old ones show any play or roughness of movement.

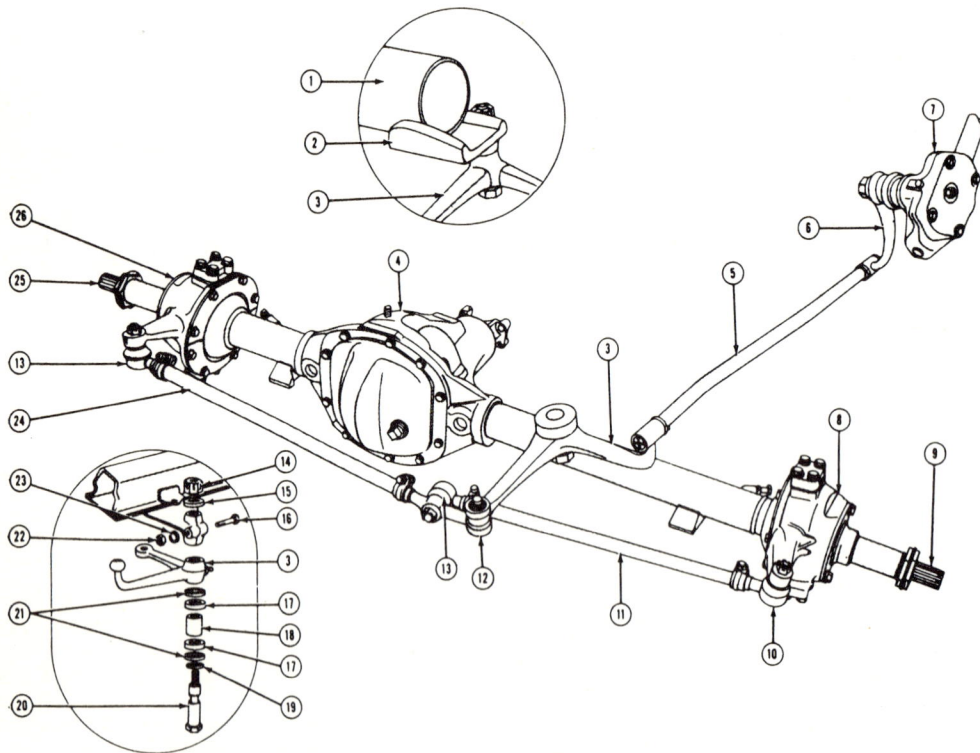

1. Frame cross tube (CJ-3B)
2. Steering bellcrank bracket (CJ-3B)
3. Steering bellcrank
4. Front axle
5. Connecting rod (drag link)
6. Steering gear arm
7. Steering gear
8. Left steering knuckle and arm
9. Left shaft and U-joint
10. Left tie rod socket
11. Left tie rod
12. Left tie rod socket
13. Right tie rod socket
14. Bellcrank nut
15. Washer
16. Bolt
17. Bellcrank bearing
18. Bearing spacer (early models)
19. Washer
20. Bellcrank shaft
21. Bearing seal
22. Nut
23. Lockwasher
24. Right steering tie rod
25. Right shaft and U-joint
26. Right steering knuckle and arm

1971 steering linkage

1. Cotter pin
2. Nut
3. Dust cover
4. Left socket
5. Nut
6. Lockwasher
7. Left tie rod
8. Lubrication fitting
9. Left socket (for right tie rod)
10. Right tie rod
11. Right socket
12. Bolt
13. Tie rod clamp

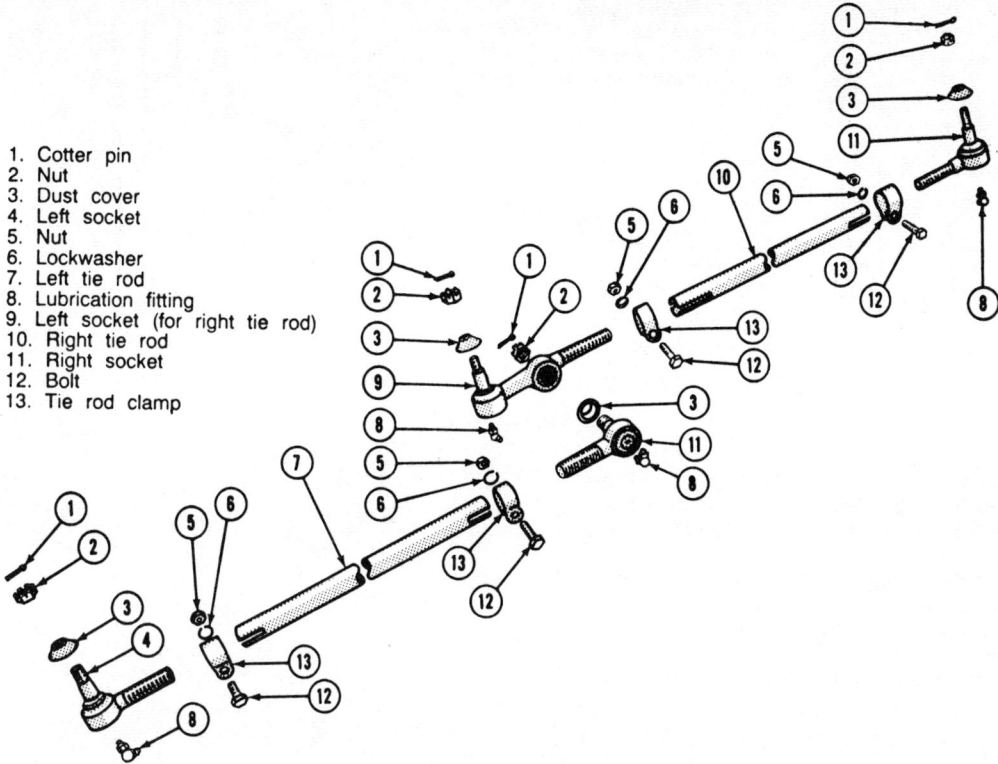

1971 tie rod assembly

7. Install the new tie rod ends, leaving the exact number of threads exposed, as previously noted. Torque all nuts to 38-42 ft. lbs.

1972-79

1. Remove the cotter pins and retaining nuts at both ends of the tie rod and from the end of the connecting rod where it attaches to the tie rod.
2. Remove the nut attaching the steering damper push rod to the tie rod bracket and move the damper aside.
3. Remove the tie rod ends from the steering arms and connecting rod with a puller.
4. Count the number of threads showing on the tie rod before removing the ends, as a guide to installation.
5. Loosen the adjusting tube clamp bolts and unthread the ends.

6. Installation is the reverse of removal. Torque the connecting rod-to-tie rod nut to 70 ft. lbs.
7. Adjust toe-in, if necessary.

1980-86

1. Raise and support the front end on jackstands.
2. Remove the cotter pins and nuts attaching the tie rod to the knuckles.
3. Remove the steering damper to tie rod nut and move the damper aside.
4. Using a separator, remove the tie rod from the knuckles.

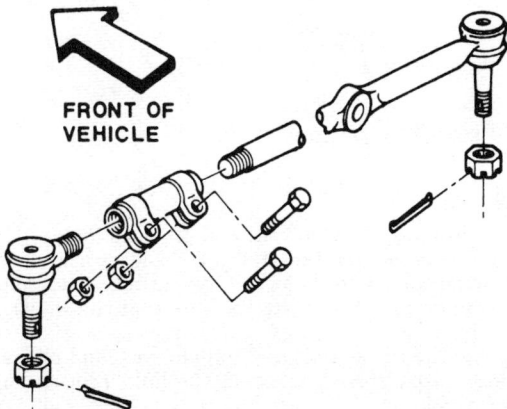

FRONT OF VEHICLE

1972–86 tie rod assembly

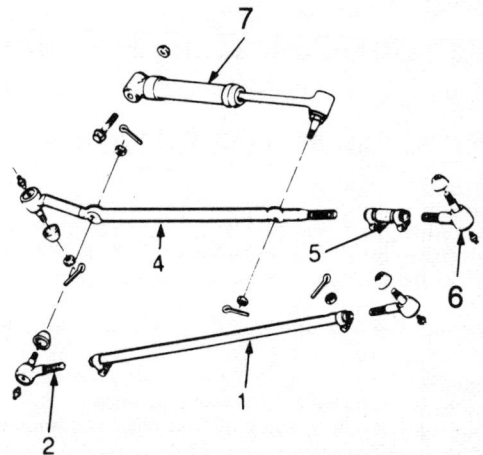

1. Connecting rod
2. Connecting rod ball stud
4. Tie rod
5. Sleeve
6. Tie rod end
7. Steering damper

1987–90 steering linkage

8 SUSPENSION AND STEERING

1. Cotter pin
2. Adjusting plug
3. Ball seat
4. Ball seat spring
5. Plug spring
6. Draglink
7. Adjusting plug
8. Dust cover
9. Dust cover shield
10. Lubricating fitting

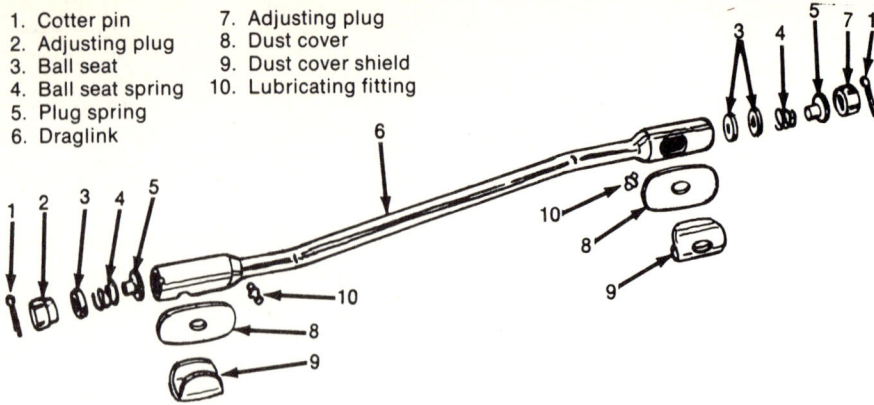

1971 drag link

5. The tie rod ends can be removed from the tie rod by loosening the clamp and unscrewing them. Before unscrewing them, note the exact number of threads visible for installation reference.

6. Install the new ends leaving the correct number of threads visible. Torque the clamp bolts to 12 ft. lbs.

7. Connect the tie rod to the knuckles and torque the nuts to 50 ft. lbs. Install new cotter pins.

8. Connect the steering damper and torque the nut to 22 ft. lbs.

1987-90

1. Remove the cotter pins and retaining nuts at both ends of the tie rod and from the end of the connecting rod where it attaches to the tie rod.

2. Remove the nut attaching the steering damper push rod to the tie rod bracket and move the damper aside.

3. Remove the tie rod ends from the steering arms and connecting rod with a puller.

4. Count the number of threads showing on the tie rod before removing the ends, as a guide to installation.

5. Loosen the adjusting tube clamp bolts and unthread the ends.

6. Installation is the reverse of removal. Torque the steering damper-to-tie rod nut to 53 ft. lbs.; all other nuts to 35 ft. lbs.

7. Adjust toe-in, if necessary.

Center Link/Connecting Rod/Drag Link

REMOVAL AND INSTALLATION

1971

1. Raise and support the front end on jackstands.
2. Remove the cotter pins at each end of the link.
3. Remove the adjusting plugs, ball seats and spring from each end.
4. Disconnect the link and remove the dust cover and dust shield.
5. Replacement kits are available which contain all the above parts. Its best to replace all these parts at once.
6. Install the dust cover and shield at each end and connect the link at the steering arm and bell crank.
7. Install the ball seat and spring at each end and turn the adjusting plugs in until they firmly contact the ball. At the front end, back off the adjusting plug ½ turn and insert a new cotter pin. At the steering arm end, back off the plug one full turn and insert a new cotter pin.

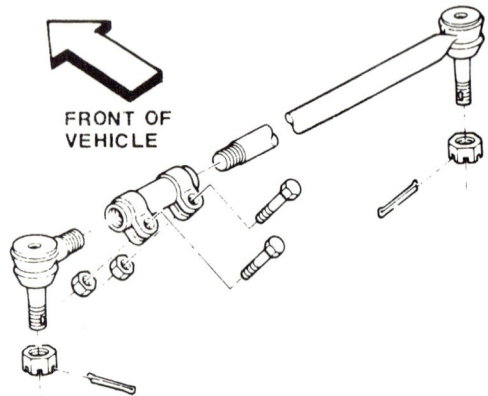

FRONT OF VEHICLE

1972–86 connecting rod

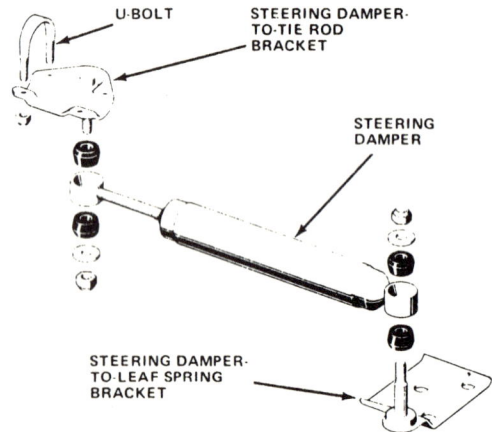

1972–79 steering damper

1972-79

1. Raise and support the front end on jackstands.
2. Place the wheels in the straight ahead position with the pitman arm parallel with the vehicle centerline. Matchmark the pitman arm and gear housing to be sure that the wheels don't move.
3. Remove the cotter pins and nuts at each end of the link.
4. Using a separator, disconnect the link from the pitman arm and tie rod.
5. With everything aligned, install the link. Torque the nuts to 70 ft. lbs., minimum, and install new cotter pins.

8-46

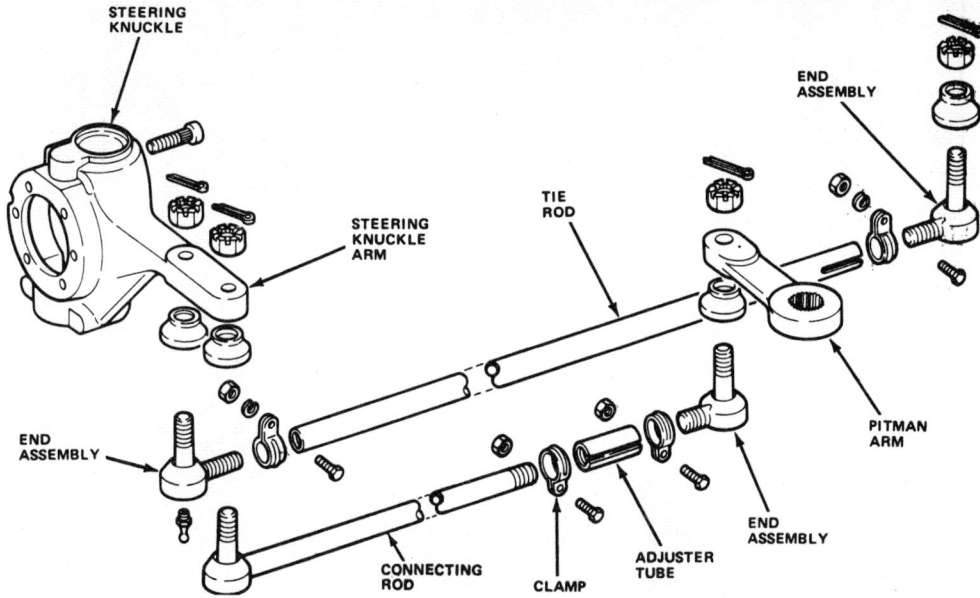

1972–86 steering linkage

1980-86

1. Raise and support the front end on jackstands.
2. Place the wheels in the straight ahead position with the pitman arm parallel with the vehicle centerline. Matchmark the pitman arm and gear housing to be sure that the wheels don't move.
3. Remove the cotter pins and nuts at each end of the link.
4. Using a separator, disconnect the link from the knuckle and pitman arm.
5. With everything aligned, install the link. Torque the nuts to 60 ft. lbs., minimum, and install new cotter pins.

1987-90

1. Raise and support the front end on jackstands.
2. Place the wheels in the straight ahead position with the pitman arm parallel with the vehicle centerline. Matchmark the pitman arm and gear housing to be sure that the wheels don't move.
3. Remove the cotter pins and nuts at each end of the link.
4. Using a separator, disconnect the link from the pitman arm and tie rod.
5. With everything aligned, install the link. Torque the nuts to 35 ft. lbs., minimum, and install new cotter pins.

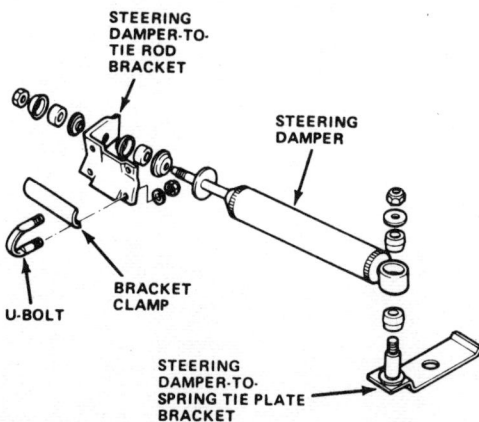

1980–86 steering damper

Steering Damper

REMOVAL AND INSTALLATION

1. Raise and support the front end on jackstands.
2. Place the wheels in a straight ahead position.
3. Remove the attaching nut at each end of the damper and remove the damper.
4. Install the damper, making sure that the wheels are still in the straight ahead position. Torque the nuts as follows.
- Through 1986
  $\frac{3}{8}$ in. nut: 22 ft. lbs.
  $\frac{7}{16}$ in. nut: 30 ft. lbs.
- 1987-90
  Tie rod end: 53 ft. lbs.
  Axle end: 55 ft. lbs.

Pitman Arm

REMOVAL AND INSTALLATION

1. Raise and support the front end on jackstands.
2. Place the wheels in a straight ahead position.
3. Matchmark the pitman arm and gear housing.
4. Disconnect the connecting rod/drag link from the arm.
5. Matchmark the pitman arm and shaft.
6. Remove the pitman arm nut and washer.
7. Using a puller, remove the pitman arm from the gear. Never hammer on the arm or use a wedge tool to remove it!
8. Install the pitman arm aligning the matchmarks on the arm and shaft.
9. Install the washer and nut. Torque the nut to:
- 1971 w/Ross gear: 70-90 ft. lbs
- 1971 w/Saginaw gear: 120-160 ft. lbs.
- 1972-90: 185 ft. lbs.
10. Connect the connecting rod/drag link, or knuckle, to the pitman arm. Observe the following torques:
- Pitman arm-to-knuckle nut: 65 ft. lbs.
- Pitman arm-to-link nut: 70 ft. lbs. on models through 1979; 60 ft. lbs. on 1980-86 models; 35 ft. lbs. on 1987-90 models.

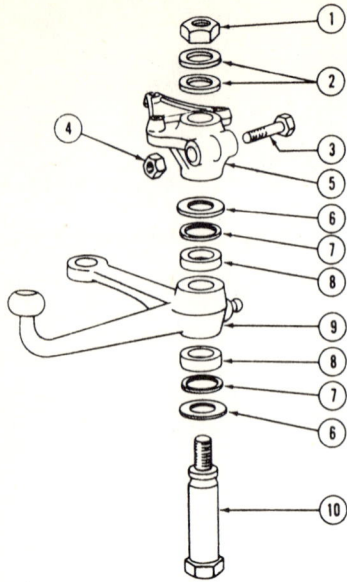

1. ⅝"-18 Stollock locknut
2. Plain washer
3. ⁷/₁₆"-20 x 2½" bolt
4. ⁷/₁₆"-20 Stollock locknut
5. Bellcrank support
6. Special ground washer
7. Seal
8. Bearing
9. Bellcrank
10. Bellcrank shaft
11. Seals
12. Bearings
13. ⅛"

Early model bellcrank

Bellcrank

REMOVAL AND INSTALLATION

1971

1. Raise and support the front end on jackstands.
2. Place the wheels in a straight ahead position.
3. Disconnect the connecting rod and tir rod from the bellcrank.
4. Remove the bellcrank-to-support bracket nut and washers.
5. Remove the bellcrank from the support bracket. It may be necessary to drive it out of the bracket with a soft drift.
6. To disassemble the bellcrank, drive out the pin and remove the parts. Service kits are available for rebuiding the bellcrank.

7. When assembling the parts, make sure that the new bearings in the bellcrank are installed ⅛ in. (3mm) below the surface of the bellcrank face. When installing the washers, make sure that the chamfer on the washers face the bellcrank.
8. After assembling the parts, install the bellcrank in the Jeep, but don't connect the linkage. Torque the bellcrank pin nut to 14-19 ft. lbs.
9. Loosen the ⁷/₁₆ in. clamp bolt and adjust the locknut on the end of the bellcrank shaft until the bellcrank just rotates freely, without binding.
10. Torque the ⁷/₁₆ in. clamp nut to 50-70 ft. lbs.
11. Connect the tie rod to the bellcrank and torque the nut to 38-45 ft. lbs.
12. Connect the connecting rod to the bellcrank and adjust is as explained above.

SPECIAL TOOLS

J-34503

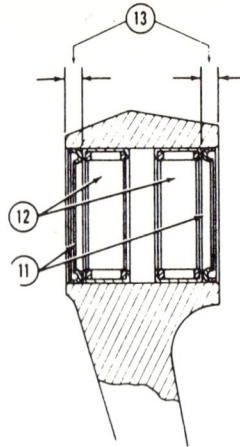

J-23653

J-21232

9
Brakes

BRAKE SPECIFICATIONS

Years	Master Cylinder Bore	Brake Disc				Brake Drum		Wheel Cyl. or Caliper Bore	
		Original Thickness	Minimum Thickness	Maximum Run-out	Diameter	Orig. Inside Dia.	Max. Wear Limit	Front	Rear
1971	1.000	—	—	—	—	9.000	9.060	1.000	0.750
1972–73	1.000	—	—	—	—	11.000	11.060	1.125	0.975
1974–76 ①	1.000	—	—	—	—	11.000	11.060	1.125	0.975
1977	1.000	1.200	1.120	0.005	12.000	11.000	11.060	3.100	0.975
1978	1.000	1.200	1.120	0.005	12.000	11.000	11.060	3.100	0.875
1979–83	1.000	1.000	0.815	0.005	12.000	10.000	10.060	2.600	0.875
1984–86	1.000	1.000	0.815	0.005	11.690	10.000	10.060	2.600	0.875
1987–90	0.937	0.884	0.815	0.004	11.040	10.000	10.060	2.600	0.875

① Includes 1977 models with front drum brakes

Troubleshooting the Brake System

Problem	Cause	Solution
Low brake pedal (excessive pedal travel required for braking action.)	• Excessive clearance between rear linings and drums caused by inoperative automatic adjusters	• Make 10 to 15 alternate forward and reverse brake stops to adjust brakes. If brake pedal does not come up, repair or replace adjuster parts as necessary.
	• Worn rear brakelining	• Inspect and replace lining if worn beyond minimum thickness specification
	• Bent, distorted brakeshoes, front or rear	• Replace brakeshoes in axle sets
	• Air in hydraulic system	• Remove air from system. Refer to Brake Bleeding.
Low brake pedal (pedal may go to floor with steady pressure applied.)	• Fluid leak in hydraulic system	• Fill master cylinder to fill line; have helper apply brakes and check calipers, wheel cylinders, differential valve tubes, hoses and fittings for leaks. Repair or replace as necessary.
	• Air in hydraulic system	• Remove air from system. Refer to Brake Bleeding.
	• Incorrect or non-recommended brake fluid (fluid evaporates at below normal temp).	• Flush hydraulic system with clean brake fluid. Refill with correct-type fluid.
	• Master cylinder piston seals worn, or master cylinder bore is scored, worn or corroded	• Repair or replace master cylinder
Low brake pedal (pedal goes to floor on first application—o.k. on subsequent applications.)	• Disc brake pads sticking on abutment surfaces of anchor plate. Caused by a build-up of dirt, rust, or corrosion on abutment surfaces	• Clean abutment surfaces

Troubleshooting the Brake System (cont.)

Problem	Cause	Solution
Fading brake pedal (pedal height decreases with steady pressure applied.)	• Fluid leak in hydraulic system	• Fill master cylinder reservoirs to fill mark, have helper apply brakes, check calipers, wheel cylinders, differential valve, tubes, hoses, and fittings for fluid leaks. Repair or replace parts as necessary.
	• Master cylinder piston seals worn, or master cylinder bore is scored, worn or corroded	• Repair or replace master cylinder
Decreasing brake pedal travel (pedal travel required for braking action decreases and may be accompanied by a hard pedal.)	• Caliper or wheel cylinder pistons sticking or seized	• Repair or replace the calipers, or wheel cylinders
	• Master cylinder compensator ports blocked (preventing fluid return to reservoirs) or pistons sticking or seized in master cylinder bore	• Repair or replace the master cylinder
	• Power brake unit binding internally	• Test unit according to the following procedure: (a) Shift transmission into neutral and start engine (b) Increase engine speed to 1500 rpm, close throttle and fully depress brake pedal (c) Slow release brake pedal and stop engine (d) Have helper remove vacuum check valve and hose from power unit. Observe for backward movement of brake pedal. (e) If the pedal moves backward, the power unit has an internal bind—replace power unit
Grabbing brakes (severe reaction to brake pedal pressure.)	• Brakelining(s) contaminated by grease or brake fluid	• Determine and correct cause of contamination and replace brakeshoes in axle sets
	• Parking brake cables incorrectly adjusted or seized	• Adjust cables. Replace seized cables.
	• Incorrect brakelining or lining loose on brakeshoes	• Replace brakeshoes in axle sets
	• Caliper anchor plate bolts loose	• Tighten bolts
	• Rear brakeshoes binding on support plate ledges	• Clean and lubricate ledges. Replace support plate(s) if ledges are deeply grooved. Do not attempt to smooth ledges by grinding.
	• Incorrect or missing power brake reaction disc	• Install correct disc
	• Rear brake support plates loose	• Tighten mounting bolts

Troubleshooting the Brake System (cont.)

Problem	Cause	Solution
Spongy brake pedal (pedal has abnormally soft, springy, spongy feel when depressed.)	• Air in hydraulic system • Brakeshoes bent or distorted • Brakelining not yet seated with drums and rotors • Rear drum brakes not properly adjusted	• Remove air from system. Refer to Brake Bleeding. • Replace brakeshoes • Burnish brakes • Adjust brakes

Hard brake pedal (excessive pedal pressure required to stop vehicle. May be accompanied by brake fade.)	• Loose or leaking power brake unit vacuum hose
	• Incorrect or poor quality brakelining
	• Bent, broken, distorted brakeshoes
	• Calipers binding or dragging on mounting pins. Rear brakeshoes dragging on support plate.
	
	• Caliper, wheel cylinder, or master cylinder pistons sticking or seized
	• Power brake unit vacuum check valve malfunction
	
	• Power brake unit has internal bind
		• Tighten connections or replace leaking hose
		• Replace with lining in axle sets
		• Replace brakeshoes
		• Replace mounting pins and bushings. Clean rust or burrs from rear brake support plate ledges and lubricate ledges with molydisulfide grease.
		NOTE: If ledges are deeply grooved or scored, do not attempt to sand or grind them smooth—replace support plate.
		• Repair or replace parts as necessary
		
		• Test valve according to the following procedure:
		  (a) Start engine, increase engine speed to 1500 rpm, close throttle and immediately stop engine
		  (b) Wait at least 90 seconds then depress brake pedal
		  (c) If brakes are not vacuum assisted for 2 or more applications, check valve is faulty
		• Test unit according to the following procedure:
		  (a) With engine stopped, apply brakes several times to exhaust all vacuum in system
		  (b) Shift transmission into neutral, depress brake pedal and start engine
		  (c) If pedal height decreases with foot pressure and less pressure is required to hold pedal in applied position, power unit vacuum system is operating normally. Test power unit. If power unit exhibits a bind condition, replace the power unit.

Troubleshooting the Brake System (cont.)

Problem	Cause	Solution
Hard brake pedal (excessive pedal pressure required to stop vehicle. May be accompanied by brake fade.)	• Master cylinder compensator ports (at bottom of reservoirs) blocked by dirt, scale, rust, or have small burrs (blocked ports prevent fluid return to reservoirs). • Brake hoses, tubes, fittings clogged or restricted • Brake fluid contaminated with improper fluids (motor oil, transmission fluid, causing rubber components to swell and stick in bores • Low engine vacuum	• Repair or replace master cylinder **CAUTION:** Do not attempt to clean blocked ports with wire, pencils, or similar implements. Use compressed air only. • Use compressed air to check or unclog parts. Replace any damaged parts. • Replace all rubber components, combination valve and hoses. Flush entire brake system with DOT 3 brake fluid or equivalent. • Adjust or repair engine
Dragging brakes (slow or incomplete release of brakes)	• Brake pedal binding at pivot • Power brake unit has internal bind • Parking brake cables incorrrectly adjusted or seized • Rear brakeshoe return springs weak or broken • Automatic adjusters malfunctioning • Caliper, wheel cylinder or master cylinder pistons sticking or seized • Master cylinder compensating ports blocked (fluid does not return to reservoirs).	• Loosen and lubricate • Inspect for internal bind. Replace unit if internal bind exists. • Adjust cables. Replace seized cables. • Replace return springs. Replace brakeshoe if necessary in axle sets. • Repair or replace adjuster parts as required • Repair or replace parts as necessary • Use compressed air to clear ports. Do not use wire, pencils, or similar objects to open blocked ports.
Vehicle moves to one side when brakes are applied	• Incorrect front tire pressure • Worn or damaged wheel bearings • Brakelining on one side contaminated • Brakeshoes on one side bent, distorted, or lining loose on shoe • Support plate bent or loose on one side • Brakelining not yet seated with drums or rotors • Caliper anchor plate loose on one side • Caliper piston sticking or seized • Brakelinings water soaked • Loose suspension component attaching or mounting bolts • Brake combination valve failure	• Inflate to recommended cold (reduced load) inflation pressure • Replace worn or damaged bearings • Determine and correct cause of contamination and replace brakelining in axle sets • Replace brakeshoes in axle sets • Tighten or replace support plate • Burnish brakelining • Tighten anchor plate bolts • Repair or replace caliper • Drive vehicle with brakes lightly applied to dry linings • Tighten suspension bolts. Replace worn suspension components. • Replace combination valve

Troubleshooting the Brake System (cont.)

Problem	Cause	Solution
Chatter or shudder when brakes are applied (pedal pulsation and roughness may also occur.)	• Brakeshoes distorted, bent, contaminated, or worn • Caliper anchor plate or support plate loose • Excessive thickness variation of rotor(s)	• Replace brakeshoes in axle sets • Tighten mounting bolts • Refinish or replace rotors in axle sets
Noisy brakes (squealing, clicking, scraping sound when brakes are applied.)	• Bent, broken, distorted brakeshoes • Excessive rust on outer edge of rotor braking surface	• Replace brakeshoes in axle sets • Remove rust
Noisy brakes (squealing, clicking, scraping sound when brakes are applied.) (cont.)	• Brakelining worn out—shoes contacting drum of rotor • Broken or loose holddown or return springs • Rough or dry drum brake support plate ledges • Cracked, grooved, or scored rotor(s) or drum(s) • Incorrect brakelining and/or shoes (front or rear).	• Replace brakeshoes and lining in axle sets. Refinish or replace drums or rotors. • Replace parts as necessary • Lubricate support plate ledges • Replace rotor(s) or drum(s). Replace brakeshoes and lining in axle sets if necessary. • Install specified shoe and lining assemblies
Pulsating brake pedal	• Out of round drums or excessive lateral runout in disc brake rotor(s)	• Refinish or replace drums, re-index rotors or replace

BASIC OPERATING PRINCIPLES

Hydraulic systems are used to actuate the brakes of all modern automobiles. The system transports the power required to force the frictional surfaces of the braking system together from the pedal to the individual brake units at each wheel. A hydraulic system is used for two reasons. First, fluid under pressure can be carried to all parts of an automobile by small hoses, some of which are flexible, without taking up a significant amount of room or posing routing problems. Second, a great mechanical advantage can be given to the brake pedal end of the system, and the foot pressure required to actuate the brakes can be reduced by making the surface area of the master cylinder pistons smaller than that of any of the pistons in the wheel cylinders or calipers.

The master cylinder consists of a fluid reservoir and either a single or double cylinder and piston assembly. Double type master cylinders are designed to separate the front and rear braking systems hydraulically in case of a leak.

Steel lines carry the brake fluid to a point on the vehicle's frame near each of the vehicle's wheels. The fluid is then carried to the wheel cylinders by flexible tubes in order to allow for suspension and steering movements.

Each wheel cylinder contains two pistons, one at either end, which push outward in opposite directions. In disc brake systems, the cylinders are part of the calipers. One or four cylinders are used to force the brake pads against the disc, but all cylinders contain one piston only. All pistons employ some type of seal, usually made of rubber, to minimize fluid leakage. A rubber dust boot seals the outer end of the cylinder against dust and dirt. The boot fits around the outer end of the piston on disc brake calipers, and around the brake actuating rod on wheel cylinders.

The hydraulic system operates as follows: When at rest, the entire system, from the piston(s) in the master cylinder to those in the wheel cylinders or calipers, is full of brake fluid. Upon application of the brake pedal, fluid trapped in front of the master cylinder piston(s) is forced through the lines to the wheel cylinders. Here, it forces the pistons outward, in the case of drum brakes, and inward toward the disc, in the case of disc brakes. The motion of the pistons is opposed by return springs mounted outside the cylinders in drum brakes, and by internal springs or spring seals, in disc brakes.

Upon release of the brake pedal, a spring located inside the master cylinder immediately returns the master cylinder pistons to the normal position. The pistons contain check valves and the master cylinder has compensating ports drilled in it. These are uncovered as the pistons reach their normal position. The piston check valves allow fluid to flow toward the wheel cylinders or calipers as the pistons withdraw. Then, as the return springs force the brake pads or shoes into the released position, the excess fluid reservoir through the compensating ports. It is during the time the pedal is in the released position that any fluid that has leaked out of the system will be replaced through the compensating ports.

Dual circuit master cylinders employ two pistons, located one behind the other, in the same cylinder. The primary piston is actuated directly by mechanical linkage from the brake pedal. The secondary piston is actuated by fluid trapped between the two pistons. If a leak develops in front of the secondary piston, it moves forward until it bottoms against the front of the master cylinder, and the fluid trapped between the pistons will operate the rear brakes. If the rear brakes develop a leak, the primary piston will move forward until direct contact with the secondary piston takes place, and it will force the secondary piston to actuate the front brakes. In either case, the brake pedal moves farther when the brakes are applied, and less braking power is available.

All dual circuit systems use a switch to warn the driver when only half of the brake system is operational. This switch is located in a valve body which is mounted on the firewall or the frame below the master cylinder. A hydraulic piston receives pressure from both circuits, each circuit's pressure being applied to one end of the piston. When the pressures are in balance, the piston remains stationary. When one circuit has a leak, however, the greater pressure in that circuit during application of the brakes will push the piston to one side, closing the switch and activating the brake warning light.

In disc brake systems, this valve body also contains a metering valve and, in some cases, a proportioning valve. The metering valve keeps pressure from traveling to the disc brakes on the front wheels until the brake shoes on the rear wheels have contacted the drums, ensuring that the front brakes will never be used alone. The proportioning valve controls the pressure to the rear brakes to avoid rear wheel lock-up during very hard braking.

Warning lights may be tested by depressing the brake pedal and holding it while opening one of the wheel cylinder bleeder screws. If this does not cause the light to go on, substitute a new lamp, make continuity checks, and, finally, replace the switch as necessary.

The hydraulic system may be checked for leaks by applying pressure to the pedal gradually and steadily. If the pedal sinks very slowly to the floor, the system has a leak. This is not to be confused with a springy or spongy feel due to the compression of air within the lines. If the system leaks, there will be a gradual change in the position of the pedal with a constant pressure.

Check for leaks along all lines and at wheel cylinders. If no external leaks are apparent, the problem is inside the master cylinder.

Disc Brakes

BASIC OPERATING PRINCIPLES

Instead of the traditional expanding brakes that press outward against a circular drum, disc brake systems utilize a disc (rotor) with brake pads positioned on either side of it. Braking effect is achieved in a manner similar to the way you would squeeze a spinning phonograph record between your fingers. The disc (rotor) is a casting with cooling fins between the two braking surfaces. This enables air to circulate between the braking surfaces making them less sensitive to heat buildup and more resistant to fade. Dirt and water do not affect braking action since contaminants are thrown off by the centrifugal action of the rotor or scraped off the by the pads. Also, the equal clamping action of the two brake pads tends to ensure uniform, straightline stops. Disc brakes are inherently self-adjusting.

There are three general types of disc brake:
1. A fixed caliper.
2. A floating caliper.
3. A sliding caliper.

The fixed caliper design uses two pistons mounted on either side of the rotor (in each side of the caliper). The caliper is mounted rigidly and does not move.

The sliding and floating designs are quite similar. In fact, these two types are often lumped together. In both designs, the pad on the inside of the rotor is moved into contact with the rotor by hydraulic force. The caliper, which is not held in a fixed position, moves slightly, bringing the outside pad into contact with the rotor. There are various methods of attaching floating calipers. Some pivot at the bottom or top, and some slide on mounting bolts. In any event, the end result is the same.

Drum Brakes

BASIC OPERATING PRINCIPLES

Drum brakes employ two brake shoes mounted on a stationary backing plate. These shoes are positioned inside a circular drum which rotates with the wheel assembly. The shoes are held in place by springs. This allows them to slide toward the drums (when they are applied) while keeping the linings and drums in alignment. The shoes are actuated by a wheel cylinder which is mounted at the top of the backing plate. When the brakes are applied, hydraulic pressure forces the wheel cylinder's actuating links outward. Since these links bear directly against the top of the brake shoes, the tops of the shoes are then forced against the inner side of the drum. This action forces the bottoms of the two shoes to contact the brake drum by rotating the entire assembly slightly (known as servo action). When pressure within the wheel cylinder is relaxed, return springs pull the shoes back away from the drum.

Modern drum brakes are designed to self-adjust themselves during application when the vehicle is moving in reverse. This motion causes both shoes to rotate very slightly with the drum, rocking an adjusting lever, thereby causing rotation of the adjusting screw.

Power Boosters

Power brakes operate just as non-power brake systems except in the actuation of the master cylinder pistons. A vacuum diaphragm is located on the front of the master cylinder and assists the driver in applying the brakes, reducing both the effort and travel he must put into moving the brake pedal.

The vacuum diaphragm housing is connected to the intake manifold by a vacuum hose. A check valve is placed at the point where the hose enters the diaphragm housing, so that during periods of low manifold vacuum brake assist vacuum will not be lost.

Depressing the brake pedal closes off the vacuum source and allows atmospheric pressure to enter on one side of the diaphragm. This causes the master cylinder pistons to move and apply the brakes. When the brake pedal is released, vacuum is applied to both sides of the diaphragm, and return springs return the diaphragm and master cylinder pistons to the released position. If the vacuum fails, the brake pedal rod will butt against the end of the master cylinder actuating rod, and direct mechanical application will occur as the pedal is depressed.

The hydraulic and mechanical problems that apply to conventional brake systems also apply to power brakes, and should be checked for if the tests below do not reveal the problem.

Test for a system vacuum leak as described below:
1. Operate the engine at idle without touching the brake pedal for at least one minute.
2. Turn off the engine, and wait one minute.
3. Test for the presence of assist vacuum by depressing the brake pedal and releasing it several times. Light application will produce less and less pedal travel, if vacuum was present. If there is no vacuum, air is leaking into the system somewhere. Test for system operation as follows:
1. Pump the brake pedal (with engine off) until the supply vacuum is entirely gone.
2. Put a light, steady pressure on the pedal.
3. Start the engine, and operate it at idle. If the system is operating, the brake pedal should fall toward the floor if constant pressure is maintained on the pedal.

Power brake systems may be tested for hydraulic leaks just as ordinary systems are tested.

--- CAUTION ---

Brake linings contain asbestos. Asbestos is a known cancer-causing agent. When working on brakes, remember that the dust which accumulates on the brake parts and/or in the drum contains asbestos. Always wear a protective face covering, such as a painter's mask, when working on the brakes. NEVER blow the dust from the brakes or drum! There are solvents made for the purpose of cleaning brake parts. Use them!

Adjustments

DRUM BRAKES

The method of brake adjustment varies depending on whether the vehicle is equipped with cam adjustment brakes or star wheel adjustment brakes with self adjusters. When the brake linings become worn, effective brake pedal travel is reduced. Adjusting the brake shoes will restore the necessary travel.

Before adjusting the brakes, check the spring nuts, brake dust shield to axle flange bolts, and wheel bearing adjustments. Any looseness in these parts will cause erratic brake operation. Also on 1971 models, make sure that the brake pedal has the correct amount of free travel without moving the master cylinder piston (free play). There should be about ½ in. of free play at the master cylinder eye bolt. Turn the eye bolt to adjust free play. On models from 1972 on, the pedal free travel is determined by the pedal pushrod length and is not adjustable. If pedal free travel is less than $\frac{1}{16}$ in. replace the pushrod.

Release the parking brakes and centralize the brake shoes in the drums by depressing the brake pedal hard and then releasing it. It is best to have all four wheels off the ground when the brakes are adjusted so that you can go back to each wheel to double check your adjustments.

Initial Brake Shoe Adjustment

If the brake assemblies have been disassembled, an initial adjustment must be made before the drum is installed. It may also be necessary to back off the adjustment to remove the drums.

When the brake parts have been installed in their correct position, adjust the adjusting screw assemblies to a point where approximately ⅜ in. of threads are exposed between the star wheel and the star wheel nut.
1. Raise and support the vehicle on jackstands.
2. Remove the access slot cover and using a brake adjusting

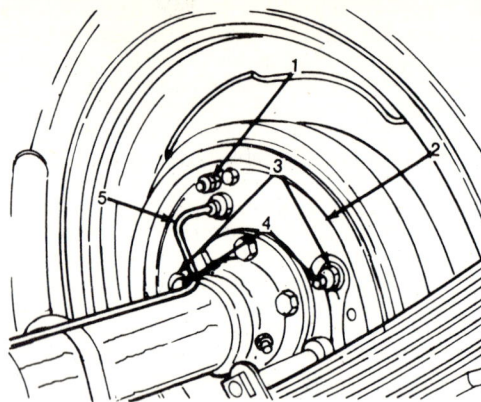

1. Bleeder screw
2. Brake backing plate
3. Eccentric lock nut
4. Eccentric adjusting screw
5. Brake fluid line

Some early 1971 CJ models had cam type adjusters

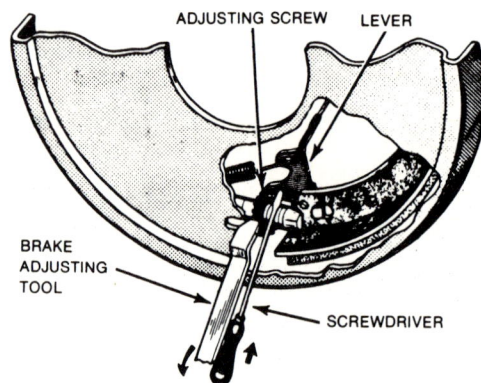

To loosen the self-adjusting drum brake, you have to hold the adjusting lever away from the star wheel

tool or screwdriver, rotate the star wheel until the wheel is locked and can't be turned in the clockwise direction.
3. Back off the star wheel until the wheel rotates freely. To back off the star wheel on the brake, insert an ice pick or thin screwdriver in the adjusting screw slot to hold the automatic adjusting lever away from the star wheel. Do not attempt to back off on the adjusting screw without holding the adjusting lever away from the star wheel as the adjuster will be damaged.

BRAKE PEDAL FREEPLAY

NOTE: Pedal freeplay is measured at the top of the pedal pad.

1971

Freeplay is controlled by the length of the master cylinder pushrod. Shorten or lengthen the pushrod to give a freeplay of ½ in.

1972-90

Proper free play should be $\frac{1}{16}$-¼ in. Free play is not adjustable. If free play is not correct, the problem is the result of worn or damaged parts.

HYDRAULIC SYSTEM

Master Cylinder

REMOVAL AND INSTALLATION

1. Disconnect and plug the brake lines.
2. Disconnect the wires from the stoplight switch.
3. Disconnect the master cylinder pushrod at the brake pedal (non-power brakes only), remove all attaching bolts and nuts and lift the assembly from the vehicle.

To install:

4. Position the master cylinder and install the mounting bolts. Torque the mounting bolts to 30 ft. lbs. on 1971–86 models; 18 ft. lbs. on 1987–90 models.
5. Connect the pushrod on non-power brakes and connect the wiring.
6. Connect the brake lines and bleed the master cylinder and brake system.

OVERHAUL

1971-75

1. Remove the filler cap and empty all the fluid.
2. The stop light switch and primary piston stop, located in the stop light switch outlet hole, must be removed before removing the snapring from the piston bore. Remove the snapring, pushrod assembly and the primary and secondary piston assemblies. Air pressure applied in the piston stop hole will help facilitate the removal of the secondary piston assembly.
3. The residual check valves are located under the front and rear fluid outlet tube seats.
4. The tube seats must be removed with self tapping screws to permit the removal of the check valves. Screw the self-tapping screws into the tube seats and place two screw driver tips under the screw head and force the screw upward.

5. Remove the expander in the rear secondary cup, secondary cups, return spring cup protector, primary cup, and washer from the secondary piston.
6. Immerse all of the metal parts in clean brake fluid and clean them. Use an air hose to blow out dirt and cleaning solvent from recesses and internal passages.
7. After cleaning, place all of the parts on clean paper or in a clean pan.

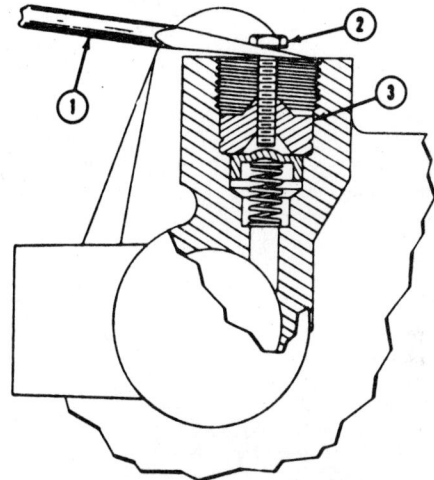

1. Screwdriver
2. Self-tapping screw
3. Tube seat

Removing tube seats

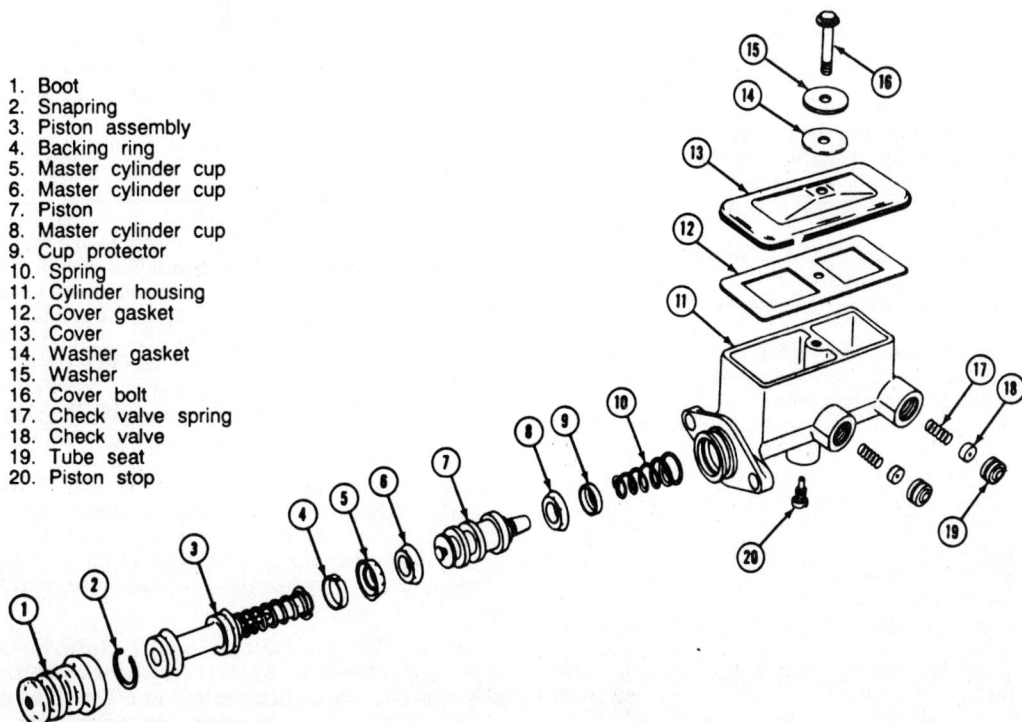

1. Boot
2. Snapring
3. Piston assembly
4. Backing ring
5. Master cylinder cup
6. Master cylinder cup
7. Piston
8. Master cylinder cup
9. Cup protector
10. Spring
11. Cylinder housing
12. Cover gasket
13. Cover
14. Washer gasket
15. Washer
16. Cover bolt
17. Check valve spring
18. Check valve
19. Tube seat
20. Piston stop

1971 master cylinder

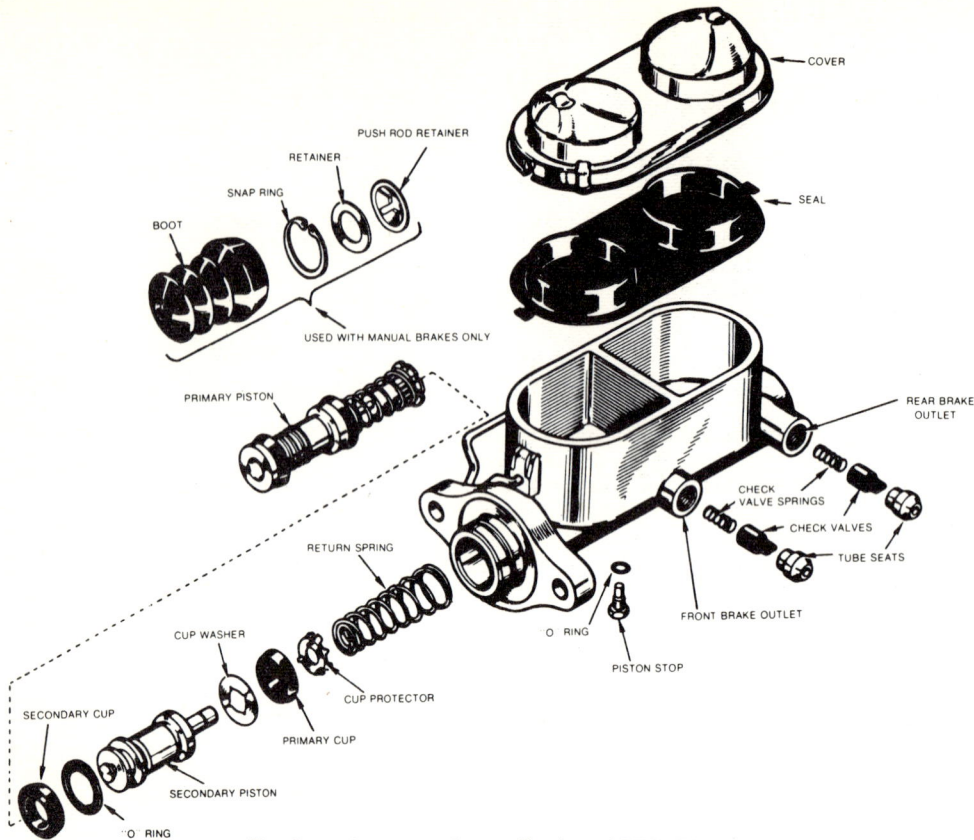

Dual system master cylinder, 1972–75

8. Inspect all parts for damage or excessive wear. Replace any damaged, worn, or chipped parts. Inspect the hydraulic cylinder bore for signs of scoring, rust, pitting, or etching. Any of these will require replacement of the hydraulic cylinder.

9. Prior to assembling the master cylinder, dip all of the components in clean brake fluid and place them on clean paper or in a clean pan.

10. Install the primary cup washer, primary cup, cup protector, and return spring in the secondary position.

11. Install the piston cups in the double groove end of the secondary piston, so the flat side of the cups face each other (lip of the cups away from each other). Install the cup expander in the lip groove of the end cup.

12. Coat the cylinder bore and piston assemblies with clean brake fluid before installing any parts in the cylinder.

13. Install the secondary piston assembly first and then the primary piston.

14. Install the pushrod assembly, which includes the pushrod, boot, and rod retainer, and secure with the snapring. Install the primary piston stop and stop light switch.

15. Place new rubber check valves over the check valve springs and install in the outlet holes, spring first.

16. Install the tube seats, flat side toward the check valve, and press in with tube nuts or the master cylinder brake tube nuts.

17. Before the master cylinder is installed on the vehicle it must be bled. Support the cylinder assembly in a vise and fill both fluid reservoirs with brake fluid.

18. Loosely install a plug in each outlet of the cylinder. Depress the push rod several times until air bubbles cease to appear in the brake fluid.

19. Tighten the plugs and attempt to depress the piston. The piston travel should be restricted after all of the air is expelled.

20. Install the master cylinder in the vehicle and bleed all the hydraulic lines at the wheel cylinders.

1976–90

1. Remove the master cylinder from the vehicle and remove the cover and diaphragm seal. Drain the brake fluid from the reservoir and mount it in a vise.

2. With non-power brakes, slide the boot back on the pushrod. Push the pushrod in and unseat the primary piston snapring. Remove the pushrod, boot, snapring, and pushrod retainer as an assembly. With power brakes, push the primary piston in with a wooden dowel and remove the piston snapring.

3. Remove the primary and secondary piston assemblies. Air pressure applied through the piston stop hole will help in the removal of the secondary piston.

4. Clamp the primary piston pushrod in a vise, with the brake pedal eyelet facing downward and the rear face of the primary piston positioned just above the jaws of the vise.

5. Remove the rubber insert from the primary piston and discard it. NEVER ATTEMPT TO REUSE THIS INSERT! A REPLACEMENT INSERT MUST BE OBTAINED!!

6. Remove the piston seal and piston cups from the secondary piston. It is not necessary to disassemble the primary piston because a new complete primary piston assembly is supplied in the rebuilding kit.

7. Clean and inspect the master cylinder. Replace the master cylinder body if the bore is severely scored, corroded, or pitted, cracked, porous, or is otherwise damaged. Check the by-pass and compensator ports to make sure that they are open and not plugged or dirty. Use brake fluid and air pressure to open these passages. Do not use wire.

NOTE: Use only clean brake fluid or an approved cleaning solvent to wash the master cylinder. Do not use any solvent containing mineral oil such as gasoline, kerosene, alcohol, or carbon tetrachloride. Mineral oil harms rubber components.

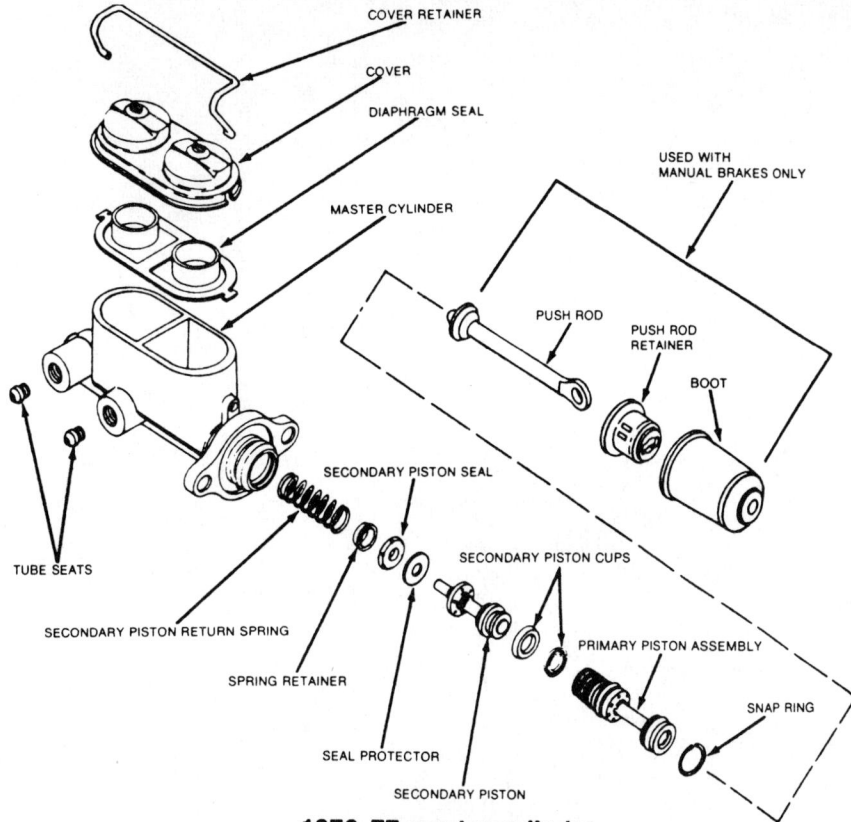

1976–77 master cylinder

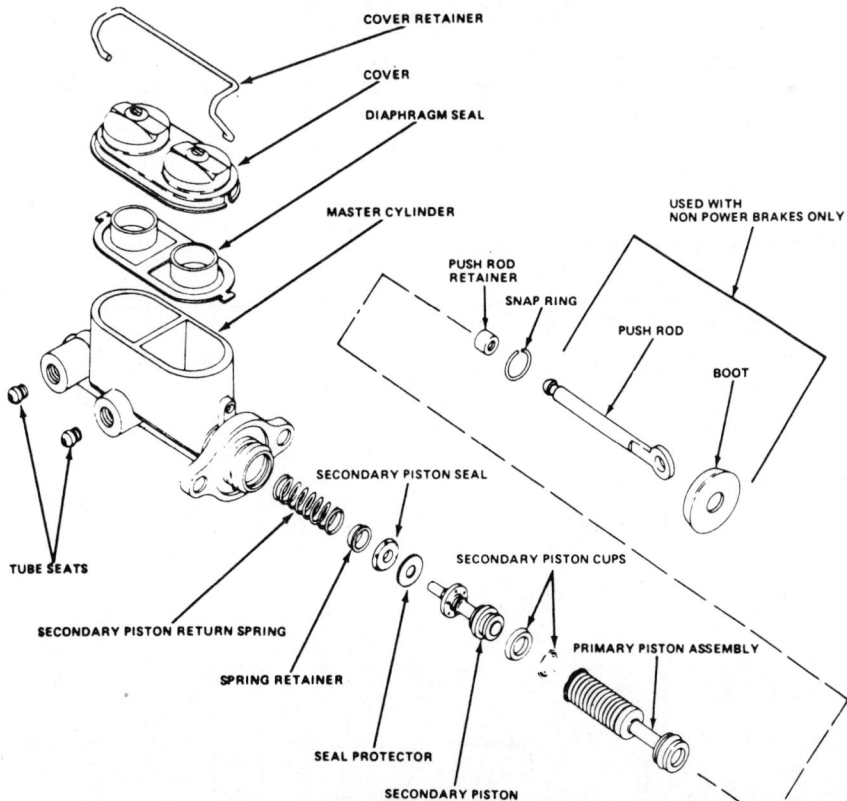

1978 master cylinder

COVER RETAINER

COVER

DIAPHRAGM SEAL

PUSH ROD RETAINER

RETAINER CLIP

WASHER

PISTON SNAP RING

USED WITH NON POWER BRAKES ONLY

MASTER CYLINDER

PUSH ROD

BOOT

USED WITH NONPOWER BRAKES ONLY

SECONDARY PISTON SEAL

TUBE SEATS

SECONDARY PISTON CUPS

SECONDARY PISTON RETURN SPRING

PRIMARY PISTON ASSEMBLY

SPRING RETAINER

SEAL PROTECTOR

SECONDARY PISTON

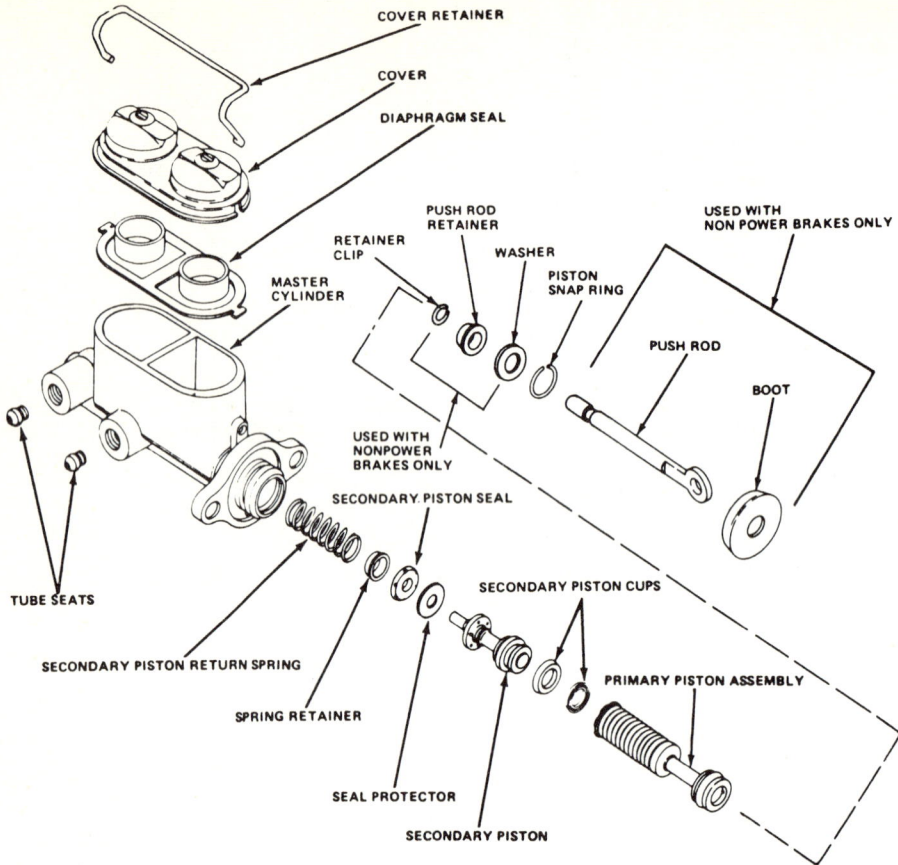

1979–81 master cylinder

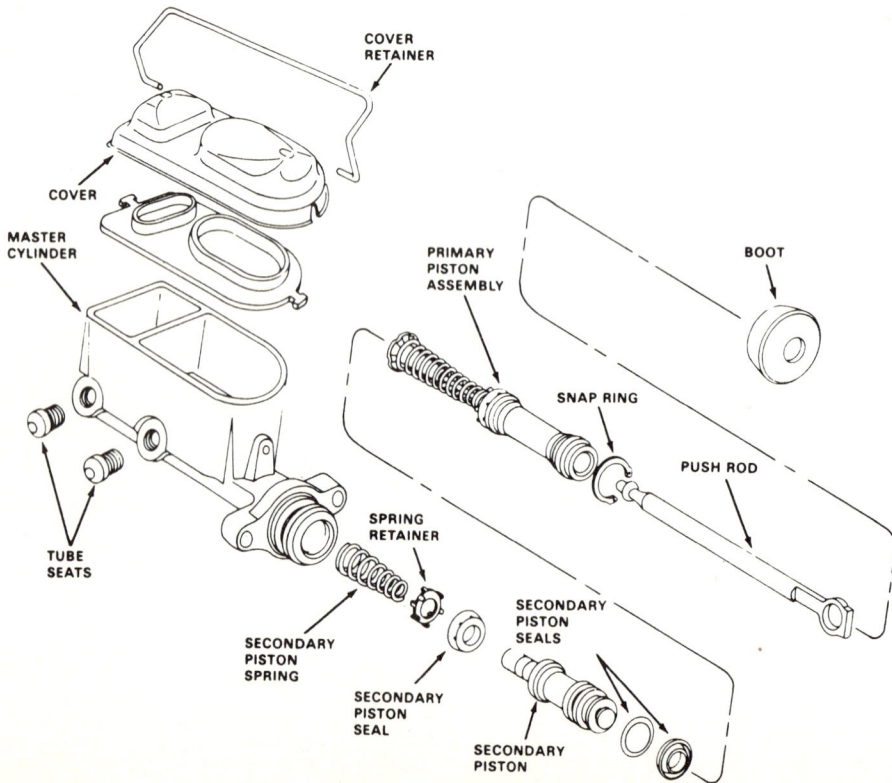

COVER RETAINER

COVER

MASTER CYLINDER

PRIMARY PISTON ASSEMBLY

BOOT

SNAP RING

PUSH ROD

TUBE SEATS

SPRING RETAINER

SECONDARY PISTON SPRING

SECONDARY PISTON SEALS

SECONDARY PISTON SEAL

SECONDARY PISTON

1982–83 master cylinder

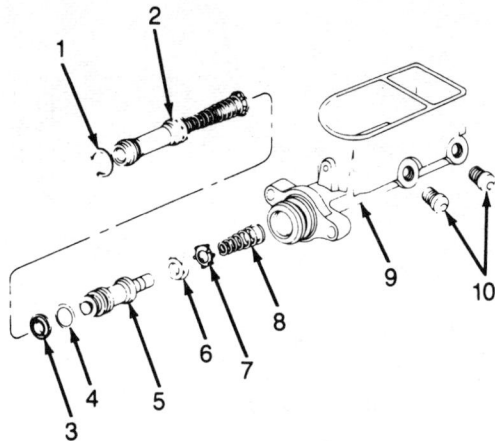

1. Snapring
2. Primary piston
3. Seal
4. Seal
5. Secondary piston
6. Rear seal
7. retainer
8. Spring
9. Master cylinder body
10. Outlet port tube seats

1984–90 master cylinder

8. Check the tube seats in the outlet ports. Replace the seats only if they are cracked, scored, cocked in the bore, or loose. Replace the tube seats as follows:

a. Enlarge the hole in the tube seat with a $^{13}/_{64}$ in. drill bit.

b. Place a flat washer on each outlet port and thread a $^{1}/_{4}$–20 x $^{3}/_{4}$ in. long screw into the seat.

c. Tighten the screw until the seat is loosened.

d. Remove the seat, screw, and washer. Flush any metal chips away with brake fluid and compressed air.

9. Install the replacement tube seats, if removed, using spare tube fitting nuts to press the seats into place. Be careful that the seats don't become cocked during installation. Make sure that the seats are bottomed. Remove the tube fitting nuts and check

for burrs or chips. Rinse the master cylinder in brake fluid and blow out all passages with compressed air.

10. Install the piston cups on the secondary piston. The piston cup installed in the groove at the end of the piston should have its lip facing away from the piston. Install the next cup so that its lip faces the piston.

11. Install the seal protector, piston seal, spring retainer, and return spring on the secondary piston. Install the piston seal so that its lip faces the interior of the master cylinder bore when the assembly is installed. Make sure that the return spring seats against the retainer and that the retainer is located inside the lip of the piston seal.

12. Lubricate the master cylinder bore and secondary piston seal and cups with brake fluid and install the secondary piston assembly in the cylinder bore.

13. Lubricate the seals on the primary piston assembly with brake fluid and install the assembly in the master cylinder bore.

14. With power brakes, push the primary piston inward with a wooden bowel and install the retaining snapring in the groove of the master cylinder bore.

15. With non-power brakes, install the pushrod, pushrod retainer, and boot. Push in and install the piston snapring.

16. Install the diaphragm seal on the master cylinder cover.

17. Install the master cylinder in the vehicle. Bleed the system.

Power Brake Booster

REMOVAL AND INSTALLATION

1. Disconnect the power unit pushrod at the pedal.

2. Disconnect the vacuum line at the power unit check valve.

3. Unbolt the master cylinder from the power unit and push the master cylinder aside carefully.

4. Unbolt the power unit bellcrank at the dash panel and remove the power unit and bellcrank as an assembly. If the power unit is being discarded, save the bellcrank for the new unit.

5. Installation is the reverse of removal. Torque the bellcrank-to-dash panel bolts to 35 ft. lbs.; the master cylinder-

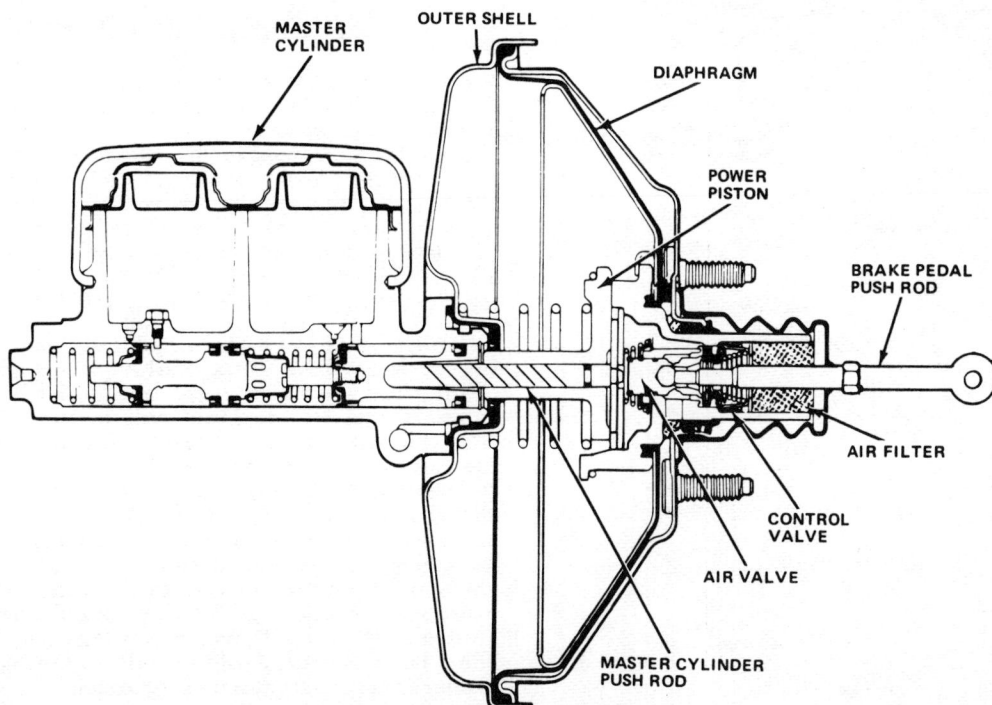

Power brake booster

to-power unit bolts to 30 ft. lbs. on models through 1986, or 18 ft. lbs. on 1987-90 models; the pushrod-to-pedal bolt and nut to 35 ft. lbs.

Proportioning Valve

The proportioning valve is actually a three part unit, containing a metering valve, pressure differential valve and brake pressure warning switch. If any of these functions fails, the unit must be replaced. It is not repairable. Two different types of valves were used. On models in the early 1970s, a D-type valve was used, in the late 1970's, a Type W was also used. They are similar in appearance and the only significant difference is in the method of holding open the valve during bleeding. Type W uses tool J-26869; type D uses tool J-23709. The valves are not interchangeable, however. In late 1983, the W-type became the only type used.

The valve is located on the frame rail, directly under the driver's position on models through 1986 and on the left fender panel just below the master cylinder, on 1987-90 models.

REMOVAL AND INSTALLATION

1. Disconnect the brake lines at the valve and plug them.
2. Unbolt and remove the valve.
3. Installation is the reverse of removal. Bleed the system.

Brake Hoses and Lines
INSPECTION

1. Inspect all connections for signs of leakage.
2. Replace any hose that shows signs of wear, scuffing or loss of rubber coating.

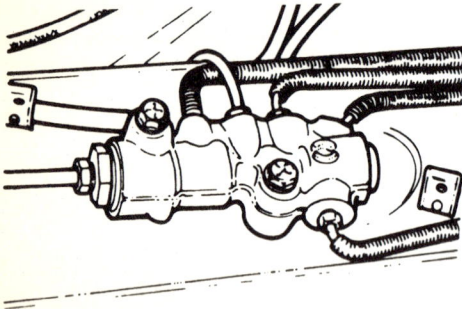

Type D combination valve

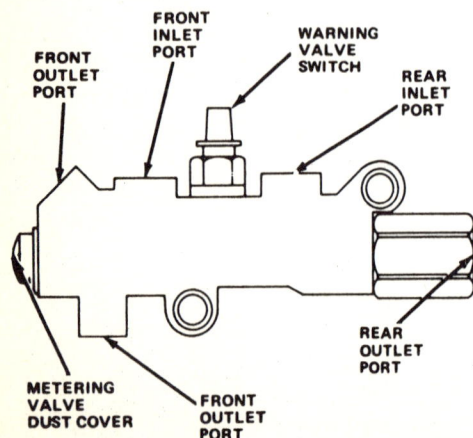

Type W combination valve

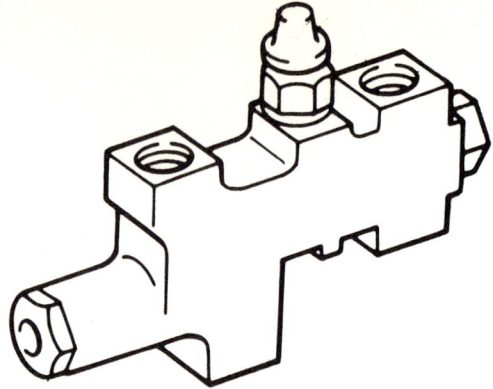

1987–90 proportioning valve

3. Replace any hose or line that becomes kinked or creased.
4. Replace any hose that shows signs of cracking or cutting.
5. Replace any hose that appears swollen.
6. Replace any line that shows signs of corrosion.

REMOVAL AND INSTALLATION

1. Clean the connections thoroughly before opening them.
2. Have handy some means of capping the line remaining in the vehicle so as to prevent excessive fluid loss.
3. Use a wrench of the exact size necessary to loosen the fitting. Avoid using an adjustable wrench or pliers, as brake line fittings are easily rounded off.
4. When replacing a brake hose connected to a brake caliper, always replace the copper gasket under the union.
5. Front brake hoses are usually connected to the metal line at a junction attached to the frame or a suspension member. In many cases, there is a U-shaped clip which must be driven out to free the junction.
6. When installing a brake hose or line, observe the following torques, where possible:

Bleeding the Brakes

The hydraulic brake system must be bled whenever a fluid line has been disconnected because air gets into the system. A leak in the system may sometimes be indicated by a spongy brake pedal. Air trapped in the system is compressible and does not permit the pressure applied to the brake pedal to be transmitted solidly through the brakes. The system must be absolutely free from air at all times. When bleeding brakes, bleed at the wheel most distant from the master cylinder first, the next most distant second, and so on. During the bleeding operation the master cylinder must be kept at least ¾ full of brake fluid.

NOTE: On 1974 and later models, there is a combination pressure differential (failure warning) and proportioning valve in the hydraulic system. It is in the engine compartment on the inner side of the left frame rail. When bleeding the brake system, the metering section of the valve must be held open. On 1974–76 models, remove the warning switch wire, switch terminal, plunger, and spring from the valve. On all 1977 models and 1978–79 models using a valve assembly with a flat exterior surface, remove the plastic dust cover at the end of the valve and hold the valve stem OUT. On 1978–79 models using a valve assembly with a rounded exterior surface, hold the valve stem IN by pressing against the dust cover boot. Tools are available to hold the valve stem in the proper location or can be fabricated.

To bleed the brakes, first carefully clean all dirt from around the master cylinder filler cap. If a bleeder tank is used follow the

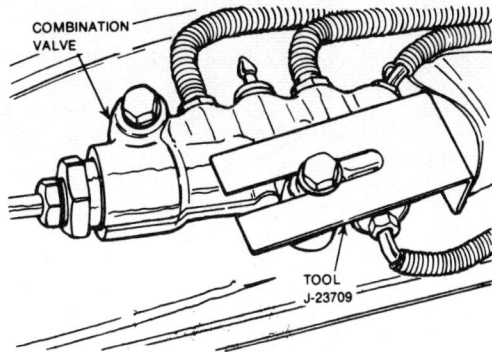

Metering valve tool installation on the type D valve

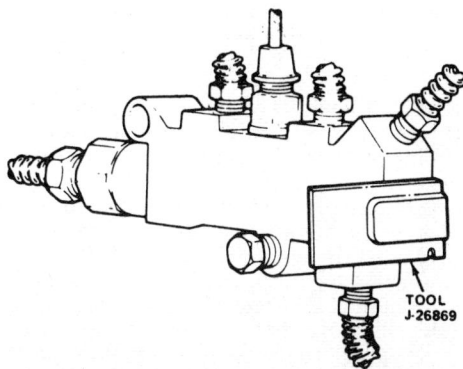

Metering valve tool installed on a type W combination valve

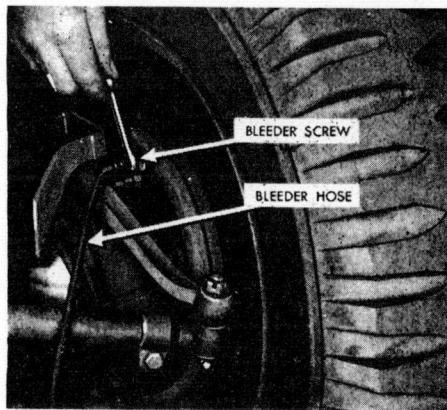

Bleeding the brakes on early models

Do not reuse the fluid which has been removed from the lines through the bleeding process because it contains air bubbles and dirt.

Stop Light Switch

Two types of switches have been used on Jeep vehicles. One type is attached to the brake pedal rod end of the push rod, and cannot be adjusted. The other type is mounted on a flange attached to the brake pedal support bracket and is held in the off position by the brake pedal's being in its released position. Upon depressing the brake pedal, the switch plunger is allowed to move outward and contact is made within the switch to allow current to pass and operate the stop lights.

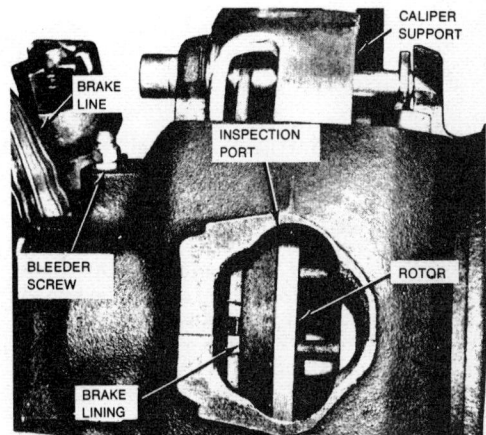

The thickness of the disc brake pad linings can be check visually. Minimum safe thickness is $1/16$ in. with bonded linings and $1/32$ in. above the rivet heads with riveted lining

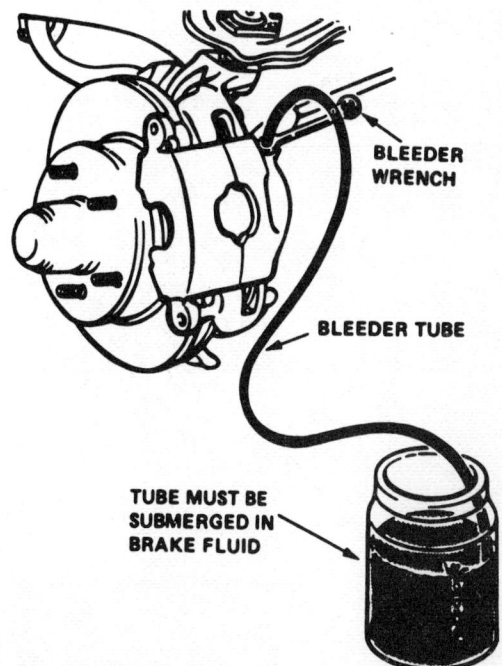

manufacturer's instructions. Remove the filler cap and fill the master cylinder to the lower edge of the filler neck. Clean off the bleeder connections at all four wheel cylinders. Attach the bleeder hose to the right rear wheel cylinder bleeder screw and place the end of the tube in a glass jar, submerged in brake fluid. Open the bleeder valve $1/2$–$3/4$ of a turn. Have an assistant depress the brake pedal slowly and allow it to return. Continue this pumping action to force any air out of the system. When bubbles cease to appear at the end of the bleeder hose, close the bleeder valve and remove the hose.

Check the level of fluid in the master cylinder reservoir and replenish as necessary.

After the bleeding operation at each wheel cylinder has been completed, fill the master cylinder reservoir and replace the filler plug.

Brake bleeding equipment

SWITCH ADJUSTMENT

NOTE: On some vehicles equipped with air conditioning, remove the screws attaching the evaporator housing to the instrument panel and move the housing away from the panel.
The switch on 1987-90 models is not adjustable.

1. Hold the brake pedal in the applied position.
2. Push the stop light switch through the mounting bracket until it stops against the brake pedal bracket. Release the pedal to set the switch in the proper position.
3. Check the position of the switch. The switch plunger should be in the ON position and activate the brake lights after a brake pedal travel of ⅜–⅝ in.

DISC BRAKES

The front disc brake consists of 3 assemblies, the caliper assembly, the hub and rotor assembly, and the support and shield assembly.

The caliper is a single piston sliding type, of one piece casting construction with the inboard side containing the single piston, piston bore and the bleeder screw and fluid inlet holes. There are two brake pads within the caliper, positioned on either side of the rotor. The brake pads take the place of brake shoes on drum brakes and the rotor takes the place of brake drums. The pads themselves actually consist of two parts: the metal shoe and the composition lining which is bonded or riveted to the shoe.

The significant operating feature of the single piston caliper is that it is free to slide laterally on the anchor plate. The pressure applied to the piston is transmitted to the inboard brake pad, forcing the lining of the pad against the inboard rotor surface. The pressure applied to the inboard end or bottom of the piston bore forces the caliper to slide toward the inboard side. This inward movement of the caliper causes the outboard section of the caliper to apply pressure against the lining of the outboard pad, forcing the lining of the outboard pad, forcing the lining against the outboard surface of the rotor. As hydraulic pressure builds within the brake lines, due to the increased application of pressure at the brake pedal, the brake pad assemblies press against the rotor surfaces with increasing force, thus slowing the rotation of the rotor.

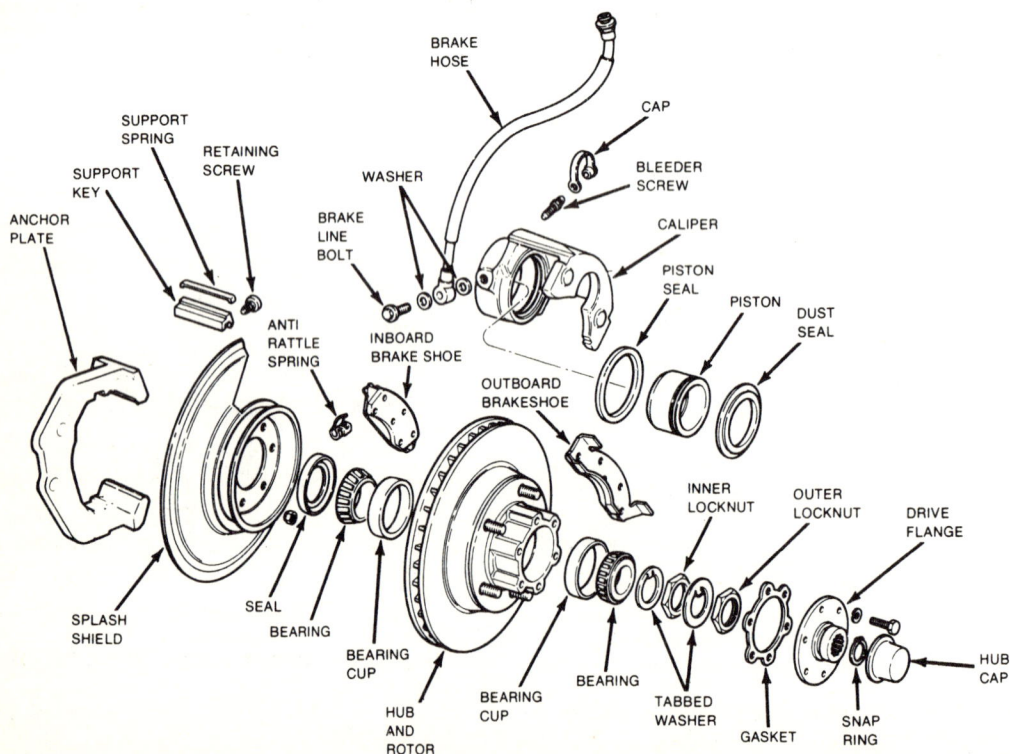

Disc brake unit used on models through 1981, without locking hubs

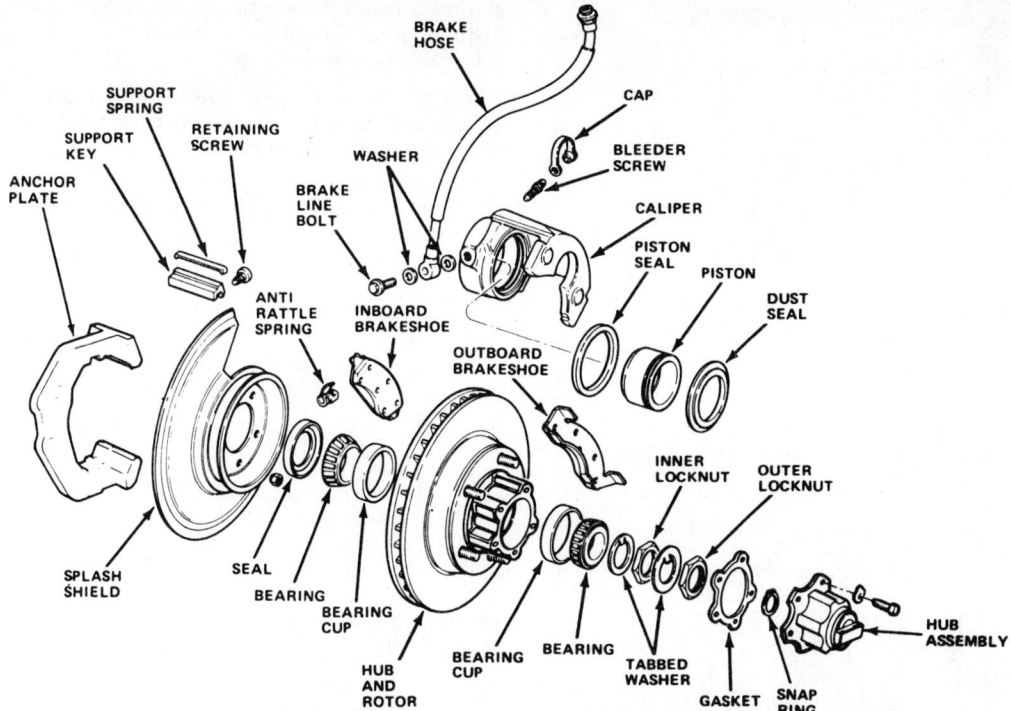

Disc brake unit used on models through 1981, with locking hubs

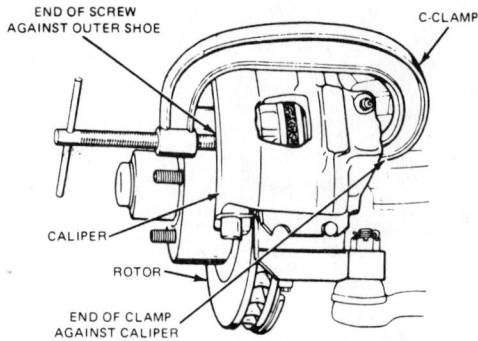

Bottoming the caliper piston on models through 1981

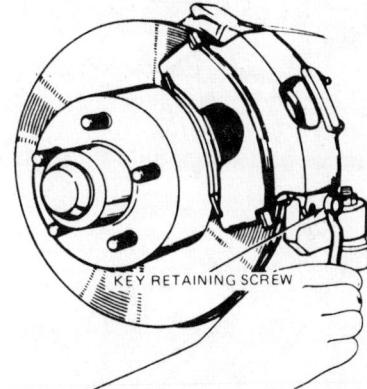

Removing the caliper retaining screw on models through 1981

Brake Pads

REMOVAL AND INSTALLATION

1977-1981

──────────── **CAUTION** ────────────

Most brake shoes contain asbestos, which has been determined to be a cancer causing agent. Never clean the brake surfaces with compressed air! Avoid inhaling any dust from any brake surface! When cleaning brake surfaces, use a commercially available brake cleaning fluid.

1. Remove ⅔ of the brake fluid from the front reservoir.
2. Raise the vehicle so that the wheel to be worked on is off the ground. Support the vehicle with jackstands.
3. Remove the front wheels.
4. Place a C-clamp on the caliper so that the solid end contacts the back of the caliper and the screw end contacts the metal part of the outboard brake pad.
5. Tighten the clamp until the caliper moves far enough to force the piston to the bottom of the piston bore. This will back

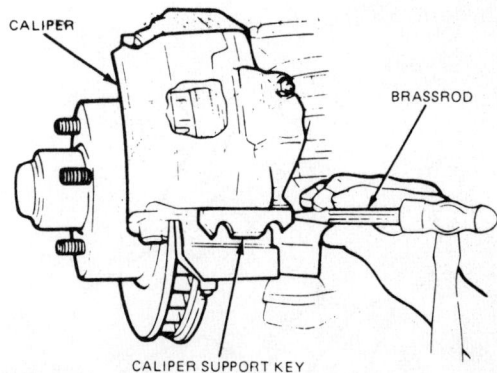

Removing the key and spring on models through 1981

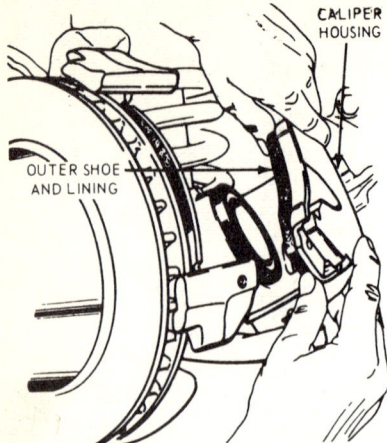

Removing the outer pad on models through 1981

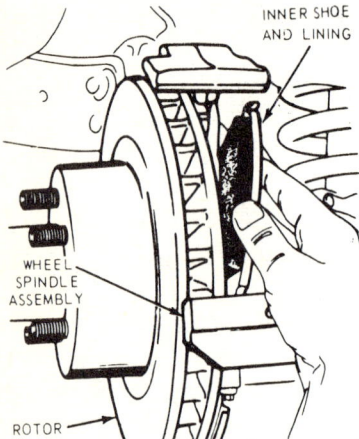

Removing the inner pad on models through 1981

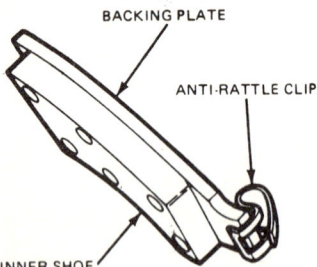

Installing the anti-rattle clip in the inner pad on models through 1981

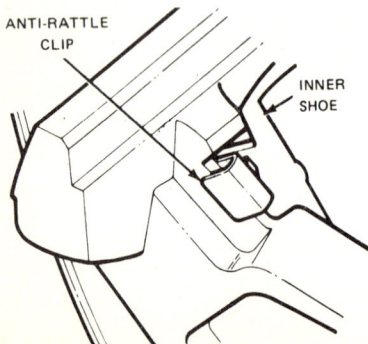

Installing the inner pad on the caliper support on models through 1981

the brake pads off the rotor surface to facilitate the removal and installation of the caliper assembly.

6. Remove the C-clamp.

NOTE: Do not push down on the brake pedal or the piston and brake pads will return to their original positions up against the rotor.

7. Remove the caliper support key retaining screw with a ¼ in. Allen wrench. Drive the support key and spring out with a punch.

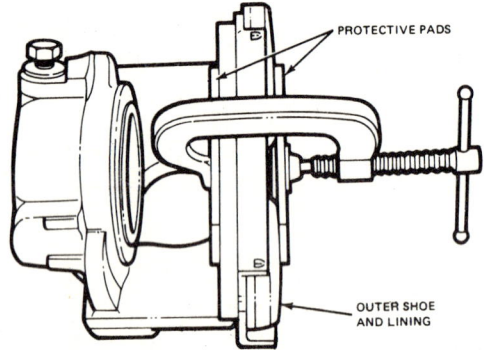

Installing the outer pad in the caliper on models through 1981

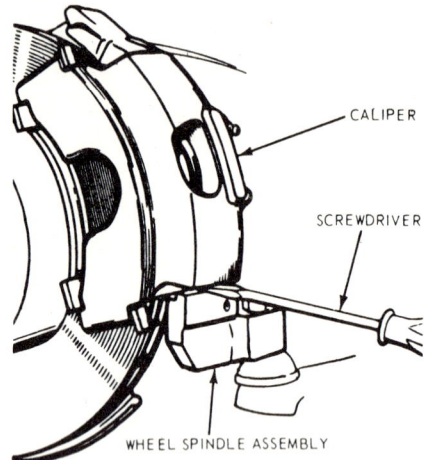

Installing the caliper on models through 1981

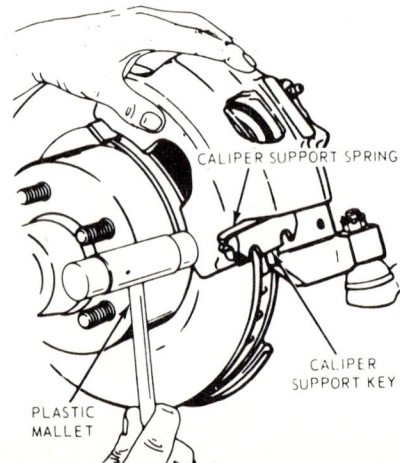

Installing the caliper support key and spring on models through 1981

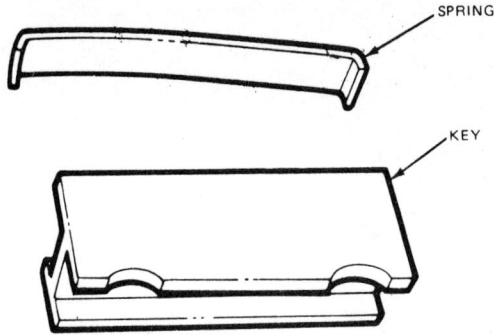

Key and spring assembly on models through 1981

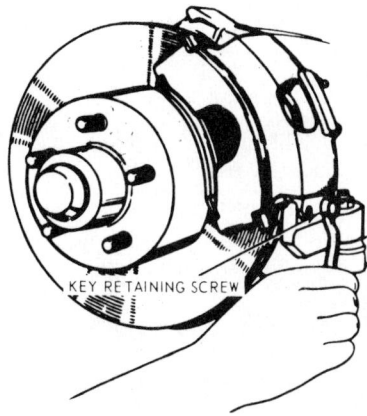

Installing the key retaining screw on models through 1981

NOTE: If just the brake pads are being replaced, it is not necessary to remove the caliper assembly entirely from the vehicle. Do not remove the brake line. Rest the caliper on the front spring or other suitable support. Do not allow the brake hose to support the weight of the caliper.

8. Remove the brake pad assemblies. Remove the anti-rattle spring from the inboard pad. Note the position of the spring before removing it for correct installation later.

9. Wipe the inside of the caliper clean, including the exterior of the dust boot. Inspect the dust boot for cuts or cracks and for proper seating in the piston bore. If evidence of fluid leakage is noted, the caliper should be rebuilt.

10. Check the sliding surface of the caliper for rust or corrosion. Clean them with a wire brush and fine sandpaper; lubricate with molybdenum disulphide grease.

11. Install the inboard pad anti-rattle spring on the rear flange of the pad. Make sure the looped section of the spring is away from the rotor.

12. Install the inboard pad with spring attached in the caliper anchor plate.

13. Install the outboard pad in the caliper.

14. Place the caliper over the rotor and on the anchor plate.

NOTE: Be careful not to damage or dislodge the dust boot.

15. Insert the support key and spring between the sliding surfaces at the rear of the caliper. Drive them into place with a punch.

16. Install the key retaining screw and torque it to 15 ft. lbs.

17. Fill the master cylinder to within ¼ in. of the rim.

18. Press firmly on the brake pedal several times till the pedal is firm.

19. Recheck the master cylinder level.

1982-86

CAUTION

Brake shoes contain asbestos, which has been determined to be a cancer causing agent. Never clean the brake surfaces with compressed air! Avoid inhaling any dust from any brake surface! When cleaning brake surfaces, use a commercially available brake cleaning fluid.

1. Remove ⅔ of the brake fluid from the front reservoir.

2. Raise the vehicle so that the wheel to be worked on is off the ground. Support the vehicle with jackstands.

3. Remove the front wheels.

4. Place a C-clamp on the caliper so that the solid end contacts the back of the caliper and the screw end contacts the metal part of the outboard brake pad.

5. Tighten the clamp until the caliper moves far enough to force the piston to the bottom of the piston bore. This will back the brake pads off the rotor surface to facilitate the removal and installation of the caliper assembly.

6. Remove the C-clamp.

NOTE: Do not push down on the brake pedal or the piston and brake pads will return to their original positions up against the rotor.

7. Remove the caliper mounting pins and lift off the caliper. Don't disconnect the brake line! Don't allow the brake line to support the weight of the caliper!

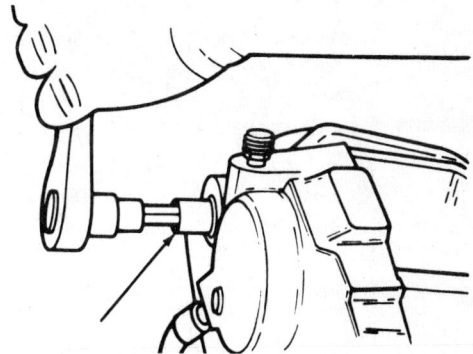

Removing the caliper pin on 1982–86 models

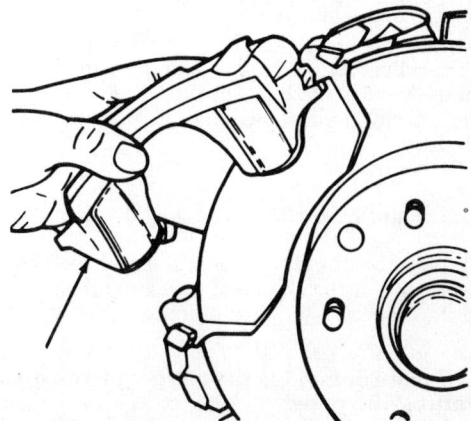

Lifting off the caliper on 1982–86 models

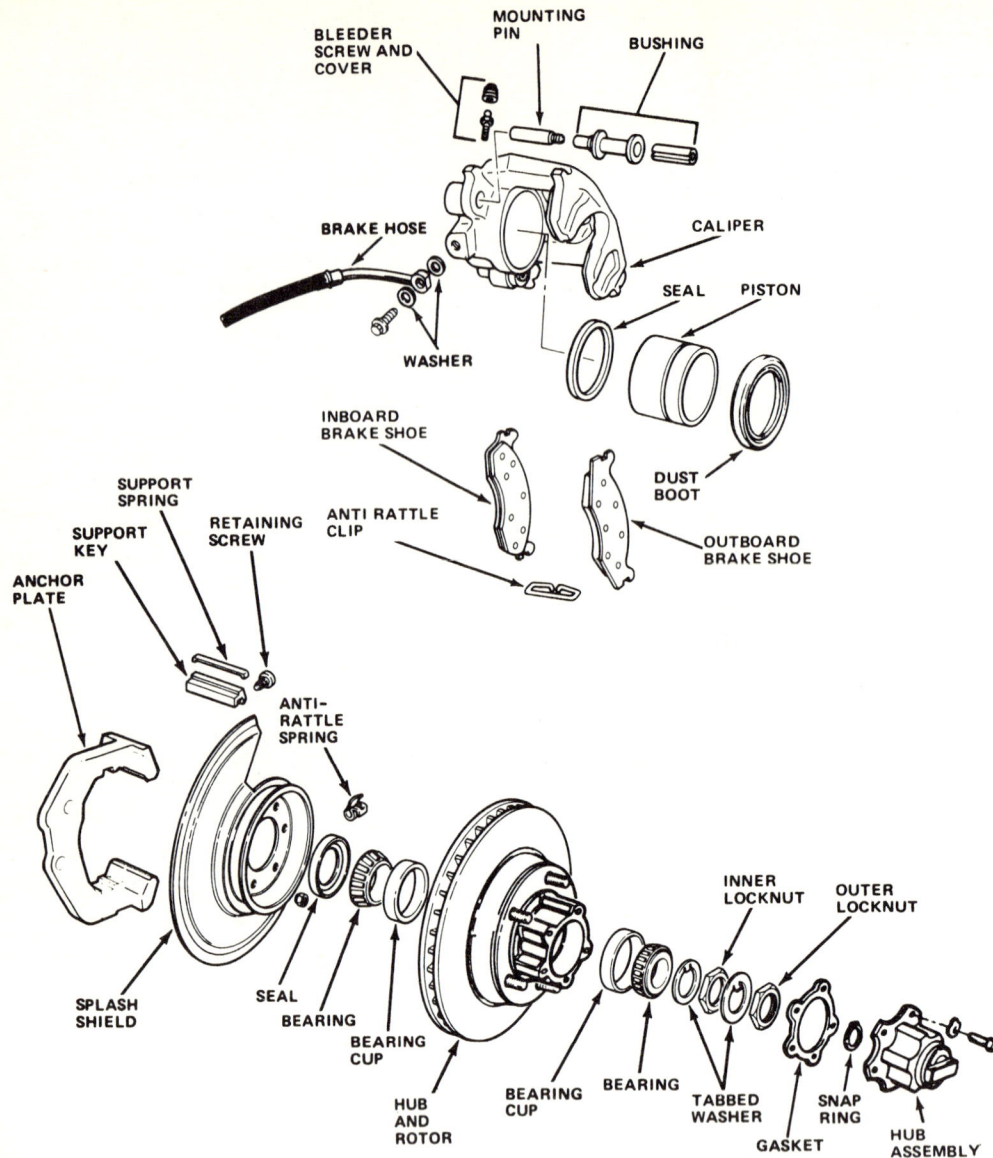

1982–86 disc brake unit

8. Hold the anti-rattle clip against the caliper anchor plate and remove the outboard pad.

9. Remove the inboard pad and anti-rattle clip.

10. Clean the caliper with a rag and solvent made for cleaning brake surfaces. Avoid disturbing the dust boot.

11. If there are any indications of fluid leakage, the caliper must be rebuilt.

12. Clean the anchor plate and caliper mounting surfaces with a wire brush.

13. Coat the mounting surfaces with a light coating of caliper lubricant.

14. Install the anti-rattle clip on the trailing edge of the anchor plate. The split end of the clip must face away from the rotor.

15. Install the inboard, then the outboard pads, while holding the clip.

WARNING: Avoid damaging the caliper piston dust boot while installing the pads!

16. Install the caliper and tighten the mounting pins to 30 ft. lbs.

17. Fill the master cylinder and press firmly on the brake pedal to seat the pads.

18. Install the wheels, and lower the truck.

1987-90

— CAUTION —

Brake shoes contain asbestos, which has been determined to be a cancer causing agent. Never clean the brake surfaces with compressed air! Avoid inhaling any dust from any brake surface! When cleaning brake surfaces, use a commercially available brake cleaning fluid.

1. Remove ⅔ of the brake fluid from the front reservoir.

2. Raise the vehicle so that the wheel to be worked on is off the ground. Support the vehicle with jackstands.

3. Remove the front wheels.

4. Place a C-clamp on the caliper so that the solid end contacts the back of the caliper and the screw end contacts the metal part of the outboard brake pad.

5. Tighten the clamp until the caliper moves far enough to

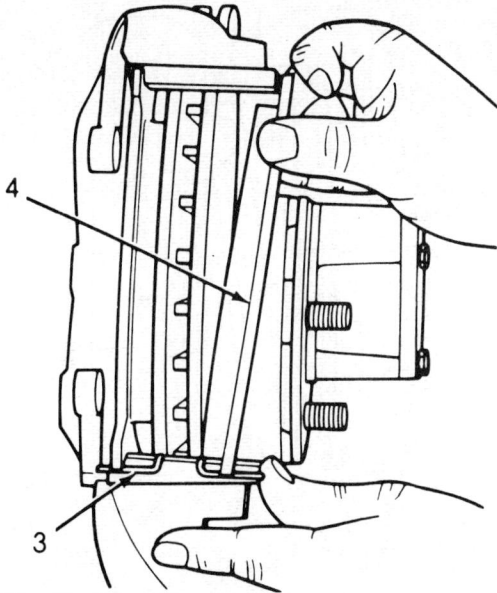

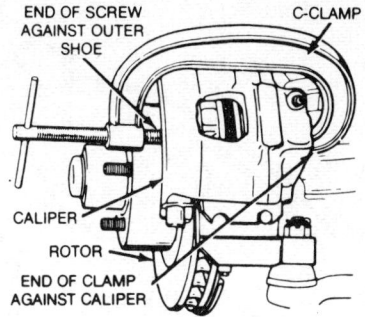

END OF SCREW
AGAINST OUTER
SHOE

C-CLAMP

CALIPER

ROTOR

END OF CLAMP
AGAINST CALIPER

Anti-rattle clip (3) and outboard pad (4) on 1982–86 models

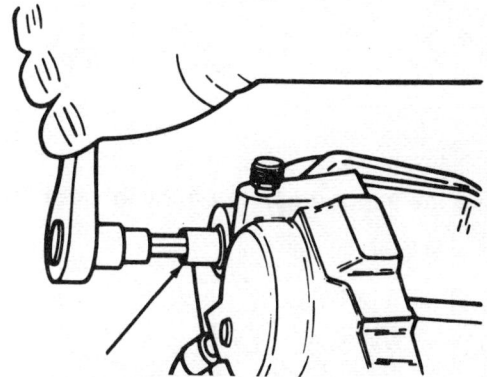

Removing the caliper mounting bolts

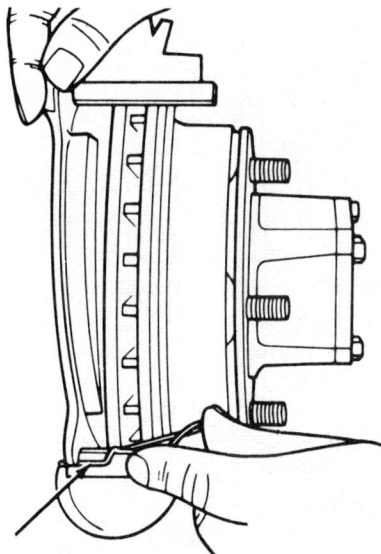

Removing the inboard pad from 1982–86 models

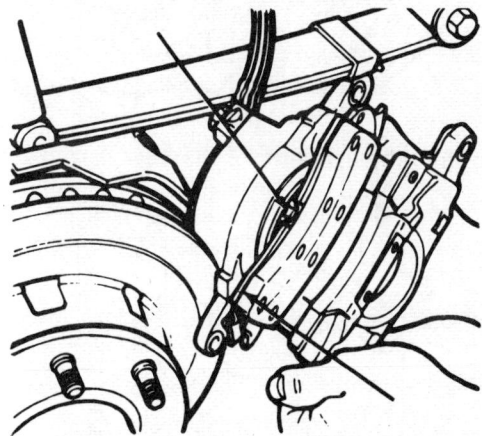

Removing the caliper from the rotor

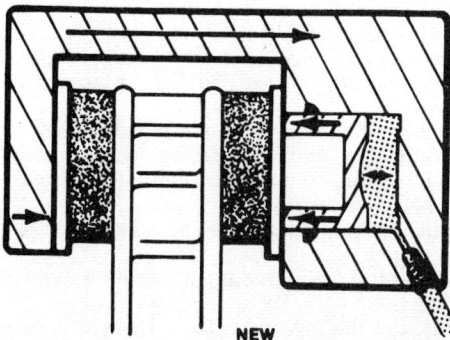

NEW

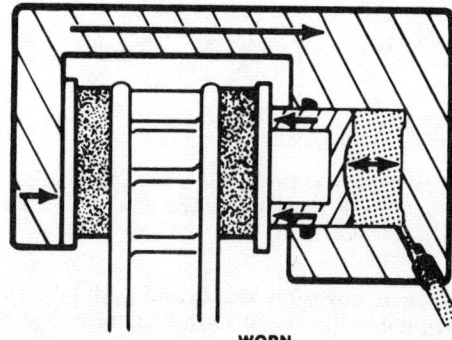

WORN

Piston extension on new and worn linings

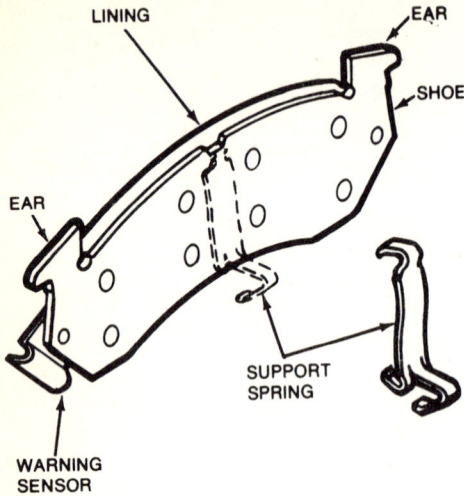

Installing the support spring on the inboard brake pad

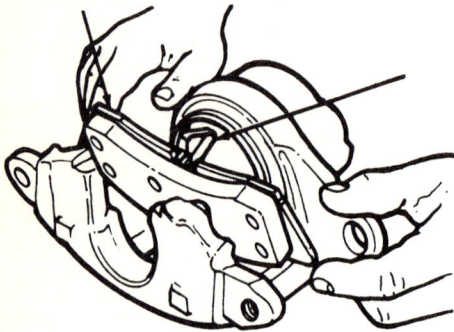

Anti-rattle clip (3) and outboard pad (4)

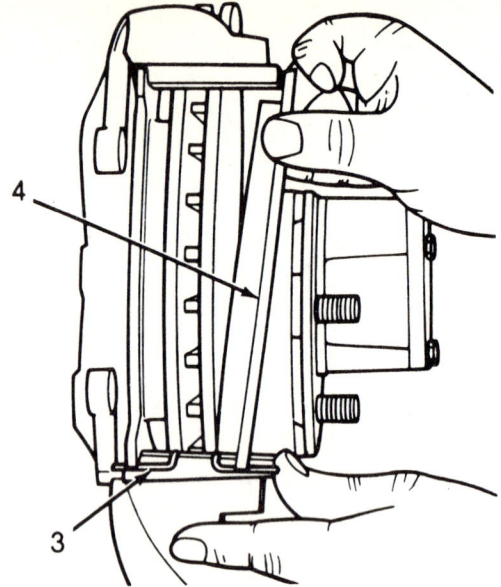

Installing the inboard pad

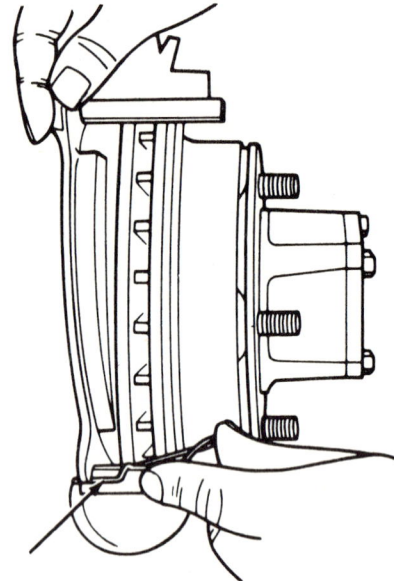

Inboard pad and anti-rattle clip

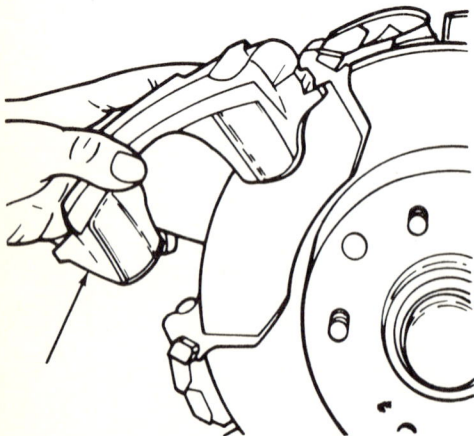

Lifting off the caliper

force the piston to the bottom of the piston bore. This will back the brake pads off the rotor surface to facilitate the removal and installation of the caliper assembly.

6. Remove the C-clamp.

NOTE: Do not push down on the brake pedal or the piston and brake pads will return to their original positions up against the rotor.

7. Remove the caliper mounting pins and lift off the caliper.

Don't disconnect the brake line! Don't allow the brake line to support the weight of the caliper!

8. Push the top anti-rattle spring aside and remove the outboard pad.

9. Remove the inboard pad and the top and bottom anti-rattle springs.

10. Clean the caliper with a rag and solvent made for cleaning brake surfaces. Avoid disturbing the dust boot.

11. If there are any indications of fluid leakage, the caliper must be rebuilt.

12. Clean the anchor plate and caliper mounting surfaces with a wire brush.

13. Coat the mounting surfaces with a light coating of caliper lubricant.

14. Install the top and bottom anti-rattle springs.

15. Install the inboard, then the outboard pads. Check the position of the springs. Make sure that they are seated properly

and that the spring ends are in contact with the pad mounting ears.

WARNING: Avoid damaging the caliper piston dust boot while installing the pads!

16. Install the caliper and tighten the mounting pins to 30 ft. lbs.

17. Fill the master cylinder and press firmly on the brake pedal to seat the pads.

18. Install the wheels, and lower the truck.

Caliper

REMOVAL AND INSTALLATION

1977-1981

──────── **CAUTION** ────────

Brake shoes contain asbestos, which has been determined to be a cancer causing agent. Never clean the brake surfaces with compressed air! Avoid inhaling any dust from any brake surface! When cleaning brake surfaces, use a commercially available brake cleaning fluid.

1. Remove ⅔ of the brake fluid from the front reservoir.

2. Raise the vehicle so that the wheel to be worked on is off the ground. Support the vehicle with jackstands.

3. Remove the front wheels.

4. Place a C-clamp on the caliper so that the solid end contacts the back of the caliper and the screw end contacts the metal part of the outboard brake pad.

5. Tighten the clamp until the caliper moves far enough to

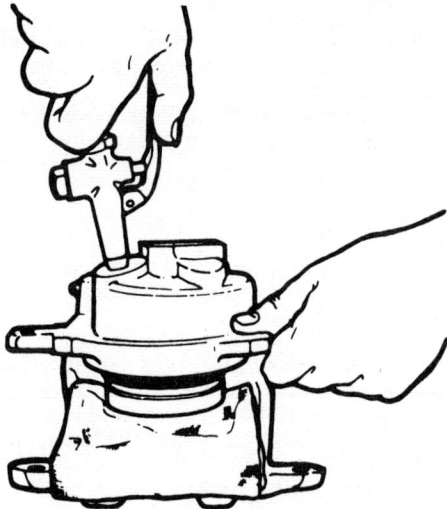

Removing the O-ring

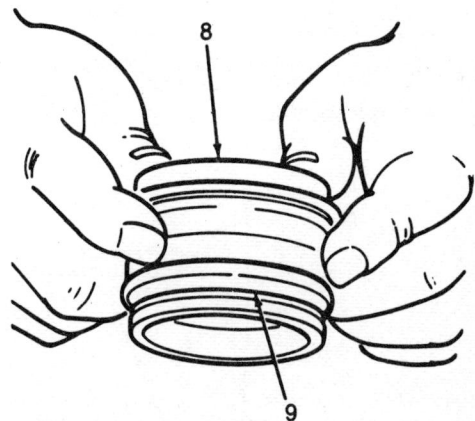

Installing the metal retainer (8) in the groove (9)

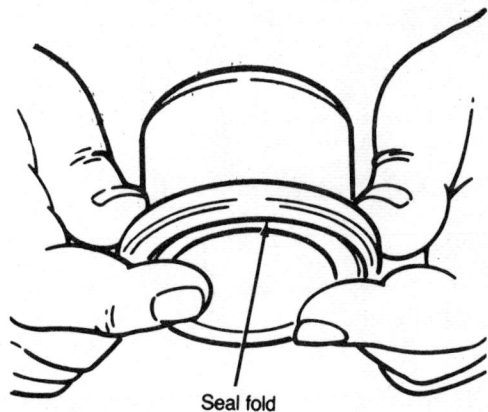

Removing the piston with compressed air

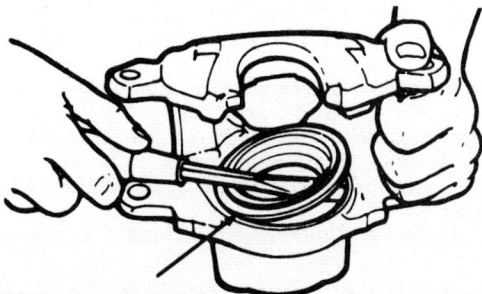
Seal fold

Dust seal installation

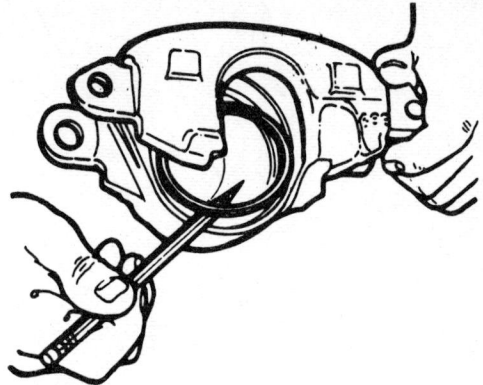

Removing the dust seal

force the piston to the bottom of the piston bore. This will back the brake pads off the rotor surface to facilitate the removal and installation of the caliper assembly.

6. Remove the C-clamp.

NOTE: Do not push down on the brake pedal or the piston and brake pads will return to their original positions up against the rotor.

7. Remove the caliper support key retaining screw with a ¼ in. Allen wrench. Drive the support key and spring out with a punch.

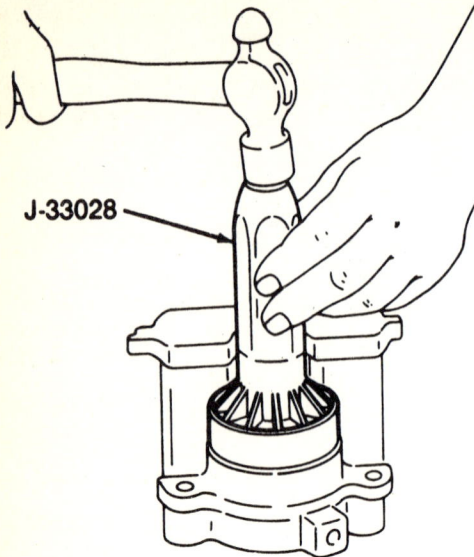

J-33028

Seating the metal retainer

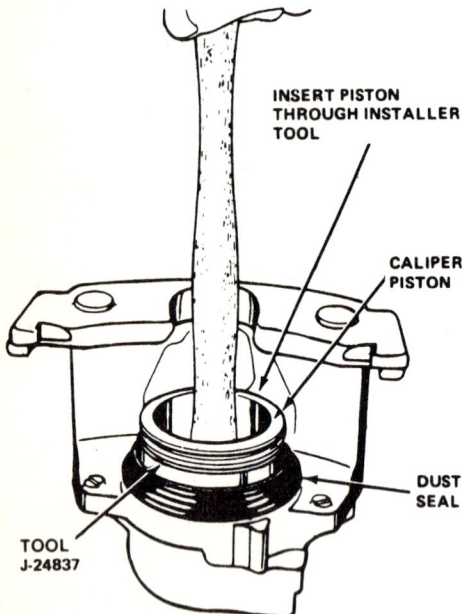

INSERT PISTON
THROUGH INSTALLER
TOOL

CALIPER
PISTON

DUST
SEAL

TOOL
J-24837

Caliper piston installation

NOTE: If just the brake pads are being replaced, it is not necessary to remove the caliper assembly entirely from the vehicle. Do not remove the brake line. Rest the caliper on the front spring or other suitable support. Do not allow the brake hose to support the weight of the caliper.

8. If the caliper is being removed in order to be rebuilt, then it is necessary to disconnect the brake fluid hose. Clean the brake fluid hose-to-caliper connection thoroughly. Remove the hose-to-caliper bolt. Cap or tape the open ends to keep dirt out. Discard the copper gaskets.
9. Wipe the inside of the caliper clean, including the exterior of the dust boot. Inspect the dust boot for cuts or cracks and for proper seating in the piston bore. If evidence of fluid leakage is noted, the caliper should be rebuilt.
10. Check the sliding surface of the caliper and anchor plate

for rust or corrosion. Clean them with a wire brush and fine sandpaper; lubricate with molybdenum disulphide grease.
11. Place the caliper over the rotor and on the anchor plate.

NOTE: Be careful not to damage or dislodge the dust boot.

12. Insert the support key and spring between the sliding surfaces at the rear of the caliper. Drive them into place with a punch.
13. Install the key retaining screw and torque it to 15 ft. lbs.
14. Connect the hose to the caliper, using a new copper gasket under the hose union. Torque the bolt to 25 ft. lbs.
15. Fill the master cylinder to within ¼ in. of the rim.
16. Press firmly on the brake pedal several times till the pedal is firm.
17. Recheck the master cylinder level.

NOTE: If the brake fluid hose was disconnected, it will be necessary to bleed the hydraulic system.

1983–90

— CAUTION —

Brake shoes contain asbestos, which has been determined to be a cancer causing agent. Never clean the brake surfaces with compressed air! Avoid inhaling any dust from any brake surface! When cleaning brake surfaces, use a commercially available brake cleaning fluid.

1. Remove ⅔ of the brake fluid from the front reservoir.
2. Raise the vehicle so that the wheel to be worked on is off the ground. Support the vehicle with jackstands.
3. Remove the front wheels.
4. Place a C-clamp on the caliper so that the solid end contacts the back of the caliper and the screw end contacts the metal part of the outboard brake pad.
5. Tighten the clamp until the caliper moves far enough to force the piston to the bottom of the piston bore. This will back the brake pads off the rotor surface to facilitate the removal and installation of the caliper assembly.
6. Remove the C-clamp.

NOTE: Do not push down on the brake pedal or the piston and brake pads will return to their original positions up against the rotor.

7. Remove the caliper mounting pins and lift off the caliper. Don't disconnect the brake line unless the caliper is being removed for service! Don't allow the brake line to support the weight of the caliper! If you disconnect the brake line, cap it to prevent fluid loss.
8. Clean the caliper with a rag and solvent made for cleaning brake surfaces. Avoid disturbing the dust boot.
9. If there are any indications of fluid leakage, the caliper must be rebuilt.
10. Clean the anchor plate and caliper mounting surfaces with a wire brush.
11. Coat the mounting surfaces with a light coating of caliper lubricant.
12. Install the caliper and tighten the mounting pins to 30 ft. lbs.
13. Connect the hose to the caliper, using a new copper gasket under the hose union. Torque the bolt to 25 ft. lbs.
14. Fill the master cylinder and press firmly on the brake pedal to seat the pads.
15. Install the wheels, and lower the truck.

NOTE: If the brake fluid hose was disconnected, it will be necessary to bleed the hydraulic system.

OVERHAUL

───── **CAUTION** ─────

Brake shoes contain asbestos, which has been determined to be a cancer causing agent. Never clean the brake surfaces with compressed air! Avoid inhaling any dust from any brake surface! When cleaning brake surfaces, use a commercially available brake cleaning fluid.

1. Remove the caliper assembly and remove the brake pads. If the pads are to be reused, mark their location in the caliper.
2. Clean the caliper exterior with clean brake fluid. Drain any residual fluid from the caliper and place it on a clean work surface.

NOTE: Removal of the caliper piston requires the use of compressed air. Do not, under any circumstances, place your fingers in front of the piston in an attempt to catch or protect it when applying compressed air to remove the piston.

3. Pad the interior of the caliper with clean cloths. Use several cloths and pad the interior well to avoid damaging the piston when it comes out of the bore.
4. Insert an air nozzle into the inlet hole in the caliper and gently apply air pressure on the piston to push it out of the bore. Use only enough air pressure to ease the piston out of the bore.
5. Pry the dust boot out of the bore with a screwdriver. Use caution during this operation to prevent scratching the bore. Discard the dust boot.
6. Remove the piston seal from the piston bore and discard the seal. Use only non-scratching implements such as a wooden stick or a piece of plastic to remove the seal. Do not use a metal tool as it could very easily scratch the bore.
7. Remove the bleeder screw. Remove and discard the sleeves and rubber bushings from the mounting ears.
8. Clean all parts with clean brake fluid. Blow out all of the passages in the caliper and bleeder valve. Use only dry and filtered compressed air.
9. Examine the piston for defects. Replace the piston if it is nicked, scratched, corroded. Examine the caliper piston bore for the same defects as the piston. Minor stains or corrosion can be polished with a fiber brush.
10. Lubricate the bore and new seal with brake fluid and install the seal in the groove in the bore.
11. Lubricate the piston with brake fluid and install the new dust boot into the piston groove so that the fold in the boot faces the open end of the piston. Slide the metal portion of the dust boot over the open end of the piston and push the retainer toward the back of the piston until the lip on the fold seats in the piston groove. Then push the retainer portion of the boot forward until the boot is flush with the rim at the open end of the piston and snaps into place.
12. Insert the piston in the bore, being careful not to unseat the piston seal. Push the piston to the bottom of the bore.
13. Install the bleeder screw.
14. Connect the brake line to the caliper using new copper gaskets.
15. Install the brake pads.
16. Install the caliper. Bleed the hydraulic system.

Rotor (Disc)
REMOVAL AND INSTALLATION

1977-1986

───── **CAUTION** ─────

Brake shoes contain asbestos, which has been determined to be a cancer causing agent. Never clean the brake surfaces with compressed air! Avoid inhaling any dust from any brake surface! When cleaning brake surfaces, use a commercially available brake cleaning fluid.

1. Loosen the front wheel lug nuts. Raise and support the front end on jackstands.
2. Remove the wheels.
3. Remove the front hub grease cap and driving hub snapring. On models equipped with locking hubs, remove the retainer knob hub ring, agitator knob, snapring, outer clutch retaining ring and actuating cam body.
4. Remove the splined driving hub and the pressure spring. This may require slight prying with a screwdriver.
5. Remove the external snapring from the spindle shaft and remove the hub shaft drive gear.
6. Remove the wheel bearing locknut, lockring, adjusting nut and inner lockring.
7. Remove the caliper and suspend it out of the way by hanging it from a suspension or frame member with a length of wire. Do not disconnect the brake hose, and be careful to avoid stretching the hose. Remove the rotor and hub assembly. The outer wheel bearing and, on vehicles with locking hubs, the spring collar, will come off with the hub.
8. If the old parts are retained, thoroughly clean them in a safe solvent and allow them to dry on a clean towel. Never spin dry them with compressed air.
9. Cover the spindle with a cloth.
10. Remove the cloth and thoroughly clean the spindle.
11. Thoroughly clean the inside of the hub.
12. Pack the inside of the hub with EP wheel bearing grease. Add grease to the hub until it is flush with the inside diameter of the bearing cup.
13. Pack the bearing with the same grease. A needle-shaped wheel bearing packer is best for this operation. If one is not available, place a large amount of grease in the palm of your hand and slide the edge of the bearing cage through the grease to pick up as much as possible, then work the grease in as best you can with your fingers.
14. If a new race is being installed, very carefully drive it into position until it bottoms all around, using a brass drift. Be careful to avoid scratching the surface.
15. Place the inner bearing in the race and install a new grease seal.
16. Place the hub assembly onto the spindle and install the inner lockring and outer bearing. Install the wheel bearing nut and torque it to 50 ft. lbs. while turning the wheel back and forth to seat the bearings. Back off the nut about ¼ turn (90°) maximum.
17. Install the lockwasher with the tab aligned with the keyway in the spindle and turn the inner wheel bearing adjusting nut until the peg on the nut engages the nearest hole in the lockwasher.
18. Install the outer locknut and torque it to 50 ft. lbs.
19. Install the spring collar, drive flange, snapring, pressure spring, and hub cap.
20. Install the caliper over the rotor.

1987-90

───── **CAUTION** ─────

Brake shoes contain asbestos, which has been determined to be a cancer causing agent. Never clean the brake surfaces with compressed air! Avoid inhaling any dust from any brake surface! When cleaning brake surfaces, use a commercially available brake cleaning fluid.

1. Raise and support the front end on jackstands.
2. Remove the wheels.
3. Remove the caliper as outlined earlier, but don't disconnect the brake line. Suspend the caliper out of the way by wiring it to the front spring.
4. Remove the rotor from the hub.
5. Inspect the rotor, clean the mounting surfaces and install it on the hub.
6. Install the caliper and wheel.

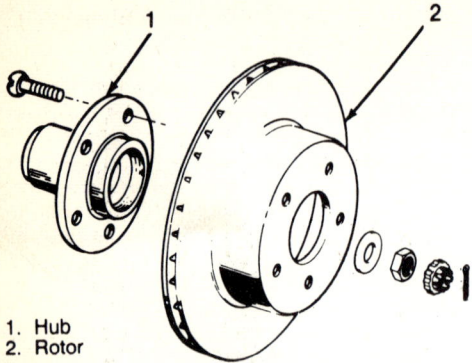

1. Hub
2. Rotor

1987–90 hub and rotor (disc)

INSPECTION AND MEASUREMENT

Check the rotor for surface cracks, nicks, broken cooling fins and scoring of both contact surfaces. Some scoring of the surfaces may occur during normal use. Scoring that is 0.009 in. (0.23mm) deep or less is not detrimental to the operation of the brakes.

If the rotor surface is heavily rusted or scaled, clean both sur-

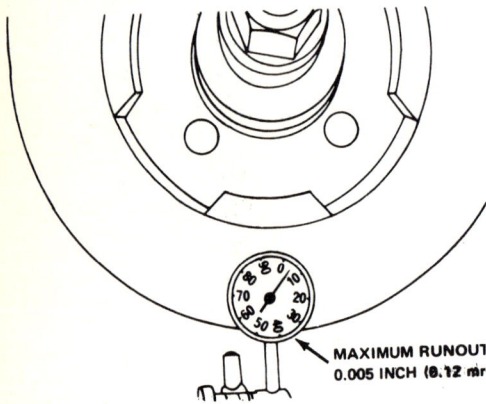

MAXIMUM RUNOUT
0.005 INCH (0.12 mm)

Checking rotor lateral runout

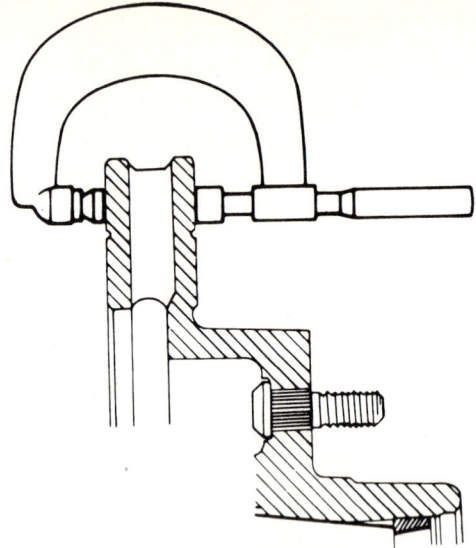

Checking rotor thickness variation

faces on a disc brake lathe using flat sanding discs before attempting any measurements.

With the hub and rotor assembly mounted on the spindle of the vehicle or a disc brake lathe and all play removed from the wheel bearings, assemble a dial indicator so that the stem contacts the center of the rotor braking surface. Zero the dial indicator before taking any measurements. Lateral runout must not exceed 0.005 in. (0.13mm) with a maximum rate of change not to exceed 0.001 in. (0.025mm) in 30 degrees of rotation. Excessive runout will cause the rotor to wobble and knock the piston back into the caliper causing increased pedal travel, noise and vibration.

After the rotor has been refinished, the minimum thickness of 1.1207 in. (28.5mm) is acceptable for models through 1978; 0.815 in. (20.7mm) for 1979–90 models. Discard the rotor if the thickness is less.

NOTE: Remember to adjust the wheel bearings after the runout measurement has been taken.

FRONT DRUM BRAKES

Front drum brakes were the only type available on Jeep through 1976 and standard on 1977 models.

Brake Drum
REMOVAL AND INSTALLATION
———————— CAUTION ————————

Brake shoes contain asbestos, which has been determined to be a cancer causing agent. Never clean the brake surfaces with compressed air! Avoid inhaling any dust from any brake surface! When cleaning brake surfaces, use a commercially available brake cleaning fluid.

The front brake drums are attached to the wheel hubs by five bolts. These bolts are also used for mounting the wheels on the hub. Press or drive out the bolts to remove the drum from the hub.

When placing the drum on the hub, make sure that the contacting surfaces are clean and flat. Line up the holes in the drum with those in the hub and put the drum over the shoulder on the hub. Insert five new bolts through the drum and hub and drive the bolts into place solidly. Place a round piece of stock approximately the diameter of the head of the bolt, in a vise. Next place the hub and drum assembly over it so that the bolt head rests on it. Then flatten the bolt head into the countersunk section of the hub with a punch.

The runout of the drum face should be within 0.030 in. (0.8mm). If the runout is found to be greater than 0.030 in. (0.8mm), it will be necessary to reset the bolts to correct the condition.

The left hand hub bolts have an **L** stamped on the head of the bolt. The left hand threaded nuts may have a groove cut around the hexagon faces, or the word **LEFT** stamped on the face.

Hubs with left hand threaded hub bolts are installed on the left hand side of the vehicle. Late production vehicles are equipped with right hand bolts and nuts on all four bolts.

INSPECTION

Using a brake drum micrometer, check all drums. Should a brake drum be scored or rough, it may be reconditioned by grinding or turning on a lathe. Do not remove more than 0.030 in. (0.8mm) thickness of metal.

Use a clean cloth to clean dirt from the brake drums. If further cleaning is required, use soap and water. Do not use brake fluid, gasoline, kerosene or any other similar solvents.

Brake Shoes

REMOVAL AND INSTALLATION
———————— CAUTION ————————

Brake shoes contain asbestos, which has been determined to be a cancer causing agent. Never clean the brake surfaces with compressed air! Avoid inhaling any dust from any brake surface! When cleaning brake surfaces, use a commercially available brake cleaning fluid.

1. Raise and support the front end on jackstands.
2. Back off the adjusters all the way.
3. Remove the drums and hubs.
4. Install wheel cylinder clamps to retain the wheel cylinder pistons in place and prevent leakage of brake fluid while replacing the shoes.
5. Remove the return springs with a brake spring removed tool.
6. Remove the adjuster cable, cable guide, adjuster lever and adjuster springs.
7. Remove the holddown clips or springs and remove the brake shoes.
8. Check the wheel cylinders for leakage. Pull back the dust covers. If there is fluid present behind the dust cover the wheel cylinder must be rebuilt or replaced.
9. Clean the backing plate with a brush or cloth. Place a dab of molybdenum disulphide grease on each spot where the shoes rub the backing plate.

NOTE: Always replace brake lining in axle sets. Never replace linings on one side.

10. Install the guide and self-adjuster cable on the anchor pin.
11. Install the shoes, one at a time, on the backing plate and secure them with the holddown springs, pins and retainers.
12. Install the cross-strut and spring.
13. Install the star wheel adjuster, spring and lever.
14. Install the return springs.
15. Make sure that the brake shoe surfaces are clean and free from any contamination.
16. Perform the initial brake shoe adjustment, outlined at the beginning of this section.

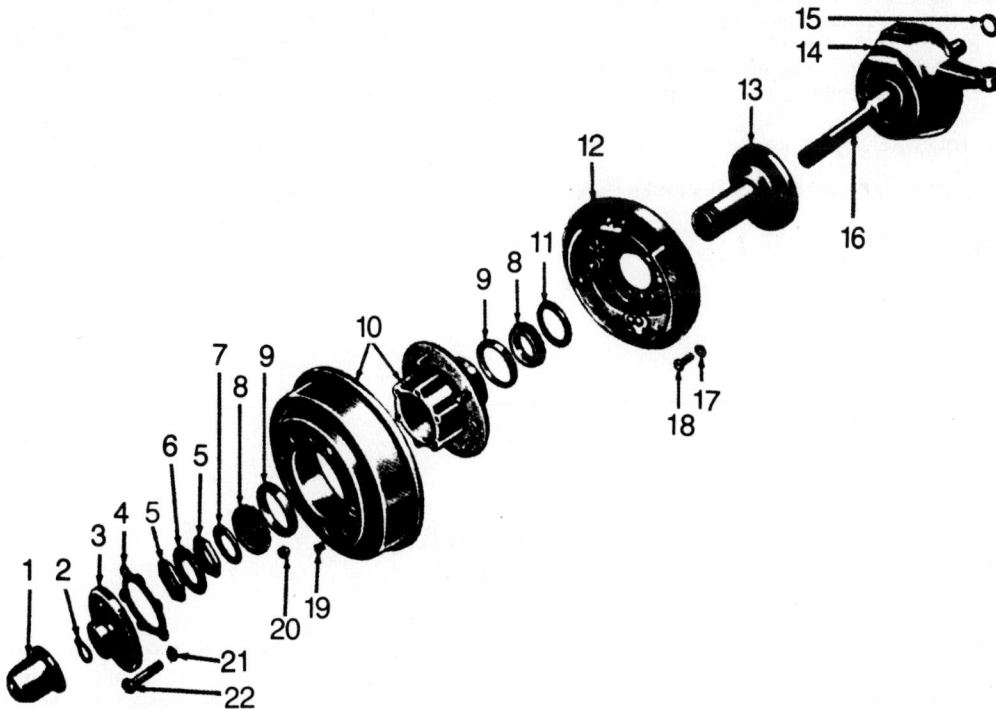

1. Hub cap	9. Cup	17. Lockwasher
2. Snap ring	10. Hub and drum	18. Bolt
3. Drive flange	11. Oil seal	19. Screw
4. Gasket	12. Left front brake	20. Nut
5. Nut	13. Spindle and bushing	21. Lockwasher
6. Lockwasher	14. Left knuckle and arm	22. Bolt
7. Lockwasher	15. Thrust washer	
8. Cone and rollers	16. Universal joint shaft	

1971 front hub, drum and brake assembly

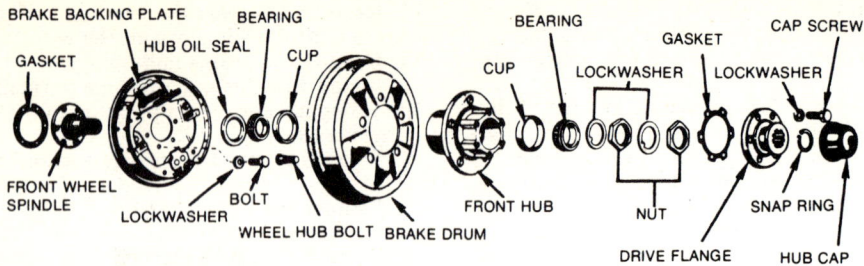

1972 and later front hub and drum brake assembly

17. Install the hubs, drums and wheels, lower the truck and check brake operation. Drive the Jeep in a series of alternating forward and reverse stops. Usually 10–15 full stops in reverse are sufficient.

Wheel Cylinders

CAUTION

Brake shoes contain asbestos, which has been determined to be a cancer causing agent. Never clean the brake surfaces with compressed air! Avoid inhaling any dust from any brake surface! When cleaning brake surfaces, use a commercially available brake cleaning fluid.

OVERHAUL

Wheel cylinder rebuilding kits are available for reconditioning wheel cylinders. The kits usually contain new cup springs, cylinder cups and in some, new boots. The most important factor to keep in mind when rebuilding wheel cylinders is cleanliness. Keep all dirt away from the wheel cylinders when you are reassembling them.

1. To remove the wheel cylinder, jack up the vehicle and remove the drum.
2. Disconnect the brake line at the fitting on the brake backing plate.
3. Remove the brake assemblies.

4. Remove the screws or nuts that hold the wheel cylinder to the backing plate and remove the wheel cylinder from the vehicle.
5. Remove the rubber dust covers on the ends of the cylinder. Remove the pistons and piston cups and the spring. Remove the bleeder screw and make sure it is not plugged.
6. Discard all of the parts that the rebuilding kit will replace.
7. Examine the inside of the cylinder. If it is severely rusted, pitted or scratched, then the cylinder must be replaced as the piston cups won't be able to seal against the walls of the cylinder.
8. Using emery cloth or crocus cloth, polish the inside of the cylinder. Do not polish in a lengthwise direction. Polish by rotating the wheel cylinder around the polishing cloth supported on your fingers. The purpose of this is to put a new surface on the inside of the cylinder. Keep the inside of the cylinder coated with brake fluid while polishing.

NOTE: Honing the wheel cylinders is not recommended due to the possibility of removing too much material from the bore, making it too large to seal.

9. Wash out the cylinder with clean brake fluid after polishing.
10. When reassembling the cylinder dip all of the parts in clean brake fluid. Reassemble in the reverse order of removal. Torque the wheel cylinder-to-backing plate fasteners to 18 ft. lbs. Torque the brake line-to-wheel cylinder connection to 160 inch lbs.

REAR DRUM BRAKES

Brake Drum

REMOVAL AND INSTALLATION

CAUTION

Most brake shoes contain asbestos, which has been determined to be a cancer causing agent. Never clean the brake surfaces with compressed air! Avoid inhaling any dust from any brake surface! When cleaning brake surfaces, use a commercially available brake cleaning fluid.

The brake drums are held in position by spring clip type locknuts or by three drum-to-hub retaining screws, depending on

the model and year. After the spring type locknuts or retaining screws are removed, the drum can be slid off the axle shaft or hub and brake shoes. It may be necessary to back off the brake shoe adjustment so that any lip on the inside of the brake drum clears the brake shoes.

INSPECTION

Using a brake drum micrometer, check all drums. Should a brake drum be scored or rough, it may be reconditioned by grinding or turning on a lathe. Do not remove more than 0.030 in. (0.8mm) thickness of metal.

Use a clean cloth to clean dirt from the brake drums. If

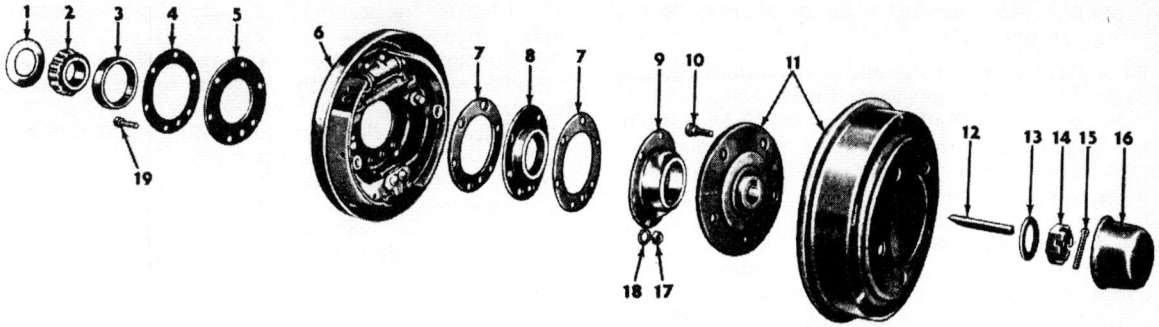

1. Oil seal
2. Cone and rollers
3. Cup
4. Shims
5. Bearing retainer

6. Brake
7. Gasket
8. Grease retainer
9. Grease protecter
10. Bolt

11. Hub and drum
12. Shaft key
13. Oil seal
14. Nut
15. Cotter pin

16. Hub cap
17. Nut
18. Lockwasher
19. Bolt

1971 rear brake and hub

further cleaning is required, use soap and water. Do not use brake fluid, gasoline, kerosene or any other similar solvents.

Brake Shoes

REMOVAL AND INSTALLATION

—————— CAUTION ——————

Brake shoes contain asbestos, which has been determined to be a cancer causing agent. Never clean the brake surfaces with compressed air! Avoid inhaling any dust from any brake surface! When cleaning brake surfaces, use a commercially available brake cleaning fluid.

1971-86

NOTE: An inexpensive brake spring removal tool, available at most good auto parts stores, will make this procedure much easier.

1. Raise and support the rear end on jackstands.
2. Back off the adjusters all the way.
3. Remove the drums.
4. Install wheel cylinder clamps to retain the wheel cylinder pistons in place and prevent leakage of brake fluid while replacing the shoes.
5. Remove the return springs with a brake spring removed tool.
6. Remove the adjuster cable, cable guide, adjuster lever and adjuster springs.
7. Remove the holddown clips or springs and remove the brake shoes.
8. Check the wheel cylinders for leakage. Pull back the dust covers. If there is fluid present behind the dust cover the wheel cylinder must be rebuilt or replaced.
9. Clean the backing plate with a brush or cloth. Place a dab of molybdenum disulphide grease on each spot where the shoes rub the backing plate.

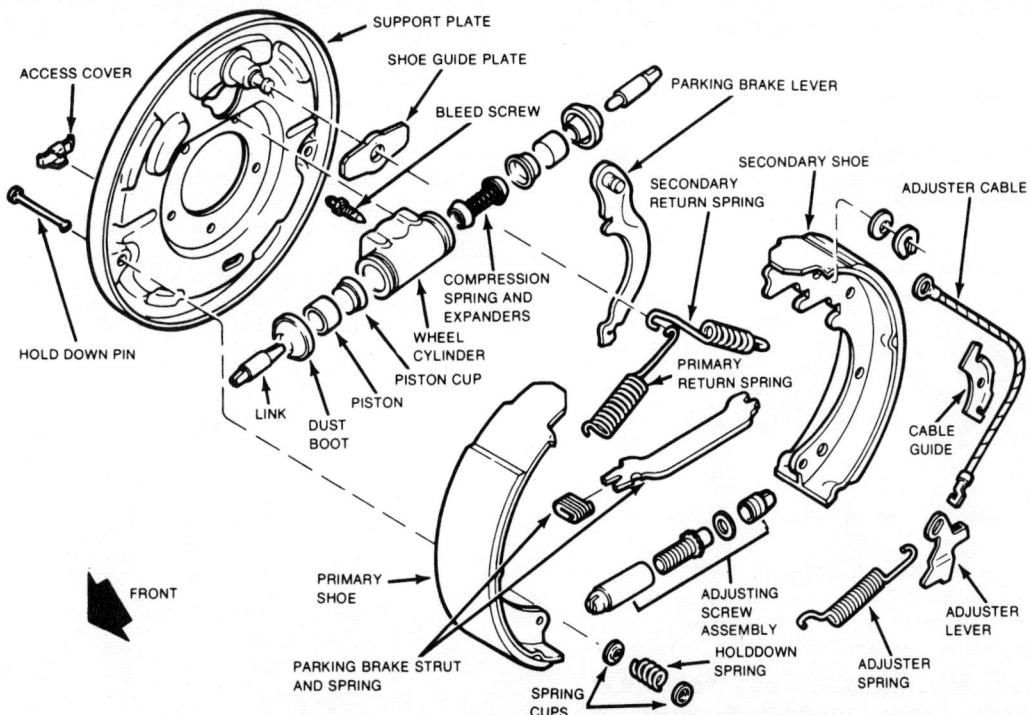

1972 and later self-adjusting rear drum brake components

NOTE: **Always replace brake lining in axle sets. Never replace linings on one side.**

10. Install the parking brake cable and lever on the secondary shoe. Secure the lever with a washer and new U-clip.

11. Install the guide and self-adjuster cable on the anchor pin.

12. Install the shoes, one at a time, on the backing plate and secure them with the holddown springs, pins and retainers.

13. Install the cross-strut and spring.

14. Install the star wheel adjuster, spring and lever.

15. Install the return springs.

16. Make sure that the brake shoe surfaces are clean and free from any contamination.

17. Perform the initial brake shoe adjustment, outlined at the beginning of this section.

18. Install the drums and wheels, lower the truck and check brake operation. Drive the Jeep in a series of alternating forward and reverse stops. Usually 10–15 full stops in reverse are sufficient.

1987-90

NOTE: **An inexpensive brake spring removal tool, available at most good auto parts stores, will make this procedure much easier.**

1. Jack the vehicle up and support it so that the wheels to be worked on are off the ground.

2. Turn the adjustment starwheel so that the brake shoes are retracted from the brake drum.

3. Remove the wheels and the drums to give access to the brake shoes.

4. Install wheel cylinder clamps to retain the wheel cylinder pistons in place and prevent leakage of brake fluid while replacing the shoes.

5. Remove the return springs with a brake spring remover tool.

6. Remove the adjuster cable, cable guide, adjuster lever and adjuster springs.

7. Remove the holddown washers and springs and remove the brake shoes.

8. Clean the backing plate with a brush or cloth. Place a dab of Lubriplate® on each spot where the brake shoes rub on the backing plate.

NOTE: **Always replace brake linings in axle sets. Never replace linings on one side or just on one wheel.**

9. Thoroughly clean the backing plate.

10. Apply a thin coat of multi-purpose chassis lube to the mounting pads on the backing plate.

11. Transfer the parking brake actuating lever to the new secondary shoe.

12. Position the brake shoes on the backing plate and install the holddown springs. Don't forget to engage the parking brake lever with the cable.

13. Install the parking brake actuating bar and spring between the parking brake lever and primary shoe.

14. Install the self-adjusting cable, cable guide and upper return springs.

15. Thoroughly clean the starwheel and lightly lubricate the threads with lithium based grease.

16. Install the starwheel.

17. Install the self-adjusting cam and lower spring. A big pair of locking pliers is good for this job.

18. Check the surface of the brake shoes for any grease that may have gotten on them.

19. Install the drum and reach through the adjusting opening in the back plate with a brake adjusting tool. Turn the starwheel outward so that the brakes lock the drum, then, holding the adjusting cam with a thin screwdriver, turn the starwheel back so that the drum is free and no drag is felt.

20. Once the wheels are on and the truck is down, Back it up several times, applying the brakes to actuate the self-adjusters.

Wheel Cylinders

—— CAUTION ——

Brake shoes contain asbestos, which has been determined to be a cancer causing agent. Never clean the brake surfaces with compressed air! Avoid inhaling any dust from any brake surface! When cleaning brake surfaces, use a commercially available brake cleaning fluid.

OVERHAUL

Wheel cylinder rebuilding kits are available for reconditioning wheel cylinders. The kits usually contain new cup springs, cylinder cups and in some, new boots. The most important factor to keep in mind when rebuilding wheel cylinders is cleanliness. Keep all dirt away from the wheel cylinders when you are reassembling them.

1. To remove the wheel cylinder, jack up the vehicle and remove the drum.

2. Disconnect the brake line at the fitting on the brake backing plate.

3. Remove the brake assemblies.

4. Remove the screws or nuts that hold the wheel cylinder to the backing plate and remove the wheel cylinder from the vehicle.

5. Remove the rubber dust covers on the ends of the cylinder. Remove the pistons and piston cups and the spring. Remove the bleeder screw and make sure it is not plugged.

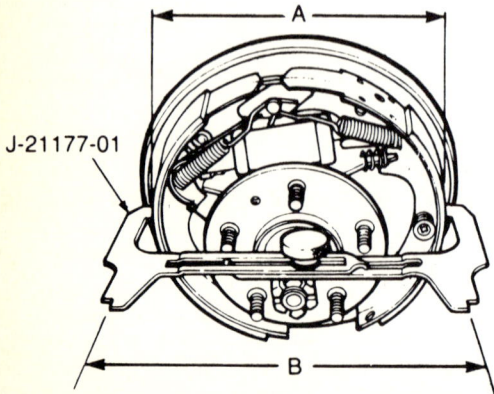

Measure the assembled shoes at A and the drum at B to obtain the initial brake shoe-to-drum clearance. The tool shown is preset for proper adjustment

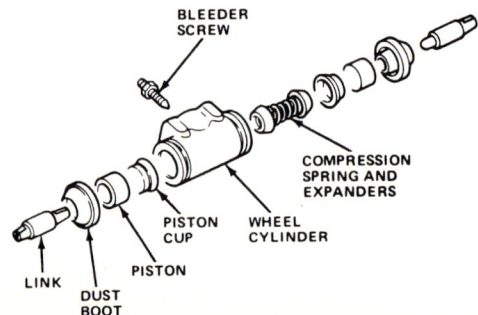

Wheel cylinder used on all Jeep vehicles

6. Discard all of the parts that the rebuilding kit will replace.

7. Examine the inside of the cylinder. If it is severely rusted, pitted or scratched, then the cylinder must be replaced as the piston cups won't be able to seal against the walls of the cylinder.

8. Using emery cloth or crocus cloth, polish the inside of the cylinder. Do not polish in a lengthwise direction. Polish by rotating the wheel cylinder around the polishing cloth supported on your fingers. The purpose of this is to put a new surface on the inside of the cylinder. Keep the inside of the cylinder coated with brake fluid while polishing.

NOTE: Honing the wheel cylinders is not recommended due to the possibility of removing too much material from the bore, making it too large to seal.

9. Wash out the cylinder with clean brake fluid after polishing.

10. When reassembling the cylinder dip all of the parts in clean brake fluid. Reassemble in the reverse order of removal. Torque the wheel cylinder-to-backing plate fasteners to 18 ft. lbs. through 1983; 15 ft. lbs. for 1984–86 models; 90 inch lbs. for 1987-90 models. Torque the brake line-to-wheel cylinder connection to 160 inch lbs.

PARKING BRAKE

Adjustment

1971 Transmission Brake

1. Make sure that the brake handle on the instrument panel is fully released.

2. Check the operating linkage and the cable to make sure that they don't bind. If necessary, free the cable and lubricate it.

3. Rotate the brake drum until one paid of the three sets of holes are over the shoe adjusting screw wheels in the brake drum.

4. Use the edge of the holes in the brake drum as a fulcrum for the brake adjusting tool or a screwdriver. Rotate each notched adjusting screw by moving the handle of the tool away from the center of the driveshaft until the shoes are snug against the drum.

5. Back off seven notches on the adjusting screw wheels to secure the proper running clearance between the shoes and the drum.

1972–86

1. Make sure that the hydraulic brakes are in satisfactory adjustment.

2. Raise the rear wheels off the ground and disengage the parking brake.

3. Loosen the locknut on the brake cable adjusting rod, located directly behind the frame center crossmember.

4. Spin the wheels and tighten the adjustment until the rear wheels drag slightly. Loosen the adjustment until there is no drag and the wheels spin freely.

5. Tighten the locknut to lock the adjusting nut.

1987-90

NOTE: This procedure requires the use of a special tool. If the special tool cannot be obtained, follow the procedure given for 1972-86 models.

1. Make sure that the hydraulic brakes are properly adjusted. Fully apply and release the parking brake five times.

2. Raise and support the truck on jackstands. Loosen the cable adjusting nuts.

3. Using an inch pound torque wrench and adjustment adapter J-34651, apply a torque of 45–50 inch lbs.

4. Adjust the equalizer adjusting nut so that the gauge point-

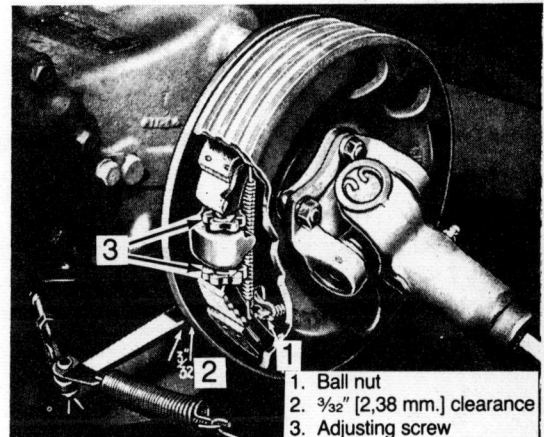

1. Ball nut
2. ³⁄₃₂″ [2.38 mm.] clearance
3. Adjusting screw

Transmission brake adjustment

er is in the blue band on the tool. Tighten the adjusting nuts.

5. Apply and release the brake lever fully, five times, and recheck the adjustment.

6. When adjustment is correct, stake the adjusting nuts.

REMOVAL AND INSTALLATION

Rear Wheel Type Hand Brake

FRONT CABLE

1. Disconnect the cable at the equalizer Disconnect the return spring, if equipped.

2. Disconnect the other end of the cable at the equalizer adjusting rod.

3. Disconnect the cable at any retaining clip on the body. On models through 1986, roll the carpet back and remove the front cable ferrule-to-lever retaining clip. On 1987-90 models, compress the lock tabs that retain the cable in the parking brake pedal. A small hose clamp is good for this purpose.

4. Remove the cable.

5. Installation is the reverse of removal. On 1987-90 models, retain the grommet from the pedal end, if the cable is being replaced.

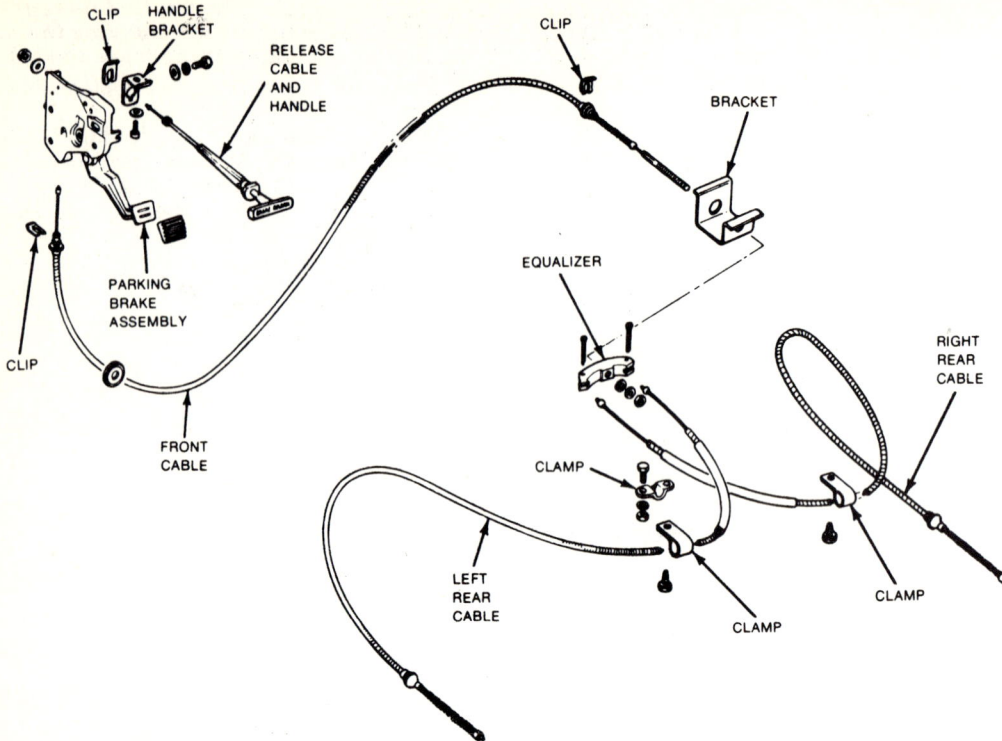

1972 and later parking brake system components

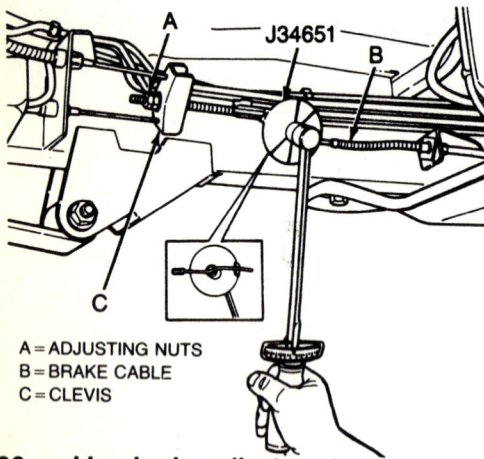

A = ADJUSTING NUTS
B = BRAKE CABLE
C = CLEVIS

1987–90 parking brake adjustment

REAR CABLES

1. Back off the adjuster at the equalizer and disconnect the cable.
2. Raise and support the rear end on jackstands. Detach the cables from any body clips.
3. Remove the wheel and brake drums.
4. Remove the brake shoes and disconnect the cable ends at the actuating levers.
5. Compress the locking tabs that secure the cables to the backing plates and slide the cable from the plates.
6. Installation is the reverse of removal. Adjust the parking brake.

Transmission Type Hand Brake

1. Fully release the hand brake.
2. Disconnect the cable at the hand brake lever.
3. Disconnect the hook from the spring at the actuating lever.
4. Installation is the reverse of removal.

TRANSMISSION BRAKE

Brake Shoe

REPLACEMENT

1971 Only

─────── **CAUTION** ───────

Brake shoes contain asbestos, which has been determined to be a cancer causing agent. Never clean the brake surfaces with compressed air! Avoid inhaling any dust from any brake surface! When cleaning brake surfaces, use a commercially available brake cleaning fluid.

1. Remove the driveshaft.
2. Remove the retracting spring clevis pin and the spring clip.
3. Remove the hub locknut, the nut and washer from the transfer case output shaft.
4. Using a puller, remove the brake drum and companion flange.
5. Remove the retracting springs and shoes from the backing plate.
6. Clean all parts with a safe solvent.
7. Installation is the reverse of removal. Replace any weak

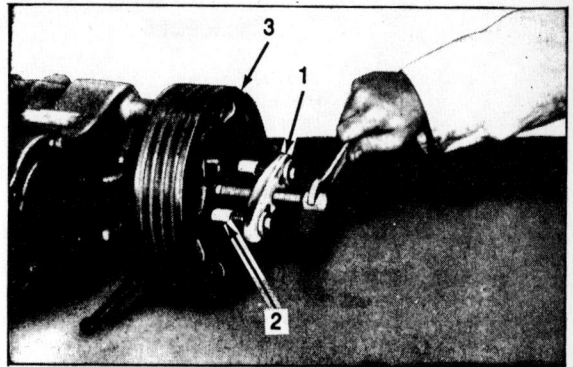

1. Tool W-172 2. Adapter 3. Brake drum

Removing transmission brake drum

springs and clean the adjuster threads. Coat the threads with a LIGHT film of oil. It will be necessary to back off the adjusting screw wheels to allow drum installation.

8. Adjust the brake.

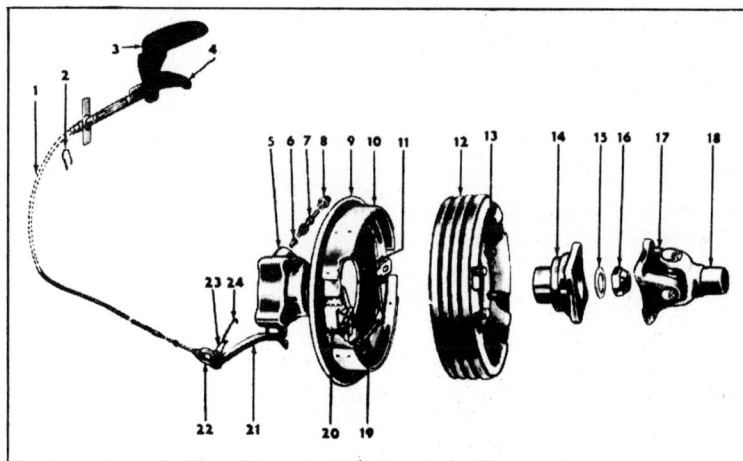

1. Cable and conduit
2. Tube fastener
3. Spring
4. Bracket
5. Rear cap
6. Bushing
7. Driven gear
8. Sleeve
9. Backing plate
10. Shoe and lining
11. Return spring
12. Drum
13. Hex bolt
14. Rear flange
15. Plain washer
16. Nut
17. Yoke
18. Yoke and plug
19. Adjusting spring
20. Bracket
21. Operating lever
22. Adjusting clevis
23. Spring clip
24. spring link

Driveshaft mounted hand brake

SPECIAL TOOLS

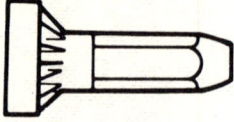

J-23709
METERING VALVE TOOL
(TYPE-D VALVE)

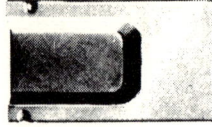

J-26869
METERING VALVE TOOL
(TYPE-W VALVE)

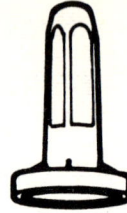

J-22904

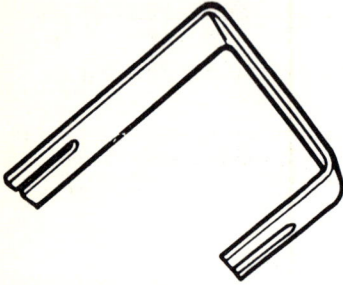

J-33028

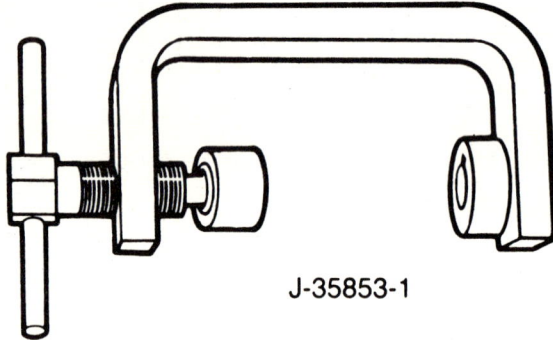

J-35853-2

J-35853-1

10 *Body*

QUICK REFERENCE INDEX

GENERAL INDEX

EXTERIOR

Windshield Frame

REMOVAL AND INSTALLATION

1971

1. On early models, disconnect the wiper vacuum hose from the vacuum motor.
2. On later models, disconnect the wiring from the electric wiper motor.
3. Unlatch the clamps on each side of the windshield.
4. Tilt the windshield forward until the slot in each hinge aligns with the flat side of the pin in the body hinges.
5. Pull the windshield frame of the pins and remove it from the body.

1972-86

1. Disengage the top from the frame.
2. Disconnect the wiper motor wiring harness from the switch.
3. Remove the hinge-to-frame bolts.
4. Remove the holddown knobs and lift off the frame.
5. Installation is the reverse of removal.

1987-90

1. Disengage the top from the frame.
2. Remove the hinge-to-frame bolts.
3. Remove the interior retainer.
4. Lift off the frame.
5. Installation is the reverse of removal.

Windshield Glass

REMOVAL AND INSTALLATION

1971-86

1. Cover all adjoining painted surfaces.
2. Remove the wiper arms.
3. Remove the rear view mirror from its bracket.
4. Remove the sun visors.
5. Disconnect the defroster ducts.
6. Have someone support the glass from the outside. Starting at the top, pull the weatherstripping away from the flange while gently pushing out on the glass. It may be necessary to break the sealer by prying with a wood spatula.
7. Work the entire weatherstripping from the flange to remove the glass.
8. Thoroughly clean and inspect the weatherstripping. Replace it if it is worn, cracked or brittle.
9. Apply a $\frac{1}{16}$ in. (1.6mm) bead of 3M Auto Bedding and Glazing Compound, or equivalent, completely around the weatherstripping in the flange cavity.
10. Install the weatherstripping on the glass. The split should be at the bottom center.
11. Position the glass and weatherstripping in the frame, and, beginning at the bottom, work the weatherstripping over the flange with a wood spatula or fiber stick.
12. When the glass is installed, apply a bead of 3M Windshield Sealer, or equivalent, between the glass and the weatherstripping on the outside. Clean off the excess.
13. Install the mirror.
14. Install the wiper arms.
15. Install the defroster ducts.
16. Install the sun visors.

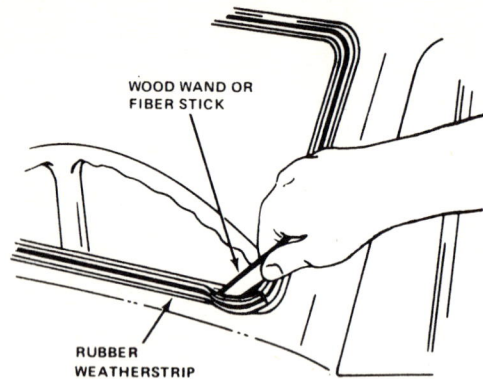

Unlocking the weatherstripping with a wood spatula

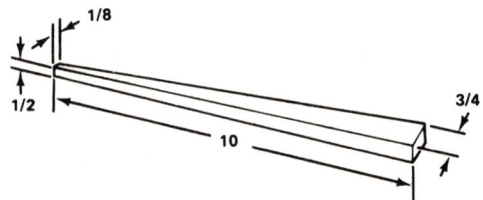

Wood spatula dimensions

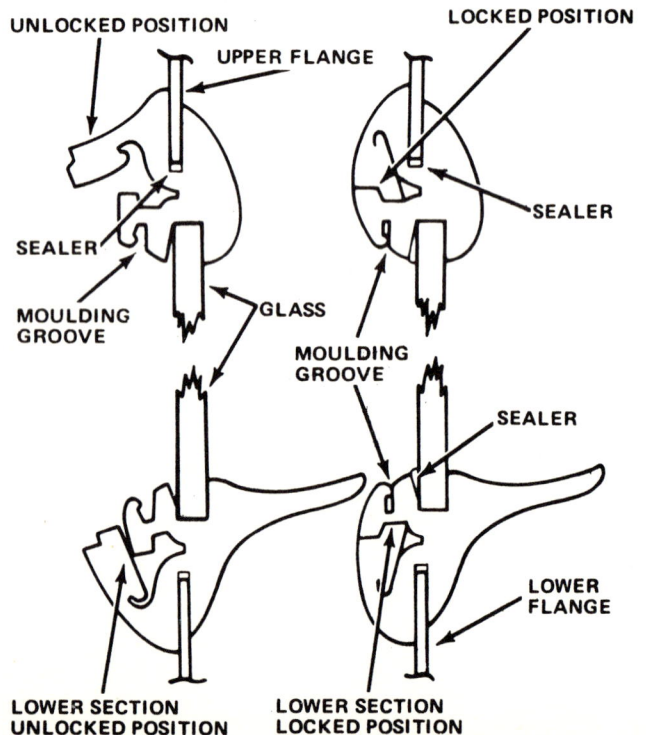

Cross-section of the windshield weatherstripping with the molding removed on 1971–86 models

1987-90

NOTE: These vehicles employ a bonded windshield system. If the windshield to be replaced has not suffered major damage, either to the glass or frame, and can be removed intact, then a do-it-yourselfer can perform the job. Even so, a special tool, called an electric hot knife must be used. If, however, the glass has been extensively damaged, or the frame and molding damaged so that a loss of adhesion has occurred, then the job should be left to a professional. The following procedure is a do-it-yourselfer procedure, assuming a simple replacement of an intact glass and good seal-to-glass adhesion.

1. Remove the wiper arms.
2. Remove the sun visors.
3. Remove the molding cap and molding.
4. Cover all adjacent painted surfaces.
5. Using an electric hot knife under the edge of the glass, cut the adhesive material as close to the inside surface of the glass as possible.

NOTE: Make sure that an even, uniform bead of sealer remains on the body for installation purposes. NEVER ALLOW THE HOT KNIFE TO STOP DURING THE CUTTING PROCESS! Allowing the knife to stop will cause permanent damage to the adhesive material.

6. Clean old sealant from the windshield opening with Kent Acrysol®, or equivalent.
7. Clean the glass with a glass cleaner that DOES NOT contain silicone.
8. Apply a smooth, continuous bead of approved urethane windshield bonding adhesive material around the entire perimeter of the windshield opening flange.
9. Carefully position the glass in the windshield opening. and press down to set the adhesive. Don't press too hard or too much adhesive will be lost.
10. Remove any sheet metal masking that you applied.
11. Install the windshield molding and cap.
12. Install the sun visors.
13. Install the wiper arms.

Hardtop

REMOVAL AND INSTALLATION

1. Remove the hardtop-to-windshield frame screws.
2. Remove the hardtop-to-rear quarter panel bolts.
3. Disconnect the dome lamp.
4. Lift the top from the Jeep, being careful to avoid damage to the foam seals.
5. Installation is the reverse of removal.

Grille

REMOVAL AND INSTALLATION

1. Remove the front crossmember cover, if so equipped.
2. Remove the screws securing the radiator and shroud to the radiator grille guard panel.

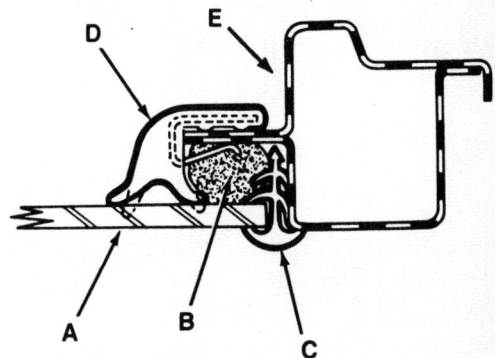

A. Glass
B. Urethane adhesive
C. Windshield reveal molding
D. Garnish molding
E. Windshield frame

Cross-section of an assembled 1987–90 windshield

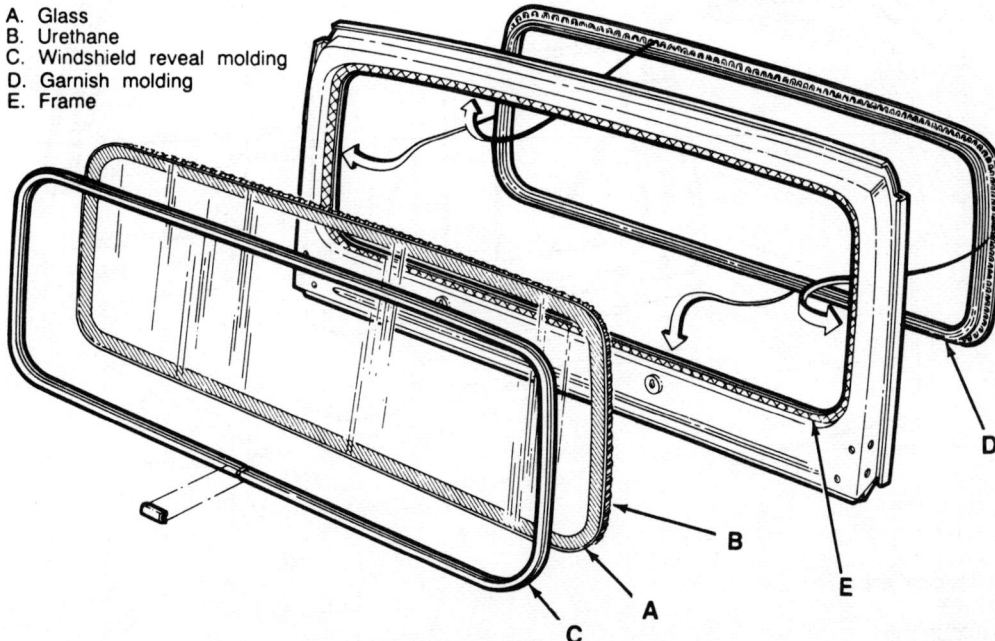

A. Glass
B. Urethane
C. Windshield reveal molding
D. Garnish molding
E. Frame

1987–90 windshield components

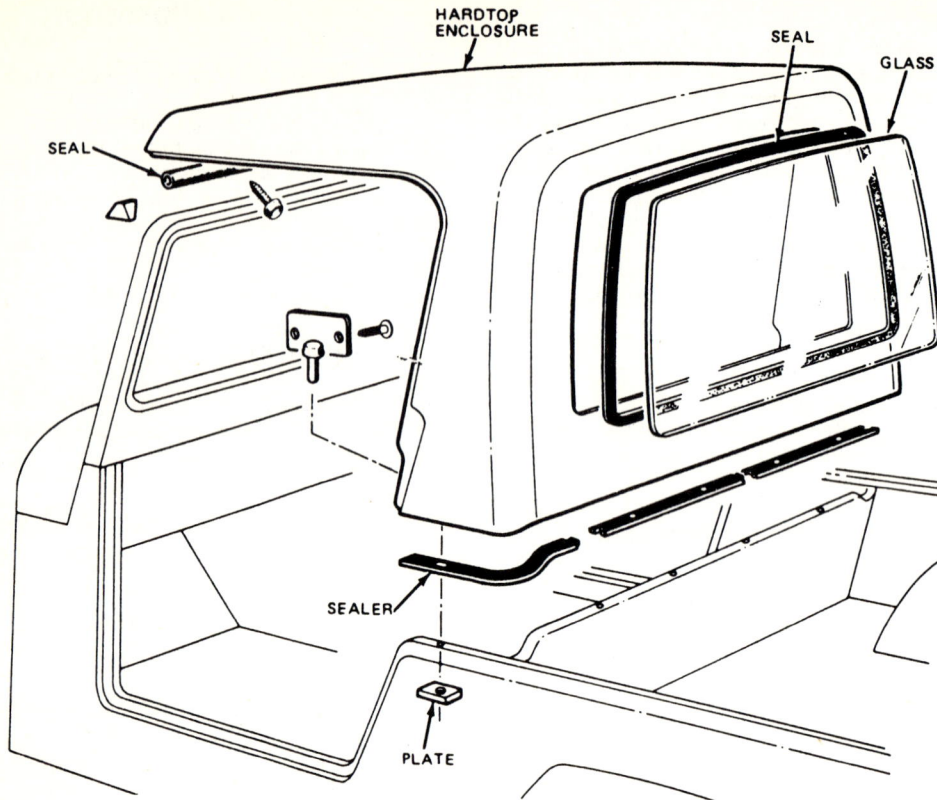

Hardtop used on the Scrambler

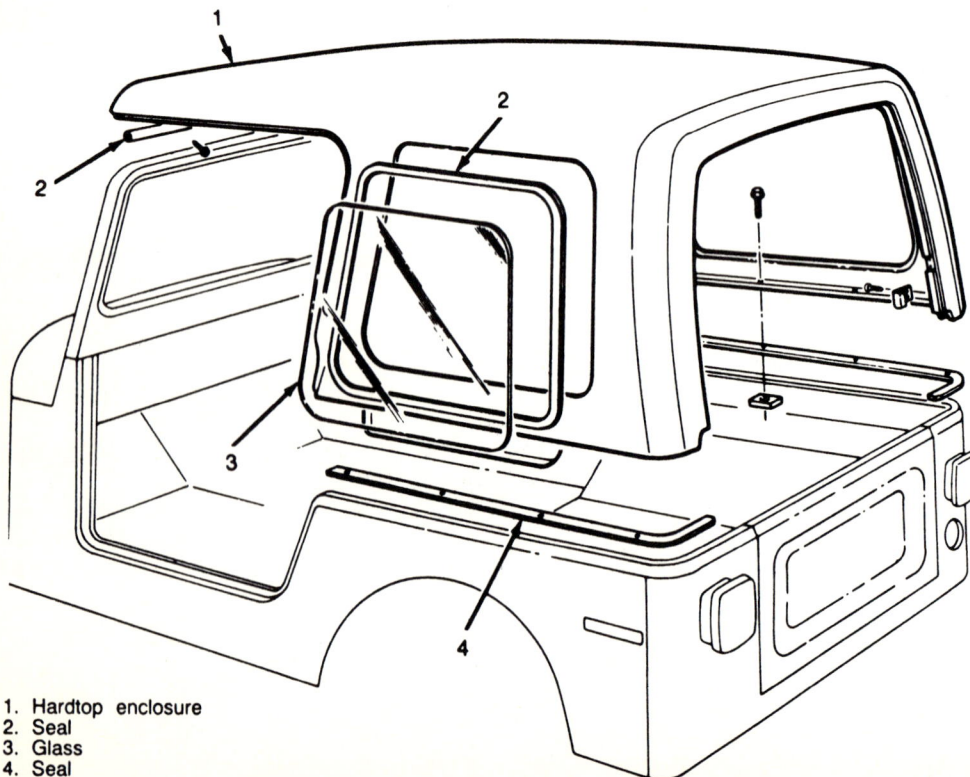

1. Hardtop enclosure
2. Seal
3. Glass
4. Seal

Hardtop used on CJ-7 and Wrangler models

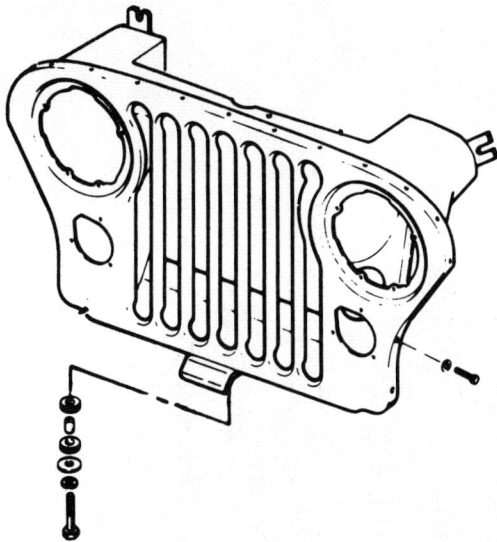

CJ grille components

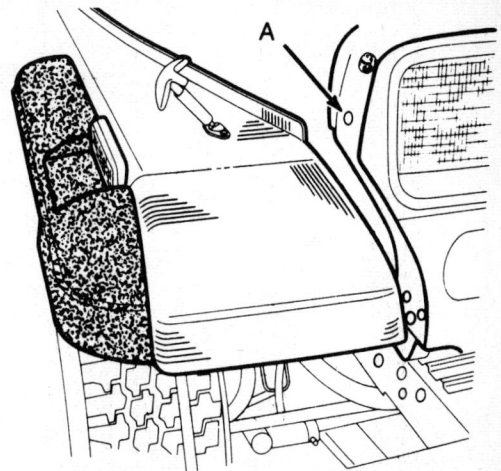

Removing the grille-to-frame crossmember holddown assembly. A are the guard panel-to-fender bolts

3. Remove the bolts securing the guard panel to the fenders.
4. Remove the grille-to-frame crossmember holddown assembly, keeping track of the sequence of parts.
5. Loosen the radiator support rods-to-grille guard support bracket nuts.
6. Remove the rods from the brackets.
7. Tilt the grille panel forward and disconnect the electrical wiring at the headlamp sealed beam unit and parking lamp assembly wiring harness at the connectors.
8. If your Jeep has air conditioning:
    a. Discharge the system as described in Section 1.
    b. Disconnect the high pressure hose at the sight glass. Immediately cap both openings!
    c. Connect the high pressure hose at the compressor. Immediately cap both openings!
9. Lift the grille from the vehicle.
10. Position the grille panel and connect the electrical wiring at the headlamp sealed beam unit and parking lamp assembly wiring harness at the connectors.
11. If your Jeep has air conditioning:
    a. Connect the high pressure hose at the compressor.
    b. Connect the high pressure hose at the sight glass.
    c. Charge the system as described in Section 1.
12. Install the rods on the brackets.
13. Tighten the radiator support rods-to-grille guard support bracket nuts.
14. Install the grille-to-frame crossmember holddown assembly.
15. Install the bolts securing the guard panel to the fenders.
16. Install the screws securing the radiator and shroud to the radiator grille guard panel.
17. Install the front crossmember cover, if so equipped.

Fenders

REMOVAL AND INSTALLATION

1. Remove or disconnect all items attached to the fender apron.
2. Disconnect the wiring at the marker light.
3. Remove the rocker panel molding on models through 1986.
4. On 1987-90 models, remove the antenna where necessary.
5. Remove the bolts attaching the fender and brace to the firewall.
6. Remove the bolts attaching the fender to the grille panel.

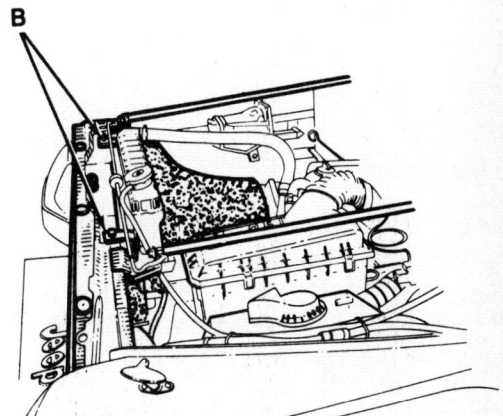

B are the radiator support rods-to-grille guard support bracket nuts

7. Pull the fender out and lift it from the Jeep.
8. Installation is the reverse of removal.

Front Bumper

REMOVAL AND INSTALLATION

1. Remove any auxiliary lighting.
2. Remove the bolts and nuts attaching the bumper to the frame extensions.
3. Remove the bumper.
4. Installation is the reverse of removal.

Rear Bumper

REMOVAL AND INSTALLATION

1. Remove the nuts and bolts attaching the bumpers to the frame.
2. Remove the bumpers.
3. Installation is the reverse of removal.

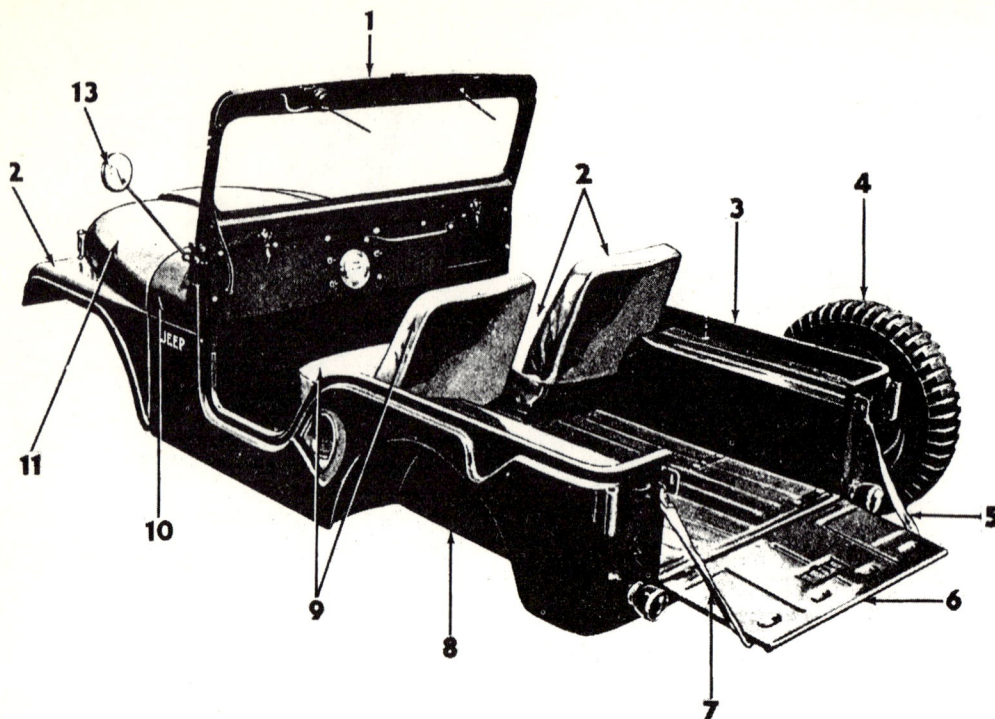

1. Windshield
2. Front passenger seat
3. Right side panel
4. Spare tire
5. Right tailgate chain
6. Tailgate
7. Left tailgate chain
8. Left side panel
9. Driver's seat
10. Cowl
11. Hood
12. Left front fender
13. Rear view mirror

Early CJ-5 body

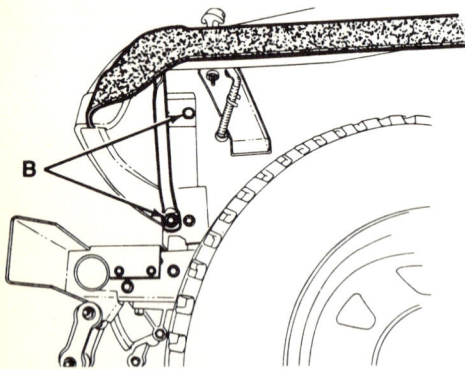

B are the radiator-to-grille panel bolts

Rear Step Bumper

REMOVAL AND INSTALLATION

Scrambler

1. Disconnect the license plate wiring at the connectors.
2. Unbolt and remove the bumper from the mounting arms.
3. Unbolt and remove the mounting arms from the frame.

Front Crossmember Cover
REMOVAL AND INSTALLATION

1. Remove the nuts and bolts attaching the crossmember to the frame.

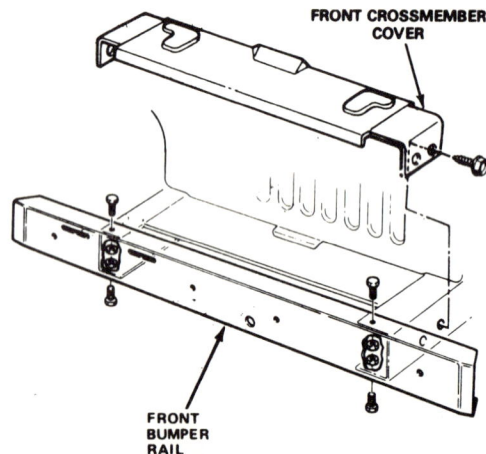

Front bumper typical of CJ models and Wrangler

2. Remove the crossmember.
3. Installation is the reverse of removal.

Swingout Spare Tire Carrier

REMOVAL AND INSTALLATION

1. Remove the tire from the carrier.
2. Remove the hinge pin nuts and bolts.
3. Support the carrier. Unlatch the handle from the latch bracket and remove the carrier and hinge spacer washers.

4. Remove the pin attaching the latch handle to the carrier and remove the handle, spring and washer.

5. Installation is the reverse of removal.

Roll Bar

REMOVAL AND INSTALLATION

NOTE: The following procedures apply only to factory-installed roll bars. Aftermarket roll bars may differ significantly!

Through 1986

NOTE: Torx® fasteners are used.

1. Remove the left front seat.
2. Remove the hardtop, or, fold the soft top back.
3. Tilt the right front seat forward.

4. Remove any carpeting.
5. Remove the roll bar-to-body bolts.
6. Using a heat gun, such as tool J-25070 or equivalent, heat the area around the mounting brackets to soften the sealer.
7. Align the bolt holes before installation. Use new sealer. Don't tighten any of the bolts until they all are installed and the assembly is aligned.

1987-90

1. Remove the bolts from the roll bar end of the extension bars.
2. Remove the bolts from the windshield end of the extension bars and remove the bars.
3. Unbolt the seatbelts from the roll bar.
4. Unbolt the roll bar from the floor pan and lift it from the Jeep.
5. Installation is the reverse of removal. Use new sealer under the roll bar legs. Torque the roll bar-to-floor pan bolts to 66

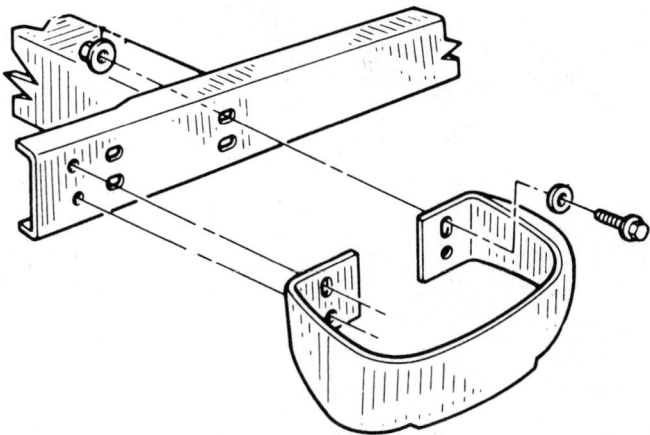

Rear bumper used on CJ and Wrangler

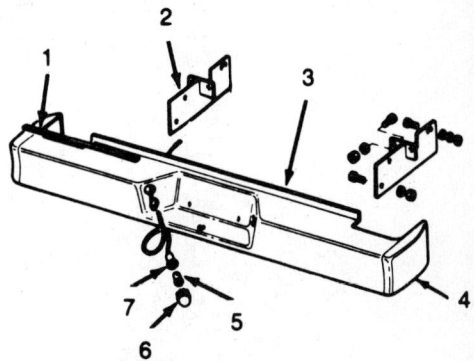

1. Antiskid tape
2. Arm
3. Bumper
4. End cap
5. Bulb
6. Lens
7. Socket

Rear step bumper used on the CJ-7

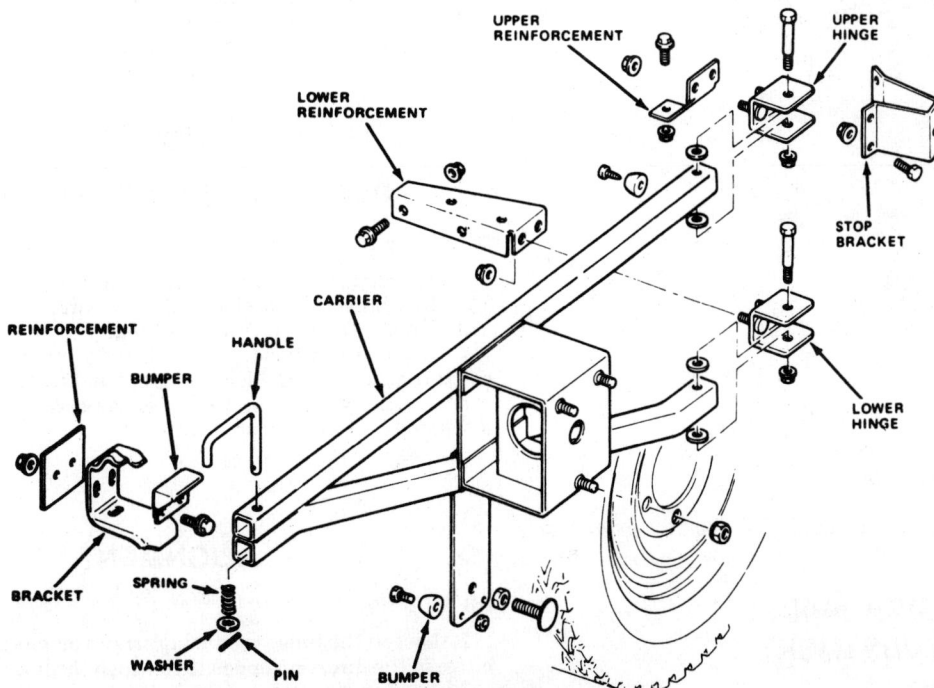

UPPER REINFORCEMENT
UPPER HINGE
LOWER REINFORCEMENT
STOP BRACKET
CARRIER
REINFORCEMENT
HANDLE
LOWER HINGE
BUMPER
BRACKET
SPRING
WASHER
PIN
BUMPER

Swingout spare tire carrier

A. Bolts
B. Bolts
C. Seat belt anchor hole
D. Floor anchor bolt
E. Rear leg bolt

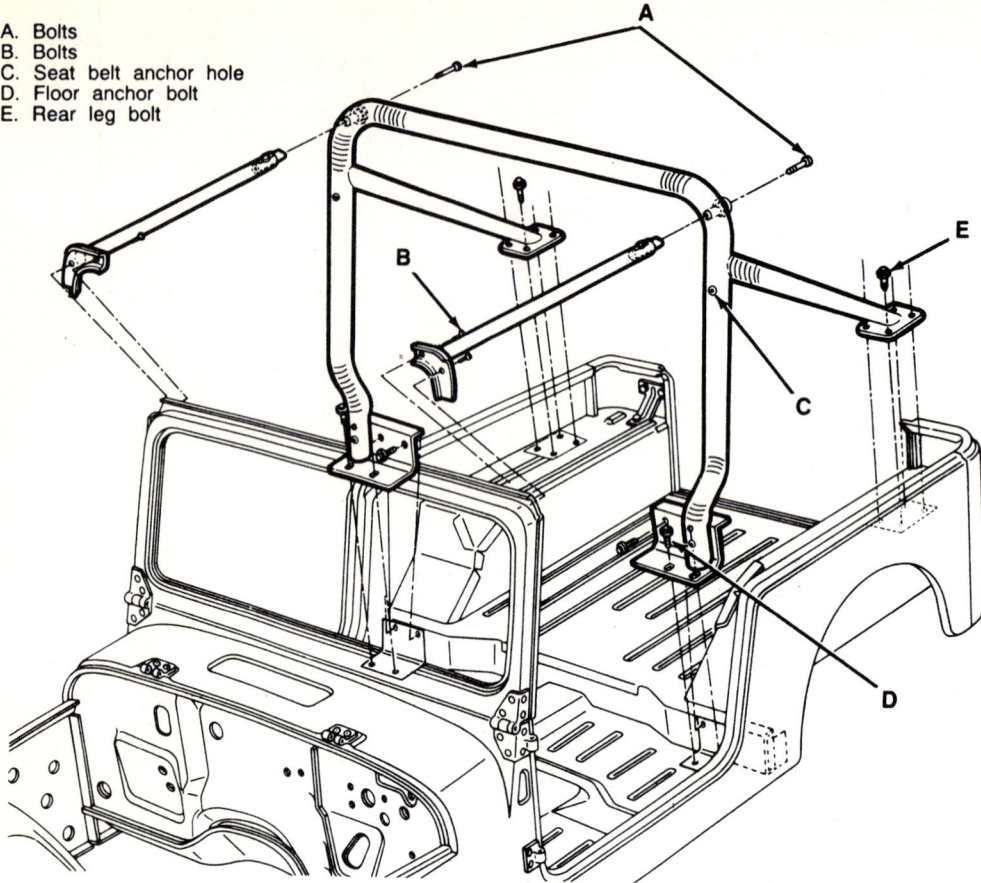

1987–90 roll bar

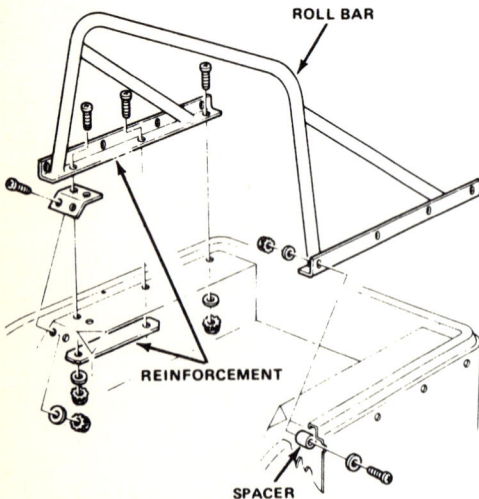

ROLL BAR

REINFORCEMENT

SPACER

Roll bar used through 1986

ft. lbs.; the seatbelt-to-roll bar bolts to 46 ft. lbs.; the extension bar bolts to 45 ft. lbs.

Soft Top
ADJUSTMENT

Tops With Metal Doors
1. Unsnap the top from the vertical support blade.

2. Loosen the adjusting screws.
3. Reposition the vertical support blade.
4. Tighten the adjusting screws.
5. Reposition and snap the soft top into place.

Hood
REMOVAL AND INSTALLATION

All Models
1. Matchmark the hinges and mounting panels.
2. Disconnect the underhood light wire.
3. Unbolt the hood from the hinges.
4. Remove the hood prop rod, prop rod retainer clip, side catch brackets, windshield bumpers and footman loop.
5. Assembly and installation is the reverse of removal and disassembly. Check hood alignment

ALIGNMENT

All Models
1. Loosen the hinge mounting screws on one side and tap the hinge in the direction opposite to which the hood is to be moved.
2. Tighten the screws.
3. Repeat this procedure on the opposite hinge.

FOOTMAN LOOP   WINDSHIELD BUMPER

HINGE

HOOD

BRACKET

CLIP

PROP ROD

SPRING

SAFETY HOOK   BRACKET   REINFORCEMENT PLATE

SIDE CATCH

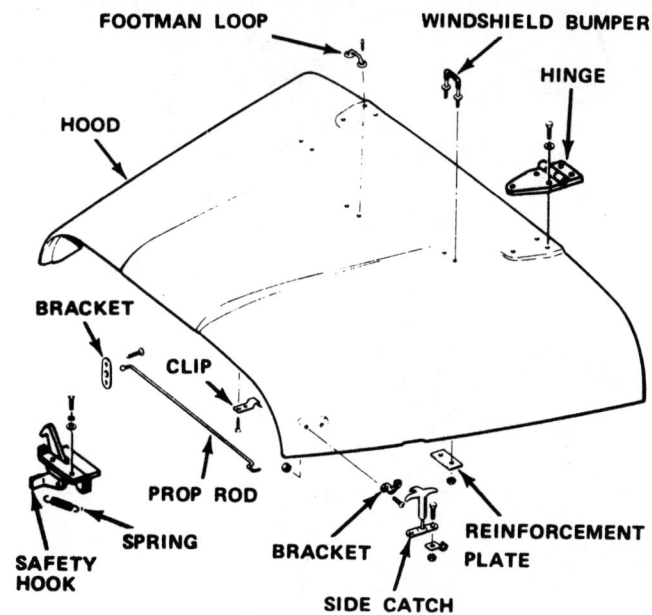

CJ hood through 1986

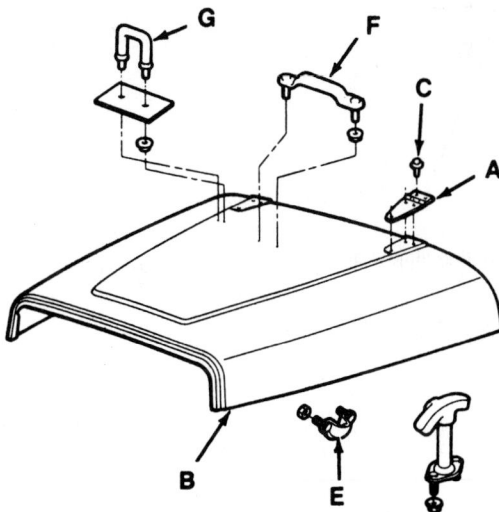

G

F

C

A

B

E

A. Hinge   E. Catch brackets
B. Hood   F. Footman loop
C. Bolts   G. Windshield bumpers
D. Light

1987–90 hood

Liftgate

REMOVAL AND INSTALLATION

CJ-7 and Wrangler

NOTE: Torx® fasteners are used in this procedure.

———————— CAUTION ————————
Never remove the liftgate supports with the liftgate closed! The supports are under considerable spring tension! After removal, never attempt to disassemble the supports!

1. Open and support the liftgate.
2. Disconnect the defogger wiring.

2. Remove the liftgate-to-supports screws and fold the supports downward.
3. Remove the hinges-to-liftgate screws and lift off the liftgate.
4. Installation is the reverse of removal. Check alignment.

ALIGNMENT

CJ-7 and Wrangler

NOTE: Torx® fasteners are used in this procedure.

———————— CAUTION ————————
Never remove the liftgate supports with the liftgate closed! The supports are under considerable spring tension! After removal, never attempt to disassemble the supports!

1. Open and support the liftgate.
2. Remove the liftgate-to-supports screws and fold the supports downward.
3. Remove the hinges-to-liftgate screws and lift off the liftgate.

NOTE: Don't disconnect the release cables from the latches.

4. Loosen the hinge attaching screws.
5. Close the liftgate and move it to obtain a satisfactory fit.
6. Open the liftgate and tighten the liftgate-to-hinge screws.
7. Install the latches and supports.

Tailgate

REMOVAL AND INSTALLATION

CJ-5, CJ-6

1. Open the tailgate to the 45° position and disengage the right hinge.
2. Open it a bit more and disengage the left hinge.
3. Installation is the reverse of removal. Alignment is accomplished by moving the hinges.

CJ-7 and Scrambler

1. Unbolt the support cables from the tailgate.
2. With the tailgate closed, remove the hinge-to-tailgate bolts and remove the tailgate.
3. Installation is the reverse of removal. Alignment is accomplished by moving the hinges.

Wrangler

1. Open the tailgate.
2. Unbolt and remove the tailgate stop swing cover.
3. Pry out the retainer spacer that secure the tailgate tension spring in the bracket.
4. Squeeze the spring and remove it from the bracket.
5. Remove the plastic isolator.
6. Close the tailgate.
7. Remove the tailgate hinge bolts.
8. Release the latch and lift out the tailgate.
9. Installation is the reverse of removal. Alignment is accomplished by loosening the hinge bolts and moving the closed tailgate to obtain a satisfactory fit.

Doors

REMOVAL AND INSTALLATION

CJ-7 and Wrangler with Hardtop

1. Matchmark the hinge-to-door position.

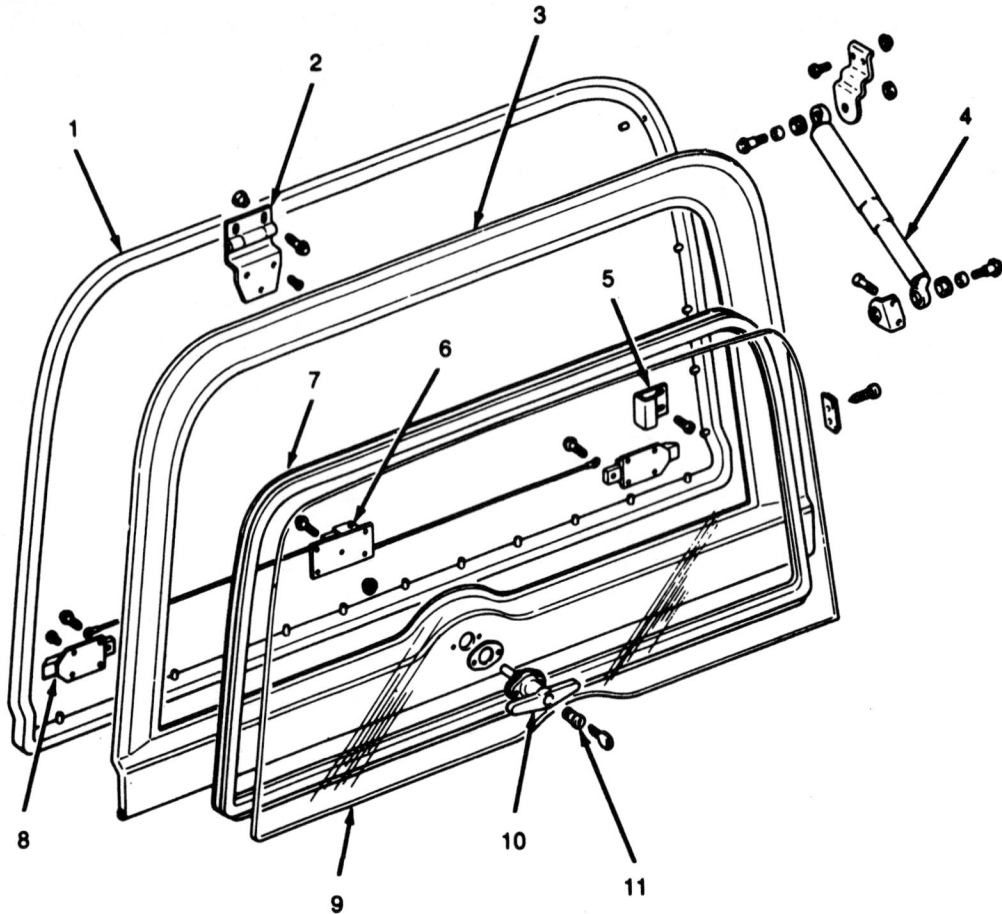

1. Rubber seal
2. Hinge
3. Liftgate
4. Support
5. Striker
6. Remote control
7. Weatherstripping
8. Latch
9. Glass
10. Outside handle
11. Lock cylinder

CJ-7 liftgate

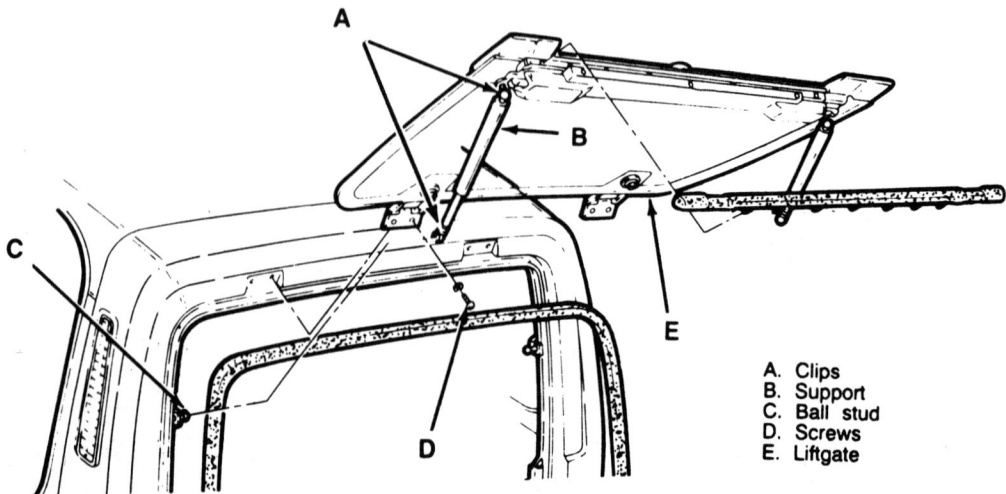

A. Clips
B. Support
C. Ball stud
D. Screws
E. Liftgate

Wrangler liftgate

Tailgate used on early model CJ-5A and CJ-6A

Tailgate used on later model CJ-5 and CJ-6

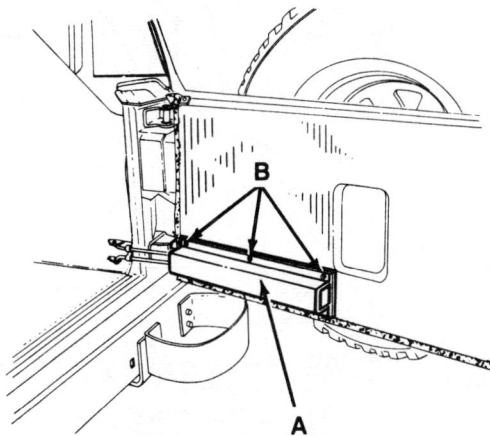
A. Shield   B. Fasteners

A is the cover; B are screws, on the Wrangler tailgate

2. Remove the hinge-to-door screws and lift off the door.
3. When installing the door, install the fasteners but don't fully tighten them. Move the door to obtain a satisfactory fit when closed, then tighten the screws.

Striker Plate

ADJUSTMENT

——— CAUTION ———

To prevent the door's opening in the safety latched position, the door

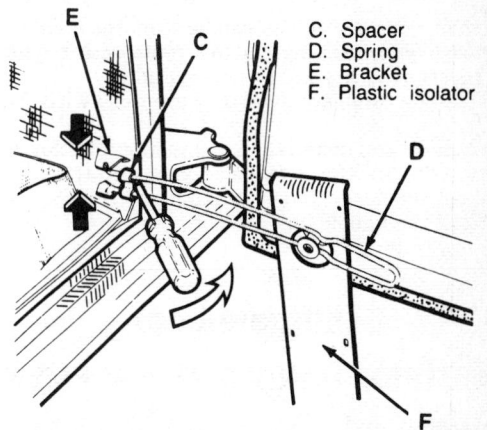
C. Spacer
D. Spring
E. Bracket
F. Plastic isolator

Removing the spring from the Wrangler tailgate

striker wedge must be properly positioned in relation to the cam surface of the lock toggle. Improper safety latch positioning will permit the toggle to override the striker pin, causing the door to open. The striker must be positioned so that the door lock toggle will be held securely in engagement with the striker pin.

1. Position a brass drift against the upper inside edge of the striker plate.
2. Using a heavy hammer, drive the striker plate outward until the wedge end of the plate firmly contacts the cam surface of the lock toggle, exerting enough pressure to prevent the toggle from overriding the striker pin.
3. Make several checks of the safety latch operation until it is satisfactory.

4. Securely tighten the striker plate-to-pillar screws after the adjustment.

Door Hinges

REMOVAL AND INSTALLATION

CJ-7 and Wrangler with Hardtop Door

NOTE: Plastic shims are used at the hinge pins. Don't lose them!

1. Matchmark the hinge-to-body position and the hinge-to-door position.
2. Remove the hinge-to-body screws and lift off the door assembly.

NOTE: The upper hinge is part of the windshield hinge assembly, so support the windshield before removing the door.

3. Remove the hinges from the door.
4. When installing the door and hinges, don't fully tighten any fasteners until door, hinge and windshield alignment is satisfactory.

Door Lock

LOCK CYLINDER AND OUTSIDE HANDLE REPLACEMENT

NOTE: Replacement outside handles come without the lock cylinder. Replacement lock cylinders come uncoded, without keys.

1. Unbolt and remove the handle from the door.
2. To code your existing key to a replacement cylinder:
   a. Insert your key in the new cylinder.
   b. File the tumblers until they are flush with the cylinder body.
   c. Remove and install the key, making sure that ll the tumblers are flush with the cylinder body with the key installed.
3. Install the new cylinder in the new handle.
4. Install the handle on the door.

Liftgate Lock

OUTSIDE HANDLE REPLACEMENT

CJ-7 Hardtop

NOTE: Replacement outside handles come without the lock cylinder. Replacement lock cylinders come uncoded, without keys.

1. Unbolt and remove the remote linkage from the inside of the liftgate.
2. Unbolt and remove the handle from the door.
3. To code your existing key to a replacement cylinder:
   a. Insert your key in the new cylinder.
   b. File the tumblers until they are flush with the cylinder body.
   c. Remove and install the key, making sure that the tumblers are flush with the cylinder body with the key installed.
4. Install the new cylinder in the new handle.
5. Install the handle on the door.
6. Install the remote linkage.

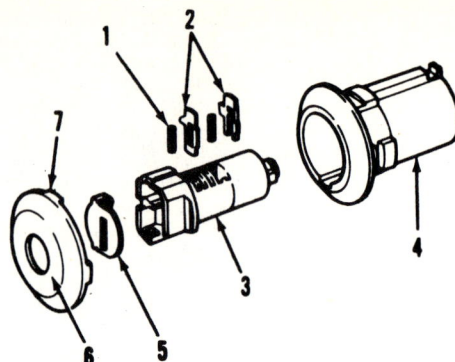

1. Spring
2. Tumblers
3. Lock cylinder
4. Lock casing
5. Door
6. Key hole
7. Bezel

CJ-7, Scrambler and Wrangler door lock, with hardtop

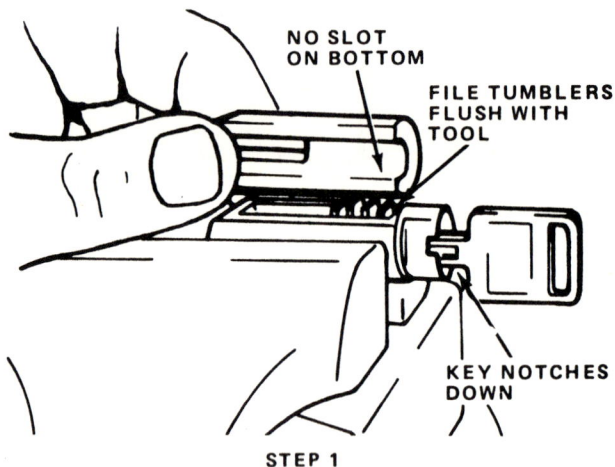

NO SLOT ON BOTTOM
FILE TUMBLERS FLUSH WITH TOOL
KEY NOTCHES DOWN
STEP 1

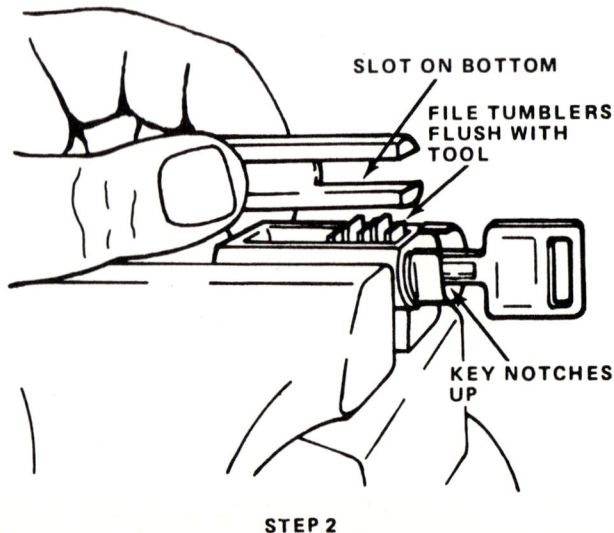

SLOT ON BOTTOM
FILE TUMBLERS FLUSH WITH TOOL
KEY NOTCHES UP
STEP 2

Filing the lock tumblers

Wrangler tailgate latch mechanism

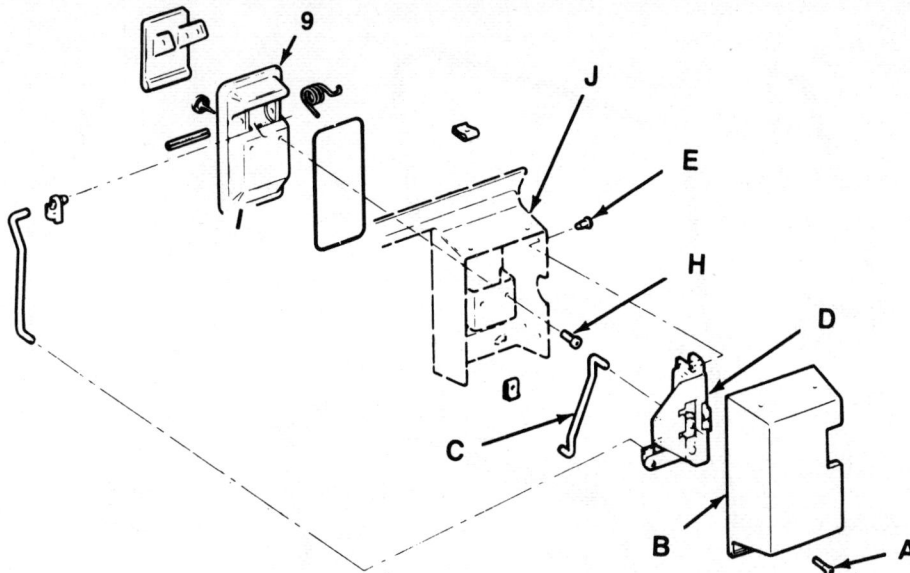

A. Screw	F. Clip
B. Latch mechanism housing	H. Screw
C. Linkage	I. Handle
D. Latch mechanism	J. Latch housing
E. Screw	

Tailgate Lock

LATCH MECHANISM REMOVAL AND INSTALLATION

Wrangler

1. Open the tailgate and remove the screw from the latch mechanism housing.
2. Disconnect the latch linkage.
3. Remove the latch-to-tailgate screws.
4. Remove the retaining clip from the lock cylinder.
5. Remove the handle retaining screw and lift out the handle.
6. Remove the latch from the tailgate.

7. Installation is the reverse of removal.

LOCK CYLINDER REMOVAL AND INSTALLATION

CJ-7

1. Remove the plastic cover.
2. Remove the retaining clip.
3. Remove the snapring.
4. Disconnect the lock cylinder from the linkage.
5. Installation is the reverse of removal.

INTERIOR

Door Trim Panel

REMOVAL AND INSTALLATION

Hardtop

1. Remove the door handle.

2. Remove the window crank handle.
3. Carefully pry the trim panel clips from the door. Be very careful! It's very easy to tear the trim panel or tear loose the clip from the panel. Inexpensive tools are made for this job and are available at most auto parts stores.
4. Remove the watershield.
5. Installation is the reverse of removal.

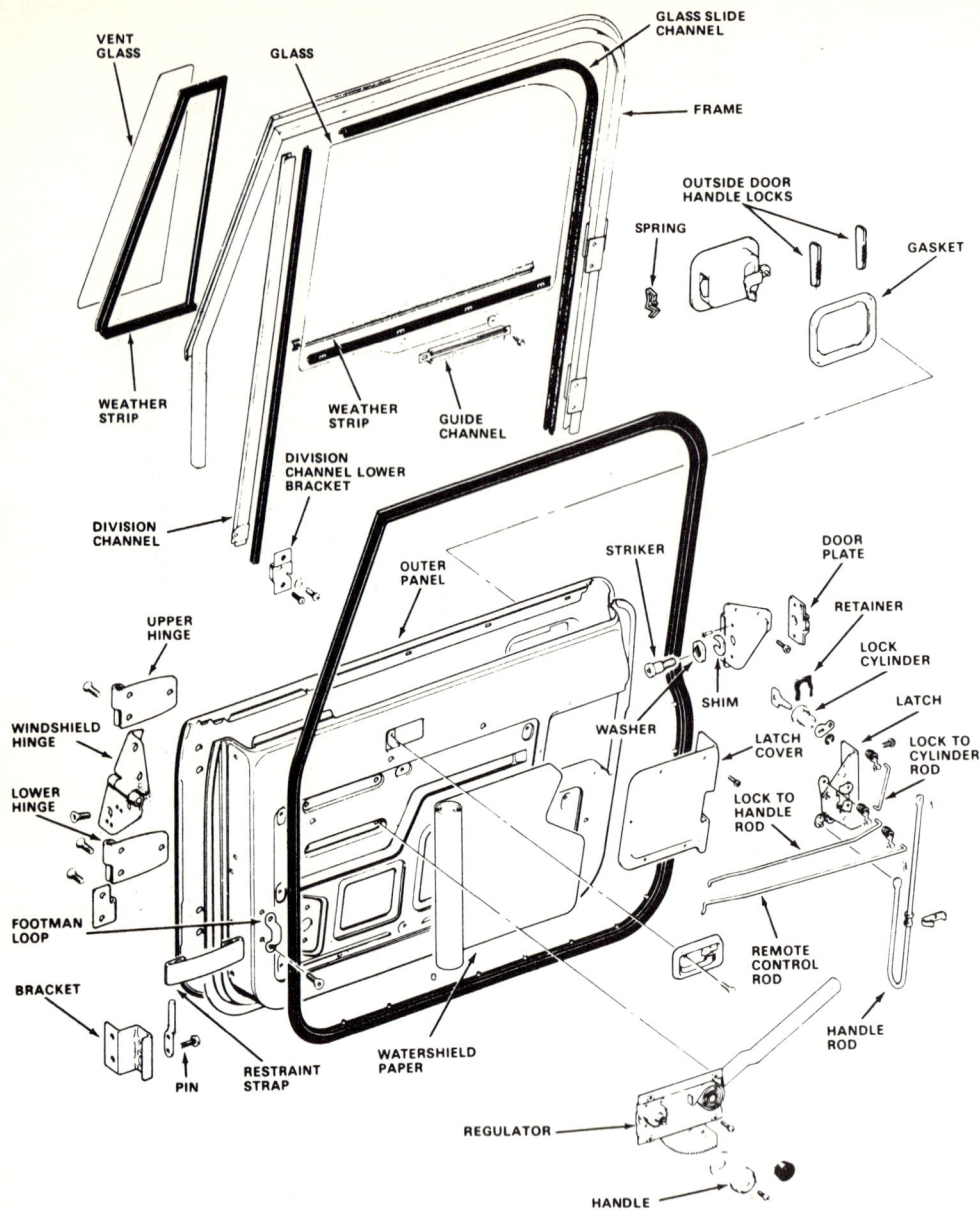

CJ-7 and Scrambler hardtop door

Door Glass

REMOVAL AND INSTALLATION

Hardtop

1. Remove the door handle.
2. Remove the window crank handle.
3. Carefully pry the trim panel clips from the door. Be very careful! It's very easy to tear the trim panel or tear loose the clip from the panel. Inexpensive tools are made for this job and are available at most auto parts stores.
4. Remove the watershield.
5. Remove the glass down-stop.
6. Remove the guide panel-to-plastic fastener screws.
7. Remove the guide channel and plastic fasteners.
8. Remove the division channel upper attaching screw and lower adjusting screw.

9. Disengage the front three inches of weatherstripping from the upper door frame.
10. Remove the division channel.
11. Tilt the glass toward the hinge side of the door and disengage it from the rear channel.
12. Pull the glass up and out of the door.
13. Insert the glass into the door with the front tilted down, while engaging the rear channel.
14. Install the plastic fasteners on the glass.
15. Lower the glass to the bottom of the door.
16. Lower the division channel into the door and position the glass in the channel.
17. Install the upper attaching screw and lower adjusting screw.
18. Install the weatherstripping.
19. Slide the guide channel on the regulator arm and position the channel on the glass. Install the screws.
20. Install the glass down-stop.

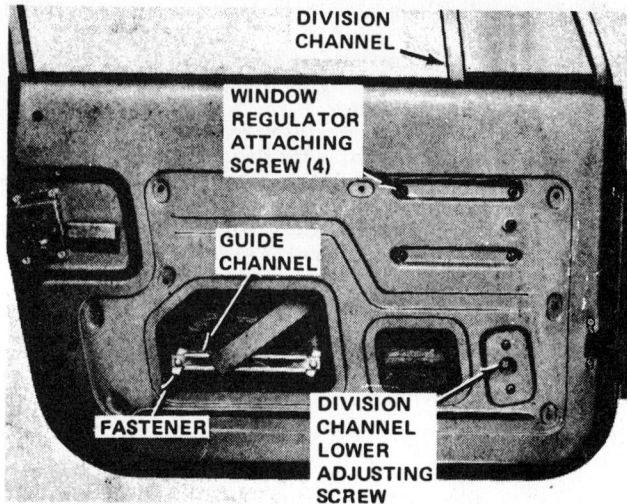

Door glass replacement on CJ-7 and Scrambler models with hardtop

21. Make sure that the window works properly.
22. Install the watershield.
23. Install the trim panel and handles.

Window Regulator

REMOVAL AND INSTALLATION

Hardtop

1. Remove the door handle.
2. Remove the window crank handle.
3. Carefully pry the trim panel clips from the door. Be very careful! It's very easy to tear the trim panel or tear loose the clip from the panel. Inexpensive tools are made for this job and are available at most auto parts stores.
4. Remove the watershield.
5. Lower the glass to gain access to the guide channel fasteners.
6. Remove the fasteners and guide channel.
7. Raise the window to thew full up position and secure it in position by using masking tape to tape it to the door frame.
8. Remove the division channel lower adjusting screw.
9. Remove the regulator attaching screws.

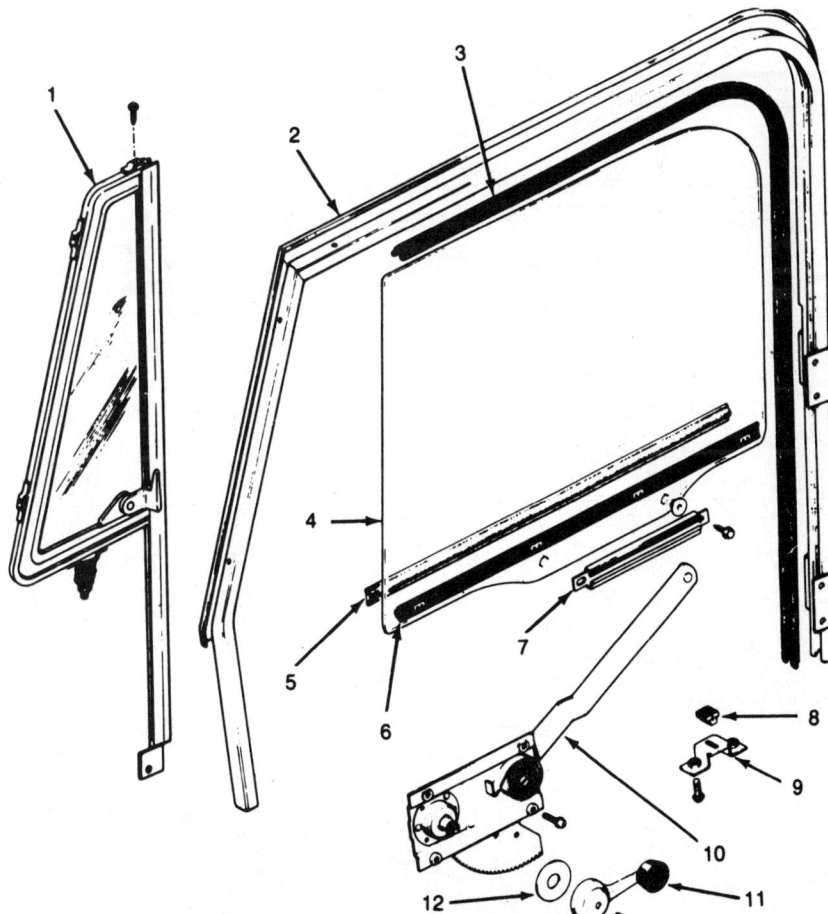

1. Ventilator assembly and glass
2. Top door frame
3. Channel
4. Glass
5. Outer weatherstripping
6. Inner weatherstripping
7. Window regulator arm
8. Stop bumper
9. Stop bracket
10. Window regulator
11. Handle
12. Washer

Door glass replacement on CJ-7 and Scrambler models with moveable vent windows

10. Push the division channel outward and remove the regulator through the access hole.

11. Installation is the reverse of removal.

Stationary Window Glass
REMOVAL AND INSTALLATION

Hardtop

1. Unlock the weatherstripping from the hardtop with a wood spatula or fiber stick.

2. Using a fiber stick, break the seal between the glass and weatherstripping.

3. Push the glass and weatherstripping outward and remove it.

4. Inspect the weatherstripping. Replace it if cracked or oth-

erwise damaged. Clean all old sealer from the glass and weatherstripping.

5. Apply a $\frac{3}{16}$ in. (5mm) bead of 3M Auto Bedding and Glazing Compound, or equivalent, in the weatherstripping glass cavity, using a pressure applicator.

6. Install the glass in the weatherstripping.

7. Place a $\frac{1}{4}$ in. (6mm) cord in the flange cavity of the weatherstripping, completely around the circumference. Let the ends of the cord hang outside the glass at the upper center.

8. Position the glass and weatherstripping in the opening. Pull on the ends of the cord to pull the lip of the weatherstripping over the hardtop flange.

9. Use a wood spatula to lock the weatherstripping.

10. Apply a bead of 3M Windshield Sealer, or equivalent, between the weatherstripping and the glass, around its entire perimeter. Clean off the excess.

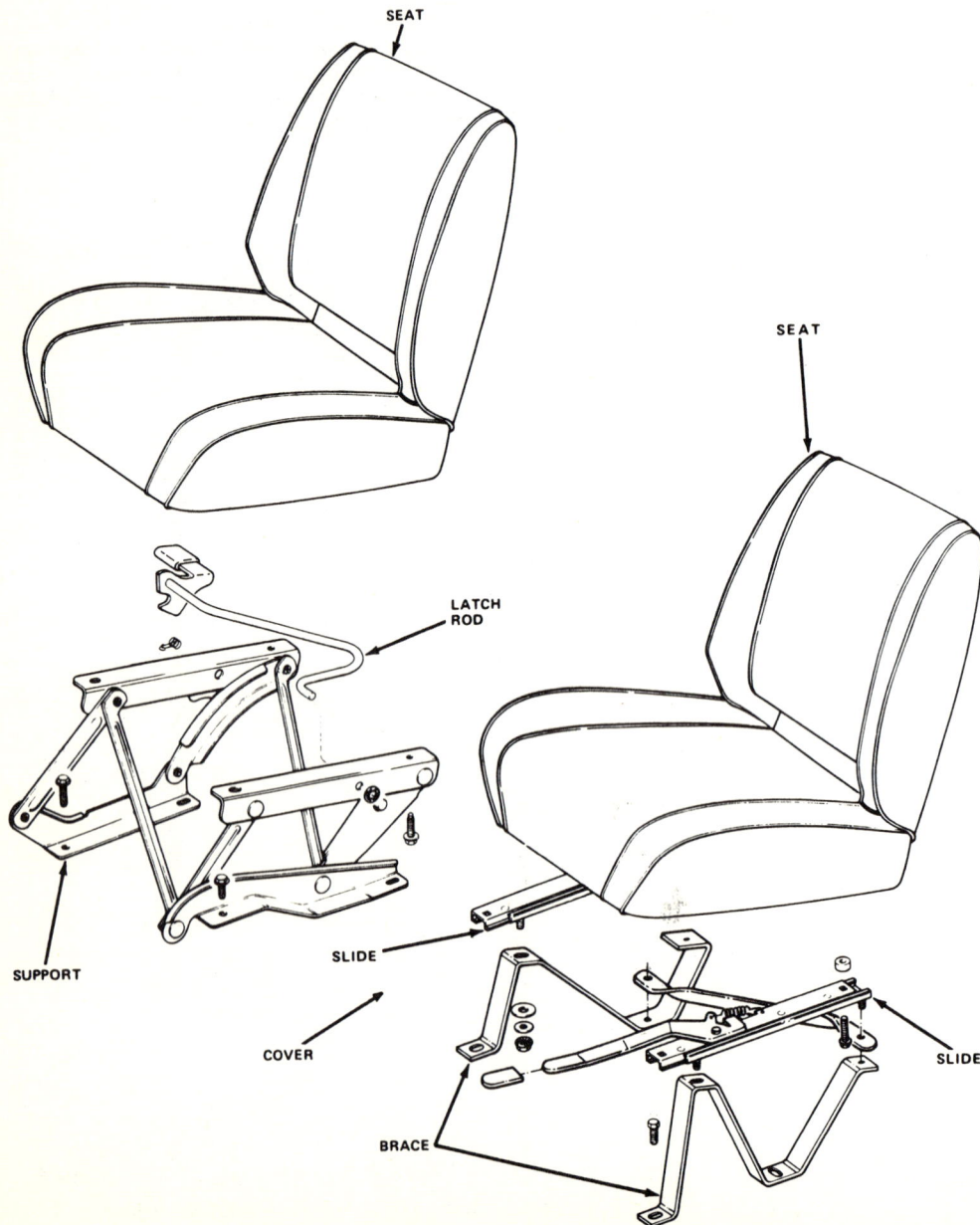

Bucket seats used on all models through 1986

Liftgate Glass

REMOVAL AND INSTALLATION

Hardtop

1. Unlock the weatherstripping from the hardtop with a wood spatula or fiber stick.

2. Using a fiber stick, break the seal between the glass and weatherstripping.

3. Push the glass and weatherstripping outward and remove it.

4. Inspect the weatherstripping. Replace it if cracked or otherwise damaged. Clean all old sealer from the glass and weatherstripping.

5. Apply a $^3/_{16}$ in. (5mm) bead of 3M Auto Bedding and Glazing Compound, or equivalent, in the weatherstripping glass cavity, using a pressure applicator.

6. Install the glass in the weatherstripping.

7. Place a ¼ in. (6mm) cord in the flange cavity of the weatherstripping, completely around the circumference. Let the ends of the cord hang outside the glass at the upper center.

8. Position the glass and weatherstripping in the opening. Pull on the ends of the cord to pull the lip of the weatherstripping over the hardtop flange.

9. Use a wood spatula to lock the weatherstripping.

10. Apply a bead of 3M Windshield Sealer, or equivalent, between the weatherstripping and the glass, around its entire perimeter. Clean off the excess.

Seats

REMOVAL AND INSTALLATION

Front or Rear

1. Unbolt the seat frame from the floor pan.
2. Remove the seat.
3. Installation is the reverse of removal. Torque the bolts to 15 ft. lbs.

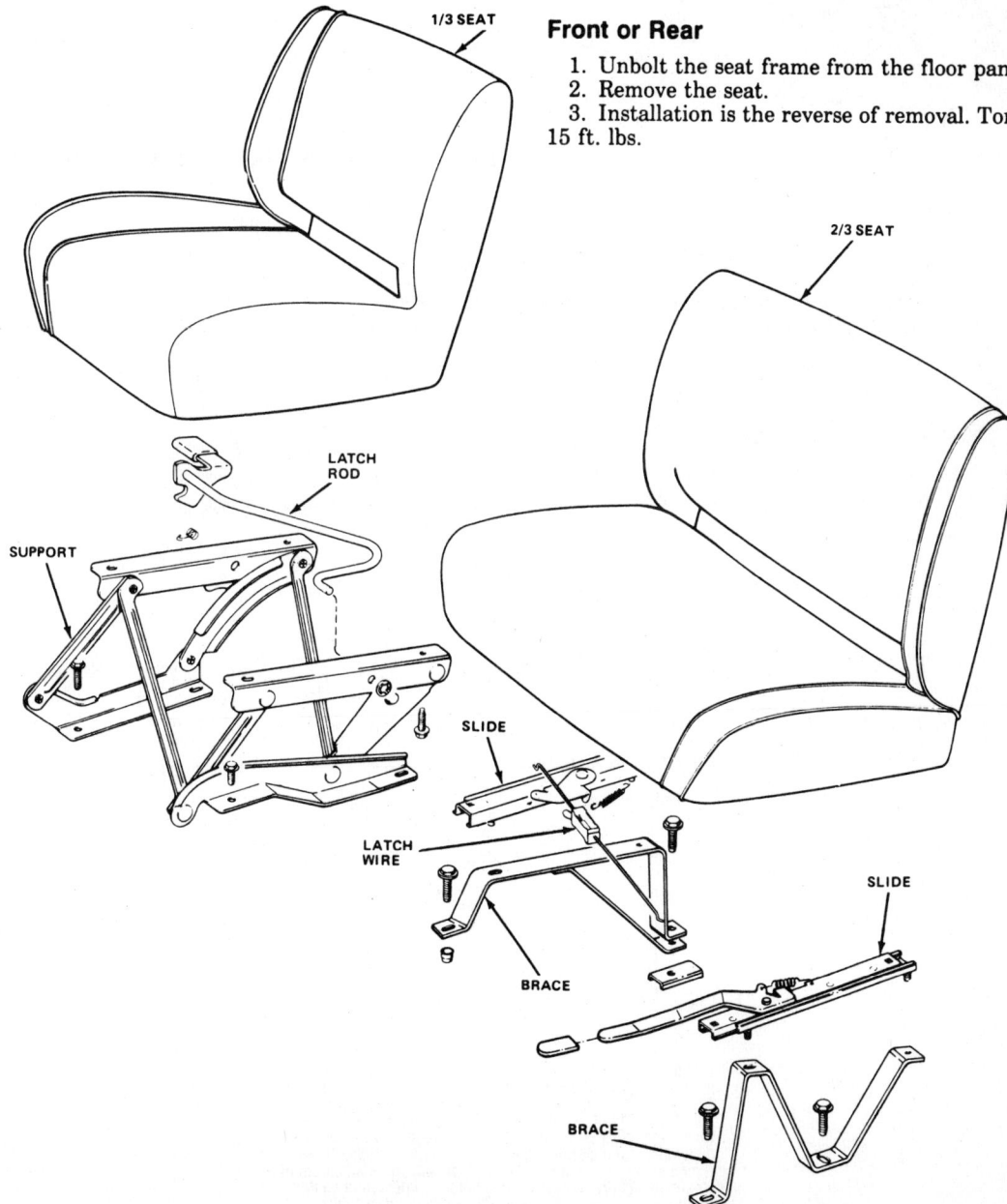

⅓–⅔ style seat, optional on CJ models

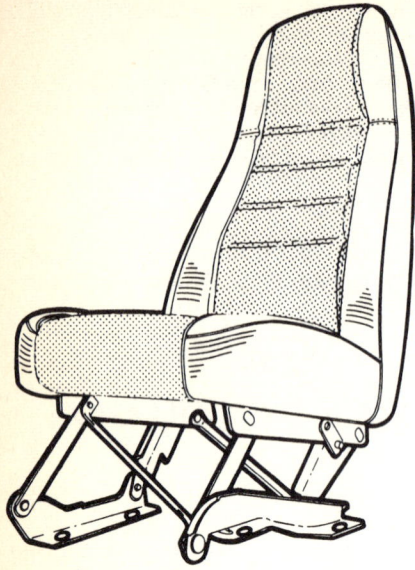

Wrangler front seat

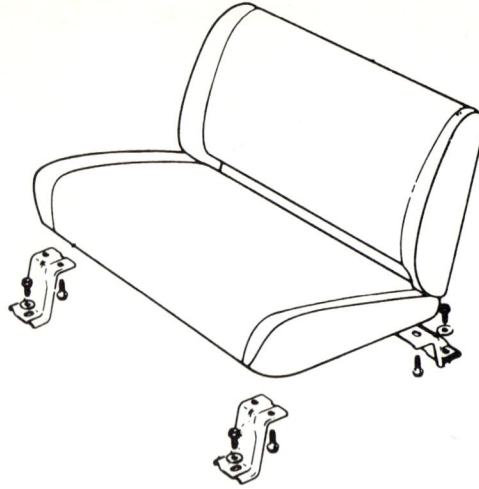

Rear seat used on all CJ and Wrangler models

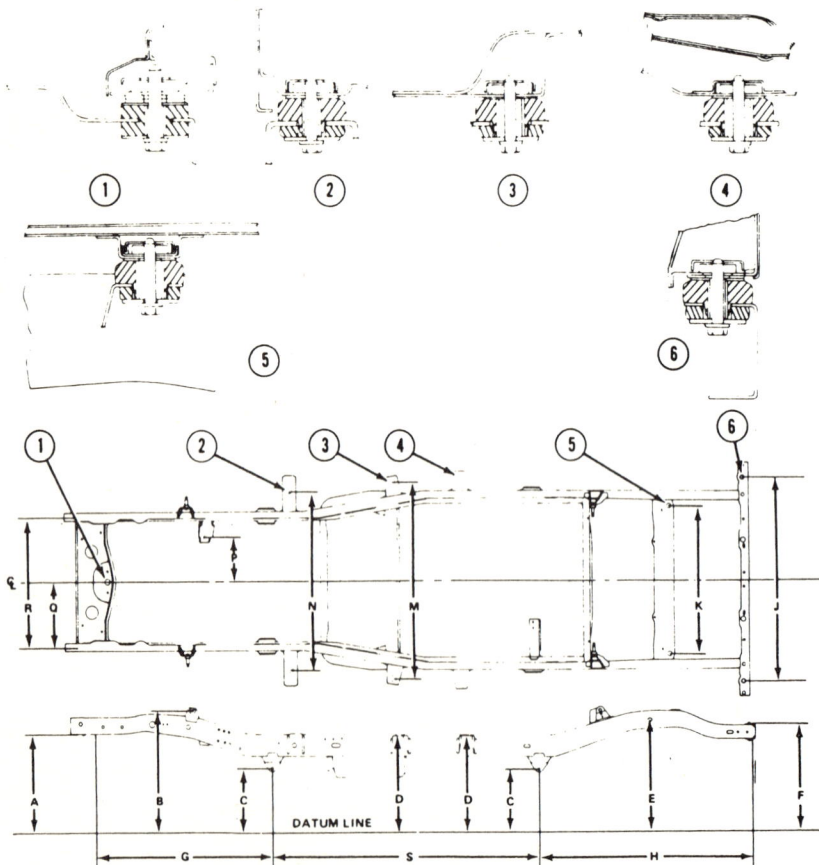

A: 12.51" (317.754mm)	J: 43.50" (1104.900mm)
B: 17.50" (444.500mm)	K: 31.26" (794.004mm)
C: 5.20" (132.080mm)	M: 41.52" (1054.608mm)
D: 12.19" (309.626mm)	N: 37.84" (961.136mm)
E: 15.73" (399.542mm)	P: 9.39" (238.506mm)
F: 14.57" (370.078mm)	Q: 13.75" (349.250mm)
G: 40.61" (1031.494mm)	R: 27.50" (698.500mm)
H: 46.16" (1172.464mm)	S: 39.06" (992.124mm)

Late model CJ-5 frame dimensions

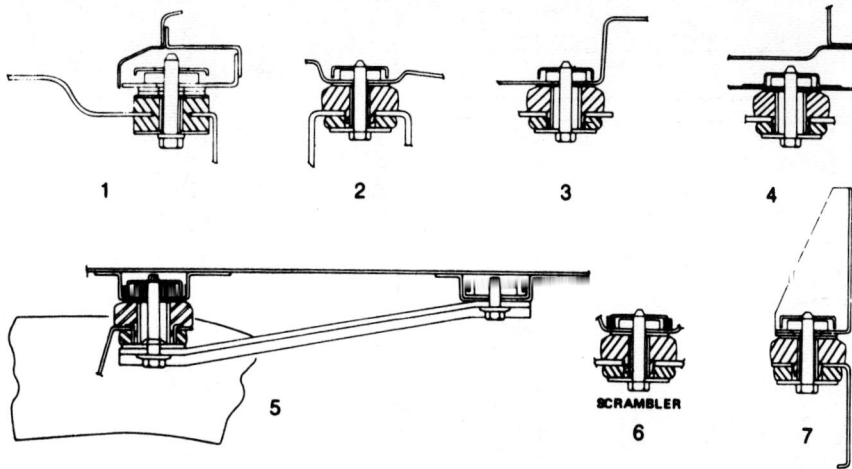

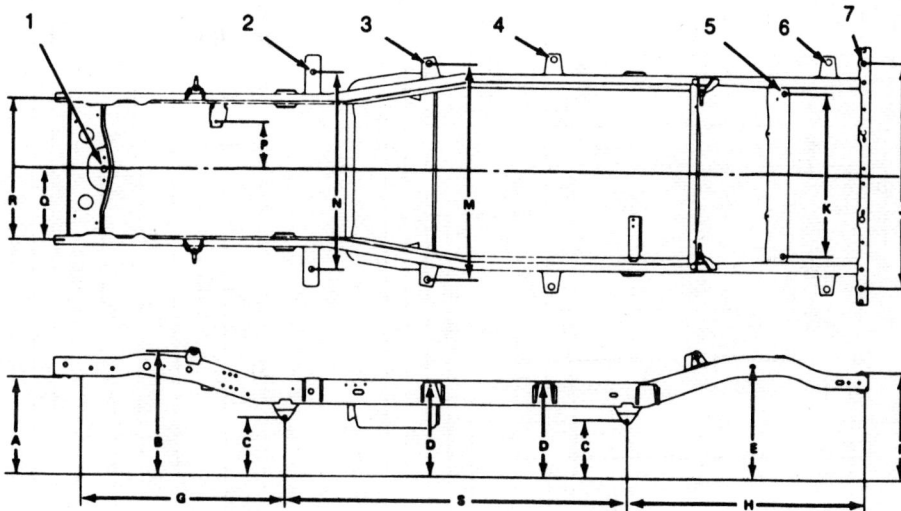

A: 12.51" (317.754mm)
B: 17.50" (444.500mm)
C: 5.20" (132.080mm)
D: 12.19" (309.626mm)
E: 15.73" (399.542mm)
F: 14.57" (370.078mm)
G: 40.61" (1031.494mm)
H: CJ-7 — 46.16" (1172.464mm); Scrambler — 70.02 (1778.508mm)
J: 43.50" (1104.900mm)
K: 31.26" (794.004mm)
M: 41.52" (1054.608mm)
N: 37.84" (961.136mm)
P: 9.39" (238.506mm)
Q: 13.75" (349.250mm)
R: 27.50" (698.500mm)
S: 49.06" (1246.124mm)

CJ-7 and Scrambler frame dimensions

Wrangler frame dimensions. All measurements in millimeters

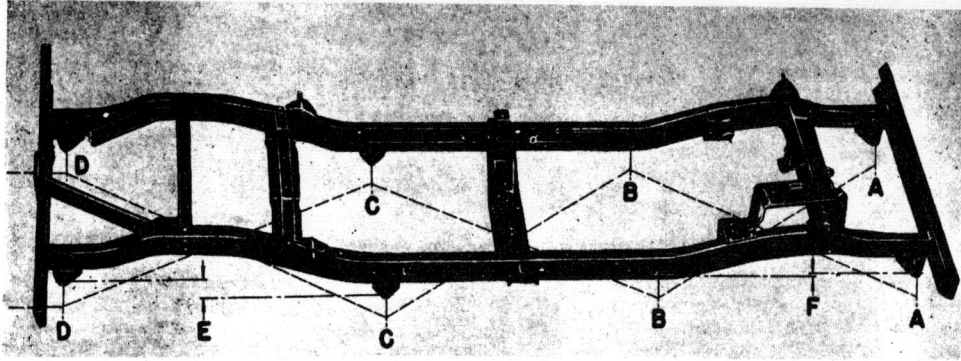

A-B: 47.08" (1195.832mm)
C-D: 52.37" (1330.198mm)
Front Width: 29.25" (742.95mm)
Rear Width: 29.25" (742.95mm)
E: 3.21875" (81.756mm)

F: 4.15625" (130.989mm)
Overall Length:
   CJ-5, CJ-5A -- 128.4375" (3262.3125mm)
   CJ-6, CJ-6A -- 148.4375" (3770.3125mm)

Frame dimensions for early model CJ-5, CJ-5A, CJ-6, CJ-6A with the F4–134 engine

How to Remove Stains from Fabric Interior

For best results, spots and stains should be removed as soon as possible. Never use gasoline, lacquer thinner, acetone, nail polish remover or bleach. Use a 3' x 3" piece of cheesecloth. Squeeze most of the liquid from the fabric and wipe the stained fabric from the outside of the stain toward the center with a lifting motion. Turn the cheesecloth as soon as one side becomes soiled. When using water to remove a stain, be sure to wash the entire section after the spot has been removed to avoid water stains. Encrusted spots can be broken up with a dull knife and vacuumed before removing the stain.

Type of Stain	How to Remove It
Surface spots	Brush the spots out with a small hand brush or use a commercial preparation such as K2R to lift the stain.
Mildew	Clean around the mildew with warm suds. Rinse in cold water and soak the mildew area in a solution of 1 part table salt and 2 parts water. Wash with upholstery cleaner.
Water stains	Water stains in fabric materials can be removed with a solution made from 1 cup of table salt dissolved in 1 quart of water. Vigorously scrub the solution into the stain and rinse with clear water. Water stains in nylon or other synthetic fabrics should be removed with a commercial type spot remover.
Chewing gum, tar, crayons, shoe polish (greasy stains)	Do not use a cleaner that will soften gum or tar. Harden the deposit with an ice cube and scrape away as much as possible with a dull knife. Moisten the remainder with cleaning fluid and scrub clean.
Ice cream, candy	Most candy has a sugar base and can be removed with a cloth wrung out in warm water. Oily candy, after cleaning with warm water, should be cleaned with upholstery cleaner. Rinse with warm water and clean the remainder with cleaning fluid.
Wine, alcohol, egg, milk, soft drink (non-greasy stains)	Do not use soap. Scrub the stain with a cloth wrung out in warm water. Remove the remainder with cleaning fluid.
Grease, oil, lipstick, butter and related stains	Use a spot remover to avoid leaving a ring. Work from the outisde of the stain to the center and dry with a clean cloth when the spot is gone.

Type of Stain	How to Remove It
Headliners (cloth)	Mix a solution of warm water and foam upholstery cleaner to give thick suds. Use only foam—liquid may streak or spot. Clean the entire headliner in one operation using a circular motion with a natural sponge.
Headliner (vinyl)	Use a vinyl cleaner with a sponge and wipe clean with a dry cloth.
Seats and door panels	Mix 1 pint upholstery cleaner in 1 gallon of water. Do not soak the fabric around the buttons.
Leather or vinyl fabric	Use a multi-purpose cleaner full strength and a stiff brush. Let stand 2 minutes and scrub thoroughly. Wipe with a clean, soft rag.
Nylon or synthetic fabrics	For normal stains, use the same procedures you would for washing cloth upholstery. If the fabric is extremely dirty, use a multi-purpose cleaner full strength with a stiff scrub brush. Scrub thoroughly in all directions and wipe with a cotton towel or soft rag.